Atomic Mass Values for Selected Elements

Aluminum	26.98
Boron	10.81
Bromine	79.90
Carbon	12.01
Chlorine	35.45
Fluorine	18.99
Hydrogen	1.008
Iodine	126.9
Lithium	6.941
Magnesium	24.30
Nitrogen	14.01
Oxygen	15.99
Phosphorus	30.97
Potassium	39.09
Silicon	28.09
Sodium	22.99
Sulfur	32.07

Concentrated Acids and Bases

Reagent	HCl	HNO_3	H_2SO_4	HCOOH	CH_3COOH	$NH_3(NH_4OH)$
Density (g/mL)	1.18	1.41	1.84	1.20	1.06	0.90
% Acid or base (by weight)	37.3	70.0	96.5	90.0	99.7	29.0
Molecular weight	36.47	63.02	98.08	46.03	60.05	17.03
Molarity of concentrated acid or base	12	16	18	23.4	17.5	15.3
Normality of concentrated acid or base	12	16	36	23.4	17.5	15.3
Volume of concentrated reagent required to prepare 1 L of 1 M solution (ml)	83	64	56	42	58	65
Volume of concentrated reagent required to prepare 1 L of 10% solution (ml)*	227	101	56	93	95	384
Molarity of a 10% solution*	2.74	1.59	1.02	2.17	1.67	5.87

*Percent solutions by weight.

Introduction to Organic Laboratory Techniques

Introduction to Organic Laboratory Techniques

A MICROSCALE APPROACH

Fourth Edition

Donald L. Pavia
Gary M. Lampman
George S. Kriz
Western Washington University
Bellingham, Washington

Randall G. Engel
North Seattle Community College
Seattle, Washington

BROOKS/COLE
CENGAGE Learning™

Australia • Brazil • Japan • Korea • Mexico • Singapore • Spain • United Kingdom • United States

BROOKS/COLE
CENGAGE Learning

**Introduction to Organic Laboratory Techniques:
A Microscale Approach, Fourth Edition**
Pavia, Lampman, Kriz, Engel

Executive Editor: David Harris

Development Editor: Peter McGahey

Editorial Assistant: Lauren Oliveira

Technology Project Manager: Donna Kelley

Marketing Manager: Amee Mosley

Marketing Assistant: Michelle Colella

Project Manager, Editorial Production: Belinda Krohmer

Creative Director: Rob Hugel

Print Buyer: Karen Hunt

Permissions Editor: Sarah Harkrader

Production Service: Interactive Composition Corporation

Text Designer: Carolyn Deacy

Copy Editor: Cara Wallace

Compositor: Interactive Composition Corporation

The cover of this book shows a cup of tea and a caffeine molecule. Traditionally, caffeine is removed from an aqueous solution of tea by extraction with an organic solvent. An alternative, green chemistry technique has been added in this edition of the book. The technique employs solid phase extraction to remove caffeine from tea or coffee, thus minimizing the use of toxic solvents. Open the book and explore the many techniques available in the organic lab.

For product information and technology assistance, contact us at
Cengage Learning Customer & Sales Support, 1-800-354-9706.

For permission to use material from this text or product, submit all requests online at
www.cengage.com/permissions.
Further permissions questions can be emailed to
permissionrequest@cengage.com.

Library of Congress Control Number: 2005938320

ISBN-13: 978-0-495-01630-4
ISBN-10: 0-495-01630-6

Brooks/Cole
20 Davis Drive
Belmont, CA 94002-3098
USA

Cengage Learning is a leading provider of customized learning solutions with office locations around the globe, including Singapore, the United Kingdom, Australia, Mexico, Brazil, and Japan. Locate your local office at: **www.cengage.com/global.**

Cengage Learning products are represented in Canada by Nelson Education, Ltd.

To learn more about Brooks/Cole, visit
www.cengage.com/brookscole.

Purchase any of our products at your local college store or at our preferred online store **www.CengageBrain.com.**

Printed in the United States of America
6 7 8 9 10 13 12 11

This book is dedicated to
our organic chemistry laboratory students

Preface

Experiments in organic chemistry can be conducted at different scales using varying amounts of chemicals and different styles of glassware. The two most common approaches are **macroscale** (also called **standard-** or **traditional-scale**) and **microscale.**

In the traditional macroscale approach, the quantities used are on the order of 5–100 grams. In another version of the macroscale approach, called **small-scale,** the experiments use smaller amounts of chemicals (1–10 grams); however, the glassware and methods used in small-scale experiments are identical to the glassware and methods used in standard-scale experiments. The small-scale approach is described in another of our textbooks, entitled *Introduction to Organic Laboratory Techniques: A Small Scale Approach, second edition.*

The microscale approach used in this textbook differs from the macroscale laboratory in that the experiments use very small amounts of chemicals (0.050–1.000 grams). In addition, some of the microscale glassware is different from standard-scale glassware, and there a few special techniques that are unique to the microscale laboratory.

In a third laboratory textbook that we have written, *Microscale and Macroscale Techniques in the Organic Laboratory*, the most important methods used in the modern organic chemistry laboratory are described. The book is written to accompany experiments using both microscale and macroscale glassware; however, no experiments or essays are included in the book.

Since the Third Edition of our microscale textbook appeared in 1999, there have been new developments in the teaching of organic chemistry laboratory. Because of these developments, this Fourth Edition includes many new experiments and significant updating of the essays and techniques chapters. We have added a special focus on new experiments that incorporate the principles of Green Chemistry. In our view, it is most timely that students begin to think about how to conduct chemical experiments in a more environmentally benign manner. In preparing this new edition, we have also attempted to incorporate the many improvements and suggestions that have been forwarded to us by the many instructors who have been using our materials over the past several years.

NEW TO THIS EDITION

To introduce environmentally responsible methods of performing laboratory experiments, we have added a new focus on Green Chemistry in this new edition. This includes several new experiments, descriptions of new "green" methods, and a new essay, *Green Chemistry*. Among these new experiments are an extraction (done as a demonstration) using supercritical carbon dioxide, an isolation experiment that utilizes solid phase extraction tubes, an experiment that can be conducted in a conventional microwave oven, and an alternative to the standard chromic acid oxidation. Other experiments have been modified to reduce their use of hazardous solvents. New experiments added for this edition include:

Experiment 7	Infrared Spectroscopy and Boiling Point Determination
Experiment 11	Isolation of Caffeine from Tea (there is a new procedure that utilizes solid phase extraction tubes)
Experiment 13	Isolation of Eugenol from Cloves (there is a new demonstration that is based on a supercritical carbon dioxide extraction)
Experiment 34	Nitration of Aromatic Compounds Using a Recyclable Catalyst
Experiment 35	An Oxidation-Reduction Scheme: Borneol, Camphor, Isoborneol (there is a new, solventless, procedure)
Experiment 39	Aqueous-Based Organozinc Reactions
Experiment 44	The Wittig Reaction: Preparation of 1,4-Diphenyl-1,3-butadiene (there is a new, solventless, procedure)
Experiment 58	Competing Nucleophiles in S_N1 and S_N2 Reactions: Investigations Using 2-Pentanol and 3-Pentanol
Experiment 62	The Use of Organozinc Reagents in Synthesis: An Exercise in Synthesis and Structure Proof by Spectroscopy
Experiment 63	Synthesis of Substituted Chalcones: A Guided Inquiry Experience
Experiment 66	An Oxidation Puzzle

Substantial revisions were made to the *Petroleum and Fossil Fuels* essay. The tables of unknowns for the qualitative analysis experiment (Experiment 54 and Appendix 1) have been greatly expanded. Chiral gas chromatography has been included in the analysis of the products of the resolution of α-phenylethylamine (Experiment 40) and the chiral reduction of ethyl acetoacetate (Experiment 33). For the qualitative analysis experiment (Experiment 54), we have added a new optional test that can be used in place of the traditional chromic acid test. This new test is safer and does not require contact with hazardous chromium compounds. The use of a conventional microwave oven to prepare organic products has been introduced. Many other experiments have been modified to improve their reliability and safety.

The Techniques chapters have been extensively reorganized, and *Laboratory Safety* can now be found as its own technique chapter. There is also a new technique chapter, *How to Find Data for Compounds: Handbooks and Catalogs*. New problems have been added to the chapters on infrared and NMR spectroscopy (Techniques 25, 26, and 27). Many of the old 60 MHz NMR spectra have been replaced by more modern 300 MHz spectra. As in previous editions, the Techniques chapters include both microscale and macroscale methods.

CUSTOMIZED OPTIONS

Because we realize that the traditional, comprehensive laboratory textbook may not fit every classroom's needs or every student's budget, we offer the opportunity to create personalized course materials. This book can be purchased in customized formats that may exclude unneeded experiments, include your local materials, and, if desired, incorporate additional content from other Cengage Learning products. For more information on custom possibilities, contact your local Cengage Learning representative. You can find contact information for your representative by visiting www.cengage.com/findrep.html and using the "Find Your Rep" link at the top of the page.

INSTRUCTOR'S MANUAL (AVAILABLE ON-LINE ONLY)

We would like to call your attention to the Instructor's Manual that accompanies our textbook and is available for download to qualified instructors as a PDF file at the book's companion website. The manual contains complete instructions for the preparation of reagents and equipment for each experiment, as well as answers to each of the questions in this textbook. In some cases, additional optional experiments are included. Other comments that should prove helpful to the instructor include the estimated time to complete each experiment—and notes for special equipment or reagent handling. There is also a new section describing recommended waste management guidelines.

We strongly recommend that instructors obtain a copy of this manual by visiting www.cengage.com/chemistry/pavia and following the instructions at the Instructor Book Companion site. You may also contact your local Cengage Learning representative for assistance. Contact information for your representative is available at www.cengage.com/findrep.html through the "Find Your Rep" link at the top of the page.

ACKNOWLEDGMENTS

We owe our sincere thanks to the many colleagues who have used our textbooks and who have offered their suggestions for changes and improvements in our laboratory procedures or discussions. Although we cannot mention everyone who has made important contributions, we must make special mention of James Patterson (North Seattle Community College), Frank Deering (North Seattle Community College), Thomas Harvey (Western Washington University), James Vyvyan (Western Washington University), and Charles Wandler (Western Washington University).

We thank all who contributed, with special thanks to our Developmental Editor, Peter McGahey and to our Project Editor, Andrea Clemente.

We are especially grateful to the students and friends who have volunteered to participate in the development of experiments or who offered their help and criticism. We thank Jeff Adamo, Carolyn Agasinski, Jeff Keverline, Sara McGarvey, Brian Michel, Liz Oxford, Emmett Richards, Jon Stewart, Sian Thornton, Dawn Wagner, and Erik Werner.

If you wish to contact us with comments, questions, or suggestions, we have a special electronic mail address for this purpose (plke@chem.wwu.edu). We encourage you to visit our home page at http://atom.chem.wwu.edu/dept/staff/org/plkhome.html. You may also wish to visit the Brooks/Cole Publishing web site at www.cengage.com/brookscole.

Finally, we wish to thank our families and special friends, especially Neva-Jean Pavia, Marian Lampman, Carolyn Kriz, and Karin Granstrom, for their encouragement, support, and patience.

Donald L. Pavia   (pavia@chem.wwu.edu)
Gary M. Lampman   (lampman@chem.wwu.edu)
George S. Kriz   (George.Kriz@wwu.edu)
Randall G. Engel   (tawnydog@earthlink.net)

January 2006

Contents

PART 1

Introduction to Basic Laboratory Techniques   1

1   Introduction to Microscale Laboratory   2

2   Solubility   13

3   Crystallization   21

   3A   Semimicroscale Crystallization—Erlenmeyer Flask and
       Hirsch Funnel   22

   3B   Microscale Crystallization—Craig Tube   25

   3C   Selecting a Solvent to Crystallize a Substance   27

   3D   Mixture Melting Points   28

   3E   Critical Thinking Application   29

4   Extraction   32

   4A   Extraction of Caffeine   33

   4B   Distribution of a Solute between Two Immiscible Solvents   36

   4C   How Do You Determine Which One Is the Organic Layer?   37

   4D   Use of Extraction to Isolate a Neutral Compound from a Mixture Containing an Acid
       or Base Impurity   38

   4E   Critical Thinking Application   40

5   Chromatography   42

   5A   Thin-Layer Chromatography   43

   5B   Selecting the Correct Solvent for Thin-Layer Chromatography   45

   5C   Monitoring a Reaction with Thin-Layer Chromatography   46

   5D   Column Chromatography   48

6   Simple and Fractional Distillation   51

   6A   Simple and Fractional Distillation (Semimicroscale Procedure)   53

   6B   Simple and Fractional Distillation (Microscale Procedure)   56

7   Infrared Spectroscopy and Boiling-Point Determination   57

Essay   Aspirin   61

8   Acetylsalicylic Acid   63

Essay   Analgesics   67

9   *Isolation of the Active Ingredient in an Analgesic Drug   71*

10 *Acetaminophen   75*

    10A *Acetaminophen (Microscale Procedure)   76*

    10B *Acetaminophen (Semimicroscale Procedure)   78*

Essay   **Identification of Drugs   80**

11 *TLC Analysis of Analgesic Drugs   82*

Essay   **Caffeine   87**

12 *Isolation of Caffeine from Tea or Coffee   90*

    12A *Extraction of Caffeine from Tea with Methylene Chloride   93*

    12B *Extraction of Caffeine from Tea or Coffee Using Solid Phase Extraction (SPE)   95*

Essay   **Esters—Flavors and Fragrances   99**

13 *Isopentyl Acetate (Banana Oil)   103*

    13A *Isopentyl Acetate (Microscale Procedure)   104*

    13B *Isopentyl Acetate (Semimicroscale Procedure)   106*

Essay   **Terpenes and Phenylpropanoids   108**

14 *Essential Oils: Extraction of Oil of Cloves by Steam Distillation and Liquid Carbon Dioxide   112*

    14A *Oil of Cloves (Microscale Procedure)   113*

    14B *Oil of Cloves (Semimicroscale Procedure)   115*

    14C *Extraction of Oil of Cloves with Liquid Carbon Dioxide—A Demonstration   116*

Essay   **Stereochemical Theory of Odor   119**

15 *Spearmint and Caraway Oil: (+)- and (−)-Carvones   124*

Essay   **The Chemistry of Vision   132**

16 *Isolation of Chlorophyll and Carotenoid Pigments from Spinach   136*

Essay   **Ethanol and Fermentation Chemistry   142**

17 *Ethanol from Sucrose   145*

PART 2

Introduction to Molecular Modeling   151

Essay   **Molecular Modeling and Molecular Mechanics   152**

18 *An Introduction to Molecular Modeling   156*

    18A *The Conformations of n-Butane: Local Minima   157*

    18B *Cyclohexane Chair and Boat Conformations   158*

    18C *Substituted Cyclohexane Rings   159*

    18D *cis- And trans-2-Butene   159*

Essay   **Computational Chemistry—*ab Initio* and Semiempirical Methods   160**

19 *Computational Chemistry   168*

    19A *Heats of Formation: Isomerism, Tautomerism, and Regioselectivity   169*

    19B *Heats of Reaction: S_N1 Reaction Rates   171*

19C *Density–Electrostatic Potential Maps—Acidities of Carboxylic Acids* 172

19D *Density–Electrostatic Potential Maps: Carbocations* 173

19E *Density–LUMO Maps: Reactivities of Carbonyl Groups* 173

PART 3

Properties and Reactions of Organic Compounds 175

20 *Reactivities of Some Alkyl Halides* 176

21 *Nucleophilic Substitution Reactions: Competing Nucleophiles* 180

    21A *Competitive Nucleophiles with 1-Butanol or 2-Butanol* 183

    21B *Competitive Nucleophiles with 2-Methyl-2-Propanol* 186

    21C *Analysis* 187

22 *Hydrolysis of Some Alkyl Chlorides* 190

23 *Synthesis of* n-*Butyl Bromide and* t-*Pentyl Chloride* 195

    23A n-*Butyl Bromide* 197

    23B n-*Butyl Bromide (Semimicroscale Procedure)* 199

    23C t-*Pentyl Chloride (Microscale Procedure)* 200

    23D t-*Pentyl Chloride (Semimicroscale Procedure)* 201

    23E t-*Pentyl Chloride (Macroscale Procedure)* 202

24 *Elimination Reactions: Dehydration and Dehydrohalogenation* 204

    24A *Dehydration of 1-Butanol and 2-Butanol* 207

    24B *Dehydrobromination of 1-Bromobutane and 2-Bromobutane* 208

25 *4-Methylcyclohexene* 211

    25A *4-Methylcyclohexene (Microscale Procedure)* 213

    25B *4-Methylcyclohexene (Semimicroscale Procedure)* 214

26 *Phase-Transfer Catalysis: Addition of Dichlorocarbene to Cyclohexene* 217

27 *Relative Reactivities of Several Aromatic Compounds* 224

28 *Nitration of Methyl Benzoate* 228

Essay **Fats and Oils** 233

29 *Methyl Stearate from Methyl Oleate* 239

Essay **Soaps and Detergents** 243

30 *Preparation of a Soap* 249

31 *Preparation of a Detergent* 251

Essay **Petroleum and Fossil Fuels** 255

32 *Gas-Chromatographic Analysis of Gasolines* 264

Essay **Green Chemistry** 268

33 *Chiral Reduction of Ethyl Acetoacetate; Optical Purity Determination* 274

    33A *Chiral Reduction of Ethyl Acetoacetate* 275

    33B *NMR Determination of the Optical Purity of Ethyl (S)-3-Hydroxybutanoate* 279

34   *Nitration of Aromatic Compounds Using a Recyclable Catalyst*   284

35   *An Oxidation–Reduction Scheme: Borneol, Camphor, Isoborneol*   288

36   *Multistep Reaction Sequences: The Conversion of Benzaldehyde to Benzilic Acid*   302
    36A   *Preparation of Benzoin by Thiamine Catalysis*   303
    36B   *Preparation of Benzil*   309
    36C   *Preparation of Benzilic Acid*   311

37   *Tetraphenylcyclopentadienone*   314

38   *Triphenylmethanol and Benzoic Acid*   317
    38A   *Triphenylmethanol*   322
    38B   *Benzoic Acid*   325

39   *Aqueous-Based Organozinc Reactions*   328

40   *Resolution of (±)-α-Phenylethylamine and Determination of Optical Purity*   331
    40A   *Resolution of (±)-α-Phenylethylamine*   334
    40B   *Determination of Optical Purity Using NMR and a Chiral Resolving Agent*   337

41   *The Aldol Condensation Reaction: Preparation of Benzalacetophenones (Chalcones)*   339

42   *Preparation of an α,β-Unsaturated Ketone via Michael and Aldol Condensation Reactions*   344

43   *Enamine Reactions: 2-Acetylcyclohexanone*   348

44   *1,4-Diphenyl-1,3-butadiene*   359
    44A   *Benzyltriphenylphosphonium Chloride (Wittig Salt)*   362
    44B   *Preparation of 1,4-Diphenyl-1,3-Butadiene Using Sodium Ethoxide*
        to Generate the Ylide   362
    44C   *Preparation of 1,4-Diphenyl-1,3-Butadiene Using Potassium Phosphate*
        to Generate the Ylide   364

Essay   Local Anesthetics   367

45   *Benzocaine*   370

46   *Methyl Salicylate (Oil of Wintergreen)*   373

Essay   Pheromones: Insect Attractants and Repellents   377

47   *N,N-Diethyl-m-Toluamide: The Insect Repellent "OFF"*   383

Essay   Sulfa Drugs   388

48   *Sulfa Drugs: Preparation of Sulfanilamide*   392

Essay   Polymers and Plastics   397

49   *Preparation and Properties of Polymers: Polyester, Nylon, and Polystyrene*   406
    49A   *Polyesters*   407
    49B   *Polyamide (Nylon)*   409
    49C   *Polystyrene*   410
    49D   *Infrared Spectra of Polymer Samples*   412

Essay   Diels–Alder Reaction and Insecticides   414

50   *The Diels–Alder Reaction of Cyclopentadiene with Maleic Anhydride*   419

51   *Photoreduction of Benzophenone and Rearrangement of Benzpinacol to Benzopinacolone*   424
    51A   *Photoreduction of Benzophenone*   425
    51B   *Synthesis of β-Benzopinacolone: The Acid-Catalyzed Rearrangement of Benzpinacol*   431

Essay Fireflies and Photochemistry 433

52 *Luminol* 437

Essay The Chemistry of Sweeteners 441

53 *Analysis of a Diet Soft Drink by HPLC* 444

PART 4

Identification of Organic Substances 447

54 *Identification of Unknowns* 448
    54A *Solubility Tests* 454
    54B *Tests for the Elements (N, S, X)* 460
    54C *Tests for Unsaturation* 466
    54D *Aldehydes and Ketones* 470
    54E *Carboxylic Acids* 477
    54F *Phenols* 479
    54G *Amines* 482
    54H *Alcohols* 486
    54I *Esters* 491

PART 5

Project-Based Experiments 495

55 *A Separation and Purification Scheme* 496

56 *Preparation of a C-4 or C-5 Acetate Ester* 498

57 *Extraction of Essential Oils from Caraway, Cinnamon, Cloves, Cumin, or Fennel by Steam Distillation* 502
    57A *Isolation of Essential Oils by Steam Distillation* 504
    57B *Identification of the Constituents of Essential Oils by Gas Chromatography–Mass Spectrometry* 507
    57C *Investigation of the Essential Oils of Herbs and Spices—A Mini–Research Project* 508

58 *Competing Nucleophiles in S_N1 and S_N2 Reactions: Investigations Using 2-Pentanol and 3-Pentanol* 510

59 *Friedel–Crafts Acylation* 515

60 *The Analysis of Antihistamine Drugs by Gas Chromatography–Mass Spectrometry* 522

61 *The Use of Organozinc Reagents in Synthesis: An Exercise in Synthesis and Structure Proof by Spectroscopy* 524

62 *The Aldehyde Enigma* 527

63 *Synthesis of Substituted Chalcones: A Guided-Inquiry Experience* 529

64 *Michael and Aldol Condensation Reactions* 533

65 *Esterification Reactions of Vanillin: The Use of NMR to Solve a Structure Proof Problem* 536

66 *An Oxidation Puzzle* 538

PART 6

The Techniques   541

1 *Laboratory Safety*   542

2 *The Laboratory Notebook, Calculations, and Laboratory Records*   558

3 *Laboratory Glassware: Care and Cleaning*   566

4 *How to Find Data for Compounds: Handbooks and Catalogs*   574

5 *Measurement of Volume and Weight*   581

6 *Heating and Cooling Methods*   589

7 *Reaction Methods*   598

8 *Filtration*   616

9 *Physical Constants of Solids: The Melting Point*   627

10 *Solubility*   637

11 *Crystallization: Purification of Solids*   647

12 *Extractions, Separations, and Drying Agents*   669

13 *Physical Constants of Liquids: The Boiling Point and Density*   694

14 *Simple Distillation*   703

15 *Fractional Distillation, Azeotropes*   715

16 *Vacuum Distillation, Manometers*   732

17 *Sublimation*   745

18 *Steam Distillation*   750

19 *Column Chromatography*   756

20 *Thin-Layer Chromatography*   777

21 *High-Performance Liquid Chromatography (HPLC)*   792

22 *Gas Chromatography*   797

23 *Polarimetry*   818

24 *Refractometry*   827

25 *Infrared Spectroscopy*   833

26 *Nuclear Magnetic Resonance Spectroscopy (Proton NMR)*   868

27 *Carbon-13 Nuclear Magnetic Resonance Spectroscopy*   906

28 *Mass Spectrometry*   924

29 *Guide to the Chemical Literature*   942

APPENDICES   **957**

1 *Tables of Unknowns and Derivatives*   958

2 *Procedures for Preparing Derivatives*   971

3 *Index of Spectra*   976

INDEX   **979**

PART 1

Introduction to Basic Laboratory Techniques

1   **EXPERIMENT 1**

Introduction to Microscale Laboratory

This textbook discusses the important laboratory techniques of organic chemistry and illustrates many important reactions and concepts. In the traditional approach to teaching this subject, the quantities of chemicals used were on the order of 5–100 grams, and glassware was designed to contain up to 500 mL of liquid. This scale of experiment we might call a **macroscale** experiment. The approach used here, a **microscale** approach, differs from the traditional laboratory course in that nearly all the experiments use small amounts of chemicals. Quantities of chemicals used range from about 50 to 1000 *milligrams* (0.050–1.000 g), and glassware is designed to contain less than 25 mL of liquid. The advantages include improved safety in the laboratory, reduced risk of fire and explosion, and reduced exposure to hazardous vapors. This approach decreases the need for hazardous waste disposal, leading to reduced contamination of the environment. You will learn to work with the same level of care and neatness that has previously been confined to courses in analytical chemistry.

This experiment introduces the equipment and shows how to construct some of the apparatus needed to carry out further experiments. Detailed discussion of how to assemble apparatus and how to practice the techniques is found in Part Six ("The Techniques") of this textbook. This experiment provides only a brief introduction, sufficient to allow you to begin working. You will need to read the techniques chapters for more complete discussions.

Microscale organic experiments require you to develop careful laboratory techniques and to become familiar with apparatus that is somewhat unusual, compared with traditional glassware. We strongly recommend that each student do Laboratory Exercises 1 and 2. These exercises will acquaint you with the most basic microscale techniques. To provide a strong foundation, we further recommend that each student complete most of Experiments 2 through 17 in Part One of this textbook before attempting any other experiments in the textbook.

READ Technique 1 "Laboratory Safety," pp. 542–558.

HEATING METHODS

Aluminum Block

The most convenient means of heating chemical reactions on a small scale is to use an **aluminum block.** An aluminum block consists of a square of aluminum that has holes drilled into it. The holes are sized to correspond to the diameters of the most common vials and flasks that are likely to be heated. Often there is also a hole intended to accept the bulb of a thermometer, so that the temperature of the block can be monitored. However, this practice is not recommended. The aluminum block is heated by placing it on a hot plate. An aluminum block is shown in Figure 1. Note that the thermometer in this figure is not used to monitor the temperature of the block.

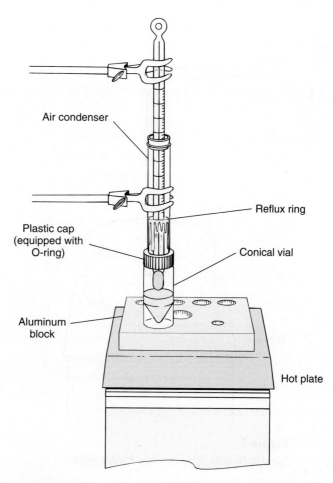

Figure 1
Aluminum block with hot plate and reflux apparatus.

> **CAUTION**
>
> **You should not use a *mercury* thermometer in direct contact with an aluminum block. If it breaks, the mercury will vaporize on the hot surface. Instead, use a nonmercury thermometer, a metal dial thermometer, or a digital electronic temperature-measuring device. See page 590.**

It is recommended that an equipment kit contain two aluminum blocks, one drilled with small holes and able to accept the conical vials found in the glassware kit and another drilled with larger holes and able to accept small round-bottom flasks. The aluminum blocks can be made from inexpensive materials in a small mechanical shop, or they can be purchased from a glassware supplier.

Sand Baths

Another commonly used means of heating chemical reactions on a small scale is to use a **sand bath.** The sand bath consists of a Petri dish or a small crystallizing dish that has been filled to a depth of about 1 cm with sand. The sand bath is also heated by placing it on a hot plate. The temperature of the sand bath may be monitored by clamping a thermometer in position so that the bulb of the thermometer is buried in the sand. A sand bath, with thermometer, is shown in Figure 2.

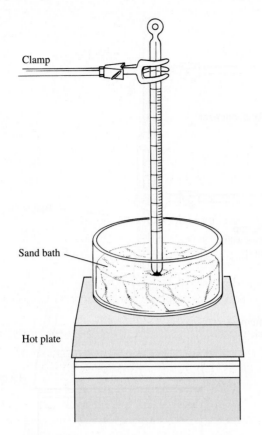

Clamp

Sand bath

Hot plate

Figure 2
Sand bath with hot plate and thermometer.

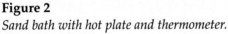

We recommend that an aluminum block, rather than a sand bath, be used as a heating source whenever possible. The aluminum block can be heated and cooled quickly, it is indestructible, and there are no problems with spillage of sand.

Water Bath

When precise control at lower temperatures (below about 80°C) is desired, a suitable alternative is to prepare a **water bath.** The water bath consists of a beaker filled to the required depth with water. The hot plate is used to heat the water bath to the desired temperature. The water in the water bath can evaporate during heating. It is useful to cover the top of the beaker with aluminum foil to diminish this problem.

CONICAL REACTION VIALS

One of the most versatile pieces of glassware contained in the microscale organic glassware kit is the **conical reaction vial.** This vial is used as a vessel in which organic reactions are performed. It may serve as a storage container. It is also used for extractions (see Technique 12). A reaction vial is shown in Figure 3.

The flat base of the vial allows it to stand upright on the laboratory bench. The interior of the vial tapers to a narrow bottom. This shape makes it possible to withdraw liquids completely from the vial, using a disposable Pasteur pipet. The vial has a screw cap, which tightens by means of threads

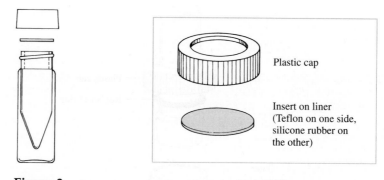

Plastic cap

Insert on liner
(Teflon on one side,
silicone rubber on
the other)

Figure 3
*A conical reaction vial. (The inset shows an expanded view of the cap
with its Teflon insert.)*

cast into the top of the vial. The top also has a ground-glass inner surface. This ground-glass joint allows you to assemble components of glassware tightly.

The plastic cap that fits the top of the conical vial has a hole in the top. This hole is large enough to permit the cap to fit over the inner joints of other components of the glassware kit (see Fig. 4). A Teflon insert, or liner, fits inside the cap to cover the hole when the cap is used to seal a vial tightly. Notice that only one side of the liner is coated with Teflon; the other side is coated with a silicone rubber. The Teflon side generally is the harder side of the insert, and it will feel more slippery. The Teflon side should always face toward the inside of the vial. An O-ring fits inside the cap when the cap is used to fasten pieces of glassware together. The cap and its Teflon insert are shown in the expanded view in Figure 3.

NOTE: Do not use the O-ring when the cap is used to seal the vial.

You can assemble the components of the glassware kit into one unit that holds together firmly and clamps easily to a ring stand. Slip the cap from the conical vial over the inner (male) joint of the upper piece of glassware and fit a rubber O-ring over the inner joint. Then assemble the apparatus by fitting the inner ground-glass joint into the outer (female) joint of the reaction vial and tighten the screw cap to attach the entire apparatus firmly together. The assembly is illustrated in Figure 4.

The walls of the conical vials are made of thick glass. Heat does not transfer through these walls very quickly. This means that if the vial is subjected to rapid changes in temperature, strain building up within the glass walls of the vial may cause the glass to crack. For this reason, do not attempt to cool these vials quickly by running cold water on them. It is safer to allow them to cool naturally by allowing them to stand.

Although the conical vials have flat bottoms intended to allow them to stand up on the laboratory bench, this does not prevent them from falling over.

NOTE: It is good practice to store the vials standing upright inside small beakers.

The vials are somewhat top-heavy, and it is easy to upset them. The beaker will prevent the vial from falling over onto its side.

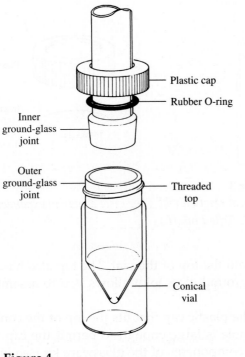

Figure 4
Assembling glassware components.

MEASUREMENT OF SOLIDS

Weighing substances to the nearest milligram requires that the weighings be done on a sensitive **top-loading balance** or an **analytical balance.**

NOTE: You must not weigh chemicals directly on balance pans.

Many chemicals can react with the metal surface of the balance pan and thus ruin it. All weighings must be made into a container that has been weighed previously (**tared**). This tare weight is subtracted from the total weight of container plus sample to give the weight of the sample. Some balances have a built-in compensating feature, the tare button, that allows you to subtract the tare weight of the container automatically, thus giving the weight of the sample directly. A top-loading and an analytical balance are shown in Figure 5.

Balances of this type are quite sensitive and expensive. Take care not to spill chemicals on the balance. It is also important to make certain that any spilled materials are cleaned up immediately.

MEASUREMENT OF LIQUIDS

In microscale experiments, liquid samples are measured using a pipet. When small quantities are used, graduated cylinders do not provide the accuracy needed to give good results. There are two common methods of delivering known amounts of liquid samples, **automatic pipets** and **graduated pipets.** When accurate quantities of liquid reagents are required, the best technique

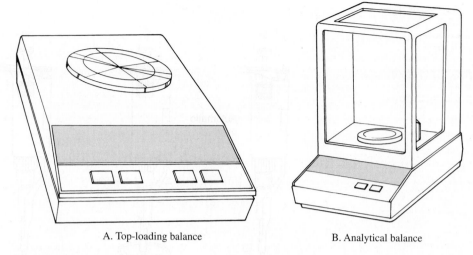

A. Top-loading balance   B. Analytical balance

Figure 5
Laboratory balances.

is to deliver the desired amount of liquid reagent from the pipet into a container whose tare weight has been determined previously. The container, with sample, is then weighed a second time in order to obtain a precise value of the amount of reagent.

Automatic Pipets

Automatic pipets may vary in design, according to the manufacturer. The following description, however, should apply to most models. The automatic pipet consists of a handle that contains a spring-loaded plunger and a micrometer dial. The dial controls the travel of the plunger and is the means used to select the amount of liquid that the pipet is intended to dispense. Automatic pipets are designed to deliver liquids within a particular range of volumes. For example, a pipet may be designed to cover the range from 10 to 100 μL (0.010 to 0.100 mL) or from 100 to 1000 μL (0.100 to 1.000 mL).

Automatic pipets must never be dipped directly into the liquid sample without a plastic tip. The pipet is designed so that the liquid is drawn only into the tip. The liquids are never allowed to come in contact with the internal parts of the pipet. The plunger has two **detent**, or "stop," positions used to control the filling and dispensing steps. Most automatic pipets have a stiffer spring that controls the movement of the plunger from the first to the second detent position. You will find a greater resistance as you press the plunger past the first detent.

To use the automatic pipet, follow the steps as outlined here. These steps are also illustrated in Figure 6.

1. Select the desired volume by adjusting the micrometer control on the pipet handle.

2. Place a plastic tip on the pipet. Be certain that that tip is attached securely.

3. Push the plunger down to the first detent position. Do not press the plunger to the second position. If you press the plunger to the second detent, an incorrect volume of liquid will be delivered.

4. Dip the tip of the pipet into the liquid sample. Do not immerse the entire length of the plastic tip in the liquid. It is best to dip the tip only to a depth of about 1 cm.

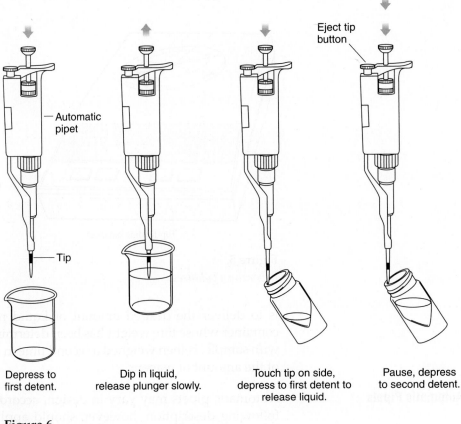

Figure 6
Use of an automatic pipet.

5. Release the plunger *slowly*. Do not allow the plunger to snap back, or liquid may splash up into the plunger mechanism and ruin the pipet. Furthermore, rapid release of the plunger may cause air bubbles to be drawn into the pipet. At this point, the pipet has been filled.

6. Move the pipet to the receiving vessel. Touch the tip of the pipet to an interior wall of the container.

7. Slowly push the plunger down to the first detent. This action dispenses the liquid into the container.

8. Pause 1–2 seconds and then depress the plunger to its second detent position to expel the last drop of liquid. The action of the plunger may be stiffer in this range than it was up to the first detent.

9. Withdraw the pipet from the receiver. If the pipet is to be used with a different liquid, remove the pipet tip and discard it.

Automatic pipets are designed to deliver aqueous solutions with an accuracy of within a few percentage points. The amount of liquid actually dispensed varies, however, depending on the viscosity, surface tension, and vapor pressure of the liquid. The typical automatic pipet is very accurate with aqueous solutions but is not always as accurate with other liquids.

Dispensing Pumps

Some scientific supply catalogs offer a series of dispensing pumps. These pumps are useful in a microscale organic laboratory because they are simple to operate, easy to clean, chemically inert, and quite accurate. The interior

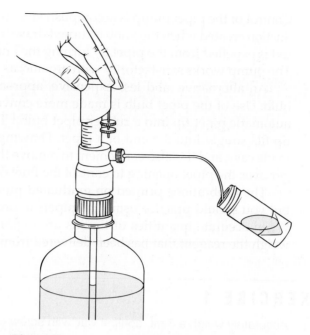

Figure 7
Use of a dispensing pump.

parts of dispensing pumps are made of Teflon, which renders them inert to most organic solvents and reagents. A dispensing pump is illustrated in Figure 7.

The first step in using a dispensing pump is to adjust the pump so that it dispenses the desired volume of liquid. Normally, the instructor will make this adjustment. Once the pump is adjusted correctly, it is a simple matter to dispense a liquid. Simply lift the head of the pump as far as it will travel. When you release the head, it will fall, and the liquid will issue from the spout. With viscous liquids, the head of the pump may not fall by itself. In such an instance, gently guide the head downward. After the liquid has been dispensed, you should touch the tip of the dispensing tube to an interior wall of the container in order to remove the last drop of liquid.

As with automatic pipets, dispensing pumps are designed to deliver aqueous solutions with an accuracy of within a few percentage points. The amount of liquid actually dispensed will vary, however, depending on the viscosity, surface tension, and vapor pressure of the liquid. You should always weigh the liquid to determine the amount accurately.

Graduated Pipets

A less-expensive means of delivering known quantities of liquid is to use a graduated pipet. Graduated pipets should be familiar to those of you who have taken general chemistry or quantitative analysis courses. Because they are made of glass, they are inert to most organic solvents and reagents. Disposable serological pipets may be an attractive alternative to standard graduated pipets. The 2-mL size of disposable pipet represents a convenient size for the organic laboratory.

Never draw liquids into the pipets using mouth suction. A pipet bulb or a pipet pump, not a rubber dropper bulb, must be used to fill pipets. We recommend the use of a pipet pump. A pipet fits snugly into the pipet pump, and the pump can be controlled to deliver precise volumes of liquids.

Control of the pipet pump is accomplished by rotating a knob on the pump. Suction created when the knob is turned draws the liquid into the pipet. Liquid is expelled from the pipet by turning the knob in the opposite direction. The pump works satisfactorily with organic, as well as aqueous, solutions.

An alternative, and less expensive, approach is to use a rubber pipet bulb. Use of the pipet bulb is made more convenient by inserting a plastic automatic pipet tip into a rubber pipet bulb.[1] The tapered end of the pipet tip fits snugly into the end of a pipet. Drawing the liquid into the pipet is made easy, and it is also convenient to remove the pipet bulb and place a finger over the pipet opening to control the flow of liquid.

The calibrations printed on graduated pipets are reasonably accurate, but you should practice using the pipets in order to achieve this accuracy. When accurate quantities of liquids are required, the best technique is to weigh the reagent that has been delivered from the pipet.

LABORATORY EXERCISE 1

Option A, Automatic Pipet

Accurately weigh a 3-mL conical vial, with screw cap and Teflon insert, on a balance. Determine its weight to the nearest milligram (nearest 0.001 g). Using the automatic pipet, dispense 0.500 mL of water into the vial, replace the cap assembly (with the insert arranged Teflon side down), and weigh the vial a second time. Determine the weight of water dispensed. Calculate the density of water from your results. Repeat the experiment using 0.500 mL of hexane. Dispose of any excess hexane in a designated waste container. Calculate the density of hexane from your data. Record the results in your notebook, along with your comments on any deviations from literature values that you may have noticed. At room temperature, the density of water is 0.997 g/mL, and the density of hexane is 0.660 g/mL.

Option B, Dispensing Pump

Accurately weigh a 3-mL conical vial, with screw cap and Teflon insert, on a balance. Determine its weight to the nearest milligram (nearest 0.001 g). Using a dispensing pump that has been adjusted to deliver 0.500 mL, dispense 0.500 mL of water into the vial, replace the cap assembly, and weigh the vial a second time. Determine the weight of water dispensed. Calculate the density of water from your results. Repeat the experiment using 0.500 mL of hexane. Dispose of any excess hexane in a designated waste container. Calculate the density of hexane from your data. Record the results in your notebook, along with your comments on any deviations from literature values that you may have noticed. See Option A for the density of water and of hexane.

Option C, Graduated Pipet

Accurately weigh a 3-mL conical vial, with screw cap and Teflon insert, on a balance. Determine its weight to the nearest milligram (nearest 0.001 g). Using a 1.0-mL graduated pipet, dispense 0.50 mL of water into the vial, replace the cap assembly, and weigh the vial a second time. Determine the weight of water dispensed. Calculate the density of water from your results. Repeat the experiment using 0.50 mL of hexane. Dispose of any excess hexane in a designated waste container. Calculate the density of hexane from your data. Record the results in your notebook along with your comments on any deviations from literature values that you may have noticed. See Option A for the density of water and of hexane.

---

[1] This technique was described in Deckey, G. "A Versatile and Inexpensive Pipet Bulb." *Journal of Chemical Education*, 57 (July 1980): 526.

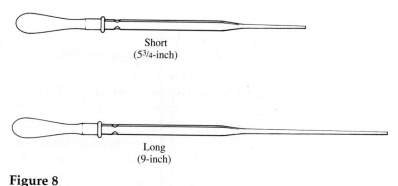

Short
(5¾-inch)

Long
(9-inch)

Figure 8
Disposable Pasteur pipets.

Disposable (Pasteur) Pipets

A convenient way of dispensing liquids when a great deal of accuracy is not required is to use a disposable pipet, or Pasteur pipet. Two sizes of Pasteur pipets are shown in Figure 8. Even though accurate calibration may not be required when these pipets are used, it is nevertheless handy to have some idea of the volume contained in the pipet. A crude calibration is, therefore, recommended.

LABORATORY EXERCISE 2

Pipet Calibration

On a balance, weigh 0.5 grams (0.5 mL) of water into a 3-mL conical vial. Select a short (5¾-inch) Pasteur pipet and attach a rubber bulb. Squeeze the rubber bulb before inserting the tip of the pipet into the water. Try to control how much you depress the bulb, so that when the pipet is placed into the water and the bulb is completely released, only the desired amount of liquid is drawn into the pipet. (This skill may take some time to acquire, but it will facilitate your use of a Pasteur pipet.) When the water has been drawn up, place a mark with an indelible marking pen at the position of the meniscus. A more durable mark can be made by scoring the pipet with a file. Repeat this procedure with 1.0 gram of water, and make a 1-mL mark on the same pipet.

Additional Pasteur pipets can be calibrated easily by holding them next to the pipet calibrated in Laboratory Exercise 2 and scoring a new mark on each pipet at the same level as the mark placed on the calibrated pipet. We recommend that several Pasteur pipets be calibrated at one time for use in future experiments.

Extraction

A technique frequently applied in purifying organic reaction products is **extraction.** In this method, a solution is mixed thoroughly with a second solvent. The second solvent is not miscible with the first solvent. When the two solvents are mixed, the dissolved substances (solutes) distribute themselves between the two solvents until an equilibrium is established. When the mixing is stopped, the two immiscible solvents separate into two distinct layers. The solutes are distributed between the two solvents so that each solute is found in greater concentration in that solvent in which it is more soluble. Separation of the two immiscible solvent layers thus becomes a means of separating solutes from one another based on their relative solubilities in the two solvents.

In a common application, an aqueous solution may contain both inorganic and organic products. An organic solvent that is immiscible with water is added, and the mixture is shaken thoroughly. When the two solvent layers are allowed to form again, on standing, the organic solutes are

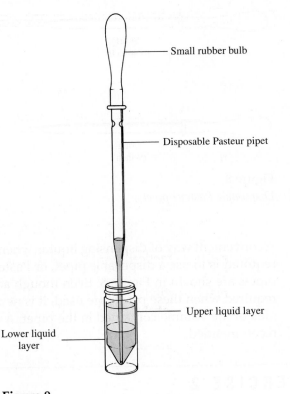

— Small rubber bulb

— Disposable Pasteur pipet

— Upper liquid layer

Lower liquid
layer

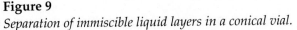

Figure 9
Separation of immiscible liquid layers in a conical vial.

transferred to the organic solvent while the inorganic solutes remain in the aqueous layer. When the two layers separate, the organic and inorganic products also separate from one another. The separation, as described here, may not be complete. The inorganic materials may be somewhat soluble in the organic solvent, and the organic products may retain some water solubility. Nevertheless, reasonably complete separations of reaction products can be achieved by the extraction method.

For microscale experiments, the conical reaction vial is the glassware item used for extractions. Place the two immiscible liquid layers in the vial, and seal the top with a screw cap and a Teflon insert (Teflon side toward the inside of the vial). Shake the vial to provide thorough mixing between the two liquid phases. As the shaking continues, vent the vial periodically by loosening the cap and then tightening it again. After about 5 or 10 seconds of shaking, loosen the cap to vent the vial, retighten it, and allow the vial to stand upright in a beaker until the two liquid layers separate completely.

The two liquid layers are separated by withdrawing the *lower* layer using a disposable Pasteur pipet. This separation technique is illustrated in Figure 9. Take care not to disturb the liquid layers by allowing bubbles to issue from the pipet. Squeeze the pipet bulb to the required amount before introducing the pipet into the vial. Also take care not to allow any of the upper liquid layer to enter the pipet. The pointed shape of the interior of the conical vial makes it easy to remove all the lower layer without allowing it be contaminated by some of the upper liquid layer. More precise control in the separation can be achieved by using a filter-tip pipet (see Technique 8, Section 8.6, p. 625).

Other Useful Techniques   The practice of organic chemistry requires you to master many more techniques than the ones described in this experiment. Those techniques included here are only the most elementary ones, those needed to get you

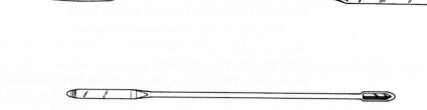

Figure 10
Microspatulas.

started in the laboratory. Additional techniques are described fully in Part Six of this textbook, and Experiments 2 through 17 expose you to the most important of them.

Some other practical hints need to be introduced at this point. The first of these involves manipulating small amounts of solid substances. The efficient transfer of solids requires a small spatula. We recommend that you have two **microspatulas,** similar to those shown in Figure 10, as part of your standard desk stock. The design of these spatulas permits the handling of milligram quantities of substances without undue spillage or waste. The larger style (see Fig. 10) is more useful when relatively large quantities of solid must be dispensed.

A clean work area is of utmost importance when working in the laboratory. The need for cleanliness is particularly great when working with the small amounts of materials characteristic of microscale laboratory experiments.

NOTE: You must read Technique 1 "Laboratory Safety." In preventing accidents, there is no substitute for preparation and care.

With this final word of caution and advice, we hope you enjoy the learning experience you are about to begin. Learning the care and precision that microscale experiments require may seem difficult at first, but before long you will be comfortable with the scale of the experiments. You will develop much better laboratory technique as a result of microscale practice, and this added skill will serve you well.

EXPERIMENT 2

Solubility

Solubility

Polarity

Acid-base chemistry

Critical thinking application

Having a good comprehension of solubility behavior is essential for understanding many procedures and techniques in the organic chemistry laboratory. For a thorough discussion of solubility, read the chapter on this concept

(Technique 10, p. 637) before proceeding because an understanding of this material is assumed in this experiment.

In Parts A and B of this experiment, you will investigate the solubility of various substances in different solvents. As you are performing these tests, it is helpful to pay attention to the polarities of the solutes and solvents and to even make predictions based on this (see "Guidelines for Predicting Polarity and Solubility," p. 639). The goal of Part C is similar to that of Parts A and B, except that you will be looking at miscible and immiscible pairs of liquids. In Part D, you will investigate the solubility of organic acids and bases. Section 10.2B on page 642 will help you understand and explain these results.

REQUIRED READING

New: Technique 5   Measurement of Volume and Weight

Technique 10   Solubility

SUGGESTED WASTE DISPOSAL

Dispose of all wastes containing methylene chloride into the container marked for halogenated waste. Place all other organic wastes into the non-halogenated organic waste container.

NOTES TO THE INSTRUCTOR

In Part A of the procedure, it is important that students follow the instructions carefully. Otherwise, the results may be difficult to interpret. It is particularly important that consistent stirring be done for each solubility test. This can be done most easily by using the larger-style microspatula shown in Figure 10 on page 13.

We have found that some students have difficulty performing Critical Thinking Application 2 (p. 17) on the same day that they complete the rest of this experiment. Many students need time to assimilate the material in this experiment before they can complete this exercise successfully. One approach is to assign Critical Thinking Applications from several technique experiments (for example, Experiments 1–3) to a laboratory period after students complete the individual technique experiments. This provides an effective way of reviewing some of the basic techniques.

PROCEDURE

NOTE: It is very important that you follow these instructions carefully and that consistent stirring be done for each solubility test.

Part A. Solubility of Solid Compounds

Place about 40 mg (0.040 g) of benzophenone into each of four *dry* test tubes.[1] (Don't try to be exact: You can be 1–2 mg off and the experiment will still work.) Label the test tubes and then add 1 mL of water to the first tube, 1 mL of methyl

---

[1] *Note to the instructor:* Grind up the benzophenone flakes into a powder.

alcohol to the second tube, and 1 mL of hexane to the third tube. The fourth tube will serve as a control. Determine the solubility of each sample in the following way: Using the rounded end of a microspatula (the larger style on p. 13), stir each sample continuously for 60 seconds by twirling the spatula rapidly. If a solid dissolves completely, note how long it takes for the solid to dissolve. *After 60 seconds* (do not stir longer), note whether the compound is soluble (dissolves completely), insoluble (none of it dissolves), or partially soluble. You should compare each tube with the control in making these determinations. You should state that a sample is partially soluble only if a significant amount (at least 50%) of the solid has dissolved. If it is not clear that a significant amount of solid has dissolved, then state that the sample is insoluble. If all but a couple of granules have dissolved, state that the sample is soluble. An additional hint for determining partial solubility is given in the next paragraph. Record these results in your notebook in the form of a table, as shown on this page. For those substances that dissolve completely, note how long it took for the solid to dissolve.

Although the instructions just given should enable you to determine whether a substance is partially soluble, you may use the following procedure to confirm this. Using a Pasteur pipet, carefully remove most of the solvent from the test tube *while leaving the solid behind.* Transfer the liquid to another test tube and then evaporate the solvent by heating the tube in a hot water bath. Directing a stream of air or nitrogen gas into the tube will speed up the evaporation (see Technique 7, Section 7.10, p. 611). When the solvent has completely evaporated, examine the test tube for any remaining solid. If there is solid in the test tube, the compound is partially soluble. If there is no, or very little, solid remaining, you can assume that the compound is insoluble.

Now repeat the directions just given, substituting malonic acid and biphenyl for benzophenone. Record these results in your notebook.

Solvents

Organic Compounds	Water (highly polar)	Methyl Alcohol (intermediate polarity)	Hexane (nonpolar)
Benzophenone			
Malonic acid			
Biphenyl			

Part B. Solubility of Different Alcohols

For each solubility test (see table on p. 16), add 1 mL of solvent (water or hexane) to a test tube. Then add one of the alcohols, *dropwise*. Carefully observe what happens as you add each drop. If the liquid solute is soluble in the solvent, you may see tiny horizontal lines in the solvent. These mixing lines indicate that solution is taking place. *Shake the tube after adding each drop*. While you shake the tube, the liquid that was added may break up into small balls that disappear in a few seconds. This also indicates that solution is taking place. Continue adding the alcohol with shaking until you have added a total of 20 drops. If an alcohol is partially soluble, you will observe that at first the drops will dissolve, but eventually a second layer of liquid (undissolved alcohol) will form in the test tube. Record your results (soluble, insoluble, or partially soluble) in your notebook in table form.

Solvents

Alcohols	Water	Hexane
1-Octanol $CH_3(CH_2)_6CH_2OH$		
1-Butanol $CH_3CH_2CH_2CH_2OH$		
Methyl alcohol CH_3OH		

Part C. Miscible or Immiscible Pairs

For each of the following pairs of compounds, add 1 mL of each liquid to the same test tube. Use a different test tube for each pair. Shake the test tube for 10–20 seconds to determine whether the two liquids are miscible (form one layer) or immiscible (form two layers). Record your results in your notebook.

Water and ethyl alcohol

Water and diethyl ether

Water and methylene chloride

Water and hexane

Hexane and methylene chloride

Part D. Solubility of Organic Acids and Bases

Place about 30 mg (0.030 g) of benzoic acid into each of three *dry* test tubes. Label the test tubes and then add 1 mL of water to the first tube, 1 mL of 1.0 *M* NaOH to the second tube, and 1 mL of 1.0 *M* HCl to the third tube. Stir the mixture in each test tube with a microspatula for 10–20 seconds. Note whether the compound is soluble (dissolves completely) or is insoluble (none of it dissolves). Record these results in table form. Now take the tube containing benzoic acid and 1.0 *M* NaOH. With stirring add 6 *M* HCl dropwise until the mixture is acidic. Test the mixture with litmus or pH paper to determine when it is acidic.[2] When it is acidic, stir the mixture for 10–20 seconds and note the result (soluble or insoluble) in the table.

Repeat this experiment using ethyl 4-aminobenzoate and the same three solvents. Record the results. Now take the tube containing ethyl 4-aminobenzoate and 1.0 *M* HCl. With stirring, add 6 *M* NaOH dropwise until the mixture is basic. Test the mixture with litmus or pH paper to determine when it is basic. Stir the mixture for 10–20 seconds and note the result.

---

[2] Do not place the litmus or pH paper into the sample; the dye will dissolve. Instead, place a drop of solution from your spatula onto the test paper. With this method, several tests can be performed using a single strip of paper.

Solvents

Compounds	Water	1.0 M NaOH	1.0 M HCl
Benzoic acid O ‖ ⬡—C—OH		Add 6.0 *M* HCl	
Ethyl 4-aminobenzoate O ‖ H₂N—⬡—C—OCH₂CH₃			Add 6.0 *M* NaOH

Part E. Critical Thinking Applications

1. Determine by experiment whether each of the following pairs of liquids is miscible or immiscible.

   Acetone and water

   Acetone and hexane

   How can you explain these results, given that water and hexane are immiscible?

2. You will be given a test tube containing two immiscible liquids and a solid organic compound that is dissolved in one of the liquids.[3] You will be told the identity of the two liquids and the solid compound, but you will not know the relative positions of the two liquids or in which liquid the solid is dissolved. Consider the following example, in which the liquids are water and hexane and the solid compound is biphenyl.

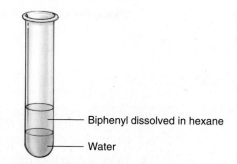

— Biphenyl dissolved in hexane

— Water

   a. Without doing any experimental work, predict where each liquid is (top or bottom) and in which liquid the solid is dissolved. Justify your prediction. You may want to consult a handbook such as *The Merck Index* or the *CRC Handbook of Chemistry and Physics* to determine the molecular structure of a compound or to find any other relevant information. Note that dilute solutions such as 1 *M* HCl are composed mainly of water.

---

[3] The sample you are given may contain one of the following combinations of solid and liquids (the solid is listed first): fluorene, methylene chloride, water; triphenylmethanol, diethyl ether, water; salicylic acid, methylene chloride, 1 *M* NaOH; ethyl 4-aminobenzoate, diethyl ether, 1 *M* HCl; naphthalene, hexane, water; benzoic acid, diethyl ether, 1 *M* NaOH; *p*-aminoacetophenone, methylene chloride, 1 *M* HCl.

b. Now try to prove your prediction experimentally. That is, demonstrate which liquid the solid compound is dissolved in and the relative positions of the two liquids. You may use any experimental technique discussed in this experiment or any other technique that your instructor will let you try. In order to perform this part of the experiment, it may be helpful to separate the two layers in the test tube. This can be done easily and effectively with a Pasteur pipet. Squeeze the bulb on the Pasteur pipet and then place the tip of the pipet on the bottom of the test tube. Now withdraw only the bottom layer and transfer it to another test tube. Note that evaporating the water from an aqueous sample takes a very long time; therefore, this may not be a good way to show that an aqueous solution contains a dissolved compound. However, other solvents may be evaporated more easily (see p. 643). Explain what you did and whether or not the results of your experimental work were consistent with your prediction.

3. Add 0.025 g of tetraphenylcyclopentadienone to a dry test tube. Add 1 mL of methyl alcohol to the tube and shake for 60 seconds. Is the solid soluble, partially soluble, or insoluble? Explain your answer.

REPORT

Part A

1. Summarize your results in table form.

2. Explain the results for all the tests done. In explaining these results, you should consider the polarities of the compound and the solvent and the potential for hydrogen bonding. For example, consider a similar solubility test for *p*-dichlorobenzene in hexane. The test indicates that *p*-dichlorobenzene is soluble in hexane. This result can be explained by stating that hexane is nonpolar, whereas *p*-dichlorobenzene is slightly polar. Because the polarities of the solvent and solute are similar, the solid is soluble. (Remember that the presence of a halogen does not significantly increase the polarity of a compound.)

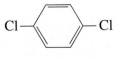

p-Dichlorobenzene

3. There should be a difference in your results between the solubilities of biphenyl and benzophenone in methyl alcohol. Explain this difference.

4. There should be a difference in your results between the solubilities of benzophenone in methyl alcohol and benzophenone in hexane. Explain this difference.

Part B

1. Summarize your results in table form.

2. Explain the results for the tests done in water. In explaining these results, you should consider the polarities of the alcohols and water.

3. Explain, in terms of polarities, the results for the tests done in hexane.

Part C

1. Summarize your results in table form.

2. Explain the results in terms of polarities and/or hydrogen bonding.

Part D

1. Summarize your results in table form.

2. Explain the results for the tube in which 1.0 *M* NaOH was added to benzoic acid. Write an equation for this. Now describe what happened when 6.0 *M* HCl was added to this same tube and explain this result.

3. Explain the results for the tube in which 1.0 *M* HCl was added to ethyl 4-aminobenzoate. Write an equation for this. Now describe what happened when 6.0 *M* NaOH was added to this same tube and explain.

Part E Give the results for any Critical Thinking Applications completed and answer all questions given in the Procedure for these exercises.

QUESTIONS

1. For each of the following pairs of solutes and solvent, predict whether the solute would be soluble or insoluble. After making your predictions, you can check your answers by looking up the compounds in *The Merck Index* or the *CRC Handbook of Chemistry and Physics*. Generally, *The Merck Index* is the easier reference book to use. If the substance has a solubility greater than 40 mg/mL, you may conclude that it is soluble.

a. Malic acid in water

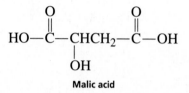

Malic acid

b. Naphthalene in water

Naphthalene

c. Amphetamine in ethyl alcohol

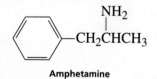

Amphetamine

d. Aspirin in water

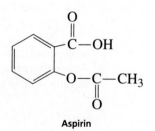

Aspirin

e. Succinic acid in hexane (*Note:* The polarity of hexane is similar to petroleum ether.)

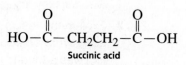

Succinic acid

f. Ibuprofen in diethyl ether

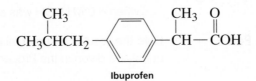

Ibuprofen

g. 1-Decanol (*n*-decyl alcohol) in water

$$CH_3(CH_2)_8CH_2OH$$
1-Decanol

2. Predict whether the following pairs of liquids would be miscible or immiscible:
   a. Water and methyl alcohol
   b. Hexane and benzene
   c. Methylene chloride and benzene
   d. Water and toluene

Toluene

   e. Ethyl alcohol and isopropyl alcohol

Isopropyl alcohol

3. Would you expect ibuprofen (see 1f) to be soluble or insoluble in 1.0 *M* NaOH? Explain.

4. Thymol is very slightly soluble in water and very soluble in 1.0 *M* NaOH. Explain.

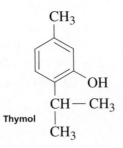

Thymol

5. Although cannabinol and methyl alcohol are both alcohols, cannabinol is very slightly soluble in methyl alcohol at room temperature. Explain.

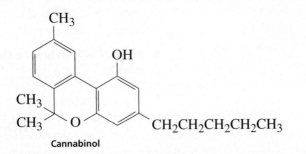

Cannabinol

3 EXPERIMENT 3

Crystallization

Crystallization
Vacuum filtration
Melting point
Finding a crystallization solvent
Mixture melting point
Critical thinking application

The purpose of this experiment is to introduce the technique of crystallization, the most common procedure used to purify crude solids in the organic laboratory. For a thorough discussion of crystallization, read the chapter on this technique (pp. 647–668) before proceeding because an understanding of this material is assumed in this experiment.

In Experiments 3A and 3B, you will carry out a crystallization of impure sulfanilamide using 95% ethyl alcohol as the solvent. The impurity is acetanilide, which is often used as a starting material for synthesizing sulfanilamide (see Experiment 48, p. 392). Sulfanilamide is one of the sulfa drugs, the first generation of antibiotics to be used in successfully treating many major diseases, such as malaria, tuberculosis, and leprosy (see the essay "Sulfa Drugs," p. 388).

In Experiments 3A and 3B, and in most of the experiments in this textbook, you are told what solvent to use for the crystallization procedure. Some of the factors involved in selecting a crystallization solvent for sulfanilamide are discussed in Section 11.5 on page 660. The most important consideration is the shape of the solubility curve for the solubility vs. temperature data. As can be seen in Figure 11.2 on page 648, the solubility curve for sulfanilamide in 95% ethyl alcohol indicates that ethyl alcohol is an ideal solvent for crystallizing sulfanilamide.

The purity of the final material after crystallization will be determined by performing a melting point on your sample. You will also weigh your sample and calculate the percentage recovery. It is impossible to obtain a 100% recovery. This is true for several reasons: There will be some experimental loss, the original sample is not 100% sulfanilamide, and some sulfanilamide is soluble in the solvent even at 0°C. Because of this last fact, some sulfanilamide will remain dissolved in the **mother liquor** (the liquid remaining after crystallization has taken place). Sometimes it is worth isolating a second crop of crystals from the mother liquor, especially if you have performed a synthesis requiring many hours of work and the amount of product is relatively small. This can be accomplished by heating the mother liquor to evaporate some of the solvent and then cooling the resultant solution to induce a second crystallization. The purity of the second crop will not be as good as the first crop, however, because the concentration of the impurities will be greater in the mother liquor after some of the solvent has been evaporated.

Two procedures are given here for crystallizing sulfanilamide: a semimicroscale procedure using an Erlenmeyer flask and a Hirsch funnel (Experiment 3A) and a microscale procedure with a Craig tube (Experiment 3B). Your instructor may assign both or just one of these procedures.

In Experiment 3C you will be given an impure sample of the organic compound fluorene (see structure that follows). You will use an experimental procedure for determining which one of three possible solvents is the most appropriate. The three solvents will illustrate three very different solubility behaviors: One of the solvents will be an appropriate solvent for crystallizing fluorene. In a second solvent, fluorene will be highly soluble, even at room temperature. Fluorene will be relatively insoluble in the third solvent, even at the boiling point of the solvent. Your task will be to find the appropriate solvent for crystallization and then perform a crystallization on this sample.

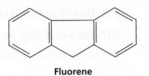

Fluorene

You should be aware that not all crystallizations will look the same. Crystals have many different shapes and sizes, and the amount of mother liquor visible at the end of the crystallization may vary significantly. The crystallizations of sulfanilamide and fluorene will appear significantly different even though the purity of the crystals in each case should be very good.

In Experiment 3D of this experiment, you will determine the identity of an unknown using the melting point technique. The **mixture melting point** technique is introduced in this part.

REQUIRED READING

Review: Technique 10   Solubility

New: Technique 8   Filtration, Sections 8.3 and 8.5
         Technique 9   Physical Constants of Solids: The Melting Point
         Technique 11   Crystallization: Purification of Solids

SUGGESTED WASTE DISPOSAL

Dispose of all organic wastes into the nonhalogenated organic waste container.

EXPERIMENT 3A

Semimicroscale Crystallization— Erlenmeyer Flask and Hirsch Funnel

This experiment assumes a familiarity with the general semimicroscale crystallization procedure (Technique 11, Section 11.3, p. 650). In this experiment, Step 2 in Figure 11.3 (removal of insoluble impurities) will not be required. Although the impure sample may have a slight color, it will also not be necessary to use a decolorizing agent (Section 11.7, p. 663). Leaving out these steps makes the crystallization easier to perform. Furthermore, very few experiments in this textbook require either of these techniques. If a filtration or decolorizing step is ever required, Technique 11 describes these procedures in detail.

Pre-Lab Calculations

1. Calculate how much 95% ethyl alcohol will be required to dissolve 0.3 g of sulfanilamide at 78°C. Use the data from the introduction to this experiment to make this calculation. The reason for making this calculation is so that you will know ahead of time the approximate amount of hot solvent you will be adding.

2. Using the volume of solvent calculated in Step 1, calculate how much sulfanilamide will remain dissolved in the mother liquor after the mixture is cooled to 0°C.

To dissolve the sulfanilamide in the minimum of hot (boiling or almost boiling) solvent, you must keep the mixture at (or near) the boiling point of 95% ethyl alcohol during the entire procedure. You will likely add more solvent than the amount you calculated because some solvent will evaporate. The amount of solvent is calculated only to indicate the approximate amount of solvent required. You should follow the procedure to determine the correct amount of solvent needed.

PROCEDURE

Preparations

Weigh 0.30 g of impure sulfanilamide[1] and transfer this solid to a 10-mL Erlenmeyer flask. To a second Erlenmeyer flask, add about 6 mL of 95% ethyl alcohol and a boiling stone. Heat the solvent on a *warm* hot plate until it is boiling.[2] Because 95% ethyl alcohol boils at a relatively low temperature (78°C), it evaporates rapidly. Setting the temperature of the hot plate too high will result in too much loss of solvent through evaporation.

Dissolving the Sulfanilamide

Before heating the flask containing the sulfanilamide, add enough hot solvent with a Pasteur pipet to barely cover the crystals. Then heat the flask containing the sulfanilamide until the solvent is boiling. At first, this may be difficult to see because so little solvent is present. Add another small portion of solvent (several drops, or about 0.25 mL), continue to heat the flask, and swirl the flask frequently. You may swirl the flask while it is on the hot plate or, for more vigorous swirling, remove it from the hot plate for a few seconds while you swirl it. When you have swirled the flask for 10–15 seconds, check to see if the solid has dissolved. If it has not, add another portion of solvent. Heat the flask again with swirling until the solvent boils. Then swirl the flask for 10–15 seconds, frequently returning the flask to the hot plate so that the temperature of the mixture does not drop. Continue repeating the process of adding solvent, heating, and swirling until all the solid has

---

[1] The impure sulfanilamide contains 5% acetanilide as the impurity.

[2] To prevent bumping in the boiling solvent, you may want to place a Pasteur pipet in the flask. Use a 25-mL flask so that the Pasteur pipet does not tip the flask over. This is a convenient method because a Pasteur pipet will also be used to transfer the solvent.

dissolved completely. Note that it is essential to add just enough solvent to dissolve the solid—neither too much nor too little. Because 95% ethyl alcohol is very volatile, you need to perform this entire procedure fairly rapidly. Otherwise, you may lose solvent nearly as quickly as you are adding it, and this procedure will take a very long time. The time from the first addition of solvent until the solid dissolves completely should not be longer than 10–15 minutes.

Crystallization

Remove the flask from the heat and allow the solution to cool *slowly* (see Section 11.3, Part C, p. 655, for suggestions). Cover the flask with a small watch glass or stopper the flask. Crystallization should begin by the time the flask has cooled to room temperature. If it has not, scratch the inside surface of the flask with a glass rod (not fire-polished) to induce crystallization (Technique 11, Section 11.8, Part A, p. 664). When it appears that no further crystallization is occurring at room temperature, place the flask in an ice-water bath using a beaker (Technique 11, Section 11.8, p. 664). Be sure that both water and ice are present and that the beaker is small enough to prevent the flask from tipping over.

Filtration

When crystallization is complete, vacuum filter the crystals using a Hirsch funnel (see Technique 8, Section 8.3, and Figure 8.5, p. 622). (If you will be performing the Optional Exercise at the end of this procedure, you must save the mother liquor from this filtration procedure. Therefore, the filter flask should be clean and dry.) Moisten the filter paper with a few drops of 95% ethyl alcohol and turn on the vacuum (or aspirator) to the fullest extent. Use a spatula dislodge the crystals from the bottom of the flask before transferring the material to the Hirsch funnel. Swirl the mixture in the flask and pour the mixture into the funnel, attempting to transfer both crystals and solvent. You need to pour the mixture quickly, before the crystals have completely resettled on the bottom of the flask. (You may need to do this in portions, depending on the size of your Hirsch funnel.) When the liquid has passed through the filter, repeat this procedure until you have transferred all the liquid to the Hirsch funnel. At this point, there will usually be some crystals remaining in the flask. Using your spatula, scrape out as many of the crystals as possible from the flask. Add about 1 mL of *ice-cold* 95% ethyl alcohol (measured with a calibrated Pasteur pipet) to the flask. Swirl the liquid in the flask and then pour the remaining crystals and alcohol into the Hirsch funnel. Not only does this additional solvent help transfer the remaining crystals to the funnel but the alcohol also rinses the crystals already on the funnel. This washing step should be done whether or not it is necessary to use the wash solvent for transferring crystals. If necessary, repeat with another 1-mL portion of ice-cold alcohol. Wash the crystals with a total of about 2 mL of ice-cold solvent.

Continue drawing air through the crystals on the Hirsch funnel by suction for about five minutes. Transfer the crystals onto a preweighed watch glass for air-drying. (Save the mother liquor in the filter flask if you will be doing the Optional Exercise.) Separate the crystals as much as possible with a spatula. The crystals should be completely dried within 10–15 minutes. You can usually determine if the crystals are still wet by observing whether or not they stick to a spatula or stay together in a clump. Weigh the dry crystals and calculate the percent recovery. Determine the melting point of the pure sulfanilamide and the original impure material. At the option of the instructor, turn in your crystallized material in a properly labeled container.

Comments on the Crystallization Procedure

1. Do not heat the crude sulfanilamide until you have added some solvent. Otherwise, the solid may melt and possibly form an oil, which may not crystallize easily.

2. When you are dissolving the solid in hot solvent, the solvent should be added in small portions with swirling and heating. The procedure calls for a specific amount (about 0.25 mL), which is appropriate for this experiment. However, the actual amount you should add each time you perform a crystallization may vary, depending on the size of your sample and the nature of the solid and solvent. You will need to make this judgment when you perform this step.

3. One of the most common mistakes is to add too much solvent. This can happen most easily if the solvent is not hot enough or if the mixture is not stirred sufficiently. If too much solvent is added, the percent recovery will be reduced; it is even possible that no crystals will form when the solution is cooled. If too much solvent is added, you must evaporate the excess by heating the mixture. Using a nitrogen or air stream directed into the container will accelerate the evaporation process (see Technique 7, Section 7.10, p. 611).

4. Sulfanilamide should crystallize as large, beautiful needles. However, this will not always happen. If the crystals form too rapidly or if there is not enough solvent, they will tend to be smaller, perhaps even appearing as a powder. Furthermore, many substances crystallize in other characteristic shapes, such as plates or prisms.

5. When the solvent is water or when the crystals form as a powder, it will be necessary to dry the crystals longer than 10–15 minutes. Overnight drying may be necessary, especially with water.

Optional Exercise

Transfer the mother liquor to a tared (preweighed) test tube. Place the test tube in a hot water bath and evaporate all the solvent from the mother liquor. Use a stream of nitrogen or air directed into the test tube to speed up the rate of evaporation (see Technique 7, Section 7.10, p. 611). Cool the test tube to room temperature and dry the outside. Weigh the test tube with solid. Compare this to the weight calculated in the Pre-Lab Calculations. Determine the melting point of this solid and compare it to the melting point of the crystals obtained by crystallization.

3B   EXPERIMENT 3B

Microscale Crystallization—Craig Tube

This experiment assumes familiarity with the general microscale crystallization procedure (Technique 11, Section 11.4, p. 656). In this experiment, Step 2 in Figure 11.6 (removal of insoluble impurities) will not be required. Although the impure sample may have a slight color, it also will not be necessary to use a decolorizing agent (Section 11.7, p. 663). Leaving out these steps makes the crystallization easier to perform. Furthermore, very few experiments in this textbook require either of these techniques. When a filtration or decolorizing step is required, Technique 11 describes these procedures in detail.

Pre-Lab Calculations

1. Calculate how much 95% ethyl alcohol will be required to dissolve 0.1 g of sulfanilamide at 78°C. Use the data from the introduction to this

experiment to make this calculation. Make this calculation so that you will know the approximate amount of hot solvent you will be adding.

2. Using the volume of solvent calculated in Step 1, calculate how much sulfanilamide will remain dissolved in the mother liquor after the mixture is cooled to 0°C.

To dissolve the sulfanilamide in the minimum of hot (boiling or almost boiling) solvent, you must keep the mixture at (or near) the boiling point of 95% ethyl alcohol during the entire procedure. You will likely add more solvent than the amount you calculated because some solvent will evaporate. Use this calculated amount only as a guide: you should follow the procedure to determine the correct amount of solvent needed.

PROCEDURE

Preparations

Weigh 0.10 g of impure sulfanilamide[3] and transfer this solid to a Craig tube. To a small test tube, add 2–3 mL of 95% ethyl alcohol and a boiling stone. Heat the solvent on a *warm (not hot)* hot plate with an aluminum block until the solvent is boiling.[4] Setting the temperature of the hot plate too high will result in too much loss of solvent through evaporation.

CAUTION ⚠

In performing the following procedure, keep in mind that the mixture in the Craig tube may erupt out of the tube if it becomes superheated. You can prevent this by stirring the mixture constantly with the spatula and by avoiding overheating the mixture.

Dissolving the Sulfanilamide

Before heating the Craig tube containing the sulfanilamide, add enough hot solvent with a Pasteur pipet to barely cover the crystals. Then heat the Craig tube containing the sulfanilamide until the solvent is boiling. At first, this may be difficult to see because so little solvent is present. Add another small portion of solvent (one or two drops), continue to heat the Craig tube, and stir the mixture by rapidly twirling a microspatula between your fingers. When you have stirred the mixture for 10–15 seconds, check to see whether the solid has dissolved. If it has not, add another portion (one or two drops) of solvent. Heat the Craig tube again with stirring until the solvent boils. Then stir the tube for 10–15 seconds. Continue repeating this process of adding solvent, heating, and stirring until all the solid has dissolved completely. Note that is it essential to add just enough solvent to dissolve the solid—neither too much nor too little. Because 95% ethyl alcohol is very volatile, you need to perform this entire procedure fairly rapidly. Otherwise, you may lose solvent nearly as rapidly as you are adding it, and this procedure will take a very long time. The time from the first addition of solvent until the solid dissolves completely should be no longer than 10–15 minutes.

---

[3] See footnote 1, p. 23.

[4] You may also use a hot water bath to heat the solvent in the test tube and to heat the Craig tube. The temperature of the water bath should be about 80°C.

Crystallization

Remove the Craig tube from the heat and insert the inner plug into the opening. Allow the Craig tube to cool slowly to room temperature by placing it into a 10-mL Erlenmeyer flask (see Section 11.4C, p. 658). Crystallization should begin by the time the Craig tube has cooled to room temperature. If it has not, *gently* scratch the inside surface of the tube with a glass rod (not fire-polished) to induce crystallization (Technique 11, Section 11.7, Part B, p. 664).[5] When it appears that no further crystallization is occurring at room temperature, place the Craig tube in an ice-water bath using a beaker (Technique 6, Section 6.5, p. 595). Be sure that both water and ice are present and that the beaker is small enough to prevent the Craig tube from tipping over.

Isolation of Crystals

When crystallization is complete, place the Craig tube in a centrifuge tube and separate the crystals from the mother liquor by centrifugation. Follow the procedure in Technique 11, Section 11.7, p. 663.

Using the copper wire, pull the Craig tube out of the centrifuge tube. If the crystals collected on the end of the inner plug, remove the plug and scrape the crystals with a spatula onto a preweighed watch glass for air-drying. Otherwise, it will be necessary to scrape the crystals from the inside surface of the outer part of the Craig tube. If you will be doing the Optional Exercise, save the mother liquor in the centrifuge tube. Separate the crystals as much as possible with a spatula. The crystals should be completely dried within 5–10 minutes. You can determine if the crystals are still wet by observing whether or not they stick to a spatula or stay together in a clump. Weigh the dry crystals and calculate the percent recovery. Determine the melting point of both the pure sulfanilamide and the original impure material. At the option of the instructor, turn in your crystallized material in a properly labeled container.

For additional information about crystallization, see "Comments on the Crystallization Procedure," p. 24.

Optional Exercise

See "Optional Exercise" on p. 25.

3C   **E X P E R I M E N T   3 C**

Selecting a Solvent to Crystallize a Substance

In this experiment you will be given an impure sample of fluorene.[6] Your goal will be to find a good solvent for crystallizing the sample. You should try water, methyl alcohol, and toluene. After you have determined which is the best solvent, crystallize the remaining material. Finally, determine the melting point of the purified compound and of the impure sample.

---

[5] An alternative method for inducing crystallization is to dip a microspatula into the solution. Then allow the solvent to evaporate so that a small amount of solid will form on the surface of the spatula. When placed back into the solution, the solid will seed the solution.

[6] The impure fluorene contains 5% fluorenone, a yellow compound.

PROCEDURE

Selecting a Solvent

Perform the procedure given in Section 11.6 on p. 662 with three separate samples of impure fluorene. Use the following solvents: methyl alcohol, water, and toluene.

Crystallizing the Sample

After you have found a good solvent, crystallize the impure fluorene using a semi-microscale (Erlenmeyer flask and Hirsch funnel) or a microscale (Craig tube) procedure. Use 0.3 g of impure fluorene if you follow the semimicroscale procedure, or use 0.05 g if you follow the microscale procedure. Weigh the impure sample carefully, and be sure to keep a little of the impure sample on which to perform a melting point. If you perform a semimicroscale crystallization, you may need to use a size of Erlenmeyer flask different from the one specified in the procedure. This decision should be made based on the amount of sample you will be crystallizing and how much solvent you think will be needed. Transfer the crystals to a preweighed watch glass and allow them to air-dry. If water was used as the solvent, you may need to let the crystals sit out overnight for drying because water is less volatile than most organic solvents. Weigh the dried sample and calculate the percent recovery. Determine the melting point of both the pure sample and the original impure material. At the option of the instructor, turn in your crystallized material in a properly labeled container.

3D   EXPERIMENT 3D

Mixture Melting Points

In Experiments 3A and 3B of this experiment, the melting point was used to determine the purity of a known substance. In some situations the melting point can also be used to determine the identity of an unknown substance.

In Experiment 3D, you will be given a pure sample of an unknown from the following list:

Compound	Melting Point (°C)
Acetylsalicylic acid	138–140
Benzoic acid	121–122
Benzoin	135–136
Dibenzoyl ethylene	108–111
Succinimide	122–124
o-Toluic acid	108–110

Your goal is to determine the identity of the unknown using the melting-point technique. If all of the compounds in the list had distinctly different melting points, it would be possible to determine the identity of the unknown by just taking its melting point. However, each of the compounds in this list has a melting point that is close to the melting point of another compound in the list. Therefore, the melting point of the unknown will allow you to narrow down the choices to two compounds. To determine the identity of your compound, you must perform mixture melting points of your unknown and each of the two compounds with similar melting points. A mixture melting point that is depressed and has a wide range indicates that the two compounds in the mixture are different.

PROCEDURE

Obtain an unknown sample and determine its melting point. Determine mixture melting points (see Technique 9, Section 9.4, p. 630) of your unknown and all compounds from the previous list that have similar melting points. To prepare a sample for a mixture melting point, use a spatula or a glass stirring rod to grind equal amounts of your unknown and the known compound in a watch glass. Record all melting points and state the identity of your unknown.

3E   EXPERIMENT 3E

Critical Thinking Application

The goal of the exercise is to find an appropriate solvent to crystallize a given compound. Rather than doing this experimentally, you will try to predict which one of three given solvents is the best. For each compound, one of the solvents has the desired solubility characteristics to be a good solvent for crystallization. In a second solvent, the compound will be highly soluble, even at room temperature. The compound will be relatively insoluble in the third solvent, even at the boiling point of the solvent. After making your predictions, you will check them by looking up the appropriate information in *The Merck Index*.

For example, consider naphthalene, which has the following structure:

Naphthalene

Consider the three solvents ether, water, and toluene. (Look up their structures if you are unsure. Remember that ether is also called diethyl ether.) Based on your knowledge of polarity and solubility behavior, make your predictions. It should be clear that naphthalene is insoluble in water because naphthalene is a hydrocarbon that is nonpolar and water is very polar. Both toluene and ether are relatively nonpolar, so naphthalene is most likely soluble in both of them. One would expect naphthalene to be more soluble in toluene because both naphthalene and toluene are hydrocarbons. In addition, they both contain benzene rings, which means that their structures are very similar. Therefore, according to the solubility rule "Like dissolves like," one would predict that naphthalene is very soluble in toluene. Perhaps it is too soluble in toluene to be a good crystallizing solvent. If so, then ether would be the best solvent for crystallizing naphthalene.

These predictions can be checked with information from *The Merck Index*. Finding the appropriate information can be somewhat difficult, especially for beginning organic chemistry students. Look up *naphthalene* in *The Merck Index*. The entry for *naphthalene* states, "Monoclinic prismatic plates from ether." This statement means that naphthalene can be crystallized from ether. It also gives the type of crystal structure. Unfortunately, sometimes the crystal structure is given without reference to the solvent. Another way to determine the best solvent is by looking at solubility-vs.-temperature data. A good solvent is one in which the solubility of the compound increases significantly as the temperature increases. To determine whether the solid is too soluble in the solvent, check the solubility at room temperature. In Technique 11 on page 647, you were instructed to add 0.5 mL of solvent to 0.05 g of compound. If the solid completely dissolved, then the solubility at room temperature was too great. Follow this same guideline here. For naphthalene, the solubility in toluene is given as 1 g in 3.5 mL. When no temperature is given, room temperature is understood. By comparing this to the 0.05 g in 0.5 mL ratio, it is clear that naphthalene is too soluble in toluene at room temperature for toluene to be a good crystallizing solvent. Finally, *The Merck Index* states that naphthalene is insoluble in water. Sometimes no information is given about solvents in which the compound is insoluble. In that case, you would rely on your understanding of solubility behavior to confirm your predictions.

When using *The Merck Index*, you should be aware that alcohol is listed frequently as a solvent. This generally refers to 95% or 100% ethyl alcohol. Because 100% (absolute) ethyl alcohol is more expensive than 95% ethyl alcohol, the cheaper grade is usually used in the chemistry lab. Finally, benzene is frequently listed as a solvent. Because benzene is a known carcinogen, it is rarely used in student labs. Toluene is a suitable substitute; the solubility behavior of a substance in benzene and toluene is so similar that you may assume any statement made about benzene also applies to toluene.

Exercise. For each of the following sets of compounds (the solid is listed first, followed by the three solvents), use your understanding of polarity and solubility to predict

1. The best solvent for crystallization

2. The solvent in which the compound is too soluble

3. The solvent in which the compound is not sufficiently soluble

Then check your predictions by looking up each compound in *The Merck Index*.

1. Phenanthrene; toluene, 95% ethyl alcohol, water

Phenanthrene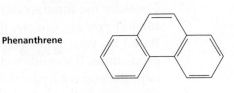

2. Cholesterol; ether, 95% ethyl alcohol, water

Cholesterol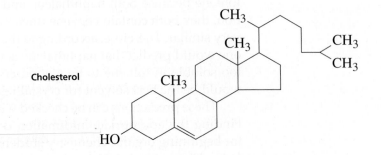

3. Acetaminophen; toluene, 95% ethyl alcohol, water

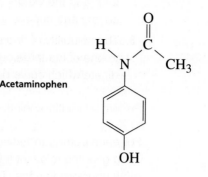

Acetaminophen

4. Urea; hexane, 95% ethyl alcohol, water

Urea

REPORT

Experiments 3A and 3B

1. Report the melting points for both the impure sulfanilamide and the crystallized sulfanilamide and comment on the differences. Also, compare these to the literature value. Look up the literature value and include this in your report. Report both the original weight of the impure sulfanilamide and the weight of the crystallized sulfanilamide. Calculate the percentage recovery and comment on several sources of loss.

2. If you completed the Optional Exercise (isolating the solid dissolved in the mother liquor), do the following:
    a. Make a table with the following information:
        i. Weight of impure sulfanilamide used in the crystallization procedure
        ii. Weight of pure sulfanilamide after crystallization
        iii. Weight of sulfanilamide plus impurity recovered from the mother liquor (see p. 25 or p. 27)
        iv. Total of items ii and iii (total weight of sulfanilamide plus impurity isolated)
        v. Calculated weight of sulfanilamide in the mother liquor (see p. 23)
    b. Comment on any differences between the values in items i and iv. Should they be the same? Explain.
    c. Comment on any differences between items iii and v. Should they be the same? Explain.
    d. Report the melting point of the solid recovered from the mother liquor. Compare this to the melting points of the crystallized sulfanilamide. Should they be the same? Explain.

Experiment 3C

1. For each of the three solvents (methyl alcohol, water, and toluene), describe the results from the tests for selecting a good crystallizing solvent for fluorene. Explain these results in terms of polarity and solubility predictions (see "Guidelines for Predicting Polarity and Solubility," p. 639).

2. Report the melting points for both the impure fluorene and the crystallized fluorene and comment on the differences. What is the literature value for the

melting point of fluorene? Report the original weight of both the impure fluorene and the weight of the crystallized fluorene. Calculate the percentage recovery and comment on several sources of loss.

3. The solubility of fluorene in each solvent used in Experiment 3B corresponds to one of the three curves shown in Figure 11.1 (p. 648). For each solvent, indicate which curve best describes the solubility of fluorene in that solvent.

Experiment 3D   Record all melting points and state the identity of your unknown.

Experiment 3E   For each compound assigned, state your predictions, along with an explanation. Then give the relevant information from *The Merck Index* that supports or contradicts your predictions. Try to explain any differences between your predictions and information found in *The Merck Index*.

QUESTIONS

1. Consider a crystallization of sulfanilamide in which 10 mL of hot 95% ethyl alcohol is added to 0.10 g of impure sulfanilamide. After the solid has dissolved, the solution is cooled to room temperature and then placed in an ice-water bath. No crystals form, even after scratching with a glass rod. Explain why this crystallization failed. What would you have to do at this point to make the crystallization work? (You may need to refer to Figure 11.2 on p. 648.)

2. Benzyl alcohol (bp 205°C) was selected by a student to crystallize fluorenol (mp 153–154°C) because the solubility characteristics of this solvent are appropriate. However, this solvent is not a good choice. Explain.

3. A student performs a crystallization on an impure sample of biphenyl. The sample weighs 0.5 g and contains about 5% impurity. Based on his knowledge of solubility, the student decides to use benzene as the solvent. After crystallization, the crystals are dried and the final weight is found to be 0.02 g. Assume that all steps in the crystallization are performed correctly, there are no spills, and the student lost very little solid on any glassware or in any of the transfers. Why is the recovery so low?

---

4   EXPERIMENT 4

Extraction

Extraction

Critical thinking application

Extraction is one of the most important techniques for isolating and purifying organic substances. In this method, a solution is mixed thoroughly with a second solvent that is **immiscible** with the first solvent. (Remember that immiscible liquids do not mix; they form two phases, or layers.) The solute is extracted from one solvent into the other because it is more soluble in the second solvent than in the first.

   The theory of extraction is described in detail in Technique 12, Sections 12.1–12.2, pp. 669–671. You should read these sections before continuing this experiment. Because solubility is the underlying principle of extraction, you may also wish to reread the introduction to the experiment on solubility.

Extraction is not only a technique used by organic chemists but it is also used to produce common products with which you are familiar. For example, vanilla extract, the popular flavoring agent, was originally extracted from vanilla beans using alcohol as the organic solvent. Decaffeinated coffee is made from coffee beans that have been decaffeinated by an extraction technique (see essay on caffeine, p. 87). This process is similar to the procedure in Experiment 4A of this experiment, in which you will extract caffeine from an aqueous solution.

The purpose of this experiment is to introduce the microscale technique for performing extractions and allow you to practice this technique. This experiment also demonstrates how extraction is used in organic experiments.

REQUIRED READING

New: Technique 12   Extraction

Essay   Caffeine (p. 87)

Review: Technique 10   Solubility

SPECIAL INSTRUCTIONS

Be careful when handling methylene chloride. It is a toxic solvent, and you should not breathe its fumes excessively or spill it on yourself.

In Experiment 4B, it is advisable to pool the data for the distribution coefficients and calculate class averages. This will compensate for differences in the values due to experimental error.

SUGGESTED WASTE DISPOSAL

You must dispose of all methylene chloride in a waste container marked for the disposal of halogenated organic wastes. Place all other organic wastes into the nonhalogenated organic waste container. The aqueous solutions obtained after the extraction steps must be disposed of in the container designated for aqueous waste.

4A   EXPERIMENT 4A

Extraction of Caffeine

One of the most common extraction procedures involves using an organic solvent (nonpolar or slightly polar) to extract an organic compound from an aqueous solution. Because water is highly polar, the mixture will separate into two layers, or phases: an aqueous layer and an organic (nonpolar) layer.

In this experiment, you will extract caffeine from an aqueous solution using methylene chloride. You will perform the extraction step three times using three separate portions of methylene chloride. Because methylene chloride is more dense than water, the organic layer (methylene chloride) will be on the bottom. After each extraction, you will remove the organic layer. The organic layers from all three extractions will be combined and dried over anhydrous sodium sulfate. After transferring the dried solution to a preweighed container, you will evaporate the methylene chloride and determine the weight of caffeine extracted from the aqueous solution. This extraction procedure succeeds because caffeine is much more soluble in methylene chloride than in water.

Pre-Lab Calculation

In this experiment, 0.070 g of caffeine is dissolved in 4.0 mL of water. The caffeine is extracted from the aqueous solution three times with 2.0-mL portions of methylene chloride. Calculate the total amount of caffeine that can be extracted into the three portions of methylene chloride (see Technique 12, Section 12.2, p. 670). Caffeine has a distribution coefficient of 4.6, between methylene chloride and water.

PROCEDURE

Preparation

Before beginning this experiment, check your screw-cap centrifuge tube for leaks.[1] Add exactly 0.070 g of caffeine to the centrifuge tube. Then add 4.0 mL of water to the tube. Cap the tube and shake it vigorously for several minutes until the caffeine dissolves completely. It may be necessary to heat the mixture slightly to dissolve all the caffeine.

Extraction

Add 2.0 mL of methylene chloride to the tube. The two layers must be mixed thoroughly so that as much caffeine as possible is transferred from the aqueous layer to the methylene chloride layer. However, if the mixture is mixed too vigorously, it may form an emulsion. Emulsions look like a third frothy layer between the other two layers, and they can make it difficult for the layers to separate. The best way to prevent an emulsion is to shake gently at first and observe whether the layers separate. If they separate quickly, continue to shake, but now more vigorously. The correct way to shake is to invert the tube and right it in a rocking motion. A good rate of shaking is about one rock per second. When it is clear that an emulsion is not forming, you may shake it more vigorously, perhaps two to three times per second. (Note that it is usually not prudent to shake the heck out of it!) Shake the tube for about one minute.

After shaking, place the tube in a test tube rack or beaker and let it stand until the layers separate completely.[2] It may be necessary to tap the sides of

---

[1] Place about 2 mL of water in the tube. Cap it and shake vigorously. It if leaks, try screwing the cap on more tightly or use a different cap. Sometimes you may need to replace the centrifuge tube itself. Discard the water in the tube.

[2] If an emulsion has formed, the two layers may not separate on standing. If they do not separate after about 1–2 minutes, it will be necessary to centrifuge the mixture to break the emulsion. Remember to balance the centrifuge by placing a tube of equal weight on the opposite side.

the tube to force all the methylene chloride layer to the bottom of the vial. Occasionally, a drop of water will get stuck in the very bottom part of the tube, below the methylene chloride layer. If this happens, depress the bulb slightly and try to draw the water drop into a Pasteur pipet. Transfer this drop to the upper layer.

Using a Pasteur pipet, you should now transfer the organic (bottom) layer into a test tube. Ideally, the goal is to remove all the organic layer without transferring any of the aqueous layer. However, this is difficult to do. Try to squeeze the bulb so that when it is released completely, you will draw up the amount of liquid that you desire. If you have to hold the bulb in a partially depressed position while making a transfer, it is likely that you will spill some liquid. It is also best to transfer the liquid in two steps. First, depress the bulb so that most (about 75%) of the bottom layer will be drawn into the pipet. Place the tip of the pipet squarely in the *V* at the bottom of the centrifuge tube and release the bulb slowly. When making the transfer, it is essential that the centrifuge tube and the test tube be held next to each other. A good technique for this is illustrated in Figure 12.6, p. 675. After transferring the first portion, depress the bulb partially, just enough to draw up the remaining liquid in the bottom layer, and place the tip of the pipet in the bottom of the tube. Draw the liquid into the pipet and transfer this liquid to the test tube.

Repeat this extraction two more times using 2 mL of fresh methylene chloride each time. Combine the organic layer from each of these extractions with the methylene chloride solution from the first extraction.

Drying the Organic Layers

Dry the combined organic layers over granular anhydrous sodium sulfate, following the instructions given in Technique 12, Section 12.9, "Drying Procedure with Anhydrous Sodium Sulfate," p. 681. Read these instructions carefully and complete Steps 1–3 in the "Microscale Drying Procedure." Step 4 (p. 683) is described in the next section, "Evaporation of Solvent."

Evaporation of Solvent

Transfer the dried methylene chloride solution with a clean, dry Pasteur pipet to a dry, preweighed 10-mL Erlenmeyer flask or test tube while leaving the drying agent behind.[3] (If you had to add more than 3–4 microspatulafuls of anhydrous sodium sulfate, rinse the sodium sulfate with about 0.5 mL of fresh methylene chloride. Stir this with a dry spatula and then transfer this solution to the same preweighed flask.) Evaporate the methylene chloride by heating the flask in a hot water bath at about 45°C. This should be done in a hood and can be accomplished more rapidly if a stream of dry air or nitrogen gas is directed at the surface of the liquid (see Technique 7, Section 7.10, p. 611). When the solvent is evaporated, remove the flask from the bath and dry the outside of the flask. When the flask has cooled to room temperature, weigh it to determine the amount of caffeine that was in the methylene chloride solution. Compare this weight with the amount of caffeine calculated in the Pre-Lab Calculation.

---

[3] It is easier to avoid transferring any drying agent if you use a filter-tip pipet (Section 8.6, p. 625).

EXPERIMENT 4B

Distribution of a Solute between Two Immiscible Solvents

In this experiment, you will investigate how several different organic solids distribute themselves between water and methylene chloride. A solid compound is mixed with the two solvents until equilibrium is reached. The organic layer is removed, dried over anhydrous sodium sulfate, and transferred to a tared container. After evaporating the methylene chloride, the weight of the organic solid that was in the organic layer is determined. By finding the difference, the amount of solute in the aqueous layer can also be determined. The distribution coefficient of the solid between the two layers can then be calculated and related to the polarity of the solid and the polarities of the two liquids.

Three different compounds will be used: benzoic acid, succinic acid, and sodium benzoate. Their structures are given below. You should perform this experiment on one of the solids and share your data with two other students who worked with the other two solids. Alternatively, data from the entire class may be pooled and averaged.

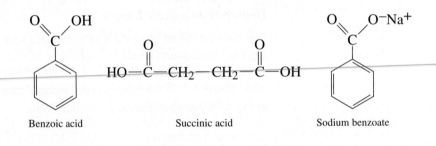

| Benzoic acid | Succinic acid | Sodium benzoate |

PROCEDURE

Place 0.050 g of one of the solids (benzoic acid, succinic acid, or sodium benzoate) into a 5-mL conical vial. Add 2.0 mL of methylene chloride and 2.0 mL of water to the vial. Cap the vial and shake it as described in Experiment 4A for about 1 minute. Check for undissolved solid. Continue shaking the vial until all the solid is dissolved. After the layers have separated, transfer the bottom organic layer to another vial or a small test tube. Using the same procedure just described in Experiment 4A (see the section on "Drying the Organic Layers"), dry this organic layer over granular anhydrous sodium sulfate.

Transfer the dried methylene chloride solution with a clean, dry Pasteur pipet to a dry, preweighed conical vial, leaving the drying agent behind. Evaporate the methylene chloride by heating the vial in a hot water bath while directing a stream of dry air or nitrogen gas at the surface of the liquid. When the solvent is evaporated, remove the vial from the bath and dry the outside of the vial. When the vial has cooled to room temperature, weigh the vial to determine the amount of solid solute that was in the methylene chloride layer. Determine by difference the amount of the solid that was dissolved in the aqueous layer. Calculate the distribution coefficient for the solid between methylene chloride and water. Because the volume of methylene chloride and water was the same, the distribution coefficient can be calculated by dividing the weight of solute in methylene chloride by the weight of solute in water.

Optional Exercise

Repeat the preceding procedure using 0.050 g of caffeine, 2.0 mL of methylene chloride, and 2.0 mL of water. Determine the distribution coefficient for caffeine between methylene chloride and water. Compare this to the literature value of 4.6.

4C   EXPERIMENT 4C

How Do You Determine Which One Is the Organic Layer?

A common problem that you might encounter during an extraction procedure is not knowing for sure which layer is organic and which is the aqueous one. Although the procedures in this textbook often indicate the expected relative positions of the two layers, not all procedures will give this information, and you should be prepared for surprises. Sometimes knowing the densities of the two solvents is not sufficient, because dissolved substances can significantly increase the density of a solution. It is very important to know the location of the two layers because usually one layer contains the desired product and the other layer is discarded. A mistake at this point in an experiment would be disastrous!

The purpose of this experiment is to give you some practice in determining which layer is aqueous and which layer is organic (see Technique 12, Section 12.8, p. 679). As described in Section 12.8, one effective technique is to add a few drops of water to each layer after the layers have been separated. If the layer is water, then the drops of added water will dissolve in the aqueous layer and increase its volume. If the added water forms droplets or a new layer, then it is the organic layer.

PROCEDURE

Obtain three test tubes, each containing two layers.[4] For each tube, you will be told the identity of the two layers, but you will not be told their relative positions. Determine experimentally which layer is organic and which layer is aqueous. Dispose of all these mixtures into the waste container designated for halogenated organic wastes. After determining the layers experimentally, look up the densities of the various liquids in a handbook to see if there is a correlation between the densities and your results.

---

[4] The three mixtures will likely be (1) water and *n*-butyl chloride, (2) water and *n*-butyl bromide, and (3) *n*-butyl bromide and saturated aqueous sodium bromide.

4D   EXPERIMENT 4D

Use of Extraction to Isolate a Neutral Compound from a Mixture Containing an Acid or Base Impurity

In this experiment you will be given a solid sample containing an unknown neutral compound and an acid or base impurity. The goal is to remove the acid or base by extraction and isolate the neutral compound. By taking the melting point of the neutral compound, you will identify it from a list of possible compounds. There are many organic reactions in which the desired product, a neutral compound, is contaminated by an acid or base impurity. This experiment illustrates how extraction is used to isolate the product in this situation.

In Technique 10, "Solubility," you learned that organic acids and bases can become ions in acid–base reactions (see "Solutions in Which the Solute Ionizes and Dissociates," p. 642). Before reading on, review this material if necessary. Using this principle, it is possible to separate an acid or base impurity from a neutral compound. The following scheme, which shows how both an acid and a base impurity are removed from the desired product, illustrates how this is accomplished:

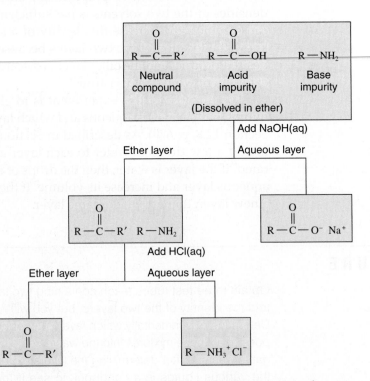

Flow chart showing how acid and base impuities are removed from the desired product.

The neutral compound can now be isolated by removing the water dissolved in the ether and evaporating the ether. Because ether dissolves a relatively large quantity of water (1.5%), the water must be removed in two steps: In the first step, the ether solution is mixed with a saturated aqueous NaCl solution. Most of the water in the ether layer will be transferred to the aqueous layer in this step (see Technique 12, Section 12.9, p. 680). Finally, the remainder of the water is removed by drying the ether layer over anhydrous sodium sulfate. The neutral compound can then be isolated by evaporating the ether. In most organic experiments that use a separation scheme such as this, it would be necessary to perform a crystallization step to purify the neutral compound. However, in this experiment the neutral compound should be sufficiently pure at this point to identify it by melting point.

The organic solvent used in this experiment is ether. Recall that the full name for ether is diethyl ether. Because ether is less dense than water, this experiment will give you practice in performing extractions where the nonpolar solvent is less dense than water.

The following procedure details the removal of an acid impurity from a neutral compound and isolating the neutral compound. It contains an additional step that is not normally part of this kind of separation scheme: The aqueous layers from each extraction are segregated and acidified with aqueous HCl. The purpose of this step is to verify that the acid impurity has been removed completely from the ether layer. In the Optional Exercise, the sample contains a neutral compound with a base impurity; however, a detailed procedure is not given. If you are assigned this exercise, you must create a procedure by using the principles discussed in this introduction and by studying the following procedure for isolating the neutral compound from an acid impurity.

PROCEDURE

Isolating a Neutral Compound from a Mixture Containing an Acid Impurity. Add 0.150 g of an unknown mixture[5] to a screw-cap centrifuge tube. Add 4.0 mL of ether to the tube and cap it. Shake the tube until all the solid dissolves completely.

Add 2.0 mL of 1.0 *M* NaOH to the tube and shake for 30 seconds. Let the layers separate. Remove the bottom (aqueous) layer, and place this in a test tube labeled "1st NaOH extract." Add another 2.0-mL portion of 1.0 *M* NaOH to the centrifuge tube and shake for 30 seconds. When the layers have separated, remove the aqueous layer and put this in a test tube labeled "2nd NaOH extract."

With stirring, add 6 *M* HCl dropwise to each of the two test tubes containing the NaOH extracts until the mixture is acidic. Test the mixture with litmus or pH paper to determine when it is acidic. Observe the amount of precipitate that forms. What is the precipitate? Does the amount of precipitate in each tube indicate that all the acid impurity has been removed from the ether layer containing the unknown neutral compound?

The drying procedure for an ether layer requires the following additional step compared to the procedure for drying a methylene chloride layer (see Technique 12, Section 12.9, "Saturated Salt Solution," p. 680). To the ether layer in the

---

[5] The mixture contains 0.100 g of one of the neutral compounds given in the list on page 39 and 0.050 g of benzoic acid, the acid impurity.

centrifuge tube, add 2.0 mL of saturated aqueous sodium chloride. Shake for 30 seconds and let the layers separate. Remove and discard the aqueous layer. With a clean, dry Pasteur pipet, transfer the ether layer (without any water) to a clean, dry test tube. Now dry the ether layer over granular anhydrous sodium sulfate (see Technique 12, Section 12.9, "Drying Procedure with Anhydrous Sodium Sulfate," p. 681).

Transfer the dried ether solution with a clean, dry Pasteur pipet to a dry, preweighed test tube, leaving the drying agent behind. Evaporate the ether by heating the tube in a hot water bath. This should be done in a hood and can be accomplished more rapidly if a stream of dry air or nitrogen gas is directed at the surface of the liquid (see Technique 7, Section 7.10, p. 611). When the solvent has evaporated, remove the test tube from the bath and dry the outside of the tube. Once the tube has cooled to room temperature, weigh it to determine the amount of solid solute that was in the ether layer. Obtain the melting point of the solid and identify it from the following list:

	Melting Point
Fluorenone	82–85°C
Fluorene	116–117°C
1,2,4,5-Tetrachlorobenzene	139–142°C
Triphenylmethanol	162–164°C

Optional Exercise: Isolating a Neutral Compound from a Mixture Containing a Base Impurity. Obtain 0.150 g of an unknown mixture containing a neutral compound and a base impurity.[6] Develop a procedure for isolating the neutral compound, using the preceding procedure as a model. After isolating the neutral compound, obtain the melting point and identify it from the list of compounds given above.

4E   **EXPERIMENT 4E**

Critical Thinking Application

PROCEDURE

1. Add 4 mL of water and 2 mL of methylene chloride to a screw-capped centrifuge tube.
2. Add 4 drops of solution A to the centrifuge tube. Solution A is a dilute aqueous solution of sodium hydroxide containing an organic compound.[7] Shake the mixture for about 30 seconds, using a rapid rocking motion. Describe the color of each layer (see the following table).

---

[6] The mixture contains 0.100 g of one of the neutral compounds given in the list above and 0.050 g of ethyl 4-aminobenzoate, a base impurity.

[7] Solution A: Mix 25 mg of 2,6-dichloroindophenol (sodium salt) with 50 mL of water and 1 mL of 1 *M* NaOH. This solution should be prepared the same day it is used.

3. Add 2 drops of 1 *M* HCl. Let the solution sit for 1 minute and note the color change. Then shake for about 1 minute, using a rapid rocking motion. Describe the color of each layer.

4. Add 4 drops of 1 *M* NaOH and shake again for about 1 minute. Describe the color of each layer.

Color

		Color
Step 2	**Aqueous**	
	Methylene chloride	
Step 3	**Aqueous**	
	Methylene chloride	
Step 4	**Aqueous**	
	Methylene chloride	

REPORT

Experiment 4A

1. Show your calculations for the amount of caffeine that should be extracted by the three 2.0-mL portions of methylene chloride (see Pre-Lab Calculation).

2. Report the amount of caffeine isolated. Compare this weight with the amount of caffeine calculated in the Pre-Lab Calculation. Comment on the similarity or difference.

Experiment 4B

1. Report in table form the distribution coefficients for the three solids: benzoic acid, succinic acid, and sodium benzoate.

2. Is there a correlation between the values of the distribution coefficients and the polarities of the three compounds? Explain.

3. If you completed the Optional Exercise, compare the distribution coefficient you obtained for caffeine with the corresponding literature value. Comment on the similarity or difference.

Experiment 4C

1. For each of the three mixtures, report which layer was on the bottom and which one was on the top. Explain how you determined this for each mixture.

2. Record the densities for the liquids given in a handbook.

3. Was there a correlation between the densities and your results? Explain.

Experiment 4D

1. Answer the following questions about the first and second NaOH extracts.
   a. Comment on the amount of precipitate for both extracts when HCl is added.
   b. What is the precipitate formed when HCl is added?
   c. Does the amount of precipitate in each tube indicate that all the acid impurity has been removed from the ether layer containing the unknown neutral compound?

2. Report the melting point and weight of the neutral compound you isolated.

3. Based on the melting point, what is the identity of this compound?

4. Calculate the percent recovery for the neutral compound. List possible sources of loss.

If you completed the Optional Exercise, complete Steps 1–4 for Experiment 4D.

Experiment 4E   Describe fully what occurred in Steps 2, 3, and 4. For each step, include (1) the nature (cation, anion, or neutral species) of the organic compound, (2) an explanation for all the color changes, and (3) an explanation for why each layer is colored as it is. Your explanation for (3) should be based on solubility principles and the polarities of the two solvents. (*Hint:* It may be helpful to review the sections in your general chemistry textbook that deal with acids, bases, and acid–base indicators.)

REFERENCE

Kelly, T. R. "A Simple, Colorful Demonstration of Solubility and Acid/Base Extraction." *Journal of Chemical Education, 70* (1993): 848.

QUESTION

1. Caffeine has a distribution coefficient of 4.6 between methylene chloride and water. If 52 mg of caffeine are added to a conical vial containing 2 mL of water and 2 mL of methylene chloride, how much caffeine would be in each layer after the mixture had been mixed thoroughly?

5   EXPERIMENT 5

Chromatography

Thin-layer chromatography

Column chromatography

Following a reaction with thin-layer chromatography

Chromatography is perhaps the most important technique used by organic chemists to separate the components of a mixture. This technique involves the distribution of the different compounds or ions in the mixture between two phases, one of which is stationary and the other moving. Chromatography works on much the same principle as solvent extraction. In extraction, the components of a mixture are distributed between two solvents according to their relative solubilities in the two solvents. The separation process in chromatography depends on differences in how strongly the components of the mixture are adsorbed to the stationary phase and how soluble they are in the moving phase. These differences depend primarily on the relative polarities of the components in the mixture.

There are many types of chromatographic techniques, ranging from thin-layer chromatography, which is relatively simple and inexpensive, to high-performance liquid chromatography, which is very sophisticated and expensive. In this experiment, you will use two of the most widely used chromatographic techniques: thin-layer and column chromatography. The purpose of this experiment is to give you practice in performing these two techniques, to illustrate the principles of chromatographic separations, and to demonstrate how thin-layer and column chromatography are used in organic chemistry.

REQUIRED READING

New: Technique 19   Column Chromatography
Technique 20   Thin-Layer Chromatography

SPECIAL INSTRUCTIONS

Many flammable solvents are used in this experiment. Use Bunsen burners for making micropipets in a part of the lab that is separate from where the solvents are being used. The thin-layer chromatography should be performed in the hood.

SUGGESTED WASTE DISPOSAL

Dispose of methylene chloride in the container designated for halogenated organic wastes. Dispose of all other organic solvents in the container for nonhalogenated organic solvents. Place the alumina in the container designated for wet alumina.

NOTES TO THE INSTRUCTOR

The column chromatography should be performed with activated alumina from EM Science (No. AX0612-1). The particle sizes are 80–200 mesh, and the material is Type F-20. The alumina should be dried overnight in an oven at 110°C and stored in a tightly sealed bottle. Alumina more than several years old may need to be dried for a longer time at a higher temperature.

For thin-layer chromatography (TLC), use flexible silica gel plates from Whatman with a fluorescent indicator (No. 4410 222). If the TLC plates have not been purchased recently, they should be placed in an oven at 100°C for 30 minutes and stored in a desiccator until used. If you use different alumina or different thin-layer plates, try out the experiment before using it with a class. Other materials than those specified here may give different results from those indicated in this experiment.

Grind up the fluorenone flakes into smaller pieces for easier dispensing.

5A EXPERIMENT 5A

Thin-Layer Chromatography

In this experiment, you will use thin-layer chromatography (TLC) to separate a mixture of three compounds: fluorene, fluorenol, and fluorenone:

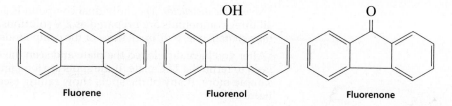

Fluorene   Fluorenol   Fluorenone

Based on the results with known samples of these compounds, you will determine which compounds are found in an unknown sample. Using TLC to identify the components in a sample is a common application of this technique.

PROCEDURE

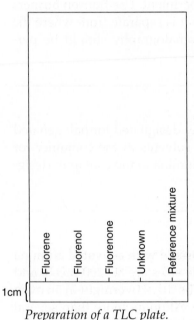

1cm {

Preparation of a TLC plate.

Preparing the TLC Plate

Technique 20 describes the procedures used for thin-layer chromatography. Use an 10 cm × 5.3 cm TLC plate (Whatman Silica Gel Plates No. 4410 222). These plates have a flexible backing but should not be bent excessively. They should be handled carefully or the adsorbent may flake off them. Also, they should be handled only by the edges; the surface should not be touched. Using a lead pencil (not a pen), lightly draw a line across the plate (short dimension) about 1 cm from the bottom (see figure). Using a centimeter ruler, move its index about 0.6 cm in from the edge of the plate and lightly mark off five 1-cm intervals on the line. These are the points at which the samples will be spotted.

Prepare five micropipets to spot the plate. The preparation of these pipets is described and illustrated in Technique 20, Section 20.4, page 782. Prepare a TLC development chamber with methylene chloride (see Technique 20, Section 20.5, p. 784). A beaker covered with aluminum foil or a wide-mouth, screw-cap bottle is a suitable container to use (see Fig. 20.4, p. 783). The backing on the TLC plates is thin, so if it touches the filter paper liner of the development chamber at any point, solvent will begin to diffuse onto the absorbent surface at that point. To avoid this, be sure that the filter paper liner does not go completely around the inside of the container. A space about 2.5 inches wide must be provided. (*Note:* This development chamber will also be used for Parts C and D in this experiment.)

On the plate, starting from left to right, spot fluorene, fluorenol, fluorenone, the unknown mixture, and the standard reference mixture, which contains all three compounds.[1] For each of the five samples, use a different micropipet to spot the sample on the plate. The correct method of spotting a TLC plate is described in Technique 20, Section 20.4, page 782. Take up part of the sample in the pipet (don't use a bulb; capillary action will draw up the liquid). Apply the sample by touching the pipet *lightly* to the thin-layer plate. The spot should be no larger than 2 mm in diameter. It will usually be sufficient to spot each sample once or twice. If you need to spot the sample more than once, allow the solvent to evaporate completely between successive applications and spot the plate in exactly the same position each time. Save the samples in case you need to repeat the TLC.[2]

Developing the TLC Plate

Place the TLC plate in the development chamber, making sure that the plate does not come in contact with the filter paper liner. Remove the plate when the solvent

---

[1] *Note to the instructor:* The individual compounds and the reference mixture containing all three compounds are prepared as 2% solutions in acetone. The unknown mixture may contain one, two, or all three of the compounds dissolved in acetone.

[2] After you have developed the plate and seen the spots, you will be able to tell if you need to rerun the TLC plate. If the spots are too faint to see clearly, you need to spot the sample more. If any of the spots show tailing (see pp. 772–773), then less sample is needed.

front is 1–2 cm from the top of the plate. Using a lead pencil, mark the position of the solvent front. Set the plate on a piece of paper towel to dry. When the plate is dry, place the plate in a jar containing a few iodine crystals, cap the jar, and warm it *gently* on a hot plate until the spots begin to appear. Remove the plate from the jar and lightly outline all the spots that became visible with the iodine treatment. Using a ruler marked in millimeters, measure the distance that each spot has traveled relative to the solvent front. Calculate the R_f values for each spot (see Technique 20, Section 20.9, p. 787). Explain the relative positions of the three compounds in terms of their polarities. Identify the compound or compounds that are found in the unknown mixture. At the instructor's option, submit the TLC plate with your report.

5B   EXPERIMENT 5B

Selecting the Correct Solvent for Thin-Layer Chromatography

In Experiment 5A, you were told what solvent to use for developing the TLC plate. In some experiments, however, it will be necessary to determine an appropriate development solvent by experimentation (Technique 20, Section 20.6, p. 785). In this experiment, you will be instructed to try three solvents for separating a pair of related compounds that differ slightly in polarity. Only one of these solvents will separate the two compounds enough so that they can be easily identified. For the other two solvents, you will be asked to explain, in terms of their polarities, why they failed.

PROCEDURE

Preparation

Your instructor will assign you a pair of compounds to run on TLC, or you will select your own pair.[3] You will need to obtain about 0.5 mL of three solutions: one solution of each of the two individual compounds and a solution containing both compounds. Prepare three thin-layer plates in the same way as you did in Experiment 5A, except that each plate should be 10 cm × 3.3 cm. When you mark them with a pencil for spotting, make three marks 1 cm apart. Prepare three micropipets to spot the plates. Prepare three TLC development chambers as you did in Experiment 5A, with each chamber containing one of the three solvents suggested for your pair of compounds.

---

[3] *Note to the instructor:* Possible pairs of compounds are given in the following list. The two compounds to be resolved are given first, followed by the three developing solvents to try: (1) benzoin and benzil; acetone, methylene chloride, hexane; (2) vanillin and vanillyl alcohol; acetone, 50% toluene–50% ethyl acetate, hexane; (3) diphenylmethanol and benzophenone; acetone, 70% hexane–30% acetone, hexane. Each compound in a pair should be prepared individually and as a mixture of the two compounds. Prepare all of them as 1% solutions in acetone.

Developing the TLC Plate

On each plate, spot the two individual compounds and the mixture of both compounds. For each of the three samples, use a different micropipet to spot the sample on the plates. Place each TLC plate in one of the three development chambers, making sure that the plate does not come in contact with the filter paper liner. Remove each plate when the solvent front is 1–2 cm from the top of the plate. Using a lead pencil, mark the position of the solvent front. Set the plate on a piece of paper towel to dry. When the plate is dry, observe it under a short-wavelength UV lamp, preferably in a darkened hood or a darkened room. With a pencil, lightly outline any spots that appear. Next, place the plate in a jar containing a few iodine crystals, cap the jar, and warm it *gently* on a hot plate until the spots begin to appear. Remove the plate from the jar and lightly outline all the spots that became visible with the iodine treatment. Using a ruler marked in millimeters, measure the distance that each spot has traveled relative to the solvent front. Calculate the R_f values for each spot. At the instructor's option, submit the TLC plates with your report.

Which of the three solvents resolved the two compounds successfully? For the two solvents that did not work, explain, in terms of their polarities, why they failed.

5C   EXPERIMENT 5C

Monitoring a Reaction with Thin-Layer Chromatography

Thin-layer chromatography is a convenient method for monitoring the progress of a reaction (Technique 20, Section 20.10, p. 789). This technique is especially useful when the appropriate reaction conditions have not yet been worked out. By using TLC to follow the disappearance of a reactant and the appearance of a product, it is relatively easy to decide when the reaction is complete. In this experiment, you will monitor the reduction of fluorenone to fluorenol:

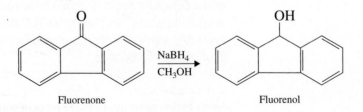

Fluorenone   Fluorenol

Although the appropriate reaction conditions for this reaction are already known, using TLC to monitor the reaction will demonstrate how to use this technique.

Then place the flask in an ice-water bath for several minutes to complete crystallization. Collect the crystals by vacuum filtration, using a small Hirsch funnel (Technique 8, Section 8.3, p. 621). Wash the crystals with three 1.0-mL portions of an ice-cold mixture of 80% methanol and 20% water. After the crystals are dry, weigh them and determine their melting point (literature, 153–154°C).

5D   EXPERIMENT 5D

Column Chromatography

The principles of column chromatography are similar to those of thin-layer chromatography. The primary difference is that the moving phase in column chromatography travels downward, whereas in TLC the solvent ascends the plate. Column chromatography is used more often than TLC to separate relatively large amounts of compounds. With column chromatography, it is possible to collect pure samples of the separated compounds and perform additional tests on them.

In this experiment, fluorene and fluorenone will be separated by column chromatography using alumina as the adsorbent. Because fluorenone is more polar than fluorene, fluorenone will be absorbed to the alumina more strongly. Fluorene will elute off the column with a nonpolar solvent hexane, whereas fluorenone will not come off until a more polar solvent (30% acetone–70% hexane) is put on the column. The purities of the two separated compounds will be tested by TLC and melting points.

PROCEDURE

Advance Preparation

Before running the column, assemble the following glassware and liquids. Obtain four dry test tubes (16 × 100 mm) and number them 1 through 4. Prepare two dry Pasteur pipets with bulbs attached. Place 9.0 mL of hexane, 2.0 mL of acetone, and 2.0 mL of a solution of 70% hexane–30% acetone (by volume) into three Erlenmeyer flasks. Clearly label and stopper each flask. Place 0.3 mL of a solution containing fluorene and fluorenone into a small test tube.[5] Stopper the test tube. Prepare one 10 cm × 3.3 cm TLC plate with four marks for spotting. Use the same TLC development chamber with methylene chloride that you used in Part A. Prepare four micropipets to spot the plates.

Prepare a chromatography column packed with alumina. Place a loose plug of cotton in a Pasteur pipet (5¾-inch) and push it gently into position using a glass rod (see figure for the correct position of the cotton). *Do not ram the cotton tightly, because this may result in the solvent flowing through the column too slowly.* Using

---

[5] *Note to the instructor:* This solution should be prepared for the entire class by dissolving 0.3 g of fluorene and 0.3 g of fluorenone in 9.0 mL of a mixture of 5% methylene chloride–95% hexane. Store this solution in a closed container to prevent evaporation of solvent. This will provide enough solution for 20 students, assuming little spillage or other types of waste.

PROCEDURE

Preparation

Work with a partner on this part of the experiment. Prepare two thin-layer
in the same way as you did in Experiment 5A, except that one plate sho
10 cm × 5.3 cm and the other one, 10 cm × 4.3 cm. When you mark them
pencil for spotting, make five marks 1 cm apart on the first plate and four ma
the second plate. During the reaction, you will be taking five samples fro
reaction mixture at 0, 15, 30, 60, and 120 seconds. Three of these samples s
be spotted on the larger plate and two of them on the smaller one. In addition
plate should be spotted with two reference solutions, one containing fluore
and the other fluorenol. Using a pencil to make very light marks, indicate at th
of each plate where each sample will be spotted so that you can keep track of
Write the number of seconds and an abbreviation for the two reference compo
Use the same TLC development chamber with methylene chloride that you us
Experiment 5A. Prepare seven micropipets to spot the plates.

Running the Reaction

Once sodium borohydride has been added to the reaction mixture (see next
graph), take samples at the times just indicated. Because this must be do
such a short time, you must be well prepared before starting the reaction.
person should be the timekeeper and the other person should take the sam
and spot the plates. Spot each sample once, using a different pipet for
sample.

Place a magnetic spin vane (Figure 7.4A, p. 603) into a 5-mL conical vial.
0.20 g of fluorenone and 4 mL of methanol to the vial. Place the vial on a mag
stirrer, using either an aluminum block or a clamp to hold the vial in place. Sti
mixture until all the solid has dissolved. Now take the first sample (the "0 sec
sample) and spot the plate. Using smooth weighing paper, weigh 0.020
sodium borohydride[4] and immediately add it to the reaction mixture. If you wai
long to add it, the sodium borohydride will become sticky because it absorbs m
ture from the air. Begin timing the reaction as soon as the sodium borohydrid
added. Use the micropipets to remove samples of the reaction mixture at the
lowing times: 15, 30, 60, and 120 seconds. Use a different micropipet each t
and spot a TLC plate with each sample. On each plate, also spot the two re
ence solutions of fluorenone and fluorenol in acetone. After developing
plates and allowing them to dry, visualize the spots with iodine, as described
Part A. Make a sketch of your plates and record the results in your noteboo
Do these results indicate that the reaction went to completion? In addition to t
TLC results, what other visible evidence indicated that the reaction went
completion? Explain.

Isolation of Fluorenol (optional procedure)

Using a Pasteur pipet, transfer the reaction mixture to a 10-mL Erlenmeyer flask
Add 1 mL of water and heat the mixture almost to boiling for about 2 minutes. Allo
the flask to cool slowly to room temperature in order to crystallize the product

---

[4] *Note to the Instructor:* The sodium borohydride should be checked to see whether it is
active: Place a small amount of powdered material in some methanol and heat it gently.
If the hydride is active, the solution should bubble vigorously.

a file, score the Pasteur pipet about 1 cm below the cotton plug. To break the tip off the pipet, put your thumbs together at the place on the pipet that you scored and push quickly with both thumbs.

CAUTION ⚠

Wear gloves or use a towel to protect your hands from being cut while breaking the pipet.

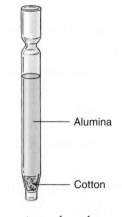

— Alumina

— Cotton

Chromatography column.

Add 1.25 g of alumina (EM Science, No. AX0612-1) to the pipet while tapping the column gently with your finger.[6] When all the alumina has been added, tap the column with your finger for several seconds to ensure that the alumina is tightly packed. Clamp the column in a vertical position so that the bottom of the column is just above the height of the test tubes you will be using to collect the fractions. Place test tube 1 under the column.

Running the Column

Using a Pasteur pipet, add 3 mL of hexane to the column. The column must be completely moistened by the solvent. Drain the excess hexane until the level of hexane reaches the top of the alumina. Once hexane has been added to the alumina, the top of the column must not be allowed to run dry. If necessary, add more hexane.

NOTE: It is essential that the liquid level not be allowed to drain below the surface of the alumina at any point in this procedure.

When the level of the hexane reaches the top of the alumina, add the solution of fluorene and fluorenone to the column using a Pasteur pipet. Begin collecting the eluent in test tube 2. Just as the solution penetrates the column, add 1 mL of hexane and drain until the surface of the liquid has reached the alumina. Add another 5 mL of hexane. As fluorene elutes off the column, some solvent will evaporate, leaving solid fluorene on the tip of the pipet. Using a Pasteur pipet, dissolve this solid off the column with a few drops of acetone. It may be necessary to do this several times, and the acetone solution is also collected in tube 2.

After you have added all the hexane, change to the more polar solvent (70% hexane–30% acetone).[7] When changing solvents, do not add the new solvent until the last solvent has nearly penetrated the alumina. The yellow band (fluorenone) should now move down the column. Just before the yellow band reaches the bottom of the column, place test tube 3 under the column. When the eluent becomes colorless again, place test tube 4 under the column and stop the procedure.

Tube 2 should contain fluorene and tube 3, fluorenone. Test the purities of these two samples using TLC. You must spot the solution from tube 2 several times in order to apply enough sample on the plate to be able to see the spots. On the plate, also spot a reference solution containing fluorene and fluorenone. After developing the plate and allowing it to dry, visualize the spots with iodine. What do the TLC results indicate about the purities of the two samples?

---

[6] As an option, students may prepare a microfunnel from a 1-mL disposable plastic pipet. The microfunnel is prepared by (1) cutting the bulb in half with a scissors and (2) cutting the stem at an angle about ½ inch below the bulb. This funnel can be placed in the top of the column (Pasteur pipet) to aid in filling the column with alumina or with the solvents (see page 764).

[7] Sometimes the fluorenone also moves through the column with hexane. Therefore, be sure to change to test tube 3 if the yellow band starts to emerge from the column.

Using a warm water bath (40–60°C) and a stream of nitrogen gas or air, evaporate the solvent from test tubes 2 and 3. As soon as all the solvent has evaporated from each of the tubes, remove them from the water bath. There may be a yellow oil in tube 3, but it should solidify when the tube cools to room temperature. If it does not, cool the tube in an ice-water bath and scratch the bottom of the test tube with a glass stirring rod or a spatula. Determine the melting points of the fluorene and fluorenone. The melting point of fluorene is 116–117°C and of fluorenone is 82–85°C.

REPORT

Experiment 5A

1. Calculate the R_f values for each spot. Include the actual plate or a sketch of the plate with your report.
2. Explain the relative R_f values for fluorene, fluorenol, and fluorenone in terms of their polarities and structures.
3. Give the composition of the unknown that you were assigned.

Experiment 5B

1. Record the names and structures of the two compounds that you ran on TLC.
2. Which solvent resolved the two compounds successfully?
3. For the other two solvents, explain, in terms of their polarities, why they failed.

Experiment 5C

1. Make a sketch of the TLC plate or include the actual plate with your report. Interpret the results. When was the reaction complete?
2. What other visible evidence indicated that the reaction went to completion?
3. If you isolated the fluorenol, record the melting point and the weight of this product.

Experiment 5D

1. Describe the TLC results on the samples in test tubes 2 and 3. What does this indicate about the purities of the two samples?
2. Record the melting points for the dried solids found in tubes 2 and 3. What do they indicate about the purities of the two samples?

QUESTIONS

1. Each of the solvents given should effectively separate one of the following mixtures by TLC. Match the appropriate solvent with the mixture that you would expect to separate well with that solvent. Select your solvent from the following: hexane, methylene chloride, or acetone. You may need to look up the structures of solvents and compounds in a handbook.
   a. 2-Phenylethanol and acetophenone
   b. Bromobenzene and *p*-xylene
   c. Benzoic acid, 2,4-dinitrobenzoic acid, and 2,4,6-trinitrobenzoic acid

2. The following questions relate to the column chromatography experiment performed in Experiment 5D.
   a. Why does the fluorene elute first from the column?
   b. Why was the solvent changed in the middle of the column procedure?

3. Consider the following errors that could be made when running TLC. Indicate what should be done to correct the error.
   a. A two-component mixture containing 1-octene and 1,4-dimethylbenzene gave only one spot with an R_f value of 0.95. The solvent used was acetone.
   b. A two-component mixture containing a dicarboxylic acid and tricarboxylic acid gave only one spot with an R_f value of 0.05. The solvent used was hexane.
   c. When a TLC plate was developed, the solvent front ran off the top of the plate.

EXPERIMENT 6[1]

Simple and Fractional Distillation

Simple distillation

Fractional distillation

Gas chromatography

Distillation is a technique frequently used to separate and purify a liquid component from a mixture. Simply stated, distillation involves heating a liquid mixture to its boiling point, where liquid is rapidly converted to vapor. The vapors, richer in the more volatile component, are then condensed into a separate container. When the components in the mixture have sufficiently different vapor pressures (or boiling points), they can be separated by distillation.

The purpose of this experiment is to illustrate the use of distillation for separating a mixture of two volatile liquids with different boiling points. Each mixture, which will be issued as an unknown, will consist of two liquids from the following table.

Compound	Boiling Point (°C)
Hexane	69
Cyclohexane	80.7
Heptane	98.4
Toluene	110.6
Ethylbenzene	136

The liquids in the mixture will be separated by two distillation techniques: simple and fractional distillation. The results of these two methods will be compared by analyzing the composition of the **distillate** (the distilled liquid) using gas chromatography. You will also construct a graph of the distillation temperature versus the total volume of distillate collected. This graph will allow you to determine the approximate boiling points of the two liquids and to make a graphical comparison of the two distillation methods.

Experiment 6A is designed to be performed with semimicroscale glassware using a conventional distillation apparatus. A microscale alternative with a Hickman head is given in Experiment 6B; however, the scale is the same in both cases, and the experiment can be performed more easily with the semimicroscale glassware.

REQUIRED READING

New: Technique 14   Simple Distillation

Technique 15   Fractional Distillation

Technique 22   Gas Chromatography

---

[1] This experiment is based on a similar one developed by James Patterson, University of Washington, Seattle.

SPECIAL INSTRUCTIONS

Many flammable solvents are used in this experiment; therefore, do not use any flames in the laboratory.

Work in pairs on this experiment. Each pair of students will be assigned an unknown containing two liquids found in the table above. One student in the pair should perform a simple distillation and the other student, a fractional distillation. The results from these two methods will be compared.

SUGGESTED WASTE DISPOSAL

Dispose of all organic liquids for nonhalogenated organic solvents.

NOTES TO THE INSTRUCTOR

One method of insulating the air condenser used for the fractional distillation column is provided by employing two layers of clear flexible tubing (PVC, polyvinyl chloride) over the air condenser. For a ½-in. diameter column, use ½-in. × ⅝-in. outer-diameter plastic tubing on the inside and ⅝-in. I.D. × ⅞-in. O.D. tubing on the outside. Cut the tubing into 3½-in. lengths. Make a slit from end to end so that the lengths can slip over the column. Slit the tubing using sharp scissors or a razor knife with a proper handle. Do not use a razor blade, or you may get badly cut. The clear tubing lets you see what is going on in the column and also provides some insulation. Another method of insulating the fractionating column is to wrap the air condenser with a cotton pad about 3½ inches square. Prepare the cotton pad by covering both sides of one layer of cotton with aluminum foil. Wrap this pad entirely with duct tape to hold the cotton in place and to make a more durable pad. Hold the pad in place with tape or twist ties.

The temperature measured during the distillation will be most accurate if a partial immersion mercury thermometer is used. See the Instructor's Manual for additional comments about the use of other kinds of thermometers in this experiment.

Prepare unknown mixtures consisting of the following pairs of liquids: hexane–heptane, hexane–toluene, cyclohexane–toluene, and heptane–ethylbenzene. For each mixture, use an equal volume of both liquids. Distillation of these mixtures should provide a good contrast between the two distillation methods. It is *important* that you read the Instructor's Manual for helpful hints about these mixtures.

The gas chromatograph is prepared as follows: column temperature, 140°C; injection temperature, 150°C; detector temperature, 140°C; carrier gas flow rate, 100 mL/min. The recommended column is 8 feet long, with a stationary phase such as Carbowax 20M.

You should determine response factors for the five liquids given in the table on p. 49. Because the data in this experiment are expressed as volume, the response factors should also be based on volume. Inject a mixture containing equal volumes of all five compounds and determine the relative peak areas. Choose one compound as the standard and define its response factor to be equal to 1.00. Calculate the other response factors based on this reference. Typical response factors are given in Footnote 2 on p. 56.

6A   EXPERIMENT 6A

Simple and Fractional Distillation (Semimicroscale Procedure)

PROCEDURE

You should work in pairs on this experiment. Each pair of students will be assigned an unknown mixture containing equal volumes of two of the liquids from the table on p. 49. One student should perform a simple distillation on the mixture, and the other person should perform a fractional distillation.

Apparatus

Assemble the appropriate distillation apparatus (see figures). Carefully notice the position of the thermometer in these figures. The bulb of the thermometer must be placed well below the sidearm or it will not read the temperature correctly. (In order to monitor the temperature most accurately, use a partial immersion mercury thermometer.) If performing the fractional distillation, pack the air condenser uniformly with 0.8–0.9 g of stainless steel cleaning pad material. Do not pack the material too tightly at any one place in the condenser.

CAUTION

You should wear heavy cotton gloves when handling the stainless steel cleaning pad. The edges are very sharp and can easily cut into the skin.

Wrap the glass section of the air condenser between the two plastic caps with plastic tubing as described in Notes to the Instructor. Alternatively, use the method with a cotton pad (see Notes to the Instructor). Hold the pad in place with tape or twist ties.

For either the simple or fractional distillation, place a boiling stone into the 10-mL round-bottom flask. Also add 7.0 mL of the unknown mixture (measured with a 10-mL graduated cylinder) to the flask. Use a hot plate and an aluminum block for heating.

Distillation

These instructions apply to both the simple and fractional distillations. Start circulating the cooling water in the condenser and adjust the heat so that the liquid boils rapidly. During the initial stages of the distillation, continue to maintain a rapid boiling rate. As the hot vapors rise, they will gradually heat up the glassware and, in the case of the fractional distillation, the fractionating column as well. Because the mass of glass and other materials is fairly large, it will take 10–20 minutes of heating before the distillation temperature begins to rise rapidly and approaches the boiling point of the distillate. (Note that this may take longer for the fractional distillation.) When the temperature begins to level off, you should soon see drops of distillate falling into the graduated cylinder.

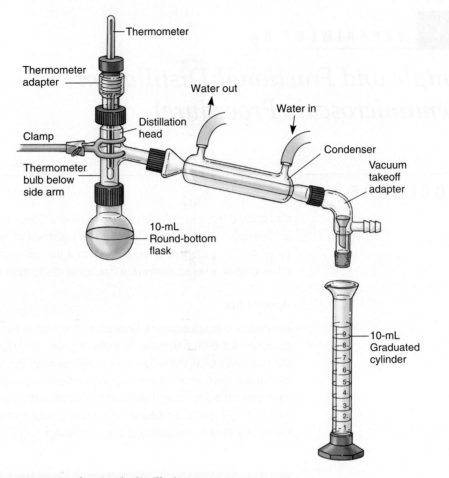

Apparatus for simple distillation.

NOTE: For the remainder of the distillation, it is very important to regulate the temperature of the hot plate so that the distillation occurs at a rate of 1 drop per 5 seconds. If the distillation is performed more rapidly than this, you may not achieve good separation between the liquids. On the other hand, if the distillation is performed less rapidly than the suggested rate, the distillation temperature may be lower than it should be.

Now you will probably need to turn down the heat control to achieve the desired rate of distillation. In addition, it may be helpful to raise the round-bottom flask slightly above the aluminum block for a minute or so to cool the mixture more quickly. You should also begin recording the distillation temperature as a function of the total volume of distillate collected. Beginning at a volume of 0.5 mL, record the temperature at every 0.5-mL interval, as determined by the volume of distillate in the 10-mL graduated cylinder. After you have collected 1.0 mL of distillate, remove the 10-mL graduated cylinder and collect the next two drops of distillate in a 3-mL conical vial. Label the vial "1-mL sample." Cap the vial tightly; otherwise, the more volatile component will evaporate more rapidly, and the composition of the mixture will change. Resume collecting the distillate in the graduated cylinder. As the distillation temperature increases, you may need to turn up the control to maintain the same rate of distillation. Continue to record the temperature and volume data. When you have collected a total of 4.5 mL of distillate, take another small sample of distillate in a second 3-mL conical vial. (If the total volume of distillate that you can collect is less than 4.5 mL, take the last two drops.) Cap the vial

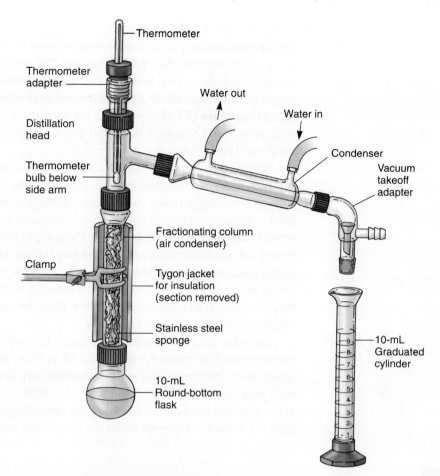

Thermometer

Thermometer adapter

Distillation head

Thermometer bulb below side arm

Clamp

Water out

Water in

Condenser

Vacuum takeoff adapter

Fractionating column (air condenser)

Tygon jacket for insulation (section removed)

Stainless steel sponge

10-mL Round-bottom flask

10-mL Graduated cylinder

Apparatus for fractional distillation.

and label it "4.5-mL sample." Then continue the distillation until there is a small amount (about 0.5 mL) of liquid remaining in the distilling flask.

NOTE: Do not distill to dryness! A dry flask may crack if it is heated too hot.

The best way to stop the distillation is to turn off the hot plate and immediately raise the entire distillation apparatus off the aluminum block.

Analysis

Distillation Curve

Using the data you collected for the distillation temperature and the total volume of distillate, construct separate graphs for the simple and fractional distillations. Plot the volume in 0.5-mL increments on the *x*-axis and the temperature on the *y*-axis. Comparing the two graphs should make clear that the fractional distillation resulted in a better separation of the two liquids. Using the graph for the fractional distillation, estimate the boiling points of the two components in your mixture by noting the two regions on the graph where the temperatures leveled off. From these approximate boiling points, try to identify the two liquids in your mixture (see table on p. 49). Note that the observed boiling point for the first component may be somewhat higher than the actual boiling point, and the observed boiling point for the second component may be somewhat lower than the actual boiling point. The reason is that the fractionating column may not be efficient enough to completely separate all of the pairs of liquids in this experiment. Therefore, it may be easier to identify the two liquids in your mixture from the gas chromatograph, as described in the next section.

Gas Chromatography

Gas chromatography is an instrumental method that separates the components of a mixture based on their boiling points. The lower-boiling component passes through the column first, followed by the higher-boiling components. The actual length of time required for a compound to pass through the column is called the **retention time** of that compound. As each component comes off the column, it is detected, and a peak is recorded that is proportional in size to the amount of the compound that was put on the column.

Gas chromatography can be used to determine the compositions of the two samples that you collected in 3-mL conical vials. The instructor or a laboratory assistant may either make the sample injections or allow you to make them. In the latter case, your instructor will give you adequate instructions beforehand. A reasonable sample size is 2.5 μL. Inject the sample into the gas chromatograph and record the gas chromatogram. Depending on how effectively the two compounds were separated by the distillation, you may see one or two peaks. The lower-boiling component has a shorter retention time than the higher-boiling one. Your instructor may provide you with the actual retention times for each compound so that you can identify each peak with more certainty.

Once the gas chromatogram has been obtained, determine the relative areas of the two peaks (Technique 22, Section 22.12, p. 810). You can calculate this by triangulation, or the instrument may do this electronically. In either case, you should divide each area by a response factor to account for differences in how the detector responds to the different compounds.[2] Calculate the percentages of the two compounds in both samples. Compare these results for the simple and fractional distillations.

6B   EXPERIMENT 6B

Simple and Fractional Distillation (Microscale Procedure)

PROCEDURE

This experiment can also be performed using a Hickman head, although it is not as easy to monitor the volume of distillate. To perform a simple distillation, refer to Figure 14.7B, page 710. For a fractional distillation, see Figure 15.2, p. 717. For both distillations, use a 10-mL round-bottom flask and a Hickman head with a side port (Fig. 14.4B, p. 707). Attach a water-cooled condenser on top of the Hickman head. It is helpful to tilt the apparatus slightly (5–10 degrees) in the direction of the side port so that the liquid in the reservoir of the Hickman head will flow toward the side port.

Follow the procedure given in Experiment 6A, except that it will be necessary to transfer the distillate from the Hickman head to a 10-mL graduated cylinder to collect data for the distillation temperature and total volume of distillate. This must

---

[2] Because response factors are instrument specific, you will be given the response factors for your instrument. Typical response factors obtained on a GowMac 69-350 gas chromatograph are hexane (1.50), cyclohexane (1.80), heptane (1.63), toluene (1.41), and ethylbenzene (1.00). These response factors were determined by injecting a mixture of equal volumes of the five liquids and determining the relative peak areas.

be done frequently so that data can be taken at 0.5-mL intervals, as indicated in the procedure. Because you will not be able to count drops, you should try to distill at a rate of three to four minutes per mL distillate. It is important to remove as much distillate as possible each time you make a transfer. Otherwise, the next sample of distillate will be contaminated by the leftover liquid.

REPORT

Distillation Curve

Record the data for the distillation temperature as a function of the volume of distillate. Construct a graph for these data (see "Analysis," p. 53). Compare the graphs for simple and fractional distillations of the same mixture. Which distillation resulted in a better separation? Explain. Report the approximate boiling points for the two compounds in your mixture and identify the compounds.

Gas Chromatography

For both the 1-mL sample and the 4.5-mL sample, determine the relative areas of the two peaks, unless there is only one peak. Divide the areas by the appropriate response factors and calculate the percentage composition of the two compounds in each sample. Compare these results for the simple and fractional distillations of the same mixture. Which distillation resulted in a better separation? Explain. Identify the two compounds in your mixture. At your instructor's option, turn in the gas chromatograms with your report.

7	**EXPERIMENT 7**

Infrared Spectroscopy and Boiling-Point Determination

Infrared spectroscopy
Boiling-point determination
Organic nomenclature
Critical thinking application

The ability to identify organic compounds is an important skill that is frequently used in the organic laboratory. Although there are several spectroscopic methods and many chemical and physical tests that can be used for identification, the goal of this experiment is to identify an unknown liquid using infrared spectroscopy and a boiling-point determination. Both methods are introduced in this experiment.

REQUIRED READING

New:	Technique 4	How to Find Data for Compounds: Handbooks and Catalogs
	Technique 13	Physical Constants of Liquids: The Boiling Point and Density, Part A. "Boiling Points and Thermometer Correction"
	Technique 25	Infrared Spectroscopy

SPECIAL INSTRUCTIONS

Many of the unknown liquids used for this experiment are flammable; therefore, do not use any flames in the laboratory. Also, be careful when handling all of the liquids because many of them are potentially toxic.

This experiment can be performed individually, with each student working on one unknown. However, the opportunity to learn is greater if students work in groups of three. In this case, each group is assigned three unknowns. Each student in the group obtains an infrared spectrum and performs a boiling-point determination on one of the unknowns. Subsequently, the student shares this information with the other two students in the group. Then each student analyzes the collective results for the three unknowns and writes a laboratory report based on all three unknowns. Your instructor will inform you whether you should work alone or in groups.

SUGGESTED WASTE DISPOSAL

If you have not identified the unknown by the end of the laboratory period, you should return the unknown liquid to your instructor in the original container in which it was issued to you. If you have identified the compound, dispose of it in either the container for halogenated waste or the one for nonhalogenated waste, whichever is appropriate.

NOTES TO THE INSTRUCTOR

If you choose to have students work in groups of three, be sure to assign unknowns that differ both in structure *and* functional group, with at least one aromatic compound in each set. If the experiment is performed early in the year, students may have some difficulty in finding the structures of the compounds that are in the list of possible unknowns (see p. 58), and they will need help. For each unknown, structures will be needed for several of the possible compounds. In fact, compounds with boiling points as much as 5°C higher should be considered because student-determined boiling points are frequently low. This will depend on the method used and the skill of the person performing the technique. *The Merck Index,* the *CRC Handbook of Chemistry and Physics,* and the lecture textbook can all be helpful in determining these structures. Technique 4, "How to Find Data for Compounds: Handbooks and Catalogs," provides helpful information for students just beginning to use handbooks. The nuclear magnetic resonance (NMR) portion of the experiment is optional. We suggest that access to the NMR be granted only after a plausible solution has been tendered. If you do not have an NMR, there are several online databases where you can obtain a printed copy of the spectrum to hand to students.

For the boiling-point determination, it is best to use partial immersion mercury thermometers for most methods. The results are generally better with mercury thermometers than with nonmercury ones, and if you use partial immersion mercury thermometers, you do not need to perform a stem correction.

PROCEDURE

Part A. Infrared Spectrum

Obtain the infrared spectrum of your unknown liquid (Technique 25, Section 25.2, p. 834). If you are working in a group, provide copies of your spectrum for everyone in your group. Identify the significant absorption peaks by labeling them *right on the spectrum,* and include the spectrum in your laboratory report. Absorption peaks corresponding to the following groups should be identified:

C—H (sp^3)

C—H (sp^2)

C—H (aldehyde)

O—H

C=O

C=C (aromatic)

aromatic substitution pattern

C—O

C—X (if applicable)

N—H

Part B. Boiling-Point Determination

Perform a boiling-point determination on your unknown liquid (Technique 13, Section 13.2, p. 695). Your instructor will indicate which method to use. Depending on the method used and the skill of the person performing the technique, boiling points can sometimes be slightly inaccurate. When experimental boiling points are inaccurate, it is more common for them to be lower than the literature value. The difference may be as much as 5°C, especially for higher-boiling liquids. Your instructor may be able to give you more guidance about what level of accuracy you can expect.

Part C. Analysis and Report

Using the structural information from the infrared spectrum and the boiling point of your unknown, identify this liquid from the list of compounds on page 60. If you are working in a group, you will need to do this for all three compounds. In order to make use of the structural information determined from the infrared spectrum, you will need to know the structures of the compounds that have boiling points close to the value you experimentally determined. You may need to consult *The Merck Index* or the *CRC Handbook of Chemistry and Physics*. It may also be helpful to look up these compounds in the index of your lecture textbook. If there is more than one compound that fits the infrared spectrum and is within a few degrees of the experimental boiling point, you should list all of these in your laboratory report.

In your laboratory report, include (1) the infrared spectrum with the significant absorption peaks identified *right on the spectrum,* (2) the experimental boiling point for your unknown, and (3) your identification of the unknown. Explain your justifications for making this identification and write out the structure of this compound.

Optional Exercise: NMR Spectrum

At the option of your instructor, you may be asked to determine the nuclear magnetic resonance spectrum of your unknown liquid (Technique 26, Section 26.1, p. 870). Alternatively, your instructor may issue you a previously run spectrum of your compound. You should provide structural assignments for all of the groups of hydrogens that are present. Do this *right on the spectrum*. If you have correctly determined the identity of your unknown, all the groups of hydrogens (and their chemical shifts) should fit your structure. Include the properly labeled spectrum in your report and explain why it fits the suggested structure.

List of possible unknown liquids

Compound	BP (°C)	Compound	BP (°C)
Acetone	56	Butyl acetate	127
2-Methylpentane	62	2-Hexanone	128
sec-Butylamine	63	Morpholine	129
Isobutyraldehyde	64	3-Methyl-1-butanol	130
Methanol	65	Hexanal	130
Isobutylamine	69	Chlorobenzene	132
Hexane	69	2,4-Pentanedione	134
Vinyl acetate	72	Cyclohexylamine	135
1,3,5-Trifluorobenzene	75	Ethylbenzene	136
Butanal	75	p-Xylene	138
Ethyl acetate	77	1-Pentanol	138
Butylamine	78	Propionic acid	141
Ethanol	78	Pentyl acetate	142
2-Butanone	80	4-Heptanone	144
Cyclohexane	81	2-Ethyl-1-butanol	146
Isopropyl alcohol	82	N-Methylcyclohexylamine	148
Cyclohexene	83	2,2,2-Trichloroethanol	151
Isopropyl acetate	85	2-Heptanone	151
Triethylamine	89	Heptanal	153
3-Methylbutanal	92	Isobutyric acid	154
3-Methyl-2-butanone	94	Bromobenzene	156
1-Propanol	97	Cyclohexanone	156
Heptane	98	Dibutylamine	159
tert-Butyl acetate	98	Cyclohexanol	160
2,2,4-Trimethylpentane	99	Butyric acid	162
2-Butanol	99	Furfural	162
Formic acid	101	Diisobutyl ketone	168
2-Pentanone	101	Furfuryl alcohol	170
2-Methyl-2-butanol	102	Octanal	171
Pentanal	102	Decane	174
3-Pentanone	102	Isovaleric acid	176
Propyl acetate	102	Limonene	176
Piperidine	106	1-Heptanol	176
2-Methyl-1-propanol	108	Benzaldehyde	179
1-Methylcyclohexene	110	Cycloheptanone	181
Toluene	111	1,4-Diethylbenzene	184
sec-Butyl acetate	111	Iodobenzene	186
Pyridine	115	1-Octanol	195
4-Methyl-2-pentanone	117	Methyl benzoate	199
2-Ethylbutanal	117	Methyl phenyl ketone	202
Methyl 3-methylbutanoate	117	Benzyl alcohol	204
Acetic acid	118	4-Methylbenzaldehyde	204
1-Butanol	118	Ethyl benzoate	212
Octane	126		

Aspirin

Aspirin is one of the most popular cure-alls of modern life. Even though its curious history began more than 200 years ago, we still have much to learn about this enigmatic remedy. No one yet knows exactly how or why it works, yet more than 15 billion aspirin tablets are consumed each year in the United States.

The history of aspirin began on June 2, 1763, when Edward Stone, a clergyman, read a paper to the Royal Society of London entitled "An Account of the Success of the Bark of the Willow in the Cure of Agues." By *ague*, Stone was referring to what we now call malaria, but his use of the word *cure* was optimistic; what his extract of willow bark actually did was to reduce the feverish symptoms of the disease. Almost a century later, a Scottish physician was to find that extracts of willow bark would also alleviate the symptoms of acute rheumatism. This extract was ultimately found to be a powerful **analgesic** (pain reliever), **antipyretic** (fever reducer), and **anti-inflammatory** (reduces swelling) drug.

Soon thereafter, organic chemists working with willow bark extract and flowers of the meadowsweet plant (which gave a similar compound) isolated and identified the active ingredient as salicylic acid (from *salix*, the Latin name for the willow tree). The substance could then be chemically produced in large quantities for medical use. It soon became apparent that using salicylic acid as a remedy was severely limited by its acidic properties. The substance irritated the mucous membranes lining the mouth, gullet, and stomach. The first attempts at circumventing this problem by using the less acidic sodium salt (sodium salicylate) were only partially successful. This substance was less irritating but had such an objectionable sweetish taste that most people could not be induced to take it. The breakthrough came at the turn of the century (1893) when Felix Hofmann, a chemist for the German firm of Bayer, devised a practical route for synthesizing acetylsalicylic acid, which was found to have all the same medicinal properties without the highly objectionable taste or the high degree of mucosal-membrane irritation. Bayer called its new product "aspirin," a name derived from *a-* for acetyl, and the root *-spir*, from the Latin name for the meadowsweet plant, *spirea*.

Salicylic acid   Sodium salicylate   Acetylsalicylic acid
(aspirin)

The history of aspirin is typical of many of the medicinal substances in current use. Many began as crude plant extracts or folk remedies whose active ingredients were isolated, and their structure was determined by chemists, who then improved on the original.

In the last few years, the mode of action of aspirin has just begun to be explained. A whole new class of compounds called **prostaglandins** has been found to be involved in the body's immune responses. Their synthesis is provoked by interference with the body's normal functioning by foreign substances or unaccustomed stimuli.

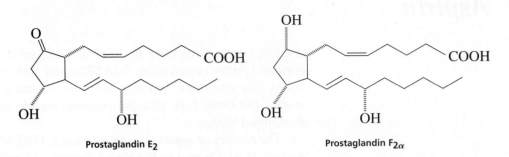

Prostaglandin E$_2$ **Prostaglandin F$_{2\alpha}$**

These substances are involved in a wide variety of physiological processes and are thought to be responsible for evoking pain, fever, and local inflammation. Aspirin has recently been shown to prevent bodily synthesis of prostaglandins and thus to alleviate the symptomatic portion (fever, pain, inflammation, menstrual cramps) of the body's immune responses (that is, the ones that let you know something is wrong). One report suggests that aspirin may inactivate one of the enzymes responsible for the synthesis of prostaglandins. The natural precursor for prostaglandin synthesis is **arachidonic acid.** This substance is converted to a peroxide intermediate by an enzyme called **cyclo-oxygenase,** or prostaglandin synthase. This intermediate is converted further to prostaglandin. The apparent role of aspirin is to attach an acetyl group to the active site of cyclo-oxygenase, thus rendering it unable to convert arachidonic acid to the peroxide intermediate. In this way, prostaglandin synthesis is blocked.

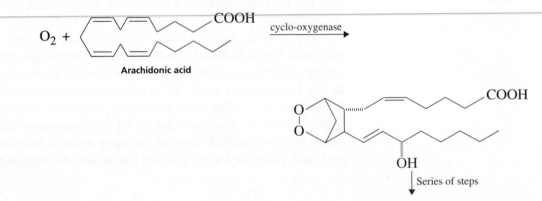

Arachidonic acid

cyclo-oxygenase

Prostaglandins

Series of steps

Aspirin tablets (5-grain size) are usually compounded of about 0.32 g of acetylsalicylic acid pressed together with a small amount of starch, which binds the ingredients. Buffered aspirin usually contains a basic buffering agent to reduce the acidic irritation of mucous membranes in the stomach because the acetylated product is not totally free of this irritating effect. Bufferin contains 0.325 g of aspirin, together with calcium carbonate, magnesium oxide, and magnesium carbonate as buffering agents. Combination pain relievers usually contain aspirin, acetaminophen, and caffeine. Extra-Strength Excedrin, for instance, contains 0.250 g aspirin, 0.250 g acetaminophen, and 0.065 g caffeine.

REFERENCES

"Aspirin Cuts Deaths after Heart Attacks." *New Scientist*, 118 (April 7, 1988): 22.

Collier, H. O. J. "Aspirin." *Scientific American*, 209 (November 1963): 96.

Collier, H. O. J. "Prostaglandins and Aspirin." *Nature*, 232 (July 2, 1971): 17.

Disla, E., Rhim, H. R., Reddy, A., and Taranta, A. "Aspirin on Trial as HIV Treatment." *Nature*, 366 (November 18, 1993): 198.

Kingman, S. "Will an Aspirin a Day Keep the Doctor Away?" *New Scientist*, 117 (February 11, 1988): 26.

Kolata, G. "Study of Reye's–Aspirin Link Raises Concerns." *Science*, 227 (January 25, 1985): 391.

Macilwain, C. "Aspirin on Trial as HIV Treatment." *Nature*, 364 (July 29, 1993): 369.

Nelson, N. A., Kelly, R. C., and Johnson, R. A. "Prostaglandins and the Arachidonic Acid Cascade." *Chemical and Engineering News* (August 16, 1982): 30.

Pike, J. E. "Prostaglandins." *Scientific American*, 225 (November 1971): 84.

Roth, G. J., Stanford, N., and Majerus, P. W. "Acetylation of Prostaglandin Synthase by Aspirin." *Proceedings of the National Academy of Science of the U.S.A.*, 72 (1975): 3073.

Street, K. W. "Method Development for Analysis of Aspirin Tablets." *Journal of Chemical Education*, 65 (October 1988): 914.

Vane, J. R. "Inhibition of Prostaglandin Synthesis as a Mechanism of Action for Aspirin-Like Drugs." *Nature–New Biology*, 231 (June 23, 1971): 232.

Weissmann, G. "Aspirin." *Scientific American*, 264 (January 1991): 84.

8   EXPERIMENT 8

Acetylsalicylic Acid

Crystallization

Vacuum filtration

Melting point

Esterification

Aspirin (acetylsalicylic acid) can be prepared by the reaction between salicylic acid and acetic anhydride:

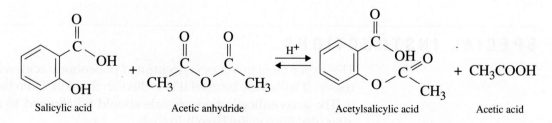

In this reaction, the **hydroxyl group** (—OH) on the benzene ring in salicylic acid reacts with acetic anhydride to form an **ester** functional group. Thus, the formation of acetylsalicylic acid is referred to as an **esterification** reaction. This reaction requires the presence of an acid catalyst, indicated by the H^+ above the equilibrium arrows.

When the reaction is complete, some unreacted salicylic acid and acetic anhydride will be present, along with acetylsalicylic acid, acetic acid, and the catalyst. The technique used to purify the acetylsalicylic acid from the other substances is called **crystallization.** This technique, which was introduced in Experiment 3, will be studied in more detail in Experiment 10. The basic principle is quite simple. At the end of this reaction, the reaction mixture will be hot, and all substances will be in solution. As the solution is allowed to cool, the solubility of acetylsalicylic acid will decrease, and it will gradually come out of solution, or crystallize. Because the other substances are either liquids at room temperature or are present in much smaller amounts, the crystals formed will be composed mainly of acetylsalicylic acid. Thus, a separation of acetylsalicylic acid from the other materials will have been accomplished. The purification process is facilitated by the addition of water after the crystals have formed. The water decreases the solubility of acetylsalicylic acid and dissolves some of the impurities.

The most likely impurity in the final product is salicylic acid itself, which can arise from incomplete reaction of the starting materials or from **hydrolysis** (reaction with water) of the product during the isolation steps. The hydrolysis reaction of acetylsalicylic acid produces salicylic acid. Salicylic acid and other compounds that contain a hydroxyl group on the benzene ring are referred to as **phenols.** Phenols form a highly colored complex with ferric chloride (Fe^{3+} ion). Aspirin is not a phenol, because it does not possess a hydroxyl group directly attached to the ring. Because aspirin will not give the color reaction with ferric chloride, the presence of salicylic acid in the final product is easily detected. The purity of your product will also be determined by obtaining the melting point.

REQUIRED READING

Review:	Introduction to Microscale Laboratory (pp. 2–13)	
	Technique 8	Filtration, Sections 8.1–8.6
	Technique 9	Physical Constants, Melting Points
New:	Technique 5	Measurement of Volume and Weight
	Technique 6	Heating and Cooling Methods
	Technique 7	Reaction Methods, Sections 7.1–7.4
	Essay	Aspirin

SPECIAL INSTRUCTIONS

This experiment involves concentrated phosphoric acid, which is highly corrosive. It will cause burns if it is spilled on the skin. Exercise care in handling it. The acetylsalicylic acid crystals should be allowed to air-dry overnight after filtration on the Hirsch funnel.

SUGGESTED WASTE DISPOSAL

Dispose of the aqueous filtrate in the container for aqueous waste.

PROCEDURE

Preparation of Acetylsalicylic Acid (Aspirin)

Prepare a hot water bath using a 250-mL beaker and a hot plate. Use about 100 mL of water and adjust the temperature to about 50°C. Weigh 0.210 g of salicylic acid (*MW* = 138.1) and place this in a **dry** 5-mL conical vial. It is not necessary for you to weigh exactly 0.210 g of salicylic acid. Try to obtain a weight within about 0.005 g of the indicated weight without spending excessive time at the balance. Record the actual weight in your notebook, and use this weight in any subsequent calculations. Using an automatic pipet or a dispensing pump, add 0.480 mL of acetic anhydride (*MW* = 102.1, *d* = 1.08 g/mL), followed by exactly one drop of concentrated phosphoric acid from a Pasteur pipet.

> **CAUTION**   ⚠
>
> **Concentrated phosphoric acid is highly corrosive. You must handle it with great care.**

Add a magnetic spin vane (Fig. 7.8a, p. 607) and attach an air condenser to the vial. Clamp this assembly so that the vial is partially submerged in the hot water bath (Fig. 6.6, p. 594). Stir the mixture with the spin vane until the salicylic acid dissolves. (If the spin vane becomes stuck in the solid salicylic acid, insert a microspatula through the air condenser into the conical vial and gently push the spin vane until it begins spinning.) Heat the mixture for 8–10 minutes after the solid dissolves to complete the reaction.

Crystallization of Acetylsalicylic Acid

Remove the vial from the water bath and allow it to cool. After the vial has cooled enough for you to handle it, detach the air condenser and remove the spin vane with forceps or a magnetic stirring bar. (If you use forceps, be sure to clean them.) Place the conical vial in a small beaker and allow the vial to cool to room temperature, during which time the acetylsalicylic acid should begin to crystallize from the reaction mixture. If it does not crystallize, scratch the walls of the vial with a glass rod (not fire-polished) and cool the mixture slightly in an ice-water bath (Technique 11, Section 11.3C, p. 655) until crystallization has occurred. (Scratching the inside walls of the container often helps to initiate crystallization.) After crystal formation is complete (usually when the product appears as a solid mass), add 3.0 mL of water (measured with a 10-mL graduated cylinder) and stir thoroughly with a microspatula.

Vacuum Filtration

Set up a Hirsch funnel for vacuum filtration (see Technique 8, Section 8.3, and Fig. 8.5, p. 622). Moisten the filter paper with a few drops of water and turn on the vacuum (or aspirator) to the fullest extent. Transfer the mixture in the conical vial to the Hirsch funnel. When you have removed as much product as possible from the vial, add about 1 mL of cold water to the vial using a calibrated Pasteur pipet (p. 11). Stir the mixture and transfer the remaining crystals and water to the Hirsch funnel. When all the crystals have been collected in the funnel, rinse them with several 0.5-mL portions of cold water. Continue drawing air through the crystals on the Hirsch funnel by suction until the crystals are nearly dry (5–10 minutes). Remove the crystals for air-drying on a watch glass or clay plate. It is convenient to hold the filter paper disc with forceps while *gently* scraping the crystals off the filter paper with a microspatula. If the paper is scraped too hard, small pieces of paper will be removed along with the crystals. To dry the crystals completely, you must set the crystals aside overnight. Weigh the dry product and calculate the percentage yield of acetylsalicylic acid (*MW* = 180.2).

Test for Purity

Ferric Chloride Test

You can perform this test on a sample of your product that is not completely dry. To determine if there is any salicylic acid remaining in your product, carry out the following procedure. Obtain three small test tubes. Add 0.5 mL of water to each test tube. Dissolve a small amount of salicylic acid in the first tube. Add a similar amount of your product to the second tube. The third test tube, which contains only solvent, will serve as the control. Add one drop of 1% ferric chloride solution to each tube and note the color after shaking. Formation of an iron–phenol complex with Fe(III) gives a definite color ranging from red to violet, depending on the particular phenol present.

Melting Point

As an additional test for purity, determine the melting point of your product (see Technique 9, Sections 9.5–9.8, pp. 631–636). The melting point must be obtained with a completely dried sample. Pure aspirin has a melting point of 135–136°C.

Place your product in a small vial, label it properly (p. 565), and submit it to your instructor.

Aspirin Tablets

Aspirin tablets are acetylsalicylic acid pressed together with a small amount of inert binding material. Common binding substances include starch, methylcellulose, and microcrystalline cellulose. You can test for the presence of starch by boiling approximately one-fourth of an aspirin tablet with 2 mL of water. Cool the liquid and add a drop of iodine solution. If starch is present, it will form a complex with the iodine. The starch–iodine complex is deep blue–violet. Repeat this test with a commercial aspirin tablet and with the acetylsalicylic acid prepared in this experiment.

QUESTIONS

1. What is the purpose of the concentrated phosphoric acid used in the first step?
2. What would happen if the phosphoric acid were left out?
3. If you used 250 mg of salicylic acid and excess acetic anhydride in the preceding synthesis of aspirin, what would be the theoretical yield of acetylsalicylic acid in moles? In milligrams?
4. What is the equation for the decomposition reaction that can occur with aspirin in water?
5. Most aspirin tablets contain five grains of acetylsalicylic acid. How many milligrams is this? (*Hint:* See the essay "Aspirin.")
6. A student performed the reaction in this experiment using a water bath at 90°C instead of 50°C. The final product was tested for the presence of phenols with ferric chloride. This test was negative (no color observed); however, the melting point of the dry product was 122–125°C. Explain these results as completely as possible.
7. If the aspirin crystals were not completely dried before the melting point was determined, what effect would this have on the observed melting point?

Analgesics

Acylated aromatic amines (those having an acyl group, $R-\overset{\overset{\displaystyle O}{\|}}{C}-$, substituted on nitrogen) are important in over-the-counter headache remedies. Over-the-counter drugs are those you may buy without a prescription. Acetanilide, phenacetin, and acetaminophen are mild analgesics (relieve pain) and antipyretics (reduce fever) and are important, along with aspirin, in many nonprescription drugs.

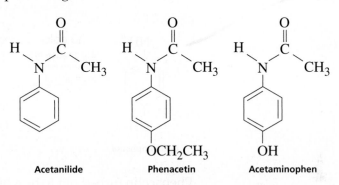

Acetanilide **Phenacetin** **Acetaminophen**

The discovery that acetanilide was an effective antipyretic came about by accident in 1886. Two doctors, Cahn and Hepp, had been testing naphthalene as a possible **vermifuge** (an agent that expels worms). Their early results on simple worm cases were discouraging, so Dr. Hepp decided to test the compound on a patient with a larger variety of complaints, including worms—a sort of shotgun approach. A short time later, Dr. Hepp excitedly reported to his colleague, Dr. Cahn, that naphthalene had miraculous fever-reducing properties.

In trying to verify this observation, the doctors discovered that the bottle they thought contained naphthalene had apparently been mislabeled. In fact, the bottle brought to them by their assistant had a label so faint as to be illegible. They were sure that the sample was not naphthalene, because it had no odor. Naphthalene has a strong odor reminiscent of mothballs. So close to an important discovery, the doctors were nevertheless stymied; they appealed to a cousin of Hepp, who was a chemist in a nearby dye factory, to help them identify the unknown compound. This compound turned out to be acetanilide, a compound with a structure not at all like that of naphthalene. Certainly, Hepp's unscientific and risky approach would be frowned on by doctors today; and to be sure, the Food and Drug Administration (FDA) would never allow human testing before extensive animal testing (consumer protection has progressed). Nevertheless, Cahn and Hepp made an important discovery.

Naphthalene

In another instance of serendipity, the publication of Cahn and Hepp, describing their experiments with acetanilide, caught the attention of

Carl Duisberg, director of research at the Bayer Company in Germany. Duisberg was confronted with the problem of profitably getting rid of nearly 50 tons of *p*-aminophenol, a by-product of the synthesis of one of Bayer's other commercial products. He immediately saw the possibility of converting *p*-aminophenol to a compound similar in structure to acetanilide by putting an acyl group on the nitrogen. It was then believed, however, that all compounds having a hydroxyl group on a benzene ring (that is, phenols) were toxic. Duisberg devised a scheme of structural modification of *p*-aminophenol to synthesize the compound phenacetin. The reaction scheme is shown here.

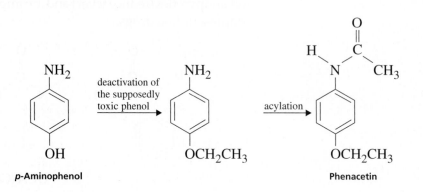

p-Aminophenol   **Phenacetin**

Phenacetin turned out to be a highly effective analgesic and antipyretic. A common form of combination pain reliever, called an APC tablet, was once available. An APC tablet contained **A**spirin, **P**henacetin, and **C**affeine (hence, **APC**). Phenacetin is no longer used in commercial pain-relief preparations. It was later found that not all aromatic hydroxyl groups lead to toxic compounds, and today the compound acetaminophen is widely used as an analgesic in place of phenacetin.

Another analgesic, structurally similar to aspirin, that has found some application is **salicylamide.** Salicylamide is found as an ingredient in some pain-relief preparations, although its use is declining.

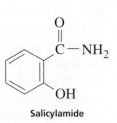

Salicylamide

On continued or excessive use, acetanilide can cause a serious blood disorder called **methemoglobinemia,** in which the central iron atom in hemoglobin is converted from Fe(II) to Fe(III) to give methemoglobin. Methemoglobin will not function as an oxygen carrier in the bloodstream. The result is a type of anemia (deficiency of hemoglobin or lack of red blood cells). Phenacetin and acetaminophen cause the same disorder but to a much lesser degree. Because they are also more effective as antipyretic and analgesic drugs than acetanilide, they are preferred remedies. Acetaminophen is marketed under a variety of trade names, including Tylenol, Datril, and Panadol, and is often successfully used by people who are allergic to aspirin.

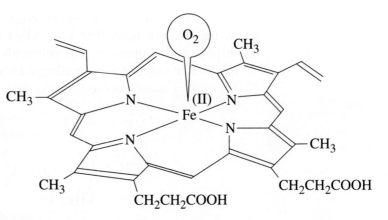

Heme portion of blood-oxygen carrier, hemoglobin.

More recently, a new drug has appeared in over-the-counter preparations. This drug is **ibuprofen,** which is marketed as a prescription drug in the United States under the name Motrin. Ibuprofen was first developed in England in 1964. United States marketing rights were obtained in 1974. Ibuprofen is now sold without prescription under brand names, which include Advil, Motrin, and Nuprin. Ibuprofen is principally an anti-inflammatory drug, but it is also effective as an analgesic and an antipyretic. It is particularly effective in treating the symptoms of rheumatoid arthritis and menstrual cramps. Ibuprofen appears to control the production of prostaglandins, which parallels the mode of action of aspirin. An important advantage of ibuprofen is that it is a powerful pain reliever. One 200-mg tablet is as effective as two tablets (650 mg) of aspirin. Furthermore, ibuprofen has a more advantageous dose–response curve, which means that taking two tablets of this drug is approximately twice as effective as one tablet for certain types of pain. Aspirin and acetaminophen reach their maximum effective dose at two tablets. Little additional relief is gained at doses above

Analgesics and caffeine in some common preparations

	Aspirin	Acetaminophen	Caffeine	Salicylamide	Ibuprofen	Ketoprofen	Naproxen
Aspirin*	0.325 g	—	—	—	—	—	—
Anacin	0.400 g	—	0.032 g	—	—	—	—
Bufferin	0.325 g	—	—	—	—	—	—
Cope	0.421 g	—	0.032 g	—	—	—	—
Excedrin (Extra-Strength)	0.250 g	0.250 g	0.065 g	—	—	—	—
Tylenol	—	0.325 g	—	—	—	—	—
B. C. Tablets	0.325 g	—	0.016 g	0.095 g	—	—	—
Advil	—	—	—	—	0.200 g	—	—
Aleve	—	—	—	—	—	—	0.220 g
Orudis	—	—	—	—	—	0.0125 g	—

Note: Nonanalgesic ingredients (e.g., buffers) are not listed.

*Five-grain tablet (1 grain = 0.0648 g).

that level. Ibuprofen, however, continues to increase its effectiveness up to the 400-mg level (the equivalent of four tablets of aspirin or acetaminophen). Ibuprofen is a relatively safe drug, but its use should be avoided in cases of aspirin allergy, kidney problems, ulcers, asthma, hypertension, or heart disease.

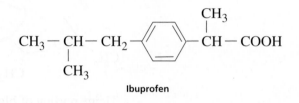

Ibuprofen

The Food and Drug Administration has also approved two other drugs with similar structures to ibuprofen for over-the-counter use as pain relievers. These new drugs are known by their generic names, **naproxen** and **ketoprofen.** Naproxen is often administered in the form of its sodium salt. Naproxen and ketoprofen can be used to alleviate the pain of headaches, toothaches, muscle aches, backaches, arthritis, and menstrual cramps, and they can also be used to reduce fever. They appear to have a longer duration of action than the older analgesics.

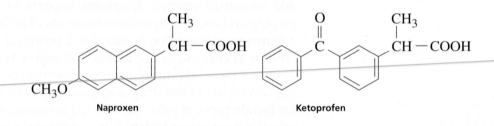

Naproxen   **Ketoprofen**

REFERENCES

Barr, W. H., and Penna, R. P. "O-T-C Internal Analgesics." In G. B. Griffenhagen, ed., *Handbook of Non-Prescription Drugs*, 7th ed. Washington, DC: American Pharmaceutical Association, 1982.

Bugg, C. E., Carson, W. M., and Montgomery, J. A. "Drugs by Design." *Scientific American*, 269 (December 1993): 92.

Flower, R. J., Moncada, S., and Vane, J. R. "Analgesic-Antipyretics and Anti-inflammatory Agents; Drugs Employed in the Treatment of Gout." In A. G. Gilman, L. S. Goodman, T. W. Rall, and F. Murad, eds., *The Pharmacological Basis of Therapeutics*, 7th ed. New York: Macmillan, 1985.

Hansch, C. "Drug Research or the Luck of the Draw." *Journal of Chemical Education*, 51 (1974): 360.

"The New Pain Relievers." *Consumer Reports*, 49 (November 1984): 636–638.

Ray, O. S. "Internal Analgesics." *Drugs, Society, and Human Behavior*, 2nd ed. St. Louis: C. V. Mosby, 1978.

Senozan, N. M. "Methemoglobinemia: An Illness Caused by the Ferric State." *Journal of Chemical Education*, 62 (March 1985): 181.

9   **EXPERIMENT 9**

Isolation of the Active Ingredient in an Analgesic Drug

Extraction

Filtration

Melting point

Most analgesic (pain-relieving) drugs found on the shelves of any drug or grocery store generally fall into one of four categories. These drugs may contain **acetylsalicylic acid, acetaminophen,** or **ibuprofen** as the active ingredient, or some **combination** of these compounds may be used in a single preparation. All tablets, regardless of type, contain a large amount of starch or other inert substance. This material acts as a binder to keep the tablet from falling apart and to make it large enough to handle. Some analgesic drugs also contain caffeine or buffering agents. In addition, many tablets are coated to make them easier to swallow and to prevent users from experiencing the unpleasant taste of the drugs.

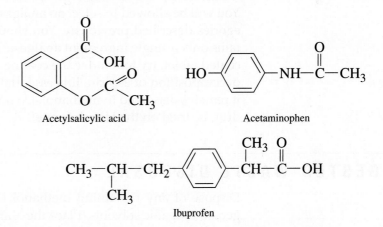

The three drugs, along with their melting points (MP) and common brand names, follow:

Drug	MP	Brand Names
Acetylsalicylic acid	135–136°C	Aspirin, ASA, acetylsalicylic acid, generic aspirin, Empirin
Acetaminophen	169–170.5°C	Tylenol, Datril, Panadol, nonaspirin pain reliever (various brands)
Ibuprofen	75–77°C	Advil, Brufen, Motrin, Nuprin

The purpose of this experiment is to demonstrate some important techniques that are applied throughout this textbook and to allow you to become accustomed to working in the laboratory at the microscale level. More specifically, you will extract (dissolve) the active ingredient of an analgesic drug by mixing the powdered tablet with a solvent, methanol. Two steps are required to remove the fine particles of binder, which remain suspended in the solvent. First, you will use centrifugation to remove most of the binder. The second step will be a filtration technique using a Pasteur pipet packed with alumina (finely ground aluminum oxide). The solvent will then be evaporated to yield the solid analgesic, which will be collected by filtration on a Hirsch funnel. Finally, you will test the purity of the drug by doing a melting-point determination.

REQUIRED READING

Review: Experiment 1   Introduction to Microscale Laboratory (pp. 2–13)

New: Technique 7   Reaction Methods, Section 7.9
Technique 8   Filtration, Sections 8.1–8.6
Technique 9   Physical Constants of Solids: The Melting Point

SPECIAL INSTRUCTIONS

You will be allowed to select an analgesic that is a member of one of the categories described previously. You should use an uncoated tablet that contains only a single ingredient analgesic and binder. If it is necessary to use a coated tablet, try to remove the coating when the tablet is crushed. To avoid decomposition of aspirin, it is essential to minimize the length of time that it remains dissolved in methanol. Do not stop this experiment until after the drug is dried on the Hirsch funnel.

SUGGESTED WASTE DISPOSAL

Dispose of any remaining methanol in the waste container for nonhalogenated organic solvents. Place the alumina in the container designated for wet alumina.

PROCEDURE

Extraction of Active Ingredient

If you are isolating aspirin or acetaminophen, use *one* tablet in this procedure. If you are isolating ibuprofen, use *two* tablets. Using a pestle, crush the tablet (or tablets) between two pieces of weighing paper. If the tablet is coated, try to remove fragments of the coating material with forceps after the tablet is first crushed. Add all the powdered material to a 3-mL conical vial. Using a calibrated Pasteur pipet (p. 11), add about 2 mL of methanol to the vial. Cap the vial and mix thoroughly by shaking. Loosen the cap at least once during the mixing process to release any pressure that may build up in the vial.

Allow the undissolved portion of the powder to settle in the vial. A cloudy suspension may remain even after 5 minutes or more. You should wait only until it is obvious that the larger particles have settled completely. Using a filter-tip pipet

(Fig. 8.9, p. 625), transfer the liquid phase to a centrifuge tube. Add a second 2-mL portion of methanol to the conical vial and repeat the shaking process described previously. After the solid has settled, transfer the liquid phase to the centrifuge tube containing the first extract.

Place the tube in a centrifuge along with another centrifuge tube of equal weight on the opposite side. Centrifuge the mixture for two to three minutes. The suspended solids should collect on the bottom of the tube, leaving a clear or nearly clear **supernatant liquid,** the liquid above the solid. If the liquid is still quite cloudy, repeat the centrifugation for a longer period or at a higher speed. Being careful not to disturb the solid at the bottom of the tube, transfer the supernatant liquid with a Pasteur pipet to a test tube or small beaker.

Column Chromatography

Prepare an alumina column using a Pasteur pipet, as shown in the figure. Insert a small ball of cotton into the top of the column. Using a long, thin object such as a glass stirring rod or a wooden applicator stick, push the cotton down so that it fits into the Pasteur pipet where the constriction begins. Add about 0.5 g of alumina to the pipet and tap the column with your finger to pack the alumina. Clamp the pipet in a vertical position so that the liquid can drain from the column into a small beaker or a 5-mL conical vial. Place a small beaker under the column. With a calibrated Pasteur pipet, add about 2 mL of methanol to the column and allow the liquid to drain until the level of the methanol just reaches the top of the alumina. Once methanol has been added to the alumina, the top of the alumina in the column should not be allowed to run dry. If necessary, add more methanol.

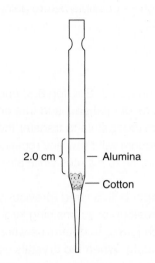

2.0 cm — Alumina

— Cotton

Column for purifying an analgesic drug.

NOTE: It is essential that the methanol not be allowed to drain below the surface of the alumina.

When the level of the methanol reaches the surface of the alumina, transfer the solution containing the drug from the beaker or test tube to the column using a Pasteur pipet. Collect the liquid that passes through the column into a 5-mL conical vial. When all the liquid from the beaker has been added to the column and has penetrated the alumina, add an additional 1 mL of methanol to the column and allow it to drain. This ensures that all the analgesic drug has been eluted from the column.

Evaporation of Solvent

If you are isolating aspirin, it is essential that the following evaporation procedure be completed in 10–15 minutes. Otherwise, the aspirin may partially decompose. Using a Pasteur pipet, transfer about half the liquid in the 5-mL conical vial to another small container. Evaporate the methanol in the 5-mL conical vial using a water bath at about 50°C (Technique 7, Section 7.10, p. 611).[1] To speed evaporation, direct a gentle stream of dry air or nitrogen into the vial containing the liquid. Evaporate the solvent until the volume is less than about 1 mL. Then add the remainder of the liquid and continue evaporation.

When the solvent has completely evaporated or it is apparent that the remaining liquid is no longer evaporating, remove the vial from the water bath (or sand bath)

---

[1] As an alternative, you may use a sand bath at about 50°C.

and allow it to cool to room temperature. (The volume of liquid should be less than 0.5 mL when you discontinue evaporation.) If liquid remains, which is likely with the ibuprofen- or acetaminophen-containing analgesics, place the cool vial in an ice-water bath for 10–15 minutes. Prepare the ice-water bath in a small beaker, using both ice and water. Be sure that the vial cannot tip over. Crystallization of the product may occur more readily if you scrape the inside of the vial with a microspatula or a glass rod (not fire-polished). If the solid is hard and clumped, you should use a microspatula to break up the solid as much as possible before going on to the next step.

Vacuum Filtration

Set up a Hirsch funnel for vacuum filtration (see Technique 8, Section 8.3, and Fig. 8.5, p. 622). Moisten the filter paper with a few drops of methanol and turn on the vacuum (or aspirator) to the fullest extent. Use a microspatula to transfer the material in the conical vial to the Hirsch funnel. The vacuum will draw any remaining solvent from the crystals. Allow the crystals to dry for 5–10 minutes while air is drawn through the crystals in the Hirsch funnel.

Carefully scrape the dried crystals from the filter paper onto a tared (previously weighed) watch glass. If necessary, use a spatula to break up any remaining large pieces of solid. Allow the crystals to air-dry on the watch glass. To determine when the crystals are dry, move them around with a dry spatula. When the crystals no longer clump or cling to the spatula, they should be dry. If you are working with ibuprofen, the solid will be slightly sticky even when it is completely dried. Weigh the watch glass with the crystals to determine the weight of analgesic drug that you have isolated. Use the weight of the active ingredient specified on the label of the container as a basis for calculating the weight percentage recovery.

Use a small sample of the crystals to determine the melting point (see Technique 9, Sections 9.5–9.8, pp. 631–636). Crush the crystals into a powder, using a stirring rod, in order to determine their melting point. You may observe some "sweating" or shrinkage (see Technique 9, Section 9.7, p. 632) before the substance actually begins to melt. The beginning of the melting-point range is when actual melting is observed, not when the solid takes on a slightly wet or shiny appearance or when shrinkage occurs. If you have isolated ibuprofen, the melting point may be somewhat lower than that given on page 69.

At the instructor's option, place your product in a small vial, label it properly (p. 565), and submit it to your instructor.

QUESTIONS

1. Why was the percentage recovery less than 100%? Give several reasons.

2. Why was the tablet crushed?

3. What was the purpose of the centrifugation step?

4. What was the purpose of the alumina column?

5. If 185 mg of acetaminophen were obtained from a tablet containing 350 mg of acetaminophen, what would be the weight percentage recovery?

6. A student who was isolating aspirin stopped the experiment after the filtration step with alumina. One week later, the methanol was evaporated and the experiment was completed. The melting point of the aspirin was found to be 110–115°C. Explain why the melting point was low and why the melting range was so wide.

10 **EXPERIMENT 10**

Acetaminophen

Decolorization

Filtration

Crystallization

Use of a Craig tube or Hirsch funnel

Preparation of an amide

Preparation of acetaminophen involves treating an amine with an acid anhydride to form an amide. In this case, *p*-aminophenol, the amine, is treated with acetic anhydride to form acetaminophen (*p*-acetamidophenol), the amide.

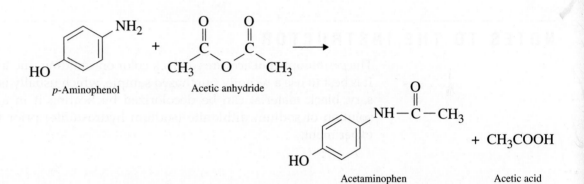

The crude solid acetaminophen contains dark impurities carried along with the *p*-aminophenol starting material. These impurities, which are dyes of unknown structure, are formed from oxidation of the starting phenol. Although the amount of the dye impurity is small, it is intense enough to impart color to the crude acetaminophen. Most of the colored impurity is destroyed by heating the crude product with sodium dithionite (sodium hydrosulfite $Na_2S_2O_4$). The dithionite reduces double bonds in the colored dye to produce colorless substances.

The decolorized acetaminophen is collected on a Hirsch funnel. It is further purified by crystallization from a methanol/water mixture. There are two procedures given in this experiment. Experiment 10A involves crystallization using a Craig tube, whereas Experiment 10B is a larger-scale reaction involving an Erlenmeyer flask and Hirsch funnel for crystallization.

REQUIRED READING

Review: Experiment 1   Introduction to Microscale Laboratory (pp. 2–13)

Techniques 5 and 6

Technique 7   Reaction Methods, Sections 7.1–7.3

Technique 8   Filtration, Sections 8.1–8.6

Technique 9   Physical Constants, Melting Points

New:	Technique 8	Filtration, Section 8.7
	Technique 11	Crystallization
	Essay	Analgesics

SPECIAL INSTRUCTIONS

Acetic anhydride can cause irritation of tissue, especially in nasal passages. Avoid breathing the vapor, and avoid contact with skin and eyes. *p*-Aminophenol is a skin irritant and is toxic.

WASTE DISPOSAL

Aqueous solutions obtained from filtration operations should be poured into the container designated for aqueous wastes. This includes the filtrates from the methanol and water crystallization steps.

NOTES TO THE INSTRUCTOR

The *p*-aminophenol acquires a black color on standing due to air oxidation. It is best to use a recently purchased sample, which usually is gray. If necessary, black material can be decolorized by heating it in a 10% aqueous solution of sodium dithionite (sodium hydrosulfite) prior to starting the experiment.

10A   **EXPERIMENT 10A**

Acetaminophen (Microscale Procedure)

PROCEDURE

Reaction Mixture

Weigh about 0.150 g of *p*-aminophenol (*MW* = 109.1) and place this in a 5-mL conical vial. Using an automatic pipet (or a dispensing pump or a graduated pipet), add 0.450 mL of water and 0.165 mL of acetic anhydride (*MW* = 102.1, *d* = 1.08 g/mL). Place a spin vane in the conical vial and attach an air condenser.

Heating

Heat the reaction mixture with an aluminum block or sand bath at about 120°C (see inset in Fig. 6.2A, p. 591) and stir gently. If you are using a sand bath, the conical vial should be partially buried in the sand so that the vial is nearly at the bottom of the sand bath. After the solid has dissolved (it may dissolve, precipitate, and redissolve), heat the mixture for an additional 20 minutes to complete the reaction.

Isolation of Crude Acetaminophen

Remove the vial from the heat and allow it to cool. When the vial has cooled to the touch, detach the air condenser and remove the spin vane with clean forceps or a magnet. Rinse the spin vane with two or three drops of warm water, allowing the water to drop into the conical vial. Place the conical vial in a small beaker and let it cool to room temperature. If crystallization has not occurred, scratch the inside of the vial with a glass stirring rod to initiate crystallization. Cool the mixture thoroughly in an ice bath for 15–20 minutes and collect the crystals by vacuum filtration on a Hirsch funnel (see Technique 8, Section 8.3, and Fig. 8.5, p. 622). Rinse the vial with about 0.5 mL of ice water and transfer this mixture to the Hirsch funnel. Wash the crystals on the funnel with two additional 0.5-mL portions of ice water. Dry the crystals for 5–10 minutes by allowing air to be drawn through them while they remain on the Hirsch funnel. Transfer the product to a watch glass or clay plate and allow the crystals to dry in air. It may take several hours for the crystals to dry completely, but you may go on to the next step before they are totally dry. Weigh the crude product and set aside a small sample for a melting-point determination and a color comparison after the next step. Calculate the percentage yield of crude acetaminophen (*MW* = 151.2). Record the appearance of the crystals in your notebook.

Decolorization of Crude Acetaminophen

Dissolve 0.2 g of sodium dithionite (sodium hydrosulfite) in 1.5 mL of water in a 5-mL conical vial. Add your crude acetaminophen to the vial. Heat the mixture at about 100°C for 15 minutes, with occasional stirring with a microspatula. Some of the acetaminophen will dissolve during the decolorization process. Cool the mixture thoroughly in an ice bath for about 10 minutes to reprecipitate the decolorized acetaminophen (scratch the inside of the vial, if necessary, to induce crystallization). Collect the purified material by vacuum filtration on a Hirsch funnel, using small portions (about 0.5 mL total) of ice water to aid the transfer. Dry the crystals for 5–10 minutes by allowing air to be drawn through them while they remain on the Hirsch funnel. You may go on to the next step before the material is totally dry. Weigh the purified acetaminophen and compare the color of the purified material to that obtained in the preceding paragraph.

Crystallization of Acetaminophen

Place the purified acetaminophen in a Craig tube. Crystallize the material from a solvent mixture composed of 50% water and 50% methanol by volume (aluminum block or sand bath set at about 100°C). Follow the crystallization procedure described in Technique 11, Section 11.4, and Figure 11.6, page 657. The solubility of acetaminophen in this hot (nearly boiling) solvent is about 0.2 g/mL. Although you can use this as a rough indication of how much solvent is required to dissolve the solid, you should still use the technique shown in Figure 11.6 to determine how much solvent to add. Add small portions (several drops) of hot solvent until the solid dissolves. Step 2 in Figure 11.6 (removal of insoluble impurities) should not be required in this crystallization. When the solid has dissolved, place the Craig tube in a 10-mL Erlenmeyer flask, insert the inner plug of the Craig tube, and let the solution cool.

When the mixture has cooled to room temperature, place the Craig tube in an ice-water bath for several minutes. If necessary, induce crystallization by gently scratching the inside of the Craig tube with your microspatula (Technique 11, Section 11.8B, p. 665). Because acetaminophen may crystallize *slowly* from the solvent, continue to cool the Craig tube in an ice bath for at least 10 minutes. Collect the crystals using the apparatus shown in Figure 8.11 on page 626. Place the assembly in a centrifuge (be sure it is balanced by a centrifuge tube filled with water so that both tubes contain the same weight) and turn on the centrifuge for several minutes. Collect the crystals on a watch glass or piece of smooth paper, as shown in Figure 11.6 on page 657. Set the crystals aside to air-dry. Very little additional time should be required to complete the drying.

Yield Calculation and Melting-Point Determination

Weigh the crystallized acetaminophen ($MW = 151.2$) and calculate the percentage yield. This calculation should be based on the number of moles of the limiting reagent used at the beginning of this procedure. Determine the melting point of the product. Compare the melting point of the final product with that of the crude acetaminophen. Also compare the colors of the crude, decolorized, and pure acetaminophen. Pure acetaminophen melts at 169.5–171°C. Place your product in a properly labeled vial and submit it to your instructor.

10B EXPERIMENT 10B

Acetaminophen (Semimicroscale Procedure)

PROCEDURE

Reaction Mixture

Weigh about 0.400 g of p-aminophenol ($MW = 109.1$) and place this in a 5-mL conical vial. Using an automatic pipet (or a dispensing pump or a graduated pipet), add 1.20 mL of water and 0.450 mL of acetic anhydride ($MW = 102.1$, $d = 1.08$ g/mL). Place a spin vane in the conical vial and attach an air condenser.

Heating

Heat the reaction mixture with an aluminum block or a sand bath at about 120°C (see inset in Fig. 7.6A, p. 601) and stir gently. If you are using an aluminum block, position the vial so that it is touching the surface of the hot plate and place aluminum collars around the vial. If you are using a sand bath, the conical vial should be partially buried in the sand so that the vial is nearly at the bottom of the sand bath. After the solid has dissolved (it may dissolve, precipitate, and redissolve), heat the mixture for an additional 20 minutes to complete the reaction.

Isolation of the Crude Acetaminophen

Remove the vial from the heat and allow it to cool. When the vial has cooled to the touch, detach the air condenser and remove the spin vane with clean forceps or a magnet. Rinse the spin vane with two or three drops of warm water, allowing the water to drop into the conical vial. Place the conical vial in a small beaker and let it cool to room temperature. If crystallization has not occurred, scratch the inside of the vial with a glass stirring rod to initiate crystallization. Cool the mixture thoroughly in an ice bath for 15–20 minutes and collect the crystals by vacuum filtration on a Hirsch funnel (see Technique 8, Section 8.3, and Fig. 8.5, p. 622). Rinse the vial with about 0.5 mL of ice-cold water and transfer this mixture to the Hirsch funnel. Repeat this rinsing with two additional 0.5-mL portions of ice-cold water. Dry the crystals for 5–10 minutes by allowing air to be drawn through them while they remain on the Hirsch funnel. Transfer the product to a watch glass or a clay plate and allow the crystals to dry in air. It may take several hours for the crystals to dry completely, but you may go on to the next step before they are totally dry. Weigh the crude product and set aside a small sample for a melting point determination and a color comparison after the next step. Calculate the percentage yield of crude acetaminophen (*MW* = 151.2). Record the appearance of the crystals in your notebook.

Decolorization of Crude Acetaminophen

Dissolve 0.5 g of sodium dithionite (sodium hydrosulfite) in 4.0 mL of water in a small Erlenmeyer flask. Add your crude acetaminophen to the flask. Heat the mixture at about 100°C for 15 minutes, with occasional swirling. Some of the acetaminophen will dissolve during the decolorization process. Cool the mixture thoroughly in an ice-water bath for about 10 minutes to reprecipitate the decolorized acetaminophen (scratch the inside of the flask, if necessary, to induce crystallization). Collect the purified material by vacuum filtration on a Hirsch funnel using small portions (about 1.0 mL total) of ice-cold water to aid the transfer. Dry the crystals for 5–10 minutes by allowing air to be drawn through them while they remain on the Hirsch funnel. You may go on to the next step before the material is totally dry. Weigh the purified acetaminophen and compare the color of the purified material to that obtained earlier.

Crystallization of Acetaminophen

Follow the semimicroscale crystallization procedure described in Technique 11, Section 11.3, pages 650–656 and shown in Figure 11.4, page 651. Step 2 in Figure 11.4 (removal of insoluble impurities) will not be required in this crystallization. Place all your acetaminophen in a 10-mL Erlenmeyer flask. In another flask, place about 3 mL of a solvent mixture composed of 50% water and 50% methanol, *with a boiling stone,* and put it on the hot plate. When the solvent begins to boil, start adding the hot solvent slowly to the acetaminophen using a Pasteur pipet. At this point, place both flasks on the hot plate to keep them hot. Continue to add the boiling solvent to the flask containing the acetaminophen until the solid *just dissolves.* Because the solubility of acetaminophen in this nearly boiling solvent is only about 0.2 g/mL, you will likely not use all the boiling solvent. The idea is to add the *minimum* amount of boiling solvent that *just dissolves* the solid.

Once the solid is dissolved, cork the flask and allow the contents of the flask to cool slowly to room temperature. Pure acetaminophen should crystallize out of the solvent. If solid does not form, scratch the inside of the flask with your microspatula. Place the flask in an ice bath to complete the crystallization for at least 10 minutes. Transfer the solid from the flask to a Hirsch funnel (Technique 8, Section 8.3, and Fig. 8.5, p. 622). Rinse the flask with about 0.5 mL of ice-cold solvent (50% methanol/50% water) and transfer this mixture to the Hirsch funnel. Repeat this rinsing with an additional 0.5-mL portion of ice-cold solvent. Dry the crystals for 5–10 minutes by allowing air to be drawn through them while they remain on the Hirsch funnel. Transfer the product to a watch glass or a clay plate and allow the crystals to dry in air. Let the crystals dry until the next laboratory period.

Yield Calculation and Melting Point Determination

Weigh the crystallized acetaminophen and calculate the percentage yield (*MW* = 151.2). This calculation should be based on the number of moles of the limiting reagent used at the beginning of this procedure. Determine the melting point of the product. Compare the melting point of the final product with that of the crude acetaminophen. Also compare the colors of the crude, decolorized, and pure acetaminophen. Pure acetaminophen melts at 169.5–171°C. Place your product in a properly labeled vial and submit it to your instructor.

QUESTIONS

1. During the crystallization of acetaminophen, why was the mixture cooled in an ice bath?

2. In the reaction between *p*-aminophenol and acetic anhydride to form acetaminophen, 0.450 mL of water was added. What was the purpose of the water?

3. Why should you use a minimum amount of water to rinse the conical vial while transferring the purified acetaminophen to the Hirsch funnel?

4. If 0.130 g of *p*-aminophenol is allowed to react with excess acetic anhydride, what is the theoretical yield of acetaminophen in moles? In grams?

5. Give two reasons, discussed in Experiments 8 and 10, why the crude product in most reactions is not pure.

6. Phenacetin has the structure shown. Write an equation for its preparation, starting from 4-ethoxyaniline.

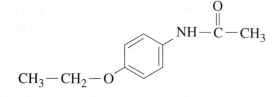

ESSAY

Identification of Drugs

Frequently, a chemist is called on to identify a particular unknown substance. If there is no prior information to work from, this can be a formidable task. There are several million known compounds, both inorganic and organic. For a completely unknown substance, the chemist must often use every available method. If the unknown substance is a mixture, then the mixture must be separated into its components and each component identified separately. A pure compound can often be identified from its physical properties (melting point, boiling point, density, refractive index, and so on) and a knowledge of its functional groups. These can be identified by the reactions that the compound is observed to undergo or by spectroscopy (infrared, ultraviolet, nuclear magnetic resonance, and mass spectroscopy). The techniques necessary for this type of identification are introduced in a later section.

A somewhat simpler situation often arises in drug identification. The scope of drug identification is more limited, and the chemist working in a hospital trying to identify the source of a drug overdose or the law enforcement officer trying to identify a suspected illicit drug or a poison usually has some clues to work from. So does the medicinal chemist working for a pharmaceutical manufacturer who might be trying to discover why a competitor's product may be better.

Consider a drug overdose case as an example. The patient is brought into the emergency ward of a hospital. This person may be in a coma or a hyperexcited state, have an allergic rash, or clearly be hallucinating. These physiological symptoms are themselves a clue to the nature of the drug. Samples of the drug may be found in the patient's possession. Correct medical treatment may require a rapid and accurate identification of a drug powder or capsule. If the patient is conscious, the necessary information can be elicited orally; if not, the drug must be examined. If the drug is a tablet or a capsule, the process is often simple because many drugs are coded by a manufacturer's trademark or logo, by shape (round, oval, bullet), by formulation (tablet, gelatin capsule, time-release microcapsules), and by color. Some drugs bear an imprinted number or code.

It is more difficult to identify a powder, but under some circumstances such identification may be easy. Plant drugs are often easily identified because they contain microscopic bits and pieces of the plant from which they are obtained. This cellular debris is often characteristic for certain types of drugs, and they can be identified on this basis alone. A microscope is all that is needed. Sometimes chemical color tests can be used as confirmation. Certain drugs give rise to characteristic colors when treated with special reagents. Other drugs form crystalline precipitates of characteristic color and crystal structure when treated with appropriate reagents.

If the drug itself is not available and the patient is unconscious (or dead), identification may be more difficult. It may be necessary to pump the stomach or bladder contents of the patient (or corpse) or to obtain a blood sample and work on these. These samples of stomach fluid, urine, or blood would be extracted with an appropriate organic solvent, and the extract would be analyzed.

Often the final identification of a drug, as an extract of urine, serum, or stomach fluid, hinges on some type of **chromatography.** Thin-layer chromatography (TLC) is often used. Under specified conditions, many drug substances can be identified by their R_f values and by the colors that their TLC spots turn when treated with various reagents or when they are observed under certain visualization methods. In the experiment that follows, TLC is applied to the analysis of an unknown analgesic drug.

REFERENCES

Keller, E. "Origin of Modern Criminology." *Chemistry, 42* (1969): 8.

Keller, E. "Forensic Toxicology: Poison Detection and Homicide." *Chemistry, 43* (1970): 14.

Lieu, V. T. "Analysis of APC Tablets." *Journal of Chemical Education, 48* (1971): 478.

Neman, R. L. "Thin Layer Chromatography of Drugs." *Journal of Chemical Education, 49* (1972): 834.

Rodgers, S. S. "Some Analytical Methods Used in Crime Laboratories." *Chemistry, 42* (1969): 29.

Tietz, N. W. *Fundamentals of Clinical Chemistry.* Philadelphia: W. B. Saunders, 1970.

Walls, H. J. *Forensic Science.* New York: Praeger, 1968.

A collection of articles on forensic chemistry can be found in

Berry, K., and Outlaw, H. E., eds. "Forensic Chemistry—A Symposium Collection." *Journal of Chemical Education, 62* (December 1985): 1043–1065.

EXPERIMENT 11

TLC Analysis of Analgesic Drugs

THIN-LAYER CHROMATOGRAPHY

In this experiment, thin-layer chromatography (TLC) will be used to determine the composition of various over-the-counter analgesics. If the instructor chooses, you may also be required to identify the components and actual identity (trade name) of an unknown analgesic. You will be given either two or three commercially prepared TLC plates with a flexible backing and a silica gel coating with a fluorescent indicator. On the first TLC plate, a reference plate, you will spot five standard compounds often used in analgesic formulations. In addition, a standard reference mixture containing four of these same compounds will be spotted. Ibuprofen is omitted from this standard mixture because it would overlap with salicylamide after the plate is developed. If your instructor wishes, you will spot five additional reference substances on a second reference plate, including the newest analgesic drugs. On the final plate (the sample plate) you will spot several commercial analgesic preparations in order to determine their composition. At your instructor's option, one or more of these may be an unknown.

The standard compounds will all be available as solutions of 1 g of each dissolved in 20 mL of a 50:50 mixture of methylene chloride and ethanol. The purpose of the first reference plate is to determine the order of elution (R_f values) of the known substances and to index the standard reference mixture. On the second reference plate (optional), several of the substances have similar R_f values, but you will note a different behavior for each spot with the visualization methods. On the sample plate, the standard reference mixture will be spotted, along with several solutions that you will prepare from commercial analgesic tablets. These tablets will each be crushed and dissolved in a 50:50 methylene chloride–ethanol mixture for spotting.

Reference Plate 1		*Reference Plate 2 (optional)*		*Sample Plate*
Acetaminophen	(Ac)	Aspirin	(Asp)	Five commercial preparations (or unknowns) plus the reference mixture
Aspirin	(Asp)	Ibuprofen	(Ibu)	
Caffeine	(Cf)	Ketoprofen	(Kpf)	
Ibuprofen	(Ibu)	Naproxen sodium	(Nap)	
Salicylamide	(Sal)	Salicylamide	(Sal)	
Reference mixture 1	(Ref-1)	Reference mixture 2	(Ref-2)	

Two methods of visualization will be used to observe the positions of the spots on the developed TLC plates. First, the plates will be observed while under illumination from a short-wavelength ultraviolet (UV) lamp. This is best done in a darkened room or in a fume hood that has been darkened by taping butcher paper or aluminum foil over the lowered glass cover. Under these conditions, some of the spots will appear as dark areas on the plate, whereas others will fluoresce brightly. This difference in appearance under UV illumination will help to distinguish the substances from one another. You will find it convenient to outline very lightly in *pencil* the spots observed and to place a small **x** inside those spots that fluoresce. For a second means of visualization, iodine vapor will be used. Not all the spots will become visible when treated with iodine, but some will turn yellow, tan, or deep brown. The differences in the behaviors of the various spots with iodine can be used to further differentiate among them.

It is possible to use several developing solvents for this experiment, but ethyl acetate with 0.5% glacial acetic acid added is preferred. The small amount of glacial acetic acid supplies protons and suppresses ionization of aspirin, ibuprofen, naproxen sodium, and ketoprofen, allowing them to travel upward on the plates in their protonated form. Without the acid, these compounds do not move.

In some analgesics, you may find ingredients besides the five mentioned previously. Some include an antihistamine and some, a mild sedative. For instance, Midol contains N-cinnamylephedrine (cinnamedrine), an antihistamine, and Excedrin PM contains the sedative methapyrilene hydrochloride. Cope contains the related sedative methapyrilene fumarate. Some tablets may be colored with a chemical dye.

REQUIRED READING

Review:	Essay	Analgesics
New:	Technique 19	Column Chromatography, Sections 19.1–19.3
	Technique 20	Thin-Layer Chromatography
	Essay	Identification of Drugs

SPECIAL INSTRUCTIONS

You must examine the developed plates under ultraviolet light first. After comparisons of *all* plates have been made with UV light, iodine vapor can be used. The iodine permanently affects some of the spots, making it impossible to go back and repeat the UV visualization. Take special care to notice those substances that have similar R_f values; these spots each have a different appearance when viewed under UV illumination or a different staining color with iodine, allowing you to distinguish among them.

Aspirin presents some special problems because it is present in a large amount in many of the analgesics and because it hydrolyzes easily. For these reasons, the aspirin spots often show excessive tailing.

SUGGESTED WASTE DISPOSAL

Dispose of all development solvent in the container for nonhalogenated organic solvents. Dispose of the ethanol–methylene chloride mixture in the container for halogenated organic solvents. The micropipets used for spotting the solution should be placed in a special container labeled for that purpose. The TLC plates should be stapled in your lab notebook.

NOTES TO THE INSTRUCTOR

If you wish, students may work in pairs on this experiment, each one preparing one of the two reference plates and then cooperating on the third plate. Alternatively, especially if students are to work alone, you may wish to omit plate 2 and forgo identifying the two newest analgesics (ketoprofen and naproxen sodium).

Perform the thin-layer chromatography with flexible Silica Gel 60 F-254 plates (EM Science, No. 5554-7). If the TLC plates have not been purchased recently, you should place them in an oven at 100°C for 30 minutes and store them in a desiccator until used. If you use different thin-layer plates, try out the experiment before using them with a class. Other plates may not resolve all five substances.

Ibuprofen and salicylamide have approximately the same R_f value, but they show up differently under the detection methods. For reasons that are not yet clear, ibuprofen sometimes gives two or even three spots. Naproxen sodium and ketoprofen have approximately the same R_f as aspirin. Once again, however, they show up differently under the detection methods. Fortunately, none of the new drugs appears in combination, either together or with aspirin or ibuprofen, in any current commercial product.

PROCEDURE

Initial Preparations

You will need at least 12 capillary micropipets (18 if both reference plates are prepared) to spot the plates. The preparation of these pipets is described and illustrated in Technique 20, Section 20.4, page 782. A common error is to pull the center section out too far when making these pipets, so that too little sample is applied to the plate. If this happens, you won't see *any* spots. Follow the directions carefully.

After preparing the micropipets, obtain two (or three) 10-cm × 6.6-cm TLC plates (EM Science Silica Gel 60 F-254, No. 5554-7) from your instructor. These plates have a flexible backing, but they should not be bent excessively. Handle them carefully or the adsorbent may flake off. Also, you should handle them only by the edges; the surface should not be touched. Using a lead pencil (not a pen), *lightly* draw a line across the plates (short dimension) about 1 cm from the bottom. Using a centimeter ruler, move its index about 0.6 cm in from the edge of the plate and lightly mark off six 1-cm intervals on the line (see figure). These are the points at which the samples will be spotted. If you are preparing two reference plates, it would be a good idea to mark a small number **1** or **2** in the upper right-hand corner of each plate to allow easy identification.

Spotting the First Reference Plate

On the first plate (marked 1), starting from left to right, spot acetaminophen, aspirin, caffeine, ibuprofen, and salicylamide. This order is alphabetic and will avoid any further memory problems or confusion. Solutions of these compounds will be found in small bottles on the side shelf. The standard reference mixture (Ref-1), also found on the side shelf, is spotted in the last position. The correct method of spotting a TLC plate is described in Technique 20, Section 20.4, page 782. It is important that the spots be made as small as possible but not too small. With too much sample, the spots will tail and will overlap one another after development. With too little sample, no spots will be observed after development. The optimum applied spot should be about 1–2 mm (1/16 in.) in diameter. If scrap pieces of the TLC plates are available, it would be a good idea to practice spotting on these before preparing the actual sample plates.

Spotting the Second Reference Plate (Optional)

On the second plate (marked 2), starting from left to right, spot aspirin, ibuprofen, ketoprofen, naproxen sodium, salicylamide, and the reference mixture (Ref-2).

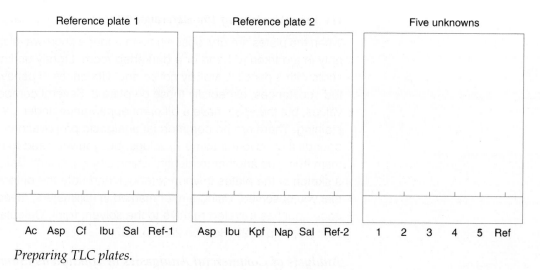

Preparing TLC plates.

Follow the same procedure and take the precautions noted above for the first plate.

Preparing the Development Chamber

When the reference plate (or plates) has been spotted, obtain a 16-oz wide-mouth, screw-cap jar (or other suitable container) for use as a development chamber. The preparation of a development chamber is described in Technique 20, Section 20.5, page 784. Because the backing on the TLC plates is thin, if they touch the filter paper liner of the development chamber *at any point,* solvent will begin to diffuse onto the absorbent surface at that point. To avoid this, you may either omit the liner or make the following modification.

If you wish to use a liner, use a narrow strip of filter paper (approximately 5 cm wide). Fold it into an L shape that is long enough to traverse the bottom of the jar and extend up the side to the top of the jar. TLC plates placed in the jar for development should *straddle* this liner strip but not touch it.

When the development chamber has been prepared, obtain a small amount of the development solvent (0.5% glacial acetic acid in ethyl acetate). Your instructor should prepare this mixture; it contains such a small amount of acetic acid that small individual portions are difficult to prepare. Fill the chamber with the development solvent to a depth of about 0.5–0.7 cm. If you are using a liner, be sure it is saturated with the solvent. Recall that the solvent level must not be above the spots on the plate or the samples will dissolve off the plate into the reservoir instead of developing.

Development of the Reference TLC Plates

Place the spotted plate (or plates) in the chamber (straddling the liner if one is present) and allow the spots to develop. If you are doing two reference plates, both plates may be placed in the same development jar. Be sure the plates are placed in the developing jar so that their bottom edge is parallel to the bottom of the jar (straight, not tilted); if not, the solvent front will not advance evenly, increasing the difficulty of making good comparisons. The plates should face each other and slant or lean back in opposite directions. When the solvent has risen to a level about 0.5 cm from the top of the plate, remove each plate from the chamber (in the hood) and, using a lead pencil, mark the position of the solvent front. Set the plate on a piece of paper towel to dry. It may be helpful to place a small object under one end to allow optimum air flow around the drying plate.

UV Visualization of the Reference Plates

When the plates are dry, observe them under a short-wavelength UV lamp, preferably in a darkened hood or a darkened room. Lightly outline all of the observed spots with a pencil. Carefully notice any differences in behavior between the spotted substances, especially those on plate 2. Several compounds have similar R_f values, but the spots have a different appearance under UV illumination or iodine staining. There are no commercial analgesic preparations containing any compounds that have the same R_f values, but you will need to be able to distinguish them from one another to identify which one is present. Before proceeding, make a sketch of the plates in your notebook and note the differences in appearance that you observed. Using a ruler marked in millimeters, measure the distance that each spot has traveled relative to the solvent front. Calculate R_f values for each spot (Technique 20, Section 20.9, p. 787).

Analysis of Commercial Analgesics or Unknowns (Sample Plate)

Next, obtain half a tablet of each of the analgesics to be analyzed on the final TLC plate. If you were issued an unknown, you may analyze four other analgesics of your choice; if not, you may analyze five. The experiment will be most interesting if you make your choices in a way that gives a wide spectrum of results. Try to pick at least one analgesic each containing aspirin, acetaminophen, ibuprofen, a newer analgesic, and, if available, salicylamide. If you have a favorite analgesic, you may wish to include it among your samples. Take each analgesic half-tablet, place it on a smooth piece of notebook paper, and crush it well with a spatula. Transfer each crushed half-tablet to a labeled test tube or a small Erlenmeyer flask. Using a graduated cylinder, mix 15 mL of absolute ethanol and 15 mL of methylene chloride. Mix the solution well. Add 5 mL of this solvent to each of the crushed half-tablets and then heat each of them *gently* for a few minutes on a steam bath or sand bath at about 100°C. Not all the tablet will dissolve, because the analgesics usually contain an insoluble binder. In addition, many of them contain inorganic buffering agents or coatings that are insoluble in this solvent mixture. After heating the samples, allow them to settle and then spot the clear liquid extracts on the sample plate. At the sixth position, spot the standard reference solution (Ref-1 or Ref-2). Develop the plate in 0.5% glacial acetic acid–ethyl acetate as before. Observe the plate under UV illumination and mark the visible spots as you did for the first plate. Sketch the plate in your notebook and record your conclusions about the contents of each tablet. This can be done by directly comparing your plate to the reference plate(s)—they can all be placed under the UV light at the same time. If you were issued an unknown, try to determine its identity (trade name).

Iodine Analysis

Do not perform this step until UV comparisons of all the plates are complete. When ready, place the plates in a jar containing a few iodine crystals, cap the jar, and warm it gently on a steam bath or warm hot plate until the spots begin to appear. Notice which spots become visible and note their relative colors. You can directly compare colors of the reference spots to those on the unknown plate(s). Remove the plates from the jar and record your observations in your notebook.

QUESTIONS

1. What happens if the spots are made too large when preparing a TLC plate for development?

2. What happens if the spots are made too small when preparing a TLC plate for development?

3. Why must the spots be above the level of the development solvent in the developing chamber?

4. What would happen if the spotting line and positions were marked on the plate with a ballpoint pen?

5. Is it possible to distinguish two spots that have the same R_f value but represent different compounds? Give two different methods.

6. Name some advantages of using acetaminophen (Tylenol) instead of aspirin as an analgesic.

ESSAY

Caffeine

The origins of coffee and tea as beverages are so old that they are lost in legend. Coffee is said to have been discovered by an Abyssinian goatherd who noticed an unusual friskiness in his goats when they consumed a certain little plant with red berries. He decided to try the berries himself and discovered coffee. The Arabs soon cultivated the coffee plant, and one of the earliest descriptions of its use is found in an Arabian medical book circa A.D. 900. The great systematic botanist Linnaeus named the tree *Coffea arabica.*

One legend of the discovery of tea—from the Orient, as you might expect—attributes the discovery to Daruma, the founder of Zen. Legend has it that he inadvertently fell asleep one day during his customary meditations. To be assured that this indiscretion would not recur, he cut off both eyelids. Where they fell to the ground, a new plant took root that had the power to keep a person awake. Although some experts assert that the medical use of tea was reported as early as 2737 B.C. in the pharmacopeia of Shen Nung, an emperor of China, the first indisputable reference is from the Chinese dictionary of Kuo P'o, which appeared in A.D. 350. The nonmedical, or popular, use of tea appears to have spread slowly. Not until about A.D. 700 was tea widely cultivated in China. Tea is native to upper Indochina and upper India, so it must have been cultivated in these places before its introduction to China. Linnaeus named the tea shrub *Thea sinensis;* however, tea is more properly a relative of the camellia, and botanists have renamed it *Camellia thea.*

The active ingredient that makes tea and coffee valuable to humans is **caffeine.** Caffeine is an **alkaloid,** a class of naturally occurring compounds containing nitrogen and having the properties of an organic amine base (alkaline, hence, *alkaloid*). Tea and coffee are not the only plant sources of caffeine. Others include kola nuts, maté leaves, guarana seeds, and in small amount, cocoa beans. The pure alkaloid was first isolated from coffee in 1821 by the French chemist Pierre Jean Robiquet.

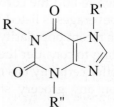

XANTHINES
Xanthine R = R' = R" = H
Caffeine R = R' = R" = CH₃
Theophylline R = R" = CH₃, R' = H
Theobromine R = H, R' = R" = CH₃

Caffeine belongs to a family of naturally occurring compounds called **xanthines.** The xanthines, in the form of their plant progenitors, are possibly the oldest known stimulants. They all, to various extents, stimulate the central nervous system and the skeletal muscles. This stimulation results in an increased alertness, the ability to put off sleep, and an increased capacity for thinking. Caffeine is the most powerful xanthine in this respect. It is the main ingredient of the popular No-Doz keep-alert tablets. Although caffeine has a powerful effect on the central nervous system, not all xanthines are as effective. Thus, theobromine, the xanthine found in cocoa, has fewer central nervous system effects. It is, however, a strong **diuretic** (induces urination) and is useful to doctors in treating patients with severe water-retention problems. Theophylline, a second xanthine found in tea, also has fewer central nervous system effects but is a strong **myocardial** (heart muscle) stimulant; it **dilates** (relaxes) the coronary artery that supplies blood to the heart. Its most important use is in the treatment of bronchial asthma because it has the properties of a **bronchodilator** (relaxes the bronchioles of the lungs). Because it is also a **vasodilator** (relaxes blood vessels), it is often used in treating hypertensive headaches. It is also used to alleviate and to reduce the frequency of attacks of **angina pectoris** (severe chest pain). In addition, it is a more powerful diuretic than theobromine.

One can develop both a tolerance for the xanthines and a dependence on them, particularly caffeine. The dependence is real, and a heavy user (>5 cups of coffee per day) will experience lethargy, headache, and perhaps nausea after about 18 hours of abstinence. An excessive intake of caffeine may lead to restlessness, irritability, insomnia, and muscular tremor. Caffeine can be toxic, but to achieve a lethal dose of caffeine, one would have to drink about 100 cups of coffee over a relatively short period.

Caffeine is a natural constituent of coffee, tea, and kola nuts (*Kola nitida*). Theophylline is found as a minor constituent of tea. The chief constituent of cocoa is theobromine. The amount of caffeine in tea varies from 2% to 5%. In one analysis of black tea, the following compounds were found: caffeine, 2.5%; theobromine, 0.17%; theophylline, 0.013%; adenine, 0.014%; and guanine and xanthine, traces. Coffee beans can contain up to 5% by weight of caffeine, and cocoa contains around 5% theobromine. Commercial cola is a beverage based on a kola nut extract. We cannot easily get kola nuts in this country, but we can get the ubiquitous commercial extract as a syrup. The syrup can be converted into "cola." The syrup contains caffeine, tannins, pigments, and sugar. Phosphoric acid is added, and caramel is added to give the syrup a deep color. The final drink is prepared by adding water and carbon dioxide under pressure to give the bubbly mixture. Before decaffeination, the Food and Drug Administration required a "cola" to contain some caffeine (about 0.2 mg per ounce). In 1990, when new nutrition labels were adopted, this requirement was dropped. The Food and Drug Administration currently requires that a "cola" contain *some* caffeine but limits this amount to a maximum of 5 milligrams per ounce. To achieve a regulated level of caffeine, most manufacturers remove all caffeine from the kola extract and then re-add the correct amount to the syrup. The caffeine content of various beverages is listed in the accompanying table.

With the recent popularity of gourmet coffee beans and espresso stands, it is interesting to consider the caffeine content of these specialty beverages. Gourmet coffee certainly has more flavor than the typical ground coffee you may find on any grocery store shelf, and the concentration of brewed gourmet coffee tends to be higher than ordinary drip-grind coffee. Brewed gourmet coffee probably contains something on the order of 20–25 mg of caffeine per ounce of liquid. Espresso coffee is a very concentrated, dark-brewed coffee. Although the darker roasted beans used for espresso actually contain less caffeine per gram than regularly roasted beans, the method of preparing espresso (extraction using pressurized steam) is more efficient, and a higher percentage of the total caffeine in the beans is extracted. The caffeine content per ounce of liquid, therefore, is substantially higher than in most brewed coffees. The serving size for espresso coffee, however, is much smaller than for ordinary coffee (about 1.5–2 oz per serving), so the total caffeine available in a serving of espresso turns out to be about the same as in a serving of ordinary coffee.

Amount of caffeine (mg/oz) found in beverages

Brewed coffee	12–30	Tea	4–20
Instant coffee	8–20	Cocoa (but 20 mg/oz theobromine)	0.5–2
Espresso (1 serving = 1.5–2 oz)	50–70		
Decaffeinated coffee	0.4–1.0	Coca-Cola	3.75

Note: The average cup of coffee or tea contains about 5–7 oz of liquid. The average bottle of cola contains about 12 oz of liquid.

Because of the central nervous system effects from caffeine, many people prefer **decaffeinated** coffee. The caffeine is removed from coffee by extracting the whole beans with an organic solvent. Then the solvent is drained off, and the beans are steamed to remove any residual solvent. The beans are dried and roasted to bring out the flavor. Decaffeination reduces the caffeine content of coffee to the range of 0.03% to 1.2% caffeine. The extracted caffeine is used in various pharmaceutical products, such as APC tablets.

Among coffee lovers, there is some controversy about the best method to remove the caffeine from coffee beans. **Direct contact** decaffeination uses an organic solvent (usually methylene chloride) to remove the caffeine from the beans. When the beans are subsequently roasted at 200°C, virtually all traces of the solvent are removed, because methylene chloride boils at 40°C. The advantage of direct contact decaffeination is that the method removes only the caffeine (and some waxes) but leaves the substances responsible for the flavor of the coffee intact in the bean. A disadvantage of this method is that all organic solvents are toxic to some extent.

Water process decaffeination is favored among many drinkers of decaffeinated coffee because it does not use organic solvents. In this method, hot water and steam are used to remove caffeine and other soluble substances from the coffee. The resulting solution is then passed through activated charcoal filters to remove the caffeine. Although this method does not use organic solvents, the disadvantage is that water is not a very selective decaffeinating agent. Many of the flavor oils in the coffee are removed at the same time, resulting in a coffee with a somewhat bland flavor.

A third method, the **carbon dioxide decaffeination process,** is being used with increasing frequency. The raw coffee beans are moistened with steam and water, and they are then placed into an extractor where they are treated with carbon dioxide gas under very high temperature and pressure. Under these conditions, the carbon dioxide gas is in a **supercritical** state, which means that it takes on the characteristics of both a liquid and a gas. The supercritical carbon dioxide acts as a selective solvent for caffeine, thus extracting it from the beans.

Caffeine has always been a controversial compound. Medically, its actions are suspect. It definitely has strong effects on the heart and blood vessels, causing an increase in blood pressure. It stimulates the central nervous system, making a person more alert but also more jittery. Many people consider caffeine to be a dangerous and addictive drug, and some religions forbid the use of beverages containing caffeine for this very reason.

Another problem, not related to caffeine but rather to the beverage tea, is that in some cases persons who consume high quantities of tea may show symptoms of Vitamin B_1 (thiamine) deficiency. It is suggested that the tannins in the tea may complex with the thiamine, rendering it unavailable for use. An alternative suggestion is that caffeine may reduce the levels of the enzyme transketolase, which depends on the presence of thiamine for its activity. Lowered levels of transketolase would produce the same symptoms as lowered levels of thiamine.

REFERENCES

Emboden, W. "The Stimulants." *Narcotic Plants,* rev. ed. New York: Macmillan, 1979.

Ray, O. S. "Caffeine." *Drugs, Society and Human Behavior,* 7th ed. St. Louis: C. V. Mosby, 1996.

Ritchie, J. M. "Central Nervous System Stimulants. II: The Xanthines." In L. S. Goodman and A. Gilman, eds., *The Pharmacological Basis of Therapeutics,* 8th ed. New York: Macmillan, 1990.

Taylor, N. *Plant Drugs That Changed the World.* New York: Dodd, Mead, 1965. Pp. 54–56.

Taylor, N. "Three Habit-Forming Nondangerous Beverages." In *Narcotics—Nature's Dangerous Gifts.* New York: Dell, 1970. (Paperbound revision of *Flight from Reality.*)

12   EXPERIMENT 12

Isolation of Caffeine from Tea or Coffee

Isolation of a natural product

Extraction

Sublimation

Green chemistry

Solid phase extraction (optional)

Gas chromatography (optional)

Infrared spectroscopy (optional)

In Experiment 12A, caffeine is isolated from tea leaves. The chief problem with the isolation is that caffeine does not exist alone in tea leaves but is accompanied by other natural substances from which it must be separated.

The main component of tea leaves is cellulose, which is the principal structural material of all plant cells. Cellulose is a polymer of glucose. Because cellulose is virtually insoluble in water, it presents no problems in the isolation procedure. Caffeine, on the other hand, is water soluble and is one of the main substances extracted into the solution called tea. Caffeine constitutes as much as 5% by weight of the leaf material in tea plants.

Tannins also dissolve in the hot water used to extract tea leaves. The term **tannin** does not refer to a single homogeneous compound or even to substances that have similar chemical structure. It refers to a class of compounds that have certain properties in common. Tannins are phenolic compounds having molecular weights between 500 and 3000. They are widely used to tan leather. They precipitate alkaloids and proteins from aqueous solutions. Tannins are usually divided into two classes: those that can be **hydrolyzed** (react with water) and those that cannot. Tannins of the first type that are found in tea generally yield glucose and gallic acid when they are hydrolyzed. These tannins are esters of gallic acid and glucose. They represent structures in which some of the hydroxyl groups in glucose have been esterified by digalloyl groups. The nonhydrolyzable tannins found in tea are condensation polymers of catechin. These polymers are not uniform in structure; catechin molecules are usually linked at ring positions 4 and 8.

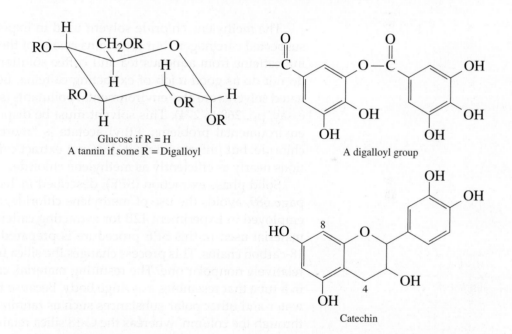

Glucose if R = H
A tannin if some R = Digalloyl

A digalloyl group

Catechin

When tannins are extracted into hot water, some of these compounds are partially hydrolyzed to form free gallic acid. The tannins, because of their phenolic groups, and gallic acid, because of its carboxyl groups, are both acidic. If sodium carbonate, a base, is added to tea water, these acids are converted to their sodium salts that are highly soluble in water.

Although caffeine is soluble in water, it is much more soluble in the organic solvent methylene chloride. Caffeine can be extracted from the basic tea solution with methylene chloride, but the sodium salts of gallic acid and the tannins remain in the aqueous layer.

The brown color of a tea solution is due to flavonoid pigments and chlorophylls and to their respective oxidation products. Although chlorophylls are soluble in methylene chloride, most other substances in tea are not. Thus, the methylene chloride extraction of the basic tea solution removes nearly pure caffeine. The methylene chloride is easily removed by evaporation (bp 40°C) to leave the crude caffeine. The caffeine is then purified by sublimation at reduced pressure to prevent decomposition.

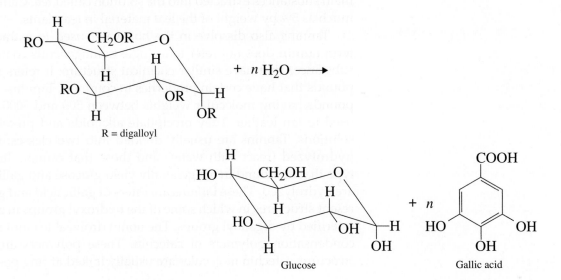

The methylene chloride solvent used in Experiment 12A is toxic and a suspected carcinogen, but it remains as one of the best solvents for extracting caffeine from aqueous tea and coffee solutions. Other solvents simply do not do as good a job of extracting caffeine. Besides being toxic, chlorinated solvents are also environmental pollutants (see the "Green Chemistry" essay, pp. 268 to 274). This solvent must be disposed of correctly to avoid environmental problems. Ethyl acetate is "more Green" than methylene chloride, but unfortunately it does not extract caffeine from aqueous solutions nearly as efficiently as methylene chloride.

Solid phase extraction (SPE), described in Technique 12, Section 12.14, page 689, avoids the use of methylene chloride, and the technique will be employed in Experiment 12B for extracting caffeine from tea or coffee. The material used in this SPE procedure is prepared by alkylating silica with 18-carbon chains. This process changes the silica from a polar substance to a relatively nonpolar one. The resulting material, called C-18 silica, is placed in a tube that resembles a syringe body. Because the silica is now nonpolar, water and other polar substances such as tannins and gallic acid will pass through the column, whereas the C-18 silica retains caffeine. To remove the caffeine from the column, ethyl acetate is added.

REQUIRED READING

Review: Techniques 5 and 6

	Technique 7	Reaction Methods, Section 7.10
	Technique 9	Physical Constants of Solids: The Melting Point
New:	Technique 12	Extractions, Separations, and Drying Agents
	Technique 17	Sublimation
	Essay	Caffeine

Essay	Green Chemistry
Technique 22	Gas Chromatography (optional)
Technique 25	Infrared Spectroscopy (optional)
Technique 28	Mass Spectrometry (optional)

SPECIAL INSTRUCTIONS

Be careful when handling methylene chloride. It is a toxic solvent and suspected carcinogen. You should not breathe it or spill it on yourself. Wear gloves and work in a fume hood. It is recommended that you use glass centrifuge tubes with screw caps. Check them first with water to make sure that they will not leak while you shake them. Plastic centrifuge tubes with screw caps can be used instead of glass ones. However, there is some tendency for the plastic to dissolve in methylene chloride, and they should not be in contact with methylene chloride for long periods (more than 1 hour).

SUGGESTED WASTE DISPOSAL

You must dispose of methylene chloride in a waste container marked for the disposal of halogenated organic waste. Dispose of the tea bag in a trash can. The aqueous solutions obtained after the extraction steps must be disposed of in a waste container labeled for aqueous waste.

12A   **EXPERIMENT 12A**

Extraction of Caffeine from Tea with Methylene Chloride

PROCEDURE

Preparing the Tea Solution

Place 20 mL of water in a 50-mL beaker. Cover the beaker with a watch glass and heat the water on a hot plate until it is almost boiling. From information provided by your instructor, determine the weight of the tea present in the tea bag.[1] Place the tea bag into the hot water so that it lies flat on the bottom of the beaker and is covered as completely as possible with water. Replace the watch glass and continue heating for about 15 minutes. During this heating period, it is important to push down *gently* on the tea bag with a test tube so that all the tea leaves are in constant contact with water. As the water evaporates during this heating step, replace it by adding water from a Pasteur pipet.

---

[1] The weight of the tea in the bag varies, whereas the weight of the bag, string and tag are relatively constant. Your instructor may provide you with the weight of everything except the tea itself. You can then determine the weight of the tea by directly weighing the tea bag provided you and subtracting the weight of the empty bag, string, and tag to determine the weight of the tea in the bag.

Using the Pasteur pipet, transfer the concentrated tea solution to two centrifuge tubes fitted with screw caps. Try to keep the liquid volume in each centrifuge tube approximately equal. To squeeze additional liquid out of the tea bag, hold the tea bag on the inside wall of the beaker and roll a test tube back and forth while exerting *gentle* pressure on the tea bag. Press out as much liquid as possible without breaking the bag. Combine this liquid with the solution in the centrifuge tubes. Place the tea bag on the bottom of the beaker again and pour 2 mL of hot water over the bag. Squeeze the liquid out, as just described, and transfer this liquid to the centrifuge tubes. Add 0.5 g of sodium carbonate to the hot liquid in each centrifuge tube. Cap the tubes and shake the mixture until the solid dissolves.

Extraction and Drying

Cool the tea solution to room temperature. Using a calibrated Pasteur pipet (p. 11), add 3 mL of methylene chloride to each centrifuge tube to extract the caffeine (Technique 12, Section 12.4, p. 674). Cap the centrifuge tubes and gently shake the mixture for several seconds. Vent the tubes to release the pressure, being careful that the liquid does not squirt out toward you. Shake the mixture for an additional 30 seconds with occasional venting. To separate the layers and break the emulsion (see Technique 12, Section 12.10, p. 684), centrifuge the mixture for several minutes (be sure to balance the centrifuge by placing the two centrifuge tubes on opposite sides). If an emulsion still remains (indicated by a green-brown layer between the clear methylene chloride layer and the top aqueous layer), centrifuge the mixture again.

Remove the lower organic layer with a Pasteur pipet and transfer it to a dry 25-mL Erlenmeyer flask. Be sure to squeeze the bulb before placing the tip of the Pasteur pipet into the liquid and try not to transfer any of the dark aqueous solution along with the methylene chloride layer. Add a fresh 3-mL portion of methylene chloride to the aqueous layer remaining in each centrifuge tube, cap the centrifuge tubes, and shake the mixture in order to carry out a second extraction. Separate the layers by centrifugation, as described previously. Combine the organic layers from each extraction into the flask containing the first extraction. If there are visible drops of the dark aqueous solution in the flask, transfer the methylene chloride solution to another flask using a clean, dry Pasteur pipet. If necessary, leave a small amount of the methylene chloride solution behind in order to avoid transferring any of the aqueous mixture. Add granular anhydrous sodium sulfate to dry the organic layer (Technique 12, Section 12.9, p. 680). If all the sodium sulfate clumps together when the mixture is stirred with a spatula, add some additional drying agent. Allow the mixture to stand for 10–15 minutes. Stir occasionally with a spatula.

Evaporation

Transfer the dry methylene chloride solution with a Pasteur pipet to a dry, preweighed 25-mL Erlenmeyer flask, while leaving the drying agent behind. Evaporate the methylene chloride by heating the flask in a hot water bath (Technique 7, Section 7.10, p. 611). This should be done in a hood and can be accomplished more rapidly if a stream of dry air or nitrogen gas is directed at the surface of the liquid. When the solvent is evaporated, the crude caffeine will coat the bottom of the flask. Do not heat the flask after the solvent has evaporated or you may sublime some of the caffeine. Weigh the flask and determine the weight of crude caffeine. Calculate the weight percentage recovery (see p. 565) of caffeine from tea leaves, using the weight of tea that you started with. You may store the caffeine by simply placing a stopper firmly into the flask.

Sublimation of Caffeine

Caffeine can be purified by sublimation (Technique 17, Section 17.5, p. 749). Assemble a sublimation apparatus as shown in Figure 17.2A, p. 748.[2] Add approximately 0.5 mL of methylene chloride to the Erlenmeyer flask and transfer the solution to a clean, 5-mL, thin-walled, conical vial, using a clean and dry Pasteur pipet. Add a few more drops of methylene chloride to the flask in order to rinse the caffeine out completely. Transfer this liquid to the conical vial. Evaporate the methylene chloride from the conical vial by gentle heating in a warm water bath under a stream of dry air or nitrogen.

Insert the cold finger into the sublimation apparatus. If you are using the subli-mator with the multipurpose adapter, adjust it so that the tip of the cold finger will be positioned about 1 cm above the bottom of the conical vial. Be sure that the inside of the assembled apparatus is clean and dry. If you are using an aspirator, install a trap between the aspirator and the sublimation apparatus. Turn on the vacuum and check to make sure that all joints in the apparatus are sealed tightly. Place *ice-cold* water in the inner tube of the apparatus. Heat the sample gently and carefully with a microburner to sublime the caffeine. Hold the burner in your hand (hold it at its base, *not* by the hot barrel), and apply the heat by moving the flame back and forth under the conical vial and up the sides. If the sample begins to melt, remove the flame for a few seconds before you resume heating. When sublimation is complete, discontinue heating. Remove the cold water and remaining ice from the inner tube and allow the apparatus to cool while continuing to apply the vacuum.

When the apparatus is at room temperature, remove the vacuum and *carefully* remove the inner tube. If this operation is done carelessly, the sublimed crystals may be dislodged from the inner tube and fall back into the conical vial. Scrape the sublimed caffeine onto a tared piece of smooth paper and determine the weight of caffeine recovered. Calculate the weight percentage recovery (see p. 565) of caffeine after the sublimation. Compare this value to the percentage recovery determined after the evaporation step. Determine the melting point of the purified caffeine. The melting point of pure caffeine is 236°C; however, the observed melting point will be lower. Submit the sample to the instructor in a labeled vial.

12B   **EXPERIMENT 12B (OPTIONAL)**

Extraction of Caffeine from Tea or Coffee Using Solid Phase Extraction (SPE)

In this experiment, caffeine is isolated from brewed tea or coffee using solid phase extraction (Technique 12, Section 12.14, p. 689). It is offered as a Green Chemistry alternative to the procedure given in Experiment 12A in which methylene chloride is used for the extraction of caffeine from tea.

---

[2] If you are using another type of sublimation apparatus, your instructor will provide you with specific instructions on how to assemble it correctly.

The solid phase extraction (SPE) columns may be obtained commercially.[3] The manufacturers covalently bond nonpolar alkyl groups to silica. This process converts the polar silica to a relatively nonpolar material. Often, 18-carbon alkyl groups are bonded to the silica creating a material that is referred to as C-18 silica.

In "normal" chromatography using nontreated silica, one expects polar substances to be attracted to the polar surface and to move more slowly through the column compared to nonpolar substances. This more traditional form of chromatography is referred to as **normal phase chromatography.** With C-18-treated silica, however, the polar substances will come off first, and the relatively nonpolar substance will be retained. This type of chromatography is referred to as **reverse phase chromatography.**

Experiment 12B makes use of reverse phase chromatography to remove caffeine from a tea or coffee solution. The technique is simple. See page Figure 12.14, page 692, for a diagram of the apparatus that you will use. The procedure involves the following steps:

1. Condition the SPE reverse phase column with methanol and water.

2. Use a vacuum to draw an aqueous solution of brewed tea or coffee through the SPE reverse phase column.

3. Caffeine is retained by the solid phase column.

4. Water, tannins, gallic acid, and other polar substances pass through the column.

5. Remove the caffeine from the column with ethyl acetate.

6. Evaporate the ethyl acetate to yield the crude caffeine.

7. Purify the caffeine by sublimation.

SUGGESTED WASTE DISPOSAL

This experiment approaches closely a high Green standard. Get into the spirit of the Green movement! Throw the tea bag into a trash can. Pour the aqueous solution from the column down the drain in a sink. Any remaining ethyl acetate should be given to another student who needs it or, as a last resort, poured into a nonhalogenated waste container. Remove the traces of ethyl acetate from the sodium sulfate drying agent with a blast of air and then dissolve the remaining solid inorganic material with water and pour it into an aqueous waste container (or pour it down the sink drain). Place the SPE column into a beaker that has been supplied for your use. Eventually, the tubes will be emptied and reused.

PROCEDURE

Preparing the Tea or Coffee Solution

Prepare a tea solution as described in Experiment 12A, page 93 (first paragraph), and footnote 1 on page 93. (If you desire to extract caffeine from coffee, dissolve

---

[3] We recommend a 6-mL column with 1000 mg of C18E sorbent (No. 8B-S001-JCH-S) for Experiment 12B. These columns are available from Phenomenex, 411 Madrid Avenue, Torrance, CA 90501-1430, phone: (310) 212-0555. About 25 mg of caffeine can be isolated from one batch of tea or coffee solution.

about 1.5 g of instant coffee in 20 mL of boiling water.[4]) Using the Pasteur pipet, transfer the concentrated tea solution into two centrifuge tubes fitted with screw caps. Try to keep the liquid volume in each centrifuge tube approximately equal. To squeeze additional liquid out of the tea bag, hold the tea bag on the inside wall of the beaker and roll a test tube back and forth while exerting gentle pressure on the tea bag. Press out as much liquid as possible without breaking the bag. Combine this liquid with the solution in the centrifuge tubes. Place the tea bag on the bottom of the beaker again and pour 2 mL of hot water over the tea bag. Squeeze the liquid out as just described and transfer this liquid to the centrifuge tubes. Let the solution in the centrifuge tube cool to room temperature. When the solution is room temperature, cap the centrifuge tubes and centrifuge the samples for 4 to 5 minutes. When the centrifugation step is complete, a solid will be present at the bottom of the tubes. Transfer the solution with a Pasteur pipet into a small beaker so that the thick solid is left behind in the centrifuge tubes.

Preparing the SPE Column

The apparatus is shown in Figure 12.14, page 692. Insert a No. 1 neoprene adapter inside a No. 2 neoprene adapter and place both adapters in the top of a 50-mL filter flask. Clamp the flask securely and then apply a vacuum source. Place the SPE column[5] inside the No. 1 neoprene adapter and condition the column under vacuum in the following way. Add 2 mL of methanol to the column, 1 mL at a time. Wait for the methanol to draw through the column completely before adding the second mL. Next, add 2 mL of room-temperature deionized (DI) water, 1 mL at a time. Proceed to the next step as quickly as possible. Don't let the column dry out.

Filtering the Tea Solution

While still drawing a vacuum, transfer the tea (or coffee) solution in the small beaker to the SPE column in approximately 1-mL portions, using a Pasteur pipet. If some solid is present in the beaker, be careful to avoid drawing up this solid at the bottom of the beaker (this solid will plug the SPE column). The SPE column should still be under vacuum. When all the solution has been drawn through the column, add 3 mL of room-temperature DI water in 1-mL portions. Continue applying the vacuum for 1 minute to remove as much water from the column as possible. Much of the caffeine is retained in the SPE column.

Removing the Caffeine from the SPE Column

Turn off the vacuum and carefully and slowly remove the hose from the side arm of the filter flask so that the liquid in the flask will not bump up onto the adapters and SPE column. Discard the filtrate in the filter flask. Clean and dry the filter flask that you used. Reconnect the SPE column to the apparatus and reapply the vacuum. Add a total of 7 mL of ethyl acetate to the column in approximately 1-mL portions. When all the ethyl acetate has been added, turn off the vacuum source and carefully remove the hose from the filter flask. Place the SPE column in the beaker supplied by the instructor.[6] Add 1.6 g of granular anhydrous sodium sulfate to the ethyl acetate solution in the filter flask. This drying operation removes any remaining water from the ethyl acetate solution. Allow the mixture to stand for 15 minutes.

---

[4] Tea yields better-quality caffeine than that obtained from coffee. Caffeine from tea is relatively colorless, whereas the caffeine extracted from coffee is highly colored. About 25 mg of caffeine is isolated in either case. Sublimation removes much of the color from the tea and coffee samples.

[5] See footnote 3 on page 96 for the SPE column required for this experiment.

[6] *Note to the instructor:* It is recommended that the C-18 silica in the SPE columns not be reused.

Stir the solution occasionally with a spatula to complete the drying step (Technique 12, Section 12.9, p. 680).

Isolation of the Caffeine

Transfer the dry ethyl acetate solution with a Pasteur pipet to a *dry, preweighed* 25-mL Erlenmeyer flask while leaving the drying agent behind. Rinse the drying agent with about 1 mL of fresh ethyl acetate and transfer it to the flask with your Pasteur pipet. Evaporate the ethyl acetate by heating the solution on a hot plate. This must be done in a hood and can be accomplished more rapidly if a stream of dry air or nitrogen gas is *carefully* directed at the surface of the liquid. Take the flask off the hot plate while some liquid remains and then use a gentle stream of air to remove the remaining traces of ethyl acetate from your sample. When the solvent has evaporated, crude solid caffeine will coat the bottom of the flask. Reweigh the flask to determine the weight of crude caffeine. Calculate the weight percentage recovery of caffeine obtained from your tea or coffee sample.[7] You may store the caffeine by placing a stopper firmly into the flask.

Sublimation of Caffeine

The caffeine obtained from tea or coffee can be purified by sublimation using the procedure described in Experiment 12A, page 95. At your instructor's option, you may combine your sample with another student's sample for sublimation. After sublimation, determine the weight of caffeine recovered and calculate the weight percentage recovery of the caffeine. Compare this value to the amount of crude sample obtained. At the instructor's option, determine the melting point of the purified caffeine. The melting point of pure caffeine is 236°C; however, the observed melting point will be lower. Submit the sample to the instructor in a labeled vial unless it is to be used for infrared spectroscopy (recommended) or mass spectroscopy (also recommended).

Analysis of Caffeine

At the option of your instructor, obtain the infrared spectrum of the caffeine as a dry film by dissolving a sample in *dry* acetone (Technique 25, Section 25.4, p. 838). Alternatively, a KBr pellet of the caffeine can be prepared to obtain the spectrum (Technique 25, Section 25.5). Include the infrared spectrum with your laboratory report, along with an interpretation of the principal peaks. A spectrum is included for comparison purposes.

At the instructor's option, you may determine the mass spectrum of the caffeine by gas chromatography/mass spectrometry (GC-MS) (Technique 28). At the same time, you can determine the purity of your sample.[8]

Report

Attach your infrared spectra to your report and label the major peaks. If you determined the mass spectrum, try to identify the important fragment ion peaks (Technique 28). Include the melting point, if it was required. Report the weight percentage of the caffeine recovered from the tea or coffee sample before and after sublimation.

---

[7] See footnote 3 on page 96 for information on the expected yield of caffeine from tea and coffee.

[8] The GC/MS may be obtained on the caffeine sample using a Varian 3800 GC with Saturn 2000 MS equipped with a capillary column that is 30 m long and 0.25 mm ID containing a 0.25-μm film of Varian CP-Sil 8CB. The injector was set at 275°C, with He gas flow at 1 mL/min. The temperature of the oven was increased from 75 to 280°C at 20°/min and held at 280°C for 1 minute. The retention time for the caffeine is about 10 minutes under these conditions. No impurities were detected.

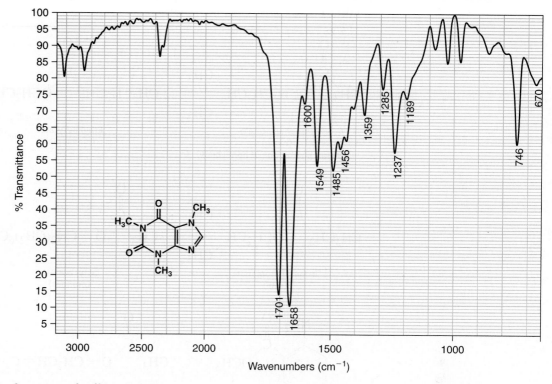

Spectrum of caffeine in KBr.

QUESTIONS

1. Outline a separation scheme for isolating caffeine from tea. Use a flowchart similar in format to that shown in Technique 2, Section 2.2, page 559.

2. Why was the sodium carbonate added in Experiment 12A?

3. The crude caffeine isolated from tea often has a green tinge. Why?

4. What are some possible explanations for why the melting point of your isolated caffeine was lower than the literature value (236°C)?

5. An alternative procedure for removing the tannins and gallic acid is to heat the tea leaves in an aqueous mixture containing calcium carbonate. Calcium carbonate reacts with the tannins and gallic acid to form insoluble calcium salts of these acids. If this procedure were used, what additional step (not done in this experiment) would be needed in order to obtain an aqueous tea solution?

6. What would happen to the caffeine if the sublimation step were performed at atmospheric pressure?

ESSAY

Esters—Flavors and Fragrances

Esters are a class of compounds widely distributed in nature. They have the general formula

Table 1 *Ester flavors and fragrances*

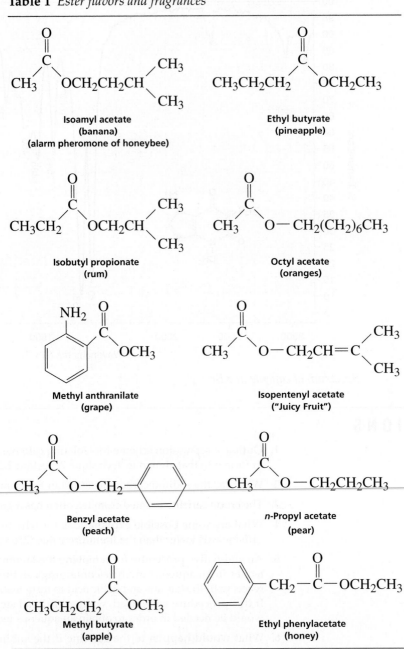

The simple esters tend to have pleasant odors. In many cases, although not exclusively so, the characteristic flavors and fragrances of flowers and fruits are due to compounds with the ester functional group. An exception is the case of the essential oils. The **organoleptic** qualities (odors and flavors) of fruits and flowers may often be due to a single ester, but more often, the flavor or the aroma is due to a complex mixture in which a single ester predominates. Some common flavor principles are listed in Table 1. Food and beverage manufacturers are familiar with these esters and often use them as additives to spruce up the flavor or odor of a dessert or beverage. Many times, such flavors or odors do not even have a natural basis, as is the case with the "juicy fruit" principle, isopentenyl acetate. An instant

Table 2 *Artificial pineapple flavor*

Pure Compounds	%	Essential Oils	%
Allyl caproate	5	Oil of sweet birch	1
Isoamyl acetate	3	Oil of spruce	2
Isoamyl isovalerate	3	Balsam Peru	4
Ethyl acetate	15	Volatile mustard oil	1
Ethyl butyrate	22	Oil cognac	5
Terpinyl propionate	3	Concentrated orange oil	4
Ethyl crotonate	5	Distilled oil of lime	2
Caproic acid	8		19
Butyric acid	12		
Acetic acid	5		
	81		

pudding that has the flavor of rum may never have seen its alcoholic namesake; this flavor can be duplicated by the proper admixture, along with other minor components, of ethyl formate and isobutyl propionate. The natural flavor and odor are not exactly duplicated, but most people can be fooled. Often, only a trained person with a high degree of gustatory perception, a professional taster, can tell the difference.

A single compound is rarely used in good-quality imitation flavoring agents. A formula for an imitation pineapple flavor that might fool an expert is listed in Table 2. The formula includes 10 esters and carboxylic acids that can easily be synthesized in the laboratory. The remaining seven oils are isolated from natural sources.

Flavor is a combination of taste, sensation, and odor transmitted by receptors in the mouth (taste buds) and nose (olfactory receptors). The stereochemical theory of odor is discussed in the essay that precedes Experiment 15. The four basic tastes (sweet, sour, salty, and bitter) are perceived in specific areas of the tongue. The sides of the tongue perceive sour and salty tastes, the tip is most sensitive to sweet tastes, and the back of the tongue detects bitter tastes. The perception of flavor, however, is not so simple. If it were, it would require only the formulation of various combinations of four basic substances—a bitter substance (a base), a sour substance (an acid), a salty substance (sodium chloride), and a sweet substance (sugar)—to duplicate any flavor! In fact, we cannot duplicate flavors in this way. The human possesses 9,000 taste buds. The combined response of these taste buds is what allows perception of a particular flavor.

Although the "fruity" tastes and odors of esters are pleasant, they are seldom used in perfumes or scents that are applied to the body. The reason for this is chemical. The ester group is not as stable under perspiration as the ingredients of the more expensive essential-oil perfumes. The latter are usually hydrocarbons (terpenes), ketones, and ethers extracted from natural sources. Esters, however, are used only for the cheapest toilet waters because on contact with sweat they undergo hydrolysis, giving organic acids. These acids, unlike their precursor esters, generally do not have a pleasant odor.

Butyric acid, for instance, has a strong odor like that of rancid butter (of which it is an ingredient) and is a component of what we normally call body odor. It is this substance that makes foul-smelling humans so easy for an animal to detect when downwind of them. It is also of great help to the bloodhound, which is trained to follow small traces of this odor.

$$R-\overset{\overset{\displaystyle O}{\|}}{C}-OR' + H_2O \longrightarrow R-\overset{\overset{\displaystyle O}{\|}}{C}-OH + R'OH$$

Ethyl butyrate and methyl butyrate, however, which are the *esters* of butyric acid, smell like pineapple and apple, respectively.

A sweet, fruity odor also has the disadvantage of possibly attracting fruit flies and other insects in search of food. Isoamyl acetate, the familiar solvent called banana oil, is particularly interesting. It is identical to a component of the alarm **pheromone** of the honeybee. Pheromone is the name applied to a chemical secreted by an organism that evokes a specific response in another member of the same species. This kind of communication is common among insects who otherwise lack means of intercourse. When a honeybee worker stings an intruder, an alarm pheromone, composed partly of isoamyl acetate, is secreted along with the sting venom. This chemical causes aggressive attack on the intruder by other bees, who swarm after the intruder. Obviously, it wouldn't be wise to wear a perfume compounded of isoamyl acetate near a beehive. Pheromones are discussed in more detail in the essay preceding Experiment 47.

REFERENCES

Bauer, K., and Garbe, D. *Common Fragrance and Flavor Materials.* Weinheim: VCH Publishers, 1985.

The Givaudan Index. New York: Givaudan-Delawanna, 1949. (Gives specifications of synthetics and isolates for perfumery.)

Gould, R. F., ed. *Flavor Chemistry, Advances in Chemistry,* No. 56. Washington, DC: American Chemical Society, 1966.

Layman, P. L. "Flavors and Fragrances Industry Taking on New Look." *Chemical and Engineering News* (July 20, 1987): 35.

Moyler, D. "Natural Ingredients for Flavours and Fragrances." *Chemistry and Industry* (January 7, 1991): 11.

Rasmussen, P. W. "Qualitative Analysis by Gas Chromatography—G.C. versus the Nose in Formulation of Artificial Fruit Flavors." *Journal of Chemical Education, 61* (January 1984): 62.

Shreve, R. N., and Brink, J. *Chemical Process Industries,* 4th ed. New York: McGraw-Hill, 1977.

Welsh, F. W., and Williams, R. E. "Lipase Mediated Production of Flavor and Fragrance Esters from Fusel Oil." *Journal of Food Science, 54* (November/December 1989): 1565.

EXPERIMENT 13

Isopentyl Acetate (Banana Oil)

Esterification

Heating under reflux

Extraction

Simple distillation

Microscale boiling point

In this experiment you will prepare an ester, isopentyl acetate. This ester is often referred to as banana oil because it has the familiar odor of this fruit.

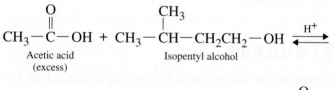

Isopentyl acetate is prepared by the direct esterification of acetic acid with isopentyl alcohol. Because the equilibrium does not favor the formation of the ester, it must be shifted to the right, in favor of the product, by using an excess of one of the starting materials. Acetic acid is used in excess because it is less expensive than isopentyl alcohol and more easily removed from the reaction mixture.

In the isolation procedure, much of the excess acetic acid and the remaining isopentyl alcohol are removed by extraction with sodium bicarbonate and water. After drying with anhydrous sodium sulfate, the ester is purified by distillation. The purity of the liquid product is analyzed by performing a microscale boiling point determination or infrared spectroscopy.

REQUIRED READING

Review: Experiment 1   Introduction to Microscale Laboratory (pp. 2–13)

Techniques 5 and 6

New: Technique 7   Reaction Methods, Sections 7.2–7.4 and 7.6

Technique 13   Physical Constants, Boiling Points

Technique 12   Extractions, Separations, and Drying Agents

Technique 14   Simple Distillation

Essay   Esters—Flavors and Fragrances

If performing the optional infrared spectroscopy, also read:

Technique 25   Preparation of Samples for Spectroscopy

SPECIAL INSTRUCTIONS

Be careful when dispensing sulfuric and glacial acetic acids. They are corrosive and will attack your skin if you make contact with them. If you get one of these acids on your skin, wash the affected area with copious quantities of running water for 10–15 minutes.

Because a 1-hour reflux is required, you should start the experiment at the beginning of the laboratory period. During the reflux period, you may perform other work.

SUGGESTED WASTE DISPOSAL

Any aqueous solutions should be placed in a container specially designated for dilute aqueous waste. Place any excess ester in the nonhalogenated organic waste container.

NOTES TO THE INSTRUCTOR

Choose either Experiment 13A or Experiment 13B, but not both. The semi-microscale procedure requires the use of equipment not found in the typical microscale kit: a 20-mL round-bottom flask, a distillation head, and a vacuum takeoff adapter. The purpose of Experiment 13B is to allow an alternative to the use of a Hickman head for the distillation step.

This experiment has been carried out successfully using Dowex 50×2-100 ion-exchange resin instead of the sulfuric acid. Amberlyst-15 resin will also work.

13A   EXPERIMENT 13A

Isopentyl Acetate (Microscale Procedure)

PROCEDURE

Apparatus

Using a 5-mL conical vial, assemble a reflux apparatus using a water-cooled condenser (Fig. 7.2A, p. 599). Top the condenser with a drying tube (Fig. 7.10B, p. 599) that contains a loose plug of glass wool. The purpose of the drying tube is to control odors rather than to protect the reaction from water. Use a hot plate and an aluminum block for heating.

Preparation

Remove the empty 5-mL conical vial, weigh it, and record its weight. Place approximately 1.0 mL of isopentyl alcohol ($MW = 88.2$, $d = 0.813$ g/mL) in the vial using an automatic pipet or a dispensing pump. Reweigh the vial containing the alcohol and subtract the tare weight to obtain an accurate weight for the alcohol. Add 1.5 mL of glacial acetic acid (*MW* = 60.1, *d* = 1.06 g/mL) using an automatic pipet or dispensing pump. Using a disposable Pasteur pipet, add two to three drops of concentrated sulfuric acid. Swirl the liquid to mix. Add a small boiling stone (or a magnetic spin vane) and reattach the vial to the apparatus.

Reflux

Bring the mixture to a boil (aluminum block at about 150–160°C). Be sure to stir the mixture if you are using a spin vane instead of a boiling stone. Continue heating under reflux for 60–75 minutes. Remove the heating source and allow the mixture to cool to room temperature.

Workup

Disassemble the apparatus and, using a forceps, remove the boiling stone (or spin vane). Using a calibrated Pasteur pipet (p.11), slowly add 1.0 mL of 5% aqueous sodium bicarbonate to the cooled mixture in the conical vial. Stir the mixture in the vial with a microspatula until carbon dioxide evolution is no longer vigorous. Then cap the vial and shake *gently* with venting until the evolution of gas is complete. Using a Pasteur pipet, remove the lower aqueous layer and discard it. Repeat the extraction two more times, as outlined previously, using a fresh 1.0-mL portion of 5% sodium bicarbonate solution each time.

If droplets of water are evident in the vial containing the ester, transfer the ester to a dry conical vial using a dry Pasteur pipet. Dry the ester over granular anhydrous sodium sulfate (see Technique 12, Section 12.9, p. 680). Allow the capped solution to stand for 10–15 minutes. Transfer the dry ester with a Pasteur pipet into a 3-mL conical vial while leaving the drying agent behind. If necessary, pick out any granules of sodium sulfate with the end of a spatula.

Distillation

Add a boiling stone (or a magnetic spin vane) to the dry ester. Clamping the glassware, assemble a distillation apparatus using a Hickman still and a water-cooled condenser on top of a hot plate with an aluminum heating block (Fig. 14.5, p. 708). In order to control odors, rather than to keep the reaction dry, top the apparatus with a drying tube packed loosely with a small amount of calcium chloride held in place by bits of cotton or glass wool. Begin the distillation by turning on the hot plate (about 180°C). Stir the mixture if you are using a spin vane instead of a boiling stove. Continue the distillation until only one or two drops of liquid remain in the distilling vial. If the Hickman head fills before the distillation is complete, it may be necessary to empty it using a Pasteur pipet (see Fig. 14.6A, p. 709) and transfer the distillate to a tared (preweighed) conical vial. Unless you have a side-ported Hickman still, it will be necessary to remove the condenser in order to perform the transfer. When the distillation is complete, transfer the final portion of the distillate to this same vial.

Determination of Yield

Weigh the product and calculate the percentage yield of the ester. Determine its boiling point (bp 142°C) using a microscale boiling-point determination (Technique 13, Section 13.2, p. 695). See page 107 for a spectral analysis.

13B   EXPERIMENT 13B

Isopentyl Acetate (Semimicroscale Procedure)

PROCEDURE

Apparatus

Assemble a reflux apparatus on top of your hot plate using a 20- or 25-mL round-bottom flask and a water-cooled condenser (refer to Fig. 7.6A, p. 601, but use a round-bottom flask instead of the conical vial). To control vapors, place a drying tube packed with calcium chloride on top of the condenser. Use a hot plate and the aluminum block with the larger set of holes for heating.

Reaction Mixture

Weigh (tare) an empty 10-mL graduated cylinder and record its weight. Place approximately 2.5 mL of isopentyl alcohol in the graduated cylinder and reweigh it to determine the weight of the alcohol. Disconnect the round-bottom flask from the reflux apparatus and transfer the alcohol into it. Do not clean or wash the graduated cylinder. Using the same graduated cylinder, measure approximately 3.5 mL of glacial acetic acid ($MW = 60$, $d = 1.06$ g/mL) and add it to the alcohol already in the flask. Using a calibrated Pasteur pipet, add 0.5 mL of concentrated sulfuric acid, mixing *immediately* (swirl), to the reaction mixture contained in the flask. Add a corundum (black) boiling stone or stirring bar and reconnect the flask. Do not use a calcium carbonate (white, marble) boiling stone, because it will dissolve in the acidic medium.

Reflux

Start water circulating in the condenser and bring the mixture to a boil. Continue heating under reflux for at least 60 minutes. Be sure to stir the mixture if you are using a stirring bar instead of a boiling stone. When the reflux period is complete, disconnect or remove the heating source and let the mixture cool to room temperature.

Extractions

Disassemble the apparatus and transfer the reaction mixture to a 15-mL capped centrifuge tube. Avoid transferring the boiling stone or stirring bar. Add 5 mL of water, cap the centrifuge tube, and mix the phases by careful shaking and venting. Allow the phases to separate and then open the cap and remove the lower aqueous layer (see a similar procedure for a conical vial in Workup on page 105). Next, extract the organic layer with 2.5 mL of aqueous sodium bicarbonate, just as you did previously with water. Extract the organic layer once again, this time with 2.5 mL of saturated aqueous sodium chloride.

Drying

Transfer the crude ester to a clean, dry, 25-mL Erlenmeyer flask and add approximately 0.5 g of anhydrous sodium sulfate. Cork the mixture and let it stand for about 10 minutes while you prepare the apparatus for distillation. If the mixture does not appear dry (the drying agent clumps and does not "flow," the solution is cloudy, or drops of water are obvious), transfer the ester to a new, clean, dry,

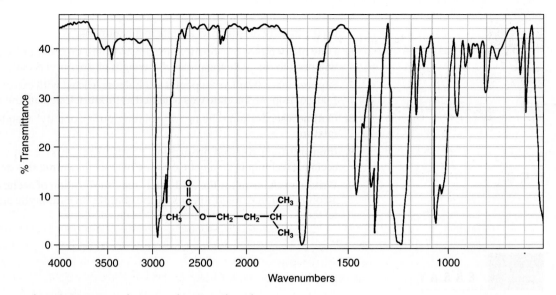

Infrared spectrum of isopentyl acetate (neat).

25-mL Erlenmeyer flask and add a new 0.25-g portion of anhydrous sodium sulfate to complete the drying.

Distillation

Assemble a distillation apparatus using your smallest round-bottom flask to distill from (Fig. 14.10, p. 713, but insert a water condenser as shown on p. 54). Use a hot plate with an aluminum block to heat. Preweigh (tare) and use a 5-mL conical vial to collect the product. Immerse the collection vial in a beaker of ice to ensure condensation and to reduce odors. Distill your ester and record its boiling-point *range* in your notebook.

Yield Determination

Weigh the product and calculate the percentage yield of the ester. At the option of your instructor, determine the boiling point using one of the methods described in Section 13.2, page 695. See below for a spectral analysis.

Infrared Spectroscopy

At your instructor's option, obtain an infrared spectrum using salt plates (Technique 25, Section 25.2, p. 834). Compare the spectrum with the one reproduced in this experiment and include it with your report to the instructor. If any of your sample remains after performing the determination of the infrared spectrum, submit it in a properly labeled vial along with your report.

QUESTIONS

1. One method for favoring the formation of an ester is to add excess acetic acid. Suggest another method, involving the right-hand side of the equation, that will favor the formation of the ester.

2. Why is it easier to remove excess acetic acid from the products than excess isopentyl alcohol?

3. Why is the reaction mixture extracted with sodium bicarbonate? Give an equation and explain its relevance.

4. Which starting material is the limiting reagent in this procedure? Which reagent is used in excess? How great is the molar excess (how many times greater)?

5. How many grams are there in 1.00 mL of isopentyl acetate? You will need to look up the density of isopentyl acetate in a handbook.

6. How many moles of isopentyl acetate are there in 1.00 g of isopentyl acetate? You will need to calculate the molecular weight of isopentyl acetate.

7. Suppose that 1.00 mL of isopentyl alcohol was reacted with excess acetic acid and that 1.00 g of isopentyl acetate was obtained as product. Calculate the percentage yield.

8. Outline a separation scheme for isolating pure isopentyl acetate from the reaction mixture.

9. Interpret the principal absorption bands in the infrared spectrum of isopentyl acetate. (Technique 25 may be of some help in answering this question.)

10. Write a mechanism for the acid-catalyzed esterification of acetic acid with isopentyl alcohol. You may need to consult the chapter on carboxylic acids in your lecture textbook.

ESSAY

Terpenes and Phenylpropanoids

Anyone who has walked through a pine or cedar forest, or anyone who loves flowers and spices, knows that many plants and trees have distinctively pleasant odors. The essences or aromas of plants are due to volatile or **essential oils,** many of which have been valued since antiquity for their characteristic odors (frankincense and myrrh, for example). A list of the commercially important essential oils would run to more than 200 entries. Allspice, almond, anise, basil, bayberry, caraway, cinnamon, clove, cumin, dill, eucalyptus, garlic, jasmine, juniper, orange, peppermint, rose, sandalwood, sassafras, spearmint, thyme, violet, and wintergreen are but a few familiar examples of such valuable essential oils. Essential oils are used for their pleasant odors in perfumes and incense. They are also used for their taste appeal as spices and flavoring agents in foods. A few are valued for antibacterial and antifungal action. Some are used medicinally (camphor and eucalyptus) and others as insect repellents (citronella). Chaulmoogra oil is one of the few known curative agents for leprosy. Turpentine is used as a solvent for many paint products.

Essential oil components are often found in the glands or intercellular spaces in plant tissue. They may exist in all parts of the plant but are often concentrated in the seeds or flowers. Many components of essential oils are steam-volatile and can be isolated by steam distillation. Other methods of isolating essential oils include solvent extraction and pressing (expression) methods. Esters (see essay, p. 99) are frequently responsible for the characteristic odors and flavors of fruits and flowers, but other types of substances may also be important components of odor or flavor principles. Besides the esters, the ingredients of essential oils may be complex mixtures of hydrocarbons, alcohols, and carbonyl compounds. These other components usually belong to one of the two groups of natural products called **terpenes** or **phenylpropanoids.**

Terpenes

Chemical investigations of essential oils in the 19th century found that many of the compounds responsible for the pleasant odors contained exactly 10 carbon atoms. These 10-carbon compounds came to be known as terpenes if they were hydrocarbons and as **terpenoids** if they contained oxygen and were alcohols, ketones, or aldehydes.

Eventually, it was found that there are also minor and less volatile plant constituents containing 15, 20, 30, and 40 carbon atoms. Because compounds of 10 carbons were originally called terpenes, they came to be called **monoterpenes.** The other terpenes were classified in the following way.

Class	No. of Carbons	Class	No. of Carbons
Hemiterpenes	5	Diterpenes	20
Monoterpenes	10	Triterpenes	30
Sesquiterpenes	15	Tetraterpenes	40

Further chemical investigations of the terpenes, all of which contain multiples of five carbons, showed them to have a repeating structural unit based on a five-carbon pattern. This structural pattern corresponds to the arrangement of atoms in the simple five-carbon compound isoprene. Isoprene was first obtained by the thermal "cracking" of natural rubber.

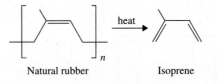

Natural rubber   Isoprene

As a result of this structural similarity, a diagnostic rule for terpenes, called the **isoprene rule,** was formulated. This rule states that a terpene should be divisible, at least formally, into **isoprene units.** The structures of a number of terpenes, along with a diagrammatic division of their structures into isoprene units, is shown in the figure on next page that accompanies this essay. Many of these compounds represent odors or flavors that should be familiar to you.

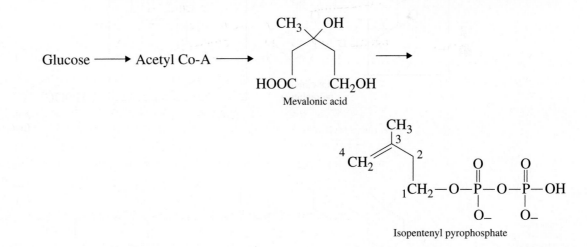

Modern research has shown that terpenes do not arise from isoprene; it has never been detected as a natural product. Instead, the terpenes arise from an important biochemical precursor compound called **mevalonic acid** (see above). This compound arises from acetyl coenzyme A, a product of the biological degradation of glucose (glycolysis) and is converted to a compound called isopentenyl pyrophosphate. Isopentenyl pyrophosphate and its isomer 3,3-dimethylallyl pyrophosphate (double bond moved to the second position) are the five-carbon building blocks used by nature to construct all the terpene compounds.

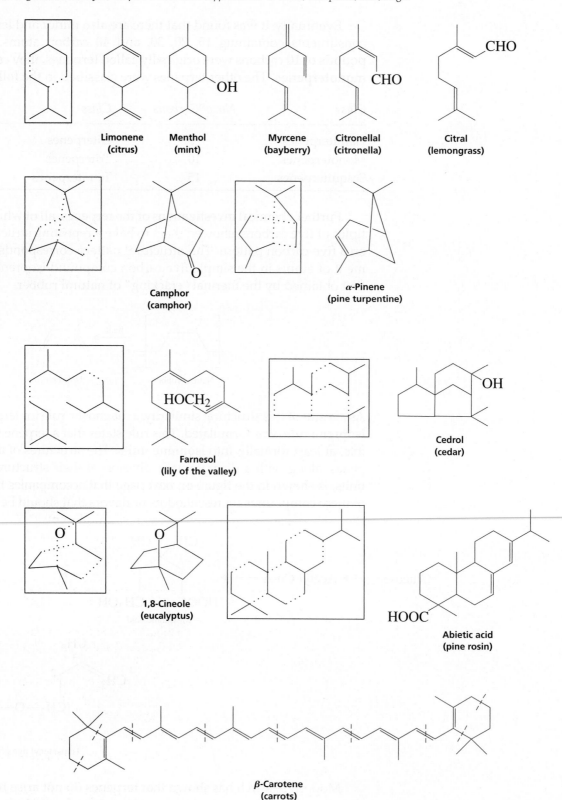

Limonene
(citrus)

Menthol
(mint)

Myrcene
(bayberry)

Citronellal
(citronella)

Citral
(lemongrass)

Camphor
(camphor)

α-Pinene
(pine turpentine)

Farnesol
(lily of the valley)

Cedrol
(cedar)

1,8-Cineole
(eucalyptus)

Abietic acid
(pine rosin)

β-Carotene
(carrots)

Selected terpenes.

Phenylpropanoids

Aromatic compounds, those containing a benzene ring, are also a major type of compound found in essential oils. Some of these compounds, like *p*-cymene, are actually cyclic terpenes that have been aromatized (had their ring converted to a benzene ring), but most are of a different origin.

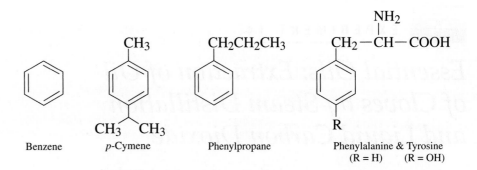

Benzene   *p*-Cymene   Phenylpropane   Phenylalanine & Tyrosine
(R = H)   (R = OH)

Many of these aromatic compounds are **phenylpropanoids,** compounds based on a phenylpropane skeleton. Phenylpropanoids are related in structure to the common amino acids phenylalanine and tyrosine, and many are derived from a biochemical pathway called the **shikimic acid pathway.**

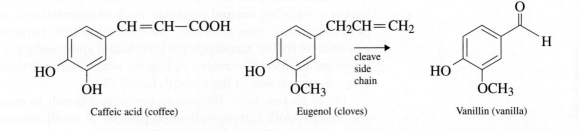

Caffeic acid (coffee)   Eugenol (cloves)   Vanillin (vanilla)

It is also common to find compounds of phenylpropanoid origin that have had the three-carbon side chain cleaved. As a result, phenylmethane derivatives, such as vanillin, are also common in plants.

REFERENCES

Cane, D. E., ed. "Isoprenoids Including Carotenoids and Steroids." In Barton, D., Nakanishi, K., and Meth-Cohn, O., eds. *Comprehensive Natural Products Chemistry.* New York: Elsevier, 1999. Vol. 2.

Cornforth, J. W. "Terpene Biosynthesis." *Chemistry in Britain,* 4 (1968): 102.

Geissman, T. A., and Crout, D. H. G. *Organic Chemistry of Secondary Plant Metabolism.* San Francisco: Freeman, Cooper and Co., 1969.

Hendrickson, J. B. *The Molecules of Nature.* New York: W. A. Benjamin, 1965.

Leeper, F. J., and Vederas, J. C., eds. *Biosynthesis: Aromatic Polyketides, Isoprenoids, Alkaloids.* New York: Springer, 2000.

Newman, A. A. *Chemistry of Terpenes and Terpenoids: A Survey for Advanced Students and Research Workers.* New York: Academic Press, 1972.

Pinder, A. R. *The Terpenes.* New York: John Wiley & Sons, 1960.

Ruzicka, L. "History of the Isoprene Rule." *Proceedings of the Chemical Society (London)* (1959): 341.

Sterret, F. S. "The Nature of Essential Oils, Part I. Production." *Journal of Chemical Education,* 39 (1962): 203.

Sterret, F. S. "The Nature of Essential Oils, Part II. Chemical Constituents. Analysis." *Journal of Chemical Education,* 39 (1962): 246.

14 EXPERIMENT 14

Essential Oils: Extraction of Oil of Cloves by Steam Distillation and Liquid Carbon Dioxide

Isolation of a natural product

Steam distillation

Green chemistry

Essential oils are volatile compounds responsible for the aromas commonly associated with many plants (see essay "Terpenes and Phenylpropanoids"). The chief constituent of the essential oil from cloves is aromatic and volatile with steam. In this experiment, you will isolate the main component derived from this spice by steam distillation. Steam distillation provides a means of isolating natural products, such as essential oils, without the risk of decomposing them thermally. Identification and characterization of this essential oil will be accomplished by infrared spectroscopy. Experiment 14C provides a Green Chemistry option in which the instructor may demonstrate the extraction of the oil with liquid CO_2.

Oil of cloves (from *Eugenia caryophyllata*) is rich in **eugenol** (4-allyl-2-methoxyphenol). Caryophyllene is present in small amounts, along with other terpenes. Eugenol (bp 250°C) is a phenol, or an aromatic hydroxy compound.

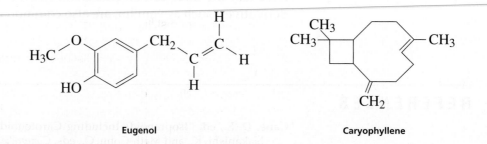

Eugenol   Caryophyllene

REQUIRED READING

Review:	Techniques 5 and 6	
	Technique 7	Reaction Methods, Section 7.10
	Technique 12	Extractions, Separations and Drying Agents, Sections 12.4 and 12.9
	Technique 25	Infrared Spectroscopy
New:	Technique 18	Steam Distillation
	Essay	Terpenes and Phenylpropanoids

If performing the optional proton NMR analysis, also read

| | Technique 26 | Nuclear Magnetic Resonance Spectroscopy |

SPECIAL INSTRUCTIONS

Be careful when handling methylene chloride. It is a toxic solvent, and you should not breathe it excessively or spill it on yourself. To complete the distillation in a reasonable time, boil the mixture as rapidly as possible without allowing the boiling mixture to rise above the neck of the Hickman head. This requires that you work with careful attention during the distillation procedure. The distillation requires 1–2 hours.

SUGGESTED WASTE DISPOSAL

You must dispose of methylene chloride in a waste container marked for the disposal of halogenated organic waste. Any residue from the ground cloves can be disposed of in an ordinary trash can. Any aqueous solutions should be placed in the container specially designated for aqueous wastes.

14A   EXPERIMENT 14A

Oil of Cloves (Microscale Procedure)

PROCEDURE

Assemble a steam distillation apparatus, as shown in Figure 18.3, page 753. Be sure to include the water condenser, as shown in the illustration. Use a 20- or 25-mL round-bottom flask as a distillation flask and either an aluminum block or a sand bath to heat the distillation flask. If you use a sand bath, you may need to cover the sand bath and distillation flask with aluminum foil.

Weight approximately 1.0 g of ground cloves or clove buds onto a weighing paper and record the exact weight. If your spice is already ground, you may proceed without grinding it; if you use clove buds, cut the buds into small pieces. Mix the spice with 12–15 mL of water in the round-bottom distillation flask, add a magnetic stirring bar, and attach the flask to the distillation apparatus. Allow the spice to soak in the water for about 15 minutes before beginning the heating. Be sure that all the spice is thoroughly wetted. Swirl the flask gently, if needed.

Steam Distillation

Turn on the cooling water in the condenser, begin stirring the mixture in the distillation flask, and begin heating the mixture to provide a steady rate of distillation. The temperature for the heating device should be about 130°C. If you approach the boiling point too quickly, you may have difficulty with frothing or bump-over. You need to find the amount of heating that provides a steady rate of distillation but avoids frothing or bumping. A good rate of distillation would be to have one drop of liquid collected every 2–5 seconds.

As the distillation proceeds, use a Pasteur pipet (5¾-inch) to transfer the distillate from the reservoir of the Hickman head to a 15-mL screw-cap centrifuge tube. If you are using a Hickman head with a side port, you can easily remove the distillate by opening the side port and withdrawing the liquid. If your Hickman head does not have a side port, you need to remove the condenser from the top of the distillation apparatus to remove the distillate. In this case, the transferring operation is best accomplished if the Pasteur pipet is bent slightly at the end. Continue distillation until 5–8 mL of distillate has been collected.

Normally in a steam distillation, the distillate is somewhat cloudy owing to separation of the essential oil as the vapors cool. You may not notice this, but you will still obtain satisfactory results.

Extraction of the Essential Oil

Collect all the distillate in a 15-mL screw-cap centrifuge tube. Using a calibrated Pasteur pipet (p. 11), add 2.0 mL of methylene chloride (dichloromethane) to extract the distillate. Cap the tube securely and shake it vigorously with frequent venting. Allow the layers to separate. Using a Pasteur pipet, transfer the lower methylene chloride layer to a clean, dry, 5-mL conical vial. Repeat this extraction procedure two more times with fresh 1.0-mL portions of methylene chloride and combine all the methylene chloride extracts in the same 5-mL conical vial that you used for the first extraction. If there are drops of water in the vial, it will be necessary to transfer the methylene chloride solution with a dry Pasteur pipet to another dry conical vial.

Drying

Dry the methylene chloride solution by adding granular anhydrous sodium sulfate to the conical vial (see Technique 12, Section 12.9, p. 680). Let the solution stand for 10–15 minutes with occasional stirring.

Evaporation

While the organic solution is being dried, clean and dry a 5-mL conical vial and weigh (tare) it accurately. With a clean, dry Pasteur pipet, transfer the dried organic layer to the tared vial, leaving the drying agent behind. Use small amounts of methylene chloride to rinse the solution completely into the tared vial. Be careful to keep any of the sodium sulfate from being transferred. Working in a hood, evaporate the methylene chloride from the solution by using a gentle stream of dry air or nitrogen while heating the vial in a warm water bath (temperature about 40°C). (See Technique 7, Section 7.10, p. 611). It is important that the stream of air or nitrogen be gentle or you will force your solution out of the conical vial. In addition, be careful not to overheat the sample. Be careful not to continue the evaporation beyond the point where all the methylene chloride has evaporated. Your product is a volatile oil (i.e., a liquid), and if you continue to heat and evaporate the liquid beyond the point where the solvent has been removed, you will likely lose your sample.

Yield Determination

When the solvent has been removed, weigh the conical vial. Calculate the weight percentage recovery (see p. 565) of the oil from the original amount of spice used. See page 115 for a spectral analysis.

14B   EXPERIMENT 14B

Oil of Cloves (Semimicroscale Procedure)

PROCEDURE

Apparatus

Assemble a semimicroscale distillation apparatus, as shown in Figure 14.10, page 713. Use a 20- or 25-mL round-bottom flask as the distillation flask and either an aluminum block or a sand bath to heat the distillation flask. If you use a sand bath, you may need to cover the sand bath and distillation flask with aluminum foil.

Preparation

Use the amounts of cloves and water described in Experiment 14A, page 113.

Steam Distillation

Proceed with the distillation as described in Experiment 14A. Note, however, that you will not have to remove distillate during the course of the distillation. Continue with the extraction, drying, evaporation, and yield determination, as described on pages 113–114.

**Spectroscopy
(Experiment 14A and 14B)**

Infrared Spectrum

Obtain the infrared spectrum of the oil as a pure liquid sample (Technique 25, Section 25.2, p. 834). Small amounts of water will damage the salt plates that are used as cells in infrared spectroscopy.

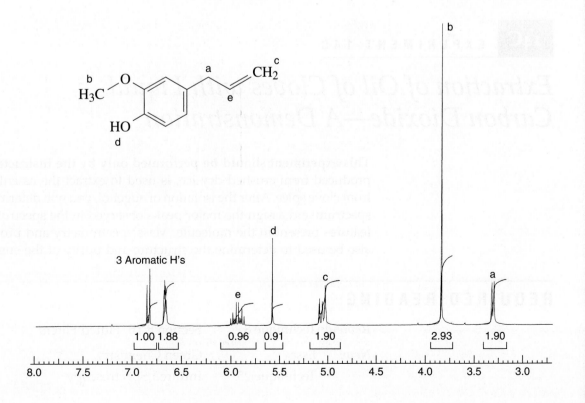

NOTE: Before proceeding with infrared spectroscopy, check with your instructor to make sure your sample is acceptable.

It may be necessary to use a capillary tube to transfer a sufficient amount of liquid to the salt plates. If the amount of liquid is too small to transfer, add one or two drops of methylene chloride to aid in the transfer. Gently blow on the plate to evaporate the solvent. Include the infrared spectrum in your laboratory report, along with an interpretation of the principal absorption peaks.

NMR Spectrum

At the instructor's option, determine the nuclear magnetic resonance spectrum of the oil (Technique 26, Section 26.1, p. 870).

QUESTIONS

1. Why is eugenol steam-distilled rather than purified by simple distillation?
2. A natural product ($MW = 150$) distills with steam at a boiling temperature of 99°C at atmospheric pressure. The vapor pressure of water at 99°C is 733 mm Hg.
   (a) Calculate the weight of the natural product that codistills with each gram of water at 99°C.
   (b) How much water must be removed by steam distillation to recover this natural product from 0.5 g of a spice that contains 10% of the desired substance?
3. In a steam distillation, the amount of water actually distilled is usually greater than the amount calculated, assuming that both water and organic substance exert the same vapor pressure when they are mixed that they exert when each is pure. Why does one recover more water in the steam distillation than was calculated? (*Hint:* Are the organic compound and water truly immiscible?)
4. Explain how caryophyllene fits the isoprene rule (see essay, "Terpenes and Phenyl-propanoids," p. 106).

14C   EXPERIMENT 14C

Extraction of Oil of Cloves with Liquid Carbon Dioxide—A Demonstration

This experiment should be performed only by the instructor. Liquid CO_2, produced from crushed dry ice, is used to extract the essential oil eugenol from clove spice. After the isolation of eugenol, you will determine its infrared spectrum and assign the major peaks observed in the spectrum to structural features present in the molecule. Mass spectrometry and proton NMR may also be used to determine the structure and purity of the eugenol obtained.

REQUIRED READING

Review: Technique 8   Section 8.1B, Fluted Filters

New: Essay   Green Chemistry

          Technique 25   Infrared Spectroscopy

Technique 26	Nuclear Magnetic Resonance Spectroscopy (Optional)
Technique 28	Mass Spectrometry (Optional)

SPECIAL INSTRUCTIONS AND NOTES TO THE INSTRUCTOR

It is recommended that you begin with clove buds and finely grind them to a powder in a mortar and pestle to improve yield. Crush the dry ice with a hammer and store in a Styrofoam container or use immediately.

Special care needs to be taken as to the type of filter paper used in this experiment. It is recommended that fastest filter paper be available to increase yield. Whatman #4 (11 cm) filter paper may be used, but paper used in "Mr Coffee" type coffeemakers works even better as long as it is a low-quality basket-type filter paper. The basket-type filter paper should be cut to 11 cm in diameter.[1] Do not use the coffee filters that have been sealed on the edges to resemble a cone.

Use a **new** 50-mL plastic centrifuge tube made of durable polypropylene provided by VWR (catalog number 20171-029). The centrifuge tubes may be reused for up to 5 times, but then they must be replaced because they will not hold the pressure required for the carbon dioxide to be maintained in the liquid state. Unused centrifuge tubes and lids will assure you that a satisfactory seal will be maintained so that you will be able to attain the pressure required to form liquid CO_2.

SAFETY PRECAUTIONS

This demonstration must be performed with a new centrifuge tube in a hood, with the safety glass pulled down. The 1000-mL graduated cylinder must be made of plastic. Special care is required when the tube is in the water bath, because the centrifuge tube could explode or the lid could fly off due to the high pressure created in the tube during the extraction. See the Instructor's Manual for additional safety discussion. Use only the equipment and methods described here. **Do not under any circumstances use glass products or other plastic centrifuge tubes than the one specified here.**

SUGGESTED WASTE DISPOSAL

Solid waste from the spice should be disposed of in a garbage can.

PROCEDURE

Apparatus

Obtain a 50-mL *plastic* centrifuge tube and cap. **Perform this experiment in a hood, with the safety glass pulled down.**

> **⚠ CAUTION**
>
> **Do not under any circumstances use glass centrifuge tubes as a substitute.**

---

[1] Costco sells bulk basket-type coffee filters designed for 8–12 cups. This type of filter fits Mr. Coffee machines. The best paper is thin so that you can see light through the paper.

Coil a 15-cm piece of copper wire twice at one end so that it will fit inside the centrifuge tube. The wire should be long enough so that you will have a "handle" that can be used to lower the spice into the tube and will allow you to easily remove the filter cone after the extraction. The coil prevents the filter paper cone containing the spice from falling to the bottom of the tube into the extracted oil (see figures).

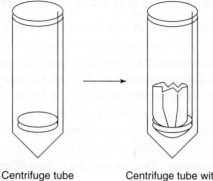

Centrifuge tube
with copper coil

Centrifuge tube with
copper coil and filter paper

Preparing the Spice

Grind about 2.0 g of clove buds to a fine powder in a mortar and pestle. Flute an 11-cm piece of filter paper by the procedure shown in Figure 8.3, page 619 (see "Special Instructions" and "Notes to the Instructor" for the type of paper required). Adjust the filter paper so that it will fit into the centrifuge tube and into the copper coil, but make sure that the tip of the filter paper doesn't project down too far into the tapered end of the centrifuge tube (see figure). It is critical that the filter paper does not touch the bottom of the centrifuge tube. Now transfer the powdered cloves to the preweighed fluted filter paper and record the exact weight of spice.

Extraction

Fill a 1000-mL **plastic** graduated cylinder two-thirds full with hot tap water (40–45°C) and place it in a hood. It is best to use a clear graduated cylinder so that you can observe the extraction as it occurs **but it must be made of plastic.** Working as quickly as possible, place some crushed dry ice *only* in the *tapered part* of the centrifuge tube at the bottom. Insert the fluted filter paper cone containing the ground spice into the copper coil and gently push the coil into the tube, sliding it down to the level of the dry ice. You will need to push the filter cone down into the tube with your index finger as you push the copper coil into the tube. Fill the remainder of the tube with the finely crushed dry ice, tapping the bottom of the tube on the counter and adding more dry ice until the tube is full. Then cap the centrifuge tube as tightly as you can. All of these operations must be carried out as quickly as possible. Proceed to the next step *immediately* after sealing the centrifuge tube.

CAUTION ⚠

Do not leave the centrifuge tube out in the open once you have twisted the cap onto the centrifuge tube. Immediately place it into the warm water in the hood.

Quickly lower the capped centrifuge tube, tapered end first, into the graduated cylinder containing warm water in a hood. Pull the hood safety glass down as soon as you place the centrifuge tube into the warm water. The dry ice should liquefy after a minute and remain in the liquid state for about 9 minutes. If the dry ice has not entered the liquid state after about 2 minutes, it is likely that the centrifuge tube leaked from around the cap and did not have an opportunity to build up sufficient pressure. You should replace the cap and tube and try again. Also, if you wait too long before sealing the centrifuge tube, the dry ice can fail to liquefy.

After about 9 minutes, all the liquid CO_2 should have vaporized, and the gas will have escaped from around the cap on the centrifuge tube. Take the centrifuge tube out of the graduated cylinder and remove the cap. Lift the filter cone out of the centrifuge tube using the wire handle on the copper coil. Dispose of the waste spice in a trash can. Remove the clove oil from the bottom of the centrifuge tube with a Pasteur pipet and transfer the clove oil to a small preweighed sample vial. Reweigh the vial to determine the weight of eugenol obtained from the extraction. Calculate the weight percentage recovery of the eugenol from the original weight of spice used.

Spectroscopy

Infrared Spectroscopy

Determine the infrared spectrum of the oil as a neat liquid sample. (Technique 25, Section 25.2) and provide copies for all students. Include the infrared spectrum with your laboratory report, along with an interpretation of the principal peaks.[2]

Nuclear Magnetic Resonance Spectroscopy (Optional)

At the instructor's option, determine the proton NMR spectrum of the eugenol. Assign the peaks in the spectrum to the structure of eugenol.

Mass Spectrometry (Optional)

At the instructor's option, determine the mass spectrum of the eugenol sample (Technique 28). Try to assign as many of the fragments in the spectrum as possible.

REPORT

Attach your infrared spectrum to your report and label the major peaks with the type of bond or group of atoms that is responsible for the absorption. If you have obtained the proton NMR spectrum of the sample, include the spectrum with your report. Be sure to interpret your spectrum fully. If you determined a mass spectrum, identify the important fragment ion peaks. Be sure to also include the weight percentage recovery calculation.

REFERENCE

McKenzie, L. C., Thompson, J. E., Sullivan, R., and Hutchison, J. E. "Green Chemical Processing in the Teaching Laboratory: A Convenient Liquid CO_2 Extraction of Natural Products." *Green Chemistry, 6* (2004): 355–358.

ESSAY

Stereochemical Theory of Odor

The human nose has an almost unbelievable ability to distinguish odors. Just consider for a few moments the different substances you can recognize by odor alone. Your list should be long. A person with a trained nose, a perfumer, for instance, can often recognize even individual components

---

[2] Extraction of eugenol with liquid CO_2 yields an extra peak at 1764 cm^{-1} that is attributed to eugenol acetate, a by-product of the extraction; GC-MS will be useful as an optional experiment in identifying the impurity.

in a mixture. Who has not met at least one cook who could sniff almost any culinary dish and identify the seasonings and spices that were used? The olfactory centers in the nose can identify odorous substances even in small amounts. Studies have shown that with some substances, as little as one 10 millionth of a gram (10^{-7} g) can be perceived. Many animals, for example, dogs and insects, have an even lower threshold of smell than humans do (see essay on pheromones that precedes Experiment 45).

There have been many theories of odor, but few have persisted. Strangely enough, one of the oldest theories, although in modern dress, is still the most current theory. Lucretius, one of the early Greek atomists, suggested that substances having odor gave off a vapor of tiny "atoms," all of the same shape and size, and that they gave rise to the perception of odor when they entered pores in the nose. The pores would have to be of various shapes, and the odor perceived would depend on which pores the atoms were able to enter. We now have many similar theories about the action of drugs (receptor-site theory) and the interaction of enzymes with their substrates (the lock-and-key hypothesis).

A substance must have certain physical characteristics to have the property of odor. First, it must be volatile enough to give off a vapor that can reach the nostrils. Second, once it reaches the nostrils, it must be somewhat water soluble, even if only to a small degree, so that it can pass through the layer of moisture (mucus) that covers the nerve endings in the olfactory area. Third, it must have lipid solubility to allow it to penetrate the lipid (fat) layers that form the surface membranes of the nerve cell endings.

Once we pass these criteria, we come to the heart of the question. Why do substances have different odors? In 1949, R. W. Moncrieff, a Scot, resurrected Lucretius' hypothesis. He proposed that in the olfactory area of the nose is a system of receptor cells of several types and shapes. He further suggested that each receptor site corresponded to a different type of primary odor. Molecules that would fit these receptor sites would display the characteristics of that primary odor. It would not be necessary for the entire molecule to fit into the receptor, so for larger molecules, any portion might fit into the receptor and activate it. Molecules having complex odors would presumably be able to activate several types of receptors.

Moncrieff's hypothesis was strengthened substantially by the work of J. E. Amoore, who began studying the subject as an undergraduate at Oxford in 1952. After an extensive search of the chemical literature, Amoore concluded that there were only seven basic primary odors. By sorting molecules with similar odor types, he even formulated possible shapes for the seven necessary receptors. For instance, from the literature he culled more than 100 compounds that were described as having a "camphoraceous" odor. Comparing the sizes and shapes of all these molecules, he postulated a three-dimensional shape for a camphoraceous receptor site. Similarly, he derived shapes for the other six receptor sites. The seven primary receptor sites he formulated are shown in the figure (p. 119), along with a typical prototype molecule of the appropriate shape to fit the receptor. The shapes of the sites are shown in perspective. Pungent and putrid odors were not thought to require a particular shape in the odorous molecules but rather to need a particular type of charge distribution.

You can verify quickly that compounds with molecules of roughly similar shape have similar odors if you compare nitrobenzene and acetophenone with benzaldehyde or *d*-camphor and hexachloroethane with

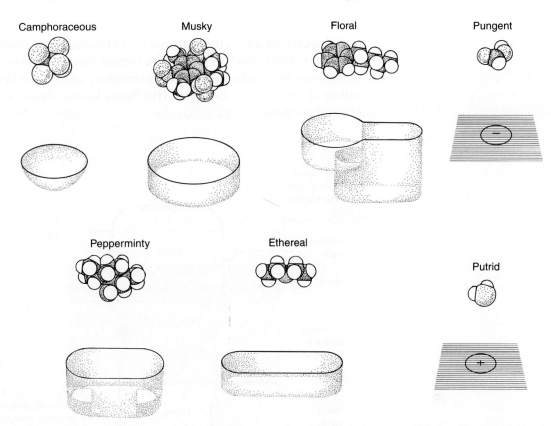

From "The Stereochemical Theory of Odor," by J. E. Amoore, J. W. Johnston Jr., and M. Rubin. Copyright © 1964 by Scientific American, Inc. All rights reserved.

cyclooctane. Each group of substances has the same basic odor *type* (primary), but the individual molecules differ in the *quality* of the odor. Some of the odors are sharp, some pungent, others sweet, and so on. The second group of substances all have a camphoraceous odor, and the molecules of these substances all have approximately the same shape.

An interesting corollary to the Amoore theory is the postulate that if the receptor sites are chiral, then optical isomers (enantiomers) of a given substance might have *different* odors. This circumstance proves true in several cases. It is true for (+)- and (−)-carvone; we investigate the idea in Experiment 15 in this textbook.

The theory changed dramatically in 1991 because of the biochemical research of Richard Axel and Linda Buck, who was a postdoctoral student in Axel's research group. Subsequently, Buck founded her own group that also continued research on the nature of the sense of smell. In 2004, Axel and Buck won the Nobel Prize in Physiology or Medicine for their combined work during the previous decade.

The 1991 paper, working with mice, described a family of membrane-spanning receptor proteins found in a small area of the upper nose called the olfactory epithelium. Mice have genes that can encode as many as 1000 types of receptor proteins. Subsequent work has estimated that humans, who have a lesser-developed sense of smell than mice, encode only about 350 of these receptor proteins. Each of these protein receptors is located on the surface of the olfactory epithelium and is connected to a single nerve cell (neuron) located in the epithelium. The neuron "fires" or sends a signal when an odorant molecule binds to the active site of the protein. The signal is carried across the bones of the skull and into a node in an area of the brain called the olfactory bulb. The signals from all receptors are processed in the olfactory bulb and sent to the memory area of the brain where recognition of the odor takes place. The figure below shows a schematic of the olfactory region.

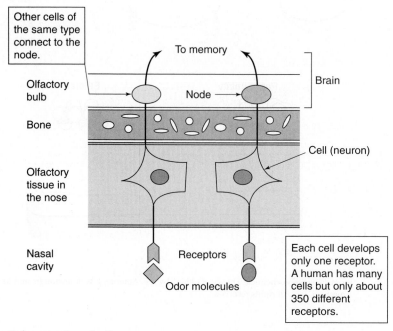

Odor receptors in the nose.

The signals from all of the types of protein receptors are collected, or integrated, in the olfactory bulb. The node (a postulated feature) is a common connection where the signals from each type of cell are collected and sent to memory, each with an intensity proportional to the numbers of cells that were stimulated by the odorant molecules. Because a given odorant molecule should be capable of binding to more than one type of receptor and because many odors are composed of more than one type of molecule, the signal sent to memory should be a complex combinatorial pattern consisting of contributions from several nodes, each with a different intensity value. This system should allow a human to recognize as many as 10,000 odors and for mice to recognize many more. The memory region in the brain can also make associations based on a given pattern. For instance, cinnamaldehyde can be recognized as the odor of the spice cinnamon, but it can also be associated with other items such as apple pie, cinnamon rolls, apple strudel, spiced cider, and, of course, pleasure. A figure showing these associations, but with a limitation of only a few receptors represented, is shown in the figure on page 123.

Although our modern understanding of the detection of odor has evolved to become a more highly detailed theory than the one proposed by Lucretius, it would appear that his fundamental hypothesis was correct and has even withstood the scrutiny of modern science.

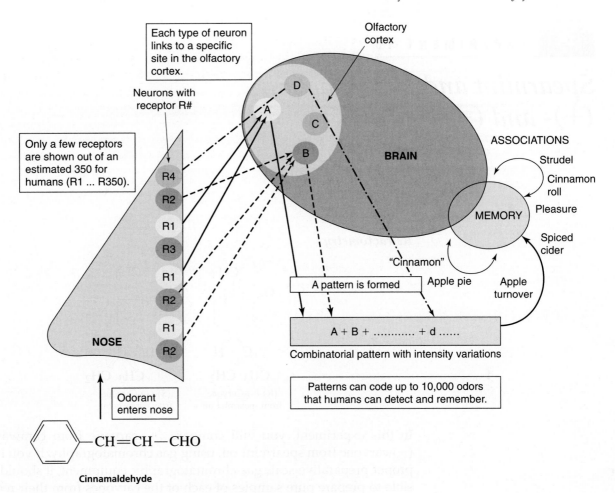

Nobel prize theory of the detection of odors (Axel and Buck, 2004).

REFERENCES

Amoore, J. E., Johnson, J. W., Jr., and Rubin, M. "The Stereochemical Theory of Odor." *Scientific American, 210* (February 1964): 1.

Amoore, J. E., Johnson, J. W., Jr., and Rubin, M. "The Stereochemical Theory of Olfaction." *Proceedings of the Scientific Section of the Toilet Goods Association* (Special Supplement to No. 37) (October 1962): 1–47.

Buck, L. "The Molecular Architecture of Odor and Pheromone Sensing in Mammals." *Cell, 100(6),* (March 2000): 611–618.

Buck, L., and Axel, R. "A Novel Multigene Family May Encode Odorant Receptors: A Molecular Basis for Odor Recognition. "*Cell, 65(1)* (April 1991): 175–187.

Lipkowitz, K. B. "Molecular Modeling in Organic Chemistry: Correlating Odors with Molecular Structure." *Journal of Chemical Education, 66* (April 1989): 275.

Malnic, B., Hirono, J., Sato, T., and Buck, L. "Combinatorial Receptor Codes for Odors." *Cell, 96(5)* (March 1999): 713–723.

Moncrieff, R. W. *The Chemical Senses.* London: Routledge & Kegan Paul, 1976.

Roderick, W. R. "Current Ideas on the Chemical Basis of Olfaction." *Journal of Chemical Education, 43* (October 1966): 510–519.

Zou, Z., Horowitz, L., Montmayeur, J., Snapper, S., and Buck, L. "Genetic Tracing Reveals a Stereotyped Sensory Map in the Olfactory Cortex." *Nature, 414(6843)* (November 2001): 173–179.

15 **EXPERIMENT 15**

Spearmint and Caraway Oil: (+)- and (−)-Carvones

Stereochemistry
Gas chromatography
Polarimetry
Spectroscopy
Refractometry

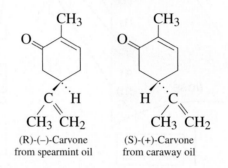

(R)-(−)-Carvone
from spearmint oil

(S)-(+)-Carvone
from caraway oil

In this experiment, you will compare (+)-carvone from caraway oil to (−)-carvone from spearmint oil, using gas chromatography. If you have the proper preparative-scale gas-chromatographic equipment, it should be possible to prepare pure samples of each of the carvones from their respective oils. If this equipment is not available, the instructor will provide pure samples of the two carvones obtained from a commercial source, and any gas-chromatographic work will be strictly analytical.

The odors of the two enantiomeric carvones are distinctly different from each other. The presence of one or the other isomer is responsible for the characteristic odors of each oil. The difference in the odors is to be expected because the odor receptors in the nose are chiral (see essay, "Stereochemical Theory of Odor," p. 119). This phenomenon, in which a chiral receptor interacts differently with each enantiomer of a chiral compound, is called **chiral recognition.**

Although we should expect the optical rotations of the isomers (enantiomers) to be of opposite sign, the other physical properties should be identical. Thus, for both (+)- and (−)-carvone, we predict that the infrared and nuclear magnetic resonance spectra, the gas-chromatographic retention times, the refractive indices, and the boiling points will be identical. Hence, the only differences in properties you will observe for the two carvones are the odors and the signs of rotation in a polarimeter.

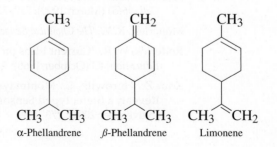

α-Phellandrene   β-Phellandrene   Limonene

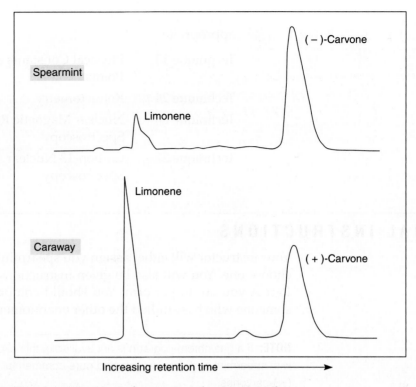

Gas chromatograms of caraway and spearmint oil.

Caraway oil contains mainly limonene and (+)-carvone. The gas chromatogram for this oil is shown in the figure. The (+)-carvone (bp 203°C) can easily be separated from the lower-boiling limonene (bp 177°C) by gas chromatography, as shown in the figure. If one has a preparative gas chromatograph, the (+)-carvone and limonene can be collected separately as they elute from the gas chromatography column. **Spearmint oil** contains mainly (−)-carvone with a smaller amount of limonene and very small amounts of the lower-boiling terpenes, α- and β-phellandrene. The gas chromatogram for this oil is also shown in the figure. With preparative equipment, you can easily collect the (−)-carvone as it exits the column. It is more difficult, however, to collect limonene in a pure form. It is likely to be contaminated with the other terpenes because they all have similar boiling points.

REQUIRED READING

Review: Experiment 1   Introduction to Microscale Laboratory (pp. 2–13)

          Technique 25   Infrared Spectroscopy

New: Technique 22   Gas Chromatography

          Technique 23   Polarimetry

          Essay   Stereochemical Theory of Odor

If performing any of the optional procedures, read as appropriate:

Technique 13   Physical Constants of Liquids, Boiling Points

Technique 24   Refractometry

Technique 26   Nuclear Magnetic Resonance Spectroscopy

Technique 27   Carbon-13 Nuclear Magnetic Resonance Spectroscopy

SPECIAL INSTRUCTIONS

Your instructor will either assign you spearmint or caraway oil or have you choose one. You will also be given instructions on which procedures from Part A you are to perform. You should compare your data with those of someone who has studied the other enantiomer.

NOTE: If a gas chromatograph is not available, this experiment can be performed with spearmint and caraway oils and pure commercial samples of the (+)- and (−)-carvones.

If the proper equipment is available, your instructor may require you to perform a gas-chromatographic analysis. If preparative gas chromatography is available, you will be asked to isolate the carvone from your oil (Part B). Otherwise, if you are using analytical equipment, you will be able to compare only the retention times and integrals from your oil to those of the other essential oil.

Although preparative gas chromatography will yield enough sample to do spectra, it will not yield enough material to do the polarimetry. Therefore, if you are required to determine the optical rotation of the pure samples, whether or not you perform preparative gas chromatography, your instructor will provide a prefilled polarimeter tube for each sample.

NOTES TO THE INSTRUCTOR

This experiment may be scheduled along with another experiment. It is best if students work in pairs, each student using a different oil. An appointment schedule for using the gas chromatograph should be arranged so that students are able to make efficient use of their time. You should prepare chromatograms using both carvone isomers and limonene as reference standards. Appropriate reference standards include a mixture of (+)-carvone and limonene and a second mixture of (−)-carvone and limonene. The chromatograms should be posted with retention times, or each student should be provided with a copy of the appropriate chromatogram.

The gas chromatograph should be prepared as follows: column temperature, 200°C; injection and detector temperature, 210°C; carrier gas flow rate, 20 mL/min. The recommended column is 8 feet long, with a stationary phase such as Carbowax 20M. It is convenient to use a Gow-Mac 69-350 instrument with the preparative accessory system for this experiment.

You should fill polarimeter cells (0.5 dm) in advance with the undiluted (+)- and (−)-carvones. There should also be four bottles containing spearmint and caraway oils and (+)- and (−)-carvone. Both enantiomers of carvone are commercially available.

PROCEDURE

Part A. Analysis of the Carvones

The samples (either those obtained from gas chromatography, Part B, or commercial samples) should be analyzed by the following methods. The instructor will indicate which methods to use. Compare your results with those obtained by someone who used a different oil. In addition, measure the observed rotation of the commercial samples of (+)-carvone and (−)-carvone. The instructor will supply prefilled polarimeter tubes.

Analyses to Be Performed on Spearmint and Caraway Oils:

Odor Carefully smell the containers of spearmint and caraway oil and of the two carvones. About 8–10% of the population cannot detect the difference in the odors of the optical isomers. Most people, however, find the difference quite obvious. Record your impressions.

Analytical Gas Chromatography If you separated your sample by preparative gas chromatography in Part B, you should already have your chromatogram. In this case, you should compare it to one done by someone using the other oil. Be sure to obtain retention times and integrals or obtain a copy of the other person's chromatogram.

If you did not perform Part B, obtain the analytical gas chromatograms of your assigned oil—spearmint or caraway—and obtain the result from the other oil from someone else. The instructor may prefer to perform the sample injections or have a laboratory assistant perform them. The sample injection procedure requires careful technique, and the special microliter syringes that are required are delicate and expensive. If you are to perform the injections yourself, your instructor will give you adequate instruction beforehand.

For both oils, determine the retention times of the components (see Technique 22, Section 22.7, p. 806). Calculate the percentage composition of the two essential oils by one of the methods explained in the same section.

Analyses to Be Performed on the Purified Carvones:

Polarimetry With the help of the instructor or assistant, obtain the observed optical rotation α of the pure (+)-carvone and (−)-carvone samples. These are provided in prefilled polarimeter tubes. The specific rotation $[\alpha]_D$ is calculated from the relationship given on page 821 of Technique 23. The concentration c will equal the density of the substances analyzed at 20°C. The values, obtained from actual commercial samples, are 0.9608 g/mL for (+)-carvone and 0.9593 g/mL for (−)-carvone. The literature values for the specific rotations are as follows: $[\alpha]_D^{20} = +61.7°$ for (+)-carvone and $-62.5°$ for (−)-carvone. These values are not identical, because trace amounts of impurities are present.

Polarimetry does not work well on the crude spearmint and caraway oils, because large amounts of limonene and other impurities are present.

Infrared Spectroscopy Obtain the infrared spectrum of the (−)-carvone sample from spearmint or of the (+)-carvone sample from caraway (see Technique 25, Section 25.2, p. 834). Compare your result with that of a person working with the other isomer. At the option of the instructor, obtain the infrared spectrum of the

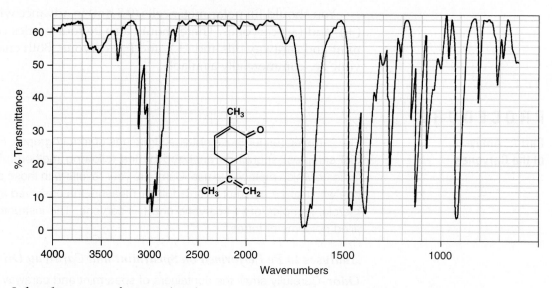

Infrared spectrum of carvone (neat).

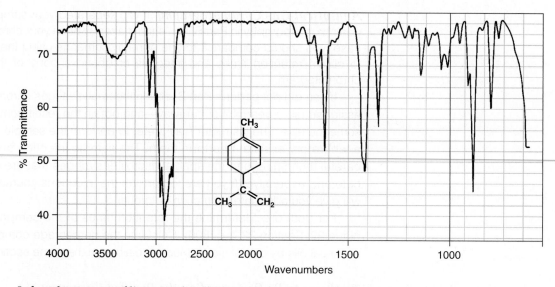

Infrared spectrum of limonene (neat).

(+)-limonene, which is found in both oils. If possible, determine all spectra using neat samples. If you isolated the samples by preparative gas chromatography, it may be necessary to add one to two drops of carbon tetrachloride to the sample. Thoroughly mix the liquids by drawing the mixture into a Pasteur pipet and expelling several times. It may be helpful to draw the end of the pipet to a narrow tip in order to withdraw all the liquid in the conical vial. As an alternative, use a microsyringe. Obtain a spectrum on this solution, as described in Technique 25, Section 25.2, page 834.

Nuclear Magnetic Resonance Spectroscopy Using an NMR instrument, obtain a proton NMR spectrum of your carvone. Compare your spectrum with the NMR spectra for (−)-carvone and (+)-limonene shown in this experiment. Attempt to

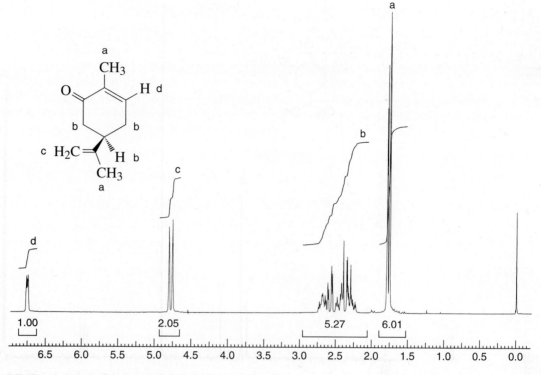

NMR spectrum of (−)-carvone from spearmint oil.

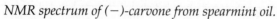

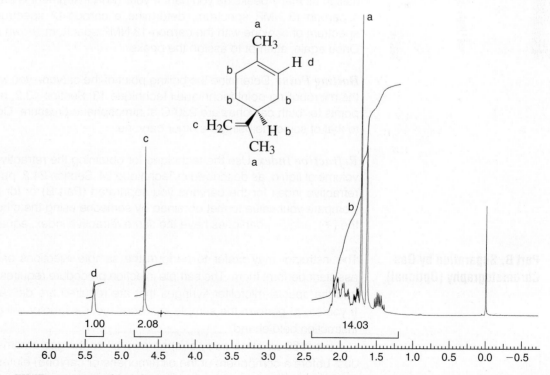

NMR spectrum of (+)-limonene.

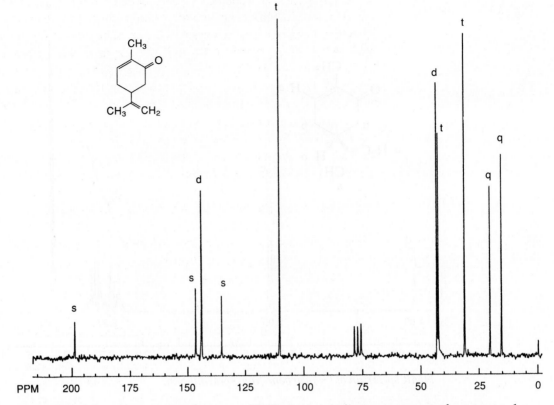

Decoupled carbon-13 spectrum of carvone, CDCl₃. Letters indicate appearance of spectrum when carbons are coupled to protons (s = singlet, d = doublet, t = triplet, q = quartet).

assign as many peaks as you can. If your NMR instrument is capable of obtaining a carbon-13 NMR spectrum, determine a carbon-13 spectrum. Compare your spectrum of carvone with the carbon-13 NMR spectrum shown in this experiment. Once again, attempt to assign the peaks.

Boiling Point Determine the boiling point of the carvone you were assigned. Use the microboiling-point technique (Technique 13, Section 13.2, p. 695). The boiling points for both carvones are 230°C at atmospheric pressure. Compare your result to that of someone using the other carvone.

Refractive Index Use the technique for obtaining the refractive index on a small volume of liquid, as described in Technique 24, Section 24.2, page 828. Obtain the refractive index for the carvone you separated (Part B) or for the one assigned. Compare your value to that obtained by someone using the other isomer. At 20°C, the (+)- and (−)-carvones have the same refractive index, equal to 1.4989.

Part B. Separation by Gas Chromatography (Optional)

The instructor may prefer to perform the sample injections or have a laboratory assistant perform them. The sample injection procedure requires careful technique, and the special microliter syringes that are required are delicate and expensive. If you are to perform the sample injections, your instructor will give you adequate instruction beforehand.

Inject 50 μL of caraway or spearmint oil on the gas-chromatography column. Just before a component of the oil (limonene or carvone) elutes from the column, install a gas-collection tube at the exit port, as described in Technique 22, Section 22.11, page 809. To determine when to connect the gas-collection tube, refer to the chromatograms prepared by your instructor. These chromatograms have been run on the same instrument you are using under the same conditions. Ideally, you should connect the gas-collection tube just before the limonene or carvone elutes from the column and remove the tube as soon as all the component has been collected but before any other compound begins to elute from the column. You can accomplish this most easily by watching the recorder as your sample passes through the column. The collection tube is connected (if possible) just before a peak is produced or as soon as a deflection is observed. When the pen returns to the baseline, remove the gas collection tube.

This procedure is relatively easy for collecting the carvone component of both oils and for collecting the limonene in caraway oil. Because of the presence of several terpenes in spearmint oil, it is somewhat more difficult to isolate a pure sample of limonene from spearmint oil (see chromatogram in figure, p. 125). In this case, you must try to collect only the limonene component and not any other compounds, such as the terpene, which produces a shoulder on the limonene peak in the chromatogram for spearmint oil.

After collecting the samples, insert the ground joint of the collection tube into a 0.1-mL conical vial, using an O-ring and screw cap to fasten the two pieces together securely. Place this assembly into a test tube, as shown in Figure 22.11, page 810. Put cotton on the bottom of the tube and use a rubber septum cap to hold the assembly in place and to prevent breakage. Balance the centrifuge by placing a tube of equal weight on the opposite side (this could be your other sample or someone else's sample). During centrifugation, the sample is forced into the bottom of the conical vial. Disassemble the apparatus, cap the vial, and perform the analyses described in Part A. You should have enough sample to perform the infrared and NMR spectroscopy, but your instructor may need to provide additional sample to perform the other procedures.

REFERENCES

Friedman, L., and Miller, J. G. "Odor, Incongruity and Chirality." *Science, 172* (1971): 1044.

Murov, S. L., and Pickering, M. "The Odor of Optical Isomers." *Journal of Chemical Education, 50* (1973): 74.

Russell, G. F., and Hills, J. I. "Odor Differences Between Enantiomeric Isomers." *Science, 172* (1971): 1043.

QUESTIONS

1. Interpret the infrared spectra for carvone and limonene and the proton and carbon-13 NMR spectra of carvone.

2. Identify the chiral centers in α-phellandrene, β-phellandrene, and limonene.

3. Explain how carvone fits the isoprene rule (see essay, "Terpenes and Phenylpropanoids," p. 109).

4. Using the Cahn–Ingold–Prelog sequence rules, assign priorities to the groups around the chiral carbon in carvone. Draw the structural formulas for (+)- and (−)-carvone with the molecules oriented in the correct position to show the R and S configurations.

5. Explain why limonene elutes from the column before either (+)- or (−)-carvone.

6. Explain why the retention times for both carvone isomers are the same.

7. The toxicity of (+)-carvone in rats is about 400 times greater than that of (−)-carvone. How do you account for this?

The Chemistry of Vision

An interesting and challenging topic for chemists to investigate is how the eye functions. What chemistry is involved in detection of light and transmission of that information to the brain? The first definitive studies on how the eye functions were begun in 1877 by Franz Boll. Boll demonstrated that the red color of the retina of a frog's eye could be bleached yellow by strong light. If the frog was then kept in the dark, the red color of the retina slowly returned. Boll recognized that a bleachable substance had to be connected somehow with the ability of the frog to perceive light.

Most of what is now known about the chemistry of vision is the result of the elegant work of George Wald, Harvard University; his studies, which began in 1933, ultimately resulted in his receiving the Nobel Prize in biology. Wald identified the sequence of chemical events during which light is converted into some form of electrical information that can be transmitted to the brain. Here is a brief outline of that process.

The retina of the eye is made up of two types of photoreceptor cells: **rods** and **cones.** The rods are responsible for vision in dim light, and the cones are responsible for color vision in bright light. The same principles apply to the chemical functioning of the rods and the cones; however, the details of functioning are less well understood for the cones than for the rods.

Each rod contains several million molecules of **rhodopsin.** Rhodospin is a complex of a protein, **opsin,** and a molecule derived from Vitamin A, 11-*cis*-retinal (sometimes called **retinene**). Little is known about the structure of opsin. The structure of 11-*cis*-retinal is shown here.

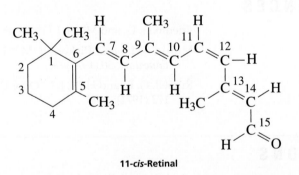

11-*cis*-Retinal

The detection of light involves the initial conversion of 11-*cis*-retinal to its all-*trans* isomer. This is the only obvious role of light in this process. The high energy of a quantum of visible light promotes the fission of the π bond between carbons 11 and 12. When the π bond breaks, free rotation about the σ bond in the resulting radical is possible. When the π bond re-forms after such rotation, all-*trans*-retinal results. All-*trans*-retinal is more stable than 11-*cis*-retinal, which is the reason the isomerization proceeds spontaneously in the direction shown.

The two molecules have different shapes because of their different structures. The 11-*cis*-retinal has a fairly curved shape, and the parts of the molecule on either side of the *cis* double bond tend to lie in different planes. Because proteins have complex and specific three-dimensional shapes (tertiary structures), 11-*cis*-retinal associates with the protein opsin in a particular manner. All-*trans*-retinal has an elongated shape, and the entire molecule tends to lie in a single plane. This different shape for the molecule, compared with that for the 11-*cis* isomer, means that all-*trans*-retinal will have a different association with the protein opsin.

In fact, all-*trans*-retinal associates weakly with opsin because its shape does not fit the protein. Consequently, the next step after the isomerization of retinal is the dissociation of all-*trans*-retinal from opsin. The opsin protein undergoes a simultaneous change in conformation as the all-*trans*-retinal dissociates.

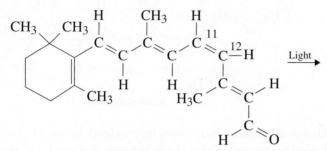

11-*cis*-Retinal

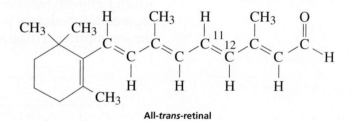

All-*trans*-retinal

At some time after the 11-*cis*-retinal–opsin complex receives a photon, a message is received by the brain. It was originally thought that either the isomerization of 11-*cis*-retinal to all-*trans*-retinal or the conformational change of the opsin protein was an event that generated the electrical message sent to the brain. Current research, however, indicates that both these events occur too slowly relative to the speed with which the brain receives the message.

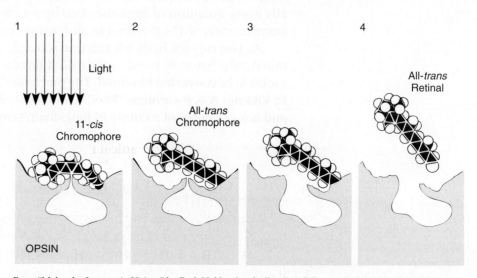

From "Molecular Isomers in Vision," by Ruth Hubbard and Allen Kropf. Copyright © 1967 by Scientific American, Inc. All rights reserved.

Current hypotheses invoke involved quantum mechanical explanations, which hold it significant that the chromophores (light-absorbing groups) are arranged in a precise geometrical pattern in the rods and cones, allowing the signal to be transmitted rapidly through space. The main physical and chemical events Wald discovered are illustrated in the figure for easy visualization. The question of how the electrical signal is transmitted still remains unsolved.

Wald was also able to explain the sequence of events by which the rhodopsin molecules are regenerated. After dissociation of all-*trans*-retinal from the protein, the following enzyme-mediated changes occur. All-*trans*-retinal is reduced to the alcohol all-*trans*-retinol, also called all-*trans*-Vitamin A.

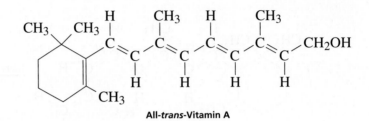

All-*trans*-Vitamin A

All-*trans*-Vitamin A is then isomerized to its 11-*cis*-Vitamin A isomer. After the isomerization, the 11-*cis*-Vitamin A is oxidized back to 11-*cis*-retinal, which forthwith recombines with the opsin protein to form rhodopsin. The regenerated rhodopsin is then ready to begin the cycle anew, as illustrated in the accompanying diagram.

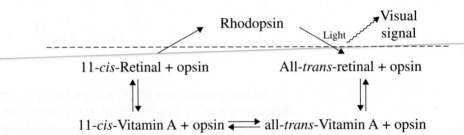

By this process, as little light as 10^{-14} of the number of photons emitted from a typical flashlight bulb can be detected. The conversion of light into isomerized retinal exhibits an extraordinarily high quantum efficiency. Virtually every quantum of light absorbed by a molecule of rhodopsin causes the isomerization of 11-*cis*-retinal to all-*trans*-retinal.

As you can see from the reaction scheme, the retinal derives from Vitamin A, which merely requires the oxidation of a —CH_2OH group to a —CHO group to be converted to retinal. The precursor in the diet that is transformed to Vitamin A is β-carotene. The β-carotene is the yellow pigment of carrots and is an example of a family of long-chain polyenes called **carotenoids.**

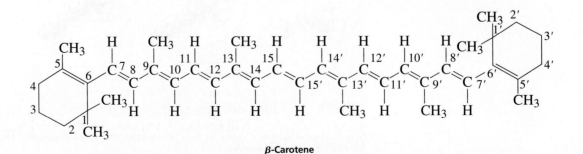

β-Carotene

In 1907, Willstätter established the structure of carotene, but it was not known until 1931–1933 that there were actually three isomers of carotene. The α-carotene differs from β-carotene in that the α isomer has a double bond between C_4 and C_5 rather than between C_5 and C_6, as in the β isomer. The γ isomer has only one ring, identical to the ring in the β isomer, whereas the other ring is opened in the γ form between C_1' and C_6'. The β isomer is by far the most common of the three.

The substance β-carotene is converted to Vitamin A in the liver. Theoretically, one molecule of β-carotene should give rise to two molecules of the vitamin by cleavage of the C_{15}–C_{15}', double bond, but actually only one molecule of Vitamin A is produced from each molecule of the carotene. The Vitamin A thus produced is converted to 11-*cis*-retinal within the eye.

Along with the problem of how the electrical signal is transmitted, color perception is also currently under study. In the human eye, there are three kinds of cone cells, which absorb light at 440, 535, and 575 nm, respectively. These cells discriminate among the primary colors. When combinations of them are stimulated, full color vision is the message received in the brain.

Because all these cone cells use 11-*cis*-retinal as a substrate trigger, it has long been suspected that there must be three different opsin proteins. Recent work has begun to establish how the opsins vary the spectral sensitivity of the cone cells, even though all of them have the same kind of light-absorbing chromophore.

Retinal is an aldehyde, and it binds to the terminal amino group of a lysine residue in the opsin protein to form a Schiff base, or imine linkage ($>C=N-$). This imine linkage is believed to be protonated (with a plus charge) and to be stabilized by being located near a negatively charged amino acid residue of the protein chain. A second negatively charged group is thought to be located near the 11-*cis* double bond. Researchers have recently shown, from synthetic models that use a simpler protein than opsin itself, that forcing these negatively charged groups to be located at different distances from the imine linkage causes the absorption maximum of the 11-*cis*-retinal chromophore to be varied over a wide enough range to explain color vision.

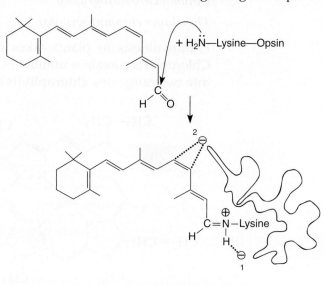

Rhodopsin.

Whether there are actually three different opsin proteins or whether there are just three different conformations of the same protein in the three types of cone cells will not be known until further work is completed on the structure of the opsin or opsins.

REFERENCES

Borman, S. "New Light Shed on Mechanism of Human Color Vision." *Chemical and Engineering News* (April 6, 1992): 27.

Fox, J. L. "Chemical Model for Color Vision Resolved." *Chemical and Engineering News,* 57(46) (November 12, 1979): 25. A review of articles by Honig and Nakanishi in the *Journal of the American Chemical Society, 101* (1979): 7082, 7084, 7086.

Hubbard, R., and Kropf, A. "Molecular Isomers in Vision." *Scientific American, 216* (June 1967): 64.

Hubbard, R., and Wald, G. "Pauling and Carotenoid Stereochemistry." In A. Rich and N. Davidson, eds., *Structural Chemistry and Molecular Biology.* San Francisco: W. H. Freeman, 1968.

MacNichol, E. F., Jr. "Three Pigment Color Vision." *Scientific American, 211* (December 1964): 48.

"Model Mechanism May Detail Chemistry of Vision." *Chemical and Engineering News* (January 7, 1985): 40.

Rushton, W. A. H. "Visual Pigments in Man." *Scientific American, 207* (November 1962): 120.

Wald, G. "Life and Light." *Scientific American, 201* (October 1959): 92.

Zurer, P. S. "The Chemistry of Vision." *Chemical and Engineering News, 61* (November 28, 1983): 24.

16   EXPERIMENT 16

Isolation of Chlorophyll and Carotenoid Pigments from Spinach

Isolation of a natural product

Extraction

Column chromatography

Thin-layer chromatography

Photosynthesis in plants takes place in organelles called **chloroplasts.** Chloroplasts contain a number of colored compounds (pigments) that fall into two categories, **chlorophylls** and **carotenoids.**

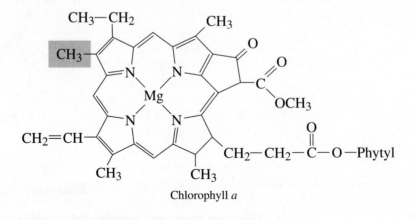

Chlorophyll *a*

$$Phytyl = -CH_2-CH=C-CH_2-(CH_2-CH_2-CH-CH_2)_2-CH_2-CH_2-CH-CH_3$$

with CH_3 groups on the designated carbons

Carotenoids are yellow pigments that are also involved in the photo-synthetic process. The structures of α- and β-**carotene** are given in the essay preceding this experiment. In addition, chloroplasts also contain several oxygen-containing derivatives of carotenes, called **xanthophylls.**

In this experiment, you will extract the chlorophyll and carotenoid pigments from spinach leaves using acetone as the solvent. The pigments will be separated by column chromatography using alumina as the adsorbent. Increasingly polar solvents will be used to elute the various components from the column. The colored fractions collected will then be analyzed using thin-layer chromatography. It should be possible for you to identify most of the pigments already discussed on your thin-layer plate after development.

Chlorophylls are the green pigments that act as the principal photoreceptor molecules of plants. They are capable of absorbing certain wavelengths of visible light that are then converted by plants into chemical energy. Two forms of these pigments found in plants are **chlorophyll** *a* and **chlorophyll** *b.* The two forms are identical, except that the methyl group that is shaded in the structural formula of chlorophyll *a* is replaced by a —CHO group in chlorophyll *b.* **Pheophytin** *a* and **pheophytin** *b* are identical to chlorophyll *a* and chlorophyll *b*, respectively, except that in each case the magnesium ion Mg^{2+} has been replaced by two hydrogen ions $2H^+$.

REQUIRED READING

Review:	Techniques 5 and 6	
	Technique 7	Reaction Methods, Section 7.9
	Technique 12	Extractions, Separations, and Drying Agents, Sections 12.5 and 12.9
	Technique 20	Thin-Layer Chromatography
New:	Technique 19	Column Chromatography
	Essay	The Chemistry of Vision

SPECIAL INSTRUCTIONS

Hexane and acetone are both highly flammable. Avoid using flames while working with these solvents. Perform the thin-layer chromatography in the hood. The procedure calls for a centrifuge tube with a tight-fitting cap. If this is not available, you can use a vortex mixer for mixing the liquids. Another alternative is to use a cork to stopper the tube; however, the cork will absorb some liquid.

Fresh spinach is preferable to frozen spinach. Because of handling, frozen spinach contains additional pigments that are difficult to identify. Because the pigments are light-sensitive and can undergo oxidation, you should work quickly. Samples should be stored in closed containers and kept in the dark when possible. The column chromatography procedure takes less than 15 minutes to perform and cannot be stopped until it is completed. It is important, therefore, that you have all the materials needed for this part of the experiment prepared in advance and that you be thoroughly familiar with the procedure before running the column. If you need to prepare the 70% hexane–30% acetone solvent mixture, be sure to mix it thoroughly before using.

SUGGESTED WASTE DISPOSAL

Dispose of all organic solvents in the container for nonhalogenated organic solvents. Place the alumina in the container designated for wet alumina.

NOTES TO THE INSTRUCTOR

The column chromatography should be performed with activated alumina from EM Science (No. AX0612-1). The particle sizes are 80–120 mesh, and the material is Type F-20. Dry the alumina overnight in an oven at 110°C and store it in a tightly sealed bottle. Alumina older than several years may need to be dried for a longer time at a higher temperature. Depending on how dry the alumina is, solvents of different polarity will be required to elute the components from the column.

For thin-layer chromatography, use flexible silica gel plates from Whatman with a fluorescent indicator (No. 4410 222). If the TLC plates have not been purchased recently, place them in an oven at 100°C for 30 minutes and store them in a desiccator until use.

If you use different alumina or different thin-layer plates, try the experiment before using it in class. Materials other than those specified here may give different results than indicated in this experiment.

PROCEDURE

Part A. Extraction of the Pigments

Weigh about 0.5 g of fresh (or 0.25 g of frozen) spinach leaves (avoid using stems or thick veins). Fresh spinach is preferable, if available. If you must use frozen spinach, dry the thawed leaves by pressing them between several layers of paper towels. Cut or tear the spinach leaves into small pieces and place them in a mortar along with 1.0 mL of cold acetone. Grind with a pestle until the spinach leaves have been broken into particles too small to be seen clearly. If too much acetone has evaporated, you may need to add an additional portion of acetone (0.5–1.0 mL) to perform the following step. Using a Pasteur pipet, transfer the mixture to a centrifuge tube. Rinse the mortar and pestle with 1.0 mL of cold acetone and transfer the remaining mixture to the centrifuge tube. Centrifuge the mixture (be sure to balance the centrifuge). Using a Pasteur pipet, transfer the liquid to a centrifuge tube with a tight-fitting cap (see "Special Instructions" if one is not available).

Add 2.0 mL of hexane to the tube, cap the tube, and shake the mixture thoroughly. Then add 2.0 mL of water and shake thoroughly with occasional venting. Centrifuge the mixture to break the emulsion, which usually appears as a cloudy green layer in the middle of the mixture. Remove the bottom aqueous layer with a Pasteur pipet. Using a Pasteur pipet, prepare a column containing anhydrous sodium sulfate to dry the remaining hexane layer, which contains the dissolved pigments. Put a plug of cotton into a Pasteur pipet (5¾-inch) and tamp it into position using a glass rod. The correct position of the cotton is shown in the figure. Add about 0.5 g of powdered or granular anhydrous sodium sulfate and tap the column with your finger to pack the material.

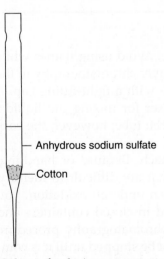

— Anhydrous sodium sulfate

— Cotton

Column for drying extract.

Clamp the column in a vertical position and place a dry test tube (13 × 100-mm) under the bottom of the column. Label this test tube with an **E** for *extract* so that you don't confuse it with the test tubes you will be working with later in this experiment. With a Pasteur pipet, transfer the hexane layer to the column. When all the solution has drained, add 0.5 mL of hexane to the column to extract all the pigments from the drying agent. Evaporate the solvent by placing the test tube in a warm water bath (40–60°C) and directing a stream of nitrogen gas (or dry air) into the test tube. Dissolve the residue in 0.5 mL of hexane. Stopper the test tube and place it in your drawer until you are ready to run the alumina chromatography column.

Part B. Column Chromatography

Introduction

The pigments are separated on a column packed with alumina. Although there are many components in your sample, they usually separate into two main bands on the column. The first band to pass through the column is yellow and consists of the carotenes. This band may be less than 1 mm wide and it may pass through the column rapidly. It is easy to miss seeing the band as it passes through the alumina. The second band consists of all the other pigments discussed in the introduction to this experiment. Although it consists of both green and yellow pigments, it appears as a green band on the column. The green band spreads out on the column more than the yellow band, and it moves more slowly. Occasionally, the yellow and green components in this band will separate as the band moves down the column. If this begins to occur, you should change to a solvent of higher polarity so that they come out as one band. As the samples elute from the column, collect the yellow band (carotenes) in one test tube and the green band in another test tube.

Because the moisture content of the alumina is difficult to control, different samples of alumina may have different activities. The activity of the alumina is an important factor in determining the polarity of the solvent required to elute each band of pigments. Several solvents with a range of polarities are used in this experiment. The solvents and their relative polarities follow:

Hexane	increasing
70% hexane–30% acetone	polarity
Acetone	↓
80% acetone–20% methanol	

A solvent of lower polarity elutes the yellow band; a solvent of higher polarity is required to elute the green band. In this procedure, you first try to elute the yellow band with hexane. If the yellow band does not move with hexane, you then add the next more polar solvent. Continue this process until you find a solvent that moves the yellow band. When you find the appropriate solvent, continue using it until the yellow band is eluted from the column. When the yellow band is eluted, change to the next more polar solvent. When you find a solvent that moves the green band, continue using it until the green band is eluted. Remember that occasionally a second yellow band will begin to move down the column before the green band moves. This yellow band will be much wider than the first one. If this occurs, change to a more polar solvent. This should bring all the components in the green band down at the same time.

Advance Preparation

Before running the column, assemble the following glassware and liquids. Obtain five dry test tubes (16 × 100-mm), and number them 1 through 5. Prepare two dry Pasteur pipets with bulbs attached. Calibrate one of them to deliver a volume of about 0.25 mL (see p. 11). Place 10.0 mL hexane, 6.0 mL 70% hexane–30% acetone solution (by volume), 6.0 mL acetone, and 6.0 mL 80% acetone–20% methanol (by volume) into four separate containers. Clearly label each container.

Prepare a chromatography column packed with alumina. Place a *loose* plug of cotton in a Pasteur pipet (5¾-inch), and push it *gently* into position using a glass rod (see figure on p. 138 for the correct position of the cotton). Add 1.25 g of alumina (EM Science, No. AX0612-1) to the pipet[1] while tapping the column gently with your finger. When all the alumina has been added, tap the column with your finger for several seconds to ensure that the alumina is tightly packed. Clamp the column in a vertical position so that the bottom of the column is just above the height of the test tubes you will be using to collect the fractions. Place test tube 1 under the column.

NOTE: Read the following procedure on running the column. The chromatography procedure takes less than 15 minutes, and you cannot stop until all the material is eluted from the column. You must have a good understanding of the whole procedure before running the column.

Running the Column

Using a Pasteur pipet, slowly add about 3.0 mL of hexane to the column. The column must be completely moistened by the solvent. Drain the excess hexane until the level of hexane reaches the top of the alumina. Once you have added hexane to the alumina, the top of the column must not be allowed to run dry. If necessary, add more hexane.

NOTE: It is essential that the liquid level not be allowed to drain below the surface of the alumina at any point during the procedure.

When the level of the hexane reaches the top of the alumina, add about half (0.25 mL) of the dissolved pigments to the column. Leave the remainder in the test tube for the thin-layer chromatography procedure. (Put a stopper on the tube and place it back in your drawer.) Continue collecting the eluent in test tube 1. Just as the pigment solution penetrates the column, add 1 mL of hexane and drain until the surface of the liquid has reached the alumina.

Add about 4 mL of hexane. If the yellow band begins to separate from the green band, continue to add hexane until the yellow band passes through the column. If the yellow band does not separate from the green band, change to the next more polar solvent (70% hexane–30% acetone). When changing solvents, do not add the new solvent until the last solvent has nearly penetrated the alumina. When the appropriate solvent is found, add this solvent until the yellow band passes through the column. Just before the yellow band reaches the bottom of the column, place test tube 2 under the column. When the eluent becomes colorless again (the total volume of the yellow material should be less than 2 mL), place test tube 3 under the column.

Add several mL of the next more polar solvent when the level of the last solvent is almost at the top of the alumina. If the green band moves down the column, continue to add this solvent until the green band is eluted from the column. If the

---

[1] As an option, students may prepare a microfunnel from a 1-mL disposable plastic pipet. The microfunnel is prepared by (1) cutting the bulb in half with a scissors and (2) cutting the stem at an angle about ½-inch below the bulb. This funnel can be placed in the top of the column (Pasteur pipet) to aid in filling the column with alumina or with the solvents (see p. 764).

green band does not move or if a diffuse yellow band begins to move, change to the next more polar solvent. Change solvents again if necessary. Collect the green band in test tube 4. When there is little or no green color in the eluent, place test tube 5 under the column and stop the procedure.

Using a warm water bath (40–60°C) and a stream of nitrogen gas, evaporate the solvent from the tube containing the yellow band (tube 2), the tube containing the green band (tube 4), and the tube containing the original pigment solution (tube E). As soon as all the solvent has evaporated from each of the tubes, remove them from the water bath. Do not allow any of the tubes to remain in the water bath after the solvent has evaporated. Stopper the tubes and place them in your drawer.

Part C. Thin-Layer Chromatography

Preparing the TLC Plate

Technique 20 describes the procedures for thin-layer chromatography. Use a 10-cm × 3.3-cm TLC plate (Whatman Silica Gel Plates No. 4410 222). These plates have a flexible backing but should not be bent excessively. Handle them carefully or the adsorbent may flake off them. Also, you should handle them only by the edges; the surface should not be touched. Using a lead pencil (not a pen), *lightly* draw a line across the plate (short dimension) about 1 cm from the bottom (see figure). Using a centimeter ruler, move its index about 0.6 cm in from the edge of the plate and lightly mark off three 1-cm intervals on the line. These are the points at which the samples will be spotted.

Prepare three micropipets to spot the plate. The preparation of these pipets is described and illustrated in Technique 20, Section 20.4, page 782. Prepare a TLC development chamber with 70% hexane–30% acetone (see Technique 20, Section 20.5, p. 784). A beaker covered with aluminum foil or a wide-mouth screw-cap bottle is a suitable container to use (see Figure 20.5, p. 784). The backing on the TLC plates is thin, so if they touch the filter paper liner of the development chamber *at any point,* solvent will begin to diffuse onto the absorbent surface at that point. To avoid this, be sure that the filter paper liner does not go completely around the inside of the container. A space about 2 inches wide must be provided.

Using a Pasteur pipet, add two drops of 70% hexane–30% acetone to each of the three test tubes containing dried pigments. Swirl the tubes so that the drops of solvent dissolve as much of the pigments as possible. The TLC plate should be spotted with three samples: the extract, the yellow band from the column, and the green band. For each of the three samples, use a different micropipet to spot the sample on the plate. The correct method of spotting a TLC plate is described in Technique 20, Section 20.4, page 782. Take up part of the sample in the pipet (don't use a bulb; capillary action will draw up the liquid). For the extract (tube labeled E) and the green band (tube 4), touch the plate once *lightly* and let the solvent evaporate. The spot should be no larger than 2 mm in diameter and should be a fairly dark green. For the yellow band (tube 2), repeat the spotting technique 5–10 times until the spot is a definite yellow. Let the solvent evaporate completely between successive applications and spot the plate in exactly the same position each time. Save the liquid samples in case you need to repeat the TLC.

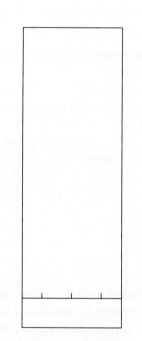

Preparing the TLC plate.

Developing the TLC Plate

Place the TLC plate in the development chamber, making sure that the plate does not come in contact with the filter paper liner. Remove the plate when the solvent front is 1–2 cm from the top of the plate. Using a lead pencil, mark the position of the solvent front. As soon as the plates have dried, outline the spots with a pencil and indicate the colors. This is important to do soon after the plates have dried because some of the pigments will change color when exposed to the air.

Analysis of the Results

In the crude extract, you should be able to see the following components (in order of decreasing R_f values):

> Carotenes (1 spot) (yellow-orange)
>
> Pheophytin a (gray, may be nearly as intense as chlorophyll b)
>
> Pheophytin b (gray, may not be visible)
>
> Chlorophyll a (blue-green, more intense than chlorophyll b)
>
> Chlorophyll b (green)
>
> Xanthophylls (possibly three spots: yellow)

Depending on the spinach sample, the conditions of the experiment, and how much sample was spotted on the TLC plate, you may observe other pigments. These additional components can result from air oxidation, hydrolysis, or other chemical reactions involving the pigments discussed in this experiment. It is common to observe other pigments in samples of frozen spinach. It is also common to observe components in the green band that were not present in the extract.

Identify as many of the spots in your samples as possible. Determine which pigments were present in the yellow band and which were present in the green band. Draw a picture of the TLC plate in your notebook. Label each spot with its color and its identity, where possible. Calculate the R_f values for each spot produced by chromatography of the extract (see Technique 20, Section 20.9, p. 787). At the instructor's option, submit the TLC plate with your report.

QUESTIONS

1. Why are the chlorophylls less mobile on column chromatography, and why do they have lower R_f values than the carotenes?

2. Propose structural formulas for pheophytin a and pheophytin b.

3. What would happen to the R_f values of the pigments if you were to increase the relative concentration of acetone in the developing solvent?

4. Using your results as a guide, comment on the purity of the material in the green and yellow bands; that is, did each band consist of a single component?

■ **ESSAY**

Ethanol and Fermentation Chemistry

The fermentation processes involved in making bread, making wine, and brewing are among the oldest chemical arts. Even though fermentation had been known as an art for centuries, not until the nineteenth century did chemists begin to understand this process from the point of view of science. In 1810, Gay-Lussac discovered the general chemical equation for the breakdown of sugar into ethanol and carbon dioxide. The manner in which the process took place was the subject of much conjecture until Louis Pasteur began his thorough examination of fermentation. Pasteur demonstrated that yeast was required in the fermentation. He was also able to identify other factors that controlled the action of the yeast cells. His results were published in 1857 and 1866.

For many years, scientists believed that the transformation of sugar into ethanol and carbon dioxide by yeasts was inseparably connected with the life process of the yeast cell. This view was abandoned in 1897, when Büchner demonstrated that yeast extract would bring about alcoholic fermentation in the absence of any yeast cells. The fermenting activity of yeast is due to a remarkably active catalyst of biochemical origin, the enzyme zymase. It is now recognized that most of the chemical transformations that go on in living cells of plants and animals are brought about by enzymes. The enzymes are organic compounds, generally proteins, and establishment of structures and reaction mechanisms of these compounds is an active field of present-day research. Zymase is now known to be a complex of at least 22 enzymes, each of which catalyzes a specific step in the fermentation reaction sequence.

Enzymes show an extraordinary specificity—a given enzyme acts on a specific compound or a closely related group of compounds. Thus, zymase acts on only a few select sugars and not on all carbohydrates; the digestive enzymes of the alimentary tract are equally specific in their activity.

The chief sources of sugars for fermentation are the various starches and the molasses residue obtained from refining sugar. Corn (maize) is the chief source of starch in the United States, and ethyl alcohol made from corn is commonly known as **grain alcohol.** In preparing alcohol from corn, the grain, with or without the germ, is ground and cooked to give the **mash.** The enzyme diastase is added in the form of **malt** (sprouted barley that has been dried in air at 40°C and ground to a powder) or of a mold such as *Aspergillus oryzae.* The mixture is kept at 40°C until all the starch has been converted to the sugar **maltose** by hydrolysis of ether and acetal bonds. This solution is known as the **wort.**

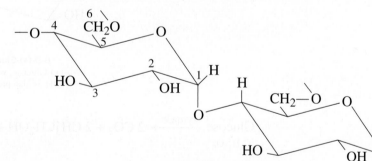

Starch

This is a glucose polymer with 1,4- and
1,6- glycosidic linkages. The linkages
at C-1 are α.

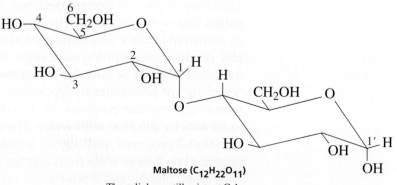

Maltose (C$_{12}$H$_{22}$O$_{11}$)

The α linkage still exists at C-1.
The —OH is shown α at the 1' postion
(axial), but it can also be ß (equatorial).

The wort is cooled to 20°C and diluted with water to 10% maltose, and a pure yeast culture is added. The yeast culture is usually a strain of *Saccharomyces cerevisiae* (or *ellipsoidus*). The yeast cells secrete two enzyme systems: maltase, which converts the maltose into glucose, and zymase, which converts the glucose into carbon dioxide and alcohol. Heat is liberated, and the temperature must be kept below 35°C by cooling to prevent destruction of the enzymes. Oxygen in large amounts is initially necessary for the optimum reproduction of yeast cells, but the actual production of alcohol is anaerobic. During fermentation, the evolution of carbon dioxide soon establishes anaerobic conditions. If oxygen were freely available, only carbon dioxide and water would be produced.

After 40–60 hours, fermentation is complete, and the product is distilled to remove the alcohol from solid matter. The distillate is fractionated by means of an efficient column. A small amount of acetaldehyde (bp 21°C) distills first and is followed by 95% alcohol. Fusel oil is contained in the higher-boiling fractions. The fusel oil consists of a mixture of higher alcohols, chiefly 1-propanol, 2-methyl-1-propanol, 3-methyl-1-butanol, and 2-methyl-1-butanol. The exact composition of fusel oil varies considerably; it particularly depends on the type of raw material that is fermented. These higher alcohols are not formed by fermentation of glucose. They arise from certain amino acids derived from the proteins present in the raw material and the yeast. These fusel oils cause the headaches associated with drinking alcoholic beverages.

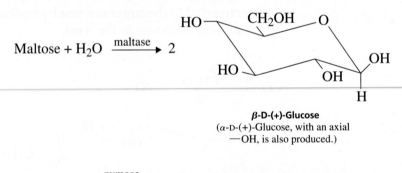

Maltose + H$_2$O $\xrightarrow{\text{maltase}}$ 2

β-D-(+)-Glucose
(α-D-(+)-Glucose, with an axial
—OH, is also produced.)

Glucose $\xrightarrow{\text{zymase}}$ 2 CO$_2$ + 2 CH$_3$CH$_2$OH + 26 Kcal
C$_6$H$_{12}$O$_6$

Industrial alcohol is ethyl alcohol used for nonbeverage purposes. Most commercial alcohol is denatured to avoid payment of taxes, the biggest cost in the price of liquor. The denaturants render the alcohol unfit for drinking. Methanol, aviation fuel, and other substances are used for this purpose. The difference in price between taxed and nontaxed alcohol is more than $20 a gallon. Before efficient synthetic processes were developed, the chief source of industrial alcohol was fermented blackstrap molasses, the noncrystallizable residue from refining cane sugar (sucrose). Most industrial ethanol in the United States is now manufactured from ethylene, a product of the "cracking" of petroleum hydrocarbons. By reaction with concentrated sulfuric acid, ethylene becomes ethyl hydrogen sulfate, which is hydrolyzed to ethanol by dilution with water. The alcohols 2-propanol, 2-butanol, and 2-methyl-2-propanol and higher secondary and tertiary alcohols also are produced on a large scale from alkenes derived from cracking.

Yeasts, molds, and bacteria are used commercially for the large-scale production of various organic compounds. An important example, in addition to ethanol production, is the anaerobic fermentation of starch by certain bacteria to yield 1-butanol, acetone, ethanol, carbon dioxide, and hydrogen.

REFERENCES

Amerine, M. A. "Wine." *Scientific American, 211* (August 1964): 46.

Hallberg, D. E. "Fermentation Ethanol." *ChemTech, 14* (May 1984): 308.

Ough, C. S. "Chemicals Used in Making Wine." *Chemical and Engineering News, 65* (January 5, 1987): 19.

Van Koevering, T. E., Morgan, M. D., and Younk, T. J. "The Energy Relationships of Corn Production and Alcohol Fermentation." *Journal of Chemical Education, 64* (January 1987): 11.

Webb, A. D. "The Science of Making Wine." *American Scientist, 72* (July–August 1984): 360.

Students wanting to investigate alcoholism and possible chemical explanations for alcohol addiction may consult the following references:

Cohen, G., and Collins, M. "Alkaloids from Catecholamines in Adrenal Tissue: Possible Role in Alcoholism." *Science, 167* (1970): 1749.

Davis, V. E., and Walsh, M. J. "Alcohol Addiction and Tetrahydropapaveroline." *Science, 169* (1970): 1105.

Davis, V. E., and Walsh, M. J. "Alcohols, Amines, and Alkaloids: A Possible Biochemical Basis for Alcohol Addiction." *Science, 167* (1970): 1005.

Seevers, M. H., Davis, V. E., and Walsh, M. J. "Morphine and Ethanol Physical Dependence: A Critique of a Hypothesis." *Science, 170* (1970): 1113.

Yamanaka, Y., Walsh, M. J., and Davis, V. E. "Salsolinol, an Alkaloid Derivative of Dopamine Formed in Vitro during Alcohol Metabolism." *Nature, 227* (1970): 1143.

17 EXPERIMENT 17

Ethanol from Sucrose

Fermentation

Fractional distillation

Azeotropes

Either sucrose or maltose can be used as the starting material for making ethanol. Sucrose is a disaccharide with the formula $C_{12}H_{22}O_{11}$. It has one glucose molecule combined with fructose. Maltose consists of two glucose molecules. The enzyme **invertase** is used to catalyze the hydrolysis of sucrose. **Maltase** is more effective in catalyzing the hydrolysis of maltose. The hydrolysis of maltose is discussed in the essay on ethanol and fermentation (p. 140). **Zymase** is used to convert the hydrolyzed sugars to alcohol and carbon dioxide. Pasteur observed that growth and fermentation were promoted by adding small amounts of mineral salts to the nutrient medium. Later, it was found that before fermentation actually begins, the hexose sugars combine with phosphoric acid, and the resulting hexose–phosphoric acid combination is then degraded into carbon dioxide and ethanol. The carbon dioxide is not wasted in the commercial process, because it is converted to dry ice.

The fermentation is inhibited by its end product ethanol; it is not possible to prepare solutions containing more than 10–15% ethanol by this method. More concentrated ethanol can be isolated by fractional distillation. Ethanol and water form an azeotropic mixture consisting of 95% ethanol and 5% water by weight, which is the most concentrated ethanol that can be obtained by fractionation of dilute ethanol–water mixtures.

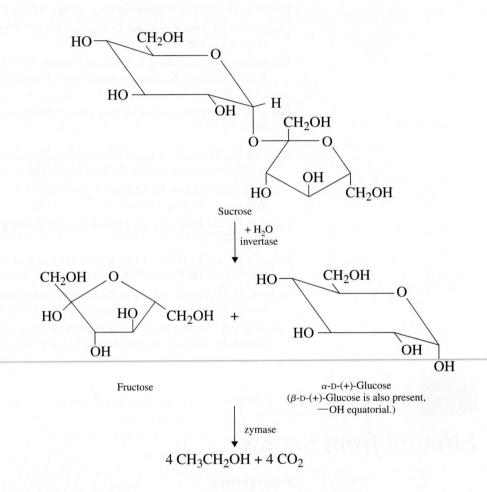

Sucrose

$+ H_2O$
invertase

Fructose

α-D-(+)-Glucose
(β-D-(+)-Glucose is also present,
—OH equatorial.)

zymase

$$4\ CH_3CH_2OH + 4\ CO_2$$

REQUIRED READING

Review:	Technique 8	Sections 8.3 and 8.4
	Technique 13	Physical Constants of Liquids
New:	Technique 15	Fractional Distillation, Azeotropes
	Essay	Ethanol and Fermentation Chemistry

SPECIAL INSTRUCTIONS

Start the fermentation at least 1 week before the period in which the ethanol will be isolated. When the aqueous ethanol solution is to be separated from the yeast cells, it is important to transfer carefully as much of the clear, supernatant liquid as possible without agitating the mixture.

NOTES TO THE INSTRUCTOR

Because the volume of the fermentation mixture is only about 20 mL, it is necessary to use an external heat source to maintain a temperature of 30–35°C. An incubator will provide the necessary temperature control. One can make a simple incubator by placing a cardboard box over a light bulb that is turned on during the fermentation. Be sure that the box does not touch the light bulb and has adequate clearance. Use aluminum foil to seal any openings and to help reflect the heat inward.

The balloons should be big enough and of sufficient quality to withstand the stretching required to be attached to the Erlenmeyer flask for 1 week.

One method of insulating the air condenser used for the fractional distillation column is provided by using two layers of clear flexible tubing (PVC) over the air condenser. For a ½-inch diameter column, use ½-inch I.D. × ⅝-inch O.D. plastic tubing on the inside and ⅝-inch I.D. × ⅞-inch O.D. tubing on the outside. Cut the tubing into 3½-inch lengths. Make a slit from end to end so that they can slip over the column. Slit the tubing using a sharp scissors or a razor knife with a proper handle.

CAUTION

Do not use a razor blade or you may get badly cut.

The clear tubing allows you to see what is going on in the column and also provides some insulation. Another method of insulating the fractionating column is to wrap the air condenser with a cotton pad about 3 inches square. Prepare the cotton pad by covering both sides of one layer of cotton with aluminum foil. Wrap this entirely with duct tape to hold the cotton in place and to make a more durable pad.

This experiment can also be performed without doing the fermentation. Provide each student with 20 mL of a 10% ethanol solution. This solution is used in place of the fermentation mixture in the Fractional Distillation section of the Procedure.

PROCEDURE

Fermentation apparatus.

Fermentation

Place 2.00 g of sucrose in a 50-mL Erlenmeyer flask. Add 18.0 mL of water warmed to 25–30°C, 2.0 mL of Pasteur's salts,[1] and 0.2 g of *dried* baker's yeast. Shake the contents vigorously to mix them, and then attach the balloon directly to the Erlenmeyer flask, as shown in the figure. The gas will cause the balloon to expand as the fermentation continues. Oxygen from the atmosphere is excluded from the chemical reaction by this technique. If oxygen were allowed to continue in contact with the fermenting solution, the ethanol could be further oxidized to acetic acid or even all the way to carbon dioxide and water. As long as carbon dioxide continues to be liberated, ethanol is being formed.

---

[1] A solution of Pasteur's salts consists of potassium dihydrogen phosphate, 1.0 g; calcium phosphate (monobasic), 0.10 g; magnesium sulfate, 0.10 g; and ammonium tartrate (diammonium salt), 5.0 g, dissolved in 430 mL water.

Allow the mixture to stand at about 30–35°C for 1 week.[2] After this time, *carefully* move the flask away from the heat source and remove the balloon. Without disturbing the sediment, transfer the clear, supernatant liquid solution to another container with a Pasteur pipet. Try to avoid drawing any of the sediment into the pipet.

If it is not possible to remove the solution completely without drawing up sediment, remove the sediment by centrifugation. Pour equal amounts of the liquid into two centrifuge tubes. After centrifugation for several minutes, decant the liquid away from the solid into another container. The liquid contains ethanol in water, plus smaller amounts of dissolved metabolites (fusel oils) from the yeast. The mixture will be subjected to fractional distillation.

Fractional Distillation

Assemble the apparatus shown in Figure 15.2, page 717; use a 20-mL or 25-mL round-bottom flask in place of the 10-mL flask. Use an aluminum block, if available, for the heat source. Pack the air condenser *uniformly* with about 1 g of stainless steel cleaning pad material. (*Use a pad that does not contain soap.*)

CAUTION   ⚠

You should wear heavy cotton gloves when handling the stainless steel cleaning pad. The edges are sharp and can easily cut into the skin.

Wrap the glass section of the air condenser between the two plastic caps with plastic tubing as described in Notes to the Instructor. Alternatively, use the method with a cotton pad (see Notes to the Instructor). Hold the pad in place with tape or twist ties. Place a boiling stone and the fermentation mixture in the round-bottom flask. If a sand bath is used, the apparatus should be clamped so that the bottom half of the flask is buried in the sand. Use a thermometer in the Hickman head to monitor the temperature of vapors rising from the liquid. Insert the thermometer so that the bulb is level with or slightly below the cap connecting the Hickman head to the air condenser. Also use a thermometer to monitor the temperature of the heat source. Cover the top of the sand bath (if used) with a square of aluminum foil with a tear from the center of one edge to the middle.

The temperature of the heat source should be adjusted to about 150–200°C. Adjust the actual temperature to achieve a rapid boiling rate in the flask. It may be necessary to increase the temperature of the heat source as the distillation proceeds. However, if liquid begins to fill the column, remove the heat source for a short time so that the liquid drains back into the flask. Once distillation begins, the temperature in the Hickman head will increase to about 78°C and remain at this temperature until the ethanol fraction is distilled. As distillate condenses in the Hickman head, transfer the liquid from the reservoir to a preweighed 3-mL conical vial. If your Hickman head does not have a side port, it will be necessary to use a 9-inch Pasteur pipet. In the latter case, it is helpful to bend the tip of the pipet slightly by heating it in a flame. The distillate can then be removed without removing the thermometer. Be sure to cap the conical vial used for storage each time after you transfer the distillate. Continue to distill the mixture, and transfer the distillate to the vial until the temperature in the Hickman head increases above 78°C or until the temperature in the Hickman head drops several degrees below 78°C

---

[2] It is typical for the balloon to expand to a volume of 100–200 cm^3. However, even when the balloon expands very little, good results are usually obtained.

and remains at this lower temperature for 10 minutes or more. You should collect about 0.4 mL of distillate. The distillation should then be interrupted by removing the apparatus from the heat source.

Analysis of Distillate

Determine the total weight of the distillate. Determine the approximate density of the distillate by transferring a known volume of the liquid with an automatic pipet or graduated pipet to a tared vial. Reweigh the vial and calculate the density. This method is good to two significant figures. Using the following table, determine the percentage composition by weight of ethanol in your distillate from the density of your sample. The extent of purification of the ethanol is limited because ethanol and water form a constant-boiling mixture, an azeotrope, with a composition of 95% ethanol and 5% water.

Percentage Ethanol by Weight	Density at 20°C (g/mL)	Density at 25°C (g/mL)
75	0.856	0.851
80	0.843	0.839
85	0.831	0.827
90	0.818	0.814
95	0.804	0.800
100	0.789	0.785

Calculate the percentage yield of alcohol. At the option of the instructor, determine the boiling point of the distillate using a microboiling-point method (Technique 13, Section 13.2, p. 695). The boiling point of the azeotrope is 78.1°C. Submit the ethanol to the instructor in a labeled vial.[3]

QUESTIONS

1. Write a balanced equation for the conversion of sucrose into ethanol.

2. By doing some library research, see whether you can find the commercial method or methods used to produce **absolute ethanol.**

3. Why is the balloon necessary in the fermentation?

4. How does acetaldehyde impurity arise in the fermentation?

5. Diethylacetal can be detected by gas chromatography. How does this impurity arise in fermentation?

6. Calculate how many milliliters of carbon dioxide would be produced theoretically from 2.0 g of sucrose at 25°C and 1 atmosphere pressure.

---

[3] A careful analysis by flame-ionization gas chromatography on a typical student-prepared ethanol sample provided the following results:

Acetaldehyde	0.060%
Diethylacetal of acetaldehyde	0.005%
Ethanol	88.3% (by hydrometer)
1-Propanol	0.031%
2-Methyl-1-propanol	0.092%
5-Carbon and higher alcohols	0.140%
Methanol	0.040%
Water	11.3% (by difference)

PART 2

Introduction to Molecular Modeling

ESSAY

Molecular Modeling and Molecular Mechanics

Since the beginnings of organic chemistry, somewhere in the middle of the nineteenth century, chemists have sought to visualize the three-dimensional characteristics of the all-but-invisible molecules that participate in chemical reactions. Concrete models that could be held in the hand were developed. Many kinds of model sets, such as framework, ball-and-stick, and space-filling models, were devised to allow people to visualize the spatial and directional relationships within molecules. These handheld models were interactive, and they could be readily manipulated in space.

Today, we can also use the computer to help us visualize these molecules. The computer images are also completely interactive, allowing us to rotate, scale, and change the type of model viewed at the press of a button or the click of a mouse. In addition, the computer can rapidly calculate many properties of the molecules that we view. This combination of visualization and calculation is often called **computational chemistry** or, more colloquially, **molecular modeling.**

There are two distinct methods of molecular modeling commonly used by organic chemists today. The first of these is **quantum mechanics,** and it involves the calculation of orbitals and their energies using solutions of the Schrödinger equation. The second method is not based on orbitals at all but is founded on our knowledge of the way in which the bonds and angles in a molecule behave. Classical equations that describe the stretching of bonds and the bending of angles are used. This second approach is called **molecular mechanics.** The two types of calculation are used for different purposes and do not calculate the same types of molecular properties. In this essay, molecular mechanics will be discussed.

Molecular Mechanics

Molecular mechanics (MM) was first developed in the early 1970s by two groups of chemical researchers: the Engler, Andose, and Schleyer group and the Allinger group. In molecular mechanics, a mechanical **force field** is defined that is used to calculate an energy for the molecule under study. The energy calculated is often called the **strain energy** or **steric energy** of the molecule. The force field is composed of several components, such as bond-stretching energy, angle-bending energy, and bond-torsion energy. A typical force field expression might be represented by the following composite expression:[1]

$$E_{strain} = E_{stretch} + E_{angle} + E_{torsion} + E_{oop} + E_{vdW} + E_{dipole}$$

To calculate the final strain energy for a molecule, the computer systematically changes every bond length, bond angle, and torsional angle in the molecule, recalculating the strain energy each time, keeping each change that minimizes the total energy and rejecting those that increase the energy. In other words, all the bond lengths and angles are changed until the energy of the molecule is *minimized.*

---

[1] Other force fields may be found that include more terms than this one and that contain more sophisticated calculational methods than those shown here.

Table 1 *Some of the factors contributing to a molecular force field*

Type of Contribution	Illustration	Typical Equation
$E_{stretch}$ (bond stretching)		$E_{stretch} = \sum\limits_{i=1}^{n\_bonds} (k_i/2)(x_i - x_0)^2$
E_{angle} (angle bending)		$E_{angle} = \sum\limits_{j=1}^{n\_angles} (k_j/2)(\theta_j - \theta_0)^2$
$E_{torsion}$ (bond torsion)		$E_{torsion} = \sum\limits_{k=1}^{n\_torsions} (k_k/2)[1 + sp_k(\cos p_k\theta)]$
E_{oop} (out of plane bending)		$E_{oop} = \sum\limits_{m=1}^{n\_oops} (k_m/2)d_m^{\,2}$
E_{vdW} (van der Waals repulsion)		$E_{vdW} = \sum\limits_{i=1}^{n\_atoms} \sum\limits_{j=1}^{n\_atoms} (E_iE_j)^{1/2}\left[\dfrac{1}{a_{ij}^{12}} - \dfrac{2}{a_{ij}^{6}}\right]$ $a_{ij} = r_{ij}/(R_i + R_j)$
E_{dipole} (electric dipole repulsion or attraction)		$E_{vdW} = K\sum\limits_{i=1}^{n\_atoms} \sum\limits_{j=i+1}^{n\_atoms} Q_iQ_j/r_{ij}^{2}$

Note: The factors selected here are similar to those in the "Tripos force field" used in the Alchemy III molecular modeling program.

Each term contained in the composite expression (E_{strain}) is defined in Table 1. All these terms come from classical physics, not quantum mechanics. We will not discuss every term but take $E_{stretch}$ as an illustrative example. Classical mechanics says that a bond behaves like a spring. Each type of bond in a molecule can be assigned a normal bond length, x_0. If the bond is stretched or compressed, its potential energy will increase, and there will be a restoring force that attempts to restore the bond to its normal length. According to Hooke's Law, the restoring force is proportional to the size of the displacement

$$F - k_i(x_f - x_0) \quad \text{or} \quad F - k_i(\Delta x)$$

where k_i is the **force constant** of the bond being studied (that is, the "stiffness" of the spring) and Δx is the change in bond length from the bond's normal length (x_0). The actual energy term that is minimized is given in Table 1. This equation indicates that all the bonds in the molecule contribute to the strain; it is a sum (Σ) starting with the first bond's contribution ($n = 1$) and proceeding through the contributions of all the other bonds (*n_bonds*).

These calculations are based on empirical data. To perform these calculations, the system must be **parameterized** with experimental data. To parameterize, a table of the normal bond lengths (x_0) and force constants (k_i) for every type of bond in the molecule has to be created. The program uses these experimental parameters to perform its calculations. The quality of the results from any molecular mechanics approach directly depends on how well the parameterization has been performed for each type of atom and bond that has to be considered. The MM procedure requires each of the factors in Table 1 to have its own parameter table.

Each of the first four terms in Table 1 is treated as a spring in the same manner as discussed for bond stretching. For instance, an angle also has a force constant k that resists a change in the size of the angle θ. In effect, in the first four terms the molecule is treated as a collection of interacting springs, and the energy of this collection of springs must be minimized. In contrast, the last two terms are based on electrostatic or "coulomb" repulsions. Without going into detail for these terms, it should be understood that these terms must also be minimized.

Minimization and Conformation

The object of minimizing the strain energy is to find the lowest energy *conformation* of a molecule. Molecular mechanics does a very good job of finding conformations because it varies bond distances, bond angles, torsional angles, and the positions of atoms in space. However, most minimizers have some limitations that users must be aware of. Many of the programs use a minimization procedure that will locate a local minimum in the energy but will not necessarily find a global minimum. The figure below illustrates the problem.

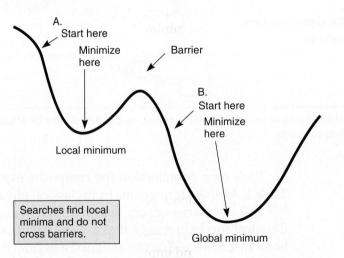

Global and local energy minima.

In the figure, the molecule under consideration has two conformations that represent energy minima for the molecule. Many minimizers will not automatically find the lowest energy conformation, the **global minimum.** The global minimum will be found only when the structure of your starting molecule is already close to the global minimum's conformation. For instance, if the starting structure corresponds to point B on the curve in the figure, then the global minimum will be found. However, if your starting molecule is not close to the global minimum in structure, a **local minimum** (one nearby) may be found. In the figure, if your starting structure corresponds to point A, then a local minimum will be found instead of the global minimum. Some of the more expensive programs always find the global minimum because they use more sophisticated minimization procedures that depend on random (Monte Carlo) changes instead of sequential ones. However, unless the program has specifically dealt with this problem, the user must be careful to avoid finding a false local minimum when the global minimum is expected. It may be necessary to use several different starting structures to discover the global minimum for a given molecule.

Limitations of Molecular Mechanics

From our discussion thus far, it should be obvious that molecular mechanics was developed to find the lowest energy conformation of a given molecule or to compare the energies of several conformations of the same molecule. Molecular mechanics calculates a "strain energy," not a thermodynamic energy such as a heat of formation. Procedures based on quantum mechanics and statistical mechanics are required to calculate thermodynamic energies. Therefore, it is dangerous to compare the strain energies of two *different* molecules. For instance, molecular mechanics can make a good evaluation of the relative energies of *anti-* and *gauche*-butane conformations, but it cannot fruitfully compare butane and cyclobutane. Isomers can be compared only if they are closely related. The *cis-* and *trans*-isomers of 1,2-dimethylcyclohexane, or those of 2-butene, can be compared. However, the isomers 1-butene and 2-butene cannot be compared; one is a monosubstituted alkene, whereas the other is disubstituted.

Molecular mechanics will perform the following tasks well:

1. It will give good estimates for the actual bond lengths and angles in a molecule.

2. It will find the best conformation for a molecule, but you must watch out for local minima!

Molecular mechanics will not calculate the following properties:

1. It will not calculate thermodynamic properties such as the heat of formation of a molecule.[2]

2. It will not calculate electron distributions, charges, or dipole moments.

3. It will not calculate molecular orbitals or their energies.

4. It will not calculate infrared, NMR, or ultraviolet spectra.

Current Implementations

With time, the most popular version of molecular mechanics has become that developed by Norman Allinger and his research group. The original program from this group was called MM1. The program has undergone constant revisions and improvements, and the current Allinger versions are now designated MM2 and MM3. However, many other versions of molecular mechanics are now available from private and commercial sources. Some popular commercial programs that now incorporate their own force fields and parameters include Alchemy III, Alchemy 2000, CAChe, Personal CAChe,

---

[2] Some of the latest versions are now parameterized to give heats of formation.

HyperChem, Insight II, PC Model, MacroModel, Spartan, PC Spartan, MacSpartan, and Sybyl. You should also realize, however, that there are many modeling programs that do not have molecular mechanics or minimization. These programs will "clean up" a structure that you create by attempting to make every bond length and angle "perfect." With these programs, every sp^3 carbon will have 109° angles, and every sp^2 carbon will have perfect 120° angles. Using one of these programs is equivalent to using a standard model set that has connectors and bond with perfect angles and lengths. If you intend to find a molecule's preferred conformation, be sure you use a program that has a force field and performs a true minimization procedure. Also remember that you may have to control the starting structure's geometry in order to find the correct result.

REFERENCES

Casanova, J. "Computer-Based Molecular Modeling in the Curriculum." Computer Series 155. *Journal of Chemical Education, 70* (November 1993): 904.

Clark, T. *A Handbook of Computational Chemistry—A Practical Guide to Chemical Structure and Energy Calculations.* New York: John Wiley & Sons, 1985.

Lipkowitz, K. B. "Molecular Modeling in Organic Chemistry—Correlating Odors with Molecular Structure." *Journal of Chemical Education, 66* (April 1989): 275.

Tripos Associates. *Alchemy III—User's Guide.* St. Louis: Tripos Associates, 1992.

Ulrich, B., and Allinger, N. L. *Molecular Mechanics,* ACS Monograph 177. Washington, DC: American Chemical Society, 1982.

18　　**EXPERIMENT 18**

An Introduction to Molecular Modeling

Molecular modeling
Molecular mechanics

REQUIRED READING

Review: The sections of your lecture textbook dealing with

　　1. Conformation of cyclic and acyclic compounds

　　2. The energies of alkenes with respect to degree of substitution

　　3. The relative energies of *cis-* and *trans-*alkenes

New:　　Essay: Molecular Modeling and Molecular Mechanics

SPECIAL INSTRUCTIONS

To perform this experiment, you must use computer software that has the ability to perform molecular mechanics (MM2 or MM3) calculations with minimization of the strain energy. Either your instructor will provide instruction for using the software or you will be provided with a handout giving instructions.

NOTES TO THE INSTRUCTOR

This molecular mechanics experiment was devised using the modeling program Alchemy III; however, it should be possible to use many other implementations of molecular mechanics. Some of the other capable programs available are Alchemy 2000, Spartan, PC Spartan, MacSpartan, HyperChem, CAChe and Personal CAChe, PCModel, Insight II, Nemesis, and Sybyl. You will have to provide your students with an introduction to your specific implementation. The introduction should show students how to build a molecule, how to minimize its energy, and how to load and save files. Students will also need to be able to measure bond lengths and bond angles.

18A EXPERIMENT 18A

The Conformations of n-Butane: Local Minima

The acyclic butane molecule has several conformations derived by rotation about the C2—C3 bond. The relative energies of these conformations have been well established experimentally and are listed in the following table.

Conformation	Torsional Angle	Relative Energy (kcal/mol)	Relative Energy (kJ/mol)	Types of Strain
Syn	0°	6.0	25.0	Steric/torsional
Gauche	60°	1.0	4.2	Steric
Eclipsed-120	120°	3.4	14.2	Torsional
Anti	180°	0	0	No strain

In this section, we will show that, although molecular mechanics does not calculate the precise thermodynamic energies for the conformations of butane, it will give strain energies that predict the *order* of stability correctly. We will also investigate the difference between a local minimum and a global minimum.

When you construct a butane skeleton, you might expect the minimizer to always arrive at the *anti* conformation (lowest energy). In fact, for most molecular mechanics programs, this will happen only if you bias the minimizer by starting with a butane skeleton that closely resembles the *anti* conformation. If this is done, the minimizer will find the *anti* conformation (the *global* minimum). However, if a skeleton is constructed that does not closely resemble the *anti* conformation, the butane will usually minimize to the *gauche* conformation (the nearest *local* minimum) and not proceed to the global minimum. For the two staggered conformations, you will begin by constructing your starting butane molecules with torsional angles slightly removed from the two minima. The eclipsed conformations, however, will be set on the exact angles to see whether they will minimize. Your data should be recorded in a table with the following headings: *Starting Angle, Minimized Angle, Final Conformation,* and *Minimized Energy.*

Your program should have a feature that allows you to set bond lengths, bond angles, and torsional angles.[1] If it does, you can merely select the torsion angle C1—C2—C3—C4 and specify 160° to set the first starting shape. Select the minimizer and allow it to run until it stops. Did it find the *anti* conformation (180°)? Record the energy. Repeat the process, starting with torsion angles of 0°, 45°, and 120° for the butane skeleton. Record the strain energies and report the final conformations that are formed in each case. What are your conclusions? Do your final results agree with those in the table?

If your minimizer rotated the two-eclipsed conformations (0° and 120°) to their closest staggered minima, you may have to restrict the minimizer to a single iteration in order to calculate their energies. This restriction calculates a **single-point energy,** and the energy of the structure is not minimized. If necessary, calculate the single-point energies of the eclipsed conformations and record your results.

The lesson here is that you may have to try several starting points to find the correct structure for the lowest energy conformation of a molecule! Do not blindly accept your first result but look at it with the skeptical eye of a practiced chemist and test it further.

Optional

Record the single-point energies for every 30-degree rotation, starting at 0° and ending at 360°. When these energies are plotted against their angle, the plot should resemble the rotational energy curve shown for butane in most organic textbooks.

18B   **EXPERIMENT 18B**

Cyclohexane Chair and Boat Conformations

In this exercise, we will investigate the chair and boat conformations of cyclohexane. Many programs will have these stored on disk as templates or fragments. If they are available as templates or fragments, you will only have to add hydrogens to the template. The chair is not difficult to build if you construct your cyclohexane on the screen in such a way that it suggests a chair (that is, just as you might draw it on paper). This crude construct will usually minimize to a chair. The boat is more difficult to construct. When you draw a crude boat on the screen, it will minimize to a *twist* boat instead of the desired symmetrical boat.

Before you construct any cyclohexanes, construct a propane molecule. Minimize it and measure the CH and CC bond lengths and the CCC bond angle. Record these values; we will use them for reference.

Now construct a cyclohexane chair and minimize it. Measure the CH and CC bond lengths and the CCC angle in the ring. Compare these values to those of propane. What do you conclude? Rotate the molecule so that you

---

[1] If your program does not have this feature, you can approximate the angles specified by constructing your starting molecules on the screen in a Z-shape for one and in a U-shape for the other.

view it end-on, looking down two of the bonds simultaneously (as in a Newman projection). Are all the hydrogens staggered? Rotate the chair and look at it from a different end-on angle. Are all the hydrogens still staggered? The van der Waals radius of a hydrogen atom is 1.20 Ångstroms. Hydrogen atoms that are closer than 2.40 Ångstroms apart will "touch" each other and create steric strain. Are any of the hydrogens in the cyclohexane chair close enough to cause steric strain? What are your conclusions?

Now construct a cyclohexane boat (from a template) and do not minimize it.[1] Measure the CH and CC bond lengths and the CCC bond angles at both the peaks and the lower corner of the ring. Compare these values to those of propane. Rotate the molecule so that you view it end-on looking down the two parallel bonds on the sides of the boat. Are the hydrogens eclipsed or staggered? Now measure the distances between the various hydrogens on the ring, including the bowsprit–flagpole hydrogens and the axial and equatorial hydrogens on the side of the ring. Are any of the hydrogens generating steric strain?

Now minimize the boat to a twist boat and repeat all the measurements. Write all of your conclusions about chairs, boats, and twist boats in your report.

18C EXPERIMENT 18C

Substituted Cyclohexane Rings

Using a cyclohexane template, construct *cis*(a,a)-1,3-dimethylcyclohexane, *cis*(e,e)-1,3-dimethylcyclohexane, and *trans*(a,e)-1,3-dimethylcyclohexane and measure their energies. In the diaxial isomer, measure the distance between the two diaxial methyl groups. What do you conclude?

Optional

Similar comparisons can be made for the *cis*- and *trans*-1,2-dimethylcyclohexanes and the *cis*- and *trans*-1,4-dimethylcyclohexanes.

Now construct *cis*(e,a)-1,4-di-*tert*-butylcyclohexane in a chair conformation, minimize it, and record its energy. Now compare it to *cis*(e,e)-1,4-di-*tert*-butylcyclohexane in a boat conformation. Minimize this isomer to a twist boat and record its energy.

18D EXPERIMENT 18D

cis- *And* trans-2-Butene

Heats of hydrogenation for the three isomers of butene are given in the following table. Construct both *cis*- and *trans*-2-butene, minimize them, and report their energies. Which of these isomers has the lowest energy? Can you determine why?

---

[1] A single-point energy (see page 158) may be obtained, if you desire.

Compound	ΔH (kcal/mol)	ΔH (kJ/mol)
trans-2-Butene	−27.6	−115
cis-2-Butene	−28.6	−120
1-Butene	−30.3	−126

Now construct and minimize 1-butene. Record its energy. Obviously, 1-butene does not fit with the hydrogenation data. Molecular mechanics works quite well for *cis*- and *trans*-2-butene because they are similar isomers. Both are 1,2-disubstituted alkenes. However, 1-butene is a monosubstituted alkene, and direct comparison to the 2-butenes cannot be made. The differences in the stability of mono- and di-substituted alkenes require that factors other than those used in molecular mechanics be used. These factors are caused by electronic and resonance differences. The molecular orbitals of the methyl groups interact with the pi bonds of the disubstituted alkenes (hyperconjugation) and help to stabilize them. Two such groups (as in 2-butene) are better than one (as in 1-butene). Therefore, although the bond lengths and angles come out pretty well for 1-butene, the energy derived for 1-butene does not directly compare to the energies of the 2-butenes. Molecular mechanics does not include terms that allow these factors to be included; it is necessary to use either semiempirical or *ab initio* quantum mechanical methods, which are based on molecular orbitals.

ESSAY

Computational Chemistry—ab Initio and Semiempirical Methods

Introduction to Terms and Methods

In an earlier essay (p. 152), the application of **molecular mechanics** to solving chemical problems was discussed. Molecular mechanics is good at giving estimates of the bond lengths and angles in a molecule. It can find the best geometry or conformation of a molecule. However, it requires the application of **quantum mechanics** to find good estimates of the thermodynamic, spectroscopic, and electronic properties of a molecule. In this essay, we will discuss the application of quantum mechanics to organic molecules.

Quantum mechanics computer programs can calculate heats of formation and the energies of transition states. The shapes of orbitals can be displayed in three dimensions. Important properties can be mapped onto the surface of a molecule. With these programs, the chemist can visualize concepts and properties in a way that the mind cannot readily imagine. Often, this visualization is the key to understanding or to solving a problem.

For you to solve the electronic structure and energy of a molecule, quantum mechanics requires that you formulate a wavefunction Ψ (psi) that describes the distribution of all the electrons within the system. The nuclei are assumed to have relatively small motions and to be essentially fixed in their equilibrium positions (Born–Oppenheimer approximation). The average energy of the system is calculated by using the Schrödinger equations as

$$E = \int \Psi^* H \Psi \, d\tau \, / \int \Psi^* \Psi \, d\tau$$

where H, the Hamiltonian operator, is a multiterm function that evaluates all the potential energy contributions (electron–electron repulsions and nuclear–electron attractions) and the kinetic energy terms for each electron in the system.

Because we can never know the true wavefunction Ψ for the molecule, we must guess at the nature of this function. According to the **Variation Principle,** a cornerstone idea in quantum mechanics, we can continue to guess at this function forever and never reach the true energy of the system, which will always be lower than our best guess. Because of the Variation Principle, we can formulate an approximate wavefunction and then consistently vary it until we minimize the energy of the system (as calculated using the Schrödinger equation). When we reach the variational minimum, the resulting wavefunction is often a good approximation of the system we are studying. Of course, you can't just make any guess and get good results. It has taken theoretical chemists quite a few years to learn how to formulate both wavefunctions and Hamiltonian operators that yield results that agree closely with experiment. Today, however, most methods for performing these calculations have been well established, and computational chemists have devised easy-to-use computer programs, which can be used by any chemist to calculate molecular wavefunctions.

Molecular quantum-mechanical calculations can be divided into two classes: *ab initio* (Latin: "from the beginning" or "from first principles") and *semiempirical.*

1. *Ab initio* **calculations** use the fully correct Hamiltonian for the system and attempt a complete solution without using any experimental parameters.

2. **Semiempirical calculations** generally use a simplified Hamiltonian operator and incorporate experimental data or a set of parameters that can be adjusted to fit experimental data.

Ab initio calculations require a great deal of computer time and memory because every term in the calculations is evaluated explicitly. Semiempirical calculations have more modest computer requirements, allowing the calculations to be completed faster and making it possible to treat larger molecules. Chemists generally use semiempirical methods whenever possible, but it is useful to understand both methods when solving a problem.

Solving the Schrödinger Equation

The Hamiltonian

The exact form of the Hamiltonian operator, which is a collection of potential energy (electrostatic attraction and repulsion terms) and kinetic energy terms, is standardized now and need not concern us here. However, all the programs require the **Cartesian coordinates** (locations in three-dimensional space) of all the atoms and a **connectivity matrix** that specifies which atoms are bonded and how (single, double, triple, H-bond, and so on). In modern programs, the user draws or constructs the molecule on the computer screen, and the program automatically constructs the atomic-coordinate and connectivity matrices.

The Wavefunction

It is not necessary for the user to construct or guess at a trial wavefunction—the program will do this. However, it is important to understand how the wavefunctions are formulated because the user frequently has a choice of methods. The complete molecular wavefunction is made up of a determinant of molecular orbitals:

$$\Psi = \begin{vmatrix} \phi_1(1) & \phi_2(1) & \phi_3(1) & \dots & \phi_n(1) \\ \phi_1(2) & \phi_2(2) & \phi_3(2) & \dots & \phi_n(2) \\ & & & \vdots & \\ \phi_1(n) & \phi_2(n) & \phi_3(n) & \dots & \phi_n(n) \end{vmatrix}$$

The molecular orbitals $\phi_i(n)$ must be made up from some type of mathematical function. They are usually made up by a **linear combination of atomic orbitals** χ_j (LCAO) from each of the atoms that make up the molecule.

$$\phi_i(n) = \Sigma_j \, c_{ji} \, \chi_j = c_1 \, \chi_1 + c_2 \, \chi_2 + c_3 \, \chi_3 \dots$$

This combination includes all the orbitals in the *core* and the *valence shell* of each atom in the molecule. The complete set of orbitals χ_j is called the **basis set** for the calculation. When an *ab initio* calculation is performed, most programs require the user to choose the basis set.

Basis Set Orbitals

It should be apparent that the most obvious basis set to use for an *ab initio* calculation is the set of hydrogen-like atomic orbitals 1*s*, 2*s*, 2*p*, and so on that we are all familiar with from atomic structure and bonding theory. Unfortunately, these "actual" orbitals present computational difficulties because they have radial nodes when they are associated with the higher shells of an atom. As a result, a more convenient set of functions was devised by Slater. These **Slater-type orbitals** (STOs) differ from the hydrogen-like orbitals in that they have no radial nodes, but they have the same angular terms and overall shape. More important, they give good results (those that agree with experiment) when used in semiempirical and *ab initio* calculations.

Slater-Type Orbitals

The radial term of an STO is an exponential function with the form $R_{nl} = r^{(n-1)}e^{[-(-Z-s)r/n]}$, where Z is the nuclear charge of the atom, and s is a "screening constant" that reduces the nuclear charge Z that is "seen" by an electron. Slater formulated a set of rules to determine the values of s that are required to produce orbitals that agree in shape with the customary hydrogen-like orbitals.

Radial Expansion and Contraction

A problem with simple STOs is that they do not have the ability to vary their radial size. Today it is common to use two or more simpler STOs so that expansion and contraction of the orbitals can occur during the calculation. For instance, if we take two functions such as $R(r) = re^{(-\zeta r)}$ with different values of ζ, the larger value of ζ gives an orbital more contracted around the nucleus (an inner STO), and the smaller value of ζ gives an orbital extended further out from the nucleus (an outer STO). By using these two functions in different combinations, any size STO can be generated.

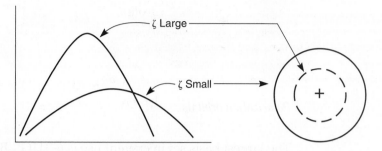

Variation of the radial size of an STO with the value of the exponent ζ (zeta).

Gaussian-Type Orbitals

Eventually, the original Slater-type orbitals were abandoned, and *simulated* STOs built from Gaussian functions were used. The most common basis set of this kind is the **STO-3G basis set,** which uses three Gaussian functions (3G) to simulate each one-electron orbital. A Gaussian function is of the type $R(r) = re^{(-\alpha r^2)}$.

In the STO-3G basis set, the coefficients of the Gaussian functions are selected to give the best fit to the corresponding Slater-type orbitals. In this formulation, for instance, a hydrogen electron is represented by a single STO (a $1s$ type orbital) that is simulated by a combination of three Gaussian functions. An electron on any period 2 element (Li to Ne) will be represented by five STOs ($1s$, $2s$, $2p_x$, $2p_y$, $2p_z$), each simulated by three Gaussian functions. Each electron in a given molecule will have its own STO. (The molecule is literally built up by a series of one-electron orbitals. A spin function is also included so that no two of the one-electron orbitals are exactly the same.)

Split-Valence Basis Sets

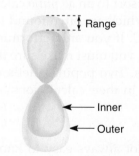

Split-valence orbitals.

In a further step of evolution, it is now common to abandon trying to simulate the hydrogen-like orbitals with STOs and to simply use an optimized combination of the Gaussian functions themselves for the basis set. The 3-21G basis set has largely replaced the STO-3G basis set for all but the largest molecules. The 3-21G symbolism means that three Gaussian functions are used for the wavefunction of each core electron, but the wavefunctions of the valence electrons are "split" two to one (21) between inner and outer Gaussian functions, allowing the valence shell to expand or contract in size.

A larger basis set (and one that requires more calculation time) is 6-31G, which uses six Gaussian "primitives" and a three-to-one split in the valence shell orbitals.

Polarization Basis Sets

Both the 3-21G and 6-31G basis sets can be extended to 3-21G* and 6-31G*. The star (*) indicates that these are **polarization sets,** where the next higher type of orbital is included (for instance, a p orbital can be polarized by adding a d orbital function). Polarization allows deformation of the orbital toward the bond on one side of the atom.

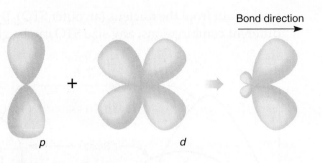

Polarization orbitals.

The largest basis set in current use is 6-311G*. Because it is computationally intensive, it is used only for **single-point calculations** (a calculation on a fixed geometry—no minimization performed). Other basis sets include the 6-31G** (which includes six *d* orbitals per atom instead of the usual five) and the 6-31+G* or 6-31++G* sets, which include diffuse *s* functions (electrons at a larger distance from the nucleus) to better deal with anions.

Semiempirical Methods

It would be impossible to give a short and complete overview of the various semiempirical methods that have evolved over time. One must really get into the mathematical details of the method to understand what approximations have been made in each case and what kinds of empirical data have been included. In many of these methods, it is common to omit integrals that are expected (either from experience or for theoretical reasons) to have negligible values. Certain integrals are stored in a table and are not calculated each time the program is applied. For instance, the **frozen core approximation** is often used. This approximation assumes that the *completed shells* of the atom do not differ from one atom to another in the same period. All the core calculations are stored in a table, and they are simply looked up when needed. This makes the computation much easier to perform.

One of the more popular semiempirical methods in use today is AM-1. The parameters in this method work especially well for organic molecules. In fact, whenever possible, you should try to solve your problem using a semiempirical method such as AM-1 before you resort to an *ab initio* calculation. Also popular are MINDO/3 and MNDO, which are often found together in a computational package called MOPAC. If you are performing semiempirical calculations on inorganic molecules, you must make sure the method you use is optimized for transition metals. Two popular methods for inorganic chemists wishing to involve metals in their calculations are PM-3 and ZINDO.

Picking a Basis Set for *ab Initio* Calculations

When you perform an *ab initio* calculation, it is not always easy to know which basis set to use. Normally, you should not use more complexity than is needed to answer your question or solve the problem. In fact, it may be desirable to determine the approximate geometry of the molecule using *molecular mechanics*. Many programs will allow you to use the result of a molecular mechanics **geometry optimization** as a starting point for an *ab initio* calculation. If possible, you should do so to save computational time.

Most of the time 3-21G is a good starting point for an *ab initio* calculation, but if you have a very large molecule, you may wish to use STO-3G, a simpler basis set. Avoid doing geometry optimizations with the larger basis sets. Often, you can do the geometry optimization first with 3-21G (or a semi-empirical method) and then polish up the result with a **single-point energy** calculation with a larger basis set, such as 6-31G. You should "move up the ladder": AM1 to STO-3G to 3-21G to 6-31G, and so on. If you don't see any change in the results as you move up to successively more complex basis sets, it is generally fruitless to continue. If you include elements beyond period 2, use polarization sets (PM3 for semiempirical). Some programs have special sets for cations and anions or for radicals. If the result doesn't match experimental results, you may not have used the correct basis set.

Heats of Formation

In classical thermodynamics, the **heat of formation,** ΔH_f, is defined as the energy consumed (endothermic reaction) or released (exothermic reaction) when the molecule is formed from its elements at standard conditions of pressure and temperature. The elements are assumed to be in their standard states.

$$2\ \text{C (graphite)} + 3\ \text{H}_2\ \text{(g)} \longrightarrow \text{C}_2\text{H}_6\ \text{(g)} + \Delta H_f \qquad (25°\text{C})$$

Both *ab initio* and semiempirical programs calculate the energy of a molecule as its "heat of formation." This heat of formation, however, is not identical to the thermodynamic function, and it is not always possible to make direct comparisons.

Heats of formation in semiempirical calculations are generally calculated in kcal/mole (1 kcal = 4.18 kJ) and are similar but not identical to the thermodynamic function. The AM1, PM3, and MNDO methods are parameterized by fitting them to a set of experimentally determined enthalpies. They are calculated from the binding energy of the system. The **binding energy** is the energy released when molecules are formed from their separated electrons and nuclei. The semiempirical heat of formation is calculated by subtracting atomic heats of formation from the binding energy. For most organic molecules, AM1 will calculate the heat of formation correctly to within a few kilocalories per mole.

In *ab initio* calculations, the heat of formation is given in **hartrees** (1 hartree = 627.5 kcal/mole = 2625 kJ/mole). In the *ab initio* calculation, the heat of formation is best defined as total energy. Like the binding energy, the **total energy** is the energy released when molecules are formed from their separated electrons and nuclei. This "heat of formation" always has a large negative value and does not relate well to the thermodynamic function.

Although these values do not relate directly to the thermodynamic values, they can be used to compare the energies of isomers (molecules of the same formula), such as *cis-* and *trans-*2-butene, or of tautomers, such as acetone in its enol and keto forms.

$$\Delta E = \Delta H_f(\text{isomer 2}) - \Delta H_f(\text{isomer 1})$$

It is also possible to compare the energies of balanced chemical equations by subtracting the energies of the products from the reactants.

$$\Delta E = [\Delta H_f(\text{product 1}) + \Delta H_f(\text{product 2})] - [\Delta H_f(\text{reactant 1}) + \Delta H_f(\text{reactant 2})]$$

Graphic Models and Visualization

Although the solution of the Schrödinger equation minimizes the *energy* of the system and gives a heat of formation, it also calculates the shapes and energies of all the molecular orbitals in the system. A big advantage of semiempirical and *ab initio* calculations, therefore, is the ability to determine the energies of the individual molecular orbitals and to plot their shapes in three dimensions. For chemists investigating chemical reactions, two molecular orbitals are of paramount interest: the HOMO and the LUMO.

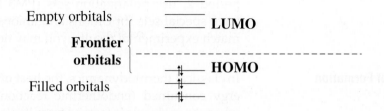

The **HOMO,** the highest occupied molecular orbital, is the last orbital in the molecule to be filled with electrons. The **LUMO,** the lowest unoccupied molecular orbital, is the first empty orbital in the molecule. These two orbitals are often called **frontier orbitals.**

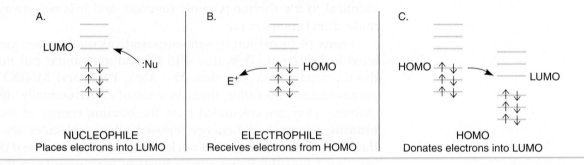

| A. NUCLEOPHILE Places electrons into LUMO | B. ELECTROPHILE Receives electrons from HOMO | C. HOMO Donates electrons into LUMO |

The frontier orbitals are similar to the valence shell of the molecule. They are where most of the chemical reactions occur. For instance, if a reagent is going to react with a Lewis base, the electron pair of the base must be placed into an empty orbital of the acceptor molecule. The most available orbital is the LUMO. By examining the structure of the LUMO, one can determine the most likely spot where the addition will take place—usually at the atom where the LUMO has its biggest lobe. Conversely, if a Lewis acid attacks a molecule, it will bond to electrons that already exist in the molecule under attack. The most likely spot for this attack would be the atom where the HOMO has its biggest lobe (the electron density should be greatest at that site). Where it is not obvious which molecule is the electron pair donor, the HOMO that has the highest orbital energy will usually be the electron pair donor, placing electrons into the LUMO of the other molecule. The frontier orbitals, HOMO and LUMO, are where most chemical reactions occur.

Surfaces

Chemists use many kinds of handheld models to visualize molecules. A framework model best represents the angles, lengths, and directions of bonds. A molecule's size and shape are probably best represented by a space-filling model. In quantum mechanics, a model similar to the space-filling model can be generated by plotting a surface that represents all the points where the electron density of the molecule's wavefunction has a constant value. If this value is picked correctly, the resulting surface will resemble the surface of a space-filling model. This type of surface is called an **electron-density surface.** The electron-density surface is useful for visualizing the size and shape of the molecule, but it does not reveal the position of the nuclei, bond lengths, or angles, because you cannot see inside the surface. The electron-density value used to define this surface will be low because electron density falls off with increasing distance from the nucleus. If you choose a higher value of electron density when you plot this surface, a **bond-density surface** will be obtained. This surface will not give you an idea of the size or shape of the molecule, but it will reveal where the bonds are located because the electron density will be higher where bonding is taking place.

Cyclopentane

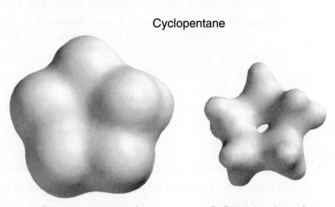

A. Electron-density surface   B. Bond-density surface

Mapping Properties onto a Density Surface

It is also possible to map a calculated property onto an electron-density surface. Because all three Cartesian coordinates are used to define the points on the surface, the property must be mapped in color, with the colors of the spectrum, red–orange–yellow–green–blue, representing a range of values. In effect, this is a four-dimensional plot (x, y, z, + property mapped). One of the most common plots of this type is the **density–electrostatic potential,** or **density–elpot,** plot. The electrostatic potential is determined by placing a unit positive charge at each point on the surface and measuring the interaction energy of this charge with the nuclei and electrons in the molecule. Depending on the magnitude of the interaction, that point on the surface is painted one of the colors of the spectrum. In the Spartan program, areas of high electron density are painted red or orange, and areas of lower electron density are plotted blue or green. When you view such a plot, the polarity of the molecule is immediately apparent.

Allyl cation

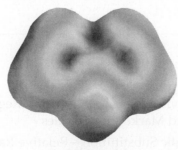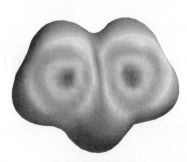

A. Density–elpot   B. LUMO   C. Density–LUMO

The second common type of mapping plots values of one of the frontier orbitals (either the HOMO or the LUMO) in color on the density surface. The color values plotted correspond to the value of the orbital where it intersects the surface. For a density–LUMO plot for instance, the "hot spot" would be where the LUMO has its largest lobe. Because the LUMO is empty, this would be a bright blue area. In a density–HOMO plot, a bright red area would be the "hot spot."

REFERENCES

Introductory

Hehre, W. J., Burke, L. D., Shusterman, A. J., and Pietro, W. J. *Experiments in Computational Organic Chemistry.* Irvine, CA: Wavefunction, Inc., 1993.

Hehre, W. J., Shusterman, A. J., and Nelson, J. E. *The Molecular Modeling Workbook for Organic Chemistry.* Irvine, CA: Wavefunction, Inc., 1998.

Hypercube, Inc. *HyperChem Computational Chemistry.* Waterloo, Ontario, Canada: HyperCube, Inc., 1996.

Shusterman, G. P., and Shusterman, A. J. "Teaching Chemistry with Electron Density Models." *Journal of Chemical Education, 74* (July 1997): 771.

Wavefunction, Inc. "PC-Spartan—Tutorial and User's Guide." Irvine, CA: Wavefunction, Inc., 1996.

Advanced

Clark, T. *Computational Chemistry.* New York: Wiley-Interscience, 1985.

Fleming, I. *Frontier Orbitals and Organic Chemical Reactions.* New York: John Wiley & Sons, 1976.

Fukui, K. "Recognition of Stereochemical Paths by Orbital Interaction." *Accounts of Chemical Research, 4* (1971): 57.

Hehre, W. J., Random, L., Schleyer, P. v. R., and Pople, J. A. Ab Initio *Molecular Orbital Theory.* New York: Wiley-Interscience, 1986.

Woodward, R. B., and Hoffmann, R. "The Conservation of Orbital Symmetry." *Accounts of Chemical Research, 1* (1968): 17.

Woodward, R. B., and Hoffmann, R. *The Conservation of Orbital Symmetry.* Weinheim: Verlag Chemie, 1970.

19 EXPERIMENT 19

Computational Chemistry

Semiempirical methods

Heats of formation

Mapped surfaces

REQUIRED READING

Review: The sections of your textbook dealing with

19A   Alkene Isomers, Tautomerism, and Regioselectivity—The Zaitsev and Markovnikoff Rules

19B   Nucleophilic Substitution—Relative Rates of Substrates in S_N1 Reactions

19C　　Acids and Bases—Inductive Effects

19D　　Carbocation Stability

19E　　Carbonyl Additions—Frontier Molecular Orbitals

New:　Essay　Computational Chemistry—*ab Initio* and Semiempirical Methods

SPECIAL INSTRUCTIONS

To perform this experiment, you must use computer software that can perform semiempirical molecular orbital calculations at the AM1 or MNDO level. In addition, the later experiments require a program that can display orbital shapes and map various properties onto an electron density surface. Either your instructor will provide instruction in the use of the software or you will be provided with a handout giving instructions.

NOTES TO THE INSTRUCTOR

This series of computational experiments was devised using the programs PC Spartan and MacSpartan; however, it should be possible to use many other implementations of semiempirical molecular orbital theory. Some of the other capable programs for the PC and the MacIntosh include HyperChem Release 5 and CAChe Workstation. You will have to provide your students with an introduction to your specific implementation. The introduction should show students how to build a molecule, how to select and submit calculations and surface models, and how to load and save files.

It is not intended that all these experiments be performed in a single session. They are intended to illustrate what you can do with computational chemistry but are not comprehensive. You may wish to assign them with either specific lecture topics or a particular experiment. Alternatively, you may wish to use them as patterns that students can use to devise their own computational procedures to solve a new problem.

For Experiments 19A and 19B, if your software will not only perform AM1 (or a similar MNDO procedure) but will also perform calculations that include the effect of aqueous solvation (such as AM1-SM2), it may be instructive to have the students work in pairs. One student can perform gasphase calculations and the other can perform the same calculations, including the solvent effect. They can then compare results in their reports.

19A　EXPERIMENT　19A

Heats of Formation: Isomerism, Tautomerism, and Regioselectivity

Part One: Isomerism

The stability of isomers may be directly compared by examining their heats of formation. In separate calculations, build models of *cis*-2-butene, *trans*-2-butene, and 1-butene. Submit each of these to AM1 calculation of the energy (heat of formation). Use the geometry optimization option in each case to find the best possible energy for each one. What do your results suggest? Do they agree with the experimental data given in Experiment 18D (p. 159)?

Part Two: Acetone and Its Enol

In this exercise, we will compare the energies of a pair of tautomers using the heats of formation calculated by the semiempirical AM1 method. These two tautomers can be directly compared because they have the same molecular formula: C_3H_6O. Most organic textbooks discuss the relative stability of ketones and their tautomeric enol forms. For acetone, there are two tautomers in equilibrium:

$$CH_3-\overset{\overset{\displaystyle O}{\|}}{C}-CH_3 \rightleftharpoons CH_3-\overset{\overset{\displaystyle OH}{|}}{C}=CH_2$$

Keto   Enol

In separate calculations, build models of acetone and its enol. Submit each model to AM1 calculation of the energy (heat of formation). Use the geometry optimization option in each case to find the best possible energy for each one.

Experimental results indicate that there is little enol (<0.0002%) in equilibrium with acetone. Do your calculations suggest a reason?

Part Three: Regioselectivity

Ionic addition reactions of alkenes are regioselective. For instance, adding concentrated HCl to 2-methylpropene produces largely 2-chloro-2-methylpropane and a much smaller amount of 1-chloro-2-methylpropane. This can be explained by examining the energies of the two carbocation intermediates that can be formed by adding a proton in the first step of the reaction:

$$\overset{\overset{\displaystyle CH_3}{|}}{\underset{\underset{\displaystyle H}{|}}{H_3C-C-CH_2}} \overset{+H^+}{\longleftarrow} \overset{\overset{\displaystyle CH_3}{|}}{H_3C-C=CH_2} \overset{+H^+}{\longrightarrow} \overset{\overset{\displaystyle CH_3}{|}}{\underset{\underset{\displaystyle +}{}}{H_3C-C-CH_3}}$$

This first step (adding a proton) is the rate-determining step of the reaction, and it is expected that the activation energies for forming these two intermediates will reflect their relative energies. That is, the activation energy leading to the lower energy intermediate will be lower than the activation energy leading to the intermediate that has higher energy. Because of this energy difference, the reaction will predominantly follow the pathway that passes through the lower energy intermediate. Because the two carbocations are isomers, and because both are formed from the same starting material, a direct comparison of their energies (heats of formation) will determine the main course of the reaction.

In separate calculations, build models of the two carbocations and submit them to AM1 calculations of their energies. Use a geometry optimization. When you build the model, most programs will require you to build the skeleton of the hydrocarbon that is closest in structure to the carbocation and then to delete the required hydrogen *and its free valence*.

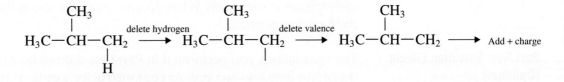

Remember also to assign a positive charge to the molecule before submitting it to calculation. This is usually done in the menus where you select the type of calculation. Compare your results for the two calculations. Which carbocation will lead to the major product? Do your results agree with the prediction made by the Markovnikoff rule?

19B EXPERIMENT 19B

Heats of Reaction: S_N1 *Reaction Rates*

In this experiment, we will attempt to determine the relative rates of selected substrates in the S_N1 reaction. The effect of the degree of substitution will be examined for the following compounds:

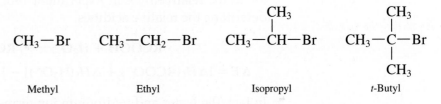

Because the four carbocations are not isomers, we cannot compare their heats of formation directly. To determine the relative rates at which these compounds react, we must determine the *activation energy* required to form the carbocation intermediate in each case. Ionization is the rate-determining step, and we will assume that the activation energy for each ionization should be *similar in magnitude* (Hammond Postulate) to the calculated energy difference between the alkyl halide and the two ions that it forms.

$$R\text{-Br} \longrightarrow R^+ + Br^- \qquad [1]$$

$$\Delta E_{activation} \cong \Delta H_f(products) - \Delta H_f(reactants) \qquad [2]$$

$$\Delta E_{activation} \cong \Delta H_f(R^+) + \Delta H_f(Br^-) - \Delta H_f(RBr) \qquad [3]$$

Because the energy of the bromide ion is a constant, it could be omitted from the calculation, but we will include it because it must be computed only once.

Part One: Ionization Energies Using the AM1 semiempirical level of calculation, compute the energies (heats of formation) of each of the starting materials and record them. Next, compute the energies of each of the carbocations that would result from the ionization of each substrate—follow the instructions given in Part Three of Experiment 19A—and record the results. Be sure to add the positive charge. Finally, compute the energy of the bromide ion, remembering to delete the free valence and add a negative charge. Once all the calculations have been performed, use equation 3 to calculate the energy required to form the carbocation in each case. What do you conclude about the relative rates of the four compounds?

Part Two: Solvation Effects (Optional)

The calculations you performed in Part One did not take the effect of solvation of the ions into account. At your instructor's option (and if you have the correct software), you may be required to repeat your calculations using a computational method that includes stabilization of the ions by solvation. Will solvation increase or decrease the ionization energies? Which will be solvated more, the reactants or the products of the ionization step? What do you conclude from your results?

19C   **E X P E R I M E N T   1 9 C**

Density–Electrostatic Potential Maps— Acidities of Carboxylic Acids

In this experiment, we will compare the acidities of acetic, chloroacetic, and trichloroacetic acid. This experiment could be approached in the same fashion as the relative rates in Experiment 19B, using the ionization energies to determine the relative acidities.

$$RCOOH + H_2O \longrightarrow RCOO^- + H_3O^+$$

$$\Delta E = [\Delta H_f(RCOO^-) + \Delta H_f(H_3O^+)] - [\Delta H_f(RCOOH) + \Delta H_f(H_2O)]$$

In fact, the water and hydronium ion terms could be omitted because they would be constant in each case.

Instead of calculating the ionization energies, we will use a more visual approach involving a property map. Set up an AM1 geometry optimization calculation for each of the acids. In addition, request that an electron-density surface be calculated with the electrostatic potential mapped onto this surface in color. In this procedure, the program plots the density surface and determines the electron density at each point by placing a test positive charge there and determining the coulomb interaction. The surface is colored using the colors of the spectrum—blue is used for positive areas (low electron density), and red is used for more negative areas (higher electron density). This plot will show the polarization of the molecule.

When you have finished the calculations, display all three maps on the screen at the same time. To compare them, you must adjust them all to the same set of color values. This can be done by observing the maximum and minimum values for each map in the surface display menus. Once you have all six values (save them), determine which two numbers give you the maximum and minimum values. Return to the surface plot menu for each of the molecules and readjust the limits of the color values to the same maximum and minimum values. Now the plots will all be adjusted to identical color scales. What do you observe for the carboxyl protons of acetic acid, chloroacetic acid, and trichloroacetic acid? The three minimum values that you saved can be compared to determine the relative electron density at each proton.

19D EXPERIMENT 19D

Density–Electrostatic Potential Maps: Carbocations

Part One: Increasing Substitution

In this experiment, we will use a density map to determine how well a series of carbocations disperses the positive charge. According to theory, increasing the number of alkyl groups attached to the carbocation center helps to spread out the charge (through hyperconjugation) and lowers the energy of the carbocation. We approached this problem from a computational (numerical) angle in Experiment 19B and now will prepare a visual solution to the problem.

Begin by performing an AM1 geometry optimization on methyl, ethyl, isopropyl, and *tert*-butyl carbocations. These carbocations are built as described in Part Three of Experiment 19A. Don't forget to specify that each one has a positive charge. Also ask for a density surface for each one, with the electrostatic potential mapped onto the surface.

When the calculations are completed, display all four density–electrostatic potential maps on the same screen and adjust the color values to the same range as described in Experiment 19C. What do you observe? Is the positive charge as localized in the *tert*-butyl carbocation as in its methyl counterpart?

Part Two: Resonance

Repeat the computational experiment described in Part One, using density–electrostatic potential maps for the allyl and benzyl carbocations. These two experiments can be performed without displaying them both on the same screen. What do you observe about the charge distribution in these two carbocations?

19E EXPERIMENT 19E

Density–LUMO Maps: Reactivities of Carbonyl Groups

In this experiment, we shall investigate how frontier molecular orbital theory applies to the reactivity of a carbonyl compound. Consider the reaction of a nucleophile such as hydride or cyanide with a carbonyl compound.

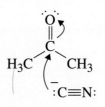

According to frontier molecular orbital theory (see the essay, page 160) the nucleophile, which is donating electrons, must place them in an empty orbital of the carbonyl. Logically, this empty orbital would be the LUMO—the **L**owest (energy) **U**noccupied **M**olecular **O**rbital.

Make a model of acetone and submit it to an AM1 calculation with geometry optimization. Also select two surfaces to display, the LUMO and a mapping of the LUMO on a density surface.

When the calculations are finished, display both surfaces on the screen at the same time. Where is the biggest lobe of the LUMO, on carbon or on oxygen? Where does the nucleophile attack? The density–LUMO surface displays the same thing but with color coding. This plot shows the spot on the surface where the LUMO has its greatest density (largest lobe) as a blue spot on the surface.

Next, continue this experiment by calculating the LUMO and the density–LUMO plots for the ketones 2-cyclohexenone and norbornanone.

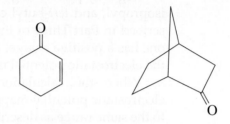

Where are the reactive sites in cyclohexenone? According to the literature, strong bases, such as Grignard reagents, attack the carbonyl, whereas weaker bases or better nucleophiles, such as amines, attack the beta carbon of the double bond, performing a conjugate addition. Can you explain this? Will a nucleophile attack norbornanone from the exo (top) or the endo (bottom) face of the molecule? See Experiment 35 for an answer.

PART 3

Properties and Reactions of Organic Compounds

20 **EXPERIMENT 20**

Reactivities of Some Alkyl Halides

S_N1/S_N2 reactions

Relative rates

Reactivities

The reactivities of alkyl halides in nucleophilic substitution reactions depend on two important factors: reaction conditions and substrate structure. The reactivities of several substrate types will be examined under both S_N1 and S_N2 reaction conditions in this experiment.

Sodium Iodide or Potassium Iodide in Acetone

A reagent composed of sodium iodide or potassium iodide dissolved in acetone is useful in classifying alkyl halides according to their reactivity in an S_N2 reaction. Iodide ion is an excellent nucleophile, and acetone is a nonpolar solvent. The tendency to form a precipitate increases the completeness of the reaction. Sodium iodide and potassium iodide are soluble in acetone, but the corresponding bromides and chlorides are not soluble. Consequently, as bromide ion or chloride ion is produced, the ion is precipitated from the solution. According to LeChâtelier's Principle, the precipitation of a product from the reaction solution drives the equilibrium toward the right; such is the case in the reaction described here:

$$R\text{—}Cl + Na^+I^- \longrightarrow RI + NaCl \text{ (s)}$$

$$R\text{—}Br + Na^+I^- \longrightarrow RI + NaBr \text{ (s)}$$

Silver Nitrate in Ethanol

A reagent composed of silver nitrate dissolved in ethanol is useful in classifying alkyl halides according to their reactivity in an S_N1 reaction. Nitrate ion is a poor nucleophile, and ethanol is a moderately powerful ionizing solvent. The silver ion, because of its ability to coordinate the leaving halide ion to form a silver halide precipitate, greatly assists the ionization of the alkyl halide. Again, a precipitate as one of the reaction products also enhances the reaction.

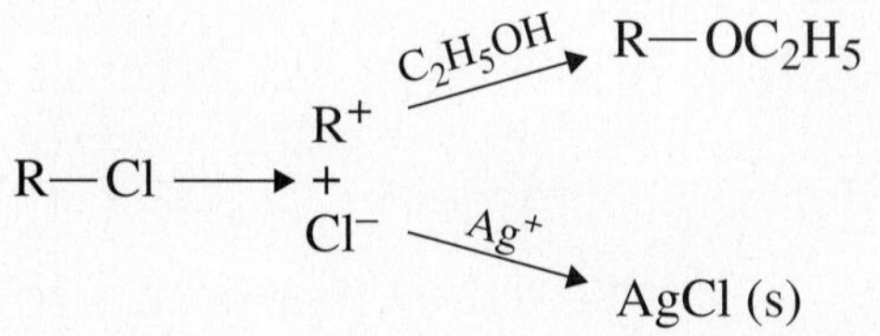

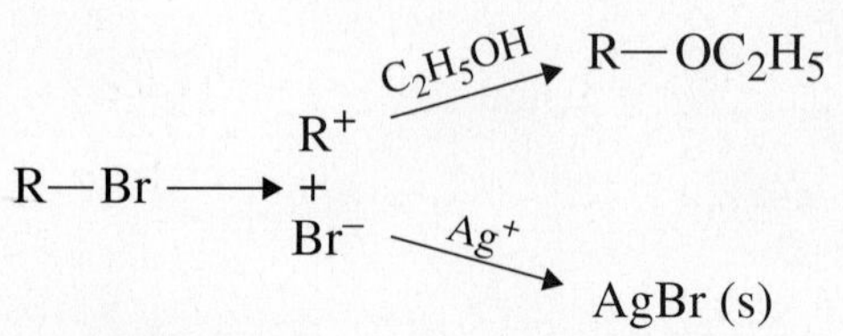

REQUIRED READING

Before beginning this experiment, review the chapters dealing with nucleophilic substitution in your lecture textbook.

SPECIAL INSTRUCTIONS

Some compounds used in this experiment, particularly crotyl chloride and benzyl chloride, are powerful lachrymators. **Lachrymators** cause eye irritation and the formation of tears.

CAUTION

Because some of these compounds are lachrymators, perform these tests in a hood. Be careful to dispose of the test solutions in a waste container marked for halogenated organic waste. After testing, rinse the test tubes with acetone and pour the contents into the same waste container.

SUGGESTED WASTE DISPOSAL

Dispose of all the halide wastes into the container marked for halogenated waste. Any acetone washings should also be placed in the same container.

NOTES TO THE INSTRUCTOR

Each of the halides should be checked with NaI/acetone and AgNO$_3$/ethanol to test for their purity before the class performs this experiment. If molecular modeling software is available, you may wish to assign the exercises included at the end of this experiment.

PROCEDURE

Part A. Sodium Iodide in Acetone

The Experiment

Label a series of ten clean and dry test tubes (10 × 75-mm test tubes may be used) from 1 to 10. In each test tube place 2 mL of a 15% NaI-in-acetone solution. Now add four drops of one of the following halides to the appropriate test tube: (1) 2-chlorobutane, (2) 2-bromobutane, (3) 1-chlorobutane, (4) 1-bromobutane, (5) 2-chloro-2-methylpropane (*t*-butyl chloride), (6) crotyl chloride CH$_3$CH=CHCH$_2$Cl (see Special Instructions), (7) benzyl chloride (α-chlorotoluene) (see Special Instructions), (8) bromobenzene, (9) bromocyclohexane, and (10) bromocyclopentane. Make certain you return the dropper to the proper container to avoid cross-contaminating these halides.

Reaction at Room Temperature

After adding the halide, shake the test tube[1] well to ensure adequate mixing of the alkyl halide and the solvent. Record the times needed for any precipitate or cloudiness to form.

---

[1] Do not use your thumb or a stopper. Instead, hold the top of the test tube between the thumb and index finger of one hand and "flick" the bottom of the test tube using the index finger of your other hand.

Reaction at Elevated Temperature

After about 5 minutes, place any test tubes that do not contain a precipitate in a 50°C water bath. Be careful not to allow the temperature of the water bath to exceed 50°C, because the acetone will evaporate or boil out of the test tube. After about 1 minute of heating, cool the test tubes to room temperature and note whether a reaction has occurred. Record the results.

Observations

Generally, reactive halides give a precipitate within 3 minutes at room temperature, moderately reactive halides give a precipitate when heated, and unreactive halides do not give a precipitate, even after being heated. Ignore any color changes.

Report

Record your results in tabular form in your notebook. Explain why each compound has the reactivity you observed. Explain the reactivities in terms of structure. Compare relative reactivities for compounds of similar structure.

Part B. Silver Nitrate in Ethanol

The Experiment

Label a series of ten clean and dry test tubes from 1 to 10, as described in the previous section. Add 2 mL of a 1% ethanolic silver nitrate solution to each test tube. Now add 4 drops of the appropriate halide to each test tube, using the same numbering scheme indicated for the sodium iodide test. To avoid cross-contaminating these halides, return the dropper to the proper container.

Reaction at Room Temperature

After adding the halide, shake the test tube well to ensure adequate mixing of the alkyl halide and the solvent. After thoroughly mixing the samples, record the times needed for any precipitate or cloudiness to form. Record your results as dense precipitate, cloudiness, or no precipitate/cloudiness.

Reaction at Elevated Temperature

After about 5 minutes, place any test tubes that do not contain a precipitate or cloudiness in a hot water bath at about 100°C. After about 1 minute of heating, cool the test tubes to room temperature and note whether a reaction has occurred. Record your results as dense precipitate, cloudiness, or no precipitate/ cloudiness.

Observations

Reactive halides give a precipitate (or cloudiness) within 3 minutes at room temperature, moderately reactive halides give a precipitate (or cloudiness) when heated, and unreactive halides do not give a precipitate, even after being heated. Ignore any color changes.

Report

Record your results in tabular form in your notebook. Explain why each compound has the reactivity that you observed. Explain the reactivities in terms of structure. Compare relative reactivities for compounds of similar structure.

MOLECULAR MODELING (OPTIONAL)

Many points developed in this experiment can be confirmed through the use of molecular modeling. The following experiments were developed with PC Spartan. It should be possible to use other software, but the instructor may have to make some modifications.

S_N1 Reactivites

Part One

The rate of an S_N1 reaction is related to the energy of the carbocation intermediate that is formed in the rate-determining ionization step of the reaction. It is expected that the activation energy required to form an intermediate is close to the energy of the intermediate. When two intermediates are compared, the activation energy leading to the intermediate of lower energy is expected to be lower than the activation energy leading to the intermediate of higher energy. The easier it is to form the carbocation, the faster the reaction will proceed. An AM1 semiempirical method for determining the approximate energies of carbocation intermediates is described in Experiment 19B, page 171. Complete the computational exercises in Experiment 19B, and compare the calculated results to the experimental results you obtained in this experiment. Do the experimental results parallel the calculated results?

Part Two

Using the density–elpot surface plot described in Experiment 19D, it is possible to compare the amount of charge delocalization in various carbocations through a visualization of the ions. Complete Experiment 19D, and determine whether the charge distributions (delocalization) are what you would expect for the series of carbocations studied.

Part Three

The benzyl (and allyl) halides are a special case; they have resonance. To see how the charge is delocalized in the benzyl carbocation, request two plots: the electrostatic potential mapped onto a density surface and the LUMO mapped onto a density surface. Submit these for calculation at the AM1 semiempirical level. On a piece of paper, draw the resonance-contributing structures for the benzyl cation. Do the computational results agree with the conclusions you draw from your resonance hybrid?

Part Four

Repeat the calculation outlined in Part Three for the benzyl cation; however, in this calculation turn the CH_2 group so that its hydrogens are perpendicular to the plane of the benzene ring. Compare your results to those obtained in Part Three.

S_N2 Reactivites

The problem in the S_N2 reaction is not an electric one but rather a steric problem. Using the AM1 semiempirical method, request a LUMO surface and a density surface for each substrate. The simplest way to visualize the steric problem is to plot the LUMO inside a density surface mapped as a net or a transparent surface. Now imagine having to attack the back lobe of the LUMO. Compare bromomethane, 2-bromo-2-methylpropane (*tert*-butyl bromide), and 1-bromo-2,2-dimethylpropane (neopentyl bromide). Is there any electron density (atoms) in the way of the nucleophile? Request and calculate another surface, mapping the LUMO onto the density surface. What are your conclusions? Can you find the "hot spot" where the nucleophile will attack? Is there any steric hindrance?

QUESTIONS

1. In the tests with sodium iodide in acetone and silver nitrate in ethanol, why should 2-bromobutane react faster than 2-chlorobutane?

2. Why is benzyl chloride reactive in both tests, whereas bromobenzene is unreactive?

3. When benzyl chloride is treated with sodium iodide in acetone, it reacts much faster than 1-chlorobutane, even though both compounds are primary alkyl chlorides. Explain this rate difference.

4. 2-Chlorobutane reacts much more slowly than 2-chloro-2-methylpropane in the silver nitrate test. Explain this difference in reactivity.

5. Bromocyclopentane is more reactive than bromocyclohexane when heated with sodium iodide in acetone. Explain this difference in reactivity.

6. How do you expect the following series of compounds to compare in behavior in the two tests?

$$CH_3-CH=CH-CH_2-Br \qquad CH_3-\underset{\underset{Br}{|}}{C}=CH-CH_3 \qquad CH_3-CH_2-CH_2-CH_2-Br$$

---

21 **EXPERIMENT 21**

Nucleophilic Substitution Reactions: Competing Nucleophiles

Nucleophilic substitution

Heating under reflux

Extraction

Refractometry

Gas chromatography

NMR spectroscopy

The purpose of this experiment is to compare the relative nucleophilicities of chloride ions and bromide ions toward each of the following alcohols: 1-butanol (*n*-butyl alcohol), 2-butanol (*sec*-butyl alcohol), and 2-methyl-2-propanol (*t*-butyl alcohol). The two nucleophiles will be present at the same time in each reaction, in equimolar concentrations, and they will be competing for substrate.

In general, alcohols do not react readily in simple nucleophilic displacement reactions. If they are attacked by nucleophiles directly, hydroxide ion, a strong base, must be displaced. Such a displacement is not energetically favorable and cannot occur to any reasonable extent:

$$X^- + ROH \xrightarrow{\quad\quad} R-X + OH^-$$

To avoid this problem, you must carry out nucleophilic displacement reactions on alcohols in acidic media. In a rapid initial step, the alcohol is protonated; then water, a stable molecule, is displaced. This displacement is energetically favorable, and the reaction proceeds in high yield:

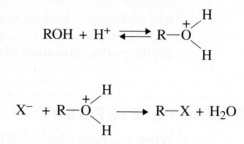

Once the alcohol is protonated, it reacts by either the S_N1 or the S_N2 mechanism, depending on the structure of the alkyl group of the alcohol. For a brief review of these mechanisms, consult the chapters on nucleophilic substitution in your lecture textbook.

You will analyze the products of the three reactions in this experiment by a variety of techniques to determine the relative amounts of alkyl chloride and alkyl bromide formed in each reaction. That is, using equimolar concentrations of chloride ions and bromide ions reacting with 1-butanol, 2-butanol, and 2-methyl-2-propanol, you will try to determine which ion is the better nucleophile. In addition, you will try to determine for which of the three substrates (reactions) this difference is important and whether an S_N1 or S_N2 mechanism predominates in each case.

REQUIRED READING

Review: Techniques 1 through 6

Technique 7	Reaction Methods, Section 7.2, 7.4, 7.5, and 7.8
Technique 12	Extractions, Separations, and Drying Agents, Sections 12.5, 12.9, and 12.11
Technique 22	Gas Chromatography
Technique 24	Refractometry
Technique 21	Nuclear Magnetic Resonance Spectroscopy

Before beginning this experiment, review the appropriate chapters on nucleophilic substitution in your lecture textbook.

SPECIAL INSTRUCTIONS

Each student will carry out the reaction with 2-methyl-2-propanol. Your instructor will also assign you either 1-butanol or 2-butanol. By sharing your results with other students, you will be able to collect data for all three alcohols. You should begin this experiment with Experiment 21A. During the lengthy reflux period, you will be instructed to go on to Experiment 21B. When you have prepared the product of that experiment, you will return to complete Experiment 21A. To analyze the results of both experiments, your instructor will assign specific analysis procedures in Experiment 21C that the class will accomplish.

The solvent–nucleophile medium contains a high concentration of sulfuric acid. Sulfuric acid is corrosive; be careful when handling it.

In each experiment, the longer your product remains in contact with water or aqueous sodium bicarbonate, the greater the risk that your product will decompose, leading to errors in your analytical results. Before coming to class, prepare so that you know exactly what you are supposed to do during the purification stage of the experiment.

SUGGESTED WASTE DISPOSAL

When you have completed the three experiments and all the analyses have been completed, discard any remaining alkyl halide mixture in the organic waste container marked for the disposal of halogenated substances. All aqueous solutions produced in this experiment should be disposed of in the container for aqueous waste.

NOTES TO THE INSTRUCTOR

The solvent–nucleophile medium must be prepared in advance for the entire class. Use the following procedure to prepare the medium.

This procedure will provide enough solvent–nucleophile medium for about 10 students (assuming no spillage or other types of waste). Place 100 g of ice in a 500-mL Erlenmeyer flask and carefully add 76 mL concentrated sulfuric acid. Carefully weigh 19.0 g ammonium chloride and 35.0 g ammonium bromide into a beaker. Crush any lumps of the reagents to powder and then, using a powder funnel, transfer the halides to an Erlenmeyer flask. Carefully add the sulfuric acid mixture to the ammonium salts a little at a time. Swirl the mixture vigorously to dissolve the salts. It will probably be necessary to heat the mixture on a steam bath or a hot plate to achieve total solution. Keep a thermometer in the mixture and make sure that the temperature does not exceed 45°C. If necessary, you may add as much as 10 mL of water at this stage. Do not worry if a few small granules do not dissolve. When solution has been achieved, pour the solution into a container that can be kept warm until all students have taken their portions. The temperature of the mixture must be maintained at about 45°C to prevent precipitation of the salts. Be careful that the solution temperature does not exceed 45°C, however. Place a 20-ml calibrated pipet fitted with a pipet helper in the mixture. The pipet is always left in the mixture to keep it warm.

Be certain that the *tert*-butyl alcohol has been melted before the beginning of the laboratory period.

The gas chromatograph should be prepared as follows: column temperature, 100°C; injection and detector temperature, 130°C; carrier gas flow rate, 50 mL/min. The recommended column is 8 feet long, with a stationary phase such as Carbowax 20M. If you wish to analyze the products from the reaction of *tert*-butyl alcohol (Exp. 21B) by gas chromatography, be sure that the *tert*-butyl halides do not undergo decomposition under the conditions set for the gas chromatograph. *tert*-Butyl bromide is susceptible to elimination.

When analyzing the product from the reaction of *tert*-butyl alcohol by refractometry, it is easy for students to make mistakes in reading the refractive index. It is therefore advisable for students to practice first by analyzing a known liquid.

21A EXPERIMENT 21A

Competitive Nucleophiles
with 1-Butanol or 2-Butanol

PROCEDURE

Apparatus

Assemble an apparatus for reflux using a 20-mL round-bottom flask, a reflux condenser, and a drying tube, as shown in the figure. Loosely insert dry glass wool into the drying tube and then add water dropwise onto the glass wool until it is partially moistened. The moistened glass wool will trap the hydrogen chloride and hydrogen bromide gases produced during the reaction. As an alternative, you can use an external gas trap as described in Technique 7, Section 7.7, Part B, page 608. Do not place the round-bottom flask into the aluminum block until the reaction mixture has been added to the flask. Six Pasteur pipets, two 3-mL conical vials with Teflon cap liners, and a 5-mL conical vial with a Teflon liner should also be assembled for use. All pipets and vials should be clean and dry.

CAUTION

The solvent–nucleophile medium contains a high concentration of sulfuric acid. This liquid will cause severe burns if it touches your skin.

Preparation of Reagents

If a calibrated pipet fitted with a pipet helper is provided, you may adjust the pipet to 10 mL and deliver the solvent–nucleophile medium directly into your 20-mL round-bottom flask (temporarily placed in a beaker for stability). Alternatively, you may use a warm 10-mL graduated cylinder to obtain 10.0 mL of the solvent–nucleophile medium. The graduated cylinder must be warm in order to prevent precipitation of the salts. Heat it by running hot water over the outside of the cylinder or by putting it in the oven for a few minutes. Immediately pour the mixture into the round-bottom flask. With either method, a small portion of the salts in the flask

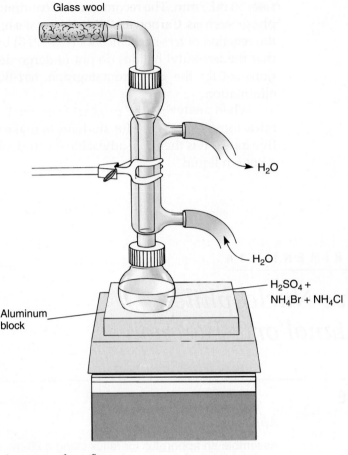

Glass wool

H₂O

H₂O

H₂SO₄ + NH₄Br + NH₄Cl

Aluminum block

Apparatus for reflux.

may precipitate as the solution cools. Do not worry about this; the salts will redissolve during the reaction.

Reflux

Assemble the apparatus shown in the figure. Using the following procedure, add 0.75 mL of 1-butanol (*n*-butyl alcohol) or 0.75 mL of 2-butanol (*sec*-butyl alcohol), depending on which alcohol you were assigned, to the solvent–nucleophile mixture contained in the reflux apparatus. Dispense the alcohol from the automatic pipet or dispensing pump into a 10-mL beaker. Remove the drying tube and, with a 9-inch Pasteur pipet, dispense the alcohol directly into the round-bottom flask by inserting the Pasteur pipet into the opening of the condenser. Also add an inert boiling stone.[1] Replace the drying tube and start circulating the cooling water. Lower the reflux apparatus so that the round-bottom flask is in the aluminum block, as shown in the figure. Adjust the heat so that this mixture maintains a *gentle* boiling action. For 1-butanol, the aluminum block temperature should be about 140°C, and for 2-butanol, the temperature should be about 120°C. Be careful to adjust the reflux ring, if one is visible, so that it remains in the lower fourth of the condenser. Violent boiling will cause loss of product. Continue heating the reaction mixture containing 1-butanol for 75 minutes. Heat the

---

[1] Do not use calcium carbonate–based stones or Boileezers, because they will partially dissolve in the highly acidic reaction mixture.

mixture containing 2-butanol for 60 minutes. During this heating period, go on to Experiment 21B and complete as much of it as possible before returning to this procedure.

Purification

When the period of reflux has been completed, discontinue heating, lift the apparatus out of the aluminum block, and allow the reaction mixture to cool. Do not remove the condenser until the flask is cool. Be careful not to shake the hot solution as you lift it from the heating block or a violent boiling and bubbling action will result; this could allow material to be lost out the top of the condenser. After the mixture has cooled for about 5 minutes, immerse the round-bottom flask (with condenser attached) in a beaker of cold tap water (no ice) and wait for this mixture to cool down to room temperature.

There should be an organic layer present at the top of the reaction mixture. Add 0.75 mL of pentane to the mixture and *gently* swirl the flask. The purpose of the pentane is to increase the volume of the organic layer so that the following operations are easier to accomplish. Using a Pasteur pipet, transfer most (about 7 mL) of the bottom (aqueous) layer to another container. Be careful that all of the top organic layer remains in the boiling flask. Transfer the remaining aqueous layer and the organic layer to a 3-mL conical vial, taking care to leave behind any solids that may have precipitated. Allow the phases to separate and remove the bottom (aqueous) layer using a Pasteur pipet.

NOTE: For the following sequence of steps, be certain to be well prepared. If you find that you are taking longer than 5 minutes to complete the entire extraction sequence, you probably have affected your results adversely!

Add 1.0 mL of water to the vial and *gently* shake this mixture. Allow the layers to separate and remove the aqueous layer, which is still on the bottom. Extract the organic layer with 1–2 mL of saturated sodium bicarbonate solution and remove the bottom aqueous layer.

Drying

Using a clean dry Pasteur pipet, transfer the remaining organic layer into a small test tube (10 × 75 mm) and dry over anhydrous granular sodium sulfate (see Technique 12, Section 12.9, page 680). Transfer the dry halide solution with a clean, dry Pasteur pipet to a small, dry conical vial, taking care not to transfer any solid. The preferred method of storage is to use a Teflon stopper that has been secured *tightly* with a plastic cap. It is helpful to cover the stopper and cap with Parafilm (on the outside of the stopper and cap). Alternatively, you may use a screw-cap vial with a Teflon liner. *Be sure the cap is screwed on tightly.* Again, it is a good idea to cover the cap with Parafilm. Do not store the liquid in a container with a cork or a rubber stopper, because these will absorb the halides. If it is necessary to store the sample overnight, store it in a refrigerator. This sample can now be analyzed by as many of the methods in Experiment 21C as your instructor indicates. However, it cannot be analyzed by refractometry because of the presence of pentane.

21B   **EXPERIMENT 21B**

Competitive Nucleophiles with 2-Methyl-2-Propanol

PROCEDURE

Place 6.0 mL of the solvent–nucleophile medium into a 15-mL centrifuge tube, using the same procedure described in the "Preparation of Reagents" section at the beginning of Experiment 21A. Place the centrifuge tube in cold tap water and wait until a few crystals of ammonium halide salts just begin to appear. Using an automatic pipet or dispensing pump, transfer 1.0 mL of 2-methyl-2-propanol (*tert*-butyl alcohol, mp 25°C) to the 15-mL centrifuge tube. Replace the cap and make sure that it doesn't leak.

CAUTION   ⚠

The solvent–nucleophile mixture contains concentrated sulfuric acid.

Shake the tube vigorously, venting occasionally, for 5 minutes (use gloves). Any solids that were originally present in the centrifuge tube should dissolve during this period. After shaking, allow the layer of alkyl halides to separate (10–15 minutes at most). A fairly distinct top layer containing the products should have formed by this time.

CAUTION   ⚠

***tert*-Butyl halides are volatile and should not be left in an open container any longer than necessary.**

Slowly remove most of the bottom aqueous layer with a Pasteur pipet and transfer it to a beaker. After waiting 10–15 seconds, remove the remaining lower layer in the centrifuge tube, including a small amount of the upper organic layer to be certain that the organic layer is not contaminated by any water.

NOTE: For the following purification sequence, be certain to be well prepared. If you find that you are taking longer than 5 minutes to complete the entire sequence, you probably have affected your results adversely!

Using a dry Pasteur pipet, transfer the remainder of the alkyl halide layer into a small test tube (10 × 75 mm) containing about 0.05 g of solid sodium bicarbonate. As soon as the bubbling stops and a clear liquid is obtained, transfer it with a Pasteur pipet into a small, dry conical vial, taking care not to transfer any solid. The preferred method of storage is to use a Teflon stopper that has been secured *tightly* with a plastic cap. It is helpful to cover the stopper and cap with Parafilm (on the outside of the stopper and cap). Alternatively, you may use a screw-cap vial with a Teflon liner. *Be sure the cap is screwed on tightly.* Again, it is a good idea to cover the cap with Parafilm. Do not store the liquid in a container with a cork or a rubber stopper, because these will absorb the halides. If it is necessary to store the sample overnight, store it in a refrigerator. This sample can now be analyzed by as many of the methods in Experiment 21C as your instructor indicates. When you have finished this procedure, return to Experiment 21A.

21C   EXPERIMENT 21C

Analysis

PROCEDURE

The ratio of 1-chlorobutane to 1-bromobutane, 2-chlorobutane to 2-bromobutane, or *tert*-butyl chloride to *tert*-butyl bromide must be determined. At your instructor's option, you may do this by one of three methods: gas chromatography, refractive index, or NMR spectroscopy. The products obtained from the reactions of 1-butanol and 2-butanol, however, cannot be analyzed by the refractive index method (they contain pentane). The products obtained from the reaction of *tert*-butyl alcohol may be difficult to analyze by gas chromatography because the *tert*-butyl halides sometimes undergo elimination in the gas chromatograph.[1]

Gas Chromatography[2]

The instructor or a laboratory assistant may either make the sample injections or allow you to make them. In the latter case, your instructor will give you adequate instruction beforehand. A reasonable sample size is 2.5 μL. Inject the sample into the gas chromatograph and record the gas chromatogram. The alkyl chloride, because of its greater volatility, has a shorter retention time than the alkyl bromide.

Once the gas chromatogram has been obtained, determine the relative areas of the two peaks (Technique 22, Section 22.12, p. 810). If the gas chromatograph has an integrator, it will report the areas. Triangulation is the preferred method of determining areas, if an integrator is not available. Record the percentages of alkyl chloride and alkyl bromide in the reaction mixture.

Refractive Index

Measure the refractive index of the product mixture (Technique 24). It is easy to make mistakes in reading the scale on some refractometers. Therefore, it is advisable to measure the refractive index of a known liquid before analyzing your mixture. To determine the composition of the mixture, assume a linear relation between

---

[1] *Note to the Instructor:* If pure samples of each product are available, check the assumption used here that the gas chromatograph responds equally to each substance. Response factors (relative sensitivities) are easily determined by injecting an equimolar mixture of products and comparing the peak areas.

[2] *Note to the Instructor:* To obtain reasonable results for the gas chromatographic analysis of the *tert*-butyl halides, it may be necessary to supply the students with response factor correction (Technique 22, Section 22.13, p. 813).

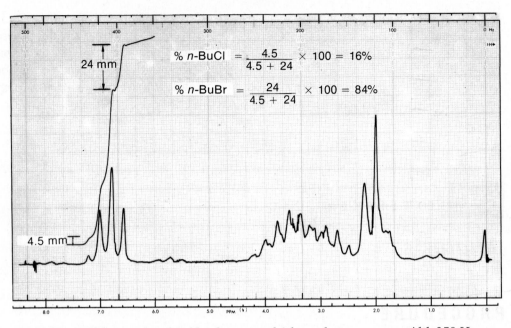

A 60-MHz NMR spectrum of 1-chlorobutane and 1-bromobutane, sweep width 250 Hz (no pentane in sample).

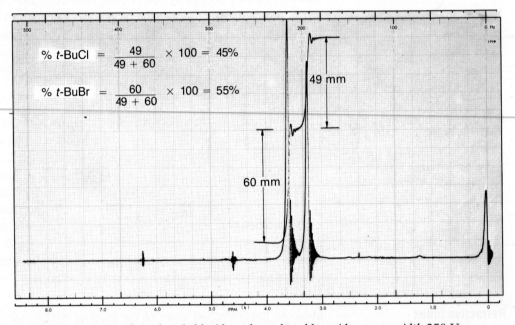

A 60-MHz spectrum of tert-butyl chloride *and* tert-butyl bromide, *sweep width 250 Hz.*

the refractive index and the molar composition of the mixture. At 20°C the refractive indices of the alkyl halides are

tert-butyl chloride	1.3877
tert-butyl bromide	1.4280

If the temperature of the laboratory room is not 20°C, the refractive index must be corrected. Add 0.0004 refractive index unit to the observed reading for each degree above 20°C and subtract the same amount for each degree below this temperature. Record the percentages of alkyl chloride and alkyl bromide in the reaction mixture.

Nuclear Magnetic Resonance Spectroscopy

The instructor or a laboratory assistant will record the NMR spectrum of the reaction mixture.[3] Submit a sample vial containing the mixture for this spectral determination. The spectrum will also contain integration of the important peaks (Technique 21, Nuclear Magnetic Resonance Spectroscopy).

If the substrate alcohol was 1-butanol, the resulting halide and pentane mixture will give rise to a complicated spectrum. Each alkyl halide will show a downfield triplet caused by the CH_2 group nearest the halogen. This triplet will appear farther downfield for the alkyl chloride than for the alkyl bromide. In a 60-MHz spectrum, these triplets will overlap, but one branch of each triplet will be available for comparison. Compare the integral of the *downfield* branch of the triplet for 1-chlorobutane with the *upfield* branch of the triplet for 1-bromobutane. The upper spectrum on the previous page provides an example. The relative heights of these integrals correspond to the relative amounts of each halide in the mixture.

If the substrate alcohol was 2-methyl-2-propanol, the resulting halide mixture will show two peaks in the NMR spectrum. Each halide will show a singlet because all the CH_3 groups are equivalent and are not coupled. In the reaction mixture, the upfield peak is due to *tert*-butyl chloride, and the downfield peak is caused by *tert*-butyl bromide. Compare the integrals of these peaks. The lower spectrum on page 188 provides an example. The relative heights of these integrals correspond to the relative amounts of each halide in the mixture.

REPORT

Record the percentages of alkyl chloride and alkyl bromide in the reaction mixture for each of the three alcohols. You need to share your data from the reaction with 1-butanol or 2-butanol with other students in order to do this. The report must include the percentages of each alkyl halide determined by each method used in this experiment for the two alcohols you studied. On the basis of product distribution, develop an argument for which mechanism (S_N1 or S_N2) predominated for each of the three alcohols studied. The report should also include a discussion of which is the better nucleophile, chloride ion or bromide ion, based on the experimental results. All gas chromatograms, refractive index data, and spectra should be attached to the report.

QUESTIONS

1. Draw complete mechanisms that explain the resultant product distributions observed for the reactions of *tert*-butyl alcohol and 1-butanol under the reaction conditions of this experiment.

2. Which is the better nucleophile, chloride ion or bromide ion? Try to explain this in terms of the nature of the chloride ion and the bromide ion.

3. What is the principal organic by-product of these reactions?

4. A student left some alkyl halides (RCl and RBr) in an open container for several minutes. What happened to the composition of the halide mixture during that time? Assume that some liquid remains in the container.

5. What would happen if all the solids in the nucleophile medium were not dissolved? How might this affect the outcome of the experiment?

6. What might have been the product ratios observed in this experiment if an aprotic solvent such as dimethyl sulfoxide had been used instead of water?

---

[3] It is difficult to determine the ratio of 2-chlorobutane to 2-bromobutane using nuclear magnetic resonance. This method requires at least a 90-MHz instrument. At 300 MHz, all downfield peaks are fully resolved.

7. Explain the order of elution you observed while performing the gas chromatography for this experiment. What property of the product molecules seems to be the most important in determining relative retention times?

8. Does it seem reasonable to you that the refractive index should be a temperature-dependent parameter? Try to explain.

9. When you calculate the percentage composition of the product mixture, exactly what kind of "percentage" (i.e., volume percent, weight percent, mole percent) are you dealing with?

10. When a pure sample of *tert*-butyl bromide is analyzed by gas chromatography, two components are usually observed. One of them is *tert*-butyl bromide and the other one is a decomposition product. As the temperature of the injector is increased, the amount of the decomposition product increases and the amount of *tert*-butyl bromide decreases.

    (a) What is the structure of the decomposition product?

    (b) Why does the amount of decomposition increase with increasing temperature?

    (c) Why does *tert*-butyl bromide decompose much more easily than *tert*-butyl chloride?

---

22 | **EXPERIMENT 22**

Hydrolysis of Some Alkyl Chlorides

Synthesis of an alkyl halide

Use of a separatory funnel

Titration

Kinetics

Two chemical reactions are of interest in this experiment. The first is the preparation of the alkyl chlorides whose hydrolysis rates are to be measured. The chloride formation is a simple nucleophilic substitution reaction carried out in a separatory funnel. Because the concentration of the initial alkyl chloride does not need to be determined for the kinetic experiment, isolation and purification of the alkyl chloride are not required.

$$\text{ROH} + \text{HCl} \longrightarrow \text{R—Cl} + \text{H}_2\text{O}$$

The second reaction is the actual hydrolysis, and the rate of this reaction will be measured. Under the conditions of this experiment, the reaction proceeds by an S_N1 pathway. The reaction rate is monitored by measuring the rate of appearance of hydrochloric acid. The concentration of hydrochloric acid is determined by titration with aqueous sodium hydroxide.

$$\text{R—Cl} + \text{H}_2\text{O} \xrightarrow{\text{solvent}} \text{R—OH} + \text{HCl}$$

The rate equation for the S_N1 hydrolysis of an alkyl chloride is

$$+ \frac{d[\text{HCl}]}{dt} = k[\text{RCl}]$$

Let c equal the initial concentration of RCl. At some time, t, x moles per liter of alkyl chloride will have decomposed, and x moles per liter of HCl will have been produced. The remaining concentration of alkyl chloride at that value of time equals $c - x$. The rate equation becomes

$$+ \frac{dx}{dt} = k(c - x)$$

On integration, this becomes

$$\ln\left(\frac{c}{c - x}\right) = kt$$

which, converted to base 10 logarithms, is

$$2.303 \log\left(\frac{c}{c - x}\right) = kt$$

This equation is of the form appropriate for a straight line $y = mx + b$ with slope m and with intercept b equal to zero. If the reaction is indeed first order, a plot of log $[c/(c - x)]$ versus t will provide a straight line whose slope is $k/2.303$.

Evaluation of the term $c/(c - x)$ remains a problem because it is experimentally difficult to determine the concentration of alkyl chloride. We can, however, determine the concentration of hydrochloric acid produced by titrating it with base. Because the stoichiometry of the reaction indicates that the number of moles of alkyl chloride consumed equals the number of moles of hydrochloric acid produced, c must also equal the number of moles of HCl produced when the reaction has gone to completion (the so-called infinity concentration of HCl), and x equals the number of moles of HCl produced at some particular value at time t. From these equalities, we can rewrite the integrated rate expression in terms of volume of base used in the titration. At the end point of the titration,

Number of moles HCl = number of moles of NaOH

or

x = number of moles NaOH at time t

and

c = number of moles NaOH at time ∞

Number of moles NaOH = [NaOH]V

where V is the volume. Substituting and canceling gives

$$\left(\frac{c}{c - x}\right) = \frac{[\cancel{NaOH}]V_\infty}{[\cancel{NaOH}](V_\infty - V_t)}$$

where V_∞ is the volume of NaOH used when the reaction is complete and V_t is the volume of NaOH used at time t. This integrated rate equation becomes

$$2.303 \log\left(\frac{V_\infty}{V_\infty - V_t}\right) = kt$$

The concentration of the base used in the titration cancels out of this equation so that it is necessary to know neither the concentration of base nor the amount of alkyl chloride used in the experiment.

A plot of log $V_\infty(V_\infty - V_t)$ versus t will provide a straight line whose slope equals $k/2.303$. The slope is determined according to the figure on page 192. If the time is measured in minutes, the units of k are min^{-1}. The experimental points plotted on the graph may contain a certain amount of scatter, but the

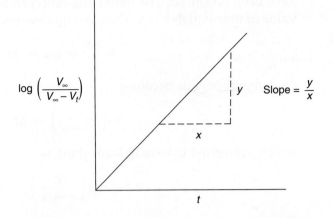

$$\log\left(\frac{V_\infty}{V_\infty - V_t}\right) \qquad \text{Slope} = \frac{y}{x}$$

A plot of log $V_\infty/V_\infty - V_t$ vs. t provides a straight line with a slope equal to k/2.303.

line drawn is the best *straight* line. The line should pass through the origin of the graph. With some reactions, competing processes may cause the line to contain a certain amount of curvature. In these cases, the slope of the initial portion of the line is used before the curvature becomes too important.

One other value often cited in kinetic studies is the **half-life** of the reaction τ. The half-life is the time required for one-half of the reactant to undergo conversion to products. During the first half-life, 50% of the available reactant is consumed. At the end of the second half-life, 75% of the reactant has been consumed. For a first-order reaction, the half-life is calculated by

$$\tau = \frac{\ln 2}{k} = \frac{0.69315}{k}$$

Two alkyl chlorides will be studied by the class in a variety of solvents. The class data will be compared so that relative reactivities of the alkyl chlorides can be determined.

REQUIRED READING

Review: Before beginning this experiment, you should read the material dealing with the methods of kinetics in your lecture textbook.

New: Technique 12 Extractions, Separations, and Drying Agents, especially Section 12.7, page 677

SPECIAL INSTRUCTIONS

You must work in pairs in this experiment in order to make the measurements rapidly. You should alternate jobs on each run, one partner doing the titrations and the other reading the stopwatch and recording the data.

CAUTION

Concentrated hydrochloric acid is corrosive; avoid any direct contact, and avoid breathing the vapors.

SUGGESTED WASTE DISPOSAL

Dispose of any unused acetone solutions of alkyl chlorides in the water container marked for the disposal of halogenated organic waste. All aqueous solutions produced in this experiment should be collected and discarded into a waste container designated for aqueous waste. The concentration of unreacted alkyl chloride that may be present in the reaction mixture (after titration) is too low to warrant any special disposal procedure.

PROCEDURE

Part A. Preparation of the Alkyl Chlorides

Select an alcohol. The choices include *t*-butyl alcohol (2-methyl-2-propanol) and α-phenylethyl alcohol (1-phenylethanol). Place the alcohol (11 mL) in a 125-mL separatory funnel, along with 25 mL of cold, concentrated hydrochloric acid (specific gravity, 1.18; 37.3% hydrogen chloride). Shake the separatory funnel vigorously, venting frequently to relieve any excess pressure, during 30 minutes. Remove the aqueous layer. Wash the organic phase quickly with three 5-mL portions of cold water, followed by a washing with 5 mL of 5% sodium bicarbonate solution. Place the organic product in a small Erlenmeyer flask over 3–4 g of anhydrous calcium chloride. Shake the flask occasionally during 5 minutes. Carefully decant the alkyl chloride from the drying agent into a small Erlenmeyer flask, which can then be stoppered tightly. The alkyl chloride is used in this experiment without prior distillation. Because the true concentration of alkyl chloride introduced into the hydrolysis reaction is determined by titration, it is not necessary to purify the product prepared in this part of the experiment.

Part B. Kinetic Study of the Hydrolysis of an Alkyl Chloride

Because the chlorides hydrolyze rapidly under the conditions used in this experiment, work in pairs to perform the kinetic studies. One partner performs the titrations while the other measures the time and records the data.

Prepare a stock solution of alkyl chloride by dissolving about 0.6 g of alkyl chloride in 50 mL of dry, reagent-grade acetone. Store this solution in a stoppered container to protect it from moisture. Use a 125-mL Erlenmeyer flask to carry out the hydrolysis. The flask should contain a magnetic stirring bar, 50 mL of solvent (see Table 1 for the appropriate solvent), and two to three drops of bromthymol blue indicator. Use absolute ethanol in preparing the aqueous ethanol solvent. Do not use denatured ethanol, because the denaturing agents may interfere with the reactions being studied. Bromthymol blue is yellow in acid solution and blue in alkaline solution.

Place a 50-mL buret filled with approximately 0.01*N* sodium hydroxide above the flask. The exact concentration of sodium hydroxide does not need to be known. Record the initial volume of sodium hydroxide at time *t* equal to 0.0 minutes. Add about 2 mL of sodium hydroxide from the buret to the Erlenmeyer flask and precisely record the new volume in the buret. Start the stirrer. At time 0.0 minutes, *rapidly* add 1.0 mL of the acetone solution of the alkyl chloride from a pipet. Start the timer when the pipet is about half empty. The indicator will undergo a color change, passing from blue through green to yellow when enough hydrogen chloride has been formed in the reaction to neutralize the sodium hydroxide in the flask. Record the time at which the color changed. This color change may not be rapid. Try to use the same color as the end point each time. Add another 2 mL of sodium hydroxide from the buret and precisely record the volume and the time at which this second volume of sodium hydroxide is consumed. Repeat the sodium hydroxide addition twice more (four total). Finally, allow the reaction to go to completion for an hour without excess sodium hydroxide present. Stopper the Erlenmeyer flask during this period.

Table 1 *Experimental conditions*

Compound	Solvent Mixtures (Volume Percentage of Organic Phase in Water)
t-Butyl chloride	40% Ethanol
	25% Acetone
	10% Acetone
α-Phenylethyl chloride	50% Ethanol
	40% Ethanol
	35% Ethanol

Table 2 *Hydrolysis of α-phenylethyl chloride in 50% ethanol*

Time (min)	Vol. NaOH Recorded	Vol. NaOH Used	$V_\infty - V_t$	$\dfrac{V_\infty}{V_\infty - V_t}$	$\ln\left(\dfrac{V_\infty}{V_\infty - V_t}\right)$
0.00	0.2	0.0	6.9	1.00	0.000
8.46	2.2	2.0	4.9	1.41	0.343
18.25	4.2	4.0	2.9	2.37	0.863
31.80	5.9	5.7	1.2	5.75	1.750
47.72	6.8	6.6	0.3	23.00	3.136
100 (∞)	7.1	6.9	0.0	. . .	. . .

After the reaction has gone to completion, *accurately* titrate the amount of hydrogen chloride in solution to the end point. The end point is reached when the color of the solution remains constant for at least 30 seconds. The time corresponding to this final volume is infinity ($t = \infty$). Repeat this process in the other two solvent mixtures indicated in Table 1. These experiments can be carried out while you are waiting for the infinity titration of the previous experiments, provided that a separate buret is used for each run, so that the infinity concentrations of hydrogen chloride produced can be accurately determined.

REPORT

Plot the data according to the method described in the introductory section of this experiment. The rate constant k and the half-life τ must be reported. The report to the instructor should include the plot of the data as well as a table of data. A sample table of data is shown in Table 2. Explain your results, especially the effect of changing the water content of the solvent on the rate of the reaction. If the instructor so desires, the results from the entire class may be compared.

QUESTIONS

1. Plot the data given in Table 2. Determine the rate constant and the half-life for this example.

2. What are the principal by-products of these reactions? Give the rate equations for these competing reactions. Should the production of these by-products go on at the same rate as the hydrolysis reactions? Explain.

3. Compare the energy diagrams for the S_N1 reaction in solvents with two different percentages of water. Explain any differences in the diagrams and their effect on the reaction rate.

4. Compare the expected rates of hydrolysis of *t*-cumyl chloride (2-chloro-2-phenyl-propane) and α-phenylethyl chloride (1-chloro-1-phenylethane) in the same solvent. Explain any differences that might be expected.

---

23 **EXPERIMENT 23**

Synthesis of n-Butyl Bromide and t-Pentyl Chloride

Synthesis of alkyl halides

Extraction

Simple distillation

The synthesis of two alkyl halides from alcohols is the basis for these experiments. In the first experiment, a primary alkyl halide *n*-butyl bromide is prepared as shown in Equation 1.

$$CH_3CH_2CH_2CH_2OH + NaBr + H_2SO_4 \longrightarrow CH_3CH_2CH_2CH_2Br + NaHSO_4 + H_2O \quad [1]$$

n-Butyl alcohol   *n*-Butyl bromide

In the second experiment, a tertiary alkyl halide *t*-pentyl chloride (*t*-amyl chloride) is prepared as shown in Equation 2.

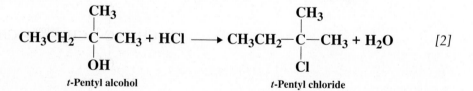

$$[2]$$

t-Pentyl alcohol   *t*-Pentyl chloride

These reactions provide an interesting contrast in mechanisms. The *n*-butyl bromide synthesis proceeds by an S_N2 mechanism, whereas *t*-pentyl chloride is prepared by an S_N1 reaction.

n-Butyl Bromide

The primary alkyl halide *n*-butyl bromide can be prepared easily by allowing *n*-butyl alcohol to react with sodium bromide and sulfuric acid by Equation 1. The sodium bromide reacts with sulfuric acid to produce hydrobromic acid.

$$2\ NaBr + H_2SO_4 \longrightarrow 2\ HBr + Na_2SO_4$$

Excess sulfuric acid serves to shift the equilibrium and thus to speed the reaction by producing a higher concentration of hydrobromic acid. The sulfuric acid also protonates the hydroxyl group of *n*-butyl alcohol so that water is displaced. The acid also protonates the water as it is produced in the reaction and deactivates it as a nucleophile. Deactivation of water keeps the alkyl halide from being converted back to the alcohol by nucleophilic attack of water. The reaction of the primary substrate proceeds via an S_N2 mechanism.

$$CH_3CH_2CH_2CH_2-O-H + H^+ \xrightarrow{\text{fast}} CH_3CH_2CH_2CH_2-\overset{+}{O}-H$$
$$\underset{H}{|}$$

$$CH_3CH_2CH_2CH_2-\overset{+}{\underset{\underset{H}{|}}{O}}-H + Br^- \xrightarrow[S_N2]{\text{slow}} CH_3CH_2CH_2CH_2-Br + H_2O$$

During the isolation of the *n*-butyl bromide, the crude product is washed with sulfuric acid, water, and sodium bicarbonate to remove any remaining acid or *n*-butyl alcohol.

t-Pentyl Chloride

The tertiary alkyl halide can be prepared by allowing *t*-pentyl alcohol to react with concentrated hydrochloric acid according to Equation 2. The reaction is accomplished simply by shaking the two reagents in a sealed conical vial. As the reaction proceeds, the insoluble alkyl halide product forms on upper phase. The reaction of the tertiary substrate occurs via an S_N1 mechanism.

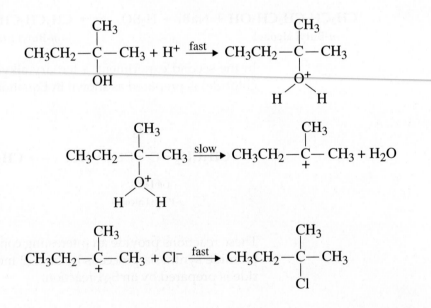

A small amount of alkene, 2-methyl-2-butene, is produced as a by-product in this reaction. If sulfuric acid had been used as it was for *n*-butyl bromide, a much larger amount of this alkene would have been produced.

REQUIRED READING

Review: Techniques 1 through 7

Techniques 12, 13, and 14

SPECIAL INSTRUCTIONS

As your instructor indicates, perform either the *n*-butyl bromide or the *t*-pentyl chloride procedure or both.

SUGGESTED WASTE DISPOSAL

Dispose of all aqueous solutions produced in this experiment in the container for aqueous waste.

EXPERIMENT 23A

n-*Butyl Bromide*

PROCEDURE

Preparation of n-*Butyl Bromide*

Using an automatic pipet or a dispensing pump, place 1.4 mL of *n*-butyl alcohol (1-butanol, *MW* = 74.1) in a preweighed 10-mL round-bottom flask. Reweigh the flask to determine the exact weight of the alcohol. Add 2.4 g of sodium bromide and 2.4 mL of water. Cool the mixture in an ice bath and slowly add 2.0 mL of concentrated sulfuric acid dropwise using a Pasteur pipet. Add a magnetic stirring bar and assemble the reflux apparatus and trap shown in the figure. The trap absorbs the hydrogen bromide gas evolved during the reaction period. While stirring, heat the mixture to a gentle boil (aluminum block temperature about 145°C) for 60–75 minutes.

Extraction

Remove the heat source and allow the apparatus to cool until you can disconnect the round-bottom flask without burning your fingers.

NOTE: Do not let the reaction mixture cool to room temperature. Complete the operations in this paragraph as quickly as possible. Otherwise, salts may precipitate, making this procedure more difficult to perform.

The *n*-butyl bromide layer should be on top. If the reaction is not yet complete, the remaining *n*-butyl alcohol will sometimes form a *second organic layer* on top of the *n*-butyl bromide layer. Treat both organic layers as if they were one. Remove and discard as much of the aqueous (bottom) layer as possible using a Pasteur pipet, but do not remove any of the organic layer (or layers). Ignore the salts during this separation. If they are drawn into the pipet, treat them as part of the aqueous layer. Transfer the remaining liquid to a 5-mL conical vial. Remove and discard any aqueous layer remaining in the conical vial.

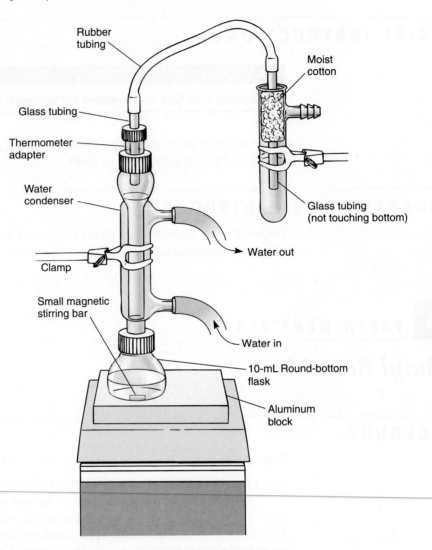

Apparatus for Experiment 23A, n-butyl bromide.

The organic and aqueous layers should separate as described in the following instructions. However, to make sure that you do not discard the wrong layer, it would be a good idea to add a drop of water to any aqueous layer you plan to discard. If a drop of water dissolves in the liquid, you can be confident that it is an aqueous layer. Add 2 mL of $9M$ H_2SO_4 to the conical vial. Cap the vial and shake it *gently*, venting occasionally. Allow the layers to separate. Because any remaining *n*-butyl alcohol is extracted by the H_2SO_4 solution, there should now be only one organic layer. The organic layer should be the top layer. Remove and discard the aqueous (bottom) layer.

Add 2 mL of water to the vial. Cap the vial and shake it *gently*, venting occasionally. Allow the layers to separate. This time, the organic layer should be the bottom layer. The bottom layer may form into a globule (ball) instead of separating cleanly. Use a microspatula to prod the ball gently into the bottom of the vial. Using a Pasteur pipet, transfer the bottom layer (or globule) to a clean 5-mL conical vial. Add 2 mL of saturated aqueous sodium bicarbonate solution, a little at a time, while stirring. Cap the vial and shake it vigorously for 1 minute, venting frequently to relieve any pressure that is produced. Allow the layers to separate and then carefully transfer the lower alkyl halide layer to a dry 3-mL conical vial using a dry Pasteur pipet. Dry the liquid over granular anhydrous sodium sulfate (Technique 12, Section 12.9, page 680).

Distillation

When the solution is dry, transfer it to a clean, dry, 3-mL vial using a Pasteur pipet and distill it (aluminum block about 140°C) using a clean, dry Hickman still (Figure 14.5, p. 708). Each time the Hickman head becomes full, transfer the distillate to a preweighed conical vial using a Pasteur pipet.

When the distillation is complete (one or two drops remaining), weigh the vial, calculate the percentage yield, and determine a microscale boiling point (Technique 13, Section 13.2, p. 695). Determine the infrared spectrum of the product using salt plates (Technique 25, Section 25.2, p. 834). Submit the remainder of your sample in a properly labeled vial, along with the infrared spectrum, when you submit your report to the instructor.

23B EXPERIMENT 23B

n-*Butyl Bromide (Semimicroscale Procedure)*

PROCEDURE

Follow the procedure given in Experiment 23A, except double the amounts of *all* reagents. Use a 25-mL round-bottom flask rather than a 10-mL round-bottom flask for running the reaction. For the separation and extraction procedures, use a screw-cap centrifuge tube in place of a 5-mL conical vial. Distill the crude *n*-butyl bromide using an apparatus similar to the semimicroscale apparatus for a simple distillation (see Figure 14.10). Make the following changes: Use a 5-mL conical vial as the distilling flask and collect the distillate in a preweighed 3-mL conical vial rather than a graduated cylinder. The bulb of the thermometer must be placed below the side arm or it will not read the temperature correctly. All the glassware must be dry. Use a boiling stone or magnetic spin vane to prevent bumping. Collect the material that boils between 85 and 102°C.

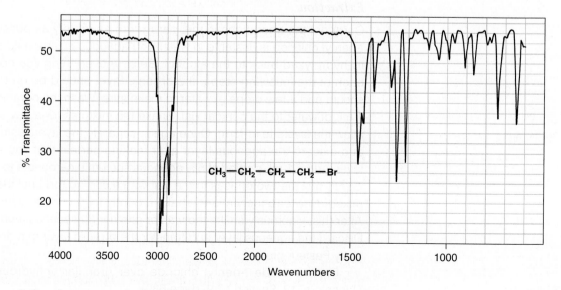

Infrared spectrum of n-*butyl bromide (neat).*

EXPERIMENT 23C

t-Pentyl Chloride (Microscale Procedure)

PROCEDURE

NOTE: In the following procedures, it may be difficult to see the interfaces between layers because the refractive index of the product will be similar to the refractive indices of the extraction solvents.

Preparation of t-Pentyl Chloride

Using an automatic pipet or a dispensing pump, place 1.0 mL of *t*-pentyl alcohol (2-methyl-2-butanol, *MW* = 88.2) in a preweighed 5-mL conical vial. Reweigh the vial to determine the exact weight of alcohol delivered.

NOTE: Before shaking the conical vial vigorously in the next step, be sure that the capped vial does not leak. If it does leak, use a Pasteur pipet to mix the two layers. Draw up as much liquid as possible into the Pasteur pipet and then expel the liquid rapidly back into the conical vial. Continue this mixing for 3–4 minutes.

Add 2.5 mL of concentrated hydrochloric acid, cap the vial, and shake it vigorously for 1 minute. After shaking the vial, loosen the cap and vent the vial. Recap the vial and shake it for 3 minutes more, venting occasionally. Allow the mixture to stand in the vial until the layer of alkyl halide product separates. The *t*-pentyl chloride (*d* = 0.865 g/mL) should be the top layer, but be sure to verify this carefully by observation as you add a few drops of hydrochloric acid.

With a Pasteur pipet, separate the layers by placing the tip of the pipet squarely into the bottom of the vial and removing the lower (aqueous) layer. Discard the aqueous layer. (Are you sure which one it is?)

Extraction

Carry out the operations in this paragraph as rapidly as possible because the *t*-pentyl chloride is unstable in water and aqueous bicarbonate solution. It is easily hydrolyzed back to the alcohol. Be sure everything you need is at hand. In each of the following steps, the organic layer should be on top; however, you should add a few drops of water to make sure. Wash the organic layer by adding 1 mL of water to the conical vial. Shake the mixtures for a few seconds and then allow the layers to re-form. Once again, separate the layers using a Pasteur pipet and discard the aqueous layer after making certain that you saved the proper layer. Add a 1-mL portion of 5% aqueous sodium bicarbonate to the organic layer. *Gently* mix the two phases in the vial with a stirring rod until they are thoroughly mixed. Now cap the vial and shake it gently for 1 minute, venting occasionally. After this, vigorously shake the vial for another 30 seconds, venting occasionally. Discard the aqueous layer and transfer the organic layer to a dry conical vial with a dry Pasteur pipet.

Dry the crude *t*-pentyl chloride over granular anhydrous sodium sulfate (Technique 12, Section 12.9, page 680).

Distillation

When the solution is dry (it should be clear), carefully separate the alkyl halide from the drying agent with a Pasteur pipet and transfer it to a clean, dry 3-mL conical vial. Add a microporous boiling stone and distill the crude *t*-pentyl chloride (Fig. 14.5, p. 708 or, if possible, Figure 14.7B, p. 710).

Using a Pasteur pipet, transfer the product to a dry, preweighed conical vial, weigh it, and calculate the percentage yield. Determine a boiling point for the product using a microscale boiling-point determination (Technique 13, Section 13.2, p. 695). Determine the infrared spectrum of the alkyl halide using salt plates (Technique 25, Section 25.2, p. 834). Submit the remainder of your sample in a properly labeled vial, along with the infrared spectrum, when you submit your report to the instructor.

23D   EXPERIMENT 23D

t-Pentyl Chloride (Semimicroscale Procedure)

PROCEDURE

Follow the procedure as written in Experiment 23C, except double the amounts of *all* reagents. Use a 15-mL screw-cap centrifuge tube instead of a 5-mL conical vial for running the reaction and performing the extractions. Distill the crude *t*-pentyl chloride using an apparatus similar to the semimicroscale apparatus for a simple distillation (see Figure 14.10). Make the following changes: Use a 5-mL

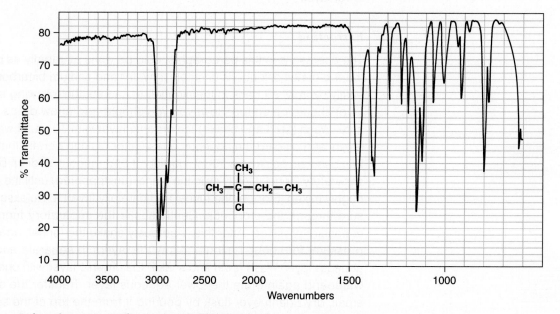

Infrared spectrum of **tert***-pentyl chloride (neat).*

conical vial as the distilling flask and collect the distillate in a preweighed 3-mL conical vial rather than a graduated cylinder. The bulb of the thermometer must be placed below the side arm or it will not read the temperature correctly. All the glassware must be dry. Use a boiling stone or magnetic spin vane to prevent bumping. Collect the material that boils between 80 and 84°C.

EXPERIMENT 23E

t-Pentyl Chloride
(Macroscale Procedure)

PROCEDURE

Preparation of t-Pentyl Chloride

In a 125-mL separatory funnel, place 10.0 mL of *tert*-pentyl alcohol (2-methyl-2-butanol, *MW* = 88.2, *d* = 0.805 g/mL) and 25 mL of concentrated hydrochloric acid (*d* = 1.18 g/mL). Do not stopper the funnel. Gently swirl the mixture in the separatory funnel for about 1 minute. After this period of swirling, stopper the separatory funnel and carefully invert it. Without shaking the separatory funnel, immediately open the stopcock to release the pressure. Close the stopcock, shake the funnel several times, and again release the pressure through the stopcock (Technique 12, Section 12.7, p. 677). Shake the funnel for 2–3 minutes, with occasional venting. Allow the mixture to stand in the separatory funnel until the two layers have completely separated. The *tert*-pentyl chloride (*d* = 0.865 g/mL) should be the top layer, but be sure to verify this by adding a few drops of water. The water should dissolve in the lower (aqueous) layer. Drain and discard the lower layer.

Extraction

The operations in this paragraph should be done as rapidly as possible because the *tert*-pentyl chloride is unstable in water and sodium bicarbonate solution. It is easily hydrolyzed back to the alcohol. In each of the following steps, the organic layer should be on top; however, you should add a few drops of water to make sure. Wash (swirl and shake) the organic layer with 10 mL of water. Separate the layers and discard the aqueous phase after making certain that the proper layer has been saved. Add a 10-mL portion of 5% aqueous sodium bicarbonate to the separatory funnel. Gently swirl the funnel (unstoppered) until the contents are thoroughly mixed. Stopper the funnel and carefully invert it. Release the excess pressure through the stopcock. Gently shake the separatory funnel, with frequent release of pressure. After this, vigorously shake the funnel, again with release of pressure, for about 1 minute. Allow the layers to separate and drain the lower aqueous layer. Wash (swirl and shake) the organic layer with one 10-mL portion of water and again drain the lower aqueous layer. Transfer the organic layer to a small, dry Erlenmeyer flask by pouring it from the top of the separatory funnel.

Dry the crude *tert*-pentyl chloride over 1.0 g of anhydrous calcium chloride until it is clear (Technique 12, Section 12.9, p. 680). Swirl the alkyl halide with the drying agent to aid the drying.

Distillation

Transfer the clear liquid to a dry 25-mL round-bottom flask using a Pasteur pipet. Add a boiling stone and distill the crude *tert*-pentyl chloride in a dry apparatus (Technique 14, Section 14.4, Figure 14.11, p. 714). Collect the pure *tert*-pentyl chloride in a receiver cooled in ice. Collect the material that boils between 78°C and 84°C. Weigh the product, calculate the percentage yield, and determine the boiling point (Technique 13, Section 13.2, pp. 695–699). Determine the infrared spectrum of the product using salt plates (Technique 25, Section 25.2, p. 834). Submit the remainder of your sample in a properly labeled vial, along with the infrared spectrum, when you submit your report to the instructor.

QUESTIONS

n-Butyl Bromide

1. What are the formulas of the salts that precipitate when the reaction mixture is cooled?

2. Why does the alkyl halide layer switch from the top layer to the bottom layer at the point where water is used to extract the organic layer?

3. An ether and an alkene are formed as by-products in this reaction. Draw the structures of these by-products and give mechanisms for their formation.

4. Aqueous sodium bicarbonate was used to wash the crude *n*-butyl bromide.

   (a) What was the purpose of this wash? Give equations.

   (b) Why would it be undesirable to wash the crude halide with aqueous sodium hydroxide?

5. Look up the density of *n*-butyl chloride (1-chlorobutane). Assume that this alkyl halide was prepared instead of the bromide. Decide whether the alkyl chloride would appear as the upper or the lower phase at each stage of the separation procedure: after the reflux, after the addition of water, and after the addition of sodium bicarbonate.

6. Why must the alkyl halide product be dried carefully with anhydrous sodium sulfate before the distillation? (*Hint:* See Technique 15, Section 15.7.)

t-Pentyl Chloride

1. Aqueous sodium bicarbonate was used to wash the crude *t*-pentyl chloride.

   (a) What was the purpose of this wash? Give equations.

   (b) Why would it be undesirable to wash the crude halide with aqueous sodium hydroxide?

2. Some 2-methyl-2-butene may be produced in the reaction as a by-product. Give a mechanism for its production.

3. How is unreacted *t*-pentyl alcohol removed in this experiment? Look up the solubility of the alcohol and the alkyl halide in water.

4. Why must the alkyl halide product be dried carefully with anhydrous sodium sulfate before the distillation? (*Hint:* See Technique 15, Section 15.7.)

5. Will *t*-pentyl chloride (2-chloro-2-methylbutane) float on the surface of water? Look up its density in a handbook.

Elimination Reactions: Dehydration and Dehydrohalogenation

Dehydration
Dehydrobromination
Collection of gaseous products
Gas chromatography
Regiochemistry
Zaitsev's Rule

Alkenes can be prepared by elimination reactions such as the dehydration of alcohols and the dehydrobromination of alkyl bromides. Dehydration reactions follow an E1 mechanistic pathway, whereas dehydrobromination occurs by an E2 mechanism. In this experiment, you will study both methods for preparing alkenes.

In the first reaction, an alcohol undergoes dehydration in the presence of strong acid to form an alkene. In many cases, alcohols give a mixture of alkenes, including *cis* and *trans* isomers. For example, 2-butanol gives a mixture of 1-butene, *cis*-2-butene, and *trans*-2-butene. The class will study the dehydration of 1-butanol and 2-butanol. The general reaction for dehydration is the following:

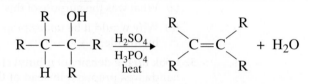

The second reaction is an example of a dehydrobromination reaction conducted in the presence of potassium hydroxide dissolved in ethanol. You will study the dehydrobromination of 1-bromobutane and 2-bromobutane. The general reaction for dehydrobromination is the following:

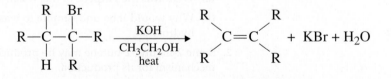

The products of the dehydration and dehydrobromination reactions, which are gases at room temperature, can be analyzed by gas chromatography. For each reaction, the relative percentages of the alkenes can then be calculated. Using your knowledge of the mechanisms of dehydration and dehydrobromination, and by applying Zaitsev's Rule, you should be able to explain the regiochemistry of these reactions.

REQUIRED READING

Review: Techniques 5, 6, and 22

New: Technique 7 Reaction Methods, Section 7.9

Before beginning this experiment, review the appropriate chapters on elimination reactions in your lecture textbook. Pay special attention to dehydration of alcohols, dehydrohalogenation of alkyl halides, E1 and E2 reactions, and Zaitsev's Rule.

SPECIAL INSTRUCTIONS

These experiments can be conveniently scheduled with another experiment because the time required for each experiment is 30–45 minutes. By scheduling these experiments over two laboratory periods, the waiting time for the gas chromatograph will also be minimized. You may be given a choice of doing one of the reactions in this experiment, or your instructor may assign one to you. In either case, you will need to share your results with other students so that you can write the laboratory report.

SUGGESTED WASTE DISPOSAL

Dispose of all halide wastes in the halogenated waste container. The alcohol mixtures from the dehydration reactions should be placed in the container designated for that purpose.

NOTES TO THE INSTRUCTOR

Because of the logistics involved, consider pairing students for this experiment. This experiment was designed to use a specific apparatus for collecting the products (see figure in this experiment). Depending on the type of glassware used by your students, it may be necessary to modify the apparatus described here. See Technique 7, Section 7.9, page 610, for possible modifications. You might find it useful to prepare the gas-collection apparatus in advance of the class and to reuse it in future classes. It is particularly difficult to insert sections of Pasteur pipets into the plastic tubing, and this should be prepared for the students in order to avoid accidents. It is also recommended to keep the 16 × 90-mm sections of glass tubing used for collecting gases.

A Gow-Mac model 69-350 or another traditional type of thermal conductivity gas chromatograph may be prepared as follows: column, injector, detector, and outlet should be at room temperature; carrier gas flow rate, 20 mL/min. An 8-foot column containing 20% DC-710 gives good separation. Retention times: 1-butene, 7 min; *trans*-2-butene, 8 min; and *cis*-2-butene, 9 min.

Conditions for a Hewlett Packard model 5890 with thermal conductivity detector are as follows: 6-foot × ⅛-inch column filled with 3% SP-2100 (a methyl silicone) on 100/120 Supelcoport, available from Supelco; oven temperature cryogenically cooled to −20°C with CO_2; helium carrier gas flow rate, 14.1 mL/min; detector temperature, 101°C; injector temperature, 40°C; column head pressure, 17 psi. A capillary gas-chromatographic column is not recommended. Packed columns, such as the one indicated, work better. A Hewlett Packard model 3393A Integrator may be set at attenuation 3; chart speed 2; area rejection 0; threshold 0; peak width 0.04. Retention times: 1-butene, 1.4 min; *trans*-2-butene, 1.6 min; and *cis*-2-butene, 1.8 min. The percentages of alkenes obtained from each reaction are listed in the Instructor's Manual.

Determine in advance how much gaseous sample to inject into your gas chromatograph. The syringe mentioned in the experimental procedure works well with the Gow-Mac chromatograph. With the Hewlett Packard chromatograph, a 5-μL sample works well. It is recommended that you use a gas-tight syringe for small samples.

PROCEDURE

Apparatus

Assemble the apparatus shown in the figure (p. 207), but do not connect the conical vial to the thermometer adapter (see individual experiments for the size of the conical vial needed). The section of the Pasteur pipet that fits into the thermometer adapter is prepared from a 5¾-inch Pasteur pipet. Your instructor may have prepared the gas-collecting equipment for you. If a collection device is unavailable, then break off the wide end of the pipet about 1 inch from where the constriction begins. Connect the narrow end of the Pasteur pipet to the ½₂-inch flexible plastic tubing. Heat the plastic tubing briefly above a flame before attaching it to the Pasteur pipet. This helps to soften and expand the tubing. While the tubing is still soft, twist it until it is firmly attached to the Pasteur pipet.

CAUTION

Accidental cuts or gouges to the skin may occur if the plastic tubing is not softened prior to attachment.

Insert the wide end of the Pasteur pipet into the thermometer adapter. Use a tight-fitting O-ring to create a better seal around the section of the Pasteur pipet and tighten the cap on the thermometer adapter. Be sure the connection between the pipet and the thermometer adapter is gas-tight. If there is a leak, you will not collect any gaseous products.

Make a mark on both the test tube and the glass tube corresponding to a volume of 4 mL. This can be done by inserting a rubber septum into the 16-mm (O.D.) glass tube and then filling each tube with 4 mL of water. Using a water-resistant marking pen, mark each tube at the level of the water. Fill a 400-mL beaker with water. Place one end of the flexible plastic tubing into the beaker so that it points slightly upward. (You may want to hold it in position by wrapping the plastic tubing with wire or by inserting the plastic tubing into a section of glass tubing bent into a U shape.) Now fill the test tube completely with water and, while holding your thumb over the opening, invert the test tube and place the open end into the beaker. Once the tube is in the water, you can release your thumb and allow the test tube to rest on the bottom of the beaker. Lift the test tube slightly and position it over the end of the plastic tubing without allowing air to enter the test tube. Repeat this filling operation with the glass tube sealed by the rubber septum, but do not insert the flexible tubing into it. The test tube

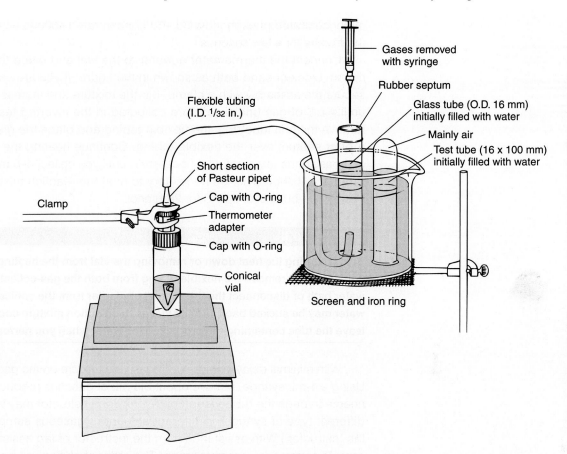

Apparatus for collecting products.

will be used to collect the first 2–4 mL of gas, which will consist mainly of air. Then insert the flexible tubing into the open end of the glass tube to collect the gaseous products.

Proceed to Experiment 24A or 24B and adjust the temperature of the aluminum block or sand bath to the temperatures listed in that experiment.

24A EXPERIMENT 24A

Dehydration of 1-Butanol and 2-Butanol

PROCEDURE

Adjust the aluminum block or sand bath to the following initial temperatures, depending on which alcohol you are using:

140°C for 1-butanol

80°C for 2-butanol

To a 3-mL conical vial, add 0.20 mL of either 1-butanol or 2-butanol and a magnetic spin vane. Using the graduated pipet provided, add 0.30 mL of the mixture of concentrated phosphoric acid and concentrated sulfuric acid[1] to the vial. Stir the mixture for a few seconds.

Connect the thermometer adapter to the vial and place the vial in the aluminum block or sand bath as shown in the figure. If you are using a sand bath, secure the assembly with a clamp. Stir the mixture and increase the heat slowly until 4 mL of gas (mainly air) are collected in the inverted test tube. Carefully remove the test tube from the flexible tubing and place the glass tube with the rubber septum over the flexible tubing. Continue heating the reaction mixture (increasing the temperature if necessary) until you collect 4–5 mL of the gaseous products. It should not be necessary to heat the reaction mixture much above 170°C for 1-butanol or 100°C for 2-butanol.

CAUTION

Before turning the heat down or removing the vial from the heating device, you must first remove the flexible tubing from both the gas-collection tube and water bath or disconnect the thermometer adapter from the conical vial. Otherwise, water may be sucked back into the vial as the reaction mixture cools. Be sure to leave the tube containing the product in the water when you perform this operation.

With minimal delay, analyze your gaseous mixture on the gas chromatograph. Using a 1-mL syringe, remove about 0.5 mL of gaseous product by injecting the needle through the rubber septum. (*Note:* Your instructor may want you to use a different type of syringe or different amount of gaseous sample; see Notes to the Instructor.) With assistance from the instructor or lab assistant, analyze this sample on the gas chromatograph. The order of elution will be 1-butene, *trans*-2-butene, and *cis*-2-butene.

Once the gas chromatograph has been obtained, determine the relative amounts of the products (see Technique 22, Section 22.12, p. 810). Triangulation is the preferred method of determining the relative areas under the peaks (use a millimeter ruler). Record the percentage of the three alkenes in the product.

24B   EXPERIMENT 24B

Dehydrobromination of 1-Bromobutane and 2-Bromobutane

PROCEDURE

Adjust the aluminum block or sand bath to the following initial temperatures, depending on which bromoalkane you are using:

> 90°C for 1-bromobutane
>
> 80°C for 2-bromobutane

---

[1] The acid mixture should be prepared for the entire class by mixing 6.0 mL of concentrated phosphoric acid and 3.0 mL of concentrated sulfuric acid. This mixture will provide enough acid for 20 students, assuming little spillage or other types of waste. Dispense this mixture with a graduated pipet and bulb.

Dehydrobromination of 1-Bromobutane

Using the graduated pipet provided, add 3.0 mL of an ethanolic potassium hydroxide solution[2] to a 5-mL conical vial. Avoid getting any of the base on the ground-glass joint. Put a thin layer of stopcock grease on the ground-glass joint of the thermometer adapter. Add 0.32 mL of 1-bromobutane and a spin vane to the vial. (Alternatively, at your instructor's option, use 3.0 mL of 1*M* potassium *tert*-butoxide solution and 0.32 mL of 1-bromobutane.[3]) Attach the thermometer adapter to the vial and place this assembly in an aluminum block or sand bath as shown in the figure. If using a sand bath, clamp the apparatus to hold it more securely.

While stirring, slowly increase the temperature of the heating source until gas begins to collect in the test tube. Continue heating the reaction mixture at this temperature or slightly higher until about 2 mL of gas (mainly air) are collected in the test tube. Carefully remove the test tube from the flexible tubing and place the glass tube with the rubber septum over the flexible tubing. Continue to heat the reaction mixture until you collect 4 mL of gaseous product in the tube or until gas evolution ceases. Ethanolic potassium hydroxide produces less gaseous product than the reaction with potassium *tert*-butoxide. With minimal delay, analyze your gaseous mixture on the gas chromatograph, as described in Experiment 24A.

CAUTION

Before turning down the heat or removing the heating source, you must first remove the flexible tubing from both the gas-collection tube and water bath or disconnect the thermometer adapter from the conical vial. Otherwise, water may be sucked back into the vial. Be sure to leave the tube containing the product in the water when you perform this operation.

Dehydrobromination of 2-Bromobutane

Using the graduated pipet provided, add 2.0 mL of an ethanolic potassium hydroxide solution[2] to a 3-mL conical vial. Put a thin layer of stopcock grease on the ground-glass joint of the thermometer adapter. Add 0.16 mL of 2-bromobutane and a spin vane to the vial. Attach the thermometer adapter to the vial and place this assembly in an aluminum block or sand bath as shown in the figure. If using a sand bath, clamp the apparatus to hold it more securely.

While stirring, slowly increase the temperature of the heating source until gas begins to collect in the test tube. Continue heating the reaction mixture at this temperature or slightly higher until 2 mL of gas (mainly air) are collected in the tube. Carefully remove the test tube from the flexible tubing and place the glass tube with the rubber septum over the flexible tubing. Continue to heat the reaction mixture until 4–5 mL of gaseous product are collected in the tube.

CAUTION

Read the caution statement on page 208 before turning down the heat.

---

[2] To prepare enough solution for 20 students (assuming little spillage or other types of waste), add 15.0 g of potassium hydroxide to 50.0 mL of 95% ethanol. Stir the mixture until the potassium hydroxide is completely dissolved. Dispense this solution with a graduated pipet and bulb.

[3] The 1*M* potassium *tert*-butoxide in 2-methyl-2-propanol is available from Aldrich Chemical Co., catalog number 33,134-1. Dispense this solution with a graduated pipet and bulb. The reagent is extremely moisture-sensitive and must be kept tightly stoppered when not in use.

With minimal time delay, analyze your gaseous mixture on the gas chromatograph, as described in Experiment 24A. Determine the percentages of three alkenes produced by the dehydrobromination of 2-bromobutane.

REPORT (EXPERIMENTS 24A AND 24B)

The data from all students will be collected and shared with everyone so that your laboratory report will reflect the results obtained by the class. Record your percentage composition and the average percentages of the isomeric butenes obtained by the class for 1-butanol, 2-butanol, 1-bromobutane, and 2-bromobutane. Using your knowledge of the mechanisms of dehydration and dehydrobromination, and by applying Zaitsev's Rule, you should be able to explain the regiochemistry of these reactions. Compare the amount of *trans*-2-butene to *cis*-2-butene and compare the total amount of the 2-butenes to 1-butene. Are these relative amounts consistent with Zaitsev's Rule and the mechanisms of the reactions? Make this comparison for all four compounds and then, in a general way, compare the results obtained from the dehydration of 1-butanol and 2-butanol with the results obtained from the dehydrobromination of 1-bromobutane and 2-bromobutane. Provide mechanisms for each reaction.

REFERENCES

Leone, S. A., and Davis, J. D. "The Dehydrohalogenation of 2-Bromobutane: A Simple Illustration of Anti-Saytzeff Elimination as a Laboratory Experiment for Organic Chemistry." *Journal of Chemical Education, 69* (1992): A175.

Gilow, H. M. "Microscale Elimination Reactions." *Journal of Chemical Education, 69* (1992): A265.

QUESTIONS

1. Give the mechanism for the dehydration of 1-butanol. Why might you expect that the dehydration of 2-butanol would produce a similar composition of alkenes?

2. Give the mechanism for the dehydrobromination of 1-bromobutane in the presence of either potassium *tert*-butoxide or alcoholic potassium hydroxide.

3. Why is there a big difference in the regioselectivity of the dehydration of 1-butanol compared with the dehydrobromination of 1-bromobutane?

4. Explain the order of elution you observed while performing the gas chromatography for this experiment. What property of the product molecules seems to be the most important in determining the relative retention times?

5. What alkenes (including *cis* and *trans* isomers) would be produced by the dehydration of the following alcohols? Where possible, predict the relative amounts of each product according to Zaitsev's Rule.

   (a) 3-Pentanol

   (b) 2-Methyl-2-butanol

   (c) 1-Butanol

   (d) 2-Methyl-1-butanol

6. What alkenes (including *cis* and *trans* isomers) would be produced by the dehydrobromination of the following alkyl bromides? Where possible, predict the relative amounts of each product according to Zaitsev's rule.

   (a) 2-Bromo-2-methylbutane

   (b) 1-Bromobutane

   (c) 2-Bromo-2,3-dimethylbutane

   (d) 3-Bromopentane

EXPERIMENT 25

4-Methylcyclohexene

Preparation of an alkene
Dehydration of an alcohol
Distillation
Bromine and permanganate tests for unsaturation

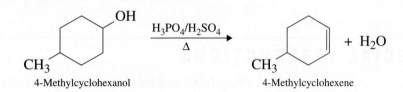

4-Methylcyclohexanol   4-Methylcyclohexene

Alcohol dehydration is an acid-catalyzed reaction performed by strong, concentrated mineral acids such as sulfuric and phosphoric acids. The acid protonates the alcoholic hydroxyl group, permitting it to dissociate as water. Loss of a proton from the intermediate (elimination) brings about an alkene. Because sulfuric acid often causes extensive charring in this reaction, phosphoric acid, which is comparatively free of this problem, is a better choice. To make the reaction proceed faster, however, a minimal amount of sulfuric acid will also be used.

The equilibrium that attends this reaction will be shifted in favor of the product by distilling it from the reaction mixture as it is formed. The 4-methylcyclohexene (bp 101–102°C) will codistill with the water that is also formed. By continuously removing the products, one can obtain a high yield of 4-methylcyclohexene. Because the starting material, 4-methylcyclohexanol, also has a somewhat low boiling point (bp 171– 173°C), the distillation must be done carefully so that the alcohol does not also distill.

Unavoidably, a small amount of phosphoric acid codistills with the product. It is removed by washing the distillate mixture with a saturated sodium chloride solution. This step also partially removes the water from the 4-methylcyclohexene layer; the drying process will be completed by allowing the product to stand over anhydrous sodium sulfate.

Compounds containing double bonds react with a bromine solution (red) to decolorize it. Similarly, they react with a solution of potassium permanganate (purple) to discharge its color and produce a brown precipitate (MnO_2). These reactions are often used as qualitative tests to determine the presence of a double bond in an organic molecule (see Experiment 54). Both tests will be performed on the 4-methylcyclohexene formed in this experiment.

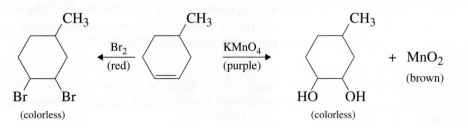

REQUIRED READING

Review: Techniques 1, 2, 3, 5, and 6

Technique 12   Extractions, Separations, and Drying Agents, Sections 12.5 and 12.9

New: Technique 14   Simple Distillation

If performing the optional boiling point or infrared spectroscopy, also read:

Technique 13   Physical Constants of Liquids: The Boiling Point and Density

Technique 25   Infrared Spectroscopy

SPECIAL INSTRUCTIONS

Phosphoric and sulfuric acids are corrosive. Do not allow either acid to touch your skin.

If you must store the 4-methylcyclohexene, place it in a conical vial sealed with a glass or Teflon stopper, an O-ring, and a compression cap. Store it in a freezer. The product is extremely volatile and is easily lost. When you remove the vial from the freezer, loosen the stopper immediately or it may expand and be difficult to remove. WARNING: It is not adequate to store the sample in a conical vial with a septum liner and cap. Conical vials typically have chips on the top edge, and the septum liners, which rest on this edge, do not make an adequate seal to contain the vapors.

SUGGESTED WASTE DISPOSAL

Any organic residues should be discarded in an organic waste container designated for the disposal of *nonhalogenated* wastes. Discard the solutions that remain after the bromine test for unsaturation in an organic waste container designated for the disposal of *halogenated* wastes. The solutions that remain after the potassium permanganate test should be discarded into a waste container specifically marked for the disposal of heavy-metal wastes. Aqueous solutions should be placed in the container designated for that purpose.

NOTE TO THE INSTRUCTOR

Amberlyst-15 ion exchange resin (sulfonic acid groups) may be used in place of the phosphoric and sulfuric acids.[1] There is less charring with the resin. Use 0.2 g of the resin and heat more slowly, increasing the reaction time to about 45 minutes. When measuring the resin (little balls), use a measuring spoon with a depression; the spheres roll off a flat spatula, and static charges sometimes complicate the weighing problem. Provide a waste container for the spent resin.

---

[1] Moeur, H. P., Swatik, S. A., and Pinnell, R. P. "Microscale Dehydration of Cyclohexanol Using a Macroreticular Cation Exchange Resin." *Journal of Chemical Education, 74* (July 1997): 833.

EXPERIMENT 25A

4-Methylcyclohexene
(Microscale Procedure)

PROCEDURE

Apparatus Assembly

Place 1.5 mL of 4-methycyclohexanol ($MW = 114.2$) in a tared 5-mL conical vial and reweigh the vial to determine an accurate weight for the alcohol. Add 0.40 mL of 85% phosphoric acid and six drops of concentrated sulfuric acid to the vial. Mix the liquids thoroughly using a glass stirring rod and add a boiling stone or a magnetic spin vane. Assemble a distillation apparatus as shown in Figure 14.5, page 708 and use a water-cooled condenser. It is recommended that you include the drying tube, filled with calcium chloride, as an odor-control measure.

Dehydration

Start circulating the cooling water in the condenser and heat the mixture until the product begins to distill (aluminum block or sand bath set to about 160–180°C). If you use an aluminum block for heating, place aluminum collars around the conical vial. Stir the mixture if you are using a spin vane instead of a boiling stone. The heating should be regulated so that the distillation requires about 30–45 minutes, heating slowly at the beginning.

During the distillation, use a Pasteur pipet to remove the distillate from the well of the Hickman head when necessary. You must remove the condenser when performing this experiment, unless you have a Hickman head with a side port. In that case, you can remove the distillate through the side port without removing the condenser. Transfer the distillate to a clean, dry, 3-mL conical vial, which should be capped except when liquid is being added. Continue the distillation until no more liquid is being distilled. This can be best determined by observing when boiling in the conical vial has ceased. Also, the volume of liquid remaining in the vial should be about 0.5 mL when distillation is complete.

When distillation is complete, remove as much distillate as possible from the Hickman head and transfer it to the 3-mL conical vial. Then, using a Pasteur pipet with the tip slightly bent, rinse the sides of the inside wall of the Hickman head with 1.0 mL of saturated sodium chloride. Do this thoroughly so that as much liquid as possible is washed down into the well of the Hickman head. Transfer this liquid to the 3-mL conical vial.

Isolation and Drying of Product

Allow the layers to separate and remove the bottom aqueous layer. Using a dry Pasteur pipet, transfer the organic layer to a small test tube, and dry it over granular anhydrous sodium sulfate (Technique 12, Section 12.9, page 680). Place a stopper in the test tube and set it aside for 10–15 minutes to remove the last traces of water. Carefully transfer as much distillate as possible to a small tared conical vial with a cap. Weigh the product ($MW = 96.2$) and calculate the percentage yield.

Boiling-Point Determination and Spectroscopy[2]

At the instructor's option, determine a more accurate boiling point on your sample using the microboiling-point method (Technique 13, Section 13.2, p. 695), and obtain the infrared spectrum of 4-methylcyclohexene (Technique 25, Section 25.2, p. 834 or Section 25.3, p. 836). Because 4-methylcyclohexene is so volatile, you must work quickly to obtain a good spectrum using sodium chloride plates. A better method is to use silver chloride plates with the orientation shown in Figure 25.3B, page 837. Compare the spectrum with the one shown in this experiment. After performing the tests on page 216, submit your sample, along with the report, to the instructor.

25B   **EXPERIMENT 25B**

4-Methylcyclohexene
(Semimicroscale Procedure)

PROCEDURE

Apparatus Assembly

Assemble a distillation apparatus as shown in Figure 14.10, page 713, but insert a water condenser as shown in Figure 14.11, page 714 and insert the thermometer a bit lower than shown so that the mercury bulb is inside the ground-glass joint. Place 4.0 mL of 4-methylcyclohexanol ($MW = 114.2$) in a 10-mL graduated cylinder and weigh it. Transfer the alcohol to the 10-mL round-bottom flask and reweigh the graduated cylinder to determine an accurate weight for the alcohol. Add 1.0 mL of 85% phosphoric acid and 16 drops of concentrated sulfuric acid to the alcohol already in the round-bottom flask. Without delay, mix the contents of the round-bottom flask thoroughly by swirling the liquids in the flask. Add a corundum (black) boiling stone or a stirring bar and reconnect the flask to the distillation apparatus.

Dehydration

Start circulating cooling water in the condenser and heat the mixture until the product begins to distill (aluminum block approximately 180°C). If you are using a stirring bar, stir the mixture. The heating should be regulated so that the distillation requires about 45 minutes to be completed. Distill slowly and monitor (record) the temperature while the liquid distills; the products should distill over a range of 100–105°C. Continue distillation until no more liquid is collected. This can best be

---

[2] An option is to submit the student samples to analysis by gas chromatography–mass spectrometry. If GC-MS is used, students can observe that several products besides 4-methylcyclohexene are formed in the reaction. Discussion of the mechanism of rearrangement should then be included.

determined by observing when the boiling in the round-bottom flask has ceased. Approximately 1 mL of a dark brown liquid will remain undistilled in the round-bottom flask when the distillation is complete.

When the distillation is complete, open the joint between the distillation head and the condenser, and then tilt the condenser so that any liquid trapped in the joints of the condenser will drip into the collection vial. If droplets of distillate are evident clinging to the inside tube of the condenser, wash them down with a small amount of saturated aqueous sodium chloride solution.

Isolation and Drying of the Product

Using a Pasteur pipet, transfer the distillate from the conical vial to a 15-mL capped centrifuge tube, and add approximately 2.0 mL of a saturated sodium chloride solution. Cap the tube and gently invert it several times with venting. Allow the layers to separate and remove the lower aqueous layer with a Pasteur pipet. Transfer the organic layer to a small test tube or Erlenmeyer flask, and dry it over granular anhydrous sodium sulfate (Technique 12, Section 12.9, page 680). Place a stopper in the test tube (or Erlenmeyer flask), and set it aside for 10–15 minutes to remove the last traces of water. Transfer the dry distillate to a small tared conical vial with a cap. Weigh the product ($MW = 96.2$) and calculate the percentage yield.

Boiling-Point Determination and Spectroscopy

At the instructor's option, determine a more accurate boiling point on your sample using the micro boiling-point method (Technique 13, Section 13.2, p. 695) and obtain the infrared spectrum of 4-methylcyclohexene (Technique 25, Section 25.2, p. 834 or Section 25.3, p. 836). Because 4-methylcyclohexene is so volatile, you must work quickly to obtain a good spectrum using sodium chloride plates. A better method is to use silver chloride plates with the orientation shown in Figure 25.3B, page 837. Compare the spectrum with the one shown in this experiment. After performing the following tests, submit your sample, along with the report, to the instructor.

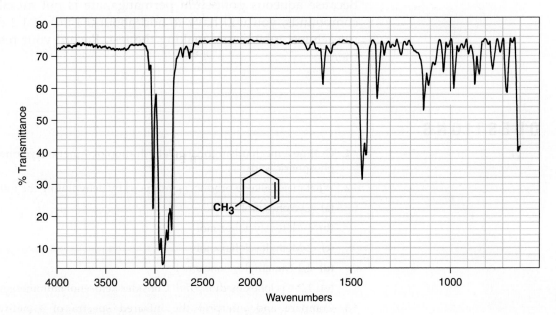

Infrared spectrum of 4-methylcyclohexene (neat).

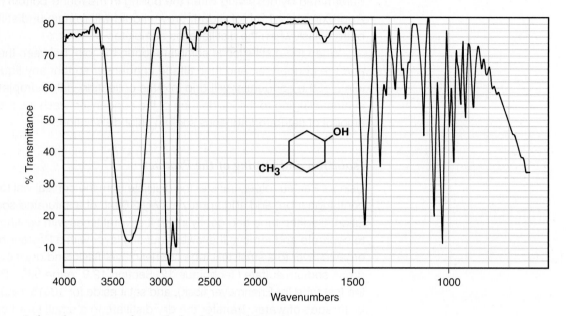

Infrared spectrum of 4-methylcyclohexanol (neat).

UNSATURATION TESTS

Place four to five drops of 4-methycyclohexanol in each of two small test tubes. In each of another pair of small test tubes, place four to five drops of the 4-methylcyclohexene you prepared. Do not confuse the test tubes. Take one test tube from each group, and add a solution of bromine in carbon tetrachloride or methylene chloride, drop by drop, to the contents of the test tube, until the red color is no longer discharged. Record the result in each case, including the number of drops required. Test the remaining two test tubes in a similar fashion with a solution of potassium permanganate. Because aqueous potassium permanganate is not miscible with organic compounds, you will have to add about 0.3 mL of 1,2-dimethoxyethane to each test tube before making the test. Record your results and explain them.

QUESTIONS

1. Outline a mechanism for the dehydration of 4-methylcyclohexanol catalyzed by phosphoric acid.

2. What major alkene product is produced by dehydrating the following alcohols?

   (a) Cyclohexanol

   (b) 1-Methylcyclohexanol

   (c) 2-Methylcyclohexanol

   (d) 2,2-Dimethylcyclohexanol

   (e) 1,2-Cyclohexanediol (*Hint:* Consider keto–enol tautomerism.)

3. Compare and interpret the infrared spectra of 4-methylcyclohexene and 4-methylcyclohexanol.

4. Identify the C-H out-of-plane bending vibrations in the infrared spectrum of 4-methylcyclohexene. What structural information can be obtained from these bands?

5. In this experiment, 1.0 mL of saturated sodium chloride is used to rinse the Hickman head after the initial distillation. Why is saturated sodium chloride, rather than pure water, used for this procedure and the subsequent washing of the organic layer?

26   EXPERIMENT 26

Phase-Transfer Catalysis: Addition of Dichlorocarbene to Cyclohexene

Carbene formation

Phase-transfer catalysis

It has long been known that a haloform CHX_3 will react with a strong base to give a highly reactive carbene species CX_2 by Reactions 1 and 2. In the presence of an alkene, this carbene adds to the double bond to produce a cyclopropane ring (Reaction 3).

$$X\!-\!\overset{\displaystyle X}{\underset{\displaystyle X}{C}}\!-\!H + {}^-:\!\ddot{O}\!-\!H \rightleftharpoons X\!-\!\overset{\displaystyle X}{\underset{\displaystyle X}{C}}\!:{}^- + H\ddot{O}H \qquad [1]$$

$$X\!-\!\overset{\displaystyle X}{\underset{\displaystyle X}{C}}\!:{}^- \overset{\text{slow}}{\rightleftharpoons} X\!-\!\overset{\displaystyle X}{C}\!: + :\ddot{X}\!:{}^- \qquad [2]$$

$$X\!-\!\overset{\displaystyle }{\underset{\displaystyle X}{C}}\!: + \!\!\diagup\!\!\overset{}{C}\!=\!\overset{}{C}\!\!\diagdown\!\! \longrightarrow \!\!-\!\overset{|}{\underset{\diagdown}{C}}\!\!-\!\!\underset{\diagup}{\overset{|}{C}}\!\!-\!\!\underset{\underset{X \quad X}{C}}{} \qquad [3]$$

Traditionally, the reaction has been carried out in *one homogeneous phase* in anhydrous *t*-butyl alcohol solvent, using *t*-butoxide ion as the base [*t*-Bu = $C(CH_3)_3$].

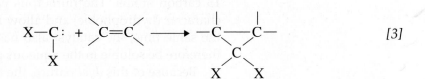

$$HCX_3 + {}^-Ot\text{-}Bu + \!\!\diagup\!\!\overset{}{C}\!=\!\overset{}{C}\!\!\diagdown\!\! \xrightarrow[\text{(solvent)}]{t\text{-BuOH}} \!\!-\!\overset{|}{\underset{\diagdown}{C}}\!\!-\!\!\underset{\diagup}{\overset{|}{C}}\!\!-\!\! + HOt\text{-}Bu + X^-$$

Haloform

Unfortunately, this technique requires time and effort to give good results. In addition, water must be avoided to prevent conversion of the haloform and carbene to formate ion and carbon monoxide by the undesirable base-catalyzed Reactions 4 and 5.

$$\begin{array}{c} \text{X} \\ | \\ \text{H—C—X} \\ | \\ \text{X} \end{array} + 2\ \text{H}_2\text{O} \longrightarrow \begin{array}{c} \text{H—C—OH} \\ \| \\ \text{O} \end{array} + 3\ \text{HX} \qquad [4]$$

$$\begin{array}{c} \text{H—C—OH} \\ \| \\ \text{O} \end{array} + {}^-\text{OH} \longrightarrow \begin{array}{c} \text{H—C—O}^- \\ \| \\ \text{O} \end{array} + \text{H}_2\text{O}$$

$$:CX_2 + H_2O \xrightarrow{\ ^-OH\ } :C\!\!\equiv\!\!O + 2\ HX \qquad [5]$$

Quaternary Ammonium Salt Catalysis

As an alternative to a homogeneous reaction, a *two-phase* reaction can be considered when the organic phase contains the alkene and a haloform CHX_3 and the aqueous phase contains the base OH^-. Unfortunately, under these conditions the reaction will be slow because the two primary reactants, CHX_3 and OH^-, are in different phases. The reaction rate can be substantially increased, however, by adding a quarternary ammonium salt such as benzyltriethylammonium chloride as a **phase-transfer catalyst.**

$$\underset{\text{benzene ring}}{\bigotimes}\!-\text{CH}_2\!-\!{}^+\!\overset{\displaystyle \text{CH}_2\text{CH}_3}{\underset{\displaystyle \text{CH}_2\text{CH}_3}{\text{N}}}\!-\text{CH}_2\text{CH}_3 \ + \ \text{Cl}^-$$

A phase-transfer catalyst: Benzyltriethylammonium chloride

Other common catalysts are tetrabutylammonium bisulfate, triocetyl-methylammonium chloride, and cetyltrimethylammonium chloride. All these catalysts, including benzyltriethylammonium chloride, have at least 13 carbon atoms. The numerous carbon atoms give the catalyst organic character (hydrophobic) and allow it to be soluble in the organic phase. At the same time, the catalyst also has ionic character (hydrophilic) and can therefore be soluble in the aqueous phase.

Because of this *dual* nature, the large cation can cross the phase boundary efficiently and transport a hydroxide ion from the aqueous phase to the organic phase (see figure). Once in the organic phase, the hydroxide ion will react with the haloform to give dihalocarbene by Reactions 1 and 2. Water, a product of the reaction, will move from the organic phase to the aqueous phase, thus keeping the water concentration in the organic phase at a low level. Because the water content in the organic phase is low, it will not interfere with the desirable reaction of the carbene with an alkene by Reaction 3. Thus, the undesirable side Reactions 4 and 5 are minimized. Finally, the halide ion, which is also produced in Reactions 1 and 2, is transported to the aqueous phase by the tetraalkylammonium cation. In this way, electrical neutrality is maintained, and the phase-transfer catalyst, R_4N^+, is returned to the aqueous phase to repeat the whole procedure. The figure shown on the next page summarizes the overall process. This process probably goes on at the interface rather than in the bulk, organic phase.

There are numerous examples of other reactions that might be effectively accelerated by a quaternary ammonium salt or other phase-transfer catalyst (see References). These reactions often involve simple experimental techniques, give shorter reaction times than noncatalyzed reactions, and avoid relatively expensive aprotic solvents that have been widely used to give one phase. Examples of reactions are shown.

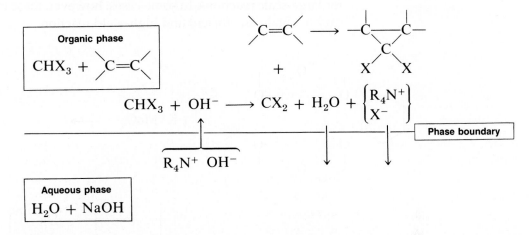

A two-phase reaction. The organic phase contains the alkene and the haloform, CHX₃, whereas the aqueous phase contains the base, OH⁻.

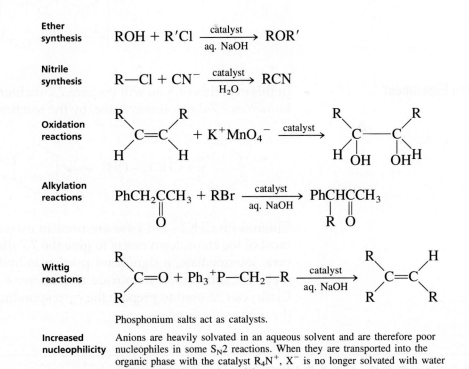

Phosphonium salts act as catalysts.

| Increased nucleophilicity | Anions are heavily solvated in an aqueous solvent and are therefore poor nucleophiles in some S_N2 reactions. When they are transported into the organic phase with the catalyst R_4N^+, X^- is no longer solvated with water and may have increased reactivity. |

Crown Ether Catalysis

Another important class of phase-transfer catalysts includes the crown ethers (not used in this experiment). Crown ethers are used to dissolve organic and inorganic alkali metal salts in organic solvents. The crown ether complexes the cation and provides it with an organic exterior (hydrophobic) so that it is soluble in organic solvents. The anion is carried along into solution as the counterion. One example of a crown ether is dicyclohexyl-18-crown-6. Potassium permanganate $KMnO_4$ complexed to the crown ether is soluble in benzene and is known as purple benzene. It is useful in various oxidation reactions.

The crown ethers catalyze many of the same types of reactions listed in the preceding section on quaternary ammonium salt catalysis. Crown ethers are expensive relative to ammonium salts and are not used as widely for large-scale reactions. In some cases, however, these ethers may be necessary to obtain an efficient and high-yield reaction.

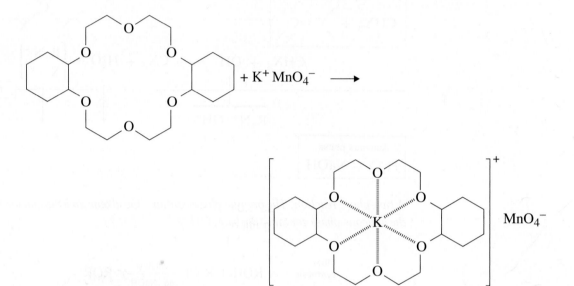

The Experiment

In this experiment, you will prepare 7,7-dichlorobicyclo[4.1.0]heptane, also known as 7,7-dichloronorcarane, by the reaction

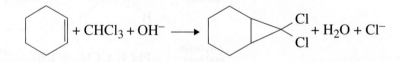

Chloroform $CHCl_3$ and base are used in excess in this reaction. Although most of the chloroform reacts to give the 7,7-dichloronorcarane via the carbene intermediate, a significant portion is hydrolyzed by the base to formate ion and carbon monoxide (Equations 4 and 5, p. 218). Bromoform $CHBr_3$ can be used to prepare the corresponding 7,7-dibromonorcarane via the dibromocarbene.

REQUIRED READING

Review:	Technique 12	Sections 12.4, 12.9, and 12.10
	Technique 25	Sections 25.2, 25.9, and 25.10
New:	Technique 26	Nuclear Magnetic Resonance Spectroscopy
	Technique 27	Carbon-13 Nuclear Magnetic Resonance Spectroscopy

SPECIAL INSTRUCTIONS

> **CAUTION** ⚠
>
> **Chloroform is a suspected carcinogen; therefore, do not let it touch your skin and avoid breathing the vapor.**

This entire experiment must be performed in a hood. Avoid contact with the caustic 50% aqueous sodium hydroxide.

SUGGESTED WASTE DISPOSAL

Dispose of all aqueous solutions produced in this experiment in the container for aqueous waste.

PROCEDURE

Reaction Mixture

Preweigh a 5-mL conical vial with cap and transfer 0.40 mL of cyclohexene ($MW = 82.2$) to the vial in a hood. Cap the vial and reweigh it to determine the weight of cyclohexene. Add 1.0 mL of 50% aqueous sodium hydroxide[1] to the vial, being careful to avoid getting any solution on the glass joint. In a hood, add 1.0 mL of chloroform ($MW = 119.4$, $d = 1.49$ g/mL) to the conical vial. Add a magnetic spin vane to the vial. Weigh 0.040 g of the phase-transfer catalyst, benzyltriethylammonium chloride, on a smooth piece of paper and *reclose the bottle* immediately. (It is hygroscopic!)[2] Add the catalyst to the vial and cap it.

> **CAUTION** ⚠
>
> **Chloroform and cyclohexene should be kept in a hood. Do your measuring and transferring operations in the hood. Avoid contact with these substances. Do not breathe the vapors.**

Reaction Period

Prepare a hot water bath at 40°C using a 250-mL beaker and hot plate. Attach an air condenser to the vial containing the reaction mixture. Clamp the condenser so that the vial is immersed in the water bath. Stir the mixture as *rapidly* as possible for 1 hour. An emulsion forms during this time.

Extraction of Product

After this reaction time, remove the vial from the water bath and allow the reaction mixture to cool to room temperature. Using a Pasteur pipet, transfer the reaction mixture to a centrifuge tube with a screw cap (you do not need to remove the spin vane). Add 1.5 mL of water and 1.0 mL of methylene chloride to the mixture. Cap the tube and shake the mixture *gently* for about 30 seconds. Allow the layers

---

[1] This reagent should be prepared by the instructor. Dissolve 15 g of sodium hydroxide in 15 mL of water. Cool the solution to room temperature and store it in a plastic bottle.

[2] *Note to the Instructor:* The activity of benzyltriethylammonium chloride varies, depending on the source of the catalyst. Try the reaction in advance of the laboratory session to make sure it works properly. We use Aldrich Chemical Co., #14,655-2.

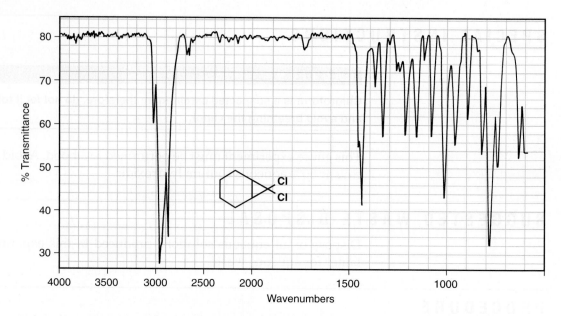

Infrared spectrum of 7,7-dichloronorcarane (neat).

to separate. Swirl the tube gently to help break up the emulsion. Remove the lower methylene chloride layer with a Pasteur pipet and transfer it to a 5-mL conical vial. The small amount of emulsion that forms at the interface should be left behind with the aqueous layer. Add another 1.0-mL portion of methylene chloride, shake the mixture for 30 seconds, remove the lower organic layer with the Pasteur pipet, and transfer the organic layer to the same storage vial. Discard the remaining aqueous layer in the centrifuge tube (avoid contact with the liquid because it is very basic). Discard it into the container for aqueous waste. Clean the centrifuge tube.

Transfer the methylene chloride extracts from the 5-mL conical vial to the centrifuge tube. Add 1.0 mL of saturated aqueous sodium chloride to the tube and shake the mixture for 30 seconds. Using a dry Pasteur pipet, transfer the lower organic layer to a dry conical vial and dry it over granular anhydrous sodium sulfate (Technique 12, Section 12.9, page 680). Cap the vial and swirl it occasionally for at least 10 minutes to dry the organic layer.

Evaporation of Solvent

Transfer the dried organic layer with a dry Pasteur pipet to a dried preweighed conical vial. Evaporate the methylene chloride, together with any remaining cyclohexene and chloroform, in a hood in a warm water bath.[3] Use a stream of dry air or nitrogen to aid the evaporation process. *Be careful or you may also evaporate the product!* The product, 7,7-dichloronorcarane, is a liquid. Continue the evaporation until the level of the liquid is no longer changing (about 0.2 mL).

Analysis of Product

After removal of methylene chloride, you are left with 7,7-dichloronorcarane of sufficient purity for spectroscopy. Weigh the conical vial, determine the weight of product, and calculate the percentage yield. Obtain the infrared spectrum (Technique 25, Section 25.2, p. 834). At the option of the instructor, obtain the decoupled and coupled carbon-13 spectra of your product. Submit any remaining product in a labeled vial with your laboratory report.

---

[3] Alternatively, you may set the uncapped vial in a hood overnight. There is less risk of evaporating the product with this method.

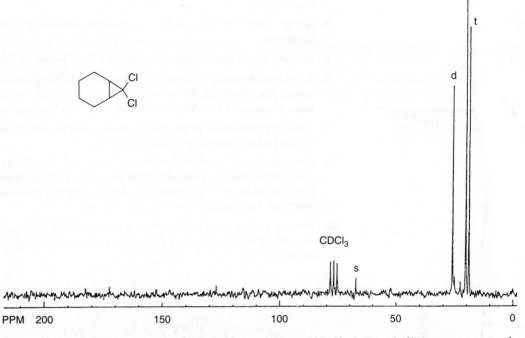

Decoupled carbon-13 spectrum of 7,7-dichloronorcarane, CDCl₃. Letters indicate appearance of the spectrum when carbons are coupled to protons (s = singlet, d = doublet, t = triplet.)

REFERENCES

Dehmlow, E. V. "Phase-Transfer Catalyzed Two-Phase Reactions in Preparative Organic Chemistry." *Angewandte Chemie, International Edition in English, 13* (1974): 170.

Dehmlow, E. V. "Advances in Phase-Transfer Catalysis." *Angewandte Chemie, International Edition in English, 16* (1977): 493.

Dehmlow, E. V., and Dehmlow, S. S. *Phase-Transfer Catalysis,* 3rd ed. Weinheim: VCH Publishers, 1992.

Gokel, G. W., and Weber, W. P. "Phase Transfer Catalysis." *Journal of Chemical Education, 55* (1978): 350, 429.

Lee, A. W. M., and Yip, W. C. "Fabric Softeners as Phase-Transfer Catalysts in Organic Synthesis." *Journal of Chemical Education, 68* (1991): 69.

Makosza, M., and Wawrzyniewicz, M. "Catalytic Method for Preparation of Dichloro-cyclopropane Derivatives in Aqueous Medium." *Tetrahedron Letters* (1969): 4659.

Starks, C. M. "Phase-Transfer Catalysis." *Journal of the American Chemical Society, 93* (1971): 195.

Starks, C. M., Liotta, C. L., and Halpern, M. *Phase-Transfer Catalysis.* New York: Chapman and Hall, 1994.

Weber, W. P., and Gokel, G. W. *Phase Transfer Catalysis in Organic Synthesis.* Berlin: Springer-Verlag, 1977.

QUESTIONS

1. Why did you need to stir the mixture vigorously during reaction?

2. Why did you wash the organic phase with saturated sodium chloride solution?

3. What short chemical test could you make on the product to indicate whether cyclohexene is present or absent?

4. Would you expect 7,7-dichloronorcarane to give a positive sodium iodide in acetone test?

5. Assign the C—H stretch for the cyclopropane ring hydrogens in the infrared spectrum.

6. Suggest why it may be necessary to use a large excess of chloroform in this reaction.

7. A student obtained a proton NMR spectrum of the product isolated in this experiment. The spectrum shows peaks at about 7.3 and 5.6 ppm. What do you think these peaks indicate? Are they part of the 7,7-dichloronorcarane spectrum?

8. Draw the structures of the products that you would expect from the reactions of *cis-* and *trans*-2-butene with dichlorocarbene.

9. Draw the structure of the expected dichlorocarbene adduct of methyl methacrylate (methyl 2-methylpropenoate). With compounds of this type, another product could have been obtained. It is the chloroform adduct to the double bond (Michael-type reaction). What would this structure look like?

10. Provide mechanisms for the following abnormal dichlorocarbene addition reactions. In both cases, the usual adduct is first obtained, and then a subsequent reaction occurs.

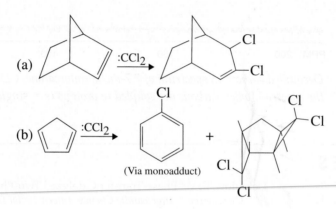

EXPERIMENT 27

Relative Reactivities of Several Aromatic Compounds

Aromatic substitution

Relative activating ability of aromatic substituents

Crystallization

When substituted benzenes undergo electrophilic aromatic substitution reactions, both the reactivity and the orientation of the electrophilic attack are affected by the nature of the original group attached to the benzene ring. Substituent groups that make the ring more reactive than benzene are called **activators.** Such groups are also said to be **ortho, para** directors because the products formed are those in which substitution occurs either ortho or para to the activating group. Various products may be formed, depending on whether substitution occurs at the ortho or para position and the number of times substitution occurs on the same molecule. Some groups may activate the benzene ring so strongly that multiple substitution consistently occurs, whereas other groups may be moderate activators, and benzene rings containing such groups may undergo only a single substitution. The purpose of this experiment is to determine the relative activating effects of several substituent groups.

In this experiment, you will study the bromination of acetanilide, aniline, and anisole:

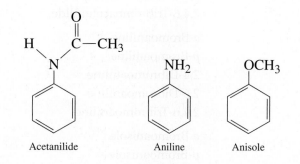

Acetanilide   Aniline   Anisole

The acetamido group, —NHCOCH₃, the amino group, —NH₂, and the methoxy group, —OCH₃, are all activators and ortho, para directors. Each student will carry out the bromination of one of these compounds and determine its melting point. By sharing your data, you will have information on the melting points of the brominated products for acetanilide, aniline, and anisole. Using the table on page 226, it will then be possible for you to rank the three substituents in order of activating strength.

The classical method of brominating an aromatic compound is to use Br_2 and a catalyst such as $FeBr_3$, which acts as a Lewis acid. The first step is the reaction between bromine and the Lewis acid:

$$Br_2 + FeBr_3 \longrightarrow [FeBr_4^- \; Br^+]$$

The positive bromine ion then reacts with the benzene ring in an aromatic electrophilic substitution reaction:

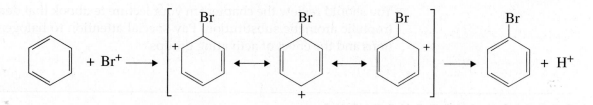

Aromatic compounds that contain activating groups can be brominated without the use of the Lewis acid catalyst because the π electrons in the benzene ring are more available and polarize the bromine molecule sufficiently to produce the required electrophile Br^+. This is illustrated by the first step in the reaction between anisole and bromine:

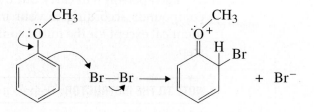

Melting points of relevant compounds

Compound	Melting Points (°C)
o-Bromoacetanilide	99
p-Bromoacetanilide	168
2,4-Dibromoacetanilide	145
2,6-Dibromoacetanilide	208
2,4,6-Tribromoacetanilide	232
o-Bromoaniline	32
p-Bromoaniline	66
2,4-Dibromoaniline	80
2,6-Dibromoaniline	87
2,4,6-Tribromoaniline	122
o-Bromoanisole	3
p-Bromoanisole	13
2,4-Dibromoanisole	60
2,6-Dibromoanisole	13
2,4,6-Tribromoanisole	87

In this experiment, the brominating mixture consists of bromine, hydrobromic acid HBr, and acetic acid. The presence of bromide ion from the hydrobromic acid helps to solubilize the bromine and increase the concentration of the electrophile.

REQUIRED READING

Review: Technique 11

You should review the chapters in your lecture textbook that deal with electrophilic aromatic substitution. Pay special attention to halogenation reactions and the effect of activating groups.

SPECIAL INSTRUCTIONS

Bromine is a skin irritant, and its vapors cause severe irritation to the respiratory tract. It will also oxidize many pieces of jewelry. Hydrobromic acid may cause skin or eye irritation. Aniline is highly toxic and a suspected teratogen. All bromoanilines are toxic. This experiment should be carried out in a fume hood or in a well-ventilated laboratory.

Each person will carry out the bromination of only one of the aromatic compounds according to your instructor's directions. The procedures are identical except for the initial compound used and the final recrystallization step.

NOTE TO THE INSTRUCTOR: Prepare the brominating mixture in advance.

SUGGESTED WASTE DISPOSAL

Dispose of the filtrate from the Hirsch funnel filtration of the crude product into a container specifically designated for this mixture. Place all other filtrates into the container for halogenated organic solvents.

PROCEDURE

Running the Reaction

To a tared 5-mL conical vial with a cap, add the given amount of one of the following compounds: 0.090 g of acetanilide, 0.060 mL of aniline, or 0.070 mL of anisole. Reweigh the conical vial to determine the actual weight of the aromatic compound. Add 0.5 mL of glacial acetic acid and a spin vane to the conical vial. Attach an air condenser and place the conical vial in a water bath at 23–27°C, as shown in Figure 6.6, page 594. Stir the mixture until the aromatic compound is completely dissolved. While the compound is dissolving, pack a drying tube loosely with glass wool. Add about 0.5 mL of 1*M* sodium bisulfite dropwise to the glass wool until it is moistened but not soaked. This apparatus will capture any bromine given off during the following reaction.

Under the hood, obtain 1.0 mL of the bromine/hydrobromic acid mixture[1] in a 3-mL conical vial. Place the cap on the vial before returning to your lab bench. While stirring, add all the bromine/hydrobromic acid mixture through the top of the air condenser, using a Pasteur pipet.

> **CAUTION** ⚠
>
> **Be careful not to spill any of this mixture.**

Attach the drying tube prepared above. Continue stirring the reaction mixture for 20 minutes.

Crystallization and Isolation of Product

When the reaction is complete, transfer the mixture to a 10-mL Erlenmeyer flask containing 5 mL of water and 0.5 mL of saturated sodium bisulfite solution. Stir this mixture with a glass stirring rod until the red color of bromine disappears.[2] If an oil has formed, it may be necessary to stir the mixture for several minutes. Place the Erlenmeyer flask in an ice bath for 10 minutes. If the product does not solidify, scratch the bottom of the flask with a glass stirring rod to induce crystallization. It may take 10–15 minutes to induce crystallization of the brominated anisole product.[3] Filter the product on a Hirsch funnel with suction and rinse with several 1-mL portions of cold water. Air-dry the product on the funnel for about 5 minutes with the vacuum on.

---

[1] *Note to the Instructor:* The brominating mixture is prepared by adding 2.6 mL of bromine to 17.4 mL of 48% hydrobromic acid. This will provide enough solution for 20 students, assuming no waste of any type. This solution should be stored in the hood.

[2] If the color of bromine is still present, add a few more drops of saturated sodium bisulfite and stir the mixture for a few more minutes. The entire mixture, including liquid and solid (or oil), should be colorless.

[3] If crystals fail to form after 15 minutes, it may be necessary to seed the mixture with a small crystal of product.

Recrystallization and Melting Point of Product

If you started with **aniline,** transfer the solid to a 10-mL Erlenmeyer flask and re-crystallize the product from 95% ethanol (see Technique 11, Section 11.3, and Fig. 11.4, p. 651). Filter the crystals on a Hirsch funnel and dry them for several minutes with suction. The brominated products from either **acetanilide** or **anisole** should be crystallized using a Craig tube (Technique 11, Section 11.4, and Fig. 11.6, p. 657). Use 95% ethanol to crystallize the acetanilide product and hexane to crystallize the brominated anisole compound. Allow the crystals to air-dry and determine the weight and melting point.

According to the melting point and the preceding table, you should be able to identify your product. Calculate the percentage yield and submit your product, along with your report, to your instructor.

REPORT

By collecting data from other students, you should be able to determine which product was obtained from the bromination of each of the three aromatic compounds. Using this information, arrange the three substituent groups (acetamido, amino, and methoxy) in order of decreasing ability to activate the benzene ring.

REFERENCE

Zaczek, N. M., and Tyszklewicz, R. B. "Relative Activating Ability of Various Ortho, Para-Directors." *Journal of Chemical Education, 63* (1986): 510.

QUESTIONS

1. Using resonance structures, show why the amino group is activating. Consider an attack by the electrophile E^+ at the *para* position.

2. For the substituent in this experiment that was found to be least activating, explain why bromination took place at the position on the ring indicated by the experimental results.

3. What other experimental techniques (including spectroscopy) might be used to identify the products in this experiment?

28 EXPERIMENT 28

Nitration of Methyl Benzoate

Aromatic substitution

Crystallization

The nitration of methyl benzoate to prepare methyl *m*-nitrobenzoate is an example of an electrophilic aromatic substitution reaction, in which a proton of the aromatic ring is replaced by a nitro group:

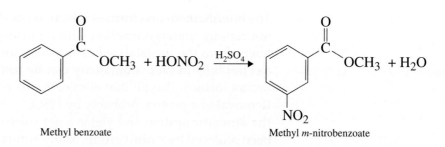

Methyl benzoate   Methyl *m*-nitrobenzoate

Many such aromatic substitution reactions are known to occur when an aromatic substrate is allowed to react with a suitable electrophilic reagent, and many other groups besides nitro may be introduced into the ring.

You may recall that alkenes (which are electron-rich due to an excess of electrons in the π system) can react with an electrophilic reagent. The intermediate formed is electron-deficient. It reacts with the nucleophile to complete the reaction. The overall sequence is called **electrophilic addition.** Addition of HX to cyclohexene is an example.

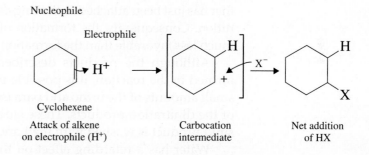

Nucleophile

Electrophile

Cyclohexene

Attack of alkene   Carbocation   Net addition
on electrophile (H⁺)   intermediate   of HX

Aromatic compounds are not fundamentally different from cyclohexene. They can also react with electrophiles. However, because of resonance in the ring, the electrons of the π system are generally less available for addition reactions because an addition would mean the loss of the stabilization that resonance provides. In practice, this means that aromatic compounds react only with **powerfully electrophilic reagents,** usually at somewhat elevated temperatures.

Benzene, for example, can be nitrated at 50°C with a mixture of concentrated nitric and sulfuric acids; the electrophile is NO_2^+ (nitronium ion), whose formation is promoted by action of the concentrated sulfuric acid on nitric acid:

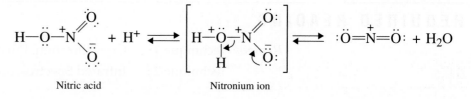

Nitric acid   Nitronium ion

The nitronium ion thus formed is sufficiently electrophilic to add to the benzene ring, *temporarily* interrupting ring resonance:

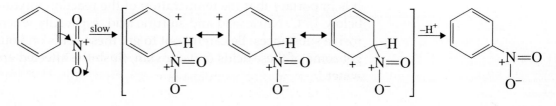

The intermediate first formed is somewhat stabilized by resonance and does not rapidly undergo reaction with a nucleophile; in this behavior, it is different from the unstabilized carbocation formed from cyclohexene plus an electrophile. In fact, aromaticity can be restored to the ring if **elimination** occurs instead. (Recall that elimination is often a reaction of carbocations.) Removal of a proton, probably by HSO_4^-, from the sp^3-ring carbon **restores the aromatic system** and yields a net **substitution** wherein a hydrogen has been replaced by a nitro group. Many similar reactions are known, and they are called **electrophilic aromatic substitution reactions.**

The substitution of a nitro group for a ring hydrogen occurs with methyl benzoate in the same way it does with benzene. In principle, one might expect that any hydrogen on the ring could be replaced by a nitro group. However, for reasons beyond our scope here (see your lecture textbook), the carbomethoxy group directs the aromatic substitution preferentially to those positions that are *meta* to it. As a result, methyl *m*-nitrobenzoate is the principal product formed. In addition, one might expect the nitration to occur more than once on the ring. However, both the carbomethoxy group and the nitro group that has just been attached to the ring *deactivate* the ring against further substitution. Consequently, the formation of a methyl dinitrobenzoate product is much less favorable than the formation of the mononitration product.

Although the products described previously are the principal ones formed in the reaction, it is possible to obtain as impurities in the reaction small amounts of the ortho and para isomers of methyl *m*-nitrobenzoate and of the dinitration products. These side products are removed when the desired product is washed with methanol and purified by crystallization.

Water has a retarding effect on the nitration because it interferes with the nitric acid–sulfuric acid equilibria that form the nitronium ions. The smaller the amount of water present, the more active the nitrating mixture. Also, the reactivity of the nitrating mixture can be controlled by varying the amount of sulfuric acid used. This acid must protonate nitric acid, which is a *weak* base, and the larger the amount of acid available, the more numerous the protonated species (and hence NO_2^+) in the solution. Water interferes because it is a stronger base than H_2SO_4 or HNO_3. Temperature is also a factor in determining the extent of nitration. The higher the temperature, the greater will be the amounts of dinitration products formed in the reaction.

REQUIRED READING

Review: Technique 11 Crystallization: Purification of Solids

Technique 25 Infrared Spectroscopy, Sections 25.4 and 25.5

SPECIAL INSTRUCTIONS

It is important that the temperature of the reaction mixture be maintained below 15°C. Nitric acid and sulfuric acid, especially when mixed, are corrosive substances. Be careful not to get these acids on your skin. If you do get some of these acids on your skin, flush the affected area liberally with water.

SUGGESTED WASTE DISPOSAL

The filtrate from the Hirsch funnel filtration should be placed in the designated container.

PROCEDURE

Add 0.210 mL of methyl benzoate to a tared 3-mL conical vial and determine the actual weight of methyl benzoate. Add 0.45 mL of concentrated sulfuric acid to the methyl benzoate, along with a magnetic spin vane. Attach an air condenser to the conical vial. The purpose of the air condenser is to make it easier to hold the conical vial in place. Prepare an ice bath in a 250-mL beaker using both ice and water. Clamp the air condenser so that the conical vial is immersed in the ice bath as shown in Figure 6.6, page 594. (Note that in Figure 6.6 a water bath is shown rather than an ice bath.) While stirring, *very slowly* add a cool mixture of 0.15 mL of concentrated sulfuric acid and 0.15 mL of concentrated nitric acid throughout a period of about 15 minutes. The acid mixture should be added with a 9-inch Pasteur pipet through the top of the air condenser. If the addition is too fast, the formation of by-product increases rapidly, reducing the yield of the desired product.

After you have added all the acid, warm the mixture to room temperature by replacing the ice water in the 250-mL beaker with water at room temperature. Let the reaction mixture stand for 15 more minutes without stirring. Then, using a Pasteur pipet, transfer the reaction mixture to a 20-mL beaker containing 2.0 g of crushed ice. After the ice has melted, isolate the product by vacuum filtration using a Hirsch funnel and wash it with two 1.0-mL portions of cold water and then with two 0.3-mL portions of ice-cold methanol. Weigh the crude, dry product and recrystallize it from methanol using a Craig tube (see Technique 11, Section 11.4, p. 656).

Determine the melting point of the product. The melting point of the recrystallized product should be 78°C. Obtain the infrared spectrum using the dry film method (Technique 25, Section 25.4, p. 838) or as a KBr pellet (Technique 25, Section 25.5, p. 838). Compare your infrared spectrum with the one reproduced on page 233. Calculate the percentage yield and submit the product to the instructor in a labeled vial.

MOLECULAR MODELING (OPTIONAL)

If you are working alone, complete Part One. If you have a partner, one of you should complete Part One and the other complete Part Two. If you work with a partner, you should combine results at the end of the experiment.

Part One: Nitration of Methyl Benzoate

In this exercise, we will try to explain the observed outcome of the nitration of methyl benzoate. The major product of this reaction is methyl *m*-nitrobenzoate, where the nitro group has been added to the *meta* position of the ring. The rate-determining step of this reaction is the attack of the nitronium ion on the benzene ring. Three benzenium ion intermediates (*ortho*, *meta*, and *para*) are possible:

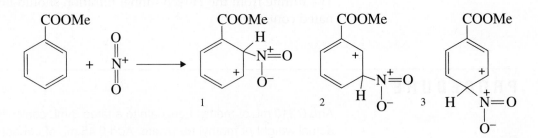

We will calculate the heats of formation for these intermediates to determine which of the three has the lowest energy. Assume that the activation energies are similar to the energies of the intermediates themselves. This is an application of the Hammond Postulate, which states that the activation energy leading to an intermediate of higher energy will be higher than the activation energy leading to an intermediate of lower energy, and vice versa. Although there are prominent exceptions, this postulate is generally true.

Make models of each of the three benzenium ion intermediates (separately), and calculate their heats of formation using an AM1-level calculation with geometry optimization. Don't forget to specify a positive charge when you submit the calculation. What do you conclude?

Now take a piece of paper and draw the resonance structures that are possible for each intermediate. Do not worry about structures involving the nitro group; only consider where the charge in the ring may be delocalized. Also note the polarity of the carbonyl group by placing a $\delta+$ symbol on the carbon and a $\delta-$ symbol on the oxygen. What do you conclude from your resonance analysis?

Part Two: Nitration of Anisole

For this computation, you will analyze the three benzenium ions formed from anisole (methoxybenzene) and the nitronium ion (see Part One). Calculate the heats of formation using AM1-level calculations with geometry optimization. Don't forget to specify a positive charge. What do you conclude for anisole? How do the results compare to those for methyl benzoate?

Now take a piece of paper and draw the resonance structures that are possible for each intermediate. Do not worry about structures involving the nitro group; only consider where the charge in the ring may be delocalized. Do not forget that the electrons on the oxygen can participate in the resonance. What do you conclude from your resonance analysis?

QUESTIONS

1. Why is methyl *m*-nitrobenzoate formed in this reaction instead of the ortho or para isomers?

2. Why does the amount of the dinitration increase at high temperatures?

3. Why is it important to add the nitric acid–sulfuric acid mixture slowly during a 15-minute period?

4. Interpret the infrared spectrum of methyl *m*-nitrobenzoate.

5. Indicate the product formed on nitration of each of the following compounds: benzene, toluene, chlorobenzene, and benzoic acid.

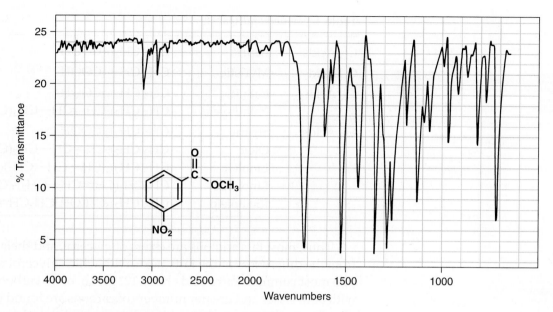

Infrared spectrum of methyl m-*nitrobenzoate, KBr.*

ESSAY

Fats and Oils

In the normal human diet, about 25 to 50% of the caloric intake consists of fats and oils. These substances are the most concentrated form of food energy in our diet. When metabolized, fats produce about 9.5 kcal of energy per gram. Carbohydrates and proteins produce less than half this amount. For this reason, animals tend to build up fat deposits as a reserve source of energy. They do this, of course, only when their food intake exceeds their energy requirements. In times of starvation, the body metabolizes these stored fats. Even so, some fats are required by animals for bodily insulation and as a protective sheath around some vital organs.

The constitution of fats and oils was first investigated by the French chemist Chevreul from 1810 to 1820. He found that when fats and oils were hydrolyzed, they gave rise to several "fatty acids" and the trihydroxylic alcohol glycerol. Thus, fats and oils are **esters** of glycerol, called **glycerides** or **acylglycerols.** Because glycerol has three hydroxyl groups, it is possible to have mono-, di-, and triglycerides. Fats and oils are predominantly triglycerides (triacylglycerols), constituted as follows:

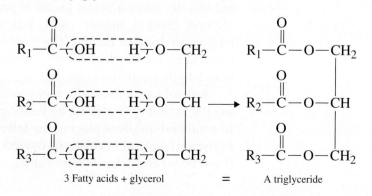

3 Fatty acids + glycerol   =   A triglyceride

Table 1 *Common fatty acids*

C_{12} Acids	Lauric	$CH_3(CH_2)_{10}COOH$
C_{14} Acids	Myristic	$CH_3(CH_2)_{12}COOH$
C_{16} Acids	Palmitic	$CH_3(CH_2)_{14}COOH$
	Palmitoleic	$CH_3(CH_2)_5CH{=}CH{-}CH_2(CH_2)_6COOH$
C_{18} Acids	Stearic	$CH_3(CH_2)_{16}COOH$
	Oleic	$CH_3(CH_2)_7CH{=}CH{-}CH_2(CH_2)_6COOH$
	Linoleic	$CH_3(CH_2)_4(CH{=}CH{-}CH_2)_2(CH_2)_6COOH$
	Linolenic	$CH_3CH_2(CH{=}CH{-}CH_2)_3(CH_2)_6COOH$
	Ricinoleic	$CH_3(CH_2)_5CH(OH)CH_2CH{=}CH(CH_2)_7COOH$

Thus, most fats and oils are esters of glycerol, and their differences result from the differences in the fatty acids with which glycerol may be combined. The most common fatty acids have 12, 14, 16, or 18 carbons, although acids with both lesser and greater numbers of carbons are found in several fats and oils. These common fatty acids are listed in Table 1 along with their structures. As you can see, these acids are both saturated and unsaturated. The saturated acids tend to be solids, whereas the unsaturated acids are usually liquids. This circumstance also extends to fats and oils. Fats are made up of fatty acids that are most saturated, whereas oils are primarily composed of fatty acid portions that have greater numbers of double bonds. In other words, unsaturation lowers the melting point. Fats (solids) are usually obtained from animal sources, whereas oils (liquids) are commonly obtained from vegetable sources. Therefore, vegetable oils usually have a higher degree of unsaturation.

About 20 to 30 fatty acids are found in fats and oils, and it is not uncommon for a given fat or oil to be composed of as many as 10 to 12 (or more) fatty acids. Typically, these fatty acids are randomly distributed among the triglyceride molecules, and the chemist cannot identify anything more than an average composition for a given fat or oil. The average fatty acid composition of some selected fats and oils is given in Table 2 on page 235. As indicated, all the values in the table may vary in percentage, depending, for instance, on the locale in which the plant was grown or on the particular diet on which the animal subsisted. Thus, perhaps there is a basis for the claims that corn-fed hogs or cattle taste better than animals maintained on other diets.

Vegetable fats and oils are usually found in fruits and seeds and are recovered by three principal methods. In the first method, **cold pressing,** the appropriate part of the dried plant is pressed in a hydraulic press to squeeze out the oil. The second method is **hot pressing,** which is the same as the first method but done at a higher temperature. Of the two methods, cold pressing usually gives a better grade of product (more bland); the hot pressing method gives a higher yield, but with more undesirable constituents (stronger odor and flavor). The third method is **solvent extraction.** Solvent extraction gives the highest recovery of all and can now be regulated to give bland, high-grade food oils.

Animal fats are usually recovered by **rendering,** which involves cooking the fat out of the tissue by heating it to a high temperature. An alternative method involves placing the fatty tissue in boiling water. The fat floats to the surface and is easily recovered. The most common animal fats, lard (from hogs) and tallow (from cattle), can be prepared in either way.

Table 2 *Average fatty acid composition (by percentage) of selected fats and oils*

	Saturated Fatty Acids (No Double Bonds)						Unsaturated (1 Double Bond)			Unsaturated (>1 Double Bond)			Unsaturated
	C_{10} C_8 C_6 C_4	C_{12} Lauric	C_{14} Myristic	C_{16} Palmitic	C_{18} Stearic	C_{20} C_{22} C_{24}	C_{16} Palmitoleic	C_{18} Oleic	C_{18} Ricinoleic	C_{18} Linoleic (2)	C_{18} Linolenic (3)	C_{18} Eleostearic (3)	C_{20} C_{22} C_{24}
Animal fats													
Tallow			2–3	24–32	14–32		1–3	35–48		2–4			
Butter	7–10	2–3	7–9	23–26	10–13		5	30–40		4–5			2
Lard			1–2	28–30	12–18		1–3	41–48		6–7			2
Animal oils													
Neat's foot				17–18	2–3			74–77					17–31
Whale			4–5	11–18	2–4		13–18	33–38					12–19
Sardine			6–8	10–16	1–2		6–15			24–30			
Vegetable oils													
Corn			0–2	7–11	3–4		0–2	43–49		34–42			
Olive			0–1	5–15	1–4		0–1	69–84		4–12			
Peanut				6–9	2–6	3–10	0–1	50–70		13–26			
Soybean			0–1	6–10	2–6			21–29		50–59	4–8		
Safflower				6–10	1–4			8–18		70–80	2–4		
Castor bean				0–1				0–9	80–92	3–7			
Cottonseed			0–2	19–24	1–2		0–2	23–33		40–48			
Linseed				4–7	2–5			9–38		3–43	25–58		
Coconut	10–22	45–51	17–20	4–10	1–5			2–10		0–2			
Palm			1–3	34–43	3–6			38–40		5–11			
Tung					2–6			4–16		0–1		74–91	

235

Many triglyceride fats and oils are used for cooking. We use them to fry meats and other foods and to make sandwich spreads. Almost all commercial cooking fats and oils, except lard, are prepared from vegetable sources. Vegetable oils are liquids at room temperature. If the double bonds in a vegetable oil are hydrogenated, the resultant product becomes solid. Manufacturers, in making commercial cooking fats (Crisco, Spry, Fluffo, etc.), hydrogenate a liquid vegetable oil until the desired degree of consistency is achieved. This makes a product that still has a high degree of unsaturation (double bonds) left. The same technique is used for margarine. "Polyunsaturated" oleomargarine is produced by the partial hydrogenation of oils from corn, cottonseed, peanut and soybean sources. The final product has a yellow dye (β-carotene) added to make it look like butter; milk, about 15% by volume, is mixed into it to form the final emulsion. Vitamins A and D are also commonly added. Because the final product is tasteless (try Crisco), salt, acetoin, and biacetyl are often added. The latter two additives mimic the characteristic flavor of butter.

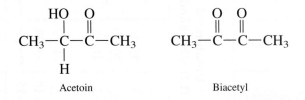

Acetoin   Biacetyl

Many producers of margarine claim it to be more beneficial to health because it is "high in polyunsaturates." Animal fats are low in unsaturated fatty acid content and are generally excluded from the diets of people who have high cholesterol levels in the blood. Such people have difficulty in metabolizing saturated fats correctly and should avoid them because they encourage cholesterol deposits to form in the arteries. This ultimately leads to high blood pressure and heart trouble. People who pay close attention to their intake of fats tend to avoid consuming large quantities of saturated fats, knowing that eating these fats increases the risk of heart disease. Diet-conscious people try to limit their fat consumption to unsaturated fats, and they make use of the current mandatory food labeling to obtain information on the fat content of the food they eat.

Unfortunately, not all the unsaturated fats appear to be equally safe. When we eat partially hydrogenated fats, we increase our consumption of **trans-fatty acids.** These acids, which are isomers of the naturally occurring *cis*-fatty acids, have been implicated in a variety of conditions, including heart disease, cancer, and diabetes. The strongest evidence that *trans*-fatty acids may be harmful comes in studies of the incidence of coronary heart disease. Ingestion of *trans*-fatty acids appears to increase blood cholesterol levels, in particular the ratio of low-density lipoproteins (LDL, or "bad" cholesterol) to high-density lipoproteins (HDL, or "good" cholesterol). The *trans*-fatty acids appear to exhibit harmful effects on the heart that are similar to those shown by saturated fatty acids.

The *trans*-fatty acids do not occur naturally to any significant extent. Rather, they are formed during the partial hydrogenation of vegetable oils to make margarine and solid forms of shortening. For a small percentage of *cis*-fatty acids subjected to hydrogenation, only one hydrogen atom is added to the carbon chain. This process forms an intermediate free radical, which is able to rotate its conformation by 180 degrees before it releases the extra hydrogen atom back to the reaction medium. The result is an isomerization of the double bond.

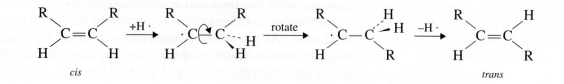

Concern over the health and nutrition of the public, particularly over the average fat intake of most Americans, has prompted food chemists and technologists to develop a variety of **fat replacers.** The objective has been to discover substances that have the taste and mouth-feel of a real fat but do not have deleterious effects on the cardiovascular system. One product that has recently appeared in certain snack foods is **olestra** (marketed under the trade name **Olean,** by the Proctor and Gamble Company). Olestra is not an acylglycerol; rather, it is composed of a molecule of **sucrose** that has been substituted by long-chain fatty acid residues. It is a **polyester,** and the body's enzyme systems are not capable of attacking it and catalyzing its breakdown into smaller molecules.

Because the body's enzyme systems are unable to break this molecule down, it does not contain any usable dietary calories. Furthermore, it is heat stable, which makes it ideal for frying and other cooking. Unfortunately, for some individuals there may be harmful or unpleasant side effects. The use

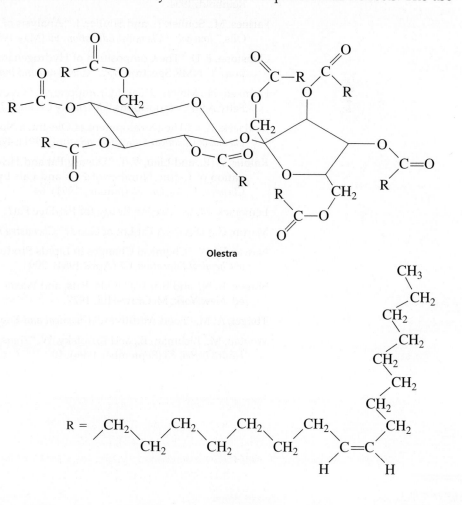

Olestra

of olestra has been reported to deplete certain fat-soluble vitamins, particularly Vitamins A, D, E, and K. For this reason, products prepared with olestra have these vitamins added to offset this effect. Also, some people have reported diarrhea and abdominal cramps.

Is the development of fat replacers such as olestra part of the wave of the future? As the average American's appetite for snack foods continues to grow and as health problems arising from obesity also increase, a demand for satisfying foods that are less fattening will always be strong. In the long run, however, it would probably be better if we all learned to curtail our appetite for fatty foods and, instead, tried to increase our intake of fruits, vegetables, and other healthful foods. At the same time, a change from a sedentary lifestyle to one that included regular exercise would be much more beneficial to our health than a search for nonfattening food additives.

REFERENCES

Dawkins, M. J. R., and Hull, D. "The Production of Heat by Fat." *Scientific American, 213* (August 1965): 62.

Dolye, E. "Olestra? The Jury's Still Out." *Journal of Chemical Education, 74* (April 1997): 370.

Eckey, E. W., and Miller, L. P. *Vegetable Fats and Oils.* ACS Monograph No. 123. New York: Reinhold, 1954.

Farines, M., Soulier, F., and Soulier, J. "Analysis of the Triglycerides of Some Vegetable Oils." *Journal of Chemical Education, 65* (May 1988): 464.

Gunstone, F. D. "The Composition of Hydrogenated Fats Determined by High Resolution ^{13}C NMR Spectroscopy." *Chemistry and Industry* (November 4, 1991): 802.

Heinzen, H., Moyna, P., and Grompone, A. "Gas Chromatographic Determination of Fatty Acid Compositions." *Journal of Chemical Education, 62* (May 1985): 449.

Jandacek, R. J. "The Development of Olestra, a Noncaloric Substitute for Dietary Fat." *Journal of Chemical Education, 68* (June 1991): 476.

Kalbus, G. E., and Lieu, V. T. "Dietary Fat and Health: An Experiment on the Determination of Iodine Number of Fats and Oils by Coulometric Titration." *Journal of Chemical Education, 68* (January 1991): 64.

Lemonick, M. D. "Are We Ready for Fat-Free Fat?" *Time, 147* (January 8, 1996): 52.

Martin, C. "TFA's—A Fat Lot of Good?" *Chemistry in Britain, 32* (October 1996): 34.

Nawar, W. W. "Chemical Changes in Lipids Produced by Thermal Processing." *Journal of Chemical Education, 61* (April 1984): 299.

Shreve, R. N., and Brink, J. "Oils, Fats, and Waxes." *The Chemical Process Industries,* 4th ed. New York: McGraw-Hill, 1977.

Thayer, A. M. "Food Additives." *Chemical and Engineering News, 70* (June 15, 1992): 26.

Wootan, M., Liebman, B., and Rosofsky, W. "*Trans:* The Phantom Fat." *Nutrition Action Health Letter, 23* (September 1996): 10.

EXPERIMENT 29

Methyl Stearate from Methyl Oleate

Catalytic hydrogenation

Recrystallization

Unsaturation tests

In this experiment, you will convert the liquid methyl oleate, an "unsaturated" fatty acid ester, to solid methyl stearate, a "saturated" fatty acid ester, by catalytic hydrogenation.

$$CH_3(CH_2)_7-CH=CH-(CH_2)_7-\overset{\overset{\textstyle O}{\|}}{C}-OCH_3 \xrightarrow[\text{H}_2]{\text{Pd/C}}$$

Methyl oleate
(methyl *cis*-9-octadecenoate)

$$CH_3(CH_2)_7-\underset{\underset{\textstyle H}{|}}{CH}-\underset{\underset{\textstyle H}{|}}{CH}-(CH_2)_7-\overset{\overset{\textstyle O}{\|}}{C}-OCH_3$$

Methyl stearate
(methyl octadecanoate)

By commercial methods such as those described in this experiment, the unsaturated fatty acids of vegetable oils are converted to margarine (see the essay "Fats and Oils"). However, rather than using the mixture of triglycerides that would be present in a cooking oil such as Mazola (corn oil), we use as a model the pure chemical methyl oleate.

For this procedure, a chemist would usually use a cylinder of hydrogen gas. Because many students will be following the procedure simultaneously, however, we use the simpler expedient of causing zinc metal to react with dilute sulfuric acid:

$$Zn + H_2SO_4 \xrightarrow{\text{H}_2\text{O}} H_2(g) + ZnSO_4$$

The hydrogen so generated will be passed into a solution containing methyl oleate and the palladium on carbon catalyst (10% Pd/C).

REQUIRED READING

Review: Techniques 1–6

New: Technique 8   Filtration, Sections 8.3–8.5

Technique 9   Physical Constants of Solids: The Melting Point

Essay   Fats and Oils

You should also read those sections in your lecture textbook that deal with catalytic hydrogenation. If the instructor indicates that you should perform the optional unsaturation tests on your starting material and product, read the descriptions of the Br_2/CH_2Cl_2 test on page 466 and in Experiment 25 on page 216.

SPECIAL INSTRUCTIONS

Because this experiment calls for generating hydrogen gas, no flames will be allowed in the laboratory.

> ### ⚠ CAUTION
>
> **No flames allowed.**

Because a buildup of hydrogen is possible within the apparatus, it is especially important to remember to wear your safety goggles; you can thus protect yourself against the possibility of minor "explosions" from joints popping open, from fires, or from any glassware accidentally cracking under pressure.

> ### ⚠ CAUTION
>
> **Wear safety goggles.**

When you operate the hydrogen generator, be sure to add sulfuric acid at a rate that does not cause hydrogen gas to evolve too rapidly. The hydrogen pressure in the vial should not rise much above atmospheric pressure. Neither should the hydrogen evolution be allowed to stop. If this happens, your reaction mixture may be "sucked back" into your hydrogen generator.

SUGGESTED WASTE DISPOSAL

Dispose of the sulfuric acid (from the hydrogen generator) by pouring it into the waste container designated for acid waste. Place any leftover zinc in the container designated for this material. After centrifugation, transfer the Pd/C catalyst to a container designated for this material. After collecting the methyl stearate by filtration, place the methanol filtrate in the nonhalogenated organic waste container. Discard the solutions that remain after the bromine test for unsaturation into a waste container designated for the disposal of halogenated organic solvents.

NOTES TO THE INSTRUCTOR

Use methyl oleate that is 100% or nearly 100% pure. We use Aldrich Chemical Co. No. 31,111-1.

PROCEDURE

Apparatus

Assemble the apparatus as illustrated in the figure on page 241. The apparatus can be simplified by using the multipurpose adapter (Fig. 14.9, p. 712) in place of the Claisen head and both thermometer adapters. The apparatus consists of basically three parts:

1. Hydrogen generator
2. Reaction flask
3. Mineral oil bubbler trap

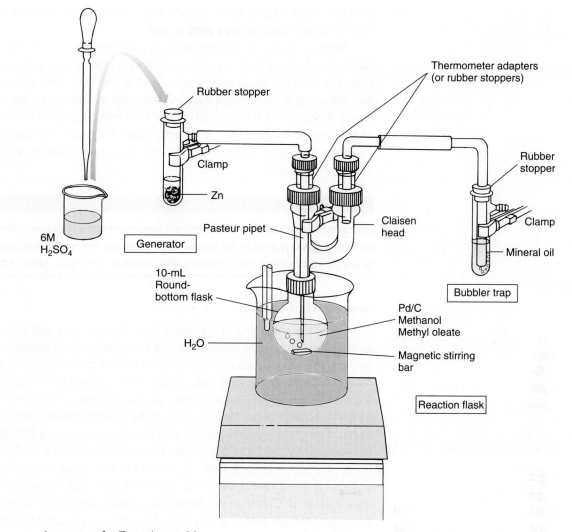

Apparatus for Experiment 29.

The mineral oil bubbler trap has two functions. First, it allows you to keep a pressure of hydrogen within the system that is slightly above atmospheric. Second, it prevents back-diffusion of air into the system. The functions of the other two units are self-explanatory.

So that hydrogen leakage is prevented, the tubing used to connect the various subunits of the apparatus should be either relatively new rubber tubing, without cracks or breaks, or Tygon tubing. The tubing can be checked for cracks or breaks simply by stretching and bending it before use. It should be of such a size that it will fit onto all connections tightly. Similarly, if any rubber stoppers are used, they should be fitted with a size of glass tubing that fits firmly through the holes in their centers. If the seal is tight, it will not be easy to slide the glass tubing up and down in the hole. The inlet tube (Pasteur pipet) in the round-bottom flask should reach almost to the bottom of the flask. Hydrogen must bubble *through* the solution.

Preparing for the Reaction

Fill the bubbler trap (second side-arm test tube) about one-third full with mineral oil. The end of the glass tube should be submerged below the surface of the oil.

To charge the hydrogen generator, weigh out about 2 g of mossy zinc and place it in the side-arm test tube. Seal the large opening at its top using a rubber stopper. Obtain about 10 mL of 6*M* sulfuric acid and place it in a small Erlenmeyer flask or beaker **but do not add it yet.**

Weigh a 10-mL round-bottom flask and then place 1.00 mL of methyl oleate into it. Reweigh the flask in order to obtain the exact amount of methyl oleate used. After this, add 6.0 mL of methanol solvent to the flask. Also place a magnetic stirring bar into the flask. Place the flask into a small beaker. Using smooth weighing paper, weigh about 0.030 g (30 mg) of 10% Pd/C. Carefully place about one-third of the catalyst into the flask and gently swirl the liquid until the solid catalyst has sunk into the liquid. Repeat this with the rest of the catalyst, adding one-third of the original amount each time.

CAUTION

Be careful when adding the catalyst; sometimes it will cause a flame. Do not hold onto the flask; it should be in a small beaker on the lab bench. Have a watch glass handy to cover the opening and smother the flame should this occur.

Running the Reaction

Complete the assembly of the apparatus, making sure that all the seals are gas tight. Place the round-bottom flask in a warm water bath maintained at 40°C. This will help to keep the product dissolved in the solution throughout the course of the reaction. If the temperature rises above 40°C, you will lose a significant amount of the methanol solvent. If this occurs, do not hesitate to add more methanol to the reaction flask through the side arm of the Claisen head, using a Pasteur pipet. Begin stirring the reaction mixture with the magnetic stirring bar. Start the evolution of hydrogen by removing the rubber stopper and adding a portion of the 6*M* sulfuric acid solution (about 6 mL) to the hydrogen generator. Replace the rubber stopper. A good rate of bubbling in the reaction flask is about three to four bubbles a second. Continue the evolution of hydrogen for 45–60 minutes. If necessary, open the generator, empty it, and refresh the zinc and sulfuric acid. (Keep in mind that the acid is used up and becomes more dilute as the zinc reacts.)

Stopping the Reaction

After the reaction is complete, stop the reaction by disconnecting the reaction flask. Decant the acid in the side-arm test tube into a designated waste container, being careful not to transfer any zinc metal. Rinse the zinc in the test tube several times with water and then place any unreacted zinc in a waste container provided for this purpose.

Keep the temperature of the reaction mixture at about 40°C until you perform the centrifugation; otherwise, the methyl stearate may crystallize and interfere with removal of the catalyst. There should not be any white solid (product) in the round-bottom flask. If there is a white solid, add more methanol and stir until the solid dissolves.

Removal of the Catalyst

Pour the reaction mixture into a centrifuge tube. Place the centrifuge tube into the water bath at 40°C until just before you are ready to centrifuge the mixture. Centrifuge the mixture for several minutes. After centrifugation, the black catalyst should be at the bottom of the tube. If some of the catalyst is still suspended in the liquid, heat the mixture to 40°C and centrifuge the mixture again. Carefully pour the supernatant liquid (leaving the black catalyst in the centrifuge tube) into a small beaker and cool to room temperature.

Crystallization and Isolation of Product

Place the beaker in an ice bath to induce crystallization. If crystals do not form or if only a few crystals form, you may need to reduce the volume of solvent. Do this by heating the beaker in a water bath and directing a slow stream of air into the beaker, using a Pasteur pipet for a nozzle (Fig. 7.12A, p. 608). If crystals begin to form while you are evaporating the solvent, remove the beaker from the water bath. If crystals do not form, reduce the volume of the solvent by about one third. Allow the solution to cool and then place it in an ice bath.

Collect the crystals by vacuum filtration, using a small Hirsch funnel (Technique 8, Section 8.3, p. 621). Save both the crystals and the filtrate for the tests later. After the crystals are dry, weigh them and determine their melting point (literature value, 39°C). Calculate the percentage yield. Submit your remaining sample to your instructor in a properly labeled container, along with your report.

Unsaturation Tests (Optional)

Number three test tubes (1, 2, 3) and place one of the following samples into each test tube:

1. About 0.1 mL of methyl oleate dissolved in a small amount of methylene chloride

2. A small spatulaful of your methyl stearate product dissolved in a small amount of methylene chloride

3. About 0.1 mL of the filtrate that you saved as directed earlier

To each test tube, add a solution of bromine in methylene chloride, drop by drop, to the contents of the test tubes until the red color is no longer discharged. Record the results in each case, including the number of drops required. What do these results indicate about the presence of unsaturated compounds in each sample?

QUESTIONS

1. Using the information in the essay on fats and oils, draw the structure of the triacylglycerol (triglyceride) formed from oleic acid, linoleic acid, and stearic acid. Give a balanced equation and show how much hydrogen would be needed to reduce the triacylglycerol completely; show the product.

2. A 0.150-g sample of a pure compound subjected to catalytic hydrogenation takes up 25.0 mL of H_2 at 25°C and 1 atm pressure. Calculate the molecular weight of the compound, assuming that it has only one double bond.

3. A compound with the formula C_5H_6 takes up 2 moles of H_2 on catalytic hydrogenation. Give one possible structure that would fit the information given.

4. A compound of formula C_6H_{10} takes up 1 mole of H_2 on reduction. Give one possible structure that would fit the information.

ESSAY

Soaps and Detergents

Soaps as we know them today were virtually unknown before the first century A.D. Clothes were cleaned primarily by the abrasive action of rubbing them on rocks in water. Somewhat later, it was discovered that certain types of leaves, roots, nuts, berries, and barks formed soapy lathers that solubilized and removed dirt from clothes. We now refer to these natural materials that lather as **saponins**. Many saponins contain pentacyclic triterpene carboxylic acids, such as oleanolic acid or ursolic acid, chemically combined with a sugar molecule. These acids also appear in the uncombined state. Saponins were probably the first known "soaps." They may have also been an early source of pollution in that they are known to be toxic to fish. The pollution problem associated with the development of soap and detergents has been long and controversial.

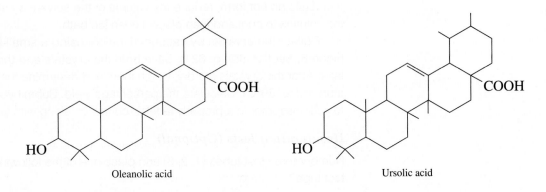

Oleanolic acid   Ursolic acid

Soap as we know it today has evolved over many centuries from experimentation with crude mixtures of alkaline and fatty materials. Pliny the Elder described the manufacture of soap during the first century A.D. A modest soap factory was even built in Pompeii. During the Middle Ages, cleanliness of the body or clothing was not considered important. Those who could afford perfumes used them to hide their body odor. Perfumes, like fancy clothes, were status symbols for the rich. An interest in cleanliness again emerged during the eighteenth century, when disease-causing microorganisms were discovered.

Soaps

The process of making soap has remained practically unchanged for 2,000 years. The procedure involves the basic hydrolysis or **saponification** of an animal fat or a vegetable oil. Chemically, fats and oils are referred to as **triglycerides** or **triacylglycerols.** They contain ester functional groups. Saponification involves heating a fat or oil with an alkaline solution. This alkaline solution was originally obtained by leaching wood ashes or from the evaporation of natural alkaline waters. Today, lye (sodium hydroxide) is used as the source of the alkali. The alkaline solution hydrolyzes the fat or oil into its component parts, the sodium salt of a long-chain carboxylic acid (soap) and an alcohol (glycerol). When common salt is added, the soap precipitates. The soap is washed free of unreacted sodium hydroxide and molded into bars. The following equation shows how soap is produced from a fat or oil.

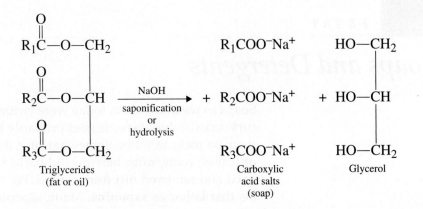

Triglycerides (fat or oil)   Carboxylic acid salts (soap)   Glycerol

The carboxylic acids represented in soap are rarely of a single type in any given fat or oil. In fact, a single triglyceride (triacylglycerol) molecule in a fat may contain three different acid residues (R_1COOH, R_2COOH, R_3COOH), and not every triglyceride in the substance will be identical. Each fat or oil, however, has a characteristic *statistical distribution* of the various types of acids possible. The carboxylic acid salts of soap usually contain 12–18 carbons arranged in a straight chain. The carboxylic acids containing even numbers of carbon atoms predominate, and the chains may contain unsaturation. The composition of the common fats and oils is given in the essay "Fats and Oils" (p. 233).

The fats and oils that are most common in soap preparations are lard and tallow from animal sources and coconut, palm, and olive oils from vegetable sources. The length of the hydrocarbon chain and the number of double bonds in the carboxylic acid portion of the fat or oil determine the properties of the resulting soap. For example, a salt of a saturated long-chain acid makes a harder, more insoluble soap. Chain length also affects solubility.

Tallow is the principal fatty material used in making soap. The solid fats of cattle are melted with steam, and the tallow layer formed at the top is removed. Soap makers usually blend tallow with coconut oil and saponify this mixture. The resulting soap contains mainly the salts of palmitic, stearic, and oleic acids from the tallow and the salts of lauric and myristic acids from the coconut oil. The coconut oil is added to produce a softer, more soluble soap. Lard (from hogs) differs from tallow (from cattle or sheep) in that lard contains more oleic acid.

Pure coconut oil yields a soap that is soluble in water. The soap contains essentially the salt of lauric acid, with some myristic acid. It is so soft (soluble) that it will lather even in seawater. Palm oil contains mainly two acids, palmitic acid and oleic acid, in about equal amounts. Saponification of this oil yields a soap that is an important constituent of toilet soaps. Olive oil contains mainly oleic acid. It is used to prepare Castile soap, named after the region in Spain in which it was first made.

Tallow	$CH_3(CH_2)_{14}COOH$	$CH_3(CH_2)_{16}COOH$
	Palmitic acid	Stearic acid

$$CH_3(CH_2)_7CH{=}CH(CH_2)_7COOH$$
Oleic acid

Coconut oil	$CH_3(CH_2)_{10}COOH$	$CH_3(CH_2)_{12}COOH$
	Lauric acid	Myristic acid

Toilet soaps generally have been carefully washed free of any alkali remaining from the saponification. As much glycerol as possible is usually left in the soap, and perfumes and medicinal agents are sometimes added. Floating soaps are produced by blowing air into the soap as it solidifies. Soft soaps are made by using potassium hydroxide, yielding potassium salts rather than the sodium salts of the acids. They are used in shaving creams and liquid soaps. Scouring soaps have abrasives added, such as fine sand or pumice.

A disadvantage of soap is that it is an ineffective cleanser in hard water. Hard water contains salts of magnesium, calcium, and iron in solution. When soap is used in hard water, "calcium soap," the insoluble calcium salts of the fatty acids, and other precipitates are deposited as **curds.**

This precipitate, or curd, is referred to as "bathtub ring." Although soap is a poor cleanser in hard water, it is an excellent cleanser in soft water.

$$2\ R\overset{\overset{\displaystyle O}{\|}}{-C}-O^-Na^+ + Ca^{2+} \longrightarrow \left(R\overset{\overset{\displaystyle O}{\|}}{-C}-\bar{O}\right)_2 Ca^{2+}$$

Soap   Curd

Water softeners are added to soaps to help remove the troublesome hard-water ions so that the soap will remain effective in hard water. Sodium carbonate or trisodium phosphate will precipitate the ions as the carbonate or phosphate. Unfortunately, the precipitate may become lodged in the fabric of items being laundered, causing a grayish or streaked appearance.

$$Ca^{2+} + CO_3^{2-} \longrightarrow CaCO_3\ (s)$$

$$3\ Ca^{2+} + 2\ PO_4^{3-} \longrightarrow Ca_3(PO_4)_2\ (s)$$

An important advantage of soap is that it is **biodegradable.** Microorganisms can consume the linear soap molecules and convert them to carbon dioxide and water. The soap is thus eliminated from the environment.

Action of Soap in Cleaning

Dirty clothes, skin, or other surfaces have particles of dirt suspended in a layer of oil or grease. Polar water molecules cannot remove the dirt embedded in nonpolar oil or grease. One can remove the dirt with soap, however, because of its dual nature. The soap molecule has a polar, *water-soluble* head (carboxylate salt) and a long, *oil-soluble* tail (the hydrocarbon chain). The hydrocarbon tail of soap dissolves in the oily substance, but the ionic end remains outside the oily surface. When enough soap molecules have oriented themselves around an oil droplet with their hydrocarbon ends dissolved in the oil, the oil droplet, together with the suspended dirt particles, is removed from the surface of the cloth or skin. The oil droplet is removed because the heavily negatively charged oil droplet is now strongly attracted to water and solvated by the water. The solvated oil droplet is called a **micelle.**

Detergents

Detergents are synthetic cleaning compounds, often referred to as **syndets.** They were developed as an alternative to soaps because they are effective in *both* soft and hard water. No precipitates form when calcium, magnesium, or iron ions are present in a detergent solution. One of the earliest detergents developed was sodium lauryl sulfate. It is prepared by the action of sulfuric acid or chlorosulfonic acid on lauryl alcohol (1-dodecanol). This detergent is relatively expensive, however. The following reactions show one industrial method of preparation:

$$CH_3(CH_2)_{10}CH_2OH + H_2SO_4 \longrightarrow CH_3(CH_2)_{10}CH_2OSO_3H + H_2O$$

Lauryl alcohol

$$CH_3(CH_2)_{10}CH_2OSO_3H + Na_2CO_3 \longrightarrow CH_3(CH_2)_{10}CH_2OSO_3^-Na^+ + NaHCO_3$$

Sodium lauryl sulfate

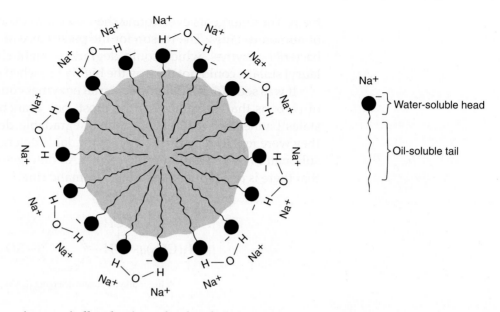

A soap micelle solvating a droplet of oil. (From W. W. Linstromberg, Organic Chemistry: A Brief Course, *D. C. Heath, 1978.)*

The first of the inexpensive detergents appeared about 1950. These detergents, called alkylbenzenesulfonates (ABS), can be prepared from inexpensive petroleum sources by the following set of reactions:

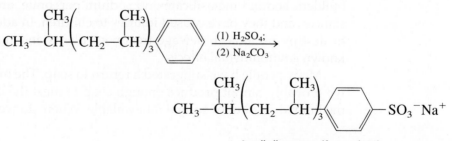

An alkylbenzenesulfonate (ABS)

Detergents became popular because they could be used effectively in all types of water and were cheap. They rapidly displaced soap as the most popular cleaning agent. A problem with the detergents was that they passed through sewage-treatment plants without being degraded by the microorganisms present, a process necessary for the full sewage treatment. Rivers and streams in many sections of the country became polluted with detergent foam. The detergents even found their way into the drinking water supplies of numerous cities. The reason for the persistence of the detergents was that bacterial enzymes, which could degrade straight-chain soaps and sodium lauryl sulfate, could not destroy the highly branched detergents such as ABS.

It was soon found that the bacterial enzymes could degrade only a chain of carbons that contained, at the most, one branch. Numerous cities and states banned the sale of the nonbiodegradable detergents, and by 1966, they were replaced by the new biodegradable detergents called linear alkylsulfonates (LAS). One example of an LAS detergent is shown here. Notice that there is one branch next to the aromatic ring.

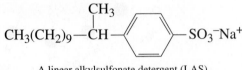

A linear alkylsulfonate detergent (LAS)

New Problems with Detergents

Detergents (also soaps) are not sold as pure compounds. A typical heavy-duty, controlled **sudser** may contain only 8–20% of the linear alkylsulfonate. A large quantity (30–50%) of a **builder** such as sodium tripolyphosphate, $Na_5P_3O_{10}$ may be present. Other additives include corrosion inhibitors, antideposition agents, and perfumes. Optical brighteners are also added. Brighteners absorb invisible ultraviolet light and reemit it as visible light, so laundry appears white and thus "clean." The phosphate builder is added to complex the hard-water ions, calcium and magnesium, and keep them in solution. Builders seem to enhance the washing ability of the LAS and also act as a cheap filler.

Unfortunately, phosphates speed the **eutrophication** of lakes and other bodies of water. The phosphates, along with other substances, are nutrients for algae. When algae begin to die and decompose, they consume so much dissolved oxygen from the water that no other life can exist in that water. The lake rapidly "dies." This is eutrophication.

Because phosphates have this undesirable effect, a search was initiated for a replacement for the phosphate builders. Some replacements have been made, but most also have problems associated with them. Two replacement builders, sodium metasilicate and sodium perborate, are highly basic substances, and they have caused injuries to children. In addition, they appear to destroy bacteria in sewage-treatment plants and may have other unknown environmental effects.

Many people have suggested a return to soap. The main problem is that we probably cannot produce enough soap to meet the demand because of the limited amount of animal fat available. Where do we go from here?

REFERENCES

Davidson, A., and Milwidsky, B. M. *Synthetic Detergents*, 6th ed. New York: Wiley, 1978.

Greek, B. F. "Detergent Components Become Increasingly Diverse, Complex." *Chemical and Engineering News, 66* (January 25, 1988): 21.

Greek, B. F., and Layman, P. L. "Higher Costs Spur New Detergent Formulations." *Chemical and Engineering News, 67* (January 23, 1989): 29.

Kirschner, E. M. "Soaps and Detergents." *Chemical and Engineering News, 76* (January 26, 1998): 39.

"LAS Detergents End Stream Foam." *Chemical and Engineering News, 45* (1967): 29.

Levey, M. "The Early History of Detergent Substances." *Journal of Chemical Education, 31* (1954): 521.

Meloan, C. E. "Detergents—Soaps and Syndets." *Chemistry, 49* (September 1976): 6.

Newell, J. "Why Plants Make Soap." *Chemistry in Britain, 34* (April 1998): 19.

Rosen, M. J. "Geminis: A New Generation of Surfactants." *Chemtech, 23* (March 1993): 30.

Rosen, M. J. *Surfactants and Interfacial Phenomena,* 2nd ed. New York: Wiley, 1989.

Rosen, M. J. "Surfactants: Designing Structure for Performance." *Chemtech, 15* (May 1985): 292.

Snell, F. D. "Soap and Glycerol." *Journal of Chemical Education, 19* (1942): 172.

Snell, F. D., and Snell, C. T. "Syndets and Surfactants." *Journal of Chemical Education, 35* (1958): 271.

Stinson, S. C. "Consumer Preferences Spur Innovation in Detergents." *Chemical and Engineering News, 65* (January 26, 1987): 21.

30   **EXPERIMENT 30**

Preparation of a Soap

Hydrolysis of a fat (ester)

Filtration

In this experiment, we prepare soap from animal fat (lard). Animal fats and vegetable oils are esters of carboxylic acids; they have a high molecular weight and contain the alcohol glycerol. Chemically, these fats and oils are called **triglycerides.** The principal acids in animal fats and vegetable oils can be prepared from the natural triglycerides by alkaline hydrolysis (saponification). See the preceding essay for a complete discussion of soaps and detergents.

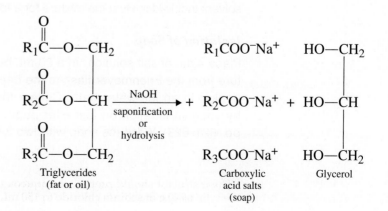

| Triglycerides (fat or oil) | | Carboxylic acid salts (soap) | Glycerol |

$$\text{R}_1\overset{\overset{\displaystyle O}{\|}}{\text{C}}\!-\!\text{O}\!-\!\text{CH}_2$$
$$\text{R}_2\overset{\overset{\displaystyle O}{\|}}{\text{C}}\!-\!\text{O}\!-\!\text{CH} \xrightarrow[\substack{\text{saponification}\\\text{or}\\\text{hydrolysis}}]{\text{NaOH}}$$
$$\text{R}_3\overset{\overset{\displaystyle O}{\|}}{\text{C}}\!-\!\text{O}\!-\!\text{CH}_2$$

$$+ \quad \text{R}_1\text{COO}^-\text{Na}^+ \qquad \text{HO}\!-\!\text{CH}_2$$
$$\text{R}_2\text{COO}^-\text{Na}^+ + \text{HO}\!-\!\text{CH}$$
$$\text{R}_3\text{COO}^-\text{Na}^+ \qquad \text{HO}\!-\!\text{CH}_2$$

REQUIRED READING

Review: Technique 7   Sections 7.1–7.3

Technique 8   Section 8.3

New: Essay   Soaps and Detergents

Essay   Fats and Oils

SPECIAL INSTRUCTIONS

This experiment is short and can easily be scheduled with another experiment. Avoid contact with sodium hydroxide. It is caustic.

SUGGESTED WASTE DISPOSAL

Dispose of all filtrates obtained in this experiment in the waste container provided for aqueous solutions.

PROCEDURE

Reaction Mixture

Prepare a solution of about 0.25 g of sodium hydroxide (two to three pellets) dissolved in a mixture of 1.0 mL of distilled water and 1.0 mL of 95% ethanol.

CAUTION

Do not touch the sodium hydroxide pellets, because they are caustic. Use a scoop. The pellets should be weighed as rapidly as possible because they tend to draw moisture from air and become sticky.

You may use a calibrated Pasteur pipet to measure and transfer the water and ethanol. Place about 0.25 g of lard in a 10-mL Erlenmeyer flask and add the sodium hydroxide solution to the flask. Heat the mixture to boiling on a hot plate. Place an inverted 20-mL beaker over the neck of the flask to help to reduce evaporation. Swirl the Erlenmeyer flask every few minutes.

The soap often begins to precipitate from the boiling mixture within about 20 minutes. If it appears that some of the alcohol and water is evaporating from the flask, you may add up to 0.4 mL of a 50% water–alcohol mixture to replace the solvent that is lost. Heat the mixture for a total of 25 minutes.

Isolation of Soap

Place 4 mL of salt solution[1] in a 20-mL beaker and transfer the saponified mixture from the Erlenmeyer flask to the beaker. Stir the mixture while cooling the beaker in an ice-water bath. Collect the prepared soap on a Hirsch funnel by vacuum filtration on fast filter paper (Technique 8, Sections 8.2 and 8.3, pp. 621–623). Wash the soap with two 3-mL portions of ice-cold distilled water

---

[1] The instructor should prepare the aqueous sodium chloride solution for the class in the ratio of 40 g of sodium chloride to 150 mL of distilled water.

to remove any excess sodium hydroxide. Continue to draw air through the filter for a few minutes to partially dry the product. **Test your soap while it is still damp** using the procedure given in the next section. Allow the remaining sample to dry until the next period. Weigh the product. Submit the sample to your instructor in a labeled vial.

Tests on Soaps and Detergents

Soap

After rinsing your product, remove a 4-mm piece (about 0.01 g) from the filter paper and place it in a clean 10-mL graduated cylinder. Add 3 mL of distilled water, place your thumb over the opening of the cylinder, and shake the mixture vigorously for about 15 seconds. After about 30 seconds, observe the level of the foam. Add two drops of 4% calcium chloride solution to the soap mixture from a Pasteur pipet. Shake the mixture for 15 seconds and let it stand for 30 seconds. Observe the effect of the calcium chloride on the foam. Do you observe anything else? Add 0.5 g of trisodium phosphate, and shake the mixture again for 15 seconds. After 30 seconds, again observe the results. Explain the results of these tests in your laboratory notebook.

Detergent

Place a 2-mm piece (about 0.005 g) of sodium lauryl sulfate (sodium dodecyl sulfate or dodecyl sodium sulfate) in a 10-mL graduated cylinder. Add 3 mL of distilled water to the sample. Place your thumb over the opening of the cylinder and shake the mixture vigorously for about 15 seconds. Allow the mixture to stand for about 30 seconds and observe the level of the foam. Add two drops of 4% calcium chloride solution. Shake the mixture for 15 seconds. After about 30 seconds, observe the effect of the calcium chloride on the foam. Explain the results of these tests in your laboratory notebook.

QUESTIONS

1. Why should the potassium salts of fatty acids yield soft soaps?
2. Why is the soap derived from coconut oil so soluble?
3. Why do you suppose a mixture of ethanol and water instead of simply water itself is used for saponification?
4. Sodium acetate and sodium propanoate are poor soaps. Why?
5. Why does adding a salt solution cause soap to precipitate?

---

31 **EXPERIMENT 31**

Preparation of a Detergent

Preparation of a sulfonate ester
Properties of soaps and detergents

In this experiment, you will prepare the detergent sodium lauryl sulfate. A detergent is usually defined as a synthetic cleaning agent, whereas a soap is derived from a natural source—a fat or an oil.

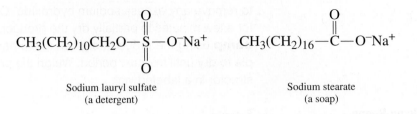

Sodium lauryl sulfate
(a detergent)

Sodium stearate
(a soap)

The differences between the two basic types of cleaning agents are discussed in the essay "Soaps and Detergents," which precedes Experiment 30. After the preparation of sodium lauryl sulfate, you will compare the properties of soap with the properties of the prepared detergent.

In the first step of the synthesis, lauryl alcohol is allowed to react with chlorosulfonic acid to give the lauryl ester of sulfuric acid. In the second step, aqueous sodium carbonate is added to produce the sodium salt (detergent).

$$CH_3(CH_2)_{10}CH_2OH + Cl-\!\!\overset{\overset{\displaystyle O}{\|}}{\underset{\underset{\displaystyle O}{\|}}{S}}\!\!-OH \longrightarrow CH_3(CH_2)_{10}CH_2O-\!\!\overset{\overset{\displaystyle O}{\|}}{\underset{\underset{\displaystyle O}{\|}}{S}}\!\!-OH + HCl \quad [1]$$

Lauryl alcohol   Chlorosulfonic acid   Lauryl ester of sulfuric acid

$$2\ CH_3(CH_2)_{10}CH_2-O-\!\!\overset{\overset{\displaystyle O}{\|}}{\underset{\underset{\displaystyle O}{\|}}{S}}\!\!-OH + Na_2CO_3 \longrightarrow$$

$$2\ CH_3(CH_2)_{10}CH_2O-\!\!\overset{\overset{\displaystyle O}{\|}}{\underset{\underset{\displaystyle O}{\|}}{S}}\!\!-O^-Na^+ + H_2O + CO_2 \quad [2]$$

Sodium lauryl sulfate

The aqueous mixture is saturated with solid sodium carbonate and extracted with 1-butanol. Sodium carbonate must be added to give phase separation; otherwise, 1-butanol would be soluble in water. The sodium salt (detergent) is more soluble in 1-butanol than in the aqueous layer because of the long hydrocarbon chain, which gives the salt considerable organic (nonpolar) character.

REQUIRED READING

Review: Technique 7   Section 7.8A

            Technique 12   Section 12.4

New: Essay   Soaps and Detergents

SPECIAL INSTRUCTIONS

Chlorosulfonic acid must be handled with care because it is a corrosive liquid and reacts violently with water. Be certain to use dry glassware.

SUGGESTED WASTE DISPOSAL

You may dispose of the aqueous layers by placing them into the aqueous waste bottle.

NOTES TO THE INSTRUCTOR

The 1-dodecanol (lauryl alcohol) is best handled as a liquid. If necessary, melt the alcohol (mp 24–27°C) and pour the liquid into a small container. Keep the alcohol in the liquid state by placing the container in a warm sand bath or on a hot plate. In this way, the alcohol will be available to the class as a liquid.

PROCEDURE

Part A. Sodium Lauryl Sulfate

Preparation of Chlorosulfonic Acid/Acetic Acid Solution

Transfer 0.100 mL of concentrated (glacial) acetic acid into a *dry* 5-mL conical vial. Cap the vial and cool it in a small beaker with ice for about 5 minutes. In a hood, remove 0.035 mL of chlorosulfonic acid ($d = 1.77$ g/mL) using the graduated pipet provided for you and add it *dropwise* to the conical vial containing the acetic acid in the ice bath. (Use safety glasses!)

> ### CAUTION ⚠
>
> **Use chlorosulfonic acid with extreme care. Avoid getting water or ice in the vial. Chlorosulfonic acid reacts violently with water to form hydrochloric acid. Transfer the material directly into your vial without dripping the liquid. Chlorosulfonic acid is an extremely strong acid similar to concentrated sulfuric acid. It will cause immediate burns.**

Reaction of Chlorosulfonic Acid with 1-Dodecanol

The following operations may be conducted at your laboratory bench if you are careful. Prepare a gas trap by placing some cotton into a drying tube and adding a few drops of water to moisten the cotton (Technique 7, Section 7.8A, p. 607). Avoid excess water or it may accidentally run down into the vial. Remove the vial from the ice bath, place a magnetic spin vane in the vial, and add 0.12 mL of 1-dodecanol (lauryl alcohol, $d = 0.831$ g/mL) into the conical vial. Place the drying tube (gas trap) on the vial, clamp the vial securely, and stir the mixture for 15 minutes at room temperature. After this time, *carefully* add 20 drops of ice-cold water to the vial with a Pasteur pipet throughout a period of 2 minutes. Continue to stir the mixture while adding the water.

> ### CAUTION ⚠
>
> **Excess chlorosulfonic acid will react violently with water. Replace the gas trap (drying tube) after each addition of water.**

Extraction of the Detergent with 1-Butanol

Add 0.30 mL of 1-butanol to the conical vial and stir the mixture with the spin vane for 5 minutes. While stirring, slowly add 0.15 g of sodium carbonate (anhydrous) to neutralize the acids and to aid in the separation of layers. The sodium carbonate will dissolve. After stirring the mixture, cap the vial and shake the conical vial so that the 1-butanol will extract the detergent from the aqueous layer. Allow the layers to separate for 5–10 minutes or until a complete separation has been achieved. The 1-butanol layer will be on top. Remove the magnetic spin vane from the vial with forceps. With care, remove the lower aqueous layer with a Pasteur pipet (Technique 12, Section 12.4, p. 674), and place it in a 3-mL conical vial. Save the organic layer (1-butanol) in the original 5-mL vial because it contains your detergent product.

Reextract the aqueous layer. To do this, add 0.3 mL of 1-butanol to the vial, cap the vial, and shake it. Let the vial stand for about 10 minutes or until a complete separation has been achieved. Remove the lower aqueous phase with a Pasteur pipet and discard it.

Combine the *two* 1-butanol organic phases in one of the vials. Allow these combined phases to stand for a few minutes to see if any further separation of layers occurs. If some further separation is observed, remove the lower aqueous layer and discard it. Otherwise, transfer the 1-butanol extracts into a preweighed 10-mL beaker.

Evaporation of 1-Butanol

At the option of your instructor, either store the 1-butanol solution of your detergent in your locker or place it in a hood until the next period. During this time, the 1-butanol should evaporate to give the detergent. If an odor of 1-butanol still remains, place the beaker in an oven maintained at about 80°C until the solid is thoroughly dry and odor-free. Use your spatula to break up the solid. Weigh the product and calculate the percentage yield (*MW* = 288.4). If the detergent is not totally free of 1-butanol, the apparent yield may exceed 100%. If necessary, continue to dry the sample. After doing the tests that follow, submit the remaining detergent to the instructor in a labeled vial.

Part B. Tests on Soaps and Detergents

Soap

Pour 3 mL of soap solution[1] into a 10-mL graduated cylinder. Place your thumb over the opening of the cylinder and shake it vigorously for about 15 seconds. Allow the solution to stand for 30 seconds and observe the level of the foam. Add 2 drops of 4% calcium chloride solution from a Pasteur pipet. Shake the mixture for 15 seconds and allow it to stand for about 30 seconds. Observe the effect of the calcium chloride on the foam. Do you observe anything else? Add 0.3 g of trisodium phosphate and shake the mixture again for about 15 seconds. Allow the solution to stand for 30 seconds. What do you observe? Explain these tests in your laboratory report.

Detergent

Place a 2-mm piece (about 0.005 g) of your prepared detergent in a 10-mL graduated cylinder and add 3 mL of distilled water. Hold your thumb over the opening and shake the graduated cylinder vigorously for 30 seconds. Allow the solution to stand for about 30 seconds and observe the level of the foam. Add 2 drops of 4% calcium chloride solution. Shake the mixture for 15 seconds and let it stand for 30 seconds. What do you observe? Explain the results of these tests in your laboratory report.

---

[1] A large batch of soap solution should be prepared by the instructor, as follows: Add one bar of Ivory soap to 1 L of distilled water. Stir the solution occasionally and let the mixture stand overnight. Remove the rest of the bar. The mixture can be used directly. Alternatively, a 0.5 g sample of soap prepared in Experiment 30 can be added to 10 mL of distilled water.

QUESTIONS

1. Draw a mechanism for the reaction of lauryl alcohol with chlorosulfonic acid.
2. Why do you suppose sodium carbonate, instead of some other base, is used for neutralization?
3. Propose a model to explain how a cationic detergent works. A cationic detergent has its polar end positively charged.
4. Sodium methyl sulfate $CH_3OSO_2^- Na^+$ is a poor detergent. Why?
5. Sodium lauryl sulfate can be prepared by replacing chlorosulfonic acid with another reagent. What could be used? Show the equations.
6. Suggest a method for synthesizing the linear alkyl sulfonate detergent shown on page 248, starting with lauryl alcohol, benzene, and any needed inorganic compounds.

ESSAY

Petroleum and Fossil Fuels

Crude petroleum is a liquid that consists of hydrocarbons, as well as some related sulfur, oxygen, and nitrogen compounds. Other elements, including metals, may be present in trace amounts. Crude oil is formed by the decay of marine animal and plant organisms that lived millions of years ago. Over many millions of years, under the influence of temperature, pressure, catalysts, radioactivity, and bacteria, the decayed matter was converted into what we now know as crude oil. The crude oil is trapped in pools beneath the ground by various geological formations.

Most crude oils have a specific gravity between 0.78 and 1.00 g/mL. As a liquid, crude oil may be as thick and black as melted tar or as thin and colorless as water. Its characteristics depend on the particular oil field from which it comes. Pennsylvania crude oils are high in straight-chain alkane compounds (called **paraffins** in the petroleum industry); those crude oils are therefore useful in the manufacture of lubricating oils. Oil fields in California and Texas produce crude oil with a higher percentage of cycloalkanes (called **naphthenes** by the petroleum industry). Some Middle East fields produce crude oil containing up to 90% cyclic hydrocarbons. Petroleum contains molecules in which the number of carbons ranges from 1 to 60.

When petroleum is refined to convert it into a variety of usable products, it is initially subjected to a fractional distillation. Table 1 lists the

Table 1 *Fractions obtained from the distillation of crude oil*

Petroleum Fraction	Composition	Commercial Use
Natural gas	C_1 to C_4	Fuel for heating
Gasoline	C_5 to C_{10}	Motor fuel
Kerosene	C_{11} to C_{12}	Jet fuel and heating
Light gas oil	C_{13} to C_{17}	Furnaces, diesel engines
Heavy gas oil	C_{18} to C_{25}	Motor oil, paraffin wax, petroleum jelly
Residuum	C_{26} to C_{60}	Asphalt, residual oils, waxes

various fractions obtained from fractional distillation. Each of these fractions has its own particular uses. Each fraction may be subjected to further purification, depending on the desired application.

The gasoline fraction obtained directly from the distillation of crude oil is called **straight-run gasoline.** An average barrel of crude oil will yield about 19% straight-run gasoline. This yield presents two immediate problems. First, there is not enough gasoline contained in crude oil to satisfy current needs for fuel to power automobile engines. Second, the straight-run gasoline obtained from crude oil is a poor fuel for modern engines. It must be "refined" at a chemical refinery.

The initial problem of the small quantity of gasoline available from crude oil can be solved by **cracking** and **polymerization.** Cracking is a refinery process by which large hydrocarbon molecules are broken down into smaller molecules. Heat and pressure are required for cracking, and a catalyst must be used. Silica–alumina and silica–magnesia are among the most effective cracking catalysts. A mixture of saturated and unsaturated hydrocarbons is produced in the cracking process. If gaseous hydrogen is also present during the cracking, only saturated hydrocarbons are produced. The hydrocarbon mixtures produced by these cracking processes tend to have a fairly high proportion of branched-chain isomers. These branched isomers improve the quality of the fuel.

$$C_{16}H_{34} + H_2 \xrightarrow[\text{heat}]{\text{catalyst}} 2\ C_8H_{18} \qquad \text{Cracking}$$

In the polymerization process, also carried out at a refinery, small molecules of alkenes are caused to react with one another to form larger molecules, which are also alkenes.

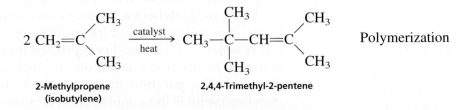

2-Methylpropene (isobutylene)   **2,4,4-Trimethyl-2-pentene**   Polymerization

The newly formed alkenes may be hydrogenated to form alkanes. The reaction sequence shown here is a common and important one in petroleum refining because the product, 2,2,4-trimethylpentane (or "isooctane"), forms the basis for determining the quality of gasoline. By these refining methods, the percentage of gasoline that can be obtained from a barrel of crude oil may rise to as much as 45% or 50%.

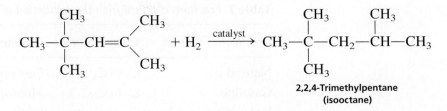

2,2,4-Trimethylpentane (isooctane)

The internal combustion engine, as it is found in most automobiles, operates in four cycles, or **strokes.** They are illustrated in the figure. The power stroke is of greatest interest from the chemical point of view because combustion occurs during this stroke.

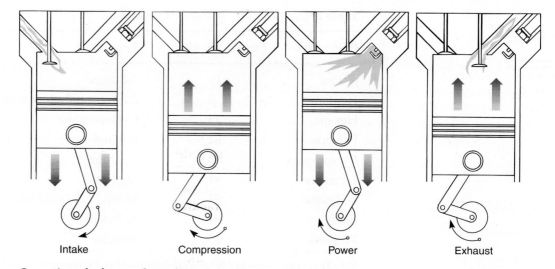

Intake   Compression   Power   Exhaust

Operation of a four-cycle engine.

When the air–fuel mixture is ignited, it does not explode. Rather, it burns at a controlled, uniform rate. The gases closest to the spark are ignited first; then they in turn ignite the molecules farther from the spark, and so on. The combustion proceeds in a wave of flame, or a **flame front,** which starts at the spark plug and proceeds uniformly outward from that point until all the gases in the cylinder have been ignited. Because a certain time is required for this burning, the initial spark is timed to ignite just before the piston has reached the top of its travel. In this way, the piston will be at the very top of its travel at the precise instant that the flame front and the increased pressure that accompanies it reach the piston. The result is a smoothly applied force to the piston, driving it downward.

If heat and compression should cause some of the air–fuel mixture to ignite before the flame front has reached it or to burn faster than expected, the timing of the combustion sequence is disturbed. The flame front arrives at the piston before the piston has reached the very top of its travel. When the combustion is not perfectly coordinated with the motion of the piston, we observe **knocking,** or **detonation** (sometimes called "pinging"). The transfer of power to the piston under these conditions is much less effective than in normal combustion. The wasted energy is merely transferred to the engine block in the form of additional heat. The opposing forces that occur in knocking may eventually damage the engine.

The tendency of a fuel to knock is a function of the structures of the molecules composing the fuel. Normal hydrocarbons, those with straight carbon chains, have a greater tendency to lead to knocking than do alkanes with highly branched chains. A fuel can be classified according to its antiknock characteristics. The most important rating system is the octane rating of gasoline. In this method of classification, the antiknock properties of a fuel are compared in a test engine with the antiknock properties of a standard mixture of heptane and 2,2,4-trimethylpentane. This latter compound is called "isooctane"; hence, the name *octane rating*. A fuel that has the same antiknock properties as a given mixture of heptane and isooctane has an octane rating numerically equal to the percentage of isooctane in that reference mixture. Today's 87-octane unleaded gasoline is a mixture of compounds that have, taken together, the same antiknock characteristics as a test fuel composed of 13% heptane and 87% isooctane. Other substances besides

Table 2 *Octane ratings of organic compounds*

Compound	Octane Number	Compound	Octane Number
Octane	−19	1-Butene	97
Heptane	0	2,2,4-Trimethylpentane	100
Hexane	25	Cyclopentane	101
Pentane	62	Ethanol	105
Cyclohexane	83	Benzene	106
1-Pentene	91	Methanol	106
2-Hexene	93	Methyl *tert*-butyl ether	116
Butane	94	*m*-Xylene	118
Propane	97	Toluene	120

Note: The octane values in this table are determined by the **research method.**

hydrocarbons may also have high resistance to knocking. Table 2 presents a list of organic compounds with their octane ratings.

Several chemical refining processes are used to improve the octane rating of gasoline and to increase the percentage of gasoline that can be obtained from petroleum. Some of these reactions, collectively known as **reforming,** are dehydrogenation, dealkylation, cyclization, and isomerization. The products of these reactions, sometimes referred to as **reformates,** contain many branched alkanes and aromatic compounds. Several examples of reforming reactions are:

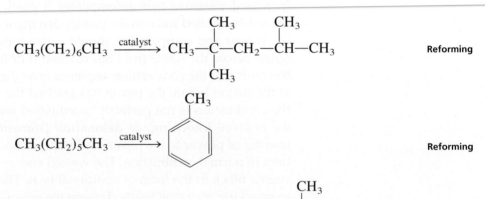

Other chemical reactions, referred to as **alkylation,** can also be used to increase the octane rating. Alkylation involves the catalytic addition of an alkane to an alkene, such as 2-methylpropane to propene or butene. The products of these reactions are sometimes referred to as **alkylates.** Another refining process, called **hydrocracking** (cracking in the presence of hydrogen gas), also produces hydrocarbons that reduce knocking.

None of these processes converts all the normal hydrocarbons into branched-chain isomers; consequently, additives are put into gasoline to improve the octane rating of the fuel. Before 1996, the most common additive used to reduce knocking was **tetraethyllead.** Gasoline that contains tetraethyllead is called **leaded** gasoline, whereas gasoline produced without tetraethyllead is sometimes called **unleaded** gasoline. Because of concern over the possible health hazard associated with emission of lead into the atmosphere, the Environmental Protection Agency began in 1973 to limit the amount of tetraethyllead in gasoline. In 1996, the Clean Air Act completely banned the sale of leaded gasoline for use in all on-road vehicles. Many other countries have followed with similar bans; however, some countries in Eastern Europe, the Middle East, and Africa continue to use leaded gasoline.

$$CH_3-CH_2-Pb-CH_2-CH_3$$

with CH_2-CH_3 groups above and below the Pb

Tetraethyllead

To replace tetraethyllead, oil companies have developed other additives and strategies that will improve the octane rating of gasoline without producing harmful emissions. One approach is to increase the quantities of hydrocarbons that have high antiknock properties themselves. Typical are the aromatic hydrocarbons, including benzene, toluene, and xylene. Such compounds are natural components of most crude petroleum, and additional aromatic compounds can be added to gasoline to improve the quality. Increasing the proportion of aromatic hydrocarbons brings with it certain hazards, however. These substances are toxic, and benzene is considered a serious carcinogenic hazard. The risk that illness will be contracted by workers in refineries, and especially by persons who work in service stations, is increased. A safer approach is to increase the amount of alkylates (see p. 258).

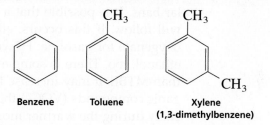

| Benzene | Toluene | Xylene (1,3-dimethylbenzene) |

Research has also been directed toward development of nonhydrocarbon compounds that can improve the quality of unleaded gasoline. To this end, compounds such as methyl *tert*-butyl ether (MTBE), ethanol, and other **oxygenates** (oxygen-containing compounds) are added to improve the octane rating of fuels. Ethanol is attractive because it is formed by fermentation of living material, a renewable resource (see essay "Ethanol and Fermentation Chemistry," p. 142). Ethanol not only would improve the antiknock properties of gasolines but also would potentially help the country to reduce its dependence on imported petroleum. Substituting ethanol for hydrocarbons in petroleum would have the effect of increasing the "yield" of fuel produced from a barrel of crude oil. As in many stories that are too good to be true, it is not clear that the energy needed to produce the ethanol by fermentation and distillation is significantly smaller than the amount of energy that is produced when the ethanol is burned in an engine!

$$CH_3-O-\underset{\underset{CH_3}{|}}{\overset{\overset{CH_3}{|}}{C}}-CH_3 \qquad CH_3-CH_2-OH$$

Methyl *tert*-butyl ether   **Ethanol**

In an effort to improve air quality in urban areas, the Clean Air Act in 1990 mandated the addition of oxygen-containing compounds in many urban areas during the winter (November to February). These compounds are expected to reduce carbon monoxide emissions produced when the gasoline burns in cold engines by helping to oxidize carbon monoxide to carbon dioxide. They also help to reduce the amount of ozone created by emission products reacting in sunlight (see p. 261), and they increase the octane rating. Refineries add "oxygenates," such as ethanol or methyl *tert*-butyl ether, to the gasoline sold in the carbon monoxide–containment areas. By law, gasoline must contain at least 2.7% oxygen by weight, and the areas must use it for a minimum of the four winter months. The 2.7% oxygen content corresponds to a 15%-by-volume content of MTBE in gasoline. In 1995, the Clean Air Act also required that **reformulated gasoline** (RFG) be sold year round in sites with the worst ground-level ozone concentrations. RFG must contain a minimum of 2% oxygen by weight.

Although methyl *tert*-butyl ether is still a widely used oxygenate additive, the use of ethanol is becoming more common. There are several reasons for an increasing preference for ethanol. First, ethanol is cheaper than MTBE because of special tax breaks and subsidies that have been granted to producers of ethanol formed by fermentation. Second, there has been some concern that MTBE may cause health problems, and there have been some widely publicized occurrences of groundwater contamination by MTBE. Furthermore, people notice the odor of gasoline more easily when MTBE is present in the fuel. Because of these concerns, the use of MTBE was outlawed by California in January 2004, and other states have issued similar bans. It is possible that a complete ban on MTBE in the United States will follow. If this occurs, ethanol will most likely become the preferred oxygenate for gasoline. However, there are disadvantages with the use of ethanol, too. There is some evidence that because ethanol is more volatile than MTBE, it may increase the emission of chemicals such as volatile organic compounds (VOCs) that contribute to smog. This is a concern especially during the warmer months. In addition, studies have suggested that fuel with ethanol increases the formation of atmospheric acetaldehyde. Because acetaldehyde is a precursor of peroxyacetyl nitrate (see p. 261), it is possible that *increased* air pollution results from the use of ethanol as an oxygenate. Because MTBE is less toxic than many other ingredients in gasoline, good arguments can be made for the continuation of its use. In fact, many states in the United States and many other countries are still using MTBE. Other oxygenates such as ethyl *tert*-butyl ether and methanol are also being considered.

The number of grams of air required for the complete combustion of one mole of gasoline (assuming the formula C_8H_{18}) is 1.735 g. This gives rise to a theoretical air–fuel ratio of 15.1:1 for complete combustion. For several reasons, however, it is neither easy nor advisable to supply each cylinder with a theoretically correct air–fuel mixture. The power and performance of an engine improve with a slightly richer mixture (lower air–fuel ratio). Maximum power is obtained from an engine when the air–fuel ratio is near 12.5:1, and maximum economy is obtained when the air–fuel ratio is near 16:1. Under conditions of idling or full load (that is, acceleration), the air–fuel ratio is lower than what would be theoretically correct. As a result, complete combustion does not take place in an internal combustion engine, and carbon monoxide (CO) is produced in the exhaust gases. Other types of nonideal combustion behavior give rise to the presence of unburned hydro-carbons in the exhaust. The high combustion temperatures cause the nitro-gen and oxygen of the air to react, forming a variety of nitrogen oxides in the exhaust. Each of these materials contributes to air pollution. Under the influence of sunlight, which has enough energy to break covalent bonds, these materials may react with each other and with air to produce **smog,** which contributes to many health problems. Smog consists of **ozone,** which deteriorates rubber and damages plant life; **particulate matter,** which pro-duces haze; **oxides of nitrogen,** which produce a brownish color in the at-mosphere; and a variety of eye irritants, such as **peroxyacetyl nitrate** (PAN). Sulfur compounds in the gasoline may lead to the production of noxious sulfur-containing gases in the exhaust.

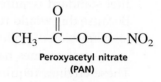

Peroxyacetyl nitrate (PAN)

Efforts to reverse the trend of deteriorating air quality caused by auto-motive exhaust have taken many forms. The advent of catalytic converters, which are mufflerlike devices containing catalysts that can convert carbon monoxide, unburned hydrocarbons, and nitrogen oxides into harmless gases, has resulted from such efforts. Some success in reducing exhaust emis-sions has been attained by modifying the design of combustion chambers of internal combustion engines. Additionally, the use of computerized control of ignition systems has achieved positive results.

It should be obvious from this discussion that there are many factors to consider in the formulation of gasoline. The gasoline produced today consists of several hundred compounds! There is substantial variation in the actual composition, depending on the local climate and regional environmental regulations. The approximate composition is 15% C_4–C_8 straight-chain alkanes, 25% C_4–C_{10} branched alkanes, 10% cycloalkanes, less than 25% aromatic compounds, and 10% straight-chain and cyclic alkenes.

Although much has been accomplished in terms of making it safer to use gasoline, there is another looming problem having to do with the sup-ply of petroleum. The amount of petroleum and other fossil fuels in the world is finite. In 1956, Marion King Hubbert, a Shell Oil geophysicist, pre-dicted that the U.S. production of oil would reach a peak around 1970, and from then on the amount extracted would decline significantly. Although most people ignored his warning, it did peak at 9 million barrels per day in 1970 and has been declining ever since, with about 6 million barrels per day being produced in 2004. Many experts have used similar methods of analy-sis to make predictions about when the world's supply of oil will peak; and although there is much variation in the actual year predicted, most experts agree that the peak will occur within the next decade. Because the demand for petroleum continues to increase every year, it is clear that declining petroleum production will have a dramatic effect on how we live. Not only is petroleum the main source of fuel used for transportation but also it provides the raw materials for a wide variety of other products, including plastics, drugs, and pesticides. Although it is possible that the decrease in production of petroleum may be partially offset by more dependence on natural gas and coal, the amount of these fossil fuels is also finite, and it seems inevitable that major adjustments will need to be made as the availability of fossil fuels declines.

There have been many developments in recent years that address some of the emission problems associated with burning gasoline and the need to stretch the supply of fossil fuels. These developments involve changes in the design of automobile engines and in the use of different fuels.

Some of the success in reducing exhaust emission has been attained by modifying the design of combustion chambers of internal combustion engines. Additionally, the use of computerized control of ignition systems has helped to reduce the level of pollutants emitted. Another strategy that could be implemented without any technological changes would be to increase fuel standard requirements, thus improving the average miles per gallon. Because this would result in less gasoline consumption, there would also be less emission of pollutants.

Diesel engines have been used in automobiles for more than 20 years. These engines require a different fraction of crude oil (see Table 1, p. 255) than gasoline, and they have been improved significantly since the initial highly polluting diesel vehicles. The diesel engine has the advantage of producing only small quantities of carbon monoxide and unburned hydrocarbons. It does, however, produce significant amounts of nitrogen oxides, soot (containing polynuclear aromatic hydrocarbons), and odor-causing compounds. Presently, the emission standards for diesel automobiles are more lenient than for those burning gasoline. More stringent standards are scheduled to be implemented in 2006 and 2009. Diesel automobiles yield higher fuel mileage than gasoline engines of a similar size; however, more oil must be refined in order to produce diesel fuel compared to gasoline. In the United States, about 3% of all new automobiles have diesel engines, whereas in Europe, about 40% of the new automobiles sold are diesel. Biodiesel, which is a chemically altered vegetable oil and can even be produced in one's garage using discarded cooking oil, can also be used in today's diesel engines, and there are fewer harmful pollutants compared to using regular diesel fuel. However, the mileage is slightly less, and it would not be possible to produce enough of this fuel for more than a small percentage of the cars on the road today.

Another possible fuel is methanol, which is produced from natural gas, coal, or biomass. Studies indicate that the amount of principal pollutants in automobiles is lowered when methanol is used instead of gasoline, but methanol is more corrosive, and extensive engine modifications must be made. Other fuels that show promise are hydrogen, methane (natural gas) and propane; however, storage and delivery of these fuels, which are gases at room temperature, are more difficult, and other significant technical problems also remain to be solved.

Hybrid-electric automobiles have become an attractive alternative to the standard automobile in the United States. Hybrid cars combine a small fuel-efficient combustion engine with an electric motor and battery. The electric motor can assist the gas engine when more power is needed, and the battery is recharged while the car is slowing down or coasting. This results in greater fuel efficiency and a drastic reduction (as much as 90%) in smog-forming pollutants. Even greater fuel efficiency is possible with diesel hybrid cars that are now being developed.

Another recent development that holds promise is the use of fuel cells that can produce electrical energy from hydrogen. This electrical energy is then used by an electric motor to propel the automobile. Although there are many proponents of hydrogen fuel cells who believe that this technology can play a major role in reducing our dependency on fossil fuels, there are significant technological challenges that must first be overcome, and the task of developing a hydrogen energy infrastructure would be costly. Furthermore, most of the hydrogen now produced comes from natural gas or coal, and this process also requires energy.

It should be clear that there are many challenges and opportunities with the use of fossil fuels in our lives. How we utilize fossil fuels will be changing in the next few decades, and chemistry will play a significant role in these changes.

REFERENCES

Ashley, S. "On the Road to Fuel-Cell Cars." *Scientific American, 292* (March 2005): 62.

Chen, C. T. "Understanding the Fate of Petroleum Hydrocarbons in the Subsurface Environment." *Journal of Chemical Education, 69* (May 1992): 357.

Goodstein, D. *Out of Gas: The End of the Age of Oil.* New York: W. W. Norton & Company, Inc., 2004.

Hogue, C. "Rethinking Ethanol." *Chemical and Engineering News, 81* (July 28, 2003): 28.

Illman, D. "Oxygenated Fuel Cost May Outweigh Effectiveness." *Chemical and Engineering News, 71* (April 12, 1993): 28.

Jacoby, M. "Fuel Cells Move Closer to Market." *Chemical and Engineering News, 81* (January 20, 2003): 328.

Kimmel, H. S., and Tomkins, R. P. T. "A Course on Synthetic Fuels." *Journal of Chemical Education, 62* (March 1985): 249.

Mielke, H. W. "Lead in the Inner Cities." *American Scientist, 87* (January 1999): 63–73.

Monahan, P., and Friedman, D. "Diesel or Gasoline? Fuel for Thought." *Catalyst, 3* (Spring 2004): 1.

Ritter, S. "Gasoline." *Chemical and Engineering News, 83* (February 21, 2005): 37.

Schriescheim, A., and Kirschenbaum, I. "The Chemistry and Technology of Synthetic Fuels." *American Scientist, 69* (September–October 1981): 536.

Seyferth, D. "The Rise and Fall of Tetraethyllead. 1. Discovery and Slow Development in European Universities, 1853–1920." *Organometallics, 22* (June 9, 2003): 2346–2357.

Seyferth, D. "The Rise and Fall of Tetraethyllead. 2." *Organometallics, 22* (December 8, 2003), 5154–5178.

Shreve, R. N., and Brink, J. "Petrochemicals." *The Chemical Process Industries*, 4th ed. New York: McGraw-Hill, 1977.

Shreve, R. N., and Brink, J. "Petroleum Refining." *The Chemical Process Industries*, 4th ed. New York: McGraw-Hill, 1977.

U.S. Environmental Protection Agency. "EPA Takes Final Step in Phaseout of Leaded Gasoline." (January 29, 1996). Available at http://www.epa.gov/history/topics/lead/0.2.htm.

U.S. Environmental Protection Agency. "MTBE in Fuels." http://www.epa.gov/mtbe/gas.htm. Accessed June 26, 2005.

Vartanian, P. F. "The Chemistry of Modern Petroleum Product Additives." *Journal of Chemical Education, 68* (December 1991): 1015.

Wald, M. L. "Questions About a Hydrogen Economy." *Scientific American, 291* (May 2004): 62.

Gas-Chromatographic Analysis of Gasolines

Gasoline

Gas chromatography

In this experiment, you will analyze samples of gasoline by gas chromatography. From your analysis, you should learn something about the composition of these fuels. Although all gasolines are compounded from the same basic hydrocarbon components, each company blends these components in different proportions to obtain a gasoline with properties similar to those of competing brands.

Sometimes the composition of the gasoline may vary, depending on the composition of the crude petroleum from which the gasoline was derived. Frequently, refineries vary the composition of gasoline in response to differences in climate or seasonal changes or environmental concerns. In the winter or in cold climates, the relative proportion of butane and pentane isomers is increased to improve the volatility of the fuel. This increased volatility permits easier starting. In the summer or in warm climates, the relative proportion of these volatile hydrocarbons is reduced. The decreased volatility reduces the possibility of forming a vapor lock. Occasionally, differences in composition can be detected by examining the gas chromatograms of a particular gasoline over several months. In this experiment, we do not try to detect such small differences.

There are different octane rating requirements for "regular" and "premium" gasolines. You may be able to observe differences in the composition of these two types of fuels. You should pay particular attention to increases in the proportions of those hydrocarbons that raise octane ratings in the premium fuels.

In some areas of the country, manufacturers are required from November to February to control the amounts of carbon monoxide produced when the gasoline burns. To do this, they add oxygenates, such as ethanol or methyl *tert*-butyl ether (MTBE), to the gasoline. You should try to observe the presence of these oxygenates, which may be observed in gasolines produced in carbon monoxide-containment areas.

The class will analyze samples of regular unleaded and premium unleaded gasolines. If available, the class will analyze oxygenated fuels. If different brands are analyzed, equivalent grades from the different companies should be compared.

Discount service stations usually buy their gasoline from one of the large petroleum-refining companies. If you analyze gasoline from a discount service station, you may find it interesting to compare that gasoline with an equivalent grade from a major supplier, noting particularly the similarities.

REQUIRED READING

New: Technique 22   Gas Chromatography
       Essay   Petroleum and Fossil Fuels

SPECIAL INSTRUCTIONS

Your instructor may want each student in the class to obtain a sample of gasoline from a service station. The instructor will compile a list of the different gasoline companies represented in the nearby area. Each student will then be assigned to collect a sample from a different company. You should collect the gasoline sample in a labeled screw-cap jar. An easy way to collect a gasoline sample for this experiment is to drain the excess gasoline from the nozzle and hose into the jar after the gasoline tank of a car has been filled. The collection of gasoline in this manner must be done *immediately after* the gas pump has been used. If not, the volatile components of the gasoline may evaporate, thus changing the composition of the gasoline. Only a small sample (a few *milli*liters) is required because the gas-chromatographic analysis requires no more than a few *micro*liters (μL) of material. Be certain to close the cap of the sample jar tightly to prevent the selective evaporation of the most volatile components. The label on the jar should list the brand of gasoline and the grade (unleaded regular, unleaded premium, oxygenated unleaded, etc.). Alternatively, your instructor may supply samples for you.

CAUTION	

Gasoline contains many highly volatile and flammable components. Do not breathe the vapors, and do not use open flames near gasoline.

This experiment may be assigned along with another short one because it requires only a few minutes of each student's time to carry out the actual gas chromatography. For this experiment to work as efficiently as possible, you may be asked to schedule an appointment for using the gas chromatograph.

SUGGESTED WASTE DISPOSAL

Dispose of all gasoline samples in the container designated for nonhalogenated wastes.

NOTES TO THE INSTRUCTOR

You need to adjust your gas chromatograph to the proper conditions for the analysis. We recommend that you prepare and analyze the reference mixture listed in the Procedure section. Most chromatographs will be able to separate this mixture cleanly, with the possible exception of the xylenes. One possible set of conditions for a Gow-Mac model 69-350 chromatograph is the following: column temperature, 110–115°C; injection port temperature, 110–115°C; carrier gas flow rate, 40–50 mL/min; column length, approximately 12 ft. The column should be packed with a nonpolar stationary phase similar to silicone oil (SE-30) on Chromosorb W or with some other stationary phase that separates components principally according to boiling point.

The chromatograms shown in this experiment were obtained on a Hewlett Packard model 5890 gas chromatograph. A 30-meter, DB 5 capillary column (0.32 mm, with 0.25 micron film) was used. A temperature program was run starting at 5°C and ramping to 105°C. Each run took about 8 minutes.

A flame-ionization detector was used. The conditions are given in the Instructor's Manual. Superior separations are obtained using capillary columns and they are recommended. Even better results are obtained with longer columns.

PROCEDURE

Reference Mixture

First, analyze a standard mixture that includes pentane, hexane (or hexanes), benzene, heptane, toluene, and xylenes (a mixture of *meta, para,* and *ortho* isomers). Inject a 0.5-μL sample into the gas chromatograph or an alternative sample size as indicated by your instructor. Measure the retention time of each component in the reference mixture on your chromatogram (Technique 22, Section 22.7, p. 806). The previously listed compounds elute in the order given (pentane first and xylenes last). Compare your chromatogram to the one posted near the gas chromatograph or the one reproduced in this experiment.

Your instructor or a laboratory assistant may prefer to perform the sample injections. The special microliter syringes used in the experiment are delicate and expensive. If you are performing the injections yourself, be sure to obtain instruction beforehand.

Oxygenated Fuel Reference Mixture

Oxygenated compounds are added to gasolines in carbon monoxide-containment areas during November through February. Currently, ethanol and methyl *tert*-butyl ether are in most common use. Your instructor may have available a reference mixture that includes all the previously listed compounds and either ethanol or methyl *tert*-butyl ether. Again, you need to inject a sample of this mixture and analyze the chromatogram to obtain the retention times for each component in this mixture.

Gasoline Samples

Inject a sample of a regular unleaded, premium unleaded, or oxygenated gasoline into the gas chromatograph and wait for the gas chromatogram to be recorded. Compare the chromatogram to the reference mixture. Determine the retention times for the major components and identify as many of the components as possible. For comparison, gas chromatograms of a premium unleaded gasoline and the reference mixture are shown on pages 267–268. On the list of the major components in gasolines, notice that the oxygenate methyl *tert*-butyl ether appears in the C_6 region. Does your oxygenated fuel show this component? See if you can notice a difference between regular and premium unleaded gasolines.

Analysis

Be certain to compare carefully the retention times of the components in each fuel sample with the standards in the reference mixture. Retention times of compounds vary with the conditions under which they are determined. It is best to analyze the reference mixture and each of the gasoline samples in succession to reduce the variations in retention times that may occur over time. Compare the gas chromatograms with those of students who have analyzed gasolines from other dealers.

Report

The report to the instructor should include the actual gas chromatograms, as well as an identification of as many of the components in each chromatogram as possible.

Major components in gasolines*

C₄ Compounds	Isobutane
	Butane
C₅ Compounds	Isopentane
	Pentane
C₆ Compounds and oxygenates	2,3-Dimethylbutane
	2-Methylpentane
	3-Methylpentane
	Hexane
	Methyl *tert*-butyl ether (oxygenate)
C₇ Compounds and aromatics (benzene)	2,4-Dimethylpentane
	Benzene (C_6H_6)
	2-Methylhexane
	3-Methylhexane
	Heptane
C₈ Compounds and aromatics (toluene, ethylbenzene, and xylenes)	2,2,4-Trimethylpentane (isooctane)
	2,5-Dimethylhexane
	2,4-Dimethylhexane
	2,3,4-Trimethylpentane
	2,3-Dimethylhexane
	Toluene (C_7H_8)
	Ethylbenzene (C_8H_{10})
	m-, *p*-, *o*-Xylenes (C_8H_{10})
C₉ Aromatic compounds	1-Ethyl-3-methylbenzene
	1,3,5-Trimethylbenzene
	1,2,4-Trimethylbenzene
	1,2,3-Trimethylbenzene

**Approximate order of elution.*

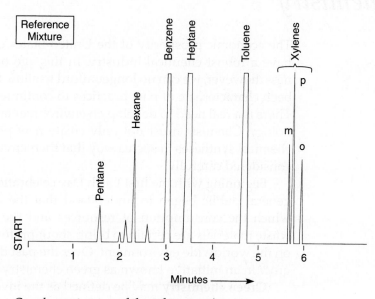

Gas chromatogram of the reference mixture.

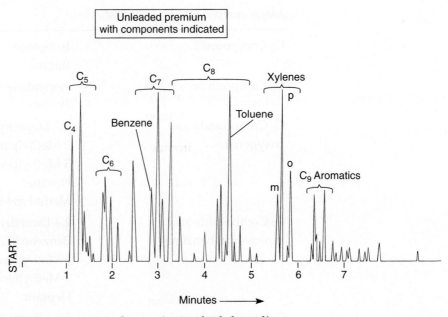

Gas chromatogram of a premium unleaded gasoline.

QUESTIONS

1. If you had a mixture of benzene, toluene, and *m*-xylene, what would be the expected order of retention times? Explain.

2. If you were a forensic chemist working for the police department and the fire marshal brought you a sample of gasoline found at the scene of an arson attempt, could you identify the service station at which the arsonist purchased the gasoline? Explain.

3. How could you use infrared spectroscopy to detect the presence of ethanol in an oxygenated fuel?

ESSAY

Green Chemistry

The economic prosperity of the United States demands that it continue to have a robust chemical industry. In this age of environmental consciousness, however, we can no longer afford to allow the type of industry that has been characteristic of past practices to continue operating as it always has. There is a real need to develop an environmentally benign, or "green," technology. Chemists must not only create new products but also design the chemical syntheses in such a way that their environmental ramifications are considered carefully.

Beginning with the first Earth Day celebration in 1970, scientists and the general public began to understand that the earth is a closed system in which the consumption of resources and the indiscriminate disposal of waste materials are certain to bring about profound and long-lasting effects on the worldwide environment. Over the past decade, interest has begun to grow in an initiative known as green chemistry.

Green chemistry may be defined as the invention, design, and application of chemical products and processes to reduce or eliminate the use and generation of hazardous substances. Practitioners of green chemistry

The twelve principles of green chemistry

1. It is better to prevent waste than to treat or clean up waste after it is formed.
2. Synthetic methods should be designed to maximize the incorporation of all materials used in the process into the final product.
3. Wherever practicable, synthetic methodologies should be designed to use and generate substances that possess little or no toxicity to human health and the environment.
4. Chemical products should be designed to preserve efficacy of function while reducing toxicity.
5. The use of auxiliary substances (solvents, separation agents, etc.) should be made unnecessary whenever possible and innocuous when used.
6. Energy requirements should be recognized for their environmental and economic impacts and should be minimized. Synthetic methods should be conducted at ambient temperature and pressure.
7. A raw material or feedstock should be renewable rather than depleting whenever technically and economically practicable.
8. Unnecessary privatization (blocking group, protection/deprotection, temporary modification of physical/chemical processes) should be avoided whenever possible.
9. Catalytic reagents (as selective as possible) are superior to stoichiometric reagents.
10. Chemical products should be designed so that at the end of their function they do not persist in the environment and do break down into innocuous degradation products.
11. Analytical methodologies need to be further developed to allow for real-time, in-process monitoring and control before the formation of hazardous substances.
12. Substances and the form of a substance used in a chemical process should be chosen to minimize the potential for chemical accidents, including releases, explosions, and fires.

Source: P. T. Anastas and J. C. Warner, *Green Chemistry: Theory and Practice.* New York: Oxford University Press, 1998. Reprinted by permission of the publisher.

strive to protect the environment by cleaning up toxic waste sites and by inventing new chemical methods that do not pollute and that minimize the consumption of energy and natural resources. Guidelines for developing green chemistry technologies are summarized in the "Twelve Principles of Green Chemistry," shown in the table.

The green chemistry program was begun shortly after the passage of the Pollution Prevention Act of 1990 and is the central focus of the Environmental Protection Agency's Design for the Environment Program. As a stimulus for research in the area of reducing the impact of chemical industry on the environment, the Presidential Green Chemistry Challenge Award was begun in 1995. The theme of the Green Chemistry Challenge is "Chemistry is not the problem; it's the solution." Since 1995, award winners have been responsible for the elimination of more than 460 million pounds of hazardous chemicals and have saved more than 440 million gallons of water and 26 million barrels of oil.

Winners of the Green Chemistry Challenge Award have developed foam fire retardants that do not use halons (compounds containing fluorine, chlorine, or bromine), cleaning agents that do not use tetrachloroethylene, methods that facilitate the recycling of polyethylene terephthalate soft-drink bottles, a method of synthesizing ibuprofen that minimizes the use of solvents and the generation of wastes, and a formulation that promotes the efficient release of ammonia from urea-based fertilizers. This latter contribution allows a more environmentally friendly means of applying fertilizers without the need for tilling or disturbing (and losing) precious topsoil.

Green syntheses of the future will require making choices about reactants, solvents, and reaction conditions that are designed to reduce resource consumption and waste production. We need to think about performing a synthesis in a way that will not consume excessive amounts of resources (and thus use less energy and be more economical), that will not produce excessive amounts of toxic or harmful by-products, and that will require milder reaction conditions.

The application of green-chemistry principles in an organic synthesis begins with the selection of the starting materials, called **feedstock.** Most organic compounds used as feedstock are derived from petroleum, a nonrenewable resource (see essay "Petroleum and Fossil Fuels," p. 255). A green approach is to replace these petrochemicals with chemicals derived from biological sources such as trees, corn or soybeans. Not only is this approach more sustainable but also the refining of organic compounds from these plant-derived materials, sometimes called **biomass,** is less polluting than the refining process for petrochemicals. Many pharmaceuticals, plastics, agricultural chemicals, and even transportation fuels can now be produced from chemicals derived from biomass. A good example of this is adipic acid, an organic chemical widely used in the production of nylon and lubricants. Adipic acid can be produced from benzene, a toxic petrochemical, or from glucose, which is found in plant sources.

Industrial processes are being designed that are based on the concept of **atom economy.** Atom economy means that close attention is paid to the design of chemical reactions so that all or most of the atoms that are starting materials in the process are converted into molecules of the desired product rather than into wasted by-products. Atom economy in the industrial world is the equivalent of ensuring that a chemical reaction proceeds with a high percentage yield in a classroom laboratory experiment.

The atom economy for a reaction can be calculated using the following equation:

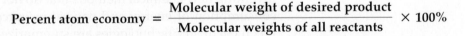

$$\text{Percent atom economy} = \frac{\text{Molecular weight of desired product}}{\text{Molecular weights of all reactants}} \times 100\%$$

For example, consider the reaction for the synthesis of aspirin (Experiment 8, "Acetylsalicylic Acid"):

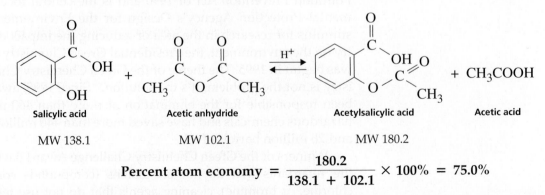

| Salicylic acid | Acetic anhydride | Acetylsalicylic acid | Acetic acid |
| MW 138.1 | MW 102.1 | MW 180.2 | |

$$\text{Percent atom economy} = \frac{180.2}{138.1 + 102.1} \times 100\% = 75.0\%$$

This calculation assumes the complete conversion of reactants into product and 100% recovery of the product, which is not possible. Furthermore, the calculation does not take into account that often an excess of one reactant is used to drive the reaction to completion. In this reaction, acetic anhydride is used in large excess to ensure the production of more acetylsalicylic acid. Nonetheless, the atom economy calculation is a good way to compare different possible pathways to a given product.

To illustrate the benefits of atom economy, consider the synthesis of ibuprofen, mentioned earlier, that won the Presidential Green Chemistry Challenge Award in 1997. In the former process, developed in the 1960s, only 40% of the reactant atoms were incorporated into the desired ibuprofen product; the remaining 60% of the reactant atoms found their way into unwanted by-products or wastes that required disposal. The new method requires fewer reaction steps and recovers 77% of the reactant atoms in the desired product. This "green" process eliminates millions of pounds of waste chemical by-products every year, and it reduces by millions of pounds the amount of reactants needed to prepare this widely used analgesic.

Another green chemistry approach is to select safer reagents that are used to carry out the synthesis of a given organic compound. In one example of this, milder or less toxic oxidizing reagents may be selected to carry out a conversion that is normally done in a less green way. For example, sodium hypochlorite (bleach) can be used in some oxidation reactions instead of the highly toxic dichromate/sulfuric acid mixture. In some reactions, it is possible to use biological reagents, such as enzymes, to carry out a transformation. Another approach in green chemistry is to use a reagent that can promote the formation of a given product in less time and with greater yield. Finally, some reagents, especially catalysts, can be recovered at the end of the reaction period and recycled for use again in the same conversion.

Many solvents used in traditional organic syntheses are highly toxic. The green chemistry approach to the selection of solvents has resulted in several strategies. One method that has been developed is to use supercritical carbon dioxide as a solvent. Supercritical carbon dioxide is formed under conditions of high pressure in which the gas and liquid phases of carbon dioxide combine to a single-phase compressible fluid that becomes an environmentally benign solvent (temperature 31°C; pressure 7280 kPa, or 72 atmospheres). Supercritical CO_2 has remarkable properties. It behaves as a material whose properties are intermediate between those of a solid and those of a liquid. The properties can be controlled by manipulating temperature and pressure. Supercritical CO_2 is environmentally benign because of its low toxicity and easy recyclability. Carbon dioxide is not added to the atmosphere; rather, it is removed from the atmosphere for use in chemical processes. It is used as a medium to carry out a large number of reactions that would otherwise have many negative environmental consequences. It is even possible to perform stereoselective synthesis in supercritical CO_2.

Some reactions can be carried out in ordinary water, the most green solvent possible. Recently, there has been much success in using near-critical water at higher temperatures where water behaves more like an organic solvent. Two of the award winners of the 2004 Green Chemistry Award, Charles Eckert and Charles Liotta, have advanced our understanding of supercritical CO_2 and near-critical water as solvents. One example of their work takes advantage of the dissociation of water that takes place under near-critical conditions, leading to a high concentration of hydronium and hydroxide ions. These ions can serve as self-neutralizing catalysts, and they can replace catalysts that must normally be added to the reaction mixture. Eckert and Liotta were able to run Friedel-Crafts reactions (Experiment 60, "Friedel-Crafts Acylation") in near-critical water without the need for the acid catalyst AlCl₃, which is normally used in large amounts in these reactions.

Research has also focused on **ionic liquids,** salts that are liquid at room temperature and do not evaporate. Ionic liquids are excellent solvents for many materials, and they can be recycled. An example of an ionic liquid is

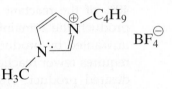

Even though many of the ionic liquids are expensive, their high initial cost is mitigated because, through recycling, they are not consumed or discarded. In addition, product recovery is often easier than with traditional solvents. In the past 5 years, many new ionic liquids have been developed with a broad range of properties. By selecting the appropriate ionic liquid, it is now possible to carry out many types of organic reactions in these solvents. In some reactions, a well-designed ionic solvent can lead to better yields under milder conditions than is possible with traditional solvents. Recently, researchers have developed ionic liquids made from artificial "sweeteners" that are nontoxic and extend even further the concept of green chemistry.

It is possible in some organic syntheses to completely eliminate the need for any solvent! Some reactions that are traditionally carried out in solvents can be carried out either in the solid or gas phases without the presence of any solvent.

Another approach to making organic chemistry greener involves the way in which a reaction is carried out, rather than in the selection of starting material, reagents, or solvents. Microwave technology (see Technique 6, Section 6.7, p. 596) can be used in some reactions to provide the heat energy required to make the transformation go to completion. With microwave technology, reactions can take place with less toxic reagents and in a shorter time, with fewer side reactions, all goals of green chemistry. Microwave technology has also been used to create supercritical water that behaves more like an organic solvent and could replace more toxic solvents in carrying out organic reactions.

Another green approach involving technology is the use of solid-phase extraction (SPE) columns (see Technique 12, Section 12.14, p. 689). Using SPE columns, extractions such as removing caffeine from tea can be carried out more quickly and with less toxic solvents. In other applications, SPE columns can be used to carry out the synthesis of organic compounds more efficiently with less use of toxic reagents.

Industry has discovered that environmental stewardship makes good economic sense, and there is a renewed interest in cleaning up manufacturing processes and products. In spite of the continuing adversarial nature of relations between industry and environmentalists, companies are discovering that preventing pollution in the first place, using less energy, and developing atom-economic methods makes as much sense as spending less money on raw materials or capturing a greater share of the market for their product. Although U.S. chemical industries are by no means near their stated goal of reducing the emission of toxic substances to zero or near-zero levels, significant progress is being made.

The teaching of the principles of green chemistry is beginning to find its way into the classroom. In this textbook, we have attempted to improve the green qualities of some of the experiments and have added several green experiments. The following table lists the experiments in this textbook that have a significant green component, along with the primary aspect of the experiment that makes it green.

Experiment	Green Aspect
Experiment 12B "Extraction of Caffeine from Tea or Coffee by Solid Phase Extraction (SPE)"	SPE column
Experiment 14C "Extraction of Eugenol from Cloves with Liquid Carbon Dioxide"	Supercritical CO_2 used in extraction
Experiment 33 "Chiral Reduction of Ethyl Acetoacetate"	Biological reagent, baker's yeast
Experiment 34 "Nitration of Aromatic Compounds Using a Recyclable Catalyst"	Recyclable catalyst
Experiment 35 "An Oxidation-Reduction Scheme: Borneol, Camphor, Isoborneol"	Less toxic oxidizing agent, sodium hypochlorite
Experiment 36A "Coenzyme Synthesis of Benzoin"	Thiamine, a biological reagent
Experiment 39 "Aqueous-Based Organozinc Reaction"	Water used as the solvent
Experiment 44B "Solvent-Free Wittig Synthesis of 1,4-Diphenyl-1,3-Butadiene"	No solvent used

It should also be mentioned that the microscale approach used in this textbook achieves many of the goals of green chemistry. By making a dramatic reduction in the scale of the reactions, the microscale approach consumes fewer chemicals and generates much less waste than the traditional macroscale approach.

Certainly, there are enormous challenges remaining. Generations of new scientists must be taught that it is important to consider the environmental impact of any new methods that are introduced. Industry and business leaders must learn to appreciate that adopting an atom-economic approach to the development of chemical processes makes good long-term economic sense and is a responsible means of conducting business. Political leaders must also develop an understanding of what the benefits of a green technology can be and why it is responsible to encourage such initiatives. In this light, it should be mentioned that even President George W. Bush has embraced the green chemistry initiative!

REFERENCES

Amato, I. "Green Chemistry Proves It Pays Companies to Find New Ways to Show That Preventing Pollution Makes More Sense Than Cleaning Up Afterward." *Fortune* (July 24, 2000). Available at www.fortune.com/fortune/articles/0.15114.368198.00.html.

Freemantle, M. "Ionic Liquids in Organic Synthesis." *Chemical and Engineering News, 82* (November 8, 2004): 44.

Jacoby, M. "Making Olefins from Soybeans." *Chemical and Engineering News, 83* (January 3, 2005): 10.

Mark, V. "Riding the Microwave." *Chemical and Engineering News, 82* (December 13, 2004): 14.

Matlack, A. "Some Recent Trends and Problems in Green Chemistry." *Green Chemistry* (February 2003): G7–G11.

Mullin, R. "Sustainable Specialties." *Chemical and Engineering News, 82* (November 8, 2004): 29.

Oakes, R. S., Clifford, A. A., Bartle, K. D., Pett, M. T., and Rayner, C. M. "Sulfur Oxidation in Supercritical Carbon Dioxide: Dramatic Pressure Dependent Enhancement of Diastereo-selectivity for Sulfoxidation of Cysteine Derivatives." *Chemical Communications* (1999): 247–248.

Ritter, S. K. "Green Innovations." *Chemical and Engineering News, 82* (July 12, 2004): 25.

33 **EXPERIMENT 33**

Chiral Reduction of Ethyl Acetoacetate; Optical Purity Determination

Green chemistry

Stereochemistry

Reduction with yeast

Use of a separatory funnel

Chiral gas chromatography

Polarimetry

Optical purity (enantiomeric excess) determination

Nuclear magnetic resonance (optional)

Chiral chemical shift reagents (optional)

The experiment described in Experiment 33A uses common baker's yeast as a chiral reducing medium to transform an achiral starting material, ethyl acetoacetate, into a chiral product. When a single stereoisomer is formed in a chemical reaction from an achiral starting material, the process is said to be **enantiospecific.** In other words, one stereoisomer (enantiomer) is formed in preference to its mirror image. In the present experiment, the ethyl (*S*)-3-hydroxybutanoate stereoisomer is formed preferentially. In actual fact,

We are grateful to Dr. Snorri Sigurdsson and James Patterson, University of Washington, Seattle, for suggested improvements.

however, some of the (R)-enantiomer is also formed in the reaction. The reaction, therefore, is described as an **enantioselective** process because the reaction does not produce one stereoisomer exclusively. Chiral gas chromatography and polarimetry will be employed to determine the percentages of each of the enantiomers. Generally, the chiral reduction produces less than 8% of the ethyl (R)-3-hydroxybutanoate.

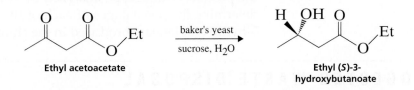

Ethyl acetoacetate   **Ethyl (S)-3-hydroxybutanoate**

In contrast, when ethyl acetoacetate is reduced with sodium borohydride in methanol, the reaction yields a 50:50 mixture of the (R) and (S)-stereoisomers. A racemic mixture is formed because the reaction is not being conducted in a chiral medium.

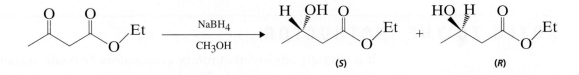

In Experiment 33B (optional), you may use nuclear magnetic resonance spectroscopy to determine the relative amounts of (R) and (S) enantiomers produced in the chiral reduction of ethyl acetoacetate. This part of the experiment requires the use of a chiral shift reagent.

33A EXPERIMENT 33A

Chiral Reduction of Ethyl Acetoacetate

REQUIRED READING

Review: Technique 8   Filtration, Section 8.4
Technique 12   Extractions, Separations, and Drying Agents, Section 12.7

Techniques 22 and 25
Techniques 26 and 27 (optional)

New: Technique 23   Polarimetry
Essay   Green Chemistry

SPECIAL INSTRUCTIONS

Day 1 of the experiment involves setting up the reaction. Another experiment can be conducted concurrently with this experiment. Part of this first laboratory period is used to mix the yeast, sucrose, and ethyl acetoacetate in a 500-mL Erlenmeyer flask. It is stirred during part of that first period.

The mixture is then covered and stored until the next period. The reduction requires at least 2 days.

Day 2 of the experiment is used to isolate the chiral ethyl 3-hydroxybutanoate. After this has been isolated, each student's product is analyzed by chiral gas chromatography and polarimetry to determine the percentages of each of the enantiomers. As an optional experiment (Experiment 33B), the products can also be analyzed by NMR using a chiral shift reagent to determine the percentages of each of the enantiomers present in the ethyl 3-hydroxybutanoate produced in the chiral reduction.

SUGGESTED WASTE DISPOSAL

The Celite, residual yeast, and cheesecloth from the reduction can be disposed of in the trash. The aqueous solutions and emulsion left from the extraction with methylene chloride should be placed in the aqueous waste container. Methylene chloride waste should be poured into the waste container designated for halogenated waste.

NOTES TO THE INSTRUCTOR

It is strongly advised that rotary evaporators be made available for this experiment. Approximately 90 mL of methylene chloride are used for each student. The experiment will be more "green" if the solvent can be recovered. The instructor will need to make available to each student a large Büchner funnel (10 cm), 500-mL filter flask, 500-mL Erlenmeyer flask, 1.5- or 2-inch magnetic stir bar, and a 500-mL separatory funnel. It is advised that packaged dry yeast be used. We suggest Fleischmann's Rapid Rise (baker's) Yeast, which contains 7 g of yeast per package. We purchase packages of 100% cotton cheesecloth that consists of three layers (do not separate the layers from each other). Cut the three-ply cheesecloth into 4-inch × 8-inch strips to be folded into 4-inch × 4-inch sections for the Büchner funnel. In some cases, the yeast does not grow substantially during the first half hour. It is best to discard the mixture and start the reaction again if it appears that the yeast is not growing. In most cases, the temperature may not have been controlled carefully. It is recommended that the flasks containing the reaction mixture be stored in an area where the temperature is maintained at about 25°C, if possible. The optimal reduction period is 4 days. A small amount of unreduced ethyl acetoacetate remains after a 2-day reduction (less than 1%), and a 4-day reduction yields no remaining ethyl acetoacetate. The expected yield of chiral hydroxyester should be around 65%, consisting of 92–94% of ethyl (*S*)-3-hydroxybutanoate.

PROCEDURE

Yeast Reduction

To a 500-mL Erlenmeyer flask, add 150 mL of deionized (DI) water and a 1.5- or 2-inch stir bar. Warm the water to about 35–40°C using a hot plate set on low. Add 7 g of sucrose and 7 g of Fleischmann's Rapid Rise (dry baker's) Yeast to the flask. Swirl the contents of the flask in order to distribute the yeast into the aqueous solution; otherwise it will remain at the top of the solution. Stir the mixture for 15 minutes while maintaining the temperature at 35°C. During this time, the yeast will become activated and will grow substantially. Add 3.0 g of ethyl acetoacetate and 8 mL of hexane to the yeast mixture. Stir the mixture with a magnetic stirrer for 1.5 hours. Because the mixture may become thick, check periodically to see whether the mixture is being stirred. The reaction is somewhat exothermic, so you may not need to heat the mixture. Nevertheless, you should monitor the temperature to make sure that it remains near 30°C. Adjust the temperature to 30°C if the temperature falls below this value.

Label the Erlenmeyer flask with your name and ask your instructor to store the flask. Cover the top of the flask loosely with aluminum foil so that carbon dioxide can be expelled during the reduction. The mixture will stand, without stirring, until the next laboratory period (2–4 days). At some point during the laboratory period, obtain the infrared spectrum of ethyl acetoacetate for the purpose of comparison to the reduced product.

CAUTION ⚠️

Do not breathe the Celite powder.

Isolation of the Alcohol Product

Obtain a 500-mL separatory funnel, a large Büchner (10 cm) funnel, and a 500-mL filter flask from your instructor. To the yeast solution, add 5 g of Celite and stir the mixture magnetically for 1 minute (Technique 7, Section 7.3, p. 602). Allow the solid to settle as much as possible (at least 5 minutes). Set up a vacuum-filtration apparatus using the large Büchner funnel (Technique 8, Section 8.3, p. 621). Wet one piece of filter paper with water and place it into the funnel. Obtain a 4-inch × 8-inch strip of cheesecloth and fold it over to make a 4-inch × 4-inch square. Wet it with water and place it on top of the filter paper so that it completely covers the filter paper and is partly up the side of the Büchner funnel. You are now ready to filter your solution. Turn on the vacuum source (aspirator or the house vacuum). Decant the clear supernatant liquid slowly into the Büchner funnel. If you do this slowly, you may avoid plugging the filter paper with small particles. Once the supernatant liquid has been poured into the funnel, add the Celite slurry to the Büchner funnel. Rinse the flask with 20 mL of water and pour the remaining Celite–yeast mixture into the Büchner funnel. Discard the Celite, yeast, and cheesecloth waste into the trash. The Celite helps to trap the tiny yeast particles. Some of the yeast and Celite will pass through the filter into the filter flask. This is unavoidable.

Add 20 g of sodium chloride to the filtrate in the filter flask and swirl the flask gently until the sodium chloride dissolves. If an emulsion forms, you may be swirling the flask too vigorously. Pour the filtrate into a 500-mL separatory funnel. Add 30 mL of methylene chloride to the funnel and stopper the funnel (Technique 12, Section 12.7, p. 677). In order to avoid a difficult emulsion, do not shake the separatory funnel; instead, slowly invert the funnel and bring it back to the upright position. Repeat this motion over a period of 5 minutes. Vent the funnel occasionally to relieve pressure. Drain the lower methylene chloride layer from the separatory funnel into a 250-mL Erlenmeyer flask, leaving behind a small amount of emulsion and the aqueous layer in the separatory funnel. Add another 30-mL portion of methylene chloride to the separatory funnel and repeat the extraction procedure. Drain the lower methylene chloride layer into the same Erlenmeyer flask holding the first methylene chloride extract. Repeat the extraction process a third time with a 30-mL portion of methylene chloride. Discard the emulsion and aqueous layer remaining in the separatory funnel into a suitable aqueous waste container.

Dry the three combined methylene chloride extracts over about 3 g of anhydrous granular sodium sulfate for at least 5 minutes. Occasionally, swirl the contents of the flask to help dry the solution. Decant the liquid into a 250-mL beaker and evaporate the solvent using an air or nitrogen stream until the volume of liquid remains constant (approximately 1–2 mL). (Alternatively, a rotary evaporator or distillation may be used to remove the methylene chloride from the product).[1] Often the remaining liquid contains some water. To remove the water, add 10 mL of methylene chloride to dissolve the product and add 0.5 g of anhydrous granular sodium sulfate to the solution. Decant the methylene chloride solution away from the drying agent into a preweighed 50-mL beaker. Evaporate the solvent using an air or nitrogen stream until the volume of liquid remains constant. The liquid contains the ethyl (S)-3-hydroxybutanoate that has been produced by chiral reduction of ethyl acetoacetate. A small amount of ethyl acetoacetate may remain unreduced in the sample. Reweigh the beaker to determine the weight of the product. Calculate the percentage yield of product.

Infrared Spectroscopy

Determine the spectrum of your isolated product. The infrared spectrum provides the best direct evidence for the reduction of ethyl acetoacetate. Look for presence of a hydroxyl group (about 3440 cm^{-1}) that was produced in the reduction of the carbonyl group. Compare the spectrum of the product, ethyl 3-hydroxybutanoate, to the starting material, ethyl acetoacetate. What differences do you notice in the two spectra? Label the two spectra with peak assignments and include them with your laboratory report.

Chiral Gas Chromatography

Chiral gas chromatography will provide a direct measure of amounts of each stereoisomer present in your chiral ethyl 3-hydroxybutanoate sample. A Varian CP-3800 equipped with J & W (Agilent) Cyclosil B capillary column (30 m, 0.25-mm ID, 0.25 μm) provides an excellent separation of (R) and (S)-enantiomers. Set the flame ionization detector (FID) detector at 270°C and the injector temperature at 250°C, with a 50:1 split ratio. Set the column oven temperature at 90°C and hold at that temperature for 20 minutes. The helium flow rate is 1 mL/min. The compounds elute in the following order: ethyl (S)-3-hydroxybutanoate (14.3 min) and the (R)-enantiomer (15.0 min). Any remaining ethyl acetoacetate present in the sample will produce a peak with a retention time of 14.1 minutes. Your observed retention times may vary from those given here, but the order of elution will be the same. Calculate the percentages of each of the enantiomers from the chiral gas chromatography results. Usually, about 92–94% of the (S)-enantiomer is obtained from the reduction.

Polarimetry

Fill a 0.5-dm polarimeter cell with your chiral hydroxyester (about 2 mL required). You may need to combine your product with material obtained by one other student in order to have enough material to fill the cell. Determine the observed optical rotation for the chiral material. Your instructor will show you how to use the polarimeter. Calculate the specific rotation for your sample using the equation

---

[1] Pour the dry methylene chloride extracts into a round-bottom flask and remove the solvent with a rotary evaporator or by distillation. After removing the solvent, add 10 mL of fresh methylene chloride and 0.5 g of anhydrous granular sodium sulfate to the round-bottom flask. Decant the solution away from the drying agent into a preweighed beaker as indicated in the procedure.

provided in Technique 23 (p. 821). The concentration value, *c*, in the equation is 1.02 g/mL. Using the published value for the *specific* rotation of ethyl (*S*)-(+)-3-hydroxybutanoate of $[\alpha_D^{25}] = +43.5°$, calculate the optical purity (enantiomeric excess) for your sample (Technique 23, Section 23.5, p. 826). Report the observed rotation, the calculated specific rotation, the optical purity (enantiomeric excess), and the percentages of each of the enantiomers to the instructor. How do the percentages of each of the enantiomers calculated from the polarimeter measurement compare to the values obtained from chiral gas chromatography?[2]

Proton and Carbon NMR Spectroscopy (Optional)

At the option of the instructor, you may obtain the proton spectrum (shown in Figures 1 and 2 and interpreted in Experiment 33B) and the carbon NMR spectra of the product. The carbon NMR spectrum shows peaks at 14.3, 22.6, 43.1, 60.7, 64.3, and 172.7 ppm.

33B EXPERIMENT 33B (OPTIONAL)

NMR Determination of the Optical Purity of Ethyl (S)-3-Hydroxybutanoate

In Experiment 33A, the yeast reduction of ethyl acetoacetate forms a product that is predominantly the (*S*)-enantiomer of ethyl 3-hydroxybutanoate. In this part of the experiment, we will use NMR to determine the percentages of each of the enantiomers in the product. The 300-MHz proton NMR spectrum of racemic ethyl 3-hydroxybutanoate is shown in Figure 1. The expansions of the individual patterns from Figure 1 are shown in Figure 2. The methyl protons (H_a) appear as a doublet at 1.23 ppm, and the methyl protons (H_b) appear as a triplet at 1.28 ppm. The methylene protons (H_c and H_d) are diastereotopic (nonequivalent) and appear at 2.40 and 2.49 ppm (each a doublet of doublets). The hydroxyl group appears at about 3.1 ppm. The quartet at 4.17 ppm results from the methylene protons (H_e) split by the protons (H_b). The methine proton (H_f) is buried under the quartet at about 4.2 ppm.

$$
\begin{array}{c}
\overset{\displaystyle OH}{|} \qquad \overset{\displaystyle O}{\|} \\
CH_3\text{-}CH\text{-}CH_2\text{-}C\text{-}O\text{-}CH_2\text{-}CH_3 \\
H_a \;\; H_f \;\; H_c \qquad\quad H_e \;\; H_b \\
H_d
\end{array}
$$

Although the normal proton NMR spectrum for the racemic ethyl 3-hydroxybutanoate is not expected to be any different from the proton

---

[2] The percentages calculated from polarimetry may vary considerably from those obtained by chiral gas chromatography. Often, the samples contain some solvent and other impurities that reduce the observed optical rotation value. The solvent and impurities do not influence the more accurate percentages obtained directly by chiral gas chromatography.

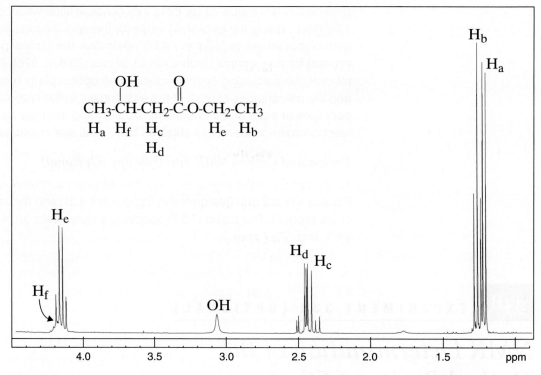

Figure 1
NMR spectrum (300 MHz) of racemic ethyl 3-hydroxybutanoate with no chiral shift reagent present.

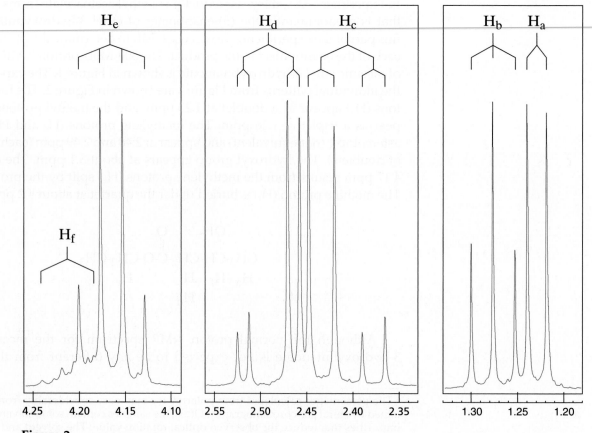

Figure 2
Expansions of the NMR spectrum of racemic ethyl 3-hydroxybutanoate.

NMR spectra of each of the enantiomers in an **achiral environment,** the introduction of a chiral shift reagent creates a **chiral environment.** This chiral environment allows the two enantiomers to be distinguished from each other. A general discussion of nonchiral chemical shift reagents is found in Technique 26, Section 26.15, p. 896. These reagents spread out the resonances of the compound with which they are used, increasing by the largest amount the chemical shifts of the protons that are nearest the center of the metal complex. Because the spectra of both enantiomers are identical under these conditions, the usual chemical shift reagent would not help our analysis. However, if we use a chemical shift reagent that is itself chiral, we can distinguish the two enantiomers by their NMR spectra. The two enantiomers, which are each chiral, will interact differently with the chiral shift reagent. The complexes formed from the (*R*)-, and (*S*)-enantiomers and with the chiral shift reagent will be diastereomers. Diastereomers usually have different physical properties, and the NMR spectra are no exception. The two complexes will be formed with slightly differing geometries. Although the effect may be small, it is large enough to see differences in the NMR spectra of the two enantiomers.

The chiral shift reagent used in this experiment is *tris*-[3-(heptafluoropropylhydroxymethylene)-(+)-camphorato]europium (III), or Eu(hfc)₃. In this complex, the europium is in a chiral environment because it is complexed to camphor, which is a chiral molecule. Eu(hfc)₃ has the structure shown here.

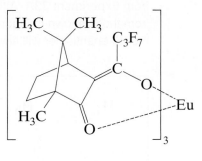

REQUIRED READING

New: Technique 26   Nuclear Magnetic Resonance Spectroscopy, Section 26.15

SPECIAL INSTRUCTIONS

This experiment requires the use of a high-field NMR spectrometer in order to obtain sufficient separation of peaks for the two enantiomers. The chiral shift reagent does cause some peak broadening, so care should be taken not to add too much of this reagent to the chiral ethyl 3-hydroxybutanoate sample. A 0.035-g sample of the chiral material and 8–11 mg of chiral shift reagent should be sufficient to give good results.

SUGGESTED WASTE DISPOSAL

Discard the remaining solution from your NMR tube into the container designated for the disposal of halogenated organic waste.

PROCEDURE

Using a Pasteur pipet to aid the transfer, weigh 0.035 g of chiral ethyl 3-hydroxy-butanoate from Experiment 33A directly into an NMR tube. Weigh 8–11 mg of *tris* [3-(heptafluoropropylhydroxymethylene)-(+)-camphorato]europium(III) chiral shift reagent on a piece of weighing paper and add the chiral shift reagent to the chiral hydroxyester in the NMR tube. Take care to avoid chipping the fragile NMR tube while adding the shift reagent with a microspatula. Add $CDCl_3$ solvent to the NMR tube until the level reaches 50 mm. Cap the tube and invert the tube to mix the sample. Allow the NMR sample to stand for a minimum of about 5–8 minutes before determining the NMR spectrum. Record in your notebook the exact weights of sample and chiral shift reagent that you used.

Determine the NMR spectrum of the sample. The peaks of interest are the methyl protons, H_a (doublet) and H_b (triplet). Notice in Figure 3 that the doublet and triplet peaks for the two methyl groups in the **racemic** ethyl 3-hydroxybutanoate are doubled. The downfield doublet (1.412 and 1.391 ppm) and triplet (1.322, 1.298, and 1.274 ppm) peaks are assigned to the (S)-enantiomer. The upfield doublet (1.405 and 1.384 ppm) and triplet (1.316, 1.293, and 1.269 ppm) peaks are assigned to the (R)-enantiomer. Your expansion of this area of the NMR spectrum should also show a doubling of the peaks as in Figure 3, but the upfield doublet for the (R)-enantiomer will be smaller. The same will be true for the (R)-enantiomer in the triplet pattern. By integration, determine the percentages of the (S)- and (R)-enantiomers in the chiral ethyl 3-hydroxybutanoate from Experiment 33A. Although the positions of the peaks may vary somewhat from those shown in Figure 3, you should still find that the doublet and triplet for the (S)-enantiomer will always be downfield relative to the (R)-enantiomer.

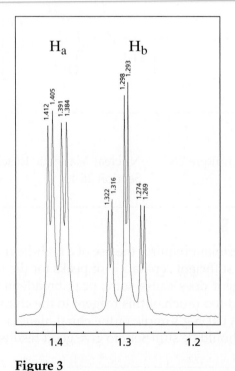

Figure 3

NMR spectrum (300 MHz) of racemic ethyl 3-hydroxybutanoate, with chiral shift reagent added.

Note: H_a for the (S)-enantiomer = 1.412, 1.391; H_b for the (S)-enantiomer = 1.322, 1.298, 1.274; H_a for the (R)-enantiomer = 1.405, 1.384; H_b for the (R)-enantiomer = 1.316, 1.293, 1.269.

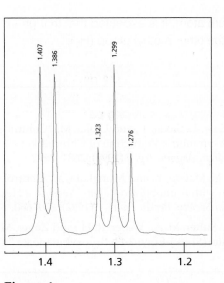

Figure 4
NMR spectrum (300 MHz) of ethyl (S)-3-hydroxybutanoate, with chiral shift reagent added.

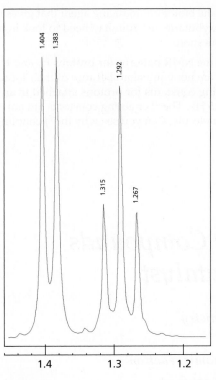

Figure 5
NMR spectrum (300 MHz) of ethyl (R)-3-hydroxybutanoate, with chiral shift reagent added.

The assignments for the (*S*)- and (*R*)-enantiomers shown in Figure 3 were determined by obtaining the NMR spectrum of pure samples of each enantiomer in the presence of the chiral shift reagent (Figures 4 and 5). You may have noticed that the doublet has moved further downfield relative to the triplet (compare Figures 2 and 3). The reason for this is that the complexation of the chiral shift reagent occurs at the hydroxyl group. Because the methyl group (H_a) is closer to the europium atom, it is expected that that group will be shifted further downfield relative to the other methyl group (H$_b$).

REFERENCES

Cui, J.-N., Ema, T., Sakai, T., and Utaka, M. "Control of Enantioselectivity in the Baker's Yeast Asymmetric Reduction of Chlorodiketones to Chloro (*S*)-Hydroxyketones." *Tetrahedron: Asymmetry, 9* (1998): 2681–2692.

Naoshima, Y., Maeda, J., and Munakata, Y. "Control of the Enantioselectivity of Bioreduction with Immobilized Bakers' Yeast in a Hexane Solvent System." *Journal of the Chemical Society, Perkin Trans. 1* (1992): 659–660.

Seebach, D., Sutter, M. A., Weber, R. H., and Züger, M. F. "Yeast Reduction of Ethyl Acetoacetate: (*S*)-(+)-Ethyl 3-Hydryoxybutanoate." *Organic Syntheses, 63* (1984): 1–9.

QUESTIONS

1. Would you expect to see a difference in retention times for the ethyl (*S*)-3-hydroxybutanoate and the (*R*)-enantiomer on gas chromatography columns described in Technique 22?

2. What is the biological reducing agent that gives rise to the formation of chiral ethyl 3-hydroxybutanoate? You may need to look in a reference book to find an answer to this question.

3. Explain the NMR patterns for protons H$_c$ and H$_d$ shown in Figure 2. (*Hints:* These protons are nonequivalent because of their location adjacent to a stereocenter. The ^2J coupling constants for protons attached to an sp^3 carbon are very large—in this case, 16.5 Hz. The ^3J coupling constants are not equal. Draw a sawhorse projection for the molecule. Can you see why the ^3J coupling constants might be different?)

---

34 EXPERIMENT 34

Nitration of Aromatic Compounds Using a Recyclable Catalyst

Green chemistry

Nitration

Atom-economic reaction

Recyclable catalyst

Rotary evaporator (optional)

Mass spectrometry

Gas chromatography

Chemists in academia and industry are attempting to make chemical reactions more environmentally friendly (see the essay "Green Chemistry"). One way to accomplish this is to use exact (stoichiometric) amounts of starting reagents so that no excess material need be thrown away, thus contributing to a higher atom economy. Another aspect of green chemistry is that chemists should use catalysts. These materials have the advantage of allowing reactions to occur under milder conditions, and catalysts can also be reused. Thus, green chemistry helps keep the environment clean while producing useful products.

In the present experiment, we employ a Lewis acid, ytterbium (III) trifluoromethanesulfonate, as a catalyst for the nitration of a series of aromatic substrates with nitric acid. This catalyst will be recycled (recovered) and reused.

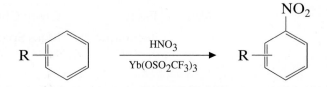

The solvent used in this reaction, 1,2-dichloroethane, is not environmentally friendly, but the solvent can be recovered using a rotary evaporator.

A proposed mechanism for this reaction involves the following three steps to generate the nitronium ion.[1] The trifluoromethanesulfonate (triflate) ions act as spectators. The ytterbium cation is believed to be hydrated by the water present in the aqueous nitric acid solution. Nitric acid binds strongly to the hydrated ytterbium cation, as shown in equation 1. A proton is generated, as shown in equation 2, by the strong polarizing effect of the metal. Nitronium ion is then formed by the process shown in equation 3. Although the nitronium ion may serve as the active electrophilic species, it is more likely that a nitronium carrier, such as the intermediate formed in equation 2, may serve as the electrophile. In any case, the reaction yields a nitrated aromatic compound.

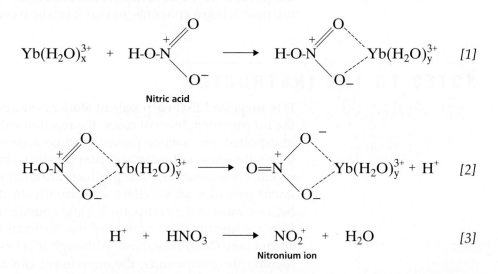

In this experiment, you will nitrate an aromatic substrate and analyze the composition of the mixture obtained by gas chromatography/mass spectrometry (GC/MS). In some cases, starting material will also be present in the mixture. You should be able to explain, mechanistically, why the observed products are obtained from the reaction.

---

[1] Braddock, C., "Novel Recyclable Catalysts for Atom Economic Aromatic Nitration," *Green Chemistry*, 3 (2001): G26–G32.

REQUIRED READING

Review: Technique 7   Reaction Methods, Sections 7.2 and 7.10

Technique 7   Section 7.11 (optional)

Technique 12   Extractions, Separations, and Drying Agents, Section 12.4

Technique 22   Gas Chromatography

New: Essay   Green Chemistry

Technique 28   Mass Spectrometry

SPECIAL INSTRUCTIONS

Some of the nitrated products may be toxic. All work should be conducted in a fume hood. Wear protective gloves to avoid skin contact with the nitrated products.

SUGGESTED WASTE DISPOSAL

The aqueous layer contains the catalyst, ytterbium triflate. Do not discard it. Instead, recycle the catalyst for future use by evaporating water on a hot plate. Transfer the colorless solid to a storage container or submit it to the instructor. If the material is highly colored, ask your instructor for advice. If the solvent, 1,2-dichloroethane, has been recovered using a rotary evaporator, pour it into a container so that it can be recycled.

NOTES TO THE INSTRUCTOR

It is suggested that each pair of students select a different substrate from the list provided. In most cases, the reaction will not go to completion, and as expected, the reaction provides isomeric products. For example, toluene yields the expected *ortho* and *para* products, but a small amount of *meta* product is also formed. The products are analyzed by GC/MS. This experiment provides an excellent opportunity to discuss mass spectrometry because most of the compounds yield abundant molecular ions. The products are identified by searching the National Institute of Standards and Technology (NIST) database. Although it is best to search the database to identify the compounds, the experiment can also be conducted with gas chromatography. If this is done, one can usually assume that the nitro compounds will emerge in the following order: *ortho*, *meta*, and *para*. Adequate separations can be achieved on a GC/MS instrument using a J & W DB-5MS or Varian CP-Sil 5CB capillary column (30 m, 0.25-mm ID, 0.25 μm). Set the injector temperature at 260°C. The column oven conditions are the following: start at 60°C (hold for 1 min), increase to 280°C at 20°C/min (12 min), and then hold at 280°C (4.5 min). Each run takes about 17 minutes. The helium flow rate is 1 mL/min. The mass range is set for 40 to 400 m/e.

PROCEDURE

Select one of the following aromatic substrates:

Toluene	Biphenyl
Butylbenzene	4-Methylbiphenyl
Isopropylbenzene	Diphenylmethane
tert-Butylbenzene	Phenylacetic acid
ortho-Xylene	Iodobenzene
meta-Xylene	Fluorobenzene
para-Xylene	Naphthalene
Anisole	Fluorene
1,2-Dimethoxybenzene (Veratrole)	Acetanilide
1,3-Dimethoxybenzene	Phenol
1,4-Dimethoxybenzene	α-Naphthol
4-Methoxytoluene	β-Naphthol

Place 0.375 g ytterbium (III) trifluoromethanesufonate hydrate catalyst (ytterbium triflate) into a 25-mL round-bottom flask. Add 10 mL of 1,2-dichloroethane solvent, followed by 0.400 mL of concentrated nitric acid (automatic pipet). Add two boiling stones to the flask. To this solution, weigh out and add approximately 6 mmol of the aromatic substrate. Connect the round-bottom flask to a reflux condenser and clamp it into place on a ring stand. Use a slow flow of water through the condenser. With a hot plate, heat the mixture to reflux for 1 hour.

After refluxing the mixture for 1 hour, allow the mixture to cool to room temperature and add 8 mL of water. Transfer the mixture into a separatory funnel. Shake the mixture gently and allow the two phases to separate. Drain the organic layer (bottom layer) into a 25-mL Erlenmeyer flask. Dry the organic layer with a small scoop of anhydrous magnesium sulfate (about 0.5 g). If a rotary evaporator is available, transfer the organic layer to a preweighed 50-mL round-bottom flask for removal of solvent. The apparatus allows the possibility of recovering most of the 1,2-dichloroethane. When the solvent has been removed, remove the flask and weigh it.

Alternatively, the solvent may be removed by using the apparatus shown in Figure 7.17C, page 612. Transfer the dried organic layer to a preweighed 125-mL filter flask. Add a melting-point capillary tube to the flask (open end down) and then cork the top. The melting-point capillary helps speed the evaporation process. Connect the side arm of the filter flask to the house vacuum system or aspirator, using a trap cooled in ice. There will be a cooling effect while the evaporation takes place, so you will need to heat the flask gently (lowest setting on a hot plate). Most of your solvent should be evaporated in less than 1 hour, under vacuum and with gentle heating. Weigh the filter flask.

The aqueous layer remaining in the separatory funnel contains the ytterbium catalyst. Pour the aqueous layer from the top of the separatory funnel into a preweighed 50-mL Erlenmeyer flask. Completely evaporate the water on a hot plate. Weigh the flask to determine how much catalyst you were able to recover. Place the catalyst in a container that holds the recycled catalyst that will be reused in other classes.

Unless instructed otherwise, prepare a sample for analysis by gas chromatography/mass spectrometry (GC/MS) by dissolving 2 drops of the mixture of nitrated aromatic compounds in about 1 mL of methylene chloride. These samples will be run using automation software on the GC/MS system.

When your sample has been run, you will have an opportunity to search the NIST mass spectral library to determine the structures of the product(s) of the nitration. Determine the structures of the product(s) and the percentages of each component. There will likely be starting material left in the reaction mixture. It would be of interest to see how your product ratios compare to the values obtained from the literature (see References).

REFERENCES

Braddock, C. "Novel Recyclable Catalysts for Atom Economic Aromatic Nitration." *Green Chemistry*, 3 (2001): G26–G32.

Schofield, K. "Aromatic Nitration." London: Cambridge University Press, 1980.

Waller, F. J., Barrett, G. M., Braddock, D. C., and Ramprasad, D. "Lanthanide (III) Triflates as Recyclable Catalysts for Atom Economic Aromatic Nitration." *Chem Communications* (1997): 613–614.

QUESTIONS

1. Interpret the mass spectrum of the compounds formed in the nitration of your aromatic substrate.

2. Draw a mechanism that explains how the nitro-substituted aromatic products observed in your reaction were formed.

---

35   **EXPERIMENT 35**

An Oxidation–Reduction Scheme: Borneol, Camphor, Isoborneol

Green chemistry

Hypochlorite (bleach) oxidation

Sodium borohydride reduction

Monitoring reactions by thin-layer chromatography (TLC)

Stereochemistry

Spectroscopy (infrared, proton NMR, carbon-13 NMR)

Gas chromatography

Computational chemistry (optional)

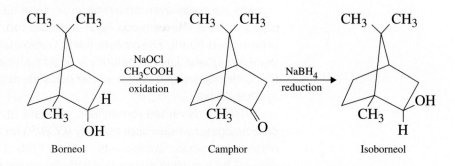

Borneol   Camphor   Isoborneol

This experiment will illustrate the use of an oxidizing agent (hypochlorous acid) for converting a secondary alcohol (borneol) to a ketone (camphor). The camphor is then reduced by sodium borohydride to give the **isomeric** alcohol isoborneol. The spectra of borneol, camphor, and isoborneol will be compared to detect structural differences and to determine the extent to which the final step produces a pure alcohol isomeric with the starting material.

Oxidation of Borneol with Hypochlorite

Sodium hypochlorite, bleach, can be used to oxidize secondary alcohols to ketones. Because this reaction occurs more rapidly in an acidic environment, it is likely that the actual oxidizing agent is hypochlorous acid HOCl. This acid is generated by the reaction between sodium hypochlorite and acetic acid.

$$\text{NaOCl} + \text{CH}_3\text{COOH} \longrightarrow \text{HOCl} + \text{CH}_3\text{COONa}$$

Although the mechanism is not fully understood, there is evidence that an alkyl hypochlorite intermediate is produced, which then gives the product via an E2 elimination:

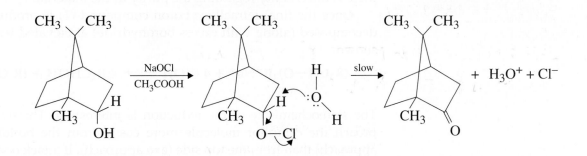

Reduction of Camphor with Sodium Borohydride

Metal hydrides (sources of H:$^-$) of the Group III elements, such as lithium aluminum hydride LiAlH$_4$ and sodium borohydride NaBH$_4$, are widely used in reducing carbonyl groups. Lithium aluminum hydride, for example, reduces many compounds containing carbonyl groups, such as aldehydes, ketones, carboxylic acids, esters, or amides, whereas sodium borohydride reduces only aldehydes and ketones. The reduced reactivity of borohydride allows it to be used even in alcohol and water solvents, whereas lithium aluminum hydride reacts violently with these solvents to produce hydrogen gas and thus must be used in nonhydroxylic solvents. In the present experiment, sodium borohydride is used because it is easily handled, and the results of reductions using either of the two reagents are essentially the same. The same care need not be taken in keeping sodium borohydride away from water, as is required with lithium aluminum hydride.

The mechanism of action of sodium borohydride in reducing a ketone is as follows:

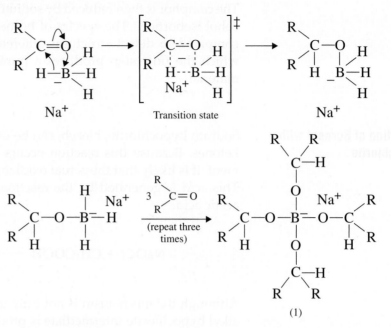

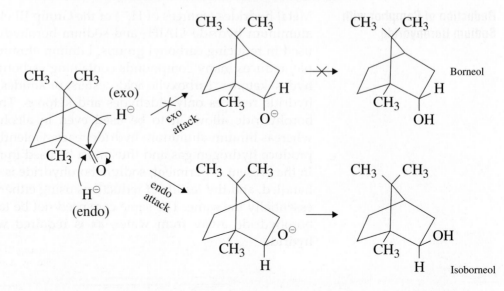

(1)

Note in this mechanism that all four hydrogen atoms are available as hydrides ($H:^-$), and thus 1 mole of borohydride can reduce 4 moles of ketones. All the steps are irreversible. Usually excess borohydride is used because there is uncertainty regarding the purity of the material.

Once the final tetraalkoxyboron compound (1) is produced, it can be decomposed (along with excess borohydride) at elevated temperatures as shown:

$$(R_2CH-O)_4B^-Na^+ + 4\ R'OH \longrightarrow 4\ R_2CHOH + (R'O)_4B^-Na^+$$
(1)

The stereochemistry of the reduction is interesting. The hydride can approach the camphor molecule more easily from the bottom side (**endo** approach) than from the top side (**exo** approach). If attack occurs at the top, a large steric repulsion is created by one of the two **geminal** methyl groups. Geminal methyl groups are groups that are attached to the same carbon. Attack at the bottom avoids this steric interaction.

It is expected, therefore, that **isoborneol,** the alcohol produced from the attack at the *least*-hindered position, will *predominate but will not be the exclusive product* in the final reaction mixture. The percentage composition of the mixture can be determined by spectroscopy.

It is interesting to note that when the methyl groups are removed (as in 2-norbornanone), the top side (**exo** approach) is favored, and the opposite stereochemical result is obtained. Again, the reaction does not give exclusively one product.

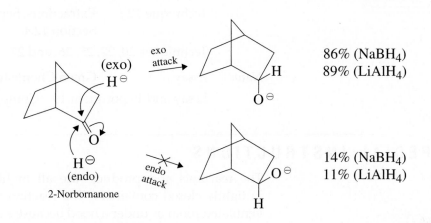

Bicyclic systems such as camphor and 2-norbornanone react predictably according to steric influences. This effect has been termed **steric approach control.** In the reduction of simple acyclic and monocyclic ketones, however, the reaction seems to be influenced primarily by thermodynamic factors. This effect has been termed **product development control.** In the reduction of 4-*t*-butylcyclohexanone, the thermodynamically more stable product is produced by product development control.

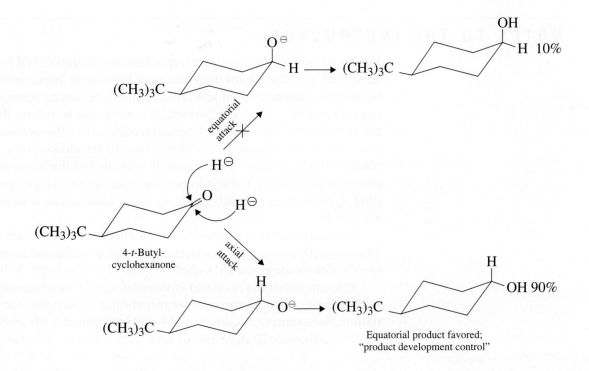

Equatorial product favored;
"product development control"

REQUIRED READING

Review: Technique 6   Heating and Cooling Methods,
Sections 6.1–6.3

Technique 7   Reaction Methods, Sections 7.1–7.4 and 7.10

Technique 8   Filtration, Section 8.3

Technique 9   Physical Constants of Solids: The Melting
Point, Sections 9.7 and 9.8

Technique 12   Extractions, Separations, and Drying Agents,
Section 12.4

Techniques 20, 22, 25, 26, and 27

New: Essay   Green Chemistry

Essay and Experiment 19   Computational Chemistry (optional)

SPECIAL INSTRUCTIONS

The reactants and products are all highly volatile and must be stored in tightly closed containers. The reaction should be carried out in a well-ventilated room or under a hood because a small amount of chlorine gas will be emitted from the reaction mixture. The reduction of camphor to isoborneol involves diethyl ether, which is extremely flammable. Be certain that no open flames of any sort are in your vicinity when you are using ether.

SUGGESTED WASTE DISPOSAL

The aqueous solutions obtained from the extraction steps should be placed in the aqueous waste container. Any leftover methanol or ether solution may be placed in the nonhalogenated organic waste container.

NOTES TO THE INSTRUCTOR

We use commercial 6% sodium hypochlorite solution (VWR Scientific Products, No. VW3248-1) for this reaction because it more reliably oxidizes borneol to camphor. Even with this solution, however, some students may not completely oxidize the borneol. It is advisable to follow the progress of the reaction by TLC. If some borneol remains after the normal reaction period, additional sodium hypochlorite should be added. Some students will obtain a liquid product. In this case, it is likely that the borneol has not been completely oxidized. If the infrared spectrum shows the presence of borneol (OH stretch), then it is advisable to use a commercial source of camphor for Part B.

The sodium borohydride should be checked to see whether it is active. Place a small amount of powdered material in some methanol and heat it gently. The solution should bubble vigorously if the hydride is active.

Percentages of borneol and isoborneol can be determined by gas chromatography. Any gas chromatograph should be suitable for this determination. For example, a Gow-Mac 69-930 instrument with an 8-foot column of 10% Carbowax 20M, at 180°C, and with a 40 mL/min helium flow rate will give a suitable separation. The compounds elute in the following order: camphor (8 min), isoborneol (10 min), and borneol (11 min). A Varian CP-3800 with autosampler equipped with a J & W DB-5 or Varian CP-Sil 5CB capillary column (30 m, 0.25-mm ID, 0.25 μm) also provides a good separation. Set the injector temperature at 250°C. The column oven conditions are the following: start at 75°C (hold for 10 min), increase to 200°C at 35°C/min, and then hold at 200°C (1 min). Each run takes about 15 minutes. The helium flow rate is 1 mL/min. The compounds elute in the following order: camphor (12.9 min), isoborneol (13.1 min), and borneol (13.2 min). An optional procedure is provided that involves computational chemistry.

PROCEDURE

Part A. Oxidation of Borneol to Camphor

Assemble the Apparatus

To a 10-mL round-bottom flask, add 0.360 g of racemic borneol, 1.0 mL of acetone, and 0.30 mL of glacial acetic acid. After adding a magnetic stir bar to the flask, attach a water condenser and place the round-bottom flask in a water bath at about 50°C, as shown in Figure 6.6, page 594. Use a 150-mL beaker for the water bath. It is important that the temperature of the water bath remain at 50°C during the entire reaction period. Stir the mixture until the borneol is dissolved. If the borneol does not dissolve, add 0.50 mL of acetone.

Addition of Sodium Hypochlorite

Measure out 6 mL of 6% sodium hypochlorite solution[1] in a 10-mL graduated cylinder. Using one of the plastic pipets, add **dropwise** 0.5 mL of the sodium hypochlorite solution **every 4 minutes** through the top of the water condenser. The addition will take 48 minutes to complete. Continue to stir and heat the mixture during the 48-minute period. After the addition, heat and stir the mixture for an additional 15 minutes.

Monitoring the Oxidation with Thin-Layer Chromatography (TLC)

The reaction progress can be monitored by TLC (Technique 20, Section 20.10, Figure 20.10, p. 790). Remove about 0.5 mL of reaction mixture with a Pasteur pipet and place it into a 3-mL conical vial. Add about 0.5 mL of methylene chloride, cap the vial, and shake the vial for a few minutes. Remove the lower, methylene chloride, layer with a Pasteur pipet to avoid drawing up any aqueous layer. Place the methylene chloride extract into a dry test tube.

Prepare a 30-mm × 70-mm silica gel TLC plate (Whatman Silica Gel plate with aluminum backing, No. 4420-222) that will be spotted with three solutions using micropipets (Technique 20, Section 20.4, p. 782). Borneol (2% in methylene chloride) is spotted in lane 1, camphor (2% in methylene chloride) in lane 2, and the reaction mixture dissolved in methylene chloride in lane 3. Spot each solution five or six times, each time spotting it on top of the previous spot (allow the previous spot to dry before applying the next one). Prepare a developing chamber from a 4-oz wide-mouth, screw-cap jar (as described in Technique 20, Section 20.5, p. 784) using methylene chloride as the solvent. Put the plate into the developing

---

[1] *Note to the Instructor:* We use sodium hypochlorite (6%) from VWR (Catalogue Number VW 3248-1). It should be stored in the refrigerator.

chamber. When the solvent front has traveled about 5 cm, remove the plate, evaporate the solvent, and place the plate into another jar that contains a few crystals of iodine (Technique 20, Section 20.7, p. 786). Heat the jar on a hot plate. The iodine vapors will visualize the spots. Camphor will have a larger R_f value than borneol. Unfortunately, camphor and borneol do not give intense spots with iodine, but you should be able to see them. The relative amounts of borneol and camphor can be determined by the relative intensity of the spots on the plates. The reaction will be judged to be complete if the borneol spot in lane 3 is not visible. If some borneol remains as determined by the TLC method, reattach the water condenser, reheat the reaction mixture in the round-bottom flask, and then add 1 mL more of sodium hypochlorite solution dropwise to the reaction mixture over a 15-minute period. Check the mixture again using the previous procedure and a new TLC plate. Ideally, borneol should not be visible on the plate, and camphor should be visible.

Extraction of Camphor

When the reaction time is complete, allow the mixture to cool to room temperature. Remove the water condenser, transfer the mixture to a screw-cap centrifuge tube, and add 2.0 mL of methylene chloride to extract the camphor (Technique 12, Section 12.4, p. 674). Cap the centrifuge tube and shake well, venting occasionally. If the layers do not separate cleanly, add another 2 mL of methylene chloride and briefly shake the mixture. Using a Pasteur pipet, transfer the lower methylene chloride layer into another centrifuge tube. Extract the aqueous layer with a second 2.0-mL portion of methylene chloride and combine it with the first methylene chloride solution.

Wash the combined methylene chloride layers with 2.0 mL of saturated aqueous sodium bicarbonate solution. Stir the liquid with a stirring rod or spatula until the bubbling produced by the formation of carbon dioxide ceases. Cap the tube and shake with frequent venting to release any pressure produced. Transfer the lower methylene chloride layer to a beaker or Erlenmeyer flask and discard the aqueous layer in the centrifuge tube.

Return the methylene chloride layer to the centrifuge tube and wash the methylene chloride with 2.0 mL of 5% aqueous sodium bisulfite. Transfer the lower methylene chloride layer to another centrifuge tube and discard the remaining aqueous layer. Finally, wash the methylene chloride with 2.0 mL of water, remove the lower methylene chloride layer with a dry Pasteur pipet, and transfer it to a dry test tube or conical vial.

Isolation of Product

Dry the methylene chloride layer over granular anhydrous sodium sulfate (see Technique 12, Section 12.9, p. 680). Transfer the dried methylene chloride layer to a preweighed 10-mL Erlenmeyer flask. Evaporate the solvent in the hood with a gentle stream of dry air or nitrogen gas while heating the Erlenmeyer flask in a water bath at 40–50°C (see Fig. 7.17, p. 612). When all the liquid has evaporated and a solid has appeared, remove the flask from the heat source. If the crystals appear wet with solvent, apply a vacuum for a few minutes to remove any residual solvent.

Analysis of Camphor

Weigh the flask to determine the weight of your product and calculate the percentage yield. At the option of the instructor, determine the melting point; the melting point of pure racemic camphor is 174°C.[2] Before proceeding to Part B, you should verify that the oxidation was successful. This can be done by determining the infrared spectrum of camphor. The spectrum is best obtained by employing the dry film method (see Technique 25, Section 25.4, p. 838). By examination of the infrared peaks, determine if the borneol (an alcohol OH stretch) is absent or nearly absent and if the borneol has been oxidized to camphor (a ketone C=O stretch). For comparison, an infrared spectrum of camphor is shown on page 296. If your oxidation was not totally successful, consult your instructor for your options. The remainder of the camphor is reduced in the next step to isoborneol, which will be carried out in the same flask. Store the camphor with the flask tightly sealed until needed.

Part B. Reduction of Camphor to Isoborneol

Reduction

The camphor obtained in Part A should not contain borneol. If it does, show your infrared spectrum to your instructor and ask for advice. Reweigh the 10-mL Erlenmeyer flask to determine the weight of camphor remaining. If the amount is less than 0.100 g, obtain some camphor from the supply shelf to supplement your yield. If it is more than 0.100 g, scale up the reagents appropriately from the following amounts. Add 1.0 mL of methanol to the camphor in the 10-mL Erlenmeyer flask. Stir with a glass stirring rod until the camphor has dissolved. In portions, cautiously and intermittently add 0.10 g of sodium borohydride to the solution. (Replace the cap on the bottle of sodium borohydride immediately.) When all the borohydride is added, boil the contents of the flask on a steam bath or a warm hot plate (low setting) for 2 minutes.

Isolation and Analysis of Product

Allow the reaction mixture to cool for a couple of minutes and carefully add 3.5 mL of ice-cold water. Collect the white solid by filtering on a Hirsch funnel and, by using suction, allow to dry for several minutes. Transfer the solid to a 10-mL Erlenmeyer flask.

Add 4 mL of methylene chloride to dissolve the product. Once the product has dissolved (add more solvent, if necessary), dry the solution over granular anhydrous sodium sulfate (see Technique 12, Section 12.9, page 680). Transfer the dried solution into a preweighed 10-mL Erlenmeyer flask. Evaporate the solvent in a hood, as described previously. Determine the weight of the product and calculate the percentage yield. Determine the melting point; pure racemic isoborneol melts at 212°C. Determine the infrared spectrum of the product by the dry film method used previously with camphor. Compare it with the spectra for borneol and isoborneol shown in the figures.

Part C. Percentages of Isoborneol and Borneol Obtained from the Reduction of Camphor

NMR Determination

The percentage of each of the isomeric alcohols in the borohydride mixture can be determined from the NMR spectrum (see Technique 26, Section 26.1, p. 870). The NMR spectra of the alcohols are shown on page 298. The hydrogen atom on the carbon atom bearing the hydroxyl group appears at 4.0 ppm for borneol and 3.6 ppm for isoborneol. To obtain the product ratio, integrate these peaks (using an expanded presentation) in the NMR spectrum of the sample obtained from borohydride reduction. In the spectrum shown on page 300, the isoborneol–borneol ratio of 6:1 was obtained. The percentages obtained are 85% isoborneol and 15% borneol.

---

[2] The observed melting point of camphor is often low. A small amount of impurity drastically reduces the melting point and increases the range (see Question 4, p. 301).

Gas Chromatography

The isomer ratio and percentages can also be obtained by gas chromatography. Your instructor will provide instructions for preparing your sample. A Gow-Mac 69-360 instrument fitted with an 8-foot column of 10% Carbowax 20M, in an oven set at 180°C, and with a 40 mL/min helium flow rate will completely separate isoborneol and borneol from each other. In addition, any residual camphor can be observed. The retention times for camphor, isoborneol, and borneol are 8, 10, and 11 minutes, respectively. Other instrument conditions are provided in the "Notes to the Instructor."

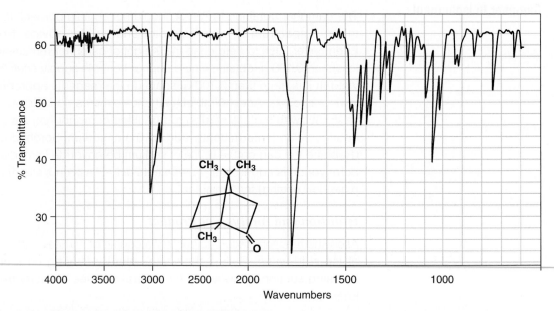

Infrared spectrum of camphor (KBr pellet).

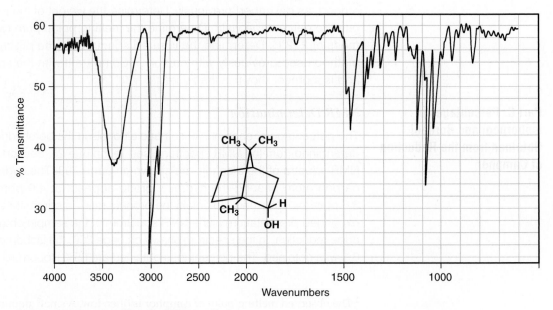

Infrared spectrum of borneol (KBr pellet).

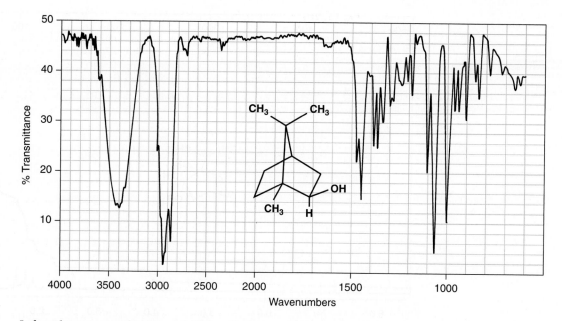

Infrared spectrum of isoborneol (KBr pellet).

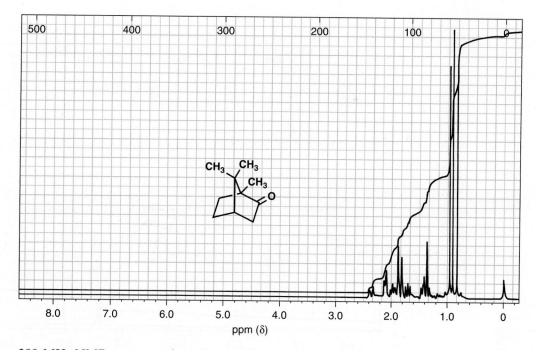

300-MHz NMR spectrum of camphor, CDCl₃.

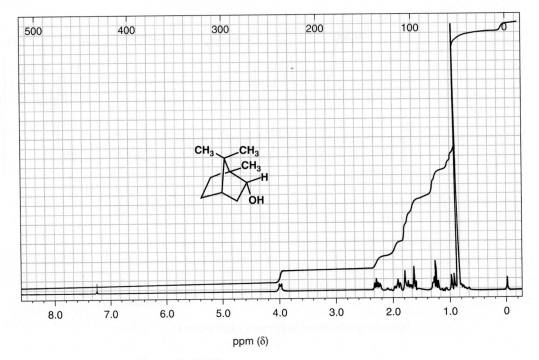

300-MHz spectrum of borneol, CDCl₃.

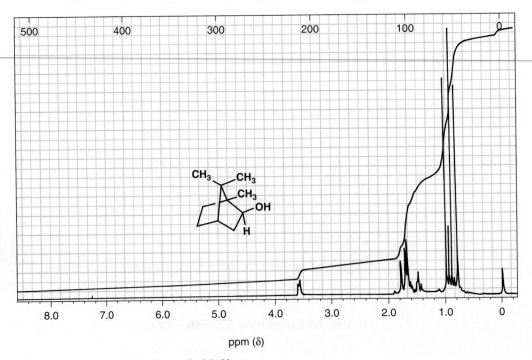

300-MHz spectrum of isoborneol, CDCl₃.

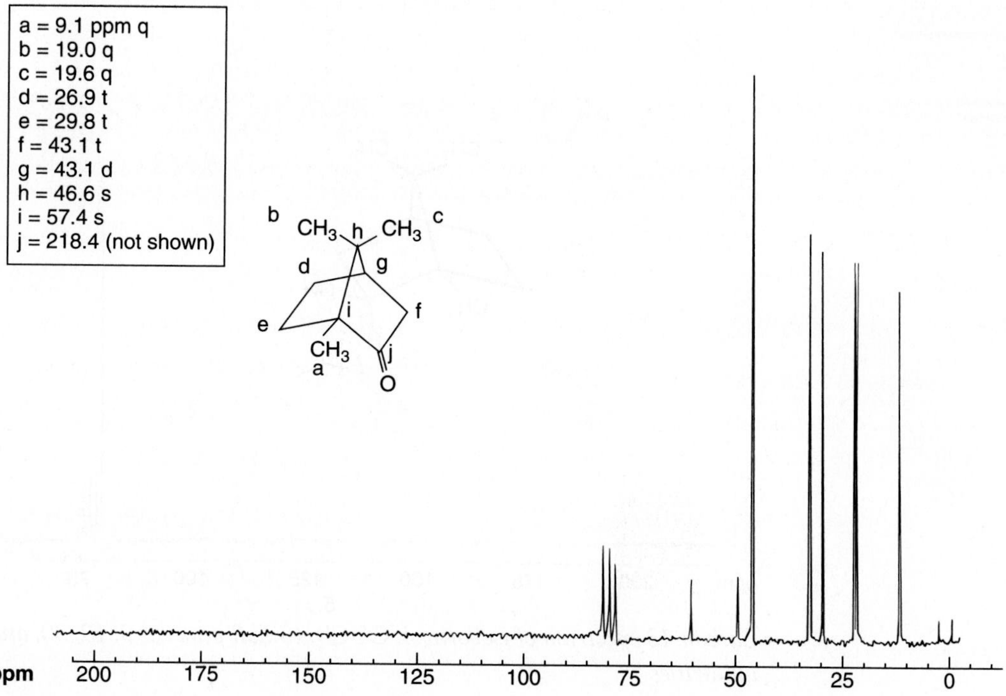
a = 9.1 ppm q
b = 19.0 q
c = 19.6 q
d = 26.9 t
e = 29.8 t
f = 43.1 t
g = 43.1 d
h = 46.6 s
i = 57.4 s
j = 218.4 (not shown)
b
CH_3
h
CH_3
c
d
g
i
f
e
CH_3
a
j
O
ppm
200
175
150
125
100
75
50
25
0

Carbon-13 spectrum of camphor, $CDCl_3$.

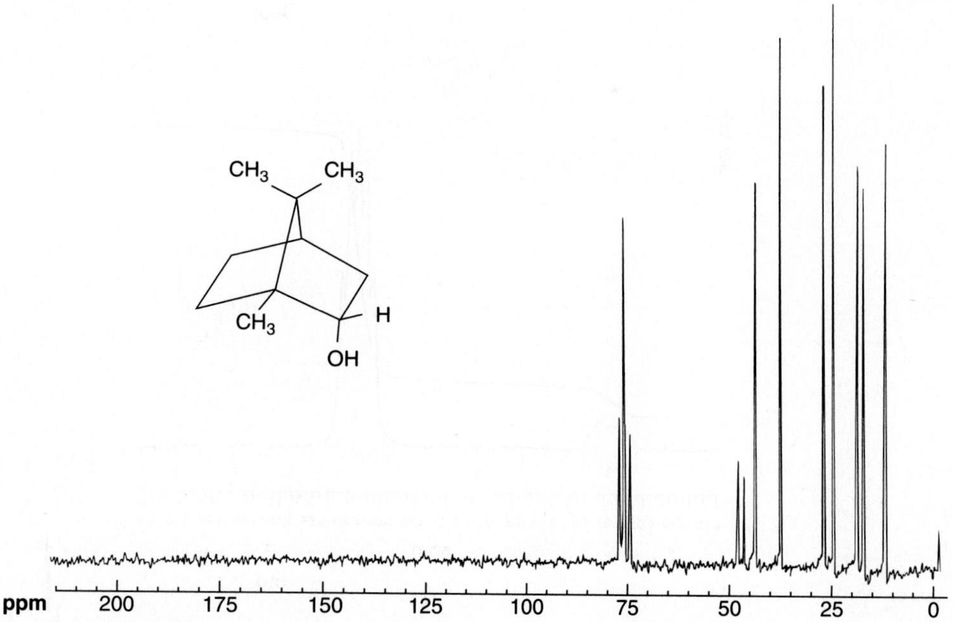
CH_3
CH_3
H
CH_3
OH
ppm
200
175
150
125
100
75
50
25
0

Carbon-13 spectrum of borneol, $CDCl_3$.

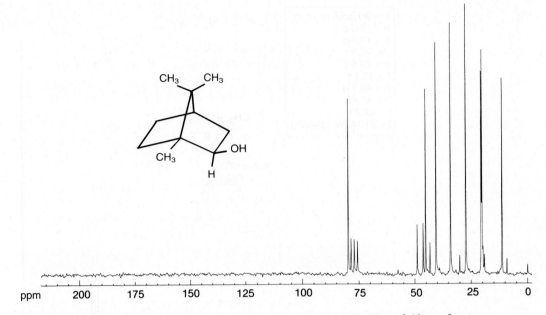

Carbon-13 spectrum of isoborneol, CDCl₃. (Small peaks at 9, 19, 30, and 43 are due to impurities.)

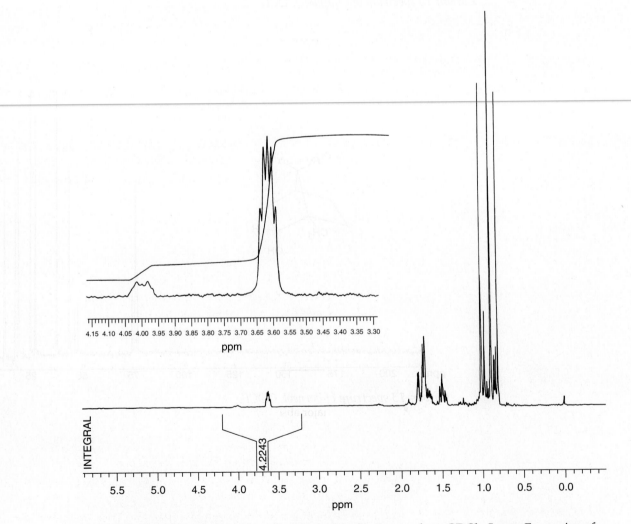

300-MHz Proton NMR spectrum of borohydride reduction product, CDCl₃. Inset: *Expansion of the 3.5–4.1 ppm region.*

MOLECULAR MODELING (OPTIONAL)

In this exercise, we will seek to understand the experimental results obtained in the borohydride reduction of camphor and compare them to the results for the simpler norbornanone system (no methyl groups). Because the hydride ion is an electron donor, it must place its electrons into an empty substrate orbital to form a new bond. The most logical orbital for this action is the LUMO (lowest unoccupied molecular orbital). Accordingly, the focus of our calculations will be the shape and location of the LUMO.

Part One   Build a model of norbornanone (page 291) and submit it to an AM1-level calculation of its energy, using a geometry optimization. Also request that density and LUMO surfaces be calculated, along with a density–LUMO surface (a mapping of the LUMO onto the density surface).

When the calculation is complete, display the LUMO on the norbornanone skeleton. Where is the size of the LUMO (its density) the largest? Which atom is this? This is the expected site of addition. Now map a density surface onto the same norbornanone surface. When you consider the approach of the borohydride ion, which face is less hindered? Is an *exo* or *endo* approach favored? An easier way to decide is to view the density–LUMO surface. On this surface, the intersection of the LUMO with the density surface is color-coded. The spot where the access to the LUMO is easiest (the location of its largest value) will be coded blue. Is this spot on the *endo* or on the *exo* face? Do your modeling results agree with observed reaction percentages (see above)?

Part Two   Follow the same instructions given earlier for norbornanone using camphor (p. 288)—that is, calculate and view density, LUMO, and density–LUMO surfaces. Do you reach the same conclusions as for norbornanone? Are there new stereochemical considerations? Do your conclusions agree with experimental results (the borneol/isoborneol ratio) you obtained in this experiment? In your report, discuss your modeling results and how they relate to your experimental results.

REFERENCES

Brown, H. C., and Muzzio, J. "Rates of Reaction of Sodium Borohydride with Bicyclic Ketones." *Journal of the American Chemical Society, 88* (1966): 2811.

Dauben, W. G., Fonken, G. J., and Noyce, D. S. "Stereochemistry of Hydride Reductions." *Journal of the American Chemical Society, 78* (1956): 2579.

Markgraf, J. H. "Stereochemical Correlations in the Camphor Series." *Journal of Chemical Education, 44* (1967): 36.

QUESTIONS

1. Interpret the major absorption bands in the infrared spectra of camphor, borneol, and isoborneol.

2. Explain why the *gem*-dimethyl groups appear as separate peaks in the proton NMR spectrum of isoborneol although they almost overlap in borneol.

3. A sample of isoborneol prepared by reduction of camphor was analyzed by infrared spectroscopy and showed a band at 1750 cm^{-1}. This result was unexpected. Why?

4. The observed melting point of camphor is often low. Look up the molal freezing-point-depression constant K for camphor and calculate the expected depression of

the melting point of a quantity of camphor that contains 0.5 molal impurity. *Hint:* Look in a general chemistry book under "freezing-point depression" or "colligative properties of solutions."

5. Why was the methylene chloride layer washed with sodium bicarbonate in the procedure for preparing camphor?

6. Why was the methylene chloride layer washed with sodium bisulfite in the procedure for preparing camphor?

7. The peak assignments are shown on the carbon-13 NMR spectrum of camphor. Using these assignments as a guide, assign as many peaks as possible in the carbon-13 spectra of borneol and isoborneol.

---

36   EXPERIMENT 36

Multistep Reaction Sequences: The Conversion of Benzaldehyde to Benzilic Acid

Green chemistry

Multistep reactions

Thiamine-catalyzed reaction

Oxidation with nitric acid

Rearrangement

Crystallization

Computational chemistry (optional)

The experiment demonstrates multistep synthesis of benzilic acid starting from benzaldehyde. In Experiment 36A, benzaldehyde is converted to benzoin using a thiamine-catalyzed reaction. This part of the experiment demonstrates how a "green" reagent can be utilized in organic chemistry. In Experiment 36B, nitric acid oxidizes benzoin to benzil. Finally, in Experiment 36C, benzil is rearranged to benzilic acid. The scheme on p. 303 shows the reactions.

REQUIRED READING

Review: Technique 6   Heating and Cooling Methods, Sections 6.1–6.3

Technique 7   Reaction Methods, Sections 7.1–7.4

Technique 8   Filtration, Section 8.3

Technique 9   Physical Constants of Solids: The Melting Point, Sections 9.7 and 9.8

Technique 11   Crystallization: Purification of Solids, Section 11.3

Technique 12   Extractions, Separations, and Drying Agents, Section 12.4

Technique 25   Infrared Spectroscopy, Section 25.4

New: Essay and Experiment 19   Computational Chemistry (Optional)

NOTES TO THE INSTRUCTOR

Although this experiment is intended to illustrate a multistep synthesis to the students, each part may be done separately, or two out of the three reactions can be linked together. The sections on "Special Instructions" and "Suggested Waste Disposal" are included in each part of this experiment. You may also create another multistep synthesis by linking together benzoin (Experiment 36A), benzil (Experiment 36B), and tetraphenylcyclopentadienone (Experiment 37).

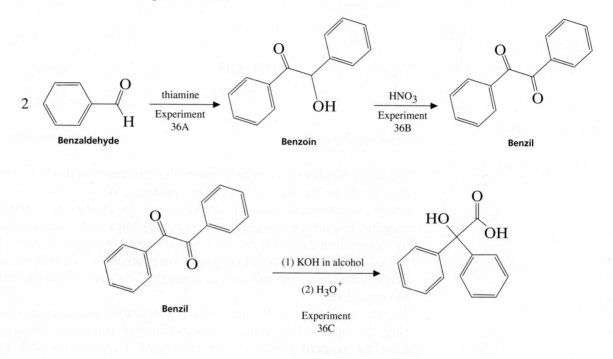

36A   EXPERIMENT 36A

Preparation of Benzoin by Thiamine Catalysis

In this experiment, two molecules of benzaldehyde will be converted to benzoin using the catalyst thiamine hydrochloride. This reaction is known as a benzoin condensation reaction:

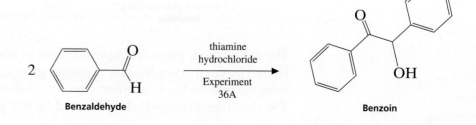

Thiamine hydrochloride is structurally similar to thiamine pyrophosphate (TPP). TPP is a coenzyme universally present in all living systems. It catalyzes several biochemical reactions in natural systems. It was originally discovered as a required nutritional factor (vitamin) in humans by its link with the disease beriberi. **Beriberi** is a disease of the peripheral nervous system caused by a deficiency of Vitamin B_1 in the diet. Symptoms include pain and paralysis of the extremities, emaciation, and swelling of the body. The disease is most common in Asia.

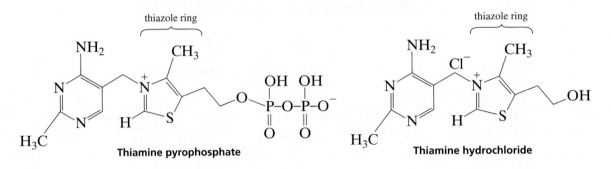

Thiamine binds to an enzyme before the enzyme is activated. The enzyme also binds to the substrate (a large protein). Without the coenzyme thiamine, no chemical reaction would occur. The coenzyme is the **chemical reagent.** The protein molecule (the enzyme) helps and mediates the reaction by controlling stereochemical, energetic, and entropic factors, but it is nonessential to the overall course of reactions that its catalyzes. A special name, vitamins, is given to coenzymes that are essential to the nutrition of the organism.

The most important part of the entire thiamine molecule is the central ring, the thiazole ring, which contains nitrogen and sulfur. This ring constitutes the **reagent** portion of the coenzyme. Experiments with the model compound 3,4-dimethylthiazolium bromide have explained how thiamine-catalyzed reactions work. It was found that this model thiazolium compound rapidly exchanged the C-2 proton for deuterium in D_2O solution. At a pD of 7 (no pH here), this proton was completely exchanged in seconds!

This indicates that the C-2 proton is more acidic than one would have expected. It is apparently easily removed because the conjugate base is a highly stabilized ylide. An **ylide** is a compound or intermediate with positive and negative formal charges on adjacent atoms.

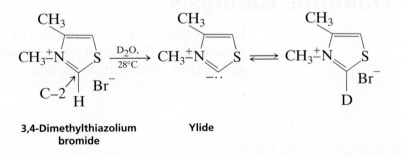

The sulfur atom plays an important role in stabilizing this ylide. This was shown by comparing the rate of exchange of 1,3-dimethyl-imidazolium ion with the rate for the thiazolium compound shown in the previous equation. The dinitrogen compound exchanged its C-2 proton more slowly than the sulfur-containing ion. Sulfur, being in the third row of the periodic chart, has *d* orbitals available for bonding to adjacent atoms. Thus, it has fewer geometrical restrictions than carbon and nitrogen atoms do and can form carbon–sulfur multiple bonds in situations in which carbon and nitrogen normally would not.

1,3-Dimethylimidazolium bromide

In Experiment 36A, we will utilize thiamine hydrochloride rather than TPP to catalyze the benzoin condensation. The mechanism is shown on page 306. For simplicity, only the thiazole ring is shown.

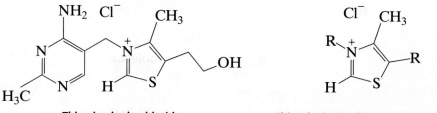

Thiamine hydrochloride **Thiazole ring in thiamine hydrochloride**

The mechanism involves the removal of the proton at C-2 from the thiazole ring with a weak base to give the ylide (Step 1). The ylide acts as a nucleophile that adds to the carbonyl group of benzaldehyde forming an intermediate (Step 2). A proton is removed to yield a new intermediate with a double bond (Step 3). Notice that the nitrogen atom helps to increase the acidity of that proton. This intermediate can now react with a second benzaldehyde molecule to yield a new intermediate (Step 4). A base removes a proton to produce benzoin and also regenerates the ylide (Step 5). The ylide reenters the mechanism to form more benzoin by the condensation of two more molecules of benzaldehyde.

SPECIAL INSTRUCTIONS

This experiment may be conducted concurrently with another experiment. It involves a few minutes at the beginning of a laboratory period for mixing reagents. The remaining portion of the period may be used for another experiment.

SUGGESTED WASTE DISPOSAL

Pour all of the aqueous solutions produced in this experiment into a waste container designated for aqueous waste. The ethanolic mixtures obtained from the crystallization of crude benzoin should be poured into a waste container designated for nonhalogenated waste.

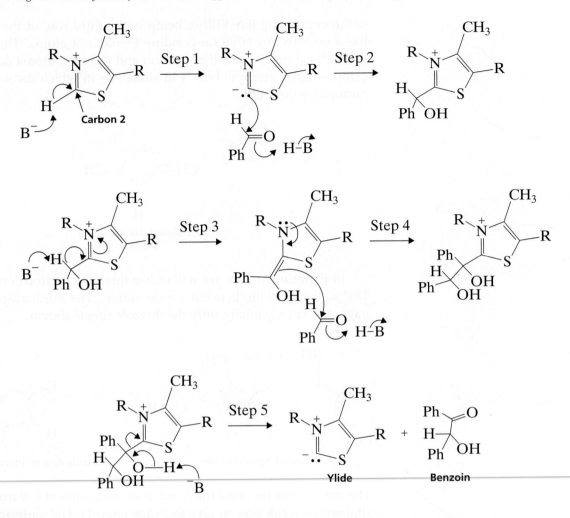

NOTES TO THE INSTRUCTOR

It is essential that the benzaldehyde used in this experiment be *pure*. Benzaldehyde is easily oxidized in air to benzoic acid. Even when benzaldehyde *appears* free of benzoic acid by infrared spectroscopy, you should check the purity of your benzaldehyde and thiamine by following the instructions given in the first paragraph of the Procedure ("Reaction Mixture"). When the benzaldehyde is pure, the solution will be nearly filled with solid benzoin after 2 days (you may need to scratch the inside of the flask to induce crystallization). If no solid appears or very little appears, then there is a problem with the purity of the benzaldehyde. If possible, use a newly opened bottle that has been purchased recently. However, it is essential that you check both the old **and** new benzaldehyde before doing the laboratory experiment.

We have found that the following procedure does an adequate job of purifying benzaldehyde. The procedure does not require distillation of benzaldehyde. Shake the benzaldehyde in a separatory funnel with an equal volume of 5% aqueous sodium carbonate solution. Shake gently and occasionally open the stopcock of the funnel to vent carbon dioxide gas. An emulsion forms that may take 2–3 hours to separate. It is helpful to stir the mixture occasionally during this period to help break the emulsion. Remove the lower sodium carbonate layer, including any remaining emulsion. Add about ¼ volume of water to the benzaldehyde and shake the mixture gently to avoid an emulsion. Remove the cloudy lower organic layer and dry the benzaldehyde with calcium chloride until the next day. Any remaining cloudiness is removed by gravity filtration through fluted filter paper. The resulting clear, purified benzaldehyde should be suitable for this experiment without vacuum distillation. You *must* check the purified benzaldehyde to see if it is suitable for the experiment by following the instructions in the first paragraph of the Procedure.

It is advisable to use a fresh bottle of thiamine hydrochloride, which should be stored in the refrigerator. Fresh thiamine does not seem to be as important as pure benzaldehyde for success in this experiment.

PROCEDURE

Reaction Mixture

Add 0.30 g thiamine hydrochloride to a 25-mL Erlenmeyer flask. Dissolve the solid in 0.45 mL of water by swirling the flask. Add 3.0 mL of 95% ethanol and swirl the solution until it is homogeneous. To this solution, add 0.90 mL of an aqueous sodium hydroxide[1] solution and swirl the flask until the bright yellow color fades to a pale yellow. Weigh the flask and solution, add 0.90 mL of benzaldehyde, and reweigh the flask to determine an accurate weight of benzaldehyde introduced to the flask. Swirl the contents of the flask until it is homogeneous. Stopper the flask and let it stand in a dark place for at least 2 days.

Isolation of Crude Benzoin

If crystals have not formed after 2 days, initiate crystallization by scratching the inside of the flask with a glass stirring rod. Allow about 5 minutes for the crystals of benzoin to form fully. Place the flask, with crystals, into an ice bath for 5–10 minutes.

If for some reason the product separates as an oil, it may be helpful to scratch the flask with a glass rod or seed the mixture by allowing a small amount of solution to dry on the end of a glass rod and then placing this into the mixture. Cool the mixture in an ice bath before filtering.

Break up the crystalline mass with a spatula, swirl the flask rapidly, and quickly transfer the benzoin to a Hirsch funnel under vacuum (Technique 8, Section 8.3, and Figure 8.5, p. 622). Wash the crystals with three 1-mL portions of ice-cold water. Allow the benzoin to dry in the Hirsch funnel by drawing air through the crystals for about 5 minutes. Transfer the benzoin to a watch glass and allow it to dry in air until the next laboratory period. The product may also be dried in a few minutes in an oven set at about 100°C.

Yield Calculation and Melting-Point Determination

Weigh the benzoin and calculate the percentage yield based on the amount of benzaldehyde used initially. Determine the melting point (pure benzoin melts between 134°C and 135°C). Because your crude benzoin will normally melt between 129°C and 132°C, the benzoin should be crystallized before the conversion to benzil (Experiment 36B).

Crystallization of Benzoin

Purify the crude benzoin by crystallization from hot 95% ethanol (use 0.8 mL of alcohol/0.1 g of crude benzoin) using a 10-mL Erlenmeyer flask for the crystallization (Technique 11, Section 11.3, p. 650; omit step 2 shown in Figure 11.4). After the

---

[1] Dissolve 8.0 g of NaOH in 100 mL water.

crystals have cooled in an ice bath, collect the crystals on a Hirsch funnel. The product may be dried in a few minutes in an oven set at about 100°C. Determine the melting point of the purified benzoin. If you are not scheduled to perform Experiment 36B, submit the sample of benzoin, along with your report, to the instructor.

Spectroscopy

Determine the infrared spectrum of the benzoin by the dry film method (Technique 25, Section 25.4, p. 838). A spectrum is shown here for comparison.

QUESTIONS

1. The infrared spectra of benzoin and benzaldehyde are given in this experiment. Interpret the principal peaks in the spectra.

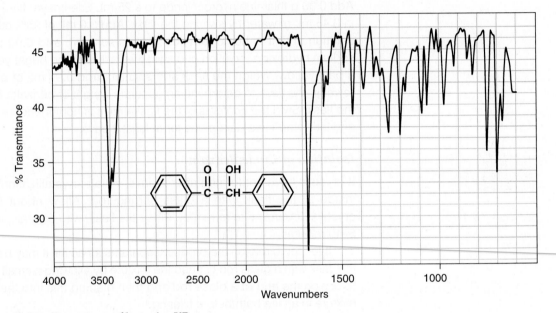

Infrared spectrum of benzoin, KBr.

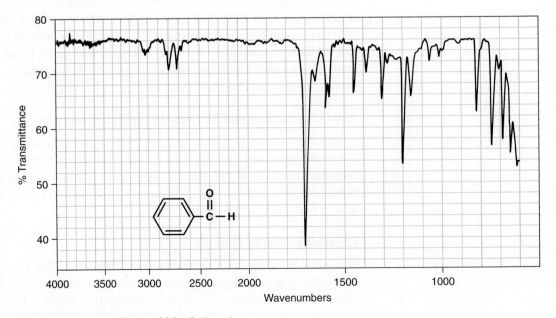

Infrared spectrum of benzaldehyde (neat).

2. How do you think the appropriate enzyme would have affected the reaction (degree of completion, yield, stereochemistry)?

3. What modifications of conditions would be appropriate if the enzyme were to be used?

4. Draw a mechanism for the cyanide-catalyzed conversion of benzaldehyde to benzoin. The intermediate, shown in brackets, is thought to be involved in the mechanism.

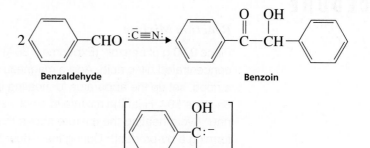

36B EXPERIMENT 36B

Preparation of Benzil

In this experiment, benzil is prepared by the oxidation of an α-hydroxyketone, benzoin. This experiment uses the benzoin prepared in Experiment 36A and is the second step in the multistep synthesis. This oxidation can be done easily with mild oxidizing agents such as Fehling's solution (alkaline cupric tartrate complex) or copper sulfate in pyridine. In this experiment, the oxidation is performed with nitric acid.

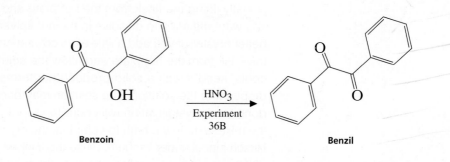

SPECIAL INSTRUCTIONS

Nitric acid should be dispensed in a good hood to avoid the choking odor of this substance. The vapors will irritate your eyes. Avoid contact with your skin. During the reaction, considerable amounts of noxious nitrogen oxide gases are evolved. Be sure to run the reaction in a good fume hood.

SUGGESTED WASTE DISPOSAL

The aqueous nitric acid wastes should be poured into a container designated for nitric acid wastes. Do not put them into the aqueous waste container. The ethanolic wastes from the crystallization should be poured into the nonhalogenated waste container.

PROCEDURE

Reaction Mixture

Place 0.30 g of benzoin (Experiment 36A) into a 5-mL conical vial and add 1.5 mL of concentrated nitric acid. Add a magnetic spin vane and attach an air condenser. In a hood, set up the apparatus for heating in a hot water bath, as shown in Figure 6.6 on page 594. Heat the mixture in a hot water bath at about 70°C for 1 hour, with stirring. Avoid heating the mixture above this temperature to reduce the possibility of forming a by-product.[1] During the 1-hour heating period, nitrogen oxide gases (red) will be evolved. If it appears that gases are still being evolved after 1 hour, continue heating for another 15 minutes but then discontinue heating at that time.

Isolation of Crude Benzil

Cool the mixture for a few minutes and detach the air condenser. With a Pasteur pipet, transfer the reaction mixture to a beaker containing 4 mL of ice-cold water. Rinse the conical vial and spin vane with a small amount of water. Cool the mixture in an ice bath until the crystals have formed. If the material oils out rather than crystallizes, scratch the oil vigorously with a spatula until it does crystallize completely. Collect the crude product on a Hirsch funnel under vacuum. Wash it well with cold water (about 5 mL) to remove the nitric acid. Continue drawing air through the solid mass on the Hirsch funnel to help dry the solid. Weigh the solid.

Crystallization of Product

Purify the solid by dissolving it in hot 95% ethanol in a small Erlenmeyer flask (about 5 mL per 0.5 g of product) using a hot plate as the heating source. Be careful not to melt the solid on the hot plate. You can avoid melting the benzil by occasionally lifting the flask from the hot plate and swirling the contents of the flask. You want the solid to dissolve in the hot solvent rather than melt. You will obtain better crystals if you add a little extra solvent after it dissolves completely. Remove the flask from the hot plate and allow the solution to cool slowly. As the solution cools, seed it with a solid product that forms on a spatula after the spatula is dipped into the solution. The solution may become supersaturated unless this is done, and crystallization will occur too rapidly. Yellow crystals are formed. Cool the mixture in an ice bath to complete the crystallization. Collect the product on a Hirsch funnel under vacuum. Rinse the flask with small amounts (about 1 mL total) of ice-cold 95% ethanol to complete the transfer of product to the Hirsch funnel. Continue drawing air through the crystals on the Hirsch funnel by suction for about 5 minutes. Then remove the crystals and air-dry them.

Yield Calculation and Melting-Point Determination

Weigh the dry benzil and calculate the percentage yield. Determine the melting point. The melting point of pure benzil is 95°C. Submit the benzil to the instructor

---

[1] At higher temperatures, some 4-nitrobenzil will be formed along with benzil.

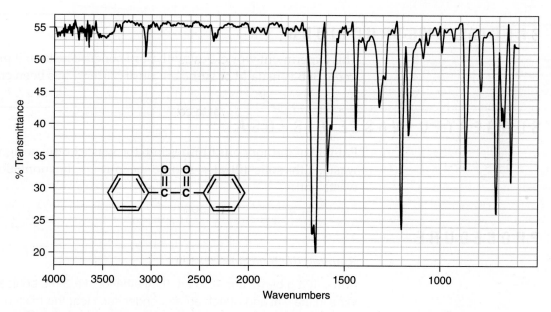

Infrared spectrum of benzil, KBr.

unless it is to be used to prepare benzilic acid (Experiment 36C) or tetraphenyl-cyclopentadienone (Experiment 37). Obtain the infrared spectrum of benzil using the dry film method. Compare the spectrum to the one shown. Also compare the spectrum with that of benzoin shown on page 308. What differences do you notice?

36C EXPERIMENT 36C

Preparation of Benzilic Acid

In this experiment, benzilic acid will be prepared by causing the rearrangement of the α-diketone benzil. Preparation of benzil is described in Experiment 36B. The rearrangement of benzil proceeds in the following way:

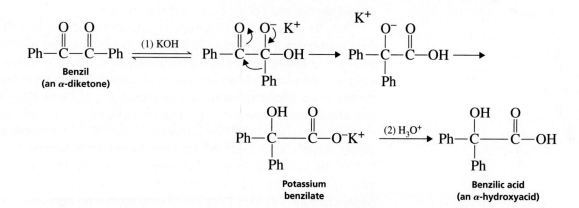

The driving force for the reaction is provided by the formation of a stable carboxylate salt (potassium benzilate). Once this salt is produced, acidification yields benzilic acid. The reaction can generally be used to convert aromatic α-diketones to aromatic α-hydroxyacids. Other compounds, however, also will undergo benzilic acid–type of rearrangement (see questions).

SPECIAL INSTRUCTIONS

This experiment works best with pure benzil. The benzil prepared in Experiment 36B is usually of sufficient purity after it has been crystallized.

SUGGESTED WASTE DISPOSAL

Pour all of the aqueous filtrates into the waste bottle designated for aqueous waste. Ethanolic filtrates should be put in the nonhalogenated organic waste bottle.

PROCEDURE

Running the Reaction

Add 0.100 g benzil and 0.30 mL 95% ethanol to a 3-mL conical vial. Place a spin vane in the vial and attach an air condenser. Heat the mixture with an aluminum block (90–100°C) while stirring until the benzil has dissolved (see inset in Figure 6.1A, p. 590). Using a 9-inch Pasteur pipet, add dropwise 0.25 mL of an aqueous potassium hydroxide solution[1] downward through the condenser into the vial. Gently boil the mixture (aluminum block about 110°C) while stirring for 15 minutes. The mixture will be blue-black. As the reaction proceeds, the color will turn to brown, and the solid should dissolve completely. Solid potassium benzilate may form during the reaction period. At the end of the heating period, remove the assembly from the aluminum block and allow it to cool for a few minutes.

Crystallization of Potassium Benzilate

Detach the air condenser when the apparatus is cool enough to handle. Transfer the reaction mixture, which may contain some solid, with a Pasteur pipet into a 10-mL beaker. Allow the mixture to cool to room temperature and then cool in an ice-water bath for about 15 minutes until crystallization is complete. It may be necessary to scratch the inside of the beaker with a glass stirring rod to induce crystallization. Crystallization is complete when virtually the entire mixture has solidified. Collect the crystals on a Hirsch funnel by vacuum filtration and wash the crystals thoroughly with three 1-mL portions of ice-cold 95% ethanol. The solvent should remove most of the color from the crystals.

Transfer the solid, which is mainly potassium benzilate, to a 10-mL Erlenmeyer flask containing 3 mL of hot (70°C) water. Stir the mixture until all solid has dissolved or until it appears that the remaining solid will not dissolve. Any remaining solid will likely form a fine suspension. If solid does remain in the flask, filter the mixture in the following manner. Place about 0.5 g of Celite (Filter Aid) in a beaker with about 5 mL of water. Stir the mixture vigorously and then pour the contents into a Hirsch funnel (with filter paper) or a small Büchner funnel while applying a *gentle* vacuum, as in a vacuum filtration (Technique 8, Section 8.3, and Figure 8.5, p. 622). Be careful not to let the Celite dry completely. This procedure will cause a thin layer of Celite to be deposited on the filter paper. Discard the water that passes through this filter. Pass the mixture containing potassium benzilate through

---

[1] The aqueous potassium hydroxide solution should be prepared for the class by dissolving 2.75 g of potassium hydroxide in 6.0 mL of water. This will provide enough solution for 20 students, assuming little solution is wasted.

this filter, using *gentle* suction. The filtrate should be clear. Transfer the filtrate to a 10-mL Erlenmeyer flask. If no solid remains in the flask, the filtration step may be omitted. In either case, proceed to the next step.

Formation of Benzilic Acid

With swirling of the flask, add dropwise 0.50 mL of 1*M* hydrochloric acid to the solution of potassium benzilate. As the solution becomes acidic, solid benzilic acid will begin to precipitate. The pH should be about 2; if it is higher than this, add a few more drops of acid and check the pH again. Allow the mixture to cool to room temperature and then complete the cooling in an ice bath. Collect the benzilic acid by vacuum filtration using a Hirsch funnel. Wash the crystals with 3–4 mL of ice-cold water to remove potassium chloride salt that sometimes coprecipitates with benzilic acid during the neutralization with hydrochloric acid. Remove the wash water by drawing air through the filter. Dry the product thoroughly by allowing it to stand until the next laboratory period.

Melting Point and Crystallization of Benzilic Acid

Weigh the dry benzilic acid and determine the percentage yield. Determine the melting point of the dry product. Pure benzilic acid melts at 150°C. If necessary, crystallize the product from hot water using a Craig tube (Technique 11, Section 11.4, and Figure 11.6, p. 657). If some impurities remain undissolved, filter the mixture using the following procedure. It will be necessary to keep the mixture hot during this filtration step. Transfer the hot mixture to a test tube with a Pasteur pipet. Clean the Craig tube and filter the mixture by transferring it back to the Craig tube with a filter-tip pipet. Cool the solution and induce crystallization, if necessary. Allow the mixture to stand at room temperature until crystallization is complete (about 15 minutes). Cool the mixture in an ice bath and collect the crystals by centrifugation. Determine the melting point of the crystallized product after it is thoroughly dry.

At the instructor's option, determine the infrared spectrum of the benzilic acid in potassium bromide (Technique 25, Section 25.5, p. 838). Calculate the percentage yield. Submit the sample to your laboratory instructor in a labeled vial.

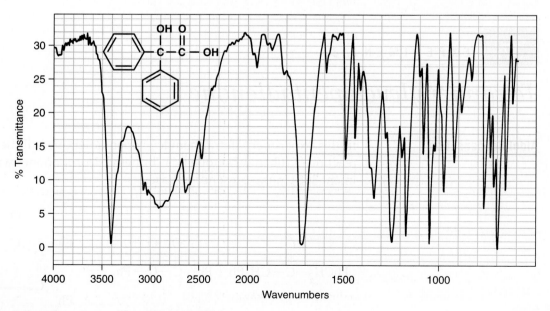

Infrared spectrum of benzilic acid, KBr.

QUESTIONS

1. Show how to prepare the following compounds, starting from the appropriate aldehyde.

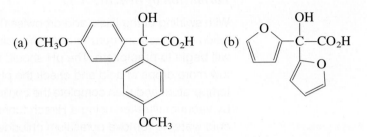

(a) CH_3O— ... —$\overset{\underset{|}{OH}}{C}$—$CO_2H$ (b) ... —$\overset{\underset{|}{OH}}{C}$—$CO_2H$

2. Give the mechanisms for the following transformations:

(a)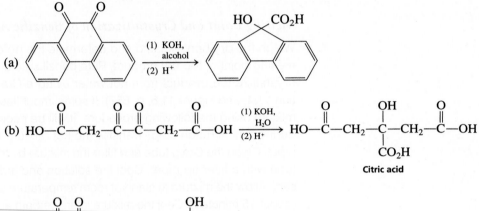

$\xrightarrow[\text{(2) H}^+]{\text{(1) KOH, alcohol}}$

(b) $HO-\overset{O}{\overset{\|}{C}}-CH_2-\overset{O}{\overset{\|}{C}}-\overset{O}{\overset{\|}{C}}-CH_2-\overset{O}{\overset{\|}{C}}-OH \xrightarrow[\text{(2) H}^+]{\substack{\text{(1) KOH,} \\ \text{H}_2\text{O}}} HO-\overset{O}{\overset{\|}{C}}-CH_2-\underset{\underset{CO_2H}{|}}{\overset{\underset{|}{OH}}{C}}-CH_2-\overset{O}{\overset{\|}{C}}-OH$

Citric acid

(c) $Ph-\overset{O}{\overset{\|}{C}}-\overset{O}{\overset{\|}{C}}-Ph \xrightarrow[\text{CH}_3\text{OH}]{^-\text{OCH}_3} Ph-\underset{\underset{Ph}{|}}{\overset{\underset{|}{OH}}{C}}-\overset{O}{\overset{\|}{C}}OCH_3$

3. Interpret the infrared spectrum of benzilic acid.

---

37   **EXPERIMENT 37**

Tetraphenylcyclopentadienone

Aldol Condensation

In this experiment, tetraphenylcyclopentadienone will be prepared by the reaction of dibenzyl ketone (1,3-diphenyl-2-propanone) with benzil (Experiment 36B) in the presence of base.

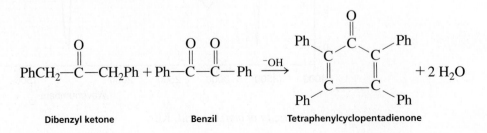

$PhCH_2-\overset{O}{\overset{\|}{C}}-CH_2Ph + Ph-\overset{O}{\overset{\|}{C}}-\overset{O}{\overset{\|}{C}}-Ph \xrightarrow{^-OH}$ ... $+ 2 H_2O$

Dibenzyl ketone   Benzil   Tetraphenylcyclopentadienone

This reaction proceeds via an aldol condensation reaction, with dehydration giving the purple unsaturated cyclic ketone. A stepwise mechanism for the reaction may proceed as follows:

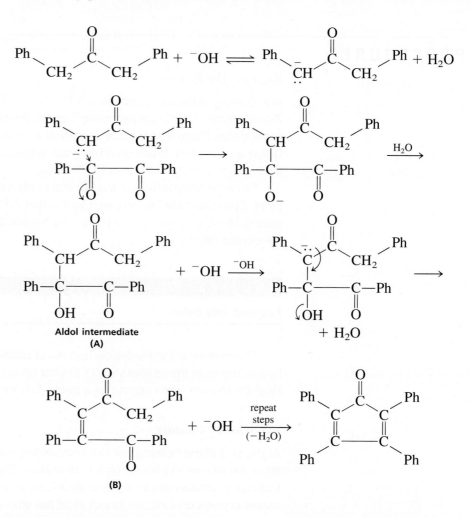

Aldol intermediate (A)

(B)

The aldol intermediate **A** readily loses water to give the highly conjugated system **B,** which reacts further to form a ring by an intramolecular aldol condensation. After a dehydration step (loss of water), the dienone product is formed.

REQUIRED READING

Review: Technique 8   Filtration, Section 8.3

Technique 11   Crystallization: Purification of Solids, Section 11.3

SPECIAL INSTRUCTIONS

This reaction can be completed in 1 hour. The ethanolic potassium hydroxide should be prepared in advance by the instructor. Another multistep synthesis can be achieved by linking Experiments 36A and 36B with Experiment 37.

SUGGESTED WASTE DISPOSAL

Place all organic wastes in the nonhalogenated organic waste container.

PROCEDURE

Running the Reaction

Add 0.100 g of benzil (Experiment 36B), 0.100 g of dibenzyl ketone (1,3-diphenyl-2-propanone, 1,3-diphenylacetone), and 0.80 mL of absolute ethanol to a 3-mL conical vial. Place a spin vane in the vial and attach a water-cooled condenser. Using an aluminum block on a hot plate that is adjusted to about 80°C, heat the mixture with stirring until the solids dissolve.

Raise the temperature of the hot plate until the mixture is just below its boiling point. Continue to stir the mixture. Using a 9-inch Pasteur pipet, carefully add drop-wise 0.15 mL of ethanolic potassium hydroxide solution[1] downward through the condenser into the vial.

> ### CAUTION ⚠
>
> **Foaming may occur.**

The mixture will immediately turn deep purple. Once you add the potassium hydroxide, raise the temperature of the hot plate until the mixture is boiling gently. Heat the mixture, while stirring, at a gentle boil for 15 minutes.

Isolation of Product

At the end of the heating period, remove the vial from the aluminum block and allow the mixture to cool to room temperature. Then place the vial in an ice-water bath for 5 minutes to complete crystallization of the product. Collect the deep-purple crystals on a Hirsch funnel. Hold the spin vane with forceps and scrape off as much solid as possible. Wash the crystals with three 0.5-mL portions of cold 95% ethanol. The rinse solvent can also be used to aid in transferring crystals from the conical vial to the Hirsch funnel and in removing the product from the spin vane. Dry the tetraphenylcyclopentadienone in an oven for 30 minutes or in air overnight.

Yield Calculation and Melting-Point Determination

Weigh the product and calculate the percentage yield. Determine the melting point (mp 218–220°C). At the option of the instructor, a small portion may be crystallized from a 1:1 mixture of 95% ethanol and toluene (12 mL/0.5 g; mp 219–220°C). Use a Craig tube for crystallization. At the instructor's option, determine the infrared spectrum of tetraphenylcyclopentadienone in potassium bromide (Technique 25, Section 25.5, page 838). Submit the product to the instructor in a labeled vial.

---

[1] *Note to the Instructor:* This solution is prepared by dissolving 0.40 g of potassium hydroxide in 4.0 mL of absolute ethanol. It will take about 30 minutes, with vigorous stirring, for the solid to dissolve. As the solid dissolves, crush the pieces with a spatula to aid in the solution process. This will provide enough solution for 20 students, assuming little material is wasted.

QUESTIONS

1. Draw the structure of the product you would expect from the reaction of benzaldehyde and acetophenone with base.

2. Suggest several possible by-products of this reaction.

---

38   **EXPERIMENT 38**

Triphenylmethanol and Benzoic Acid

Grignard reactions

Extractions

Crystallization

Use of a separatory funnel

In this experiment, you will prepare a Grignard reagent, or organomagnesium reagent. The reagent is phenylmagnesium bromide.

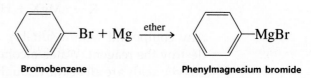

Bromobenzene   **Phenylmagnesium bromide**

This reagent will be converted to a tertiary alcohol or a carboxylic acid, depending on the experiment selected.

**Triphenylmethanol
(Experiment 38A)**

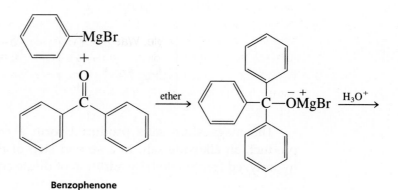

Benzophenone

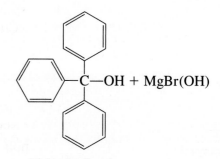

Triphenylmethanol

Benzoic Acid (Experiment 38B)

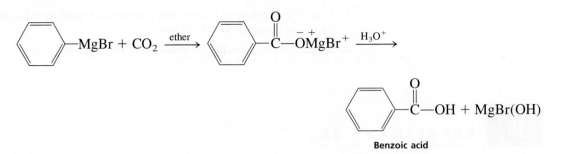

Benzoic acid

The alkyl portion of the Grignard reagent behaves as if it had the characteristics of a **carbanion.** We may write the structure of the reagent as a partially ionic compound:

$$\overset{\delta-}{R}\cdots\overset{\delta+}{MgX}$$

This partially bonded carbanion is a Lewis base. It reacts with strong acids, as you would expect, to give an alkane:

$$\overset{\delta-}{R}\cdots\overset{\delta+}{MgX} + HX \longrightarrow R—H + MgX_2$$

Any compound with a suitably acidic hydrogen will donate a proton to destroy the reagent. Water, alcohols, terminal acetylenes, phenols, and carboxylic acids are all acidic enough to bring about this reaction.

The Grignard reagent also functions as a good nucleophile in nucleophilic addition reactions of the carbonyl group. The carbonyl group has electrophilic character at its carbon atom (due to resonance), and a good nucleophile seeks out this center for addition.

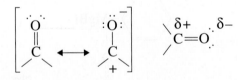

The magnesium salts produced form a complex with the addition product, an alkoxide salt. In a second step of the reaction, these must be hydrolyzed (protonated) by addition of dilute aqueous acid:

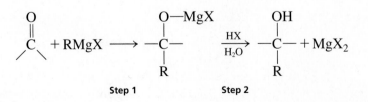

The Grignard reaction is used synthetically to prepare secondary alcohols from aldehydes and tertiary alcohols from ketones. The Grignard reagent will react with esters twice to give tertiary alcohols. Synthetically, it also can be allowed to react with carbon dioxide to give carboxylic acids and with oxygen to give hydroperoxides:

$$RMgX + O{=}C{=}O \longrightarrow R{-}\underset{\underset{O}{\|}}{C}{-}OMgX \xrightarrow[H_2O]{HX} R{-}\underset{\underset{O}{\|}}{C}{-}OH$$

Carboxylic acid

$$RMgX + O_2 \longrightarrow ROOMgX \xrightarrow[H_2O]{HX} ROOH$$

Hydroperoxide

Because the Grignard reagent reacts with water, carbon dioxide, and oxygen, it must be protected from air and moisture when it is used. The apparatus in which the reaction is to be conducted must be scrupulously dry (recall that 18 mL of H_2O is 1 mole), and the solvent must be free of water, or anhydrous. During the reaction, the flask must be protected by a calcium chloride drying tube. Oxygen should also be excluded. In practice, this can be done by allowing the solvent ether to reflux. This blanket of solvent vapor keeps air from the surface of the reaction mixture.

In the experiment described here, the principal impurity is **biphenyl,** which is formed by a heat- or light-catalyzed coupling reaction of the Grignard reagent and unreacted bromobenzene. A high reaction temperature favors the formation of this product. Biphenyl is highly soluble in petroleum ether, and it is easily separated from triphenylmethanol. Biphenyl can be separated from benzoic acid by extraction.

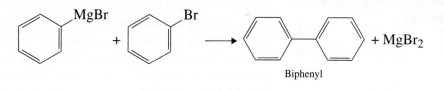

Biphenyl

REQUIRED READING

Review: Technique 8   Section 8.3

Technique 11   Section 11.3

Technique 12   Sections 12.5, 12.7, 12.9, 12.11

Technique 25   Section 25.5

SPECIAL INSTRUCTIONS

This experiment must be conducted in one laboratory period either to the point after which benzophenone is added (Experiment 38A) or to the point after which the Grignard reagent is poured over dry ice (Experiment 38B). The Grignard reagent cannot be stored. This reaction involves the use of diethyl ether, which is extremely flammable. Be certain that no open flames of any sort are in your vicinity when you are using ether.

During this experiment, you will need to use *anhydrous* diethyl ether, which is usually contained in metal cans with a screw cap. You are instructed in the experiment to transfer a small portion of this solvent to an Erlenmeyer flask. Be certain to minimize exposure to atmospheric water. Always recap the container after use. Do not use solvent-grade ether because it may contain some water.

All students will prepare the Grignard reagent, phenylmagnesium bromide. At the option of the instructor, you should proceed to either Experiment 38A (triphenylmethanol) or Experiment 38B (benzoic acid).

SUGGESTED WASTE DISPOSAL

All aqueous solutions should be placed in the aqueous waste container. Be sure to decant these solutions away from any magnesium chips. The unreacted magnesium chips should be placed in a solid waste container.

Place all ether solutions in the container for nonhalogenated liquid wastes. Likewise, the mother liquor from the crystallization using isopropyl alcohol (Experiment 38A) should also be placed in the container for nonhalogenated liquid wastes.

PROCEDURE

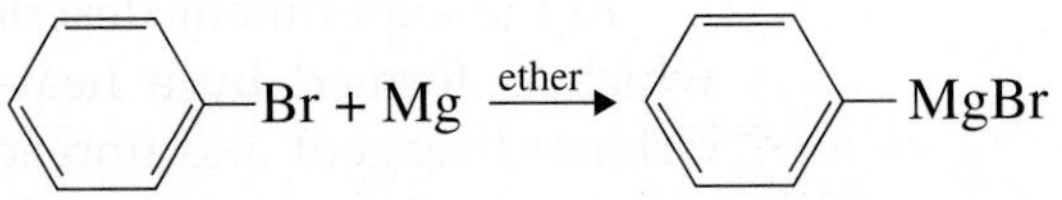

Preparation of the Grignard Reagent: Phenylmagnesium Bromide

Preparation of Glassware

All glassware used in a Grignard reaction must be *scrupulously* dried. Surprisingly large amounts of water adhere to the walls of glassware, even glassware that is apparently dry. For this experiment, make sure all pieces of glassware have been rinsed with acetone and allowed to dry for at least two days. *If the equipment has been dried in this manner, then it is not necessary to dry the equipment in an oven.* Dry the following pieces of equipment before doing this experiment: a 20-mL round-bottom flask, two 5-mL conical vials, a 50-mL Erlenmeyer flask, a Claisen head, a syringe, and a calibrated Pasteur pipet (0.5-mL and 1.0-mL calibration marks) for use in dispensing ether. If, after drying as described, signs of water are still visible in the apparatus, dry the equipment in an oven. Prepare a drying tube with anhydrous calcium chloride.

Obtain 0.15 g of *shiny* magnesium turnings and place them in the *dry* round-bottom flask. Place a small *dry* magnetic stirring bar into the flask. Assemble the remainder of the apparatus, as shown in the figure. Seal off the open end of the Claisen head with a rubber septum.

Formation of the Grignard Reagent

Transfer about 20 mL of *anhydrous diethyl ether* into a dry 50-mL Erlenmeyer flask and stopper the flask. Use the flask to store your dry ether during the course of this experiment. During the experiment, remove the ether from this flask with a dry calibrated Pasteur pipet.

Place 0.70 mL of bromobenzene ($MW = 157.0$) into a preweighed 5-mL conical vial and determine the weight of the material transferred. Add 4.0 mL of *anhydrous* ether to the vial. After the bromobenzene dissolves, withdraw about 0.8 mL of this solution into the syringe and cap the vial. You will need to save the

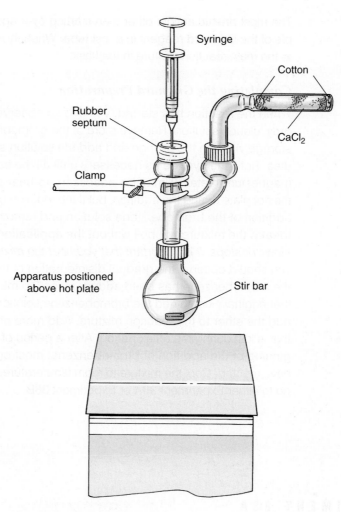

Apparatus for Experiment 38.

remainder of the bromobenzene/ether solution for later use; recap the vial between uses. After inserting the syringe needles through the rubber septum, add 0.8 mL of the bromobenzene solution to the magnesium in the round-bottom flask. Position the apparatus just above the hot plate (about 60°C) and stir the mixture gently to avoid throwing the magnesium onto the side of the flask. You should begin to notice the evolution of bubbles, from the metal surface, that signals that the reaction is starting. It will probably be necessary to heat the mixture to start the reaction. Because ether has a low boiling point (35°C), it may be sufficient to heat the flask by placing it just above the hot plate. Check to see if the bubbling action continues after the apparatus is removed from the heat. The reaction should start, but if you experience difficulty, proceed to the next paragraph.

Optional Steps

You may need to employ one or more of the following procedures if heating fails to start the reaction. If you are experiencing difficulty, remove the syringe and rubber septum. Place a *dry* glass stirring rod into the flask and gently twist the stirring rod to crush the magnesium against the glass surface. Reattach the rubber septum and again heat the mixture. Repeat the crushing procedure several times, if necessary, to start the reaction. If the crushing procedure fails to start the reaction, then add one small crystal of iodine to the flask. Again, heat the mixture gently.

The most drastic action, other than starting over again, is to prepare a small sample of the Grignard reagent in a test tube. When this reaction is started, it is added to the main reaction mixture in the flask.

Completing the Grignard Preparation

When the reaction has started, you should observe the formation of a brownish-gray, cloudy solution. Remove more of the bromobenzene/ether solution from the storage vial with the syringe and add the solution slowly over a period of 15 minutes. Refill the syringe as necessary until all the solution has been added to the magnesium metal. It may be necessary to heat the mixture occasionally with the hot plate during the addition, but if the reaction becomes too vigorous, slow the addition of the bromobenzene solution and remove the flask from the hot plate. Ideally, the mixture will boil without the application of external heat. *If the reflux slows or stops, it is important that you heat the mixture.* As the reaction proceeds, you should observe the gradual disintegration of the magnesium metal. When all the bromobenzene has been added, place 2.0 mL of *anhydrous* ether in the vial that originally contained the bromobenzene solution, draw it into the syringe, and add the ether to the reaction mixture. Add more anhydrous ether to replace any that is lost during the reflux period. After a period of about 30 minutes from the beginning of the addition of bromobenzene, most or all of the magnesium should have reacted. Cool the mixture to room temperature. As your instructor designates, go to either Experiment 38A or Experiment 38B.

38A   EXPERIMENT 38A

Triphenylmethanol

PROCEDURE

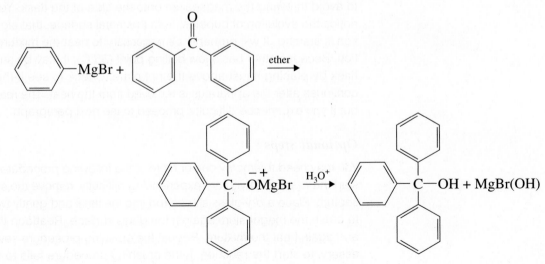

Adduct

Addition of Benzophenone

While the phenylmagnesium bromide solution is being heated and stirred under reflux, make a solution of 1.09 g benzophenone in 2 mL of *anhydrous* ether in a 5-mL conical vial. Cap the vial until the reflux period is over. Once the Grignard reagent is cooled to room temperature, draw some of the benzophenone solution into the syringe. Add this solution as rapidly (but not all at once) as possible to the stirred Grignard reagent. Do not add the solution so rapidly that the solution begins to boil. Add the remainder of the benzophenone solution with the syringe. Once the addition has been completed, cool the mixture to room temperature. The solution turns red and then gradually solidifies as the adduct is formed. When stirring is no longer effective, remove the syringe and septum and stir the mixture with a spatula. Rinse the vial that contained the benzophenone solution with about 1 mL of anhydrous ether and add it to the mixture. Remove the reaction flask from the apparatus and cap it. Occasionally stir the contents of this flask. Recap the flask when it is standing to avoid contact with water vapor. The adduct should be fully formed after about 15 minutes. You may stop here.

Hydrolysis

Add 6.0 mL of *6M* hydrochloric acid (*dropwise at first*) to neutralize the reaction mixture. The acid converts the adduct to triphenylmethanol and inorganic compounds (MgX_2). Any unreacted magnesium will react with the acid to evolve hydrogen gas. Use a spatula to break up the solid while adding the hydrochloric acid. You may need to cap the flask and shake it vigorously to dissolve the solid. Because the neutralization procedure evolves heat, some ether will be lost due to evaporation. You should add enough additional ether to maintain at least a 10-mL volume in the upper organic phase. Eventually, you should obtain two distinct layers: The upper ether layer will contain triphenylmethanol; the lower aqueous hydrochloric acid layer will contain the inorganic compounds. Make sure you have two distinct liquid layers, with no sign of any solid, before separating the layers. More ether or hydrochloric acid may be added, if necessary, to dissolve any remaining solid.

Transfer the entire contents of the reaction flask to a small separatory funnel, leaving the stirring bar behind. Use a small amount of ether to rinse the reaction flask and add this ether to the separatory funnel. If some solid material appears or if there are three layers, add more ether and hydrochloric acid to the separatory funnel and shake it. Continue adding small portions of ether and hydrochloric acid to the separatory funnel and shake it until everything dissolves. In some cases, it may be necessary to add more water instead of hydrochloric acid. Ultimately, you should have two distinct liquid layers with no sign of any solid, except possibly some magnesium. If a small amount of unreacted magnesium metal is present, you will observe bubbles of hydrogen being formed. You may remove the aqueous layer from the separatory funnel even though the magnesium is still producing hydrogen.

Separation and Drying

Drain the lower aqueous layer into a beaker. Pour the remaining ether layer that contains the triphenylmethanol product into a dry Erlenmeyer flask. Pour the aqueous layer back into the separatory funnel and reextract it with 5 mL of ether. Remove the lower aqueous phase and discard it. Combine the remaining ether phase with the first ether extract. Dry the ether solution with granular anhydrous sodium sulfate (Technique 12, Section 12.9, p. 680).

Evaporation

Remove the dried ether solution from the drying agent by decanting it into a small Erlenmeyer flask and rinse the drying agent with more diethyl ether. Evaporate the solvent in a hood by heating the flask in a hot water bath at about 50°C (use an air or nitrogen stream to aid the evaporation process). After removal of the solvent, an oily solid should be left. This crude mixture contains the desired triphenylmethanol and the by-product, biphenyl. Most of the biphenyl can be removed by adding about 3 mL of *petroleum ether (30 to 60°C)*. Petroleum ether is a mixture of hydrocarbons that easily dissolves the hydrocarbon biphenyl and leaves behind the alcohol triphenylmethanol. Do not confuse this solvent with diethyl ether ("ether"). Heat the mixture slightly, stir it, and then cool the mixture to room temperature. Collect the triphenylmethanol by vacuum filtration on a Hirsch funnel and rinse it with small portions of petroleum ether (Technique 8, Section 8.3 and Figure 8.5, pp. 621–622). Air-dry the solid, weigh it, and calculate the percentage yield of the crude triphenylmethanol (*MW* = 260.3).

Crystallization

Crystallize all your product from *hot* isopropyl alcohol in an Erlenmeyer flask using a hot plate as the heating source. Be sure to add the hot alcohol in small portions to the crude product. Add the hot solvent until the solid just dissolves. Then allow the flask to cool slowly. When it has cooled, place the flask in an ice bath to complete the crystallization. Collect the solid on a Hirsch funnel and wash it with a small amount of *cold* isopropyl alcohol. Set the crystals aside to air-dry. Report the melting point of the purified triphenylmethanol (literature value, 162°C) and recovered yield in grams. Submit the sample to the instructor in a properly labeled vial.

Spectroscopy

At the option of the instructor, determine the infrared spectrum of the purified material in a KBr pellet (Technique 25, Section 25.5, p. 838). Your instructor may assign certain tests on the product you prepared. These tests are described in the Instructor's Manual.

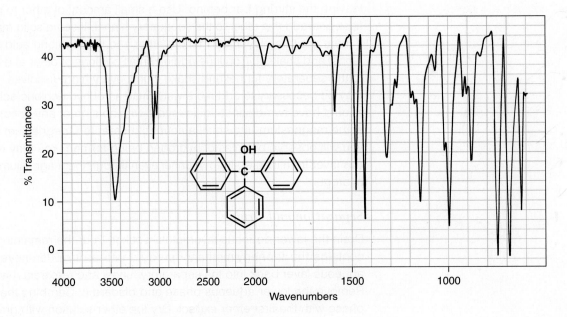

Infrared spectrum of triphenylmethanol, KBr.

EXPERIMENT 38B

Benzoic Acid

PROCEDURE

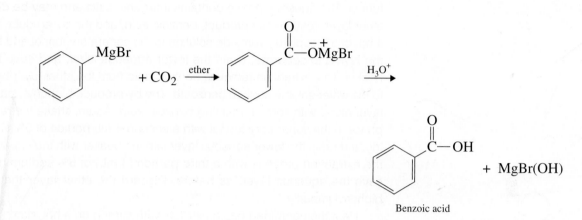

Addition of Dry Ice

When the phenylmagnesium bromide has cooled to room temperature, use a Pasteur pipet to transfer this reagent as quickly as possible to 4 g of crushed dry ice contained in a beaker. The dry ice should be weighed as quickly as possible to avoid contact with atmospheric moisture. It need not be weighed precisely. Rinse the flask with 2 or 3 mL of anhydrous ether and add it to the beaker.

> **CAUTION** ⚠
>
> **Exercise caution in handling dry ice. Contact with the skin can cause severe frostbite. Always use gloves or tongs. The dry ice is best crushed by wrapping large pieces in a clean, dry towel and striking them with a hammer or a wooden block. It should be used as soon as possible after crushing it to avoid contact with atmospheric water.**

Cover the reaction mixture with a watch glass and let it stand until the excess dry ice has completely sublimed. The Grignard addition compound will appear as a viscous glassy mass.

Hydrolysis

Hydrolyze the Grignard adduct by slowly adding 10 mL of 6*M* hydrochloric acid, with stirring, to the beaker. Any remaining magnesium chips will react with acid to evolve hydrogen. If you have solid present (other than magnesium), try adding a little more ether. If the solid is insoluble in ether, try adding a little 6*M* hydrochloric acid solution. If neither seems to dissolve the solid, try adding a little water. Benzoic acid is soluble in ether, whereas the inorganic compounds (MgX_2) are soluble in the acid solution. Ultimately, you should have two distinct liquid layers in the beaker, with no sign of any solid, except possibly some magnesium. Transfer the liquid layers to a separatory funnel with a Pasteur pipet, leaving behind any residual magnesium.[1] If a separatory funnel is not available, you may use a centrifuge tube to separate the mixture. Add more ether to the beaker to rinse the beaker. Again, transfer the ether solution to the separatory funnel.

Isolation of the Product

Drain the lower aqueous layer and keep the upper ether layer in the separatory funnel. The aqueous phase contains inorganic salts and may be discarded. The ether layer contains the product, benzoic acid, and the by-product, biphenyl. Add 4 mL of 5% sodium hydroxide solution to the separatory funnel and shake it. Allow the layers to separate, drain the lower aqueous layer, and save this layer in a beaker. This extraction removes benzoic acid from the ether layer by converting it to the water-soluble sodium benzoate. The by-product, biphenyl, stays in the ether layer along with some remaining benzoic acid. Again, shake the remaining ether phase in the separatory funnel with a second 4-mL portion of 5% sodium hydroxide and drain the lower aqueous layer into the beaker with the first extract. Repeat the extraction process with a third portion (4 mL) of 5% sodium hydroxide and save the aqueous layer, as before. Discard the ether layer that contains the biphenyl impurity.

Heat the combined basic extracts with stirring on a hot plate (hot enough to boil the aqueous mixture) for about 5 minutes to remove any ether that may be dissolved in this aqueous phase. Stir the mixture as it is being heated. Ether is soluble in water to the extent of 7%. During this heating period, you may observe slight bubbling, but the volume of liquid will not decrease substantially. Unless the ether is removed before the benzoic acid is precipitated, the product may appear as a waxy solid instead of crystals.

Cool the alkaline solution and precipitate the benzoic acid by adding 5 mL of 6*M* hydrochloric acid with stirring. Cool the mixture in an ice bath. Collect the solid by vacuum filtration on a Hirsch funnel (Technique 8, Section 8.3, and Figure 8.5, p. 622). The transfer may be aided and the solid washed with several small portions of cold water (total volume, 4 mL). Allow the crystals to dry thoroughly at room temperature at least overnight in an open container. Weigh the solid and calculate the percentage yield of benzoic acid (*MW* = 122.1).

Crystallization

Crystallize all your product from *hot* water in an Erlenmeyer flask using a hot plate as the heating source. Be sure to add the hot water in small portions to the crude product. Add the hot water until the solid just dissolves. Then allow the flask to cool slowly. After the flask cools to room temperature, place it in an ice bath to complete the crystallization. Collect the solid on a small Hirsch funnel and wash it with a small amount of *cold* water. Set the crystals aside to air-dry at room temperature until the next laboratory period before determining the melting point of the purified benzoic acid (literature value, 122°C). Also determine the recovered yield in grams. Submit the sample to the instructor in a properly labeled vial.

----

[1] If it is necessary to set this experiment aside overnight, do not transfer the solution to the separatory funnel. Instead, transfer the solution to a 25-mL Erlenmeyer flask. Stopper the flask tightly. When you resume the experiment, transfer this solution to the separatory funnel using about 4 mL of ether to aid the transfer and proceed as instructed.

Spectroscopy

At the option of the instructor, determine the infrared spectrum of the purified material in a KBr pellet (Technique 25, Section 25.5, p. 838). Your instructor may assign certain tests on the product you prepared. These tests are described in the Instructor's Manual.

QUESTIONS

1. Benzene is often produced as a side product during Grignard reactions using phenylmagnesium bromide. How can its formation be explained? Give a balanced equation for its formation.

2. Write a balanced equation for the reaction of benzoic acid with hydroxide ion. Why is it necessary to extract the ether layer with sodium hydroxide?

3. Interpret the principal peaks in the infrared spectrum of either triphenylmethanol or benzoic acid, depending on the procedure used in this experiment.

4. Outline a separation scheme for isolating either triphenylmethanol or benzoic acid from the reaction mixture, depending on the procedure used in this experiment.

5. Provide methods for preparing the following compounds by the Grignard method:

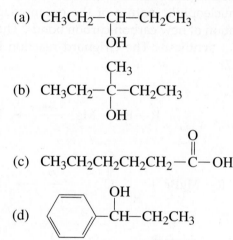

(a) $CH_3CH_2{-}CH{-}CH_2CH_3$
            |
           OH

(b) $CH_3CH_2{-}\overset{\displaystyle CH_3}{\underset{\displaystyle OH}{C}}{-}CH_2CH_3$

(c) $CH_3CH_2CH_2CH_2CH_2{-}\overset{\displaystyle O}{C}{-}OH$

(d)   phenyl$-\overset{\displaystyle OH}{CH}{-}CH_2CH_3$

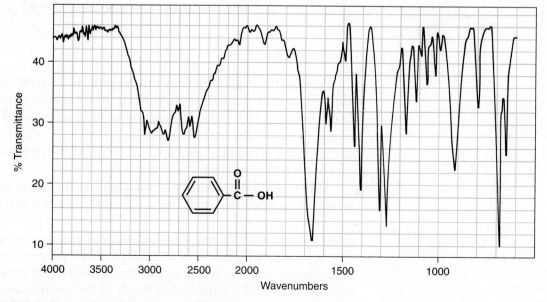

Infrared spectrum of benzoic acid, KBr.

Aqueous-Based Organozinc Reactions

Organometallic reactions

Green chemistry

Extractions

Use of a separatory funnel

Gas chromatography

Spectroscopy

One of the most important categories of reactions in organic synthesis is the class of reactions that result in the formation of a carbon-carbon bond. Among these, one of the best-known reactions is the Grignard reaction, where an organomagnesium reagent is formed from an alkyl halide and then allowed to react with a variety of substances to form new molecules. The nucleophilic nature of the organomagnesium reagent is used in the formation of new carbon-carbon bonds. The equation shown illustrates this type of synthesis. The Grignard reaction is introduced in Experiment 38 (p. 317).

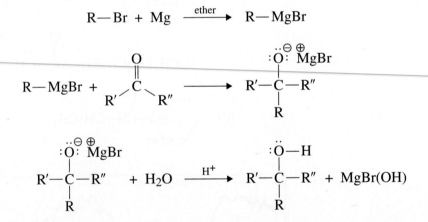

Because the organomagnesium reagent reacts with water, carbon dioxide, and oxygen, it must be protected from air and moisture when it is used. The apparatus in which the reaction is to be conducted must be scrupulously dry, and the solvent must be completely anhydrous. In addition, diethyl ether is required as a solvent; without the presence of an ether, the organomagnesium reagent will not form.

This experiment presents a variation on the basic idea of a Grignard synthesis, but one that does not use magnesium and that can be conducted in a mixed organic-aqueous solution. The reaction presented in this experiment is a variation on the Barbier-Grignard reaction, where zinc is used as the metal. A small amount of an ether, in this case tetrahydrofuran (THF), is still required for this reaction, but the principal component of the solvent system is water. By eliminating much of the organic solvent, this method can be used to illustrate some of the principles of "Green Chemistry," in which reactions are conducted under conditions that are less harmful to the environment than traditional chemical methods.

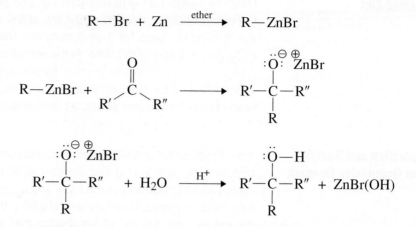

Although this organozinc method of synthesis is very similar to the Grignard reaction, there are also some interesting differences. The organozinc reagent is much more selective than the organomagnesium reagent, and rearrangements of the alkyl group attached to the metal are also possible. Whereas the formation of Grignard reagents from allylic halides is notoriously difficult, the formation of organozinc reagents seems to require that one begin with an allylic halide. A comparison of the structure of the products of this reaction with the structure of the starting alkyl halide can reveal some of this interesting chemistry.

REQUIRED READING

Review: Technique 8   Section 8.3

Technique 7   Section 7.10

Technique 12   Sections 12.5, 12.8, 12.9, 12.11

Technique 22

Technique 25   Sections 25.2, 25.4

Technique 26   Section 26.1

Technique 27   Section 27.1

SPECIAL INSTRUCTIONS

This reaction involves the use of allyl bromide, a substance that is volatile and may also be a **lachrymator.** Be certain to dispense this material under the hood. Do not attempt to weigh this substance; determine the approximate volume of allyl bromide needed using the specific gravity provided in this experiment, and dispense the allyl bromide by volume using a calibrated pipet. Students should work in pairs for this experiment.

SUGGESTED WASTE DISPOSAL

All aqueous solutions should be placed in a waste container designated for the disposal of aqueous wastes.

PROCEDURE

Activated Zinc

Carefully weigh 1.31 g (0.02 moles) of zinc powder and add it to a small (10-mL) Erlenmeyer flask or beaker. Add 1 mL of 5% aqueous hydrochloric acid and allow the mixture to stand for 1 to 2 minutes. There will be a noticeable evolution of hydrogen gas during this time. At the end of this period, pour the entire mixture into a Hirsch funnel and isolate the zinc by vacuum filtration. Rinse the zinc with 1 mL of water, followed by 1 mL of ethanol and 1 mL of diethyl ether. The zinc should be ready to use for the procedure, as described below.

Preparation and Reaction of the Organozinc Reagent

Add 10 mL of saturated aqueous ammonium chloride solution to a 25-mL round-bottom flask. Add 1.31 g zinc powder (0.02 moles) and a stirring bar to the flask. Attach an air condenser to the flask and begin continuous stirring while adding the remaining reagents. Carefully weigh 0.86 g (0.01 moles) of the 3-pentanone. Add the ketone and 1.6 mL of tetrahydrofuran to a test tube and add this solution dropwise to the zinc/NH$_4$Cl solution. The rate of addition should be about one drop per second. Note that this addition can be made by dropping the solution carefully down the opening in the air condenser; use a Pasteur pipet to add the solution. Allow the solution to stir for 10 to 15 minutes, giving time for the carbonyl compound to form a complex with the zinc. Add 2.4 g (0.02 moles—use the specific gravity 1.398 g/mL to estimate the volume required) of allyl bromide (3-bromopropene) to the stirring solution. *Be sure to dispense this reagent in the hood!* The rate of addition should be about one drop per second. Add the halide carefully by dropping it down the opening in the air condenser. Allow the reaction mixture to stir for 1 hour.

Set up a vacuum filtration apparatus with a Hirsch funnel. Decant the liquid from the reaction mixture through the Hirsch funnel. Rinse the round-bottom flask with approximately 1 mL of diethyl ether and pour the liquid into the Hirsch funnel. Using a second 1-mL portion of diethyl ether, rinse the solid that has collected in the Hirsch funnel. Discard the solid. Prepare a filter-tip pipet and transfer the liquid that was collected in the vacuum filtration into a separatory funnel. Use 1 mL of diethyl ether to rinse the inside of the filter flask and use the filter-tip pipet to transfer this liquid to the separatory funnel. Shake the separatory funnel gently to extract the organic material from the aqueous layer to the ether layer. Drain the lower (aqueous) layer into a 50-mL Erlenmeyer flask. *Do not discard this aqueous layer.* Collect the upper (organic) layer from the separatory funnel into a 25-mL Erlenmeyer flask (remember to collect the upper layer by pouring it from the top of the separatory funnel). Replace the aqueous layer in the separatory funnel and wash it with a 2-mL portion of ether. Separate the layers, save the aqueous layer in the same 50-mL Erlenmeyer flask as before, and combine the ether layer with the ether solution collected in the previous extraction. Repeat this extraction process of the aqueous phase one more time using a fresh 2-mL portion of ether. Dry the combined ether extracts with 3–4 microspatulafuls of anhydrous sodium sulfate (see Technique 12, Section 12.9, page 680). Stopper the Erlenmeyer flask with a cork and allow it to stand for at least 15 minutes (or overnight).

Use a filter-tip pipet to transfer the dried liquid to a clean, preweighed Erlenmeyer flask. Use a small amount of ether to rinse the inside of the original flask and add this ether to the dried liquid. Evaporate the ether with a rotary evaporator or under a gentle stream of air. When the ether has evaporated completely, reweigh the flask to determine the yield of product. If it should be necessary to store your final product, use Parafilm® to seal the container.

Prepare a sample of your final product for analysis by gas chromatography. Determine the infrared spectrum and both proton and ^{13}C NMR spectrum of your product. Use these spectra to determine the structure of your product. In your laboratory report, include an interpretation of each spectrum, identifying the principal absorption bands and demonstrating how the spectrum corresponds to the structure of your compound. Submit your sample in a labeled vial with your laboratory report.

QUESTIONS

1. Write *balanced* chemical equations for the formation of a substance that you prepared in this experiment.

2. Outline a series of chemical equations to show how your product could have been prepared using a Grignard reaction. Be sure to show the structures of all starting materials and intermediates.

3. Draw the structure of the product that would have been formed if benzaldehyde had been used in place of 3-pentanone in this experiment.

4. When benzaldehyde is used as the carbonyl compound in this experiment, the CH_2 peak in the proton NMR spectrum appears as *two* separate, complex resonances. Explain why this is observed.

---

40 **EXPERIMENT 40**

Resolution of (±)-α-Phenylethylamine and Determination of Optical Purity

Resolution of enantiomers

Use of a separatory funnel

Polarimetry

Chiral gas chromatography

NMR spectroscopy

Chiral resolving agent

Diastereomeric methyl groups

Although racemic (±)-α-phenylethylamine is readily available from commercial sources, the pure enantiomers are more difficult to obtain. In this experiment, you will isolate one of the enantiomers, the levorotatory one, in a high state of optical purity (large enantiomeric excess). A **resolution,** or separation, of enantiomers will be performed, using (+)-tartaric acid as the resolving agent.

Resolution of Enantiomers

The resolving agent to be used is (+)-tartaric acid, which forms diastereomic salts with racemic α-phenylethylamine. The important reactions for this experiment follow.

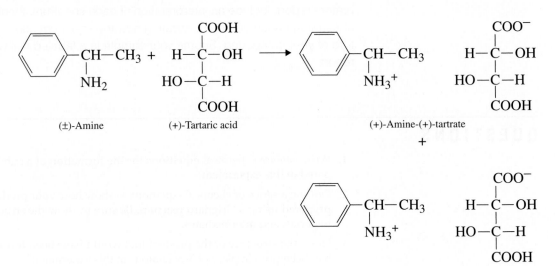

Optically pure (+)-tartaric acid is abundant in nature. It is frequently obtained as a by-product of winemaking. The separation depends on the fact that diastereomers usually have different physical and chemical properties. The (−)-amine-(+)-tartrate salt has a lower solubility than its diastereomeric counterpart, the (+)-amine-(+)-tartrate salt. With some care, the (−)-amine-(+)-tartrate salt can be induced to crystallize, leaving (+)-amine-(+)-tartrate in solution. The crystals are removed by filtration and purified. The (−)-amine can be obtained from the crystals by treating them with base. This breaks apart the salt by removing the proton, and it regenerates the free, unprotonated (−)-amine.

A polarimeter will be used to measure the observed rotation, α, of the resolved amine sample. From this value, you will calculate the specific rotation $[\alpha]_D$ and the optical purity (enantiomeric excess) of the amine. You will then calculate the percentages of each of the enantiomers present in the resolved sample. The (S)-α-phenylethylamine predominates in the sample. An optional chiral gas chromatographic method may be used to directly determine the percentages of each of the enantiomers in the sample.

NMR Determination of Optical Purity

An alternate means of determining the optical purity of the sample makes use of NMR spectroscopy (Experiment 40B). A group attached to a sterogenic (chiral) carbon normally has the same chemical shift whether that carbon has either R or S configuration. However, that group can be made diastereomeric in the NMR spectrum (have different chemical shifts) when the racemic parent compound is treated with an optically pure chiral resolving agent to produce diastereomers. In this case, the group is no longer found in two enantiomers but rather in two different diastereomers, and its chemical shift will be different in each environment.

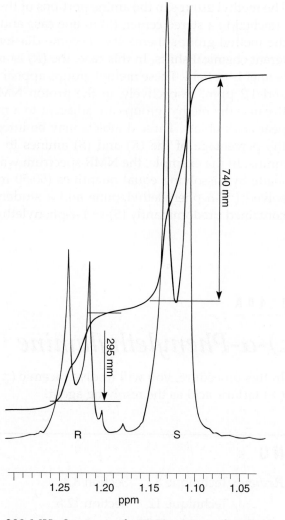

300-MHz Spectrum of a 50:50 mixture of resolved and unresolved α-phenylethylamine in CDCl₃. The chiral resolving agent (S)-(+)-O-acetylmandelic acid was added.

In this experiment, the partly resolved amine (containing both *R* and *S* enantiomers) is mixed with optically pure (S)-(+)-O-acetylmandelic acid in an NMR tube containing CDCl₃. Two diastereomers are formed:

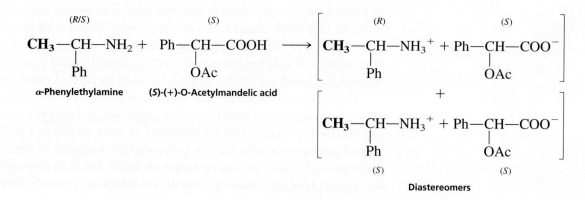

The methyl groups in the amine portions of the two diastereomeric salts are attached to a stereo center, (S) in one case and (R) in the other. As a result, the methyl groups themselves become diastereomeric, and they have different chemical shifts. In this case, the (R) isomer is downfield, and the (S) isomer is upfield. These methyl groups appear at approximately (varies) 1.1 and 1.2 ppm, respectively, in the proton NMR spectrum of the mixture. Because the methyl groups are adjacent to a methine (CH) group, they appear as doublets. These doublets may be integrated in order to determine the percentage of the (R) and (S) amines in the resolved α-phenylethylamine. In the example, the NMR spectrum was determined with a mixture made by dissolving equal quantities (50:50 mixture) of the original unresolved (±)-α-phenylethylamine and a student's resolved product, which contained predominantly (S)-(−)-α-phenylethylamine.

40A   EXPERIMENT 40A

Resolution of (±)-α-Phenylethylamine

In this procedure, you will resolve racemic (±)-α-phenylethylamine, using (+)-tartaric acid as the resolving agent.

REQUIRED READING

Review: Technique 8   Section 8.3

Technique 12   Section 12.8

Technique 23

Technique 22

SPECIAL INSTRUCTIONS

α-Phenylethylamine readily reacts with carbon dioxide in the air to form a white solid, the N-carboxyl amine derivative. Every effort should be taken to avoid prolonged exposure of the amine to air. Be sure to close the bottle tightly after you have measured the rotation of your amine and be sure to place your sample quickly into the flask where you will perform the resolution. This flask should also be stoppered. Use a cork stopper because a rubber stopper will dissolve somewhat and discolor your solution. The crystalline salt will not react with carbon dioxide until you decompose it to recover the resolved amine. Then you must be careful once again.

The observed rotation for a sample isolated by a single student may be only a few degrees, which limits the precision of the optical purity determination. Better results can be obtained if four students combine their resolved amine products for the polarimetric analysis. If you have allowed your amine to have excessive exposure to air, the polarimetry solution may be cloudy. This will make it difficult to obtain an accurate determination of the optical rotation.

SUGGESTED WASTE DISPOSAL

Place the mother liquor solution from the crystallization, which contains (+)-α-phenylethylamine, (+)-tartaric acid, and methanol, in the special container provided for this purpose. Aqueous extracts will contain tartaric acid, dilute base, and water; they should be placed in the container designated for aqueous wastes. When you are finished with polarimetry, depending on the wishes of your instructor, you should either place your resolved (S)-(−)-α-phenylethylamine in a special container marked for this purpose or you should submit it to your instructor in a suitably labeled container that includes the names of those people who have combined their samples.

PROCEDURE

NOTE TO THE INSTRUCTOR: This experiment is designed for students to work individually but to combine their products with three other students for polarimetry.

Preparations

Place 7.8 g of L-(+)-tartaric acid and 125 mL of methanol in a 250-mL Erlenmeyer flask. Heat this mixture on a hot plate until the solution is nearly boiling. Slowly add 6.25 g of racemic α-phenylethylamine (α-methylbenzylamine) to this hot solution.

CAUTION

At this step, the mixture is likely to froth and boil over.

Crystallization

Stopper the flask and let it stand overnight. The crystals that form should be prismatic. If needles form, they are not optically pure enough to give a complete resolution of the enantiomers; *prisms must form*. Needles should be dissolved (by careful heating) and cooled slowly to crystallize once again. When you recrystallize, you can "seed" the mixture with a prismatic crystal, if one is available. If it appears that you have prisms but that they are overgrown (covered) with needles, the mixture may be heated until *most* of the solid has dissolved. The needle crystals dissolve easily, and usually a small amount of the prismatic crystals remains to seed the solution. After dissolving the needles, allow the solution to cool slowly and form prismatic crystals from the seeds.

Workup

Filter the crystals, using a Büchner funnel (Technique 8, Section 8.3, and Fig. 8.5, p. 622), and rinse them with a few portions of cold methanol. Partially dissolve the crystalline amine-tartrate salt in 25 mL of water, add 4 mL of 50% sodium hydroxide, and extract this mixture with three 10-mL portions of methylene chloride using a separatory funnel (Technique 12, Section 12.7, p. 677). Combine the organic layers from each extraction in a stoppered flask and dry them over about 1 g of anhydrous sodium sulfate for about 10 minutes. During this waiting period, preweigh a clean, dry, 50-mL Erlenmeyer flask. Decant the dried solution into the flask and evaporate the methylene chloride on a hot plate (about 60°C) in the hood.

A stream of nitrogen or air should be directed into the flask to increase the rate of evaporation. When the volume of liquid reaches about 2 or 3 mL total, you should carefully insert a hose attached to the house vacuum or aspirator system to remove any remaining methylene chloride. The hose should be inserted into the neck of the flask. Note that the desired product is a **liquid.** Some solid amine carbonate may start to form on the sides of the flask during the course of the evaporation. This undesired solid is more likely to form if you prolong the heating operation. You will want to take care to avoid the formation of this white solid if at all possible. If you do obtain a cloudy solution or there are solids present, transfer the material to a centrifuge tube and centrifuge the sample. Then remove the clear liquid for the polarimetry part of this experiment.

Yield Calculation and Storage

Stopper the flask and weigh it to determine the yield. Also calculate the percentage yield of the (S)-$(-)$-amine based on the amount of the racemic amine you started with.

Polarimetry

Combine your product with the products obtained by three other students. If anyone's product is cloudy or highly colored, do not use it. If there is floating solid in anyone's sample, try to avoid transferring it. Be sure to keep the pure amine tightly stoppered. Mix the combined liquids and use them to fill a preweighed 10-mL volumetric flask. Weigh the flask to determine the weight of amine and calculate the density (concentration) in grams per milliliter. You should obtain a value of about 0.94 g/mL. This should give you a sufficient amount of material to proceed with the polarimetry measurements that follow without diluting your sample. If, however, your combined products do not amount to more than 10 mL of the amine, you may have to dilute your sample with methanol (check with your instructor).

If you have less than 10 mL of product, weigh the flask to determine the amount of the amine present. Then fill the volumetric flask to the mark with absolute methanol and mix the solution thoroughly by inverting 10 times. The concentration of your solution in grams per milliliter is easily calculated.

Transfer the solution to a 0.5-dm polarimeter tube and determine its observed rotation. Your instructor will show you how to use the polarimeter. Report the values of the observed rotation, specific rotation, and optical purity (enantiomeric excess) to the instructor. The published value for the specific rotation is $[\alpha]_D^{22} = -40.3°$. Calculate the percentage of *each* of the enantiomers in the sample (Technique 23, Section 23.5, p. 826), and include the figures in your report.

Because of the presence of some methylene chloride in the sample of the chiral amine, you may obtain low rotation values from polarimetry. Because of this, your calculated value of the optical purity (enantiomeric excess) and percentages of the enantiomers will be in error. The percentages of the enantiomers obtained from the optional chiral gas chromatography experiment, below, should provide more accurate percentages of each of the stereoisomers.

Chiral Gas Chromatography (Optional)

Chiral gas chromatography will provide a direct measure of amounts of each stereoisomer present in your resolved α-phenylethylamine sample. A Varian CP-3800 equipped with an J & P (Agilent) Cyclosil B capillary column (30 m, 0.25-mm ID, 0.25 μm) provides an excellent separation of (R) and (S)-enantiomers.

Set the FID detector at 270°C and the injector temperature at 250°C. The initial split ratio should be set at 150:1 and then changed to 10:1 after 1.5 minutes. Set the oven temperature at 100°C and hold at that temperature for 25 minutes. The helium flow rate is 1 mL/min. The compounds elute in the following order: (R)-α-phenylethylamine (17.5 min) and (S)-enantiomer (18.1 min). Your observed retention times may vary from those given here, but the order of elution will be the same. Because the peaks overlap slightly, you may not observe a distinct peak for the (R)-enantiomer. Instead, what you may observe is a shoulder for the (R)-enantiomer peak on the side of the large peak for the (S)-enantiomer. If you are able to see the (R)-enantiomer, integrate the area under the peaks to obtain the percentages of each of the enantiomers in your sample and compare your results to those obtained with the polarimeter. It should be noted the resolution process used in this experiment is highly selective for the (S)-enantiomer. That is the good news; the bad news is that you may have such a pure (S)-α-phenylethylamine sample that you will not be able to obtain percentages from the analysis on the chiral column!

40B EXPERIMENT 40B

Determination of Optical Purity Using NMR and a Chiral Resolving Agent

In this procedure, you will use NMR spectroscopy with the chiral resolving agent (S)-(+)-O-acetylmandelic acid to determine the optical purity of the (S)-(−)-α-phenylethylamine you isolated in Experiment 40A.

REQUIRED READING

New: Technique 26   Nuclear Magnetic Resonance Spectroscopy

SPECIAL INSTRUCTIONS

Be sure to use a clean Pasteur pipet whenever you remove $CDCl_3$ from its supply bottle. Avoid contaminating the stock of NMR solvent! Also be sure to fill and empty the pipet several times before attempting to remove the solvent from the bottle. If you bypass this equilibration technique, the volatile solvent may squirt out of the pipet before you can transfer it successfully to another container.

SUGGESTED WASTE DISPOSAL

When you dispose of your NMR sample, which contains $CDCl_3$, place it in the container designated for halogenated wastes.

PROCEDURE

Using a small test tube, weigh approximately 0.05 mmole (0.006 g, *MW* = 121) of your resolved amine by adding it from a Pasteur pipet. Cork the test tube to protect it from atmospheric carbon dioxide. Carbon dioxide reacts with the amine to form an amine carbonate (white solid). Using a weighing paper, weigh approximately 0.06 mmole (0.012 g, *MW* = 194) of (*S*)-(+)-O-acetylmandelic acid and add it to the amine in the test tube. Using a clean Pasteur pipet, add about 0.25 mL of $CDCl_3$ to dissolve everything. If the solid does not completely dissolve, you can mix the solution by drawing it several times into your Pasteur pipet and redelivering it back into the test tube. When everything is dissolved, transfer the mixture to an NMR tube using a Pasteur pipet. Using a clean Pasteur pipet, add enough $CDCl_3$ to bring the total height of the solution in the NMR tube to 50 mm.

Determine the proton NMR spectrum, preferably at 300 MHz, using a method that expands and integrates the peaks of interest. Using the integrals, calculate the percentages of the *R* and *S* isomers in the sample, and its optical purity.[1] Compare your results from this NMR determination to those you obtained by polarimetry (Experiment 40A).

REFERENCES

Ault, A. "Resolution of D,L-α-Phenylethylamine." *Journal of Chemical Education, 42* (1965): 269.

Jacobus, J., and Raban, M. "An NMR Determination of Optical Purity." *Journal of Chemical Education, 46* (1969): 351.

Parker, D., and Taylor, R. J. "Direct ^{1}H NMR Assay of the Enantiomeric Composition of Amines and β-Amino Alcohols Using O-Acetyl Mandelic Acid as a Chiral Solvating Agent." *Tetrahedron, 43*, No. 22 (1987): 5451.

QUESTIONS

1. Using a reference textbook, find examples of reagents used in performing chemical resolutions of acidic, basic, and neutral racemic compounds.

2. Propose methods of resolving each of the following racemic compounds

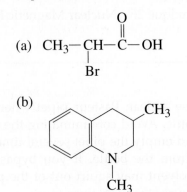

$$
\text{(a)} \quad CH_3-\underset{\underset{Br}{|}}{CH}-\overset{\overset{O}{||}}{C}-OH
$$

(b)

---

[1] *Note to the Instructor:* In some cases, the resolution is so successful that it is difficult to detect the doublet arising from the (*R*)-(+)-α-phenylethylamine + (*S*)-(+)-O-acetylmandelic acid diastereomer. If this occurs, it is useful to have the students add a single drop of *racemic* α-phenylethylamine to the NMR tube and redetermine the spectrum. In this way, both disasteromers can be clearly seen.

3. Explain how you would proceed to isolate (R)-$(+)$-α-phenylethylamine from the *mother liquor* that remained after you crystallized (S)-$(-)$-α-phenylethylamine.

4. What is the white solid that forms when α-phenylethylamine comes in contact with carbon dioxide? Write an equation for its formulation.

5. Which method, polarimetry or NMR spectroscopy, gives the more accurate results in this experiment? Explain.

6. Draw the three-dimensional structure of (S)-$(-)$-α-phenylethylamine.

7. Draw the three-dimensional structure of the diastereomer formed when (S)-$(-)$-α-phenylethylamine is reacted with (S)-$(+)$-O-acetylmandelic acid.

| 41 | **E X P E R I M E N T   4 1** |

The Aldol Condensation Reaction: Preparation of Benzalacetophenones (Chalcones)

Aldol condensation

Crystallization

Benzaldehyde reacts with a ketone in the presence of base to give α,β-unsaturated ketones. This reaction is an example of a crossed aldol condensation where the intermediate undergoes dehydration to produce the resonance-stabilized unsaturated ketone.

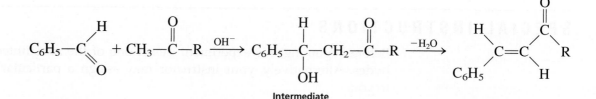

Intermediate

Crossed aldol condensations of this type proceed in high yield because benzaldehyde cannot react with itself by an aldol condensation reaction because it has no α-hydrogen. Likewise, ketones do not react easily with themselves in aqueous base. Therefore, the only possibility is for a ketone to react with benzaldehyde.

In this experiment, procedures are given for preparing benzalacetophenones (chalcones). You should choose one of the substituted benzaldehydes and react it with the ketone, acetophenone. All the products are solids that can be recrystallized easily.

Benzalacetophenones (chalcones) are prepared by the reaction of a substituted benzaldehyde with acetophenone in aqueous base. Piperonaldehyde, p-anisaldehyde, and 3-nitrobenzaldehyde are used.

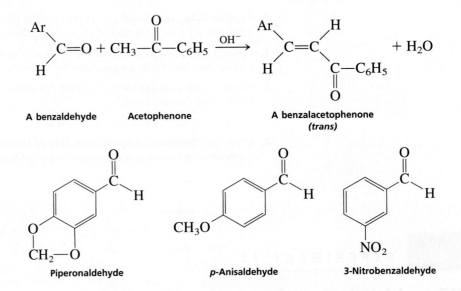

A benzaldehyde   **Acetophenone**   **A benzalacetophenone**
                                                                  (trans)

Piperonaldehyde   **p-Anisaldehyde**   **3-Nitrobenzaldehyde**

An optional molecular modeling exercise is provided in this experiment. We will examine the reactivity of the enolate ion of a ketone to see which atom, oxygen or carbon, is more nucleophilic. The molecular modeling part of this experiment will help you to rationalize the experimental results of this experiment. It would be helpful to look at Experiment 19E, starting on page 173, in addition to the material given in this experiment.

REQUIRED READING

Review: Technique 8   Section 8.7

Technique 11   Sections 11.3 and 11.4

SPECIAL INSTRUCTIONS

Before beginning this experiment, select one of the substituted benzaldehydes. Alternatively, your instructor may assign a particular compound to you.

SUGGESTED WASTE DISPOSAL

All filtrates should be poured into a waste container designated for non-halogenated organic waste.

PROCEDURE

Running the Reaction

Choose one of three aldehydes for this experiment: piperonaldehyde (solid), 3-nitrobenzaldehyde (solid), or *p*-anisaldehyde (liquid). Place 0.150 g of piperonaldehyde (3,4-methylenedioxybenzaldehyde, *MW* = 150.1) or 0.151 g of

3-nitrobenzaldehyde ($MW = 151.1$) into a 5-mL conical vial. Alternatively, transfer 0.13 mL of *p*-anisaldehyde (4-methoxybenzaldehyde, $MW = 136.2$) to a *tared* conical vial and reweigh the vial to determine the weight of material transferred.

Add 0.12 mL of acetophenone ($MW = 120.2$, $d = 1.03$ g/mL) and 0.80 mL of 95% ethanol to the vial containing your choice of aldehyde. Place the conical vial into a 50-mL beaker. Stir the mixture with a microspatula to dissolve any solids present. You may need to warm the mixture on a hot plate to dissolve the solids. If this is necessary, then cool the solution to room temperature before proceeding with the next step.

Add 0.10 mL of aqueous sodium hydroxide solution[1] to the aldehyde/acetophenone mixture. Stir the mixture with your microspatula until it solidifies. Before the mixture solidifies, you may observe some cloudiness. *Wait until the cloudiness has been replaced with an obvious precipitate settling out to the bottom of the conical vial before proceeding to the next paragraph.* Continue stirring with your microspatula until a solid forms (approximately 3 to 5 minutes).[2] Scratching the inside of the conical vial with your microspatula may help to crystallize the chalcone.

Isolation of the Crude Chalcone

Add 2 mL of ice water to the vial *after a solid has formed, as indicated in the previous paragraph.* Stir the solid in the mixture with a spatula to break up the solid mass.

Transfer the mixture to a small beaker with 3 mL of ice water. Stir the precipitate to break it up and then collect the solid on a Hirsch funnel. Wash the product with cold water. Let the solid air-dry for about 30 minutes. Weigh the solid and determine the percentage yield.

Crystallization of the Chalcone

Purify all of the crude chalcone from hot 95% ethanol or hot methanol using the semimicroscale crystallization procedure (Technique 11, Section 11.3, p. 650). Alternatively, purify part of the chalcone using a Craig tube (Technique 8, Section 8.7, p. 625 and Technique 11, Section 11.4, p. 656), as follows:

3,4-Methylenedioxychalcone (from piperonaldehyde) Crystallize a 0.040-g sample from about 0.5 mL of hot 95% ethanol; literature melting point is 122°C.

4-Methoxychalcone (from p-anisaldehyde) Crystallize a 0.075-g sample from about 0.3 mL of hot 95% ethanol. Scratch the tube to induce crystallization while cooling; literature melting point is 74°C.

3-Nitrochalcone (from 3-nitrobenzaldehyde) Crystallize a 0.025-g sample from about 1 mL of hot methanol. Scratch the tube gently to induce crystallization while cooling; literature melting point is 146°C.

---

[1] The instructor should prepare the concentrated aqueous sodium hydroxide in advance, in the ratio of 0.60 g of sodium hydroxide to 1 mL water.

[2] In some cases, the chalcone may not precipitate. If this is the case, cap the conical vial and allow it to stand until the next laboratory period. Usually, the chalcone will precipitate during that time. An additional portion of base will sometimes be helpful, as will *gentle* warming.

Laboratory Report

Determine the melting point of your purified product. At the option of the instructor, obtain the proton and/or carbon-13 NMR spectrum. Include a balanced equation for the reaction in your report. Submit the crude and purified samples to the instructor in labeled vials.

MOLECULAR MODELING (OPTIONAL)

In this exercise we will examine the enolate ion of acetone and determine which atom, oxygen or carbon, is the more nucleophilic site. Two resonance structures can be drawn for the enolate ion of acetone, one with the negative charge on oxygen, structure **A,** and one with the negative charge on carbon, structure **B.**

$$H_2C=\overset{\overset{\displaystyle :\ddot{O}:^-}{|}}{C}-CH_3 \quad\longleftrightarrow\quad \bar{H}_2\ddot{C}-\overset{\overset{\displaystyle \cdot\ddot{O}\cdot}{\|}}{C}-CH_3$$
$$\text{A} \qquad\qquad\qquad \text{B}$$

The enolate ion is an **ambident nucleophile,** a nucleophile that has two possible nucleophilic sites. Resonance theory indicates that structure **A** should be the major contributing structure because the negative charge is better accommodated by oxygen, a more electronegative atom than carbon. However, the reactive site of this ion is carbon, not oxygen. Aldol condensations, brominations, and alkylations take place at carbon, not oxygen. In frontier molecular orbital terms (see the essay on p. 160), the enolate ion is an electron pair donor, and we would expect the pair of electrons donated to be those in the highest occupied molecular orbital, the HOMO.

In the structure-building editor of your modeling program, build structure **A.** Be sure to delete an unfilled valence from oxygen and to place a –1 charge on the molecule. Request a geometry optimization at the AM1 semiempirical level. Also request the HOMO surface and maps of the HOMO and the electrostatic potential onto the electron density surface. Submit your selections for computation. Plot the HOMO on the screen. Where are the biggest lobes of the HOMO, on carbon or on oxygen? Now map the HOMO onto the electron density surface. The "hot spot," the place where the HOMO has the highest density at the point where it intersects the surface, will be bright blue. What do you conclude from this mapping? Finally, map the electrostatic potential onto the electron density. This shows the electron distribution in the molecule. Where is the overall electron density highest, on oxygen or on carbon?

Finally, build structure **B** and calculate the same surfaces as requested for structure **A.** Do you obtain the same surfaces as for structure **A,** or are they different? What do you conclude? Include your results, along with your conclusions, in your report on this experiment.

QUESTIONS

1. Give a mechanism for the preparation of the appropriate benzalacetophenone using the aldehyde and ketone that you selected in this experiment.

2. Draw the structure of the *cis* and *trans* isomers of the compound that you prepared. Why did you obtain the *trans* isomer?

3. Using proton NMR, how could you experimentally determine that you have the *trans* isomer rather than the *cis* one? (*Hint:* See Technique 25, p. 854.)

4. Provide the starting materials needed to prepare the following compounds:

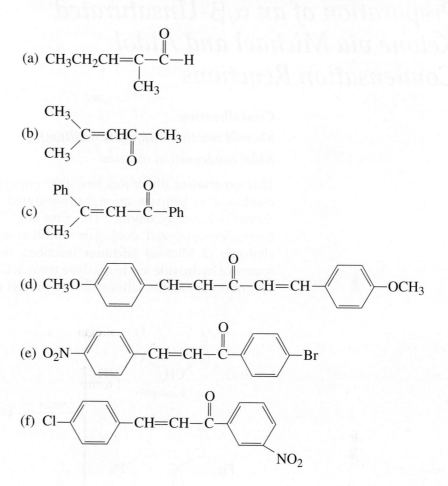

5. Prepare the following compounds starting from benzaldehyde and the appropriate ketone. Provide reactions for preparing the ketones starting from aromatic hydrocarbon compounds (see Experiment 59).

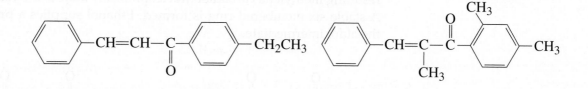

EXPERIMENT 42

Preparation of an α,β-Unsaturated Ketone via Michael and Aldol Condensation Reactions

Crystallization

Michael reaction (conjugate addition)

Aldol condensation reaction

This experiment illustrates how two important synthetic reactions can be combined to prepare an α,β-unsaturated ketone, 6-ethoxycarbonyl-3,5-diphenyl-2-cyclohexenone. The first step in this synthesis is a sodium hydroxide–catalyzed conjugate addition of ethyl acetoacetate to *trans*-chalcone (a Michael addition reaction). Sodium hydroxide serves as a source of hydroxide ion to catalyze the reaction.[1] In the reactions that follow, Ph and Et are abbreviations for the phenyl and ethyl groups, respectively.

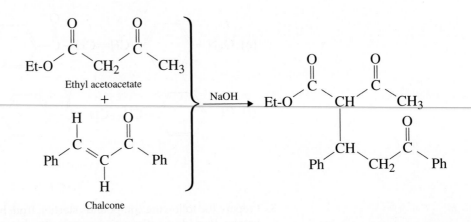

The second step of the synthesis is a base-catalyzed aldol condensation reaction. The methyl group loses a proton in the presence of base, and the resulting methylene carbanion nucleophilically attacks the carbonyl group. A stable six-membered ring is formed. Ethanol supplies a proton to yield the aldol intermediate.

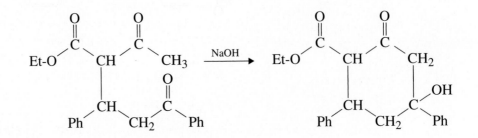

---

[1] Barium hydroxide has also been used as a catalyst (see References).

Finally, the aldol intermediate is dehydrated to form the final product, 6-ethoxycarbonyl-3,5-diphenyl-2-cyclohexenone. The α,β-unsaturated ketone that is formed is stable because of the conjugation of the double bond with both the carbonyl group and a phenyl group.

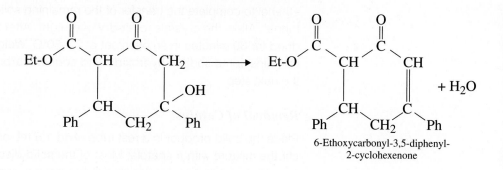

6-Ethoxycarbonyl-3,5-diphenyl-
2-cyclohexenone

REQUIRED READING

Review: Techniques 7, 8, 11, 12

SUGGESTED WASTE DISPOSAL

Dispose of all aqueous wastes containing ethanol in the bottle designated for aqueous wastes. Ethanolic filtrates from the crystallization of the product should be poured into the nonhalogenated organic waste container.

NOTES TO THE INSTRUCTOR

The *trans*-chalcone (Aldrich Chemical Co., #13,612-3) should be finely ground for use by the class.

PROCEDURE

Assembling the Apparatus

To a 10-mL round-bottom flask, add 0.24 g of finely ground *trans*-chalcone, 0.15 g of ethyl acetoacetate, and 5 mL of absolute ethanol. Swirl the flask until all or most of the solid dissolves and place a boiling stone in the flask. Add 0.25 mL of 2.2*M* NaOH to the mixture. Attach a water-jacketed condenser to the round-bottom flask and heat the mixture to reflux using an aluminum block and hot plate. Once the mixture has been brought to a gentle boil, continue to reflux the mixture for at least 1 hour. During this reflux, the mixture will become very cloudy.

Isolation of the Crude Product

After the end of the reflux period, let the mixture cool to room temperature. Add 2 mL of water and scratch the inside of the flask with a glass stirring rod to induce crystallization (an oil may form; scratch vigorously). Place the flask in an ice bath for a minimum of 30 minutes. It is essential to cool the mixture thoroughly in order to completely crystallize the product. Because the product may precipitate slowly, you should also scratch the inside of the flask occasionally over the 30-minute period, as well as cool it in an ice bath.

Vacuum filter the crystals on a Hirsch funnel, using 1 mL of ice-cold water to aid in the transfer. Then rinse the round-bottom flask with 1 mL of ice-cold 95% ethanol to complete the transfer of the remaining solid from the flask to the Hirsch funnel. Allow the crystals to air-dry overnight. Alternatively, the crystals may be dried for 30 minutes in an oven set at 75–80°C. Weigh the dry product. The solid contains some sodium hydroxide and sodium carbonate, which are removed in the next step.

Removal of Catalyst

Place the solid product in a test tube. Add 1.5 mL of reagent-grade acetone and stir the mixture with a spatula. Most of the solid dissolves in acetone, but do not expect all of it to dissolve. Using a Pasteur pipet, remove the liquid and transfer it to a glass centrifuge tube, leaving as much solid as possible behind in the test tube. It is impossible to avoid drawing some solid up into the pipet, so the transferred liquid will contain suspended solids and the solution will be very cloudy. You should not be concerned about the suspended solids in the cloudy acetone extract, because the centrifugation step will clear the liquid completely. Centrifuge the acetone extract for approximately 2–3 minutes or until the liquid clears. Using a clean, dry Pasteur pipet, transfer the *clear* acetone extract from the centrifuge tube to a *dry, preweighed* test tube. If the transfer operation is done carefully, you should be able to leave the solid behind in the centrifuge tube. The solids left behind in the test tube and centrifuge tube are inorganic materials related to the sodium hydroxide originally used as the catalyst.

Evaporate the acetone solvent by carefully heating the test tube in a hot water bath while directing a light stream of dry air or nitrogen in the tube. Use a *slow* stream of gas to avoid blowing your product out of the tube. When the acetone has evaporated, you may be left with an oily solid in the bottom of the tube. Scratch the oily product with a spatula to induce crystallization. You may need to redirect air or nitrogen in the test tube to remove all traces of acetone. Reweigh the test tube to determine the yield of this partially purified product.

Crystallization of Product

Crystallize the product in a 10-mL Erlenmeyer flask using a minimum amount (approximately 2 mL) of hot 95% ethanol. After the solid has dissolved, allow the flask to cool slightly. Scratch the inside of the flask with a glass stirring rod until crystals appear. Allow the flask to sit undisturbed at room temperature for a few minutes. Then place the flask in an ice-water bath for at least 15 minutes.

Collect the crystals by vacuum filtration on a Hirsch funnel. Use two 0.5-mL portions of ice-cold 95% ethanol to aid in the transfer. Allow the crystals to dry overnight or dry them for 30 minutes in a 75–80°C oven. Weigh the dry 6-ethoxycarbonyl-3,5-diphenyl-2-cyclohexenone and calculate the percentage yield. Determine the melting point of the product (literature value, 111–112°C). Submit the sample to the instructor in a labeled vial.

Spectroscopy

At the option of the instructor, obtain the infrared spectrum using the dry film method (Technique 25, Section 25.4, p. 838). You should observe absorbances at 1734 and 1660 cm^{-1} for the ester carbonyl and enone groups, respectively.

Compare your spectrum to that shown in this experiment. Your instructor may also want you to determine the proton and carbon NMR spectra. These may be run in CDCl$_3$ solvent.[2]

REFERENCES

García-Raso, A., García-Raso, J., Campaner, B., Mestres, R., and Sinisterra, J. V. "An Improved Procedure for the Michael Reaction of Chalcones." *Synthesis*, 1982: 1037.

García-Raso, A., García-Raso, J., Sinisterra, J. V., and Mestres, R. "Michael Addition and Aldol Condensation: A Simple Teaching Model for Organic Laboratory." *Journal of Chemical Education, 63* (May 1986): 443.

QUESTIONS

1. Why was it possible to separate the product from sodium hydroxide using acetone?

2. The white solid that remains in the centrifuge tube after acetone extraction fizzes when hydrochloric acid is added, suggesting that sodium carbonate is present. How did this substance form? Give a balanced equation for its formation. Also give an equation for the reaction of sodium carbonate with hydrochloric acid.

3. Draw a mechanism for each of the three steps in the preparation of the 6-ethoxycarbonyl-3,5-diphenyl-2-cyclohexenone. You may assume that sodium hydroxide functions as a base and ethanol serves as a proton source.

4. Indicate how you could synthesize *trans*-chalcone.

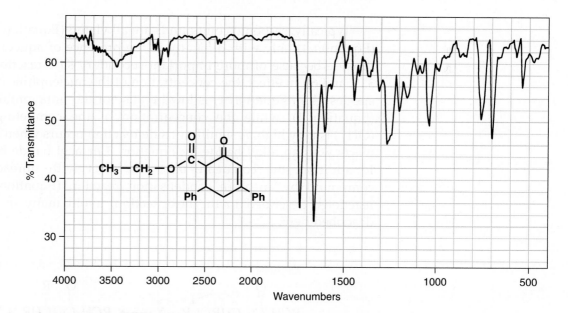

Infrared spectrum of 6-ethoxycarbonyl-3,5-diphenyl-2-cyclohexenone, KBr.

---

[2] Proton NMR determined at 300 MHz: 1.05 ppm (triplet, 3 H, J = 7.1 Hz), 2.95–3.05 ppm (multiplet, 1 H), 3.05–3.15 ppm (multiplet, 1 H), 3.80 ppm (multiplet, 2 H), 4.05 ppm (quartet, 2 H, J = 7.1 Hz), 6.57 ppm (doublet, 1 H, J = 2.0 Hz), and 7.30–7.45 (multiplets, 10 H). Carbon NMR determined at 75 MHz: 17 peaks; 14.1, 36.3, 44.3, 59.8, 61.1, 124.3, 126.4, 127.5, 127.7, 129.0, 129.1, 130.7, 137.9, 141.2, 158.8, 169.5, and 194.3 ppm.

43 EXPERIMENT 43

Enamine Reactions:
2-Acetylcyclohexanone

Enamine reaction
Azeotropic distillation
Column chromatography
Keto-enol tautomerism
Infrared and NMR spectroscopy

Hydrogens on the α-carbon of ketones, aldehydes, and other carbonyl compounds are weakly acidic and are removed in a basic solution (Equation 1). Although resonance stabilizes the conjugate base **A** in such a reaction, the equilibrium is still unfavorable because of the high pK_a (about 20) of a carbonyl compound.

$$R-CH_2\overset{O}{\overset{\|}{C}}CH_2R + {}^-OH \rightleftharpoons \left[RCH_2-\overset{O}{\overset{\|}{C}}-\overset{..}{\overset{..}{C}}HR \longleftrightarrow RCH_2-\overset{O^-}{\overset{|}{C}}=CHR\right] + H_2O \quad [1]$$

$$A$$

Typically, carbonyl compounds are alkylated (Equation 2) or acylated (Equation 4) only with difficulty in the presence of aqueous sodium hydroxide because of more important secondary side reactions (Equations 3, 5, and 6). In effect, the concentration of the nucleophilic conjugate base species (**A** in Equation 1) is low because of the unfavorable equilibrium (Equation 1), while the concentration of the competing nucleophile (OH^-) is very high. A significant side reaction occurs when hydroxide ion reacts with an alkyl halide by Equation 3 or acyl halide by Equation 5. In addition, the conjugate base can react with unreacted carbonyl compound by an aldol condensation reaction (Equation 6). Enamine reactions, described in the next section, avoid many of the problems described here.

Alkylation

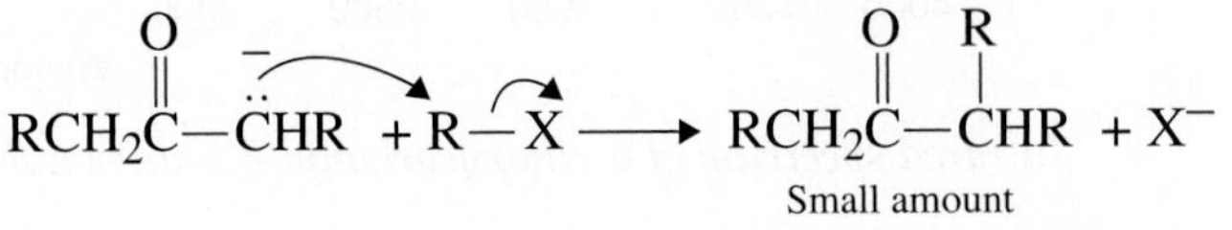

$$RCH_2\overset{O}{\overset{\|}{C}}-\overset{..}{\overset{..}{C}}HR + R-X \longrightarrow RCH_2\overset{O}{\overset{\|}{C}}-\overset{R}{\overset{|}{C}}HR + X^- \quad [2]$$

Small amount

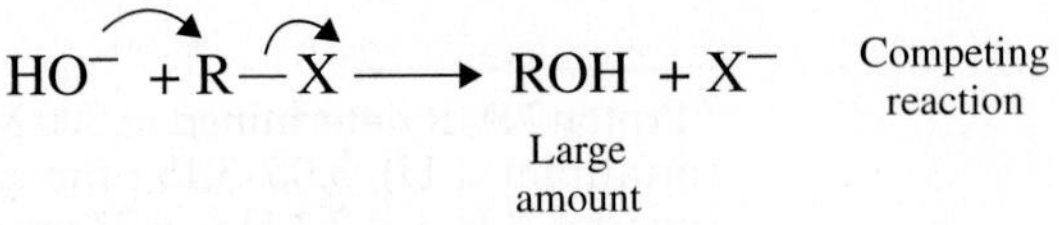

$$HO^- + R-X \longrightarrow ROH + X^- \quad \begin{array}{l}\text{Competing}\\\text{reaction}\end{array} \quad [3]$$

Large
amount

Acylation

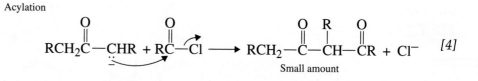

[4]

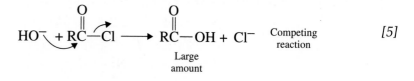

[5]

Aldol condensation

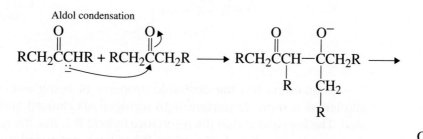

[6]

Formation and Reactivity of Enamines

Enamines are prepared easily from carbonyl compounds (for example, cyclohexanone) and a secondary amine (for example, pyrrolidine) by an acid-catalyzed addition–elimination reaction. Water, the other product of the reaction, is removed by azeotropic distillation with toluene, which drives the equilibrium to the right:

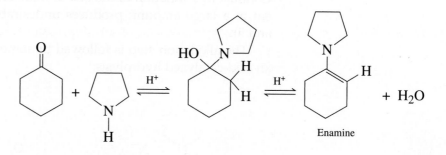

If the water were not removed, the equilibrium would be unfavorable, and only a small amount of enamine would be produced. Azeotropic distillation of water is an important "trick" used in organic chemistry to produce desired products in spite of an unfavorable equilibrium.

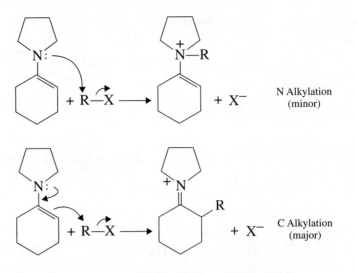

An enamine has the desirable property of being nucleophilic (carbon alkylation is more important than nitrogen alkylation) and is easily alkylated. The *key point* is that the resonance hybrid **B** is like the resonance hybrid **A** shown in Equation 1. However, **B** has been produced under nearly neutral conditions so that it is the *only* nucleophile present.

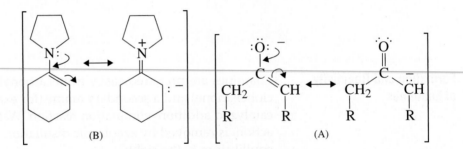

Contrast this situation to the one in Equation 1 where hydroxide ion, present in a large amount, produces undesirable side reactions (Equations 3 and 5).

The alkylation step is followed by removal of the secondary amine by an acid-catalyzed hydrolysis:

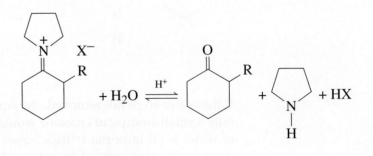

Examples of Enamine Reactions

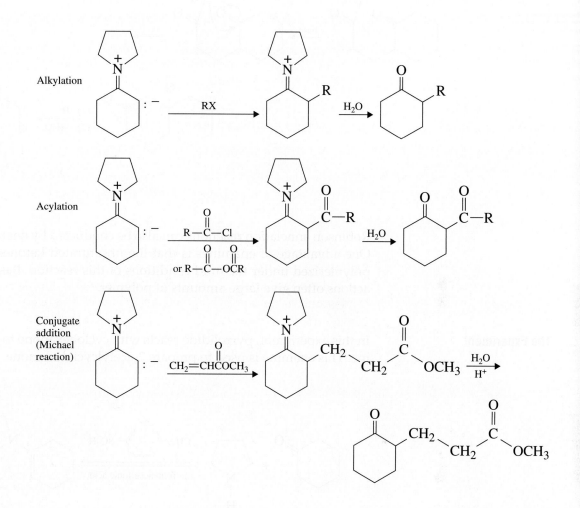

Alkylation

Acylation

Conjugate addition (Michael reaction)

Robinson Annelation (Ring-Formation) Reaction

Reactions that combine the Michael addition reaction and aldol condensation to form a six-member ring fused on another ring are well known in the steroid field. These reactions are known as **Robinson annelation reactions.** An example is the formation of $\Delta^{1,9}$-2-octalone.

Michael addition (conjugate addition)

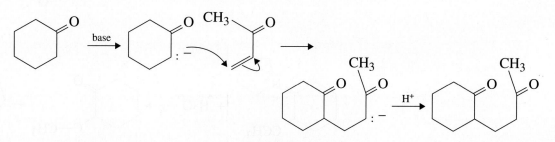

Aldol condensation

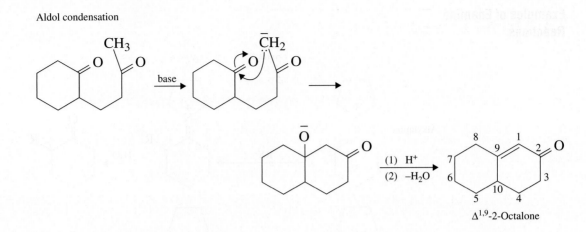

$\Delta^{1,9}$-2-Octalone

Robinson annelation reactions can also be conducted by enamine chemistry. One advantage of enamines is that the unsaturated ketones are not easily polymerized under the mild conditions of this reaction. Base-catalyzed reactions often give large amounts of polymer.

The Experiment

In this experiment, pyrrolidine reacts with cyclohexanone to give the enamine. This enamine is used to prepare 2-acetylcyclohexanone.

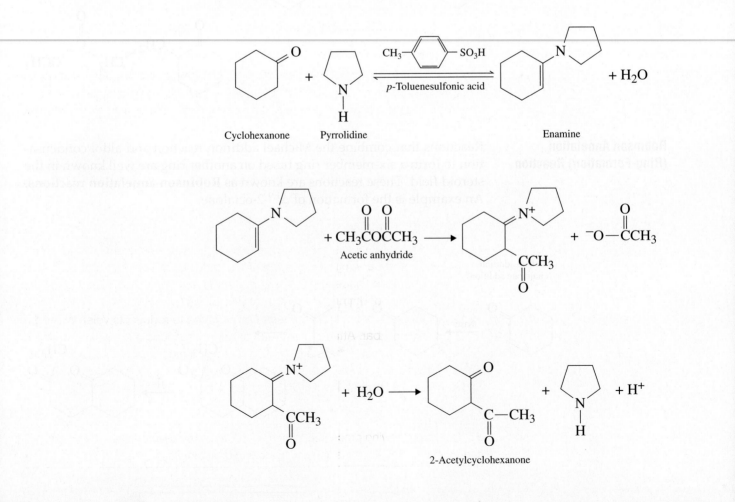

REQUIRED READING

Review: Technique 12   Sections 12.x, 12.x, and 12.x

Technique 14   Section 14.1

Technique 19   Sections 19.6, 19.8, 19.9

New: Technique 15   Fractional Distillation, Azeotropes, Part B, Azeotropes, Sections 15.7, and 15.8

SPECIAL INSTRUCTIONS

Pyrrolidine and acetic anhydride are toxic and noxious. You must measure and transfer these substances in a hood. If you are not careful, the entire room will be filled with vapors of pyrrolidine, and it will not be pleasant to work in the laboratory.

The enamine should be made during the first part of the laboratory period and used as soon as possible. Once the acetic anhydride has been added, the reaction mixture must be allowed to stand in your drawer for at least 48 hours to complete the reaction. The second period is used for the workup and column chromatography. The yields in these reactions are low (less than 20%), partly due to reduced reaction periods necessary to fit the experiment into convenient 3-hour laboratory periods.

SUGGESTED WASTE DISPOSAL

Dispose of the toluene–water azeotrope into the waste container designated for nonhalogenated organic solvents. All aqueous solutions produced in this experiment should be disposed of by pouring them into the container designated for aqueous waste.

PROCEDURE

Part A. Preparation of Enamine

Running the Reaction

Place 0.64 mL of cyclohexanone ($MW = 98.1$) into a preweighed 10-mL round-bottom flask and determine the weight of the material transferred. Add 4.0 mL of toluene to the flask. Place about 20 mg of *p*-toluenesulfonic acid monohydrate in the mixture. In a hood, transfer 0.54 mL of pyrrolidine ($MW = 71.1$, $d = 0.85$ g/mL) to this flask from a bottle that has been cooled in ice to reduce its volatility. Add a magnetic stirring bar. Attach a Hickman head, a water-cooled condenser, and a drying tube that contains moistened cotton. The apparatus is shown in Figure 15.15, page 730, and Figure 14.5, page 708. If a ported Hickman head is not available, you may use an unported Hickman head (Fig. 14.4A, p. 707).

NOTE: In the following procedure, toluene and water are collected in the Hickman head, whereas the enamine remains in the round-bottom flask.

Distillation

While stirring, distill the mixture with an aluminum block adjusted to at least 140°C. Adjust the rate of distillation so that it takes about 30 minutes. Collect the distillate in the Hickman head (Technique 14, Section 14.3, p. 707, and Fig. 14.6, p. 709). You will need to remove 2 mL of distillate. Because some Hickman stills have capacities of less than 1 mL, you should remove the distillate each time the reservoir is filled. Detach the condenser (or open the port), remove the distillate, and transfer it to a conical vial for storage. Continue to remove the distillate until 2 mL of liquid has been removed from the Hickman head and transferred to the conical vial (use the graduations on the vial for measurement).

Water is formed in the reaction as the enamine is produced, and it azeotropes with the solvent toluene and collects in the Hickman head. Only a small amount of water is produced in this reaction, and it is soluble at elevated temperatures in the Hickman head (no cloudiness). When you remove the distillate from the Hickman head, the liquid cools rapidly, and the mixture becomes cloudy as water separates from the toluene. Discard the azeotrope that you collected in the conical vial.

After the distillation has been completed, allow the reaction mixture to cool to room temperature. Remove the flask and prepare 2-acetylcyclohexanone as described in the next section. Proceed to the next step during this laboratory period.

Part B. Preparation of 2-Acetylcyclohexanone

Running the Reaction

In a hood, dissolve 0.64 mL of acetic anhydride ($MW = 102.1$, $d = 1.08$ g/mL) in 1.0 mL of toluene in a conical vial. Using a Pasteur pipet, add this solution to the enamine that you already have prepared. Cap the round-bottom flask, swirl it for a few minutes at room temperature, and allow the mixture to stand for at least 48 hours.

After this period, add 1.0 mL of water. Attach a water-cooled condenser (without the Hickman head) and boil the mixture with stirring for 30 minutes in an aluminum block at about 120°C. Cool the flask to room temperature. Using a Pasteur pipet, transfer the liquid to a 15-mL centrifuge tube with a screw cap. Add another 1.0 mL of water, cap the centrifuge tube, shake it, and allow the layers to separate. The 2-acetylcyclohexane is contained in the upper toluene layer. Remove the lower aqueous layer and discard it.

Extraction

Add 2 mL of 6*M* hydrochloric acid to the toluene layer remaining in the tube and shake the mixture to extract any nitrogen-containing contaminants from the organic phase. After allowing the layers to separate, remove the lower aqueous layer and discard it. Finally, shake the organic phase with 1.0 mL of water, remove the lower aqueous layer, and discard it. Using a dry Pasteur pipet, transfer the organic layer to a dry conical vial and add granular anhydrous sodium sulfate (four microspatulafuls measured in the V-grooved end) to dry the organic layer. Using a *dry* Pasteur pipet, transfer the dried organic phase from the drying agent and place it in a *dry* 5-mL conical vial. Rinse the drying agent with a minimum amount of fresh toluene and add this to the vial.

Evaporate the toluene in a water bath at about 70°C, using a stream of dry air or nitrogen. *Watch the liquid carefully during this procedure or your product may evaporate.* When the toluene has all been removed, the volume of liquid will remain constant (0.3 to 0.5 mL). Save the yellow liquid residue for purification by column chromatography.

Column Chromatography

Prepare a column for column chromatography using a 5¾-inch Pasteur pipet as a column (Technique 19, Section 19.6, p. 764). Use alumina as the absorbent and methylene chloride as the eluent. Place a small piece of cotton in the pipet and gently push it into the constriction. Pour 1.0 g of alumina[1] into the column. Obtain 5 mL of methylene chloride in a graduated cylinder. Use the methylene chloride to prepare the column, dissolve the crude product, and elute the purified product as described in the next paragraph.

Dissolve the crude product in 0.5 mL of methylene chloride. Clamp the column above a 10-mL Erlenmeyer flask. Then add about 1 mL of the methylene chloride to the column and let it percolate through the alumina. Allow the solvent to drain until the solvent surface just begins to enter the alumina. Add the crude product to the top of the column and let the mixture pass onto the column. Use about 0.5 mL of methylene chloride to rinse the vial that contained the crude product. When the first batch of crude product has drained so that the surface of the liquid just begins to enter the alumina, add the methylene chloride rinse to the column.

When the solvent level has again reached the top of the alumina, add more methylene chloride with a Pasteur pipet to elute the product into the flask. Continue adding methylene chloride to the column until all the colored material has eluted off the column. Collect all the liquid that passes through the column as one fraction.

Evaporation of Solvent

Preweigh a 5-mL conical vial and transfer about half the liquid in the Erlenmeyer flask to the conical vial. Place the conical vial in a warm water bath (about 50°C) and evaporate the methylene chloride with a light stream of air or nitrogen in a hood until the volume is about 0.5 mL. Transfer the remaining liquid in the Erlenmeyer flask to the conical vial and continue evaporating the methylene chloride to give the 2-acetylcyclohexanone as a yellow liquid. When the solvent has been removed, reweigh the vial to determine the weight of product. Calculate the percentage yield ($MW = 140.2$).

At the option of the instructor, determine the infrared spectrum, the NMR spectrum, or both. The NMR spectrum may be used to determine the percentage enol content for 2-acetylcyclohexanone. This compound is highly enolic, giving calculated values between 25% and 70%. The enol content depends on the time delay between when the compound was synthesized and when the enol content is measured. Solvent effects also influence the enol content. Submit the remaining sample to the instructor in a labeled vial with your laboratory report.

NOTE: The percentage enol content can be calculated using the 60-MHz NMR spectrum reproduced in this experiment. The offset peak is assigned to the enolic hydrogen (integral height, 10 mm). The remaining absorptions at 1.5 to 2.85 ppm (integral height, 155 mm) are assigned to the 11 protons remaining in the enol structure and the 12 protons in the keto structure. Thus, 110 mm (10 × 11) of the

---

[1] EM Science (No. AX 0612-1). The particle sizes are 80–200 mesh, and the material is Type F-20.

155-mm integral height is assigned to the enol structure. Enol % = 110/155 = 71; keto % = 45/155 = 29. At 300 MHz, one can integrate the enol hydrogen at 16 ppm and compare it to the methyl hydrogens at 2.15 ppm. At 60 MHz, the methyl groups are not clearly resolved.

REFERENCES

Augustine, R. L., and Caputa, J. A. "$\Delta^{1,9}$-2-Octalone." *Organic Syntheses,* Coll. Vol. 5 (1973): 869.

Cook, A. G., ed. *Enamines: Synthesis, Structure, and Reactions.* New York: Marcel Dekker, 1969.

Dyke, S. F. *The Chemistry of Enamines.* London: Cambridge University Press, 1973.

Mundy, B. P. "The Synthesis of Fused Cycloalkenones via Annelation Methods." *Journal of Chemical Education,* 50 (1973): 110.

Stork, G., Brizzolara, A., Landesman, H., Szmuszkovicz, J., and Terrell, R. "The Enamine Alkylation and Acylation of Carbonyl Compounds." *Journal of the American Chemical Society,* 85 (1963): 207.

QUESTIONS

1. Draw a mechanism for the enamine synthesis of $\Delta^{1,9}$-2-octalone. Why is this octalone rather than the $\Delta^{9,10}$-2-octalone the main product in the reaction? On the other hand, why is there a substantial amount of the $\Delta^{9,10}$-2-octalone produced in the reaction?

2. **(a)** The enamine formed from pyrrolidine and 2-methylcyclohexanone has the structure **A**. What reason can you give for the less substituted enamine being formed instead of the more substituted enamine **B**? (*Hint:* Consider steric effects.)

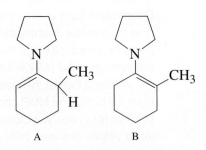

        A   B

  (b) Draw the structure of the product that would result from the reaction of enamine **A** with methyl vinyl ketone. Compare its structure with the product obtained in Question 3.

3. **(a)** The enolate formed from 2-methylcyclohexanone has the following structure. What is the structure of the other possible enolate, and why is it not as stable as the one shown here?

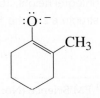

(b) Draw the structure of the product that would result from the reaction with methyl vinyl ketone. Compare its structure with the product obtained in Question 2.

4. Draw the structures of the Robinson annelation products that would result from the following reactions:

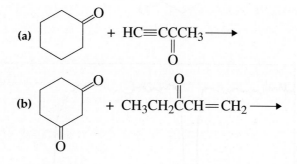

5. Draw the structures of the products that would result from the following enamine reactions. Use pyrrolidine as the amine and write equations for the reaction sequence.

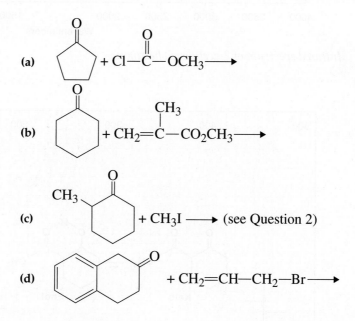

6. Interpret the infrared spectrum of 2-acetylcyclohexanone, especially in the O—H and C=O stretch regions of the spectrum.

7. Calculate the amount of water produced during the formation of the enamine in this experiment.

8. Write the equations showing how one could carry out the following multistep transformation, starting from the indicated materials. One need not use an enamine synthesis.

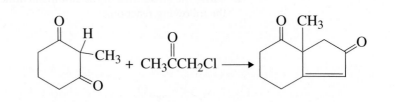

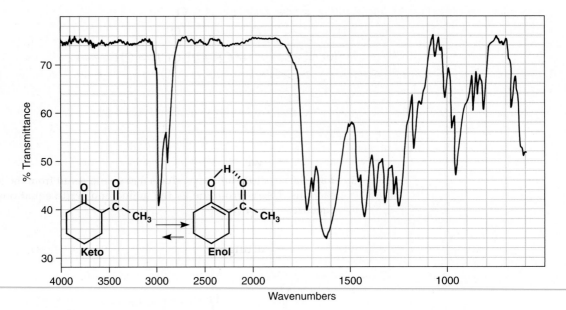

Infrared spectrum of 2-acetylcyclohexanone.

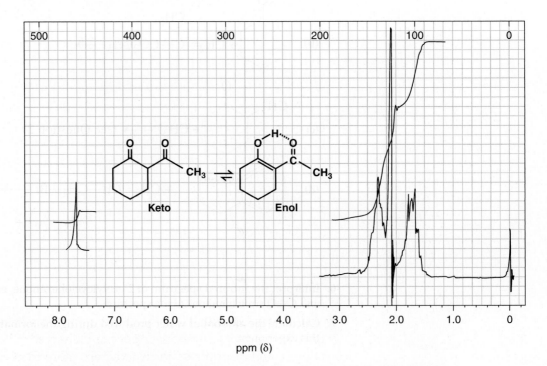

ppm (δ)

NMR spectrum of 2-acetylcyclohexanone, CDCl$_3$, offset peak shifted by 500 Hz.

EXPERIMENT 44

1,4-Diphenyl-1,3-butadiene

Wittig reaction
Working with sodium ethoxide
Thin-layer chromatography
UV/NMR spectroscopy
Solventless Wittig reactions

The Wittig reaction is often used to form alkenes from carbonyl compounds. In this experiment, the isomeric dienes *cis,trans-*, and *trans,trans-*1,4-diphenyl-1,3-butadiene will be formed from cinnamaldehyde and benzyltriphenyl-phosphonium chloride.

Two procedures are provided for preparing *trans,trans-*1,4-diphenyl-1,3-butadiene. In Experiment 44B, the reaction uses sodium ethoxide in ethanol solvent as the base, whereas in Experiment 44C, a green chemistry alternative is provided whereby potassium phosphate is employed as the base that is conducted without any solvent. The mechanism of the Wittig reaction in the presence of either sodium ethoxide or potassium phosphate is essentially identical. Sodium ethoxide is shown as the base in the mechanism that follows. In Experiment 44A, an optional procedure is provided for preparing one of the starting materials for the Wittig reaction.

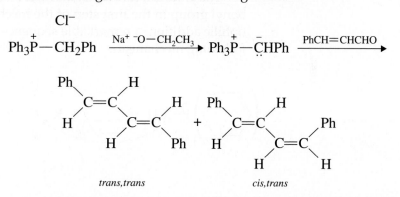

The reaction is carried out in two steps. First, the phosphonium salt is formed by the reaction of triphenylphosphine with benzyl chloride in Experiment 44A. The reaction is a simple nucleophilic displacement of chloride ion by triphenylphosphine. The salt that is formed is called the "Wittig reagent" or "Wittig salt."

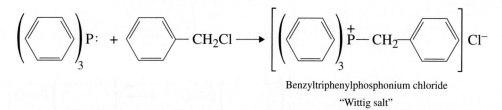

Benzyltriphenylphosphonium chloride
"Wittig salt"

When treated with base, the Wittig salt forms an **ylide.** An ylide is a species having adjacent atoms oppositely charged. The ylide is stabilized due to the ability of phosphorus to accept more than eight electrons in its valence shell. Phosphorus uses its 3d orbitals to form the overlap with the

2p orbital of carbon that is necessary for resonance stabilization. Resonance stabilizes the carbanion.

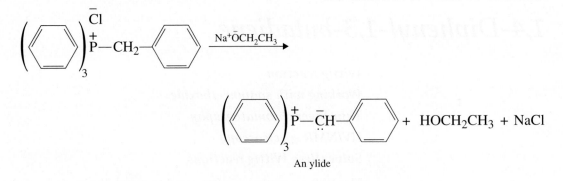

An ylide

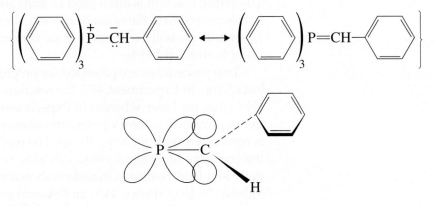

The ylide is a carbanion that acts as a nucleophile, and it adds to the carbonyl group in the first step of the mechanism. Following the initial nucleophilic addition, a remarkable sequence of events occurs, as outlined in the following mechanism:

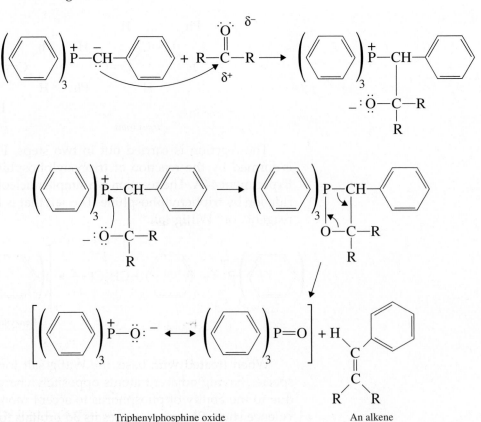

Triphenylphosphine oxide   An alkene

The addition intermediate, formed from the ylide and the carbonyl compound, cyclizes to form a four-membered-ring intermediate. This new intermediate is unstable and fragments into an alkene and triphenylphosphine oxide. Notice that the ring breaks open differently from the way it was formed. The driving force for this ring-opening process is the formation of a very stable substance, triphenylphosphine oxide. A large decrease in potential energy is achieved on the formation of this thermodynamically stable compound.

In this experiment, cinnamaldehyde is used as the carbonyl compound and yields mainly the *trans,trans*-1,4-diphenyl-1,3-butadiene, which is obtained as a solid. The *cis,trans* isomer is formed in smaller amounts, but it is a liquid that is not isolated in this experiment. The *trans,trans* isomer is the more stable isomer and is formed preferentially.

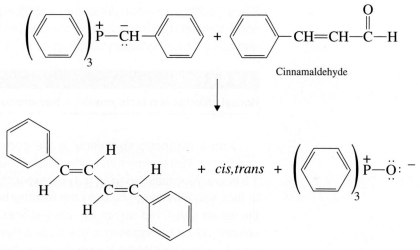

Cinnamaldehyde

trans,trans-1,4-Diphenyl-1,3-butadiene   Triphenylphosphine oxide

---

REQUIRED READING

Review: Technique 8   Section 8.3

Technique 20

SPECIAL INSTRUCTIONS

Your instructor may ask you to prepare 1,4-diphenyl-1,3-butadiene starting with commercially available benzyltriphenylphosphonium chloride. If so, start with Experiment 44B or 44C. The sodium ethoxide solution used in Experiment 44B must be kept tightly stoppered when not in use because it reacts readily with atmospheric water. Fresh cinnamaldehyde must be used in this experiment. Old cinnamaldehyde should be checked by infrared spectroscopy to be certain that it does not contain any cinnamic acid.

If your instructor asks you to prepare benzyltriphenylphosphonium chloride in Experiment 44A, you can conduct another experiment concurrently during the 1.5-hour reflux period. Triphenylphosphine is rather toxic. Be careful not to inhale the dust. Benzyl chloride is a skin irritant and a lachrymator. It should be handled in the hood with care.

SUGGESTED WASTE DISPOSAL

If you conducted Experiment 44A, place the wastes in the nonhalogenated waste container. For Experiment 44B and 44C, dispose of organic wastes in the nonhalogenated waste container. Place the aqueous waste into the bottle designated for aqueous wastes.

44A EXPERIMENT 44A (OPTIONAL)

Benzyltriphenylphosphonium Chloride (Wittig Salt)

Place 0.550 g of triphenylphosphine ($MW = 262.3$) into a 5-mL conical vial. In a hood, transfer 0.36 mL of benzyl chloride ($MW = 126.6$, $d = 1.10$ g/mL) to the vial and add 2.0 mL of xylenes (mixture of *o*-, *m*-, and *p*-isomers).

> **CAUTION** ⚠
>
> **Benzyl chloride is a lachrymator, a tear-producing substance.**

Add a magnetic spin vane to the conical vial and attach a water-cooled condenser. Boil the mixture using an aluminum block at about 165°C for at least 1.5 hours. An increased yield may be expected when the mixture is heated longer. In fact, you may begin heating the mixture before the temperature has reached the values given, but do not include this time in the 1.5-hour reaction period. The solution will be homogeneous at first, and then the Wittig salt will begin to precipitate. Maintain the stirring during the entire heating period or bumping may occur. Remove the apparatus from the aluminum block and allow it to cool for a few minutes. Remove the vial and cool it thoroughly in an ice bath for about 5 minutes.

Collect the Wittig salt by vacuum filtration using a Hirsch funnel. Use three 1-mL portions of cold petroleum ether (bp 60–90°C) to aid the transfer and to wash the crystals free of the xylene solvent. Dry the crystals, weigh them, and calculate the percentage yield of the Wittig salt. At the option of the instructor, obtain the proton NMR spectrum of the salt in CDCl$_3$. The methylene group appears as a doublet (J = 14 Hz) at 5.5 ppm because of ^{1}H-^{31}P coupling.

44B EXPERIMENT 44B

Preparation of 1,4-Diphenyl-1,3-Butadiene Using Sodium Ethoxide to Generate the Ylide

In the following operations, cap the 5-mL conical vial whenever possible to avoid contact with moisture from the atmosphere. If you prepared your own benzyltriphenylphosphonium chloride in Experiment 44A, you may need to supplement your yield in this part of the experiment.

Preparation of the Ylide

Place 0.480 g of benzyltriphenylphosphonium chloride (*MW* = 388.9) in a *dry* 5-mL conical vial. Add a magnetic spin vane. Transfer 2.0 mL of absolute (anhydrous) ethanol to the vial and stir the mixture to dissolve the phosphonium salt (Wittig salt). Add 0.75 mL of sodium ethoxide solution[1] to the vial using a *dry* pipet while stirring continuously. Cap the vial and stir this mixture for 15 minutes. During this period, the cloudy solution acquires the characteristic yellow color of the ylide.

Reaction of the Ylide with Cinnamaldehyde

Measure 0.15 mL of *pure* cinnamaldehyde (*MW* = 132.2, *d* = 1.11 g/mL) and place it in another small conical vial. Add 0.50 mL of absolute ethanol to the cinnamaldehyde. Cap the vial until it is needed. After the 15-minute period, use a Pasteur pipet to mix the cinnamaldehyde with the ethanol and add this solution to the ylide in the reaction vial. A color change should be observed as the ylide reacts with the aldehyde and the product precipitates. Stir the mixture for 10 minutes.

Separation of the Isomers of 1,4-Diphenyl-1,3-Butadiene

Cool the vial thoroughly in an ice-water bath (10 min), stir the mixture with a spatula, and transfer the material from the vial to a Hirsch funnel under vacuum. Use two 1-mL portions of ice-cold absolute ethanol to aid the transfer and to rinse the product. Dry the crystalline *trans,trans*-1,4-diphenyl-1,3-butadiene by drawing air through the solid. The product has a small amount of sodium chloride that is removed as described in the next paragraph. The cloudy material in the filter flask contains triphenylphosphine oxide, the *cis,trans*-isomer, and some *trans,trans* product. Pour the filtrate into a beaker and save it for the thin-layer chromatography experiment described in the next section.

Remove the *trans,trans*-1,4-diphenyl-1,3-butadiene from the filter paper, place the solid in a 10-mL beaker, and add 3 mL of water. Stir the mixture and filter it on a Hirsch funnel, under vacuum, to collect the nearly colorless crystalline *trans,trans* product. Use about 1 mL of water to aid the transfer. Allow the solid to dry thoroughly.

Thin-Layer Chromatography

Use thin-layer chromatography to analyze the filtrate that you saved in the previous section. This mixture must be analyzed as soon as possible so that the *cis,trans* isomer will not be photochemically converted to the *trans,trans* compound. Use a 2 × 8-cm silica gel TLC plate that has a fluorescent indicator

---

[1] This reagent is prepared in advance by the instructor. Carefully dry a 250-mL Erlenmeyer flask and insert a drying tube filled with calcium chloride into a one-hole rubber stopper. Obtain a large piece of sodium, clean it by cutting off the oxidized surface, weigh out a 2.30-g piece, cut it into 20 smaller pieces, and store it under xylene. Using tweezers, remove each piece, wipe off the xylene, and add the sodium slowly over a period of about 30 minutes to 40 mL of absolute (anhydrous) ethanol in the 250-mL Erlenmeyer flask. After the addition of each piece, replace the stopper. The ethanol will warm as the sodium reacts, but do not cool the flask. After the sodium has been added, warm the solution and shake it *gently* until all the sodium reacts. Cool the sodium ethoxide solution to room temperature. This reagent may be prepared in advance of the laboratory period, but it must be stored in a refrigerator between laboratory periods. When it is stored in a refrigerator, it may be kept for about 3 days. Before using this reagent, bring it to room temperature and swirl it gently in order to redissolve any precipitated sodium ethoxide. Keep the flask stoppered between each use.

(Eastman Chromatogram Sheet, No. EK 1224294). At one position on the TLC plate, spot the filtrate, as is, without dilution. Dissolve a few crystals of the *trans,trans*-1,4-diphenyl-1,3-butadiene in a few drops of acetone and spot it at another position on the plate. Use hexane as a solvent to develop (run) the plate.

Visualize the spots with a UV lamp using both the long and short wavelength settings. The order of increasing R_f values is as follows: triphenylphosphine oxide, *trans,trans*-diene, and *cis,trans*-diene. It is easy to identify the spot for the *trans,trans* isomer because it fluoresces brilliantly. What conclusion can you make about the contents of the filtrate and the purity of the *trans,trans* product? Report the results that you obtain, including R_f values and the appearance of the spots under illumination. Discard the filtrate in the container designated for non-halogenated waste.

Yield Calculation and Melting-Point Determination

When the *trans,trans*-1,4-diphenyl-1,3-butadiene is dry, determine the melting point (literature, 151°C). Weigh the solid and determine the percentage yield. If the melting point is below 145°C, recrystallize a portion of the compound from hot 95% ethanol (20 mg/1.3 mL ethanol) in a Craig tube. Redetermine the melting point.

  EXPERIMENT 44C

Preparation of 1,4-Diphenyl-1,3-Butadiene Using Potassium Phosphate to Generate the Ylide

Experiment 44C provides an alternative green chemistry method for preparing 1,4-diphenyl-1,3-butadiene by the Wittig reaction. No solvent is used in this experiment. Instead, the starting materials are ground together with potassium phosphate in a mortar and pestle. This experiment will demonstrate to students a more environmentally friendly method for carrying out a reaction that might be performed on a larger scale in industry.

The reaction will be accomplished by grinding cinnamaldehyde with benzyltriphenylphosphonium chloride and potassium phosphate (tribasic, K_3PO_4). This is done using a mortar and pestle. TLC will be used to analyze the crystallized *trans,trans*-1,4-diphenyl-1,3-butadiene product, as well as the filtrate from the crystallization procedure that contains both the *cis,trans* and *trans,trans*-1,4-diphenyl-1,3,-butadiene isomers.

Reaction

Using an analytical balance, weigh out 309 mg of benzyltriphenylphosphonium chloride and 656 mg of potassium phosphate (tribasic, K_3PO_4) and place the solids into a clean and dry 6-cm (inside diameter) porcelain mortar with a pour lip. Using an automatic pipet, measure and add 100 μL of cinnamaldehyde to the mixture in the mortar. Grind the mixture together for a total of 20 minutes. It is much easier to use a pestle that is long enough to grip securely in your hand, thus saving one's fingers from getting sore or tired. At the beginning of the grinding operation, the mixture will act like putty and will have a definite yellow color. After a few minutes of grinding, the mixture starts to turn into a thick paste that adheres to the inside of the mortar and the edges of the pestle. Bend the end of a spatula as shown in the figure. This bent spatula is useful for scraping the material off of the inside of the mortar and pestle and directing the mass into the center of the mortar. Repeat the scraping operation after every 1 to 2 minutes of grinding. Include that time in the total of 20 minutes of grinding time.

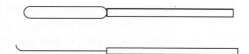

Isolation of Crude 1,4-Diphenyl-1,3-Butadiene

After 20 minutes, add a few milliliters of deionized water to the material in the mortar. Scrape the mortar and pestle a final time to loosen all of the product from the mortar. Pour the mixture into a Hirsch funnel inserted into a filter flask under vacuum. Use a squirt bottle with deionized water to transfer any remaining off-white product into the Hirsch funnel. Discard the filtrate that contains potassium phosphate and some triphenylphosphine oxide. The off-white solid consists of mainly of the *trans,trans* isomer, but some of the *cis,trans* isomer will be present, as well.

Crystallization

Purify the off-white solid by crystallization from absolute ethanol in a small test tube using the standard technique of adding hot solvent until the solid dissolves. A small amount of impurity might not dissolve. If this is the case, use a Pasteur pipet to *rapidly* remove the hot solution away from the impurity and transfer the hot solution to another test tube. Cork the test tube and place it in a warm 25-mL Erlenmeyer flask. Allow the solution to cool slowly. Once the test tube has cooled and crystals have formed, place the test tube in an ice bath for at least 10 minutes to complete the crystallization process. Place 2 mL of absolute ethanol in another test tube and cool the solvent in the ice bath (this solvent will be used to aid the transfer of the product). Loosen the crystals in the test tube with a microspatula and pour the contents of the test tube into a Hirsch funnel under vacuum. Remove the remaining crystals from the test tube using the chilled ethanol and a spatula. Dry the colorless crystalline (plates) of *trans,trans*-1,4-diphenyl-1,3-butadiene on the Hirsch funnel for about 5 minutes to completely dry them. Save the filtrate from the crystallization for analysis by thin-layer chromatography. The *cis,trans*-1,4-diphenyl-1,3-butadiene, which is also formed in the Wittig reaction, is a liquid, and crystallization effectively removes the isomer from the solid *trans,trans* product.

Yield Calculation and Melting Point Determination

Weigh the purified *trans,trans* product and calculate the percentage yield. Determine the melting point of the product (literature, 151°C).

Thin-Layer Chromatography

Following the procedure in Experiment 44B, analyze the filtrate from the crystallization and the purified solid product by thin-layer chromatography. Develop the plate with hexane. This solvent will separate the *cis,trans*-diene from the *trans,trans* isomer. The order of increasing R_f values is as follows: triphenylphosphine oxide, *trans,trans*-diene, and *cis-trans*-diene. Triphenylphosphine oxide is so polar that the R_f value will be nearly zero. After developing the plate in hexane, as indicated in Experiment 44B, use the short and long wavelength settings with a UV lamp to visualize the spots. Calculate the R_f values and record them in your notebook.

Spectroscopy (Optional)

Prepare an NMR sample by dissolving at least 20 mg of crystallized product in about 1 ml of $CDCl_3$ in a small test tube. Transfer the solution to an NMR tube and add more solvent until the level of the solution is about 50 mm in the tube. Run the proton and carbon NMR spectra on the sample. At 300 MHz, the proton spectrum shows multiplets at 6.68 ppm and 6.95 ppm for the vinyl protons and 7.24 ppm (triplet, 2 H), 7.34 ppm (triplet, 4 H), and 7.44 ppm (doublet, 4 H) for the aromatic protons. The carbon spectrum shows peaks at 125.4, 126.6, 127.7, 128.3, 131.8, and 136.4 ppm. To determine the UV spectrum of the product, dissolve a 10-mg sample in 100 mL of hexane in a volumetric flask. Remove 10 mL of this solution and dilute it to 100 mL in another volumetric flask. This concentration should be adequate for analysis. The *trans,trans* isomer absorbs at 328 nm and possesses fine structure, whereas the *cis,trans* isomer absorbs at 313 nm and has a smooth curve.[2] See if your spectrum is consistent with these observations. Submit the spectral data with your laboratory report.

QUESTIONS

1. There is an additional isomer of 1,4-diphenyl-1,3-butadiene (mp 70°C), which has not been shown in this experiment. Draw the structure and name it. Why is it not produced in this experiment? (*Hint:* The cinnamaldehyde has *trans* stereochemistry).

2. Why should the *trans,trans* isomer be the thermodynamically most stable one?

3. A lower yield of phosphonium salt is obtained in refluxing benzene than in xylene (Experiment 44A). Look up the boiling points for these solvents, and explain why the difference in boiling points might influence the yield.

4. Outline a synthesis for *cis* and *trans* stilbene (the 1,2-diphenylethenes) using the Wittig reaction.

5. The sex attractant of the female housefly (*Musca domestica*) is called **muscalure,** and its structure follows. Outline a synthesis of muscalure, using the Wittig reaction. Will your synthesis lead to the required *cis* isomer?

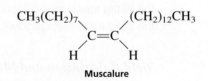

Muscalure

---

[2] The comparative study of the stereoisometric 1,4-diphenyl-1,3-butadienes has been published: Pinkard, J. H., Wille, B., and Zechmeister, L. *Journal of the American Chemical Society, 70* (1948): 1938.

Local Anesthetics

Local anesthetics, or "painkillers," are a well-studied class of compounds. Chemists have shown their ability to study the essential features of a naturally occurring drug and to improve on them by substituting totally new, synthetic surrogates. Often such substitutes are superior in desired medical effects and have fewer unwanted side effects or hazards.

The coca shrub (*Erythroxylon coca*) grows wild in Peru, specifically in the Andes Mountains, at elevations of 1,500 to 6,000 feet above sea level. The natives of South America have long chewed these leaves for their stimulant effects. Leaves of the coca shrub have even been found in pre-Inca Peruvian burial urns. Chewing the leaves brings about a definite sense of mental and physical well-being and the power to increase endurance. For chewing, the Indians smear the coca leaves with lime and roll them. The lime, $Ca(OH)_2$, apparently releases the free alkaloid components; it is remarkable that the Indians learned this subtlety long ago by some empirical means. The pure alkaloid responsible for the properties of the coca leaves is **cocaine.**

The amounts of cocaine the Indians consume in this way are extremely small. Without such a crutch of central-nervous-system stimulation, the natives of the Andes would probably find it more difficult to perform the nearly Herculean tasks of their daily lives, such as carrying heavy loads over the rugged mountainous terrain. Unfortunately, overindulgence can lead to mental and physical deterioration and eventually an unpleasant death.

The pure alkaloid in large quantities is a common drug of addiction. Sigmund Freud first made a detailed study of cocaine in 1884. He was particularly impressed by the ability of the drug to stimulate the central nervous system, and he used it as a replacement drug to wean one of his addicted colleagues from morphine. This attempt was successful, but unhappily, the colleague became the world's first known cocaine addict.

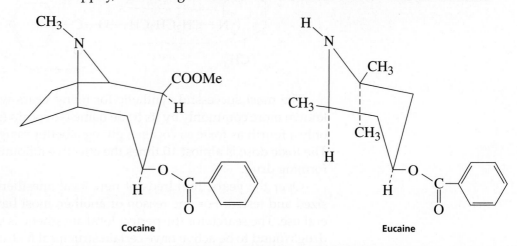

Cocaine   Eucaine

An extract from coca leaves was one of the original ingredients in Coca-Cola. However, early in the present century, government officials, with much legal difficulty, forced the manufacturer to omit coca from its beverage. The company has managed to this day to maintain the *coca* in its trademarked title, even though "Coke" contains none.

Our interest in cocaine lies in its anesthetic properties. The pure alkaloid was isolated in 1862 by Niemann, who noted that it had a bitter taste and produced a queer numbing sensation on the tongue, rendering it almost devoid of sensation. (Oh, those brave, but foolish chemists of yore who used to taste everything!) In 1880, Von Anrep found that the skin was made numb and insensitive to the prick of a pin when cocaine was injected subcutaneously. Freud and his assistant Karl Koller, having failed at attempts to rehabilitate morphine addicts, turned to a study of the anesthetizing properties of cocaine. Eye surgery is made difficult by involuntary reflex movements of the eye in response to even the slightest touch. Koller found that a few drops of a solution of cocaine would overcome this problem. Not only can cocaine serve as a local anesthetic but also it can be used to produce **mydriasis** (dilation of the pupil). The ability of cocaine to block signal conduction in nerves (particularly of pain) led to its rapid medical use in spite of its dangers. It soon found use as a "local" in both dentistry (1884) and in surgery (1885). In this type of application, it was injected directly into the particular nerves it was intended to deaden.

Soon after the structure of cocaine was established, chemists began to search for a substitute. Cocaine has several drawbacks for wide medical use as an anesthetic. In eye surgery, it also produces mydriasis. It can also become a drug of addiction. Finally, it has a dangerous effect on the central nervous system.

The first totally synthetic substitute was eucaine. It was synthesized by Harries in 1918 and retains many of the essential skeletal features of the cocaine molecule. The development of this new anesthetic partly confirmed the portion of the cocaine structure essential for local anesthetic action. The advantage of eucaine over cocaine is that it does not produce mydriasis and is not habit forming. Unfortunately, it is highly toxic.

A further attempt at simplification led to piperocaine. The molecular portion common to cocaine and eucaine is outlined by dotted lines in the structure shown here. Piperocaine is only a third as toxic as cocaine itself.

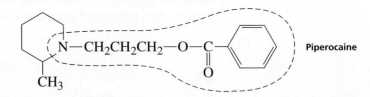

Piperocaine

The most successful synthetic for many years was the drug procaine, known more commonly by its trade name Novocain (see table). Novocain is only a fourth as toxic as cocaine, giving a better margin of safety in its use. The toxic dose is almost 10 times the effective amount, and it is not a habit-forming drug.

Over the years, hundreds of new local anesthetics have been synthesized and tested. For one reason or another, most have not come into general use. The search for the perfect local anesthetic is still under way. All the drugs found to be active have certain structural features in common. At one end of the molecule is an aromatic ring. At the other is a secondary or tertiary amine. These two essential features are separated by a central chain of atoms usually one to four units long. The aromatic part is usually an ester of an aromatic acid. The ester group is important to the bodily detoxification of these compounds. The first step in deactivating them is a hydrolysis of

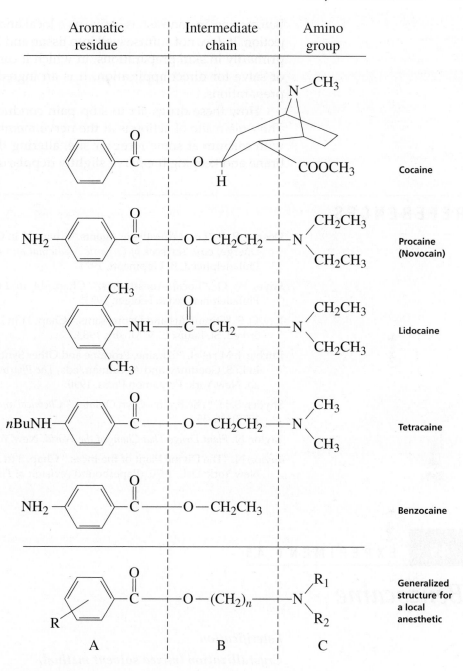

Aromatic residue	Intermediate chain	Amino group	
A	B	C	

Local anesthetics.

this ester linkage, a process that occurs in the bloodstream. Compounds that do not have the ester link are both longer lasting in their effects and generally more toxic. An exception is lidocaine, which is an amide. The tertiary amino group is apparently necessary to enhance the solubility of the compounds in the injection solvent. Most of these compounds are used in their hydrochloride salt forms, which can be dissolved in water for injection.

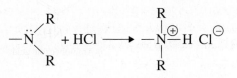

Benzocaine, in contrast, is active as a local anesthetic but is not used for injection. It does not suffuse well into tissue and is not water soluble. It is used primarily in skin preparations, in which it can be included in an ointment or salve for direct application. It is an ingredient of many sunburn-relief preparations.

How these drugs act to stop pain conduction is not well understood. Their main site of action is at the nerve membrane. They seem to compete with calcium at some receptor site, altering the permeability of the membrane and keeping the nerve slightly depolarized electrically.

REFERENCES

Doerge, R. F. "Local Anesthetic Agents." Chap. 22 in C. O. Wilson, O. Gisvold, and R. F. Doerge, eds. *Textbook of Organic Medicinal and Pharmaceutical Chemistry*, 6th ed. Philadelphia: J. B. Lippincott, 1971.

Foye, W. O. "Local Anesthetics." Chap. 14 in *Principles of Medicinal Chemistry*. Philadelphia: Lea & Febiger, 1974.

Ray, O. S. "Stimulants and Depressants." Chap. 11 in *Drugs, Society, and Human Behavior*, 3rd ed. St. Louis: C. V. Mosby, 1983.

Ritchie, J. M., et al. "Cocaine, Procaine and Other Synthetic Local Anesthetics." Chap. 15 in L. S. Goodman and A. Gilman, eds. *The Pharmacological Basis of Therapeutics*, 8th ed. New York: Pergamon Press, 1990.

Snyder, S. H. "The Brain's Own Opiates." *Chemical and Engineering News* (November 28, 1977): 26–35.

Taylor, N. *Plant Drugs That Changed the World*. New York: Dodd, Mead, 1965. Pp. 14–18.

Taylor, N. "The Divine Plant of the Incas." Chap. 3 in *Narcotics: Nature's Dangerous Gifts*. New York: Dell, 1970. (Paperbound revision of *Flight from Reality*.)

45 EXPERIMENT 45

Benzocaine

Esterification

Crystallization (mixed solvent method)

In this experiment, a procedure is given for the preparation of a local anesthetic, benzocaine, by the direct esterification of *p*-aminobenzoic acid with ethanol. At the instructor's option, you may test the prepared anesthetic on a frog's leg muscle.

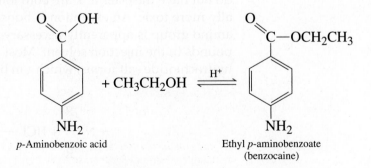

p-Aminobenzoic acid   Ethyl *p*-aminobenzoate (benzocaine)

REQUIRED READING

Review: Technique 8   Filtration, Section 8.3

Technique 11   Crystallization, Sections 11.4 and 11.10

New: Essay   Local Anesthetics

SPECIAL INSTRUCTIONS[1]

Sulfuric acid is corrosive. Do not allow it to touch your skin.

SUGGESTED WASTE DISPOSAL

Dispose of all filtrates into the container designated for nonhalogenated organic solvents.

PROCEDURE

Running the Reaction

Place 0.120 g of *p*-aminobenzoic acid and 1.20 mL of absolute ethanol into a 3-mL conical vial. Add a magnetic spin vane and stir the mixture until the solid dissolves completely. While stirring, add 0.10 mL of concentrated sulfuric acid dropwise. A large amount of precipitate forms when you add the sulfuric acid, but this solid slowly dissolves during the reflux that follows. Attach a water-cooled condenser and heat the mixture at a gentle boil for 60–75 minutes with an aluminum block at about 105°C. Stir the mixture during this heating period.

Precipitation of Benzocaine

At the end of the reaction time, remove the apparatus from the aluminum block and allow the reaction mixture to cool for several minutes. Using a Pasteur pipet, transfer the contents of the vial to a small beaker containing 3.0 mL of water. When the liquid has cooled to room temperature, add a 10% sodium carbonate solution (about 1 mL needed) dropwise to neutralize the mixture. Stir the contents of the beaker with a stirring rod or spatula. After each addition of the sodium carbonate solution, extensive gas evolution (frothing) will be perceptible until the mixture is nearly neutralized. As the pH increases, a white precipitate of benzocaine is produced. When gas no longer evolves as you add a drop of sodium carbonate, check the pH of the solution and add further portions of sodium carbonate until the pH is about 8.

Collect the benzocaine by vacuum filtration, using a Hirsch funnel. Use three 1-mL portions of water to aid in the transfer and to wash the product in the funnel. Be sure that the solid is rinsed thoroughly with the water. After the product has dried overnight, weigh it, calculate the percentage yield, and determine its melting point. The melting point of pure benzocaine is 92°C.

Recrystallization and Characterization of Benzocaine

Although the product should be fairly pure, it may be recrystallized by the mixed solvent method using methanol and water (Technique 11, Section 11.10, p. 667). Place the product in a Craig tube, add several drops of methanol, and, while heating the Craig tube in an aluminum block (60–70°C) and stirring the mixture with a

---

[1] *Note to the Instructor:* Benzocaine may be tested for its effect on a frog's leg muscle. See Instructor's Manual for instructions.

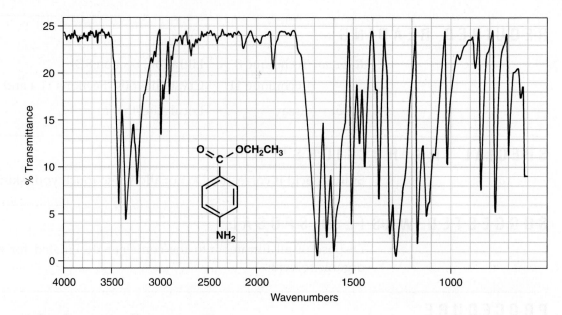

Infrared spectrum of benzocaine, KBr

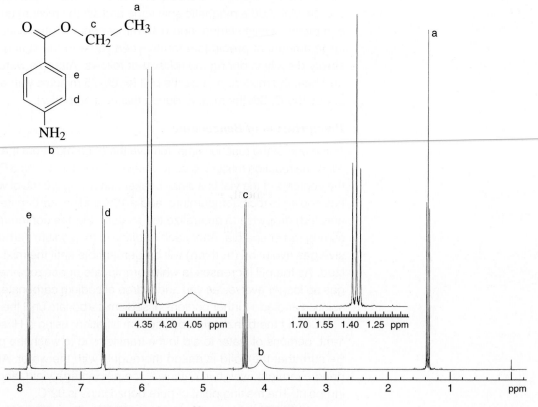

300-MHz proton NMR spectrum of benzocaine (CDCl₃).

microspatula, add methanol dropwise until all the solid dissolves. Add two to three additional drops of methanol and then add hot water dropwise until the mixture turns cloudy or a white precipitate forms. Add methanol again until the solid dissolves completely. Insert the inner plug of the Craig tube and allow the solution to cool slowly to room temperature. Complete the crystallization by cooling the mixture in an ice bath and collect the crystals by centrifugation (Technique 8, Section 8.8, p. 627). Weigh the purified benzocaine and determine its melting point.

At the option of the instructor, obtain the infrared spectrum using the dry film method (Technique 25, Section 25.4, p. 838) or as a KBr pellet (Technique 25, Section 25.5, p. 838) and the NMR spectrum in carbon tetrachloride or CDCl₃ (Technique 26, Section 26.1, p. 870). Submit the sample in a labeled vial to the instructor.

QUESTIONS

1. Interpret the infrared and NMR spectra of benzocaine.

2. What is the structure of the precipitate that forms after the sulfuric acid has been added?

3. When 10% sodium carbonate solution is added, a gas evolves. What is the gas? Give a balanced equation for this reaction.

4. Explain why benzocaine precipitates during the neutralization.

5. Refer to the structure of procaine in the table in the essay "Local Anesthetics." Using *p*-aminobenzoic acid, give equations showing how procaine and procaine monohydrochloride could be prepared. Which of the two possible amino functional groups in procaine will be protonated first? Defend your choice. (*Hint:* Consider resonance.)

---

46 · EXPERIMENT 46

Methyl Salicylate (Oil of Wintergreen)

Synthesis of an ester
Heating under reflux
Extraction
Vacuum distillation

In this experiment, you will prepare a familiar-smelling organic ester, oil of wintergreen. Methyl salicylate was first isolated in 1843 by extraction from the wintergreen plant (*Gaultheria*). It was soon found that this compound had analgesic and antipyretic character almost identical to that of salicylic acid (see the essay "Aspirin") when taken internally. This medicinal character probably derives from the ease with which methyl salicylate is hydrolyzed to salicylic acid under the alkaline conditions found in the intestinal tract. Salicylic acid is known to have analgesic and antipyretic properties. Methyl salicylate can be taken internally or absorbed through the skin; thus, it finds much use in liniment preparations. Applied to the skin, it produces a mild tingling or soothing sensation, which probably comes from the action of its phenolic hydroxyl group. This ester also has a pleasant odor, and it is used to a small extent as a flavoring principle.

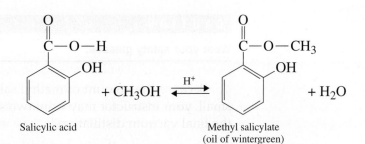

Salicylic acid   Methyl salicylate
                                  (oil of wintergreen)

Methyl salicylate will be prepared from salicylic acid, which is esterified at the carboxyl group with methanol. You should recall from your organic chemistry lecture course that esterification is an acid-catalyzed equilibrium reaction. The equilibrium does not lie far enough to the right to favor the formation of the ester in high yield. More product can be formed by increasing the concentrations of one of the reactants. In this experiment, a large excess of methanol will shift the equilibrium to favor a more complete formation of the ester.

This experiment also illustrates the use of distillation under reduced pressure for purifying high-boiling liquids. Distillation of high-boiling liquids at atmospheric pressure is often unsatisfactory. At the high temperatures required, the material being distilled (the ester, in this case) may partially or even completely decompose, causing loss of product and contamination of the distillate. When the total pressure inside the distillation apparatus is reduced, however, the boiling point of the substance is lowered. In this way, the substance can be distilled without being decomposed.

REQUIRED READING

Review: Techniques 5 and 6

Technique 13   Physical Constants of Liquids, Part A, Boiling Points and Thermometer Correction

New: Technique 16   Vacuum Distillation

Technique 25   Preparation of Samples for Spectroscopy

Essay   Esters—Flavors and Fragrances

SPECIAL INSTRUCTIONS

The experiment must be started at the beginning of the laboratory period because a long reflux time is needed to esterify salicylic acid and obtain a respectable yield. Perform a supplementary experiment during the reaction period or complete work that is pending from previous experiments. Enough time should remain at the end of the period to perform the extractions, place the product over the drying agent, assemble the apparatus, and perform the vacuum distillation.

CAUTION

Handle the concentrated sulfuric acid carefully; it can cause severe burns.

When a distillation is conducted under reduced pressure, it is important to guard against the dangers of an implosion. Inspect the glassware for flaws and cracks and replace any that is defective.

CAUTION

Wear your safety glasses.

Because the amount of methyl salicylate obtained in this experiment is small, your instructor may want two students to combine their products for the final vacuum distillation.

SUGGESTED WASTE DISPOSAL

The aqueous extracts from this experiment should be placed in the container designated for this purpose. Place any remaining methylene chloride in the container designated for halogenated waste.

PROCEDURE

Assemble equipment for reflux using a 5-mL conical vial and a water-cooled condenser (Fig. 7.6A, p. 601). Top the apparatus with a calcium chloride drying tube. Use a hot plate with an aluminum block. Place 0.65 g of salicylic acid, 2.0 mL of methanol ($d = 0.792$ g/mL), and a spin vane in the vial. Stir the mixture until the salicylic acid dissolves. Carefully add 0.75 mL of concentrated sulfuric acid, *in small portions,* to the mixture in the vial while stirring. A white precipitate may form, but it will redissolve during the reflux period. Complete assembly of the apparatus and, while stirring, gently boil the mixture (aluminum block 80°C) for 60–75 minutes.

After the mixture has cooled, extract it with three 1-mL portions of methylene chloride (Technique 12, Section 12.4, p. 674). Add the methylene chloride, cap the vial, shake it, and then loosen the cap. When the layers separate, transfer the lower layer with a filter-tip pipet to another container. After completing the three extractions, discard the aqueous layer and return the three methylene chloride extracts to the vial. Extract the methylene chloride layers with a 1-mL portion of 5% aqueous sodium bicarbonate. Transfer the lower organic layer to a clean, dry conical vial. Discard the aqueous layer. Dry the organic layer over anhydrous sodium sulfate (see Technique 12, Section 12.9, p. 680). When the solution is dry, transfer it to a clean, dry, 3-mL, conical vial with a filter-tip pipet. Evaporate the methylene chloride using a warm waterbath (40–50°C) in the hood. A stream of nitrogen or air will accelerate the evaporation (Fig. 7.17A, p. 612). The product may be stored in the capped vial and saved for the next period, or it may be distilled under vacuum during the same period.

Vacuum Distillation

Using the procedure described in Technique 16, Section 16.4, page 737, distill the product by vacuum distillation using an apparatus fitted with a Hickman still and a water-cooled condenser (Fig. 16.5, p. 736). Place a small piece of a stainless steel sponge in the lower stem of the Hickman still to prevent bumpover and stir vigorously with a magnetic spin vane. Use an aspirator for the vacuum source and attach a manometer if one is available (see Fig. 16.10, p. 742). You may use an aluminum block to heat the distillation mixture. The aluminum block temperature will be about 130°C (with 20 mm Hg vacuum). If you have less than 0.75 mL, you should combine your product with that of another student.

When the distillation is complete, transfer the distillate to a tared 3-mL conical vial with a Pasteur pipet and weigh it to determine the percentage yield. Determine a microscale boiling point (Technique 13, Section 13.2, pp. 695) for your product.

Spectroscopy

At your instructor's option, obtain an infrared spectrum using salt plates (Technique 25, Section 25.2, p. 834). Compare your spectrum with the one reproduced on page 376. Interpret the spectrum and include it in your report to the instructor. You may also be required to determine and interpret the proton and carbon-13 NMR spectra (Technique 26, Part A, p. 870, and Technique 27, p. 906). Submit your sample in a properly labeled vial with your report.

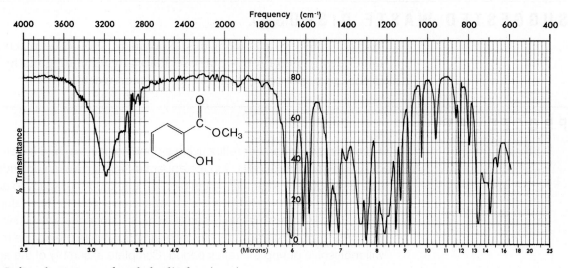

Infrared spectrum of methyl salicylate (neat).

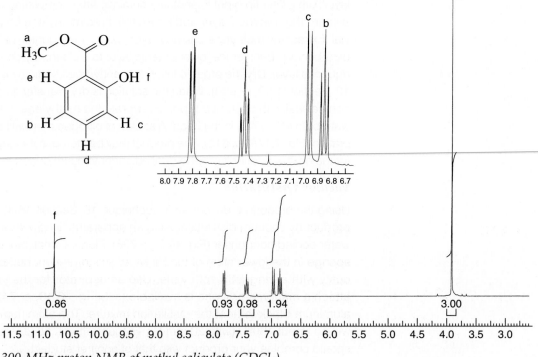

300-MHz proton NMR of methyl salicylate (CDCl₃).

QUESTIONS

1. Write a mechanism for the acid-catalyzed esterification of salicylic acid with methanol. You may need to consult the chapter on carboxylic acids in your lecture textbook.

2. What is the function of the sulfuric acid in this reaction? Is it consumed in the reaction?

3. In this experiment, excess methanol was used to shift the equilibrium toward the formation of more ester. Describe other methods for achieving the same result.

4. How are sulfuric acid and the excess methanol removed from the crude ester after the reaction has been completed?

5. Why was 5% $NaHCO_3$ used in the extraction? What would have happened if 5% NaOH had been used?

6. Interpret the principal absorption bands in the infrared spectrum of methyl salicylate. Also interpret the proton NMR spectrum shown on page 376.

ESSAY

Pheromones: Insect Attractants and Repellents

It is difficult for humans, who are accustomed to heavy reliance on visual and verbal forms of communication, to imagine that there are forms of life that depend primarily on the release and perception of *odors* to communicate with one another. Among insects, however, this is perhaps the chief form of communication. Many species of insects have developed a virtual "language" based on the exchange of odors. These insects have well-developed scent glands, often of several types, which have as their sole purpose the synthesis and release of chemical substances. When these chemical substances, known as **pheromones,** are secreted by insects and detected by other members of the same species, they induce a specific and characteristic response. Pheromones are usually of two distinct types: releaser pheromones and primer pheromones. **Releaser pheromones** produce an immediate *behavioral* response in the recipient insect; **primer pheromones** trigger a series of *physiological* changes in the recipient. Some pheromones, however, combine both releaser and primer effects.

Sex Attractants

Among the most important types of releaser pheromones are the sex attractants. **Sex attractants** are pheromones secreted by either the female or, less commonly, the male of the species to attract the opposite member for the purpose of mating. In large concentrations, sex pheromones also induce a physiological response in the recipient (for example, the changes necessary to the mating act) and thus have a primer effect and so are misnamed.

Anyone who has owned a female cat or dog knows that sex pheromones are not limited to insects. Female cats or dogs widely advertise, by odor, their sexual availability when they are "in heat." This type of pheromone is not uncommon to mammals. Some persons even believe that there are human pheromones responsible for attracting certain sensitive males and females to one another. This idea is, of course, responsible for many of the perfumes now widely available. Whether or not the idea is correct cannot yet be established, but there are proven sexual differences in the ability of humans to smell certain substances. For instance, Exaltolide, a synthetic lactone of 14-hydroxytetradecanoic acid, can be perceived only by females, or by males after they have been injected with an estrogen. Exaltolide is very similar in overall structure to civetone (civet cat) and muskone (musk deer), which are two naturally occurring compounds believed to be mammalian sex pheromones.

Whether or not humans use pheromones as a means of attracting the opposite sex has never been completely established, although it is an active area of research. Humans, like other animals, emit odors from many parts of their bodies. Body odor consists of secretions from several types of skin glands, most of which are concentrated in the underarm region of the body. Do these secretions contain substances that might act as human sex attractants?

Research has shown that a mother can correctly identify the odor of her newborn infant or older child by smelling clothing worn previously by the child and can distinguish the clothing from that worn by another child of the same age. Studies conducted over 30 years ago showed that the menstrual cycles of women who are roommates or close friends tend to converge over time. These and other similar investigations suggest that some forms of pheromone-like communication are possible in humans.

Recent studies have clearly identified a specialized structure, called the **vomeronasal organ,** in the nose. This organ appears to respond to a variety of chemical stimuli. In a recent article, researchers at the University of Chicago reported that when they wiped human body-odor secretions from one group of women under the noses of other women, the second group showed changes in their menstrual cycles. The cycles grew either longer or shorter, depending on where the donors were in their own menstrual cycles. The affected women claimed that they did not smell anything except the alcohol on the cotton pads. Alcohol alone had no effect on the women's menstrual cycles. The timing of ovulation for the female test subjects was affected in a similar manner. Although the nature of substances responsible for these effects has not yet been identified, clearly the potential for chemical communication regulating sexual function has been established in humans.

One of the first identified insect attractants belongs to the gypsy moth, *Lymantria dispar.* This moth is a common agricultural pest, and it was hoped that the sex attractant that females emitted could be used to lure and trap males. Such a method of insect control would be preferable to inundating large areas with DDT and would be species-specific. Nearly 50 years of work were expended in identifying the chemical substance responsible. Early in this period, researchers found that an extract from the tail sections of female gypsy moths would attract males, even from a great distance. In experiments with the isolated gypsy moth pheromone, it was found that the male gypsy moth has an almost unbelievable ability to detect extremely small amounts of the substance. He can detect it in concentrations lower than a few hundred *molecules* per cubic centimeter (about 10^{-19}–10^{-20} g/cc)! When a male moth encounters a small concentration of pheromone, he immediately heads into the wind and flies upward in search of higher concentrations and the female. In only a mild breeze, a continuously emitting female can activate a space 300 ft high, 700 ft wide, and almost 14,000 feet (nearly 3 miles) long!

In subsequent work, researchers isolated 20 mg of a pure chemical substance from solvent extracts of the two extreme tail segments collected from each of 500,000 female gypsy moths (about 0.1 μg/moth). This emphasizes that pheromones are effective in very minute amounts and that chemists must work with very small amounts to isolate them and prove their structures. It is not unusual to process thousands of insects to get even a minute sample of these substances. Extremely sophisticated analytical and instrumental methods, such as spectroscopy, must be used to determine the structure of a pheromone.

In spite of these techniques, the original researchers assigned an incorrect structure to the gypsy moth pheromone and proposed for it the name **gyplure.** Because of its great promise as a method of insect control, gyplure was soon synthesized. The synthetic material turned out to be totally inactive. After some controversy about why the synthetic material was incapable of luring male gypsy moths (see the References for the complete story), it was finally shown that the proposed structure for the pheromone (that is, the gyplure structure) was incorrect. The actual pheromone was found to be *cis*-7,8-epoxy-2-methyloctadecane, also named (7*R*,8*S*)-epoxy-2-methyloctadecane. This material was soon synthesized, found to be active, and given the name **disparlure.** In recent years, disparlure traps have been found to be a convenient and economical method for controlling the gypsy moth.

A similar story of mistaken identity can be related for the structure of the pheromone of the pink bollworm, *Pectinophora gossypiella.* The originally proposed structure was called **propylure.** Synthetic propylure turned out to be inactive. Subsequently, the pheromone was shown to be a mixture of two isomers of 7,11-hexadecadien-1-yl acetate, the *cis,cis* (7Z,11Z) isomer and the *cis,trans* (7Z,11E) isomer. It turned out to be quite easy to synthesize a 1:1 mixture of these two isomers, and the 1:1 mixture was named **gossyplure.** Curiously, adding as little as 10% of either of the other two possible isomers, *trans,cis* (7E,11Z) or *trans,trans* (7E,11E), to the 1:1 mixture greatly diminishes its activity, apparently masking it. Geometric isomerism can be important! The details of the gossyplure story can also be found in the References.

Both these stories have been partly repeated here to point out the difficulties of research on pheromones. The usual method is to propose a structure determined by work on *tiny* amounts of the natural material. The margin for error is great. Such proposals are usually not considered "proved" until synthetic material is shown to be as biologically effective as the natural pheromone.

Other Pheromones

The most important example of a primer pheromone is found in honeybees. A bee colony consists of one queen bee, several hundred male drones, and thousands of worker bees, or undeveloped females. It has recently been found that the queen, the only female that has achieved full development and reproductive capacity, secretes a primer pheromone called the **queen substance.** The worker females, while tending the queen bee, continuously ingest quantities of the queen substance. This pheromone, which is a mixture of compounds, prevents the workers from rearing any competitive queens and prevents the development of ovaries in all other females in the hive. The substance is also active as a sex attractant; it attracts drones to the queen during her "nuptial flight." The major component of queen substance is shown in the figure.

Honeybees also produce several other important types of pheromones. It has long been known that bees will swarm after an intruder. It has also been known that isopentyl acetate induces a similar behavior in bees. Isopentyl acetate (Experiment 13) is an **alarm pheromone.** When an angry

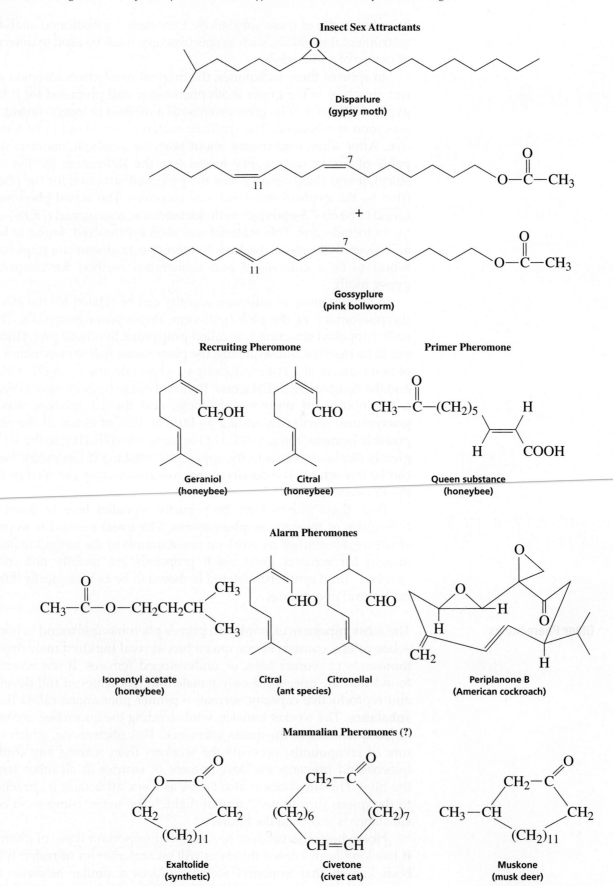

Insect Sex Attractants

Disparlure
(gypsy moth)

Gossyplure
(pink bollworm)

Recruiting Pheromone

Geraniol
(honeybee)

Citral
(honeybee)

Primer Pheromone

Queen substance
(honeybee)

Alarm Pheromones

Isopentyl acetate
(honeybee)

Citral   Citronellal
(ant species)

Periplanone B
(American cockroach)

Mammalian Pheromones (?)

Exaltolide
(synthetic)

Civetone
(civet cat)

Muskone
(musk deer)

worker bee stings an intruder, she discharges, along with the sting venom, a mixture of pheromones that incites the other bees to swarm on and attack the intruder. Isopentyl acetate is an important component of the alarm pheromone mixture. Alarm pheromones have also been identified in many other insects. In insects less aggressive than bees or ants, the alarm pheromone may take the form of a **repellent,** which induces the insects to go into hiding or leave the immediate vicinity.

Honeybees also release **recruiting** or **trail pheromones.** These pheromones attract others to a source of food. Honeybees secrete recruiting pheromones when they locate flowers in which large amounts of sugar syrup are available. Although the recruiting pheromone is a complex mixture, both geraniol and citral have been identified as components. In a similar fashion, when ants locate a source of food, they drag their tails along the ground on their way back to the nest, continuously secreting a trail pheromone. Other ants follow the trail to the source of food.

In some species of insects, **recognition pheromones** have been identified. In carpenter ants, a caste-specific secretion has been found in the mandibular glands of the males of five species. These secretions have several functions, one of which is to allow members of the same species to recognize one another. Insects not having the correct recognition odor are immediately attacked and expelled from the nest. In one species of carpenter ant, the recognition pheromone has been shown to have methyl anthranilate as an important component.

We do not yet know all the types of pheromones that any given species of insect may use, but it seems that as few as 10 or 12 pheromones could constitute a "language" that could adequately regulate the entire life cycle of a colony of social insects.

Insect Repellents

Currently, the most widely used **insect repellent** is the synthetic substance *N,N*-diethyl-*m*-toluamide (Experiment 47), also called Deet. It is effective against fleas, mosquitoes, chiggers, ticks, deerflies, sandflies, and biting gnats. A specific repellent is known for each of these types of insects, but none has the wide spectrum of activity that this repellent has. Exactly why these substances repel insects is not yet fully understood. The most extensive investigations have been carried out on the mosquito.

Originally, many investigators thought that repellents might simply be compounds that provided unpleasant or distasteful odors to a wide variety of insects. Others thought that they might be alarm pheromones for the species affected or that they might be the alarm pheromones of a hostile species. Early research with the mosquito indicates that at least for several varieties of mosquitoes, none of these is the correct answer.

Mosquitoes seem to have hairs on their antennae that are receptors enabling them to find a warm-blooded host. These receptors detect the convection currents arising from a warm and moist living animal. When a mosquito encounters a warm and moist convection current, the mosquito moves steadily forward. If it passes out of the current into dry air, it turns until it finds the current again. Eventually, it finds the host and lands. Repellents cause a mosquito to turn in flight and become confused. Even if it should land, it becomes confused and flies away again.

Researchers have found that the repellent prevents the moisture receptors of the mosquito from responding normally to the raised humidity of the subject. At least two sensors are involved, one responsive to carbon dioxide and the other responsive to water vapor. The carbon dioxide sensor is activated by the repellent, but if exposure to the chemical continues, adaptation occurs, and the sensor returns to its usual low output of signal. The moisture sensor, on the other hand, simply seems to be deadened, or turned off, by the repellent. Therefore, mosquitoes have great difficulty in finding and interpreting a host when they are in an environment saturated with repellent. They fly right through warm and humid convection currents as if the currents did not exist. Only time will tell if other biting insects respond likewise.

REFERENCES

Agosta, W. C. "Using Chemicals to Communicate." *Journal of Chemical Education, 71* (March 1994): 242.

Batra, S. W. T. "Polyester-Making Bees and Other Innovative Insect Chemists." *Journal of Chemical Education, 62* (February 1985): 121.

Katzenellenbogen, J. A. "Insect Pheromone Synthesis: New Methodology." *Science, 194* (October 8, 1976): 139.

Leonhardt, B. A. "Pheromones." *ChemTech, 15* (June 1985): 368.

Prestwick, G. D. "The Chemical Defenses of Termites." *Scientific American, 249* (August 1983): 78.

Silverstein, R. M. "Pheromones: Background and Potential Use for Insect Control." *Science, 213* (September 18, 1981): 1326.

Stine, W. R. "Pheromones: Chemical Communication by Insects." *Journal of Chemical Education, 63* (July 1986): 603.

Villemin, D. "Olefin Oxidation: A Synthesis of Queen Bee Pheromone." *Chemistry and Industry* (January 20, 1986): 69.

Wilson, E. O. "Pheromones." *Scientific American, 208* (May 1963): 100.

Winston, M. L., and Slessor, K. N. "The Essence of Royalty: Honey Bee Queen Pheromone." *American Scientist, 80* (July–August 1992): 374.

Wood, W. F. "Chemical Ecology: Chemical Communication in Nature." *Journal of Chemical Education, 60* (July 1983): 531.

Wright, R. H. "Why Mosquito Repellents Repel." *Scientific American, 233* (July 1975): 105.

Yu, H., Becker, H., and Mangold, H. K. "Preparation of Some Pheromone Bouquets." *Chemistry and Industry* (January 16, 1989): 39.

Gypsy Moth

Beroza, M., and Knipling, E. F. "Gypsy Moth Control with the Sex Attractant Pheromone." *Science, 177* (1972): 19.

Bierl, B. A., Beroza, M., and Collier, C. W. "Potent Sex Attractant of the Gypsy Moth: Its Isolation, Identification, and Synthesis." *Science, 170* (1970): 87.

Pink Bollworm

Anderson, R. J., and Henrick, C. A. "Preparation of the Pink Bollworm Sex Pheromone Mixture, Gossyplure." *Journal of the American Chemical Society, 97* (1975): 4327.

Hummel, H. E., Gaston, L. K., Shorey, H. H., Kaae, R. S., Byrne, K. J., and Silverstein, R. M. "Clarification of the Chemical Status of the Pink Bollworm Sex Pheromone." *Science, 181* (1973): 873.

American Cockroach

Adams, M. A., Nakanishi, K., Still, W. C., Arnold, E. V., Clardy, J., and Persoon, C. J. "Sex Pheromone of the American Cockroach: Absolute Configuration of Periplanone-B." *Journal of the American Chemical Society, 101* (1979): 2495.

Still, W. C. "(±)-Periplanone-B: Total Synthesis and Structure of the Sex Excitant Pheromone of the American Cockroach." *Journal of the American Chemical Society,* *101* (1979): 2493.

Stinson, S. C. "Scientists Synthesize Roach Sex Excitant." *Chemical and Engineering News,* *57* (April 30, 1979): 24.

Spider   Schulz, S., and Toft, S. "Identification of a Sex Pheromone from a Spider." *Science, 260* (June 11, 1993): 1635.

Silkworm   Emsley, J. "Sex and the Discerning Silkworm." *New Scientist, 135* (July 11, 1992): 18.

Aphids   Coghlan, A. "Aphids Fall for Siren Scent of Pheromones." *New Scientist, 127* (July 21, 1990): 32.

Snakes   Mason, R. T., Fales, H. M., Jones, T. H., Pannell, L. K., Chinn, J. W., and Crews, D. "Sex Pheromones in Snakes." *Science, 245* (July 21, 1989): 290.

Oriental Fruit Moth   Mithran, S., and Mamdapur, V. R. "A Facile Synthesis of the Oriental Fruit Moth Sex Pheromone." *Chemistry and Industry* (October 20, 1986): 711.

Human   Stern, K., and McClintock, M. K. "Regulation of Ovulation by Human Pheromones." *Nature, 392* (March 12, 1998): 177.

Weller, A. "Communication through Body Odour." *Nature, 392* (March 12, 1998): 126.

47   **EXPERIMENT 47**

N,N-*Diethyl-m-Toluamide:*
The Insect Repellent "OFF"

Preparation of an amide

Extraction

Column chromatography

In this experiment, you will synthesize the active ingredient of the insect repellent "OFF," *N,N*-diethyl-*m*-toluamide. This substance belongs to the class of compounds called **amides.** Amides have the generalized structure

$$\overset{\displaystyle O}{\underset{\displaystyle R-C-NH_2}{\|}}$$

The amide to be prepared in this experiment is a disubstituted amide. That is, two of the hydrogens on the amide —NH_2 group have been replaced with ethyl groups. Amides cannot be prepared directly by mixing a carboxylic acid with an amine. If an acid and an amine are mixed, an acid–base reaction occurs, giving the conjugate base of the acid, which will not react further while in solution:

$$RCOOH + R_2NH \longrightarrow [RCOO^-R_2NH_2^+]$$

However, if the amine salt is isolated as a crystalline solid and strongly heated, the amide can be prepared:

$$[RCOO^- R_2NH_2^+] \xrightarrow{\text{heat}} [RCONR_2 + H_2O]$$

Because of the high temperature required for this reaction, this is not a convenient laboratory method.

Amides are usually prepared via the acid chloride, as in this experiment. In Step 1, *m*-toluic acid is converted to its acid chloride derivative using thionyl chloride ($SOCl_2$).

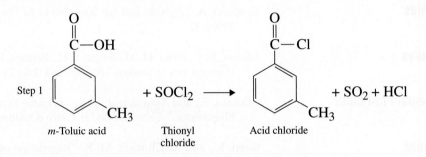

Step 1

m-Toluic acid   Thionyl chloride   Acid chloride

The acid chloride is not isolated or purified, and it is allowed to react directly with diethylamine in Step 2. An excess of diethylamine is used in this experiment to react with the hydrogen chloride produced in Step 2.

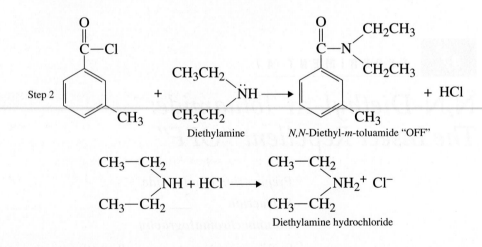

Step 2

Diethylamine   *N,N*-Diethyl-*m*-toluamide "OFF"

Diethylamine hydrochloride

REQUIRED READING

Review: Technique 7   Reaction Methods, Section 7.10

Technique 12   Extractions

Technique 19   Column Chromatography, Sections 19.6 and 19.8

New: Essay   Pheromones: Insect Attractants and Repellents

SPECIAL INSTRUCTIONS

All equipment used in this experiment should be dry because thionyl chloride reacts with water to liberate HCl and SO_2. Likewise, *anhydrous* ether should be used, because water reacts with both thionyl chloride and the intermediate acid chloride.

Thionyl chloride is a noxious and corrosive chemical and should be handled with care. If it is spilled on the skin, serious burns will result. Thionyl chloride and diethylamine must be dispensed *in the hood* from bottles that should be kept tightly closed when not in use. Diethylamine is also noxious and corrosive. In addition, it is quite volatile (bp 56°C) and must be cooled in a hood prior to use.

SUGGESTED WASTE DISPOSAL

All aqueous extracts should be poured into the waste bottle designated for aqueous waste.

PROCEDURE

Apparatus Assembly

Assemble the apparatus as shown in the figure, except for the syringe. The drying tube is packed with glass wool, and a few drops of water are added to the drying tube. Excess water should be avoided so that water does not get into the flask. The moistened glass wool traps the hydrogen chloride and sulfur dioxide that are evolved in the reaction. You can save time by setting the dial on your hot plate to give an aluminum block temperature of about 90°C prior to measuring reagents.

Preparation of the Acid Chloride

Place 0.272 g of *m*-toluic acid (3-methylbenzoic acid, *MW* = 136.1) into the dry 10-mL round-bottom flask. In a hood, transfer 0.30 mL of thionyl chloride (*MW* = 118.9, *d* = 1.64 g/mL) into the flask with the dry graduated pipet provided.

> **CAUTION** ⚠️
>
> **The thionyl chloride is kept in a hood. Do not breathe the vapors of this noxious and corrosive chemical. Use dry equipment when handling this material because it reacts violently with water. Do not get it on your skin. Once the drying tube containing the moistened glass wool has been attached, the apparatus may be taken to your desk.**

Add a magnetic spin bar, start the circulation of water in the reflux condenser, and heat the mixture while stirring in an aluminum block at about 90°C. Boil the mixture gently for 15 minutes.

Preparation of the Amide

Raise the apparatus from the aluminum block and let the flask cool to *room temperature*. Remove the aluminum block from the hot plate. Turn off the heater and allow the unit to cool. You may need to place an insulating pad between the flask and the stirring unit to avoid heating the flask with residual heat from the hot plate. The next part of this reaction sequence is conducted at room temperature.

Inject 4.0 mL of *anhydrous* ether into the reaction flask using a syringe and stir the mixture at room temperature until a homogeneous solution is obtained. In a hood, place 0.66 mL of ice-cold diethylamine (*MW* = 73.1, *d* = 0.71 g/mL) in a small conical vial and dissolve it in 1.33 mL of *anhydrous* ether. Draw this solution into the syringe and insert the needle through the rubber septum of your apparatus. While stirring the mixture, add the solution of diethylamine and ether *dropwise* over

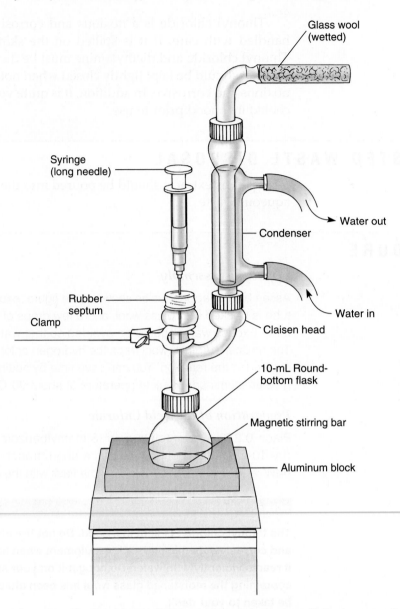

Apparatus for Experiment 47. **Note:** *A long syringe needle is recommended.*

a 10- to 15-minute period to the round-bottom flask. As the solution is added, a voluminous white cloud of diethylamine hydrochloride will form in the flask.

After adding the diethylamine, stir the mixture for 10 minutes at room temperature. After this time, inject 2 mL of a 10% aqueous sodium hydroxide solution into the conical vial and stir the mixture for 15 minutes. During this time, the sodium hydroxide converts any remaining acid chloride to the sodium salt of *m*-toluic acid. This salt is soluble in the aqueous layer. Diethylamine hydrochloride is also water soluble. Any remaining thionyl chloride is destroyed by water. The desired amide is soluble in ether.

Extraction of Product

Remove the drying tube (gas trap), the condenser, and the Claisen head. Using a Pasteur pipet, transfer all the liquid to a centrifuge tube with a cap. After the two layers separate, draw out the lower aqueous layer with a Pasteur pipet so that the desired ether layer remains in the centrifuge tube. Discard the aqueous layer.

Add another 2-mL portion of 10% sodium hydroxide to the remaining ether layer and cap the tube. Shake the mixture occasionally over a period of 5 minutes, allow the layers to separate, and again remove the lower aqueous layer. Discard the aqueous layer. Add additional ether to replace solvent lost by evaporation during the extractions.

Now extract the ether layer with a 2-mL portion of 10% hydrochloric acid to remove any remaining diethylamine as its hydrochloride salt. Finally, wash the ether layer with a 2-mL portion of water. Each time, shake the mixture vigorously, allow time for the phases to separate, and remove the lower aqueous layer with a pipet. Discard all the aqueous phases and keep the ether layer.

Transfer the ether layer containing the amide product with a dry Pasteur pipet to a dry conical vial and dry the ether phase over granular anhydrous sodium sulfate (see Technique 12, Section 12.9, p. 680). Remove the solution from the drying agent with a dry Pasteur pipet, and transfer it to another dry vial. A small amount of additional ether may be used to aid in a complete transfer. Place the vial in a warm waterbath (about 50°C) and evaporate the ether using a stream of air or nitrogen under a hood to give the crude dark-brown amide, which is a liquid (Technique 7, Section 7.10, p. 611). Column chromatography is used to remove much of the dark color from the product.

Column Chromatography

Preweigh a 5-mL conical vial for use in collecting the material eluted from the column. Prepare a column for column chromatography using a 5¾-inch Pasteur pipet as a column (Technique 19, Section 19.6B, p. 765). Place a small piece of cotton in the pipet and gently push it into the constriction. Pour 1.0 g of alumina[1] into the column. Obtain 4 mL of hexane in a graduated cylinder. The hexane will be used to prepare the column, dissolve the crude product, and elute the purified product as described in the next paragraph.

Dissolve the crude product in 10 drops of hexane. Clamp the column above the preweighed 5-mL conical vial. Then add about 1 mL of the hexane to the column and let it percolate through the alumina. Allow the solvent to drain until the solvent level just begins to enter the alumina. Add the crude product to the top of the column and allow the mixture to pass onto the column. Use about 0.5 mL of hexane to rinse the vial that contained the crude product. When the first batch of crude product has drained so that the surface of the liquid just begins to enter the top of the alumina, place the hexane rinse on the column.

When the solvent level has again reached the top of the alumina, add more hexane with a Pasteur pipet to elute the product into the conical vial. You should place 2 mL of hexane, in portions, on the column to elute the product. Collect all the liquid that passes through the column as one fraction (yellow material). Place the conical vial in a warm waterbath (about 50°C) and evaporate the hexane with a light stream of air or nitrogen in a hood to give the *N,N*-diethyl-*m*-toluamide as a light tan liquid. If necessary, use a few drops of hexane to rinse the product from the side of the vial into the bottom. Evaporate this solvent.

Analysis of Product

Reweigh the vial to determine the weight of product. Calculate the percentage yield (*MW* = 193.1). Determine the infrared spectrum of your product. Submit the remaining sample to the instructor. The infrared spectrum can be compared to the one reproduced as follows.

---

[1] EM Science (No. AX0612-1). The particle sizes are 80-200 mesh, and the material is Type F-20.

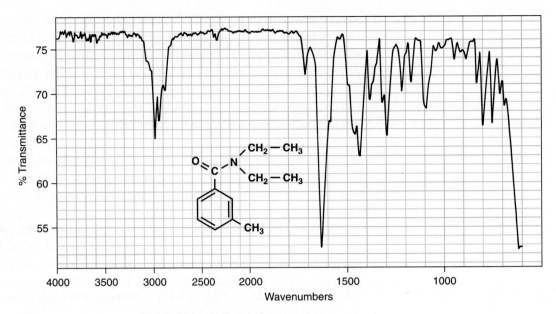

Infrared spectrum of N,N-*diethyl-m-toluamide, neat.*

REFERENCE

Wang, B. J-S. "An Interesting and Successful Organic Experiment." *Journal of Chemical Education,* 51 (October 1974): 631. (The synthesis of *N,N*-diethyl-*m*-toluamide.)

QUESTIONS

1. Write an equation that describes the reaction of thionyl chloride with water.

2. What reaction would take place if the acid chloride of *m*-toluic acid were mixed with water?

3. Why is the reaction mixture extracted with 10% aqueous sodium hydroxide? Write an equation.

4. Write a mechanism for each step in the preparation of *N,N*-diethyl-*m*-toluamide.

5. Interpret each of the principal peaks in the infrared spectrum of *N,N*-diethyl-*m*-toluamide.

6. A student determined the infrared spectrum of the product and found an absorption at 1785 cm^{-1}. The rest of the spectrum resembled the one given in this experiment. Assign this peak and provide an explanation for this unexpected result. (See Technique 25, Section 25.14, page 853).

ESSAY

Sulfa Drugs

The history of chemotherapy extends as far back as 1909, when Paul Ehrlich first used the term. Although Ehrlich's original definition of chemotherapy was limited, he is recognized as one of the giants of medicinal chemistry. **Chemotherapy** might be defined as "the treatment of disease by chemical reagents." It is preferable that these chemical reagents exhibit a toxicity toward only the pathogenic organism and not toward both the organism and the host. A chemotherapeutic agent would not be useful if it poisoned the patient at the same time that it cured the patient's disease!

In 1932, the German dye manufacturing firm I. G. Farbenindustrie patented a new drug, Prontosil. Prontosil is a red azo dye, and it was first prepared for its dye properties. Remarkably, it was discovered that Prontosil showed antibacterial action when it was used to dye wool. This discovery led to studies of Prontosil as a drug capable of inhibiting the growth of bacteria. The following year, Prontosil was successfully used against staphylococcal septicemia, a blood infection. In 1935, Gerhard Domagk published the results of his research, which indicated that Prontosil was capable of curing streptococcal infections of mice and rabbits. Prontosil was shown to be active against a wide variety of bacteria in later work. This important discovery, which paved the way for a tremendous amount of research on the chemotherapy of bacterial infections, earned for Domagk the 1939 Nobel Prize in medicine, but an order from Hitler prevented Domagk from accepting the honor.

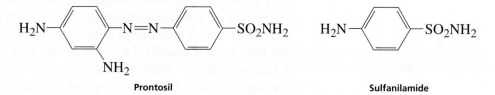

Prontosil

Sulfanilamide

Prontosil is an effective antibacterial substance **in vivo,** that is, when injected into a living animal. Prontosil is not medicinally active when the drug is tested **in vitro,** that is, on a bacterial culture grown in the laboratory. In 1935, the research group at the Pasteur Institute in Paris headed by J. Tréfouël learned that Prontosil is metabolized in animals to **sulfanilamide.** Sulfanilamide had been known since 1908. Experiments with sulfanilamide showed that it had the same action as Prontosil in vivo and that it was also active in vitro, where Prontosil was known to be inactive. It was concluded that the active portion of the Prontosil molecule was the sulfanilamide moiety. This discovery led to an explosion of interest in sulfonamide derivatives. Well over a thousand sulfonamide substances were prepared within a few years of these discoveries.

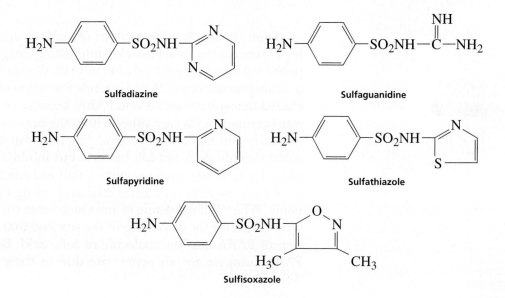

Sulfadiazine

Sulfaguanidine

Sulfapyridine

Sulfathiazole

Sulfisoxazole

Although many sulfonamide compounds were prepared, only a relative few showed useful antibacterial properties. As the first useful antibacterial drugs, these few medicinally active sulfonamides, or **sulfa drugs,** became the wonder drugs of their day. An antibacterial drug may be either **bacteriostatic** or **bactericidal.** A bacteriostatic drug suppresses the growth of bacteria; a bactericidal drug kills bacteria. Strictly speaking, the sulfa drugs are bacteriostatic. The structures of some of the most common sulfa drugs are shown here. These more complex sulfa drugs have various important applications. Although they do not have the simple structure characteristic of sulfanilamide, they tend to be less toxic than the simpler compound.

Sulfa drugs began to lose their importance as generalized antibacterial agents when production of antibiotics in large quantity began. In 1929, Sir Alexander Fleming made his famous discovery of **penicillin.** In 1941, penicillin was first used successfully on humans. Since that time, the study of antibiotics has spread to molecules that bear little or no structural similarity to the sulfonamides. Besides penicillin derivatives, antibiotics that are derivatives of **tetracycline,** including Aureomycin and Terramycin, were also discovered. These newer antibiotics have high activity against bacteria, and they do not usually have the severe unpleasant side effects of many of the sulfa drugs. Nevertheless, the sulfa drugs are still widely used in treating malaria, tuberculosis, leprosy, meningitis, pneumonia, scarlet fever, plague, respiratory infections, and infections of the intestinal and urinary tracts.

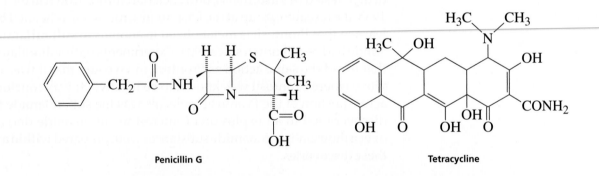

Penicillin G   Tetracycline

Even though the importance of sulfa drugs has declined, studies of how these materials act provide interesting insights into how chemotherapeutic substances might behave. In 1940, Woods and Fildes discovered that *p*-aminobenzoic acid (PABA) inhibits the action of sulfanilamide. They concluded that sulfanilamide and PABA, because of their structural similarity, must compete with each other within the organism, even though they cannot carry out the same chemical function. Further studies indicated that sulfanilamide does not kill bacteria but inhibits their growth. In order to grow, bacteria require an enzyme-catalyzed reaction that uses **folic acid** as a cofactor. Bacteria synthesize folic acid, using PABA as one of the components. When sulfanilamide is introduced into the bacterial cell, it competes with PABA for the active site of the enzyme that carries out the incorporation of PABA into the molecule of folic acid. Because sulfanilamide and PABA compete for an active site due to their structural similarity and because sulfanilamide cannot carry out the chemical transformations characteristic of PABA once it has formed a complex with the enzyme, sulfanilamide is called a **competitive inhibitor** of the enzyme. The enzyme, once it has formed a complex with sulfanilamide, is incapable of catalyzing the reaction required for the synthesis of folic acid. Without folic acid, the bacteria cannot synthesize the nucleic acids required for growth. As a result, bacterial growth is arrested until the body's immune system can respond and kill the bacteria.

One might well ask the question, Why, when someone takes sulfanilamide as a drug, doesn't it inhibit the growth of *all* cells, bacterial and human alike? The answer is simple. Animal cells cannot synthesize folic acid. Folic acid must be a part of the diet of animals and is therefore an essential vitamin. Because animal cells receive their fully synthesized folic acid molecules through the diet, only the bacterial cells are affected by the sulfanilamide, and only their growth is inhibited.

For most drugs, a detailed picture of their mechanism of action is unavailable. The sulfa drugs, however, provide a rare example from which we can theorize how other therapeutic agents carry out their medicinal activity.

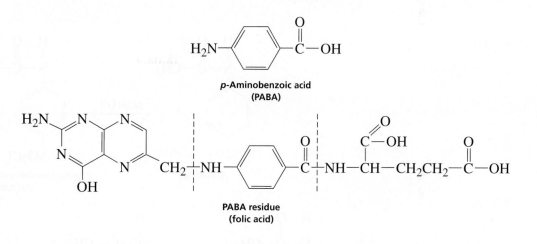

p-Aminobenzoic acid
(PABA)

PABA residue
(folic acid)

REFERENCES

Amundsen, L. H. "Sulfanilamide and Related Chemotherapeutic Agents." *Journal of Chemical Education, 19* (1942): 167.

Evans, R. M. *The Chemistry of Antibiotics Used in Medicine.* London: Pergamon Press, 1965.

Fieser, L. F., and Fieser, M. "Chemotherapy." Chap. 7 in *Topics in Organic Chemistry.* New York: Reinhold, 1963.

Garrod, L. P., and O'Grady, F. *Antibiotic and Chemotherapy.* Edinburgh: E. and S. Livingstone, Ltd., 1968.

Mandell, G. L., and Sande, M. A. "The Sulfonamides." Chap. 45 in L. S. Goodman and A. Gilman, eds., *The Pharmacological Basis of Therapeutics,* 8th ed. New York: Pergamon Press, 1990.

Sementsov, A. "The Medical Heritage from Dyes." *Chemistry, 39* (November 1966): 20.

Zahner, H., and Maas, W. K. *Biology of Antibiotics.* Berlin: Springer-Verlag, 1972.

Sulfa Drugs: Preparation of Sulfanilamide

Crystallization

Protecting groups

Testing the action of drugs on bacteria

Preparation of a sulfonamide

Aromatic substitution

In this experiment, you will prepare the sulfa drug sulfanilamide by the following synthetic scheme. The synthesis involves converting acetanilide to the intermediate *p*-acetamidobenzenesulfonyl chloride in Step 1. This intermediate is converted to sulfanilamide by way of *p*-acetamidobenzenesulfonamide in Step 2.

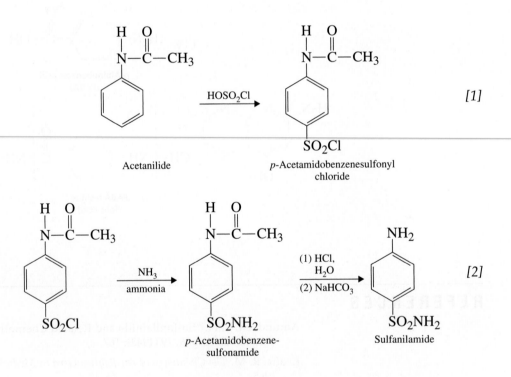

Acetanilide, which can easily be prepared from aniline, is allowed to react with chlorosulfonic acid to yield *p*-acetamidobenzenesulfonyl chloride. The acetamido group directs substitution almost totally to the *para* position. The reaction is an example of an electrophilic aromatic substitution reaction. Two problems would result if aniline itself were used in the reaction. First, the amino group in aniline would be protonated in strong acid to become a *meta* director; and, second, the chlorosulfonic acid would react with the amino group rather than with the ring, to give C_6H_5—$NHSO_3H$. For these reasons, the amino group has been "protected" by acetylation.

The acetyl group will be removed in the final step, after it is no longer needed, to regenerate the free amino group present in sulfanilamide.

p-Acetamidobenzenesulfonyl chloride is isolated by adding the reaction mixture to ice water, which decomposes the excess chlorosulfonic acid. This intermediate is fairly stable in water; nevertheless, it is converted slowly to the corresponding sulfonic acid (Ar—SO₃H). Thus, it should be isolated as soon as possible from the aqueous medium by filtration.

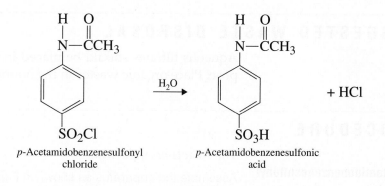

p-Acetamidobenzenesulfonyl   *p*-Acetamidobenzenesulfonic
           chloride   acid

The intermediate sulfonyl chloride is converted to *p*-acetamidobenzene-sulfonamide by a reaction with aqueous ammonia (Step 2). Excess ammonia neutralizes the hydrogen chloride produced. The only side reaction is the hydrolysis of the sulfonyl chloride to *p*-acetamidobenzenesulfonic acid.

The protecting acetyl group is removed by acid-catalyzed hydrolysis to generate the hydrochloride salt of the product, sulfanilamide. Note that of the two amide linkages present, only the carboxylic acid amide (acetamido group) was cleaved, not the sulfonic acid amide (sulfonamide). The salt of the sulfa drug is converted to sulfanilamide when the base, sodium bicarbonate, is added.

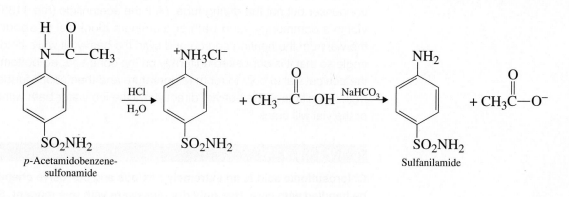

p-Acetamidobenzene-   Sulfanilamide
    sulfonamide

REQUIRED READING

Review: Technique 7   Reaction Methods

Technique 8   Filtration, Sections 8.3 and 8.7

Technique 11   Crystallization, Section 11.4

Technique 25   Infrared Spectroscopy, Sections 25.4 and 25.5

New: Essay   Sulfa Drugs

SPECIAL INSTRUCTIONS

Chlorosulfonic acid must be handled with care because it is a corrosive liquid and reacts violently with water. The *p*-acetamidobenzenesulfonyl chloride should be used during the same laboratory period in which it is prepared. It is unstable and will not survive long storage. The sulfa drug may be tested on several kinds of bacteria (Instructor's Manual).

SUGGESTED WASTE DISPOSAL

Aqueous filtrates should be placed in the container provided for this purpose. Place organic wastes in the nonhalogenated waste container.

PROCEDURE

Part A.
p-Acetamidobenzenesulfonyl Chloride

The Reaction Apparatus

Assemble the apparatus as shown in Figure 7.6A (inset) on page 601 using dry glassware. You will need a 5-mL conical vial, an air condenser, and a drying tube, which will be used as a gas trap. Prepare the drying tube for use as a gas trap by packing the tube loosely with dry glass wool (Technique 7, Section 7.8A, p. 607). Moisten the glass wool slightly with several drops of water. The moistened glass wool traps the hydrogen chloride that is evolved in the reaction. Attach the 5-mL conical vial after the acetanilide and chlorosulfonic acid have been added, as directed in the following paragraph. You should adjust the temperature of the aluminum block to about 110°C for use later in the experiment.

Reaction of Acetanilide with Chlorosulfonic Acid

Place 0.18 g of acetanilide in the dry 5-mL conical vial and connect the air condenser but not the drying tube. Melt the acetanilide (mp 113°C) by heating the vial in a community sand bath or aluminum block set to about 160°C. Remove the vial from the heating source and swirl the heavy oil while holding the vial at an angle so that it is deposited uniformly on the cone-shaped bottom of the vial. Allow the conical vial to cool to room temperature and then cool it further in an ice-water bath. (Don't place the hot vial directly into the ice-water bath without prior cooling, or the vial will crack.)

> **CAUTION**
>
> **Chlorosulfonic acid is an extremely noxious and corrosive chemical and should be handled with care. Use only dry glassware with this reagent. Should the chlorosulfonic acid be spilled on your skin, wash it off immediately with water. Be very careful when washing any glassware that has come in contact with chlorosulfonic acid. Even a small amount of the acid will react vigorously with water and may splatter. Wear safety glasses.**

Remove the air condenser. In a hood, transfer 0.50 mL of chlorosulfonic acid $ClSO_2OH$ ($MW = 116.5$, $d = 1.77$ g/mL) to the acetanilide in the conical vial using the graduated pipet provided. Reattach the air condenser and drying tube. Allow the mixture to stand for 5 minutes and then heat the reaction vial in the aluminum block at about 110°C for 10 minutes to complete the reaction. Remove the vial from the aluminum block. Allow the vial to cool to the touch and then cool it in an ice-water bath.

Isolation of p-Acetamidobenzenesulfonyl Chloride

The operations described in this paragraph should be conducted as rapidly as possible because the *p*-acetamidobenzenesulfonyl chloride reacts with water. Add 3 g of crushed ice to a 20-mL beaker. In a hood, transfer the cooled reaction mixture dropwise (it may splatter somewhat) with a Pasteur pipet onto the ice while stirring the mixture with a glass stirring rod. (The remaining operations in this paragraph may be completed at your laboratory bench.) Rinse the conical vial with a few drops of cold water and transfer the contents to the beaker containing the ice. Stir the precipitate to break up the lumps and then filter the *p*-acetamidobenzenesulfonyl chloride on a Hirsch funnel (Technique 8, Section 8.3, p. 621, and Fig. 8.5, p. 622). Rinse the conical vial and beaker with two 1-mL portions of ice water. Use the rinse water to wash the crude product on the funnel. Any remaining solid in the conical vial should be left there because this vial is used again in the next section. Do not stop here. Convert the solid into *p*-acetamidobenzenesulfonamide in the same laboratory period.

Part B. Sulfanilamide

Preparation of p-Acetamidobenzenesulfonamide

Prepare a hot waterbath at 70°C. Place the crude *p*-acetamidobenzenesulfonyl chloride into the original 5-mL conical vial and add 1.1 mL of dilute ammonium hydroxide solution.[1] Stir the mixture well with a spatula and reattach the air condenser and drying tube (gas trap) using fresh, moistened glass wool. Heat the mixture in the hot waterbath for 10 minutes. Allow the conical vial to cool to the touch and place it in an ice-water bath for several minutes. Collect the *p*-acetamidobenzenesulfonamide on a Hirsch funnel and rinse the vial and product with a small amount of ice water. You may stop here.

Hydrolysis of p-Acetamidobenzenesulfonamide

Transfer the solid into the conical vial and add 0.53 mL of dilute hydrochloric acid solution.[2] Attach the air condenser and heat the mixture in an aluminum block at about 130°C until all the solid has dissolved. Then heat the solution for an additional 5 minutes. Allow the mixture to cool to room temperature. If a solid (unreacted starting material) appears, heat the mixture for several minutes at 130°C. When the vial has cooled to room temperature, no further solids should appear.

Isolation of Sulfanilamide

With a Pasteur pipet, transfer the solution to a 20-mL beaker. While stirring with a glass rod, cautiously add dropwise a slurry of 0.5 g of sodium bicarbonate in about 1 mL of water to the mixture in the beaker. Foaming will occur after each addition of the bicarbonate solution because of carbon dioxide evolution. Allow gas evolution to cease before making the next addition. Eventually, sulfanilamide will begin to precipitate. At this point, begin to check the pH of the solution. Add the aqueous sodium bicarbonate until the pH of the solution is between 4 and 6. Cool the mixture thoroughly in an ice-water bath. Collect the sulfanilamide on a Hirsch funnel and rinse the beaker and solid with about 0.5 mL of cold water. Allow the solid to air dry on the Hirsch funnel for several minutes using suction.

---

[1] Prepared by mixing 11.0 mL of concentrated ammonium hydroxide with 11.0 mL of water.

[2] Prepared by mixing 7.0 mL of water with 3.6 mL of concentrated hydrochloric acid.

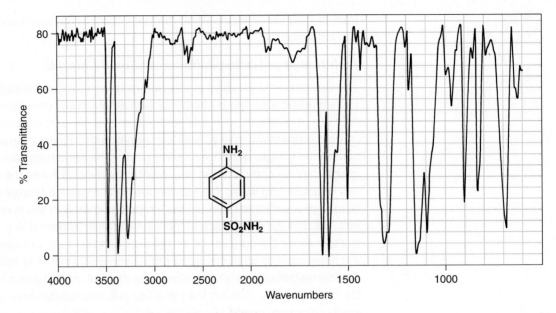

Infrared spectrum of sulfanilamide, KBr.

Crystallization of Sulfanilamide

Weigh the crude product and crystallize it from hot water (use 1.0 to 1.2 mL water/ 0.1 g) using a Craig tube (Technique 11, Section 11.4, p. 656, and Fig. 11.6, p. 657). Step 2 in Figure 11.6 (removal of insoluble impurities) should not be required in this crystallization. Let the purified product dry until the next laboratory period.

Yield Calculation, Melting Point, and Infrared Spectrum

Weigh the dry sulfanilamide and calculate the percentage yield ($MW = 172.2$). Determine the melting point (pure sulfanilamide melts at 163–164°C). At the option of the instructor, obtain the infrared spectrum using the dry film method (Technique 25, Section 25.4, p. 838) or as a KBr pellet (Technique 25, Section 25.5, p. 838). Compare your infrared spectrum with the one reproduced here. Submit the sulfanilamide to the instructor in a labeled vial or save it for the tests with bacteria (see Instructor's Manual).

QUESTIONS

1. Write an equation showing how excess chlorosulfonic acid is decomposed in water.
2. In the preparation of sulfanilamide, why was aqueous sodium bicarbonate, rather than aqueous sodium hydroxide, used to neutralize the solution in the final step?
3. At first glance, it might seem possible to prepare sulfanilamide from sulfanilic acid by the set of reactions shown here.

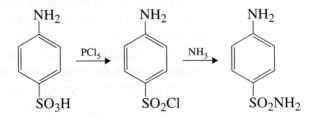

When the reaction is conducted in this way, however, a polymeric product is produced after Step 1. What is the structure of the polymer? Why does *p*-acetamidobenzenesulfonyl chloride not produce a polymer?

Polymers and Plastics

Chemically, plastics are composed of chainlike molecules of high molecular weight called **polymers.** Polymers have been built up from simpler chemicals called **monomers.** The word *poly* is defined as "many," *mono* means "one," and *mer* indicates "units." Thus, many monomers are combined to give a polymer. A different monomer or combination of monomers is used to manufacture each type or family of polymers. There are two broad classes of polymers: addition and condensation. Both types are described here.

Many polymers (plastics) produced in the past were of such low quality that they gained a bad reputation. The plastics industry now produces high-quality materials that are increasingly replacing metals in many applications. They are used in many products such as clothes, toys, furniture, machine components, paints, boats, automobile parts, and even artificial organs. In the automobile industry, metals have been replaced with plastics to help reduce the overall weight of the car and to help reduce corrosion. This reduction in weight helps improve gas mileage. Epoxy resins can even replace metal in engine parts.

Chemical Structures of Polymers

Basically, a polymer is made up of many repeating molecular units formed by sequential addition of monomer molecules to one another. Many monomer molecules of A, say 1000 to 1 million, can be linked to form a gigantic polymeric molecule:

Many A ⟶ etc. **—A-A-A-A-A—**etc.   or $\left(\!\!+\text{A}\!\!+\!\right)_n$

Monomer molecules   Polymer molecule

Monomers that are different can also be linked to form a polymer with an alternating structure. This type of polymer is called a **copolymer.**

Many A + many B ⟶ etc. **—A-B-A-B-A-B—**etc.   or $\left(\!\!+\text{A-B}\!\!+\!\right)_n$

Monomer molecules   Polymer molecule

Types of Polymers

For convenience, chemists classify polymers in several main groups, depending on the method of synthesis.

1. **Addition polymers** are formed by a reaction in which monomer units simply add to one another to form a long-chain (generally linear or branched) polymer. The monomers usually contain carbon–carbon double bonds. Examples of synthetic addition polymers include polystyrene (Styrofoam), polytetrafluoroethylene (Teflon), polyethylene, polypropylene, polyacrylonitrile (Orlon, Acrilan, Creslan), poly(vinyl chloride)

(PVC), and poly(methyl methacrylate) (Lucite, Plexiglas). The process can be represented as follows:

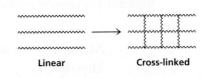

2. **Condensation polymers** are formed by the reaction of bifunctional or polyfunctional molecules, with the elimination of some small molecule (such as water, ammonia, or hydrogen chloride) as a by-product. Familiar examples of synthetic condensation polymers include polyesters (Dacron, Mylar), polyamides (nylon), polyurethanes, and epoxy resin. Natural condensation polymers include polyamino acids (protein), cellulose, and starch. The process can be represented as follows:

$$\text{H}-\boxed{}-\text{X} + \text{H}-\boxed{}-\text{X} \longrightarrow \text{H}-\boxed{}-\boxed{}-\text{X} + \text{HX}$$

3. **Cross-linked polymers** are formed when long chains are linked in one gigantic, three-dimensional structure with tremendous rigidity. Addition and condensation polymers can exist with a cross-linked network, depending on the monomers used in the synthesis. Familiar examples of cross-linked polymers are Bakelite, rubber, and casting (boat) resin. The process can be represented as follows:

Linear   Cross-linked

Linear and crossed-linked polymers.

Thermal Classification of Polymers

Industrialists and technologists often classify polymers as either thermoplastics or thermoset plastics rather than as addition or condensation polymers. This classification takes into account their thermal properties.

1. **Thermal properties of thermoplastics.** Most addition polymers and many condensation polymers can be softened (melted) by heat and reformed (molded) into other shapes. Industrialists and technologists often refer to these types of polymers as **thermoplastics.** Weaker, noncovalent bonds (dipole–dipole and London dispersion) are broken during the heating. Technically, thermoplastics are the materials we call plastics. Thermoplastics may be repeatedly melted and recast into new shapes. They may be recycled as long as degradation does not occur during reprocessing.

Some addition polymers, such as poly(vinyl chloride), are difficult to melt and process. Liquids with high boiling points, such as dibutyl phthalate, are added to the polymer to separate the chains from each other. These compounds are called **plasticizers.** In effect, they act as lubricants that neutralize the attractions that exist between chains. As a result, the polymer can be melted at a lower temperature to aid in processing. In addition, the polymer becomes more flexible at room temperature. By varying the amount of plasticizer, poly(vinyl chloride) can range from a very flexible, rubberlike material to a very hard substance.

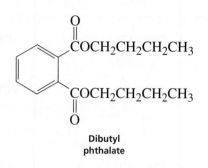

Dibutyl phthalate

Phthalate plasticizers are volatile compounds of low molecular weight. Part of the new car smell comes from the odor of these materials as they evaporate from the vinyl upholstery. The vapor often condenses on the windshield as an oily film. After some time, the vinyl material may lose enough plasticizer to cause it to crack.

2. **Thermal properties of thermoset plastics.** Industrialists use the term **thermoset** plastics to describe materials that melt initially but on further heating become permanently hardened. Once formed, thermoset materials cannot be softened and remolded without destruction of the polymer, because covalent bonds are broken. Thermoset plastics cannot be recycled. Chemically, thermoset plastics are cross-linked polymers. They are formed when long chains are linked in one gigantic, three-dimensional structure with tremendous rigidity.

Polymers can also be classified in other ways; for example, many varieties of rubber are often referred to as *elastomers*, Dacron is a fiber, and poly(vinyl acetate) is an adhesive. The addition and condensation classifications are used in this essay.

Addition Polymers

By volume, most of the polymers prepared in industry are of the addition type. The monomers generally contain a carbon–carbon double bond. The most important example of an addition polymer is the well-known polyethylene, for which the monomer is ethylene. Countless numbers (n) of ethylene molecules are linked in long-chain polymeric molecules by breaking the pi bond and creating two new single bonds between the monomer units. The number of recurring units may be large or small, depending on the polymerization conditions.

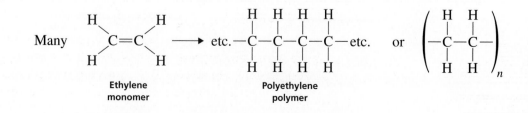

Ethylene monomer   Polyethylene polymer

Table 1 *Addition polymers*

Example	Monomer(s)	Polymer	Uses
Polyethylene	$CH_2{=}CH_2$	$-CH_2-CH_2-$	Most common and important polymer; bags, insulation for wires, squeeze bottles
Polypropylene	$CH_2{=}CH$ $\quad\ \ \vert$ $\quad\ \ CH_3$	$-CH_2-CH-$ $\qquad\ \ \vert$ $\qquad\ \ CH_3$	Fibers, indoor–outdoor carpets, bottles
Polystyrene	$CH_2{=}CH$ $\quad\ \ \vert$ (phenyl)	$-CH_2-CH-$ $\qquad\ \ \vert$ (phenyl)	Styrofoam, inexpensive house-hold goods, inexpensive molded objects
Poly(vinyl chloride) (PVC)	$CH_2{=}CH$ $\quad\ \ \vert$ $\quad\ \ Cl$	$-CH_2-CH-$ $\qquad\ \ \vert$ $\qquad\ \ Cl$	Synthetic leather, clear bottles, floor covering, water pipe
Polytetrafluoroethylene (Teflon)	$CF_2{=}CF_2$	$-CF_2-CF_2-$	Nonstick surfaces, chemically resistant films
Poly(methyl methacrylate) (Lucite, Plexiglas)	$CH_2{=}C$ $\quad\ \ \vert$ CO_2CH_3, CH_3	$-CH_2-C-$ $\qquad\ \ \vert$ CO_2CH_3, CH_3	Unbreakable "glass," latex paints
Polyacrylonitrile (Orlon, Acrilan, Creslan)	$CH_2{=}CH$ $\quad\ \ \vert$ $\quad\ \ CN$	$-CH_2-CH-$ $\qquad\ \ \vert$ $\qquad\ \ CN$	Fiber used in sweaters, blankets, carpets
Poly(vinyl acetate) (PVA)	$CH_2{=}CH$ $\quad\ \ \vert$ $\quad\ \ OCCH_3$ $\qquad\ \ \Vert$ $\qquad\ \ O$	$-CH_2-CH-$ $\qquad\ \ \vert$ $\qquad\ \ OCCH_3$ $\qquad\qquad\ \Vert$ $\qquad\qquad\ O$	Adhesive, latex paints, chewing gum, textile coatings
Natural rubber	$\qquad\quad\ CH_3$ $\qquad\quad\ \vert$ $CH_2{=}CCH{=}CH_2$	$\qquad\qquad\ CH_3$ $\qquad\qquad\ \vert$ $-CH_2-C{=}CH-CH_2-$	Polymer cross-linked with sulfur (vulcanization)
Polychloroprene (neoprene rubber)	$\qquad\quad\ Cl$ $\qquad\quad\ \vert$ $CH_2{=}CCH{=}CH_2$	$\qquad\qquad\ Cl$ $\qquad\qquad\ \vert$ $-CH_2-C{=}CH-CH_2-$	Cross-linked with ZnO; resistant to oil and gasoline
Styrene butadiene rubber (SBR)	$CH_2{=}CH$ $\quad\ \ \vert$ (phenyl) $CH_2{=}CHCH{=}CH_2$	$-CH_2CH-CH_2CH{=}CHCH_2-$ $\qquad\ \vert$ (phenyl)	Cross-linked with peroxides; most common rubber; used for tires; 25% styrene, 75% butadiene

This reaction can be promoted by heat, pressure, and a chemical catalyst. The molecules produced in a typical reaction vary in the number of carbon atoms in their chains. In other words, a mixture of polymers of varying length, rather than a pure compound, is produced.

Polyethylenes with linear structures can pack together easily and are referred to as high-density polyethylenes. They are fairly rigid materials. Low-density polyethylenes consist of branched-chain molecules, with some cross-linking in the chains. They are more flexible than the high-density polyethylenes. The reaction conditions and the catalysts that produce polyethylenes of low and high density are quite different. The monomer, however, is the same in each case.

Another example of an addition polymer is polypropylene. In this case, the monomer is propylene. The polymer that results has a branched methyl on alternate carbon atoms of the chain.

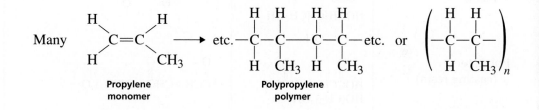

Propylene
monomer

Polypropylene
polymer

A number of common addition polymers are shown in Table 1. Some of their principal uses are also listed. The last three entries in the table all have a carbon–carbon double bond remaining after the polymer is formed. These bonds activate or participate in a further reaction to form cross-linked polymers called *elastomers;* this term is almost synonymous with *rubber,* because elastomers are materials with common characteristics.

Condensation Polymers

Condensation polymers, for which the monomers contain more than one type of functional group, are more complex than addition polymers. In addition, most condensation polymers are copolymers made from more than one type of monomer. Recall that addition polymers, in contrast, are all prepared from substituted ethylene molecules. The single functional group in each case is one or more double bonds, and a single type of monomer is generally used.

Dacron, a polyester, can be prepared by causing a dicarboxylic acid to react with a bifunctional alcohol (a diol):

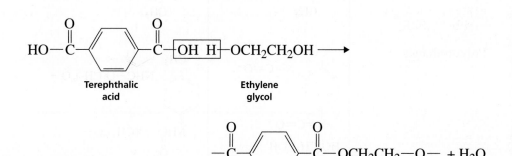

Terephthalic
acid

Ethylene
glycol

Dacron

Table 2 *Condensation polymers*

Example	Monomers	Polymer	Uses
Polyamides (nylon)	$HOC(CH_2)_nCOH$ $H_2N(CH_2)_nNH_2$	$-C(CH_2)_nC-NH(CH_2)_nNH-$	Fibers, molded objects
Polyesters (Dacron, Mylar, Fortrel)	$HOC-\!\!\!\!\!\!\bigcirc\!\!\!\!\!\!-COH$ $HO(CH_2)_nOH$	$-C-\!\!\!\!\!\!\bigcirc\!\!\!\!\!\!-C-O(CH_2)_nO-$	Linear polyesters, fibers, recording tape
Polyesters (Glyptal resin)	phthalic anhydride; $HOCH_2CHCH_2OH$ with OH	$-COCH_2CHCH_2O-$	Cross-linked polyester, paints
Polyesters (casting resin)	$HOCCH=CHCOH$ $HO(CH_2)_nOH$	$-CCH=CHC-O(CH_2)_nO-$	Cross-linked with styrene and peroxide, fiberglass boat resin
Phenol-formaldehyde resin (Bakelite)	OH (phenol) $CH_2=O$	$-CH_2-\!\!\!\bigcirc(OH)\!\!\!-CH_2-\!\!\!\bigcirc(OH)\!\!\!-CH_2-$ with CH_2 cross-links	Mixed with fillers, molded electrical goods, adhesives, laminates, varnishes
Cellulose acetate*	glucose (CH_2OH, OH) CH_3COOH	cellulose (CH_2OAc, OAc)	Photographic film
Silicones	CH_3 $Cl-Si-Cl$ H_2O CH_3	CH_3 $-O-Si-O-$ CH_3	Water-repellent coatings, temperature-resistant fluids and rubbers (CH_3SiCl_3 cross-links in water)
Polyurethanes	toluene diisocyanate (CH_3, $N=C=O$, $N=C=O$) $HO(CH_2)_nOH$	CH_3, $NHC-O(CH_2)_nO-$, $NHC-O(CH_2)_nO-$	Rigid and flexible foams, fibers

*Cellulose, a polymer of glucose, is used as the monomer.

Nylon 6-6, a polyamide, can be prepared by causing a dicarboxylic acid to react with a bifunctional amine.

$$HO-\overset{\overset{O}{\|}}{C}(CH_2)_4\overset{\overset{O}{\|}}{C}-\boxed{OH\ \ H}-\overset{\overset{H}{|}}{N}(CH_2)_6\overset{\overset{H}{|}}{N}H \longrightarrow -\overset{\overset{O}{\|}}{C}(CH_2)_4\overset{\overset{O}{\|}}{C}-\underset{\underset{H}{|}}{N}(CH_2)_6\underset{\underset{H}{|}}{N}- + H_2O$$

Adipic   **Hexamethylene-**   **Nylon**
acid   **diamine**

Notice in each case that a small molecule, water, is eliminated as a product of the reaction. Several other condensation polymers are listed in Table 2. Linear (or branched) chain polymers, as well as cross-linked polymers, are produced in condensation reactions.

The nylon structure contains the amide linkage at regular intervals:

$$-\overset{\overset{O}{\|}}{C}-\overset{\overset{H}{|}}{N}-$$

This type of linkage is extremely important in nature because of its presence in proteins and polypeptides. Proteins are gigantic polymeric substances made up of monomer units of amino acids. They are linked by the peptide (amide) bond.

Other important natural condensation polymers are starch and cellulose. They are polymeric materials made up of the sugar monomer glucose. Another important natural condensation polymer is the DNA molecule. A DNA molecule is made up of the sugar deoxyribose linked with phosphates to form the backbone of the molecule.

Disposability Problems

What do we do with all our wastes? Currently, the most popular method is to bury our garbage in sanitary landfills. However, as we run out of good places to bury our garbage, incineration has become a more attractive method for solving the solid waste problem. Plastics, which compose about 2% of our garbage, burn readily. The new high-temperature incinerators are extremely efficient and can be operated with little air pollution. It should also be possible to burn our garbage and generate electrical power from it.

Ideally, we should either recycle all our wastes or not produce the waste in the first place. Plastic waste consists of about 55% polyethylene and polypropylene, 20% polystyrene, and 11% PVC. All these polymers are thermoplastics and can be recycled. They can be resoftened and remolded into new goods. Unfortunately, thermosetting plastics (cross-linked polymers) cannot be remelted. They decompose on high-temperature heating. Thus, thermosetting plastics should not be used for "disposable" purposes. To recycle plastics effectively, we must sort the materials according to the various types. The plastics industry has introduced a code system consisting of seven categories for the common plastics used in packaging. The code is conveniently stamped on the bottom of the container. Using these codes,

Table 3 *Code system for plastic materials*

Code	Polymer	Uses
1 PETE	Poly(ethylene terephthlate) (PET) $-O-CH_2-CH_2-O-C-\text{(benzene ring)}-C-$ with O double bonds	Soft-drink bottles
2 HDPE	High-density polyethylene $-CH_2-CH_2-CH_2-CH_2-$	Milk and beverage containers, products in squeeze bottles
3 V	Vinyl/poly(vinyl chloride) (PVC) $-CH_2-CH-CH_2-CH-$ with Cl substituents	Some shampoo containers, bottles with cleaning materials in them
4 LDPE	Low-density polyethylene $-CH_2-CH_2-CH_2-CH_2-$ with some branches	Thin plastic bags, some plastic wrap
5 PP	Polypropylene $-CH_2-CH-CH_2-CH-$ with CH_3 substituents	Heavy-duty, microwaveable containers used in kitchens
6 PS	Polystyrene $-CH_2-CH-CH_2-CH-$ with phenyl substituents	Beverage/foam cups, window in envelopes
7 Other	All other resins, layered multimaterials, containers made of different materials	Some ketchup bottles, snack packs, mixture where top differs from bottom

consumers can separate the plastics into groups for recycling purposes. These codes are listed in Table 3, together with the most common uses around the home. Notice that the seventh category is a miscellaneous one, called "Other."

It is amazing that so few plastics are used in packaging. The most common ones are polyethylene (low and high density), polypropylene, polystyrene, and poly(ethylene terephthlate). All these materials can easily be recycled because they are thermoplastics. Incidentally, vinyls (polyvinyl chloride) are becoming less common in packaging. The "Other" category, Code 7, is virtually nonexistent and usually consists of packaging where the top is made of a material different from that of the bottom. This dilemma should be easy to solve by placing the appropriate code on each part of the container.

Polymers, if they are well made, will not corrode or rust, and they last almost indefinitely. Unfortunately, these desirable properties also lead to a problem when plastics are buried in a landfill or thrown on the landscape—they do not decompose. Research is being undertaken to discover plastics that are biodegradable or photodegradable so that either microorganisms or light from the sun can decompose our litter and garbage. Although there are some advantages to this approach, it is probably better to eliminate packaging at the source or to engage in an effective recycling program. We must learn to use plastics wisely.

REFERENCES

Ainsworth, S. J. "Plastics Additives." *Chemical and Engineering News, 70* (August 31, 1992): 34–55.

Burfield, D. R. "Polymer Glass Transition Temperatures." *Journal of Chemical Education, 64* (1987): 875.

Carraher, C. E., Jr., Hess, G., and Sperling, L. H. "Polymer Nomenclature—or What's in a Name?" *Journal of Chemical Education, 64* (1987): 36.

Carraher, C. E., Jr., and Seymour, R. B. "Physical Aspects of Polymer Structure: A Dictionary of Terms." *Journal of Chemical Education, 63* (1986): 418.

Carraher, C. E., Jr., and Seymour, R. B. "Polymer Properties and Testing—Definitions." *Journal of Chemical Education, 64* (1987): 866.

Carraher, C. E., Jr., and Seymour, R. B. "Polymer Structure—Organic Aspects (Definitions)." *Journal of Chemical Education, 65* (1988): 314.

Fried, J. R. "The Polymers of Commercial Plastics." *Plastics Engineering* (June 1982): 49–55.

Fried, J. R. "Polymer Properties in the Solid State." *Plastics Engineering* (July 1982): 27–37.

Fried, J. R. "Molecular Weight and Its Relation to Properties." *Plastics Engineering* (August 1982): 27–33.

Fried, J. R. "Elastomers and Thermosets." *Plastics Engineering* (March 1983): 67–73.

Fried, J. R., and Yeh, E. B. "Polymers and Computer Alchemy." *Chemtech, 23* (March 1993): 35–40.

Goodall, B. L. "The History and Current State of the Art of Propylene Polymerization Catalysts." *Journal of Chemical Education, 63* (1986): 191.

Harris, F. W., et al. "State of the Art: Polymer Chemistry." *Journal of Chemical Education, 58* (November 1981). (This issue contains 17 papers on polymer chemistry. The series covers structures, properties, mechanisms of formation, methods of preparation, stereochemistry, molecular weight distribution, rheological behavior of polymer melts, mechanical properties, rubber elasticity, block and graft copolymers, organometallic polymers, fibers, ionic polymers, and polymer compatibility.)

Jordan, R. F. "Cationic Metal–Alkyl Olefin Polymerization Catalysts." *Journal of Chemical Education, 65* (1988): 285.

Kauffman, G. B. "Wallace Hume Carothers and Nylon, the First Completely Synthetic Fiber." *Journal of Chemical Education, 65* (1988): 803.

Kauffman, G. B. "Rayon: The First Semi-Synthetic Fiber Product." *Journal of Chemical Education, 70* (1993): 887.

Kauffman, G. B., and Seymour, R. B. "Elastomers I. Natural Rubber." *Journal of Chemical Education, 67* (1990): 422.

Kauffman, G. B., and Seymour, R. B. "Elastomers II. Synthetic Rubbers." *Journal of Chemical Education, 68* (1991): 217.

Morse, P. M. "New Catalysts Renew Polyolefins." *Chemical and Engineering News, 76* (July 6, 1998): 11–16.

Seymour, R. B. "Polymers Are Everywhere." *Journal of Chemical Education, 65* (1988): 327.

Seymour, R. B. "Alkenes and Their Derivatives: The Alchemists' Dream Come True." *Journal of Chemical Education, 66* (1989): 670.

Seymour, R. B., and Kauffman, G. B. "Polymer Blends: Superior Products from Inferior Materials." *Journal of Chemical Education, 69* (1992): 646.

Seymour, R. B., and Kauffman, G. B. "Polyurethanes: A Class of Modern Versatile Materials." *Journal of Chemical Education, 69* (1992): 909.

Seymour, R. B., and Kauffman, G. B. "The Rise and Fall of Celluloid." *Journal of Chemical Education, 69* (1992): 311.

Seymour, R. B., and Kauffman, G. B. "Thermoplastic Elastomers." *Journal of Chemical Education, 69* (1992): 967.

Stevens, M. P. "Polymer Additives: Chemical and Aesthetic Property Modifiers." *Journal of Chemical Education, 70* (1993): 535.

Stevens, M. P. "Polymer Additives: Mechanical Property Modifiers." *Journal of Chemical Education, 70* (1993): 444.

Stevens, M. P. "Polymer Additives: Surface Property and Processing Modifiers." *Journal of Chemical Education, 70* (1993): 713.

Thayer, A. M. "Metallocene Catalysts Initiate New Era in Polymer Synthesis." *Chemical and Engineering News, 73* (September 11, 1995): 15–20.

Waller, F. J. "Fluoropolymers." *Journal of Chemical Education, 66* (1989): 487.

Webster, O. W. "Living Polymerization Methods." *Science, 251* (1991): 887.

---

49 | **EXPERIMENT 49**

Preparation and Properties of Polymers: Polyester, Nylon, and Polystyrene

Condensation polymers

Addition polymers

Cross-linked polymers

Infrared spectroscopy

In this experiment, the syntheses of two polyesters (Experiment 49A), nylon (Experiment 49B), and polystyrene (Experiment 49C) will be described. These polymers represent important commercial plastics. They also represent the main classes of polymers: condensation (linear polyester, nylon), addition (polystyrene), and cross-linked (Glyptal polyester). Infrared spectroscopy is used in Experiment 49D to determine the structure of polymers.

REQUIRED READING

Review: Technique 25   Infrared Spectroscopy, Section 25B

New: Essay   Polymers and Plastics

SPECIAL INSTRUCTIONS

Experiments 49A, 49B, and 49C all involve toxic vapors. Each experiment should be conducted in a good hood. The styrene used in Experiment 49C irritates the skin and eyes. Avoid breathing its vapors. Styrene must be dispensed and stored in a hood. Benzoyl peroxide is flammable and may detonate on impact or on heating.

SUGGESTED WASTE DISPOSAL

The test tubes containing the polyester polymers from Experiment 49A should be placed in a box designated for disposal of these samples. The nylon from Experiment 49B should be washed thoroughly with water and placed in a waste basket. The liquid wastes from Experiment 49B (nylon) should be poured into a container designated for disposal of these wastes. The polystyrene prepared in Experiment 49C should be placed in the container designated for solid wastes.

49A   EXPERIMENT 49A

Polyesters

Linear and cross-linked polyesters will be prepared in this experiment. Both are examples of condensation polymers. The linear polyester is prepared as follows:

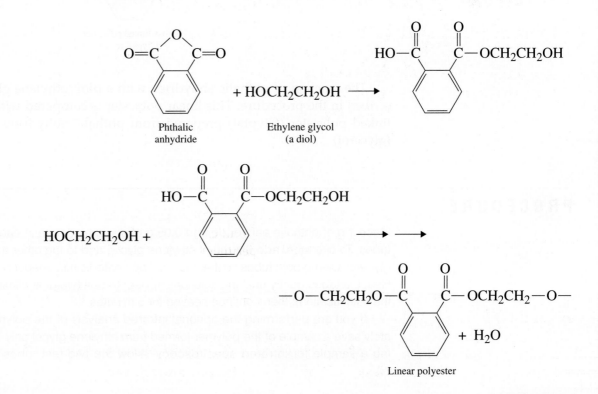

Phthalic anhydride

Ethylene glycol (a diol)

Linear polyester

This linear polyester is isomeric with Dacron, which is prepared from terephthalic acid and ethylene glycol (see the preceding essay). Dacron and the linear polyester made in this experiment are both thermoplastics.

If more than two functional groups are present in one of the monomers, the polymer chains can be linked to one another (cross-linked) to form a three-dimensional network. Such structures are usually more rigid than linear structures and are useful in making paints and coatings. They may be classified as thermosetting plastics. The polyester Glyptal is prepared as follows:

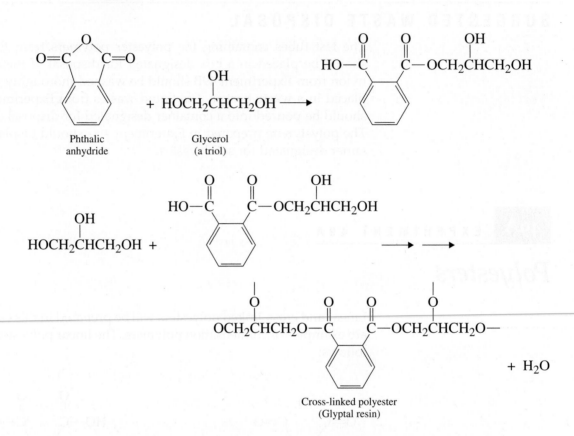

The reaction of phthalic anhydride with a diol (ethylene glycol) is described in the procedure. This linear polyester is compared with the cross-linked polyester (Glyptal) prepared from phthalic anhydride and a triol (glycerol).

PROCEDURE

Place 1 g of phthalic anhydride and 0.05 g of sodium acetate in each of two test tubes. To one tube, add 0.4 mL of ethylene glycol, and to the other add 0.4 mL of glycerol. Clamp both tubes so that they can be heated simultaneously with a flame. Heat the tubes gently until the solutions appear to boil (water is eliminated during the esterification); then continue heating for 5 minutes.

If you are performing the optional infrared analysis of the polymer, immediately save a sample of the polymer formed from ethylene glycol only. After removing a sample for infrared spectroscopy, allow the two test tubes to cool and compare the viscosity and brittleness of the two polymers. The test tubes cannot be cleaned.

Infrared Spectroscopy (Optional)

Lightly coat a watch glass with stopcock grease. Pour some of the *hot* polymer from the tube containing ethylene glycol; use a wooden applicator stick to spread the polymer on the surface so as to create a thin film of the polymer. Remove the polymer from the watch glass and save it for Experiment 49D.

49B   EXPERIMENT 49B

Polyamide (Nylon)

Reaction of a dicarboxylic acid, or one of its derivatives, with a diamine leads to a linear polyamide through a condensation reaction. Commercially, nylon 6-6 (so called because

$$\text{Cl}-\overset{\overset{\text{O}}{\|}}{\text{C}}\text{CH}_2\text{CH}_2\text{CH}_2\text{CH}_2\overset{\overset{\text{O}}{\|}}{\text{C}}-\text{Cl} + \text{H}-\overset{\overset{\text{H}}{|}}{\text{N}}\text{CH}_2\text{CH}_2\text{CH}_2\text{CH}_2\text{CH}_2\text{CH}_2\overset{\overset{\text{H}}{|}}{\text{N}}-\text{H} \longrightarrow$$

Adipoyl chloride   Hexamethylenediamine

$$-\overset{\overset{\text{O}}{\|}}{\text{C}}\text{CH}_2\text{CH}_2\text{CH}_2\text{CH}_2\overset{\overset{\text{O}}{\|}}{\text{C}}-\overset{\overset{\text{H}}{|}}{\text{N}}\text{CH}_2\text{CH}_2\text{CH}_2\text{CH}_2\text{CH}_2\text{CH}_2\overset{\overset{\text{H}}{|}}{\text{N}}-$$

Nylon 6-6

each monomer has six carbons) is made from adipic acid and hexamethylenediamine. In this experiment, you will use the acid chloride instead of adipic acid. The acid chloride is dissolved in cyclohexane, and this is added *carefully* to hexamethylenediamine dissolved in water. These liquids do not mix, so two layers will form. The polymer can then be drawn out continuously to form a long strand of nylon. Imagine how many molecules have been linked in this long strand! It is a fantastic number.

PROCEDURE

Pour 10 mL of a 5% aqueous solution of hexamethylenediamine (1,6-hexanediamine) into a 50-mL beaker. Add 10 drops of 20% sodium hydroxide solution. Carefully add 10 mL of a 5% solution of adipoyl chloride in cyclohexane to the solution by pouring it down the wall of the slightly tilted beaker. Two layers will form (see figure), and there will be an immediate formation of a polymer film at the liquid–liquid interface. Using a copper-wire hook (a 6-in. piece of wire bent at one end), gently free the walls of the beaker from polymer strings. Then hook the mass at the center and slowly raise the wire so that polyamide forms continuously, producing a rope that can be drawn out for many feet. The strand can be broken by pulling it faster. Rinse the rope several times with water and lay it on a paper

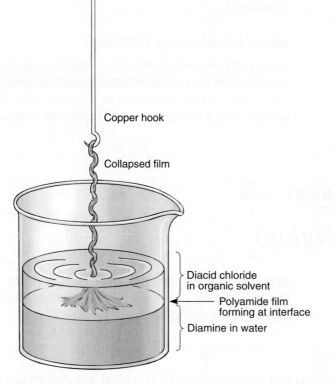

Preparation of nylon.

towel to dry. With the piece of wire, vigorously stir the remainder of the two-phase system to form additional polymer. Decant the liquid and wash the polymer thoroughly with water. Allow the polymer to dry. Do not discard the nylon in the sink; use a waste container.

49C EXPERIMENT 49C

Polystyrene

An addition polymer, polystyrene, will be prepared in this experiment. Reaction can be brought about by free-radical, cationic, or anionic catalysts, the first of these being the most common. In this experiment, polystyrene is prepared by free-radical-catalyzed polymerization.

The reaction is initiated by a free-radical source. The initiator will be benzoyl peroxide, a relatively unstable molecule, which at 80–90°C decomposes with homolytic cleavage of the oxygen–oxygen bond:

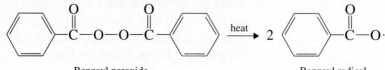

Benzoyl peroxide   Benzoyl radical

If an unsaturated monomer is present, the catalyst radical adds to it, initiating a chain reaction by producing a new free radical. If we let R stand for the catalyst radical, the reaction with styrene can be represented as

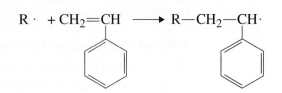

The chain continues to grow:

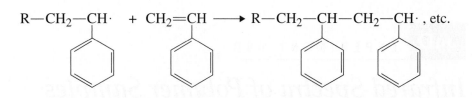

The chain can be terminated by causing two radicals to combine (either both polymer radicals or one polymer radical and one initiator radical) or by causing a hydrogen atom to become abstracted from another molecule.

PROCEDURE

Because it is difficult to clean the glassware, this experiment is best performed by the laboratory instructor. One large batch should be made for the entire class (at least 10 times the amounts given). After the polystyrene is prepared, a small amount will be dispensed to each student. The students will provide their own watch glass for this purpose. Perform the experiment in a hood. Place several thicknesses of newspaper in the hood.

CAUTION

Styrene vapor is irritating to the eyes, mucous membranes, and upper respiratory tract. Do not breathe the vapor, and do not get it on your skin. Exposure can cause nausea and headaches. All operations with styrene must be conducted in a hood.

 Benzoyl peroxide is flammable and may detonate on impact or on heating (or grinding). It should be weighed on glassine (glazed, not ordinary) paper. Clean all spills with water. Wash the glassine paper with water before discarding it.

 Place 12–15 mL of styrene monomer in a 100-mL beaker and add 0.35 g of benzoyl peroxide. Heat the mixture on a hot plate until the mixture turns yellow. When the color disappears and bubbles begin to appear, immediately take the beaker of styrene off the hot plate because the reaction is exothermic (use tongs or an insulated glove). After the reaction subsides, put the beaker of styrene back on the hot plate and continue heating it until the liquid becomes syrupy. With a stirring rod, draw out a long filament of material from the beaker. If this filament can be cleanly snapped after a few seconds of cooling, the polystyrene is ready to be poured. If the filament does not break, continue heating the mixture and repeat this process until the filament breaks easily.

If you are performing the optional infrared analysis of the polymer, immediately save a sample of the polymer. After removing a sample for infrared spectroscopy, pour the remainder of the syrupy liquid on a watch glass that has been lightly coated with stopcock grease. After being cooled, the polystyrene can be lifted from the glass surface by gently prying with a spatula.

Infrared Spectroscopy (Optional)

Pour a small amount of the *hot* polymer from the beaker onto a warm watch glass (no grease) and spread the polymer with a wooden applicator stick to create a thin film of the polymer. Peel the polymer from the watch glass and save it for Experiment 49D.

49D   EXPERIMENT 49D

Infrared Spectra of Polymer Samples

Infrared spectroscopy is an excellent technique for determining the structure of a polymer. For example, polyethylene and polypropylene have relatively simple spectra because they are saturated hydrocarbons. Polyesters have stretching frequencies associated with the C=O and C—O groups in the polymer chain. Polyamides (nylon) show absorptions that are characteristic for the C=O stretch and N—H stretch. Polystyrene has characteristic features of a monosubstituted aromatic compound (see Fig. 19.11 on p. 772). You may determine the infrared spectra of the linear polyester from Procedure 49A and polystyrene from Experiment 49C in this part of the experiment. Your instructor may ask you to analyze a sample that you bring to the laboratory or one supplied to you.

PROCEDURE

Mounting the Samples

Prepare cardboard mounts for your polymer samples. Cut 3×5-inch index cards so that they fit into the sample cell holder of your infrared spectrometer. Then cut a 0.5-inch-wide $\times$ 1-inch-high rectangular hole in the center of the card stock. Attach a polymer sample on the cardboard mount with tape.

Choices of Polymer Samples

If you have completed Experiments 49A and 49C, you can obtain the spectra of your polyester or polystyrene. Alternatively, your instructor may provide you with known or unknown polymer samples for you to analyze.

Your instructor may ask you to bring a polymer sample of your own choice. If possible, these samples should be clear and as thin as possible (similar to the thickness of plastic sandwich wrap). Good choices of plastic materials include windows from envelopes, plastic sandwich wrap, sandwich bags, soft drink bottles, milk containers, shampoo bottles, candy wrappers, and shrink-wrap. If necessary, the samples can be heated in an oven and stretched to obtain thinner samples. If you are bringing a sample cut from a plastic container, obtain the recycling code from the bottom of the container, if one is given.

Running the Infrared Spectrum

Insert the cardboard mount into the cell holder in the spectrometer so that your polymer sample is centered in the infrared beam of the instrument. Find the thinnest place in your polymer sample. Determine the infrared spectrum of your sample. Because of the thickness of your polymer sample, many absorptions are so strong that you will not be able to see individual bands. To obtain a better spectrum, try moving the sample to a new position in the beam and rerun the spectrum.

Analyzing the Infrared Spectrum

You can use the essay "Polymers and Plastics" and Technique 25 with your spectrum to help determine the structure of the polymer. Most likely, the polymers will consist of plastic materials listed in Table 3 of the essay (p. 404). This table lists the recycling codes for a number of household plastics used in packaging. Submit the infrared spectrum, along with the structure of the polymer, to your instructor. Do your spectrum and structure agree with the recycling code? Label the spectrum with the important absorption bands consistent with the structure of the polymer.

Using a Polymer Library

If your instrument has a polymer library, you can search the library for a match. Do this after you have made a preliminary "educated guess" as to the structure of the polymer. The library search should help confirm the structure you determined.

QUESTIONS

1. Ethylene dichloride $ClCH_2CH_2Cl$ and sodium polysulfide Na_2S_4 react to form a chemically resistant rubber, Thiokol A. Draw the structure of the rubber.

2. Draw the structure for the polymer produced from the monomer, vinylidene chloride ($CH_2{=}CCl_2$).

3. Draw the structure of the copolymer produced from vinyl acetate and vinyl chloride. This copolymer is employed in some paints, adhesives, and paper coatings.

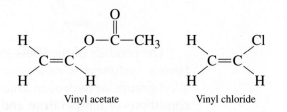

Vinyl acetate   Vinyl chloride

4. Isobutylene $CH_2{=}C(CH_3)_2$ is used to prepare cold-flow rubber. Draw a structure for the addition polymer formed from this alkene.

5. Kel-F is an addition polymer with the following partial structure. What is the monomer used to prepare it?

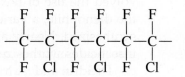

6. Maleic anhydride reacts with ethylene glycol to produce an alkyd resin. Draw the structure of the condensation polymer produced.

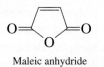

Maleic anhydride

7. Kodel is a condensation polymer made from terephthalic acid and 1,4-cyclohexane-dimethanol. Write the structure of the resulting polymer.

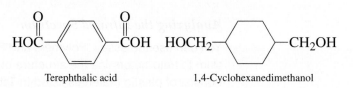

Terephthalic acid   1,4-Cyclohexanedimethanol

ESSAY

Diels–Alder Reaction and Insecticides

Since the 1930s, it has been known that the addition of an unsaturated molecule across a diene system forms a substituted cyclohexene. The original research dealing with this type of reaction was performed by Otto Diels and Kurt Alder in Germany, and the reaction is known as the **Diels–Alder reaction.** The Diels–Alder reaction is the reaction of a **diene** with a species capable of reacting with the diene, the **dienophile.**

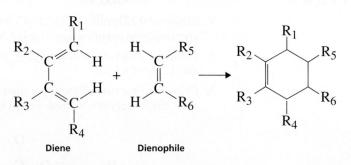

Diene   Dienophile

The product of the Diels–Alder reaction is usually a structure that contains a cyclohexene ring system. If the substituents as shown are simply alkyl groups or hydrogen atoms, the reaction proceeds only under extreme conditions of temperature and pressure. With more complex substituents, however, the Diels–Alder reaction may go on at low temperatures and under mild conditions. The reaction of cyclopentadiene with maleic anhydride (Experiment 50) is an example of a Diels–Alder reaction carried out under reasonably mild conditions.

In the past, a commercially important use of the Diels–Alder reaction involved the use of hexachlorocyclopentadiene as the diene. Depending on the dienophile, a variety of chlorine-containing addition products may be synthesized. Nearly all these products were powerful **insecticides.** Three insecticides synthesized by the Diels–Alder reaction are shown on page 415.

Dieldrin and Aldrin are named after Diels and Alder. These insecticides were once used against the insect pests of fruits, vegetables, and cotton;

against soil insects, termites, and moths; and in the treatment of seeds. Chlordane was used in veterinary medicine against insect pests of animals, including fleas, ticks, and lice. These insecticides are seldom used today.

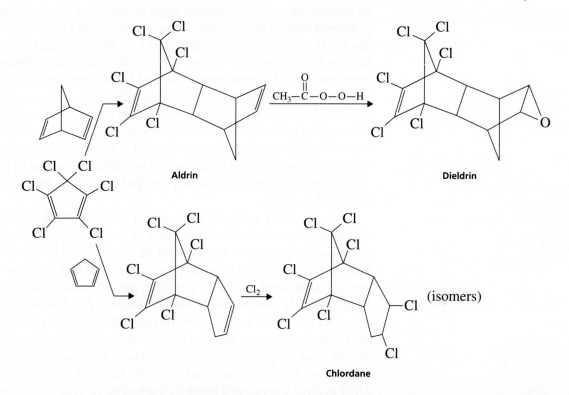

The best-known insecticide, DDT, is not prepared by the Diels–Alder reaction but is nevertheless the best illustration of the difficulties that were experienced when chlorinated insecticides were used indiscriminately. DDT was first synthesized in 1874, and its insecticidal properties were first demonstrated in 1939. It is easily synthesized commercially, with inexpensive reagents.

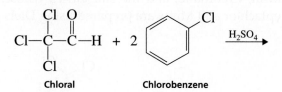

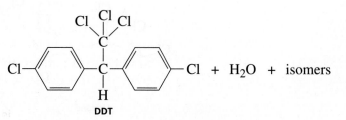

At the time DDT was introduced, it was a boon to humanity. It was effective in controlling lice, fleas, and malaria-carrying mosquitoes and thus helped control human and animal disease. The use of DDT rapidly spread to the control of hundreds of insects that damage fruit, vegetable, and grain crops.

Pesticides that persist in the environment for a long time after application are called **hard pesticides.** Beginning in the 1960s, some of the harmful effects of such hard pesticides as DDT and the other chlorocarbon materials became known. DDT is a fat-soluble material and is therefore likely to collect in the fat, nerve, and brain tissues of animals. The concentration of DDT in tissues increases in animals high in the food chain. Thus, birds that eat poisoned insects accumulate large quantities of DDT. Animals that feed on the birds accumulate even more DDT. In birds, at least two undesirable effects of DDT have been recognized. First, birds whose tissues contain large amounts of DDT have been observed to lay eggs having shells too thin to survive until young birds are hatched. Second, large quantities of DDT in the tissues seem to interfere with normal reproductive cycles. The massive destruction of bird populations that sometimes occurred after heavy spraying with DDT became an issue of great concern. The brown pelican and the bald eagle were placed in danger of extinction. The use of chlorocarbon insecticides was identified as the principal reason for the decline in the numbers of these birds.

Because DDT is chemically inert, it persists in the environment without decomposing to harmless materials. It can decompose slowly, but the decomposition products are every bit as harmful as DDT itself. Consequently, each application of DDT means that still more DDT will pass from species to species, from food source to predator, until it concentrates in the higher animals, possibly endangering their existence. Even humans may be threatened. As a result of evidence of the harmful effects of DDT, the Environmental Protection Agency banned general use of DDT in the early 1970s; it may still be used for certain purposes, although permission of the Environmental Protection Agency is required. In 1974, the EPA granted permission to use DDT against the tussock moth in the forests of Washington and Oregon.

Because the life cycles of insects are short, they can evolve an immunity to insecticides within a short period. As early as 1948, several strains of DDT-resistant insects were identified. Today, the malaria-bearing mosquitoes are almost completely resistant to DDT, an ironic development. Other chlorocarbon insecticides were developed to use as alternatives to DDT against resistant insects. Examples of these chlorocarbon materials include Dieldrin, Aldrin, Chlordane, and the substances whose structures are shown here. Heptachlor and Mirex are prepared using Diels–Alder reactions.

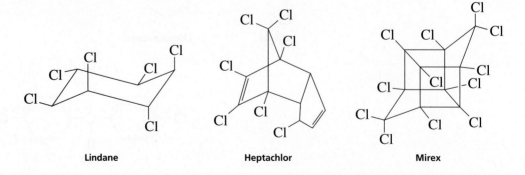

Lindane   Heptachlor   Mirex

In spite of structural similarity, Chlordane and Heptachlor behave differently than DDT, Dieldrin, and Aldrin. Chlordane, for instance, is short lived and less toxic to mammals. Nevertheless, all the chlorocarbon insecticides have been the objects of much suspicion. A ban on the use of Dieldrin and Aldrin has also been ordered by the Environmental Protection Agency. In addition, strains of insects resistant to Dieldrin, Aldrin, and other materials have been observed. Some insects become addicted to a chlorocarbon insecticide and thrive on it!

The problems associated with chlorocarbon materials have led to the development of "soft" insecticides. These usually are organophosphorus or carbamate derivatives, and they are characterized by a short lifetime before they are decomposed to harmless materials in the environment.

The organic structures of some organophosphorus insecticides are shown here.

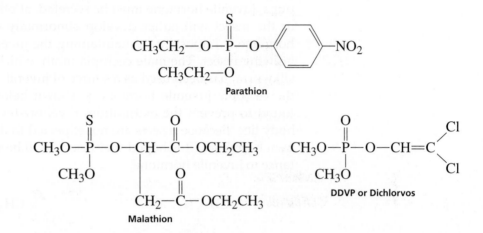

Parathion and Malathion are used widely for agriculture. DDVP is used in "pest strips," which are used for combating household insect pests. The organophosphorus materials do not persist in the environment, so they are not passed between species up the food chain, as the chlorocarbon compounds are. However, the organophosphorus compounds are highly toxic to humans. Some migrant and other agricultural workers have lost their lives because of accidents involving these materials. Stringent safety precautions must be applied when organophosphorus insecticides are being used.

The carbamate derivatives, including Carbaryl, tend to be less toxic than the organophosphorus compounds. They are also readily degraded to harmless materials. Nevertheless, insects resistant to soft insecticides have also been observed. Furthermore, the organophosphorus and carbamate derivatives destroy many more nontarget pests than the chlorocarbon compounds do. The danger to earthworms, mammals, and birds is high.

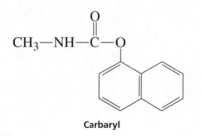

Carbaryl

Alternatives to Insecticides

Several alternatives to the massive application of insecticides have recently been explored. Insect attractants, including the pheromones (see the essay preceding Experiment 45), have been used in localized traps. Such methods have been effective against the gypsy moth. A "confusion technique," whereby a pheromone is sprayed into the air in such high concentrations that male insects are no longer able to locate females, has been studied. These methods are specific to the target pest and do not cause repercussions in the general environment.

Recent research has been focused on using an insect's own biochemical processes to control pests. Experiments with **juvenile hormone** have shown promise. Juvenile hormone is one of three internal secretions used by insects to regulate growth and metamorphosis from larva to pupa and thence to the adult. At certain stages in the metamorphosis from larva to pupa, juvenile hormone must be secreted; at other stages it must be absent or the insect will either develop abnormally or fail to mature. Juvenile hormone is important in maintaining the juvenile, or larval, stage of the growing insect. The male cecropia moth, which is the mature form of the silkworm, has been used as a source of juvenile hormone. The structure of the cecropia juvenile hormone is shown below. This material has been found to prevent the maturation of yellow-fever mosquitoes and human body lice. Because insects are not expected to develop a resistance to their own hormones, it is hoped that insects will be unlikely to develop a resistance to juvenile hormone.

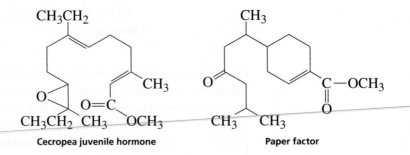

Cecropea juvenile hormone   Paper factor

Although it is difficult to get enough of the natural substance for use in agriculture, synthetic analogues have been prepared, and they have been shown to be similar in properties and effectiveness to the natural substance. A substance has been found in the American balsam fir (*Abies balsamea*), known as **paper factor.** Paper factor is active against the linden bug, *Pyrrhocoris apterus*, a European cotton pest. This substance is merely one of thousands of terpenoid materials synthesized by the fir tree. Other terpenoid substances are being investigated as potential juvenile hormone analogues.

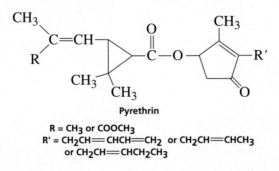

Pyrethrin

R = CH₃ or COOCH₃
R' = CH₂CH=CHCH=CH₂ or CH₂CH=CHCH₃
or CH₂CH=CHCH₂CH₃

Certain plants are capable of synthesizing substances that protect them against insects. Included among these natural insecticides are the **pyrethrins** and derivatives of **nicotine.**

The search for environmentally suitable means of controlling agricultural pests continues with a great sense of urgency. Insects cause billions of dollars of damage to food crops each year. With food becoming increasingly scarce and with the world's population growing at an exponential rate, preventing such losses to food crops becomes essential.

REFERENCES

Berkoff, C. E. "Insect Hormones and Insect Control." *Journal of Chemical Education, 48* (1971): 577.

Bowers, W. S., and Nishida, R. "Juvocimenes: Potent Juvenile Hormone Mimics from Sweet Basil." *Science, 209* (1980): 1030.

Carson, R. *Silent Spring.* Boston: Houghton Mifflin, 1962.

Keller, E. "The DDT Story." *Chemistry, 43* (February 1970): 8.

O'Brien, R. D. *Insecticides: Action and Metabolism.* New York: Academic Press, 1967.

Peakall, D. B. "Pesticides and the Reproduction of Birds." *Scientific American, 222* (April 1970): 72.

Saunders, H. J. "New Weapons against Insects." *Chemical and Engineering News, 53* (July 28, 1975): 18.

Williams, C. M. "Third-Generation Pesticides." *Scientific American, 217* (July 1967): 13.

Williams, W. G., Kennedy, G. G., Yamamoto, R. T., Thacker, J. D., and Bordner, J. "2-Tridecanone: A Naturally Occurring Insecticide from the Wild Tomato." *Science, 207* (1980): 888.

50   **EXPERIMENT 50**

The Diels–Alder Reaction of Cyclopentadiene with Maleic Anhydride

Diels–Alder reaction

Fractional distillation

Cyclopentadiene and maleic anhydride react readily in a Diels–Alder reaction to form the adduct, *cis*-norbornene-5,6-*endo*-dicarboxylic anhydride:

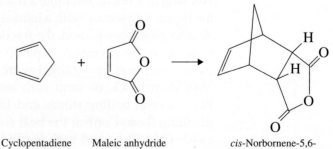

Cyclopentadiene   Maleic anhydride   *cis*-Norbornene-5,6-
                                              endo-dicarboxylic anhydride

Because two molecules of cyclopentadiene can also undergo a Diels–Alder reaction to form dicyclopentadiene, it is not possible to store cyclopentadiene in the monomeric form. Therefore, it is necessary to first "crack" dicyclopentadiene to produce cyclopentadiene for use in this experiment. This will be accomplished by heating the dicyclopentadiene to a boil and collecting the cyclopentadiene as it is formed by fractional distillation. To keep it from dimerizing, the cyclopentadiene must be kept cold and used fairly soon.

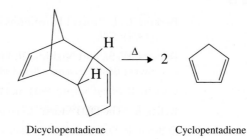

Dicyclopentadiene   Cyclopentadiene

REQUIRED READING

Review: Technique 11   Crystallization, Section 11.4

New: Essay   Diels–Alder Reaction and Insecticides

SPECIAL INSTRUCTIONS

The cracking of dicyclopentadiene should be performed by the instructor or laboratory assistant. If a flame is used for this, be sure that there are no leaks in the system, because both cyclopentadiene and the dimer are highly flammable. The procedure provides enough cyclopentadiene for about 50 students.

SUGGESTED WASTE DISPOSAL

Dispose of the mother liquor from the crystallization in the container designated for nonhalogenated organic solvents.

NOTES TO THE INSTRUCTOR

Working in a hood, assemble a fractional distillation apparatus, as shown in the figure. Glassware with a joint size of ℸ 19/22 or larger should be used. If smaller glassware is used, the fractionating column may not be long enough to achieve the necessary separation. Although the required temperature control can best be obtained with a microburner, using a heating mantle, aluminum block, or sand bath lessens the possibility of a fire occurring. Place several boiling stones and 15 mL of dicyclopentadiene in the 50-mL distilling flask. Control the heat source so that the cyclopentadiene distills at 40–43°C. (If a sand bath is used, the temperature should be 190–200°C, and it may be necessary to cover the sand bath and distilling flask with aluminum foil.) After 30–45 minutes, 6–7 mL of cyclopentadiene should be collected, and the distillation can be stopped. If the cyclopentadiene is cloudy, dry the liquid over granular anhydrous sodium sulfate. Store the product in a sealed container and keep it cooled in an ice-water bath until all students have taken their portions.

PROCEDURE

Preparation of the Adduct

To a Craig tube add 0.100 g of maleic anhydride and 0.40 mL of ethyl acetate. Without inserting the plug, shake the tube gently to dissolve the solid (slight heating in a warm waterbath may be necessary). Add 0.40 mL of ligroin (bp 60–90°C) and shake the tube gently to mix the solvents and reactant thoroughly. Add 0.10 mL of cyclopentadiene and mix thoroughly by shaking until no visible layers of liquid are present. Because this reaction is exothermic, the temperature of the mixture will likely become high enough to keep the product in solution. However, if a solid does form at this point, it will be necessary to heat the mixture gently in a warm waterbath to dissolve any solids present. If necessary, add a drop of ethyl acetate to help dissolve the solid, and, again, heat the mixture gently.

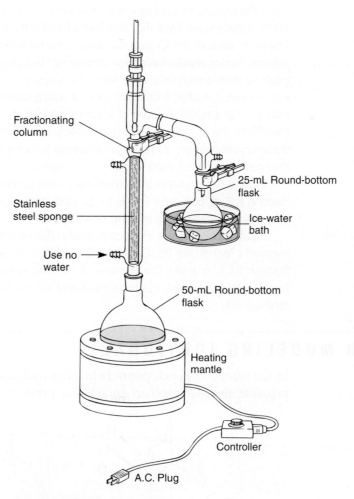

Fractional distillation apparatus for cracking dicyclopentadiene.

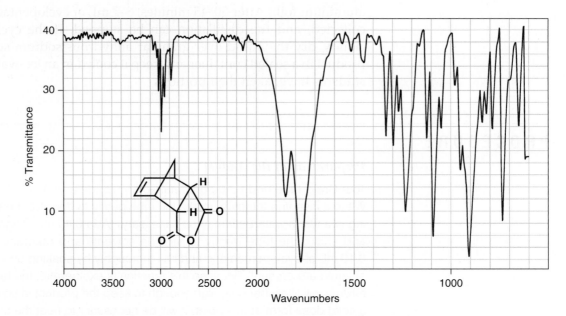

Infrared spectrum of cis-*norbornene-5,6-*endo-*dicarboxylic anhydride, KBr.*

Crystallization of Product

Allow the mixture to cool slowly to room temperature by placing the Craig tube in a 10-mL Erlenmeyer flask that has been filled with about 8 mL of water at 50–60°C. The inner plug of the Craig tube should be inserted to prevent evaporation of the solvent. Better crystal formation can be achieved by seeding the solution before it cools to room temperature. To seed the solution, dip a spatula or glass stirring rod into the solution after it has cooled for about 5 minutes. Allow the solvent to evaporate so that a small amount of solid forms on the surface of the spatula or glass rod. Place the spatula or stirring rod back into the solution for a few seconds to induce crystallization. When crystallization is complete at room temperature, cool the mixture in an ice bath for several minutes.

Isolate the crystals from the Craig tube by centrifugation (see Technique 8, Section 8.7, p. 625, and Fig. 8.11, p. 626) and allow the crystals to air-dry. Determine the weight and the melting point (164°C).

At the option of the instructor, obtain the infrared spectrum using the dry film method (Technique 25, Section 25.4, p. 838) or as a KBr pellet (Technique 25, Section 25.5, p. 838). Compare your infrared spectrum with the one reproduced here. Calculate the percentage yield and submit the product to the instructor in a labeled vial.

MOLECULAR MODELING (OPTIONAL)

In the reaction of cyclopentadiene with maleic anhydride, two products are possible: the *endo* product and the *exo* product.

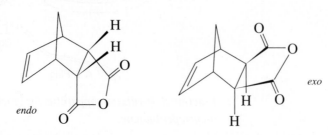

Calculate the heats of formation for both of these products to determine which is the expected **thermodynamic product** (product of lowest energy). Perform the calculations at the AM1 level with a geometry optimization. The actual product of the Diels–Alder reaction is the *endo* product; is this the thermodynamic product? Display a space-filling model for each structure. Which one appears most crowded?

Woodward and Hoffmann have pointed out that the diene is the electron donor and the dienophile the electron acceptor in this reaction. In accordance with this idea, dienes that have electron-donating groups are more reactive than those without, and dienophiles with electron-withdrawing groups are most reactive. Using the reasoning of frontier molecular orbital theory (see the essay "Computational Chemistry" on page 160), the electrons in the HOMO of the diene will be placed into the LUMO of the dienophile when reaction occurs. Using the AM1 level, calculate the HOMO surface for the diene (cyclopentadiene) and the LUMO surface for the dienophile (maleic anhydride). Display the two simultaneously on the screen in the orientations that will lead to the *endo* and *exo* products.

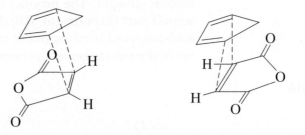

Woodward and Hoffmann suggested that the orientation that leads to the largest degree of constructive overlap between the two orbitals (HOMO and LUMO) is the orientation that would lead to the product. Do you agree?

Depending on the capability of your software, it may be possible to determine the geometrics (and energies) of the transition states that lead to each product. Your instructor will have to show you how to do this.

QUESTIONS

1. Draw a structure for the exo product formed by cyclopentadiene and maleic anhydride.

2. Because the exo form is more stable than the endo form, why is the endo product formed almost exclusively in this reaction?

3. In addition to the main product, what are two side reactions that could occur in this experiment?

4. The infrared spectrum of the adduct is given in this experiment. Interpret the principal peaks.

EXPERIMENT 51

Photoreduction of Benzophenone and Rearrangement of Benzpinacol to Benzopinacolone

Photochemistry
Photoreduction
Energy transfer
Pinacol rearrangement

This experiment consists of two parts. In the first part (Experiment 51A), benzophenone will be subjected to **photoreduction,** a dimerization brought about by exposing a solution of benzophenone in isopropyl alcohol to natural sunlight. The product of this photoreaction is benzpinacol. In the second part (Experiment 51B), benzpinacol will be induced to undergo an acid-catalyzed rearrangement called the **pinacol rearrangement.** The product of the rearrangement is benzopinacolone.

Experiment 51A

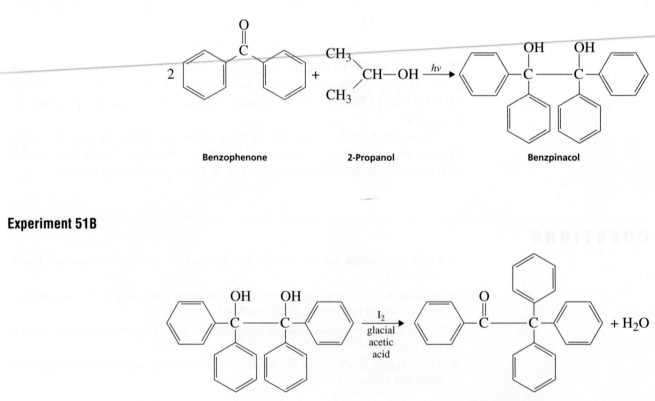

Benzophenone   **2-Propanol**   **Benzpinacol**

Experiment 51B

Benzpinacol   **Benzopinacolone**

51A EXPERIMENT 51A

Photoreduction of Benzophenone

The photoreduction of benzophenone is one of the oldest and most thoroughly studied photochemical reactions. Early in the history of photochemistry, it was discovered that solutions of benzophenone are unstable in light when certain solvents are used. If benzophenone is dissolved in a "hydrogen-donor" solvent, such as 2-propanol, and exposed to ultraviolet light, hν, an insoluble dimeric product, benzpinacol, will form.

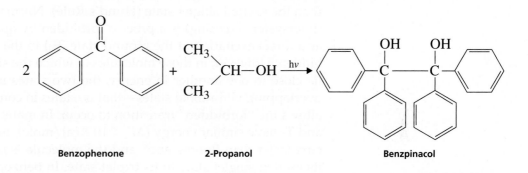

Benzophenone	2-Propanol	Benzpinacol

To understand this reaction, let's review some simple photochemistry as it relates to aromatic ketones. In the typical organic molecule, all the electrons are paired in the occupied orbitals. When such a molecule absorbs ultraviolet light of the appropriate wavelength, an electron from one of the occupied orbitals, usually the one of highest energy, is excited to an unoccupied molecular orbital, usually to the one of lowest energy. During this transition, the electron must retain its spin value, because during an electronic transition a change of spin is forbidden by the laws of quantum mechanics. Therefore, just as the two electrons in the highest occupied orbital of the molecule originally had their spins paired (opposite), so they will retain paired

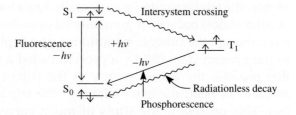

Electronic states of a typical molecule and the possible interconversions. In each state (S_0, S_1, T_1), the lower line represents the highest occupied orbital, and the upper line represents the lowest unoccupied orbital of the unexcited molecule. Straight lines represent processes in which a photon is absorbed or emitted. Wavy lines represent radiationless processes—those that occur without emission or absorption of a photon.

spins in the first electronically excited state of the molecule. This is true even though the two electrons will be in *different* orbitals after the transition. This first excited state of a molecule is called a **singlet state** (S_1) because its spin multiplicity ($2S + 1$) is 1. The original unexcited state of the molecule is also a singlet state because its electrons are paired, and it is called the **ground-state** singlet state (S_0) of the molecule.

The excited state singlet S_1 may return to the ground state S_0 by re-emission of the absorbed photon of energy. This process is called **fluorescence.** Alternatively, the excited electron may undergo a change of spin to give a state of higher multiplicity, the excited **triplet state,** so called because its spin multiplicity ($2S + 1$) is 3. The conversion from the first excited singlet state to the triplet state is called **intersystem crossing.** Because the triplet state has a higher multiplicity, it inevitably has a lower energy state than the excited singlet state (Hund's Rule). Normally, this change of spin (intersystem crossing) is a process forbidden by quantum mechanics, just as a direct excitation of the ground state (S_0) to the triplet state (T_1) is forbidden. However, in those molecules in which the singlet and triplet states lie close to one another in energy, the two states inevitably have several overlapping vibrational states—that is, states in common—a situation that allows the "forbidden" transition to occur. In many molecules in which S_1 and T_1 have similar energy ($\Delta E < 10$ Kcal/mole), intersystem crossing occurs faster than fluorescence, and the molecule is rapidly converted from its excited singlet state to its triplet state. In benzophenone, S_1 undergoes intersystem crossing to T_1 with a rate of $k_{isc} = 10^{10}$ sec^{-1}, meaning that the lifetime of S_1 is only 10^{-10} second. The rate of fluorescence for benzophenone is $k_f = 10^6$ sec^{-1}, meaning that intersystem crossing occurs at a rate that is 10^4 times faster than fluorescence. Thus, the conversion of S_1 to T_1 in benzophenone is essentially a quantitative process. In molecules that have a wide energy gap between S_1 and T_1, this situation would be reversed. As you will see shortly, the naphthalene molecule presents a reversed situation.

Because the excited triplet state is lower in energy than the excited singlet state, the molecule cannot easily return to the excited singlet state. Nor can it easily return to the ground state by returning the excited electron to its original orbital. Once again, the transition $T_1 \rightarrow S_0$ would require a change of spin for the electron, and this is a forbidden process. Hence, the triplet excited state usually has a long lifetime (relative to other excited states) because it generally has nowhere to which it can easily go. Even though the process is forbidden, the triplet T_1 may eventually return to the ground state (S_0) by a process called a **radiationless transition.** In this process, the excess energy of the triplet is lost to the surrounding solution as heat, thereby "relaxing" the triplet back to the ground state (S_0). This process is the study of much current research and is not well understood. In the second process, in which a triplet state may revert to the ground state, **phosphorescence,** the excited triplet emits a photon to dissipate the excess energy and returns directly to the ground state. Although this process is "forbidden," it nevertheless occurs when there is no other open pathway by which the molecule can dissipate its excess energy. In benzophenone, radiationless decay is the faster process, with rate $k_d = 10^5$ sec^{-1}, and phosphorescence, which is not observed, has a lower rate of $k_p = 10^2$ sec^{-1}.

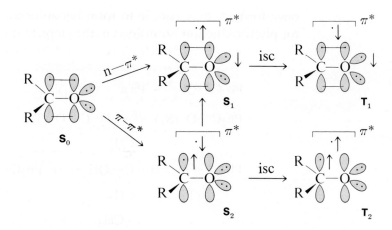

n–π* and π–π* transitions for ketones.

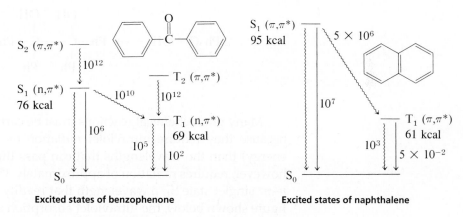

Excited energy states of benzophenone and naphthalene.

Benzophenone is a ketone. Ketones have *two* possible excited singlet states and, consequently, two excited triplet states as well. This occurs because two relatively low-energy transitions are possible in benzophenone. It is possible to excite one of the π electrons in the carbonyl π bond to the lowest-energy unoccupied orbital, a π* orbital. It is also possible to excite one of the unbonded or *n* electrons on oxygen to the same orbital. The first type of transition is called a π–π* transition, whereas the second is called an *n*–π* transition. In the figure, these transitions and the states that result are illustrated pictorially.

Spectroscopic studies show that for benzophenone and most other ketones, the *n*–π* excited states S_1 and T_1 are of lower energy than the π–π* excited states. An energy diagram depicting the excited states of benzophenone (along with one that depicts those of naphthalene) is shown.

It is now known that the photoreduction of benzophenone is a reaction of the *n*–π* triplet state (T_1) of benzophenone. The *n*–π* excited states have radical character at the carbonyl oxygen atom because of the unpaired electron in the nonbonding orbital. Thus, the radical-like and energetic T_1 excited-state species can abstract a hydrogen atom from a suitable donor molecule to form the diphenylhydroxymethyl radical. Two of these radicals, once formed, may couple to form benzpinacol. The complete mechanism for photoreduction is outlined in the steps that follow.

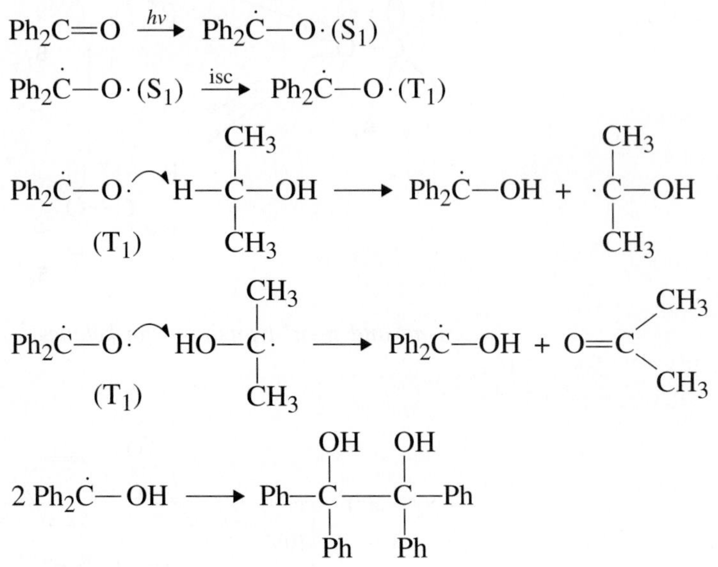

$$Ph_2C{=}O \xrightarrow{h\nu} Ph_2\dot{C}{-}O\cdot (S_1)$$

$$Ph_2\dot{C}{-}O\cdot (S_1) \xrightarrow{isc} Ph_2\dot{C}{-}O\cdot (T_1)$$

$$2\ Ph_2\dot{C}{-}OH \longrightarrow Ph\underset{\underset{Ph}{|}}{\overset{\overset{OH}{|}}{C}}{-}\underset{\underset{Ph}{|}}{\overset{\overset{OH}{|}}{C}}{-}Ph$$

Many photochemical reactions must be carried out in a quartz apparatus because they require ultraviolet radiation of shorter wavelengths (higher energy) than the wavelengths that can pass through Pyrex. Benzophenone, however, requires radiation of approximately 350 nm to become excited to its $n–\pi^*$ singlet state S_1, a wavelength that readily passes through Pyrex. In the figure shown below, the ultraviolet absorption spectra of benzophenone and naphthalene are given. Superimposed on their spectra are two curves, which show the wavelengths that can be transmitted by Pyrex and quartz, respectively. Pyrex will not allow any radiation of wavelength shorter than approximately 300 nm to pass, whereas quartz allows wavelengths as short as 200 nm to pass. Thus, when benzophenone is placed in a Pyrex flask, the only electronic transition possible is the $n–\pi^*$ transition, which occurs at 350 nm.

However, even if it were possible to supply benzophenone with radiation of the appropriate wavelength to produce the second excited singlet state of the molecule, this singlet would rapidly convert to the lowest singlet state (S_1). The state S_2 has a lifetime of less than 10^{-12} second. The conversion process $S_2 \rightarrow S_1$ is called an **internal conversion.** Internal conversions are processes of conversion between excited states of the same multiplicity (singlet–singlet or triplet–triplet), and they usually are rapid. Thus, when an S_2 or T_2 is formed, it readily converts to S_1 or T_1, respectively. As a consequence of their short lifetimes, little is known about the properties or the exact energies of S_2 and T_2 of benzophenone.

Energy Transfer

Using a simple **energy-transfer** experiment, one can show that the photoreduction of benzophenone proceeds via the T_1 excited state of benzophenone rather than the S_1 excited state. If naphthalene is added to the reaction, the photoreduction is stopped because the excitation energy of the benzophenone triplet is transferred to naphthalene. The naphthalene is said to have **quenched** the reaction. This occurs in the following way.

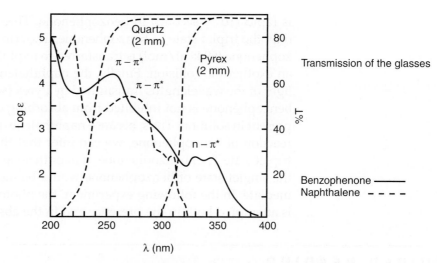

Ultraviolet absorption spectra for benzophenone and naphthalene.

When the excited states of molecules have long enough lifetimes, they often can transfer their excitation energy to another molecule. The mechanisms of these transfers are complex and cannot be explained here; however, the essential requirements can be outlined. First, for two molecules to exchange their respective states of excitation, the process must occur with an overall decrease in energy. Second, the spin multiplicity of the total system must not change. These two features can be illustrated by the two most common examples of energy transfer: singlet transfer and triplet transfer. In these two examples, the superscript 1 denotes an excited singlet state, the superscript 3 denotes a triplet state, and the subscript 0 denotes a ground-state molecule. The designations A and B represent different molecules.

$$A^1 + B_0 \longrightarrow B^1 + A_0 \qquad \textbf{Singlet energy transfer}$$
$$A^3 + B_0 \longrightarrow B^3 + A_0 \qquad \textbf{Triplet energy transfer}$$

In singlet energy transfer, excitation energy is transferred from the excited singlet state of A to a ground-state molecule of B, converting B to its excited singlet state and returning A to its ground state. In triplet energy transfer, there is a similar interconversion of excited state and ground state. Singlet energy is transferred through space by a dipole–dipole coupling mechanism, but triplet energy transfer requires the two molecules involved in the transfer to collide. In the usual organic medium, about 10^9 collisions occur per second. Thus, if a triplet state A^3 has a lifetime longer than 10^{-9} second, and if an acceptor molecule B_0, which has a lower triplet energy than that of A^3, is available, energy transfer can be expected. If the triplet A^3 undergoes a reaction (such as photoreduction) at a rate lower than the rate of collisions in the solution, and if an acceptor molecule is added to the solution, the reaction can be *quenched*. The acceptor molecule, which is called a **quencher,** deactivates, or "quenches," the triplet before it has a chance to react. Naphthalene has the ability to quench benzophenone triplets in this way and to stop the photoreduction.

Naphthalene cannot quench the excited-state singlet S_1 of benzophenone because its own singlet has an energy (95 kcal/mol) that is higher than the energy of benzophenone (76 kcal/mol). In addition, the conversion $S_1 \rightarrow T_1$

is rapid (10^{-10} second) in benzophenone. Thus, naphthalene can intercept only the triplet state of benzophenone. The triplet excitation energy of benzophenone (69 kcal/mol) is transferred to naphthalene ($T_1 = 61$ kcal/mol) in an exothermic collision. Finally, the naphthalene molecule does not absorb light of the wavelengths transmitted by Pyrex (see spectra, p. 429); therefore, benzophenone is not inhibited from absorbing energy when naphthalene is present in solution. Thus, because naphthalene quenches the photoreduction reaction of benzophenone, we can infer that this reaction proceeds via the triplet state T_1 of benzophenone. If naphthalene did not quench the reaction, the singlet state of benzophenone would be indicated as the reactive intermediate. In the following experiment, the photoreduction of benzophenone is attempted both in the presence and in the absence of added naphthalene.

REQUIRED READING

Review: Technique 8   Filtration, Section 8.3

SPECIAL INSTRUCTIONS

This experiment may be performed concurrently with some other experiment. It requires only 15 minutes during the first laboratory period and only about 15 minutes in a subsequent laboratory period about 1 week later (or at the end of the laboratory period if you use a sunlamp).

Using Direct Sunlight

It is important that the reaction mixture be left where it will receive direct sunlight. If it does not, the reaction will be slow and may need more than 1 week for completion. It is also important that the room temperature not be too low or the benzophenone will precipitate. If you perform this experiment in the winter and the laboratory is not heated at night, you must shake the solutions every morning to redissolve the benzophenone. Benzpinacol should not redissolve easily.

Using a Sunlamp

If you wish, you may use a 275-W sunlamp instead of direct sunlight. Place the lamp in a hood that has had its window covered with aluminum foil (shiny side in). The lamp (or lamps) should be mounted in a ceramic socket attached to a ring stand with a three-pronged clamp.

CAUTION

The purpose of the aluminum foil is to protect the eyes of people in the laboratory. You should not view a sunlamp directly or damage to the eyes may result. Take all possible viewing precautions.

Attach samples to a ring stand placed at least 18 inches from the sunlamp. Placing them at this distance will avoid their being heated by the lamp. Heating may cause loss of the solvent. It is a good idea to agitate the samples every 30 minutes. With a sunlamp, the reaction will be complete in 3–4 hours.

SUGGESTED WASTE DISPOSAL

Dispose of the filtrate from the vacuum filtration procedure in the container designated for nonhalogenated organic wastes.

PROCEDURE

Label two 13-mm × 100-mm test tubes near the top of the tubes. The labels should have your name and "No. 1" and "No. 2" written on them. Place 0.50 g of benzophenone in the first tube. Place 0.50 g of benzophenone and 0.05 g of naphthalene in the second tube. Add about 2 mL of 2-propanol (isopropyl alcohol) to each tube and warm them in a beaker of warm water to dissolve the solids. When the solids have dissolved, add one small drop (Pasteur pipet) of glacial acetic acid to each tube and then fill each tube nearly to the top with more 2-propanol. Stopper the tubes tightly with rubber stoppers, shake them well, and place them in a beaker on a windowsill where they will receive direct sunlight.

NOTE: You may be directed by your instructor to use a sunlamp instead of direct sunlight (see "Special Instructions").

The reaction requires about 1 week for completion (3 hours with a sunlamp). If the reaction has occurred during this period, the product will have crystallized from the solution. Observe the result in each test tube. Collect the product by vacuum filtration using a small Büchner or Hirsch funnel (Technique 8, Section 8.3, p. 621) and allow it to dry. Weigh the product and determine its melting point and percentage yield. At the option of the instructor, obtain the infrared spectrum using the dry film method (Technique 25, Section 25.4, p. 838) or as a KBr pellet (Technique 25, Section 25.5, p. 838). Submit the product to the instructor in a labeled vial along with the report.

REFERENCE

Vogler, A., and Kunkely, H. "Photochemistry and Beer." *Journal of Chemical Education,* 59 (January 1982): 25.

51B   EXPERIMENT 51B

Synthesis of β-Benzopinacolone: The Acid-Catalyzed Rearrangement of Benzpinacol

The ability of carbocations to rearrange represents an important concept in organic chemistry. In this experiment, the benzpinacol, prepared in Experiment 51A, will rearrange to **benzopinacolone (2,2,2-triphenylacetophenone)** under the influence of iodine in glacial acetic acid.

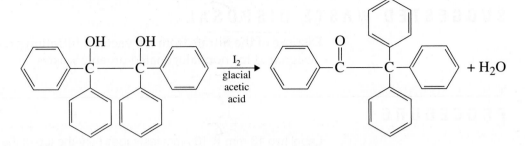

The product is isolated as a crystalline white solid. Benzopinacolone is known to crystallize in two crystalline forms, each with a different melting point. The **alpha** form has a melting point of 206–207°C, whereas the **beta** form melts at 182°C. The product formed in this experiment is the β-benzopinacolone.

REQUIRED READING

Review: Technique 7   Reaction Methods, Section 7.2

Technique 11   Crystallization: Purification of Solids, Section 11.3

Technique 25   Infrared Spectroscopy, Part B

Technique 26   Nuclear Magnetic Resonance Spectroscopy, Part B

Before beginning this experiment, you should read the material dealing with carbocation rearrangements in your lecture textbook.

SPECIAL INSTRUCTIONS

This experiment requires little time and can be coscheduled with another short experiment.

SUGGESTED WASTE DISPOSAL

All organic residues must be placed in the appropriate container for non-halogenated organic waste.

PROCEDURE

In a 25-mL round-bottom flask, add 5 mL of a 0.015 *M* solution of iodine dissolved in glacial acetic acid. Add 1 g of benzpinacol and attach a water-cooled condenser. Using a small heating mantle, allow the solution to heat under reflux for 5 minutes. Crystals may begin to appear from the solution during this heating period.

Remove the heat source and allow the solution to cool slowly. The product will crystallize from the solution as it cools. When the solution has cooled to room temperature, collect the crystals by vacuum filtration using a small Büchner funnel. Rinse the crystals with three 2-mL portions of cold, glacial acetic acid. Allow the crystals to dry in the air overnight. Weigh the dried product and determine its melting point. Pure β-benzopinacolone melts at 182°C. Obtain the infrared spectrum using the dry film method (Technique 25, Section 25.4, p. 838) or as a KBr pellet (Technique 25, Section 25.5, p. 838) and the NMR spectrum in carbon tetrachloride or CDCl$_3$ (Technique 26, Section 26.1, p. 870).

Calculate the percentage yield. Submit the product to your instructor in a labeled vial, along with your spectra. Interpret your spectra, showing how they are consistent with the rearranged structure of the product.

QUESTIONS

1. Can you think of a way to produce the benzophenone $n–\pi^*$ triplet T_1 *without* having benzophenone pass through its first singlet state? Explain.

2. A reaction similar to the one described here occurs when benzophenone is treated with the metal magnesium (pinacol reduction).

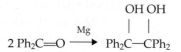

$$2 \ Ph_2C{=}O \xrightarrow{Mg} \ Ph_2\overset{\overset{\displaystyle OH}{|}}{C}{-}\overset{\overset{\displaystyle OH}{|}}{C}Ph_2$$

Compare the mechanism of this reaction with the photoreduction mechanism. What are the differences?

3. Which of the following molecules do you expect would be useful in quenching benzophenone photoreduction? Explain.

Oxygen	$(S_1 = 22 \text{ kcal/mol})$
9,10-Diphenylanthracene	$(T_1 = 42 \text{ kcal/mol})$
trans-1,3-Pentadiene	$(T_1 = 59 \text{ kcal/mol})$
Naphthalene	$(T_1 = 61 \text{ kcal/mol})$
Biphenyl	$(T_1 = 66 \text{ kcal/mol})$
Toluene	$(T_1 = 83 \text{ kcal/mol})$
Benzene	$(T_1 = 84 \text{ kcal/mol})$

ESSAY

Fireflies and Photochemistry

The production of light as a result of a chemical reaction is called **chemiluminescence.** A chemiluminescent reaction generally produces one of the product molecules in an electronically excited state. The excited state emits a photon, and light is produced. If a reaction that produces light is biochemical, occurring in a living organism, the phenomenon is called **bioluminescence.**

The light produced by fireflies and other bioluminescent organisms has fascinated observers for many years. Many different organisms have developed the ability to emit light. They include bacteria, fungi, protozoans, hydras, marine worms, sponges, corals, jellyfish, crustaceans, clams, snails, squids, fish, and insects. Curiously, among the higher forms of life, only fish are included on the list. Amphibians, reptiles, birds, mammals, and the higher plants are excluded. Among the marine species, none is a freshwater organism. The excellent *Scientific American* article by McElroy and Seliger (see References) delineates the natural history, characteristics, and habits of many bioluminescent organisms.

The first significant studies of a bioluminescent organism were performed by the French physiologist Raphael Dubois in 1887. He studied the mollusk *Pholas dactylis,* a bioluminescent clam indigenous to the Mediterranean Sea. Dubois found that a cold-water extract of the clam was able to emit light for several minutes following the extraction. When the light emission ceased, it could be restored, he found, by a material extracted from the clam by hot water. A hot-water extract of the clam alone did not produce the luminescence. Reasoning carefully, Dubois concluded that there was an enzyme in the cold-water extract that was destroyed in hot water. The luminescent compound, however, could be extracted without destruction in either hot or cold water. He called the luminescent material **luciferin,** and the enzyme that induced it to emit light, **luciferase;** both names were derived from *Lucifer,* a Latin name meaning "bearer of light." Today the luminescent materials from all organisms are called *luciferins,* and the associated enzymes are called *luciferases.*

The most extensively studied bioluminescent organism is the firefly. Fireflies are found in many parts of the world and probably represent the most familiar example of bioluminescence. In such areas, on a typical summer evening, fireflies, or "lightning bugs," can frequently be seen to emit flashes of light as they cavort over the lawn or in the garden. It is now universally accepted that the luminescence of fireflies is a mating device. The male firefly flies about 2 feet above the ground and emits flashes of light at regular intervals. The female, who remains stationary on the ground, waits a characteristic interval and then flashes a response. In return, the male reorients his direction of flight toward her and flashes a signal once again. The entire cycle is rarely repeated more than 5 to 10 times before the male reaches the female. Fireflies of different species can recognize one another by their flash patterns, which vary in number, rate, and duration among species.

Although the total structure of the luciferase enzyme of the American firefly *Photinus pyralis* is unknown, the structure of the luciferin has been established. In spite of a large amount of experimental work, however, the complete nature of the chemical reactions that produce the light is still subject to some controversy. It is possible, nevertheless, to outline the most salient details of the reaction.

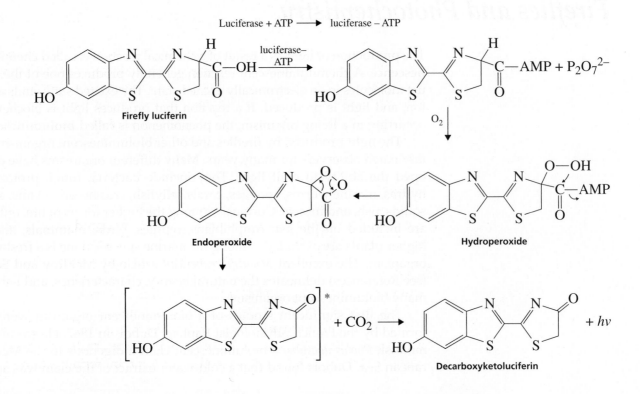

Luciferase + ATP ⟶ luciferase – ATP

Firefly luciferin ⟶ (luciferase–ATP) ⟶ ... $+ P_2O_7^{2-}$

Hydroperoxide (via O_2)

Endoperoxide

$+ CO_2 \longrightarrow$ **Decarboxyketoluciferin** $+ h\nu$

Besides the luciferin and the luciferase, other substances—magnesium(II), ATP (adenosine triphosphate), and molecular oxygen—are needed to produce the luminescence. In the postulated first step of the reaction, the luciferase complexes with an ATP molecule. In the second step, the luciferin binds to the luciferase and reacts with the already bound ATP molecule to become "primed." In this reaction, pyrophosphate ion is expelled, and AMP (adenosine monophosphate) becomes attached to the carboxyl group of the luciferin. In the third step, the luciferin–AMP complex is oxidized by molecular oxygen to form a hydroperoxide; this cyclizes with the carboxyl group, expelling AMP and forming the cyclic endoperoxide. This reaction would be difficult if the carboxyl group of the luciferin had not been primed with ATP. The endoperoxide is unstable and readily decarboxylates, producing decarboxyketoluciferin in an *electronically excited state*, which is deactivated by the emission of a photon (fluorescence). Thus, it is the cleavage of the four-membered-ring endoperoxide that leads to the electronically excited molecule and hence the bioluminescence.

That one of the two carbonyl groups, either that of the decarboxyketoluciferin or that of the carbon dioxide, should be formed in an excited state can be readily predicted from the orbital symmetry conservation principles of Woodward and Hoffmann. This reaction is formally like the decomposition of a cyclobutane ring and yields two ethylene molecules. In analyzing the forward course of that reaction, that is, 2 ethylene → cyclobutane, one can easily show that the reaction, which involves four π electrons, is forbidden for two ground-state ethylenes but allowed for only one ethylene in the ground state and the other in an excited state. This suggests that, in the reverse process, one of the ethylene molecules should be formed in an excited state. Extending these arguments to the endoperoxide also suggests that one of the two carbonyl groups should be formed in its excited state.

The emitting molecule, decarboxyketoluciferin, has been isolated and synthesized. When it is excited photochemically by photon absorption in basic solution (pH > 7.5–8.0), it fluoresces, giving a fluorescence emission spectrum that is identical to the emission spectrum produced by the interaction of firefly luciferin and firefly luciferase. The emitting form of decarboxyketoluciferin has thus been identified as the **enol dianion.** In neutral or acidic solution, the emission spectrum of decarboxyketoluciferin does not match the emission spectrum of the bioluminescent system.

The exact function of the enzyme firefly luciferase is not yet known, but it is clear that all these reactions occur while luciferin is bound to the enzyme as a substrate. Also, because the enzyme undoubtedly has several basic groups ($-COO^-$, $-NH_2$, and so on), the buffering action of those groups would easily explain why the enol dianion is also the emitting form of decarboxyketoluciferin in the biological system.

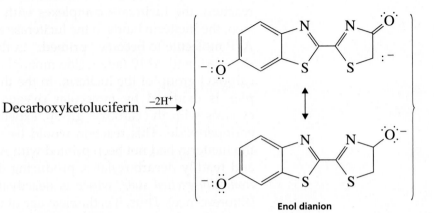

Decarboxyketoluciferin $\xrightarrow{-2H^+}$

Enol dianion

Most chemiluminescent and bioluminescent reactions require oxygen. Likewise, most produce an electronically excited emitting species through the decomposition of a **peroxide** of one sort or another. In the experiment that follows, a **chemiluminescent** reaction that involves the decomposition of a peroxide intermediate is described.

REFERENCES

Clayton, R. K. "The Luminescence of Fireflies and Other Living Things." Chap. 6 in *Light and Living Matter. Vol. 2: The Biological Part.* New York: McGraw-Hill, 1971.

Fox, J. L. "Theory May Explain Firefly Luminescence." *Chemical and Engineering News,* 56 (March 6, 1978): 17.

Harvey, E. N. *Bioluminescence.* New York: Academic Press, 1952.

Hastings, J. W. "Bioluminescence." *Annual Review of Biochemistry,* 37 (1968): 597.

McCapra, F. "Chemical Mechanisms in Bioluminescence." *Accounts of Chemical Research,* 9 (1976): 201.

McElroy, W. D., and Seliger, H. H. "Biological Luminescence." *Scientific American,* 207 (December 1962): 76.

McElroy, W. D., Seliger, H. H., and White, E. H. "Mechanism of Bioluminescence, Chemiluminescence and Enzyme Function in the Oxidation of Firefly Luciferin." *Photochemistry and Photobiology,* 10 (1969): 153.

Seliger, H. H., and McElroy, W. D. *Light: Physical and Biological Action.* New York: Academic Press, 1965.

52   EXPERIMENT 52

Luminol

Chemiluminescence

Energy transfer

Reduction of a nitro group

Amide formation

In this experiment, the chemiluminescent compound **luminol,** or **5-amino-phthalhydrazide,** will be synthesized from 3-nitrophthalic acid.

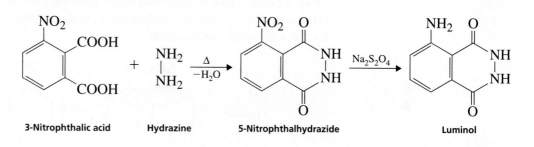

| 3-Nitrophthalic acid | Hydrazine | 5-Nitrophthalhydrazide | Luminol |

The first step of the synthesis is the simple formation of a cyclic diamide, 5-nitrophthalhydrazide, by reaction of 3-nitrophthalic acid with hydrazine. Reduction of the nitro group with sodium dithionite affords luminol.

In neutral solution, luminol exists largely as a dipolar anion (zwitterion). This dipolar ion exhibits a weak blue fluorescence after being exposed to light. However, in alkaline solution, luminol is converted to its dianion, which may be oxidized by molecular oxygen to give an intermediate that is chemiluminescent. The reaction is thought to have the following sequence:

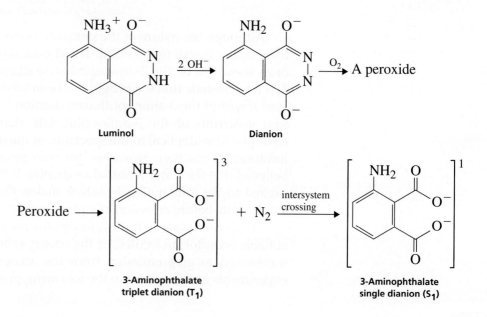

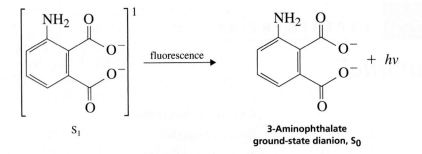

3-Aminophthalate
ground-state dianion, S_0

The dianion of luminol undergoes a reaction with molecular oxygen to form a peroxide of unknown structure. This peroxide is unstable and decomposes with the evolution of nitrogen gas, producing the 3-amino-phthalate dianion in an electronically excited state. The excited dianion emits a photon that is visible as light. One very attractive hypothesis for the structure of the peroxide postulates a cyclic endoperoxide that decomposes by the following mechanism:

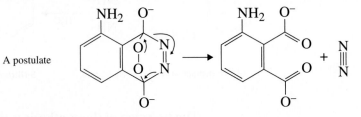

A postulate

Certain experimental facts argue against this intermediate, however. For instance, certain acyclic hydrazides that cannot form a similar intermediate have also been found to be chemiluminescent.

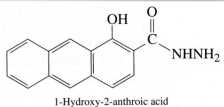

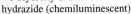

1-Hydroxy-2-anthroic acid
hydrazide (chemiluminescent)

Although the nature of the peroxide is still debatable, the remainder of the reaction is well understood. The chemical products of the reaction have been shown to be the 3-aminophthalate dianion and molecular nitrogen. The intermediate that emits light has been identified definitely as the *excited state singlet* of the 3-aminophthalate dianion.[1] Thus, the fluorescence emission spectrum of the 3-aminophthalate dianion (produced by photon absorption) is identical to the spectrum of the light emitted from the chemiluminescent reaction. However, for numerous complicated reasons, it is believed that the 3-aminophthalate dianion is formed first as a vibrationally excited triplet state molecule, which makes the intersystem crossing to the singlet state before emission of a photon.

The excited state of the 3-aminophthalate dianion may be quenched by suitable acceptor molecules, or the energy (about 50–80 Kcal/mol) may be transferred to give emission from the acceptor molecules. Several such experiments are described in the following procedure.

---

[1] The terms *singlet, triplet, intersystem crossing, energy transfer,* and *quenching* are explained in Experiment 51.

The system chosen for the chemiluminescence studies of luminol in this experiment uses dimethylsulfoxide $(CH_3)_2SO$ as the solvent, potassium hydroxide as the base required for the formation of the dianion of luminol, and molecular oxygen. Several alternative systems have been used, substituting hydrogen peroxide and an oxidizing agent for molecular oxygen. An aqueous system using potassium ferricyanide and hydrogen peroxide is an alternative system used frequently.

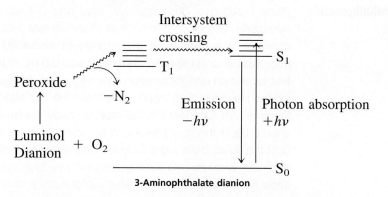

Fluorescence emission spectrum of the 3-aminophthalate dianion.

REFERENCES

Rahaut, M. M. "Chemiluminescence from Concerted Peroxide Decomposition Reactions." *Accounts of Chemical Research, 2* (1969): 80.

White, E. H., and Roswell, D. F. "The Chemiluminescence of Organic Hydrazides." *Accounts of Chemical Research, 3* (1970): 54.

REQUIRED READING

Review: Technique 3   Reaction Methods, Section 7.10
New: Essay   Fireflies and Photochemistry

SPECIAL INSTRUCTIONS

This entire experiment can be completed in about 1 hour. When you are working with hydrazine, you should remember that it is toxic and should not be spilled on the skin. It is also a suspected carcinogen. Dimethylsulfoxide may also be toxic; avoid breathing the vapors or spilling it on your skin.

A darkened room is required to observe adequately the chemiluminescence of luminol. A darkened hood that has had its window covered with butcher paper or aluminum foil also works well. Other fluorescent dyes besides those mentioned (for instance, 9,10-diphenylanthracene) can also be used for the energy-transfer experiments. The dyes selected may depend on what is immediately available. The instructor may have each student use one dye for the energy-transfer experiments, with one student making a comparison experiment without a dye.

SUGGESTED WASTE DISPOSAL

Dispose of the filtrate from the vacuum filtration of 5-nitrophthalhydrazide in the container designated for nonhalogenated organic solvents. The filtrate from the vacuum filtration of 5-aminophthalhydrazide may be diluted with water and poured into the waste container designated for aqueous waste. The mixture containing potassium hydroxide, dimethyl-sulfoxide, and luminol should be placed in the special container designated for this material.

PROCEDURE

Part A. 5-Nitrophthalhydrazide

Place 0.300 g of 3-nitrophthalic acid and 0.4 mL of a 10% aqueous solution of hydrazine (use gloves) in a small side-arm test tube.[2] At the same time, heat 4 mL of water in a beaker on a hot plate to about 80°C. Heat the test tube over a microburner until the solid dissolves. Add 0.8 mL of triethylene glycol and clamp the test tube in an upright position on a ring stand. Place a thermometer (do not seal the system) and a boiling stone in the test tube and attach a piece of pressure tubing to the side arm. Connect this tubing to an aspirator (use a trap). The thermometer bulb should be in the liquid as much as possible. Heat the solution with a microburner until the liquid boils vigorously and the refluxing water vapor is drawn away by the aspirator vacuum (the temperature will rise to about 120°C). Continue heating and allow the temperature to increase rapidly until it rises just above 200°C. This heating requires 1–2 minutes, and you must watch the temperature closely to avoid heating the mixture well above 200°C. Remove the burner briefly when this temperature has been achieved and then resume gentle heating to maintain a fairly constant temperature of 210–220°C for about 2 minutes. Allow the test tube to cool to about 100°C, add the 4.0 mL of hot water that was prepared previously, and cool the test tube to room temperature by allowing tap water to flow over the outside of the test tube. Collect the brown crystals of 5-nitrophthalhydrazide by vacuum filtration, using a small Hirsch funnel. It is not necessary to dry the product before you go on with the next reaction step.

Part B. Luminol (5-Aminophthalhydrazide)

Transfer the moist 5-nitrophthalhydrazide to a 13 × 100-mm test tube. Add 1.30 mL of a 10% sodium hydroxide solution and agitate the mixture until the hydrazide dissolves. Add 0.80 g of sodium dithionite dihydrate (sodium hydrosulfite dihydrate, $Na_2S_2O_4 \cdot 2H_2O$). Using a Pasteur pipet, add 1–2 mL of water to wash the solid from the walls of the test tube. Add a boiling stone to the test tube. Heat the test tube until the solution boils, agitate the solution, and maintain the boiling, continuing agitation, for 5 minutes. Add 0.50 mL of glacial acetic acid and cool the test tube to room temperature by allowing tap water to flow over the outside of it. Agitate the mixture during the cooling step. Collect the light yellow or gold crystals of luminol by vacuum filtration, using a small Hirsch funnel. Save a small sample of this product, allow it to dry overnight, and determine its melting point (mp 319–320°C). The remainder of the luminol may be used without drying for the chemiluminescence experiments.

Part C. Chemiluminescence Experiments

CAUTION

Be careful not to let any of the mixture touch your skin while shaking the flask. Hold the stopper securely.

Cover the bottom of a 10-mL Erlenmeyer flask with a layer of potassium hydroxide pellets. Add enough dimethylsulfoxide to cover the pellets. Add about

---

[2] A 10% aqueous solution of hydrazine can be prepared by diluting 15.6 g of a commercial 64% hydrazine solution to a volume of 100 mL using water.

0.025 g of the moist luminol to the flask, stopper it, and shake it vigorously to mix air into the solution.[3] In a dark room, a faint glow of bluish white light will be visible. The intensity of the glow will increase with continued shaking of the flask and occasional removal of the stopper to admit more air.

To observe energy transfer to a fluorescent dye, dissolve one or two crystals of the indicator dye in about 0.25 mL of water. Add the dye solution to the dimethylsulfoxide solution of luminol, stopper the flask, and shake the mixture vigorously. Observe the intensity and the color of the light produced.

The following table shows some dyes and the colors produced when they are mixed with luminol. Other dyes not included on this list may also be tested in this experiment.

Fluorescent Dye	Color
No dye	Faint bluish white
2,6-Dichloroindophenol	Blue
9-Aminoacridine	Blue-green
Eosin	Salmon pink
Fluorescein	Yellow-green
Dichlorofluorescein	Yellow-orange
Rhodamine B	Green
Phenolphthalein	Purple

ESSAY

The Chemistry of Sweeteners

Americans, as no other nationalities in the world do, possess a particularly demanding sweet tooth. Our craving for sugar, either added to food or included in candies and desserts, is astounding. Even when we are choosing a food that is not considered to be sweet, we ingest large quantities of sugar. A casual examination of the *Nutrition Facts* label and the list of ingredients of virtually any processed food will reveal that sugar is generally one of the principal ingredients.

Americans, paradoxically, are also obsessed by the need to diet. As a result, the search for noncaloric substitutes for natural sugar represents a multimillion-dollar business in this country. There is a ready market for foods that taste sweet but don't contain sugar.

For a molecule to taste sweet, it must fit into a taste bud site where a nerve impulse can carry the message of sweetness from the tongue to the brain. Not all natural sugars, however, trigger an equivalent neural response. Some sugars, such as glucose, have a relatively bland taste, and others, such as fructose, taste very sweet. Fructose, in fact, has a sweeter taste than common table sugar, or sucrose. Furthermore, individuals vary in their ability to perceive sweet substances. The relationship between perceived sweetness and molecular structure is complicated, and, to date, it is rather poorly understood.

---

[3] An alternative method for demonstrating chemiluminescence, using potassium ferricyanide and hydrogen peroxide as oxidizing agents, is described in E. H. Huntress, L. N. Stanley, and A. S. Parker, *Journal of Chemical Education, 11* (1934): 142.

The most common sweetener is, of course, common table sugar, or **sucrose.** Sucrose is a disaccharide, consisting of a unit of glucose and a unit of fructose connected by a 1,2-glycosidic linkage. Sucrose is purified and crystallized from the syrups that are extracted from such plants as sugar cane and sugar beets.

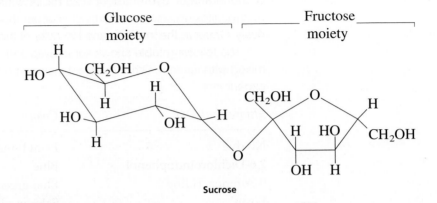

Sucrose

When sucrose is hydrolyzed, it yields one molecule of D-fructose and one of D-glucose. This hydrolysis is catalyzed by an enzyme, **invertase,** and produces a mixture known as **invert sugar.** Invert sugar derives its name from the fact that the mixture is levorotatory, whereas sucrose is dextrorotatory. Thus, the sign of the rotation has been "inverted" in the course of the hydrolysis. Invert sugar is somewhat sweeter than sucrose, owing to the presence of free fructose. **Honey** is composed mostly of invert sugar, which is the reason it has such a sweet taste.

Persons who suffer from diabetes are urged to avoid sugar in their diets. Nevertheless, those individuals also have a craving for sweet foods. A substitute sweetener that is used for food items recommended for diabetics is **sorbitol,** which is an alcohol formed by the catalytic hydrogenation of glucose. Sorbitol has about 60% the sweetness of sucrose. It is a common ingredient in products such as sugarless chewing gum. Even though sorbitol is a different substance from sucrose, it still possesses about the same number of calories per gram. Therefore, sorbitol is not a suitable sweetener for diet foods or beverages.

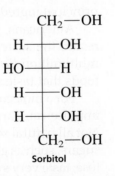

Sorbitol

As sucrose and honey are implicated in problems of tooth decay, as well as being culprits in the continuing battle against obesity, an active field of study is the search for new, noncaloric, noncarbohydrate sweeteners. Even if such a nonnutritive sweetener possessed some calories, if it were very sweet, it would not be necessary to use as much of the sweetener; therefore, the impact on dental hygiene and on diet would be less.

The first artificial sweetener to be used extensively was **saccharin,** which is used commonly as its more soluble sodium salt. Saccharin is about 300 times sweeter than sucrose. The discovery of saccharin was hailed as a great benefit for diabetics because it could be used as an alternative to sugar. As a pure substance, the sodium salt of saccharin has an intense, sweet taste, with a somewhat bitter aftertaste. Because it has such an intense taste, it can be used in minute amounts to achieve the desired effect. In some preparations, sorbitol is added to ameliorate the bitter aftertaste. Prolonged studies on laboratory animals have shown that saccharin is a possible carcinogen. In spite of this health risk, the government has permitted saccharin to be used in foods that are intended to be used by diabetics.

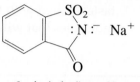

Saccharin (sodium salt)

Another artificial sweetener, which gained wide use in the 1960s and 1970s, is **sodium cyclamate.** Sodium cyclamate, which is 33 times as sweet as sucrose, belongs to the class of compounds known as the **sulfamates.** The sweet taste of many of the sulfamates has been known since 1937, when Sweda accidentally discovered that sodium cyclamate had a powerfully sweet taste. The availability of sodium cyclamate spurred the popularity of diet soft drinks. Unfortunately, in the 1970s, research showed that a metabolite of sodium cyclamate, cyclohexylamine, posed some potentially serious health risks, including a risk of cancer. This sweetener has been withdrawn from the market.

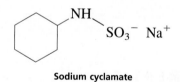

Sodium cyclamate

The most widely used artificial sweetener available today is a dipeptide, consisting of a unit of aspartic acid linked to a unit of phenylalanine. The carboxyl group of the phenylalanine moiety has been converted to the methyl ester. This substance is known commercially as **aspartame,** but it is also sold under the trade names **NutraSweet** and **Equal.** Aspartame is about 200 times more sweet than sucrose. It is found in diet soft drinks, puddings, juices, and many other foods. Unfortunately, aspartame is not stable when heated, so it is not suitable as an ingredient in cooking. Other dipeptides that have structures similar to that of aspartame are many thousands of times sweeter than sucrose.

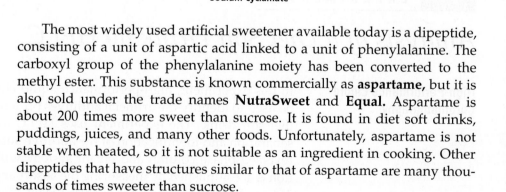

Aspartame

When aspartame was being developed as a commercial product, concern was raised over potential health hazards associated with its use. The potential cancer-causing effects of aspartame, along with other potential adverse side effects, were considered. Extensive testing of this product demonstrated that it met health-risk criteria established by the Food and Drug Administration, which granted approval for the sale of aspartame as a food additive in 1974.

The search for new substances that can serve as sweeteners continues. There is a great deal of interest in substances that are naturally occurring and that can be isolated from various plants. In addition, research, including molecular modeling studies and spectroscopic investigations, is attempting to clarify exactly what structural features are required for a sweet taste. Armed with that information, chemists will then be able to synthesize molecules that will be designed specifically for their sweet taste.

REFERENCES

Barker, S. A., Garegg, P. J., Bucke, C., Rastall, R. A., Sharon, N., Lis, H., and Hounsell, E. F. "Contemporary Carbohydrate Chemistry." *Chemistry in Britain, 26* (1990): 663. (This is a series of five articles, each written by one or two of the cited authors, compiled as part of a series of articles on carbohydrate chemistry.)

Bragg, R. W., Chow, Y., Dennis, L., Ferguson, L. N., Howell, S., Morga, G., Ogino, C., Pugh, H., and Winters, M. "Sweet Organic Chemistry." *Journal of Chemical Education, 55* (1978): 281.

Crammer, B., and Ikan, R. "Sweet Glycosides from the Stevia Plant." *Chemistry in Britain, 22* (1986): 915.

Sharon, N. "Carbohydrates." *Scientific American, 243* (November 1980): 90.

53   EXPERIMENT 53

Analysis of a Diet Soft Drink by HPLC

High-Performance Liquid Chromatography

In this experiment, high-performance liquid chromatography (HPLC) will be used to identify the artificial additives present in a sample of commercial diet soft drink. The experiment uses HPLC as an analytical tool for the separation and identification of the additive substances. The method uses a reversed-phase column and eluent system, with isochratic elution. Detection is accomplished by measuring the absorbance of ultraviolet radiation at 254 nm by the solution as it is eluted from the column. The mobile phase that will be used is a mixture of 80% $1M$ acetic acid and 20% acetonitrile, buffered to pH 4.2.

Diet soft drinks contain many chemical additives, including several substances that can be used as artificial sweeteners. Among these additives are the four substances that we will be detecting in this experiment: caffeine, saccharine, benzoic acid, and aspartame. The structure of these compounds is shown here.

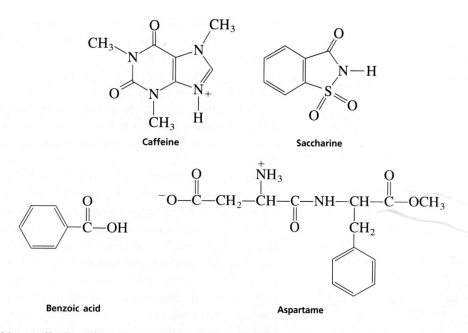

Caffeine

Saccharine

Benzoic acid

Aspartame

You will identify each compound in a sample of diet soft drink by its retention time on the HPLC column. You will be provided with data for a reference mixture of each substance in a test mixture in order to compare retention times in your test sample with a set of standards.

REQUIRED READING

New: Technique 21   High-Performance Liquid Chromatography (HPLC)

SPECIAL INSTRUCTIONS

The instructor will provide specific instruction in the operation of the particular HPLC instrument being used in your laboratory. The instructions that follow indicate the general procedure.

SUGGESTED WASTE DISPOSAL

Discard the excess acetic acid–methanol solvent in the organic waste container designated for the disposal of nonhalogenated organic wastes. The acetonitrile–acetic acid solvent mixture should be collected in a specially designated container so that it may be either safely discarded or reused.

PROCEDURE

Following your instructor's directions, form a small group of students to perform this experiment. Each small group will analyze a different diet soft drink, and the results obtained by each group will be shared among all students in the class.

The instructor will prepare a mixed standard of the four components, consisting of 200 mg of aspartame, 40 mg of benzoic acid, 40 mg of saccharine, and 20 mg of caffeine in 100 mL of solvent. The solvent for these standards is a mixture of 80% acetic acid and 20% methanol, buffered to pH 4.2 with 50% sodium hydroxide. The lab instructor will also run an HPLC of this standard mixture beforehand,

and you should obtain a copy of the results. Some of the steps described in the next two paragraphs may be completed in advance by your instructor.

You may select from a variety of diet soft drinks with different chemical compositions.[1] Select a soft drink from the supply shelf and dispense approximately 50 mL into a small flask.

Completely remove the carbon dioxide gas, which causes the bubbles in the soft drink, before examining the sample by HPLC. The bubbles will affect the retention times of the compounds and possibly cause damage to the expensive HPLC columns. Most of the gas can be eliminated by allowing the containers of soft drinks to remain open overnight. To remove the final traces of dissolved gases, set up a filtering flask with a Büchner funnel and connect it to a vacuum line. Place a 4-μm filter in the Büchner funnel. (*Note:* Be sure to use a piece of filter paper, not one of the colored spacers that are placed between the pieces of filter paper. The spacers are normally blue.) Filter the soda sample by vacuum filtration through the 4-μm filter and place the filtered sample in a *clean* 4-dram snap-cap vial.

Before using the HPLC instrument, be certain that you have obtained specific instruction in the operation of the instrument in your laboratory. Alternatively, your instructor may have someone operate the instrument for you. Before your sample is analyzed on the HPLC instrument, it should be filtered one more time, this time through a 0.2-μm filter. The recommended sample size for analysis is 10 μL. The solvent system used for this analysis is a mixture of 80% 1M acetic acid and 20% acetonitrile, buffered to pH 4.2. The instrument will be operated in an isochratic mode.

When you examine the chart obtained from the analysis, you may find that the peak corresponding to aspartame appears to be rather small. The peak is small because aspartame absorbs ultraviolet radiation most efficiently at 220 nm, whereas the detector is set to measure the absorption of light at 254 nm. Nevertheless, the observed retention time of aspartame will not depend upon the setting of the detector, and the interpretation of the results should not be affected. The expected order of elution is saccharine (first), caffeine, aspartame, and benzoic acid. Another interesting point is that the caffeine peak appears to be quite large in this analysis; it is nevertheless quite small when compared with the peak that would be obtained if you injected coffee into the HPLC. For a caffeine peak from coffee to fit onto your graph, you would have to dilute the coffee *at least* 10-fold. Even decaffeinated coffee usually has more caffeine in it than most sodas (decaffeinated coffee is required to be only 95–96% decaffeinated).

When you have completed your experiment, report your results by preparing a table showing the retention times of each of the four standard substances. In your report, be sure to identify the diet soft drink that you used and to identify the substances that you found in that sample. Also report the substances that were found in each of the other soft-drink samples that were tested by other groups in your class.

REFERENCE

Bidlingmeyer, B. A., and Schmitz, S. "The Analysis of Artificial Sweeteners and Additives in Beverages by HPLC." *Journal of Chemical Education, 68* (August 1991): A195.

---

[1] *Note to the instructor:* The experiment will be more interesting if the diet soft drink TAB is included among the choices. TAB is one of the few readily available diet soft drinks that contain substantial amounts of saccharine.

PART 4

Identification of Organic Substances

54 **EXPERIMENT 54**

Identification of Unknowns

Qualitative organic analysis, the identification and characterization of unknown compounds, is an important part of organic chemistry. Every chemist must learn the appropriate methods for establishing the identity of a compound. In this experiment, you will be issued an unknown compound and will be asked to identify it through chemical and spectroscopic methods. Your instructor may give you a general unknown or a specific unknown. With a **general unknown,** you must first determine the class of compound to which the unknown belongs, that is, identify its main functional group; then you must determine the specific compound in that class that corresponds to the unknown. With a **specific unknown,** you will know the class of compound (ketone, alcohol, amine, and so on) in advance, and it will be necessary to determine only whatever specific member of that class was issued to you as an unknown. This experiment is designed so that the instructor can issue several general unknowns or as many as six successive specific unknowns, each having a different main functional group.

Although there are millions of organic compounds that an organic chemist might be called on to identify, the scope of this experiment is necessarily limited. In this textbook, about 500 compounds are included in the tables of possible unknowns given for the experiment (see Appendix 1). Your instructor may wish to expand the list of possible unknowns, however. In such a case, you will have to consult more extensive tables, such as those found in the work compiled by Rappoport (see References). In addition, the experiment is restricted to include only seven important functional groups:

Aldehydes	Amines
Ketones	Alcohols
Carboxylic acids	Esters
Phenols	

Even though this list of functional groups omits some of the important types of compounds (alkyl halides, alkenes, alkynes, aromatics, ethers, amides, mercaptans, nitriles, acid chlorides, acid anhydrides, nitro compounds, and so on), the methods introduced here can be applied equally well to other classes of compounds. The list is sufficiently broad to illustrate all the principles involved in identifying an unknown compound.

In addition, although many of the functional groups listed as being excluded will not appear as the major functional group in a compound, several of them will frequently appear as secondary, or subsidiary, functional groups. Three examples of this are presented here.

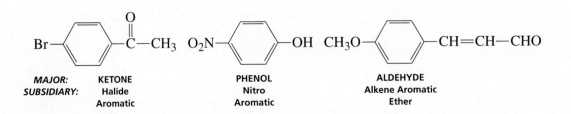

MAJOR:	**KETONE**	**PHENOL**	**ALDEHYDE**
SUBSIDIARY:	Halide	Nitro	Alkene Aromatic
	Aromatic	Aromatic	Ether

The groups included that have subsidiary status are

—Cl	Chloro	—NO$_2$	Nitro	C=C	Double Bond
—Br	Bromo	—C≡N	Cyano	C≡C	Triple Bond
—I	Iodo	—OR	Alkoxy	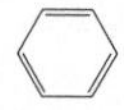 ⬡	Aromatic

The experiment presents all the chief chemical and spectroscopic methods of determining the main functional groups, and it includes methods for verifying the presence of the subsidiary functional groups as well. It will usually not be necessary to determine the presence of the subsidiary functional groups to identify the unknown compound correctly. Every piece of information helps the identification, however, and if these groups can be detected easily, you should not hesitate to determine them. Finally, complex bifunctional compounds are generally avoided in this experiment; only a few are included.

How to Proceed

Fortunately, we can detail a fairly straightforward procedure for determining all the necessary pieces of information. This procedure consists of the following steps:

Part One: Chemical Classification

1. Preliminary classification by physical state, color, and odor
2. Melting-point or boiling-point determination; other physical data
3. Purification, if necessary
4. Determination of solubility behavior in water and in acids and bases
5. Simple preliminary tests: Beilstein, ignition (combustion)
6. Application of relevant chemical classification tests
7. Inspection of tables for possible structure(s) of unknown; elimination of unlikely compounds

Part Two: Spectroscopy

8. Determination of infrared and NMR spectra

Part Three: Optional Procedures

9. Elemental analysis, if necessary
10. Preparation of derivatives, if required
11. Confirmation of identity

Each of these steps is discussed briefly in the following sections.

1. PRELIMINARY CLASSIFICATION

Note the physical characteristics of the unknown, including its color, its odor, and its physical state (liquid, solid, crystalline form). Many compounds have characteristic colors or odors, or they crystallize with a specific crystal structure. This information can often be found in a handbook and can be checked later. Compounds with a high degree of conjugation are frequently yellow to red. Amines often have a fishlike odor. Esters have a pleasant fruity or floral odor. Acids have a sharp and pungent odor. A part of the training of every good chemist includes cultivating the ability to recognize familiar or typical odors. As a note of caution, many compounds have distinctly unpleasant or nauseating odors. Some have corrosive vapors. Sniff any unknown substance with the greatest caution. As a first step, open the container, hold it away from you and, using your hand, carefully waft the vapors toward your nose. If you get past this stage, a closer inspection will be possible.

2. MELTING-POINT OR BOILING-POINT DETERMINATION

The single most useful piece of information to have for an unknown compound is its melting point or boiling point. Either piece of data will drastically limit the compounds that are possible. The electric melting-point apparatus gives a rapid and accurate measurement (see Technique 9, Sections 9.5 and 9.7). To save time, you can often determine two separate melting points. The first determination can be made rapidly to get an approximate value. Then you can determine the second melting point more carefully. Because some of the unknown solids contain traces of impurities, you may find that your observed melting point is lower than the values found in the tables in Appendix 1. This is especially true for low-melting compounds (<50°C). For these low-melting compounds, it is a good idea to look at compounds in the tables in Appendix 1 that have melting points above your observed melting-point range. The same advice may apply to other solid compounds issued to you as unknowns.

The boiling point is easily obtained by a simple distillation of the unknown (Technique 14, Section 14.4), by reflux (Technique 13, Section 13.2), or by a microboiling-point determination (Technique 13, Section 13.2). The simple distillation has the advantage in that it also purifies the compound. The smallest distilling flask available should be used if a simple distillation is performed, and you should be sure that the thermometer bulb is fully immersed in the vapor of the distilling liquid. The liquid should be distilled rapidly to determine an accurate boiling-point value. The microboiling-point method requires the least amount of unknown, but the refluxing method is more reliable and requires much less liquid than performing a distillation.

When inspecting the tables of unknowns in Appendix 1, you may find that the observed boiling point that you determined is lower than the value for the corresponding compound listed in the tables. This is especially true for compounds boiling above 200°C. It is less likely, but not impossible, that the observed boiling point of your unknown will be higher than the value given in the table. Thus, your strategy should be to look for boiling points of compounds in the tables that are nearly equal to or above the value you obtained, within a range of about ±5°C. For high-boiling liquid compounds (>200°C), you may need to apply a thermometer correction (Technique 13, Section 13.3).

3. PURIFICATION

If the melting point of a solid has a wide range (about 5°C), the solid should be recrystallized and the melting point redetermined.

If a liquid was highly colored before distillation, if it yielded a wide boiling-point range, or if the temperature did not hold constant during the distillation, it should be redistilled to determine a new temperature range. A reduced-pressure distillation is in order for high-boiling liquids or for those that show any sign of decomposition on heating.

Occasionally, column chromatography may be necessary to purify solids that have large amounts of impurities and do not yield satisfactory results on crystallization.

Acidic or basic impurities that contaminate a neutral compound may often be removed by dissolving the compound in a low-boiling solvent, such as CH_2Cl_2 or ether, and extracting with 5% $NaHCO_3$ or 5% HCl, respectively. Conversely, acidic or basic compounds can be purified by dissolving them in 5% $NaHCO_3$ or 5% HCl, respectively, and extracting them with a low-boiling organic solvent to remove impurities. After the aqueous solution has been neutralized, the desired compound can be recovered by extraction.

4. SOLUBILITY BEHAVIOR

Tests on solubility are described fully in Experiment 54A. They are extremely important. Determine the solubility of small amounts of the unknown in water, 5% HCl, 5% $NaHCO_3$, 5% NaOH, concentrated H_2SO_4, and organic solvents. This information reveals whether a compound is an acid, a base, or a neutral substance. The sulfuric acid test reveals whether a neutral compound has a functional group that contains an oxygen, a nitrogen, or a sulfur atom that can be protonated. This information allows you to eliminate or to choose various functional-group possibilities. The solubility tests must be made on *all* unknowns.

5. PRELIMINARY TESTS

The two combustion tests, the Beilstein test (Experiment 54B) and the ignition test (Experiment 54C), can be performed easily and quickly, and they often give valuable information. It is recommended that they be performed on all unknowns.

6. CHEMICAL CLASSIFICATION TESTS

The solubility tests usually suggest or eliminate several possible functional groups. The chemical classification tests listed in Experiments 54D to 54I allow you to distinguish among the possible choices. Choose only those tests that the solubility tests suggest might be meaningful. Time will be wasted performing unnecessary tests. There is no substitute for a firsthand, thorough knowledge of these tests. Study each of the sections carefully until you understand the significance of each test. Also, it is essential to actually try the tests on *known* substances. In this way, it will be easier to recognize a positive test. Appropriate test compounds are listed for many of the tests. When you are performing a test that is new to you, it is always good practice to run the test separately on both a known substance and the unknown *at the same time*. This practice lets you compare results directly.

Do not perform the chemical tests either haphazardly or in a methodical, comprehensive sequence. Instead, use the tests selectively. Solubility tests automatically eliminate the need for some of the chemical tests. Each successive test will either eliminate the need for another test or dictate its use. You should also examine the tables of unknowns in Appendix 1 carefully. The boiling point or the melting point of the unknown may eliminate the need for many of the tests. For instance, the possible compounds may simply not include one with a double bond. *Efficiency* is the key word here. Do not waste time performing nonsensical or unnecessary tests. Many possibilities can be eliminated on the basis of logic alone.

How you proceed with the following steps may be limited by your instructor's wishes. Many instructors may restrict your access to infrared and NMR spectra until you have narrowed your choices to a few compounds *all within the same class.* Others may have you determine these data routinely. Some instructors may want students to perform elemental analysis on all unknowns; others may restrict it to only the most essential situations. Again, some instructors may require derivatives as a final confirmation of the compound's identity; others may not wish to use them at all.

7. INSPECTION OF TABLES FOR POSSIBLE STRUCTURES

Once the melting or boiling point, the solubilities, and the main chemical classification tests have been made, you should be able to identify the class of compound (aldehyde, ketone, and so on). At this stage, with the melting point or boiling point as a guide, you can compile a list of possible compounds from one of the appropriate tables in Appendix 1. It is very important to draw out the structures of compounds that fit the solubility, classification tests, and melting point or boiling point that were determined. If necessary, you can look up the structures in the *CRC Handbook, The Merck Index,* or the *Aldrich Handbook.* Remember that the boiling point or melting point recorded in the table may be higher than what you obtained in the laboratory (see Section 2 on page 450).

The short list that you developed by inspection of the tables in Appendix 1 and the structures drawn should suggest that some additional tests may be needed to distinguish among the possibilities. For instance, one compound may be a methyl ketone, and the other may not. The iodoform test is called for to distinguish the two possibilities. The tests for the subsidiary functional groups may also be required. These tests are described in Experiments 54B and 54C. These tests should also be studied carefully; there is no substitute for firsthand knowledge about these tests.

8. SPECTROSCOPY

Spectroscopy is probably the most powerful and modern tool available to the chemist for determining the structure of an unknown compound. It is often possible to determine structure through spectroscopy alone. On the other hand, there are also situations for which spectroscopy may not be of much help, and the traditional methods must be relied on. For this reason, you should not use spectroscopy to the exclusion of the more traditional tests but rather as a confirmation of those results. Nevertheless, the main functional groups and their immediate environmental features can be determined quickly and accurately with spectroscopy.

9. ELEMENTAL ANALYSIS

Elemental analysis—which allows you to determine the presence of nitrogen, sulfur, or a specific halogen atom (Cl, Br, I) in a compound—is often useful; however, other information may render these tests unnecessary. A compound identified as an amine by solubility tests obviously contains nitrogen. Many nitrogen-containing groups (for instance, nitro groups) can be identified by infrared spectroscopy. Finally, it is not usually necessary to identify a specific halogen. The simple information that the compound contains a halogen (any halogen) may be enough information to distinguish between two compounds. A simple Beilstein test provides this information.

10. DERIVATIVES

One of the principal tests for the correct identification of an unknown compound comes in trying to convert the compound by a chemical reaction to another known compound. This second compound is called a **derivative.** The best derivatives are solid compounds because the melting point of a solid provides an accurate and reliable identification of most compounds. Solids are also easily purified through crystallization. The derivative provides a way of distinguishing two otherwise very similar compounds. Usually, they will have derivatives (both prepared by the same reaction) that have different melting points. Tables of unknowns and derivatives are listed in Appendix 1. Procedures for preparing derivatives are given in Appendix 2.

11. CONFIRMATION OF IDENTITY

A rigid and final test for identifying an unknown can be made if an "authentic" sample of the compound is available for comparison. One can compare infrared and NMR spectra of the unknown compound with the spectra of the known compound. If the spectra match, peak for peak, then the identity is probably certain. Other physical and chemical properties can also be compared. If the compound is a solid, a convenient test is the mixture melting point (Technique 9, Section 9.4). Thin-layer or gas-chromatographic comparisons may also be useful. For thin-layer analysis, however, it may be necessary to experiment with several different development solvents to reach a satisfactory conclusion about the identity of the substance in question.

Although we cannot be complete in this experiment in terms of the functional groups covered or the tests described, the experiment should provide a good introduction to the methods and the techniques chemists use to identify unknown compounds. Textbooks that cover the subject more thoroughly are listed in the References. You are encouraged to consult these for more information, including specific methods and classification tests.

REFERENCES

Comprehensive Textbooks

Cheronis, N. D., and Entrikin, J. B. *Identification of Organic Compounds.* New York: Wiley-Interscience, 1963.

Pasto, D. J., and Johnson, C. R. *Laboratory Text for Organic Chemistry.* Englewood Cliffs, NJ: Prentice-Hall, 1979.

Shriner, R. L., Hermann, C. K. F., Morrill, T. C., Curtin, D. Y., and Fuson, R. C. *The Systematic Identification of Organic Compounds,* 8th ed. New York: Wiley, 2003.

Spectroscopy

Bellamy, L. J. *The Infra-red Spectra of Complex Molecules,* 3rd ed. New York: Methuen, 1975.

Colthup, N. B., Daly, L. H., and Wiberly, S. E. *Introduction to Infrared and Raman Spectroscopy,* 3rd ed. San Diego, CA: Academic Press, 1990.

Lin-Vien, D., Colthup, N. B., Fateley, W. B., and Grasselli, J. G. *The Handbook of Infrared and Raman Characteristic Frequencies of Organic Molecules.* San Diego, CA: Academic Press, 1991.

Nakanishi, K. *Infrared Absorption Spectroscopy,* 2nd ed. San Francisco: Holden-Day, 1977.

Pavia, D. L., Lampman, G. M., and Kriz, G. S. *Introduction to Spectroscopy: A Guide for Students of Organic Chemistry,* 3rd ed. Fort Worth, TX: Harcourt, 2001.

Silverstein, R. M., Webster, F. X., and Kiemle, D. *Spectrometric Identification of Organic Compounds,* 7th ed. New York: Wiley, 2004.

Extensive Tables of Compounds and Derivatives

Rappoport, Z., ed. *Handbook of Tables for Organic Compound Identification,* 3rd ed. Boca Raton, FL: CRC Press, 1967.

54A   EXPERIMENT 54A

Solubility Tests

Solubility tests should be performed on *every unknown.* They are extremely important in determining the nature of the main functional group of the unknown compound. The tests are very simple and require only small amounts of the unknown. In addition, solubility tests reveal whether the compound is a strong base (amine), a weak acid (phenol), a strong acid (carboxylic acid), or a neutral substance (aldehyde, ketone, alcohol, ester). The common solvents used to determine solubility types are

5% HCl	Concentrated H_2SO_4
5% $NaHCO_3$	Water
5% NaOH	Organic solvents

The solubility chart on page 455 indicates solvents in which compounds containing the various functional groups are likely to dissolve. The summary charts in Experiments 54D through 54I repeat this information for each functional group included in this experiment. In this section, the correct procedure for determining whether a compound is soluble in a test solvent is given. Also given is a series of explanations detailing the reasons that compounds having specific functional groups are soluble in only specific solvents. This is accomplished by indicating the type of chemistry or the type of chemical interaction that is possible in each solvent.

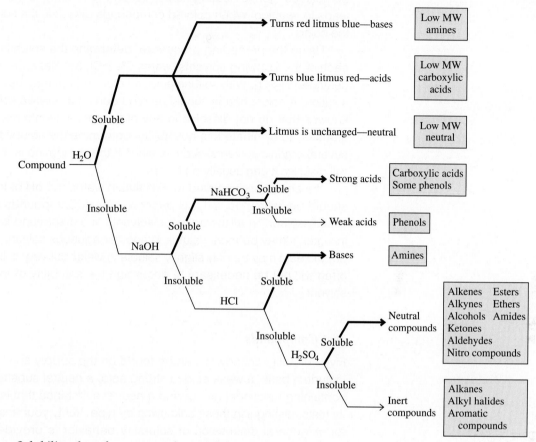

Solubility chart for compounds containing various functional groups.

SUGGESTED WASTE DISPOSAL

Dispose of all aqueous solutions in the container designated for aqueous waste. Any remaining organic compounds must be disposed of in the appropriate organic waste container.

SOLUBILITY TESTS

Procedure

Place about 2 mL of the solvent in a small test tube. Add *1 drop* of an unknown liquid from a Pasteur pipet or a few crystals of an unknown solid from the end of a spatula directly into the solvent. Gently tap the test tube with your finger to ensure mixing and then observe whether any mixing lines appear in the solution. The disappearance of the liquid or solid or the appearance of the mixing lines indicates that solution is taking place. Add several more drops of the liquid or a few more crystals of the solid to determine the extent of the compound's solubility. A common mistake in determining the solubility of a compound is testing with a quantity of the unknown too large to dissolve in the chosen solvent. Use small amounts. It may take several minutes to dissolve solids. Compounds in the form of large crystals need more time to dissolve than powders or very small crystals. In some cases, it is helpful to use a mortar and pestle to pulverize a compound with large crystals. Sometimes, gentle heating helps, but strong heating is discouraged, as it often leads to reaction. When colored compounds dissolve, the solution often assumes the color.

Using the preceding procedure, determine the solubility of the unknown in each of the following solvents: water, 5% HCl, 5% NaHCO$_3$, 5% NaOH, and concentrated H$_2$SO$_4$. With sulfuric acid, a color change may be observed rather than solution. A color change should be regarded as a positive solubility test. Solid unknowns that do not dissolve in any of the test solvents may be inorganic substances. To eliminate this possibility, determine the solubility of the unknown in several organic solvents, such as ether. If the compound is organic, a solvent that will dissolve it can usually be found.

If a compound is found to dissolve in water, the pH of the aqueous solution should be estimated with pH paper or litmus. Compounds soluble in water are usually soluble in *all* the aqueous solvents. If a compound is only slightly soluble in water, it may be *more* soluble in another aqueous solvent. For instance, a carboxylic acid may be only slightly soluble in water but very soluble in dilute base. It often will not be necessary to determine the solubility of the unknown in every solvent.

Test Compounds

Five solubility unknowns can be found on the supply shelf. The five unknowns include a base, a weak acid, a strong acid, a neutral substance with an oxygen-containing functional group, and a neutral substance that is inert. Using solubility tests, distinguish these unknowns by type. Verify your answer with the instructor. A general discussion of solubility behavior is provided in Technique 10, Section 10.2.

Discussion

Solubility in Water

Compounds that contain four or fewer carbons and also contain oxygen, nitrogen, or sulfur are often soluble in water. Almost any functional group containing these elements will lead to water solubility for low-molecular-weight (C$_4$) compounds. Compounds having five or six carbons and any of those elements are often insoluble in water or have borderline solubility. Branching of the alkyl chain in a compound lowers the intermolecular forces between its molecules. This is usually reflected in a lowered boiling point or melting point and a greater solubility in water for the branched compound than for the corresponding straight-chain compound. This occurs simply because the molecules of the branched compound are more easily separated from one another. Thus, *t*-butyl alcohol would be expected to be more soluble in water than *n*-butyl alcohol.

When the ratio of the oxygen, nitrogen, or sulfur atoms in a compound to the carbon atoms is increased, the solubility of that compound in water often increases. This is due to the increased number of polar functional groups. Thus, 1,5-pentanediol would be expected to be more soluble in water than 1-pentanol.

As the size of the alkyl chain of a compound is increased beyond about four carbons, the influence of a polar functional group is diminished, and the water solubility begins to decrease. A few examples of these generalizations are given here.

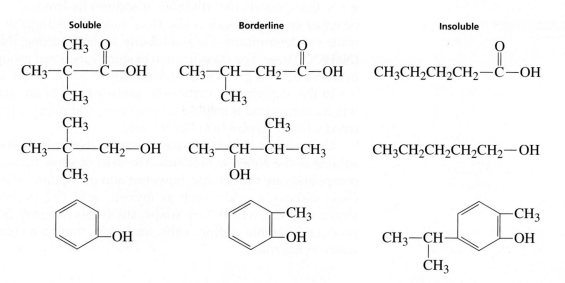

Solubility in 5% HCl

The possibility of an amine should be considered immediately if a compound is soluble in dilute acid (5% HCl). Aliphatic amines (RNH_2, R_2NH, R_3N) are basic compounds that readily dissolve in acid because they form hydrochloride salts that are soluble in the aqueous medium:

$$R—\overset{..}{N}H_2 + HCl \longrightarrow R—NH_3^+ + Cl^-$$

The substitution of an aromatic (benzene) ring Ar for an alkyl group R reduces the basicity of an amine somewhat, but the amine will still protonate, and it will still generally be soluble in dilute acid. The reduction in basicity in an aromatic amine is due to the resonance delocalization of the unshared electrons on the amino nitrogen of the free base. The delocalization is lost on protonation, a problem that does not exist for aliphatic amines. The substitution of two or three aromatic rings on an amine nitrogen reduces the basicity of the amine even further. Diaryl and triaryl amines do not dissolve in dilute HCl because they do not protonate easily. Thus, Ar_2NH and Ar_3N are insoluble in dilute acid. Some amines of very high molecular weight, such as tribromoaniline ($MW = 330$), may also be insoluble in dilute acid.

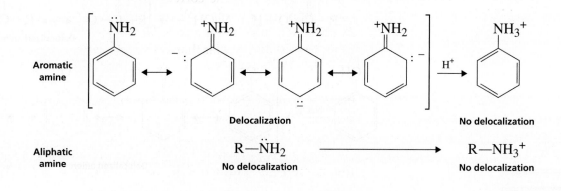

Solubility in 5% NaHCO₃ and 5% NaOH

Compounds that dissolve in sodium bicarbonate, a weak base, are strong acids. Compounds that dissolve in sodium hydroxide, a strong base, may be either strong or weak acids. Thus, one can distinguish weak and strong acids by determining their solubility in both strong (NaOH) and weak (NaHCO₃) base. The classification of some functional groups as either weak or strong acids is given in the table below.

In this experiment, carboxylic acids ($pK_a \sim 5$) are generally indicated when a compound is soluble in both bases, and phenols ($pK_a \sim 10$) are indicated when it is soluble in NaOH only.

Compounds dissolve in base because they form sodium salts that are soluble in the aqueous medium. The salts of some high-molecular-weight compounds are not soluble, however, and precipitate. The salts of the long-chain carboxylic acids, such as myristic acid C₁₄, palmitic acid C₁₆, and stearic acid C₁₈, which form soaps, are in this category. Some phenols also produce insoluble sodium salts, and often these are colored due to resonance in the anion.

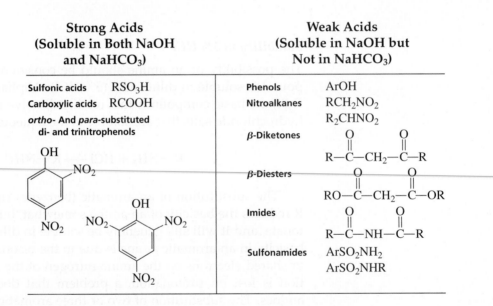

Strong Acids (Soluble in Both NaOH and NaHCO₃)		Weak Acids (Soluble in NaOH but Not in NaHCO₃)	
Sulfonic acids	RSO₃H	Phenols	ArOH
Carboxylic acids	RCOOH	Nitroalkanes	RCH₂NO₂
ortho- And para-substituted di- and trinitrophenols			R₂CHNO₂
		β-Diketones	R—C(=O)—CH₂—C(=O)—R
		β-Diesters	RO—C(=O)—CH₂—C(=O)—OR
		Imides	R—C(=O)—NH—C(=O)—R
		Sulfonamides	ArSO₂NH₂
			ArSO₂NHR

Both phenols and carboxylic acids produce resonance-stabilized conjugate bases. Thus, bases of the appropriate strength may easily remove their acidic protons to form the sodium salts.

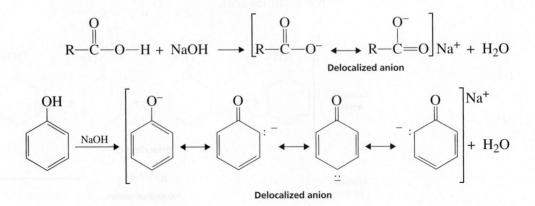

Delocalized anion

Delocalized anion

In phenols, substitution of nitro groups in the *ortho* and *para* positions of the ring increases the acidity. Nitro groups in these positions provide additional delocalization in the conjugate anion. Phenols that have two or three nitro groups in the *ortho* and *para* positions often dissolve in *both* sodium hydroxide and sodium bicarbonate solutions.

Solubility in Concentrated Sulfuric Acid

Many compounds are soluble in cold, concentrated sulfuric acid. Of the compounds included in this experiment, alcohols, ketones, aldehydes, and esters are in this category. These compounds are described as being "neutral." Other compounds that also dissolve include alkenes, alkynes, ethers, nitroaromatics, and amides. Because several different kinds of compounds are soluble in sulfuric acid, further chemical tests and spectroscopy will be needed to differentiate among them.

Compounds that are soluble in concentrated sulfuric acid but not in dilute acid are extremely weak bases. Almost any compound containing a nitrogen, an oxygen, or a sulfur atom can be protonated in concentrated sulfuric acid. The ions produced are soluble in the medium.

$$\text{R}-\overset{..}{\underset{..}{\text{O}}}-\text{H} + \text{H}_2\text{SO}_4 \longrightarrow \text{R}-\overset{\overset{+}{..}}{\underset{\underset{\text{H}}{|}}{\text{O}}}-\text{H} + \text{HSO}_4^- \longrightarrow \text{R}^+ + \text{H}_2\text{O} + \text{HSO}_4^-$$

$$\text{R}-\overset{\overset{\displaystyle :\text{O}:}{\|}}{\text{C}}-\text{R} + \text{H}_2\text{SO}_4 \longrightarrow \text{R}-\overset{\overset{\displaystyle \overset{+}{..}\text{O}-\text{H}}{\|}}{\text{C}}-\text{R} + \text{HSO}_4^-$$

$$\text{R}-\overset{\overset{\displaystyle :\text{O}:}{\|}}{\text{C}}-\text{OR} + \text{H}_2\text{SO}_4 \longrightarrow \text{R}-\overset{\overset{\displaystyle \overset{+}{..}\text{O}-\text{H}}{\|}}{\text{C}}-\text{OR} + \text{HSO}_4^-$$

$$\overset{\text{R}}{\underset{\text{R}}{}}\text{C}=\text{C}\overset{\text{R}}{\underset{\text{R}}{}} + \text{H}_2\text{SO}_4 \longrightarrow \text{R}-\overset{\overset{\text{R}}{|}}{\underset{\underset{\text{H}}{|}}{\text{C}}}-\overset{\overset{\text{R}}{|}}{\underset{+}{\text{C}}}-\text{R} + \text{HSO}_4^-$$

Inert Compounds

Compounds not soluble in concentrated sulfuric acid or any of the other solvents are said to be **inert.** Compounds not soluble in concentrated sulfuric acid include the alkanes, most simple aromatics, and the alkyl halides. Some examples of inert compounds are hexane, benzene, chlorobenzene, chlorohexane, and toluene.

54B EXPERIMENT 54B

Tests for the Elements (N, S, X)

$$-N- \qquad \overset{-Br}{} \qquad -C{\equiv}N$$

$$-Cl \qquad \overset{-NO_2}{\underset{S}{}} \qquad -I$$

Except for amines (Experiment 54G), which are easily detected by their solubility behavior, all compounds issued in this experiment will contain heteroelements (N, S, Cl, Br, or I) only as *secondary* functional group. These will be subsidiary to some other important functional group. Thus, no alkyl or aryl halides, nitro compounds, thiols, or thioethers will be issued. However, some of the unknowns may contain a halogen or a nitro group. Less frequently, they may contain a sulfur atom or a cyano group.

Consider as an example *p*-bromobenzaldehyde, an **aldehyde** that contains bromine as a ring substituent. The identification of this compound would hinge on whether the investigator could identify it as an aldehyde. It could probably be identified *without* proving the existence of bromine in the molecule. That information, however, could make the identification easier. In this experiment, methods are given for identifying the presence of a halogen or a nitro group in an unknown compound. Also given is a general method (sodium fusion) for detecting the principal heteroelements that may exist in organic molecules.

Classification tests

Halides	Nitro Groups	N, S, X (Cl, Br, I)
Beilstein test	Ferrous hydroxide	Sodium fusion
Silver nitrate		
Sodium iodide/acetone		

SUGGESTED WASTE DISPOSAL

Dispose of all solutions containing silver into a waste container designated for this purpose. Any other aqueous solutions should be disposed of in the container designated for aqueous waste. Any remaining organic compounds must be disposed of in the appropriate organic waste container under the hood. This is particularly true of any solution containing benzyl bromide, which is a lachrymator.

TESTS FOR A HALIDE

BEILSTEIN TEST

Procedure

Adjust the air and gas mixture so that the flame of a Bunsen burner or microburner is blue. Bend the end of a piece of copper wire so that a small closed loop is created. Heat the loop end of the wire in the flame until it glows brightly. After the wire has cooled, dip the wire directly into a sample of the unknown. If the unknown is a solid and won't adhere to the copper wire, place a small amount of the substance on a watch glass, wet the copper wire in distilled water, and place the wire into the sample on the watch glass. The solid should adhere to the wire. Now heat the wire in the Bunsen burner flame again. The compound will first burn. After the burning, a green flame will be produced if a halogen is present. You should hold the wire in the flame either just above the tip of the flame or at its outside edge near the bottom of the flame. You will need to experiment to find the best position to hold the copper wire to obtain the best result.

Test Compounds

Try this test on bromobenzene and benzoic acid.

Discussion

Halogens can be detected easily and reliably by the Beilstein test. It is the simplest method for determining the presence of a halogen, but it does not differentiate among chlorine, bromine, and iodine, any one of which will give a positive test. However, when the identity of the unknown has been narrowed to two choices, of which one has a halogen and one does not, the Beilstein test will often be enough to distinguish between the two.

A positive Beilstein test results from the production of a volatile copper halide when an organic halide is heated with copper oxide. The copper halide imparts a blue-green color to the flame.

This test can be sensitive to small amounts of halide impurities in some compounds. Therefore, use caution in interpreting the results of the test if you obtain only a weak color.

SILVER NITRATE TEST

Procedure

Add 1 drop of a liquid or 5 drops of a concentrated ethanolic solution of a solid unknown to 2 mL of a 2% ethanolic silver nitrate solution. If no reaction is observed after 5 minutes at room temperature, heat the solution in a hot water bath at about 100°C and note whether a precipitate forms. If a precipitate forms, add 2 drops of 5% nitric acid and note whether the precipitate dissolves. Carboxylic acids give a false test by precipitating in silver nitrate, but they dissolve when nitric acid is added. Silver halides, in contrast, do not dissolve in nitric acid.

Test Compounds

Apply this test to benzyl bromide (α-bromotoluene) and bromobenzene. Discard all waste reagents in a suitable waste container in the hood because benzyl bromide is a lachrymator.

Discussion

This test depends on the formation of a white or off-white precipitate of silver halide when silver nitrate is allowed to react with a sufficiently reactive halide.

$$RX + Ag^+NO_3^- \rightarrow AgX + R^+NO_3^- \xrightarrow{CH_3CH_2OH} R-O-CH_2CH_3$$

Precipitate

The test does not distinguish among chlorides, bromides, and iodides but does distinguish **labile** (reactive) halides from halides that are unreactive. Halides substituted on an aromatic ring will not usually give a positive silver nitrate test; however, alkyl halides of many types will give a positive test.

The most reactive compounds are those able to form stable carbocations in solution and those equipped with good leaving groups (X = I, Br, Cl). Benzyl, allyl, and tertiary halides react immediately with silver nitrate. Secondary and primary halides do not react at room temperature but react readily when heated. Aryl and vinyl halides do not react at all, even at elevated temperatures. This pattern of reactivity fits the stability order for various carbocations quite well. Compounds that produce stable carbocations react at higher rates than those that do not.

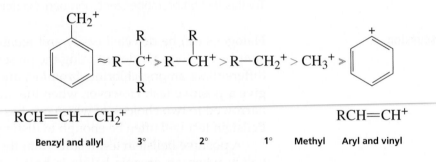

| Benzyl and allyl | 3° | 2° | 1° | Methyl | Aryl and vinyl |

The fast reaction of benzylic and allylic halides is a result of the resonance stabilization that is available to the intermediate carbocations formed. Tertiary halides are more reactive than secondary halides, which are in turn more reactive than primary or methyl halides because alkyl substituents are able to stabilize the intermediate carbocations by an electron-releasing effect. The methyl carbocations have no alkyl groups and are the least stable of all carbocations mentioned thus far. Vinyl and aryl carbocations are extremely unstable because the charge is localized on an sp^2-hybridized carbon (double-bond carbon) rather than on one that is sp^3-hybridized.

SODIUM IODIDE IN ACETONE

Procedure

This test is described in Experiment 20.

Test Compounds

Apply this test to benzyl bromide (α-bromotoluene), bromobenzene, and 2-chloro-2-methylpropane (*tert*-butyl chloride).

DETECTION OF NITRO GROUPS

Although nitro compounds will not be issued as distinct unknowns, many of the unknowns may have a nitro group as a secondary functional group. The presence of a nitro group, and hence nitrogen, in an unknown compound is determined most easily by infrared spectroscopy. However, many nitro compounds give a positive result in the following test. Unfortunately, functional groups other than the nitro group may also give a positive result. You should interpret the results of this test with caution.

FERROUS HYDROXIDE TEST

Procedure

Place 1.5 mL of freshly prepared 5% aqueous ferrous ammonium sulfate in a small test tube and add about 10 mg of a solid or 5 drops of a liquid compound. Mix the solution well and then add first 1 drop of *2M* sulfuric acid and then 1 mL of *2M* potassium hydroxide in methanol. Stopper the test tube and shake it vigorously. A positive test is indicated by the formation of a red-brown precipitate, usually within 1 minute.

Test Compound

Apply this test to 2-nitrotoluene.

Discussion

Most nitro compounds oxidize ferrous hydroxide to ferric hydroxide, which is a red-brown solid. A precipitate indicates a positive test.

$$R{-}NO_2 + 4H_2O + 6Fe(OH)_2 \longrightarrow R{-}NH_2 + 6Fe(OH)_3$$

INFRARED SPECTROSCOPY

The nitro group gives two strong bands near 1560 cm^{-1} and 1350 cm^{-1}. See Technique 25 for details.

DETECTION OF A CYANO GROUP

Although nitriles will not be given as unknowns in this experiment, the cyano group may be a subsidiary functional group whose presence or absence is important to the final identification of an unknown compound. The cyano group can be hydrolyzed in a strong base by heating vigorously to give a carboxylic acid and ammonia gas:

$$R{-}C{\equiv}N + 2\ H_2O \xrightarrow[\Delta]{NaOH} R{-}COOH + NH_3$$

The ammonia can be detected by its odor or by moist pH paper. However, this method is somewhat difficult, and the presence of a nitrile group is confirmed most easily by infrared spectroscopy. No other functional groups (except some C≡C) absorb in the same region of the spectrum as C≡N.

INFRARED SPECTROSCOPY

C≡N stretch is a sharp band of medium intensity near 2250 cm^{-1}. See Technique 25 for details.

SODIUM FUSION TESTS (DETECTION OF N, S, AND X) (OPTIONAL)

When an organic compound containing nitrogen, sulfur, or halide atoms is fused with sodium metal, there is a reductive decomposition of the compound, which converts these atoms to the sodium salts of the inorganic ions CN^-, S^{2-}, and X^-.

$$[N, S, X] \xrightarrow[\Delta]{Na} NaCN, Na_2S, NaX$$

When the fusion mixture is dissolved in distilled water, the cyanide, sulfide, and halide ions can be detected by standard qualitative inorganic tests.

CAUTION ⚠

Always remember to manipulate the sodium metal with a knife or a forceps. Do not touch it with your fingers. Keep sodium away from water. Destroy all waste sodium with 1-butanol or ethanol. Wear safety glasses.

Preparation of Stock Solution

GENERAL METHOD

Procedure

Using a forceps and a knife, take some sodium from the storage container, cut a small piece about the size of a small pea (3 mm on a side), and dry it on a paper towel. Place this small piece of sodium in a clean, dry, small test tube (10 mm × 75 mm). Clamp the test tube to a ring stand and heat the bottom of the tube with a microburner until the sodium melts and its metallic vapor can be seen to rise about a third of the way up the tube. The bottom of the tube will probably have a dull red glow. Remove the burner and *immediately* drop the sample directly into the tube. Use about 10 mg of a solid placed on the end of a spatula or 2–3 drops of a liquid. Be sure to drop the sample directly down the center of the tube so that it touches the hot sodium metal and does not adhere to the side of the test tube. If the fusion is successful, there will usually be a flash or a small explosion. If the reaction is not successful, heat the tube to red heat for a few seconds to ensure complete reaction.

Allow the test tube to cool to room temperature and then carefully add 10 drops of methanol, a drop at a time, to the fusion mixture. Using a spatula or a long glass rod, reach into the test tube and stir the mixture to ensure complete reaction of any excess sodium metal. The fusion will have destroyed the test tube for other uses. Thus, the easiest way to recover the fusion mixture is to crush the test tube into a small beaker containing 5–10 mL of *distilled* water. The tube is easily crushed if it is placed in the angle of a clamp holder. Tighten the clamp until the tube is securely held near its bottom and then—standing back from the beaker and holding the clamp at its opposite end—continue tightening the clamp until the test tube breaks and the pieces fall into the beaker. Stir the solution well, heat it to boiling, and then filter it by gravity through a fluted filter (Figure 8.3, p. 619). Portions of this solution will be used for the tests to detect nitrogen, sulfur, and the halogens.

ALTERNATIVE METHOD

Procedure

With some volatile liquids, the previous method will not work. The compounds volatilize before they reach the sodium vapors. For such compounds, place 4 or 5 drops of the pure liquid in the clean, dry test tube, clamp it, and cautiously add the small piece of sodium metal. If there is any reaction, wait until it subsides. Then heat the test tube to red heat and continue according to the instructions in the second paragraph of the preceding procedure.

NITROGEN TEST

Procedure

Using pH paper and a 10% sodium hydroxide solution, adjust the pH of about 1 mL of the stock solution to pH 13. Add 2 drops of saturated ferrous ammonium sulfate solution and 2 drops of 30% potassium fluoride solution. Boil the solution for about 30 seconds. Then acidify the hot solution by adding 30% sulfuric acid dropwise until the iron hydroxides dissolve. Avoid using excess acid. If nitrogen is present, a dark blue (not green) precipitate of Prussian blue $NaFe_2(CN)_6$ will form, or the solution will assume a dark blue color.

Reagents

Dissolve 5 g of ferrous ammonium sulfate in 100 mL of water. Dissolve 30 g of potassium fluoride in 100 mL of water.

SULFUR TEST

Procedure

Acidify about 1 mL of the test solution with acetic acid and add a few drops of a 1% lead acetate solution. The presence of sulfur is indicated by a black precipitate of lead sulfide PbS.

> **CAUTION** ⚠
>
> **Many compounds of lead(II) are suspected carcinogens (see p. 557) and should be handled with care. Avoid contact.**

HALIDE TESTS

Procedure

Cyanide and sulfide ions interfere with the test for halides. If such ions are present, they must be removed. To accomplish this, acidify the solution with dilute nitric acid and boil it for about 2 minutes. This will drive off any HCN or H_2S that is formed. When the solution cools, add a few drops of a 5% silver nitrate solution. A *voluminous* precipitate indicates a halide. A faint turbidity *does not* mean a positive test. Silver chloride is white. Silver bromide is off-white. Silver iodide is yellow. Silver chloride will readily dissolve in concentrated ammonium hydroxide, whereas silver bromide is only slightly soluble.

DIFFERENTIATION OF CHLORIDE, BROMIDE, AND IODIDE

Procedure

Acidify 2 mL of the test solution with 10% sulfuric acid and boil it for about 2 minutes. Cool the solution and add about 0.5 mL of methylene chloride. Add a few drops of chlorine water or 2–4 mg of calcium hypochlorite.[1] Check to be sure that the solution is still acidic. Then stopper the tube, shake it vigorously, and set it aside to allow the layers to separate. An orange to brown color in the methylene chloride layer indicates bromine. Violet indicates iodine. No color or a *light* yellow indicates chlorine.

---

[1] Clorox, the commercial bleach, is a permissible substitute for chlorine water, as is any other brand of bleach, provided that it is based on sodium hypochlorite.

54C   EXPERIMENT 54C

Tests for Unsaturation

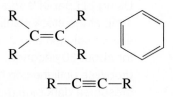

The unknowns to be issued for this experiment have neither a double bond nor a triple bond as their *only* functional group. Hence, simple alkenes and alkynes can be ruled out as possible compounds. Some of the unknowns may have a double or a triple bond, however, *in addition to* another more important functional group. The tests described allow you to determine the presence of a double bond or a triple bond (unsaturation) in such compounds.

Classification tests

Unsaturation	Aromaticity
Bromine–methylene chloride	Ignition test
Potassium permanganate	

SUGGESTED WASTE DISPOSAL

Test reagents that contain bromine should be discarded into a special waste container designated for this purpose. Methylene chloride must be placed in the organic waste container designated for the disposal of halogenated organic wastes. Dispose of all other aqueous solutions in the container designated for aqueous waste. Any remaining organic compounds must be disposed of in the appropriate organic waste container.

TESTS FOR SIMPLE MULTIPLE BONDS

BROMINE IN METHYLENE CHLORIDE

Procedure

Dissolve 50 mg of a solid unknown or 4 drops of a liquid unknown in 1 mL of methylene chloride (dichloromethane) or in 1,2-dimethoxyethane. Add a 2% (by volume) solution of bromine in methylene chloride, dropwise, with shaking. If you find that the red color remains after adding one or two drops of the bromine solution, the test is negative. If the red color disappears, continue adding the bromine in methylene chloride until the red bromine color remains. The test is positive if more than 5 drops of the bromine solution were added, with discharge of the red color of bromine. If the red color disappears, try adding more drops of the bromine solution to see how many drops are necessary before the red color persists. Usually, many drops of the bromine solution will be decolorized when an isolated double bond is present. Hydrogen bromide should not be evolved. If hydrogen bromide gas is evolved, you will note a "fog" while you blow across the mouth of the test tube. The HBr can also be detected by a moistened piece of litmus or pH paper. If hydrogen bromide is evolved, the reaction is a **substitution reaction** (see following discussion) and not an **addition reaction,** and a double or triple bond is probably not present.

Reagent

The classic method for running this test is to use bromine dissolved in carbon tetrachloride. Because of the toxic nature of this solvent, methylene chloride has been substituted for carbon tetrachloride. The instructor must prepare this reagent because of the danger associated with the very toxic bromine vapor. Be sure to work in an efficient fume hood. Dissolve 2 mL of bromine in 100 mL of methylene chloride (dichloromethane). The solvent will undergo a light-induced free-radical substitution producing hydrogen bromide over a period of time. After about 1 week, the color of a 2% solution of bromine in methylene chloride fades noticeably, and the odor of the HBr can be detected in the reagent. Although the decolorization tests still work satisfactorily, the presence of HBr makes it difficult to distinguish between addition and substitution reactions. A freshly prepared solution of bromine in methylene chloride must be used to make this distinction. Deterioration of the reagent can be forestalled by storing it in a brown glass bottle.

Test Compounds

Try this test with cyclohexene, cyclohexane, toluene, and acetone.

Discussion

A successful test depends on the addition of bromine, a red liquid, to a double or a triple bond to give a colorless dibromide:

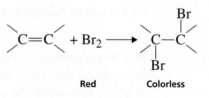

Not all double bonds react with the bromine solution. Only those that are electron rich are sufficiently reactive nucleophiles to initiate the reaction. A double bond that is substituted by electron-withdrawing groups often fails to react or reacts slowly. Fumaric acid is an example of a compound that fails to give the reaction.

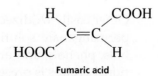

Fumaric acid

Aromatic compounds either do not react with the bromine reagent or they react by **substitution.** Only the aromatic rings that have activating groups as substituents (—OH, —OR, or —NR$_2$) give the substitution reaction.

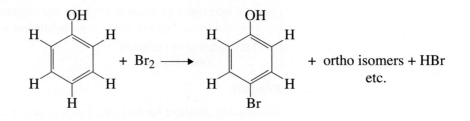

Some ketones and aldehydes react with bromine to give a **substitution product,** but this reaction is slow except for ketones that have a high enol content. When substitution occurs, not only is the bromine color discharged but hydrogen bromide gas is also evolved.

POTASSIUM PERMANGANATE (BAEYER TEST)

Procedure

Dissolve 25 mg of a solid unknown or 2 drops of a liquid unknown in 2 mL of 95% ethanol (1,2-dimethoxyethane may also be used). Slowly add a 1% aqueous solution (weight/volume) of potassium permanganate, drop by drop while shaking, to the unknown. In a positive test, the purple color of the reagent is discharged, and a brown precipitate of manganese dioxide forms, usually within 1 minute. If alcohol was the solvent, the solution should not be allowed to stand for more than 5 minutes, because oxidation of the alcohol will begin slowly. Because permanganate solutions undergo some decomposition to manganese dioxide on standing, any small amount of precipitate should be interpreted with caution.

Test Compounds

Try this test on cyclohexene and toluene.

Discussion

This test is positive for double and triple bonds but not for aromatic rings. It depends on the conversion of the purple ion MnO$_4^-$ to a brown precipitate of MnO$_2$ following the oxidation of an unsaturated compound.

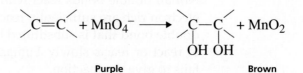

Other easily oxidized compounds also give a positive test with potassium permanganate solution. These substances include aldehydes, some alcohols, phenols, and aromatic amines. If you suspect that any of these functional groups is present, you should interpret the test with caution.

SPECTROSCOPY

Infrared

Double Bonds (C=C)

C=C stretch usually occurs near 1680–1620 cm^{-1}. Symmetrical alkenes may have no absorption.

C—H stretch of vinyl hydrogens occurs > 3000 cm^{-1} but usually not higher than 3150 cm^{-1}.

C—H out-of-plane bending occurs near 1000–700 cm^{-1}.

See Technique 25 for details.

Triple Bonds (C≡C)

C≡C stretch usually occurs near 2250–2100 cm^{-1}. The peak is usually sharp. Symmetrical alkynes show no absorption.

C—H stretch of terminal acetylenes occurs near 3310–3200 cm^{-1}.

Nuclear Magnetic Resonance

Vinyl hydrogens have resonance near 5–7 ppm and have coupling values as follows: J_{trans} = 11–18 Hz, J_{cis} = 6–15 Hz, $J_{geminal}$ = 0–5 Hz. Allylic hydrogens have resonance near 2 ppm. Acetylenic hydrogens have resonance near 2.8–3.0 ppm. See Technique 26 for details on proton NMR. Carbon NMR is described in Technique 27.

TESTS FOR AROMATICITY

None of the unknowns to be issued for this experiment will be simple aromatic hydrocarbons. All aromatic compounds will have a principal functional group as a part of their structure. Nevertheless, in many cases it will be useful to be able to recognize the presence of an aromatic ring. Although infrared and nuclear magnetic spectroscopy provide the most reliable methods of determining aromatic compounds, often they can be detected by a simple ignition test.

IGNITION TEST

Procedure

Working in a hood, place a small amount of the compound on a spatula and place it in the flame of a Bunsen burner. Observe whether a sooty flame results. Compounds giving the sooty yellow flame have a high degree of unsaturation and may be aromatic. This test should be interpreted with care because some nonaromatic compounds may produce soot. If in doubt, use spectroscopy to more reliably determine the presence or absence of an aromatic ring.

Test Compounds

Try this test with ethyl benzoate and benzoin.

Discussion

The presence of an aromatic ring will usually lead to the production of a sooty yellow flame in this test. In addition, halogenated alkanes and high-molecular-weight aliphatic compounds may produce a sooty yellow flame. Aromatic compounds with high oxygen content may burn cleaner and produce less soot even though the compound contains an aromatic ring.

This is actually a test to determine the ratio of carbon to hydrogen and oxygen in an unknown substance. If the carbon-to-hydrogen ratio is high and if little or no oxygen is present, you will observe a sooty flame. For instance, acetylene, C_2H_2 (a gas), will burn with a sooty flame unless mixed with oxygen. When the carbon-to-hydrogen ratio is nearly equal to one, you will be very likely to see a sooty flame.

SPECTROSCOPY

Infrared

C=C aromatic ring double bonds appear in the 1600–1450 cm^{-1} region. There are often four sharp absorptions that occur in pairs near 1600 cm^{-1} and 1450 cm^{-1}, which are characteristic of an aromatic ring.

Special ring absorptions: There are often weak ring absorptions around 2000–1600 cm^{-1}. These are frequently obscured, but when they can be observed, the relative shapes and numbers of these peaks can often be used to ascertain the type of ring substitution.

=C—H stretch, aromatic ring: The aromatic C—H stretch always occurs at a higher frequency than 3000 cm^{-1}.

=C—H out-of-plane bending peaks appear in the region 900–690 cm^{-1}. The number and position of these peaks can be used to determine the substitution pattern of the ring.

See Technique 25 for details.

Nuclear Magnetic Resonance

Hydrogens attached to an aromatic ring usually have resonance near 7 ppm. Monosubstituted rings not substituted by anisotropic or electronegative groups often give a single resonance for all the ring hydrogens. Monosubstituted rings with anisotropic or electronegative groups usually have the aromatic resonances split into two groups integrating either 3:2 or 2:3. A nonsymmetric, *para*-disubstituted ring has a characteristic four-peak splitting pattern (see Technique 26). Carbon NMR is described in Technique 27.

54D   EXPERIMENT 54D

Aldehydes and Ketones

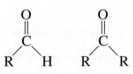

Compounds containing the carbonyl functional group $\diagdown$C=O, where it has only hydrogen atoms or alkyl groups as substituents, are called aldehydes RCHO or ketones RCOR'. The chemistry of these compounds is primarily due to the chemistry of the carbonyl functional groups. These compounds are identified by the distinctive reactions of the carbonyl function.

Solubility Characteristics					*Classification Tests*	
HCl	$NaHCO_3$	NaOH	H_2SO_4	Ether	**Aldehydes and ketones**	
(−)	(−)	(−)	(+)	(+)	2,4-Dinitrophenylhydrazine	
Water: $< C_5$ and some $C_6(+)$					**Aldehydes only**	**Methyl ketones**
$> C_5(-)$					Tollens reagent	Iodoform test
					Chromic acid	
					Compounds with high enol content	
					Ferric chloride test	

SUGGESTED WASTE DISPOSAL

Solutions containing 2,4-dinitrophenylhydrazine or derivatives formed from it should be placed in a waste container designated for these compounds. Any solution containing chromium must be disposed of in a waste container specifically identified for the disposal of chromium wastes. Dispose of all solutions containing silver by acidifying them with 5% hydrochloric acid and then placing them in a waste container designated for this purpose. Dispose of all other aqueous solutions in the container designated for aqueous waste. Any remaining organic compounds must be disposed of in the appropriate organic waste container.

CLASSIFICATION TESTS

Most aldehydes and ketones give a solid, yellow to red precipitate when mixed with 2,4-dinitrophenylhydrazine. However, only aldehydes will reduce chromium(VI) or silver(I). By this difference in behavior, you can differentiate between aldehydes and ketones.

2,4-DINITROPHENYLHYDRAZINE

Procedure

Place 1 drop of the liquid unknown in a small test tube and add 1 mL of the 2,4-dinitrophenylhydrazine reagent. If the unknown is a solid, dissolve about 10 mg (estimate) in a minimum amount of 95% ethanol or di(ethylene glycol) diethyl ether before adding the reagent. Shake the mixture vigorously. Most aldehydes and ketones will give a yellow to red precipitate immediately. However, some compounds will require up to 15 minutes, or even *gentle* heating, to give a precipitate. A precipitate indicates a positive test.

Test Compounds

Try this test on cyclohexanone, benzaldehyde, and benzophenone.

CAUTION

Many derivatives of phenylhydrazine are suspected carcinogens (see p. 557) and should be handled with care. Avoid contact.

Reagent

Dissolve 3.0 g of 2,4-dinitrophenylhydrazine in 15 mL of concentrated sulfuric acid. In a beaker, slowly add, with mixing, 23 mL of water until the solid dissolves. Add

75 mL of 95% ethanol to the warm solution, with stirring. After thorough mixing, filter the solution if any solid remains. This reagent needs to be prepared fresh each term.

Discussion

Most aldehydes and ketones give a precipitate, but esters generally do not give this result. Thus, an ester usually can be eliminated by this test. The color of the 2,4-dinitrophenylhydrazone (precipitate) formed is often a guide to the amount of conjugation in the original aldehyde or ketone. Unconjugated ketones, such as cyclohexanone, give yellow precipitates, whereas conjugated ketones, such as benzophenone, give orange to red precipitates. Compounds that are highly conjugated give red precipitates. However, the 2,4-dinitrophenylhydrazine reagent is itself orange red, and the color of any precipitate must be judged cautiously. Occasionally, compounds that are either strongly basic or strongly acidic precipitate the unreacted reagent.

Some allylic and benzylic alcohols give this test result because the reagent can oxidize them to aldehydes and ketones, which subsequently react. Some alcohols may be contaminated with carbonyl impurities, either as a result of their method of synthesis (reduction) or as a result of their becoming air-oxidized. A precipitate formed from small amounts of impurity in the solution will be formed in small amount. With some caution, a test that gives only a slight amount of precipitate can usually be ignored. The infrared spectrum of the compound should establish its identity and identify any impurities present.

TOLLENS TEST

Procedure

The reagent must be prepared immediately before use. To prepare the reagent, mix 1 mL of Tollens solution A with 1 mL of Tollens solution B. A precipitate of silver oxide will form. Add enough dilute (10%) ammonia solution (dropwise) to the mixture to dissolve the silver oxide *just barely*. The reagent so prepared can be used immediately for the following test.

Dissolve 1 drop of a liquid aldehyde or 10 mg (approximate) of a solid aldehyde in the minimum amount of di(ethylene glycol) diethyl ether. Add this solution, a little at a time, to the 2–3 mL of reagent contained in a small test tube. Shake the solution well. If a mirror of silver is deposited on the inner walls of the test tube, the test is positive. In some cases, it may be necessary to warm the test tube in a warm waterbath.

Test Compounds

Try the test on benzaldehyde, butanal (butyraldehyde), and cyclohexanone.

CAUTION ⚠️

The reagent should be prepared immediately before use and all residues disposed of immediately after use. Dispose of any residues by acidifying them with 5% hydrochloric acid and then placing them in a waste container designated for this purpose. On standing, the reagent tends to form silver fulminate, a *very explosive* substance. Solutions containing the mixed Tollens reagent should never be stored.

Reagents

Solution A: Dissolve 3.0 g of silver nitrate in 30 mL of water. *Solution B:* Prepare a 10% sodium hydroxide solution.

Discussion

Most aldehydes reduce ammoniacal silver nitrate solution to give a precipitate of silver metal. The aldehyde is oxidized to a carboxylic acid:

$$RCHO + 2\ Ag(NH_3)_2OH \longrightarrow 2\ Ag + RCOO^-NH_4^+ + H_2O + NH_3$$

Ordinary ketones do not give a positive result in this test. The test should be used only if it has already been shown that the unknown compound is either an aldehyde or a ketone.

CHROMIC ACID TEST: ALTERNATIVE TEST

CAUTION ⚠️

Many chromium (VI) compounds are suspected carcinogens. If you would like to run this test, talk to your instructor first. Most often, the Tollens test will easily distinguish between aldehydes and ketones, and you should do that test first. If you run the chromic acid test, be sure to wear gloves to avoid contact with this reagent.

Procedure

Dissolve 1 drop of a liquid or 10 mg (approximate) of a solid aldehyde in 1 mL of *reagent-grade* acetone. Add several drops of the chromic acid reagent, a drop at a time while shaking the mixture. A positive test is indicated by a green precipitate and a loss of the orange color in the reagent. With aliphatic aldehydes, RCHO, the solution turns cloudy within 5 seconds, and a precipitate appears within 30 seconds. With aromatic aldehydes, ArCHO, it generally takes 30–120 seconds for a precipitate to form, but with some it may take even longer. In some cases, however, you may find that some of the original orange color may remain together with a green or brown precipitate. This should be interpreted as a positive test. In a negative test, a nongreen precipitate may form in an orange solution.

In performing this test, make sure that the acetone used for the solvent does not give a positive test with the reagent. Add several drops of the chromic acid reagent to a few drops of the reagent acetone contained in a small test tube. Allow this mixture to stand for 3–5 minutes. If no reaction has occurred by this time, the acetone is pure enough to use as a solvent for the test. If a positive test resulted, try another bottle of acetone.

Test Compounds

Try this test on benzaldehyde, butanal (butyraldehyde), and cyclohexanone.

Reagent

Dissolve 20 g of chromium trioxide (CrO_3) in 60 mL of cold water in a beaker. Add a magnetic stir bar to the solution. With stirring, slowly and carefully add 20 mL of concentrated sulfuric acid to the solution. This reagent should be prepared fresh each term.

Discussion

This test has as its basis the fact that aldehydes are easily oxidized to the corresponding carboxylic acid by chromic acid. The green precipitate is due to chromous sulfate.

$$2\,CrO_3 + 2\,H_2O \xrightleftharpoons{H^+} 2\,H_2CrO_4 \xrightleftharpoons{H^+} H_2Cr_2O_7 + H_2O$$

$$\underset{\text{Orange}}{3\,RCHO + H_2Cr_2O_7 + 3\,H_2SO_4} \longrightarrow \underset{\text{Green}}{3\,RCOOH + Cr_2(SO_4)_3 + 4\,H_2O}$$

Primary and secondary alcohols are also oxidized by this reagent (see Experiment 54H). Therefore, this test is not useful in identifying aldehydes *unless* a positive identification of the carbonyl group has already been made. Aldehydes give a 2,4-dinitrophenylhydrazine test result, whereas alcohols do not.

There are numerous other tests used to detect the aldehyde functional group. Most are based on an easily detectable oxidation of the aldehyde to a carboxylic acid. The most common tests are the Tollens, Fehling's, and Benedict's tests. Only the Tollens test is described in this book. The Tollens test is often more reliable than the chromic acid test for aldehydes.

IODOFORM TEST

Procedure

Prepare a 60–70°C waterbath in a beaker. Using a Pasteur pipet, add 6 drops of a liquid unknown to a 15-mm × 100-mm or 15-mm × 125-mm test tube. Alternatively, 0.06 g of a solid unknown may be used. Dissolve the liquid or solid unknown compound in 2 mL of 1,2-dimethoxyethane. Add 2 mL of 10% aqueous sodium hydroxide solution and place the test tube in the hot waterbath. Next add 4 mL of iodine–potassium iodide solution in 1-mL portions to the test tube. *Cork* the test tube and shake it after adding each portion of iodine reagent. Heat the mixture in the hot waterbath for about 5 minutes, shaking the test tube occasionally. It is likely that some or all the dark color of the iodine reagent will be discharged.

If the dark color of the iodine reagent is still apparent after heating, add 10% sodium hydroxide solution until the dark color of the iodine reagent has been discharged. Shake the mixture in the test tube (corked) during the addition of the sodium hydroxide. Care need not be taken to avoid adding excess sodium hydroxide.

After the dark iodine color of the solution has been discharged, fill the test tube with water to within 2 cm of the top. Cork the test tube and shake it vigorously. Allow the tube to stand for at least 15 minutes at room temperature. The appearance of a pale yellow precipitate of iodoform CHI$_3$ constitutes a positive test, indicating that the unknown is a methyl ketone or a compound that is easily oxidized to a methyl ketone, such as a 2-alkanol. Other ketones will also decolorize the iodine solution, but they will not give a precipitate of iodoform *unless* there is an impurity of a methyl ketone present in the unknown.

The yellow precipitate usually settles out slowly onto the bottom of the test tube. Sometimes, the yellow color of iodoform is masked by a dark substance. If this is the case, cork the test tube and shake it vigorously. If the dark color persists, add more sodium hydroxide solution and shake the test tube again. Then allow the tube to stand for at least 15 minutes. If there is some doubt as to whether the solid is iodoform, collect the precipitate on a Hirsch funnel and dry it. Iodoform melts at 119–121°C.

You may find on some occasions that the methyl ketone only gives a yellow coloration to the solution rather than a distinct yellow precipitate. You should be cautious about any conclusions about this result. Therefore, you should depend on proton NMR to confirm the presence of a methyl group attached directly to a carbonyl group (singlet at about 2 ppm).

Test Compounds

Try the test on 2-heptanone, 4-heptanone (dipropyl ketone), and 2-pentanol.

Reagents

The iodine reagent is prepared by dissolving 20 g of potassium iodide and 10 g of iodine in 100 mL of water. The aqueous sodium hydroxide solution is prepared by dissolving 10 g of sodium hydroxide in 100 mL of water.

Discussion

The basis of this test is the ability of certain compounds to form a precipitate of iodoform when treated with a basic solution of iodine. Methyl ketones are the most common types of compounds that give a positive result in this test. However, acetaldehyde CH$_3$CHO and alcohols with the hydroxyl group at the 2-position of the chain also give a precipitate of iodoform. 2-Alkanols of the type described are easily oxidized to methyl ketones under the conditions of the reaction. The other product of the reaction, besides iodoform, is the sodium or potassium salt of a carboxylic acid.

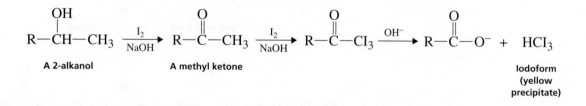

FERRIC CHLORIDE TEST

Procedure

Some aldehydes and ketones, those that have a high **enol content,** give a positive ferric chloride test, as described for phenols in Experiment 54F.

SPECTROSCOPY

Infrared

The carbonyl group is usually one of the strongest-absorbing groups in the infrared spectrum, with a very broad range: 1800–1650 cm^{-1}. The aldehyde functional group has *very characteristic* C—H stretch absorptions: two sharp peaks that lie *far outside* the usual region for —C—H, =C—H, or ≡C—H.

Aldehydes

C=O stretch at approximately 1725 cm^{-1} is normal.
1725–1685 cm^{-1}.*

C—H stretch (aldehyde–CHO) has two weak bands at about 2750 cm^{-1} and 2850 cm^{-1}.

See Technique 25 for details.

Ketones

C=O stretch at approximately 1715 cm^{-1} is normal.
1780–1665 cm^{-1}.*

Nuclear Magnetic Resonance

Hydrogens alpha to a carbonyl group have resonance in the region between 2 ppm and 3 ppm. The hydrogen of an aldehyde group has a characteristic resonance between 9 ppm and 10 ppm. In aldehydes, there is coupling between the aldehyde hydrogen and any alpha hydrogens (J = 1–3 Hz).

See Technique 26 for details on proton NMR. Carbon NMR is described in Technique 27.

DERIVATIVES

The most common derivatives of aldehydes and ketones are 2,4-dinitrophenylhydrazones, oximes, and semicarbazones. Procedures for preparing these derivatives are given in Appendix 2.

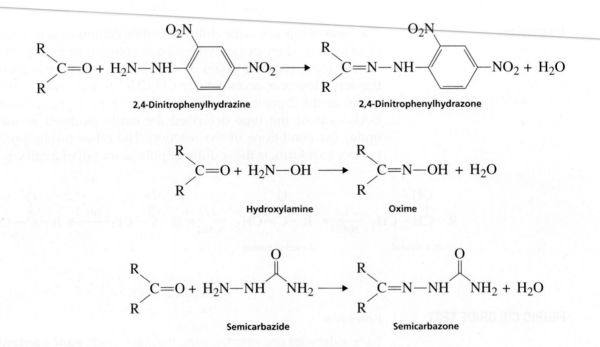

* **Conjugation** moves the absorption to lower frequencies. **Ring strain** (cyclic ketones) moves the absorption to higher frequencies.

54E EXPERIMENT 54E

Carboxylic Acids

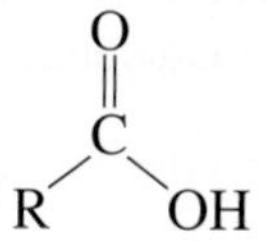

Carboxylic acids are detectable mainly by their solubility characteristics. They are soluble in *both* dilute sodium hydroxide and sodium bicarbonate solutions.

Solubility Characteristics					*Classification Tests*
HCl	NaHCO$_3$	NaOH	H$_2$SO$_4$	Ether	pH of an aqueous solution
(−)	(+)	(+)	(+)	(+)	Sodium bicarbonate
Water: $< C_6(+)$					Silver nitrate
$> C_6(−)$					Neutralization equivalent

SUGGESTED WASTE DISPOSAL

Dispose of all aqueous solutions in the container designed for aqueous waste. Any remaining organic compounds must be disposed of in the appropriate organic waste container.

CLASSIFICATION TESTS

pH OF AN AQUEOUS SOLUTION

Procedure

If the compound is soluble in water, simply prepare an aqueous solution and check the pH with pH paper. If the compound is an acid, the solution will have a low pH.

Compounds that are insoluble in water can be dissolved in ethanol (or methanol) and water. First, dissolve the compound in the alcohol and then add water until the solution *just* becomes cloudy. Clarify the solution by adding a few drops of the alcohol and then determine its pH using pH paper.

SODIUM BICARBONATE

Procedure

Dissolve a small amount of the compound in a 5% aqueous sodium bicarbonate solution. Observe the solution carefully. If the compound is an acid, you may see bubbles of carbon dioxide form. In some cases with solids, the evolution of carbon dioxide may not be that obvious.

$$RCOOH + NaHCO_3 \longrightarrow RCOO^-Na^+ + H_2CO_3 \text{ (unstable)}$$

$$H_2CO_3 \longrightarrow CO_2 + H_2O$$

SILVER NITRATE

Procedure

Acids may give a false silver nitrate test, as described in Experiment 54B.

NEUTRALIZATION EQUIVALENT (OPTIONAL)

Procedure

Accurately weigh (three significant figures) approximately 0.2 g of the acid and place in a 125-mL Erlenmeyer flask. Dissolve the acid in about 50 mL of water or aqueous ethanol (the acid need not dissolve completely, because it will dissolve as it is titrated). Titrate the acid, using a solution of sodium hydroxide of known molarity (about 0.1*M*) and a phenolphthalein indicator.

Calculate the neutralization equivalent (NE) from the equation

$$NE = \frac{\text{mg acid}}{\text{molarity of NaOH} \times \text{mL of NaOH added}}$$

The NE is identical to the equivalent weight of the acid. If the acid has only one carboxyl group, the neutralization equivalent and the molecular weight of the acid are identical. If the acid has more than one carboxyl group, the neutralization equivalent equals the molecular weight of the acid divided by the number of carboxyl groups, that is, the equivalent weight. The NE can be used much like a derivative to identify a specific acid.

Some phenols are sufficiently acidic to behave much like carboxylic acids. This is especially true of those substituted with electron-withdrawing groups at the *ortho* and *para* ring positions. These phenols, however, can usually be eliminated either by the ferric chloride test (Experiment 54F) or by spectroscopy (phenols have no carbonyl group).

SPECTROSCOPY

Infrared

C=O stretch is very strong and often broad in the region between 1725 cm^{-1} and 1690 cm^{-1}.

O—H stretch is a very broad absorption in the region between 3300 cm^{-1} and 2500 cm^{-1}; it usually overlaps the CH stretch region.

See Technique 25 for details.

Nuclear Magnetic Resonance

The acid proton of a —COOH group usually has resonance near 12.0 ppm. See Technique 26 for details. Carbon NMR is described in Technique 27.

DERIVATIVES

Derivatives of acids are usually amides. They are prepared via the corresponding acid chloride:

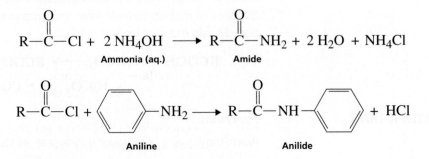

$$R{-}\overset{\overset{\text{O}}{\|}}{C}{-}OH + SOCl_2 \longrightarrow R{-}\overset{\overset{\text{O}}{\|}}{C}{-}Cl + SO_2 + HCl$$

The most common derivatives are the amides, the anilides, and the *p*-toluidides.

$$R{-}\overset{\overset{\text{O}}{\|}}{C}{-}Cl + 2\,NH_4OH \longrightarrow R{-}\overset{\overset{\text{O}}{\|}}{C}{-}NH_2 + 2\,H_2O + NH_4Cl$$

Ammonia (aq.)   Amide

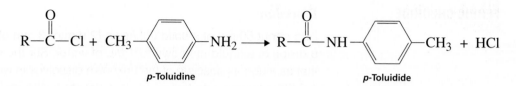

p-Toluidine → *p*-Toluidide

Procedures for the preparation of these derivatives are given in Appendix 2.

54F EXPERIMENT 54F

Phenols

Like carboxylic acids, phenols are acidic compounds. However, except for the nitro-substituted phenols (discussed in the section covering solubilities), they are not as acidic as the carboxylic acids. The pK_a of a typical phenol is 10, whereas the pK_a of a carboxylic acid is usually near 5. Hence, phenols are generally not soluble in the weakly basic sodium bicarbonate solution, but they dissolve in sodium hydroxide solution, which is more strongly basic.

Solubility Characteristics					*Classification Tests*
HCl	NaHCO$_3$	NaOH	H$_2$SO$_4$	Ether	Colored phenolate anion
(−)	(−)	(+)	(+)	(+)	Ferric chloride
Water: Most are insoluble, although phenol itself and the nitrophenols are soluble					Ce(IV) Test
					Bromine/water

SUGGESTED WASTE DISPOSAL

Dispose of all aqueous solutions in the container designated for aqueous waste. Any remaining organic compounds must be disposed of in the appropriate organic waste container.

CLASSIFICATION TESTS

SODIUM HYDROXIDE SOLUTION

With phenols that have a high degree of conjugation possible in their conjugate base (phenolate ion), the anion is often colored. To observe the color, dissolve a small amount of the phenol in 10% aqueous sodium hydroxide solution. Some phenols do not give a color. Others have an insoluble anion and give a precipitate. The more acidic phenols, such as the nitrophenols, tend more toward colored anions.

FERRIC CHLORIDE

Procedure

Add about 50 mg of a solid unknown (2 mm or 3 mm off the end of a spatula) or 5 drops of a liquid unknown to 1 mL of water. Stir the mixture with a spatula so that as much as possible of the unknown dissolves in water. Add several drops of a 2.5% aqueous solution of ferric chloride to the mixture. Most water-soluble phenols produce an intense red, blue, purple, or green color. Some colors are transient, and it may be necessary to observe the solution carefully just as the solutions are mixed. The formation of a color is usually immediate, but the color may not last over any great period. Some phenols do not give a positive result in this test, so a negative test must not be taken as significant without other adequate evidence.

Test Compound

Try this test on phenol.

Discussion

The colors observed in this test result from the formation of a complex of the phenols with Fe(III) ion. Carbonyl compounds that have a high enol content also give a positive result in this test. The ferric chloride test works best with water soluble phenols. A more reliable test, especially for water insoluble phenols, is the Ce(IV) test.

CERIUM(IV) TEST

Add 3 mL of 1,2-dimethoxyethane to 0.5 mL of Cerium(IV) reagent in a dry test tube. Gently shake the solution to thoroughly mix the solution and then add 4 drops of a liquid compound to be tested. If you have a solid, you can directly add a few milligrams of the solid to the solution. Enough will dissolve to give a test if an —OH group is present. Gently shake the mixture and look for an *immediate color change* from a yellow-orange solution to a red-orange or a deep red color indicating the presence of a phenol. The unsubstituted phenol, C_6H_5-OH, forms a dark-brown precipitate. Other phenols should yield a deep-red solution.

Test Compounds

Try this test on β-naphthol (2-naphthol).

Reagent

Prepare 2*M* nitric acid solution by diluting 12.8 mL of concentrated nitric acid to 100 mL with water. Dissolve 8 g of ceric ammonium nitrate [$Ce(NH_4)_2(NO_3)_6$] in 20 mL of the dilute nitric acid solution.

Discussion

The Ce(IV) test provides a more reliable way of detecting the presence of the hydroxyl group in water-insoluble phenols than the ferric chloride test. Since alcohols also give a color change with this reagent, you will first need to distinguish between alcohols and phenols by determining the solubility behavior of your compound. Phenols should be soluble in sodium hydroxide, whereas alcohols will not dissolve in aqueous sodium hydroxide.

BROMINE WATER

Procedure

Prepare a 1% aqueous solution of the unknown and then add a saturated solution of bromine in water to it, drop by drop while shaking, until the bromine color is no longer discharged. A positive test is indicated by the precipitation of a substitution product at the same time that the bromine color of the reagent is discharged.

Test Compound

Try this test on a 1% aqueous phenol solution.

Discussion

Aromatic compounds with ring-activating substituents give a positive test with bromine in water. The reaction is an aromatic substitution reaction that introduces bromine atoms into the aromatic ring at the positions *ortho* and *para* to the hydroxyl group. All available positions are usually substituted. The precipitate is the brominated phenol, which is generally insoluble because of its large molecular weight.

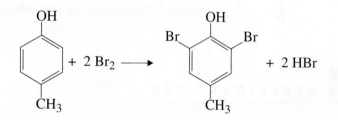

Other compounds that give a positive result with this test include aromatic compounds that have activating substituents other than hydroxyl. These compounds include anilines and alkoxyaromatics.

SPECTROSCOPY

Infrared

O—H stretch is observed near 3400 cm^{-1}.

C—O stretch is observed near 1200 cm^{-1}.

The typical aromatic ring absorptions between 1600 cm^{-1} and 1450 cm^{-1} are also found. Aromatic C—H is observed near 3100 cm^{-1}.

See Technique 25 for details.

Nuclear Magnetic Resonance

Aromatic protons are observed near 7 ppm. The hydroxyl proton has a resonance position that is concentration-dependent.

See Technique 26 for details. Carbon NMR is described in Technique 27.

DERIVATIVES

Phenols form the same derivatives as alcohols (Experiment 54H). They form urethanes by reaction with isocyanates. Phenylurethanes are used for alcohols, and the α-naphthylurethanes are more useful for phenols. Like alcohols, phenols yield 3,5-dinitrobenzoates.

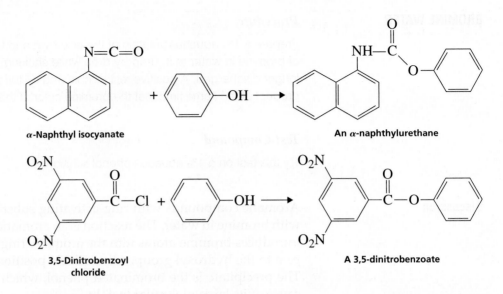

α-Naphthyl isocyanate   An α-naphthylurethane

3,5-Dinitrobenzoyl   A 3,5-dinitrobenzoate
chloride

The bromine–water reagent yields solid bromo derivatives of phenols in several cases. These solid derivatives can be used to characterize an unknown phenol. Procedures for preparing these derivatives are given in Appendix 2.

54G   EXPERIMENT 54G

Amines

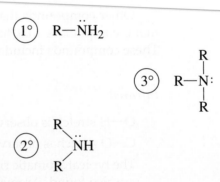

Amines are detected best by their solubility behavior and their basicity. They are the only basic compounds that will be issued for this experiment. Hence, once the compound has been identified as an amine, the main problem that remains is to decide whether it is primary (1°), secondary (2°), or tertiary (3°). This can usually be decided either by the nitrous acid tests or by infrared spectroscopy.

Solubility Characteristics					*Classification Tests*
HCl	NaHCO₃	NaOH	H₂SO₄	Ether	pH of an aqueous solution
(+)	(−)	(−)	(+)	(+)	Hinsberg test
Water: < C₆(+)					Nitrous acid test
> C₆(−)					Acetyl chloride

SUGGESTED WASTE DISPOSAL

Residues from the nitrous acid test should be poured into a waste container containing 6*M* hydrochloric acid. Dispose of all aqueous solutions in the container designated for aqueous waste. Any remaining organic compounds must be disposed of in the appropriate organic waste container.

CLASSIFICATION TESTS

NITROUS ACID TEST

Procedure

Dissolve 0.1 g of an amine in 2 mL of water to which 8 drops of concentrated sulfuric acid have been added. Use a large test tube. Often, a considerable amount of solid forms in the reaction of an amine with sulfuric acid. This solid is likely to be the amine sulfate salt. Add about 4 mL of water to help dissolve the salt. Any remaining solid will not interfere with the results of this test. Cool the solution to 5°C or less in an ice bath. Also cool 2 mL of 10% aqueous sodium nitrite in another test tube. In a third test tube, prepare a solution of 0.1 g β-naphthol in 2 mL of aqueous 10% sodium hydroxide and place it in an ice bath to cool. Add the cold sodium nitrite solution, drop by drop while shaking, to the cooled solution of the amine. Look for bubbles of nitrogen gas. Be careful not to confuse the evolution of the *colorless* nitrogen gas with an evolution of *brown* nitrogen oxide gas. Substantial evolution of gas at 5°C or below indicates a primary aliphatic amine RNH_2. The formation of a yellow oil or a yellow solid usually indicates a secondary amine R_2NH. Either tertiary amines do not react or they behave like secondary amines.

If little or no gas evolves at 5°C, take *half* the solution and warm it gently to about room temperature. Nitrogen gas bubbles at this elevated temperature indicate that the original compound was a **primary aromatic** $ArNH_2$. Take the other half of the solution and drop by drop add the solution of β-naphthol in base. If a red dye precipitates, the unknown has been conclusively shown to be a primary aromatic amine $ArNH_2$.

Test Compounds

Try this test with aniline, *N*-methylaniline, and butylamine.

CAUTION

The products of this reaction may include nitrosamines. Nitrosamines are suspected carcinogens. Avoid contact and dispose of all residues by pouring them into a waste container that contains 6*M* hydrochloric acid.

Discussion

Before you make this test, it should definitely be proved by some other method that the unknown is an amine. Many other compounds react with nitrous acid (phenols, ketones, thiols, amides), and a positive result with one of these could lead to an incorrect interpretation.

The test is best used to distinguish *primary* aromatic and *primary* aliphatic amines from secondary and tertiary amines. It also differentiates aromatic and aliphatic primary amines. It cannot distinguish between secondary and tertiary amines. You will need to use infrared spectroscopy to make the distinction between secondary and tertiary amines. Primary aliphatic amines lose nitrogen gas at low temperatures under the conditions of this test. Aromatic amines yield a more stable diazonium salt and do not lose nitrogen until the temperature is elevated. In addition, aromatic diazonium salts produce a red azo dye when β-naphthol is added. Secondary and tertiary amines produce yellow nitroso compounds, which may be soluble or may be oils or solids. Many nitroso compounds have been shown to be carcinogenic. Avoid contact and immediately dispose of all such solutions in an appropriate waste container.

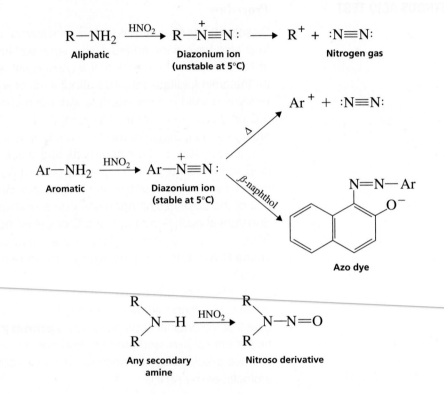

HINSBERG TEST

A traditional method for classifying amines is the **Hinsberg test.** A discussion of this test can be found in the comprehensive textbooks listed on page 454. We have found that infrared spectroscopy is a more reliable method of distinguishing between primary, secondary, and tertiary amines.

pH OF AN AQUEOUS SOLUTION

Procedure

If the compound is soluble in water, simply prepare an aqueous solution and check the pH with pH paper. If the compound is an amine, it will be basic, and the solution will have a high pH. Compounds that are insoluble in water can be dissolved in ethanol–water or 1,2-dimethoxyethane–water.

ACETYL CHLORIDE

Procedure

Primary and secondary amines give a positive acetyl chloride test result (liberation of heat). This test is described for alcohols in Experiment 54H. Cautiously add dropwise the acetyl chloride to the liquid amine. This reaction can be very exothermic and violent! When the test mixture is diluted with water, primary and secondary amines often give a solid acetamide derivative; tertiary amines do not.

Test Compounds

Try this test with aniline and butylamine.

SPECTROSCOPY

Infrared

N—H stretch. Both aliphatic and aromatic primary amines show two absorptions (doublet due to symmetric and asymmetric stretches) in the region 3500–3300 cm^{-1}. Secondary amines show a single absorption in this region. Tertiary amines have no N—H bonds.

N—H bend. Primary amines have a strong absorption at 1640–1560 cm^{-1}. Secondary amines have an absorption at 1580–1490 cm^{-1}.

Aromatic amines show bands typical for the aromatic ring in the region 1600–1450 cm^{-1}.

Aromatic C—H is observed near 3100 cm^{-1}.

See Technique 25 for details.

NUCLEAR MAGNETIC RESONANCE

The resonance position of amino hydrogens is extremely variable. The resonance may also be very broad (quadrupole broadening). Aromatic amines give resonances near 7 ppm due to the aromatic ring hydrogens.

See Technique 26 for details. Carbon NMR is described in Technique 27.

DERIVATIVES

The derivatives of amines that are most easily prepared are the acetamides and the benzamides. These derivatives work well for both primary and secondary amines but not for tertiary amines.

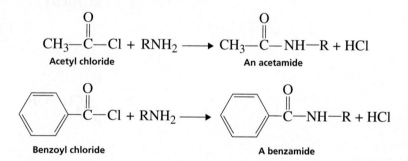

$$CH_3-\overset{\overset{\displaystyle O}{\|}}{C}-Cl + RNH_2 \longrightarrow CH_3-\overset{\overset{\displaystyle O}{\|}}{C}-NH-R + HCl$$

Acetyl chloride   **An acetamide**

$$\langle \rangle-\overset{\overset{\displaystyle O}{\|}}{C}-Cl + RNH_2 \longrightarrow \langle \rangle-\overset{\overset{\displaystyle O}{\|}}{C}-NH-R + HCl$$

Benzoyl chloride   **A benzamide**

The most general derivative that can be prepared is the picric acid salt, or picrate, of an amine. This derivative can be used for primary, secondary, and tertiary amines.

Great care must be taken when working with saturated solutions of picric acid. Picric acid may detonate when heated above 300°C! It is also known to explode when heated rapidly. For this reason, it is strongly recommended that you check with your instructor before preparing this derivative.

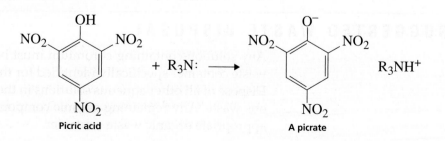

Picric acid   **A picrate**

For tertiary amines, the methiodide salt is often useful.

$$CH_3I + R_3N: \longrightarrow CH_3-NR_3^+I^-$$

A methiodide

Procedures for preparing derivatives from amines can be found in Appendix 2.

54H   EXPERIMENT 54H

Alcohols

Alcohols are neutral compounds. The only other classes of neutral compounds used in this experiment are the aldehydes and ketones and the esters. Alcohols and esters usually do not give a positive 2,4-dinitrophenylhydrazine test; aldehydes and ketones do. Esters do not react with Ce(IV) or acetyl chloride or with Lucas reagent, as alcohols do, and they are easily distinguished from alcohols on this basis. Primary and secondary alcohols are easily oxidized; esters and tertiary alcohols are not. A combination of the Lucas test and the chromic acid test will differentiate among primary, secondary, and tertiary alcohols.

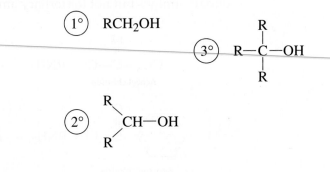

Solubility Characteristics					Classification Tests
HCl	NaHCO₃	NaOH	H₂SO₄	Ether	Cerium(IV) test
(−)	(−)	(−)	(+)	(+)	Acetyl chloride
Water: $< C_6(+)$					Lucas test
$> C_6(-)$					Chromic acid test
					Iodoform test

SUGGESTED WASTE DISPOSAL

Any solution containing chromium must be disposed of by placing it in a waste container specifically identified for the disposal of chromium wastes. Dispose of all other aqueous solutions in the container designated for aqueous waste. Any remaining organic compounds must be disposed of in the appropriate organic waste container.

CLASSIFICATION TESTS

CERIUM(IV) TEST

Procedure for Water-Soluble or Partially Soluble Compounds

Add 3 mL of water to 0.5 mL of Cerium(IV) reagent in a test tube. Gently shake the solution to thoroughly mix the solution and then add 4 drops of the compound to be tested. Gently shake the mixture and look for an immediate color change from a yellow-orange solution to a red-orange or deep red color indicating the presence of an —OH group in an alcohol or phenol. Phenol forms a dark-brown precipitate.

Test Compounds

Try this test on 1-butanol, 2-pentanol, 2-methyl-2-butanol, phenol, butanal, cyclo-hexanone and ethyl acetate.

Procedure for Water-Insoluble Compounds

Add 3 mL of 1,2-dimethoxyethane to 0.5 mL of Cerium(IV) reagent in a dry test tube. Gently shake the solution to thoroughly mix the solution and then add 4 drops of a liquid compound to be tested. If you have a solid, you can directly add a few milligrams of the solid to the solution. Enough will dissolve to give a test if an —OH group is present. Gently shake the mixture and look for an *immediate color change* from a yellow-orange solution to a reddish brown color indicating the presence of an alcohol or phenol.

Test Compounds

Try this test on 1-octanol, β-naphthol (2-naphthol) and benzoic acid.

Reagent

Prepare 2M nitric acid solution by diluting 12.8 mL of concentrated nitric acid to 100 mL with water. Dissolve 8 g of ceric ammonium nitrate [$Ce(NH_4)_2(NO_3)_6$] in 20 mL of the dilute nitric acid solution.

Discussion

Primary, secondary, and tertiary alcohols and phenols form 1:1 colored complexes with Ce(IV) and are an excellent way to detect hydroxyl groups. However, this test is limited to compounds with no more than 10 carbon atoms. Unfortunately, the test cannot distinguish between primary, secondary, or tertiary alcohols. The Lucas test or chromium oxide test will need to be used for this purpose. Esters, ketones, carboxylic acids, and simple aldehydes do not change the color of the reagent and give a negative test with the Ce(IV) reagent. Thus, esters and other neutral compounds can be distinguished from alcohols by this test. Amines produce a flocculent white precipitate with the reagent. Cerium solutions can oxidize alcohols, but this usually occurs when the solution is heated or when the alcohol is in contact with the reagent for long periods.

ACETYL CHLORIDE

Procedure

Cautiously add about 5–10 drops of acetyl chloride, drop by drop, to about 0.25 mL of the liquid alcohol contained in a small test tube. Evolution of heat and hydrogen chloride gas indicates a positive reaction. Check for the evolution of HCl with a piece of wet blue litmus paper. Hydrogen chloride will turn the litmus paper red. Adding water will sometimes precipitate the acetate.

Test Compounds

Try this test with 1-butanol.

Discussion

Acid chlorides react with alcohols to form esters. Acetyl chloride forms acetate esters.

$$CH_3-\overset{\displaystyle O}{\overset{\|}{C}}-Cl + ROH \longrightarrow CH_3-\overset{\displaystyle O}{\overset{\|}{C}}-O-R + HCl$$

Usually, the reaction is exothermic, and the heat evolved is easily detected. Phenols react with acid chlorides somewhat as alcohols do. Hence, phenols should be eliminated as possibilities before this test is attempted. Amines also react with acetyl chloride to evolve heat (see Experiment 54G). This test does not work well with solid alcohols.

LUCAS TEST

Procedure

Place 2 mL of Lucas reagent in a small test tube and add 3–4 drops of the alcohol. Stopper the test tube and shake it vigorously. Tertiary (3°), benzylic, and allylic alcohols give an immediate cloudiness in the solution as the insoluble alkyl halide separates from the aqueous solution. After a short time, the immiscible alkyl halide may form a separate layer. Secondary (2°) alcohols produce a cloudiness after 2–5 minutes. Primary (1°) alcohols dissolve in the reagent to give a clear solution (no cloudiness). Some secondary alcohols may have to be heated slightly to encourage reaction with the reagent.

NOTE: This test works only for alcohols that are soluble in the reagent. This often means that alcohols with more than six carbon atoms cannot be tested.

Test Compounds

Try this test with 1-butanol (*n*-butyl alcohol), 2-butanol (*sec*-butyl alcohol), and 2-methyl-2-proponol (*t*-butyl alcohol).

Reagent

Cool 10 mL of concentrated hydrochloric acid in a beaker, using an ice bath. While still cooling and while stirring, dissolve 16 g of anhydrous zinc chloride in the acid.

Discussion

This test depends on the appearance of an alkyl chloride as an insoluble second layer when an alcohol is treated with a mixture of hydrochloric acid and zinc chloride (Lucas reagent):

$$R-OH + HCl \xrightarrow{ZnCl_2} R-Cl + H_2O$$

Primary alcohols do not react at room temperature; therefore, the alcohol is seen simply to dissolve. Secondary alcohols react slowly, whereas tertiary, benzylic, and allylic alcohols react instantly. These relative reactivities are explained on the same basis as the silver nitrate reaction, which is discussed in Experiment 54B. Primary carbocations are unstable and do not form under the conditions of this test; hence, no results are observed for primary alcohols.

$$
\begin{array}{c}
R \\
| \\
R-C-OH \\
| \\
R
\end{array}
+ ZnCl_2 \longrightarrow
\begin{array}{c}
R \\
| \quad \delta^+ \ \delta^- \\
R-C-O\cdots ZnCl_2 \\
| \quad | \\
R \quad H
\end{array}
\longrightarrow
\left[
\begin{array}{c}
R \\
| \\
R-C^+ \\
| \\
R
\end{array}
\right]
\xrightarrow{Cl^-}
\begin{array}{c}
R \\
| \\
R-C-Cl \\
| \\
R
\end{array}
$$

The Lucas test does not work well with solid alcohols or with liquid alcohols with about six or more carbon atoms.

CHROMIC ACID TEST: ALTERNATIVE TEST

> **CAUTION**
>
> Many chromium(VI) compounds are suspected carcinogens. If you would like to run this test, talk to your instructor first. The Lucas test will distinguish between 1°, 2°, and 3° alcohols, and you should do that test first. If you run the chromic acid test, be sure to wear gloves to avoid contact with this reagent.

Procedure

Dissolve 1 drop of a liquid or about 10 mg of a solid alcohol in 1 mL of *reagent-grade* acetone. Add 1 drop of the chromic acid reagent and note the result that occurs within 2 seconds. A positive test for a primary or a secondary alcohol is the appearance of a blue-green color. Tertiary alcohols do not produce the test result within 2 seconds, and the solution remains orange. To make sure that the acetone solvent is pure and does not give a positive test result, add 1 drop of chromic acid to 1 mL of acetone that does not have an unknown dissolved in it. The orange color of the reagent should persist for *at least* 3 seconds. If it does not, a new bottle of acetone should be used.

Test Compounds

Try this test with 1-butanol (*n*-butyl alcohol), 2-butanol (*sec*-butyl alcohol), and 2-methyl-2-propanol (*t*-butyl alcohol).

Reagent

Dissolve 20 g of chromium trioxide (CrO_3) in 60 mL of cold water in a beaker. Add a magnetic stir bar to the solution. With stirring, slowly and carefully add 20 mL of concentrated sulfuric acid to the solution. This reagent should be prepared fresh each term.

Discussion

This test is based on the reduction of chromium(VI), which is orange, to chromium(III), which is green, when an alcohol is oxidized by the reagent. A change in color of the reagent from orange to green represents a positive test. Primary alcohols are oxidized by the reagent to carboxylic acids; secondary alcohols are oxidized to ketones.

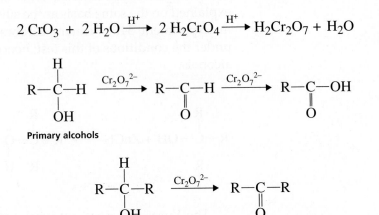

Although primary alcohols are first oxidized to aldehydes, the aldehydes are further oxidized to carboxylic acids. The ability of chromic acid to oxidize aldehydes but not ketones is taken advantage of in a test that uses chromic acid to distinguish between aldehydes and ketones (Experiment 54D). Secondary alcohols are oxidized to ketones but no further. Tertiary alcohols are not oxidized at all by the reagent; hence, this test can be used to distinguish primary and secondary alcohols from tertiary alcohols. Unlike the Lucas test, this test can be used with all alcohols regardless of molecular weight and solubility.

IODOFORM TEST

Alcohols with the hydroxyl group at the 2-position of the chain give a positive iodoform test. See the discussion in Experiment 54D.

SPECTROSCOPY

Infrared

O—H stretch. A medium to strong, and usually broad, absorption comes in the region 3600–3200 cm^{-1}. In dilute solutions or with little hydrogen bonding, there is a sharp absorption near 3600 cm^{-1}. In more concentrated solutions, or with considerable hydrogen bonding, there is a broad absorption near 3400 cm^{-1}. Sometimes both bands appear.

C—O stretch. There is a strong absorption in the region 1200–1500 cm^{-1}. Primary alcohols absorb nearer 1050 cm^{-1}; tertiary alcohols and phenols absorb nearer 1200 cm^{-1}. Secondary alcohols absorb in the middle of this range.

See Technique 25 for details.

Nuclear Magnetic Resonance

The hydroxyl resonance is extremely concentration-dependent, but it is usually found between 1 ppm and 5 ppm. Under normal conditions, the hydroxyl proton does not couple with protons on adjacent carbon atoms.

See Technique 26 for details. Carbon NMR is described in Technique 27.

DERIVATIVES

The most common derivatives for alcohols are the 3,5-dinitrobenzoate esters and the phenylurethanes. Occasionally, the α-naphthylurethanes

(Experiment 54F) are also prepared, but these latter derivatives are more often used for phenols.

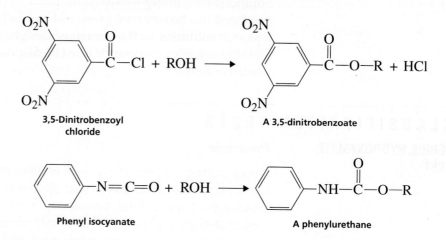

3,5-Dinitrobenzoyl chloride → A 3,5-dinitrobenzoate

Phenyl isocyanate → A phenylurethane

Procedures for preparing these derivatives are given in Appendix 2.

54I   EXPERIMENT 54I

Esters

Esters are formally considered "derivatives" of the corresponding carboxylic acid. They are frequently synthesized from the carboxylic acid and the appropriate alcohol:

$$R-COOH + R'-OH \overset{H^+}{\rightleftharpoons} R-COOR' + H_2O$$

Thus, esters are sometimes referred to as though they were composed of an acid part and an alcohol part.

Although esters, like aldehydes and ketones, are neutral compounds that have a carbonyl group, they do not usually give a positive 2,4-dinitrophenylhydrazine test result. The two most common tests for identifying esters are the basic hydrolysis and ferric hydroxamate tests.

Solubility Characteristics					*Classification Tests*
HCl	NaHCO$_3$	NaOH	H$_2$SO$_4$	Ether	Ferric hydroxamate test
(−)	(−)	(−)	(+)	(+)	Basic hydrolysis
Water: < C$_4$(+)					
> C$_5$(−)					

SUGGESTED WASTE DISPOSAL

Solutions containing hydroxylamine or derivatives formed from it should be placed in a beaker containing 6*M* hydrochloric acid. Dispose of any other aqueous solutions in the container designated for aqueous waste. Any remaining organic compounds must be disposed of in the appropriate organic waste container.

CLASSIFICATION TESTS

FERRIC HYDROXAMATE TEST

Procedure

Before starting, you must determine whether the compound to be tested already has enough enolic character in acid solution to give a positive ferric chloride test. Dissolve 1 or 2 drops of the unknown liquid or a few crystals of the unknown solid in 1 mL of 95% ethanol and add 1 mL of 1*M* hydrochloric acid. Add 1 or 2 drops of 5% ferric chloride solution. If a burgundy magenta or reddish-brown color appears, the ferric hydroxamate test cannot be used. It contains enolic character (see p. 480).

If the compound did not show enolic character, continue as follows. Dissolve 5 or 6 drops of a liquid ester, or about 40 mg of a solid ester, in a mixture of 1 mL of 0.5*M* hydroxylamine hydrochloride (dissolved in 95% ethanol) and 0.4 mL of 6*M* sodium hydroxide. Heat the mixture to boiling for a few minutes. Cool the solution and then add 2 mL of 1*M* hydrochloric acid. If the solution becomes cloudy, add 2 mL of 95% ethanol to clarify it. Add a drop of 5% ferric chloride solution and note whether a color is produced. If the color fades, continue to add ferric chloride until the color persists. A positive test should give a deep burgundy or magenta or reddish-brown color.

Test Compound

Try this test with ethyl butanoate.

Discussion

On being heated with hydroxylamine, esters are converted to the corresponding hydroxamic acids:

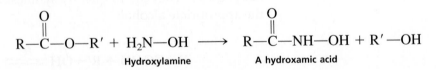

The hydroxamic acids form strong, colored complexes with ferric ion.

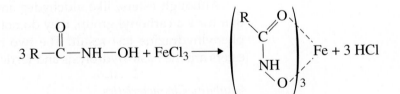

BASIC HYDROLYSIS (OPTIONAL)

Procedure

Place 0.7 g of the ester in a 10-mL round-bottom flask with 7 mL of 25% aqueous sodium hydroxide. Add a boiling stone, and attach a water condenser. Use a small amount of stopcock grease to lubricate the ground-glass joint. Boil the mixture for about 30 minutes. Stop the heating and observe the solution to determine whether the oily ester layer has disappeared or whether the odor of the ester (usually pleasant) has disappeared. Low-boiling esters (below 110°C) usually dissolve within 30 minutes if the alcohol part has a low molecular weight. If the ester has not dissolved, reheat the mixture to reflux for 1–2 hours. After that time, the oily ester layer should have disappeared, along with the characteristic odor. Esters with boiling points up to 200°C should hydrolyze during this time. Compounds remaining after this extended period of heating are either unreactive esters or are not esters at all.

For esters derived from solid acids, the acid part can, if desired, be recovered after hydrolysis. Extract the basic solution with ether to remove any unreacted ester (even if it appears to be gone), acidify the basic solution with hydrochloric acid, and extract the acidic phase with ether to remove the acid. Dry the ether layer over anhydrous sodium sulfate, decant, and evaporate the solvent to obtain the parent acid from the original ester. The melting point of the parent acid can provide valuable information in the identification process.

Discussion

This procedure converts the ester to its separate acid and alcohol parts. The ester dissolves because the alcohol part (if small) is usually soluble in the aqueous medium, as is the sodium salt of the acid. Acidification produces the parent acid:

$$R\!-\!\overset{\overset{\displaystyle O}{\|}}{C}\!-\!O\!-\!R' \xrightarrow{\text{NaOH}} R\!-\!\overset{\overset{\displaystyle O}{\|}}{C}\!-\!O^-Na^+ + R'OH \xrightarrow{\text{HCl}} R\!-\!\overset{\overset{\displaystyle O}{\|}}{C}\!-\!O\!-\!H + R'OH$$

Ester | **Salt of acid part** **Alcohol part**

All derivatives of carboxylic acids are converted to the parent acid on basic hydrolysis. Thus, amides, which are not covered in this experiment, would also dissolve in this test, liberating the free amine and the sodium salt of the carboxylic acid.

SPECTROSCOPY

Infrared

The ester–carbonyl group (C=O) peak is usually a strong absorption, as is the absorption of the carbonyl–oxygen link (C—O) to the alcohol part. C=O stretch at approximately 1735 cm^{-1} is normal.[1] C—O stretch usually gives two or more absorptions, one stronger than the others, in the region 1280–1050 cm^{-1}.

See Technique 25 for details.

Nuclear Magnetic Resonance

Hydrogens that are alpha to an ester carbonyl group have resonance in the region 2–3 ppm. Hydrogens alpha to the alcohol oxygen of an ester have resonance in the region 3–5 ppm.

See Technique 26 for details. Carbon NMR is described in Technique 27.

---

[1] Conjugation with the carbonyl group moves the carbonyl absorption to lower frequencies. Conjugation with the alcohol oxygen raises the carbonyl absorption to higher frequencies. Ring strain (lactones) moves the carbonyl absorption to higher frequencies.

DERIVATIVES

Esters present a double problem when trying to prepare derivatives. To characterize an ester completely, you need to prepare derivatives of *both* the acid part and the alcohol part.

Acid Part

The most common derivative of the acid is the *N*-benzylamide derivative.

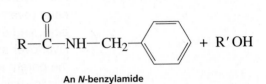

An *N*-benzylamide

The reaction does not proceed well unless R′ is methyl or ethyl. For alcohol portions that are larger, the ester must be transesterified to a methyl or an ethyl ester before preparing the derivative.

$$R-\overset{O}{\underset{||}{C}}-OR' + CH_3OH \xrightarrow{H^+} R-\overset{O}{\underset{||}{C}}-O-CH_3 + R'OH$$

Hydrazine also reacts well with methyl and ethyl esters to give acid hydrazides.

$$R-\overset{O}{\underset{||}{C}}-OR' + NH_2NH_2 \longrightarrow R-\overset{O}{\underset{||}{C}}-NHNH_2 + R'OH$$

An acid hydrazide

Alcohol Part

The best derivative of the alcohol part of an ester is the 3,5-dinitrobenzoate ester, which is prepared by an acyl interchange reaction:

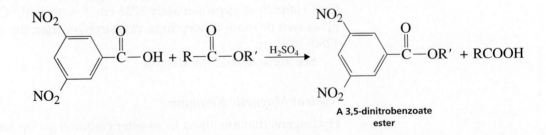

A 3,5-dinitrobenzoate ester

Most esters are composed of very simple acid and alkyl portions. For this reason, spectroscopy is usually a better method of identification than is the preparation of derivatives. Not only is it necessary to prepare two derivatives with an ester but all esters with the same acid portion, or all those with the same alcohol portion, give identical derivatives of those portions.

PART 5

Project-Based Experiments

EXPERIMENT 55

A Separation and Purification Scheme

Extraction

Crystallization

Devising a procedure

Critical thinking application

There are many organic experiments in which the components of a mixture must be separated, isolated, and purified. Although detailed procedures are usually given for carrying this out, devising your own scheme can help you understand these techniques more thoroughly. In this experiment, you will devise a separation and purification scheme for a three-component mixture that will be assigned to you. The mixture will contain a neutral organic compound and either an organic acid or base in nearly equal amounts. The third component, also a neutral compound, will be present in a much smaller amount. Your goal will be to isolate in pure form *two* of the three compounds. The components of your mixture may be separated and purified by a combination of acid–base extractions and crystallizations. You will be told the composition of your mixture well in advance of the laboratory period so that you will have time to write a procedure for this experiment.

REQUIRED READING

Review: Technique 11   Crystallization: Purification of Solids

Technique 12   Extractions, Separations, and Drying Agents

SUGGESTED WASTE DISPOSAL

Dispose of all filtrates that may contain 1,4-dibromobenzene or methylene chloride into the container designated for halogenated organic wastes. All other filtrates may be disposed of into the container for nonhalogenated organic wastes.

NOTES TO THE INSTRUCTOR

Students must be told the composition of their mixture well in advance of the laboratory period so that they have enough time to devise a procedure. It is advisable to require that students turn in a copy of their procedure at the beginning of the lab period. You may wish to allow enough time for students to repeat the experiment if their procedure doesn't work the first time or if they want to improve on their percentage recovery and purity. If you

Experiment 55 is based on a similar experiment developed by James Patterson, University of Washington, Seattle.

allow enough time for students to perform this experiment just once, it will be helpful to put out pure samples of the compounds in the mixtures so students can try out different solvents to determine a good solvent for crystallizing each compound.

PROCEDURE

Advance Preparation

Each student will be assigned a mixture of three compounds.[1] Before coming to the laboratory, you must work out a detailed procedure that can be used to separate, isolate, and purify *two* of the compounds in your mixture. You may not be able to specify all the reagents or the volumes required ahead of time, but the procedure should be as complete as possible. It will be helpful to consult the following experiments and techniques:

> Experiment 2, "Solubility," Part D, pp. 16–17
>
> Experiment 4D, "Extraction," pp. 38–40
>
> Technique 10, Section 10.2B, p. 642
>
> Technique 12, Section 12.11, p. 685

The following reagents will be available: $1M$ NaOH, $6M$ NaOH, $1M$ HCl, $6M$ HCl, $1M$ NaHCO$_3$, saturated sodium chloride, diethyl ether, 95% ethanol, methanol, isopropyl alcohol, acetone, hexane, toluene, methylene chloride, and granular anhydrous sodium sulfate. Other solvents that can be used for crystallization may also be available.

Separation

The first step in your procedure should be to dissolve about 0.5 g (record exact weight) of the mixture in the minimum amount of diethyl ether or methylene chloride. If more than about 4 mL of a solvent is required, you should use the other solvent. Most of the compounds in the mixtures are more soluble in methylene chloride than diethyl ether; however, you may need to determine the appropriate solvent by experimentation. Once you have selected a solvent, this same solvent should be used throughout the procedure when an organic solvent is required. If you use diethyl ether, you must use two steps to dry the organic layer. First, the organic layer must be mixed with saturated sodium chloride (see p. 698), and then the liquid is dried over granular anhydrous sodium sulfate.

Purification

To improve the purity of your final samples, you should include a backwashing step at the appropriate place in your procedure. See Section 12.11, page 685, for a discussion of this technique. Crystallization will be required to purify both of the compounds you isolate. To find an appropriate solvent, you should consult a handbook. You can also use the procedure in Section 11.6, page 662, to determine a

---

[1] Your mixture may be one of the following: (1) 50% benzoic acid, 40% benzoin, 10% 1,4-dibromobenzene; (2) 50% fluorene, 40% *o*-toluic acid, 10% 1,4-dibromobenzene; (3) 50% phenanthrene, 40% methyl 4-aminobenzoate, 10% 1,4-dibromobenzene; or (4) 50% 4-aminoacetophenone, 40% 1,2,4,5-tetrachlorobenzene, 10% 1,4-dibromobenzene. Other mixtures are given in the Instructor's Manual, along with some suggestions about these mixtures.

good solvent experimentally. Your procedure should include at least one method for determining if you have obtained both compounds in a pure form. Hand in each compound in a labeled vial.

When performing the laboratory work, you should strive to obtain a high recovery of both compounds in a highly pure form. If your procedure fails, modify it and repeat the experiment.

REPORT

Write out a complete procedure by which you separated and isolated pure samples of two of the compounds in your mixture. Describe how you determined that your procedure was successful and give any data or results used for this purpose. Calculate the percentage recovery for both compounds.

EXPERIMENT 56

Preparation of a C-4 or C-5 Acetate Ester

Esterification

Separatory funnel

Conventional distillation

In this experiment, we prepare an ester from acetic acid and a C-4 or a C-5 alcohol. This experiment is a conventional-scale preparation, but it is similar to the microscale preparation of isopentyl acetate, which is described in Experiment 13. However, for the experiment, either your instructor will assign, or you will pick, one of the following C-4 or C-5 alcohols to react with acetic acid:

1-Butanol (*n*-butyl alcohol)	1-Pentanol (*n*-pentyl alcohol)
2-Butanol (*sec*-butyl alcohol)	2-Pentanol
2-Methyl-1-propanol (isobutyl alcohol)	3-Pentanol
Cyclopentanol	3-Methyl-1-butanol (isopentyl alcohol)

If an NMR spectrometer is available, your instructor may wish to give you one of these alcohols as an unknown, leaving it to you to determine which alcohol was issued. For this purpose, you could use the infrared and NMR spectra, as well as the boiling points of the alcohol and its ester.

REQUIRED READING

Review: Essay   Esters—Flavors and Fragrances

Experiment 13

Techniques 12, 13, and 14

SPECIAL INSTRUCTIONS

Be careful when dispensing sulfuric and glacial acetic acids. They are corrosive and will attack your skin if you make contact with them. If you get one of these acids on your skin, wash the affected area with large amounts of running water for 10–15 minutes.

If you select 2-butanol, reduce the amount of concentrated sulfuric acid to 0.5 mL. Also reduce the heating time to 60 minutes or less. Secondary alcohols have a tendency to give a significant percentage of elimination in strongly acidic solutions. Some of the alcohols may undergo elimination, leading to the formation of some low-boiling material (alkenes). In addition, cyclopentanol forms some dicyclopentyl ether, a solid.

SUGGESTED WASTE DISPOSAL

Any aqueous solutions should be placed in the container designated for dilute aqueous waste. Place any excess ester in the nonhalogenated organic waste container.

NOTES TO THE INSTRUCTOR

The sulfuric acid used as a catalyst in this reaction may be replaced with Dowex 50WX8-100 cationic exchange resin (sulfonate groups).

The purity of the esters can be determined by gas chromatography. It is recommended that a gas chromatogram of each of the starting alcohols be performed prior to determining the gas chromatogram of the esters. In this way, the peak corresponding to the parent alcohol can be identified by its retention time and the percentage of unreacted alcohol in the sample can be obtained. Approximate gas chromatography conditions for a GowMac Series 580 instrument with an ⅛-inch OV-1 column: 0.5 μL sample; flow rate, 27 mL/min; column temperature, 82°C; injector temperature, 170°C; detector temperature, 180°C; detector current, 200 mA.

PROCEDURE

Apparatus

Assemble a reflux apparatus on top of your hot plate using a 20- or 25-mL round-bottom flask and a water-cooled condenser (refer to Fig. 7.6A, p. 601, but use a round-bottom flask instead of the conical vial). To control vapors, place a drying tube packed with calcium chloride on top of the condenser. Use a hot plate and the aluminum block with the larger set of holes for heating.

Reaction Mixture

Weigh (tare) an empty 10-mL graduated cylinder and record its weight. Place approximately 5.0 mL of your chosen alcohol in the graduated cylinder and reweigh it to determine the weight of alcohol. Disconnect the round-bottom flask from the reflux apparatus and transfer the alcohol into it. Do not clean or wash the graduated cylinder. Using the same graduated cylinder, measure approximately 7.0 mL of glacial acetic acid (*MW* = 60.1, *d* = 1.06 g/mL) and add it to the alcohol already in the flask. Using a calibrated Pasteur pipet, add 1 mL of concentrated sulfuric acid (0.5 mL if you have chosen 2-butanol), mixing *immediately* (swirl), to the reaction mixture contained in the flask. Add a corundum boiling stone or stirring bar and reconnect the flask. Do not use a calcium carbonate (marble) boiling stone, because it will dissolve in the acidic medium.

Reflux

Start water circulating in the condenser and bring the mixture to a boil. Continue heating under reflux for 60–75 minutes. Be sure to stir the mixture if you are using a stirring bar instead of a boiling stone. Then disconnect or remove the heating source and let the mixture cool to room temperature.

Extractions

Disassemble the apparatus and transfer the reaction mixture to a separatory funnel (60 or 125 mL) placed in a ring attached to a ring stand. Be sure the stopcock is closed and, using a funnel, pour the mixture into the top of the separatory funnel. Also be careful to avoid transferring the boiling stone (or stirring bar), or you will need to remove it after the transfer. Add 10 mL of water, stopper the funnel, and mix the phases by careful shaking and venting (Section 12.7 and Fig. 12.9, pp. 677–679). Allow the phases to separate and then uncap the funnel and drain the lower aqueous layer through the stopcock into a beaker or other suitable container. Next, extract the organic layer with 5 mL of 5% aqueous sodium bicarbonate just as you did previously with water. Extract the organic layer once again, this time with 5 mL of saturated aqueous sodium chloride.

Drying

Transfer the crude ester to a clean, dry, 25-mL Erlenmeyer flask and add approximately 1.0 g of anhydrous sodium sulfate. Cork the mixture and let it stand for 10–15 minutes while you prepare the apparatus for distillation. If the mixture does not appear dry (the drying agent clumps and does not "flow," the solution is cloudy, or drops of water are obvious), transfer the ester to a new, clean, dry, 25-mL Erlenmeyer flask and add a new 0.5-g portion of anhydrous sodium sulfate to complete the drying.

Distillation

Assemble a distillation apparatus using your smallest round-bottom flask to distill from (Fig. 14.10, p. 713, but insert a water condenser as shown on p. 714). Use a hot plate with an aluminum block to heat. Preweigh (tare) and use a 5-mL conical vial to collect the product. (It might be wise to have a second tared 5-mL conical vial handy in case you fill the first one.) Immerse the collection flask in a beaker of ice to ensure condensation and to reduce odors. If your alcohol is not an unknown, you can look up its boiling point in a handbook; otherwise, you can expect your ester to have a boiling point between 95 and 150°C. Continue distillation until only 1 or 2 drops of liquid remain in the distilling flask. Record the observed boiling-point *range* in your notebook.

Yield Determination

Weigh the product and calculate the percentage yield of the ester. At the option of your instructor, determine the boiling point using one of the methods described in Section 13.2, page 695.

Spectroscopy

At your instructor's option, obtain an infrared spectrum using salt plates (Technique 25, Section 25.2, p. 834). Compare the spectrum with the one reproduced in Experiment 13 (p. 107). The spectrum of your ester should have similar features to the one shown. Interpret the spectrum and include it in your report to the instructor. You may also be required to determine and interpret the proton and carbon-13 NMR spectra (Technique 26, Section 26.1, p. 870 and Technique 27, Section 27.1, page 906). Submit your sample in a properly labeled vial with your report.

Gas Chromatography (Optional)

At your instructor's option, perform a gas-chromatographic analysis of your ester. Either your instructor will provide a gas chromatogram of your starting alcohol or you will be asked to determine one at the same time that you do the analysis of your ester. Using both chromatograms, identify the alcohol and ester peaks and calculate the percentage of unreacted alcohol (if any) still remaining in your sample. Is there any evidence of a product from a competing elimination reaction? Attach the chromatograms to your notebook or your final report, and be sure to include a discussion of the results in your report.

QUESTIONS

1. One method of favoring the formation of an ester is to add excess acetic acid. Suggest another method, involving the right-hand side of the equation, that will favor the formation of the ester.

2. Why is the mixture extracted with sodium bicarbonate? Give an equation and explain its relevance.

3. Why are gas bubbles observed?

4. Using your alcohol, determine which starting material is the limiting reagent in this procedure. Which reagent is used in excess? How great is the molar excess (how many times greater)?

5. Outline a separation scheme for isolating your pure ester from the reaction mixture.

6. Interpret the principal absorption bands in the infrared spectrum of your ester or, if you did not determine the infrared spectrum of you ester, do this for the spectrum of isopentyl acetate on page 107. (Technique 25 may be of some help.)

7. Write a mechanism for the acid-catalyzed esterification that uses your alcohol and acetic acid. You may need to consult the chapter on carboxylic acids in your lecture textbook.

8. Tertiary alcohols do not work well in the procedure outlined for this experiment; they give a different product from what you might expect. Explain this and draw the expected product from *t*-butyl alcohol (2-methyl-2-propanol).

9. Why is glacial acetic acid designated as "glacial"? (*Hint:* Consult a handbook of physical properties.)

57

EXPERIMENT 57

Extraction of Essential Oils from Caraway, Cinnamon, Cloves, Cumin, or Fennel by Steam Distillation

Steam distillation

Extraction

Infrared spectroscopy

Nuclear magnetic resonance spectroscopy (optional)

High-performance liquid chromatography (optional)

Gas chromatography–mass spectrometry (optional)

Mini–research project (optional)

In Experiment 57A, you will steam distill the essential oil from a spice. You will choose, or the instructor will assign you, a spice from the following list: caraway, cinnamon, cloves, cumin, or fennel. Each spice produces a relatively pure essential oil. The structures for the major essential oil components of the spices are shown here. Your spice will yield one of these compounds. You are to determine which structure represents the essential oil that was distilled from your spice.

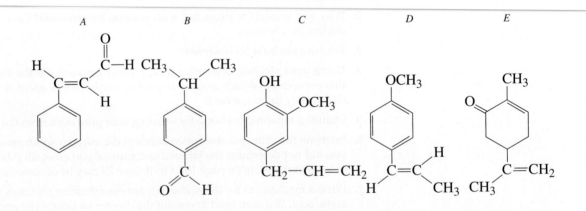

In trying the determine your structure, look for the following features (stretching frequencies) in the infrared spectrum: C=O (ketone or aldehyde), C—H (aldehyde), O—H (phenol), C—O (ether), benzene ring, and C=C (alkene). Also be sure to look for the aromatic-ring, out-of-plane bending frequencies, which may help you determine the substitution patterns of the benzene rings (see p. 856). The out-of-plane bending region may also be of help in determining the degree of substitution on the alkene double bond where it exists (see p. 854). There are enough differences in the infrared spectra of the five possible compounds that you should be able to identify your essential oil.

If NMR spectroscopy is available, it will provide a nice confirmation of your conclusions. Carbon-13 NMR would be even more informative than proton magnetic resonance. However, neither of these techniques is required for a solution.

If high-performance liquid-chromatography (HPLC) equipment is available, you can analyze the product of the steam distillation using this technique. The experiment uses HPLC as an analytical tool for separating and identifying the components of the steam distillate. The method uses a reversed-phase column and eluent system, with isochratic elution. Detection is accomplished by measuring the absorbance of ultraviolet radiation at 254 nm by the solution as it is eluted from the column. The mobile phase that will be used is a mixture of 85% methanol and 15% water.

In Experiment 57A, we have assumed that each of the spices provides *one* major product in the steam distillation. HPLC analysis lets you test whether or not this assumption is correct. You should also be able to determine the percentage of the major essential oil component in the distillate.

If gas chromatography–mass spectrometry (GC-MS) equipment is available, you can also analyze the steam distillate using this method (Experiment 57B). GC-MS is a sensitive method for determining the components in a volatile mixture. This technique is capable not only of separating the components of a mixture but also of identifying each component of the mixture. By comparing the mass spectrum of each substance eluting from the column with mass spectra from the computer-based library of spectra in the instrument's memory, you can completely identify each component of the mixture.

Finally, a variety of additional spices and herbs can be investigated by a combination of steam distillation and GC-MS analysis (Experiment 57C). This experiment is intended to be a mini–research project. Your instructor will assign you a particular spice or herb to analyze, or you will choose your own plant material. In this project, you will not have advance information about the components of the plant material that you investigate.

REQUIRED READING

Review: Technique 12

New:

	Essay	Terpenes and Phenylpropanoids
	Technique 18	Steam Distillation
	Technique 21	High-Performance Liquid Chromatography (HPLC)
	Technique 22	Section 22.14, Gas Chromatography–Mass Spectrometry (GC-MS)
	Technique 28	Mass Spectrometry

SPECIAL INSTRUCTIONS

If you use finely ground spices, foaming can be a serious problem. It is recommended that you use clove buds, or cinnamon sticks instead of ground spices. For Experiment 57A, however, be sure to cut or break up the large pieces or crush them with a mortar and pestle.

If your instructor assigns the HPLC option, you will have to determine the best operating conditions for your instrument and conditions. Your instructor should test this experiment in advance so you can have a good idea of which column to use and which flow rate of solvent works best. Your instructor will provide specific instruction in the operation of the particular HPLC instrument being used in your laboratory. The instructions that follow outline the general procedure.

For Experiment 57B, similar instructions also pertain. Your instructor will provide instruction for the operation of the specific GC-MS instrument used in your laboratory. Your instructor should also tell which column to use and which operating conditions work best. The instructions that follow outline the general procedure.

Your instructor may also assign Experiment 57C, which extends the basic techniques developed in Experiments 57A and 57B to a larger list of plant materials. For this assignment, either your instructor will assign you a particular spice or herb to analyze or you will choose your own plant material to analyze.

SUGGESTED WASTE DISPOSAL

Any aqueous solutions should be placed in the container designated for aqueous wastes. Be sure to place any solid spice residues in the garbage can because they will plug the sink. Mixed organic/aqueous solvents should be disposed of in the container designated for aqueous wastes.

NOTES TO THE INSTRUCTOR

If ground spices are used, you may want to have the students insert a Claisen head between the round-bottom flask and the distillation head to allow extra volume in case the mixture foams. Problems with foaming can be greatly ameliorated by applying an aspirator vacuum to the spice–water mixture before the steam distillation is begun.

For the HPLC option in Experiment 57A, you must determine the best operating conditions in advance of the experiment. You will also need to prepare instructions for operating your particular instrument. In a similar way, for Experiments 57B and 57C, you must test the experiment in advance on your GC-MS instrument and prepare operating instructions.

| 57A | **EXPERIMENT 57A** |

Isolation of Essential Oils by Steam Distillation

PROCEDURE

Apparatus

Using a 20- or 25-mL round-bottom flask to distill and a 10-mL round-bottom flask to collect, assemble a distillation apparatus similar to that shown in Figure 14.10, page 713. Use an aluminum block to heat and insert a water condenser as shown on page 714. The collection flask may be immersed in ice to ensure condensation of the distillate. Be careful not to assemble the apparatus permanently because you will have to open it to add the spice and prepare the distillation mixture.

Preparing the Spice

Weigh approximately 1.0 g of your spice or herb onto a piece of weighing paper and record the exact weight. If your spice or herb is already ground, you may proceed without grinding it; otherwise, break up the seeds, leaves, or roots with a mortar and pestle or cut larger pieces into smaller ones using a scissors. Mix the spice or herb with 12–15 mL of water in the 20-mL round-bottom flask and add a magnetic stirring bar or boiling stone. If the spice or herb is a powder or seed, less water is necessary; if the material is a leaf or root, more water will be necessary.

Attach the round-bottom flask to an aspirator or vacuum source. To do this, place a piece of glass tubing through a thermometer adapter, attach the adapter to the flask, and use a piece of vacuum tubing to connect the vacuum to the tubing in the thermometer adapter. Start the magnetic stirring bar and apply the vacuum. The solution will begin to foam. Be careful not to let the foaming action become vigorous enough to rise above the level of the neck of the round-bottom flask. As the solution begins to foam, reduce the vacuum so that the bubbles recede into the flask. Repeat this process until foaming action subsides or until 15 minutes have passed, whichever is longer. Disconnect the vacuum and attach the flask to the distillation apparatus.

Steam Distillation

Turn on the cooling water in the condenser, begin stirring if you are using a stirring bar, and begin heating the mixture to provide a steady rate of distillation. If you approach the boiling point too quickly, you may have difficulty with frothing or bump-over. You will need to find the amount of heating that provides a steady rate of distillation but avoids frothing or bumping. A good rate of distillation would be to have 1 drop of liquid collected every 2–5 seconds. Continue distillation until at least 5 mL of distillate has been collected.

Normally, in a steam distillation the distillate will be somewhat cloudy due to separation of the essential oil as the vapors cool. However, you may not notice this and still obtain satisfactory results.

Extraction of the Essential Oil

Transfer the distillate to a 15-mL screw-cap centrifuge tube and add 1.0 mL of methylene chloride (dichloromethane) to extract the distillate. Cap the tube securely and shake it vigorously, venting frequently. Allow the layers to separate.

If the layers do not separate well, the mixture may be spun in a centrifuge. Stirring gently with a spatula sometimes helps resolve an emulsion. It may also help to add about 1 mL of a saturated sodium chloride solution. For the following directions, however, be aware that the saturated salt solution is quite dense, and the aqueous layer may change places with the methylene chloride layer, which is normally on the bottom.

Using a Pasteur pipet, transfer the lower methylene chloride layer to a clean, dry, 5-mL conical vial. Repeat this extraction procedure two more times with fresh 1.0-mL portions of methylene chloride and place them in the same 5-mL conical vial as you placed the first extraction. If there are visible drops of water, you need to transfer the methylene chloride solution with a dry Pasteur pipet to a clean, dry, 5-mL conical vial.

Drying

Dry the methylene chloride solution by adding granular anhydrous sodium sulfate to the conical vial (see Technique 12, Section 12.9, p. 680). Let the solution stand for 10–15 minutes and stir occasionally.

Evaporation

While the organic solution is being dried, clean and dry the first 5-mL conical vial and weigh (tare) it accurately. With a clean, dry filter-tip pipet, transfer the dried organic layer to this tared vial, leaving the drying agent behind. Use small amounts of clean methylene chloride to rinse the solution completely into the tared vial. Be careful to keep any of the sodium sulfate from being transferred. Working in a hood, evaporate the methylene chloride from the solution by using a gentle stream of air or nitrogen and heating to about 40°C (see Technique 7, Section 7.10, p. 611).

> ⚠ **C A U T I O N**
>
> **The stream of air or nitrogen must be very gentle or you will blast your solution out of the conical vial. In addition, do not overheat the sample. Do not continue the evaporation beyond the point where all the methylene chloride has evaporated. Your product is a volatile oil (i.e., liquid). If you continue to heat and evaporate, you will lose it. It is better to leave some methylene chloride than to lose your sample.**

Yield Determination

When the solvent has been removed, weigh the conical vial. Calculate the weight percentage recovery of the oil from the original amount of spice used.

SPECTROSCOPY

Infrared

Obtain the infrared spectrum of the oil as a pure liquid sample (Technique 25, Section 25.2, p. 834). It may be necessary to use a microsyringe or a Pasteur pipet with a narrow tip to transfer a sufficient amount to the salt plates. Include the infrared spectrum in your laboratory report, along with an interpretation of the principal peaks.

Nuclear Magnetic Resonance

At the instructor's option, determine the nuclear magnetic resonance spectrum of the oil (Technique 26, Section 26.1, p. 870).

REPORT

From the infrared spectrum (and any other data you have used), you should determine the structure (A–E) that best matches the essential oil you isolated from your spice. Label the major peaks in the infrared spectrum and give an argument that supports your choice of structure. Also be sure to include your weight percentage recovery calculation.

HIGH-PERFORMANCE LIQUID CHROMATOGRAPHY (OPTIONAL)

Following your instructor's directions, form a small group of students to perform this experiment. Each small group will be assigned the same spice to analyze, and the results obtained will be shared among all students in the group.

Dissolve your sample of essential oil in methanol. A reasonable concentration can be obtained by dissolving 25 mg of your sample in 10 mL of methanol. To remove all traces of dissolved gases and solid impurities, set up a filtering flask with a Büchner funnel and connect it to a vacuum line. Place a 4-μm filter in the Büchner funnel. (*Note:* Be sure to use a piece of filter paper, not one of the colored spacers that are placed between the pieces of filter paper. The spacers are normally blue.) Filter the essential oil solution by vacuum filtration through the 4-μm filter and place the filtered sample in a *clean* 4-dram snap-cap vial.

Before using the HPLC instrument, be certain you have obtained specific instruction in operating the instrument in your laboratory. Alternatively, your instructor may have someone operate the instrument for you. Before your sample is analyzed on the HPLC instrument, it should be filtered one more time, this time through a 0.2-μm filter. The recommended sample size for analysis is 10-μL. The solvent system used for this analysis is a mixture of 80% methanol and 20% water. The instrument will be operated in an isochratic mode.

When you have completed your experiment, report your results by preparing a table showing the retention times of each substance identified in the analysis. Determine the relative percentages of each component and record these values in your table, along with the name of each substance identified.

REFERENCE

McKone, H. T. "High Performance Liquid Chromatography of Essential Oils." *Journal of Chemical Education*, 56 (October 1979): 698.

57B EXPERIMENT 57B

Identification of the Constituents of Essential Oils by Gas Chromatography–Mass Spectrometry

PROCEDURE

Sample Preparation

Obtain a sample of essential oil by steam distillation of the spice, according to the method shown in Experiment 57A.

Analysis by GC-MS

For the GC-MS analysis, a dilute solution (about 500 ppm) is recommended. To prepare this solution, dip an end of a length of capillary tube (ca. 1.8 mm inner diameter, open at both ends) into the sample of the essential oil. Transfer the contents of the capillary tube into a clean, calibrated, 15-mL centrifuge tube by flushing methylene chloride through the capillary tube. Note that to avoid getting solvent on your fingers, you will have to hold the capillary tube with a pair of forceps. Add additional methylene chloride to the centrifuge tube to obtain a total volume of 6 mL. Add one or two microspatulafuls of anhydrous sodium sulfate to the centrifuge tube, place a piece of aluminum foil over the top, and screw the cap over the aluminum foil.

Before injecting the solution onto the GC-MS column, it is necessary to filter the solution. Draw a portion of the solution into a clean hypodermic syringe (without needle). Attach a 0.45-μm filter cartridge to the tip of the syringe and force the solution through the filter cartridge into a clean sample vial. Cover the sample vial with aluminum foil until the solution is used.

Inject the solution onto the column of the GC-MS instrument. As each component in the solution appears on the graph, use the built-in computer library to identify each component. Use the "quality" or "confidence" indicators on the printed lists to determine whether or not the compounds suggested are plausible. In your laboratory report, identify each component in the essential oil by providing its name and structural formula.

57C EXPERIMENT 57C

Investigation of the Essential Oils of Herbs and Spices—A Mini–Research Project

PROCEDURE

Obtain a sample of essential oil by steam distillation of the spice or herb, according to the method shown in Experiment 57A. Prepare the sample for analysis by gas chromatography–mass spectrometry by the method described in Experiment 57B.

Using the results of your GC-MS analysis, prepare a brief report describing your experimental method and presenting the results of your analysis. In your report, be sure to identify each important component of the essential oil you analyzed, draw its complete structural formula, and indicate the relative percentage of that substance in the essential oil mixture.

SPECTROSCOPY

Infrared Spectroscopy

Determine the infrared spectrum of the oil as a neat liquid sample. (Technique 25, Section 25.2). It will be necessary to use a Pasteur pipet to transfer a sufficient amount to the salt plates, or a solution of the oil in methylene chloride or dry acetone can be used to aid the transfer. Blow gently on the surface of the salt plate to evaporate the solvent. Include the infrared spectrum with your laboratory report, along with an interpretation of the principal peaks.[1]

Nuclear Magnetic Resonance Spectroscopy (Optional)

At the instructor's option, determine the proton NMR spectrum of your essential oil. NMR provides a nice confirmation of your structure. In keeping with the spirit of green chemistry, deuterated acetone can be used as an alternative solvent to deuterated chloroform. You can use deuterated acetone to remove the essential oil from the centrifuge tube and directly transfer the solution to an NMR tube. Once the NMR has been determined, you can transfer some of the NMR sample to a salt plate for infrared spectral analysis. Assign the peaks in the spectrum to the structure of your essential oil.

Mass Spectrometry (Optional)

At the instructor's option, determine the mass spectrum of the essential oil sample (Technique 28). You may search the database to help in the identification of the structure. Try to assign as many of the fragments in the spectrum as possible.

REPORT

From the infrared spectrum (and any other data you have used), you should determine the structure (A-E on p. 502) that best matches the essential oil you isolated from your spice. Label the major peaks in the infrared spectrum and give an argument that supports your choice of structure. If you determined the mass spectrum, identify the important fragment ion peaks. If you determined the NMR spectrum, also include this, along with an interpretation of the peaks and splitting patterns. Be sure to also include the calculation of your weight percentage recovery.

QUESTIONS

1. Take a sheet of paper and build a matrix by drawing each of the five possible essential oil compounds shown on page 502 down the left side of the sheet and by listing each of the possible infrared spectral features given previously along the top of the sheet. Draw lines to form boxes. Inside the boxes opposite each compound, note the expected infrared observation. Is the peak expected to be present or absent? If not absent, give the expected number of peaks and the probable frequencies. A good set of correlation charts and tables will help you with this.

2. Why does the newly condensed steam distillate appear cloudy?

3. After the drying step, what observations will help you to determine if the extracted solution is "dry" (i.e., free of water)?

REFERENCE

McKenzie, L. C., Thompson, J. E., Sullivan, R., and Hutchison, J. E. "Green Chemical Processing in the Teaching Laboratory: A Convenient Liquid CO_2 Extraction of Natural Products." *Green Chemistry*, 6 (2004): 355–358.

---

[1] Extraction of cloves yields an extra peak at 1764 cm^{-1}, which is attributed to eugenol acetate, a by-product of the extraction; GC-MS will be useful as an optional experiment in identifying the impurity.

58

EXPERIMENT 58

Competing Nucleophiles in S_N1 and S_N2 Reactions: Investigations Using 2-Pentanol and 3-Pentanol

Nucleophilic substitution

Heating under reflux

Extraction

Gas chromatography

NMR spectroscopy

This experiment is based on the procedure outlined in Experiment 21. The purpose of this experiment is to examine the products formed when competing nucleophiles, equimolar concentrations of chloride ions and bromide ions, are allowed to react with either 2-pentanol or 3-pentanol. Based on the products formed in each reaction, students can advance a variety of hypotheses that account for the number and proportions of products formed.

Because the starting alcohols are secondary alcohols, one might expect that the substitution reactions will take place by a combination of S_N1 and S_N2 pathways. You will analyze the products of the three reactions in this experiment by a variety of techniques to determine the relative amounts of alkyl chloride and alkyl bromide formed in each reaction and to identify all of the products that are observed.

REQUIRED READING

Experiment 21	Nucleophilic Substitution Reactions: Competing Nucleophiles
Technique 7	Reaction Methods, Section 7.2, 7.4, 7.5, and 7.7
Technique 12	Extractions, Separations, and Drying Agents, Sections 12.5, 12.9, and 12.11
Technique 22	Gas Chromatography
Technique 26	Nuclear Magnetic Resonance Spectroscopy

Before beginning this experiment, review the appropriate chapters on nucleophilic substitution in your lecture textbook.

SPECIAL INSTRUCTIONS

Your instructor will also assign you either 2-pentanol or 3-pentanol. By sharing your results with other students, you will be able to collect data for both alcohols. To analyze the results of both experiments, your instructor will assign specific analysis procedures that the class will accomplish.

The solvent–nucleophile medium contains a high concentration of sulfuric acid. Sulfuric acid is corrosive; be careful when handling it.

During the extractions, the longer your product remains in contact with water or aqueous sodium bicarbonate, the greater the risk that your product will decompose, leading to errors in your analytical results. Before coming to class, prepare so that you know exactly what you are supposed to do during the purification stage of the experiment.

SUGGESTED WASTE DISPOSAL

When you have completed the two experiments and all the analyses have been completed, discard any remaining alkyl halide mixture in the organic waste container marked for the disposal of halogenated substances. All aqueous solutions produced in this experiment should be disposed of in the container for aqueous waste.

NOTES TO THE INSTRUCTOR

The solvent–nucleophile medium must be prepared in advance for the entire class. Use the following procedure to prepare the medium. This procedure will provide enough solvent–nucleophile medium for about 10 students (assuming no spillage or other types of waste). Place 100 g of ice in a 500-mL Erlenmeyer flask and carefully add 76 mL concentrated sulfuric acid. Carefully weigh 19.0 g ammonium chloride and 35.0 g ammonium bromide into a beaker. Crush any lumps of the reagents to powder and then, using a powder funnel, transfer the halides to an Erlenmeyer flask. Carefully add the sulfuric acid mixture to the ammonium salts a little at a time. Swirl the mixture vigorously to dissolve the salts. It will probably be necessary to heat the mixture on a steam bath or a hot plate to achieve total solution. Keep a thermometer in the mixture and make sure that the temperature does not exceed 45°C. If necessary, you may add as much as 10 mL of water at this stage. Do not worry if a few small granules do not dissolve. When solution has been achieved, pour the solution into a container that can be kept warm until all students have taken their portions. The temperature of the mixture must be maintained at about 45°C to prevent precipitation of the salts. Be careful that the solution temperature does not exceed 45°C, however. Place a 20-mL calibrated pipet fitted with a pipet helper in the mixture. The pipet is always left in the mixture to keep it warm.

The gas chromatograph should be prepared as follows: Agilent (J & W) DB5 capillary column (30 m, 0.32 mm ID, 0.25 μm). Set injector temperature at 260°C. FID detector temperature is 280°C. The column oven conditions are as follows: start at 40°C (hold 2 min.), increase to 140°C at 20°C/min. (5 min.). The helium flow rate is 1.0 mL/min. The hydrogen FID gas makeup flow is 35 mL/min.

PROCEDURE

CAUTION

The solvent–nucleophile medium contains a high concentration of sulfuric acid. This liquid will cause severe burns if it touches your skin.

Apparatus

Assemble an apparatus for reflux using a 20-mL round-bottom flask, a reflux condenser, and a drying tube, as shown in the figure that accompanies Experiment 21 (p. 184). Loosely insert dry glass wool into the drying tube and then add water dropwise onto the glass wool until it is partially moistened. The moistened glass wool will trap the hydrogen chloride and hydrogen bromide gases produced during the reaction. As an alternative, you can use an external gas trap as described in Technique 7, Section 7.8, Part B, page 607. Do not place the round-bottom flask into the aluminum block until the reaction mixture has been added to the flask. Six Pasteur pipets, two 3-mL conical vials with Teflon cap liners, and a 5-mL conical vial with a Teflon liner should also be assembled for use. All pipets and vials should be clean and dry.

Preparation of Reagents

If a calibrated pipet fitted with a pipet helper is provided, you may adjust the pipet to 10 mL and deliver the solvent-nucleophile medium directly into your 20-mL round-bottom flask (temporarily placed in a beaker for stability). Alternatively, you may use a warm 10-mL graduated cylinder to obtain 10.0 mL of the solvent–nucleophile medium. The graduated cylinder must be warm in order to prevent precipitation of the salts. Heat it by running hot water over the outside of the cylinder or by putting it in the oven for a few minutes. Immediately pour the mixture into the round-bottom flask. With either method, a small portion of the salts in the flask may precipitate as the solution cools. Do not worry about this; the salts will redissolve during the reaction.

Reflux

Assemble the apparatus shown in the figure (page 184). Using the following procedure, add 0.75 mL of either 2-pentanol or 3-pentanol, depending on which alcohol you were assigned, to the solvent–nucleophile mixture contained in the reflux apparatus. Dispense the alcohol from the automatic pipet or dispensing pump into a 10-mL beaker. Remove the drying tube and, with a 9-inch Pasteur pipet, dispense the alcohol directly into the round-bottom flask by inserting the Pasteur pipet into the opening of the condenser. Also add an inert boiling stone.[1] Replace the drying tube and start circulating the cooling water. Lower the reflux apparatus so that the round-bottom flask is in the aluminum block, as shown in the figure. Adjust the heat so that this mixture maintains a *gentle* boiling action. The aluminum block temperature should be about 140°C. Be careful to adjust the reflux ring, if one is visible, so that it remains in the lower fourth of the condenser. Violent boiling will cause loss of product. Heat the mixture for 45 minutes.

Purification

When the period of reflux has been completed, discontinue heating, lift the apparatus out of the aluminum block, and allow the reaction mixture to cool. Do not remove the condenser until the flask is cool. Be careful not to shake the hot solution

---

[1] Do not use calcium carbonate–based stones or Boileezers, because they will partially dissolve in the highly acidic reaction mixture.

as you lift it from the heating block or a violent boiling and bubbling action will result; this could allow material to be lost out the top of the condenser. After the mixture has cooled for about 5 minutes, immerse the round-bottom flask (with condenser attached) in a beaker of cold tap water (no ice) and wait for this mixture to cool down to room temperature.

There should be an organic layer present at the top of the reaction mixture. Add 0.75 mL of pentane to the mixture and *gently* swirl the flask. The purpose of the pentane is to increase the volume of the organic layer so that the following operations are easier to accomplish. Using a Pasteur pipet, transfer most (about 7 mL) of the bottom (aqueous) layer to another container. Be careful that the entire top organic layer remains in the boiling flask. Transfer the remaining aqueous layer and the organic layer to a 3-mL conical vial, taking care to leave behind any solids that may have precipitated. Allow the phases to separate and remove the bottom (aqueous) layer using a Pasteur pipet.

NOTE: For the following sequence of steps, be certain to be well prepared in advance. If you find that you are taking longer than 5 minutes to complete the entire extraction sequence, you probably have affected your results adversely!

Add 1.0 mL of water to the vial and *gently* shake this mixture. Allow the layers to separate and remove the aqueous layer, which is still on the bottom. Extract the organic layer with 1–2 mL of saturated sodium bicarbonate solution and remove the bottom aqueous layer.

Drying

Using a clean dry Pasteur pipet, transfer the remaining organic layer into a small test tube (10 × 75 mm) containing 3 to 4 microspatulafuls (using the V-grooved end) of anhydrous granular sodium sulfate. Stir the mixture with a microspatula, put a stopper on the tube, and set it aside for 10–15 minutes or until the solution is clear. If the mixture does not turn clear, add more anhydrous sodium sulfate. Transfer the halide solution with a clean, dry Pasteur pipet to a small, dry conical vial, taking care not to transfer any solid. The preferred method of storage is to use a Teflon stopper that has been secured *tightly* with a plastic cap. It is helpful to cover the stopper and cap with Parafilm (on the outside of the stopper and cap). Alternatively, you may use a screw-cap vial with a Teflon liner. *Be sure the cap is screwed on tightly.* Again, it is a good idea to cover the cap with Parafilm. Do not store the liquid in a container with a cork or a rubber stopper, because these will absorb the halides. If it is necessary to store the sample overnight, store it in a refrigerator. This sample can now be analyzed by as many of the methods as your instructor indicates.

Analysis

PROCEDURE

The ratio of secondary pentyl chlorides and bromides must be determined. At your instructor's option, you may do this by gas chromatography or NMR spectroscopy or by both methods.

GAS CHROMATOGRAPHY[2]

The instructor or a laboratory assistant may either make the sample injections or allow you to make them. In the latter case, your instructor will give you adequate instruction beforehand. A reasonable sample size is 2.5 μL. Inject the sample into the gas chromatograph and record the gas chromatogram. The alkyl chlorides, because of their greater volatility, have a shorter retention time than the alkyl bromides.

Once the gas chromatogram has been obtained, determine the relative areas of the peaks (Technique 22, Section 22.12, p. 810). If the gas chromatograph has an integrator, it will report the areas. Triangulation is the preferred method of determining areas if an integrator is not available. Record the percentages of all alkyl chloride and alkyl bromide products in the reaction mixture.

NUCLEAR MAGNETIC RESONANCE SPECTROSCOPY

The instructor or a laboratory assistant will record the NMR spectrum of the reaction mixture.[3] Submit a sample vial containing the mixture for this spectroscopic determination. The spectrum will also contain integration of the important peaks (Technique 26, "Nuclear Magnetic Resonance Spectroscopy"). Compare the integral of the *downfield* peaks of the alkyl halide multiplets. The relative heights of these integrals correspond to the relative amounts of each halide in the mixture.

REPORT

Record the percentages of all of the alkyl chloride and alkyl bromide products in the reaction mixture for each of the isomeric pentanol substrates. You will need to share your data with results obtained by someone in the class who used the other starting alcohol. Your laboratory report must include the percentages of each alkyl halide determined by each method used in this experiment for the two alcohols that were studied. On the basis of the products identified and their relative percentages, develop an argument for a mechanism that will account for all of the results obtained. All gas chromatograms and spectra should be attached to the report.

---

[2] *Note to the Instructor:* To obtain reasonable results for the gas chromatographic analysis of the pentyl halides, it may be necessary to supply the students with response-factor corrections (Technique 22, Section 22.13, p. 813). If pure samples of each product are available, check the assumption used here that the gas chromatograph responds equally to each substance. Response factors (relative sensitivities) are easily determined by injecting an equimolar mixture of products and comparing peak areas.

[3] It is difficult to determine the ratio of pentyl chlorides and pentyl bromides using nuclear magnetic resonance. This method requires at least a 90-MHz instrument. At 300 MHz, the peaks are completely resolved.

59   **E X P E R I M E N T   5 9**

Friedel–Crafts Acylation

Aromatic substitution

Directive groups

Vacuum distillation (optional)

Infrared spectroscopy

NMR spectroscopy (proton/carbon-13)

Structure proof

In this experiment, a Friedel–Crafts acylation of an aromatic compound is undertaken, using acetyl chloride:

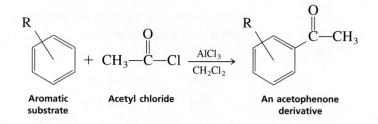

Aromatic   Acetyl chloride   An acetophenone
substrate   derivative

If benzene (R = H) were used as the substrate, the product would be a ketone, acetophenone. Instead of using benzene, however, you will perform the acylation on one of the following compounds:

Toluene	Ethylbenzene
o-Xylene ⎫	Mesitylene (1,3,5-trimethylbenzene)
m-Xylene ⎬ Dimethylbenzenes	Cumene (isopropylbenzene)
p-Xylene ⎭	Anisole (methoxybenzene)
p-Cymene	
(1-isopropyl-4-methylbenzene)	

Each of these products will give a single product, a *substituted* acetophenone. You are to isolate this product and to determine its structure by infrared and NMR spectroscopy. That is, you are to determine at which position of the original compound the new acetyl group becomes attached.

This experiment is much the same kind that professional chemists perform every day. A standard procedure, Friedel–Crafts acylation, is applied to a new compound for which the results are not known (at least not to you). A chemist who knows reaction theory well should be able to predict the result in each case. However, once the reaction is completed, it must be proved that the expected product has actually been obtained. If it has not, and sometimes surprises do occur, then the structure of the unexpected product must be determined.

To determine the position of substitution, several features of the product's spectra should be examined closely. These include the following:

Infrared Spectrum

The C—H Out-of-Plane Bending Modes Found between 900 and 690 cm⁻¹

The C—H out-of-plane absorptions (Technique 25, Figure 25.19A, p. 856) often allow us to determine the type of ring substitution by their numbers, intensities, and positions.

The Weak Combination and Overtone Absorptions That Occur between 2000 and 1667 cm^{-1}

These combination bands (Fig. 25.19B, p. 856) may not be as useful as those given here because the spectral sample must be concentrated for them to be visible. They are often weak. In addition, a broad carbonyl absorption may overlap and obscure this region, rendering it useless.

Proton NMR Spectrum

The Integral Ratio of the Downfield Peaks in the Aromatic Ring Resonances Found between 6 and 8 ppm

The acetyl group has a significant anisotropic effect, and those protons found *ortho* to this group on an aromatic ring usually have a greater chemical shift than the other ring protons (see Technique 26, Section 26.8, p. 880, and Section 26.13, p. 889).

A Splitting Analysis of the Patterns Found in the 6–8-ppm Region of the NMR Spectrum

The coupling constants for protons in an aromatic ring differ according to their positional relations:

ortho	J = 6–10 Hz
meta	J = 1–4 Hz
para	J = 0–2 Hz

If complex second-order splitting interaction does not occur, a simple splitting diagram will often suffice to determine the positions of substitution for the protons on the ring. For several of these products, however, such an analysis will be difficult. In other cases, an easily interpretable pattern such as those described in Section 26.13 (p. 889) will be found.

Carbon-13 NMR Spectrum

In *completely decoupled* carbon-13 spectra, the number of resonances for the aromatic ring carbons (at about 120–130 ppm) will give some help in deciding the substitution patterns of the ring. Ring carbons that are equivalent by symmetry will give rise to a single peak, thereby causing the number of aromatic carbon peaks to fall below the maximum of six. A *p*-disubstituted ring, for instance, will show only four resonances. Carbons that bear a hydrogen usually will have a larger intensity than "quaternary" carbons. (See Technique 27, Section 27.6, p. 913)

In *coupled* carbon-13 spectra, the ring carbons that bear hydrogen atoms will be split into doublets, allowing them to be easily recognized.[1]

As a final note, you should not eschew using the library. Technique 29 (p. 942) outlines how to find several important types of information. Once you think you know the identity of your compound, you might well try to find whether it has been reported previously in the literature and, if so, whether or not the reported data match your own findings. You may also wish to consult some spectroscopy books, such as Pavia, Lampman, and Kriz, *Introduction to Spectroscopy*, or one of the other textbooks listed at the end of either Technique 25 or Technique 26 for additional help in interpreting your spectra.

---

[1] *Note to the Instructor:* For those not equipped to perform carbon-13 NMR spectroscopy, carbon-13 NMR spectra can be found in the Instructor's Manual.

REQUIRED READING

Review: Techniques 5, 6, 12, 25, 26, and 27

Technique 7   Reaction Methods, Sections 7.5 and 7.8

Technique 13   Physical Constants of Liquids: The Boiling Point and Density

New: Technique 16   Vacuum Distillation, Manometers, Sections 16.1, 16.2, and 16.8

Before you begin this experiment, you should review the chapters in your lecture text that deal with electrophilic aromatic substitution. Pay special attention to Friedel–Crafts acylation and to the explanation of directing groups. You should also review what you have learned about the infrared and NMR spectra of aromatic compounds.

SPECIAL INSTRUCTIONS

Both acetyl chloride and aluminum chloride are corrosive reagents. You should not allow them to come in contact with your skin, nor should you breathe them, because they generate HCl on hydrolysis. They may even react explosively on contact with water. Weighing and dispensing operations should be carried out in a hood. The work-up procedure wherein excess aluminum chloride is decomposed with ice water should also be performed in the hood.

Your instructor will either assign you a compound or have you choose one yourself from the list given on page 515. Although you will acetylate only one of these compounds, you should learn much more from this experiment by comparing results with other students.

Notice that the details of the vacuum distillation are left for you to figure out on your own. However, there are two hints. First, all the products boil between 100 and 150°C at 20-mm pressure. Second, if your chosen substrate is anisole, the product will be a *solid* with a low melting point and will solidify soon after the vacuum distillation is completed. In this case, it might be worthwhile to preweigh the Hickman head itself because it will be difficult to transfer the entire solidified product to another container to determine a yield.

SUGGESTED WASTE DISPOSAL

All aqueous solutions should be collected in a container specially marked for aqueous wastes. Place organic liquids in the container designated for non-halogenated organic waste unless they contain methylene chloride. Waste materials that contain methylene chloride should be placed in the container designated for halogenated organic wastes. Note that your instructor may establish a different method of collecting wastes in this experiment.

PROCEDURE

Assemble the reaction apparatus shown in the figure. It consists of a 10-mL round-bottom flask and a Claisen head with one opening fitted with a rubber septum and the other attached to an inverted-funnel trap for acidic gases. Secure the Claisen head and the gas-trap funnel with clamps. The funnel should be about 2 mm *above* the water. Remove the Claisen head and add 2 mL of methylene chloride, 0.8 g of AlCl₃, and a magnetic stirring bar to the 10-mL round-bottom flask. Replace the Claisen head and begin stirring.

CAUTION ⚠

Both aluminum chloride and acetyl chloride are corrosive and noxious. Avoid contact and conduct all weighings in a hood. On contact with water, either compound may react violently.

Fill your 1-mL syringe (needle attached) with no less than 0.5 mL of fresh acetyl chloride. Insert the syringe through the rubber septum cap (see figure) and add the acetyl chloride slowly over a 2-minute period. (Rapid addition of the acetyl chloride may cause foaming.) Using a graduated pipet and pipet pump, transfer exactly 0.5 mL of your chosen aromatic compound to a preweighed 3-mL conical vial. Determine the weight of material delivered by weighing on a balance. Take up the aromatic compound with your syringe and slowly add it through the rubber septum over a 5-minute period. (This should not be done hastily, because the reaction is very exothermic; the mixture may boil up into the Claisen head.) When the aromatic compound has been added, rinse the vial with 1 mL of methylene chloride and, using the syringe, add this rinse to the reaction flask. Continue stirring for at least 30 minutes after the final addition has been made.

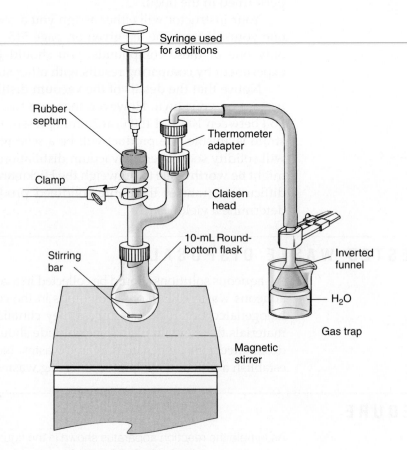

Apparatus for Friedel–Crafts reaction.

Isolation of Product

Remove the gas trap from the Claisen head and take the remaining apparatus, including the stirrer, to the hood. With your syringe, slowly add 4 mL of *ice-cold water* to the reaction mixture over a 5-minute period while stirring slowly. Next, add 4 mL of concentrated HCl with a Pasteur pipet and then stir the mixture vigorously with the magnetic stirrer until all the aluminum salts dissolve. At this point, discontinue stirring and allow the organic layer to separate. If the organic layer does not separate cleanly, add 0.5 mL of methylene chloride, stir again, and allow the organic layer to separate. You may have to add up to 1.5 mL of methylene chloride to induce the organic layer to separate cleanly.

Decant the entire mixture into a 15-mL centrifuge tube, leaving the stirring bar behind. Transfer the lower organic layer to a 5-mL conical vial with a filter-tip pipet. Avoid transferring any of the aqueous layer. If necessary, add a small amount of water and reseparate the layers that have been transferred to the conical vial. If a significant amount of the original highly acidic aqueous layer is present, violent foaming will occur in the next step. Add about 1 mL of 5% sodium bicarbonate to the conical vial containing the organic layer. Cap the vial and shake it *gently.* Carefully vent the vial by loosening the cap and resealing it after a few moments. Repeat this mixing several times until the evolution of CO_2 is no longer apparent.

Transfer the organic layer to a dry 3-mL conical vial (5-mL if necessary) and add 3 to 4 microspatulafuls of anhydrous sodium sulfate (use the V-grooved end). Cap the vial and set it aside for 10–15 minutes while the liquid is dried. If the liquid appears cloudy, shake the vial several times during the drying period or add more sodium sulfate.

The product dissolved in the methylene chloride solution is likely to be highly colored at this point. Some of the color can be removed by employing column chromatography. Gently push a small amount of cotton into the constricted end of a Pasteur pipet. Add about 3 cm of alumina to the pipet. Remove the methylene chloride solution from the drying agent with a Pasteur pipet, and add it directly to the dry alumina contained in the chromatography column. Collect the eluent in a *preweighed and dry* conical vial. After collecting the liquid, add about 1 mL of fresh methylene chloride to the column and collect this eluent in the same conical vial. In a hood, place the vial in a hot waterbath regulated to a temperature of about 40°C and direct a stream of air into the vial to evaporate the methylene chloride (Fig. 7.17A, p. 612). Do not rush this process. Allow the methylene chloride to be driven off completely. Monitor the evaporation by checking the volume markings on the side of the vial. When the volume is constant, the methylene chloride has been removed. If your instructor directs you to do a vacuum distillation, perform the optional procedure given in the next section; otherwise, skip to the "Boiling Point Determination and Spectroscopy" section. Weigh the conical vial after the methylene chloride has been removed and determine the percentage yield.

Vacuum Distillation (Optional)

If you are using a sand bath to heat, you should preheat it to about 165°C while assembling the apparatus. Assemble the apparatus *above* the sand bath; do not lower it into the sand bath until you are ready to distill. If you are using an aluminum block, preheating will not be necessary.

NOTE: Review Technique 16, Sections 16.1, 16.2, and 16.4, before proceeding.

Assemble an apparatus for vacuum distillation using an aspirator as shown in Figure 16.5, page 736. A manometer should be attached as shown in Figure 16.13, page 744. A piece of stainless steel sponge should be placed in the bottom portion of the neck of the Hickman still to protect the distilled product from any bumping action. Do not pack the stainless steel sponge too tightly. You may wish to preweigh the Hickman head (without the packing) to avoid having to transfer the product to determine the yield. This will be especially convenient if anisole was used as the substrate in the reaction. Using an *empty* conical vial, evacuate the system and check for any leaks. When there are no significant leaks, add a spin vane to the 3-mL conical vial containing the product (methylene chloride removed). Attach the vial to the distillation apparatus and reestablish the vacuum.

If using a sand bath, lower the apparatus to begin the distillation and cover the sand bath with aluminum foil. If using an aluminum block, begin heating after lowering the apparatus. Adjust the spin vane to its maximum rate of spin. If boiling, bumping, or refluxing has not occurred after 3 minutes of heating, you may increase the heat. A sand bath or aluminum block temperature in the range of 165–200°C will be required, depending on your compound. Once the distillate begins to appear on the walls of the Hickman still, the distillation proceeds rapidly. When no liquid remains in the 3-mL vial or when liquid is no longer distilling, raise the apparatus immediately to discontinue the distillation. If you overheat the vial, it may crack. Turn the hot plate off. Allow the apparatus to cool to room temperature and then vent the system. Transfer the product to a preweighed storage container and determine its weight. (If you preweighed your Hickman still, remove the stainless steel sponge and transfer the Hickman still to a beaker for weighing.) Calculate the percentage yield.

Boiling-Point Determination and Spectroscopy

At the instructor's option, determine the boiling point of your product using the microboiling-point method (Technique 13, Section 13.2, p. 695). Determine both the infrared and the NMR spectra (proton and carbon-13). The infrared spectra may be determined neat, using salt plates (Technique 25, Section 25.2, p. 834), except for the product from anisole, which is a solid. For this product, one of the solution spectrum techniques (Technique 25, Section 25.6, p. 842) should be used. The proton NMR spectra can be determined as described in Technique 26, Section 26.1, page 870. Deuteriochloroform is also an excellent solvent for all the carbon-13 samples as described in Technique 27, Section 27.1, page 906. Any residual methylene chloride appears at 5.3 ppm in the proton spectrum and at 54 ppm in the carbon spectrum.

REPORT

In the usual fashion, you should report the boiling point (or melting point) of your product, calculate the percentage yield, and construct a separation scheme diagram. You should also give the actual structure of your product. Include the infrared and NMR spectra and discuss carefully what you learned from each spectrum. If it did not help you determine the structure, explain why not. As many peaks as possible should be assigned on each spectrum and all important features explained, including the NMR splitting patterns, if possible. Discuss any literature you consulted and compare the reported results with your own.

Explain in terms of aromatic substitution theory why the substitution occurred at the position observed and why a single substitution product was obtained. Could you have predicted the result in advance?

REFERENCE

Schatz, P. F. "Friedel–Crafts Acylation." *Journal of Chemical Education, 56* (July 1979): 480.

QUESTIONS

1. The following are all relatively inexpensive aromatic compounds that could have been used as substrates in this reaction. Predict the product or products, if any, that would be obtained on acylation of each of them using acetyl chloride.

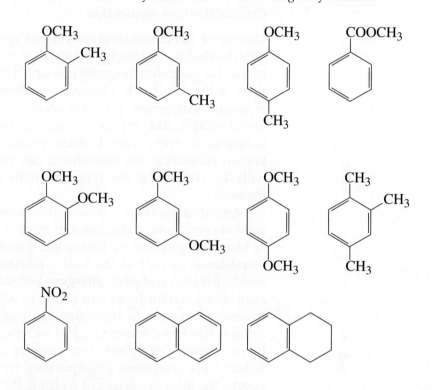

2. Why are only monosubstitution products obtained in the acylation of the substrate compounds chosen for this experiment?

3. Draw a full mechanism for the acylation of the compound you chose for this experiment. Include attention to any relevant directive effects.

4. Why do none of the substrates given as choices for this experiment include any with meta-directing groups?

5. Write equations for what happens when aluminum chloride is hydrolyzed in water. Do the same for acetyl chloride.

6. Explain carefully, with a drawing, why the protons substituted ortho to an acetyl group normally have a greater chemical shift than the other protons on the ring.

7. The compounds shown are possible acylation products from 1,2,4-trimethylbenzene (pseudocumene). Explain the only way you could distinguish these two products by NMR spectroscopy.

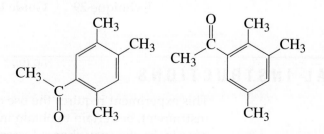

60   EXPERIMENT 60

The Analysis of Antihistamine Drugs by Gas Chromatography–Mass Spectrometry

Gas chromatography–mass spectrometry

Critical thinking application

The use of **gas chromatography–mass spectrometry (GC-MS)** as an analytical technique is growing in importance. GC-MS is a powerful technique in which a gas chromatograph is coupled to a mass spectrometer that functions as the detector. If a sample is sufficiently volatile to be injected into a gas chromatograph, the mass spectrometer can detect each component in the sample and display its mass spectrum. The user can identify the substance by comparing its mass spectrum with the mass spectrum of a known substance. The instrument can also make this comparison internally by comparing the spectrum with spectra stored in its computer memory.

Antihistamines are a class of pharmaceutical agents commonly used to combat symptoms of allergies and colds. They reduce physiological effects of histamine production. Histamine, a protein, is normally released in the bloodstream as part of the body's reaction to intrusions by pollen, dust, molds, pet hair, and other **allergens** (substances that cause an allergic reaction). Even certain foods can cause an allergic response in some people. Excessive amounts of histamine can cause various disorders, including asthma, hay fever, sneezing, nasal secretions, skin irritations and swelling, hives, digestive disorders, and watering eyes. We take antihistamines to reduce these symptoms. Unfortunately, antihistamines also have some side effects, the most important of which is that they cause drowsiness. In fact, certain antihistamines are also sold as sleep aids.

In this experiment, you will prepare solutions of common over-the-counter antihistamine and cold-remedy tablets. The samples, once prepared, will be analyzed using a GC-MS instrument, and you will use the results to identify the significant antihistamine substances that comprise the original tablet.

REQUIRED READING

New:	Technique 22	Section 22.12
	Technique 28	Mass Spectrometry
	Technique 29	Guide to the Chemical Literature

SPECIAL INSTRUCTIONS

This experiment requires the use of a GC-MS instrument. Before using this instrument, be certain to obtain instructions on its operation. As an option, your instructor may choose to perform the injections.

SUGGESTED WASTE DISPOSAL

Dispose of all solutions by discarding them in the container specified for nonhalogenated organic solvents. If your antihistamine contains either brompheniramine or chlorpheniramine, discard the solutions in the container specified for halogenated organic wastes.

PROCEDURE

Before beginning this experiment, you will need to rinse two 50-mL beakers, a syringe, and a snap-cap sample vial with HPLC-grade or spectrograde ethanol, and the glassware should be clean and dry before rinsing. Two rinsings of each item of glassware are recommended.

If your tablet has a colored coating, remove it by chipping it away from the tablet with a microspatula. Grind the tablet to a fine powder using a mortar and pestle. Weigh approximately 0.100 g of the powdered tablet into a prerinsed 50-mL beaker which has been prerinsed with ethanol. Add 10 mL of HPLC-grade ethanol to the beaker and let this solution stand, covered, for several minutes. Filter this solution by gravity through a fluted filter into a second prerinsed 50-mL beaker.

Draw the filtered solution into a prerinsed 5-mL syringe (without a needle), attach a 0.45-μm filter cartridge to the syringe, and expel the solution through the filter cartridge into a prerinsed sample vial. Repeat this process with a second syringeful of solution. Cover the top of the sample vial with a square of aluminum foil and attach the snap-cap to the vial, over the top of the foil. Label the vial and store it in the refrigerator.

Analyze the sample by gas chromatography–mass spectrometry. Your instructor or laboratory assistant may either make the sample injections or allow you to make them. In the latter case, your instructor will give you adequate instruction beforehand. A reasonable sample size is 2 μL. Inject the sample into the gas chromatograph and obtain the printout of the total ion chromatogram, along with the mass spectrum of each component. You should also obtain the results of a library search for each component.

The library search will give you a list of components detected in your sample and the retention time and relative area for each component. The results will also list possible substances that the computer has tried to match against the mass spectrum of each component. This list—often called a "hit list"—will include the name of each possible compound, its Chemical Abstracts Registry number (CAS number), and a "quality" ("confidence") measure, expressed as a percentage. The "quality" parameter estimates how closely the mass spectrum of the substance on the "hit list" fits the observed spectrum of that component in the gas chromatogram.

In your report, you should identify each significant component in the sample and provide its name and structural formula. You may have to use the CAS number as a key to look up the complete name and structure of the compound (Technique 29, Section 29.11, p. 952). You may need to search a computerized database to get the necessary information, or you may be able to find it in the *Aldrich Catalog Handbook of Fine Chemicals,* issued by the Aldrich Chemical Company. Current issues of this catalogue include listings of substances by CAS number. In your report, you should also report the relative percentage of the substance in the tablet extract. Finally, your instructor may also ask you to include the "quality" parameter from the "hit list." If possible, determine which components have antihistamine activity and which ones are present for another purpose. The *Merck Index* may provide this information.

E X P E R I M E N T   6 1

The Use of Organozinc Reagents in Synthesis: An Exercise in Synthesis and Structure Proof by Spectroscopy

Organometallic reactions

Green chemistry

Extractions

Use of a separatory funnel

Gas chromatography

Spectroscopy

A "green chemistry" alternative to the Grignard reaction was introduced in Experiment 39. This was the preparation of an organozinc reagent that was then allowed to react with a carbonyl compound. The reaction results in the formation of a new carbon–carbon bond.

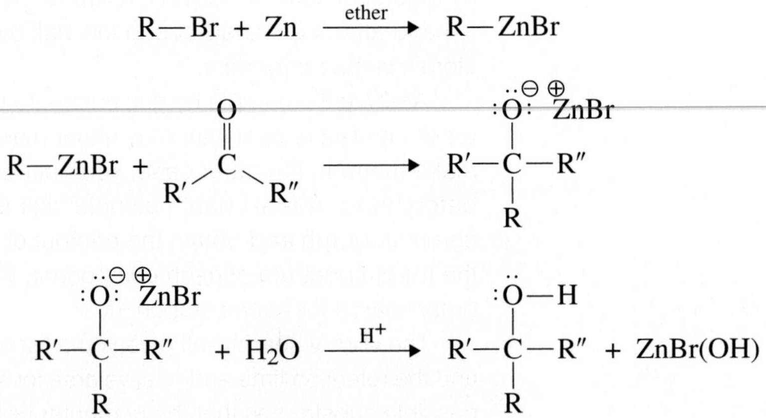

$$R—Br + Zn \xrightarrow{\text{ether}} R—ZnBr$$

$$R—ZnBr + \underset{R'\quad R''}{\overset{O}{\underset{\|}{C}}} \longrightarrow R'—\overset{:\overset{\ominus}{O}:\ \overset{\oplus}{ZnBr}}{\underset{R}{\underset{|}{\overset{|}{C}}}}—R''$$

$$R'—\overset{:\overset{\ominus}{O}:\ \overset{\oplus}{ZnBr}}{\underset{R}{\underset{|}{\overset{|}{C}}}}—R'' + H_2O \xrightarrow{H^+} R'—\overset{:\overset{..}{O}—H}{\underset{R}{\underset{|}{\overset{|}{C}}}}—R'' + ZnBr(OH)$$

The theory that lies beneath this reaction was also presented.

The procedure that was outlined in Experiment 39 involved the preparation of a relatively simple alcohol. In this experiment, we will use the same general laboratory procedure, but we will use it to prepare substances that extend the synthetic methods presented in Experiment 39 and that will form products with interesting spectroscopic properties.

Each of the recommended starting carbonyl compounds has two different substituents attached to the carbonyl carbon. As a result, the products that are formed will contain a stereocenter (chiral carbon). The presence of the stereocenter will cause neighboring protons to become **diastereotopic.** Proton NMR spectra of these products will show the added complication resulting from unequal coupling to these diastereotopic protons. One of the purposes of this experiment is to ask students to develop a complete analysis of the NMR spectra of the products formed in this reaction.

REQUIRED READING

Review: Experiment 39

Technique 8	Section 8.3
Technique 12	Sections 12.7, 12.8, 12.9, 12.11
Technique 22	
Technique 25	Sections 25.2, 25.4
Technique 26	Section 26.1
Technique 27	Section 27.1

SPECIAL INSTRUCTIONS

This reaction involves the use of alkyl halides that are volatile and may also be **lachrymators.** Be certain to dispense these materials under the hood. Do not attempt to weigh these substances; determine the approximate volume of alkyl halide needed using the table of specific gravities provided in this experiment, and dispense the alkyl halides by volume using a calibrated pipet.

Students will be allowed to choose an alkyl halide and a carbonyl compound from the list provided (see pages 526–527).

SUGGESTED WASTE DISPOSAL

All aqueous solutions should be placed in a waste container designated for the disposal of aqueous wastes.

PROCEDURE

Activated Zinc

Carefully weigh 1.31 g (0.02 moles) of zinc powder and add it to a small (10-mL) Erlenmeyer flask or beaker. Add 1 mL of 5% aqueous hydrochloric acid and allow the mixture to stand for 1 to 2 minutes. There will be a noticeable evolution of hydrogen gas during this time. At the end of this period, pour the entire mixture into a Hirsch funnel and isolate the zinc by vacuum filtration. Rinse the zinc with 1 mL of water, followed by 1 mL of ethanol and 1 mL of diethyl ether. The zinc should be ready to use for the procedure, as described below.

Reaction of the Organozinc Reagent

Add 10 mL of saturated aqueous ammonium chloride solution to a 25-mL round-bottom flask. Add 1.31 g zinc powder (0.02 moles) and a stirring bar to the flask. Attach an air condenser to the flask and begin continuous stirring while adding the remaining reagents. Carefully weigh 0.01 moles of the carbonyl compound. Add the ketone or aldehyde and 1.6 mL of tetrahydrofuran to a test tube and add this solution dropwise to the zinc/NH$_4$Cl solution. The rate of addition should be about 1 drop per second. Note that this addition can be made by dropping the solution carefully down the opening in the air condenser; use a Pasteur pipet to add the solution. Allow the solution to stir for 10 to 15 minutes, giving time for the carbonyl compound to form a complex with the zinc. Add 0.02 moles (use the specific gravity to estimate the volume required) of the assigned alkyl halide to the stirring solution. *Be sure to dispense this reagent in the hood!* The rate of addition should be about 1 drop per second. Add the halide carefully by dropping it down the opening in the air condenser. Allow the reaction mixture to stir for 1 hour.

Set up a vacuum filtration with a Hirsch funnel. Decant the liquid from the reaction mixture through the Hirsch funnel. Rinse the round-bottom flask with approximately 1 mL of diethyl ether and pour the liquid into the Hirsch funnel. Using a second 1-mL portion of diethyl ether, rinse the solid that has collected in the Hirsch funnel. Discard the solid. Prepare a filter-tip pipet and transfer the liquid that was collected in the vacuum filtration into a separatory funnel. Use 1 mL of diethyl ether to rinse the inside of the filter flask and use the filter-tip pipet to transfer this liquid to the separatory funnel. Shake the separatory funnel gently to extract the organic material from the aqueous layer to the ether layer. Drain the lower (aqueous) layer into a 50-mL Erlenmeyer flask. *Do not discard this aqueous layer.* Collect the upper (organic) layer from the separatory funnel into a 25-mL Erlenmeyer flask (remember to collect the upper layer by pouring it from the top of the separatory funnel). Replace the aqueous layer in the separatory funnel and wash it with a 2-mL portion of ether. Separate the layers, save the aqueous layer in the same 50-mL Erlenmeyer flask as before, and combine the ether layer with the ether solution collected in the previous extraction. Repeat this extraction process of the aqueous phase one more time, using a fresh 2-mL portion of ether. Dry the combined ether extracts with 3–4 microspatulafuls of anhydrous sodium sulfate. Stopper the Erlenmeyer flask with a cork and allow it to stand for at least 15 minutes (or overnight).

Use a filter-tip pipet to transfer the dried liquid to a clean, preweighed Erlenmeyer flask. Use a small amount of ether to rinse the inside of the original flask and add this ether to the dried liquid. Evaporate the ether under a gentle stream of air. When the ether has evaporated completely, reweigh the flask to determine the yield of product. If it should be necessary to store your final product, use Parafilm to seal the container.

Prepare a sample of your final product for analysis by gas chromatography. Determine the infrared spectrum and both proton and ^{13}C NMR spectrum of your product. Use these spectra to determine the structure of your product. In your laboratory report, include an interpretation of each spectrum, identifying the principal absorption bands and demonstrating how the spectrum corresponds to the structure of your compound. Submit your sample in a labeled vial with your laboratory report.

Carbonyl starting materials

Isobutryaldehyde
MW = 72.1 g/mole

Acetophenone
MW = 120.2 g/mole

Benzaldehyde
MW = 106.1 g/mole

Alkyl halide starting materials

$$CH_2{=}CH{-}CH_2{-}Br$$

Allyl bromide
MW = 121.0 g/mole
Specific gravity = 1.398 g/mL

$$CH_3{-}CH{=}CH{-}CH_2{-}Cl$$

Crotyl chloride
MW = 90.6 g/mole
Specific gravity = 0.92 g/mL

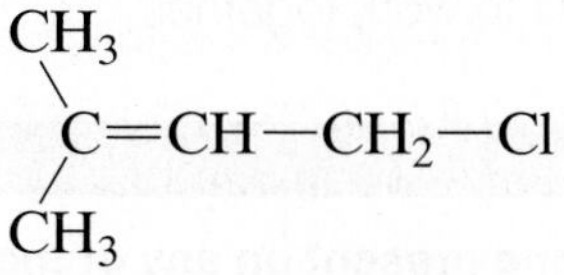

$$\underset{CH_3}{\overset{CH_3}{\diagdown}}C{=}CH{-}CH_2{-}Cl$$

1-Chloro-3-methyl-2-butene
MW = 104.6 g/mole
Specific gravity = 0.98 g/mL

| 62 | EXPERIMENT 62 |

The Aldehyde Enigma

Aldehyde chemistry

Extraction

Crystallization

Spectroscopy

Devising a procedure

Critical thinking application

The reaction mixture in this experiment contains 4-chlorobenzaldehyde, methanol, and aqueous potassium hydroxide. A reaction occurs that produces two organic compounds, Compound 1 and Compound 2. Both are solids at room temperature. Your task is to isolate, purify, and identify both compounds. A specific procedure is given for preparing the compounds, but you will need to work out the procedures for most other parts of this experiment.

SPECIAL INSTRUCTIONS

If the work on this experiment is done in pairs, work closely together as a team, dividing up the work equitably. A logical division of labor is for one student to work on Compound 1 and the other to work on Compound 2. Whether you work in pairs or not, you will need to plan your work carefully before coming to the laboratory, to make efficient use of class time.

SUGGESTED WASTE DISPOSAL

Dispose of all filtrates into the container designated for halogenated organic wastes.

PROCEDURE

This procedure should produce enough of each compound to complete the experiment; however, in some cases it may be necessary to run the reaction a second time. Although this experiment can be done individually, it works out especially well for two students to work together.

CAUTION

Be sure there is no acetone present on any of the glassware. Acteone will interfere with the desired reaction.

Running the Reaction

Add 1.50 g of 4-chlorobenzaldehyde and 4.0 mL of methanol to a 25-mL round-bottom flask. With gentle swirling, add 4.0 mL of an aqueous potassium hydroxide solution[1] with a Pasteur pipet. **Avoid getting potassium hydroxide solution on the ground-glass joint!** Add a stir bar to the flask and attach a water-cooled condenser. Using a hot waterbath, heat the reaction mixture at about 65°C with stirring for 1 hour. Cool the mixture to room temperature and add 10 mL of water to the flask. Pour the mixture into a beaker and use another 10 mL of water to aid the transfer into the beaker.

Using a separatory funnel, extract the reaction mixture with 10 mL of methylene chloride. Drain the organic layer (bottom) into another container. Extract the aqueous layer with another 10-mL portion of methylene chloride. Combine the organic layers. The organic layer contains Compound 1, and the aqueous layer contains Compound 2.

Organic Layer

Wash the organic layer two times with 10-mL portions of 5% aqueous sodium bicarbonate solution. Then wash the organic layer with an equal volume of water. If an emulsion forms, use a little saturated sodium chloride solution to break it. Dry the organic layer over anhydrous sodium sulfate for 10–15 minutes. After the dried solution is removed from the drying agent, the organic layer should contain only Compound 1 and methylene chloride. Isolate Compound 1 by removing the methylene chloride.

Purify Compound 1 by crystallization. See "Testing Solvents for Crystallization," Technique 11, Section 11.6, for instructions on how to determine an appropriate solvent. You should try 95% ethanol and xylene. After determining the best solvent, crystallize the compound using a hot waterbath at about 70°C for heating to avoid melting the solid. Identify Compound 1 using some or all of the techniques given next in the section "Identification of Compounds."

Aqueous Layer

To precipitate Compound 2, add 10 mL of cold water and acidify with 6*M* HCl. As acid is added, stir the mixture. Do not overacidify the solution; pH 3 or 4 is fine. If no precipitate is formed on acidification, add saturated NaCl to aid the process. This is called **salting out.**

---

[1] Dissolve 61.7 g of potassium hydroxide in 100 mL of water.

Isolate Compound 2 and dry it in an oven at about 110°C. Purify it by crystallization (see p. 660 for instructions on how to determine an appropriate solvent). You should try methanol and 95% ethanol. After determining the best solvent, purify the compound by crystallization and identify the purified solid using some or all of the techniques given in the next section, "Identification of Compounds."

IDENTIFICATION OF COMPOUNDS

Identify Compound 1 and Compound 2 using any of the following techniques:

1. *Melting point:* Consult a handbook for literature values.
2. *Infrared spectroscopy:* KBr pellet is preferred.
3. *Proton or carbon NMR:* Compound 1 dissolves easily in $CDCl_3$; use deuterated DMSO or Unisol to dissolve Compound 2.[2]
4. Some of the "Wet" chemical tests listed in Experiment 54 may be helpful: solubility tests, Beilstein test for halide, and others you may think appropriate.
5. Physical properties such as color and shape of crystals may also be helpful.

REPORT

Write out a complete procedure by which you synthesized and isolated Compounds 1 and 2. Describe the results of your experiments to determine a good crystallization solvent for both compounds. Draw the structures of Compounds 1 and 2. Give all melting-point data and results of other tests used to identify the two compounds. Identify significant peaks in the infrared spectrum and proton/carbon NMR spectra. Show clearly how all these results confirm the identity of the two compounds. Write a balanced equation for the synthesis of Compounds 1 and 2. What type of reaction is this? Propose a mechanism for the reaction. Determine the percentage yield of each of the compounds.

63 EXPERIMENT 63

Synthesis of Substituted Chalcones: A Guided-Inquiry Experience

Crystallization

Aldol condensation

Use of the chemical literature

Project-based experiment

In Experiment 41, you were introduced to the **aldol condensation reaction,** which you used to prepare a variety of **benzalacetophenones,** or **chalcones.** In this experiment, you will again prepare chalcones, but in this case you

[2] Unisol is a mixture of chloroform-d and DMSO-d_6 available from Norell, Inc., 120 Marlin Lane, Mays Landing, NJ 08330.

will do so in a guided-inquiry experiment that simulates some of the methodology that you are likely to use in research.

You will select from a variety of substituted benzaldehydes (1) and substituted acetophenones (2) to prepare benzalacetophenones (chalcones) (3) that bear a combination of substituents in each of the aromatic rings (see figure).

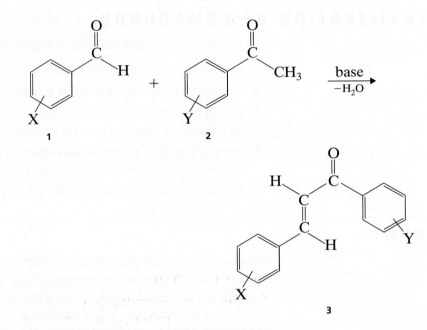

Once you have selected your starting materials, you will determine the complete structure of the condensation product that you expect to be formed in your reaction. You will also determine the molecular formula. With this information, you will be able to conduct an online literature search of *Chemical Abstracts* using **STN Easy** or **SciFinder Scholar.** From the literature search, you will be able to obtain the complete name of your target chalcone, its CAS Registry Number, and literature citations from the primary chemical literature. These literature citations should be able to afford you characterization information about your target chalcone, including melting points, infrared spectra, and NMR spectra.

After you have conducted the literature search, the final step will be to prepare your chalcone and compare its properties with those that you were able to find in the literature. The purpose of this experiment is to introduce you to many of the activities that you are likely to encounter in research. These include an examination of the target molecule, selection of appropriate starting materials, a thorough search of the primary chemical literature, laboratory synthesis of the desired compound, and characterization (including a comparison of the physical properties of the product with published values found in journal articles or other tables of data).

REQUIRED READING

Review: Technique 8   Filtration, Section 8.3

Technique 11   Crystallization: Purification of Solids, Section 11.3

New: Technique 29   Guide to the Chemical Literature

SPECIAL INSTRUCTIONS

Before beginning this experiment, you should select a substituted benzaldehyde and a substituted acetophenone. Your instructor will determine the method of assigning these reactants. You should also sign up for an STN Easy or SciFinder Scholar computer session. Your instructor will provide you with instructions on how to conduct an online computer search. Before coming to the computer session, you should work out the structure of your target compound and determine its molecular formula.

Note that sodium hydroxide solutions are caustic. Be careful when handling the substituted benzaldehydes and acetophenones. Wear personal protective equipment and work in a well-ventilated area.

SUGGESTED WASTE DISPOSAL

All filtrates should be poured into a waste container designated for non-halogenated organic waste. Note that your instructor may establish a different method of collecting wastes in this experiment.

NOTES TO THE INSTRUCTOR

It is best to introduce this project two to three weeks before the date of the actual chalcone synthesis to allow time for searching the literature. You will have to develop a method of assigning a target compound to each student. You will also have to schedule computer time for the online searching of *Chemical Abstracts*. We recommend that you prepare a handout that describes how to search *Chemical Abstracts* using STN Easy or SciFinder Scholar. The handout should guide the students through the process of finding the registry number for the target compound and for finding pertinent references, with particular attention to references describing the preparation of the compound. Finally, you will have to determine whether or not to require a formal laboratory report and what the expected format should be.

PROCEDURE

Before beginning the synthesis of your chalcone, determine its structure and molecular formula and perform the online search of *Chemical Abstracts*, following the instructions that your instructor provides.

Running the Reaction

Place 0.005 moles of your substituted benzaldehyde into a *tared* 50-mL Erlenmeyer flask and reweigh the flask to determine the weight of material transferred.

Add 0.005 moles of the substituted acetophenone and 4.0 mL of 95% ethanol to the flask that contains the substituted benzaldehyde. Add a magnetic stirring bar to the flask. Swirl the flask to mix the reagents and dissolve any solids present. It may be necessary to warm the mixture on a steam bath or hot plate to dissolve the solids. If this is necessary, the solution should be cooled to room temperature before proceeding to the next step.

Add 0.50 mL of sodium hydroxide solution to the aldehyde/acetophenone mixture.[1] Stir the mixture. Before the mixture solidifies, you may observe some cloudiness. *Wait until the cloudiness has been replaced with an obvious precipitate settling out to the bottom of the conical vial before proceeding to the next paragraph.* Continue stirring until a solid forms (approximately 3 to 5 minutes).[2] Scratching the inside of the conical vial with you microspatula may help to crystallize the chalcone.

Isolation of the Crude Product

Add 10 mL of ice water to the vial *after a solid has formed as indicated in the previous paragraph.* Stir the solid in the mixture with a spatula to break up the solid mass. Transfer the mixture to a small beaker with 5 mL of ice water. Stir the precipitate to break it up and then collect the solid on a Hirsch funnel. Wash the product with cold water. Allow the solid to air dry for about 30 minutes. Weigh the solid and determine the percentage yield.

Crystallization

Crystallize all of your sample from hot 95% ethanol. You will have to use the crystallization techniques introduced in Experiment 3 (p. 21) to crystallize the chalcone. Once the crystals have been allowed to dry thoroughly, weigh the solid, determine the percentage yield, and determine the melting point.

Laboratory Report

At the option of your instructor, obtain the infrared and proton and/or carbon-13 NMR spectra of your product. Again, at the option of your instructor, you may be required to write a formal laboratory report. If this is the case, use the format that your instructor provides or base your report on the style found in the *Journal of Organic Chemistry* (see Technique 29). Include a balanced equation for the reaction in your report. Submit the purified sample of your chalcone in a labeled vial to the instructor.

REFERENCE

Vyvyan, J. R., Pavia, D. L., Lampman, G. M., and Kriz, G. S. "Preparing Students for Research: Synthesis of Substituted Chalcones as a Comprehensive Guided-Inquiry Experience." *Journal of Chemical Education, 79* (September 2002): 1119–1121.

---

[1] This reagent should be prepared in advance by the instructor in the ratio of 6.0 g of sodium hydroxide to 10 mL of water.

[2] In some cases, the chalcone may not precipitate. If this is the case, cap the conical vial and allow it to stand until the next laboratory period. It is sometimes helpful to add an additional portion of base. Usually, the chalcone will precipitate during that time.

EXPERIMENT 64

Michael and Aldol Condensation Reactions

Aldol condensation

Michael reaction (conjugate addition)

Crystallization

Devising a procedure

Critical thinking application

In Experiment 41 ("The Aldol Condensation Reaction: Preparation of Benzalacetophenones"), substituted benzaldehydes are reacted with acetophenone in a crossed aldol condensation to prepare benzalacetophenones (chalcones). This is illustrated in the following reaction, where Ar and Ph are used as abbreviations for a substituted benzene ring and the phenyl group, respectively.

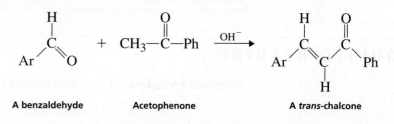

A benzaldehyde   **Acetophenone**   **A *trans*-chalcone**

Experiment 42 involves the reaction between ethyl acetoacetate and *trans*-chalcone in the presence of base. Under the conditions of this experiment, a sequence of three reactions takes place: a Michael addition followed by an internal aldol reaction and a dehydration.

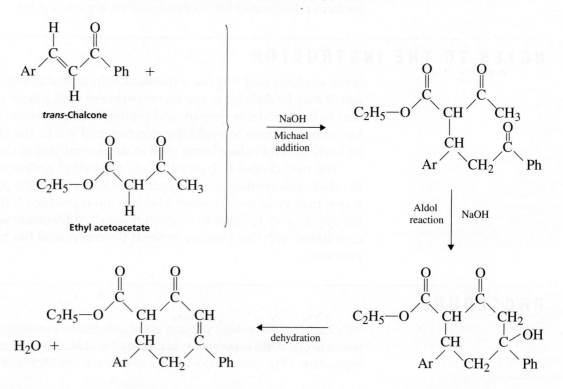

The purpose of this experiment is to combine the reactions introduced in Experiments 41 and 42 in the form of a project. Starting with one of four possible substituted benzaldehydes, you will synthesize a chalcone using the procedure given in Experiment 41. After performing a melting point to verify that this step has been completed successfully, you will perform a Michael/aldol reaction with the chalcone and ethyl acetoacetate using the procedure given in Experiment 42. The identity of this final product will be confirmed by melting point and possibly infrared and NMR spectroscopy.

You will be assigned one of the aromatic aldehydes shown in the following list. For each aldehyde, the melting points of the corresponding chalcone and the Michael/aldol product are given:

Aldehyde	Chalcone (mp, °C)	Michael/Aldol Product (mp, °C)
4-Chlorobenzaldehyde	114–115	141–143
4-Methoxybenzaldehyde	73–74	106–108
4-Methylbenzaldehyde	92–94	139–142
Piperonaldehyde	121–122	146–147

REQUIRED READING

Review: Technique 11   Crystallization: Purification of Solids

SUGGESTED WASTE DISPOSAL

If your starting compound is 4-chlorobenzaldehyde, all filtrates should be poured into a waste container designated for halogenated organic wastes. If you use one of the other three aldehydes, dispose of all filtrates in the container designated for nonhalogenated organic wastes.

NOTES TO THE INSTRUCTOR

Some students may require individual help with this experiment. As a result, it may be difficult to use this experiment with a large class. It is a good idea to have students prepare and present their procedure for approval before allowing them to begin the experimental work. The chalcones should be finely ground before being used in the second part of the experiment.

You may choose to have students react ethyl acetoacetate with one of the chalcones synthesized in Experiment 63. Because the product of the reaction may yield an unknown Michael/aldol product, a student will have the opportunity to conduct original research. A literature search may be incorporated with this exercise to see if the compound has been synthesized previously.

PROCEDURE

Your instructor will assign you one of the substituted benzaldehydes in the table above to use in this experiment. To prepare the chalcone, refer to the procedure in Experiment 41 (p. 339). To convert the chalcone to the Michael/aldol product, refer to the procedure given in Experiment 42 (p. 344). Using these procedures as a guide, devise the entire experimental procedure together with reagent quantities. The chalcone you prepare should be finely ground before using it in the second part of this experiment.

Initially, you should follow the procedures in Experiments 41 and 42 as closely as possible with appropriate adjustments in the scale. If either procedure does not work, you may need to modify the procedure and run the experiment again. An unsuccessful procedure will most likely be indicated by either the melting point or spectral data. The problem you would most likely encounter in preparing the chalcone is difficulty in getting the product to solidify from the reaction mixture. The Michael/aldol reaction is more complicated because there are two intermediate compounds that could be present in a significant amount in the final sample. If this occurs, both the melting point and the infrared spectrum may provide clues about what happened. It is possible you will need to increase the reaction time for this part of the experiment.

You must pay attention to scale so that you prepare enough of the chalcone for use in the next step and so that you finish up with a reasonable amount of the final product, about 0.1–0.2 g. It is possible, therefore, that the amounts of reagents given in Experiments 41 and 42 will need to be adjusted. If the scale needs to be changed in either experiment, be sure to adjust the amounts of all reagents proportionately and make any necessary changes in the glassware. In making your initial decision about scale, assume that the percentage yield of the chalcone after crystallization will be about 50%. Likewise, assume that the procedure in Experiment 41 will result in about a 50% yield.

To determine an accurate melting point of the chalcone or the final product, the sample must be pure and dry. In most cases, 95% ethanol can be used to crystallize these compounds. If this solvent does not work, you can use the procedure in Technique 11, Section 11.6, to find an appropriate solvent. Other solvents to try include methanol or a mixture of ethanol and water. If you are unsuccessful in finding an appropriate solvent, consult your instructor.

It is particularly important that the chalcone be highly pure before going on to the next step. If the melting point after crystallization is not within 3–4°C of the melting point given in the table on page 534, you may need to crystallize the material a second time.

SPECTROSCOPY

Infrared Spectrum

You should obtain an infrared spectrum of the chalcone and the final product to verify the identity of each product in the reaction sequence. Obtain the infrared spectrum by the dry film method (Technique 25, Section 25.4, p. 838) or as a KBr pellet (Technique 25, Section 25.5, p. 838). For the Michael/aldol product, you should observe absorbances at about 1735 and 1660 cm^{-1} for the ester carbonyl and enone groups, respectively.

NMR Spectrum

Your instructor may also want you to determine the proton and carbon NMR spectra of each product. These may be run in CDCl$_3$ solvent. Some of the expected signals can be determined by referring to the spectral data given in the footnote on page 347. Although these data are for a slightly different compound, many of the signals will have similar splitting patterns and similar chemical shifts.

REPORT

The report should include balanced equations for the preparation of the chalcone and the Michael/aldol product. You should calculate both the theoretical and percentage yields for each step. Write your complete procedure as you actually performed it. Include the actual results of your melting-point determinations and compare them to the expected results.

Include any infrared spectra obtained and interpret the major absorption peaks. If you determined NMR spectra, you should include them, along with an interpretation of the peaks and splitting patterns.

REFERENCE

Garcia-Raso, A., Garcia-Raso, J., Campaner, B., Maestres, R., and Sinisterra, J. V. "An Improved Procedure for the Michael Reaction of Chalcones." *Synthesis* (1982): 1037.

EXPERIMENT 65

Esterification Reactions of Vanillin: The Use of NMR to Solve a Structure Proof Problem[1]

Esterification

Crystallization

Use of a Craig tube

Nuclear magnetic resonance

Critical thinking application

The reaction of vanillin with acetic anhydride, in the presence of base, is an example of the esterification of a phenol. The product, which is a white solid, can be characterized easily by its infrared and NMR spectra.

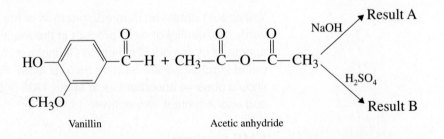

Vanillin   Acetic anhydride

---

[1] This experiment is based on a paper presented at the 12th Biennial Conference on Chemical Education, Davis, California, August 2–7, 1992, by Professor Rosemary Fowler, Cottey College, Nevada, Missouri. The authors are very grateful to Professor Fowler for her generosity in sharing her ideas.

When vanillin is esterified with acetic anhydride under acidic conditions, however, the product that is isolated has a different melting point and different spectra. The object of this experiment is to identify the products formed in each of these reactions and to propose mechanisms that will explain why the reaction proceeds differently under basic and acidic conditions.

REQUIRED READING

Review: Techniques 8, 11, 25, and 26

You should also read the sections in your lecture textbook that deal with the formation of esters and nucleophilic addition reactions of aldehydes.

SPECIAL INSTRUCTIONS

Sulfuric acid is corrosive. Do not allow it to touch your skin.

SUGGESTED WASTE DISPOSAL

All filtrates and organic residues should be disposed of into the container designated for nonhalogenated organic wastes. Dispose of solutions used for NMR spectroscopy in the waste container designated for the disposal of halogenated materials.

PROCEDURE

Preparation of 4-Acetoxy-3-Methoxybenzaldehyde (Vanillyl Acetate)

Dissolve 0.30 g of vanillin in 5 mL of 10% sodium hydroxide in a 50-mL Erlenmeyer flask. Add 6 g of crushed ice and 0.8 mL of acetic anhydride. Stopper the flask with a clean, rubber stopper and shake it several times over a 20-minute period. A cloudy, milky white precipitate will form immediately on adding the acetic anhydride. Filter the precipitate, using a Hirsch funnel, and wash the solid with three 1-mL portions of ice-cold water.

In a Craig tube, recrystallize the solid from 95% ethyl alcohol. Heat the mixture in a hot waterbath at about 60°C to avoid melting the solid. When the crystals are dry, weigh them and calculate the percentage yield. Obtain the melting point (literature value is 77–79°C). Determine the infrared spectrum of the product using the dry film method. Determine the proton NMR spectrum of the product in CDCl$_3$ solution. Using the spectral data, confirm that the structure of the product is consistent with the predicted result.

Esterification of Vanillin in the Presence of Acid

Place a magnetic spin vane in a 3-mL conical vial. Add 0.15 g of vanillin and 1.0 mL of acetic anhydride to the conical vial. Stir the mixture at room temperature until the solid dissolves. While continuing to stir the mixture, using a pasteur pipet add 1 drop of 1.0M sulfuric acid to the reaction mixture. Cap the vial and stir at room temperature for 1 hour. During this period, the solution will turn purple or purple-orange.

At the end of the reaction period, transfer the reaction mixture to a centrifuge tube with a screw cap. Cool the tube in an ice-water bath for 3–4 minutes. Add 3.5 mL of ice-cold water to the mixture in the centrifuge tube. Cap the tube and shake it vigorously—almost as hard as you can shake! Continue to cool and shake the tube to induce crystallization. Crystallization has occurred when you can see small solid clumps separating from the cloudy liquid and settling to the bottom of the tube. If crystallization does not occur after 10–15 minutes, it may be necessary to seed the mixture with a small crystal of the product. Filter the product on a Hirsch funnel and wash the solid with three 1-mL portions of ice cold water.

Using a Craig tube, recrystallize the crude product from hot 95% ethanol. Allow the crystals to dry. Weigh the dried crystals, calculate the percentage yield, and determine the melting point (literature value is 90–91°C). Determine the infrared spectrum of the product using the dry film method. Determine the proton NMR spectrum of the product in CDCl$_3$ solution.

REPORT

Compare the two sets of spectra obtained for the base- and acid-promoted reactions. Using the spectra, identify the structures of the compounds formed in each reaction. Record the melting points and compare them to the literature values. Write balanced equations for both reactions and calculate the percentage yields. Outline mechanistic pathways to account for the formation of both products isolated in this experiment.

---

66   EXPERIMENT 66 [1]

An Oxidation Puzzle

Oxidation of alcohols
Infrared spectroscopy
Critical thinking application

Sodium hypochlorite in acetic acid is an oxidizing agent capable of oxidizing alcohols to the corresponding aldehydes or ketones. In this experiment, you will oxidize a diol, 2-ethyl-1,3-hexanediol (1) and then use infrared spectroscopy to determine which of the alcohol functional groups was oxidized.

You will determine whether the oxidation occurred selectively (and which functional group was oxidized) or whether both functional groups were oxidized at the same time. The possible outcomes of the oxidation are shown in the figure. If only the primary alcohol is oxidized, the aldehyde (2) will be formed; if only the secondary alcohol is oxidized, the ketone (3) will be the product. If both alcohol functional groups are oxidized, compound

---

[1] Experiment 66 is adapted from M. W. Pelter, R. M. Macudzinski, and M. E. Passarelli, "A Microscale Oxidation Puzzle," *Journal of Chemical Education*, 77 (November 2000): 1481.

(4) will be observed. Your assignment will be to use infrared spectroscopy to determine the structure of the product and decide which of these three possible outcomes actually takes place.

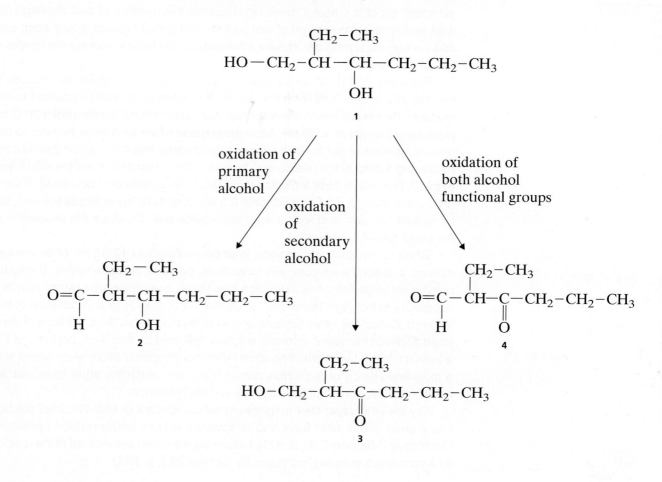

REQUIRED READING

Review: Techniques 12 and 25

SPECIAL INSTRUCTIONS

Glacial acetic acid is corrosive; it can cause burns on the skin and on mucous membranes in the nose and mouth. Its vapors are also hazardous. Dispense it in the hood and use personal protective equipment. Avoid contact with skin, eyes, and clothing. Sodium hypochlorite emits chlorine gas, which is a respiratory and eye irritant. Dispense it in a fume hood.

SUGGESTED WASTE DISPOSAL

All aqueous solutions should be collected in a container specially marked for aqueous wastes. Place organic liquids in the container designated for nonhalogenated organic waste. Note that your instructor may establish a different method of collecting wastes in this experiment.

PROCEDURE

Dispense 0.5 mL of 2-ethyl-1,3-hexanediol into a *tared* 10-mL Erlenmeyer flask. An automatic pipet is a useful device to dispense this quantity of diol. Reweigh the flask to determine the weight of diol added. Add 3 mL of glacial acetic acid; also add a magnetic stirring bar. Have a thermometer available to monitor the temperature of the reaction.

Place the mixture in an ice bath on a magnetic stirrer. While the mixture is stirring, slowly add 3 mL of a 6% aqueous sodium hypochlorite solution to the mixture.[2] Be careful not to allow the reaction temperature to rise above 30°C by controlling the rate of addition. Allow the solution to stir for 1 hour. In order to determine whether or not there is excess hypochlorite, test the solution periodically by placing a drop of the reaction mixture on a strip of potassium iodide starch test paper. A blue-black color indicates that there is an excess of hypochlorite. If there is no color change, add an additional 0.5 mL of sodium hypochlorite solution, stir for several minutes, and repeat the starch-iodide test. Continue this process until the paper turns blue-black.

When the reaction is complete, pour the mixture into 10–15 mL of an ice–salt mixture. Extract the mixture with three 5-mL portions of diethyl ether. It may be convenient to perform this extraction in a 15-mL centrifuge tube rather than in a separatory funnel (see Technique 12, Section 12.6, p. 677, for a description of this method). Collect the ether extracts and wash them with two 3-mL portions of saturated aqueous sodium carbonate solution, followed by two 3-mL portions of 5% aqueous sodium hydroxide. The ether layer should appear basic when tested with a *moistened* piece of red litmus paper. If it is not, wash the ether layer with an additional 3-mL portion of 5% aqueous sodium hydroxide.

Dry the ether layer over magnesium sulfate. Decant or filter the dried solution into a tared 25-mL filter flask and remove the solvent under reduced pressure (Technique 7, Section 7.10, p. 611). Determine the infrared spectrum of the residue as a pure liquid sample (Technique 25, Section 25.2, p. 834).

REPORT

Using your infrared spectrum, determine the structure of the oxidation product (see the structures of the possible products on p. 539). Is the oxidation selective? Did the hypochlorite oxidize both alcohol functional groups?

---

[2] Your instructor will have prepared this solution in advance.

PART 6

The Techniques

Laboratory Safety

In any laboratory course, familiarity with the fundamentals of laboratory safety is critical. Any chemistry laboratory, particularly an organic chemistry laboratory, can be a dangerous place in which to work. Understanding potential hazards will serve you well in minimizing that danger. It is ultimately your responsibility, along with your laboratory instructor's, to make sure that all laboratory work is carried out in a safe manner.

1.1 Safety Guidelines

It is vital that you take necessary precautions in the organic chemistry laboratory. Your laboratory instructor will advise you of specific rules for the laboratory in which you work. The following list of safety guidelines should be observed in all organic chemistry laboratories.

A. Eye Safety

Always Wear Approved Safety Glasses or Goggles. It is essential to wear eye protection whenever you are in the laboratory. Even if you are not actually carrying out an experiment, a person near you might have an accident that could endanger your eyes. Even dishwashing can be hazardous. We know of cases in which a person has been cleaning glassware only to have an undetected piece of reactive material explode, throwing fragments into the person's eyes. To avoid such accidents, wear your safety glasses or goggles at all times.

Learn the Location of Eyewash Facilities. If there are eyewash fountains in your laboratory, determine which one is nearest to you before you start to work. If any chemical enters your eyes, go immediately to the eyewash fountain and flush your eyes and face with large amounts of water. If an eyewash fountain is not available, the laboratory will usually have at least one sink fitted with a piece of flexible hose. When the water is turned on, this hose can be aimed upward, and the water can be directed into the face, working much as an eyewash fountain does. To avoid damaging the eyes, the water flow rate should not be set too high, and the water temperature should be slightly warm.

B. Fires

Use Care with Open Flames in the Laboratory. Because an organic chemistry laboratory course deals with flammable organic solvents, the danger of fire is frequently present. Because of this danger, DO NOT SMOKE IN THE LABORATORY. Furthermore, use extreme caution when you light matches or use any open flame. Always check to see whether your neighbors on either side, across the bench, and behind you are using flammable solvents. If so, either wait or move to a safe location, such as a fume hood, to use your open flame. Many flammable organic substances are the source of dense vapors that can travel for some distance down a bench. These vapors present a fire danger, and you should be careful because the source of those vapors may be far away from you. Do not use the bench sinks to dispose of flammable solvents. If your bench has a trough running along it, pour only *water* (no flammable solvents!) into it. The troughs and sinks are designed to carry water—not flammable materials—from the condenser hoses and aspirators.

Learn the Location of Fire Extinguishers, Fire Showers, and Fire Blankets.
For your own protection in case of a fire, you should immediately determine the location of the nearest fire extinguisher, fire shower, and fire blanket. You should learn how to operate these safety devices, particularly the fire extinguisher. Your instructor can demonstrate this.

If there is a fire, the best advice is to get away from it and let the instructor or laboratory assistant take care of it. DON'T PANIC! Time spent in thought before action is never wasted. If it is a small fire in a container, it can usually be extinguished quickly by placing a wire-gauze screen with a ceramic fiber center or, possibly, a watch glass over the mouth of the container. It is good practice to have a wire screen or watch glass handy whenever you are using a flame. If this method does not extinguish the fire and if help from an experienced person is not readily available, then extinguish the fire yourself with a fire extinguisher.

Should your clothing catch on fire, DO NOT RUN. Walk *purposefully* toward the fire shower station or the nearest fire blanket. Running will fan the flames and intensify them.

C. Organic Solvents: Their Hazards

Avoid Contact with Organic Solvents. It is essential to remember that most organic solvents are flammable and will burn if they are exposed to an open flame or a match. Remember also that on repeated or excessive exposure, some organic solvents may be toxic, carcinogenic (cancer causing), or both. For example, many chlorocarbon solvents, when accumulated in the body, result in liver deterioration similar to cirrhosis caused by excessive use of ethanol. The body does not easily rid itself of chlorocarbons, nor does it detoxify them; they build up over time and may cause future illness. Some chlorocarbons are also suspected of being carcinogens. MINIMIZE YOUR EXPOSURE. Long-term exposure to benzene may cause a form of leukemia. Do not sniff benzene, and avoid spilling it on yourself. Many other solvents, such as chloroform and ether, are good anesthetics and will put you to sleep if you breathe too much of them. They subsequently cause nausea. Many of these solvents have a synergistic effect with ethanol, meaning that they enhance its effect. Pyridine causes temporary impotence. In other words, organic solvents are just as dangerous as corrosive chemicals, such as sulfuric acid, but manifest their hazardous nature in other, more subtle ways.

If you are pregnant, you may want to consider taking this course at a later time. Some exposure to organic fumes is inevitable, and any possible risk to an unborn baby should be avoided.

Minimize any direct exposure to solvents and treat them with respect. The laboratory room should be well ventilated. Normal cautious handling of solvents should not result in any health problem. If you are trying to evaporate a solution in an open container, you must do the evaporation in the hood. Excess solvents should be discarded in a container specifically intended for waste solvents, rather than down the drain at the laboratory bench. A sensible precaution is to wear gloves when working with solvents. Gloves made from polyethylene are inexpensive and provide good protection. The disadvantage of polyethylene gloves is that they are slippery. Disposable surgical gloves provide a better grip on glassware and other equipment, but they do not offer as much protection as polyethylene gloves. Nitrile gloves offer better protection (see p. 547).

Do Not Breathe Solvent Vapors. In checking the odor of a substance, be careful not to inhale very much of the material. The technique for smelling flowers is not advisable here; you could inhale dangerous amounts of the compound. Rather, a technique for smelling minute amounts of a substance is used. Pass a stopper or spatula moistened with the substance (if it is a liquid) under your nose. Or hold the substance away from you and waft the vapors toward you with your hand. But *never* hold your nose over the container and inhale deeply!

The hazards associated with organic solvents you are likely to encounter in the organic laboratory are discussed in detail beginning on page 554. If you use proper safety precautions, your exposure to harmful organic vapors will be minimized and should present no health risk.

Safe Transportation of Chemicals. When transporting chemicals from one location to another, particularly from one room to another, it is always best to use some form of **secondary containment.** This means that the bottle or flask is carried inside another, larger container. This outer container serves to contain the contents of the inner vessel in case a leak or breakage should occur. Scientific suppliers offer a variety of chemical-resistant carriers for this purpose.

D. Waste Disposal

Do Not Place Any Liquid or Solid Waste in Sinks; Use Appropriate Waste Containers. Many substances are toxic, flammable, and difficult to degrade; it is neither legal nor advisable to dispose of organic solvents or other liquid or solid reagents by pouring them down the sink.

The correct disposal method for wastes is to put them in appropriately labeled waste containers. These containers should be placed in the hoods in the laboratory. The waste containers will be disposed of safely by qualified persons using approved protocols. Specific guidelines for disposing of waste will be determined by the people in charge of your laboratory and by local regulations. Two alternative systems for handling waste disposal are presented here. For each experiment that you are assigned, you will be instructed to dispose of all wastes according to the system that is in operation in your laboratory.

In one model of waste collection, a separate waste container for each experiment is placed in the laboratory. In some cases, more than one container, each labeled according to the type of waste that is anticipated, is set out. The containers will be labeled with a list that details each substance that is present in the container. In this model, it is common practice to use separate waste containers for aqueous solutions, organic halogenated solvents, and other organic nonhalogenated materials. At the end of the laboratory class period, the waste containers are transported to a central hazardous materials storage location. These wastes may be later consolidated and poured into large drums for shipping. Complete labeling, detailing each chemical contained in the waste, is required at each stage of this wastehandling process, even when the waste is consolidated into drums.

In a second model of waste collection, you will be instructed to dispose of all wastes in one of the following ways:

Nonhazardous solids. Nonhazardous solids such as paper and cork can be placed in an ordinary wastebasket.

Broken glassware. Broken glassware should be put into a container specifically designated for broken glassware.

Organic solids. Solid products that are not turned in or any other organic solids should be disposed of in the container designated for organic solids.

Inorganic solids. Solids such as alumina and silica gel should be put in a container specifically designated for them.

Nonhalogenated organic solvents. Organic solvents such as diethyl ether, hexane, and toluene, or any solvent that does not contain a halogen atom, should be disposed of in the container designated for nonhalogenated organic solvents.

Halogenated solvents. Methylene chloride (dichloromethane), chloroform, and carbon tetrachloride are examples of common halogenated organic solvents. Dispose of all halogenated solvents in the container designated for them.

Strong inorganic acids and bases. Strong acids such as hydrochloric, sulfuric, and nitric acid will be collected in specially marked containers. Strong bases such as sodium hydroxide and potassium hydroxide will also be collected in specially designated containers.

Aqueous solutions. Aqueous solutions will be collected in a specially marked waste container. It is not necessary to separate each type of aqueous solution (unless the solution contains heavy metals); rather, unless otherwise instructed, you may combine all aqueous solutions into the same waste container. Although many types of solutions (aqueous sodium bicarbonate, aqueous sodium chloride, and so on) may seem innocuous and it may seem that their disposal down the sink drain is not likely to cause harm, many communities are becoming increasingly restrictive about what substances they will permit to enter municipal sewage-treatment systems. In light of this trend toward greater caution, it is important to develop good laboratory habits regarding the disposal of *all* chemicals.

Heavy metals. Many heavy-metal ions such as mercury and chromium are highly toxic and should be disposed of in specifically designated waste containers.

Whichever method is used, the waste containers must eventually be labeled with a complete list of each substance that is present in the waste. Individual waste containers are collected, and their contents are consolidated and placed into drums for transport to the waste-disposal site. Even these drums must bear labels that detail each of the substances contained in the waste.

In either waste-handling method, certain principles will always apply:

- Aqueous solutions should not be mixed with organic liquids.

- Concentrated acids should be stored in separate containers; certainly they must *never* be allowed to come into contact with organic waste.

- Organic materials that contain halogen atoms (fluorine, chlorine, bromine, or iodine) should be stored in separate containers from those used to store materials that do not contain halogen atoms.

In each experiment in this textbook, we have suggested a method of collecting and storing wastes. Your instructor may opt to use another method for collecting wastes.

E. Use of Flames

Even though organic solvents are frequently flammable (for example, hexane, diethyl ether, methanol, acetone, and petroleum ether), there are certain laboratory procedures for which a flame must be used. Most often, these procedures involve an aqueous solution. In fact, as a general rule, use a flame to heat only aqueous solutions. Heating methods that do not use a flame are discussed in detail in Technique 6, starting on page 589. Most organic solvents boil below 100°C, and an aluminum block, heating mantle, sand bath, or waterbath may be used to heat these solvents safely. Common organic solvents are listed in Technique 10, Table 10.3, page 643. Solvents marked in the table with boldface type will burn. Diethyl ether, pentane, and hexane are especially dangerous because, in combination with the correct amount of air, they may explode.

Some commonsense rules apply to using a flame in the presence of flammable solvents. Again, we stress that you should check to see whether anyone in your vicinity is using flammable solvents before you ignite any open flame. If someone is using a flammable solvent, move to a safer location before you light your flame. Your laboratory should have an area set aside for using a burner to prepare micropipets or other pieces of glassware.

The drainage troughs or sinks should never be used to dispose of flammable organic solvents. They will vaporize if they are low boiling and may encounter a flame farther down the bench on their way to the sink.

F. Inadvertently Mixed Chemicals

To avoid unnecessary hazards of fire and explosion, never pour any reagent back into a stock bottle. There is always the chance that you may accidentally pour back some foreign substance that will react explosively with the chemical in the stock bottle. Of course, by pouring reagents back into the stock bottles, you may introduce impurities that could spoil the experiment for the person using the stock reagent after you. Pouring things back into bottles is not only a dangerous practice but also an inconsiderate one. Thus, you should not take more chemicals than you need.

G. Unauthorized Experiments

Never undertake any unauthorized experiments. The risk of an accident is high, particularly if the experiment has not been completely checked to reduce hazards. Never work alone in the laboratory. The laboratory instructor or supervisor must always be present.

H. Food in the Laboratory

Because all chemicals are potentially toxic, avoid accidentally ingesting any toxic substance; therefore, never eat or drink any food while in the laboratory. There is always the possibility that whatever you are eating or drinking may become contaminated with a potentially hazardous material.

I. Clothing

Always wear closed shoes in the laboratory; open-toed shoes or sandals offer inadequate protection against spilled chemicals or broken glass. Do not wear your best clothing in the laboratory because some chemicals can make holes in or permanent stains on your clothing. To protect yourself and your clothing, it is advisable to wear a full-length laboratory apron or coat.

When working with chemicals that are very toxic, wear some type of gloves. Disposable gloves are inexpensive, offer good protection, provide acceptable "feel," and can be bought in many departmental stockrooms and college bookstores. Disposable latex surgical or polyethylene gloves are the least expensive type of glove; they are satisfactory when working with inorganic reagents and solutions. Better protection is afforded by disposable nitrile gloves. This type of glove provides good protection against organic chemicals and solvents. Heavier nitrile gloves are also available.

Finally, hair that is shoulder length or longer should be tied back. This precaution is especially important if you are working with a burner.

J. First Aid: Cuts, Minor Burns, and Acid or Base Burns

If any chemical enters your eyes, immediately irrigate the eyes with copious quantities of water. Tempered (slightly warm) water, if available, is preferable. Be sure that the eyelids are kept open. Continue flushing the eyes in this way for 15 minutes.

In case of a cut, wash the wound well with water unless you are specifically instructed to do otherwise. If necessary, apply pressure to the wound to stop the flow of blood. If you are assisting someone else, prudence dictates that you wear gloves in order to avoid contact with the blood of another person. Minor burns caused by flames or contact with hot objects may be soothed by immediately immersing the burned area in cold water or cracked ice until you no longer feel a burning sensation. Applying salves to burns is discouraged. Severe burns must be examined and treated by a physician. For chemical acid or base burns, rinse the burned area with copious quantities of water for at least 15 minutes.

If you accidentally ingest a chemical, call the local poison control center for instructions. Do not drink anything until you have been told to do so. It is important that the examining physician be informed of the exact nature of the substance ingested.

1.2 Right-to-Know Laws

The federal government and most state governments now require that employers provide their employees with complete information about hazards in the workplace. These regulations are often referred to as **Right-to-Know Laws.** At the federal level, the Occupational Safety and Health Administration (OSHA) is charged with enforcing these regulations.

In 1990, the federal government extended the Hazard Communication Act, which established the Right-to-Know Laws, to include a provision that requires the establishment of a Chemical Hygiene Plan at all academic laboratories. Every college and university chemistry department should have a Chemical Hygiene Plan. Having this plan means that all the safety regulations and laboratory safety procedures should be written in a manual. The plan also provides for the training of all employees in laboratory safety. Your laboratory instructor and assistants should have this training.

One of the components of Right-to-Know Laws is that employees and students have access to information about the hazards of any chemicals with which they are working. Your instructor will alert you to dangers to which you need to pay particular attention. However, you may want to seek additional information. Two excellent sources of information are labels on the bottles that come from a chemical manufacturer and **Material Safety Data Sheets** (MSDSs). The MSDSs are also provided by the manufacturer and must be kept available for all chemicals used at educational institutions.

A. Material Safety Data Sheets

Reading an MSDS for a chemical can be a daunting experience, even for an experienced chemist. MSDSs contain a wealth of information, some of which must be decoded to understand. The MSDS for methanol is shown on pages 550–554. Only the information that might be of interest to you is described in the paragraphs that follow.

Section 1. The first part of Section 1 identifies the substance by name, formula, and various numbers and codes. Most organic compounds have more than one name. In this case, the systematic (or International Union of Pure and Applied Chemistry [IUPAC]) name is methanol, and the other names are common names or are from an older system of nomenclature. The Chemical Abstract Service Number (CAS No.) is often used to identify a substance, and it may be used to access extensive information about a substance found in many computer databases or in the library.

Section 3. The Baker SAF-T-DATA System is found on all MSDSs and bottle labels for chemicals supplied by J. T. Baker, Inc. For each category listed, the number indicates the degree of hazard. The lowest number is 0 (very low hazard), and the highest number is 4 (extreme hazard). The Health category refers to damage involved when the substance is inhaled, ingested, or absorbed. Flammability indicates the tendency of a substance to burn. Reactivity refers to how reactive a substance is with air, water, or other substances. The last category, Contact, refers to how hazardous a substance is when it comes in contact with external parts of the body. Note that this rating scale is applicable only to Baker MSDSs and labels; other rating scales with different meanings are also in common use.

Section 4. This section provides helpful information for emergency and first aid procedures.

Section 6. This part of the MSDS deals with procedures for handling spills and disposal. The information could be very helpful, particularly if a large amount of the chemical was spilled. More information about disposal is also given in Section 13.

Section 8. Much valuable information is found in Section 8. To help you understand this material, some of the more important terms used in this section are defined:

> *Threshold Limit Value (TLV).* The American Conference of Governmental Industrial Hygienists (ACGIH) developed the TLV: This is the maximum concentration of a substance in air that a person should be exposed to on a regular basis. It is usually expressed in ppm or mg/m^3. Note that this value assumes that a person is exposed to the substance 40 hours per week, on a long-term basis. This value may not be particularly applicable in the case of a student performing an experiment in a single laboratory period.

> *Permissible Exposure Limit (PEL).* This has the same meaning as TLV; however, PELs were developed by OSHA. Note that for methanol, the TLV and PEL are both 200 ppm.

Section 10. The information contained in Section 10 refers to the stability of the compound and the hazards associated with mixing of chemicals. It is important to consider this information before carrying out an experiment not previously done.

Section 11. More information about the toxicity is given in this section. Another important term must first be defined:

> ***Lethal Dose, 50% Mortality (LD₅₀).*** This is the dose of a substance that will kill 50% of the animals administered a single dose. Different means of administration are used, such as oral, intraperitoneal (injected into the lining of the abdominal cavity), subcutaneous (injected under the skin), and application to the surface of the skin. The LD_{50} is usually expressed in milligrams (mg) of substance per kilogram (kg) of animal weight. The lower the value of LD_{50}, the more toxic the substance. It is assumed that the toxicity in humans will be similar.

Unless you have considerably more knowledge about chemical toxicity, the information in Sections 8 and 11 is most useful for comparing the toxicity of one substance with another. For example, the TLV for methanol is 200 ppm, whereas the TLV for benzene is 10 ppm. Clearly, performing an experiment involving benzene would require much more stringent precautions than an experiment involving methanol. One of the LD_{50} values for methanol is 5628 mg/kg. The comparable LD_{50} value of aniline is 250 mg/kg. Clearly, aniline is much more toxic, and because it is easily absorbed through the skin, it presents a significant hazard. It should also be mentioned that both TLV and PEL ratings assume that the worker comes in contact with a substance on a repeated and long-term basis. Thus, even if a chemical has a relatively low TLV or PEL, it does not mean that using it for one experiment will present a danger to you. Furthermore, by performing experiments using small amounts of chemicals and with proper safety precautions, your exposure to organic chemicals in this course will be minimal.

Section 16. Section 16 contains the National Fire Protection Association (NFPA) rating. This is similar to the Baker SAF-T-DATA (discussed in Section 3), except that the number represents the hazards when a fire is present. The order here is Health, Flammability, and Reactivity. Often, this is presented in graphic form on a label (see figure). The small diamonds are often color coded: blue for Health, red for Flammability, and yellow for Reactivity. The bottom diamond (white) is sometimes used to display graphic symbols denoting unusual reactivity, hazards, or special precautions to be taken.

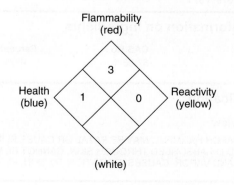

B. Bottle Labels

Reading the label on a bottle can be a helpful way of learning about the hazards of a chemical. The amount of information varies greatly, depending on which company supplied the chemical.

Apply some common sense when you read MSDSs and bottle labels. Using these chemicals does not mean you will experience the consequences that can potentially result from exposure to each chemical. For example, an MSDS for sodium chloride states, "Exposure to this product may have serious adverse health effects." Despite the apparent severity of this cautionary statement, it would not be reasonable to expect people to stop using sodium chloride in a chemistry experiment or to stop sprinkling a small amount of it (as table salt) on eggs to enhance their flavor. In many cases, the consequences described in MSDSs from exposure to chemicals are somewhat overstated, particularly for students using these chemicals to perform a laboratory experiment.

MSDS Number: M2015	Effective Date: 12/8/96

MSDS **Material Safety Data Sheet**

From: Mallinckrodt Baker, Inc.
222 Red School Lane
Phillipsburg, NJ 08865

MALLINCKRODT   **J.T.Baker**

24 Hour Emergency Telephone: 908-859-2151
CHEMTREC: 1-800-424-9300

National Response in Canada
CANUTEC: 613-996-6666

Outside U.S. and Canada
Chemtrec: 202-483-7616

NOTE: CHEMTREC, CANUTEC and National Response Center emergency numbers to be used only in the event of chemical emergencies involving a spill, leak, fire, exposure or accident involving chemicals.

All non-emergency questions should be directed to Customer Service (1-800-582-2537) for assistance.

METHYL ALCOHOL

1.   Product Identification

Synonyms: Wood alcohol; methanol; carbinol
CAS No: 67-56-1
Molecular Weight: 32.04
Chemical Formula: CH_3OH
Product Codes: **J.T. Baker:**

5217, 5370, 5794, 5807, 5811, 5842, 5869, 9049, 9063, 9066, 9067, 9069, 9070, 9071, 9073, 9075, 9076, 9077, 9091, 9093, 9096, 9097, 9098, 9263, 9893

Mallinckrodt:

3004, 3006, 3016, 3017, 3018, 3024, 3041, 3701, 4295, 5160, 8814, H080, H488, H603, V079, V571

2.   Composition/Information on Ingredients

Ingredient	CAS No.	Percent	Hazardous
Methyl Alcohol	67-56-1	100%	Yes

3.   Hazards Identification

Emergency Overview

POISON! DANGER! VAPOR HARMFUL. MAY BE FATAL OR CAUSE BLINDNESS IF SWALLOWED. HARMFUL IF INHALED OR ABSORBED THROUGH SKIN. CANNOT BE MADE NONPOISONOUS. FLAMMABLE LIQUID AND VAPOR. CAUSES IRRITATION TO SKIN, EYES AND RESPIRATORY TRACT. AFFECTS THE LIVER.

J.T. Baker SAF-T-DATA(tm) Ratings

(Provided here for your convenience)

Health:	Flammability:	Reactivity:	Contact:
3 - Severe (Poison)	4 - Extreme (Flammable)	1 - Slight	1 - Slight

Lab Protection Equip:	GOGGLES & SHIELD; LAB COAT & APRON; VENT HOOD; PROPER GLOVES; CLASS B EXTINGUISHER
Storage Color Code:	Red (Flammable)

Potential Health Effects

Inhalation:

A slight irritant to the mucous membranes. Toxic effects exerted upon nervous system, particularly the optic nerve. Once absorbed into the body, it is very slowly eliminated. Symptoms of overexposure may include headache, drowsiness, nausea, vomiting, blurred vision, blindness, coma, and death. A person may get better but then worse again up to 30 hours later.

Ingestion:

Toxic. Symptoms parallel inhalation. Can intoxicate and cause blindness. Usual fatal dose: 100-125 milliliters.

Skin Contact:

Methyl alcohol is a defatting agent and may cause skin to become dry and cracked. Skin absorption can occur; symptoms may parallel inhalation exposure.

Eye Contact:

Irritant. Continued exposure may cause eye lesions.

Chronic Exposure:

Marked impairment of vision and enlargement of the liver has been reported. Repeated or prolonged exposure may cause skin irritation.

Aggravation of Pre-existing Conditions:

Persons with pre-existing skin disorders or eye problems or impaired liver or kidney function may be more susceptible to the effects of the substance.

4. First Aid Measures

Inhalation:

Remove to fresh air. If not breathing, give artificial respiration. If breathing is difficult, give oxygen. Call a physician.

Ingestion:

Induce vomiting immediately as directed by medical personnel. Never give anything by mouth to an unconscious person.

Skin Contact:

Remove any contaminated clothing. Wash skin with soap or mild detergent and water for at least 15 minutes. Get medical attention if irritation develops or persists.

Eye Contact:

Immediately flush eyes with plenty of water for at least 15 minutes, lifting lower and upper eyelids occasionally. Get medical attention immediately.

5. Fire Fighting Measures

Fire:

Flash point: 12°C (54°F) CC
Autoignition temperature: 464°C (867°F)
Flammable limits in air % by volume:
lel: 7.3; uel: 36
Flammable.

Explosion:

Above flash point, vapor-air mixtures are explosive within flammable limits noted above. Moderate explosion hazard and dangerous fire hazard when exposed to heat, sparks or flames. Sensitive to static discharge.

Fire Extinguishing Media:

Water spray, dry chemical, alcohol foam, or carbon dioxide.

Special Information:

In the event of a fire, wear full protective clothing and NIOSH-approved self-contained breathing apparatus with full facepiece operated in the pressure demand or other positive pressure mode. Use water spray to blanket fire, cool fire exposed containers, and to flush non-ignited spills or vapors away from fire. Vapors can flow along surfaces to distant ignition source and flash back.

6. Accidental Release Measures

Ventilate area of leak or spill. Remove all sources of ignition. Wear appropriate personal protective equipment as specified in Section 8. Isolate hazard area. Keep unnecessary and unprotected personnel from entering. Contain and recover liquid when possible. Use non-sparking tools and equipment. Collect liquid in an appropriate container or absorb with an inert material (e. g., vermiculite, dry sand, earth), and place in a chemical waste container. Do not use combustible materials, such as saw dust. Do not flush to sewer! J. T. Baker SOLUSORB® solvent adsorbent is recommended for spills of this product.

7.   Handling and Storage

Protect against physical damage. Store in a cool, dry well-ventilated location, away from any area where the fire hazard may be acute. Outside or detached storage is preferred. Separate from incompatibles. Containers should be bonded and grounded for transfers to avoid static sparks. Storage and use areas should be No Smoking areas. Use non-sparking type tools and equipment, including explosion proof ventilation. Containers of this material may be hazardous when empty since they retain product residues (vapors, liquid); observe all warnings and precautions listed for the product.

8.   Exposure Controls/Personal Protection

Airborne Exposure Limits:

For Methyl Alcohol:
- OSHA Permissible Exposure Limit (PEL):
     200 ppm (TWA)
- ACGIH Threshold Limit Value (TLV):
     200 ppm (TWA), 250 ppm (STEL) skin

Ventilation System:

A system of local and/or general exhaust is recommended to keep employee exposures below the Airborne Exposure Limits. Local exhaust ventilation is generally preferred because it can control the emissions of the contaminant at its source, preventing dispersion of it into the general work area. Please refer to the ACGIH document, "Industrial Ventilation, A Manual of Recommended Practices", most recent edition, for details.

Personal Respirator (NIOSH Approved)

If the exposure limit is exceeded, wear a supplied air, full-facepiece respirator, airlined hood, or full-facepiece self-contained breathing apparatus.

Skin Protection:

Rubber or neoprene gloves and additional protection including impervious boots, apron, or coveralls, as needed in areas of unusual exposure.

Eye Protection:

Use chemical safety goggles. Maintain eye wash fountain and quick-drench facilities in work area.

9.   Physical and Chemical Properties

Appearance:	**Boiling Point:**
Clear, colorless liquid.	64.5°C (147°F)
Odor:	**Melting Point:**
Characteristic odor.	-98°C (-144°F)
Solubility:	**Vapor Density (Air=1):**
Miscible in water.	1.1
Specific Gravity:	**Vapor Pressure (mm Hg):**
0.8	97 @ 20°C (68°F)
pH:	**Evaporation Rate (BuAc=1):**
No information found.	5.9
% Volatiles by volume @ 21°C (70°F):	
100	

10.   Stability and Reactivity

Stability:

Stable under ordinary conditions of use and storage.

Hazardous Decomposition Products:

May form carbon dioxide, carbon monoxide, and formaldehyde when heated to decomposition.

Hazardous Polymerization:

Will not occur.

Incompatabilities:

Strong oxidizing agents such as nitrates, perchlorates or sulfuric acid. Will attack some forms of plastics, rubber, and coatings. May react with metallic aluminum and generate hydrogen gas.

Conditions to Avoid:

Heat, flames, ignition sources and incompatibles.

11.   Toxicological Information

Methyl Alcohol (Methanol) Oral rat LD50: 5628 mg/kg; inhalation rat LC50: 64000 ppm/4H; skin rabbit LD50: 15800 mg/kg; Irritation data-standard Draize test: skin, rabbit: 20mg/24 hr. Moderate; eye, rabbit: 100 mg/24 hr. Moderate; Investigated as a mutagen, reproductive effector.

Cancer Lists			
	---NTP Carcinogen---		
Ingredient	Known	Anticipated	IARC Category
Methyl Alcohol (67-56-1)	No	No	None

12. Ecological Information

Environmental Fate:

When released into the soil, this material is expected to readily biodegrade. When released into the soil, this material is expected to leach into groundwater. When released into the soil, this material is expected to quickly evaporate. When released into the water, this material is expected to have a half-life between 1 and 10 days. When released into water, this material is expected to readily biodegrade. When released into the air, this material is expected to exist in the aerosol phase with a short half-life. When released into the air, this material is expected to be readily degraded by reaction with photochemically produced hydroxyl radicals. When released into air, this material is expected to have a half-life between 10 and 30 days. When released into the air, this material is expected to be readily removed from the atmosphere by wet deposition.

Environmental Toxicity:

This material is expected to be slightly toxic to aquatic life.

13. Disposal Considerations

Whatever cannot be saved for recovery or recycling should be handled as hazardous waste and sent to a RCRA approved incinerator or disposed in a RCRA approved waste facility. Processing, use or contamination of this product may change the waste management options. State and local disposal regulations may differ from federal disposal regulations.

Dispose of container and unused contents in accordance with federal, state and local requirements.

14. Transport Information

Domestic (Land, D.O.T.)

Proper Shipping Name:	METHANOL		
Hazard Class:	3		
UN/NA:	UN1230	**Packing Group:**	II
Information reported for product/size:	350LB		

International (Water, I.M.O.)

Proper Shipping Name:	METHANOL		
Hazard Class:	3.2, 6.1		
UN/NA:	UN1230	**Packing Group:**	II
Information reported for product/size:	350LB		

15. Regulatory Information

Chemical Inventory Status

Ingredient	TSCA	EC	Japan	Australia	Korea	DSL	NDSL	Phil.
						---Canada---		
Methyl Alcohol (67-56-1)	Yes	Yes	Yes	Yes	Yes	Yes	No	Yes

Federal, State & International Regulations

	---SARA 302---		----------SARA 313----------			-RCRA-	-TSCA-
Ingredient	RQ	TPQ	List	Chemical Catg.	CERCLA	261.33	8(d)
Methyl Alcohol (67-56-1)	No	No	Yes	No	5000	U154	No

Chemical Weapons Convention: No   **TSCA 12(b):** No   **CDTA:** No

SARA 311/312: Acute: Yes   Chronic: Yes   Fire: Yes   Pressure: No   Reactivity: No   (Pure / Liquid)

Australian Hazchem Code: 2PE   **Australian Poison Schedule:** S6

WHMIS: This MSDS has been prepared according to the hazard criteria of the Controlled Products Regulations (CPR) and the MSDS contains all of the information required by the CPR.

16. Other Information

NFPA Ratings:

Health: 1 Flammability: 3 Reactivity: 0

Label Hazard Warning:

POISON! DANGER! VAPOR HARMFUL. MAY BE FATAL OR CAUSE BLINDNESS IF SWALLOWED. HARMFUL IF INHALED OR ABSORBED THROUGH SKIN. CANNOT BE MADE NONPOISONOUS. FLAMMABLE LIQUID AND VAPOR. CAUSES IRRITATION TO SKIN, EYES AND RESPIRATORY TRACT. AFFECTS THE LIVER.

Label Precautions:

Keep away from heat, sparks and flame.
Keep container closed.
Use only with adequate ventilation.
Wash thoroughly after handling.
Avoid breathing vapor.
Avoid contact with eyes, skin and clothing.

Label First Aid:

If swallowed, induce vomiting immediately as directed by medical personnel. Never give anything by mouth to an unconscious person. In case of contact, immediately flush eyes or skin with plenty of water for at least 15 minutes while removing contaminated clothing and shoes. Wash clothing before reuse. If inhaled, remove to fresh air. If not breathing give artificial respiration. If breathing is difficult, give oxygen. In all cases get medical attention immediately.

Product Use:

Laboratory Reagent.

Revision Information:

New 16 section MSDS format, all sections have been revised.

Disclaimer:

Mallinckrodt Baker, Inc. provides the information contained herein in good faith but makes no representation as to its comprehensiveness or accuracy. This document is intended only as a guide to the appropriate precautionary handling of the material by a properly trained person using this product. Individuals receiving the information must exercise their independent judgment in determining its appropriateness for a particular purpose. MALLINCKRODT BAKER, INC. MAKES NO REPRESENTATIONS OR WARRANTIES, EITHER EXPRESS OR IMPLIED, INCLUDING WITHOUT LIMITATION ANY WARRANTIES OR MERCHANTABILITY, FITNESS FOR A PARTICULAR PURPOSE WITH RESPECT TO THE INFORMATION SET FORTH HEREIN OR THE PRODUCT TO WHICH THE INFORMATION REFERS. ACCORDINGLY, MALLINCKRODT BAKER, INC. WILL NOT BE RESPONSIBLE FOR DAMAGES RESULTING FROM USE OF OR RELIANCE UPON THIS INFORMATION.

Prepared By: Strategic Services Division
Phone Number: (314) 539-1600 (U.S.A.)

1.3 Common Solvents

Most organic chemistry experiments involve an organic solvent at some step in the procedure. A list of common organic solvents follows, with a discussion of toxicity, possible carcinogenic properties, and precautions that you should use when handling these solvents. A tabulation of the compounds currently suspected of being carcinogens appears at the end of Technique 1.

Acetic Acid. Glacial acetic acid is corrosive enough to cause serious acid burns on the skin. Its vapors can irritate the eyes and nasal passages. Care should be exercised not to breathe the vapors and not to allow them to escape into the laboratory.

Acetone. Relative to other organic solvents, acetone is not very toxic. It is flammable, however. Do not use acetone near open flames.

Benzene. Benzene can damage bone marrow, it causes various blood disorders, and its effects may lead to leukemia. Benzene is considered a serious carcinogenic hazard. It is absorbed rapidly through the skin and also poisons the liver and kidneys. In addition, benzene is flammable. Because of its toxicity and its carcinogenic properties, benzene should not be used in the laboratory; you should use some less dangerous solvent instead. Toluene is considered a safer alternative solvent in procedures that specify benzene.

Carbon Tetrachloride. Carbon tetrachloride can cause serious liver and kidney damage, as well as skin irritation and other problems. It is absorbed rapidly through the skin. In high concentrations, it can cause death as a result of respiratory failure. Moreover, carbon tetrachloride is suspected of being a carcinogenic material. Although this solvent has the advantage of being nonflammable (in the past, it was used on occasion as a fire extinguisher), it causes health problems, so it should not be used routinely in the laboratory. If no reasonable substitute exists, however, it must be used in small quantities, as in preparing samples for infrared (IR) and nuclear magnetic resonance (NMR) spectroscopy. In such cases, you must use it in a hood.

Chloroform. Chloroform is similar to carbon tetrachloride in its toxicity. It has been used as an anesthetic. However, chloroform is currently on the list of suspected carcinogens. Because of this, do not use chloroform routinely as a solvent in the laboratory. If it is occasionally necessary to use chloroform as a solvent for special samples, then you must use it in a hood. Methylene chloride is usually found to be a safer substitute in procedures that specify chloroform as a solvent. Deuterochloroform, $CDCl_3$, is a common solvent for NMR spectroscopy. Caution dictates that you should treat it with the same respect as chloroform.

1,2-Dimethoxyethane (Ethylene Glycol Dimethyl Ether or Monoglyme). Because it is miscible with water, 1,2-dimethoxyethane is a useful alternative to solvents such as dioxane and tetrahydrofuran, which may be more hazardous. 1,2-Dimethoxyethane is flammable and should not be handled near an open flame. On long exposure of 1,2-dimethoxyethane to light and oxygen, explosive peroxides may form. 1,2-Dimethoxyethane is a possible reproductive toxin.

Dioxane. Dioxane has been used widely because it is a convenient, water-miscible solvent. It is now suspected, however, of being carcinogenic. It is also toxic, affecting the central nervous system, liver, kidneys, skin, lungs, and mucous membranes. Dioxane is also flammable and tends to form explosive peroxides when it is exposed to light and air. Because of its carcinogenic properties, it is no longer used in the laboratory unless absolutely necessary. Either 1,2-dimethoxyethane or tetrahydrofuran is a suitable, water-miscible alternative solvent.

Ethanol. Ethanol has well-known properties as an intoxicant. In the laboratory, the principal danger arises from fires because ethanol is a flammable solvent. When using ethanol, take care to work where there are no open flames.

Ether (Diethyl Ether). The principal hazard associated with diethyl ether is fire or explosion. Ether is probably the most flammable solvent found in the laboratory. Because ether vapors are much denser than air, they may travel along a laboratory bench for a considerable distance from their source before being ignited. Before using ether, it is important to be sure that no one is working with matches or any open flame. Ether is not a particularly toxic solvent, although in high enough concentrations it can cause drowsiness and perhaps nausea. It has been used as a general anesthetic. Ether can form highly explosive peroxides when exposed to air. Consequently, you should never distill it to dryness.

Hexane. Hexane may be irritating to the respiratory tract. It can also act as an intoxicant and a depressant of the central nervous system. It can cause skin irritation because it is an excellent solvent for skin oils. The most serious hazard, however, comes from its flammability. The precautions recommended for using diethyl ether in the presence of open flames apply equally to hexane.

Ligroin. See Hexane.

Methanol. Much of the material outlining the hazards of ethanol applies to methanol. Methanol is more toxic than ethanol; ingestion can cause blindness and even death. Because methanol is more volatile, the danger of fires is more acute.

Methylene Chloride (Dichloromethane). Methylene chloride is not flammable. Unlike other members of the class of chlorocarbons, it is not currently considered a serious carcinogenic hazard. Recently, however, it has been the subject of much serious investigation, and there have been proposals to regulate it in industrial situations in which workers have high levels of exposure on a day-to-day basis. Methylene chloride is less toxic than chloroform and carbon tetrachloride. It can cause liver damage when ingested, however, and its vapors may cause drowsiness or nausea.

Pentane. See Hexane.

Petroleum Ether. See Hexane.

Pyridine. Some fire hazard is associated with pyridine. However, the most serious hazard arises from its toxicity. Pyridine may depress the central nervous system; irritate the skin and respiratory tract; damage the liver, kidneys, and gastrointestinal system; and even cause temporary sterility. You should treat pyridine as a highly toxic solvent and handle it only in the fume hood.

Tetrahydrofuran. Tetrahydrofuran may cause irritation of the skin, eyes, and respiratory tract. It should never be distilled to dryness because it tends to form potentially explosive peroxides on exposure to air. Tetrahydrofuran does present a fire hazard.

Toluene. Unlike benzene, toluene is not considered a carcinogen. However, it is at least as toxic as benzene. It can act as an anesthetic and damage the central nervous system. If benzene is present as an impurity in toluene, expect the usual hazards associated with benzene. Toluene is also a flammable solvent, and the usual precautions about working near open flames should be applied.

You should not use certain solvents in the laboratory because of their carcinogenic properties. Benzene, carbon tetrachloride, chloroform, and dioxane are among these solvents. For certain applications, however, notably as solvents for infrared or NMR spectroscopy, there may be no suitable alternative. When it is necessary to use one of these solvents, use safety precautions and refer to the discussions in Techniques 25–28.

Because relatively large amounts of solvents may be used in a large organic laboratory class, your laboratory supervisor must take care to store these substances safely. Only the amount of solvent needed for a particular experiment should be kept in the laboratory. The preferred location for bottles of solvents being used during a class period is in a hood. When the solvents are not being used, they should be stored in a fireproof storage cabinet for solvents. If possible, this cabinet should be ventilated into the fume hood system.

1.4 Carcinogenic Substances

A **carcinogen** is a substance that causes cancer in living tissue. The usual procedures for determining whether a substance is carcinogenic is to expose laboratory animals to high dosages over a long period. It is not clear whether short-term exposure to these chemicals carries a comparable risk, but it is prudent to use these substances with special precautions.

Many regulatory agencies have compiled lists of carcinogenic substances or substances suspected of being carcinogenic. Because these lists are inconsistent, compiling a definitive list of carcinogenic substances is difficult. The following common substances are included in many of these lists.

Acetamide	4-Methyl-2-oxetanone (β-butyrolactone)
Acrylonitrile	1-Naphthylamine
Asbestos	2-Naphthylamine
Benzene	N-Nitroso compounds
Benzidine	2-Oxetanone (β-propiolactone)
Carbon tetrachloride	Phenacetin
Chloroform	Phenylhydrazine and its salts
Chromic oxide	Polychlorinated biphenyl (PCB)
Coumarin	Progesterone
Diazomethane	Styrene oxide
1,2-Dibromoethane	Tannins
Dimethyl sulfate	Testosterone
p-Dioxane	Thioacetamide
Ethylene oxide	Thiourea
Formaldehyde	o-Toluidine
Hydrazine and its salts	Trichloroethylene
Lead (II) acetate	Vinyl chloride

REFERENCES

Aldrich Catalog and Handbook of Fine Chemicals. Milwaukee, WI: Aldrich Chemical Co., current edition.

Armour, M. A., *Pollution Prevention and Waste Minimization in Laboratories.* Reinhardt, P. A., Leonard, K. L., Ashbrook, P. C., eds. Boca Raton, FL: Lewis Publishers, 1996.

Fire Protection Guide on Hazardous Materials, 10th ed. Quincy, MA: National Fire Protection Association, 1991.

Flinn Chemical Catalog Reference Manual. Batavia, IL: Flinn Scientific, current edition.

Gosselin, R. E., Smith, R. P., and Hodge, H. C. *Clinical Toxicology of Commercial Products,* 5th ed. Baltimore, MD: Williams & Wilkins, 1984.

Lenga, R. E., ed. *The Sigma-Aldrich Library of Chemical Safety Data.* Milwaukee, WI: Sigma-Aldrich, 1985.

Lewis, R. J. *Carcinogenically Active Chemicals: A Reference Guide.* New York: Van Nostrand Reinhold, 1990.

Lewis, R. J. *Sax's Dangerous Properties of Industrial Materials,* 8th ed. New York: Van Nostrand Reinhold, 1992.

The Merck Index, 13th ed. Rahway, NJ: Merck and Co., 2001.

Prudent Practices in the Laboratory: Handling and Disposal of Chemicals. Washington, DC: Committee on Prudent Practices for Handling, Storage, and Disposal of Chemicals in Laboratories, Board on Chemical Sciences and Technology, Commission on Physical Sciences, Mathematics, and Applications, National Research Council, National Academy Press, 1995.

Renfrew, M. M., ed. *Safety in the Chemical Laboratory.* Easton, PA: Division of Chemical Education, American Chemical Society, 1967–1991.

Safety in Academic Chemistry Laboratories, 4th ed. Washington, DC: Committee on Chemical Safety, American Chemical Society, 1985.

Sax, N. I., and Lewis, R. J. *Dangerous Properties of Industrial Materials,* 7th ed. New York: Van Nostrand Reinhold, 1988.

Sax, N. I., and Lewis, R. J., eds. *Rapid Guide to Hazardous Chemicals in the Work Place,* 2nd ed. New York: Van Nostrand Reinhold, 1990.

Useful Safety-Related Internet Addresses

Interactive Learning Paradigms, Inc.
http://www.ilpi.com/msds/   (accessed April 2005)
This is an excellent general site for MSDS sheets. The site lists chemical manufacturers and suppliers. Selecting a company will take you directly to the appropriate place to obtain an MSDS sheet. Many of the sites listed require you to register in order to obtain a MSDS sheet for a particular chemical. Ask your departmental or college safety supervisor to obtain the information for you.

Acros Chemicals and Fisher Scientific
https://www1.fishersci.com   (accessed April 2005)

Alfa Aesar
http://www.alfa.com/alf/index.htm   (accessed April 2005)

Eastman Kodak
http://msds.kodak.com/ehswww/external/index.jsp   (accessed April 2005)

EMD Chemicals (formerly EM Science) and Merck
http://www.emdchemicals.com/corporate/emd_corporate.asp   (accessed April 2005)

J. T. Baker and Mallinckrodt Laboratory Chemicals
http://www.jtbaker.com/asp/Catalog.asp   (accessed April 2005)

National Institute for Occupational Safety and Health (NIOSH) has an excellent website that includes databases and information resources, including links:
http://www.cdc.gov/niosh/topics/chemical-safety/default.html   (accessed April 2005)

Sigma, Aldrich and Fluka
http://www.sigmaaldrich.com/Area_of_Interest/The_Americas/United_States.html   (accessed April 2005)

VWR Scientific Products
http://www.vwrsp.com/search/index.cgi?tmpl5_msds&src=topnv-msds   (accessed April 2005)

2 **TECHNIQUE 2**

The Laboratory Notebook, Calculations, and Laboratory Records

It is important that you do some advance preparation for all laboratory work. Presented here are some suggestions about what specific information you should try to obtain in your advance studying. Because much of this information must be obtained while preparing your laboratory notebook, the two subjects, advance study and notebook preparation, are developed simultaneously.

An important part of any laboratory experience is learning to maintain complete records of every experiment undertaken and every item of data obtained. Far too often, careless recording of data and observations has resulted in mistakes, frustration, and lost time due to needless repetition of experiments. If reports are required, you will find that proper collection and recording of data can make your report writing much easier.

Because organic reactions are seldom quantitative, special problems result. Frequently, reagents must be used in large excess to increase the amount of product. Some reagents are expensive, and, therefore, care must be used in measuring the amounts of these substances. Often, many more reactions take place than you desire. These extra reactions, or **side reactions,** may form products other than the desired product. These are called **side products.** For all these reasons, you must plan your experimental procedure carefully before undertaking the actual experiment.

2.1 The Notebook

For recording data and observations during experiments, use a *bound notebook.* The notebook should have consecutively numbered pages. If it does not, number the pages immediately. A spiral-bound notebook or any other notebook from which the pages can be removed easily is not acceptable, because the possibility of losing the pages is great.

All data and observations must be recorded in the notebook. Paper towels, napkins, toilet tissue, or scratch paper tend to become lost or destroyed. It is bad laboratory practice to record information on such random and perishable pieces of paper. All entries must be recorded in *permanent ink.* It can be frustrating to have important information disappear from the notebook because it was recorded in washable ink or pencil and could not survive a flood caused by the student at the next position on the bench. Because you will be using your notebook in the laboratory, the book will probably become soiled or stained by chemicals, filled with scratched-out entries, or even slightly burned. That is expected and is a normal part of laboratory work.

Your instructor may check your notebook at any time, so you should always have it up to date. If your instructor requires reports, you can prepare them quickly from the material recorded in the laboratory notebook.

2.2 Notebook Format

A. Advance Preparation

Individual instructors vary greatly in the type of notebook format they prefer; such variation stems from differences in philosophies and experience. You must obtain specific directions from your own instructor for preparing a notebook. Certain features, however, are common to most notebook formats. The following discussion indicates what might be included in a typical notebook.

It will be helpful and you can save much time in the laboratory if for each experiment you know the main reactions, the potential side reactions, the mechanism, and the stoichiometry and you understand fully the procedure and the theory underlying it before you come to the laboratory. Understanding the procedure by which the desired product is to be separated from undesired materials is also important. If you examine each of these topics before coming to class, you will be prepared to do the experiment efficiently. You will have your equipment and reagents already prepared when they are to be used. Your reference material will be at hand when you need it. Finally, with your time efficiently organized, you will be able to take advantage of long reaction or reflux periods to perform other tasks, such as doing shorter experiments or finishing previous ones.

For experiments in which a compound is synthesized from other reagents, that is, **preparative experiments,** it is essential to know the main reaction. To perform stoichiometric calculations, you should balance the equation for the main reaction. Therefore, before you begin the experiment, your notebook should contain the balanced equation for the pertinent reaction. Using the preparation of isopentyl acetate, or banana oil, as an example, you should write the following:

$$CH_3-\overset{\displaystyle O}{\overset{\|}{C}}-OH \ + \ CH_3-\overset{\displaystyle CH_3}{\overset{|}{CH}}-CH_2-CH_2-OH \ \xrightarrow{\ H^+\ }$$

Acetic acid **Isopentyl alcohol**

$$CH_3-\overset{\displaystyle O}{\overset{\|}{C}}-O-CH_2-CH_2-\overset{\displaystyle CH_3}{\overset{|}{CH}}-CH_3 \ + \ H_2O$$

Isopentyl acetate

Also enter in the notebook the possible side reactions that divert reagents into contaminants (side products), before beginning the experiment. You will have to separate these side products from the major product during purification.

You should list physical constants such as melting points, boiling points, densities, and molecular weights in the notebook when this information is needed to perform an experiment or to do calculations. These data are located in sources such as the *CRC Handbook of Chemistry and Physics, The Merck Index, Lange's Handbook of Chemistry,* or *Aldrich Handbook of Fine Chemicals.* Write physical constants required for an experiment in your notebook before you come to class.

Advance preparation may also include examining some subjects, information not necessarily recorded in the notebook, that should prove useful in understanding the experiment. Included among these subjects are an understanding of the mechanism of the reaction, an examination of other methods by which the same compound might be prepared, and a detailed study of the experimental procedure. Many students find that an outline of the procedure, prepared *before* they come to class, helps them use their time more efficiently once they begin the experiment. Such an outline could well be prepared on some loose sheet of paper rather than in the notebook itself.

Once the reaction has been completed, the desired product does not magically appear as purified material; it must be isolated from a frequently complex mixture of side products, unreacted starting materials, solvents, and catalysts. You should try to outline a **separation scheme** in your notebook for isolating the product from its contaminants. At each stage, you should try to understand the reason for the particular instruction given in the experimental procedure. This not only will familiarize you with the basic separation and purification techniques used in organic chemistry but also will help you understand when to use these techniques. Such an outline might take the form of a flowchart. For example, see the separation scheme for isopentyl acetate (Figure 2.1). Careful attention to understanding the separation,

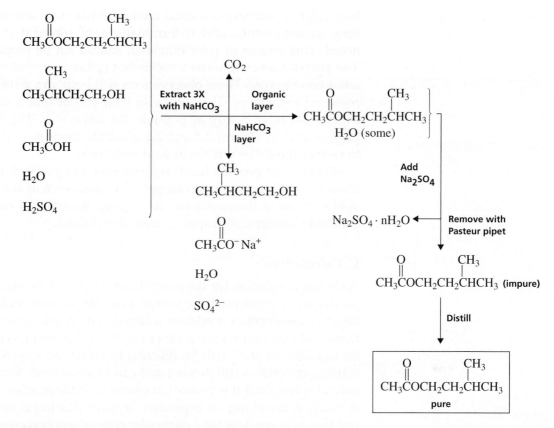

Figure 2.1
Separation scheme for isopentyl acetate.

besides familiarizing you with the procedure by which the desired product is separated from impurities in your particular experiments, may prepare you for original research in which no experimental procedure exists.

In designing a separation scheme, note that the scheme outlines those steps undertaken once the reaction period has been concluded. For this reason, the represented scheme does not include steps such as the addition of the reactants (isopentyl alcohol and acetic acid) and the catalyst (sulfuric acid) or the heating of the reaction mixture.

For experiments in which a compound is isolated from a particular source and is not prepared from other reagents, some information described in this section will not be applicable. Such experiments are called **isolation experiments.** A typical isolation experiment involves isolating a pure compound from a natural source. Examples include isolating caffeine from tea or isolating cinnamaldehyde from cinnamon. Although isolation experiments require somewhat different advance preparation, this advance study may include looking up physical constants for the compound isolated and outlining the isolation procedure. A detailed examination of the separation scheme is important here because it is the heart of such an experiment.

B. Laboratory Records

When you begin the actual experiment, keep your notebook nearby so you will be able to record those operations you perform. When working in the laboratory, the notebook serves as a place in which to record a rough transcript of your experimental method. Data from actual weighings, volume measurements, and determinations of physical constants are also noted. This section of your notebook should *not* be prepared in advance. The purpose is not to write a recipe but rather to record what you *did* and what you *observed*. These observations will help you write reports without resorting to memory. They will also help you or other workers repeat the experiment in as nearly as possible the same way. The sample notebook pages found in Figures 2.2 and 2.3 illustrate the type of data and observations that should be written in your notebook.

When your product has been prepared and purified, or isolated if it is an isolation experiment, record pertinent data such as the melting point or boiling point of the substance, its density, its index of refraction, and the conditions under which spectra were determined.

C. Calculations

A chemical equation for the overall conversion of the starting materials to products is written on the assumption of simple ideal stoichiometry. Actually, this assumption is seldom realized. Side reactions or competing reactions will also occur, giving other products. For some synthetic reactions, an equilibrium state will be reached in which an appreciable amount of starting material is still present and can be recovered. Some of the reactant may also remain if it is present in excess or if the reaction was incomplete. A reaction involving an expensive reagent illustrates another reason for needing to know how far a particular type of reaction converts reactants to products. In such a case, it is preferable to use the most efficient method for this conversion. Thus, information about the efficiency of conversion for various reactions is of interest to the person contemplating the use of these reactions.

The quantitative expression for the efficiency of a reaction is found by calculating the **yield** for the reaction. The **theoretical yield** is the number of grams of the product expected from the reaction on the basis of ideal stoichiometry, with side reactions, reversibility, and losses ignored. To calculate the theoretical yield, it is first necessary to determine the **limiting reagent.** The limiting reagent is the reagent that is not present in excess and on which the overall yield of product depends. The method for determining the limiting reagent in the isopentyl acetate experiment is illustrated in the sample notebook pages shown in Figures 2.2 and 2.3. You should consult your general chemistry textbook for more complicated examples. The theoretical yield is then calculated from the expression

Theoretical yield = (moles of limiting reagent)(ratio)(molecular weight of product)

The ratio here is the stoichiometric ratio of product to limiting reagent. In preparing isopentyl acetate, that ratio is 1:1. One mole of isopentyl alcohol, under ideal circumstances, should yield 1 mole of isopentyl acetate.

The **actual yield** is simply the number of grams of desired product obtained. The **percentage yield** describes the efficiency of the reaction and is determined by

$$\text{Percentage yield} = \frac{\text{Actual yield}}{\text{Theoretical yield}} \times 100$$

THE PREPARATION OF ISOPENTYL ACETATE (BANANA OIL)

MAIN REACTION

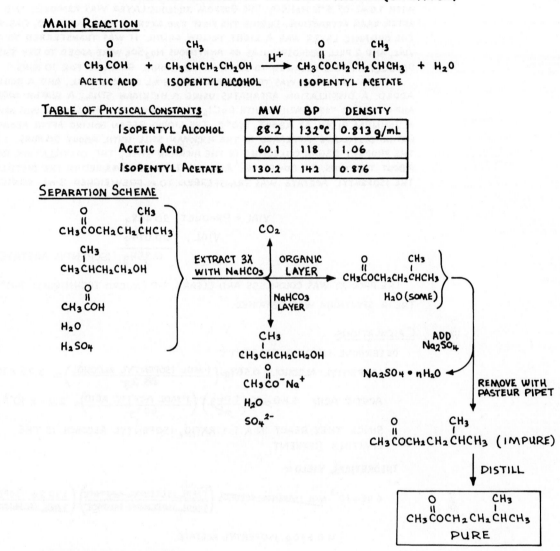

$$\underset{\text{ACETIC ACID}}{CH_3\overset{\displaystyle O}{\overset{\displaystyle \|}{C}}OH} \ + \ \underset{\text{ISOPENTYL ALCOHOL}}{CH_3\overset{\displaystyle CH_3}{\underset{\displaystyle |}{C}}HCH_2CH_2OH} \ \xrightarrow{\ H^+\ } \ \underset{\text{ISOPENTYL ACETATE}}{CH_3\overset{\displaystyle O}{\overset{\displaystyle \|}{C}}OCH_2CH_2\overset{\displaystyle CH_3}{\underset{\displaystyle |}{C}}HCH_3} \ + \ H_2O$$

TABLE OF PHYSICAL CONSTANTS

	MW	BP	DENSITY
ISOPENTYL ALCOHOL	88.2	132°C	0.813 g/mL
ACETIC ACID	60.1	118	1.06
ISOPENTYL ACETATE	130.2	142	0.876

SEPARATION SCHEME

DATA AND OBSERVATIONS

0.70 mL OF ISOPENTYL ACETATE WAS ADDED TO A PREWEIGHED
5-mL CONICAL VIAL:

VIAL + ALCOHOL	25.524 g
VIAL	24.955 g
	0.569 g ISOPENTYL ALCOHOL

ACETIC ACID (1.4 mL) AND 3 DROPS OF CONCENTRATED H₂SO₄ (USING A PASTEUR
PIPET) WERE ALSO ADDED TO THE CONICAL VIAL ALONG WITH A SMALL BOILING STONE.
A WATER-COOLED CONDENSER TOPPED WITH A DRYING TUBE CONTAINING A LOOSE
PLUG OF GLASS WOOL WAS ATTACHED TO THE VIAL. THE REACTION MIXTURE WAS
REFLUXED IN AN ALUMINUM BLOCK (ABOUT 155°) FOR 75 MIN. AND THEN COOLED TO
ROOM TEMPERATURE. THE COLOR OF THE REACTION MIXTURE WAS BROWNISH-YELLOW.

Figure 2.2
A sample notebook, page 1.

THE BOILING STONE WAS REMOVED AND THE REACTION MIXTURE WAS EXTRACTED THREE TIMES WITH 1.0mL OF 5% NaHCO₃. THE BOTTOM AQUEOUS LAYER WAS REMOVED AND DISCARDED AFTER EACH EXTRACTION. DURING THE FIRST TWO EXTRACTIONS, MUCH CO₂ GAS WAS GIVEN OFF. THE ORGANIC LAYER WAS A LIGHT YELLOW COLOR. IT WAS TRANSFERRED TO A DRY CONICAL VIAL, AND 2 FULL MICROSPATULAS OF ANHYDROUS Na₂SO₄ WERE ADDED TO DRY THE CRUDE PRODUCT. IT WAS ALLOWED TO SET WITH OCCASIONAL STIRRING FOR 10 MINS.

THE DRY PRODUCT WAS TRANSFERRED TO A 3-mL CONICAL VIAL, AND A BOILING STONE WAS ADDED. A DISTILLATION APPARATUS USING A HICKMAN STILL, A WATER-COOLED CONDENSER, AND A DRYING TUBE PACKED WITH CaCl₂ WAS ASSEMBLED. THE SAMPLE WAS HEATED IN AN ALUMINUM BLOCK AT ABOUT 180°C. THE LIQUID BEGAN BOILING AFTER ABOUT FIVE MINS, BUT NO DISTILLATE APPEARED IN THE HICKMAN STILL UNTIL ABOUT 20 MINS. LATER. ONCE THE PRODUCT BEGAN COLLECTING IN THE HICKMAN STILL, THE DISTILLATION REQUIRED ONLY ABOUT TWO MINS. TO COMPLETE. ABOUT 1-2 DROPS REMAINED IN THE DISTILLING VIAL. THE ISOPENTYL ACETATE WAS TRANSFERRED TO A PREWEIGHED 3-mL CONICAL VIAL.

$$\begin{aligned}\text{VIAL + PRODUCT} \quad & 20.428g \\ \text{VIAL} \quad & \underline{20.074g} \\ & 0.354g \quad \text{ISOPENTYL ACETATE}\end{aligned}$$

THE PRODUCT WAS COLORLESS AND CLEAR. BP (MICRO TECHNIQUE): 140°C. THE IR SPECTRUM WAS OBTAINED.

CALCULATIONS

DETERMINE LIMITING REAGENT:

ISOPENTYL ALCOHOL $\quad 0.569g \left(\dfrac{1 \text{ MOL ISOPENTYL ALCOHOL}}{88.2g}\right) = 6.45 \times 10^{-3}$ MOL

ACETIC ACID $\quad 1.40 \text{mL} \left(\dfrac{1.06g}{\text{mL}}\right)\left(\dfrac{1 \text{ MOL ACETIC ACID}}{60.1g}\right) = 2.47 \times 10^{-2}$ MOL

SINCE THEY REACT IN A 1:1 RATIO, ISOPENTYL ALCOHOL IS THE LIMITING REAGENT.

THEORETICAL YIELD =

6.45×10^{-3} MOL ISOPENTYL ALCOHOL $\left(\dfrac{1 \text{ MOL ISOPENTYL ACETATE}}{1 \text{ MOL ISOPENTYL ALCOHOL}}\right)\left(\dfrac{130.2g \text{ ISOPENTYL ACETATE}}{1 \text{ MOL ISOPENTYL ACETATE}}\right)$

$= 0.840g$ ISOPENTYL ACETATE

PERCENTAGE YIELD $= \dfrac{0.354g}{0.840g} \times 100 = 42.1\%$

Figure 2.3
A sample notebook, page 2.

Calculation of the theoretical yield and percentage yield can be illustrated using hypothetical data for the isopentyl acetate preparation:

$$\text{Theoretical yield} = (6.45 \times 10^{-3} \underline{\text{mol isopentyl alcohol}})\left(\frac{1 \text{ mol isopentyl acetate}}{1 \text{ mol isopentyl alcohol}}\right)$$

$$\times \left(\frac{(130.2 \text{ g isopentyl acetate})}{1 \text{ mol isopentyl acetate}}\right) = 0.840 \text{ g isopentyl acetate}$$

$$\text{Actual yield} = 0.354 \text{ g isopentyl acetate}$$

$$\text{Percentage yield} = \frac{0.354 \text{ g}}{0.840 \text{ g}} \times 100 = 42.1\%$$

For experiments that have the principal objective of isolating a substance such as a natural product rather than preparing and purifying some reaction product, the **weight percentage recovery** and not the percentage yield is calculated. This value is determined by

$$\text{Weight percentage recovery} = \frac{\text{Weight of substance isolated}}{\text{Weight of original material}} \times 100$$

Thus, for instance, if 0.014 g of caffeine was obtained from 2.3 g of tea, the weight percentage recovery of caffeine would be

$$\text{Weight percentage recovery} = \frac{0.014 \text{ g Caffeine}}{2.3 \text{ g Tea}} \times 100 = 0.61\%$$

2.3 Laboratory Reports

Various formats for reporting the results of the laboratory experiments may be used. You may write the report directly in your notebook in a format similar to the sample notebook pages included in this section. Alternatively, your instructor may require a more formal report that is not written in your notebook. When you do original research, these reports should include a detailed description of all the experimental steps undertaken. Frequently, the style used in scientific periodicals such as *Journal of the American Chemical Society* is applied to writing laboratory reports. Your instructor is likely to have his or her own requirements for laboratory reports and should describe the requirements to you.

2.4 Submission of Samples

In all preparative experiments and in some isolation experiments, you will be required to submit to your instructor the sample of the substance you prepared or isolated. How this sample is labeled is important. Again, learning a correct method of labeling bottles and vials can save time in the laboratory because fewer mistakes will be made. More important, learning to label properly can decrease the danger inherent in having samples of material that cannot be identified correctly at a later date.

Solid materials should be stored and submitted in containers that permit the substance to be removed easily. For this reason, narrow-mouthed bottles or vials are not used for solid substances. Liquids should be stored in containers that will not let them escape through leakage. Be careful not to store volatile liquids in containers that have plastic caps, unless the cap is lined with an inert material such as Teflon. Otherwise, the vapors from the liquid are likely to contact the plastic and dissolve some of it, thus contaminating the substance being stored.

On the label, print the name of the substance, its melting or boiling point, the actual and percentage yields, and your name. An illustration of a properly prepared label follows:

> **Isopentyl Acetate**
> **BP 140°C**
> **Yield 3.81 g (42.1%)**
> **Joe Schmedlock**

Laboratory Glassware: Care and Cleaning

Because your glassware is expensive and you are responsible for it, you will want to give it proper care and respect. If you read this chapter carefully and follow the procedures presented here, you may be able to avoid some unnecessary expense. You may also save time because cleaning problems and replacing broken glassware are time-consuming.

If you are unfamiliar with the equipment found in an organic laboratory or are uncertain about how such equipment should be treated, this chapter provides some useful information. It includes topics such as cleaning glassware, caring for glassware when using corrosive or caustic reagents, and assembling components from your organic laboratory kit. At the end of this section are illustrations and names of most of the equipment you are likely to find in your drawer or locker.

3.1 Cleaning Glassware

Glassware can be cleaned easily if you clean it immediately. It is good practice to do your "dishwashing" right away. With time, organic tarry materials left in a container begin to attack the surface of the glass. The longer you wait to clean glassware, the more extensively this interaction will have progressed. Cleaning is then more difficult because water will no longer wet the surface of the glass as effectively. If you can't wash your glassware immediately after use, soak the dirty pieces in soapy water. A half-gallon plastic container is convenient for soaking and washing glassware. Using a plastic container also helps prevent the loss of small pieces of equipment used in microscale techniques.

Various soaps and detergents are available for washing glassware. They should be tried first when washing dirty glassware. Organic solvents can also be used because the residue remaining in dirty glassware is likely to be soluble in some organic solvent. After the solvent has been used, the conical vial or flask probably will have to be washed with soap and water to remove the residual solvent. When you use solvents in cleaning glassware, use caution, because the solvents are hazardous (see Technique 1). Use fairly small amounts of a solvent for cleaning purposes. Usually 1–2 mL will be sufficient. Acetone is commonly used, but it is expensive. Your **wash acetone** can be used effectively several times before it is "spent." Once your acetone is spent, dispose of it as your instructor directs. If acetone does not work, other organic solvents such as methylene chloride or toluene can be used.

> **CAUTION** ⚠
>
> **Acetone is very flammable. Do not use it around flames.**

For troublesome stains and residues that adhere to the glass despite your best efforts, use a mixture of sulfuric acid and nitric acid. Cautiously add about 20 drops of concentrated sulfuric acid and 5 drops of concentrated nitric acid to the flask or vial.

> ### CAUTION
>
> **You must wear safety glasses when you are using a cleaning solution made from sulfuric acid and nitric acid. Do not allow the solution to come into contact with you skin or clothing. It will cause severe burns on your skin and create holes in your clothing. The acids may also react with the residue in the container.**

Swirl the acid mixture in the container for a few minutes. If necessary, place the glassware in a warm waterbath and heat cautiously to accelerate the cleaning process. Continue heating until any sign of a reaction ceases. When the cleaning procedure is completed, decant the mixture into an appropriate waste container.

> ### CAUTION
>
> **Do not pour the acid solution into a waste container that is intended for organic wastes.**

Rinse the piece of glassware thoroughly with water and then wash with soap and water. For most common organic chemistry applications, any stains that survive this treatment are not likely to cause difficulty in subsequent laboratory procedures.

If the glassware is contaminated with stopcock grease (unlikely with the glassware recommended in this book), rinse the glassware with a small amount (1–2 mL) of methylene chloride. Discard the rinse solution into an appropriate waste container. Once the grease is removed, wash the glassware with soap or detergent and water.

3.2 Drying Glassware

The easiest way to dry glassware is to let it stand overnight. Store conical vials, flasks, and beakers upside down on a piece of paper towel to permit the water to drain from them. Drying ovens can be used to dry glassware if they are available, and if they are not being used for other purposes. Rapid drying can be achieved by rinsing the glassware with acetone and air-drying it or placing it in an oven. First, thoroughly drain the glassware of water. Then rinse it with one or two *small* portions (1–2 mL) of acetone. Do not use any more acetone than is suggested here. Return the used acetone to a waste acetone container for recycling. After you rinse the glassware with acetone, dry it by placing it in a drying oven for a few minutes or allow it to air-dry at room temperature. The acetone can also be removed by aspirator suction. In some laboratories, it may be possible to dry the glassware by blowing a *gentle* stream of dry air into the container. (Your laboratory instructor will indicate if you should do this.) Before drying the glassware with air, make sure that the air line is not filled with oil. Otherwise, the oil will be blown into the container, and you will have to clean it again. It is not necessary to blast the acetone out of the glassware with a wide-open stream of air; a gentle stream of air is just as effective and will not startle other people in the room.

Do not dry your glassware with a paper towel unless the towel is lint-free. Most paper will leave lint on the glass that can interfere with subsequent procedures performed in the equipment. Sometimes it is not necessary to dry a piece of equipment thoroughly. For example, if you are going to place water or an aqueous solution in a container, it does not need to be completely dry.

3.3 Ground-Glass Joints

It is likely that the glassware in your organic kit has **standard-taper ground-glass joints.** For example, the air condenser in the figure consists of an inner (male) ground-glass joint at the bottom and an outer (female) joint at the top. Each end is ground to a precise size, which is designated by the symbol ⊺ followed by two numbers. A common joint size in microscale glassware is ⊺14/10. The first number indicates the diameter (in millimeters) of the joint at its widest point, and the second number refers to its length (see Figure 3.1). One advantage of standard-taper joints is that the pieces fit together snugly and form a good seal. In addition, standard-taper joints allow all glassware components with the same joint size to be connected, thus permitting the assembly of a wide variety of apparatus. One disadvantage of glassware with ground-glass joints, however, is that it is expensive.

Some pieces of glassware with ground-glass joints also have threads cast into the outside surface of the outer joints (see top of air condenser in Figure 3.1). The threaded joint allows the use of a plastic screw cap with a hole in the top to fasten two pieces of glassware together securely. The plastic cap is slipped over the inner joint of the upper piece of glassware, followed by a rubber O-ring (see Figure 3.2). The O-ring should be pushed down so that it fits snugly on top of the ground-glass joint. The inner ground-glass joint is then fitted into the outer joint of the bottom piece of glassware. The screw cap is tightened without excessive force to attach the entire apparatus firmly together. The O-ring provides an additional seal that makes this joint airtight. With this connecting system, it is unnecessary to use any type of grease to seal the joint. The O-ring *must be used* to obtain a good seal and to lessen the chances of breaking the glassware when you tighten the plastic cap.

It is important to make sure no solid or liquid is on the joint surfaces. Such material will lessen the efficiency of the seal, and the joints may leak. The presence of solid particles could cause the ground-glass joints to break when

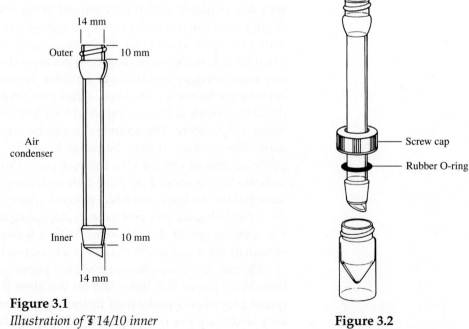

Figure 3.1

Illustration of ⊺14/10 inner and outer joints showing dimensions.

Figure 3.2

A microscale standard-taper joint assembly.

the plastic cap is tightened. Also, if the apparatus is to be heated, material caught between the joint surfaces will increase the tendency for the joints to stick. If the joint surfaces are coated with liquid or adhering solid, you should wipe them with a cloth or lint-free paper towel before assembling.

3.4 Separating Ground-Glass Joints

The most important thing you can do to prevent ground-glass joints from becoming "frozen," or stuck together, is to disassemble the glassware as soon as possible after a procedure is completed. Even when this precaution is followed, ground-glass joints may become stuck tightly together. The same is true of glass stoppers in bottles or conical vials. Because microscale glassware is small and fragile, it is relatively easy to break a piece of glassware when trying to pull two pieces apart. If the pieces do not separate easily, you must be careful when you try to pull them apart. The best way is to hold the two pieces, with your hands touching, as close as possible to the joint. With a firm grasp, try to loosen the joint with a slight twisting motion (do not twist very hard). If this does not work, try to pull your hands apart without pushing sideways on the glassware.

If it is not possible to pull the pieces apart, the following methods may help. A frozen joint can sometimes be loosened if you tap it *gently* with the wooden handle of a spatula. Then, try to pull it apart as already described. If this procedure fails, you may try heating the joint in hot water or a steam bath. If this heating fails, the instructor may be able to advise you. As a last resort, you may try heating the joint in a flame. You should not try this unless the apparatus is hopelessly stuck, because heating by flame often causes the joint to expand rapidly and crack or break. If you use a flame, make sure the joint is clean and dry. Heat the outer part of the joint slowly, in the yellow portion of a low flame, until it expands and breaks away from the inner section. Heat the joint very slowly and carefully or it may break.

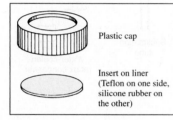

Plastic cap

Insert on liner
(Teflon on one side,
silicone rubber on
the other)

Figure 3.3
Plastic cap and Teflon insert.

3.5 Etching Glassware

Glassware that has been used for reactions involving strong bases such as sodium hydroxide or sodium alkoxides must be cleaned thoroughly *immediately* after use. If these caustic materials are allowed to remain in contact with the glass, they will etch the glass permanently. The etching makes later cleaning more difficult because dirt particles may become trapped within the microscopic surface irregularities of the etched glass. Furthermore, the glass is weakened, so the lifetime of the glassware is shortened. If caustic materials are allowed to come into contact with ground-glass joints without being removed promptly, the joints will become fused, or "frozen." It is extremely difficult to separate fused joints without breaking them.

3.6 Assembling the Apparatus

Care must be taken when assembling the glass components into the desired apparatus. Always remember that Newtonian physics applies to chemical apparatus, and unsecured pieces of glassware are certain to respond to gravity. You should always clamp the glassware securely to a ring stand. Throughout this textbook, the illustrations of the various glassware arrangements include the clamps that attach the apparatus to a ring stand. You should assemble your apparatus using the clamps as shown in the illustrations.

3.7 Capping Conical Vials or Openings

The plastic screw caps used to join two pieces of glassware together can also be used to cap conical vials (see Figure 3.3) or other openings. A Teflon insert, or liner, fits inside the cap to cover the hole when the cap is used to seal a vial. Only one side of the liner is coated with Teflon. This side should always face toward the inside of the vial. (Note that the O-ring is not used when the cap is used to seal a vial.) To seal a vial, it is necessary to tighten the cap firmly but not too tightly. It is possible to crack the vial if you apply too much force. Some Teflon liners have a soft backing materials (silicone rubber) that allows the liner to compress slightly when the cap is screwed down. It is easier to cap a vial securely with these liners without breaking the vial than with liners that have a harder backing materials.

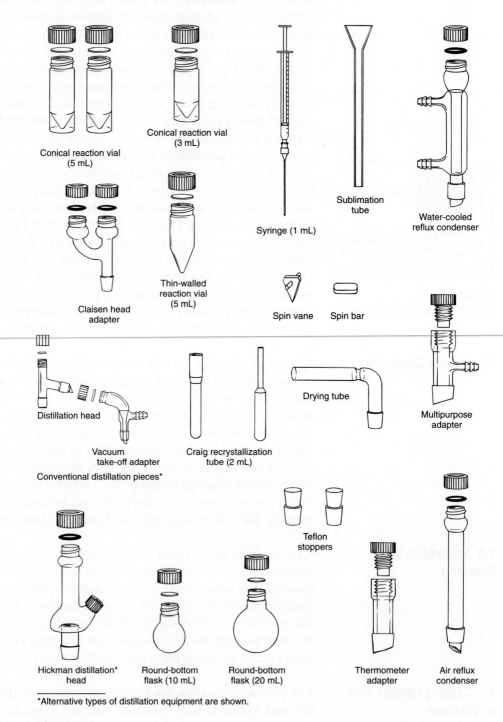

Figure 3.4
Components of a microscale organic kit.

3.8 Attaching Rubber Tubing to Equipment

When you attach rubber tubing to the glass apparatus or when you insert glass tubing into rubber stoppers, first lubricate the rubber tubing or the rubber stopper with either water or glycerin. Without such lubrication, it can be difficult to attach rubber tubing to the side arms of items of glassware such as condensers and filter flasks. Furthermore, glass tubing may break when it is inserted into rubber stoppers. Water is a good lubricant for most purposes. Do not use water as a lubricant when it might contaminate the reaction. Glycerin is a better lubricant than water and should be used when there is considerable friction between the glass and rubber. If glycerin is the lubricant, be careful not to use too much.

3.9 Description of Equipment

The components of the organic kit recommended for use in this textbook are given in Figures 3.4–3.7. Notice that most of the joints in these pieces of glassware are ℑ14/10, and all the outer joints are threaded. The organic kits used in your laboratory may have different joint sizes, or some of the outer joints may not be threaded. In particular, some older organic kits contain a number of pieces of glassware with ℑ7/10 joints. These kits will work as well with the experiments in this book as the glassware recommended in the figures. In addition, there are microscale kits containing glassware that is connected without the use of ground-glass joints. The experiments in this book can also be performed with these glassware kits. Modifications with organic kits not containing the recommended glassware are discussed in the Technique chapters and in some of the experiments.

Figures 3.4–3.7 include glassware and equipment that are commonly used in the organic laboratory. Your glassware and equipment may vary slightly from the pieces shown on this spread and on the following pages.

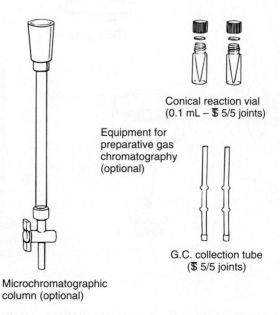

Conical reaction vial
(0.1 mL – ℑ 5/5 joints)

Equipment for
preparative gas
chromatography
(optional)

G.C. collection tube
(ℑ 5/5 joints)

Microchromatographic
column (optional)

Figure 3.5
Optional pieces of microscale glassware. Note:
*The optional pieces of equipment shown in this
figure are not part of the standard microscale kit.
They must be purchased separately.*

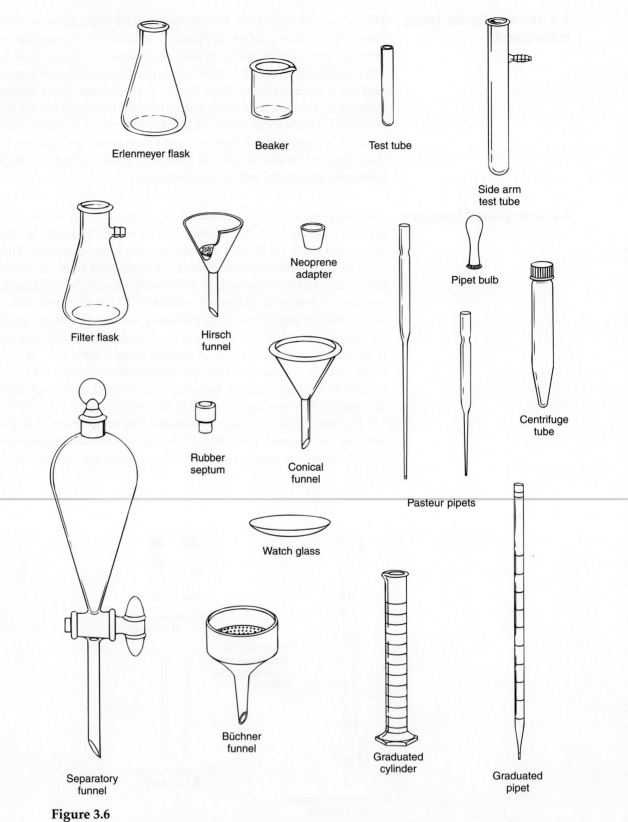

Figure 3.6
Equipment commonly used in the organic laboratory.

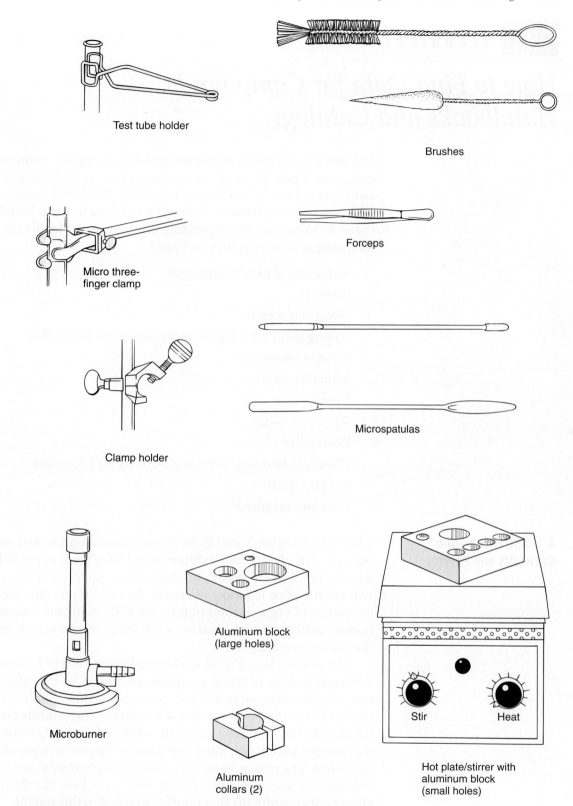

Test tube holder

Brushes

Forceps

Micro three-finger clamp

Microspatulas

Clamp holder

Microburner

Aluminum block
(large holes)

Aluminum
collars (2)

Hot plate/stirrer with
aluminum block
(small holes)

Stir   Heat

Figure 3.7
Equipment commonly used in the organic laboratory.

How to Find Data for Compounds: Handbooks and Catalogs

The best way to find information quickly on organic compounds is to consult a handbook. We will discuss the use of the *CRC Handbook of Chemistry and Physics, Lange's Handbook of Chemistry, The Merck Index,* and the *Aldrich Handbook of Fine Chemicals.* Complete citations to these handbooks are provided in Technique 29. Depending on the type of handbook consulted, the following information may be found:

Name and common synonyms

Formula

Molecular weight

Boiling point for a liquid or melting point for a solid

Beilstein reference

Solubility data

Density

Refractive index

Flash point

Chemical Abstracts Service (CAS) Registry Number

Toxicity data

Uses and synthesis

4.1 CRC Handbook of Chemistry and Physics

This is the handbook that is most often consulted for data on organic compounds. Although a new edition of the handbook is published each year, the changes that are made are often minor. An older copy of the handbook will often suffice for most purposes. In addition to the extensive tables of properties of organic compounds, the *CRC Handbook* includes sections on nomenclature and ring structures, an index of synonyms, and an index of molecular formulas.

The nomenclature used in this book most closely follows the Chemical Abstracts system of naming organic compounds. This system differs, but only slightly, from standard IUPAC nomenclature. Table 4.1 lists some examples of how some commonly encountered compounds are named in this handbook. The first thing you will notice is that this handbook is not like a dictionary. Instead, you must first identify the *parent* name of the compound of interest. The parent names are found in alphabetical order. Once the parent name is identified and found, then you look for the particular substituent or substituents that may be attached to this parent.

For most compounds, it is easy to find what you are looking for as long as you know the parent name. Alcohols are, as expected, named by IUPAC nomenclature. Notice in Table 4.1 that the branched-chain alcohol, isopentyl alcohol, is listed as 1-butanol, 3-methyl.

Esters, amides, and acid halides are usually named as derivatives of the parent carboxylic acid. Thus, in Table 4.1, you find ethyl propanoate listed

Table 4.1 *Examples of names of compounds in the* CRC Handbook

Name of Organic Compound	Location in CRC Handbook
1-Chloropentane	Pentane, 1-chloro-
1,4-Dichlorobenzene	Benzene, 1,4-dichloro-
4-Chlorotoluene	Benzene, 1-chloro-4-methyl-
Ethanoic acid	Acetic acid
tert-Butyl acetate (ethanoate)	Acetic acid, 1,1-dimethylethyl ester
Ethyl propanoate	Propanoic acid, ethyl ester
Isopentyl alcohol	1-Butanol, 3-methyl-
Isopentyl acetate (banana oil)	1-Butanol, 3-methyl-, acetate
Salicylic acid	Benzoic acid, 2-hydroxy-
Acetylsalicylic acid (aspirin)	Benzoic acid, 2-acetyloxy-

under the parent carboxylic acid, propanoic acid. If you have trouble finding a particular ester under the parent carboxylic acid, try looking under the alcohol part of the name. For example, isopentyl acetate is not listed under acetic acid, as expected, but instead is found under the alcohol part of the name (see Table 4.1). Fortunately, this handbook has a Synonym Index that nicely locates isopentyl acetate for you in the main part of the handbook.

Once you locate the compound by its name, you will find the following useful information:

CRC number	This is an identification number for the compound. You can use this number to find the molecular structure located elsewhere in the handbook. This is especially useful when the compound has a complicated structure.
Name and synonym	The Chemical Abstracts name and possible synonyms.
Mol. form.	Molecular formula for the compound.
Mol. wt.	Molecular weight.
CAS RN	Chemical Abstracts Service Registry Number. This number is useful for locating additional information on the compound in the primary chemical literature (see Technique 29, Section 29.11).
mp/°C	Melting point of the compound in degrees Celsius.
bp/°C	Boiling point of the compound in degrees Celsius. A number without a superscript indicates that the recorded boiling point was obtained at 760 mm Hg pressure (atmospheric pressure). A number with a superscript indicates that the boiling point was obtained at reduced pressure. For example, an entry of 234; 122^{16} would indicate that the compound boils at 234°C at 760 mm Hg and 122°C at 16 mm Hg pressure.
Den/g cm^{-3}	Density of a liquid. A superscript indicates the temperature in degrees Celsius at which the density was obtained.

n_D		Refractive index determined at a wavelength of 589 nm, the yellow line in a sodium lamp (D line). A superscript indicates the temperature at which the refractive index was obtained (see Technique 24).
Solubility	Solubility classification	Solvent abbreviations

Solubility classification	Solvent abbreviations
1 = insoluble	ace = acetone
2 = slightly soluble	bz = benzene
3 = soluble	chl = chloroform
4 = very soluble	EtOH = ethanol
5 = miscible	eth = ether
6 = decomposes	hx = hexane

Beil. ref.	*Beilstein* reference. An entry of 4-02-00-00157 would indicate that the compound is found in the fourth supplement in Volume 2, with no subvolume, on page 157 (see Technique 29, Section 29.10 for details on the use of *Beilstein*).
Merck No.	*Merck Index* number in the 11th edition of the handbook. These numbers change each time a new edition of *The Merck Index* is issued.

Examples of sample handbook entries for isopentyl alcohol (1-butanol, 3-methyl) and isopentyl acetate (1-butanol, 3-methyl, acetate) are shown in Table 4.2.

4.2 Lange's Handbook of Chemistry

This handbook tends not to be as available as the *CRC Handbook,* but it has some interesting differences and advantages. *Lange's Handbook* has synonyms listed at the bottom of each page, along with structures of more complicated molecules. The most noticeable difference is in how compounds are named. For many compounds, the system lists names as they would appear in a dictionary. Table 4.3 lists examples of how some commonly encountered compounds are named in this handbook. Most often, you do not need to identify the *parent* name. Unfortunately, *Lange's Handbook* frequently uses common names that are becoming obsolete. For example, propionate is used rather than propanoate. Nevertheless, this handbook often names compounds as a practicing organic chemist would tend to name them. Notice how easy it is to find the entries for isopentyl acetate and acetylsalicylic acid (aspirin) in this handbook.

Table 4.2 *Properties of isopentyl alcohol and isopentyl acetate as listed in the* CRC Handbook

No.	Name Synonym	Mol. Form. Mol. Wt.	CAS RN mp/°C	Merck No. bp/°C	Beil. Ref. den/g cm^{-3}	Solubility n_D
3627	1-Butanol, 3-methyl	$C_5H_{12}O$	123-51-3	5081	4-01-00-01677	ace 4; eth 4; EtOH 4
	Isopentyl alcohol	88.15	−117.2	131.1	0.8104[20]	1.4053[20]
3631	1-Butanol, 3-methyl, acetate	$C_7H_{14}O_2$	123-92-2	4993	4-02-00-00157	H_2O 2; EtOH 5; eth 5; ace 3
	Isopentyl acetate	130.19	−78.5	142.5	0.876[15]	1.4000[20]

Table 4.3 *Examples of names of compounds in* Lange's Handbook

Name of Organic Compound	*Location in* Lange's Handbook
1-Chloropentane	1-Chloropentane
1,4-Dichlorobenzene	1,4-Dichlorobenzene
4-Chlorotoluene	4-Chlorotoluene
Ethanoic acid	Acetic acid
tert-Butyl acetate (ethanoate)	*tert*-Butyl acetate
Ethyl propanoate	Ethyl propionate
Isopentyl alcohol	3-Methyl-1-butanol
Isopentyl acetate (banana oil)	Isopentyl acetate
Salicylic acid	2-Hydroxybenzoic acid
Acetylsalicylic acid (aspirin)	Acetylsalicylic acid

Once you locate the compound by its name, you will find the following useful information:

Lange's number	This is an identification number for the compound.
Name	See examples in Table 4.3.
Formula	Structures are drawn out. If they are complicated, then the structures are shown at the bottom of the page.
Formula weight	Molecular weight of the compound.
Beilstein reference	An entry of 2, 132 would indicate that the compound is found in Volume 2 of the main work on page 132. An entry of 3^2, 188 would indicate that the compound is found in Volume 3 of the second supplement on page 188 (see Technique 29, Section 29.10 for details on the use of *Beilstein*).
Density	Density is usually expressed in units of g/mL or g/cm^3. A superscript indicates the temperature at which the density was measured. If the density is also subscripted, usually 4°, it indicates that the density was measured at a certain temperature relative to water at its maximum density, 4°C. Most of the time you can simply ignore the subscripts and superscripts.
Refractive index	A superscript indicates the temperature at which the refractive index was determined (see Technique 24).
Melting point	Melting point of the compound in degrees Celsius. When a "d" or "dec" appears with the melting point, it indicates that the compound decomposes at the melting point. When decomposition occurs, you will often observe a change in color of the solid.
Boiling point	Boiling point of the compound in degrees Celsius. A number without a superscript indicates that the recorded boiling point was obtained at 760 mm Hg pressure (atmospheric pressure). A number with a

superscript indicates that the boiling point was obtained at reduced pressure. For example, an entry of $102^{11 \text{ mm}}$ would indicate that the compound boils at 102°C at 11 mm Hg pressure.

Flash point

This number is the temperature in degrees Celsius at which the compound will ignite when heated in air and a spark is introduced into the vapor. There are a number of different methods that are used to measure this value, so this number varies considerably. It gives a crude indication of flammability. You may need this information when heating a substance with a hot plate. Hot plates can be a serious source of trouble because of the sparking action that can occur with switches and thermostats used in hot plates.

Solubility in 100 parts solvent

Parts by weight of a compound that can be dissolved in 100 parts by weight of solvent at room temperature. In some cases, the values given are expressed as the weight in grams that can be dissolved in 100 mL of solvent. This handbook is not consistent in describing solubility. Sometimes gram amounts are provided, but in other cases the description will be more vague, using terms such as *soluble*, *insoluble*, or *slightly soluble*.

Solvent abbreviations	Solubility characteristics
acet = acetone	i = insoluble
bz = benzene	s = soluble
chl = chloroform	sls = slightly soluble
aq = water	vs = very soluble
alc = ethanol	misc = miscible
eth = ether	
HOAc = acetic acid	

Examples of sample handbook entries for isopentyl alcohol (3-methyl-1-butanol) and isopentyl acetate are shown in Table 4.4.

4.3 The Merck Index

The Merck Index is a useful book because it has additional information not found in the other two handbooks. This handbook, however, tends to emphasize medicinally related compounds, such as drugs and biological compounds, although it also lists many other common organic compounds. It is not revised each year; new editions are published in five- or six-year cycles. It does not contain all of the compounds listed in *Lange's Handbook* or the

Table 4.4 *Properties of 3-methyl-1-butanol and isopentyl acetate as listed in* Lange's Handbook

No.	Name	Formula	Formula Weight	Beilstein Reference	Density	Refractive Index	Melting Point	Boiling Point	Flash Point	Solubility in 100 Parts Solvent
m155	3-methyl-1-butanol	$(CH_3)_2CHCH_2CH_2OH$	88.15	1, 392	0.8129^{15}_4	1.4085^{15}	−117.2	132.0	45	2 aq; misc alc, bz, chl, eth, HOAc
i80	Isopentyl acetate	$CH_3COOCH_2CH_2CH(CH_3)_2$	130.19	2, 132	0.876^{15}_4	1.4007^{20}	−78.5	142.0	80	0.25 aq; misc alc, eth

CRC Handbook. However, for the compounds listed, it provides a wealth of useful information. The handbook will provide you with some or all of the following data for each entry.

Merck number, which changes each time a new edition is issued

Name, including synonyms and stereochemical designation

Molecular formula and structure

Molecular weight

Percentages of each of the elements in the compound

Uses

Source and synthesis, including references to the primary literature

Optical rotation for chiral molecules

Density, boiling point, and melting point

Solubility characteristics, including crystalline form

Pharmacology information

Toxicity data

One of the problems with looking up a compound in this handbook is trying to decide the name under which the compound will be listed. For example, isopentyl alcohol can also be named as 3-methyl-1-butanol or isoamyl alcohol. In the 12th edition of the handbook, it is listed under the name isopentyl alcohol (#5212) on page 886. Finding isopentyl acetate is even a more challenging task. It is located in the handbook under the name isoamyl acetate (#5125) on page 876. Often, it is easier to look up the name in the name index or to find it in the formula index.

The handbook has some useful appendices that include the CAS registry numbers, a biological activity index, a formula index, and a name index that also includes synonyms. When looking up a compound in one of the indexes, you need to remember that the numbers provided are compound numbers, rather than page numbers. There is also a useful section on organic name reactions that includes references to the primary literature.

4.4 Aldrich Handbook of Fine Chemicals

The *Aldrich Handbook* is actually a catalog of chemicals sold by the Aldrich Chemical Company. The company includes in its catalog a large body of useful data on each compound that it sells. Because the catalog is reissued each year at no cost to the user, you should be able to find an old copy when the new one is issued. As you are mainly interested in the data on a particular compound and not the price, an old volume is perfectly fine. Isopentyl alcohol is listed as 3-methyl-1-butanol, and isopentyl acetate is listed as isoamyl acetate in the *Aldrich Handbook.* The following are some of the properties and information listed for individual compounds.

Aldrich catalog number

Name: Aldrich uses a mixture of common and IUPAC names. It takes a bit of time to master the names. Fortunately, the catalog does a good job of cross-referencing compounds and has a very good molecular formula index.

CAS Registry Number

Structure

Synonym

Formula weight

Boiling point/melting point

Index of refraction

Density

Beilstein reference

Merck reference

Infrared spectrum reference to the Aldrich Library of FT-IR spectra

NMR spectrum reference to the Aldrich Library of ^{13}C and ^{1}H FT-NMR spectra

Literature references to the primary literature on the uses of the compound

Toxicity

Safety data and precautions

Flash point

Prices of chemicals

4.5 Strategy for Finding Information: Summary

Most students and professors find *The Merck Index* and *Lange's Handbook* easier and more "intuitive" to use than the *CRC Handbook*. You can go directly to a compound without rearranging the name according to the parent or base name followed by its substituents. Another great source of information is the *Aldrich Handbook,* which contains those compounds that are easily available from a commercial source. Many compounds are found in the *Aldrich Handbook* that you may never find in any of the other handbooks. The Sigma–Aldrich Web site (http://www.sigmaaldrich.com/) allows you to search by name, synonym, and catalog number.

PROBLEMS

1. Using *The Merck Index,* find and draw structures for the following compounds:

   a. atropine

   b. quinine

   c. saccharin

   d. benzo[*a*]pyrene (benzpyrene)

   e. itaconic acid

   f. adrenosterone

   g. chrysanthemic acid (chrysanthemumic acid)

   h. cholesterol

   i. Vitamin C (ascorbic acid)

2. Find the melting points for the following compounds in the *CRC Handbook, Lange's Handbook,* or the *Aldrich Handbook:*

   a. biphenyl

   b. 4-bromobenzoic acid

   c. 3-nitrophenol

3. Find the boiling point for each compound in the references listed in problem 2:

   a. octanoic acid at reduced pressure

   b. 4-chloroacetophenone at atmosphere and reduced pressure

   c. 2-methyl-2-heptanol

4. Find the index of refraction n_D and density for the liquids listed in problem 3.

5. Using the *Aldrich Handbook,* report the specific rotations for the enantiomers of camphor.

6. Read the section on carbon tetrachloride in *The Merck Index* and list some of the health hazards for this compound.

5   TECHNIQUE 5

Measurement of Volume and Weight

Special care must be taken when working with small amounts of liquid or solids. In the typical microscale experiment, a student will use from 10 to 1000 mg of a liquid or solid. Specially designed microscale equipment will be used for these small-scale reactions. You may not be used to working with such small quantities, but after a while you will adjust to "thinking small."

Liquids to be used for an experiment will usually be found in small containers in a hood. For experiments in this book, an automatic pipet, dispensing pump, or calibrated pipet will be used for measuring the volume of a liquid. It is critical that **limiting reactants** be weighed for accuracy purposes. *Do not calculate the weight using densities!* Measurement of a small volume of a liquid is subject to a large experimental error when converted to a weight using the density of a liquid. To determine the weight of a liquid when dealing with limiting reactants, preweigh the container before adding the liquid to the container and then reweigh the container after adding the liquid. This gives an **exact weight** and avoids the experimental error involved in using densities to calculate weights when working with smaller amounts of a liquid. For **nonlimiting** liquid reactants, you may calculate the weight of the liquid from the volume you have delivered using the density of the liquid and the following equation:

$$\textbf{Weight (g) = density (g/mL)} \times \textbf{volume (mL)}$$

Solids may be found near the balance. When an accurate measurement is required, solids must be weighed on a balance that reads to the nearest milligram (0.001 g) or tenth of a milligram (0.0001 g). To weigh a solid, place your conical vial or round-bottom flask in a small beaker and take these with you to the balance. Place a piece of paper that has been folded once on the balance pan. The folded paper will enable you to pour the solid into the conical vial or flask without spilling. Use the larger of your two spatulas (p. 49) to aid the transfer of the solid to the paper. Never weigh directly into a conical vial or flask and never pour, dump, or shake a material from a bottle. While still at the balance, carefully transfer the solid from the paper to your vial or flask. The vial or flask should be in a beaker while transferring the solid. The beaker traps any material that fails to make it into the container. It also supports the vial or flask so that it does not fall over. It is not necessary to obtain the exact amount specified in the experimental procedure, and trying to be exact requires too much time at the balance. For example, if you obtained 0.140 g of a solid, rather than the 0.136 g specified in a procedure, you could use it, but the actual amount weighed should be recorded in your notebook. Use the amount you weighed to calculate the theoretical yield, if this solid is the limiting agent.

Careless dispensing of liquids and solids is a hazard in any laboratory. When reagents are spilled, you may be subjected to an unnecessary health or fire hazard. In addition, you may waste expensive chemicals, destroy balance pans and clothing, and damage the environment. Always clean up any spills immediately.

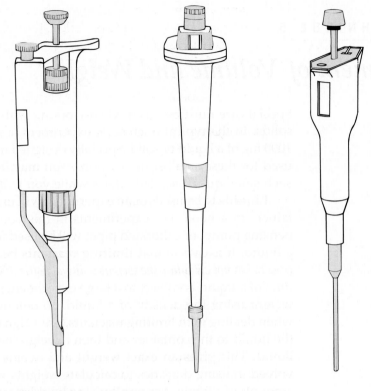

Figure 5.1
The adjustable automatic pipet.

5.1 Automatic Pipets

When available, an automatic pipet increases the speed of transfer of liquids from reagent bottles. These pipets are expensive and must be shared by the entire laboratory. A number of types of units are available commercially. We describe the use of the continuously adjustable automatic pipet. This type of pipet can be adjusted for any volume within its defined range using a three- or four-digit readout. Several types of adjustable automatic pipets are shown in Figure 5.1. The typical laboratory may have several units available: one 10–100 μL (0.01–0.10 mL) pipet for smaller volumes, and two 100–1000 μL (0.10–1.00 mL) pipets for larger volumes. Disposable tips are available for each of these units and are color coded: yellow and blue for the small and large units, respectively. The automatic pipet is accurate with aqueous solutions, but it is not as accurate with organic liquids.

In most cases, the instructor will adjust the pipet so that it will deliver the desired volume. It will be placed in a convenient location near the reagent bottle, usually in a hood, and students will reuse the tip. Your instructor will give directions for the correct use of the automatic pipet. Students must practice using the automatic pipet by following the instructions given on pages 7–8. Remember that the automatic pipet is expensive and must be handled carefully. To protect the unit, you must always use a tip on the end of the pipet. Liquid must be drawn only into this plastic tip and never up into the unit itself. If this happens, you should notify your laboratory instructor immediately. Keep the pipet upright and immerse the tip just below the surface of the liquid. Automatic pipets should never be used with corrosive liquids, such as sulfuric acid or hydrochloric acid.

5.2 Dispensing Pumps

Dispensing pumps may be used in place of automatic pipets when larger amounts (more than 0.1 mL) of liquids are being dispensed in the laboratory.

Figure 5.2
Dispensing pump.

The pumps are simple to operate, chemically inert, and accurate. Because the plunger assembly is made of Teflon, the dispensing pump may be used with most corrosive liquids and organic solvents. Dispensing pumps come in a variety of sizes, but the 1-, 2-, and 5-mL sizes are most useful in the microscale organic laboratory. The pump is attached to a bottle containing the liquid being dispensed. The liquid is drawn up from this reservoir into the pump assembly through a piece of inert plastic tubing.

Dispensing pumps are somewhat more difficult to adjust to the proper volume than automatic pipets. Normally, the instructor or assistant will carefully adjust the unit to deliver the proper amount of liquid. As shown in Figure 5.2, the plunger is pulled up as far as it will travel to draw in the liquid from the glass reservoir. To expel the liquid from the spout into a container, slowly guide the plunger down. With low-viscosity liquids, the weight of the plunger will expel the liquid. With more viscous liquids, however, you may need to push the plunger gently to deliver the liquid into a container. Remove the last drop of liquid on the end of the spout by touching the tip on the interior wall of the container. When the liquid being transferred is a limiting reagent or when you need to know the weight precisely, you should weigh the liquid to determine the amount accurately.

As you pull up the plunger, look to see if the liquid is being drawn up into the pump unit. Some volatile liquids may not be drawn up in the expected manner, and you will observe an air bubble. Air bubbles are commonly observed when the pump has not been used for a while. The air bubble can be removed from the pump by dispensing and discarding several volumes of liquid to "reprime" the dispensing pump. Also check to see if the spout is filled completely with liquid. An accurate volume will not be dispensed unless the spout is filled with liquid before you lift up the plunger.

5.3 Graduated Pipets

A suitable alternative to an automatic pipet or a dispensing pump is the graduated serological pipet. These *glass* pipets are available commercially in a number of sizes. "Disposable" pipets may be used many times and discarded only when the graduations become too faint to be seen. A good assortment of these pipets consists of the following:

0.50-mL pipets calibrated in 0.01-mL divisions (5/10 in 1/100 mL)

1.00-mL pipets calibrated in 0.01-mL divisions (1 in 1/100 mL)

2.00-mL pipets calibrated in 0.01-mL divisions (2 in 1/100 mL)

Liquids may be measured and transferred using a graduated pipet and a pipet pump. The style of pipet pump shown in Figure 5.3A is available in four sizes. The 2-mL size (blue) works well with the range of pipets previously indicated. To fill the pipet, one simply rotates the knurled wheel forward so that the piston moves upward. The liquid is discharged by slowly turning the wheel backward until the proper amount of liquid has been expelled. The top of the pipet must be inserted securely into the pump and held there with one hand to obtain an adequate seal. The other hand is used to load and release the liquid.

The pipet pump shown in Figure 5.3B may also be used with graduated pipets. The knob is turned counterclockwise to draw in the liquid, and then the liquid is released by turning the knob clockwise. With this style of pipet, the top of the pipet is held securely by a rubber O-ring, and it is easily handled with one hand. You should be certain that the pipet is held securely by

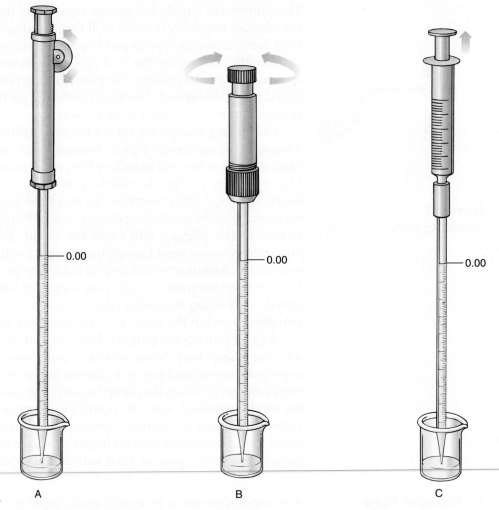

Figure 5.3
Pipet pumps.

the O-ring before using it. Disposable pipets may not fit tightly in the O-ring, because they often have smaller diameters than nondisposable pipets.

A syringe may be used as a pipet pump, as shown in Figure 5.3C. In the design shown here, a 1- or 2-mL syringe is attached to the graduated pipet using a short piece of plastic tubing. The liquid is drawn up into the pipet when the plunger is pulled up, and it is expelled when the plunger is pushed down.

Excellent results may be obtained with graduated pipets if you transfer by difference between marked calibrations and avoid transferring the entire contents of the pipet. When expelling the liquid, be sure to touch the tip of the pipet to the inside of the container before withdrawing the pipet. Graduated pipets are commonly used when dispensing corrosive liquids, such as sulfuric acid or hydrochloric acid. The pipet will be supplied with a bulb or pipet pump.

Pipets may be obtained in a number of styles, but only three types will be described here (Figure 5.4). One type of graduated pipet is calibrated "to deliver" (TD) its total capacity when the last drop is blown out. This style of pipet, shown in Figure 5.4A is probably the most common type of graduated pipet in use in the laboratory; it is designated by two rings at the top.

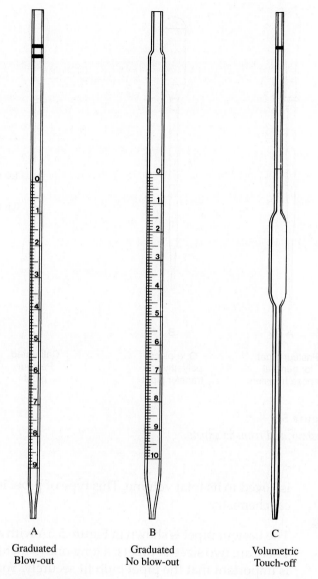

A
Graduated
Blow-out

B
Graduated
No blow-out

C
Volumetric
Touch-off

Figure 5.4
Pipets.

Of course, one does not need to transfer the entire volume to a container. To deliver a more accurate volume, you should transfer an amount less than the total capacity of the pipet using the graduations on the pipet as a guide.

Another type of graduated pipet is shown in Figure 5.4B. This pipet is calibrated to deliver its total capacity when the meniscus is located on the last graduation mark near the bottom of the pipet. For example, the pipet shown in the Figure 5.4B delivers 10.0 mL of liquid when it has been drained to the point where the meniscus is located on the 10.0-mL mark. With this type of pipet, you must not drain the entire pipet or blow it out. In contrast, notice that the pipet shown in Figure 5.4A has its last graduation at 0.90 mL. The last 0.10-mL volume is blown out to give the 1.00-mL volume.

A nongraduated volumetric pipet is shown in Figure 5.4C. It is easily identified by the large bulb in the center of the pipet. This pipet is calibrated so that it will retain its last drop after the tip is touched on the side of the container. It must not be blown out. These pipets often have a single colored band at the top that identifies it as a "touch-off" pipet. The color of the band

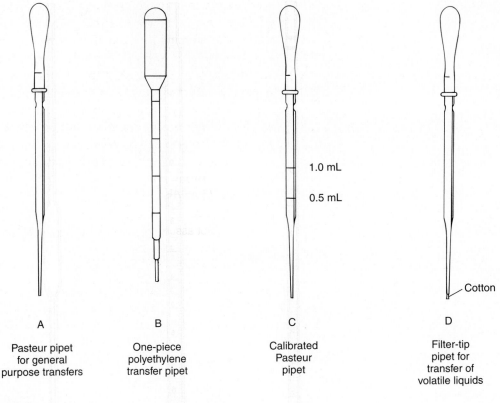

Figure 5.5
Pasteur and transfer pipets.

is keyed to its total volume. This type of pipet is commonly used in analytical chemistry.

5.4 Pasteur Pipets

The Pasteur pipet is shown in Figure 5.5A with a 2-mL rubber bulb attached. There are two sizes of pipets: a long one (9 inch) and a short one (5¾ inch). It is important that the pipet bulb fit securely. You should not use a medicine dropper bulb, because of its small capacity. A Pasteur pipet is an indispensable piece of equipment for the routine transfer of liquids. It is also used for separations (Technique 12). Pasteur pipets may be packed with cotton for use in gravity filtration (Technique 8) or packed with an adsorbent for small-scale column chromatography (Technique 19). Although they are considered disposable, you should be able to clean them for reuse as long as the tip remains unchipped.

A Pasteur pipet may be supplied by your instructor for dropwise addition of a particular reagent to a reaction mixture. For example, concentrated sulfuric acid is often dispensed in this way. When sulfuric acid is transferred, take care to avoid getting the acid into the rubber or latex dropper bulb. It is best to avoid the rubber dropper bulb entirely by using one-piece transfer pipets made entirely of polyethylene. These plastic pipets are available in 1- or 2-mL sizes. They come from the manufacturers with approximate calibration marks stamped on them (Fig. 5.5B).

Pipets may be calibrated for use in operations where the volume does not need to be known precisely. Examples include measurement of solvents needed for extraction and for washing a solid obtained following crystallization. It is suggested that you calibrate several 5¾-inch pipets following the procedure given on page 11. A calibrated Pasteur pipet is shown in Figure 5.5C. Your instructor may provide you with a calibrated Pasteur pipet and bulb for transferring liquids where an accurate volume is not required. The pipet may be used to transfer a volume of 1.5 mL or less. You may find that the instructor has taped a test tube to the side of the storage bottle. The pipet is stored in the test tube with that particular reagent.

In general, Pasteur pipets should not be used to measure volumes of reagents needed for organic reactions, because they are not accurate enough for this purpose. In some cases, however, your instructor may have available a calibrated pipet for transferring nonlimiting reagents that may damage an automatic pipet. For example, a calibrated Pasteur pipet may be used with concentrated acids.

NOTE: You should not assume that a certain number of drops equals a 1-mL volume. The common rule that 20 drops equal 1 mL does not hold true for a Pasteur pipet!

A Pasteur pipet may be packed with cotton to create a filter-tip pipet as shown in Figure 5.5D. This pipet is prepared by the instructions given in Technique 8, Section 8.6, page 625. Pipets of this type are useful in transferring volatile solvents during extractions and in filtering small amounts of solid impurities from solutions.

5.5 Syringes

Syringes may be used to add a pure liquid or a solution to a reaction mixture. They are especially useful when anhydrous conditions must be maintained. The needle is inserted through a septum, and the liquid is added to the reaction mixture. Although syringes come in a number of sizes, we will use a 1-mL unit in this textbook. Caution should be used with disposable syringes because they often use solvent-soluble rubber gaskets on the plungers. A syringe should be cleaned carefully after each use by drawing acetone or another volatile solvent into it and expelling the solvent with the plunger. Repeat this procedure several times to clean the syringe thoroughly. Draw air through the barrel with an aspirator to dry the syringe.

Syringes are usually supplied with volume graduations inscribed on the barrel. Large-volume syringes are not accurate enough to be used for measuring liquids in microscale experiments. A small microliter syringe, however, such as that used in gas chromatography, delivers a precise volume.

5.6 Graduated Cylinders

Graduated cylinders are used to measure relatively large volumes of liquids where accuracy is not required. For example, you could use a 10-mL graduated cylinder to obtain about 2 mL of a solvent for a crystallization procedure. You should use an automatic pipet, dispensing pump, or a graduated pipet for accurate transfer of liquids in microscale work. Use a *clean and dry* Pasteur pipet to transfer the liquid from the storage container into the graduated cylinder. Do not attempt to pour the liquid directly into the cylinder from the storage bottle or you may spill the fluid. Some instructors may want you to pour some of the liquid into a beaker first and then use a Pasteur pipet to transfer the liquid to a graduated cylinder. Remember that you should not take more than you need. Excess material should never be returned to the storage bottle. Unless you can convince someone else to take it, it must be poured into the appropriate waste container. You should be frugal in estimating amounts needed.

5.7 Measuring Volumes with Conical Vials, Beakers, and Erlenmeyer Flasks

Conical vials, beakers, and Erlenmeyer flasks all have graduations inscribed on them. Beakers and flasks can be used to give only a crude approximation of the volume. They are much less precise than graduated cylinders for measuring volume. In some cases, a conical vial may be used to estimate volumes. For example, the graduations are sufficiently accurate for measuring a solvent needed to wash a solid obtained on a Hirsch funnel after a crystallization. You should use an automatic pipet, dispensing pump, or graduated transfer pipet for accurate measurement of liquids.

5.8 Balances

Solids and some liquids will need to be weighed on a balance that reads to at least the nearest milligram (0.001 g). A top-loading balance (see Fig. 5.6) works well if the balance pan is covered with a plastic draft shield. The shield has a flap that opens to allow access to the balance pan. An analytical balance (see Fig. 5.7) may also be used. This type of balance will weigh to the nearest tenth of a milligram (0.0001 g) when provided with a glass draft shield.

Modern electronic balances have a tare device that automatically subtracts the weight of a container or a piece of paper from the combined weight to give the weight of the sample. With solids, it is easy to place a piece of paper on the balance pan, press the tare device so that the paper appears to have zero weight, and then add your solid until the balance gives the weight you desire. You can then transfer the weighed solid to a container. You should always use a spatula to transfer a solid and never pour material from a bottle. In addition, solids must be weighed on paper and not directly on the balance pan. Remember to clean any spills.

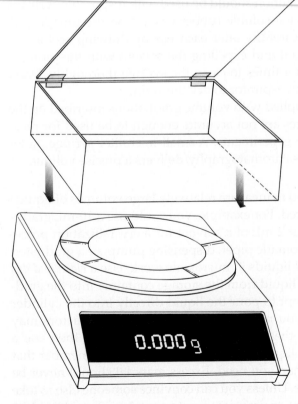

Figure 5.6
A top-loading balance with plastic draft shield.

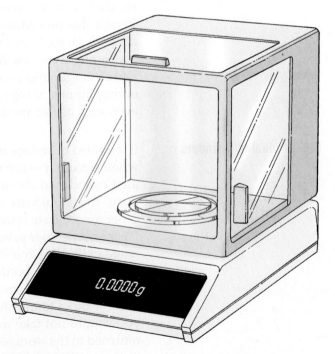

Figure 5.7
An analytical balance with glass draft shield.

With liquids, you should weigh the conical vial to determine the tare weight, transfer the liquid with an automatic pipet, dispensing pump, or graduated pipet into the vial, and then reweigh it. With liquids, it is usually necessary to weigh only the limiting reagent. The other liquids may be transferred using an automatic pipet, dispensing pump, or graduated pipet. Their weights can be calculated by knowing the volumes and densities of the liquids.

PROBLEMS

1. What measuring device would you use to measure the volume under each of the conditions described below? In some cases, there may be more than one answer to the question.

   a. 5 mL of a solvent needed for a crystallization

   b. 0.76 mL of a liquid needed for a reaction

   c. 1 mL of a solvent needed for an extraction

2. Assume that the liquid used in part (b) is a limiting reagent for a reaction. What should you do after measuring the volume?

3. Calculate the weight of a 0.25-mL sample of each of the following liquids:

   a. Diethyl ether (ether)

   b. Methylene chloride (dichloromethane)

   c. Acetone

4. A laboratory procedure calls for 0.146 g of acetic anhydride. Calculate the volume of this reagent needed in the reaction.

5. Criticize the following techniques:

   a. A 100-mL graduated cylinder is used to measure accurately a volume of 2.8 mL.

   b. A one-piece polyethylene transfer pipet (Figure 5.6B) is used to transfer precisely 0.75 mL of a liquid that is being used as the limiting reactant.

   c. A calibrated Pasteur pipet (Figure 5.6C) is used to transfer 25 mL of a solvent.

   d. The volume markings on a 100-mL beaker are used to transfer accurately 5 mL of a liquid.

   e. An automatic pipet is used to transfer 10 mL of a liquid.

   f. A graduated cylinder is used to transfer 0.126 mL of a liquid.

   g. For a small-scale reaction, the weight of a liquid limiting reactant is calculated from its density and volume.

6   TECHNIQUE 6

Heating and Cooling Methods

Most organic reaction mixtures need to be heated in order to complete the reaction. In general chemistry, you may have used a Bunsen burner for heating because nonflammable aqueous solutions were used. In an organic laboratory, however, the student must heat nonaqueous solutions that may contain *highly flammable* solvents. You *should not heat organic mixtures with a Bunsen burner* unless you are directed by your laboratory instructor. Open flames present a fire hazard. Whenever possible you should use one of the alternative heating methods, as described in the following sections.

6.1 Aluminum Block with Hot Plate/Stirrer

Most microscale organic laboratories now use an aluminum block and a hot plate, rather than a sand bath, for heating conical vials or flasks. There are several advantages to heating with an aluminum block. First, the metal will heat faster than a sand bath. Second, you can obtain a higher temperature with an aluminum block. Higher temperatures are often needed when distilling liquids with high boiling points at atmospheric pressure or under vacuum. Third, you can cool the aluminum rapidly by removing it with crucible tongs and immersing it in cold water.

Aluminum heating blocks can be fabricated readily in a machine shop or purchased from commercial suppliers.[1] The two aluminum blocks shown in Figure 6.1 will handle most heating applications in the laboratory. The block with the smaller holes will hold conical vials (Fig. 6.1A). Holes have been drilled in the block so that different-sized conical vials will fit into the holes. This aluminum block may also be used in crystallizations using a Craig tube (Techniques 8 and 11). A hole is often provided for a mercury thermometer, but we do not recommend using it (see the caution box that follows). The aluminum block with the larger holes, as shown in Figure 6.1B, is designed to hold 10-, 20-, or 25-mL round-bottom flasks, as well as a thermometer.

Figure 6.2 shows a reaction mixture being heated with an aluminum block on a hot plate/stirrer unit. Also shown in Figure 6.2 is a split aluminum collar that may be used when very high temperatures are required. The collar is split to facilitate easy placement around a 5-mL conical vial. The collar helps distribute heat farther up the wall of the vial.

Because some hot plates vary widely in the temperature achieved for a given dial setting, some instructors may ask you to calibrate the hot plate so you have an approximate idea where to set the control on the hot plate to achieve a desired temperature. Place an aluminum block on the hot plate and insert a nonmercury thermometer into the small hole in the block, as shown in Figure 6.2 (without the glassware). *Make sure the thermometer fits loosely in the hole or it may break.* Secure the thermometer with a clamp. Select five equally spaced settings on the heating control of the hot plate. Set the dial to the first of these settings and monitor the temperature recorded on the thermometer. When the thermometer reading arrives at a constant value, record this final temperature, along with the dial setting, in your notebook. Repeat this procedure with the remaining four settings and record the temperatures corresponding to the dial settings. Plot the data and keep it for future reference.

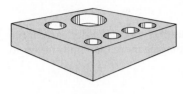

A

Aluminum block with small holes to fit Craig tube and 3-mL and 5-mL conical vials.

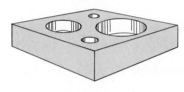

B

Aluminum block with large holes to fit 10-mL and 25-mL flasks

Figure 6.1
Aluminum heating blocks.

CAUTION

You should not use a *mercury* thermometer with an aluminum block. If it breaks, the mercury will vaporize on the hot surface. Instead, use a nonmercury glass thermometer, a metal dial thermometer, or a digital electronic temperature-measuring device.

---

[1] The use of solid aluminum heating devices was developed by Siegfried Lodwig at Centralia College, Centralia, WA: Lodwig, S. N. "The Use of Solid Aluminum Heat Transfer Devices in Organic Chemistry Laboratory Instruction and Research." *Journal of Chemical Education, 66* (1989): 77.

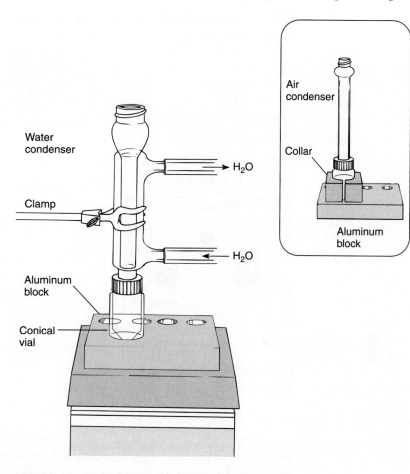

Figure 6.2
Heating with an aluminum block.

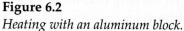

To avoid the possibility of breaking a glass thermometer, your hot plate may have a hole drilled into the metal plate so that a metal dial thermometer can be inserted into the unit (Fig. 6.3A). These metal thermometers, such as the one shown in Figure 6.3B, can be obtained in a number of temperature ranges. For example, a 0–250°C thermometer with 2-degree divisions can be obtained at a reasonable price. Also shown in Figure 6.3 (inset) is an aluminum block with a small hole drilled into it so that a metal thermometer can be inserted.[2] An alternative to the metal thermometer is a digital electronic temperature-measuring device that can be inserted into the aluminum block or hot plate. It is strongly recommended that mercury thermometers be avoided when measuring the surface temperature of the hot plate or aluminum block. If a mercury thermometer is broken on a hot surface, you will introduce toxic mercury vapors into the laboratory. Nonmercury thermometers filled with high-boiling colored liquids are available as alternatives.

It is a good idea to use the same hot plate each time. It is likely that two hot plates of the same type may give different temperatures with an identical setting. Record the identification number printed on the unit that you are using in your notebook to ensure that you always use the same hot plate.

---

[2] Garner, C. M. "A Mercury-Free Alternative for Temperature Measurement in Aluminum Blocks." *Journal of Chemical Education, 68* (1991): A244.

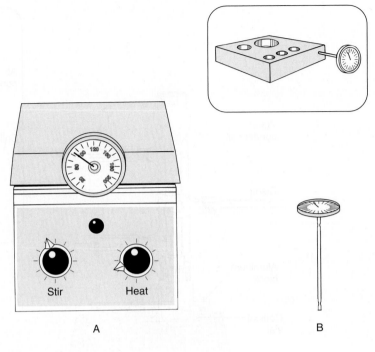

Figure 6.3
Dial thermometers.

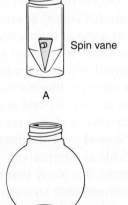

Figure 6.4

Methods of stirring in a conical vial or round-bottom flask.

Although we provide aluminum block temperatures in most of the experiments in this textbook, they should be taken as *approximate* values. You may need to adjust the temperature of the aluminum block appropriately to achieve the conditions you require. Each student must determine the actual temperature required to carry out a particular procedure. When a temperature is suggested, consider it as nothing more than a guide. Pay more attention to what is going on in your reaction vial or flask. If the temperature of your aluminum block equals the suggested temperature but the solution in the flask is not boiling (and you want it to boil), you clearly will need to increase the temperature of the aluminum block. Likewise, if the solution is boiling too rapidly, then you will need to reduce the temperature of the block.

When an aluminum block temperature is not given in the procedure and the liquid needs to be brought to a boil, you can determine the approximate setting from the boiling point of the liquid. Because the temperature inside the vial is lower than the aluminum block temperature, you should add at least 20°C to the boiling point of the liquid and set the aluminum block at this higher temperature. In fact, you may need to raise the temperature even higher than this value.

Many organic mixtures need to be stirred, as well as heated, to achieve satisfactory results. To stir a mixture, place a magnetic spin vane (Technique 7, Fig. 7.8B, p. 603) in a conical vial containing the reaction mixture as shown in Figure 6.4A. If the mixture is to be heated as well as stirred, attach a water condenser or an air condenser, as shown in Figure 6.2. With the combination stirrer/hot plate unit, it is possible to stir and heat a mixture simultaneously. Many reactions in this textbook are stirred continuously during the course of the reaction. With round-bottom flasks, a magnetic stir bar must be used to stir mixtures (Technique 7, Fig. 7.8A, p. 603). This is shown

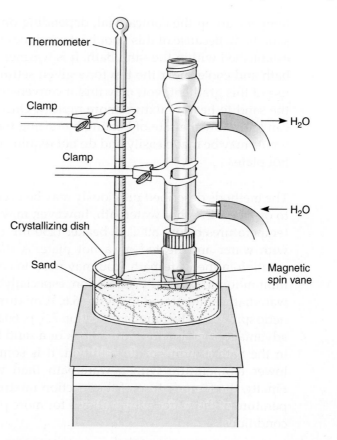

Figure 6.5
Heating with a sand bath.

in Figure 6.4B. Many laboratories will have another aluminum block drilled to accommodate 10- and 25-mL round-bottom flasks. More uniform stirring will be obtained if the vial or flask is placed in the aluminum block so that it is centered on the hot plate. Mixing may also be achieved by boiling the mixture. A boiling stone (Technique 7, Section 7.4, p. 603) should be added when a mixture is boiled without magnetic stirring.

6.2 Sand Bath with Hot Plate/Stirrer

The sand bath is used in some microscale laboratories to heat organic mixtures. Sand provides a clean way of distributing heat to a reaction mixture. To prepare a sand bath, place about a 1-cm depth of sand in a crystallizing dish or a Petri dish and then set the dish on a hot plate/stirrer unit. The apparatus is shown in Figure 6.5. Clamp the thermometer into position in the sand bath. You should calibrate the sand bath in a manner similar to that used with the aluminum block. Because sand heats more slowly than an aluminum block, you will need to begin heating the sand bath well before using it.

Do not heat the sand bath much above 200°C or you may break the dish. If you need to heat at very high temperatures, you should use an aluminum block rather than a sand bath (Section 6.1). With sand baths, it may be necessary to cover the dish with aluminum foil to achieve a temperature near 200°C. Keep in mind that the temperature obtained at a particular setting on the hot plate may vary for several reasons. First, you may place the thermometer at a different depth from time to time. Second, because of the relatively poor heat conduction of sand, you may obtain a different temperature in the conical vial, depending on the depth of the vial in the sand bath. Because of this poor heat conductivity, a temperature gradient is established within the sand bath. It is warmer near the bottom of the sand bath and cooler near the top for a given setting on the hot plate. To make use of this gradient, you may find it convenient to bury the vial or flask in the sand to heat a mixture more rapidly. Once the mixture is boiling, you can then slow the rate of heating by raising the vial or flask. These adjustments may be made easily and do not require a change in the setting on the hot plate.

6.3 Waterbath with Hot Plate/Stirrer

The methods described previously may be used over a range of about 50°C to over 200°C. A hot waterbath, however, may be a suitable alternative for temperatures below 80°C. A beaker (250 mL or 400 mL) is partially filled with water and heated on a hot plate. A thermometer is clamped into position in the waterbath. You may need to cover the waterbath with aluminum foil to prevent evaporation, especially at higher temperatures. The waterbath is illustrated in Figure 6.6. A mixture can be stirred with a magnetic spin vane (Technique 7, Section 7.3, p. 602). A hot waterbath has some advantage over an aluminum block or a sand bath in that the temperature in the bath is uniform. In addition, it is sometimes easier to establish a lower temperature with a waterbath than with other heating devices. Finally, the temperature of the reaction mixture will be closer to the temperature of the water, which allows for more precise control of the reaction conditions.

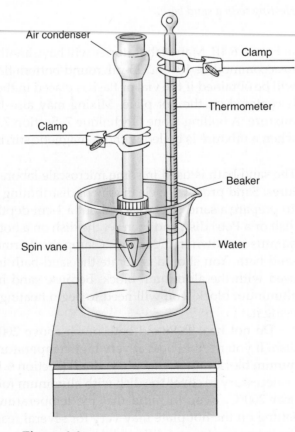

Figure 6.6
Waterbath.

6.4 Flames

The simplest technique for heating mixtures is to use a Bunsen burner. Because of the high danger of fires, however, the use of the Bunsen burner should be strictly limited to those cases for which the danger of fire is low or for which no reasonable alternative source of heat is available. A flame should generally be used only to heat aqueous solutions or solutions with very high boiling points. You should always check with your instructor about using a burner. If you use a burner at your bench, great care should be taken to ensure that others in the vicinity are not using flammable solvents.

In heating a flask with a Bunsen burner, you will find that using a wire gauze can produce more even heating over a broader area. The wire gauze, when placed under the object being heated, spreads the flame to keep the flask from being heated in one small area only.

Bunsen burners may be used to prepare capillary micropipets for thin-layer chromatography or to prepare other pieces of glassware requiring an open flame. For these purposes, burners should be used in designated areas in the laboratory and not at your laboratory bench.

6.5 Cold Baths

At times, you may need to cool a conical vial or flask below room temperature. A cold bath is used for this purpose. The most common cold bath is an **ice bath,** which is a highly convenient source of 0°C temperatures. An ice bath requires water along with ice to work well. If an ice bath is made up of only ice, it is not an efficient cooler because the large pieces of ice do not make good contact with the flask or vial. Enough water should be present with ice so that the flask is surrounded by water but not so much that the temperature is no longer maintained at 0°C. In addition, if too much water is present, the buoyancy of a flask resting in the ice bath may cause it to tip over. There should be enough ice in the bath to allow the flask to rest firmly.

For temperatures somewhat below 0°C, you may add some solid sodium chloride to the ice-water bath. The ionic salt lowers the freezing point of the ice so that temperatures from 0 to −10°C can be reached. The lowest temperatures are reached with ice-salt-water mixtures that contain relatively little water.

A temperature of −78.5°C can be obtained with solid carbon dioxide or dry ice. Large chunks of dry ice do not provide uniform contact with a flask being cooled. A liquid such as isopropyl alcohol is mixed with small pieces of dry ice to provide an efficient cooling mixture. Acetone and ethanol can be used in place of isopropyl alcohol. Be careful when handling dry ice because it can inflict severe frostbite. Extremely low temperatures can be obtained with liquid nitrogen (−195.8°C).

6.6 Steam Baths

The steam cone or steam bath is a good source of heat when temperatures less than 100°C are needed. Steam baths are used to heat reaction mixtures and solvents needed for crystallization. A steam cone and a portable steam bath are shown in Figure 6.7. These methods of heating have the disadvantage that water vapor may be introduced, through condensation of steam, into the mixture being heated. A slow flow of steam may minimize this difficulty.

Because water condenses in the steam line when it is not in use, it is necessary to purge the line of water before the steam will begin to flow. This purging should be accomplished before the flask is placed on the steam bath. The steam flow should be started with a high rate to purge the line; then the flow should be reduced to the desired rate. When using a portable

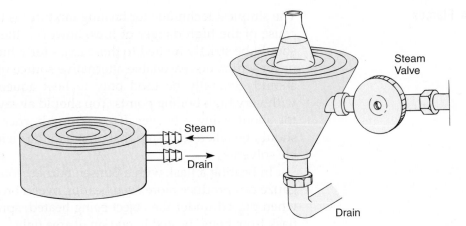

Figure 6.7
Steam bath and steam cone.

steam bath, be certain that condensate (water) is drained into a sink. Once the steam bath or cone is heated, a slow steam flow will maintain the temperature of the mixture being heated. There is no advantage to having a Vesuvius on your desk! An excessive steam flow may cause problems with condensation in the flask. This condensation problem can often be avoided by selecting the correct place at which to locate the flask on top of the steam bath.

The top of the steam bath consists of several flat concentric rings. The amount of heat delivered to the flask being heated can be controlled by selecting the correct sizes of these rings. Heating is most efficient when the largest opening that will still support the flask is used. Heating large flasks on a steam bath while using the smallest opening leads to slow heating and wastes laboratory time.

6.7 Microwave Ovens

Microwave technology is rapidly becoming a method of choice for heating and conducting chemical reactions. In the late 1990s, chemists used kitchen microwave ovens from appliance stores. Unfortunately, these units had a problem with variable field strengths and nonuniform heating capabilities. Yields tended to be variable, depending on where the reaction vessel was placed in the microwave. Worse, the reaction mixtures sometimes would blow up! Cases have even been recorded where the door of the microwave would blow off!

The good news that came out of the early use of domestic kitchen microwave ovens was that microwaves accelerate chemical reactions when compared to heating with a hot plate or other heating device. Often, the technique reduced side reactions and increased the yields of reactions. For example, a reaction that might take 25 hours to run using conventional heating devices might take only 5 or 10 minutes to run to completion using a microwave oven.

Because of the safety issues and variable results obtained with kitchen microwave ovens, companies began to produce units with more uniform fields that could be controlled more effectively. These commercial units were specially designed for chemists. The modern microwave units can run a large number of reactions on a small scale at the same time. This provides a way of doing development work in a shorter time. Many pharmaceutical and biotechnology companies make heavy use of microwave technology to speed development of new compounds. The companies might now be able to conduct reactions in minutes rather than hours, making it possible to do reactions that they might not have attempted in the past.

Currently, microwave technology is expensive and is used primarily in research laboratories at a small scale. As these ovens become less expensive, it is possible that microwave units might even replace the hot plate or oil bath (see p. 590) in the organic laboratory. Furthermore, microwave technology would help to make the organic laboratory more green (see Green Chemistry Essay, p. 268). When heating with microwaves, reactions can take place with less toxic reagents and in a shorter time, with less side reactions, all goals of green chemistry. Microwave technology has also been used to create supercritical water, which behaves more like an organic solvent and could replace toxic solvents in some organic reactions.

PROBLEMS

1. What would be the preferred heating device(s) in each of the following situations?
   a. Refluxing a solvent with a 56°C boiling point
   b. Refluxing a solvent with a 120°C boiling point
   c. Distilling a substance that boils at 220°C

2. Obtain the boiling points for the following compounds by using a handbook (Technique 4, pp. 574–580). In each case, suggest a heating device(s) that should be used for refluxing the substance.
   a. Butyl benzoate
   b. 1-Pentanol
   c. 1-Chloropropane

3. What type of bath would you use to get a temperature of −10°C?

4. Obtain the melting point and boiling point for benzene and ammonia from a handbook (Technique 4, pp. 574–580) and answer the following questions.
   a. A reaction was conducted in benzene as the solvent. Because the reaction was very exothermic, the mixture was cooled in a ice-salt-water bath. This was a bad choice. Why?
   b. What bath should be used for a reaction that is conducted in *liquid* ammonia as the solvent?

5. Criticize the following techniques:
   a. Refluxing a mixture that contains diethyl ether using a Bunsen burner
   b. Refluxing a mixture that contains a large amount of toluene using a hot water bath
   c. Using a mercury thermometer that is inserted into an aluminum block on a hot plate
   d. Running a reaction with *tert*-butyl alcohol (2-methyl-2-propanol) that is cooled to 0°C in an ice bath

7   **TECHNIQUE 7**

Reaction Methods

The successful completion of an organic reaction requires the chemist to be familiar with a variety of laboratory methods. These methods include operating safely, assembling the apparatus, heating and stirring reaction mixtures, adding liquid reagents, maintaining anhydrous and inert conditions in the reaction, and collecting gaseous products. Several techniques that are used in bringing a reaction to a successful conclusion are discussed here.

7.1 Assembling the Apparatus

Care must be taken when assembling the glass components into the desired apparatus. You should always remember that Newtonian physics applies to chemical apparatus, and unsecured pieces of glassware are certain to respond to gravity.

Assembling an apparatus in the correct manner requires that the individual pieces of glassware be connected to each other securely and the entire apparatus be held in the correct position. This can be accomplished by using **adjustable metal clamps** or a combination of adjustable metal clamps and **plastic joint clips.**

Two types of adjustable metal clamps are shown in Figure 7.1. Although these two types of clamps can usually be interchanged, the extension clamp is more commonly used to hold round-bottom flasks in place, and the three-finger clamp is frequently used to clamp condensers. Both types of clamps must be attached to a ring stand using a clamp holder, shown in Figure 7.1C.

A. Securing Macroscale Apparatus Assemblies

It is possible to assemble an apparatus using only adjustable metal clamps. An apparatus used to perform a distillation is shown in Figure 7.2. It is held together securely with three metal clamps. Because of the size of the apparatus and its geometry, the various clamps would likely be attached to three different ring stands. This apparatus would be somewhat difficult to assemble because it is necessary to ensure that the individual pieces stay together while securing and adjusting the clamps required to hold the entire apparatus in place. In addition, one must be careful not to bump any part of the apparatus or the ring stands after the apparatus is assembled.

A more convenient alternative is to use a combination of metal clamps and plastic joint clips. A plastic joint clip is shown in Figure 7.3A. These clips

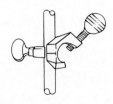

A. Extension clamp   B. Three-finger clamp   C. Clamp holder

Figure 7.1
Adjustable metal clamps.

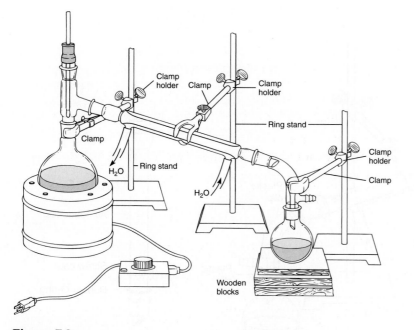

Figure 7.2
Distillation apparatus secured with metal clamps.

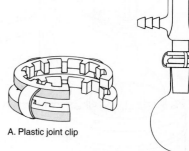

A. Plastic joint clip

B. Joint connected
by plastic clip

Figure 7.3
Plastic joint clip.

are easy to use (they just clip on), will withstand temperatures up to 140°C, and are durable. They hold together two pieces of glassware that are connected by ground-glass joints, as shown in Figure 7.3B. These clips come in different sizes to fit ground-glass joints of different sizes, and they are color coded for each size.

When used in combination with metal clamps, the plastic joint clips make it much easier to assemble most apparatus securely. There is less chance of dropping the glassware while assembling the apparatus, and once the apparatus is set up, it is more secure. Figure 7.4 shows the same distillation apparatus held in place with adjustable metal clamps and plastic joint clips.

To assemble this apparatus, first connect all of the individual pieces using the plastic clips. The entire apparatus is then connected to the ring stands using the adjustable metal clamps. Note that only two ring stands are required and the wooden blocks are not needed.

B. Securing Microscale Apparatus Assemblies

The glassware in most microscale kits is made with standard-taper ground joints. The most common joint size is ᵀ 14/10. Some microscale glassware with ground-glass joints also has threads cast into the outside surface of the outer joints (see top of air condenser in Figure 7.5). The threaded joint allows the use of a plastic screw cap with a hole in the top to fasten two pieces of glassware together securely. The plastic cap is slipped over the inner joint of the upper piece of glassware, followed by a rubber O-ring (see Figure 7.5). The O-ring should be pushed down so that it fits snugly on top of the ground-glass joint. The inner ground-glass joint is then fitted into the outer joint of the bottom piece of glassware. The screw cap is tightened, without excessive force, to attach the entire apparatus firmly together. The O-ring provides an additional seal that makes

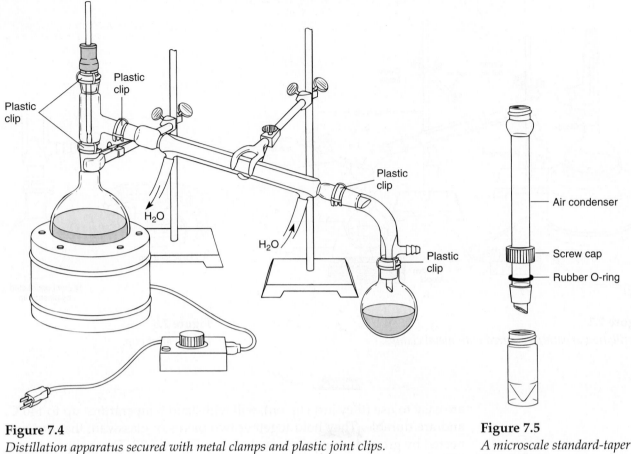

Figure 7.4
Distillation apparatus secured with metal clamps and plastic joint clips.

Figure 7.5
*A microscale standard-taper
joint assembly.*

this joint airtight. With this connecting system, it is unnecessary to use any
type of grease to seal the joint. The O-ring *must be used* to obtain a good
seal and to lessen the chances of breaking the glassware when you tighten
the plastic cap.

Microscale glassware connected in this fashion can be assembled easily.
The entire apparatus is held together securely, and usually only one metal
clamp is required to hold the apparatus onto a ring stand.

7.2 Heating under Reflux

Often we wish to heat a mixture for a long time and to leave it untended. A
reflux apparatus (see Figure 7.6) allows such heating. It also keeps the solvent from being lost by evaporation. A condenser is attached to the reaction
vial or boiling flask.

Choice of Condenser. The condenser used in a reflux apparatus can be either
of two types. An **air condenser** is simply a long tube. The surrounding air
removes heat from the vapors within the tube and condenses them to liquid. A **water-jacketed condenser** consists of two concentric tubes with the
outer cooling tube sealed onto the inner tube. The vapors rise within the
inner tube, and water circulates through the outer tube. The circulating
water removes heat from the vapors and condenses them. The air condenser
is suitable for use with high-boiling liquids or with small quantities of
material that are being heated gently. The water-jacketed condenser must
be used when the vapors are difficult to condense, usually because the

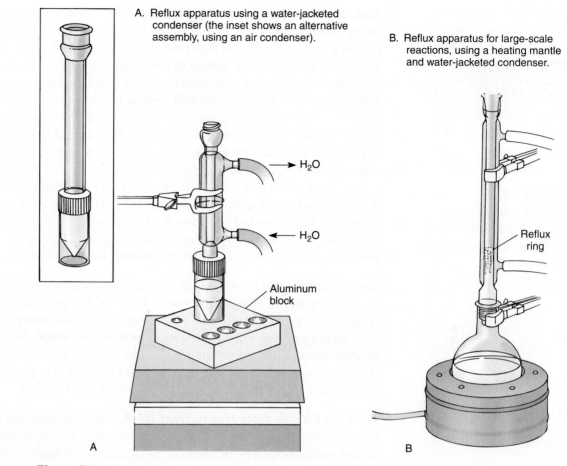

A. Reflux apparatus using a water-jacketed condenser (the inset shows an alternative assembly, using an air condenser).

B. Reflux apparatus for large-scale reactions, using a heating mantle and water-jacketed condenser.

H_2O

H_2O

Aluminum block

Reflux ring

A

B

Figure 7.6
Heating under reflux.

substance is volatile, or when vigorous boiling action is desired. In either case, the condenser prevents the vapors from escaping. Glassware assemblies using both air and water-jacketed condensers are shown in Figure 7.6A. The figure also shows a typical macroscale apparatus for heating large quantities of material under reflux (Fig. 7.6B).

When a water-jacketed condenser is used, the direction of the water flow should be such that the condenser will fill with cooling water. The water should enter the bottom of the condenser and leave from the top. The water should flow fast enough to withstand any changes in pressure in the water lines, but it should not flow any faster than absolutely necessary. An excessive flow rate greatly increases the chance of a flood, and high water pressure may force the hose from the condenser. Cooling water should be flowing before heating is begun! If the water is to remain flowing overnight, it is advisable to fasten the rubber tubing securely with wire to the condenser. If a flame is used as a source of heat, it is wise to use a wire gauze beneath the flask to provide an even distribution of heat from the flame. In most cases, an aluminum block, a sand bath, water bath, heating mantle, or steam bath is preferred over a flame.

Stirring. When heating a solution, always use a magnetic stirrer or a boiling stone (see Sections 7.3 and 7.4) to keep the solution from "bumping."

Figure 7.7
Tended reflux of small quantities on a steam cone (this can also be done with a hot plate).

Rate of Heating. If the heating rate has been correctly adjusted, the liquid being heated under reflux will travel only partway up the condenser tube before condensing. Below the condensation point, solvent will be seen running back into the flask; above it, the interior of the condenser will appear dry. The boundary between the two zones will be clearly demarcated, and a **reflux ring** or a ring of liquid will appear there. The reflux ring can be seen in Figure 7.6B. In heating under reflux, the rate of heating should be adjusted so that the reflux ring is no higher than a third to half the distance to the top of the condenser. With microscale experiments, the quantities of vapor rising in the condenser frequently are so small that a clear reflux ring cannot be seen. In those cases, the heating rate must be adjusted so that the liquid boils smoothly but not so rapidly that solvent can escape the condenser. With such small volumes, the loss of even a small amount of solvent can affect the reaction. With large-scale reactions, the reflux ring is much easier to see, and one can adjust the heating rate more easily.

Tended Reflux. It is possible to heat small amounts of a solvent under reflux in an Erlenmeyer flask. With gentle heating, the evaporated solvent will condense in the relatively cold neck of the flask and return to the solution. This technique (see Fig. 7.7) requires constant attention. The flask must be swirled frequently and removed from the heating source for a short period if the boiling becomes too vigorous. When heating is in progress, the reflux ring should not be allowed to rise into the neck of the flask.

How Do I Know How Hot to Heat It? A common problem that inexperienced students encounter when they assemble an apparatus for heating under reflux is that it is difficult to decide what temperature setting to use for heating the contents of a vial or flask to the desired temperature. This problem becomes more acute when the students attempt to reproduce the temperatures that are specified in the laboratory procedures of a textbook.

First, you should understand that the temperatures specified are only approximate suggestions. The actual temperature required to carry out a particular procedure must be determined for each student and each apparatus. When you see a temperature stipulated, consider it as nothing more than a guide. You will have to make adjustments to suit your own situation.

Second, you must always pay attention to what is going on in your reaction flask. If the temperature of your aluminum block or sand bath equals the suggested temperature, but the solution in your flask is not boiling, you clearly will have to increase the temperature of the heating device. Remember that what really matters is what is going on in the flask, not what the textbook says! The *external* temperature, as measured by a thermometer placed into the heating device, is not the important temperature. Far more critical is the temperature *inside* the flask, which may be considerably lower than the external temperature.

7.3 Stirring Methods

When a solution is heated, there is a danger that it may become superheated. When this happens, very large bubbles sometimes erupt violently from the solution; this is called **bumping.** Bumping must be avoided because of the risk that material may be lost from the apparatus, that a fire may start, or that the apparatus may break.

Magnetic stirrers are used to prevent bumping because they produce turbulence in the solution. The turbulence breaks up the large bubbles that

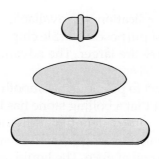

A. Standard-sized
   magnetic stirring bars

B. Microscale magnetic
   spin vane

C. Small magnetic stirring
   bar ("disposable" type)

Figure 7.8
Magnetic stirring bars.

form in boiling solutions. An additional purpose for using a magnetic stirrer is to stir the reaction to ensure that all the reagents are thoroughly mixed. A magnetic stirring system consists of a magnet that is rotated by an electric motor. The rate at which this magnet rotates can be adjusted by a potentiometric control. A small magnet, which is coated with a nonreactive material such as Teflon or glass, is placed in the flask. The magnet within the flask rotates in response to the rotating magnetic field caused by the motor-driven magnet. The result is that the inner magnet stirs the solution as it rotates. A common type of magnetic stirrer includes the stirring system within a hot plate. This type of hot plate/stirrer permits one to heat the reaction and stir it simultaneously. In order for the magnetic stirrer to be effective, the contents of the flask being stirred should be placed as close to the center of the hot plate as possible and not offset.

For macroscale apparatus, magnetic stirring bars of various sizes and shapes are available. For microscale apparatus, a **magnetic spin vane** is often used. It is designed to contain a tiny bar magnet and to have a shape that conforms to the conical bottom of a reaction vial. A small Teflon-coated magnetic stirring bar works well with very small round-bottom boiling flasks. Small stirring bars of this type (often sold as "disposable" stirring bars) can be obtained cheaply. A variety of magnetic stirring bars is illustrated in Figure 7.8.

There is also a variety of simple techniques that may be used to stir a liquid mixture in a centrifuge tube or conical vial. A thorough mixing of the components of a liquid can be achieved by repeatedly drawing the liquid into a Pasteur pipet and then ejecting the liquid back into the container by pressing sharply on the dropper bulb. Liquids can also be stirred effectively by placing the flattened end of a spatula into the container and twirling it rapidly.

7.4 Boiling Stones

A **boiling stone,** also known as a **boiling chip** or **Boileezer,** is a small lump of porous material that produces a steady stream of fine air bubbles when it is heated in a solvent. This stream of bubbles and the turbulence that accompanies it break up the large bubbles of gases in the liquid. In this way, it reduces the tendency of the liquid to become superheated, and it promotes the smooth boiling of the liquid. The boiling stone decreases the chances for bumping.

Two common types of boiling stones are carborundum and marble chips. Carborundum boiling stones are more inert, and the pieces are usually small, suitable for most applications. If available, carborundum boiling stones are preferred for most purposes. Marble chips may dissolve in strong acid solutions, and the pieces are larger. The advantage of marble chips is that they are cheaper.

Because boiling stones act to promote the smooth boiling of liquids, you should always make certain that a boiling stone has been placed in a liquid *before* heating is begun. If you wait until the liquid is hot, it may have become superheated. Adding a boiling stone to a superheated liquid will cause all the liquid to try to boil at once. The liquid, as a result, would erupt entirely out of the flask or froth violently.

As soon as boiling ceases in a liquid containing a boiling stone, the liquid is drawn into the pores of the boiling stone. When this happens, the boiling stone no longer can produce a fine stream of bubbles; it is spent. You may have to add a new boiling stone if you have allowed boiling to stop for a long period.

Wooden applicator sticks are used in some applications. They function in the same manner as boiling stones. Occasionally, glass beads are used. Their presence also causes sufficient turbulence in the liquid to prevent bumping.

7.5 Addition of Liquid Reagents

Liquid reagents and solutions are added to a reaction by several means, some of which are shown in Figure 7.9. For microscale experiments, the simplest approach is simply to add the liquid to the reaction by means of a Pasteur pipet. This method is shown in Figure 7.9A. In this technique, the system is open to the atmosphere. A second microscale method, shown in Figure 7.9B, is suitable for experiments in which the reaction should be kept isolated from the atmosphere. In this approach, the liquid is kept in a hypodermic syringe. The syringe needle is inserted through a rubber septum, and the liquid is added dropwise from the syringe. The septum seals the apparatus from the atmosphere, which makes this technique useful for reactions that are conducted under an atmosphere of inert gas or where anhydrous conditions must be maintained. As an alternative, the rubber septum may be replaced by a cap and Teflon insert or liner. A disadvantage of the Teflon insert, however, is that the insert may no longer form an effective seal after being punctured by the needle.

The most common type of assembly for macroscale experiments is shown in Figure 7.9C. In this apparatus, a separatory funnel is attached to the side arm of a three-necked round-bottom flask. The separatory funnel must be equipped with a standard-taper, ground-glass joint to be used in this manner. The liquid is stored in the separatory funnel (which is called an **addition funnel** in this application) and is added to the reaction. The rate of addition is controlled by adjusting the stopcock. When it is being used as an addition funnel, the upper opening must be kept open to the atmosphere. If the upper hole is stoppered, a vacuum will develop in the funnel and will prevent the liquid from passing into the reaction vessel. Because the funnel is open to the atmosphere, there is a danger that atmospheric moisture can contaminate the liquid reagent as it is being added. To prevent this outcome, a drying tube (see Section 7.6) is attached to the upper opening of the addition funnel. The drying tube allows the funnel to maintain atmospheric pressure without allowing the passage of water vapor into the reaction.

Figure 7.9D shows an alternative type of addition funnel that is useful for reactions that must be maintained under an atmosphere of inert gas. This is the **pressure-equalizing addition funnel.** With this glassware, the upper

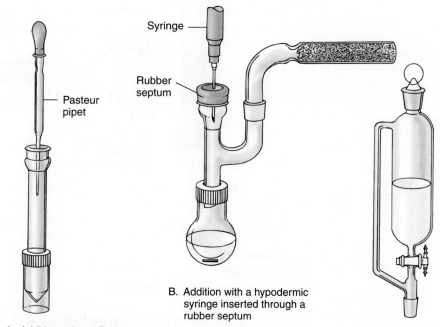

Pasteur pipet

A. Addition using a Pasteur pipet inserted into the top of an air condenser

Syringe

Rubber septum

B. Addition with a hypodermic syringe inserted through a rubber septum

D. A pressure equalizing addition funnel

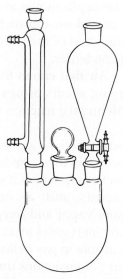

C. Macroscale equipment, using a separatory funnel as an addition funnel

Figure 7.9
Methods of adding liquid reagents to a reaction.

opening is stoppered. The side arm allows the pressure above the liquid in the funnel to be in equilibrium with the pressure in the rest of the apparatus, and it lets the inert gas flow over the top of the liquid as it is being added.

7.6 Drying Tubes

With certain reactions, atmospheric moisture must be prevented from entering the reaction vessel. A **drying tube** can be used to maintain anhydrous conditions within the apparatus. Two types of drying tubes are shown in Figure 7.10. The typical drying tube is prepared by placing a small, loose plug of glass wool or cotton into the constriction at the end of the tube nearest the ground-glass joint or hose connection. The plug is tamped gently

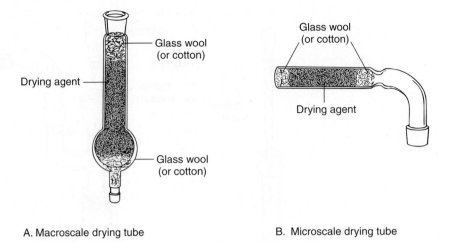

A. Macroscale drying tube   B. Microscale drying tube

Figure 7.10
Drying tubes.

with a glass rod or piece of wire to place it in the correct position. A drying agent, typically calcium sulfate ("Drierite") or calcium chloride (see Technique 12, Section 12.9, p. 680), is poured on top of the plug to the approximate depth shown in Figure 7.10. Another loose plug of glass wool or cotton is placed on top of the drying agent to prevent the solid material from falling out of the drying tube. The drying tube is then attached to the flask or condenser.

Air that enters the apparatus must pass through the drying tube. The drying agent absorbs any moisture from air passing through it so that air entering the reaction vessel has had the water vapor removed from it.

7.7 Reactions Conducted under an Inert Atmosphere

Some reactions are sensitive to oxygen and water vapor present in air and require an inert atmosphere in order to obtain satisfactory results. The usual reactions in which it is desirable to exclude air often include organometallic reagents, such as organomagnesium or organolithium reagents, where water vapor and oxygen (air) react with these compounds. The most common inert gases available in a laboratory are nitrogen and argon, which are available in gas cylinders. Nitrogen is probably the gas most often used to carry out reactions under an inert atmosphere, although argon has a distinct advantage because it is denser than air. This allows the argon to push air away from the reaction mixture.

When laboratories are not equipped with individual gas lines to benches or hoods, it is useful to supply nitrogen or argon to the reaction apparatus using a balloon assembly (shown in Figure 7.11). Your instructor will provide you with the apparatus.

Construct the balloon assembly by cutting off the top of a 3-mL disposable plastic syringe. Attach a small balloon snugly to the top of the syringe, securing it with a small rubber band that has been doubled to hold the balloon securely to the body of the syringe. Attach a needle to the syringe. Fill the balloon with the inert gas through the needle using a piece of rubber tubing attached to the gas source. When the balloon has been inflated to 2–3 inches in diameter, quickly pinch off the neck of the balloon while removing the gas source. Now push the needle into a rubber stopper to keep the balloon inflated. It is possible to keep an assembly like this filled with inert gas for several days without the balloon deflating.

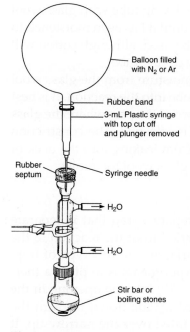

Figure 7.11
Conducting a reaction under an inert atmosphere using a balloon assembly.

Before you start the reaction, you may need to dry your apparatus thoroughly in an oven. Add all reagents carefully to avoid water. The following instructions are based on the assumption that you are using an apparatus consisting of a round-bottom flask equipped with a condenser. Attach a rubber septum to the top of your condenser. Now flush the air out of the apparatus with the inert gas. It is best not to use the balloon assembly for this purpose, unless you are using argon (see next paragraph). Instead, remove the round-bottom flask from the apparatus and, with the help of your instructor, flush it with the inert gas using a Pasteur pipet to bubble the gas through the solvent and reaction mixture in the flask. In this way, you can remove air from the reaction assembly before attaching the balloon assembly. Quickly reattach the flask to the apparatus. Pinch off the neck of the balloon between your fingers, remove the rubber stopper, and insert the needle into the rubber septum. The reaction apparatus is now ready for use.

When argon is employed as an inert gas, you can use the balloon assembly to remove air from the reaction apparatus in the following way. Insert the balloon assembly into the rubber septum as previously described. Also insert a second needle (no syringe attached) through the septum. The pressure from the balloon will force argon down the reflux condenser (argon is denser than air) and push the less dense air out through the second syringe needle. When the apparatus has been thoroughly flushed with argon, remove the second needle. Nitrogen does not work as well with this method because it is less dense than air, and it will be difficult to remove the air that is in contact with the reaction mixture in the round-bottom flask.

For reactions conducted at room temperature, you can remove the condenser shown in Figure 7.11. Attach the rubber septum directly to the round-bottom flask and insert the needle of an argon-filled balloon assembly through the rubber septum. To flush the air out of the reaction flask, insert a second syringe needle into the rubber septum. Any air present in the flask will be flushed out through this second syringe needle, and the air will be replaced with argon. Now remove the second needle, and you have a reaction mixture free of air.

7.8 Capturing Noxious Gases

Many organic reactions involve the production of a noxious gaseous product. The gas may be corrosive, such as hydrogen chloride, hydrogen bromide, or sulfur dioxide, or it may be toxic, such as carbon monoxide. The safest way to avoid exposure to these gases is to conduct the reaction in a ventilated hood where the gases can be safely drawn away by the ventilation system.

In many instances, however, it is safe and efficient to conduct the experiment on the laboratory bench, away from the hood. This is particularly true when the gases are soluble in water. Some techniques for capturing noxious gases are presented in this section.

A. Drying Tube Method

Microscale experiments have the advantage that the amounts of gases produced are small. Hence, it is easy to trap them and prevent them from escaping into the laboratory. You can take advantage of the water solubility of corrosive gases such as hydrogen chloride, hydrogen bromide, and sulfur dioxide. A simple technique is to attach the drying tube (see Fig. 7.10B) to the top of the reaction vial or condenser. The drying tube is filled with moistened glass wool. The moisture in the glass wool absorbs the gas, preventing its escape. To prepare this type of gas trap, fill the drying tube with glass wool and then add water dropwise to the glass wool until it has been moistened to the desired degree. Moistened cotton can also be used, although cotton will absorb so much water that it is easy to plug the drying tube.

When using glass wool in a drying tube, moisture from the glass wool must not be allowed to drain from the drying tube into the reaction. It is best to use a drying tube that has a constriction between the part where the glass wool is placed and the neck, where the joint is attached. The constriction acts as a partial barrier preventing the water from leaking into the neck of the drying tube. Make certain not to make the glass wool too moist.

B. External Gas Traps

Another approach to capturing gases is to prepare a trap that is separate from the reaction apparatus. The gases are carried from the reaction to the trap by means of tubing. There are several variations on this type of trap. One method that works well for microscale experiments is to place a thermometer adapter (Technique 14, Fig. 14.9A, p. 712) into the opening in the reaction apparatus. A Pasteur pipet is inserted upside down through the adapter, and a piece of fine flexible tubing is fitted over the narrow tip. It may be helpful to break the Pasteur pipet before using it for this purpose so that only the narrow tip and a short section of the barrel are used. The other end of the flexible tubing is placed through a large plug of moistened glass wool in a test tube. The water in the glass wool absorbs the water-soluble gases. This method is shown in Figure 7.12.

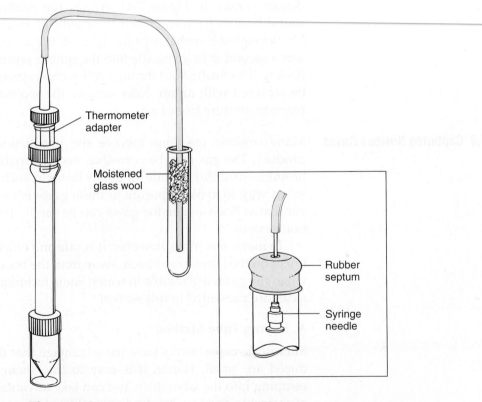

Figure 7.12

Microscale external gas trap. (The inset shows an expanded view of an alternative fitting, using a syringe needle and a rubber septum.)

A variation on the Pasteur pipet method uses a hypodermic syringe needle inserted upside down (from the inside) through a rubber septum, which has been fitted over the opening at the top of the reaction apparatus. Flexible tubing, fitted over the syringe needle, leads to a trap such as the one using wet glass wool described previously. This variation is also shown in Figure 7.12.

Another alternative to the apparatus shown in Figure 7.12 is to use a multipurpose adapter in place of the thermometer adapter (p. 712). The flexible tubing can be attached directly to the side arm of the multipurpose adapter, thus connecting the apparatus to the gas trap. If the multipurpose adapter is used for this purpose, the upper opening of the adapter must be closed; this is accomplished most easily by inserting a piece of glass rod or a short piece of glass tubing sealed at one end into the opening and tightening the fittings around it.

With large-scale reactions, a trap using an inverted funnel placed in a beaker of water is used. A piece of glass tubing, inserted through a thermometer adapter attached to the reaction apparatus, is connected to flexible tubing. The tubing is attached to a conical funnel. The funnel is clamped in place inverted over a beaker of water. The funnel is clamped so that its lip *almost touches* the water surface, but is not placed below the surface of the water. With this arrangement, water cannot be sucked back into the reaction if the pressure in the reaction vessel changes suddenly. This type of trap can also be used in microscale applications. An example of the inverted-funnel type of gas trap is shown in Figure 7.13.

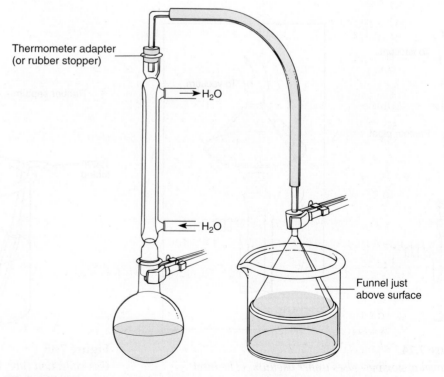

Figure 7.13
An inverted-funnel gas trap.

C. Removal of Noxious Gases Using an Aspirator

An aspirator can be used to remove noxious gases from the reaction. The simplest approach is to clamp a disposable Pasteur pipet so that its tip is placed well into the condenser atop the reaction vial. An inverted funnel clamped over the apparatus can also be used. The pipet or funnel is attached to an aspirator with flexible tubing. A trap should be placed between the pipet or funnel and the aspirator. As gases are liberated from the reaction, they rise into the condenser. The vacuum draws the gases away from the apparatus. Both types of systems are shown in Figure 7.14. In the special case in which the noxious gases are soluble in water, connecting a water aspirator to the pipet or funnel removes the gases from the reaction and traps them in the flowing water without the need for a separate gas trap.

7.9 Collecting Gaseous Products

In Section 7.8, means of removing unwanted gaseous products from the reaction system were examined. Some experiments produce gaseous products that you must collect and analyze. Methods to collect gaseous products are all based on the same principle. The gas is carried through tubing from the reaction to the opening of a flask or a test tube, which has been filled with water and is inverted in a container of water. The gas is allowed to bubble into the inverted collection tube (or flask). As the collection tube fills with gas, the water is displaced into the water container. If the collection tube is graduated, as in a graduated cylinder or a centrifuge tube, you can monitor the quantity of gas produced in the reaction.

If the inverted gas-collection tube is constructed from a piece of glass tubing, a rubber septum can be used to close the upper end of the container. This type of collection tube is shown in Figure 7.15. A sample of the gas can

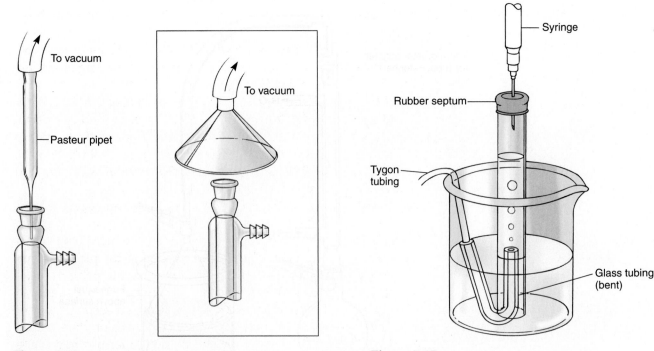

Figure 7.14

Removal of noxious gases under vacuum. (The inset shows an alternative assembly, using a funnel in place of the Pasteur pipet.)

Figure 7.15

Gas-collection tube, with rubber septum.

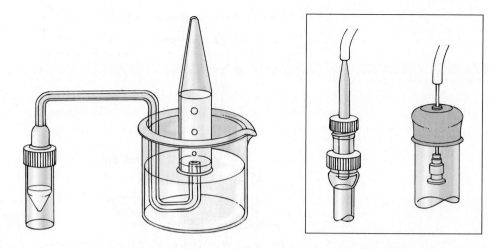

A. Apparatus using a capillary
   gas delivery tube

B. The inset shows alternative
   assemblies, using flexible
   tubing

Figure 7.16
Gas-delivery tubes.

be removed using a syringe equipped with a needle. The gas that is removed can be analyzed by gas chromatography (see Technique 22).

Many of the glassware kits for microscale experiments contain a special,
all-glass, capillary gas-delivery tube. The tube is attached to the top of the
reaction apparatus by means of a ground-glass joint, and the open end of
the capillary tubing is placed into an inverted, water-filled flask or test tube,
clamped over a waterbath. An example of a microscale kit gas-delivery tube
is shown in Figure 7.16A. This type of tube is an efficient means of collecting gases. A disadvantage, however, is that it is expensive and relatively
easy to break.

A simpler, less expensive approach is to use flexible tubing of a fine diameter to lead the gases from the reaction vessel to the collecting container.
One method is to place a hypodermic syringe needle, point upward,
through a rubber septum. The septum is attached to the top of the reaction
apparatus, and a piece of fine flexible tubing is fitted over the end of the needle. The free end of the tubing is placed in the waterbath, underneath the
opening of the water-filled collection container. The gases bubble into the
container, where they are collected. This alternative apparatus is shown in
Figure 7.15 and also as an inset in Figure 7.16B.

Another alternative, which may also be used with larger-scale experiments, is to place a piece of glass tubing or the tip of a Pasteur pipet through
a thermometer adapter. The thermometer adapter is attached to the top of
the reaction apparatus, and flexible tubing is attached to the piece of glass
tubing. The free end of the tubing is positioned in the opening of the water-
filled collection vessel, as described previously. This variation is also shown
as an inset in Figure 7.16. As an option, you may attach a second piece of
glass tubing to the free end of the flexible hose. This piece of glass tubing
sometimes makes it easier to fix the open end in the proper position in the
opening of the collection flask.

7.10 Evaporation of Solvents   In many experiments, it is necessary to remove excess solvent from a solution. An obvious approach is to allow the container to stand unstoppered in

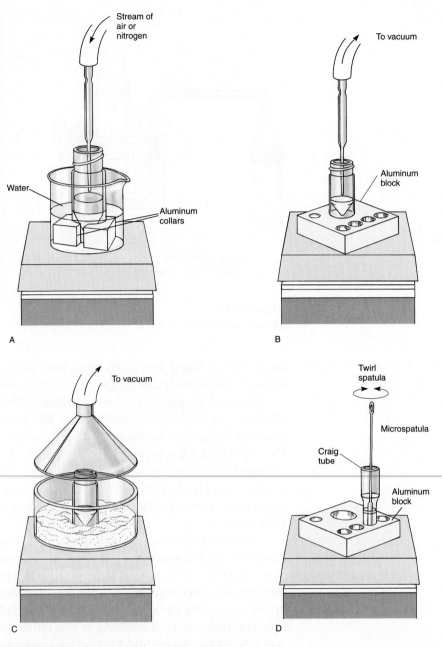

Figure 7.17
Evaporation of solvents (microscale methods).

the hood for several hours until the solvent has evaporated. This method is generally not practical, however, and a quicker, more efficient means of evaporating solvents must be used. Figures 7.17 and 7.18 show several methods of removing solvents by evaporation. Figure 7.17 depicts microscale methods; Figure 7.18 is devoted to large-scale procedures.

NOTE: It is good laboratory practice to evaporate solvents in the hood.

Microscale Methods. A simple means of evaporating a solvent is to place a conical vial in a warm waterbath or a warm sand bath. The heat from the waterbath or sand bath will warm the solvent to a temperature where it can

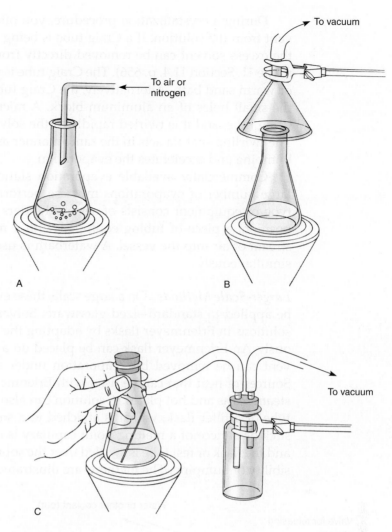

Figure 7.18
Evaporation of solvents (large-scale methods).

evaporate within a short time. The heat from the waterbath or sand bath can be adjusted to provide the best rate of evaporation, but the liquid should not be allowed to boil vigorously. The evaporation rate can be increased by allowing a stream of dry air or nitrogen to be directed into the vial (Fig. 7.17A). The moving gas stream will sweep the vapors from the vial and accelerate the evaporation. As an alternative, a vacuum can be applied above the vial to draw away solvent vapors (Fig. 7.17B and 7.17C).

A convenient waterbath suitable for microscale methods can be constructed by placing the aluminum collars, which are generally used with aluminum heating blocks into a 150-mL beaker (Fig. 7.17A). In some cases, it may be necessary to round off the sharp edges of the collars with a file in order to allow them to fit properly into the beaker. Held by the aluminum collars, the conical vial will stand securely in the beaker. This assembly can be filled with water and placed on a hot plate for use in the evaporation of small amounts of solvent.

Aluminum heating blocks placed on a hot plate can also be used for the evaporation of solvents (Fig. 7.17B). You must be careful, however, not to allow the aluminum block to become too hot, or the sample may decompose thermally.

During a crystallization procedure, you often must remove excess solvent from the solution. If a Craig tube is being used for the crystallization, the excess solvent can be removed directly from the Craig tube (see Technique 11, Section 11.4, p. 656). The Craig tube is placed in a warm waterbath or warm sand bath. Alternatively, the Craig tube can be placed into one of the small holes of an aluminum block. A microspatula is placed into the Craig tube, and it is twirled rapidly as the solvent evaporates (Fig. 7.17D). The twirling spatula acts in the same manner as a boiling stone; it prevents bumping and accelerates the evaporation.

Commercially available evaporation stations may be useful when a large number of evaporations must be performed at the same time. This type of equipment consists of several holders for vials or flasks. At each position, a piece of tubing equipped with a metal tip is used to direct a stream of air into the vessel. A waterbath is used to heat all the containers simultaneously.

Larger-Scale Methods.   On a large scale, these evaporation methods can also be applied to standard-sized glassware. Solvents can be evaporated from solutions in Erlenmeyer flasks by adapting the techniques described previously. An Erlenmeyer flask can be placed on a source of heat, and the solvent can be removed by evaporation under a gas stream or a vacuum. Sources of heat that can be used with Erlenmeyer flasks include sand and steam baths and hot plates. A solution can also be placed in a side arm test tube or a filter flask, which is attached to a source of vacuum. A wooden stick or a piece of a melting point capillary is often placed in the solution, and the flask or test tube is swirled over the source of heat to reduce the possibility of bumping. The methods are illustrated in Figure 7.18.

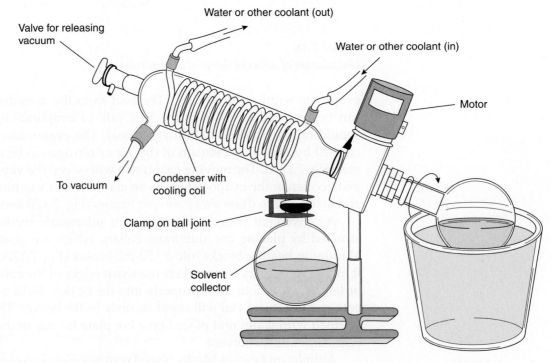

Figure 7.19
A rotary evaporator.

7.11 Rotary Evaporator

In some organic chemistry laboratories, solvents are evaporated under reduced pressure using a **rotary evaporator.** This is a motor-driven device that is designed for rapid evaporation of solvents, with heating, while minimizing the possibility of bumping. A vacuum is applied to the flask, and the motor spins the flask. The rotation of the flask spreads a thin film of the liquid over the surface of the glass, which accelerates evaporation. The rotation also agitates the solution sufficiently to reduce the problem of bumping. A waterbath can be placed under the flask to warm the solution and increase the vapor pressure of the solvent. One can select the speed at which the flask is rotated and the temperature of the waterbath to attain the desired evaporation rate. As the solvent evaporates from the rotating flask, the vapors are cooled by the condenser, and the resulting liquid collects in the flask. The product remains behind in the rotating flask. A complete rotary evaporator assembly is shown in Figure 7.19. If the coolant is sufficiently cold, virtually all of the solvent can be recovered and recycled. This is a good example of *Green Chemistry* (see Green Chemistry essay, p. 268).

PROBLEMS

1. What is the best type of stirring device to use for stirring a reaction that takes place in the following type of glassware?

    a. A conical vial

    b. A 10-mL round-bottom flask

    c. A 250-mL round-bottom flask

2. Should you use a drying tube for the following reaction? Explain.

$$CH_3-\overset{\overset{\displaystyle O}{\|}}{C}-OH + CH_3-\underset{\underset{\displaystyle CH_3}{|}}{CH}-CH_2-CH_2-OH \rightleftharpoons CH_3-\overset{\overset{\displaystyle O}{\|}}{C}-O-CH_2-CH_2-\underset{\underset{\displaystyle CH_3}{|}}{CH}-CH_3 + H_2O$$

3. For which of the following reactions should you use a trap to collect noxious gases?

    a.

    b.

    c. $C_{12}H_{22}O_{11} + H_2O \longrightarrow 4\,CH_3-CH_2-OH + 4\,CO_2$
       (Sucrose)

    d. $CH_3-\underset{\underset{\displaystyle H}{|}}{C}=NH + H_2O \xrightarrow[\text{heat}]{\text{base}} CH_3-\underset{\underset{\displaystyle H}{|}}{C}=O + NH_3$

4. Criticize the following techniques:

    a. A reflux is conducted with a stopper in the top of the condenser.

    b. Water is passed through the reflux condenser at the rate of 1 gallon per minute.

    c. No water hoses are attached to the condenser during a reflux.

    d. A boiling stone is not added to the round-bottom flask until the mixture is boiling vigorously.

e. To save money, you decide to save your boiling stones for another experiment.

f. The reflux ring is located near the top of the condenser in a reflux setup.

g. A rubber O-ring is omitted when the water condenser is attached to a conical vial.

h. A gas trap is assembled with the funnel in Figure 7.13, completely submerged in the water in the beaker.

i. Powdered drying agent is used rather than granular material.

j. A reaction involving hydrogen chloride is conducted on the laboratory bench and not in a hood.

k. An air-sensitive reaction apparatus is set up as shown in Figure 7.6.

l. Air is used to evaporate solvent from an air-sensitive compound.

8   TECHNIQUE 8

Filtration

Filtration is a technique used for two main purposes. The first is to remove solid impurities from a liquid. The second is to collect a desired solid from the solution from which it was precipitated or crystallized. Several different kinds of filtration are commonly used: two general methods include gravity filtration and vacuum (or suction) filtration. Two techniques specific to the microscale laboratory are filtration with a filter-tip pipet and filtration with a Craig tube. The various filtration techniques and their applications are summarized in Table 8.1. These techniques are discussed in more detail in the following sections.

8.1 Gravity Filtration

The most familiar filtration technique is probably filtration of a solution through a paper filter held in a funnel, allowing gravity to draw the liquid through the paper. Because even a small piece of filter paper will absorb a significant volume of liquid in most microscale procedures requiring filtration, this technique is useful only when the volume of mixture to be filtered is greater than 10 mL. For many macroscale and microscale procedures, a more suitable technique, which also makes use of gravity, is to use a Pasteur (or disposable) pipet with a cotton or glass wool plug (called a filtering pipet).

A. Filter Cones

This filtration technique is most useful when the solid material being filtered from a mixture is to be collected and used later. The filter cone, because of its smooth sides, can easily be scraped free of collected solids. Because of the many folds, fluted filter paper, described in the next section, cannot be scraped easily. The filter cone is likely to be used in microscale experiments only when a relatively large volume (greater than 10 mL) is being filtered and when a Hirsch funnel (Section 8.3) is not appropriate.

The filter cone is prepared as indicated in Figure 8.1. It is then placed into a funnel of an appropriate size. With filtrations using a simple filter cone, solvent may form seals between the filter and the funnel and between the funnel and the lip of the receiving flask. When a seal forms, the filtration

Table 8.1 *Filtration methods*

Method	Application	Section
Gravity filtration		
Filter cones	The volume of liquid to be filtered is about 10 mL or greater, and the solid collected in the filter is saved.	8.1A
Fluted filters	The volume of liquid to be filtered is greater than about 10 mL, and solid impurities are removed from a solution; often used in crystallization procedures.	8.1B
Filtering pipets	Used with volumes less than about 10 mL to remove solid impurities from a liquid.	8.1C
Decantation	Although not a filtration technique, decantation can be used to separate a liquid from large, insoluble particles.	8.1D
Vacuum filtration		
Hirsch funnels	Used in the same way as Büchner funnels, except the volume of liquid is usually smaller (1–10 mL).	8.3
Büchner funnels	Primarily used to collect a desired solid from a liquid when the volume is greater than about 10 mL; used frequently to collect the crystals obtained from crystallization.	8.3
Filtering media	Used to remove finely divided impurities.	8.4
Filter-tip pipets	May be used to remove a small amount of solid impurities from a small volume (1–2 mL) of liquid; also useful for pipetting volatile liquids, especially in extraction procedures.	8.6
Craig tubes	Used to collect a small amount of crystals resulting from crystallizations in which the volume of the solution is less than 2 mL.	8.7
Centrifugation	Although not strictly a filtration technique, centrifugation may be used to remove suspended impurities from a liquid (1–25 mL).	8.8

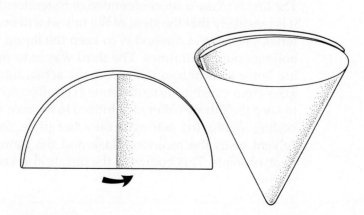

Figure 8.1
Folding a filter cone.

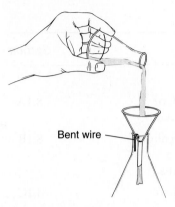

Bent wire

Figure 8.2
Gravity filtration with a filter cone.

stops because the displaced air has no possibility of escaping. To avoid the solvent seal, you can insert a small piece of paper, a paper clip, or some other bent wire between the funnel and the lip of the flask to let the displaced air escape. As an alternative, you can support the funnel by a clamp fixed *above* the flask rather than placed on the neck of the flask. A gravity filtration using a filter cone is shown in Figure 8.2.

B. Fluted Filters

This filtration method is also most useful when filtering a relatively large amount of liquid. Because a fluted filter is used when the desired material is expected to remain in solution, this filter is used to remove undesired solid materials, such as dirt particles, decolorizing charcoal, and undissolved impure crystals. A fluted filter is often used to filter a hot solution saturated with a solute during a crystallization procedure.

The technique for folding a fluted filter paper is shown in Figure 8.3. An advantage of a fluted filter is that it increases the speed of filtration in two ways. First, it increases the surface area of the filter paper through which the solvent seeps; second, it allows air to enter the flask along its sides to permit rapid pressure equalization. If pressure builds up in the flask from hot vapors, filtering slows down. This problem is especially pronounced with filter cones. The fluted filter tends to reduce this problem considerably, but it may be a good idea to clamp the funnel above the receiving flask or to use a piece of paper, paper clip, or wire between the funnel and the lip of the flask as an added precaution against solvent seals.

Filtration with a fluted filter is relatively easy to perform when the mixture is at room temperature. However, when it is necessary to filter a hot solution saturated with a dissolved solute, a number of steps must be taken to ensure that the filter does not become clogged by solid material accumulated in the stem of the funnel or in the filter paper. When the hot, saturated solution comes in contact with a relatively cold funnel (or a cold flask, for that matter), the solution is cooled and may become supersaturated. If crystallization then occurs in the filter, either the crystals will fail to pass through the filter paper or they will clog the stem of the funnel.

To keep the filter from clogging, use one of the following four methods. The first is to use a short-stemmed or a stemless funnel. With these funnels, it is less likely that the stem of the funnel will become clogged by solid material. The second method is to keep the liquid to be filtered at or near its boiling point at all times. The third way is to preheat the funnel by pouring hot solvent through it before the actual filtration. This keeps the cold glass from causing instantaneous crystallization. And fourth, it is helpful to keep the **filtrate** (filtered solution) in the receiver hot enough to continue boiling *slightly* (by setting it on a hot plate, for example). The refluxing solvent heats the receiving flask and the funnel stem and washes them clean of solids. This boiling of the filtrate also keeps the liquid in the funnel warm.

C. Filtering Pipets

A filtering pipet is a microscale technique most often used to remove solid impurities from a liquid with a volume less than 10 mL. It is important that the mixture being filtered be at or near room temperature because it is

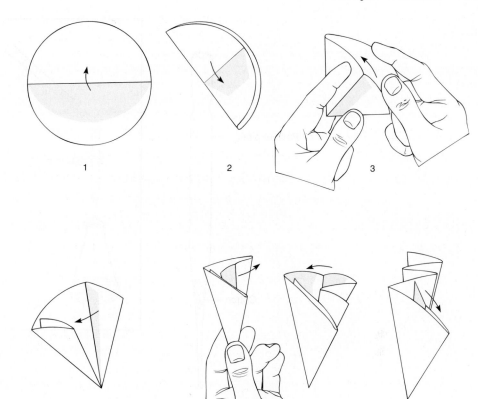

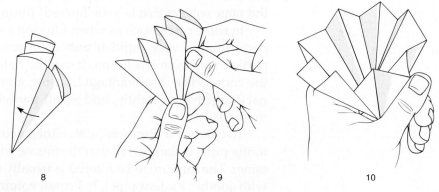

Figure 8.3
Folding a fluted filter paper, or origami at work in the organic chemistry laboratory.

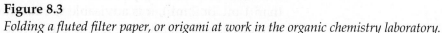

difficult to prevent premature crystallization in a hot solution saturated with a solute.

To prepare this filtration device, a small piece of cotton is inserted into the top of a Pasteur (disposable) pipet and pushed down to the beginning of the lower constriction in the pipet, as shown in Figure 8.4. It is important to use enough cotton to collect all the solid being filtered; however, the amount of cotton used should not be so large that the flow rate through the pipet is significantly restricted. For the same reason, the cotton should not be packed too tightly. The cotton plug can be pushed down gently with a long

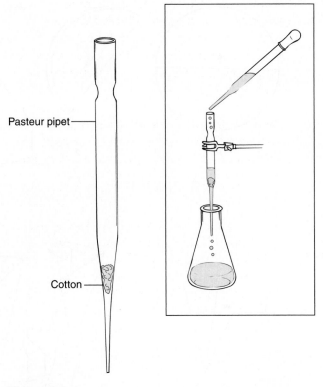

Figure 8.4
A filtering pipet.

thin object such as a glass stirring rod or a wooden applicator stick. It is advisable to wash the cotton plug by passing about 1 mL of solvent (usually the same solvent that is to be filtered) through the filter.

In some cases, such as when filtering a strongly acidic mixture or when performing a very rapid filtration to remove dirt or impurities of large particle size from a solution, it may be better to use glass wool in place of the cotton. The disadvantage in using glass wool is that the fibers do not pack together as tightly, and small particles will pass through the filter more easily.

To conduct a filtration (with either a cotton or glass wool plug), the filtering pipet is clamped so that the filtrate will drain into an appropriate container. The mixture to be filtered is usually transferred to the filtering pipet with another Pasteur pipet. If a small volume of liquid is being filtered (less than 1 mL or 2 mL), it is advisable to rinse the filter and plug with a small amount of solvent after the last of the filtrate has passed through the filter. The rinse solvent is then combined with the original filtrate. The rate of filtration can be increased by gently applying pressure to the top of the pipet using a pipet bulb, if desired.

Depending on the amount of solid being filtered and the size of the particles (small particles are more difficult to remove by filtration), it may be necessary to put the filtrate through a second filtering pipet. This should be done with a new filtering pipet rather than with the one already used.

D. Decantation

It is not always necessary to use filter paper to separate insoluble particles. If you have large, heavy, insoluble particles, with careful pouring you can decant the solution, leaving behind the solid particles that will settle to the bottom of the flask. The term *decant* means "to carefully pour out the liquid, leaving the insoluble particles behind." For example, boiling stones or sand granules in the bottom of an Erlenmeyer flask filled with a liquid can easily be separated in this way. This procedure is often preferred over filtration and usually results in a smaller loss of material. If there are a large number of particles and they retain a significant amount of the liquid, they can be rinsed with solvent and a second decantation performed. The term *decant* was coined in the wine industry, where it is often necessary to let the wine settle and then carefully pour it out of the original bottle into a clean one, leaving the "must" (insoluble particles) behind.

8.2 Filter Paper

Many kinds and grades of filter paper are available. The paper must be correct for a given application. In choosing filter paper, you should be aware of its various properties. **Porosity** is a measure of the size of the particles that can pass through the paper. Highly porous paper does not remove small particles from solution; paper with low porosity removes very small particles. **Retentivity** is a property that is the opposite of porosity. Paper with low retentivity does not remove small particles from the filtrate. The **speed** of filter paper is a measure of the time it takes a liquid to drain through the filter. Fast paper allows the liquid to drain quickly; with slow paper, it takes much longer to complete the filtration. Because all these properties are related, fast filter paper usually has a low retentivity and high porosity, and slow filter paper usually has high retentivity and low porosity.

Table 8.2 compares some commonly available qualitative filter paper types and ranks them according to porosity, retentivity, and speed. Eaton–Dikeman (E&D), Schleicher and Schuell (S&S), and Whatman are the most common brands of filter paper. The numbers in the table refer to the grades of paper used by each company.

8.3 Vacuum Filtration

Vacuum, or suction, filtration is more rapid than gravity filtration and is most often used to collect solid products resulting from precipitation or crystallization. This technique is used primarily when the volume of liquid being filtered is more than 1–2 mL. With smaller volumes, use of the Craig tube (Section 8.7) is the preferred technique. In a vacuum filtration, a receiver flask with a side arm, a **filter flask,** is used. For microscale work, the

Table 8.2 *Some common qualitative filter paper types and approximate relative speeds and retentivities*

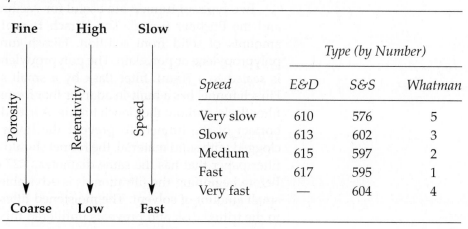

Speed	E&D	S&S	Whatman
Very slow	610	576	5
Slow	613	602	3
Medium	615	597	2
Fast	617	595	1
Very fast	—	604	4

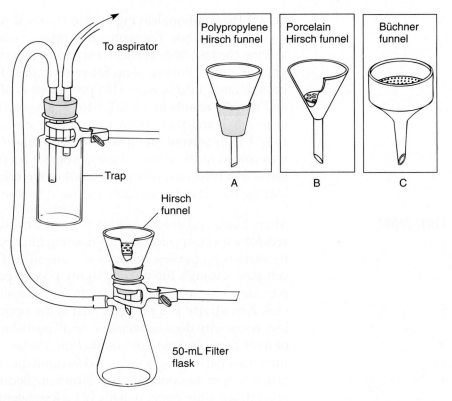

Figure 8.5
Vacuum filtration.

most useful size is a 50-mL filter flask. For macroscale laboratory work, the most useful sizes of filter flasks range from 50 mL to 500 mL, depending on the volume of liquid being filtered. The side arm is connected by *heavy-walled* rubber tubing (see Technique 16, Figure 16.3, p. 734) to a source of vacuum. Thin-walled tubing will collapse under vacuum, due to atmospheric pressure on its outside walls, and will seal the vacuum source from the flask. Because this apparatus is unstable and can tip over easily, it must be clamped, as shown in Figure 8.5.

> **CAUTION** ⚠
>
> **It is essential that the filter flask be clamped.**

Two types of funnels are useful for vacuum filtration, the Hirsch funnel and the Büchner funnel. The **Hirsch funnel** is used for filtering smaller amounts of solid from solution. Hirsch funnels are usually made from polypropylene or porcelain. The polypropylene Hirsch funnel (see Fig. 8.5A) is sealed to a 50-mL filter flask by a small section of Gooch tubing. This Hirsch funnel has a built-in adapter that forms a tight seal with some 25-mL filter flasks without the Gooch tubing. A fritted polyethylene disk fits into the bottom of the funnel. To prevent the holes in this disk from becoming clogged with solid material, the funnel should always be used with a circular filter paper that has the same diameter (1.27-cm) as the polyethylene disk. Before beginning the filtration, it is advisable to moisten the paper with a small amount of solvent. The moistened filter paper adheres more strongly to the fritted disk and prevents unfiltered mixture from passing around the edges of the filter paper. A porcelain Hirsch funnel (see Fig. 8.5B) is sealed to the filter flask by a rubber stopper or a filter (Neoprene) adapter. The flat bottom of this Hirsch funnel, which should be 1–2 cm in diameter, is covered with an unfolded piece of circular filter paper. To prevent the escape of solid materials from the funnel, you must be certain that the filter paper fits the funnel exactly. It must cover all the holes in the bottom of the funnel but not extend up the sides of the funnel. With a porcelain Hirsch funnel, it is also important to moisten the paper with a small amount of solvent before beginning the filtration.

The **Büchner funnel,** which is shown in Figure 8.5C, operates on the same principle as the Hirsch funnel, but it is usually larger, and its sides are vertical rather than sloped. It is sealed to the filter flask with a rubber stopper or a Neoprene adapter. In the Büchner funnel, the filter paper must also cover all the holes in the bottom but must not extend up the sides.

Because the filter flask is attached to a source of vacuum, a solution poured into a Hirsch funnel or Büchner Funnel is literally sucked rapidly through the filter paper. For this reason, vacuum filtration is generally not used to separate fine particles such as decolorizing charcoal, because the small particles would likely be pulled through the filter paper. However, this problem can be alleviated, when desired, by the use of specially prepared filter beds (see Section 8.4).

8.4 Filtering Media

It is occasionally necessary to use specially prepared filter beds to separate fine particles when using vacuum filtration. Often, very fine particles either pass right through a paper filter or clog it so completely that the filtering stops. This is avoided by using a substance called Filter Aid, or Celite. This material is also called **diatomaceous earth** because of its source. It is a finely divided inert material derived from the microscopic shells of dead diatoms (a type of phytoplankton that grows in the sea).

CAUTION

Diatomaceous earth is a lung irritant. When using Filter Aid, take care not to breathe the dust.

Filter Aid will not clog the fiber pores of filter paper. It is **slurried,** mixed with a solvent to form a rather thin paste, and filtered through a Hirsch or Büchner funnel (with filter paper in place) until a layer of diatoms about 2–3 mm thick is formed on top of the filter paper. The solvent in which the diatoms were slurried is poured from the filter flask, and, if necessary, the filter flask is cleaned before the actual filtration is begun. Finely divided particles can now be suction-filtered through this layer and will be caught in the Filter Aid. This technique is used for removing impurities, not for collecting a product. The filtrate (filtered solution) is the desired material in this procedure. If the material caught in the filter were the desired material, you would have to try to separate the product from all those diatoms! Filtration with Filter Aid is not appropriate when the desired substance is likely to precipitate or crystallize from solution.

In microscale work, it may sometimes be more convenient to use a column prepared with a Pasteur pipet to separate fine particles from a solution. The Pasteur pipet is packed with alumina or silica gel, as shown in Figure 8.6.

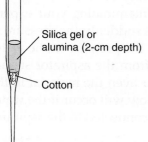

Silica gel or alumina (2-cm depth)

Cotton

Figure 8.6
A Pasteur pipet with filtering media.

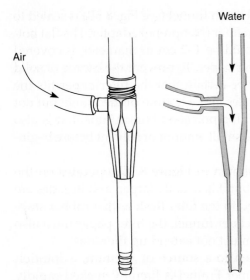

Figure 8.7
An aspirator.

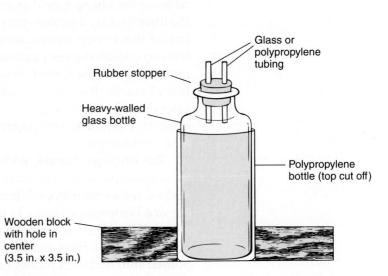

Figure 8.8
A simple aspirator trap and holder.

8.5 The Aspirator

The most common source of vacuum (approximately 10–20 mm Hg) in the laboratory is the water aspirator, or "water pump," illustrated in Figure 8.7. This device passes water rapidly past a small hole to which a side arm is attached. The water pulls air in through the side arm. This phenomenon, called the Bernoulli effect, causes a reduced pressure along the side of the rapidly moving water stream and creates a partial vacuum in the side arm.

NOTE: The aspirator works most effectively when the water is turned on to the fullest extent.

A water aspirator can never lower the pressure beyond the vapor pressure of the water used to create the vacuum. Hence, there is a lower limit to the pressure (on cold days) of 9–10 mm Hg. A water aspirator does not provide as high a vacuum in the summer as in the winter, because of this water-temperature effect.

A trap must be used with an aspirator. One type of trap is illustrated in Figure 8.5. Another method for securing this type of trap is shown in Figure 8.8. This simple holder can be constructed from readily available material and can be placed anywhere on the laboratory bench. Although not often needed, a trap can prevent water from contaminating your experiment. If the water pressure in the laboratory drops suddenly, the pressure in the filter flask may suddenly become lower than the pressure in the water aspirator. This would cause water to be drawn from the aspirator stream into the filter flask and contaminate the filtrate or even the material in the filter. The trap stops this reverse flow. A similar flow will occur if the water flow at the aspirator is stopped before the tubing connected to the aspirator side arm is disconnected.

NOTE: Always disconnect the tubing before stopping the aspirator.

If a "backup" begins, disconnect the tubing as rapidly as possible before the trap fills with water. Some chemists like to fit a stopcock into the stopper on top of the trap. A three-hole stopper is required for this purpose. With a stopcock in the trap, the system can be vented before the aspirator is shut off. Then water cannot back up into the trap.

Aspirators do not work well if too many people use the water line at the same time because the water pressure is lowered. Also, the sinks at the ends of the lab benches or the lines that carry away the water flow may have a limited capacity for draining the resultant water flow from too many aspirators. Care must be taken to avoid floods.

8.6 Filter-Tip Pipet

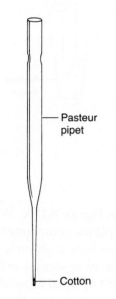

Figure 8.9
A filter-tip pipet.

The filter-tip pipet, illustrated in Figure 8.9, has two common uses. The first is to remove a small amount of solid, such as dirt or filter paper fibers, from a small volume of liquid (1–2 mL). It can also be helpful when using a Pasteur pipet to transfer a highly volatile liquid, especially during an extraction procedure (see Technique 12, Section 12.5, p. 675).

Preparing a filter-tip pipet is similar to preparing a filtering pipet, except that a much smaller amount of cotton is used. A *tiny* piece of cotton is loosely shaped into a ball and placed into the large end of a Pasteur pipet. Using a wire with a diameter slightly smaller than the inside diameter of the narrow end of the pipet, push the ball of cotton to the bottom of the pipet. If it becomes difficult to push the cotton, you have probably started with too much cotton; if the cotton slides through the narrow end with little resistance, you probably have not used enough.

To use a filter-tip pipet as a filter, the mixture is drawn up into the Pasteur pipet using a pipet bulb and then expelled. With this procedure, a small amount of solid will be captured by the cotton. However, very fine particles, such as activated charcoal, cannot be removed efficiently with a filter-tip pipet, and this technique is not effective in removing more than a trace amount of solid from a liquid.

Transferring many organic liquids with a Pasteur pipet can be a somewhat difficult procedure for two reasons. First, the liquid may not adhere well to the glass. Second, as you handle the Pasteur pipet, the temperature of the liquid in the pipet increases slightly, and the increased vapor pressure may tend to "squirt" the liquid out the end of the pipet. This problem can be particularly troublesome when separating two liquids during an extraction procedure. The purpose of the cotton plug in this situation is to slow the rate of flow through the end of the pipet so you can control the movement of liquid in the Pasteur pipet more easily.

8.7 Craig Tubes

The **Craig tube,** illustrated in Figure 8.10, is used primarily to separate crystals from a solution after a microscale crystallization procedure has been performed (Technique 11, Section 11.4, p. 656). Although it may not be a filtration procedure in the traditional sense, the outcome is similar. The outer part of the Craig tube is similar to a test tube, except that the diameter of the tube becomes wider part of the way up the tube, and the glass is ground at this point so that the inside surface is rough. The inner part (plug) of the Craig tube may be made of Teflon or glass. If this part is glass, the end of the plug is also ground. With either a glass or a Teflon inner plug, there is only a partial seal where the plug and the outer tube come together. Liquid may pass through, but solid will not. This is the place where the solution is separated from the crystals.

After crystallization has been completed in the outer Craig tube, replace the inner plug (if necessary) and connect a thin copper wire or strong thread

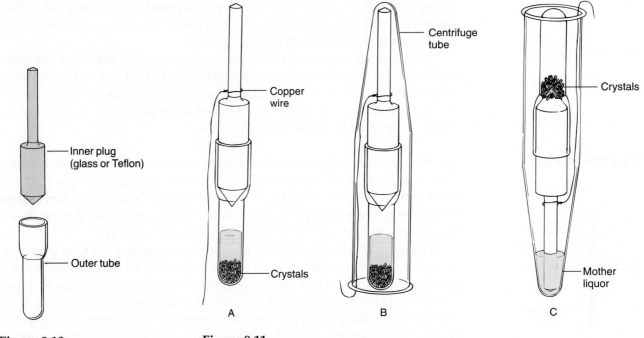

Figure 8.10
A Craig tube (2 mL).

Figure 8.11
Separation with a Craig tube.

to the narrow part of the inner plug, as indicated in Figure 8.11A. While holding the Craig tube in an upright position, place a plastic centrifuge tube over the Craig tube so that the bottom of the centrifuge tube rests on top of the inner plug, as shown in Figure 8.11B. The copper wire should extend just below the lip of the centrifuge tube and is now bent upward around the lip of the centrifuge tube. This apparatus is then turned over so that the centrifuge tube is upright. The Craig tube is spun in a centrifuge (be sure it is balanced by placing another tube filled with water on the opposite side of the centrifuge) for several minutes until the **mother liquor** (solution from which the crystals grew) goes to the bottom of the centrifuge tube and the crystals collect on the end of the inner plug (see Figure 8.11C). Depending on the consistency of the crystals and the speed of the centrifuge, the crystals may spin down to the inner plug, or (if you are unlucky) they may remain at the other end of the Craig tube.[1] If the latter situation occurs, it may be helpful to centrifuge the Craig tube longer or, if this problem is anticipated, to stir the crystal- and solution-mixture with a spatula or stirring rod before centrifugation.

Using the copper wire, then pull the Craig tube out of the centrifuge tube. If the crystals collected on the end of the inner plug, it is now a simple procedure to remove the plug and scrape the crystals with a spatula onto a watch glass, a clay plate, or a piece of smooth paper. Otherwise, it will be

---

[1] *Note to the Instructor:* In some centrifuges, the bottom of the Craig tube may be close to the center of the centrifuge when the Craig tube assembly is placed into the centrifuge. In this situation, very little centrifugal force will be applied to the crystals, and it is likely that the crystals will not spin down. It may then be helpful to use an inner plug with a shorter stem. The stem on a Teflon inner plug can be easily cut off about 0.5 inch with a pair of wire cutters. This will help to spin down the crystals to the inner plug, and also the centrifuge can be run at a lower speed, which can help prevent breakage of the Craig tube.

necessary to scrape the crystals from the inside surface of the outer part of the Craig tube.

8.8 Centrifugation

Sometimes, centrifugation is more effective in removing solid impurities than conventional filtration techniques. Centrifugation is particularly effective in removing suspended particles, which are so small that the particles would pass through most filtering devices. Centrifugation may also be useful when the mixture must be kept hot to prevent premature crystallization while the solid impurities are removed.

Centrifugation is performed by placing the mixture in one or two centrifuge tubes (be sure to balance the centrifuge) and centrifuging for several minutes. The supernatant liquid is then decanted (poured off) or removed with a Pasteur pipet.

PROBLEMS

1. In each of the following situations, what type of filtration device would you use? (Do not consider centrifugation when answering this question.)

   a. Remove powdered decolorizing charcoal from 20 mL of solution

   b. Collect crystals obtained from crystallizing a substance from about 1 mL of solution

   c. Remove a very small amount of dirt from 1 mL of liquid

   d. Isolate 0.2 g of crystals from about 5 mL of solution after performing a crystallization

   e. Remove dissolved colored impurities from about 3 mL of solution

   f. Remove solid impurities from 5 mL of liquid at room temperature

9   TECHNIQUE 9

Physical Constants of Solids: The Melting Point

9.1 Physical Properties

The physical properties of a compound are those properties that are intrinsic to a given compound when it is pure. A compound may often be identified simply by determining a number of its physical properties. The most commonly recognized physical properties of a compound include its color, melting point, boiling point, density, refractive index, molecular weight, and optical rotation. Modern chemists would include the various types of spectra (infrared, nuclear magnetic resonance, mass, and ultraviolet-visible) among the physical properties of a compound. A compound's spectra do not vary from one pure sample to another. Here, we look at methods of determining the melting point. Boiling point and density of compounds are covered in Technique 13. Refractive index, optical rotation, and spectra are also considered separately.

Many reference books list the physical properties of substances. You should consult Technique 4 for a complete discussion on how to find data for specific compounds. The works most useful for finding lists of values for the nonspectroscopic physical properties include

The Merck Index

The CRC Handbook of Chemistry and Physics

Lange's Handbook of Chemistry

Aldrich Handbook of Fine Chemicals

Complete citations for these references can be found in Technique 29 (p. 942). Although the *CRC Handbook* has good tables, it adheres strictly to IUPAC nomenclature. For this reason, it may be easier to use one of the other references, particularly *The Merck Index* or the *Aldrich Handbook of Fine Chemicals,* in your first attempt to locate information (see Technique 4).

9.2 The Melting Point

The melting point of a compound is used by the organic chemist not only to identify the compound but also to establish its purity. A small amount of material is heated *slowly* in a special apparatus equipped with a thermometer or thermocouple, a heating bath or heating coil, and a magnifying eyepiece for observing the sample. Two temperatures are noted. The first is the point at which the first drop of liquid forms among the crystals; the second is the point at which the whole mass of crystals turns to a *clear* liquid. The melting point is recorded by giving this range of melting. You might say, for example, that the melting point of a substance is 51–54°C. That is, the substance melted over a 3-degree range.

The melting point indicates purity in two ways. First, the purer the material, the higher its melting point. Second, the purer the material, the narrower its melting-point range. Adding successive amounts of an impurity to a pure substance generally causes its melting point to decrease in proportion to the amount of impurity. Looking at it another way, adding impurities lowers the freezing point. The freezing point, a colligative property, is simply the melting point (solid → liquid) approached from the opposite direction (liquid → solid).

Figure 9.1 is a graph of the usual melting-point behavior of mixtures of two substances, A and B. The two extremes of the melting range (the low and high temperature) are shown for various mixtures of the two. The upper curves indicate the temperatures at which all the sample has melted. The lower curves indicate the temperature at which melting is observed to

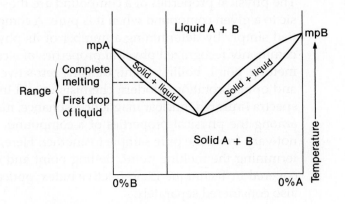

Figure 9.1

A melting-point–composition curve.

begin. With pure compounds, melting is sharp and without any range. This is shown at the left- and right-hand edges of the graph. If you begin with pure A, the melting point decreases as impurity B is added. At some point, a minimum temperature, or **eutectic,** is reached, and the melting point begins to increase to that of substance B. The vertical distance between the lower and upper curves represents the melting range. Notice that for mixtures that contain relatively small amounts of impurity (<15%) and are not close to the eutectic, the melting range increases as the sample becomes less pure. The range indicated by the lines in Figure 9.1 represents the typical behavior.

We can generalize the behavior shown in Figure 9.1. Pure substances melt with a narrow range of melting. With impure substances, the melting range becomes wider, and the entire melting range is lowered. Be careful to note, however, that at the minimum point of the melting-point–composition curves, the mixture often forms a eutectic, which also melts sharply. Not all binary mixtures form eutectics, and some caution must be exercised in assuming that every binary mixture follows the previously described behavior. Some mixtures may form more than one eutectic; others might not form even one. In spite of these variations, both the melting point and its range are useful indications of purity, and they are easily determined by simple experimental methods.

9.3 Melting-Point Theory

Figure 9.2 is a phase diagram describing the usual behavior of a two-component mixture (A + B) on melting. The behavior on melting depends on the relative amounts of A and B in the mixture. If A is a pure substance (no B), then A melts sharply at its melting point t_A. This is represented by point A on the left side of the diagram. When B is a pure substance, it melts at t_B; its melting point is represented by point B on the right side of the diagram. At either point A or point B, the pure solid passes cleanly, with a narrow range, from solid to liquid.

In mixtures of A and B, the behavior is different. Using Figure 9.2, consider a mixture of 80% A and 20% B on a mole-per-mole basis (that is, mole percentage). The melting point of this mixture is given by t_M at point M on the diagram. That is, adding B to A has lowered the melting point of A from t_A to t_M. It has also expanded the melting range. The temperature t_M corresponds to the **upper limit** of the melting range.

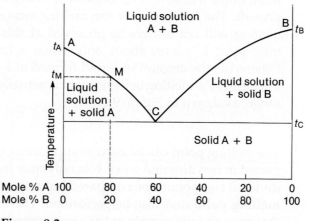

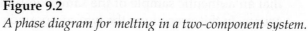

Figure 9.2

A phase diagram for melting in a two-component system.

Lowering the melting point of A by adding impurity B comes about in the following way. Substance A has the lower melting point in the phase diagram shown, and if heated, it begins to melt first. As A begins to melt, solid B begins to dissolve in the liquid A that is formed. When solid B dissolves in liquid A, the melting point is depressed. To understand this, consider the melting point from the opposite direction. When a liquid at a high temperature cools, it reaches a point at which it solidifies, or "freezes." The temperature at which a liquid freezes is identical to its melting point. Recall that the freezing point of a liquid can be lowered by adding an impurity. Because the freezing point and the melting point are identical, lowering the freezing point corresponds to lowering the melting point. Therefore, as more impurity is added to a solid, its melting point becomes lower. There is, however, a limit to how far the melting point can be depressed. You cannot dissolve an infinite amount of the impurity substance in the liquid. At some point, the liquid will become saturated with the impurity substance. The solubility of B in A has an upper limit. In Figure 9.2, the solubility limit of B in liquid A is reached at point C, the **eutectic point.** The melting point of the mixture cannot be lowered below t_C, the melting temperature of the eutectic.

Now consider what happens when the melting point of a mixture of 80% A and 20% B is approached. As the temperature is increased, A begins to "melt." This is not really a visible phenomenon in the beginning stages; it happens before liquid is visible. It is a softening of the compound to a point at which it can begin to mix with the impurity. As A begins to soften, it dissolves B. As it dissolves B, the melting point is lowered. The lowering continues until all B is dissolved or until the eutectic composition (saturation) is reached. When the maximum possible amount of B has been dissolved, actual melting begins, and one can observe the first appearance of liquid. The initial temperature of melting will be below t_A. The amount below t_A at which melting begins is determined by the amount of B dissolved in A but will never be below t_C. Once all B has been dissolved, the melting point of the mixture begins to rise as more A begins to melt. As more A melts, the semisolid solution is diluted by more A, and its melting point rises. While all this is happening, you can observe *both* solid and liquid in the melting-point capillary. Once all A has begun to melt, the composition of the mixture M becomes uniform and will reach 80% A and 20% B. At this point, the mixture finally melts sharply, giving a clear solution. The maximum melting-point range will be $t_C - t_M$, because t_A is depressed by the impurity B that is present. The lower end of the melting range will always be t_C; however, melting will not always be observed at this temperature. An observable melting at t_C comes about only when a large amount of B is present. Otherwise, the amount of liquid formed at t_C will be too small to observe. Therefore, the melting behavior that is actually observed will have a smaller range, as shown in Figure 9.1.

9.4 Mixture Melting Points

The melting point can be used as supporting evidence in identifying a compound in two different ways. Not only may the melting points of the two individual compounds be compared but a special procedure called a **mixture melting point** may also be performed. The mixture melting point requires that an authentic sample of the same compound be available from another source. In this procedure, the two compounds (authentic and suspected) are finely pulverized and mixed together in equal quantities. Then the melting point of the mixture is determined. If there is a melting-point depression or if the range of melting is expanded by a large amount compared to that of the individual substances, you may conclude that one compound has acted as an impurity toward the other and that they are not the same compound. If there is no lowering of the melting point for the mixture (the melting point is identical with those of pure A and pure B), then A and B are almost certainly the same compound.

9.5 Packing the Melting-Point Tube

Melting points are usually determined by heating the sample in a piece of thin-walled capillary tubing (1 mm × 100 mm) that has been sealed at one end. To pack the tube, press the open end gently into a *pulverized* sample of the crystalline material. Crystals will stick in the open end of the tube. The amount of solid pressed into the tube should correspond to a column no more than 1–2 mm high. To transfer the crystals to the closed end of the tube, drop the capillary tube, closed end first, down a ⅔-m length of glass tubing, which is held upright on the desktop. When the capillary tube hits the desktop, the crystals will pack down into the bottom of the tube. This procedure is repeated if necessary. Tapping the capillary on the desktop with fingers is not recommended because it is easy to drive the small tubing into a finger if the tubing should break.

Some commercial melting-point instruments have a built-in vibrating device that is designed to pack capillary tubes. With these instruments, the sample is pressed into the open end of the capillary tube, and the tube is placed in the vibrator slot. The action of the vibrator will transfer the sample to the bottom of the tube and pack it tightly.

9.6 Determining the Melting Point—the Thiele Tube

There are two principal types of melting-point apparatus available: the Thiele tube and commercially available, electrically heated instruments. The Thiele tube, shown in Figure 9.3, is the simpler device and was once widely used. It is a glass tube designed to contain a heating oil (mineral oil or silicone oil) and a thermometer to which a capillary tube containing the sample is attached. For best results use a mercury thermometer. The shape of the Thiele tube allows convection currents to form in the oil when it is heated. These currents maintain a uniform temperature distribution through the oil in the tube. The side arm of the tube is designed to generate these convection currents and thus transfer the heat from the flame evenly and rapidly throughout the oil. The sample, which is in a capillary tube attached to the thermometer, is held by a rubber band or a thin slice of rubber tubing. It is important that this rubber band be above the level of the oil (allowing for expansion of the oil on heating) so that the oil does not soften the rubber and allow the capillary tubing to fall into the oil. If a cork or a rubber stopper is used to hold the thermometer, a triangular channel should be cut into the side of it to allow pressure equalization.

The Thiele tube is usually heated by a microburner. During the heating, the rate of temperature increase should be regulated. Hold the burner by its cool base and, using a low flame, move the burner slowly back and forth along the bottom of the arm of the Thiele tube. If the heating is too fast, remove the burner for a few seconds and then resume heating. The rate of heating should be *slow* near the melting point (about 1°C per minute) to ensure that the temperature increase is not faster than the rate at which heat can be transferred to the sample being observed. At the melting point, it is necessary that the mercury in the thermometer and the sample in the capillary tube be at temperature equilibrium.

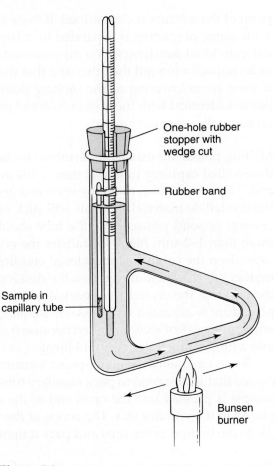

One-hole rubber
stopper with
wedge cut

Rubber band

Sample in
capillary tube

Bunsen
burner

Figure 9.3
A Thiele tube.

9.7 Determining the Melting Point—Electrical Instruments

Three types of electrically heated melting-point instruments are illustrated in Figure 9.4. In each case, the melting-point tube is filled as described in Section 9.5 and placed in a holder located just behind the magnifying eyepiece. The apparatus is operated by moving the switch to the ON position, adjusting the potentiometric control dial for the desired rate of heating, and observing the sample through the magnifying eyepiece. The temperature is read from a thermometer or, in the most modern instruments, from a digital display attached to a thermocouple. Your instructor will demonstrate and explain the type used in your laboratory.

Most electrically heated instruments do not heat or increase the temperature of the sample linearly. Although the rate of increase may be linear in the early stages of heating, it usually decreases and leads to a constant temperature at some upper limit. The upper-limit temperature is determined by the setting of the heating control. Thus, a family of heating curves is usually obtained for various control settings, as shown in Figure 9.5. The four hypothetical curves shown (1–4) might correspond to different control settings. For a compound melting at temperature t_1, the setting corresponding to curve 3 would be ideal. In the beginning of the curve, the temperature is increasing too rapidly to allow determination of an accurate melting point, but after the change in slope, the temperature increase will have slowed to a more usable rate.

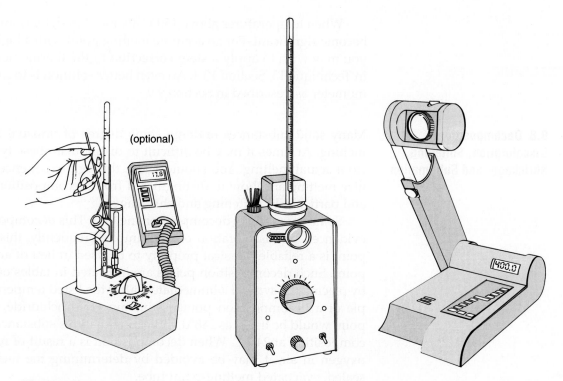

Figure 9.4
Melting-point apparatus.

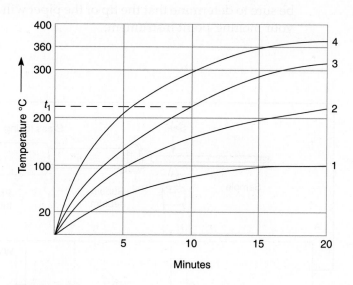

Figure 9.5
Heating-rate curves.

If the melting point of the sample is unknown, you can often save time by preparing two samples for melting-point determination. With one sample, you can rapidly determine a crude melting-point value. Then repeat the experiment more carefully using the second sample. For the second determination, you already have an approximate idea of what the melting-point temperature should be, and a proper rate of heating can be chosen.

When temperatures above 150°C are measured, thermometer errors can become significant. For an accurate melting point with a high-melting solid, you may wish to apply a **stem correction** to the thermometer as described in Technique 13, Section 13.4. An even better solution is to calibrate the thermometer as described in Section 9.9.

9.8 Decomposition, Discoloration, Softening, Shrinkage, and Sublimation

Many solid substances undergo some degree of unusual behavior before melting. At times it may be difficult to distinguish these types of behavior from actual melting. You should learn, through experience, how to recognize melting and how to distinguish it from decomposition, discoloration, and particularly, softening and shrinkage.

Some compounds decompose on melting. This decomposition is usually evidenced by discoloration of the sample. Frequently, this decomposition point is a reliable physical property to be used in lieu of an actual melting point. Such decomposition points are indicated in tables of melting points by placing the symbol *d* immediately after the listed temperature. An example of a decomposition point is thiamine hydrochloride, whose melting point would be listed as 248°d, indicating that this substance melts with decomposition at 248°C. When decomposition is a result of reaction with the oxygen in air, it may be avoided by determining the melting point in a sealed, evacuated melting-point tube.

Figure 9.6 shows two simple methods of evacuating a packed tube. Method A uses an ordinary melting-point tube, and method B constructs the melting-point tube from a disposable Pasteur pipet. Before using method B, be sure to determine that the tip of the pipet will fit into the sample holder in your melting-point instrument.

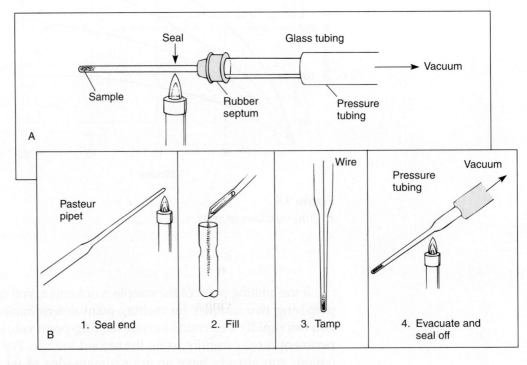

Figure 9.6
Evacuation and sealing of a melting-point capillary.

Method A

In method A, a hole is punched through a rubber septum using a large pin or a small nail, and the capillary tube is inserted from the inside, sealed end first. The septum is placed over a piece of glass tubing connected to a vacuum line. After the tube is evacuated, the upper end of the tube may be sealed by heating and pulling it closed.

Method B

In method B, the thin section of a 9-inch Pasteur pipet is used to construct the melting-point tube. Carefully seal the tip of the pipet using a flame. Be sure to hold the tip *upward* as you seal it. This will prevent water vapor from condensing inside the pipet. When the sealed pipet has cooled, the sample may be added through the open end using a microspatula. A small wire may be used to compress the sample into the closed tip. (If your melting-point apparatus has a vibrator, it may be used in place of the wire to simplify the packing.) When the sample is in place, the pipet is connected to the vacuum line with tubing and evacuated. The evacuated sample tube is sealed by heating it with a flame and pulling it closed.

Some substances begin to decompose *below* their melting points. Thermally unstable substances may undergo elimination reactions or anhydride formation reactions during heating. The decomposition products formed represent impurities in the original sample, so the melting point of the substance may be lowered due to their presence.

It is normal for many compounds to soften or shrink immediately before melting. Such behavior represents not decomposition but a change in the crystal structure or a mixing with impurities. Some substances "sweat," or release solvent of crystallization, before melting. These changes do not indicate the beginning of melting. Actual melting begins when the first drop of liquid becomes visible, and the melting range continues until the temperature is reached at which all the solid has been converted to the liquid state. With experience, you soon learn to distinguish between softening, or "sweating," and actual melting. If you wish, the temperature of the onset of softening or sweating may be reported as a part of your melting-point range: 211°C (softens), 223–225°C (melts).

Some solid substances have such a high vapor pressure that they sublime at or below their melting points. In many handbooks, the sublimation temperature is listed along with the melting point. The symbols *sub, subl,* and sometimes *s* are used to designate a substance that sublimes. In such cases, the melting-point determination must be performed in a sealed capillary tube to avoid loss of the sample. The simplest way to accomplish sealing a packed tube is to heat the open end of the tube in a flame and pull it closed with tweezers or forceps. A better way, although more difficult to master, is to heat the center of the tube in a small flame, rotating it about its axis, and keeping the tube straight, until the center collapses. If this is not done quickly, the sample may melt or sublime while you are working. With the smaller chamber, the sample will not be able to migrate to the cool top of the tube that may be above the viewing area. Figure 9.7 illustrates the method.

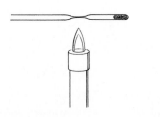

Figure 9.7

Sealing a tube for a substance that sublimes.

9.9 Thermometer Calibration

When a melting-point or boiling-point determination has been completed, you expect to obtain a result that duplicates the result recorded in a handbook or in the original literature. It is not unusual, however, to find a discrepancy of several degrees from the literature value. Such a discrepancy does not necessarily indicate that the experiment was incorrectly performed or that the material is impure; rather, it may indicate that the thermometer used for the determination was slightly in error. Most thermometers do not measure the temperature with perfect accuracy.

To determine accurate values, you must calibrate the thermometer that is used. This calibration is done by determining the melting points of a variety of standard substances with the thermometer. A plot is drawn of the observed temperature vs. the published value of each standard substance. A smooth line is drawn through the points to complete the chart. A correction chart prepared in this way is shown in Figure 9.8. This chart is used to correct any melting point determined with that particular thermometer. Each thermometer requires its own calibration curve. A list of suitable standard substances for calibrating thermometers is provided in Table 9.1. The standard substances, of course, must be pure in order for the corrections to be valid.

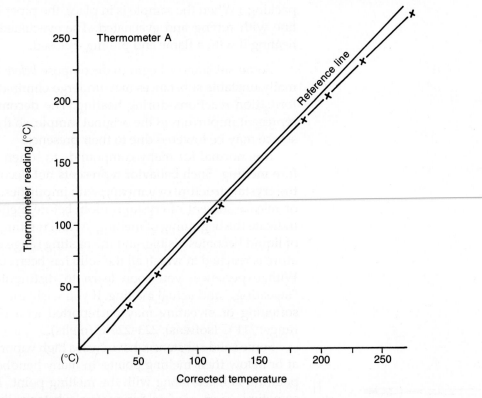

Figure 9.8
A thermometer-calibration curve.

Table 9.1 *Melting-point standards*

Compound	Melting Point (°C)
Ice (solid–liquid water)	0
Acetanilide	115
Benzamide	128
Urea	132
Succinic acid	189
3,5-Dinitrobenzoic acid	205

PROBLEMS

1. Two substances, A and B, have the same melting point. How can you determine if they are the same without using any form of spectroscopy? Explain in detail.

2. Using Figure 9.5, determine which heating curve would be most appropriate for a substance with a melting point of about 150°C.

3. What steps can you take to determine the melting point of a substance that sublimes before it melts?

4. A compound melting at 134°C was suspected to be either aspirin (mp 135°C) or urea (mp 133°C). Explain how you could determine whether one of these two suspected compounds was identical to the unknown compound without using any form of spectroscopy.

5. An unknown compound gave a melting point of 230°C. When the molten liquid solidified, the melting point was redetermined and found to be 131°C. Give a possible explanation for this discrepancy.

10   **TECHNIQUE 10**

Solubility

The solubility of a **solute** (a dissolved substance) in a **solvent** (the dissolving medium) is the most important chemical principle underlying three basic techniques you will study in the organic chemistry laboratory: crystallization, extraction, and chromatography. In this discussion of solubility, you will gain an understanding of the structural features of a substance that determine its solubility in various solvents. This understanding will help you to predict solubility behavior and to understand the techniques that are based on this property. Understanding solubility behavior will also help you understand what is going on during a reaction, especially when there is more than one liquid phase present or when a precipitate is formed.

10.1 Definition of Solubility

Although we often describe solubility behavior in terms of a substance being **soluble** (dissolved) or **insoluble** (not dissolved) in a solvent, solubility can be described more precisely in terms of the *extent* to which a substance is soluble. Solubility may be expressed in terms of grams of solute per liter (g/L) or milligrams of solute per milliliter (mg/mL) of solvent. Consider the solubilities at room temperature for the following three substances in water:

Cholesterol	0.002 mg/mL
Caffeine	22 mg/mL
Citric acid	620 mg/mL

In a typical test for solubility, 40 mg of solute is added to 1 mL of solvent. Therefore, if you were testing the solubility of these three substances, cholesterol would be insoluble, caffeine would be partially soluble, and citric acid would be soluble. Note that a small amount (0.002 mg) of cholesterol would dissolve. It is unlikely, however, that you would be able to observe this small amount dissolving, and you would report that cholesterol is insoluble. On the other hand, 22 mg (55%) of the caffeine would dissolve. It is likely that you would be able to observe this, and you would state that caffeine is partially soluble.

When the solubility of a liquid solute in a solvent is described, it is sometimes helpful to use the terms **miscible** and **immiscible.** Two liquids that are miscible will mix homogeneously (one phase) in all proportions. For example, water and ethyl alcohol are miscible. When they are mixed in any proportion, only one layer will be observed. When two liquids are miscible, it is also true that either one of them will be completely soluble in the other one. Two immiscible liquids do not mix homogeneously in all proportions, and under some conditions they will form two layers. Water and diethyl ether are immiscible. When mixed in roughly equal amounts, they will form two layers. However, each liquid is slightly soluble in the other one. Even when two layers are present, a small amount of water will be soluble in the diethyl ether, and a small amount of diethyl ether will be soluble in the water. Furthermore, if only a small amount of either one is added to the other, it may dissolve completely, and only one layer will be observed. For example, if a small amount of water (less than 1.2% at 20°C) is added to diethyl ether, the water will dissolve completely in the diethyl ether, and only one layer will be observed. When more water is added (more than 1.2%), some of the water will not dissolve, and two layers will be present.

Although the terms *solubility* and *miscibility* are related in meaning, it is important to understand that there is one essential difference. There can be different degrees of solubility, such as slightly, partially, very, and so on. Unlike solubility, miscibility does not have any degrees—a pair of liquids is either miscible or it is not.

10.2 Predicting Solubility Behavior

A major goal of this section is to explain how to predict whether a substance will be soluble in a given solvent. This is not always easy, even for an experienced chemist. However, guidelines will help you make a good guess about the solubility of a compound in a specific solvent. In discussing these guidelines, it is helpful to separate the types of solutions we will be looking at into two categories: solutions in which both the solvent and the solute are covalent (molecular) and ionic solutions, in which the solute ionizes and dissociates.

A. Solutions in Which the Solvent and Solute Are Molecular

A useful generalization in predicting solubility is the widely used rule "Like dissolves like." This rule is most commonly applied to polar and nonpolar compounds. According to this rule, a polar solvent will dissolve polar (or ionic) compounds, and a nonpolar solvent will dissolve nonpolar compounds.

The reason for this behavior involves the nature of intermolecular forces of attraction. Although we will not be focusing on the nature of these forces, it is helpful to know what they are called. The force of attraction between polar molecules is called **dipole–dipole interaction;** between nonpolar molecules, forces of attraction are called **van der Waals forces** (also called **London** or **dispersion forces**). In both cases, these attractive forces can occur between molecules of the same compound or different compounds. Consult your lecture textbook for more information on these forces.

To apply the rule "Like dissolves like," you must first determine whether a substance is polar or nonpolar. The polarity of a compound is dependent on both the polarities of the individual bonds and the shape of the molecule. For most organic compounds, evaluating these factors can become quite complicated because of the complexities of the molecules. However, it is possible to make some reasonable predictions just by looking at the types of atoms that a compound possesses. As you read the following guidelines, it is important to understand that although we often describe compounds as being polar or nonpolar, polarity is a matter of degree, ranging from nonpolar to highly polar.

Guidelines for Predicting Polarity and Solubility

1. All hydrocarbons are nonpolar.
   Examples:

$$CH_3CH_2CH_2CH_2CH_2CH_3$$

   Hexane   **Benzene**

   Hydrocarbons such as benzene are slightly more polar than hexane because of their pi (π) bonds, which allow for greater van der Waals or London attractive forces.

2. Compounds possessing the electronegative elements oxygen or nitrogen are polar.
   Examples:

$$\overset{\displaystyle O}{\overset{\|}{CH_3CCH_3}} \qquad CH_3CH_2OH \qquad \overset{\displaystyle O}{\overset{\|}{CH_3COCH_2CH_3}}$$

   Acetone   **Ethyl alcohol**   **Ethyl acetate**

$$CH_3CH_2NH_2 \qquad CH_3CH_2OCH_2CH_3 \qquad H_2O$$

   Ethylamine   **Diethyl ether**   **Water**

   The polarity of these compounds depends on the presence of polar C—O, C=O, OH, NH, and CN bonds. The compounds that are most polar are capable of forming hydrogen bonds (see Guideline 6) and have NH or OH bonds. Although all these compounds are polar, the degree of polarity ranges from slightly polar to highly polar. This is due to the effect on polarity of the shape of the molecule and size of the carbon chain and whether the compound can form hydrogen bonds.

3. The presence of halogen atoms, even though their electronegativities are relatively high, does not alter the polarity of an organic compound in a significant way. Therefore, these compounds are only slightly polar.

The polarities of these compounds are more similar to those of hydrocarbons, which are nonpolar, than to that of water, which is highly polar.

Examples:

CH_2Cl_2

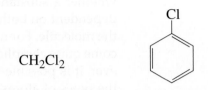

Methylene chloride (dichloromethane)   **Chlorobenzene**

4. When comparing organic compounds within the same family, note that adding carbon atoms to the chain decreases the polarity. For example, methyl alcohol (CH_3OH) is more polar than propyl alcohol ($CH_3CH_2CH_2OH$). The reason is that hydrocarbons are nonpolar, and increasing the length of a carbon chain makes the compound more hydrocarbon-like.

5. Compounds that contain four or fewer carbons and also contain oxygen or nitrogen are often soluble in water. Almost any functional group containing these elements will lead to water solubility for low-molecular-weight (up to C_4) compounds. Compounds having five or six carbons and containing one of these elements are often insoluble in water or have borderline solubility.

6. As mentioned earlier, the force of attraction between polar molecules is dipole–dipole interaction. A special case of dipole–dipole interaction is hydrogen bonding. Hydrogen bonding is a possibility when a compound possesses a hydrogen atom bonded to a nitrogen, oxygen, or fluorine atom. The bond is formed by the attraction between this hydrogen atom and a nitrogen, oxygen, or fluorine atom in another molecule. Hydrogen bonding may occur between two molecules of the same compound or between molecules of different compounds:

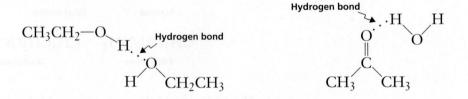

Hydrogen bonding is the strongest type of dipole–dipole interaction. When hydrogen bonding between solute and solvent is possible, solubility is greater than one would expect for compounds of similar polarity that cannot form hydrogen bonds. Hydrogen bonding is very important in organic chemistry, and you should be alert for situations in which hydrogen bonding may occur.

7. Another factor that can affect solubility is the degree of branching of the alkyl chain in a compound. Branching of the alkyl chain in a compound lowers the intermolecular forces between the molecules. This

is usually reflected in a greater solubility in water for the branched compound than for the corresponding straight-chain compound. This occurs simply because the molecules of the branched compounds are more easily separated from one another.

8. The solubility rule ("Like dissolves like") may be applied to organic compounds that belong to the same family. For example, 1-octanol (an alcohol) is soluble in the solvent ethyl alcohol. Most compounds within the same family have similar polarity. However, this generalization may not apply if there is a substantial difference in size between the two compounds. For example, cholesterol, an alcohol with a molecular weight (MW) of 386.64, is only slightly soluble in methanol (MW 32.04). The large hydrocarbon component of cholesterol negates the fact that they belong to the same family.

9. The stability of the crystal lattice also affects solubility. Other things being equal, the higher the melting point (the more stable the crystal) is, the less soluble the compound. For instance, *p*-nitrobenzoic acid (mp 242°C) is, by a factor of 10, less soluble in a fixed amount of ethanol than the *ortho* (mp 147°C) and *meta* (mp 141°C) isomers.

You can check your understanding of some of these guidelines by studying the list given in Table 10.1, which is given in order of increasing polarity. The structures of these compounds are given on pages 639–640.

This list can be used to make some predictions about solubility, based on the rule "Like dissolves like." Substances that are close to one another on this list will have similar polarities. Thus, you would expect hexane to be soluble in methylene chloride but not in water. Acetone should be soluble in ethyl alcohol. On the other hand, you might predict that ethyl alcohol would be insoluble in hexane. However, ethyl alcohol is soluble in hexane because ethyl alcohol is somewhat less polar than methyl alcohol or water.

Table 10.1 *Compounds in increasing order of polarity*

Increasing Polarity

Aliphatic hydrocarbons
  Hexane (nonpolar)
Aromatic hydrocarbons (π bonds)
  Benzene (nonpolar)
Halocarbons
  Methylene chloride (slightly polar)
Compounds with polar bonds
  Diethyl ether (slightly polar)
  Ethyl acetate (intermediate polarity)
  Acetone (intermediate polarity)
Compounds with polar bonds and hydrogen bonding
  Ethyl alcohol (intermediate polarity)
  Methyl alcohol (intermediate polarity)
  Water (highly polar)

Table 10.2 *Solvents in increasing order of polarity*

Increasing Polarity (Approximate)

RH	Alkanes (hexane, petroleum ether)
ArH	Aromatics (benzene, toluene)
ROR	Ethers (diethyl ether)
RX	Halides ($CH_2Cl_2 > CHCl_3 > CCl_4$)
RCOOR	Esters (ethyl acetate)
RCOR	Aldehydes, ketones (acetone)
RNH_2	Amines (triethylamine, pyridine)
ROH	Alcohols (methanol, ethanol)
$RCONH_2$	Amides (N,N-dimethylformamide)
RCOOH	Organic acids (acetic acid)
H_2O	Water

This last example demonstrates that you must be careful in using the guidelines on polarity for predicting solubilities. Ultimately, solubility tests must be done to confirm predictions until you gain more experience.

The trend in polarities shown in Table 10.1 can be expanded by including more organic families. The list in Table 10.2 gives an approximate order for the increasing polarity of organic functional groups. It may appear that there are some discrepancies between the information provided in these two tables. The reason is that Table 10.1 provides information about specific compounds, whereas the trend shown in Table 10.2 is for major organic families and is approximate.

B. Solutions in Which the Solute Ionizes and Dissociates

Many ionic compounds are highly soluble in water because of the strong attraction between ions and the highly polar water molecules. This also applies to organic compounds that can exist as ions. For example, sodium acetate consists of Na^+ and CH_3COO^- ions, which are highly soluble in water. Although there are some exceptions, you may assume that all organic compounds that are in the ionic form will be water soluble.

The most common way by which organic compounds become ions is in acid–base reactions. For example, carboxylic acids can be converted to water-soluble salts when they react with dilute aqueous NaOH:

$$CH_3CH_2CH_2CH_2CH_2CH_2\overset{\overset{\displaystyle O}{\|}}{C}OH + NaOH\ (aq) \longrightarrow$$
Water-insoluble carboxylic acid

$$CH_3CH_2CH_2CH_2CH_2CH_2\overset{\overset{\displaystyle O}{\|}}{C}O^-\ Na^+ + H_2O$$
Water-soluble salt

The water-soluble salt can then be converted back to the original carboxylic acid (which is insoluble in water) by adding another acid (usually aqueous HCl) to the solution of the salt. The carboxylic acid precipitates out of solution.

Amines, which are organic bases, can also be converted to water-soluble salts when they react with dilute aqueous HCl:

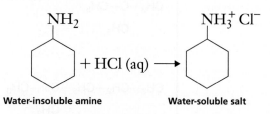

Water-insoluble amine   **Water-soluble salt**

This salt can be converted back to the original amine by adding a base (usually aqueous NaOH) to the solution of the salt.

10.3 Organic Solvents

Organic solvents must be handled safely. Always remember that organic solvents are all at least mildly toxic and that many are flammable. You should become thoroughly familiar with laboratory safety (see Technique 1, pp. 542–558).

The most common organic solvents are listed in Table 10.3 along with their boiling points. Solvents marked in boldface type will burn. Ether, pentane, and hexane are especially dangerous; if they are combined with the correct amount of air, they will explode.

The terms **petroleum ether** and **ligroin** are often confusing. Petroleum ether is a mixture of hydrocarbons with isomers of formulas C_5H_{12} and

Table 10.3 *Common organic solvents*

Solvent	Bp (°C)	Solvent	Bp (°C)
Hydrocarbons		Ethers	
Pentane	36	**Ether** (diethyl)	35
Hexane	69	Dioxane[a]	101
Benzene[a]	80	**1,2-Dimethoxyethane**	83
Toluene	111	Others	
Hydrocarbon mixtures		Acetic acid	118
Petroleum ether	30–60	Acetic anhydride	140
Ligroin	60–90	**Pyridine**	115
Chlorocarbons		**Acetone**	56
Methylene chloride	40	**Ethyl acetate**	77
Chloroform[a]	61	Dimethylformamide	153
Carbon tetrachloride[a]	77	Dimethylsulfoxide	189
Alcohols			
Methanol	65		
Ethanol	78		
Isopropyl alcohol	82		

Note: **Boldface type** indicates flammability.
[a]Suspect carcinogen (see p. 557).

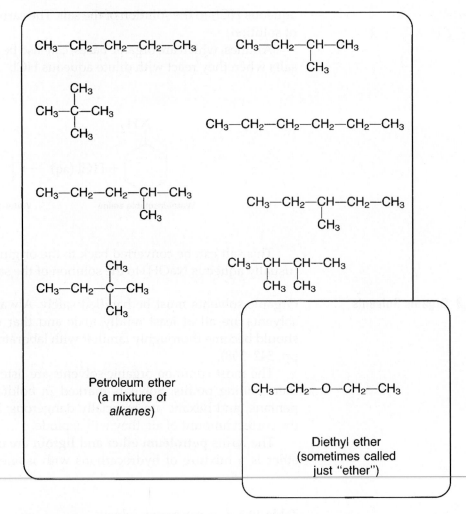

Figure 10.1
A comparison between "ether" (diethyl ether) and "petroleum ether."

C_6H_{14} predominating. Petroleum ether is not an ether at all because there are no oxygen-bearing compounds in the mixture. In organic chemistry, an ether is usually a compound containing an oxygen atom to which two alkyl groups are attached. Figure 10.1 shows some of the hydrocarbons that appear commonly in petroleum ether. It also shows the structure of ether (diethyl ether). Use special care when instructions call for either **ether** or **petroleum ether;** the two must not become accidentally confused. Confusion is particularly easy when one is selecting a container of solvent from the supply shelf.

Ligroin, or high-boiling petroleum ether, is like petroleum ether in composition except that compared with petroleum ether, ligroin generally includes higher-boiling alkane isomers. Depending on the supplier, ligroin may have different boiling ranges. Whereas some brands of ligroin have boiling points ranging from about 60°C to about 90°C, other brands have boiling points ranging from about 60°C to about 75°C. The boiling-point ranges of petroleum ether and ligroin are often included on the labels of the containers.

PROBLEMS

1. For each of the following pairs of solutes and solvent, predict whether the solute would be soluble or insoluble. After making your predictions, you can check your answers by looking up the compounds in *The Merck Index* or the *CRC Handbook of Chemistry and Physics*. Generally, *The Merck Index* is the easier reference book to use. If the substance has a solubility greater than 40 mg/mL, you may conclude that it is soluble.

   a. Malic acid in water

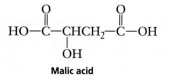

   Malic acid

   b. Naphthalene in water

   Naphthalene

   c. Amphetamine in ethyl alcohol

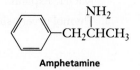

   Amphetamine

   d. Aspirin in water

   Aspirin

   e. Succinic acid in hexane (*Note:* the polarity of hexane is similar to that of petroleum ether.)

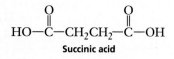

   Succinic acid

   f. Ibuprofen in diethyl ether

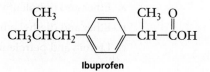

   Ibuprofen

g. 1-Decanol (*n*-decyl alcohol) in water

$$CH_3(CH_2)_8CH_2OH$$
1-Decanol

2. Predict whether the following pairs of liquids would be miscible or immiscible:

   a. Water and methyl alcohol

   b. Hexane and benzene

   c. Methylene chloride and benzene

   d. Water and toluene

Toluene

   e. Ethyl alcohol and isopropyl alcohol

OH
$$CH_3CHCH_3$$
Isopropyl alcohol

3. Would you expect ibuprofen (see problem 1f) to be soluble or insoluble in 1.0 *M* NaOH? Explain.

4. Thymol is very slightly soluble in water and very soluble in 1.0 *M* NaOH. Explain.

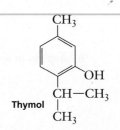
Thymol

5. Although cannabinol and methyl alcohol are both alcohols, cannabinol is very slightly soluble in methyl alcohol at room temperature. Explain.

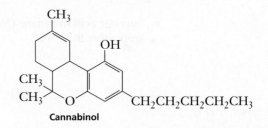
Cannabinol

6. What is the difference between the compounds in each of the following pairs?

   a. Ether and petroleum ether

   b. Ether and diethyl ether

   c. Ligroin and petroleum ether

TECHNIQUE 11

Crystallization: Purification of Solids

In most organic chemistry experiments, the desired product is first isolated in an impure form. If this product is a solid, the most common method of purification is crystallization. The general technique involves dissolving the material to be crystallized in a *hot* solvent (or solvent mixture) and cooling the solution slowly. The dissolved material has a decreased solubility at lower temperatures and will separate from the solution as it is cooled. This phenomenon is called either **crystallization,** if the crystal growth is relatively slow and selective, or **precipitation,** if the process is rapid and non-selective. Crystallization is an equilibrium process and produces very pure material. A small seed crystal is formed initially, and it then grows layer by layer in a reversible manner. In a sense, the crystal "selects" the correct molecules from the solution. In precipitation, the crystal lattice is formed so rapidly that impurities are trapped within the lattice. Therefore, any attempt at purification with too rapid a process should be avoided. Because the impurities are usually present in much smaller amounts than the compound being crystallized, most of the impurities will remain in the solvent even when it is cooled. The purified substance can then be separated from the solvent and from the impurities by filtration.

In microscale organic work, two methods are commonly used to perform crystallizations. The first method, which is carried out with an Erlenmeyer flask to dissolve the material and a Hirsch funnel to filter the crystals, is normally used when the weight of solid to be crystallized is more than 0.1 g. This technique, called **semimicroscale crystallization,** is discussed in Section 11.3. The second method is performed with a Craig tube and is used with smaller amounts of solid. Referred to as **microscale crystallization,** this technique is discussed in Section 11.4. The weight of solid to be crystallized, however, is not the only factor to consider when choosing a method for crystallization. Because the solubility of a substance in a given solvent must also be taken into account, the weight, 0.1 g, should not be adhered to rigidly in determining which method to use. In this textbook, you will usually be advised which method to use in the experimental procedure.

The method described here for semimicroscale crystallizations is nearly identical to that used for crystallizing larger amounts of materials than those encountered in this textbook. Therefore, this technique can also be used to perform crystallizations at the macroscale level (more than several grams).

PART A. THEORY

11.1 Solubility

The first problem in performing a crystallization is selecting a solvent in which the material to be crystallized shows the desired solubility behavior. In an ideal case, the material should be sparingly soluble at room temperature and yet quite soluble at the boiling point of the solvent selected. The solubility curve should be steep, as can be seen in line A of Figure 11.1. A curve with a low slope (line B) would not cause significant crystallization when the temperature of the solution was lowered. A solvent in which the

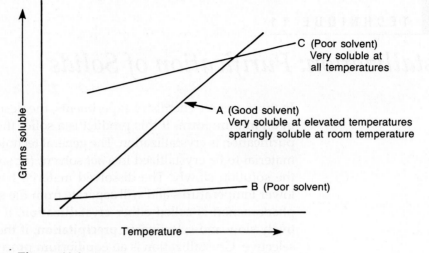

Figure 11.1
Graph of solubility vs. temperature.

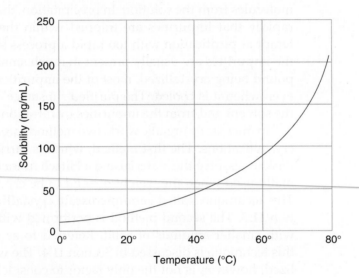

Figure 11.2
Solubility of sulfanilamide in 95% ethyl alcohol.

material is very soluble at all temperatures (line C) also would not be a suitable crystallization solvent. The basic problem in performing a crystallization is to select a solvent (or mixed solvent) that provides a steep solubility-vs.-temperature curve for the material to be crystallized. A solvent that allows the behavior shown in line A is an ideal crystallization solvent. It should also be mentioned that solubility curves are not always linear, as they are depicted in Figure 11.1. This figure represents an idealized form of solubility behavior. The solubility curve for sulfanilamide in 95% ethyl alcohol, shown in Figure 11.2, is typical of many organic compounds and shows what solubility behavior might look like for a real substance.

The solubility of organic compounds is a function of the polarities of both the solvent and the **solute** (dissolved material). A general rule is "Like dissolves like." If the solute is very polar, a very polar solvent is needed to dissolve it; if the solute is nonpolar, a nonpolar solvent is needed. Applications of this rule are discussed extensively in Technique 10, Section 10.2, page 638, and in Section 11.5, page 660.

11.2 Theory of Crystallization

A successful crystallization depends on a large difference between the solubility of a material in a hot solvent and its solubility in the same solvent when it is cold. When the impurities in a substance are equally soluble in both the hot and the cold solvent, an effective purification is not easily achieved through crystallization. A material can be purified by crystallization when both the desired substance and the impurity have similar solubilities, but only when the impurity represents a small fraction of the total solid. The desired substance will crystallize on cooling, but the impurities will not.

For example, consider a case in which the solubilities of substance A and its impurity B are both 1 g/100 mL of solvent at 20°C and 10 g/100 mL of solvent at 100°C. In the impure sample of A, the composition is 9 g of A and 2 g of B. In the calculations for this example, it is assumed that the solubilities of both A and B are unaffected by the presence of the other substance. To make the calculations easier to understand, 100 mL of solvent are used in each crystallization. Normally, the minimum amount of solvent required to dissolve the solid would be used.

At 20°C, this total amount of material will not dissolve in 100 mL of solvent. However, if the solvent is heated to 100°C, all 11 g dissolve. The solvent has the capacity to dissolve 10 g of A *and* 10 g of B at this temperature. If the solution is cooled to 20°C, only 1 g of each solute can remain dissolved, so 8 g of A and 1 g of B crystallize, leaving 2 g of material in the solution. This crystallization is shown in Figure 11.3. The solution that remains after a crystallization is called the **mother liquor.** If the process is now repeated by treating the crystals with 100 mL of fresh solvent, 7 g of A will crystallize again, leaving 1 g of A and 1 g of B in the mother liquor. As a result of these operations, 7 g of pure A are obtained, but with the loss of 4 g of material (2 g of A plus 2 g of B). Again, this second crystallization step is illustrated in Figure 11.3. The final result illustrates an important aspect of crystallization—it is wasteful. Nothing can be done to prevent this waste; some A must be lost along with the impurity B for the method to be successful. Of course, if the impurity B were *more* soluble than A in the solvent, the losses would be reduced. Losses could also be reduced if the impurity were present in *much smaller* amounts than the desired material.

Note that in the preceding case, the method operated successfully because A was present in substantially larger quantity than its impurity B. If there had been a 50-50 mixture of A and B initially, no separation would

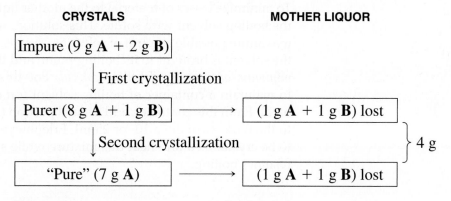

Figure 11.3
Purification of a mixture by crystallization.

have been achieved. In general, a crystallization is successful only if there is a *small* amount of impurity. As the amount of impurity increases, the loss of material must also increase. Two substances with nearly equal solubility behavior, present in equal amounts, cannot be separated. If the solubility behavior of two components present in equal amounts is different, however, a separation or purification is frequently possible.

In the preceding example, two crystallization procedures were performed. Normally, this is not necessary; however, when it is, the second crystallization is more appropriately called **recrystallization.** As illustrated in this example, a second crystallization results in purer crystals, but the yield is lower.

In some experiments, you will be instructed to cool the crystallizing mixture in an ice-water bath before collecting the crystals by filtration. Cooling the mixture increases the yield by decreasing the solubility of the substance; however, even at this reduced temperature, some of the product will be soluble in the solvent. It is not possible to recover all your product in a crystallization procedure even when the mixture is cooled in an ice-water bath. A good example of this is illustrated by the solubility curve for sulfanilamide shown in Figure 11.2. The solubility of sulfanilamide at 0°C is still significant, 14 mg/mL.

PART B. SEMIMICROSCALE CRYSTALLIZATION

11.3 Semimicroscale Crystallization—Hirsch Funnel

The crystallization technique described in this section is used when the weight of solid to be crystallized is more than 0.1 g. The four main steps in a semimicroscale crystallization are

1. Dissolving the solid
2. Removing insoluble impurities (when necessary)
3. Crystallization
4. Isolation of crystals

These steps are illustrated in Figure 11.4. It should be pointed out that a microscale crystallization with a Craig tube involves the same four steps, although the apparatus and procedures are somewhat different (see Section 11.4).

A. Dissolving the Solid

To minimize losses of material to the mother liquor, it is desirable to *saturate* the boiling solvent with solute. This solution, when cooled, will return the maximum possible amount of solute as crystals. To achieve this high return, the solvent is brought to its boiling point, and the solute is dissolved in the *minimum amount (!) of boiling solvent.* For this procedure, it is advisable to maintain a container of boiling solvent (on either a hot plate or a sand bath). From this container, add a small portion (about 0.5 mL) of the solvent to the flask[1] (usually a 10- or 25-mL Erlenmeyer flask) containing the solid to be crystallized and heat this mixture while swirling occasionally until it resumes boiling.

---

[1] A beaker should not be used, because the large opening allows the solvent to evaporate too rapidly and dust particles to get in too easily.

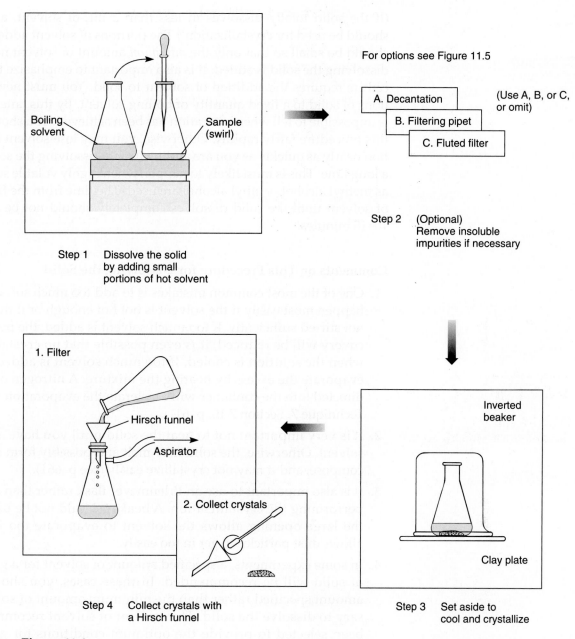

Figure 11.4

Steps in a semimicroscale crystallization (no decolorization).

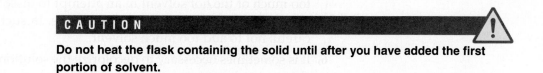

CAUTION

Do not heat the flask containing the solid until after you have added the first portion of solvent.

If the solid does not dissolve in the first portion of boiling solvent, then another small portion of boiling solvent is added to the flask. The mixture is heated again until it resumes boiling. If the solid dissolves, no more solvent is added. But if the solid has not dissolved, another portion of boiling solvent is added as before, and the process is repeated until the solid dissolves.

(If the solid totally dissolves in less than 2 mL of solvent, a Craig tube should be used for crystallization.) The portions of solvent added each time should be small so that only the *minimum* amount of solvent necessary for dissolving the solid is added. It is also important to emphasize that the procedure requires the addition of solvent to solid. You must never add portions of solid to a fixed quantity of boiling solvent. By this latter method, it is impossible to tell when saturation has been achieved. You should perform this procedure fairly rapidly. Otherwise, you may lose solvent by evaporation nearly as quickly as you are adding it, and dissolving the solid will take a long time. This is most likely to happen when highly volatile solvents such as methyl alcohol or ethyl alcohol are used. The time from the first addition of solvent until the solid dissolves completely should not be longer than 10–15 minutes.

Comments on This Procedure for Dissolving the Solid

1. One of the most common mistakes is to add too much solvent. This can happen most easily if the solvent is not hot enough or if the mixture is not stirred sufficiently. If too much solvent is added, the percentage recovery will be reduced; it is even possible that no crystals will form when the solution is cooled. If too much solvent is added, you must evaporate the excess by heating the mixture. A nitrogen or air stream directed into the container will accelerate the evaporation process (see Technique 7, Section 7.10, p. 611).

2. It is very important not to heat the solid until you have added some solvent. Otherwise, the solid may melt and possibly form an oil or decompose, and it may not crystallize easily (see p. 661).

3. It is also important to use an Erlenmeyer flask rather than a beaker for performing the crystallization. A beaker should not be used because the large opening allows the solvent to evaporate too rapidly and allows dust particles to get in too easily.

4. In some experiments, a specified amount of solvent for a given weight of solid will be recommended. In these cases, you should use the amount specified rather than the minimum amount of solvent necessary to dissolve the solid. The amount of solvent recommended has been selected to provide the optimum conditions for good crystal formation.

5. Occasionally, you may encounter an impure solid that contains small particles of insoluble impurities, pieces of dust, or paper fibers that will not dissolve in the hot crystallizing solvent. A common error is to add too much of the hot solvent in an attempt to dissolve these small particles, not realizing that they are insoluble. In such cases, you must be careful not to add too much solvent.

6. It is sometimes necessary to decolorize the solution by adding activated charcoal or by passing the solution through a column containing alumina or silica gel (see Section 11.7 and Technique 19, Section 19.14, p. 773). A decolorization step should be performed only if the mixture is *highly* colored and it is clear that the color is due to impurities and not to the actual color of the substance being crystallized. If decolorization is necessary, it should be accomplished before the following filtration step.

B. Removing Insoluble Impurities

It is necessary to use one of the following three methods only if insoluble material remains in the hot solution or if decolorizing charcoal has been used.

NOTE: Indiscriminate use of the procedure can lead to needless loss of your product.

Decantation is the easiest method of removing solid impurities and should be considered first. A filtering pipet is used when the volume of liquid to be filtered is less than 10 mL (see Technique 8, Section 8.1, Part C, p. 618), and you should use gravity filtration through a fluted filter when the volume is 10 mL or greater (see Technique 8, Section 8.1, Part B, p. 618). These three methods are illustrated in Figure 11.5.

Decantation. If the solid particles are relatively large or they easily settle to the bottom of the flask, it may be possible to separate the hot solution from the impurities by carefully pouring off the liquid, leaving the solid behind. This is accomplished most easily by holding a glass stirring rod along the top of the flask and tilting the flask so that the liquid pours out along one end of the glass rod into another container. A technique similar in principle to decantation, which may be easier to perform with smaller amounts of liquid, is to use a **preheated Pasteur pipet** to remove the hot solution. With this method, it may be helpful to place the tip of the pipet against the bottom of the flask when removing the last portion of solution. The small space between the tip of the pipet and the inside surface of the flask prevents solid material from being drawn into the pipet. An easy way to preheat the pipet is to draw up a small portion of hot *solvent* (not the *solution* being transferred) into the pipet and expel the liquid. Repeat this process several times.

Filtering Pipet. If the volume of solution after dissolving the solid in hot solvent is less than 10 mL, gravity filtration with a filtering pipet may be used to remove solid impurities. However, using a filtering pipet to filter a hot solution saturated with solute can be difficult without premature crystallization. The best way to prevent this from occurring is to add enough solvent to dissolve the desired product at room temperature (be sure not to add too much solvent) and perform the filtration at room temperature, as described in Technique 8, Section 8.1, Part C, p. 618). After filtration, the excess solvent is evaporated by boiling until the solution is saturated at the boiling point of the mixture (see Technique 7, Section 7.10, p. 611). If powdered decolorizing charcoal was used, it will probably be necessary to perform two filtrations with a filtering pipet, or else the method described next can be used.

Fluted Filter. This method is the most effective way to remove solid impurities when the volume of liquid is greater than 10 mL or when decolorizing charcoal has been used (see Technique 8, Section 8.1, Part B, p. 618). You should add a small amount of extra solvent to the hot mixture. This procedure helps to prevent crystal formation in the filter paper or the stem of the funnel during the filtration. The funnel is fitted with a fluted filter and installed at the top of the Erlenmeyer flask to be used for the actual filtration. It is advisable to place a small piece of wire between the funnel and the mouth of the flask to relieve any increase in pressure caused by hot filtrate.

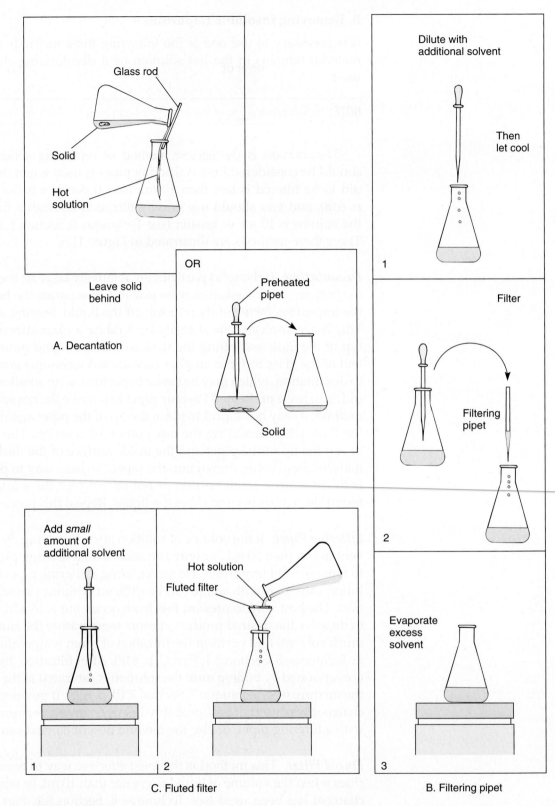

Figure 11.5

Methods for removing insoluble impurities in a semimicroscale crystallization.

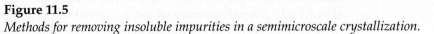

The Erlenmeyer flask containing the funnel and the fluted paper is placed on top of a sand bath or hot plate (low setting). The liquid to be filtered is brought to its boiling point and poured through the filter in portions. (If the volume of the mixture is less than 10 mL, it may be more convenient to transfer the mixture to the filter with a preheated Pasteur pipet.) It is necessary to keep the solutions in both flasks at their boiling temperatures to prevent premature crystallization. The refluxing action of the filtrate keeps the funnel warm and reduces the chance that the filter will clog with crystals that may have formed during the filtration. With low-boiling solvents, be aware that some solvent may be lost through evaporation. Consequently, extra solvent must be added to make up for this loss. If crystals begin to form in the filter during filtration, a minimum amount of boiling solvent is added to redissolve the crystals and to allow the solution to pass through the funnel. If the volume of liquid being filtered is less than 10 mL, a small amount of hot solvent should be used to rinse the filter after all the filtrate has been collected. The rinse solvent is then combined with the original filtrate.

After the filtration, it may be necessary to remove extra solvent by evaporation until the solution is saturated at the boiling point of the solvent (see Technique 7, Section 7.10, p. 611).

C. Crystallizing the Solid

An Erlenmeyer flask, not a beaker, should be used for crystallization. The large open top of a beaker makes it an excellent dust catcher. The narrow opening of the Erlenmeyer flask reduces contamination by dust and allows the flask to be stoppered if it is to be set aside for a long period. Mixtures set aside for long periods must be stoppered after cooling to room temperature to prevent evaporation of solvent. If all the solvent evaporates, no purification is achieved, and the crystals originally formed become coated with the dried contents of the mother liquor. Even if the time required for crystallization to occur is relatively short, it is advisable to cover the top of the Erlenmeyer flask with a small watch glass or inverted beaker to prevent evaporation of solvent while the solution is cooling to room temperature.

The chances of obtaining pure crystals are improved if the solution cools to room temperature slowly. When the volume of solution is 10 mL or less, the solution is likely to cool more rapidly than is desired. This can be prevented by placing the flask on a surface that is a poor heat conductor and covering the flask with a beaker to provide a layer of insulating air. Appropriate surfaces include a clay plate or several pieces of filter paper on top of the laboratory bench. It may also be helpful to use a clay plate that has been warmed slightly on a hot plate or in an oven.

After crystallization has occurred, it is sometimes desirable to cool the flask in an ice-water bath. Because the solute is less soluble at lower temperatures, this will increase the yield of crystals.

If a cooled solution does not crystallize, it will be necessary to induce crystallization. Several techniques are described in Section 11.8, Part A.

D. Collecting and Drying the Crystals

After the flask has been cooled, the crystals are collected by vacuum filtration through a Hirsch (or Büchner) funnel (see Technique 8, Section 8.3, p. 621, and Fig. 8.5). The crystals should be washed with a small amount of *cold* solvent to remove any mother liquor adhering to their surface. Hot or warm solvent will dissolve some of the crystals. The crystals should then be left for a short time in the funnel, where air, as it passes, will dry them free of most of the solvent. It is often wise to cover the Hirsch funnel with an oversize filter paper or towel during this air-drying. This precaution prevents accumulation of dust in the crystals. When the crystals are nearly dry, they should be gently scraped off (so paper fibers are not removed with the crystals) the filter paper onto a watch glass or clay plate for further drying (see Section 11.9).

PART C. MICROSCALE CRYSTALLIZATION

11.4 Microscale Crystallization—Craig Tube

In most microscale experiments, the amount of solid to be crystallized is small enough (generally less than 0.1 g) that a **Craig tube** (see Fig. 8.10, p. 626) is the preferred method for crystallization. The main advantage of the Craig tube is that it minimizes the number of transfers of solid material, thus resulting in a greater yield of crystals. Also, the separation of the crystals from the mother liquor with the Craig tube is very efficient, and little time is required for drying the crystals. The steps involved are, in principle, the same as those performed when a crystallization is accomplished with an Erlenmeyer flask and a Hirsch funnel. The steps in a microscale crystallization using a Craig tube are illustrated in Figure 11.6.

A. Dissolving the Solid

In crystallizations in which a filtration step is not required to remove insoluble impurities such as dirt or activated charcoal, this first step can be performed directly in the Craig tube. Otherwise, use a small test tube. The solid is transferred to the Craig tube, and the appropriate solvent, contained in a test tube, is heated to boiling on an aluminum block. A small portion (several drops) of hot solvent is added to the Craig tube, which is subsequently heated on the aluminum block until the solution in the Craig tube starts to boil. The hot mixture should be stirred continuously with a microspatula using a twirling motion. Stirring not only helps to dissolve the solute but also prevents the boiling liquid from bumping. Additional portions of hot solvent are added until all the solid has dissolved. In order to obtain the maximum yield, it is important not to add too much solvent, although any excess solvent can be evaporated later. You should perform this procedure fairly rapidly. Otherwise you may lose solvent by evaporation nearly as quickly as you are adding it, and dissolving all the solid will take a long time. The time required to dissolve the solid should not be longer than 15 minutes.

In many of the experiments in this textbook, a specified amount of solvent for a given weight of solid is recommended. In these cases, use the amount specified rather than the minimum amount of solvent necessary for dissolving the solid. The amount of solvent recommended has been selected to provide the optimum conditions for good crystal formation.

If the mixture is *highly* colored, and it is clear that the color is due to impurities and not to the actual color of the substance being crystallized, it will be necessary to decolorize the liquid. If decolorization is necessary, it should be accomplished before the following filtration step. Decolorizing charcoal may be used or the mixture may be passed through an alumina or silica gel column (see Section 11.7, Parts B and C, and Technique 19, Section 19.14, p. 773).

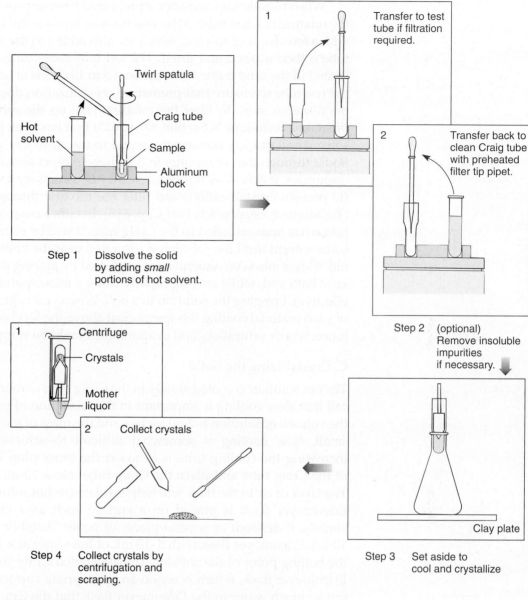

Step 1 Dissolve the solid by adding *small* portions of hot solvent.

1 Transfer to test tube if filtration required.

2 Transfer back to clean Craig tube with preheated filter tip pipet.

Step 2 (optional) Remove insoluble impurities if necessary.

Step 3 Set aside to cool and crystallize

Step 4 Collect crystals by centrifugation and scraping.

Figure 11.6
Steps in a microscale crystallization (no decolorization).

B. Removing Insoluble Impurities

You should be alert for the presence of impurities that will not dissolve in the hot solvent, no matter how much solvent is added. If it appears that most of the solid has dissolved and the remaining solid has no tendency to dissolve, or if the liquid has been decolorized with charcoal, it will be necessary to remove the solid particles. Two methods are discussed.

If the impurities are relatively large or concentrated in one part of the mixture, it may be possible to use a Pasteur pipet preheated with hot solvent to draw up the liquid without removing any solid. One way to do this is to expel the air from the pipet and then place the end of the pipet on the bottom of the tube, being careful not to trap any solid in the pipet. The small space between the pipet and the bottom of the tube should allow you to draw up the liquid without removing any solid.

When filtration is necessary, a preheated Pasteur pipet is used to transfer the mixture to a test tube. After this transfer is made the Craig tube is rinsed with a few drops of solvent, which are also added to the test tube. The Craig tube is then washed and dried. The test tube containing the mixture is also heated in the sand bath. An additional 5 to 10 drops of solvent are added to the test tube to ensure that premature crystallization does not occur during the filtration step. To filter the mixture, take up the mixture in a filter-tip pipet (see Technique 8, Section 8.6, p. 625) that has been preheated with hot solvent and quickly transfer the liquid to the clean Craig tube. Passing the liquid through the cotton plug in the filter-tip pipet should remove the solid impurities. If this does not occur, it may be necessary to add more solvent (to prevent crystallization) and filter the mixture through a filtering pipet (Technique 8, Section 8.1, Part C, p. 618). In either case, once the filtered solution has been returned to the Craig tube, it will be necessary to evaporate some solvent until the solution is saturated near the boiling point of the liquid. This is most conveniently accomplished by placing the Craig tube in the sand bath and, while stirring rapidly using a microspatula (twirling is most effective), bringing the solution to a boil. When you begin to observe a trace of solid material coating the spatula just above the level of the liquid, the solution is near saturation, and evaporation should be stopped.

C. Crystallizing the Solid

The hot solution is cooled slowly in the Craig tube to room temperature. Recall that slow cooling is important in the formation of pure crystals. When the volume of solution is 2 mL or less and the mass of glassware is relatively small, slow cooling is somewhat difficult to achieve. One method of increasing the cooling time is to insert the inner plug into the outer part of the Craig tube and place the Craig tube into a 10-mL Erlenmeyer flask. The layer of air in the flask will help insulate the hot solution as it cools. The Erlenmeyer flask is placed on a surface such as a clay plate (warmed slightly, if desired) or several pieces of paper. Another method is to fill a 10-mL Erlenmeyer flask with 8–10 mL of hot water at a temperature below the boiling point of the solvent. The assembled Craig tube is placed in the Erlenmeyer flask, which is set on an appropriate surface. Be careful not to put so much water in the Erlenmeyer flask that the Craig tube floats. After crystallization at room temperature is complete, the Craig tube can be placed in an ice-water bath to maximize the yield.

If crystals have not formed after the solution has cooled to room temperature, it will be necessary to induce crystallization. Several techniques are described in Section 11.8.

A common occurrence with crystallizations using a Craig tube is to obtain a seemingly solid mass of small crystals. This may not be a problem, but if there is very little mother liquor present or the crystals are impure, it may be necessary to repeat the crystallization. This situation may have resulted either because the cooling process occurred too rapidly or because the solubility–temperature curve was so steep for a given solvent that very little mother liquor remained after the crystallization. In either case, you may want to repeat the crystallization to obtain a better (purer) yield of crystals. Three measures may be taken to avoid this problem. A small amount of extra solvent may be added before heating the mixture again and allowing it to cool. A second measure is to cool the solution more slowly. Finally, it may be helpful to try to induce crystallization *before* the solution has cooled to room temperature.

D. Collecting and Drying the Crystals

When the crystals have formed and the mixture has cooled in an ice-water bath (if desired), the Craig tube is placed in a centrifuge tube and the crystals are separated from the mother liquor by centrifugation (see Technique 8, Section 8.8, p. 627). The crystals are then scraped off the end of the inner plug or from inside the Craig tube onto a watch glass or piece of paper. Minimal drying will be necessary (see Section 11.9)

The four steps in a semimicroscale or microscale crystallization are summarized in Table 11.1.

Table 11.1 *Steps in a semimicroscale or microscale crystallization*

A. Dissolving the Solid

1. Find a solvent with a steep solubility-vs.-temperature characteristic. (Done by trial and error using small amounts of material or by consulting a handbook.)
2. Heat the desired solvent to its boiling point.
3. Dissolve the solid in a **minimum** of boiling solvent (either in a flask or a Craig tube).
4. If necessary, add decolorizing charcoal or decolorize the solution on a silica gel or alumina column.

B. Removing Insoluble Impurities

1. Decant or remove the solution with a Pasteur pipet, or
2. Filter the hot solution through a fluted filter, a filtering pipet, or a filter tip pipet to remove insoluble impurities or charcoal.

NOTE: If no decolorizing charcoal has been added or if there are no undissolved particles, Part B should be omitted.

C. Crystallizing the Solid

1. Allow the solution to cool.
2. If crystals appear, cool the mixture in an ice-water bath (if desired) and go to Part D. If crystals do not appear, go to the next step.
3. Inducing crystallization
   (a) Scratch the flask with a glass rod; or, if using a Craig tube, dip a glass rod or spatula into the solution, let the liquid evaporate, and place the glass rod or spatula back into the solution to seed it.
   (b) Seed the solution with original solid, if available.
   (c) Cool the solution in an ice-water bath.
   (d) Evaporate excess solvent and allow the solution to cool again.

D. Collecting and Drying the Crystals

1. Collect crystals by vacuum filtration using a Hirsch funnel or by centrifugation using a Craig tube.
2. If using a Hirsch funnel, rinse crystals with a small portion of **cold** solvent.
3. Continue suction until crystals are nearly dry, if using vacuum filtration.
4. Drying.
   (a) Air-dry the crystals, or
   (b) Place the crystals in a drying oven, or
   (c) Dry the crystals *in vacuo*.

PART D. ADDITIONAL EXPERIMENTAL CONSIDERATIONS: SEMIMICROSCALE AND MICROSCALE

11.5 Selecting a Solvent

A solvent that dissolves little of the material to be crystallized when it is cold but a great deal of the material when it is hot is a good solvent for crystallization. Quite often, correct crystallization solvents are indicated in the experimental procedures that you will be following. When a solvent is not specified in a procedure, you can determine a good crystallization solvent by consulting a handbook or making an educated guess based on polarities, both discussed in this section. A third approach, involving experimentation, is discussed in Section 11.6.

With compounds that are well known, the correct crystallization solvent has already been determined through the experiments of earlier researchers. In such cases, the chemical literature can be consulted to determine which solvent should be used. Sources such as *The Merck Index* or the *CRC Handbook of Chemistry and Physics* may provide this information.

For example, consider naphthalene, which is found in *The Merck Index*, which states under the entry for naphthalene: "Monoclinic prismatic plates from ether." This statement means that naphthalene can be crystallized from ether. It also gives the type of crystal structure. Unfortunately, the crystal structure may be given without reference to the solvent. Another way to determine the best solvent is by looking at solubility-vs.-temperature data. When this is given, a good solvent is one in which the solubility of the compound increases significantly as the temperature increases. Sometimes, the solubility data will be given for only cold solvent and boiling solvent. This should provide enough information to determine whether this would be a good solvent for crystallization.

In most cases, however, the handbooks will state only whether a compound is soluble or not in a given solvent, usually at room temperature. Determining a good solvent for crystallization from this information can be somewhat difficult. The solvent in which the compound is soluble may or may not be an appropriate solvent for crystallization. Sometimes, the compound may be too soluble in the solvent at all temperatures, and you would recover little of your product if this solvent were used for crystallization. It is possible that an appropriate solvent would be the one in which the compound is nearly insoluble at room temperature because the solubility-vs.-temperature curve is steep. Although the solubility information may give you some ideas about what solvents to try, you will most likely need to determine a good crystallizing solvent by experimentation as described in Section 11.6.

When using *The Merck Index* or *Handbook of Chemistry and Physics*, you should be aware that alcohol is frequently listed as a solvent. This generally refers to 95% or 100% ethyl alcohol. Because 100% (absolute) ethyl alcohol is more expensive than 95% ethyl alcohol, the cheaper grade is usually used in the chemistry laboratory. Another solvent frequently listed is benzene. Benzene is a known carcinogen, so it is rarely used in student laboratories. Toluene is a suitable substitute; the solubility behavior of a substance in benzene and toluene is so similar that you may assume any statement made about benzene also applies to toluene.

Another way to identify a solvent for crystallization is to consider the polarities of the compound and the solvents. Generally, you would look for a solvent that has a polarity somewhat similar to that of the compound to be crystallized. Consider the compound sulfanilamide, shown in the figure. There are several polar bonds in sulfanilamide, the NH and the

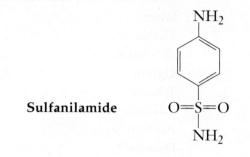

Sulfanilamide

SO bonds. In addition, the NH_2 groups and the oxygen atoms in sulfanilamide can form hydrogen bonds. Although the benzene ring portion of sulfanilamide is nonpolar, sulfanilamide has an intermediate polarity because of the polar groups. A common organic solvent of intermediate polarity is 95% ethyl alcohol. Therefore, it is likely that sulfanilamide would be soluble in 95% ethyl alcohol because they have similar polarities. (Note that the other 5% in 95% ethyl alcohol is usually a substance such as water or isopropyl alcohol, which does not alter the overall polarity of the solvent.) Although this kind of analysis is a good first step in determining an appropriate solvent for crystallization, without more information it is not enough to predict the shape of the solubility curve for the temperature-vs.-solubility data (see Figure 11.1, p. 648). Therefore, knowing that sulfanilamide is soluble in 95% ethyl alcohol does not necessarily mean that this is a good solvent for crystallizing sulfanilamide. You would still need to test the solvent to see if it is appropriate. The solubility curve for sulfanilamide (see Figure 11.2, page 648) indicates that 95% ethyl alcohol is a good solvent for crystallizing this substance.

When choosing a crystallization solvent, do not select one whose boiling point is higher than the melting point of the substance (solute) to be crystallized. If the boiling point of the solvent is too high, the substance may come out of solution as a liquid rather than a crystalline solid. In such a case, the solid may **oil out.** Oiling out occurs when, on cooling the solution to induce crystallization, the solute begins to come out of solution at a temperature above its melting point. The solute will then come out of solution as a liquid. Furthermore, as cooling continues, the substance may still not crystallize; rather, it will become a supercooled liquid. Oils may eventually solidify if the temperature is lowered, but often they will not actually crystallize. Instead, the solidified oil will be an amorphous solid or a hardened mass. In this case, purification of the substance will not have occurred as it does when the solid is crystalline. It can be difficult to deal with oils when trying to obtain a pure substance. You must try to redissolve them and hope that the substance will crystallize with slow, careful cooling. During the cooling period, it may be helpful to scratch the glass container where the oil is present with a glass stirring rod that has not been fire polished. Seeding the oil as it cools with a small sample of the original solid is another technique that is sometimes helpful in working with difficult oils. Other methods of inducing crystallization are discussed in Section 11.8.

Table 11.2 *Common solvents for crystallization*

	Boils (°C)	Freezes (°C)	Soluble in H_2O	Flammability
Water	100	0	+	−
Methanol	65	*	+	+
95% Ethanol	78	*	+	+
Ligroin	60–90	*	−	+
Toluene	111	*	−	+
Chloroform**	61	*	−	−
Acetic acid	118	17	+	+
Dioxane**	101	11	+	+
Acetone	56	*	+	+
Diethyl ether	35	*	Slightly	++
Petroleum ether	30–60	*	−	++
Methylene chloride	41	*	−	−
Carbon tetrachloride**	77	*	−	−

*Lower than 0°C (ice temperature).
**Suspected carcinogen.

One additional criterion for selecting the correct crystallization solvent is the **volatility** of that solvent. Volatile solvents have low boiling points or evaporate easily. A solvent with a low boiling point may be removed from the crystals through evaporation without much difficulty. It will be difficult to remove a solvent with a high boiling point from the crystals without heating them under vacuum.

Table 11.2 lists common crystallization solvents. The solvents used most commonly are listed in the table first.

11.6 Testing Solvents for Crystallization

When the appropriate solvent is not known, select a solvent for crystallization by experimenting with various solvents and a small amount of the material to be crystallized. Experiments are conducted on a small test tube scale before the entire quantity of material is committed to a particular solvent. Such trial-and-error methods are common when trying to purify a solid material that has not been studied.

Procedure

1. Place about 0.05 g of the sample in a test tube.

2. Add about 0.5 mL of solvent at room temperature and stir the mixture by rapidly twirling a microspatula between your fingers. If all (or almost all) of the solid dissolves at room temperature, then your solid is *probably* too soluble in this solvent and little compound would be recovered if this solvent were used. Select another solvent.

3. If none (or very little) of the solid dissolves at room temperature, heat the tube carefully and stir with a spatula. (A hot waterbath is perhaps better than an aluminum block because you can more easily control the temperature of the hot waterbath. The temperature of the hot waterbath should be slightly higher than the boiling point of the solvent.)

Add more solvent dropwise while continuing to heat and stir. Continue adding solvent until the solid dissolves, but do not add more than about 1.5 mL (total) of solvent. If all the solid dissolves, go to step 4. If all the solid has not dissolved by the time you have added 1.5 mL of solvent, this is probably not a good solvent. However, if most of the solid has dissolved at this point, you might try adding a little more solvent. Remember to heat and stir at all times during this step.

4. If the solid dissolves in about 1.5 mL or less of boiling solvent, then remove the test tube from the heat source, stopper the tube, and allow it to cool to room temperature. Then place it in an ice-water bath. If a lot of crystals come out, this is most likely a good solvent. If crystals do not come out, scratch the sides of the tube with a glass stirring rod to induce crystallization. If crystals still do not form, this is probably not a good solvent.

Comments about This Procedure

1. Selecting a good solvent is something of an art. There is no perfect procedure that can be used in all cases. You must think about what you are doing and use some common sense in deciding whether to use a particular solvent.

2. Do not heat the mixture above the melting point of your solid. This can occur most easily when the boiling point of the solvent is higher than the melting point of the solid. Normally, do not select a solvent that has a higher boiling point than the melting point of the substance. If you do, make certain that you do not heat the mixture beyond the melting point of your solid.

11.7 Decolorization

Small amounts of highly colored impurities may make the original crystallization solution appear colored; this color can often be removed by **decolorization,** either by using activated charcoal (often called Norit) or by passing the solution through a column packed with alumina or silica gel. A decolorizing step should be performed only if the color is due to impurities, not to the color of the desired product, and if the color is significant. Small amounts of colored impurities will remain in solution during crystallization, making the decolorizing step unnecessary. The use of activated charcoal is described separately for macroscale and microscale crystallizations, and then the column technique, which can be used with both crystallization techniques, is described.

A. Semimicroscale—Powdered Charcoal

As soon as the solute is dissolved in the minimum amount of boiling solvent, the solution is allowed to cool slightly, and a small amount of Norit (powdered charcoal) is added to the mixture. The Norit adsorbs the impurities. When performing a crystallization in which the filtration is performed with a fluted filter, you should add powdered Norit because it has a larger surface area and can remove impurities more effectively. A reasonable amount of Norit is what could be held on the end of a microspatula, or about 0.01–0.02 g. If too much Norit is used, it will adsorb product as well as impurities. A small amount of Norit should be used, and its use should be repeated if necessary. (It is difficult to determine if the initial amount added is sufficient until after the solution is filtered because the suspended particles of charcoal will obscure the color of the liquid.) Caution should be exercised so that the solution does not froth or erupt when the finely divided charcoal is added. The mixture is boiled with the Norit for several minutes and then filtered by gravity, using a fluted filter (see Section 11.3 and Technique 8, Section 8.1B, p. 618), and the crystallization is carried forward as described in Section 11.3.

The Norit preferentially adsorbs the colored impurities and removes them from the solution. The technique seems to be most effective with hydroxylic solvents. In using Norit, be careful not to breathe the dust. Normally, small quantities are used so that little risk of lung irritation exists.

B. Microscale—Pelletized Norit

If the crystallization is being performed in a Craig tube, it is advisable to use pelletized Norit. Although this is not as effective in removing impurities as powdered Norit, it is easier to remove, and the amount of pelletized Norit required is more easily determined because you can see the solution as it is being decolorized. Again, the Norit is added to the hot solution (the solution should not be boiling) after the solid has dissolved. This should be performed in a test tube rather than in a Craig tube. About 0.02 g is added, and the mixture is boiled for a minute or so to see if more Norit is required. More Norit is added, if necessary, and the liquid is boiled again. It is important not to add too much pelletized Norit because the Norit will also adsorb some of the desired material, and it is possible that not all the color can be removed no matter how much is added. The decolorized solution is then removed with a preheated filter-tip pipet (see Technique 8, Section 8.6, p. 625) to filter the mixture and transferred to a Craig tube for crystallization as described in Section 11.4.

C. Decolorization on a Column

The other method for decolorizing a solution is to pass the solution through a column containing alumina or silica gel. The adsorbent removes the colored impurities while allowing the desired material to pass through (see Technique 8, Figure 8.6, p. 623, and Technique 19, Section 19.14, p. 773). If this technique is used, it will be necessary to dilute the solution with additional solvent to prevent crystallization from occurring during the process. The excess solvent must be evaporated after the solution is passed through the column (Technique 7, Section 7.10, p. 611), and the crystallization procedure is continued as described in Sections 11.3 or 11.4.

11.8 Inducing Crystallization

If a cooled solution does not crystallize, several techniques may be used to induce crystallization. Although identical in principle, the actual procedures vary slightly when macroscale and microscale crystallizations are performed.

A. Semimicroscale

In the first technique, you should try scratching the inside surface of the flask vigorously with a glass rod that has not been fire polished. The motion of the rod should be vertical (in and out of the solution) and should be vigorous enough to produce an audible scratching. Such scratching often induces crystallization, although the effect is not well understood. The high-frequency vibrations may have something to do with initiating crystallization; or perhaps—a more likely possibility—small amounts of solution dry by evaporation on the side of the flask, and the dried solute is pushed into the solution. These small amounts of material would provide "seed crystals," or nuclei, on which crystallization may begin.

A second technique that can be used to induce crystallization is to cool the solution in an ice bath. This method decreases the solubility of the solute.

A third technique is useful when small amounts of the original material to be crystallized are saved. The saved material can be used to "seed" the cooled solution. A small crystal dropped into the cooled flask often will start the crystallization—this is called **seeding.**

If all these measures fail to induce crystallization, it is likely that too much solvent was added. The excess solvent must then be evaporated (Technique 7, Section 7.10, p. 611) and the solution allowed to cool.

B. Microscale

The strategy is basically the same as described for macroscale crystallizations. Scratching vigorously with a glass rod should be avoided, however, because the Craig tube is fragile and expensive. Scratching *gently* is allowed.

Another measure is to dip a spatula or glass stirring rod into the solution and allow the solvent to evaporate so that a small amount of solid will form on the surface of the spatula or glass rod. When placed back into the solution, the solid will seed the solution. A small amount of the original material, if some was saved, may also be used to seed the solution.

A third technique is to cool the Craig tube in an ice-water bath. This method may also be combined with either of the previous suggestions.

If none of these measures is successful, it is possible that too much solvent is present, and it may be necessary to evaporate some of the solvent (Technique 7, Section 7.10, p. 611) and allow the solution to cool again.

11.9 Drying Crystals

The most common method of drying crystals involves allowing them to dry in air. Several methods are illustrated in Figure 11.7, below. In all three methods, the crystals must be covered to prevent accumulation of dust particles. Note that in each method, the spout on the beaker provides an opening so that solvent vapor can escape from the system. The advantage of this method is that heat is not required, thus reducing the danger of

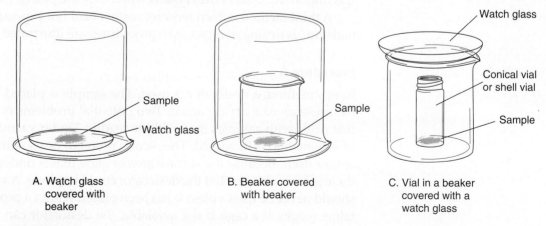

A. Watch glass covered with beaker

B. Beaker covered with beaker

C. Vial in a beaker covered with a watch glass

Figure 11.7
Methods for drying crystals in air.

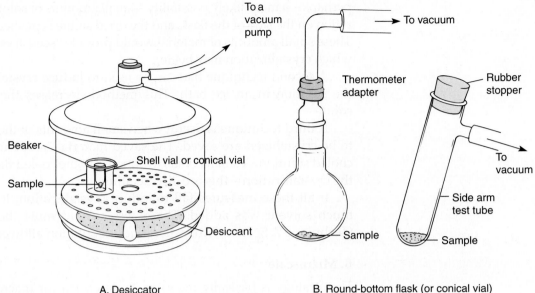

A. Desiccator

B. Round-bottom flask (or conical vial) or side arm test tube

Figure 11.8
Methods for drying crystals in a vacuum.

decomposition or melting; however, exposure to atmospheric moisture may cause the hydration of strongly hygroscopic materials. A **hygroscopic** substance is a substance that absorbs moisture from the air.

Another method of drying crystals is to place the crystals on a watch glass, a clay plate, or a piece of absorbent paper in an oven. Although this method is simple, some possible difficulties deserve mention. Crystals that sublime readily should not be dried in an oven because they might vaporize and disappear. Care should be taken that the temperature of the oven does not exceed the melting point of the crystals. Remember that the melting point of crystals is lowered by the presence of solvent; allow for this melting-point depression when selecting a suitable oven temperature. Some materials decompose on exposure to heat, and they should not be dried in an oven. Finally, when many different samples are being dried in the same oven, crystals might be lost due to confusion or reaction with another person's sample. It is important to label the crystals when they are placed in the oven.

A third method, which requires neither heat nor exposure to atmospheric moisture, is drying *in vacuo*. Two procedures are illustrated in Figure 11.8.

Procedure A

In this method, a desiccator is used. The sample is placed under vacuum in the presence of a drying agent. Two potential problems must be noted. The first deals with samples that sublime readily. Under vacuum, the likelihood of sublimation is increased. The second problem deals with the vacuum desiccator itself. Because the surface area of glass that is under vacuum is large, there is some danger that the desiccator could implode. A vacuum desiccator should never be used unless it has been placed within a protective metal container (cage). If a cage is not available, the desiccator can be wrapped with electrical or duct tape. If you use an aspirator as a source of vacuum, you should use a water trap (see Figure 8.5, p. 622).

Table 11.3 *Common solvent pairs for crystallization*

Methanol–water	Ether–acetone
Ethanol–water	Ether–petroleum ether
Acetic acid–water	Toluene–ligroin
Acetone–water	Methylene chloride–methanol
Ether–methanol	Dioxane[a]–water

[a]Suspected carcinogen.

Procedure B

This method can be accomplished with a round-bottom flask and a thermometer adapter equipped with a short piece of glass tubing, as illustrated in Figure 11.8B. In microscale work, the apparatus with the round-bottom flask can be modified by replacing the round-bottom flask with a conical vial. The glass tubing is connected by vacuum tubing to either an aspirator or a vacuum pump. A convenient alternative, using a side arm test tube, is also shown in Figure 11.8B. With either apparatus, install a water trap when an aspirator is used.

11.10 Mixed Solvents

Often, the desired solubility characteristics for a particular compound are not found in a single solvent. In these cases, a mixed solvent may be used. You simply select a first solvent in which the solute is soluble and a second solvent, miscible with the first, in which the solute is relatively insoluble. The compound is dissolved in a minimum amount of the boiling solvent in which it is soluble. Following this, the second hot solvent is added to the boiling mixture, dropwise, until the mixture barely becomes cloudy. The cloudiness indicates precipitation. At this point, more of the first solvent should be added. Just enough is added to clear the cloudy mixture. At that point, the solution is saturated, and as it cools, crystals should separate. Common solvent mixtures are listed in Table 11.3.

It is important not to add an excess of the second solvent or to cool the solution too rapidly. Either of these actions may cause the solute to oil out, or separate as a viscous liquid. If this happens, reheat the solution and add more of the first solvent.

PROBLEMS

1. Listed below are solubility-vs.-temperature data for an organic substance A dissolved in water.

Temperature (°C)	Solubility of A in 100 mL of Water (g)
0	1.5
20	3.0
40	6.5
60	11.0
80	17.0

    a. Graph the solubility of A vs. temperature. Use the data given in the table. Connect the data points with a smooth curve.

    b. Suppose 0.1 g of A and 1.0 mL of water were mixed and heated to 80°C. Would all the substance A dissolve?

    c. The solution prepared in (b) is cooled. At what temperature will crystals of A appear?

    d. Suppose the cooling described in (c) were continued to 0°C. How many grams of A would come out of solution? Explain how you obtained your answer.

2. What would likely happen if a hot saturated solution were filtered by vacuum filtration using a Hirsch funnel? (*Hint:* The mixture will cool as it comes in contact with the Hirsch funnel.)

3. A compound you have prepared is reported in the literature to have a pale yellow color. When the substance is dissolved in hot solvent to purify it by crystallization, the resulting solution is yellow. Should you use decolorizing charcoal before allowing the hot solution to cool? Explain your answer.

4. After a crude product is dissolved in 1.5 mL of hot solvent, the resulting solution is dark brown. Because the pure compound is reported in the literature to be colorless, it is necessary to perform a decolorizing procedure. Should you use pelletized Norit or powdered activated charcoal to decolorize the solution? Explain your answer.

5. While performing a crystallization, you obtain a light tan solution after dissolving your crude product in hot solvent. A decolorizing step is determined to be unnecessary, and there are no solid impurities present. Should you perform a filtration to remove impurities before allowing the solution to cool? Why or why not?

6. **a.** Draw a graph of a cooling curve (temperature vs. time) for a solution of a solid substance that shows no supercooling effects. Assume that the solvent does not freeze.

    b. Repeat the instructions in (a) for a solution for a solid substance that shows some supercooling behavior but eventually yields crystals if the solution is cooled sufficiently.

7. A solid substance A is soluble in water to the extent of 10 mg/mL of water at 25°C and 100 mg/mL of water at 100°C. You have a sample that contains 100 mg of A and an impurity B.

    a. Assuming that 2 mg of B are present along with 100 mg of A, describe how you can purify A if B is completely insoluble in water. Your description should include the volume of solvent required.

    b. Assuming that 2 mg of the impurity B are present along with 100 mg of A, describe how you can purify A if B has the same solubility behavior as A. Will one crystallization produce pure A? (Assume that the solubilities of both A and B are unaffected by the presence of the other substance.)

    c. Assume that 25 mg of the impurity B are present along with 100 mg of A. Describe how you can purify A if B has the same solubility behavior as A. Each time, use the minimum amount of water to just dissolve the solid. Will one crystallization produce absolutely pure A? How many crystallizations would be needed to produce pure A? How much A will have been recovered when the crystallizations have been completed?

8. An organic chemistry student dissolved 0.095 g of a crude product in 3.5 mL (the minimum amount required) of ethanol at 25°C. He cooled the solution in an icewater bath for 15 minutes and obtained beautiful crystals. He filtered the crystals on a Hirsch funnel and rinsed them with about 0.5 mL of ice-cold ethanol. After drying, the weight of the crystals was found to be 0.005 g. Why was the recovery so low?

12 **TECHNIQUE 12**

Extractions, Separations, and Drying Agents

PART A. THEORY

12.1 Extraction

Transferring a solute from one solvent into another is called **extraction,** or, more precisely, liquid–liquid extraction. The solute is extracted from one solvent into the other because the solute is more soluble in the second solvent than in the first. The two solvents must not be **miscible** (mix freely), and they must form two separate **phases,** or layers, in order for this procedure to work. Extraction is used in many ways in organic chemistry. Many natural products (organic chemicals that exist in nature) are present in animal and plant tissues having high water content. Extracting these tissues with a water-immiscible solvent is useful for isolating natural products. Often, diethyl ether (commonly referred to as "ether") is used for this purpose. Sometimes alternative water-immiscible solvents such as hexane, petroleum ether, ligroin, and methylene chloride are used. For instance, caffeine, a natural product, can be extracted from an aqueous tea solution by shaking it successively with several portions of methylene chloride.

A generalized extraction process that uses a conical vial is illustrated in Figure 12.1. The first solvent contains a mixture of black and white molecules (Fig. 12.1A). A second solvent that is not miscible with the first is added.

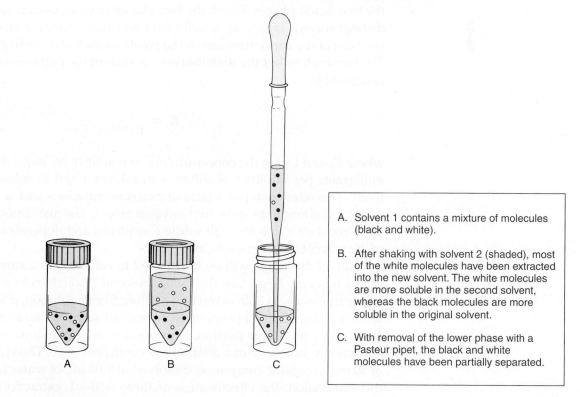

A. Solvent 1 contains a mixture of molecules (black and white).

B. After shaking with solvent 2 (shaded), most of the white molecules have been extracted into the new solvent. The white molecules are more soluble in the second solvent, whereas the black molecules are more soluble in the original solvent.

C. With removal of the lower phase with a Pasteur pipet, the black and white molecules have been partially separated.

Figure 12.1
The extraction process.

After the vial is capped and shaken, the layers separate. In this example, the second solvent is less dense, so it becomes the top layer (Fig. 12.1B). Because of differences in physical properties, the white molecules are more soluble in the second solvent, whereas the black molecules are more soluble in the first solvent. Most of the white molecules are in the upper layer, but there are some black molecules there, too. Likewise, most of the black molecules are in the lower layer. However, there are a few white molecules in this lower phase. A Pasteur pipet may be used to remove the lower layer (Fig. 12.1C). In this way, a partial separation of black and white molecules has been achieved. In this example, notice that it was not possible to effect a complete separation with one extraction. This is a common occurrence in organic chemistry. Many organic substances are soluble in both water and organic solvents.

Water can be used to extract or "wash" water-soluble impurities from an organic reaction mixture. To carry out a "washing" operation, you add water to the reaction mixture contained in a conical vial. After capping the vial and shaking it, you allow the organic layer and the aqueous (water) layer to separate from each other in the vial. A water wash removes highly polar and water-soluble materials, such as sulfuric acid, hydrochloric acid, or sodium hydroxide from the organic layer. A water wash can also be used to remove water-soluble and low-molecular-weight compounds, such as ethanol or acetic acid, from the organic layer. The washing operation helps purify the desired organic compound present in the original reaction mixture.

12.2 Distribution Coefficient

When a solution (solute A in solvent 1) is shaken with a second solvent (solvent 2) with which it is not miscible, the solute distributes itself between the two liquid phases. When the two phases have separated again into two distinct solvent layers, an equilibrium will have been achieved such that the ratio of the concentrations of the solute in each layer defines a constant. The constant, called the **distribution coefficient** (or partition coefficient) K, is defined by

$$K = \frac{C_2}{C_1}$$

where C_1 and C_2 are the concentrations at equilibrium, in grams per liter or milligrams per milliliter of solute A in solvent 1 and in solvent 2, respectively. This relationship is a ratio of two concentrations and is independent of the actual amounts of the two solvents mixed. The distribution coefficient has a constant value for each solute considered and depends on the nature of the solvents used in each case.

Not all the solute will be transferred to solvent 2 in a single extraction unless K is very large. Usually it takes several extractions to remove all the solute from solvent 1. In extracting a solute from a solution, it is always better to use several small portions of the second solvent than to make a single extraction with a large portion. Suppose that, as an illustration, a particular extraction proceeds with a distribution coefficient of 10. The system consists of 50 mg of organic compound dissolved in 1.00 mL of water (solvent 1). In this illustration, the effectiveness of three 0.50-mL extractions with ether (solvent 2) is compared with one 1.50-mL extraction with ether. In the first 0.50-mL extraction, the amount extracted into the ether layer is given by the following calculation. The amount of compound remaining in the aqueous phase is given by x.

$$K = 10 = \frac{C_2}{C_1} = \frac{\left(\dfrac{(50.0 - x)}{0.50} \dfrac{mg}{mL\ ether}\right)}{\left(\dfrac{x}{1.00} \dfrac{mg}{mL\ water}\right)} ; 10 = \frac{(50.0 - x)(1.00)}{0.50x}$$

$$5.0x = 50.0 - x$$

$$6.0x = 50.0$$

$$x = 8.3\ \text{mg remaining in the aqueous layer}$$

$$50.0 - x = 41.7\ \text{mg in the ether layer}$$

As a check on the calculation, it is possible to substitute the value 8.3 mg for x in the original equation and demonstrate that the concentration in the ether phase divided by the concentration in the water phase equals the distribution coefficient.

$$\frac{\left(\dfrac{(50.0 - x)}{0.50} \dfrac{mg}{mL\ ether}\right)}{\left(\dfrac{x}{1.00} \dfrac{mg}{mL\ water}\right)} = \frac{\dfrac{41.7}{0.50}}{\dfrac{8.3}{1.00}} = \frac{83\ mg/mL}{8.3\ mg/mL} = 10 = K$$

The second extraction with another 0.50-mL portion of fresh ether is performed on the aqueous phase, which now contains 8.3 mg of the solute. The amount of solute extracted is given by the calculation shown in Figure 12.2. Also shown in the figure is a calculation for a third extraction with another 0.50-mL portion of ether. This third extraction will transfer 1.2 mg of solute into the ether layer, leaving 0.2 mg of solute remaining in the water layer. A total of 49.8 mg of solute will be extracted into the combined ether layers, and 0.2 mg will remain in the aqueous phase.

Figure 12.3 shows the result of a *single* extraction with 1.50 mL of ether. As shown there, 46.9 mg of solute was extracted into the ether layer, leaving 3.1 mg of compound in the aqueous phase. One can see that three successive 0.50-mL ether extractions (Fig. 12.2) succeeded in removing 2.9 mg more solute from the aqueous phase than using one 1.50-mL portion of ether (Fig. 12.3). This differential represents 5.8% of the total material.

12.3 Choosing an Extraction Method and a Solvent

Three types of apparatus are used for extractions: conical vials, centrifuge tubes, and separatory funnels. These are shown in Figure 12.4. Conical vials may be used with volumes of less than 4 mL; volumes of up to 10 mL may be handled in centrifuge tubes. A centrifuge tube equipped with a screw cap is particularly useful for extractions. The separatory funnel is used in large-scale reactions. Each type of equipment is discussed in a separate section.

Before using a conical vial for an extraction, make sure that the capped conical vial does not leak when shaken. To do this, place some water in the conical vial, place the Teflon liner in the cap, and screw the cap securely onto the conical vial. Shake the vial vigorously and check for leaks. Conical vials that are used for extractions must not be chipped on the edge of the vial or they will not seal adequately. If there is a leak, try tightening the cap

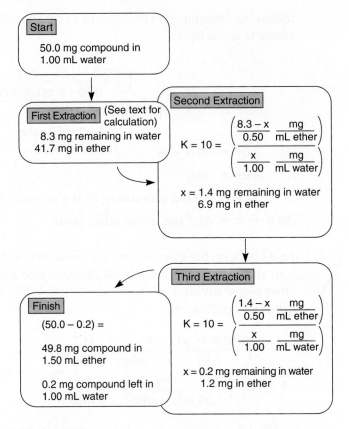

Figure 12.2
The result of extraction of 50.0 mg of compound in 1.00 mL
of water by three successive 0.50-mL portions of ether.
Compare this result with that of Figure 12.3.

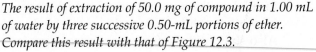

or replacing the Teflon liner with another one. Sometimes it helps to use the silicone rubber side of the liner to seal the conical vial. Some laboratories are supplied with Teflon stoppers that fit into the 5-mL conical vials. You may find that this stopper eliminates leakage.

When shaking the conical vial, do it gently at first in a rocking motion. When it is clear that an emulsion will not form (see Section 12.10, p. 684), you can shake it more vigorously.

In some cases, adequate mixing can be achieved by spinning your microspatula for at least 10 minutes in the conical vial. Another technique of mixing involves drawing up the mixture into a Pasteur pipet and squirting it rapidly back into the vial. Repeat this process for at least 5 minutes to obtain an adequate extraction.

If you are using a screw-cap centrifuge tube, put some water in the tube, cap it, and shake it vigorously to check for leaks. If the centrifuge tube leaks, try replacing the cap with another one. If available in the laboratory, a vortex mixer may be used to mix the phases. A vortex mixer works well with a variety of containers, including small flasks, test tubes, conical vials, and centrifuge tubes. You start the mixing action on a vortex mixer by holding the test tube or other container on one of the pads. The unit mixes the sample by high-frequency vibration.

Most extractions consist of an aqueous phase and an organic phase. In order to extract a substance from an aqueous phase, an organic solvent that

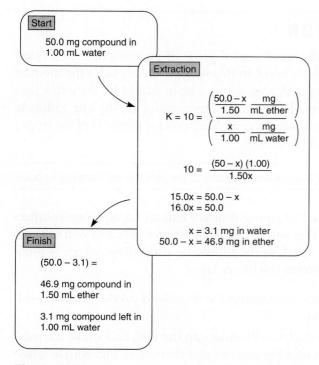

Figure 12.3
The result of extraction of 50.0 mg of compound in 1.00 mL of water with one 1.5-mL portion of ether. Compare this result with that of Figure 12.2.

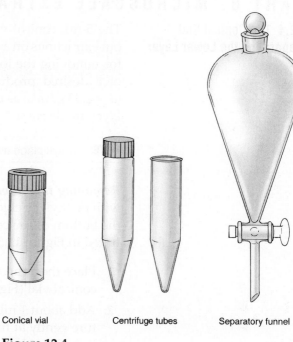

Conical vial   Centrifuge tubes   Separatory funnel

Figure 12.4
Apparatus used for extraction.

Table 12.1 *Densities of common extraction solvents*

Solvent	Density (g/mL)
Ligroin	0.67–0.69
Diethyl ether	0.71
Toluene	0.87
Water	1.00
Methylene chloride	1.330

is not miscible with water must be used. Table 12.1 lists a number of the common organic solvents that are not miscible with water and are used for extraction.

Those solvents that have a density less than that of water (1.00 g/mL) will separate as the top layer when shaken with water. Those solvents that have a greater density than water will separate into the lower layer. For instance, diethyl ether ($d = 0.71$ g/mL) when shaken with water will form the upper layer, whereas methylene chloride ($d = 1.33$ g/mL) will form the lower layer. When an extraction is performed, slightly different methods are used when you wish to separate the lower layer (whether it is the aqueous layer or the organic layer) than when you wish to separate the upper layer.

PART B. MICROSCALE EXTRACTION

12.4 The Conical Vial—Separating the Lower Layer

The 5-mL conical vial is the most useful piece of equipment for carrying out extractions on a microscale level. In this section, we consider the method for removing the lower layer. A concrete example would be the extraction of a desired product from an aqueous layer using methylene chloride ($d = 1.33$ g/mL) as the extraction solvent. Methods for removal of the upper layer are discussed in the next section.

NOTE: Always place a conical vial in a small beaker to prevent the vial from falling over.

Removing the Lower Layer. Suppose that we extract an aqueous solution with methylene chloride. This solvent is denser than water and will settle to the bottom of the conical vial. Use the following procedure, which is illustrated in Figure 12.5, to remove the lower layer.

1. Place the aqueous phase containing the dissolved product into a 5-mL conical vial (Fig. 12.5A).

2. Add about 1 mL of methylene chloride, cap the vial, and shake the mixture gently at first in a rocking motion and then more vigorously when it is clear that an emulsion will not form. Vent or unscrew the cap

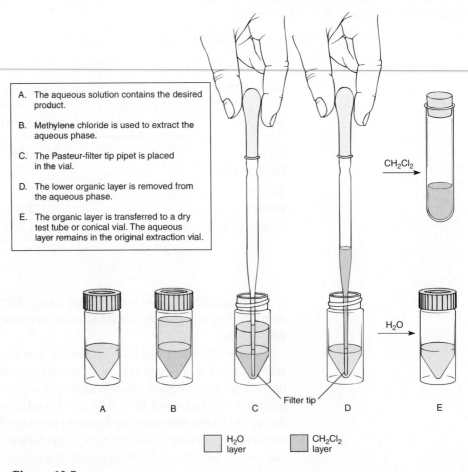

A. The aqueous solution contains the desired product.

B. Methylene chloride is used to extract the aqueous phase.

C. The Pasteur-filter tip pipet is placed in the vial.

D. The lower organic layer is removed from the aqueous phase.

E. The organic layer is transferred to a dry test tube or conical vial. The aqueous layer remains in the original extraction vial.

CH_2Cl_2

H_2O

Filter tip

A   B   C   D   E

☐ H_2O layer   ☐ CH_2Cl_2 layer

Figure 12.5

Extraction of an aqueous solution using a solvent denser than water: methylene chloride.

slightly to release the pressure in the vial. Allow the phases to separate completely so that you can detect two distinct layers in the vial. The organic phase will be the lower layer in the vial (Fig. 12.5B). If necessary, tap the vial with your finger or stir the mixture gently if some of the organic phase is suspended in the aqueous layer.

3. Prepare a Pasteur filter-tip pipet (Technique 8, Section 8.6, p. 625) using a 5¾-inch pipet. Attach a 2-mL rubber bulb to the pipet, depress the bulb, and insert the pipet into the vial so that the tip touches the bottom (Fig. 12.5C). The filter-tip pipet gives you better control in removing the lower layer. In some cases, however, you may be able to use a Pasteur pipet (no filter tip), but considerably more care must be taken to avoid losing liquid from the pipet during the transfer operation. With experience, you should be able to judge how much to squeeze the bulb to draw in the desired volume of liquid.

4. Slowly draw the lower layer (methylene chloride) into the pipet in such a way that you exclude the aqueous layer and any emulsion (Section 12.10) that might be at the interface between the layers (Fig. 12.5D). Be sure to keep the tip of the pipet squarely in the V at the bottom of the vial.

5. Transfer the withdrawn organic phase into a *dry* test tube or another *dry* conical vial if one is available. It is best to have the test tube or vial located next to the extraction vial. Hold the vials in the same hand between your index finger and thumb, as shown in Figure 12.6. This avoids messy and disastrous transfers. The aqueous layer (upper layer) is left in the original conical vial (Fig. 12.5E).

In performing an actual extraction in the laboratory, you would extract the aqueous phase with a second 1-mL portion of fresh methylene chloride to achieve a more complete extraction. Steps 2–5 would be repeated, and the organic layers from both extractions would be combined. In some cases, you may need to extract a third time with yet another 1-mL portion of methylene chloride. Again, the methylene chloride would be combined with the other extracts. The overall process would use three 1-mL portions of methylene chloride to transfer the product from the water layer into methylene chloride. Sometimes you will see the statement "extract the aqueous phase with three 1-mL portions of methylene chloride" in an experimental procedure. This statement describes in a shorter fashion the process described previously. Finally, the methylene chloride extracts will contain some water and must be dried with a drying agent as indicated in Section 12.9.

In this example, we extracted water with the heavy solvent methylene chloride and removed it as the lower layer. If you were extracting a light solvent (for instance, diethyl ether) with water, and you wished to keep the water layer, the water would be the lower layer and would be removed using the same procedure. You would not dry the water layer, however.

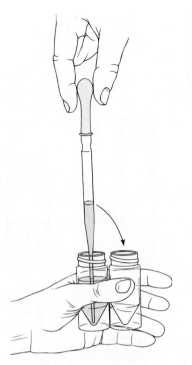

Figure 12.6
Method of holding vials while transferring liquids.

12.5 The Conical Vial— Separating the Upper Layer

In this section, we consider the method used when you wish to remove the upper layer. A concrete example would be the extraction of a desired product from an aqueous layer using diethyl ether ($d = 0.71$ g/mL) as the extraction solvent. Methods for removing the lower layer were discussed previously.

NOTE: Always place a conical vial in a small beaker to prevent the vial from falling over.

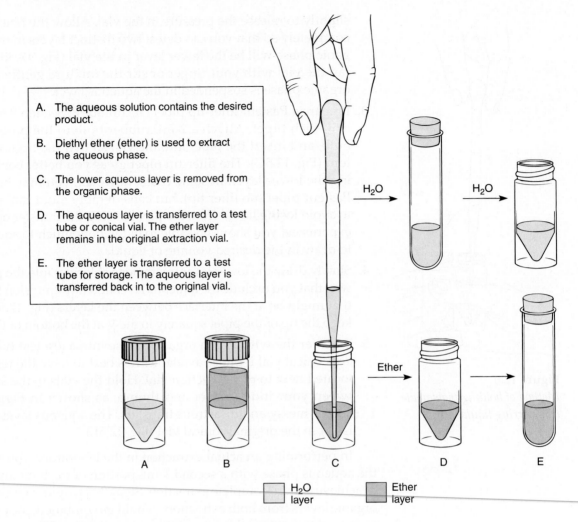

A. The aqueous solution contains the desired product.

B. Diethyl ether (ether) is used to extract the aqueous phase.

C. The lower aqueous layer is removed from the organic phase.

D. The aqueous layer is transferred to a test tube or conical vial. The ether layer remains in the original extraction vial.

E. The ether layer is transferred to a test tube for storage. The aqueous layer is transferred back in to the original vial.

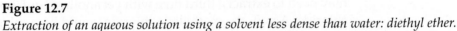

Figure 12.7
Extraction of an aqueous solution using a solvent less dense than water: diethyl ether.

Removing the Upper Layer. Suppose we extract an aqueous solution with diethyl ether (ether). This solvent is less dense than water and will rise to the top of the conical vial. Use the following procedure, which is illustrated in Figure 12.7, to remove the upper layer.

1. Place the aqueous phase containing the dissolved product in a 5-mL conical vial (Fig. 12.7A).

2. Add about 1 mL of ether, cap the vial, and shake the mixture vigorously. Vent or unscrew the cap slightly to release the pressure in the vial. Allow the phases to separate completely so that you can detect two distinct layers in the vial. The ether phase will be the upper layer in the vial (Fig. 12.7B).

3. Prepare a Pasteur filter-tip pipet (Technique 8, Section 8.6, p. 625) using a 5¾-inch pipet. Attach a 2-mL rubber bulb to the pipet, depress the bulb, and insert the pipet into the vial so that the tip touches the bottom. The filter-tip pipet gives you better control in removing the lower layer. In some cases, however, you may be able to use a Pasteur pipet

(no filter tip), but considerably more care must be taken to avoid losing liquid from the pipet during the transfer operation. With experience, you should be able to judge how much to squeeze the bulb to draw in the desired volume of liquid. Slowly draw the lower *aqueous* layer into the pipet. Be sure to keep the tip of the pipet squarely in the V at the bottom of the vial (Fig. 12.7C).

4. Transfer the withdrawn aqueous phase into a test tube or another conical vial for temporary storage. It is best to have the test tube or vial located next to the extraction vial. This avoids messy and disastrous transfers. Hold the vials in the same hand between your index finger and thumb as shown in Fig. 12.6. The ether layer is left behind in the conical vial (Fig. 12.7D).

5. The ether phase remaining in the original conical vial should be transferred with a Pasteur pipet into a test tube for storage and the aqueous phase returned to the original conical vial (Fig. 12.7E).

In performing an actual extraction, you would extract the aqueous phase with another 1-mL portion of fresh ether to achieve a more complete extraction. Steps 2–5 would be repeated, and the organic layers from both extractions would be combined in the test tube. In some cases, you may need to extract the aqueous layer a third time with yet another 1-mL portion of ether. Again, the ether would be combined with the other two layers. This overall process uses three 1-mL portions of ether to transfer the product from the water layer into ether. The ether extracts contain some water and must be dried with a drying agent as indicated in Section 12.9.

12.6 The Centrifuge Tube

A screw-cap centrifuge tube may be employed instead of a conical vial for separations (Fig. 12.4). Before using the centrifuge tube, be sure to check it for leaks as indicated in Section 12.3. You should use the same extraction and separation procedures described in Sections 12.4 and 12.5. You may also use a "regular" (nonscrew-cap) centrifuge tube for extractions, although it will be necessary to cork the tube before shaking it. Because a regular centrifuge tube will probably leak around the cork, it is best to mix the contents with a vortex mixer (Section 12.3) to avoid shaking the tube. If an emulsion has formed after mixing or shaking, you can use a centrifuge to aid in the separation of the layers (Section 12.10). Once the layers have separated, it is easy to use a Pasteur pipet to withdraw the lower layer from the tapered bottom of the centrifuge tube.

PART C. MACROSCALE EXTRACTION

12.7 The Separatory Funnel

The separatory funnel is often used in large-scale reactions. This apparatus is illustrated in Figure 12.8. To fill the separatory funnel, support it in an iron ring attached to a ring stand. Cut pieces of rubber tubing and attach them to the iron ring to cushion the separatory funnel as shown in Figure 12.8. This protects the funnel against possible breakage.

When beginning an extraction, the first step is close the stopcock. (Don't forget!) Using a powder funnel (wide bore) placed in the top of the separatory funnel, fill it with both the solution to be extracted and the extraction

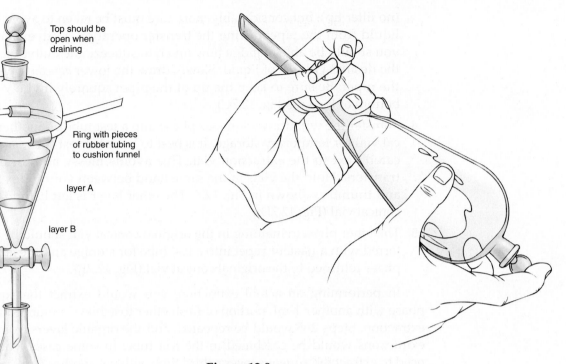

Top should be
open when
draining

Ring with pieces
of rubber tubing
to cushion funnel

layer A

layer B

Figure 12.9

Correct way of shaking and venting the separatory funnel.

Figure 12.8

The separatory funnel.

solvent. Swirl the funnel gently by holding it by its upper neck, and then stopper it. Pick up the separatory funnel with two hands and hold it as shown in Figure 12.9. Hold the stopper in place firmly because the two immiscible liquids will build pressure when they mix, and this pressure may force the stopper out of the separatory funnel. To release this pressure, vent the funnel by holding it upside down (hold the stopper securely) and slowly open the stopcock. Usually the rush of vapors out of the opening can be heard. Continue shaking and venting until the "whoosh" is no longer audible. Now continue shaking the mixture gently for about one minute. This can be done by inverting the funnel in a rocking motion repeatedly or, if the formation of an emulsion is not a problem (see Section 12.10, p. 684), by shaking the funnel more vigorously for less time.

NOTE: There is an art to shaking and venting a separatory funnel correctly, and it usually seems awkward to the beginner. The technique is best learned by observing a person, such as your instructor, who is thoroughly familiar with the separatory funnel's use.

When you have finished mixing the liquids, place the separatory funnel in the iron ring and remove the top stopper immediately. The two immiscible solvents separate into two layers after a short time, and they can be separated from each other by draining most of the lower layer through the stopcock.[1] Allow a few minutes to pass so that any of the lower phase

---

[1] A common error is to try to drain the separatory funnel without removing the top stopper. Under this circumstance, the funnel will not drain, because a partial vacuum is created in the space above the liquid.

adhering to the inner glass surfaces of the separatory funnel can drain down. Open the stopcock again and allow the remainder of the lower layer to drain until the interface between the upper and lower phases just begins to enter the bore of the stopcock. At this moment, close the stopcock and remove the remaining upper layer by pouring it from the top opening of the separatory funnel.

NOTE: To minimize contamination of the two layers, the lower layer should always be drained from the bottom of the separatory funnel and the upper layer poured out from the top of the funnel.

When methylene chloride is used as the extracting solvent with an aqueous phase, it will settle to the bottom and be removed through the stopcock. The aqueous layer remains in the funnel. A second extraction of the remaining aqueous layer with fresh methylene chloride may be needed.

With a diethyl ether (ether) extraction of an aqueous phase, the organic layer will form on top. Remove the lower aqueous layer through the stopcock and pour the upper ether layer from the top of the separatory funnel. Pour the aqueous phase back into the separatory funnel and extract it a second time with fresh ether. The combined organic phases must be dried using a suitable drying agent (Section 12.9) before the solvent is removed.

For microscale procedures, a 60- or 125-mL separatory funnel is recommended. Because of surface tension, water has a difficult time draining from the bore of smaller funnels. Funnels larger than 125 mL are simply too large for microscale experiments, and a good deal of material is lost in "wetting" their surfaces.

PART D. ADDITIONAL EXPERIMENTAL CONSIDERATIONS: MICROSCALE AND MACROSCALE

12.8 How Do You Determine Which One Is the Organic Layer?

A common problem encountered during an extraction is trying to determine which of the two layers is the organic layer and which is the aqueous (water) layer. The most common situation occurs when the aqueous layer is on the bottom in the presence of an upper organic layer consisting of ether, ligroin, petroleum ether, or hexane (see densities in Table 12.1). However, the aqueous layer will be on the top when you use methylene chloride as a solvent (again, see Table 12.1). Although a laboratory procedure may frequently identify the expected relative positions of the organic and aqueous layers, sometimes their actual positions are reversed. Surprises usually occur in situations in which the aqueous layer contains a high concentration of sulfuric acid or a dissolved ionic compound, such as sodium chloride. Dissolved substances greatly increase the density of the aqueous layer, which may lead to the aqueous layer being found on the bottom even when coexisting with a relatively dense organic layer such as methylene chloride.

NOTE: Always keep both layers until you have actually isolated the desired compound or until you are certain where your desired substance is located.

To determine if a particular layer is the aqueous one, add a few drops of water to the layer. Observe closely as you add the water to see where it goes.

If the layer is water, then the drops of added water will dissolve in the aqueous layer and increase its volume. If the added water forms droplets or a new layer, however, you can assume that the suspected aqueous layer is actually organic. You can use a similar procedure to identify a suspected organic layer. This time, try adding more of the solvent, such as methylene chloride. The organic layer should increase in size, without separation of a new layer, if the tested layer is actually organic.

When performing an extraction procedure on the microscale level, you can use the following approach to identify the layers. When both layers are present, it is always a good idea to think carefully about the volumes of materials that you have added to the conical vial. You can use the graduations on the vial to help determine the volumes of the layers in the vial. If, for example, you have 1 mL of methylene chloride in a vial and you add 2 mL of water, you should expect the water to be on top because it is less dense than methylene chloride. As you add the water, watch to see where it goes. By noting the relative volumes of the two layers, you should be able to tell which is the aqueous layer and which is the organic layer. This approach can also be used when performing an extraction procedure using a centrifuge tube. Of course, you can always test to see which layer is the aqueous layer by adding one or two drops of water, as described previously.

12.9 Drying Agents

After an organic solvent has been shaken with an aqueous solution, it will be "wet"; that is, it will have dissolved some water even though its solubility with water is not great. The amount of water dissolved varies from solvent to solvent; diethyl ether represents a solvent in which a fairly large amount of water dissolves. To remove water from the organic layer, use a **drying agent.** A drying agent is an *anhydrous* inorganic salt that acquires waters of hydration when exposed to moist air or a wet solution:

Insoluble   Insoluble

$$Na_2SO_4(s) + \text{Wet Solution } (nH_2O) \longrightarrow Na_2SO_4 \cdot nH_2O \text{ (s)} + \text{Dry Solution}$$

Anhydrous   Hydrated
drying agent   drying agent

The insoluble drying agent is placed directly into the solution, where it acquires water molecules and becomes hydrated. If enough drying agent is used, all of the water can be removed from a wet solution, making it "dry," or free of water.

The following anhydrous salts are commonly used: sodium sulfate, magnesium sulfate, calcium chloride, calcium sulfate (Drierite), and potassium carbonate. These salts vary in their properties and applications. For instance, not all will absorb the same amount of water for a given weight, nor will they dry the solution to the same extent. **Capacity** refers to the amount of water a drying agent absorbs per unit weight. Sodium and magnesium sulfates absorb a large amount of water (high capacity), but magnesium sulfate dries a solution more completely. **Completeness** refers to a compound's effectiveness in removing all the water from a solution by the time equilibrium has been reached. Magnesium ion, a strong Lewis acid, sometimes causes rearrangements of compounds such as epoxides. Calcium chloride is a good drying agent but cannot be used with many compounds containing oxygen or nitrogen because it forms complexes. Calcium chloride absorbs methanol and ethanol in addition to water, so it is useful for removing these materials when they are present as impurities. Potassium carbonate is a

Table 12.2 *Common drying agents*

	Acidity	*Hydrated*	*Capacity*[a]	*Completeness*[b]	*Rate*[c]	*Use*
Magnesium sulfate	Neutral	$MgSO_4 \cdot 7H_2O$	High	Medium	Rapid	General
Sodium sulfate	Neutral	$Na_2SO_4 \cdot 7H_2O$ $Na_2SO_4 \cdot 10H_2O$	High	Low	Medium	General
Calcium chloride	Neutral	$CaCl_2 \cdot 2H_2O$ $CaCl_2 \cdot 6H_2O$	Low	High	Rapid	Hydrocarbons Halides
Calcium sulfate (Drierite)	Neutral	$CaSO_4 \cdot \frac{1}{2}H_2O$ $CaSO_4 \cdot 2H_2O$	Low	High	Rapid	General
Potassium carbonate	Basic	$K_2CO_3 \cdot 1\frac{1}{2}H_2O$ $K_2CO_3 \cdot 2H_2O$	Medium	Medium	Medium	Amines, esters, bases, ketones
Potassium hydroxide	Basic	—	—	—	Rapid	Amines only
Molecular sieves (3 or 4 Å)	Neutral	—	High	Extremely high	—	General

[a]Amount of water removed per given weight of drying agent.
[b]Refers to amount of H_2O still in solution at equilibrium with drying agent.
[c]Refers to rate of action (drying).

base and is used for drying solutions of basic substances, such as amines. Calcium sulfate dries a solution completely but has a low capacity.

Anhydrous sodium sulfate is the most widely used drying agent. The granular variety is recommended because it is easier to remove the dried solution from it than from the powdered variety. Sodium sulfate is mild and effective. It will remove water from most common solvents, with the possible exception of diethyl ether, in which case a prior drying with saturated salt solution may be advised (see p. 684). Sodium sulfate must be used at room temperature to be effective; it cannot be used with boiling solutions. Table 12.2 compares the various common drying agents.

Drying Procedure with Anhydrous Sodium Sulfate. In experiments that require a drying step, the instructions are usually given in the following way: dry the organic layer (or phase) over granular anhydrous sodium sulfate (or some other drying agent). More specific instructions, such as the amount of drying agent to add, usually will not be given, and you will need to determine this each time that you perform a drying step. The drying procedure consists of four steps:

1. Remove the organic layer from any visible water.
2. Add the appropriate amount of granular anhydrous sodium sulfate (or other drying agent).
3. Allow a drying period during which dissolved water is removed from the organic layer by the drying agent.
4. Separate the dried organic layer from the drying agent.

More specific instructions are given below for both microscale and macroscale procedures. The only differences between these two procedures is that they are intended for different volumes of liquid and they require different glassware. The microscale procedure is generally for volumes up to about 5 mL, and the macroscale procedure is usually appropriate for volumes of 5 mL or greater.

A. Microscale Drying Procedure

Step 1. (Removal of Visible Water). Before attempting to dry an organic layer, check closely to see that there are no visible signs of water. If there is a separate layer of water (top or bottom), droplets or a globule of water floating in the organic layer, or water droplets clinging to the sides of the container, then transfer the organic layer with a *dry* Pasteur pipet to a *dry* container, usually a conical vial or test tube, before adding any drying agent. If there is any doubt about whether water is present, it is advisable to make a transfer to a dry container. Performing this step when necessary may save time later in the drying procedure and result in a greater recovery of the desired substance.

Step 2. (Addition of Drying Agent). Each time a drying procedure is performed, it is necessary to determine how much granular anhydrous sodium sulfate (or other drying agent) should be added. This will depend on the total volume of the organic phase and how much water is dissolved in the solvent. Nonpolar organic solvents such as methylene chloride or hydrocarbons (hexane, pentane, etc.) can dissolve relatively small amounts of water and generally require less drying agent, whereas more polar organic solvents such as ether and ethyl acetate can dissolve more water, and more drying agent will be required. Begin by adding one spatulaful of granular anhydrous sodium sulfate (or other drying agent) from the V-grooved end of a microspatula (smaller microspatula on page 13) into the solution. If all the drying agent "clumps," add another spatulaful of sodium sulfate. To determine if the drying agent has clumped, it is helpful to stir the mixture with a clean, dry spatula or to rapidly swirl the container. If any portion of the drying agent flows freely on the bottom of the container when stirred or swirled, then you can assume that enough of the drying agent has been added. Otherwise, you must continue adding one spatulaful of drying agent at a time until it is clear that the drying agent has stopped clumping. Stir or swirl the mixture after adding each spatulaful of the drying agent. For small amounts of liquid (less than 5 mL), about 1–6 microspatulafuls of drying agent will *usually* be required. However, the actual amount must be determined by experimentation, as just described. It is best to use a slight excess of drying agent; but if too great an excess is used, the recovery may be poor because some of the solution always adheres to the solid drying agent after the liquid is separated from the drying agent (Step 4). Take care not to add so much drying agent that all of the liquid is absorbed (disappears). If you do this you will have to add additional solvent to recover your product from the drying agent!

Step 3. (Drying Period). Stopper or cap the container and let the solution dry for at least 15 minutes.

NOTE: It is important that you stopper or cap the container to prevent evaporation and exposure to atmospheric moisture.

Stir the mixture occasionally with a spatula during the drying period. The mixture is dry if it appears clear (not cloudy) and shows the common

Table 12.3 *Common signs that indicate a solution is dry*

1. There are no visible water droplets on the side of flask or suspended in solution.
2. There is not a separate layer of liquid or a "puddle."
3. The solution is clear, not cloudy. Cloudiness indicates water is present.
4. The drying agent (or a portion of it) flows freely on the bottom of the container when stirred or swirled and does not "clump" together as a solid mass.

signs of a dry solution given in Table 12.3. Note that a "clear" solution may be colorless or colored. If the solution remains cloudy after treatment with the first batch of drying agent, add more drying agent and repeat the drying procedure. However, if a water layer forms or if drops of water are visible, transfer the organic layer to a dry container before adding fresh drying agent, as described in Step 2. It will also be necessary to repeat the 15-minute drying step described in Step 3.

Step 4. (Removal of Liquid from Drying Agent). When the organic phase is dry, use a *dry* Pasteur pipet or a *dry* filter-tip pipet (Technique 8, Section 8.6, p. 625) to remove the dried organic layer from the drying agent and transfer the solution to a *dry* conical vial or test tube. Be careful not to transfer any of the drying agent when performing this step. Rinse the drying agent with a small amount of fresh solvent and transfer this additional solvent to the vial containing the dried organic layer. To isolate the desired material, remove the solvent by evaporation using heat and a stream of air or nitrogen (Technique 7, Section 7.10, p. 611).

An alternative method of drying a small volume of organic phase is to pass it through a filtering pipet (Technique 8, Section 8.1C, p. 618) that has been packed with a small amount (about 2 cm) of drying agent. Again, the solvent is removed by evaporation.

B. Macroscale Drying Procedure

To dry a large amount of organic liquid (greater than about 5 mL), follow the same four steps just described for the "Microscale Drying Procedure." The main differences are that an Erlenmeyer flask is used rather than a test tube or conical vial and more drying agent will be required. The size of the Erlenmeyer flask is not critical, but it's best that the flask not be filled more than half full with the solution being dried.

Step 1. (Removal of Visible Water). Refer to Step 1 above for instructions. If the amount of water is large, it may be best to separate the layers using a separatory funnel. If visible water must be removed in this step, place the separated organic layer in a clean, *dry* Erlenmeyer flask.

Step 2. (Addition of Drying Agent). Refer to Step 2 in the "Microscale Drying Procedure" for the basic instructions. Read these instructions carefully. The only difference is that in this macroscale procedure, more drying agent will be required. A common guideline is to add enough granular anhydrous sodium sulfate (or other drying agent) to give a 1- to 3-mm layer on the bottom of the flask, depending on the volume of the solution. However, it is best to add the drying agent in small portions, as described above. In this procedure, use the larger microspatula shown in the figure (p. 13) to add the drying agent. Generally, an appropriate portion to add each time is about 0.5–1.0 g.

Step 3. (Drying Period). The instructions are the same as for Step 3 in the "Microscale Drying Procedure."

Step 4. (Removal of Liquid from Drying Agent). When the solution is dry, the drying agent should be removed by using decantation (pouring carefully to leave the drying agent behind). Transfer the liquid to a dry Erlenmeyer flask. If the volume of liquid is relatively small (less than 10 mL), it may be easier to complete this step by using a *dry* Pasteur pipet or a *dry* filter-tip pipet (Technique 8, Section 8.6, p. 625) to remove the dried organic layer. With granular sodium sulfate, decantation is easy to perform because of the size of the drying-agent particles. If a powdered drying agent, such as magnesium sulfate, is used, it may be necessary to use gravity filtration (Technique 8, Section 8.1B, p. 618) to remove the drying agent. Finally, to isolate the desired material, remove the solvent by distillation (Technique 14, Section 14.3, p. 707) or evaporation (Technique 7, Section 7.10, p. 611).

Saturated Salt Solution. At room temperature, diethyl ether (ether) dissolves 1.5% by weight of water, and water dissolves 7.5% of ether. Ether, however, dissolves a much smaller amount of water from a saturated aqueous sodium chloride solution. Hence, the bulk of water in ether, or ether in water, can be removed by shaking it with a saturated aqueous sodium chloride solution. A solution of high ionic strength is usually not compatible with an organic solvent and forces separation of it from the aqueous layer. The water migrates into the concentrated salt solution. The ether phase (organic layer) will be on top, and the saturated sodium chloride solution will be on the bottom ($d = 1.2$ g/mL). After removing the organic phase from the aqueous sodium chloride, dry the organic layer completely with sodium sulfate or with one of the other drying agents listed in Table 12.2.

12.10 Emulsions

An **emulsion** is a colloidal suspension of one liquid in another. Minute droplets of an organic solvent are often held in suspension in an aqueous solution when the two are mixed or shaken vigorously; these droplets form an emulsion. This is especially true if any gummy or viscous material was present in the solution. Emulsions are often encountered in performing extractions. Emulsions may require a long time to separate into two layers and are a nuisance to the organic chemist.

Fortunately, several techniques may be used to break a difficult emulsion once it has formed.

1. Often an emulsion will break up if it is allowed to stand for some time. Patience is important here. Gently stirring with a stirring rod or spatula may also be useful.

2. If one of the solvents is water, adding a saturated aqueous sodium chloride solution will help destroy the emulsion. The water in the organic layer migrates into the concentrated salt solution.

3. With microscale experiments, the mixture may be transferred to a centrifuge tube. The emulsion will often break during centrifugation. Remember to place another tube filled with water on the opposite side of the centrifuge to balance it. The two tubes should weigh the same.

4. Adding a small amount of a water-soluble detergent may also help. This method has been used in the past for combating oil spills. The detergent helps to solubilize the tightly bound oil droplets.

5. Gravity filtration (see Technique 8, Section 8.1, p. 616) may help to destroy an emulsion by removing gummy polymeric substances. With large volumes, you might try filtering the mixture through a fluted filter (Technique 8, Section 8.1B, p. 618) or a piece of cotton. With small-scale reactions, a filtering pipet may work (Technique 8, Section 8.1C, p. 618). In many cases, once the gum is removed, the emulsion breaks up rapidly.

6. If you are using a separatory funnel, you might try to use a gentle swirling action in the funnel to help break an emulsion. Gently stirring with a stirring rod may also be useful.

When you know through experience that a mixture may form a difficult emulsion, you should avoid shaking the mixture vigorously. When using conical vials for extractions, it may be better to use a magnetic spin vane for mixing and not shake the mixture at all. When using separatory funnels, extractions should be performed with gentle swirling instead of shaking or with several gentle inversions of the separatory funnel. Do not shake the separatory funnel vigorously in these cases. It is important to use a longer extraction period if the more gentle techniques described in this paragraph are being employed. Otherwise, you will not transfer all the material from the first phase to the second one.

12.11 Purification and Separation Methods

In nearly all synthetic experiments undertaken in the organic laboratory, a series of operations involving extractions is used after the actual reaction has been concluded. These extractions form an important part of the purification. Using them, you separate the desired product from unreacted starting materials or from undesired side products in the reaction mixture. These extractions may be grouped into three categories, depending on the nature of the impurities they are designed to remove.

The first category involves extracting or "washing" an organic mixture with water. Water washes are designed to remove highly polar materials, such as inorganic salts, strong acids or bases, and low-molecular-weight, polar substances including alcohols, carboxylic acids, and amines. Many organic compounds containing fewer than five carbons are water soluble. Water extractions are also used immediately following extractions of a mixture with either acid or base to ensure that all traces of acid or base have been removed.

The second category concerns extraction of an organic mixture with a dilute acid, usually 1–2 M hydrochloric acid. Acid extractions are intended to remove basic impurities, especially such basic impurities as organic amines. The bases are converted to their corresponding cationic salts by the acid used in the extraction. If an amine is one of the reactants or if pyridine or another amine is a solvent, such an extraction might be used to remove any excess amine present at the end of a reaction.

$$RNH_2 + HCl \longrightarrow RNH_3^+Cl^-$$

(water-soluble ammonium salt)

Cationic ammonium salts are usually soluble in the aqueous solution, and they are thus extracted from the organic material. A water extraction may be used immediately following the acid extraction to ensure that all traces of the acid have been removed from the organic material.

The third category is extraction of an organic mixture with a dilute base, usually 1 *M* sodium bicarbonate, although extractions with dilute sodium hydroxide can also be used. Such basic extractions are intended to convert acidic impurities, such as organic acids, to their corresponding anionic salts. For example, in the preparation of an ester, a sodium bicarbonate extraction might be used to remove any excess carboxylic acid that is present.

$$RCOOH + NaHCO_3 \longrightarrow RCOO^-Na^+ + H_2O + CO_2$$

(p$K_a \sim 5$)　　　　　(water-soluble carboxylate salt)

Anionic carboxylate salts, being highly polar, are soluble in the aqueous phase. As a result, these acid impurities are extracted from the organic material into the basic solution. A water extraction may be used after the basic extraction to ensure that all the base has been removed from the organic material.

Occasionally, phenols may be present in a reaction mixture as impurities, and removing them by extraction may be desired. Because phenols, although they are acidic, are about 10^5 times less acidic than carboxylic acids, basic extractions may be used to separate phenols from carboxylic acids by a careful selection of the base. If sodium bicarbonate is used as a base, carboxylic acids are extracted into the aqueous base, but phenols are not. Phenols are not sufficiently acidic to be deprotonated by the weak base bicarbonate. Extraction with sodium hydroxide, on the other hand, extracts both carboxylic acids and phenols into the aqueous basic solution because hydroxide ion is a sufficiently strong base to deprotonate phenols.

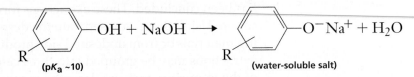

(p$K_a \sim 10$)　　　　　　　　　　(water-soluble salt)

Mixtures of acidic, basic, and neutral compounds are easily separated by extraction techniques. One such example is shown in Figure 12.10. The original compounds are dissolved in ether.

Organic acids or bases that have been extracted can be regenerated by neutralizing the extraction reagent. This would be done if the organic acid or base were a product of a reaction rather than an impurity. For example, if a carboxylic acid has been extracted with the aqueous base, the compound can be regenerated by acidifying the extract with 6 *M* HCl until the solution becomes *just* acidic, as indicated by litmus or pH paper. When the solution becomes acidic, the carboxylic acid will separate from the aqueous solution. If the acid is a solid at room temperature, it will precipitate and can be purified by filtration and crystallization. If the acid is a liquid, it will form a separate layer. In this case, it would usually be necessary to extract the mixture with ether or methylene chloride. After removing the organic layer and drying it, the solvent can be evaporated to yield the carboxylic acid.

In the example shown in Figure 12.10, you also need to perform a drying step at (3) before isolating the neutral compound. When the solvent is ether, you should first extract the ether solution with saturated aqueous sodium chloride to remove much of the water. The ether layer is then dried over a drying agent such as anhydrous sodium sulfate. If the solvent were

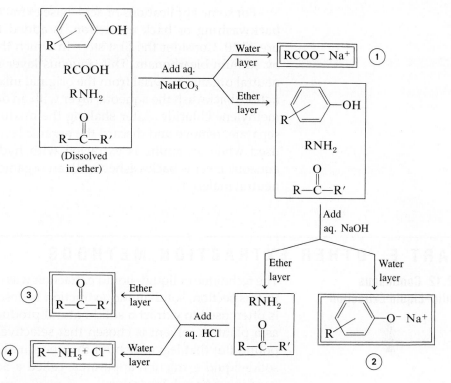

Figure 12.10
Separating a four-component mixture by extraction.

methylene chloride, it would not be necessary to do the step with saturated sodium chloride.

When acid–base extractions are performed, it is common practice to extract a mixture several times with the appropriate reagent. For example, if you were extracting a carboxylic acid from a mixture, you might extract the mixture three times with 2-mL portions of 1 *M* NaOH. In most published experiments, the procedure will specify the volume and concentration of extracting reagent and the number of times to do the extractions. If this information is not given, you must devise your own procedure. Using a carboxylic acid as an example, if you know the identity of the acid and the approximate amount present, you can actually calculate how much sodium hydroxide is needed. Because the carboxylic acid (assuming it is monoprotic) will react with sodium hydroxide in a 1:1 ratio, you would need the same number of moles of sodium hydroxide as there are moles of acid. To ensure that all the carboxylic acid is extracted, you should use about a *twofold* excess of the base. From this, you could calculate the number of milliliters of base needed. This should be divided into two or three equal portions, one portion for each extraction. In a similar fashion, you could calculate the amount of 5% sodium bicarbonate required to extract an acid or the amount of 1 *M* HCl required to extract a base. If the amount of organic acid or base is not known, then the situation is more difficult. A guideline that sometimes works is to do two or three extractions so that the total volume of the extracting reagent is approximately equal to the volume of the organic layer. To test this procedure, neutralize the aqueous layer from the last extraction. If a precipitate or cloudiness results, perform another extraction and test again. When no precipitate forms, you know that all the organic acid or base has been removed.

For some applications of acid–base extraction, **an additional step, called backwashing** or **back extraction,** is added to the scheme shown in Figure 12.10. Consider the first step, in which the carboxylic acid is extracted by sodium bicarbonate. This aqueous layer may contain some unwanted neutral organic material from the original mixture. To remove this contamination, backwash the aqueous layer with an organic solvent such as ether or methylene chloride. After shaking the mixture and allowing the layers to separate, remove and discard the organic layer. This technique may also be used when an amine is extracted with hydrochloric acid. The resulting aqueous layer is backwashed with an organic solvent to remove unwanted neutral material.

PART E. OTHER EXTRACTION METHODS

12.12 Continuous Solid–Liquid Extraction

The technique of liquid–liquid extraction was described in Sections 12.1–12.8. In this section, solid–liquid extraction is described. Solid–liquid extraction is often used to extract a solid natural product from a natural source, such as a plant. A solvent is chosen that selectively dissolves the desired compound but that leaves behind the undesired insoluble solid. A continuous solid–liquid extraction apparatus, called a Soxhlet extractor, is commonly used in a research laboratory.

As shown in Figure 12.11, the solid to be extracted is placed in a thimble made from filter paper, and the thimble is inserted into the central chamber. A low-boiling solvent, such as diethyl ether, is placed in the round-bottom distilling flask and is heated to reflux. The vapor rises through the left side arm into the condenser where it liquefies. The condensate (liquid) drips into the thimble containing the solid. The hot solvent begins to fill the thimble and extracts the desired compound from the solid. Once the thimble is filled with solvent, the side arm on the right acts as a siphon, and the solvent, which now contains the dissolved compound, drains back into the distillation flask. The vaporization–condensation– extraction–siphoning process is repeated hundreds of times, and the desired product is concentrated in the distillation flask. The product is concentrated in the flask because the product has a boiling point higher than that of the solvent or because it is a solid.

12.13 Continuous Liquid–Liquid Extraction

When a product is very soluble in water, it is often difficult to extract using the techniques described in Sections 12.4–12.7 because of an unfavorable distribution coefficient. In this case, you need to extract the aqueous solution numerous times with fresh batches of an immiscible organic solvent to remove the desired product from water. A less labor-intensive technique involves the use of a continuous liquid–liquid extraction apparatus. One type of extractor, used with solvents that are less dense than water, is shown in Figure 12.12. Diethyl ether is usually the solvent of choice.

The aqueous phase is placed in the extractor, which is then filled with diethyl ether up to the side arm. The round-bottom distillation flask is partially filled with ether. The ether is heated to reflux in the round-bottom flask, and the vapor is liquefied in the water-cooled condenser. The ether drips into the central tube, passes through the porous sintered glass tip, and flows through the aqueous layer. The solvent extracts the desired compound from the aqueous phase, and the ether is recycled back into the

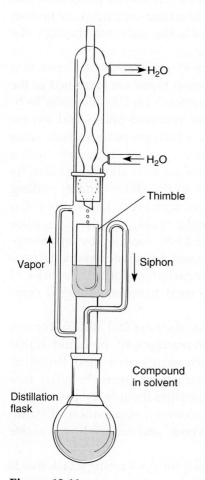

Figure 12.11
Continuous solid–liquid extraction using a Soxhlet extractor.

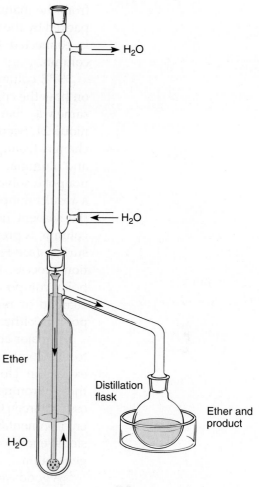

Figure 12.12
Continuous liquid–liquid extraction using a solvent less dense than water.

round-bottom flask. The product is concentrated in the flask. The extraction is rather inefficient and must be placed in operation for at least 24 hours to remove the compound from the aqueous phase.

12.14 Solid Phase Extraction

Solid phase extraction (SPE) is a relatively new technique, which is similar in appearance and function to column chromatography and high-performance liquid chromatography (Techniques 19 and 21). In some applications, SPE is also similar to liquid–liquid extraction, discussed in this technique chapter. In addition to performing separation processes, SPE can also be used to carry out reactions in which new compounds are prepared.

A typical SPE column is constructed from the body of a plastic syringe, which is packed with a **sorbent.** The term *sorbent* is used by many manufacturers as a general term for materials that can both adsorb (attract to the surface of the sorbent by a physical attraction) or absorb (penetrate into the material like a sponge). A frit is inserted at the bottom of the column to support the sorbent. After the sorbent is added, another frit is inserted on top of the sorbent to hold it in place. The remainder of the tube serves as a reservoir for the solvent. Generally, the column comes packed with the sorbent from the manufacturer, but unpacked columns can also be purchased and packed by the user for specific applications. The Luer-lock tip at the bottom is connected to a vacuum source that pulls the solvents through the column.

SPE columns can be packed with many kinds of sorbent, depending on how the column will be used. Some common types are identified in the same way that column chromatography adsorbents are classified (see Technique 21, Section 21.1, p. 793): normal-phase, reversed-phase, and ion exchange. Examples of normal-phase sorbents, which are polar, include silica and alumina. These columns are used to isolate polar compounds from a nonpolar solvent. Reversed-phase sorbents are made by alkylating silica. As a result, nonpolar alkyl groups are bonded to the silica surface, making the sorbent nonpolar. A common column of this type, known as a C_{18} column, is prepared by attaching an octadecyl ($-C_8H_{18}$) group to the silica surface (see Figure 12.13). C_{18} columns most likely function by an adsorption process. Reversed-phase sorbents are used to isolate relatively nonpolar compounds from polar solvents. Ion-exchange sorbents consist of charged or highly polar materials and are used to isolate charged compounds, either as anions or cations.

A major advantage of SPE columns is that they are fast and convenient to use compared to traditional column chromatography or liquid–liquid extraction. However, there are many other advantages that are of benefit to the environment, and their use is a good example of green chemistry (see essay "Green Chemistry," p. 268). These advantages include the use of more environmentally friendly solvents, higher recovery, elimination of emulsions, enormous decrease in the use of solvents, and reduced toxic waste generation.

A good example of the use of SPE columns for performing a task that is normally done by liquid–liquid extraction is the isolation of caffeine from tea or coffee. In this application, a C_{18} column is used. As the tea or coffee flows through the column, caffeine is attracted to the sorbent, and the polar impurities come off with water. Ethyl acetate is then used to remove the caffeine from the column. The experimental setup for this is shown in Figure 12.14. The SPE column[2] is attached to the filter flask by using two neoprene adapters (sizes #1 and #2). The filter flask is connected to either a vacuum line or a water aspirator to provide the vacuum. After each step, the solvents with impurities or desired product are drawn through the column into the filter flask using the vacuum.

The following steps are used with an SPE tube to remove caffeine from tea or coffee (see Figure 12.15):

A. Condition the C_{18} reversed-phase silica column by passing methanol and water through the tube.

B. Apply the sample of caffeinated drink to the column.

C. Wash the polar impurities from the column with water.

D. Elute the caffeine from the tube with ethyl acetate.

---

[2] This is a Strata SPE column available from Phenomenex, 411 Madrid Ave, Torrance, CA 90501-1430; phone (310)-212-0555. Part number: 8B-S001-JCH-S, Strata C-18-E, 1000 mg sorbent/6-mL tube.

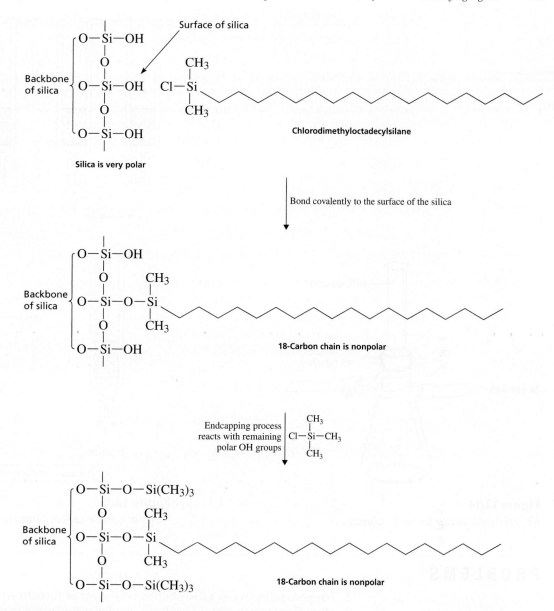

Figure 12.13

Preparation of C-18 silica for reversed-phase extractions using SPE tubes. The process changes polar silica (hydrophilic material) to nonpolar silica (hydrophobic material).

Even though Figure 12.15 is applied to the isolation of caffeine, the general scheme may be used in any application in which it is desired to separate polar substances, such as water, from a relatively nonpolar substance. Numerous applications are found in the medical field, in which analyzing body fluids is important.

There are many other diverse applications that SPE columns can be used for. By modifying the silica with specific chemical reagents, new compounds can be prepared in SPE columns. For example, oxidation reactions can be performed by mixing the silica with the appropriate oxidizing agents. Aldol condensation reactions can also be conducted in SPE columns. In another type of application, SPE has been adopted as an alternative to liquid–liquid extraction.

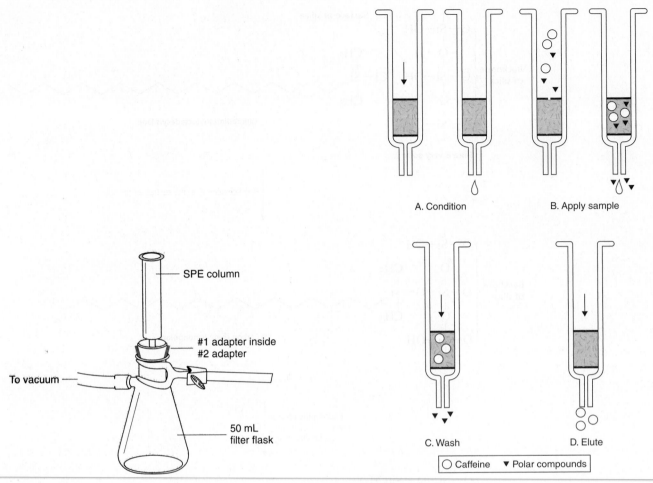

Figure 12.14
Experimental setup for SPE column.

Figure 12.15
Steps to remove caffeine from tea or coffee.

PROBLEMS

1. Suppose solute A has a distribution coefficient of 1.0 between water and diethyl ether. Demonstrate that if 4.0 mL of a solution of 0.20 g of A in water were extracted with two 1.0-mL portions of ether, a smaller amount of A would remain in the water than if the solution were extracted with one 2.0-mL portion of ether.

2. Write an equation to show how you could recover the parent compounds from their respective salts (1, 2, and 4) shown in Figure 12.10.

3. Aqueous hydrochloric acid was used *after* the sodium bicarbonate and sodium hydroxide extractions in the separation scheme shown in Figure 12.10. Is it possible to use this reagent earlier in the separation scheme to achieve the same overall result? If so, explain where you would perform this extraction.

4. Using aqueous hydrochloric acid, sodium bicarbonate, or sodium hydroxide solutions, devise a separation scheme using the style shown in Figure 12.10 to separate the following two-component mixtures. All the substances are soluble in ether. Also indicate how you would recover each of the compounds from its respective salts.

   a. Give two different methods for separating this mixture.

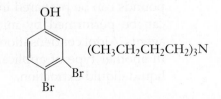

b. Give two different methods for separating this mixture.

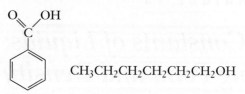

CH₃CH₂CH₂CH₂CH₂CH₂OH

c. Give one method for separating this mixture.

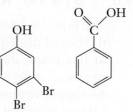

5. Solvents other than those in Table 12.1 may be used for extractions. Determine the relative positions of the organic layer and the aqueous layer in a conical vial or separatory funnel after shaking each of the following solvents with an aqueous phase. Find the densities for each of these solvents in a handbook (see Technique 4, p. 574).

   a. 1,1,1-Trichloroethane

   b. Hexane

6. A student prepares ethyl benzoate by the reaction of benzoic acid with ethanol using a sulfuric acid catalyst. The following compounds are found in the crude reaction mixture: ethyl benzoate (major component), benzoic acid, ethanol, and sulfuric acid. Using a handbook, obtain the solubility properties in water for each of these compounds (see Technique 4, p. 574). Indicate how you would remove benzoic acid, ethanol, and sulfuric acid from ethyl benzoate. At some point in the purification, you should also use an aqueous sodium bicarbonate solution.

7. Calculate the weight of water that could be removed from a wet organic phase using 50.0 mg of magnesium sulfate. Assume that it gives the hydrate listed in Table 12.2.

8. Explain exactly what you would do when performing the following laboratory instructions:

   a. "Wash the organic layer with 1.0 mL of 1 *M* aqueous sodium bicarbonate."

   b. "Extract the aqueous layer three times with 1-mL portions of methylene chloride."

9. Just prior to drying an organic layer with a drying agent, you notice water droplets in the organic layer. What should you do next?

10. What should you do if there is some question about which layer is the organic one during an extraction procedure?

11. Saturated aqueous sodium chloride (*d* = 1.2 g/mL) is added to the following mixtures in order to dry the organic layer. Which layer is likely to be on the bottom in each case?

    a. Sodium chloride layer or a layer containing a high-density organic compound dissolved in methylene chloride (*d* = 1.4 g/mL)

    b. Sodium chloride layer or a layer containing a low-density organic compound dissolved in methylene chloride (*d* = 1.1 g/mL)

Physical Constants of Liquids: The Boiling Point and Density

PART A. BOILING POINTS AND THERMOMETER CORRECTION

13.1 The Boiling Point

As a liquid is heated, the vapor pressure of the liquid increases to the point at which it just equals the applied pressure (usually atmospheric pressure). At this point, the liquid is observed to boil. The normal boiling point is measured at 760 mm Hg (760 torr), or 1 atm. At a lower applied pressure, the vapor pressure needed for boiling is also lowered, and the liquid boils at a lower temperature. The relation between applied pressure and temperature of boiling for a liquid is determined by its vapor pressure–temperature behavior. Figure 13.1 is an idealization of the typical vapor pressure–temperature behavior of a liquid.

Because the boiling point is sensitive to pressure, it is important to record the barometric pressure when determining a boiling point if the determination is being conducted at an elevation significantly above or below sea level. Normal atmospheric variations may affect the boiling point, but they are usually of minor importance. However, if a boiling point is being monitored during the course of a vacuum distillation (Technique 16) that is being performed with an aspirator or a vacuum pump, the variation from the atmospheric value will be especially marked. In these cases, it is important to know the pressure as accurately as possible.

As a rule of thumb, the boiling point of many liquids drops about 0.5°C for a 10-mm decrease in pressure when in the vicinity of 760 mm Hg. At lower pressures, a 10°C drop in boiling point is observed for each halving of the pressure. For example, if the observed boiling point of a liquid is 150°C at 10 mm pressure, then the boiling point would be about 140°C at 5 mm Hg.

A more accurate estimate of the change in boiling point with a change of pressure can be made by using a nomograph. In Figure 13.2, a nomograph

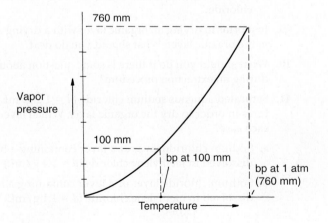

Figure 13.1

The vapor pressure–temperature curve for a typical liquid.

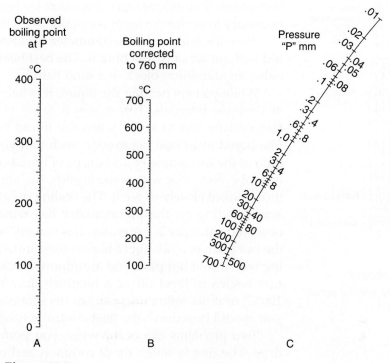

Figure 13.2

Pressure–temperature alignment nomograph. How to use the nomograph: Assume a reported boiling point of 100°C (column A) at 1 mm. To determine the boiling point at 18 mm, connect 100°C (column A) to 1 mm (column C) with a transparent plastic rule and observe where this line intersects column B (about 280°C). This value would correspond to the normal boiling point. Next, connect 280°C (column B) with 18 mm (column C) and observe where this intersects column A (151°C). The approximate boiling point will be 151°C at 18 mm. (Reprinted courtesy of EMD Chemicals, Inc.)

is given, and a method is described for using it to obtain boiling points at various pressures when the boiling point is known at some other pressure.

13.2 Determining the Boiling Point—Microscale and Macroscale Methods

Several experimental methods of determining boiling points are available. Selecting a method depends on how much liquid is available and the availability of specific apparatus. In either microscale or macroscale experiments in which 0.3–0.5 mL of liquid is available, the semimicroscale direct method is usually most reliable. If less material is available, it will be necessary to perform either the semimicroscale or microscale inverted capillary method. With practice, these methods can be reliable, too. With larger quantities in both microscale and macroscale experiments, the boiling point can be observed best while performing a distillation.

Semimicroscale Direct Method. The apparatus for this method is shown in Figure 13.3. With this method, the bulb of the thermometer can be immersed in vapor from the boiling liquid for a period long enough to allow it to equilibrate and give a good temperature reading. A 13-mm × 100-mm test tube works well in this procedure. Use 0.3–0.5 mL of liquid and a small, inert carborundum (black) boiling stone. This method works best with a partial

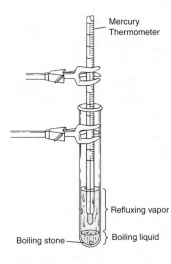

Mercury
Thermometer

Refluxing vapor

Boiling stone

Boiling liquid

Figure 13.3
*Macroscale method of
determining the boiling point.*

immersion (76 mm) mercury thermometer (see Section 13.3, p. 700). It is not necessary to perform a stem correction with this type of thermometer.

Place the bulb of the thermometer as close as possible to the boiling liquid without actually touching it. The best heating device is a hot plate with either an aluminum block or a sand bath.[1]

While you are heating the liquid, it is helpful to record the temperature at 1-minute intervals. This makes it easier to keep track of changes in the temperature and to know when the liquid has reached the boiling point. The liquid must boil vigorously, such that you see a reflux ring above the bulb of the thermometer and drops of liquid condensing on the sides of the test tube. Note that with some liquids, the reflux ring will be faint, and you must looked closely to see it. The boiling point is reached when the temperature reading on the thermometer has remained constant at its highest observed value for 2–3 minutes. It is usually best to turn the heat control on the hot plate to a relatively high setting initially, especially if you are starting with a cold hot plate and aluminum block or sand bath. If the temperature begins to level off at a relatively low temperature (less than about 100°C) or if the reflux ring reaches the immersion ring on the thermometer, you should turn down the heat-control setting immediately.

Two problems can occur when you perform this boiling-point procedure. The first is much more common and occurs when the temperature appears to be leveling off at a temperature below the boiling point of the liquid. This is more likely to happen with a relatively high-boiling liquid (boiling points greater than about 150°C) or when the sample is not heated sufficiently. The best way to prevent this problem is to heat the sample more strongly. With high-boiling liquids, it may be helpful to wait for the temperature to remain constant for 3–4 minutes to make sure that you have reached the actual boiling point.

The second problem, which is rare, occurs when the liquid evaporates completely, and the temperature inside the dry test tube may rise higher than the actual boiling point of the liquid. This is more likely to happen with low-boiling liquids (boiling point less than 100°C) or if the temperature on the hot plate is set too high for too long. To check for this possibility, observe the amount of liquid remaining in the test tube as soon as you have finished with the procedure. If there is no liquid remaining, it is possible that the highest temperature you observed is greater than the boiling point of the liquid. In this case, you should repeat the boiling-point determination, heating the sample less strongly or using more sample.

Depending on the skill of the person performing this technique, boiling points may be slightly inaccurate. When experimental boiling points are inaccurate, it is more common for them to be lower than the literature value, and inaccuracies are more likely to occur for higher-boiling liquids. With higher-boiling liquids, the difference may be as much as 5°C. Carefully following the previous instructions will make it more likely that your experimental value will be close to the literature value.

---

[1] *Note to the Instructor:* The aluminum block should have a hole drilled in it that goes all the way through the block and is just slightly larger than the outside diameter of the test tube. A sand bath can be conveniently prepared by adding 40 mL of sand to a 150-mL beaker or by using a heating mantle partially filled with sand. For additional comments about these heating methods, see the Instructor's Manual, Experiment 7, "Infrared Spectroscopy and Boiling-Point Determination."

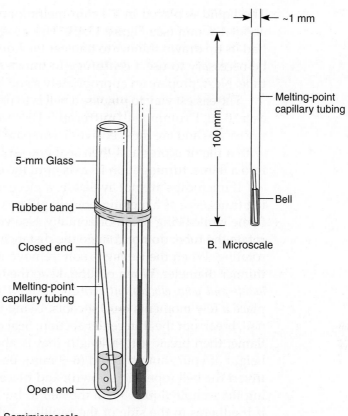

Figure 13.4
Boiling-point determinations.

With smaller amounts of material, you can carry out a microscale or semimicroscale determination of the boiling point by using the apparatus shown in Figure 13.4.

Semimicroscale Inverted Capillary Method. To carry out the semimicroscale determination, attach a piece of 5-mm glass tubing (sealed at one end) to a thermometer with a rubber band or a thin slice of rubber tubing. The liquid whose boiling point is being determined is introduced with a Pasteur pipet into this piece of tubing, and a short piece of melting-point capillary (sealed at one end) is dropped in with the open end down. The whole unit is then placed in a Thiele tube. The rubber band should be placed above the level of the oil in the Thiele tube; otherwise, the band may soften in the hot oil. When positioning the band, keep in mind that the oil will expand when heated. Next, the Thiele tube is heated in the same fashion as described in Technique 9, Section 9.6, page 631, for determining a melting point. Heating is continued until a rapid and continuous stream of bubbles emerges from the inverted capillary. At this point, you should stop heating. Soon, the stream of bubbles slows down and stops. When the bubbles stop, the liquid enters the capillary tube. The moment at which the liquid enters the capillary tube corresponds to the boiling point of the liquid, and the temperature is recorded.

Microscale Inverted Capillary Method. In microscale experiments, there often is too little product available to use the semimicroscale method just described. However, the method can be scaled down in the following manner.

The liquid is placed in a 1-mm melting-point capillary tube to a depth of about 4–6 mm (see Figure 13.4B). Use a syringe or a Pasteur pipet that has had its tip drawn thinner to transfer the liquid into the capillary tube. It may be necessary to use a centrifuge to transfer the liquid to the bottom of the tube. Next, prepare an appropriately sized inverted capillary, or **bell.**

The easiest way to prepare a bell is to use a commercial micropipet, such as a 10-μL Drummond "microcap." These are available in vials of 50 or 100 microcaps and are inexpensive. To prepare the bell, cut the microcap in half with a file or scorer and then seal one end by inserting it a small distance into a flame, turning it on its axis until the opening closes.

If microcaps are not available, a piece of 1-mm open-end capillary tubing (same size as a melting-point capillary) can be rotated along its axis in a flame while being held horizontally. Use your index fingers and thumbs to rotate the tube; do not change the distance between your two hands while rotating. When the tubing is soft, remove it from the flame and pull it to a thinner diameter. When pulling, keep the tube straight by *moving both your hands and your elbows outward* by about 4 inches. Hold the pulled tube in place a few moments until it cools. Using the edge of a file or your fingernail, break out the thin center section. Seal one end of the thin section in the flame; then break it to a length that is about one and one-half times the height of your sample liquid (6–9 mm). Be sure the break is done squarely. Invert the bell (open end down), and place it in the capillary tube containing the sample liquid. Push the bell to the bottom with a fine copper wire if it adheres to the side of the capillary tube. A centrifuge may be used if you prefer. Figure 13.5 shows the construction method for the bell and the final assembly.

Place the microscale assembly in a standard melting-point apparatus (or a Thiele tube if an electrical apparatus is not available) to determine the boiling point. Heating is continued until a rapid and continuous stream of bubbles emerges from the inverted capillary. At this point, stop heating. Soon, the stream of bubbles slows down and stops. When the bubbles stop, the liquid enters the capillary tube. The moment at which the liquid enters the capillary tube corresponds to the boiling point of the liquid, and the temperature is recorded.

Explanation of the Method.   During the initial heating, the air trapped in the inverted bell expands and leaves the tube, giving rise to a stream of bubbles. When the liquid begins boiling, most of the air has been expelled; the bubbles of gas are due to the boiling action of the liquid. Once the heating is stopped, most of the vapor pressure left in the bell comes from the vapor of the heated liquid that seals its open end. There is always vapor in equilibrium with a heated liquid. If the temperature of the liquid is above its boiling point, the pressure of the trapped vapor will either exceed or equal the atmospheric pressure. As the liquid cools, its vapor pressure decreases. When the vapor pressure drops just below atmospheric pressure (just below the boiling point), the liquid is forced into the capillary tube.

Difficulties.   Three problems are common to this method. The first arises when the liquid is heated so strongly that it evaporates or boils away. The second arises when the liquid is not heated above its boiling point before heating is discontinued. If the heating is stopped at any point below the actual boiling point of the sample, the liquid enters the bell *immediately,* giving an apparent boiling point that is too low. Be sure you observe a

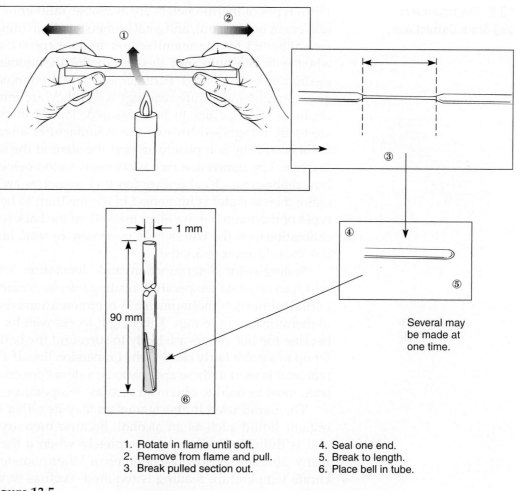

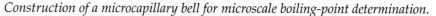

1. Rotate in flame until soft.
2. Remove from flame and pull.
3. Break pulled section out.
4. Seal one end.
5. Break to length.
6. Place bell in tube.

Figure 13.5
Construction of a microcapillary bell for microscale boiling-point determination.

continuous stream of bubbles, too fast for individual bubbles to be distinguished, before lowering the temperature. Also be sure the bubbling action decreases slowly before the liquid enters the bell. If your melting-point apparatus has fine enough control and fast response, you can actually begin heating again and force the liquid out of the bell before it becomes completely filled with the liquid. This allows a second determination to be performed on the same sample. The third problem is that the bell may be so light that the bubbling action of the liquid causes the bell to move up the capillary tube. This problem can sometimes be solved by using a longer (heavier) bell or by sealing the bell so that a larger section of solid glass is formed at the sealed end of the bell.

When measuring temperatures above 150°C, thermometer errors can become significant. For an accurate boiling point with a high-boiling liquid, you may wish to apply a **stem correction** to the thermometer, as described in Section 13.3, or to calibrate the thermometer, as described in Technique 9, Section 9.9, page 635.

Microscale or Macroscale–Distillation Method. When you have large quantities of material, you can simply record the boiling point (or boiling range) as viewed on a thermometer while performing a simple distillation (see Technique 14). When this method is used to determine a boiling point, it is best to use a partial immersion mercury thermometer for more accurate readings.

13.3 Thermometers and Stem Corrections

Three types of thermometers are available: bulb immersion, partial immersion (stem immersion), and total immersion. Bulb immersion thermometers are calibrated by the manufacturer to give correct temperature readings when only the bulb (not the rest of the thermometer) is placed in the medium to be measured. Partial immersion thermometers are calibrated to give correct temperature readings when they are immersed to a specified depth in the medium to be measured. Partial immersion thermometers are easily recognized because the manufacturer always scores a mark, or immersion ring, completely around the stem at the specified depth of immersion. The immersion ring is normally found below any of the temperature calibrations. Total immersion thermometers are calibrated when the entire thermometer is immersed in the medium to be measured. The three types of thermometer are often marked on the back (opposite side from the calibrations) by the words *bulb, immersion,* or *total,* but this may vary from one manufacturer to another.

Boiling-point determination and distillation are two techniques in which an accurate temperature reading may be obtained most easily with a partial immersion thermometer. A common immersion length for this type of thermometer is 76 mm. This length works well for these two techniques because the hot vapors are likely to surround the bottom of the thermometer up to a point fairly close to the immersion line. If a total immersion thermometer is used in these applications, a stem correction, which is described later, must be used to obtain an accurate temperature reading.

The liquid used in thermometers may be either mercury or a colored organic liquid such as an alcohol. Because mercury is highly poisonous and is difficult to clean up completely when a thermometer is broken, many laboratories now use nonmercury thermometers. When a highly accurate temperature reading is required, such as in a boiling-point determination or in some distillations, mercury thermometers may have an advantage over nonmercury thermometers for two reasons. Mercury has a lower coefficient of expansion than the liquids used in nonmercury thermometers. Therefore, a partial immersion mercury thermometer will give a more accurate reading when the thermometer is not immersed in the hot vapors exactly to the immersion line. In other words, the mercury thermometer is more forgiving. Furthermore, because mercury is a better conductor of heat, a mercury thermometer will respond more quickly to changes in the temperature of the hot vapors. If the temperature is read before the thermometer reading has stabilized, which is more likely to occur with a nonmercury thermometer, the temperature reading will be inaccurate.

Manufacturers design total immersion thermometers to read correctly only when they are immersed totally in the medium to be measured. The entire mercury thread must be covered. Because this situation is rare, a **stem correction** should be added to the observed temperature. This correction, which is positive, can be fairly large when high temperatures are being measured. Keep in mind, however, that if your thermometer has been calibrated for its desired use (such as described in Technique 9, Section 9.9 for a melting-point apparatus), a stem correction should not be necessary for any temperature within the calibration limits. You are most likely to want a stem correction when you are performing a distillation. If you determine a melting point or boiling point using an uncalibrated, total immersion thermometer, you will also want to use a stem correction.

When you wish to make a stem correction for a total immersion thermometer, the following formula may be used. It is based on the fact that the portion of the mercury thread in the stem is cooler than the portion immersed in the vapor or the heated area around the thermometer. The mercury will not have expanded in the cool stem to the same extent as in the warmed section of the thermometer. The equation used is

$$(0.000154)(T - t_1)(T - t_2) = \text{correction to be added to } T \text{ observed}$$

1. The factor 0.000154 is a constant, the coefficient of expansion for the mercury in the thermometer.

2. The term $T - t_1$ corresponds to the length of the mercury thread not immersed in the heated area. Use the temperature scale on the thermometer itself for this measurement rather than an actual length unit. T is the observed temperature, and t_1 is the *approximate* place where the heated part of the stem ends and the cooler part begins.

3. The term $T - t_2$ corresponds to the difference between the temperature of the mercury in the vapor T and the temperature of the mercury in the air outside the heated area (room temperature). The term T is the observed temperature, and t_2 is measured by hanging another thermometer so the bulb is close to the stem of the main thermometer.

Figure 13.6 shows how to apply this method for a distillation. By the formula just given, it can be shown that high temperatures are more likely to

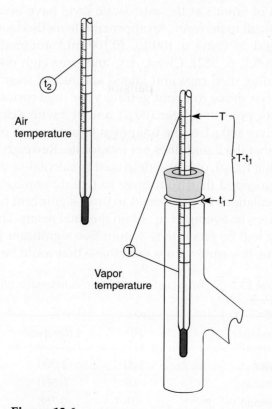

Figure 13.6
Measurement of a thermometer stem correction during distillation.

require a stem correction and that low temperatures need not be corrected. The following sample calculations illustrate this point.

Example 1	*Example 2*
$T = 200°C$	$T = 100°C$
$t_1 = 0°C$	$t_1 = 0°C$
$t_2 = 35°C$	$t_2 = 35°C$
$(0.000154)(200)(165) = 5.1°$ stem correction	$(0.000154)(100)(165) = 1.0°$ stem correction
$200°C + 5°C = 205°C$ corrected temperature	$100°C + 1°C = 101°C$ corrected temperature

PART B. DENSITY

13.4 Density

Density is defined as mass per unit volume and is generally expressed in units of grams per milliliter (g/mL) for a liquid and grams per cubic centimeter (g/cm³) for a solid.

$$\text{Density} = \frac{\text{mass}}{\text{volume}} \quad \text{or} \quad D = \frac{M}{V}$$

In organic chemistry, density is most commonly used in converting the weight of liquid to a corresponding volume, or vice versa. It is often easier to measure a volume of a liquid than to weigh it. As a physical property, density is also useful for identifying liquids in much the same way that boiling points are used.

Although precise methods that allow the measurements of the densities of liquids at the microscale level have been developed, they are often difficult to perform. An approximate method for measuring densities can be found in using a 100-μL (0.100-mL) automatic pipet (Technique 5, Section 5.1, p. 582). Clean, dry, and preweigh one or more conical vials (including their caps and liners) and record their weights. Handle these vials with a tissue to avoid getting your fingerprints on them. Adjust the automatic pipet to deliver 100 μL and fit it with a clean, new tip. Use the pipet to deliver 100 μL of the unknown liquid to each of your tared vials. Cap them so that the liquid does not evaporate. Reweigh the vials and use the weight of the 100 μL of liquid delivered to calculate a density for each case. It is recommended that from three to five determinations be performed, that the calculations be performed to three significant figures, and that all the calculations be averaged to obtain the final result. This determination of the density will be accurate to within two significant figures. Table 13.1 compares some literature values with those that could be obtained by this method.

Table 13.1 *Densities determined by the automatic pipet method (g/mL)*

Substance	*BP*	*Literature*	*100 μL*
Water	100	1.000	1.01
Hexane	69	0.660	0.66
Acetone	56	0.788	0.77
Dichloromethane	40	1.330	1.27
Diethyl ether	35	0.713	0.67

PROBLEMS

1. Using the pressure–temperature alignment chart in Figure 13.2, answer the following questions.

   a. What is the normal boiling point (at 760 mm Hg) for a compound that boils at 150°C at 10 mm Hg pressure?

   b. At what temperature would the compound in (a) boil if the pressure were 40 mm Hg?

   c. A compound was distilled at atmospheric pressure and had a boiling point of 285°C. What would be the approximate boiling range for this compound at 15 mm Hg?

2. Calculate the corrected boiling point for nitrobenzene by using the method given in Section 13.3. The boiling point was determined using an apparatus similar to that shown in Figure 13.3. Assume that a total immersion thermometer was used. The observed boiling point was 205°C. The reflux ring in the test tube just reached up to the 0°C mark on the thermometer. A second thermometer suspended alongside the test tube, at a slightly higher level than the one inside, gave a reading of 35°C.

3. Suppose that you had calibrated the thermometer in your melting-point apparatus against a series of melting-point standards. After reading the temperature and converting it using the calibration chart, should you also apply a stem correction? Explain.

4. The density of a liquid was determined by the automatic pipet method. A 100-μL automatic pipet was used. The liquid had a mass of 0.082 g. What was the density in grams per milliliter of the liquid?

5. During the microscale boiling-point determination (see p. 697) of an unknown liquid, heating was discontinued at 154°C and the liquid immediately began to enter the inverted bell. Heating was begun again at once, and the liquid was forced out of the bell. Heating was again discontinued at 165°C, at which time a rapid stream of bubbles emerged from the bell. On cooling, the rate of bubbling gradually diminished until the liquid reached a temperature of 161°C and entered and filled the bell. Explain this sequence of events. What was the boiling point of the liquid?

---

14 ■ TECHNIQUE 14

Simple Distillation

Distillation is the process of vaporizing a liquid, condensing the vapor, and collecting the condensate in another container. This technique is useful for separating a liquid mixture when the components have different boiling points or when one of the components will not distill. It is one of the principal methods of purifying a liquid. Four basic distillation methods are available to the chemist: simple distillation, vacuum distillation (distillation at reduced pressure), fractional distillation, and steam distillation. This technique chapter will discuss simple distillation. Vacuum distillation will be discussed in Technique 16. Fractional distillation will be discussed in Technique 15, and steam distillation will be discussed in Technique 18.

14.1 The Evolution of Distillation Equipment

There are probably more types and styles of distillation apparatus than exist for any other technique in chemistry. Over the centuries, chemists have devised just about every conceivable design. The earliest known types of distillation apparatus were the **alembic** and the **retort** (Fig. 14.1). They were

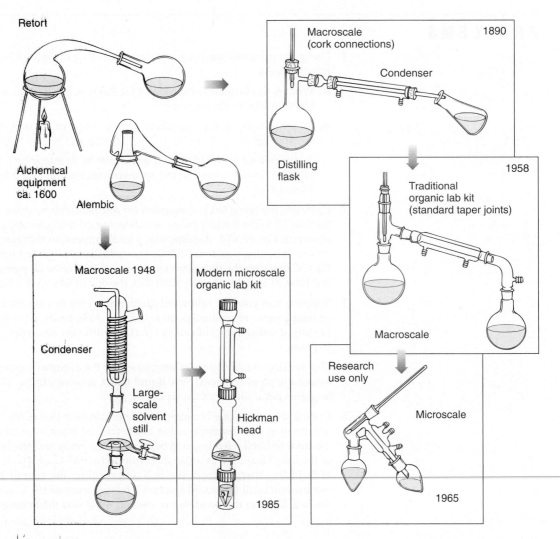

Figure 14.1

Some stages in the evolution of microscale distillation equipment from alchemical equipment (dates represent approximate time of popular use).

used by alchemists in the Middle Ages and the Renaissance and probably even earlier by Arabic chemists. Most other distillation equipment has evolved as variations on these designs.

Figure 14.1 shows several stages in the evolution of distillation equipment as it relates to the organic laboratory. It is not intended to be a complete history; rather, it is representative. Up until recent years, equipment based on the retort design was common in the laboratory. Although the retort itself was still in use early in the twentieth century, it had evolved by that time into the distillation flask and water-cooled condenser combination. This early equipment was connected with corks. By 1958, most introductory laboratories were beginning to use "organic lab kits" that included glassware connected by standard-taper glass joints. The original lab kits contained large ⊤ 24/40 joints. Within a short time, they became smaller, with ⊤ 19/22 and even ⊤ 14/20 joints. These later kits are still being used today in many organic courses. Small-scale variations of these kits are also used today by chemical researchers, but they are too expensive to use in an introductory laboratory. Instead, the "microscale" equipment you

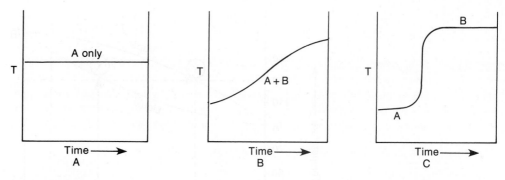

Figure 14.2
Three types of temperature behavior during a simple distillation. A. A single pure component. B. Two components of similar boiling points. C. Two components with widely differing boiling points. Good separations are achieved in A and C.

are using in this course is coming into common use. This equipment has ₮ 14/10 standard-taper joints, threaded outer joints with screwcap connectors, and an internal O-ring. The distillation apparatus in microscale kits is designed for work with small amounts of material, and it is different from the more traditional larger-scale equipment. It is perhaps more closely related to the alembic design than to that of the retort. Because both types of equipment are in use today, after we describe microscale equipment, we will also show the equivalent large-scale apparatus used to perform distillation.

14.2 Distillation Theory

In the traditional distillation of a pure substance, vapor rises from the distillation flask and comes into contact with a thermometer that records its temperature. The vapor then passes through a condenser, which reliquefies the vapor and passes it into the receiving flask. The temperature observed during the distillation of a **pure substance** remains constant throughout the distillation so long as both vapor *and* liquid are present in the system (see Fig. 14.2A). When a **liquid mixture** is distilled, often the temperature does not remain constant but increases throughout the distillation. The reason for this is that the composition of the vapor that is distilling varies continuously during the distillation (see Fig. 14.2B).

For a liquid mixture, the composition of the vapor in equilibrium with the heated solution is different from the composition of the solution itself. This is shown in Figure 14.3, which is a phase diagram of the typical vapor–liquid relationship for a two-component system (A + B).

On this diagram, horizontal lines represent constant temperatures. The upper curve represents vapor composition, and the lower curve represents liquid composition. For any horizontal line (constant temperature), like that shown at t, the intersections of the line with the curves give the compositions of the liquid and the vapor that are in equilibrium with each other at that temperature. In the diagram, at temperature t, the intersection of the curve at X indicates that liquid of composition W will be in equilibrium with vapor of composition Z, which corresponds to the intersection at Y. Composition is given as a mole percentage of A and B in the mixture. Pure A, which boils at temperature t_A, is represented at the left. Pure B, which boils at temperature t_B, is represented on the right. For either pure A or pure B, the vapor and liquid curves meet at the boiling point. Thus, either pure A or pure B will distill at a constant temperature (t_A or t_B). Both the vapor and the

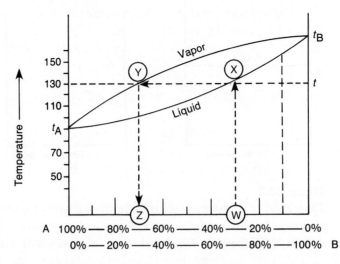

Figure 14.3
Phase diagram for a typical liquid mixture of two components.

liquid must have the same composition in either of these cases. This is not the case for mixtures of A and B.

A mixture of A and B of composition W will have the following behavior when heated. The temperature of the liquid mixture will increase until the boiling point of the mixture is reached. This corresponds to following line WX from W to X, the boiling point of the mixture t. At temperature t the liquid begins to vaporize, which corresponds to line XY. The vapor has the composition corresponding to Z. In other words, the first vapor obtained in distilling a mixture of A and B does not consist of pure A. It is richer in A than the original mixture but still contains a significant amount of the higher-boiling component B, even from the very beginning of the distillation. The result is that it is never possible to separate a mixture completely by a simple distillation. However, in two cases it is possible to get an acceptable separation into relatively pure components. In the first case, if the boiling points of A and B differ by a large amount (>100 degrees) and if the distillation is carried out carefully, it will be possible to get a fair separation of A and B. In the second case, if A contains a fairly small amount of B (<10%), a reasonable separation of A from B can be achieved. When the boiling-point differences are not large and when highly pure components are desired, it is necessary to perform a **fractional distillation.** Fractional distillation is described in Technique 15, in which the behavior during a simple distillation is also considered in detail. Note only that as vapor distills from the mixture of composition W (Fig. 14.3), it is richer in A than is the solution. Thus, the composition of the material left behind in the distillation becomes richer in B (moves to the right from W toward pure B in the graph). A mixture of 90% B (dotted line on the right side in Fig. 14.3) has a higher boiling point than at W. Hence, the temperature of the liquid in the distillation flask will increase during the distillation, and the composition of the distillate will change (as is shown in Fig. 14.3B).

When two components that have a large boiling-point difference are distilled, the temperature remains constant while the first component distills. If the temperature remains constant, a relatively pure substance is being distilled. After the first substance distills, the temperature of the vapors rises, and the second component distills, again at a constant temperature. This is shown in Figure 14.2C. A typical application of this type of distillation might be an instance of a reaction mixture containing the desired component A (bp 140°C) contaminated with a small amount of undesired component B (bp 250°C) and mixed with a solvent such as diethyl ether (bp 36°C). The ether is removed easily at low temperature. Pure A is removed at a higher temperature and collected in a separate receiver. Component B can then be distilled, but it usually is left as a residue and not distilled. This separation is not difficult and represents a case where simple distillation might be used to advantage.

14.3 Microscale Equipment

Most large-scale distillation equipment requires the distilled liquid to travel a long distance from the distillation flask, through the condenser, to the receiving flask. When working at the microscale level, a long distillation path must be avoided. With small quantities of liquid, there are too many opportunities to lose all the sample. The liquid will adhere to, or wet, surfaces and get lost in every little nook and cranny of the system. A system with a long path also has a large volume, and a small amount of liquid may not produce enough vapor to fill it. Small-scale distillation requires a "short path" distillation. In order to make the distilling path as short as possible, the **Hickman head** has been adopted as the principal receiving device for most microscale distilling operations.

The Hickman Head. Two types of Hickman head (also called a Hickman "still") are shown in Figure 14.4. One of these variations has a convenient opening, or port, in the side, making removal of liquid that has collected in it easier. In operation, the liquid to be distilled is placed in a flask or vial attached to the bottom joint of the Hickman head and heated. If desired, you can attach a condenser to the top joint. Either a magnetic spin vane or a boiling stone is used to prevent bumping. Some typical assemblies are shown in Figures 14.5 and 14.7. The vapors of the heated liquid rise upward and are cooled and condensed on either the walls of the condenser or, if no condenser is used, on the inside walls of the Hickman head itself. As liquid drains downward, it collects in the circular well at the bottom of the still.

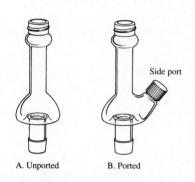

A. Unported   B. Ported

Figure 14.4
The Hickman head.

Collecting Fractions. The liquid that distills is called the **distillate.** Portions of the distillate collected during the course of a distillation are called **fractions.** A small fraction (usually discarded) collected before the distillation is begun in earnest is called a **forerun.** The well in a Hickman head can contain anywhere from 1 to 2 mL of liquid. In the style with the side port, fractions may be removed by opening the port and inserting a Pasteur pipet (Fig. 14.6C). The unported head works equally well, but the head is emptied from the top by using a Pasteur pipet (Fig. 14.6A). If a condenser or an internal thermometer is used, the distilling apparatus must be partially disassembled to remove liquid when the well fills. In some stills, the inner diameter of the head is small, and it is difficult to reach in at an angle with the pipet and make contact with the liquid. To remedy this problem, you may be able to use the longer (9-inch) Pasteur pipet instead of the shorter (5¾-inch) one. The longer pipet has a much longer narrow section (tip) and can adapt more effectively to the required angle. The disadvantage of the longer tip is that you are more likely to break it off inside the still. You may prefer to modify a short pipet by bending its tip slightly in a flame (Fig. 14.6B).

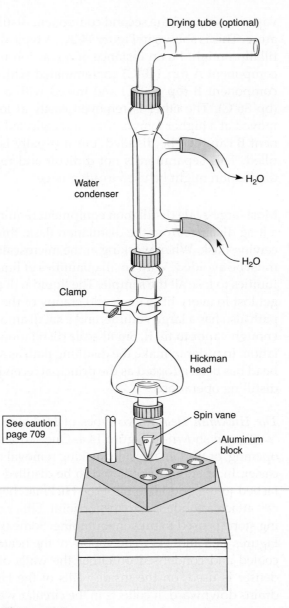

Figure 14.5
Basic microscale distillation (external monitoring of temperature).

Choice of Condenser. If you are careful (slow heating) or if the liquid to be distilled has a high boiling point, it many not be necessary to use a condenser with the Hickman head (Fig. 14.7). In this case, the liquid being distilled must condense on the cooler sides of the head itself without any being lost through evaporation. If the liquid has a low boiling point or is very volatile, a condenser must be used. With very volatile liquids, a water-cooled condenser must be used; however, an air-cooled condenser may suffice for less demanding cases. When using a water condenser, remember that water should enter the lower opening and exit from the upper one. If the hoses carrying the water in and out are connected in reverse fashion, the water jacket of the condenser will not fill completely.

Sealed Systems. Whenever you perform a distillation, be sure the system you are heating is not sealed off completely from the outside atmosphere.

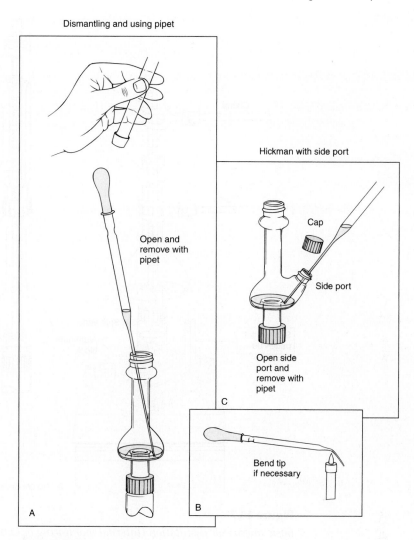

Figure 14.6
Removing fractions.

During a distillation, the air and vapors inside the system will both expand and contract. If pressure builds up inside a sealed system, the apparatus may explode. In performing a distillation, you should leave a small opening at the far end of the system. If water vapor could be harmful to the substances being distilled, a calcium chloride drying tube may be used to protect the system from moisture. Carefully examine each system discussed to see how an opening to the outside is provided.

External Monitoring of Temperature. The simple assembly using the Hickman head shown in Figure 14.5 does not monitor the temperature inside the apparatus. Instead, the temperature is monitored externally with a thermometer placed in an aluminum block.

CAUTION

You should not use a *mercury* thermometer with an aluminum block. If it breaks, the mercury will vaporize on the hot surface. Instead, use a nonmercury glass thermometer, a metal dial thermometer, or a digital electronic temperature-measuring device.

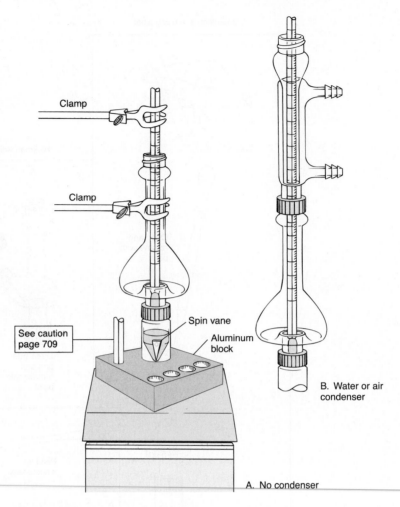

Figure 14.7

Basic microscale distillation (internal monitoring of temperature).

External monitoring of the temperature has the disadvantage that the exact temperature at which liquid distills is never known. In many cases, this does not matter or is unavoidable, and the boiling point of the distilled liquid can be checked later by performing a microboiling-point determination (Technique 13, Section 13.2, p. 695).

As a rule, there is at least a 15-degree difference in temperature between the temperature of the aluminum block or sand bath and that of the liquid in the heated distillation vial or flask. However, the magnitude of this difference cannot be relied on. Keep in mind that the liquid in the vial or flask may be at a different temperature than the vapor that is distilling. In many procedures in this text, the *approximate* temperature of the heating device will be given instead of the boiling point of the liquid involved. Because this method of monitoring the temperature is rather approximate, you will need to make the actual heater setting based on what is supposed to be occurring in the vial or flask.

Internal Monitoring of Temperature. When you wish to monitor the actual temperature of a distillation, a thermometer must be placed inside the apparatus. Figures 14.7 and 14.8 show distillation assemblies that use an internal thermometer. The apparatus in Figure 14.7A represents the simplest possible distillation assembly. It does not use a condenser, and the

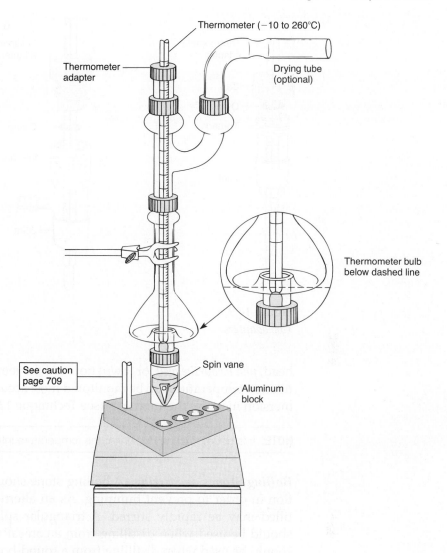

Thermometer (−10 to 260°C)

Thermometer
adapter

Drying tube
(optional)

Thermometer bulb
below dashed line

See caution
page 709

Spin vane

Aluminum
block

Figure 14.8

*Basic microscale distillation using thermometer adapter (internal monitoring
of temperature).*

thermometer is suspended from a clamp. It is possible to add either an air or
a water condenser to this basic assembly (Fig. 14.7B) and maintain internal
monitoring of the temperature.

In the arrangement shown in Figure 14.8, a thermometer adapter is
used. A thermometer adapter (Fig. 14.9A) provides a convenient way of
holding a thermometer in place. The Claisen head is used to provide an
opening to the atmosphere, thereby avoiding a sealed system. With the
Claisen head, a drying tube may be used to protect the system from atmospheric moisture.

If protection from atmospheric moisture is not required, the multipurpose adapter may be used. The multipurpose adapter (Fig. 14.9B) replaces
both the thermometer adapter and the Claisen head. With this adapter, the
necessary opening to the atmosphere is provided by the side arm. The
threaded joint holds the thermometer in place.

Carefully notice the position of the thermometer in Figures 14.7 and 14.8.
The bulb of the thermometer must be placed in the stem of the Hickman

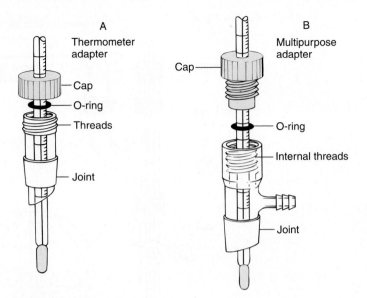

Figure 14.9
Two adapters.

head, *just below the well,* or it will not read the temperature correctly. The distillation temperature can be monitored most accurately by using a partial immersion mercury thermometer (see Technique 13, Section 13.3, p. 700).

NOTE: It is good practice to monitor the temperature internally whenever possible.

Boiling Stones or Stirring.   A boiling stone should be used during distillation in order to prevent bumping. As an alternative, the liquid being distilled may be rapidly stirred. A triangular spin vane of the correct size should be used when distilling from a conical vial, whereas a stirring bar should be used when distilling from a round-bottom flask.

Size of Distillation Flask.   As a rule, the distillation flask or vial should not be filled to more than two thirds of its total capacity. This allows room for boiling and stirring action, and it prevents contamination of the distillate by bumping. A flask that is too large should also be avoided. With too large a flask, the **holdup** is excessive; the holdup is the amount of material that cannot distill because some vapor must fill the empty flask.

Assembling the Apparatus.   You should not grease the joints when assembling the apparatus. Ungreased joints seal well enough to allow you to perform a simple distillation. Stopcock grease can introduce a serious contaminant into your product.

Rate and Degree of Heating.   Take care not to distill too quickly. If you vaporize liquid at a rate faster than it can be recondensed, some of your product may be lost by evaporation. On the other hand, you should not distill too slowly. This may also lead to loss of product because there is a longer period during which vapors can escape. Carefully examine your apparatus during distillation to monitor the position of either a reflux ring or a wet appearance on the surface of the glass. Either of these indicates the place at which condensation is occurring. The position at which condensation occurs should be well inside the Hickman head. Be sure that liquid is collecting in the well. If all the surfaces are shiny (wet) and there is no distillate, you are losing material.

NOTE: A slower rate of heating also helps to avoid bumping.

If you are using a sand bath, material may be lost because the hot sand bath radiates too much heat upward and warms the Hickman still. If you believe this to be the case, it can often be remedied by placing a small square of aluminum foil over the top of the sand bath. Make a tear from one edge to the center of the foil to wrap it around the apparatus.

14.4 Semimicroscale and Macroscale Equipment

When you wish to distill quantities of liquid that are larger than 2–3 mL, different equipment is required. Most manufacturers of microscale equipment make two pieces of conventional distillation equipment sized to work with the ⊤ 14/10 microscale kit components. These two pieces, the **distillation head** and the **bent vacuum adapter,** are not provided in student microscale kits but must be purchased separately. Figure 14.10 shows a semimicroscale assembly using these components. Note that the bulb of the thermometer must be placed *below the side arm* if it is to be bathed in vapor and give a correct temperature reading. This apparatus assumes that a condenser is not necessary; however, you could easily insert one between the distilling head and the bent vacuum adapter. This insertion would produce a completely traditional distillation apparatus but would use microscale equipment. (See Figure on p. 714.) A distillation apparatus constructed from a "macroscale" organic laboratory kit is shown in Figure 14.11. This type of equipment is being used today in organic laboratories that have not converted to microscale. Electrically regulated **heating mantles** are often used with this equipment.

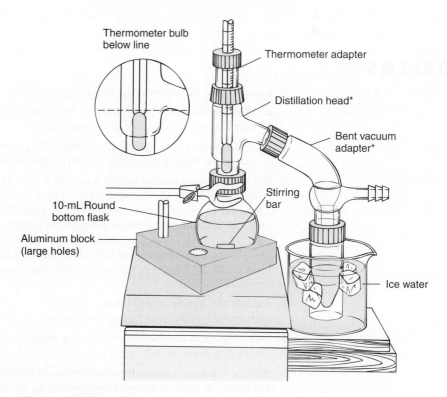

Figure 14.10
Semimicroscale distillation (*requires special pieces).

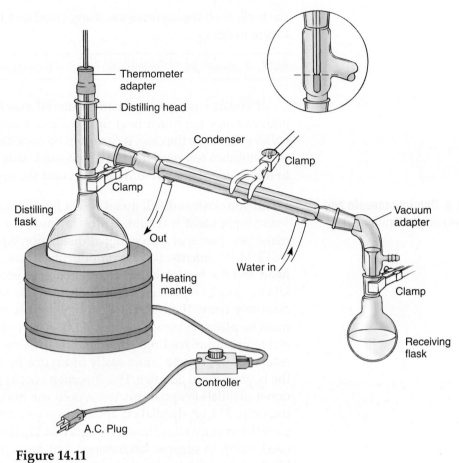

Figure 14.11
Distillation with the standard macroscale organic lab kit.

PROBLEMS

1. Using Figure 14.3, answer the following questions.

   a. What is the molar composition of the vapor in equilibrium with a boiling liquid that has a composition of 60% A and 40% B?

   b. A sample of vapor has the composition 50% A and 50% B. What is the composition of the boiling liquid that produced this vapor?

2. Use an apparatus similar to that shown in Figure 14.10 and assume that the round-bottom flask holds 10 mL and that the Claisen head has an internal volume of about 2 mL in the vertical section. At the end of a distillation, vapor would fill this volume, but it could not be forced through the system. No liquid would remain in the distillation flask. Assuming this **holdup volume** of 12 mL, use the ideal gas law and assume a boiling point of 100°C (760 mm Hg) to calculate the number of microliters of liquid ($d = 0.9$ g/mL, $MW = 200$) that would recondense into the distillation flask on cooling.

3. Explain the significance of a horizontal line connecting a point on the lower curve with a point on the upper curve (such as line XY) in Figure 14.3.

4. Using Figure 14.3, determine the boiling point of a liquid having a molar composition of 50% A and 50% B.

5. What is the approximate difference between the temperature of a boiling liquid in a conical vial and the temperature read on an *external* thermometer when both are placed on an aluminum block?

6. Where should the thermometer bulb be located for internal monitoring in

    a. a distillation apparatus using a Hickman head?

    b. a large-scale distillation using a Claisen head with a water condenser placed beyond it?

7. Under what conditions can a good separation be achieved with a simple distillation?

---

15   TECHNIQUE 15

Fractional Distillation, Azeotropes

Simple distillation, described in Technique 14, works well for most routine separation and purification procedures for organic compounds. When boiling-point differences of components to be separated are not large, however, **fractional distillation** must be used to achieve a good separation.

PART A. FRACTIONAL DISTILLATION

15.1 Differences between Simple and Fractional Distillation

When an ideal solution of two liquids, such as benzene (bp 80°C) and toluene (bp 110°C), is distilled by simple distillation, the first vapor produced will be enriched in the lower-boiling component (benzene). However, when that initial vapor is condensed and analyzed, the distillate will not be pure benzene. The boiling-point difference of benzene and toluene (30°C) is too small to achieve a complete separation by simple distillation. Following the principles outlined in Technique 14, Section 14.2 (pp. 705–707), and using the vapor–liquid composition curve given in Figure 15.1, you can see what would happen if you started with a equimolar mixture of benzene and toluene.

Following the dashed lines shows that an equimolar mixture (50 mole % benzene) would begin to boil at about 91°C and, far from being 100% benzene, the distillate would contain about 74 mole % benzene and 26 mole % toluene. As the distillation continued, the composition of the undistilled liquid would move in the direction of A' (there would be increased toluene due to removal of more benzene than toluene), and the corresponding vapor would contain a progressively smaller amount of benzene. In effect, the temperature of the distillation would continue to increase throughout the distillation (as in Figure 14.2B, p. 705), and it would be impossible to obtain any fraction that consisted of pure benzene.

Suppose, however, that we are able to collect a small quantity of the first distillate that was 74 mole % benzene, and to redistill it. Using Figure 15.1, we can see that this liquid would begin to boil at about 84°C and would give an initial distillate containing 90 mole % benzene. If we were experimentally able to continue taking small fractions at the beginning of each distillation and redistill them, we would eventually reach a liquid with a composition of nearly 100 mole % benzene. However, because we only took a small amount of material at the beginning of each distillation, we would have lost most of the material we started with. To recapture a reasonable amount of benzene, we would have to process each of the fractions left behind in the same way as our early fractions. As each of them was partially distilled, the material advanced would become progressively richer in benzene, whereas

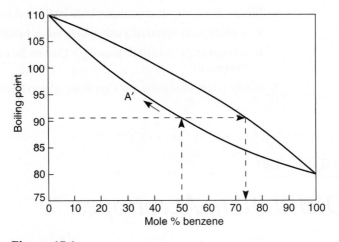

Figure 15.1
The vapor–liquid composition curve for mixtures of benzene and toluene.

that left behind would become progressively richer in toluene. It would require thousands (maybe millions) of such microdistillations to separate benzene from toluene.

Obviously, the procedure just described would be very tedious; fortunately, it need not be performed in usual laboratory practice. **Fractional distillation** accomplishes the same result. You simply have to use a column inserted between the distillation flask and the receiver (Hickman head), as shown in Figure 15.2. This **fractionating column** is filled, or **packed,** with a suitable material such as a stainless steel sponge. This packing allows a mixture of benzene and toluene to be subjected continuously to many vaporization–condensation cycles as the material moves up the column. With each cycle within the column, the composition of the vapor is progressively enriched in the lower-boiling component (benzene). Nearly pure benzene (bp 80°C) finally emerges from the top of the column, condenses, and passes into the receiving head or flask. This process continues until all the benzene is removed. The distillation must be carried out slowly to ensure that numerous vaporization–condensation cycles occur. When nearly all the benzene has been removed, the temperature begins to rise, and a small amount of a second fraction, which contains some benzene and toluene, may be collected. When the temperature reaches 110°C, the boiling point of pure toluene, the vapor is condensed and collected as the third fraction. A plot of boiling point versus volume of condensate (distillate) would resemble Figure 15.3. This separation would be much better than that achieved by simple distillation (Figure 15.1).

15.2 Vapor–Liquid Composition Diagrams

A vapor–liquid composition-phase diagram like the one in Figure 15.4 can be used to explain the operation of a fractionating column with an **ideal solution** of two liquids, A and B. An ideal solution is one in which the two liquids are chemically similar, miscible (mutually soluble) in all proportions, and do not interact. Ideal solutions obey **Raoult's Law.** Raoult's Law is explained in detail in Section 15.3.

The phase diagram relates the compositions of the boiling liquid (lower curve) and its vapor (upper curve) as a function of temperature. Any horizontal line drawn across the diagram (a constant-temperature line) intersects the diagram in two places. These intersections relate the vapor composition to

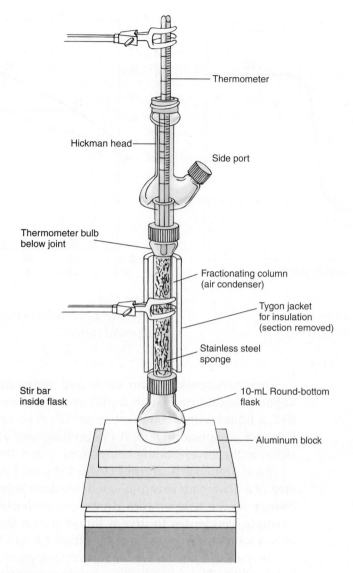

Figure 15.2
Microscale apparatus for fractional distillation.

the composition of the boiling liquid that produces that vapor. By convention, composition is expressed either in **mole fraction** or in **mole percentage**. The mole fraction is defined as follows:

$$\text{Mole fraction A} = N_A = \frac{\text{moles A}}{\text{moles A} + \text{moles B}}$$

$$\text{Mole fraction B} = N_B = \frac{\text{moles B}}{\text{moles A} + \text{moles B}}$$

$$N_A + N_B = 1$$

$$\text{Mole percentage A} = N_A \times 100$$

$$\text{Mole percentage B} = N_B \times 100$$

The horizontal and vertical lines shown in Figure 15.4 represent the processes that occur during a fractional distillation. Each of the **horizontal lines** (L_1V_1, L_2V_2, etc.) represents the **vaporization** step of a given

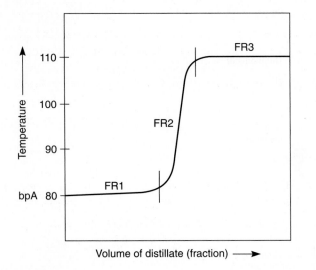

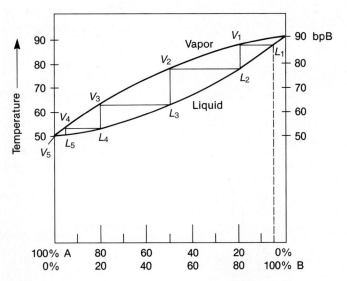

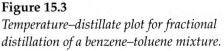

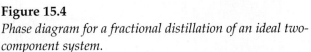

Figure 15.3

Temperature–distillate plot for fractional distillation of a benzene–toluene mixture.

Figure 15.4

Phase diagram for a fractional distillation of an ideal two-component system.

vaporization–condensation cycle and represents the composition of the vapor in equilibrium with liquid at a given temperature. For example, at 63°C a liquid with a composition of 50% A (L_3 on the diagram) would yield vapor of composition 80% A (V_3 on diagram) at equilibrium. The vapor is richer in the lower-boiling component A than the original liquid was.

Each of the **vertical lines** (V_1L_2, V_2L_3, etc.) represents the **condensation** step of a given vaporization–condensation cycle. The composition does not change as the temperature drops on condensation. The vapor at V_3, for example, condenses to give a liquid (L_4 on the diagram) of composition 80% A with a drop in temperature from 63° to 53°C.

In the example shown in Figure 15.4, pure A boils at 50°C and pure B boils at 90°C. These two boiling points are represented at the left- and right-hand edges of the diagram, respectively. Now consider a solution that contains only 5% of A but 95% of B. (Remember that these are *mole* percentages.) This solution is heated (following the dashed line) until it is observed to boil at L_1 (87°C). The resulting vapor has composition V_1 (20% A, 80% B). The vapor is richer in A than the original liquid, but it is by no means pure A. In a simple distillation apparatus, this vapor would be condensed and passed into the receiver in a very impure state. However, with a fractionating column in place, the vapor is condensed in the **column** to give liquid L_2 (20% A, 80% B). Liquid L_2 is immediately revaporized (bp 78°C) to give a vapor of composition V_2 (50% A, 50% B), which is condensed to give liquid L_3. Liquid L_3 is revaporized (bp 63°C) to give vapor of composition V_3 (80% A, 20% B), which is condensed to give liquid L_4. Liquid L_4 is revaporized (bp 53°C) to give vapor of composition V_4 (95% A, 5% B). This process continues to V_5, which condenses to give nearly pure liquid A. The fractionating process follows the stepped lines in the figure downward and to the left.

As this process continues, all of liquid A is removed from the distillation flask or vial, leaving nearly pure B behind. If the temperature is raised, liquid B may be distilled as a nearly pure fraction. Fractional distillation will have achieved a separation of A and B, a separation that would have been nearly impossible with simple distillation. Notice that the boiling point of the liquid becomes lower each time it vaporizes. Because the temperature at the bottom of a column is normally higher than the temperature at the top, successive vaporizations occur higher and higher in the column as the composition of the distillate approaches that of pure A. This process is illustrated in Figure 15.5, where the composition of the liquids, their boiling points, and the composition of the vapors present are shown alongside the fractionating column.

15.3 Raoult's Law

Two liquids (A and B) that are miscible and that do not interact form an **ideal solution** and follow Raoult's Law. The law states that the partial vapor pressure of component A in the solution (P_A) equals the vapor pressure of pure A (P_A°) times its mole fraction (N_A) (Eq. 1). A similar expression can be written for component B (Eq. 2). The mole fractions N_A and N_B were defined in Section 15.2.

$$\text{Partial vapor pressure of A in solution} = P_A = (P_A^\circ)(N_A) \qquad [1]$$

$$\text{Partial vapor pressure of B in solution} = P_B = (P_B^\circ)(N_B) \qquad [2]$$

P_A° is the vapor pressure of pure A, independent of B. P_B° is the vapor pressure of B, independent of A. In a mixture of A and B, the partial vapor pressures are added to give the total vapor pressure above the solution (Eq. 3). When the total pressure (sum of the partial pressures) equals the applied pressure, the solution boils.

$$P_{total} = P_A + P_B = P_A^\circ N_A + P_B^\circ N_B \qquad [3]$$

The composition of A and B in the vapor produced is given by Equations 4 and 5.

$$N_A \text{ (vapor)} = \frac{P_A}{P_{total}} \qquad [4]$$

$$N_B \text{ (vapor)} = \frac{P_B}{P_{total}} \qquad [5]$$

Several problems involving applications of Raoult's Law are illustrated in Figure 15.6. Note, particularly in the result from Equation 4, that the vapor is richer ($N_A = 0.67$) in the lower-boiling (higher vapor pressure) component A than it was before vaporization ($N_A = 0.50$). This proves mathematically what was described in Section 15.2.

The consequences of Raoult's Law for distillations are shown schematically in Figure 15.7. In Part A, the boiling points are identical (vapor pressures the same), and no separation is attained regardless of how the distillation is conducted. In Part B, a fractional distillation is required, whereas in Part C a simple distillation provides an adequate separation.

When a solid B (rather than another liquid) is dissolved in a liquid A, the boiling point is increased. In this extreme case, the vapor pressure of B is negligible, and the vapor will be pure A no matter how much solid B is added. Consider a solution of salt in water.

$$P_{total} = P_{water}^\circ N_{water} + P_{salt}^\circ N_{salt}$$

$$P_{salt}^\circ = 0$$

$$P_{total} = P_{water}^\circ N_{water}$$

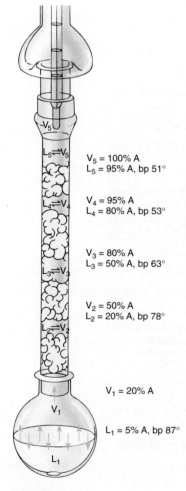

$V_5 = 100\%$ A
$L_5 = 95\%$ A, bp 51°

$V_4 = 95\%$ A
$L_4 = 80\%$ A, bp 53°

$V_3 = 80\%$ A
$L_3 = 50\%$ A, bp 63°

$V_2 = 50\%$ A
$L_2 = 20\%$ A, bp 78°

$V_1 = 20\%$ A

$L_1 = 5\%$ A, bp 87°

Figure 15.5

Vaporization–condensation in a fractionation column.

Consider a solution at 100°C where $N_A = 0.5$ and $N_B = 0.5$.

1. What is the partial vapor pressure of A in the solution if the vapor pressure of pure A at 100°C is 1020 mm Hg?

   Answer: $P_A = P_A^\circ N_A = (1020)(0.5) = 510$ mm Hg

2. What is the partial vapor pressure of B in the solution if the vapor pressure of pure B at 100°C is 500 mm Hg?

   Answer: $P_B = P_B^\circ N_B = (500)(0.5) = 250$ mm Hg

3. Would the solution boil at 100°C if the applied pressure were 760 mm Hg?

   Answer: Yes. $P_{total} = P_A + P_B = (510 + 250) = 760$ mm Hg

4. What is the composition of the vapor at the boiling point?

   Answer: The boiling point is 100°C

   $$N_A \text{ (vapor)} = \frac{P_A}{P_{total}} = 510/760 = 0.67$$

   $$N_B \text{ (vapor)} = \frac{P_B}{P_{total}} = 250/760 = 0.33$$

Figure 15.6

Sample calculations with Raoult's Law.

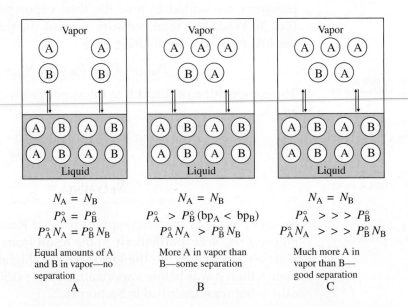

$$N_A = N_B$$
$$P_A^\circ = P_B^\circ$$
$$P_A^\circ N_A = P_B^\circ N_B$$

Equal amounts of A and B in vapor—no separation

A

$$N_A = N_B$$
$$P_A^\circ > P_B^\circ \ (bp_A < bp_B)$$
$$P_A^\circ N_A > P_B^\circ N_B$$

More A in vapor than B—some separation

B

$$N_A = N_B$$
$$P_A^\circ >>> P_B^\circ$$
$$P_A^\circ N_A >>> P_B^\circ N_B$$

Much more A in vapor than B—good separation

C

Figure 15.7

Consequences of Raoult's Law. (A) Boiling points (vapor pressures) are identical—no separation. (B) Boiling point somewhat lower for A than for B—requires fractional distillation. (C) Boiling point much lower for A than for B—simple distillation will suffice.

A solution whose mole fraction of water is 0.7 will not boil at 100°C, because $P_{total} = (760)(0.7) = 532$ mm Hg and is less than atmospheric pressure. If the solution is heated to 110°C, it will boil because $P_{total} = (1085)(0.7) = 760$ mm Hg. Although the solution must be heated to 110°C to boil it, the vapor is pure water and has a boiling-point temperature of 100°C. (The vapor pressure of water at 110°C can be looked up in a handbook; it is 1085 mm Hg.)

15.4 Column Efficiency

A common measure of the efficiency of a column is given by its number of **theoretical plates.** The number of theoretical plates in a column is related to the number of vaporization–condensation cycles that occur as a liquid mixture travels through it. Using the example mixture in Figure 15.4, if the first distillate (condensed vapor) had the composition at L_2 when starting with liquid of composition L_1, the column would be said to have *one theoretical plate.* This would correspond to a simple distillation, or one vaporization–condensation cycle. A column would have two theoretical plates if the first distillate had the composition at L_3. The two-theoretical-plate column essentially carries out "two simple distillations." According to Figure 15.4, *five theoretical plates* would be required to separate the mixture that started with composition L_1. Notice that this corresponds to the number of "steps" that need to be drawn in the figure to arrive at a composition of 100% A.

Most columns do not allow distillation in discrete steps, as indicated in Figure 15.4. Instead, the process is *continuous,* allowing the vapors to be continuously in contact with liquid of changing composition as they pass through the column. Any material can be used to pack the column as long as it can be wetted by the liquid and as long as it does not pack so tightly that vapor cannot pass.

The approximate relationship between the number of theoretical plates needed to separate an ideal two-component mixture and the difference in boiling points is given in Table 15.1. Notice that more theoretical plates are required as the boiling-point differences between the components decrease. For instance, a mixture of A (bp 130°C) and B (bp 166°C) with a boiling-point difference of 36°C would be expected to require a column with a minimum of five theoretical plates.

15.5 Types of Fractionating Columns and Packings

Several types of fractionating columns are shown in Figure 15.8. The Vigreux column, shown in Part A, has indentations that incline downward at angles of 45 degrees and are in pairs on opposite sides of the column. The projections into the column provide increased possibilities for condensation and

Table 15.1 *Theoretical plates required to separate mixtures, based on boiling-point differences of components*

Boiling-Point Difference	Number of Theoretical Plates
108	1
72	2
54	3
43	4
36	5
20	10
10	20
7	30
4	50
2	100

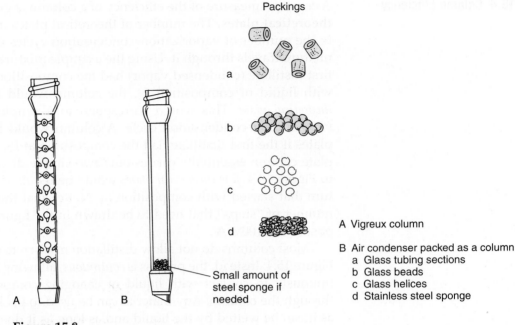

Packings

A Vigreux column

B Air condenser packed as a column
  a Glass tubing sections
  b Glass beads
  c Glass helices
  d Stainless steel sponge

Small amount of steel sponge if needed

Figure 15.8
Columns for fractional distillation.

for the vapor to equilibrate with the liquid. Vigreux columns are popular in cases where only a small number of theoretical plates are required. They are not very efficient (a 20-cm column might have only 2.5 theoretical plates), but they allow for rapid distillation and have a small **holdup** (the amount of liquid retained by the column). A column packed with a stainless steel sponge is a more effective fractionating column than a Vigreux column, but not by a large margin. Glass beads, or glass helices, can also be used as a packing material, and they have a slightly greater efficiency. The air condenser or the water condenser can be used as an improvised column if an actual fractionating column is unavailable. If a condenser is packed with glass beads, glass helices, or sections of glass tubing, the packing must be held in place by inserting a small plug of stainless steel sponge into the bottom of the condenser.

The most effective type of column is the **spinning-band column.** In the most elegant form of this device, a tightly fitting, twisted platinum screen or a Teflon rod with helical threads is rotated rapidly inside the bore of the column (Fig. 15.9). A spinning-band column that is available for microscale work is shown in Figure 15.10. This spinning-band column has a band about 2–3 cm in length and provides 4–5 theoretical plates. It can separate 1–2 mL of a mixture with a 30°C boiling-point difference. Larger research models of this spinning-band column can provide as many as 20 or 30 theoretical plates and can separate mixtures with a boiling-point difference of as little as 5–10°C.

Manufacturers of fractionating columns often offer them in a variety of lengths. Because the efficiency of a column is a function of its length, longer columns have more theoretical plates than shorter ones. It is common to express efficiency of a column in a unit called **HETP,** the Height of a column that is Equivalent to one Theoretical Plate. HETP is usually expressed in units of cm/plate. When the height of the column (in centimeters) is divided by this value, the total number of theoretical plates is specified.

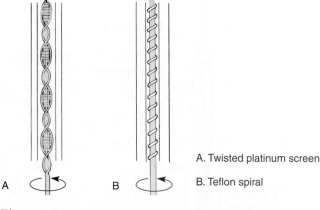

A. Twisted platinum screen

B. Teflon spiral

Figure 15.9
Bands for spinning-band columns.

15.6 Fractional Distillation: Methods and Practice

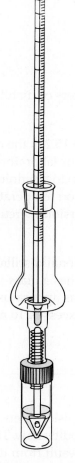

Figure 15.10
A commercially available microscale spinning-band column.

When a fractional distillation is performed, the column should be clamped in a vertical position. The distillation should be conducted as slowly as possible, but the rate of distillation should be steady enough to produce a constant temperature reading at the thermometer.

Many fractionating columns must be insulated so that temperature equilibrium is maintained at all times. Additional insulation will not be required for columns that have an evacuated outer jacket, but those that do not can benefit from being wrapped in insulation.

A microscale air condenser can be converted to a column by packing it with a piece of stainless steel sponge. The simplest form of insulation is Tygon tubing that has been split lengthwise. Select a piece with an inner diameter that just matches or is slightly smaller than the diameter of the fractionating column so that it will fit snugly.

CAUTION

Cut the tubing to the correct length and then slit it with a sharp scissors. Do not use a razor blade or knife. Tygon tubing is difficult to cut; it is a nonslip substance and will "grab" even a single-edged razor blade in a way that can give you a nasty cut. See Experiment 6, page 52, for complete instructions.

Glass wool and aluminum foil (shiny side in) are often used for insulation. You can wrap the column with glass wool and then use a wrapping of the aluminum foil to keep it in place. An especially effective method is to make an insulation blanket by placing a layer of glass wool or cotton between two rectangles of aluminum foil, placed shiny side in. The sandwich is bound together with duct tape. This blanket, which is reusable, can be wrapped around the column and held in place with twist ties or tape.

The **reflux ratio** is defined as the ratio of the number of drops of distillate that return to the distillation flask compared to the number of drops of distillate collected. In an efficient column, the reflux ratio should equal or exceed the number of theoretical plates. A high reflux ratio ensures that the column will achieve temperature equilibrium and achieve its maximum efficiency. This ratio is not easy to determine; in fact, it is impossible to determine when using a Hickman head, and it should not concern a beginning student. In some cases, the **throughput,** or **rate of takeoff,** of a column may

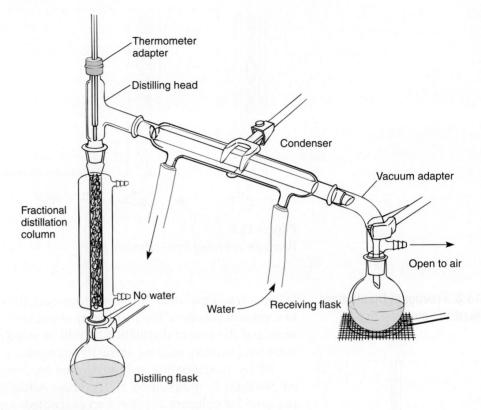

Figure 15.11
Large-scale fractional distillation apparatus.

be specified. This is expressed as the number of milliliters of distillate that can be collected per unit of time, usually as mL/min.

Microscale Apparatus. The apparatus shown in Figure 15.2 is the one you are most likely to use in the microscale laboratory. If your laboratory is one of the better-equipped ones, you may have access to spinning-band columns like those shown in Figure 15.10. The distillation temperature can be monitored most accurately by using a partial immersion mercury thermometer (see Technique 13, Section 13.3, page 700).

Macroscale Apparatus. Figure 15.11 illustrates a fractional distillation assembly that can be used for larger-scale distillations. It has a glass-jacketed column that is packed with a stainless steel sponge. This apparatus would be common in situations where quantities of liquid in excess of 10 mL were to be distilled.

PART B. AZEOTROPES

15.7 Nonideal Solutions: Azeotropes

Some mixtures of liquids, because of attractions or repulsions between the molecules, do not behave ideally; they do not follow Raoult's Law. There are two types of vapor–liquid composition diagrams that result from this non-ideal behavior: **minimum-boiling-point** and **maximum-boiling-point** diagrams. The minimum or maximum points in these diagrams correspond to a constant-boiling mixture called an **azeotrope**. An azeotrope is a mixture with a fixed composition that cannot be altered by either simple or fractional

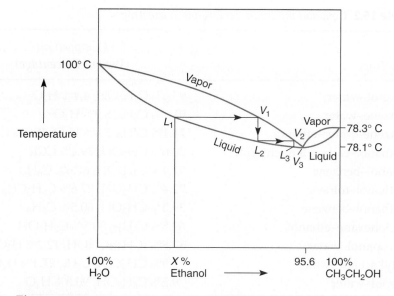

Figure 15.12
Ethanol–water minimum-boiling-point phase diagram.

distillation. An azeotrope behaves as if it were a pure compound, and it distills from the beginning to the end of its distillation at a constant temperature, giving a distillate of constant (azeotropic) composition. The vapor in equilibrium with an azeotropic liquid has the same composition as the azeotrope. Because of this, an azeotrope is represented as a *point* on a vapor–liquid composition diagram.

A. Minimum-Boiling-Point Diagrams

A minimum-boiling-point azeotrope results from a slight incompatibility (repulsion) between the liquids being mixed. This incompatibility leads to a higher-than-expected combined vapor pressure from the solution. This higher combined vapor pressure brings about a lower boiling point for the mixture than is observed for the pure components. The most common two-component mixture that gives a minimum-boiling-point azeotrope is the ethanol–water system shown in Figure 15.12. The azeotrope at V_3 has a composition of 96% ethanol–4% water and a boiling point of 78.1°C. This boiling point is not much lower than that of pure ethanol (78.3°C), but it means that it is impossible to obtain pure ethanol from the distillation of any ethanol–water mixture that contains more than 4% water. Even with the best fractionating column, you cannot obtain 100% ethanol. The remaining 4% of water can be removed by adding benzene and removing a different azeotrope, the ternary benzene–water–ethanol azeotrope (bp 65°C). Once the water is removed, the excess benzene is removed as an ethanol–benzene azeotrope (bp 68°C). The resulting material is free of water and is called "absolute" ethanol.

The fractional distillation of an ethanol–water mixture of composition X can be described as follows. The mixture is heated (follow line XL_1) until it is observed to boil at L_1. The resulting vapor at V_1 will be richer in the lower-boiling component, ethanol, than the original mixture.[1] The condensate at

_____
[1] Keep in mind that this distillate is not pure ethanol but is an ethanol–water mixture.

Table 15.2 *Common minimum-boiling-point azeotropes*

Azeotrope	Composition (Weight Percentage)	Boiling Point (°C)
Ethanol–water	95.6% C_2H_5OH, 4.4% H_2O	78.17
Benzene–water	91.1% C_6H_6, 8.9% H_2O	69.4
Benzene–water–ethanol	74.1% C_6H_6, 7.4% H_2O, 18.5% C_2H_5OH	64.9
Methanol–carbon tetrachloride	20.6% CH_3OH, 79.4% CCl_4	55.7
Ethanol–benzene	32.4% C_2H_5OH, 67.6% C_6H_6	67.8
Methanol–toluene	72.4% CH_3OH, 27.6% $C_6H_5CH_3$	63.7
Methanol–benzene	39.5% CH_3OH, 60.5% C_6H_6	58.3
Cyclohexane–ethanol	69.5% C_6H_{12}, 30.5% C_2H_5OH	64.9
2-Propanol–water	87.8% $(CH_3)_2CHOH$, 12.2% H_2O	80.4
Butyl acetate–water	72.9% $CH_3COOC_4H_9$, 27.1% H_2O	90.7
Phenol–water	9.2% C_6H_5OH, 90.8% H_2O	99.5

L_2 is vaporized to give V_2. The process continues, following the lines to the right, until the azeotrope is obtained at V_3. The liquid that distills is not pure ethanol, but it has the azeotropic composition of 96% ethanol and 4% water, and it distills at 78.1°C. The azeotrope, which is richer in ethanol than the original mixture, continues to distill. As it distills, the percentage of water left behind in the distillation flask continues to increase. When all the ethanol has been distilled (as the azeotrope), pure water remains behind in the distillation flask, and it distills at 100°C.

If the azeotrope obtained by the preceding procedure is redistilled, it distills from the beginning to the end of the distillation at a constant temperature of 78.1°C as if it were a pure substance. There is no change in the composition of the vapor during the distillation.

Some common minimum-boiling azeotropes are given in Table 15.2. Numerous other azeotropes are formed in two- and three-component systems; such azeotropes are common. Water forms azeotropes with many substances; therefore, water must be carefully removed with **drying agents** whenever possible before compounds are distilled. Extensive azeotropic data are available in references such as the *Handbook of Chemistry and Physics*.[2]

B. Maximum-Boiling-Point Diagrams

A maximum-boiling-point azeotrope results from a slight attraction between the component molecules. This attraction leads to lower combined vapor pressure than expected in the solution. The lower combined vapor pressures cause a higher boiling point than would be characteristic for the components. A two-component maximum-boiling-point azeotrope is illustrated in Figure 15.13. Because the azeotrope has a higher boiling point than

---

[2] More examples of azeotropes, with their compositions and boiling points, can be found in the *CRC Handbook of Chemistry and Physics*; also in L. H. Horsley, ed., *Advances in Chemistry Series*, no. 116. Azeotropic Data, III (Washington, DC: American Chemical Society, 1973).

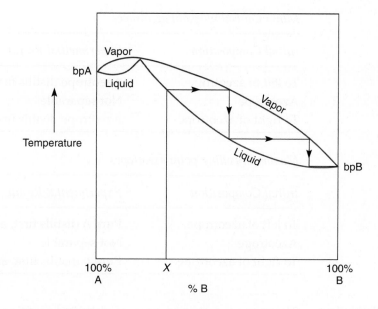

Figure 15.13
A maximum-boiling-point phase diagram.

Table 15.3 *Maximum-boiling-point azeotropes*

Azeotrope	Composition (Weight Percentage)	Boiling Point (°C)
Acetone–chloroform	20.0% CH_3COCH_3, 80.0% $CHCl_3$	64.7
Chloroform–methyl ethyl ketone	17.0% $CHCl_3$, 83.0% $CH_3COCH_2CH_3$	79.9
Hydrochloric acid	20.2% HCl, 79.8% H_2O	108.6
Acetic acid–dioxane	77.0% CH_3COCH, 23.0% $C_4H_8O_2$	119.5
Benzaldehyde–phenol	49.0% C_6H_5CHO, 51.0% C_6H_5OH	185.6

any of the components, it will be concentrated in the distillation flask as the distillate (pure B) is removed. The distillation of a solution of composition X would follow to the right along the lines in Figure 15.13. Once the composition of the material remaining in the flask has reached that of the azeotrope, the temperature will rise, and the azeotrope will begin to distill. The azeotrope will continue to distill until all the material in the distillation flask has been exhausted.

Some maximum-boiling-point azeotropes are listed in Table 15.3. They are not nearly as common as minimum-boiling-point azeotropes.

C. Generalizations

There are some generalizations that can be made about azeotropic behavior. They are presented here without explanation, but you should be able to verify them by thinking through each case using the phase diagrams given. (Note that pure A is always to the left of the azeotrope in these diagrams, whereas pure B is to the right of the azeotrope.)

Minimum-boiling-point azeotropes

Initial Composition	Experimental Result
To left of azeotrope	Azeotrope distills first, pure A second
Azeotrope	Not separable
To right of azeotrope	Azeotrope distills first, pure B second

Maximum-boiling-point azeotropes

Initial Composition	Experimental Result
To left of azeotrope	Pure A distills first, azeotrope second
Azeotrope	Not separable
To right of azeotrope	Pure B distills first, azeotrope second

15.8 Azeotropic Distillation: Applications

There are many examples of chemical reactions in which the amount of product is low because of an unfavorable equilibrium. An example is the direct acid-catalyzed esterification of a carboxylic acid with an alcohol:

$$\underset{\text{O}}{R-\overset{\text{O}}{\overset{\|}{C}}-OH} + R-O-H \underset{}{\overset{H^+}{\rightleftarrows}} R-\overset{\text{O}}{\overset{\|}{C}}-OR + H_2O$$

Because the equilibrium does not favor formation of the ester, it must be shifted to the right, in favor of the product, by using an excess of one of the starting materials. In most cases, the alcohol is the least expensive reagent and is the material used in excess. Isopentyl acetate (Experiment 13) and methyl salicylate (Experiment 46) are examples of esters prepared by using one of the starting materials in excess.

Another way of shifting the equilibrium to the right is to remove one of the products from the reaction mixture as it is formed. In the preceding example, water can be removed as it is formed by **azeotropic distillation.** A common large-scale method is to use the Dean–Stark water separator shown in Figure 15.14A. In this technique, an inert solvent, commonly benzene or toluene, is added to the reaction mixture contained in the round-bottom flask. The side arm of the water separator is also filled with this solvent. If benzene is used, as the mixture is heated under reflux, the benzene–water azeotrope (bp 69.4°C, Table 15.3) distills out of the flask.[3] When the vapor condenses, it enters the side arm directly below the condenser, and water separates from the benzene–water condensate; benzene and water mix as vapors, but they are not miscible as cooled liquids. Once the water (lower phase) separates from the benzene (upper phase), liquid benzene overflows from the side arm back into the flask. The cycle is repeated continuously until no more water forms in the side arm. You may calculate the weight of water that should theoretically be produced and compare this value with the amount of water collected in the side arm. Because the density of water is 1.0,

---

[3] Actually, with ethanol, a lower-boiling-point three-component azeotrope distills at 64.9°C (see Table 15.3). It consists of benzene–water–ethanol. Because some ethanol is lost in the azeotropic distillation, a large excess of ethanol is used in esterification reactions. The excess also helps shift the equilibrium to the right.

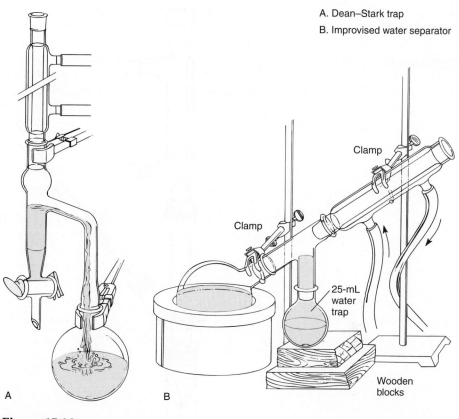

A. Dean–Stark trap

B. Improvised water separator

Clamp

Clamp

25-mL water trap

Wooden blocks

A   B

Figure 15.14
Large-scale water separators.

the volume of water collected can be compared directly with the calculated amount, assuming 100% yield.

An improvised water separator, constructed from the components found in the traditional organic kit, is shown in Figure 15.14B. Although this requires the condenser to be placed in a nonvertical position, it works quite well.

At the microscale level, water separation can be achieved using a standard distillation assembly with a water condenser and a Hickman head (Fig. 15.15). The side-ported variation of the Hickman head is the most convenient one to use for this purpose, but it is not essential. In this variation, you simply remove all the distillate (both solvent and water) several times during the course of the reaction. Use a Pasteur pipet to remove the distillate, as shown in Technique 14 (Fig. 14.6, p. 709). Because both the solvent and water are removed in this procedure, it may be desirable to add more solvent from time to time, adding it through the condenser with a Pasteur pipet.

The most important consideration in using azeotropic distillation to prepare an ester (described above) is that the azeotrope containing water must have a **lower boiling point** than the alcohol used. With ethanol, the benzene–water azeotrope boils at a much lower temperature (69.4°C) than ethanol (78.3°C), and the technique previously described works well. With higher-boiling-point alcohols, azeotropic distillation works well because of the large boiling-point difference between the azeotrope and the alcohol.

With methanol (bp 65°C), however, the boiling point of the benzene–water azeotrope is actually *higher* by about 5°C, and methanol distills first. Thus, in esterifications involving methanol, a totally different approach

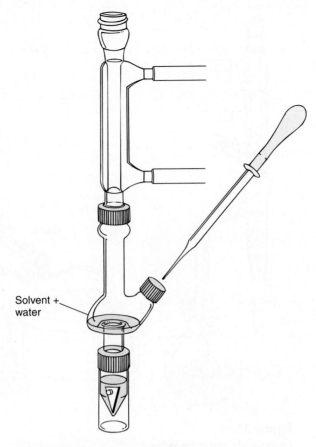

Solvent + water

Figure 15.15
Microscale water separator (both layers are removed).

must be taken. For example, you can mix the carboxylic acid, methanol, the acid catalyst, and 1,2-dichloroethane in a conventional reflux apparatus (Technique 7, Fig. 7.6, p. 601) without a water separator. During the reaction, water separates from the 1,2-dichloroethane because it is not miscible; however, the remainder of the components are soluble, so the reaction can continue. The equilibrium is shifted to the right by the "removal" of water from the reaction mixture.

Azeotropic distillation is also used in other types of reactions, such as ketal or acetal formation, and in enamine formation. The use of azeotropic distillation is illustrated in the formation of 2-acetylcyclohexanone (Experiment 43) via the enamine intermediate. Toluene is used in the azeotropic distillation of water. The Hickman head is used as a water separator.

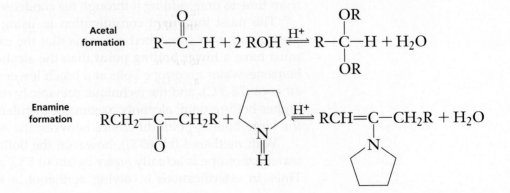

PROBLEMS

1. In the accompanying chart are approximate vapor pressures for benzene and toluene at various temperatures.

Temp (°C)	mm Hg	Temp (°C)	mm Hg
Benzene 30	120	Toluene 30	37
40	180	40	60
50	270	50	95
60	390	60	140
70	550	70	200
80	760	80	290
90	1010	90	405
100	1340	100	560
		110	760

   a. What is the mole fraction of each component if 3.9 g of benzene C_6H_6 is dissolved in 4.6 g of toluene C_7H_8?

   b. Assuming that this mixture is ideal—that is, it follows Raoult's Law—what is the partial vapor pressure of benzene in this mixture at 50°C?

   c. Estimate to the nearest degree the temperature at which the vapor pressure of the solution equals 1 atm (bp of the solution).

   d. Calculate the composition of the vapor (mole fraction of each component) that is in equilibrium in the solution at the boiling point of this solution.

   e. Calculate the composition in weight percentage of the vapor that is in equilibrium with the solution.

2. Estimate how many theoretical plates are needed to separate a mixture that has a mole fraction of B equal to 0.70 (70% B) in Figure 15.4.

3. Two moles of sucrose are dissolved in 8 moles of water. Assume that the solution follows Raoult's Law and that the vapor pressure of sucrose is negligible. The boiling point of water is 100°C. The distillation is carried out at 1 atm (760 mm Hg).

   a. Calculate the vapor pressure of the solution when the temperature reaches 100°C.

   b. What temperature would be observed during the entire distillation?

   c. What would be the composition of the distillate?

   d. If a thermometer were immersed below the surface of the liquid of the boiling flask, what temperature would be observed?

4. Explain why the boiling point of a two-component mixture rises slowly throughout a simple distillation when the boiling-point differences are not large.

5. Given the boiling points of several known mixtures of A and B (mole fractions are known) and the vapor pressures of A and B in the pure state (P_A° and P_B°) at these same temperatures, how would you construct a boiling-point-composition phase diagram for A and B? Give a stepwise explanation.

6. Describe the behavior on distillation of a 98% ethanol solution through an efficient column. Refer to Figure 15.12.

7. Construct an approximate boiling-point-composition diagram for a benzene–methanol system. The mixture shows azeotropic behavior (see Table 15.3). Include on the graph the boiling points of pure benzene and pure methanol and the boiling point of the azeotrope. Describe the behavior on distillation of a mixture that is initially rich in benzene (90%) and then for a mixture that is initially rich in methanol (90%).

8. Construct an approximate boiling-point-composition diagram for an acetone–chloroform system, which forms a maximum boiling azeotrope (Table 15.4). Describe the behavior on distillation of a mixture that is initially rich in acetone (90%), and then describe the behavior of a mixture that is initially rich in chloroform (90%).

9. Two compounds have boiling points of 130 and 150°C. Estimate the number of theoretical plates needed to separate these substances in a fractional distillation.

10. A spinning-band column has an HETP of 0.25 in./plate. If the column has 12 theoretical plates, how long is it?

---

16   TECHNIQUE 16

Vacuum Distillation, Manometers

Vacuum distillation (distillation at reduced pressure) is used for compounds that have high boiling points (above 200°C). Such compounds often undergo thermal decomposition at the temperatures required for their distillation at atmospheric pressure. The boiling point of a compound is lowered substantially by reducing the applied pressure. Vacuum distillation is also used for compounds that, when heated, might react with the oxygen present in air. It is also used when it is more convenient to distill at a lower temperature because of experimental limitations. For instance, a heating device may have difficulty heating to a temperature in excess of 250°C.

The effect of pressure on the boiling point is discussed more thoroughly in Technique 13 (Section 13.1, p. 694). A nomograph is given (Fig. 13.2, p. 695) that allows you to estimate the boiling point of a liquid at a pressure different from the one at which it is reported. For example, a liquid reported to boil at 200°C at 760 mm Hg would be expected to boil at 90°C at 20 mm Hg. This is a significant decrease in temperature, and it would be advantageous to use a vacuum distillation if any problems were to be expected. Counterbalancing this advantage, however, is the fact that separations of liquids of different boiling points may not be as effective with a vacuum distillation as with a simple distillation.

16.1 Microscale Methods

When working with glassware that is to be evacuated, you should wear safety glasses at all times. There is always danger of an implosion.

CAUTION

Safety glasses must be worn at all times during vacuum distillation.

It is a good idea to work in a hood when performing a vacuum distillation. If the experiment will involve high temperatures (>220°C) for distillation or an extremely low pressure (<0.1 mm Hg), for your own safety you should definitely work in a hood, behind a shield.

A basic apparatus similar to the one shown in Figure 16.1 (or Fig. 16.5) may be used for microscale vacuum distillations. As is the case for simple distillation, this apparatus uses the Hickman head as a means to reduce the length of the vapor path. The major difference to be found when comparing this assembly to one for simple distillation (Fig. 14.8, p. 711) is that the opening to the atmosphere has been replaced by a connection to a vacuum source (top right-hand side). The usual sources of vacuum are the aspirator (Technique 8, Section 8.5, p. 624), a mechanical vacuum pump, or a "house" vacuum line (one piped directly to the laboratory bench). The aspirator is probably the simplest of these sources and the vacuum source most likely to be available. However, if pressures below 10–20 mm Hg are required, a vacuum pump must be used.

Assembling the Apparatus. When assembling an apparatus for vacuum distillation, it is important that all joints and connections be airtight. The joints in the newest microscale kits are standard-taper ground-glass joints, with a compression cap that contains an O-ring seal. Glassware that contains this type of compression joint will hold a vacuum quite easily. Under normal conditions, it is not necessary to grease these joints.

NOTE: Normally, you should not grease joints. It is necessary to grease the joints in a vacuum distillation only if you cannot achieve the desired pressure without using grease.

If you must grease joints, take care not to use too much grease. You are working with small quantities of liquid in a microscale distillation, and the grease can become a very serious contaminant if it oozes out the bottom of the joints into your system. Apply a small amount of grease (thin film) completely around the top of the *inner* joint; then mate the joints and turn them slightly to spread the grease evenly. If you have used the correct amount of grease, it will not ooze out the bottom; rather, the entire joint will appear clear and without striations or uncovered areas.

Make doubly sure that any connections to pressure tubing are tight. The pressure tubing itself should be relatively new and without cracks. If the tubing shows cracks when you stretch or bend it, it may be old and leak air into the system. Glass tubing should fit securely into any rubber stoppers. If you can move the tubing up and down with only gentle force, it is too loose, and you should obtain a larger size. Check all glassware to be sure there are no cracks and that there are no chips in the standard-taper joints. Cracked glassware may break when evacuated.

Connecting to Vacuum. In Figure 16.1, the connection to vacuum has been made using a multipurpose adapter (see Fig. 14.9B, p. 712). If a multipurpose adapter is not available, an alternative method uses a Claisen head and two thermometer adapters (Fig. 16.2). If two thermometer adapters are not available, a #0 rubber stopper fitted with glass tubing can be used.

Whichever is used, the connection to the vacuum source is made using **pressure tubing.** Pressure tubing (also called vacuum tubing), unlike the more common thin-walled tubing used to carry water or gas, has heavy walls that will not collapse inward when it is evacuated. Compare the two types of tubing shown in Figure 16.3.

Water Trap. If an aspirator is used as a source of vacuum, a water trap must be placed between it and the distillation assembly. A commonly used type of water trap is shown at the bottom right of Figure 16.1. A different type of trap, which is also common, is shown in Figure 16.5. Variations in water pressure are to be expected when using an aspirator. If the pressure drops

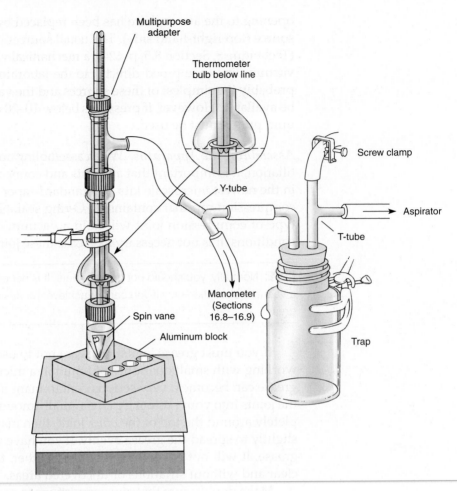

Figure 16.1
Reduced-pressure microscale distillation (internal monitoring of temperature).

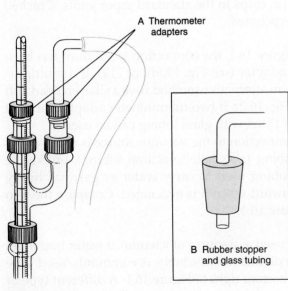

Figure 16.2
Alternative vacuum connections.

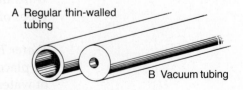

Figure 16.3
Comparison of tubing.

low enough, the vacuum in the system will draw water from the aspirator into the connecting line. The trap allows you to see this happening and take corrective action (prevent water from entering the distillation apparatus). The correct action for anything but a small amount of water is to "vent the system." This can be accomplished by opening the screw clamp at the top of the trap to let air into the system. When performing a vacuum distillation, you should also realize that the system should always be vented before stopping the aspirator. If you turn off the aspirator while the system is still under vacuum, water will be drawn into the connecting line and trap.

Manometer Connection. A Y-tube is shown in the line from the apparatus to the trap. This branching connection is optional but is required if you wish to monitor the actual pressure of the system using a manometer. The operation of manometers is discussed in Sections 16.8 and 16.9.

Thermometer Placement. If a thermometer is used, be sure that the bulb is placed in the stem of the Hickman head just below the well. If it is placed higher, it may not be surrounded by a constant stream of vapor from the material being distilled. If the thermometer is not exposed to a continuous stream of vapor, it may not reach temperature equilibrium. As a result, the temperature reading would be incorrect (low).

Preventing Bumpover. When a distillation flask is heated, there is always the possibility that the boiling action will become too vigorous (mainly due to superheating) and "bump" some of the undistilled liquid up into the Hickman head. The simplest way to prevent bumping is to stir the boiling liquid with a magnetic spin vane. Stirring rapidly will distribute the heat evenly, keep the boiling action smooth, and prevent bumping. Boiling stones cannot be used for this purpose in a vacuum distillation; they do not work in vacuum. In a conventional vacuum distillation (macroscale), it is customary to maintain smooth boiling action by using an **ebulliator tube.** The ebulliator tube agitates the boiling solution by providing a small, continuous stream of air bubbles. Figure 16.4 shows how a microscale vacuum distillation may be modified to use an ebulliator tube. The amount of air (rate of bubbles) provided by the ebulliator is adjusted by either tightening or loosening the screw clamp at the top. A Pasteur pipet makes an excellent ebulliator tube. As Figure 16.4 shows, the ebulliator tube replaces the thermometer. Hence, the ebulliator should be used only when internal monitoring of temperature is not required. In practice, although this method works satisfactorily, better results are obtained by stirring and distilling slowly.

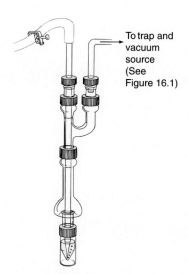

To trap and vacuum source (See Figure 16.1)

Figure 16.4
Use of ebulliator tube instead of thermometer.

NOTE: Heating slowly helps to avoid bumping.

16.2 Simplified Microscale Apparatus

The apparatus shown in Figure 16.5 will often produce very satisfactory results when internal temperature monitoring is not required. It is the apparatus we prefer for the experiments in this textbook that require vacuum distillation. Distillation (heating) should be performed slowly while stirring briskly. Just before the well begins to fill, you will see reflux action (condensation) in the stem. In many cases, this will occur even before there is any evidence of boiling in the heated liquid. In Figure 16.5 an aluminum heating block is shown. The aluminum block is an effective heat source whenever you want a fast heating response or a high temperature.

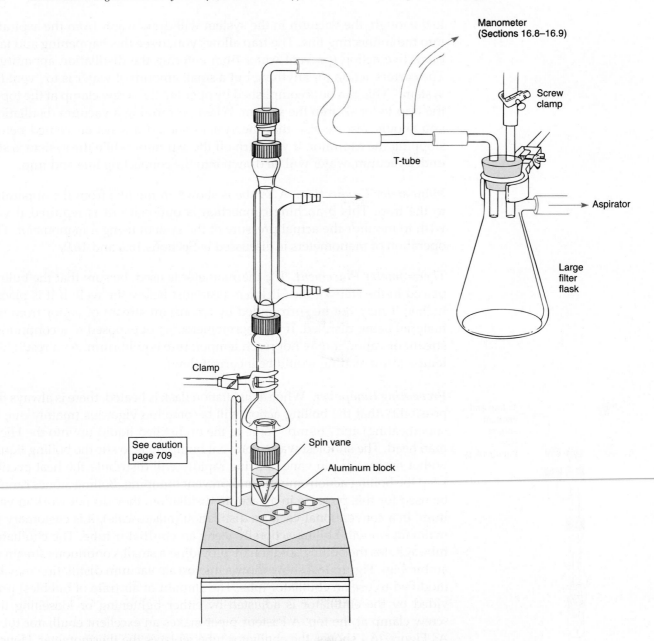

Figure 16.5
Simplified vacuum distillation apparatus (external monitoring of temperature).

16.3 Semimicroscale and Macroscale Equipment

A vacuum distillation apparatus using the components of the traditional organic laboratory kit is shown in Figure 16.6. It uses the ebulliator tube, the Claisen head, and a thermometer for internal temperature monitoring. A water condenser is shown, but with high-boiling liquids, this apparatus may be simplified by removing the water condenser. A special vacuum adapter allows connection to the manometer and vacuum source.

The Claisen head is used in larger-scale vacuum distillations because it allows the use of an ebulliator tube. The bend it provides in the distilling path helps to prevent bumpover. Because the Claisen head increases the holdup of the system, it cannot be used with very-small-scale distillations (<10 mL).

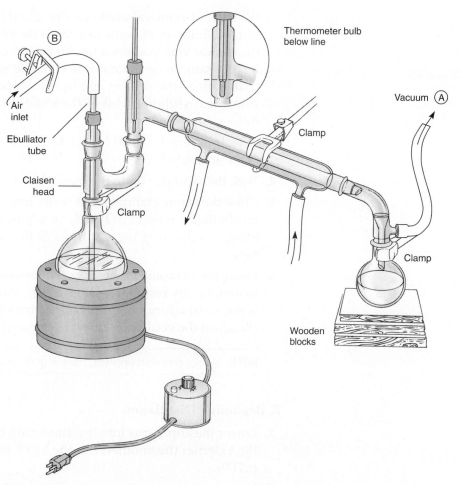

Figure 16.6
Macroscale vacuum distillation using the standard organic laboratory kit.

16.4 Stepwise Instructions for Microscale Vacuum Distillations

The following set of instructions is a step-by-step account of how to carry out a vacuum distillation. The microscale apparatus illustrated in Figure 16.1 will be used; however, the procedures apply to any vacuum distillation.

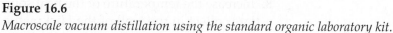

CAUTION

Safety glasses must be worn at all times during vacuum distillation.

A. Evacuating the Apparatus

1. Assemble the apparatus as shown in Figure 16.1. It should be held with a clamp attached to the top of the Hickman head and placed *above* the aluminum block.

    NOTE: If you expect the temperature of the distillation to rise above 150°C, omit the threaded cap and O-ring between the conical vial and Hickman head. They will melt at high temperature.

2. If the sample contains solvent, concentrate the sample to be distilled in the conical vial (or round-bottom flask) that you are using. Use one of

the solvent-removal methods discussed in Technique 7. If you have a large volume of solvent to evaporate and the sample does not fit in the conical vial, you must use an Erlenmeyer flask first and then transfer the sample to the conical vial. (Be sure to rinse the Erlenmeyer flask with a little solvent and then reevaporate in the conical vial.) As a rule, the distillation vial or flask should be no more than two-thirds full.

3. Attach the conical vial (or flask) to the apparatus and make sure all joints are sealed.

4. Turn the aspirator on to the maximum extent.

5. Close the screw clamp on the water trap very tightly. (If you are using an ebulliator tube as in Figure 16.4, you next regulate the rate of bubbling by adjusting the tightness of the screw clamp at the top of the tube.)

6. Using the manometer, observe the pressure. It may take a few minutes to remove any residual solvent and evacuate the system. If the pressure is not satisfactory, check all connections to see whether they are tight. (Readjust the ebulliator tube if necessary.)

NOTE: Do not proceed until you have a good vacuum.

B. Beginning Distillation

7. Lower the apparatus into the aluminum block and begin to heat. Place the external thermometer in the block now if you wish (see caution, p. 709).

8. Increase the temperature of the heat source until you begin to see distillate collect in the well of the Hickman head. (Observe very carefully; liquid may appear almost "magically" without any sign of boiling or any obvious reflux ring.)

9. If you are using a thermometer, record the temperature and pressure when distillate begins to appear. (If you are not using an internal thermometer, record the external temperature. If you have two thermometers, record both temperatures.)

C. Collecting a Fraction

10. To collect a fraction, raise the apparatus above the aluminum block and allow it to cool a bit before opening it.

11. Open the screw clamp on the water trap to allow air to enter the system. (If you are using an ebulliator tube, you also need to open the screw clamp at its top *immediately*, or the liquid in the distillation flask will be forced upward into it.)

12. Partially disassemble the apparatus and remove the fraction with a Pasteur pipet, as shown in Figure 14.6A. (If you have a Hickman head with a side port, you may simply open the side port to remove the fraction. This is shown in Figure 14.6C.)

NOTE: If you do not intend to collect a second fraction, go directly to steps 18–20.

---

13. Reassemble the apparatus (or close the side port) and tighten the clamp at the top of the ebulliator tube.

14. Tighten the screw clamp on the water trap and reestablish the desired pressure. If the pressure is not satisfactory, check all connections to make sure they are sealed.

15. Lower the apparatus back into the aluminum block and continue the distillation.

D. Shutdown

16. At the end of the distillation, raise the apparatus from the aluminum block and allow it to cool. Also let the aluminum block cool.

17. Open the screw clamp on the water trap first, and then immediately open the one at the top of the ebulliator tube.

18. Turn off the water at the aspirator. (Do not do this before step 17!)

19. Remove any distilled material by one of the methods shown in Figure 14.6.

20. Disassemble the apparatus and clean all glassware as soon as possible to prevent the joints from sticking.

NOTE: If you used grease, thoroughly clean all grease off the joints, or it will contaminate your samples in other procedures.

16.5 Rotary Fraction Collectors

With the types of apparatus we have discussed previously, the vacuum must be stopped to remove fractions when a new substance (fraction) begins to distill. Quite a few steps are required to perform this change, and it is quite inconvenient when there are several fractions to be collected. Two pieces of semimicroscale apparatus that are designed to alleviate the difficulty of collecting fractions while working under vacuum are shown in Figure 16.7. The collector, which is shown to the right, is sometimes called a "cow" because of its appearance. With these rotary fraction collecting devices, all you need to do is rotate the device to collect fractions.

16.6 Bulb-to-Bulb Distillation

The ultimate in microscale methods is to use a bulb-to-bulb distillation apparatus. This apparatus is shown in Figure 16.8. The sample to be distilled is placed in the glass container attached to one of the arms of the apparatus. The sample is frozen solid, usually by using liquid nitrogen, but dry ice in 2-propanol or an ice–salt-water mixture may also be used. The coolant container shown in the figure is a **Dewar flask.** The Dewar flask has a double wall with the space between the walls evacuated and sealed. A vacuum is a very good thermal insulator, and there is little heat loss from the cooling solution.

After the sample is frozen, the entire apparatus is evacuated by opening the stopcock. When the evacuation is complete, the stopcock is closed, and the Dewar flask is removed. The sample is allowed to thaw, and then it is frozen again. This freeze–thaw–freeze cycle removes any air or gases that were trapped in the frozen sample. Next, the stopcock is opened to evacuate the system again. When the second evacuation is complete, the stopcock is closed, and the Dewar flask is moved to the other arm to cool the empty container. As the sample warms, it will vaporize, travel to the other

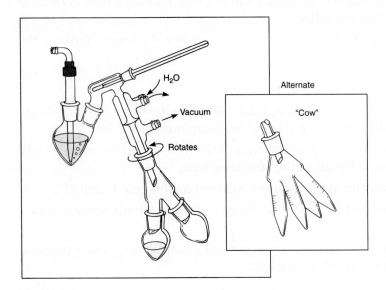

Figure 16.7
Rotary fraction collector.

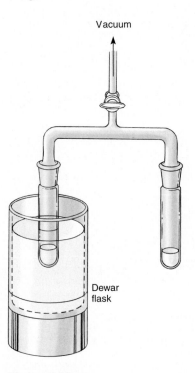

Figure 16.8
Bulb-to-bulb distillation.

side, and be frozen or liquefied by the cooling solution. This transfer of the liquid from one arm to the other may take quite a while, but no heating is required.

The bulb-to-bulb distillation is most effective when liquid nitrogen is used as coolant and when the vacuum system can achieve a pressure of 10^{-3} mm Hg or lower. This requires a vacuum pump; an aspirator cannot be used.

16.7 The Mechanical Vacuum Pump

The aspirator is not capable of yielding pressures below about 5 mm Hg. This is the vapor pressure of water at 0°C, and water freezes at this temperature. A more realistic value of pressure for an aspirator is about 20 mm Hg. When pressures below 20 mm Hg are required, a vacuum pump will have to be employed. Figure 16.9 illustrates a mechanical vacuum pump and its associated glassware. The vacuum pump operates on a principle similar to that of the aspirator, but the vacuum pump uses a high-boiling oil, rather than water, to remove air from the attached system. The oil used in a vacuum pump, a silicone oil or a high-molecular-weight hydrocarbon-based oil, has a very low vapor pressure, and very low system pressures can be achieved. A good vacuum pump, with new oil, can achieve pressures of 10^{-3} or 10^{-4} mm Hg. Instead of the oil being discarded as it is used, it is recycled continuously through the system.

A cooled trap is required when using a vacuum pump. This trap protects the oil in the pump from any vapors that may be present in the system. If vapors from organic solvents, or from the organic compounds being distilled, dissolve in the oil, the vapor pressure of the oil will increase,

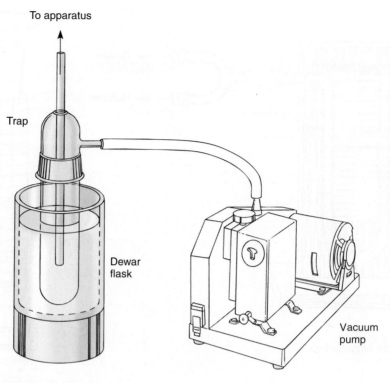

To apparatus

Trap

Dewar flask

Vacuum pump

Figure 16.9
A vacuum pump and its trap.

rendering it less effective. A special type of vacuum trap is illustrated in Figure 16.9. It is designed to fit into an insulated Dewar flask so that the coolant will last for a long period. At a minimum, this flask should be filled with ice water, but a dry ice–acetone mixture or liquid nitrogen is required to achieve lower temperatures and better protect the oil. Often, two traps are used; the first trap contains ice water and the second trap dry ice–acetone or liquid nitrogen. The first trap liquefies low-boiling vapors that might freeze or solidify in the second trap and block it.

16.8 The Closed-End Manometer

The principal device used to measure pressures in a vacuum distillation is the **closed-end manometer.** Two basic types are shown in Figures 16.10 and 16.11. The manometer shown in Figure 16.10 is widely used because it is relatively easy to construct. It consists of a U-tube that is closed at one end and mounted on a wooden support. You can construct the manometer from 9-mm glass capillary tubing and fill it, as shown in Figure 16.12.

C A U T I O N ⚠

Mercury is a very toxic metal with cumulative effects. Because mercury has a high vapor pressure, it must not be spilled in the laboratory. You must not touch it. Seek immediate help from an instructor in case of a spill or if you break a manometer. Spills must be cleaned immediately.

A small filling device is connected to the U-tube with pressure tubing. The U-tube is evacuated with a good vacuum pump; then the mercury is introduced by tilting the mercury reservoir.

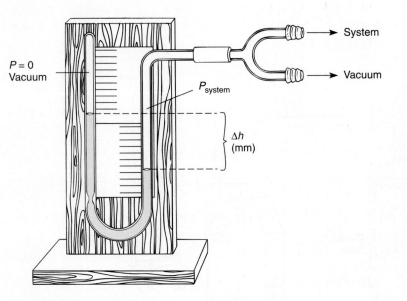

Figure 16.10
A simple U-tube manometer.

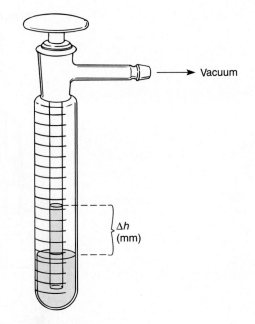

Figure 16.11
Commercial "stick" manometer.

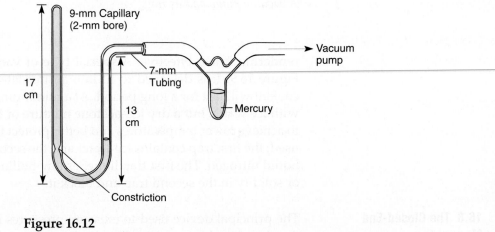

Figure 16.12
Filling a U-tube manometer.

NOTE: The entire filling operation should be conducted in a shallow pan in order to contain any spills that might occur.

Enough mercury should be added to form a column about 20 cm in total length. When the vacuum is interrupted by admitting air, the mercury is forced by atmospheric pressure to the end of the evacuated tube. The manometer is then ready for use. The constriction shown in Figure 16.12 helps to protect the manometer against breakage when the pressure is released. Be sure that the column of mercury is long enough to pass through this constriction.

When an aspirator or any other vacuum source is used, a manometer can be connected into the system. As the pressure is lowered, the mercury rises in the right tube and drops in the left tube until Δh corresponds to the approximate pressure of the system (see Fig. 16.10).

$$\Delta h = (P_{\text{system}} - P_{\text{reference arm}}) = (P_{\text{system}} - 10^{-3} \text{ mm Hg}) \approx P_{\text{system}}$$

A short piece of metric ruler or a piece of graph paper ruled in millimeter squares is mounted on the support board to allow Δh to be read. No addition or subtraction is necessary, because the reference pressure (created by the initial evacuation when filling) is approximately zero (10^{-3} mm Hg) when referred to readings in the 10- to 50-mm Hg range. To determine the pressure, count the number of millimeter squares beginning at the top of the mercury column on the left and continuing downward to the top of the mercury column on the right. This is the height difference Δh, and it gives the pressure in the system directly.

A commercial counterpart to the U-tube manometer is shown in Figure 16.11. With this manometer, the pressure is given by the difference in the mercury levels in the inner and outer tubes.

The manometers described here have a range of about 1–150 mm Hg in pressure. They are convenient to use when an aspirator is the source of vacuum. For high-vacuum systems (pressures below 1 mm Hg), a more elaborate manometer or an electronic measuring device must be used. These devices will not be discussed here.

16.9 Connecting and Using a Manometer

The most common use of a closed-end manometer is to monitor pressure during a reduced-pressure distillation. The manometer is placed in a vacuum distillation system, as shown in Figure 16.13. Generally, an aspirator is the source of vacuum. Both the manometer and the distillation apparatus should be protected by a trap from possible backups in the water line. Alternatives to the trap arrangements shown in Figure 16.13 appear in Figures 16.1 and 16.5. Notice in each case that the trap has a device (screw clamp or stopcock) for opening the system to the atmosphere. This is especially important in using a manometer because you should always make pressure changes slowly. If this is not done, there is a danger of spraying mercury throughout the system, breaking the manometer, or spurting mercury into the room. In the closed-end manometer, if the system is opened suddenly, the mercury rushes to the closed end of the U-tube. The mercury rushes with such speed and force that the end will be broken out of the manometer. Air should be admitted *slowly* by opening the valve cautiously. In a similar fashion, the valve should be closed slowly when the vacuum is being started, or mercury may be forcefully drawn into the system through the open end of the manometer.

If the pressure in a reduced-pressure distillation is lower than that desired, it is possible to adjust it by means of a **bleed valve.** The stopcock can serve this function in Figure 16.13 if it is opened only a small amount. In those systems with a screw clamp on the trap (Figs. 16.1 and 16.5), remove the screw clamp from the trap valve and attach the base of a Tirrill-style Bunsen burner. The needle valve in the base of the burner can be used to adjust precisely the amount of air that is admitted (bled) to the system and hence control the pressure.

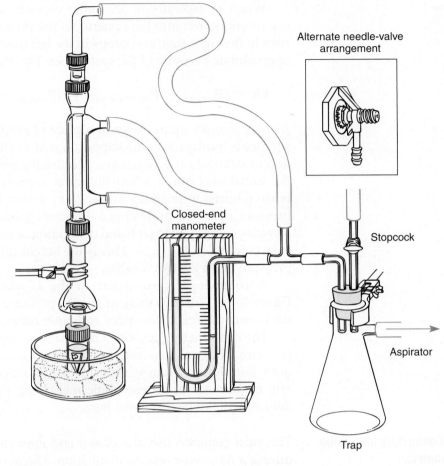

Figure 16.13

Connecting a manometer to the system. To construct a "bleed," the needle valve may replace the stopcock.

PROBLEMS

1. Give some reasons that would lead you to purify a liquid by using vacuum distillation rather than by using simple distillation.

2. When using an aspirator as a source of vacuum in a vacuum distillation, do you turn off the aspirator before venting the system? Explain.

3. A compound was distilled at atmospheric pressure and had a boiling range of 310–325°C. What would be the approximate boiling range of this liquid if it were distilled under vacuum at 20 mm Hg?

4. Boiling stones generally do not work when a vacuum distillation is performed. What substitutes may be used?

5. What is the purpose of the trap that is used during a vacuum distillation performed with an aspirator?

17 TECHNIQUE 17

Sublimation

17.1 Vapor-Pressure Behavior of Solids and Liquids

In Technique 13, the influence of temperature on the change in vapor pressure of a liquid was considered (see Fig. 13.1, p. 694). It was shown that the vapor pressure of a liquid increases with temperature. Because the boiling point of a liquid occurs when its vapor pressure is equal to the applied pressure (normally atmospheric pressure), the vapor pressure of a liquid equals 760 mmHg at its boiling point. The vapor pressure of a solid also varies with temperature. Because of this behavior, some solids can pass directly into the vapor phase without going through a liquid phase. This process is called **sublimation.** Because the vapor can be resolidified, the overall vaporization–solidification cycle can be used as a purification method. The purification can be successful only if the impurities have significantly lower vapor pressures than the material being sublimed.

In Figure 17.1, vapor-pressure curves for solid and liquid phases for two different substances are shown. Along lines *AB* and *DF*, the sublimation curves, the solid and vapor are at equilibrium. To the left of these lines, the solid phase exists, and to the right of these lines, the vapor phase is present. Along lines *BC* and *FG*, the liquid and vapor are at equilibrium. To the left of these lines, the liquid phase exists, and to the right, the vapor is present. The two substances vary greatly in their physical properties, as shown in Figure 17.1.

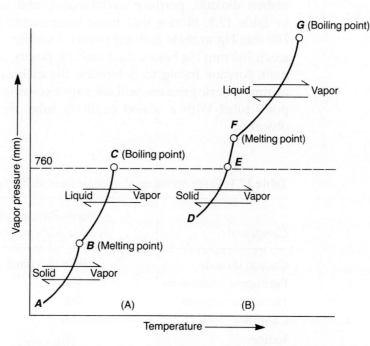

Figure 17.1

Vapor-pressure curves for solids and liquids. (A) Substance shows normal solid to liquid to gas transitions at 760 mm Hg pressure.
(B) Substance shows a solid to gas transition at 760 mm Hg pressure.

In the first case (Fig. 17.1A), the substance shows normal change-of-state behavior on being heated, going from solid to liquid to gas. The dashed line, which represents an atmospheric pressure of 760 mm Hg, is located *above* the melting point *B* in Figure 17.1A. Thus, the applied pressure (760 mm Hg) is *greater* than the vapor pressure of the solid–liquid phase at the melting point. Starting at *A*, as the temperature of the solid is raised, the vapor pressure increases along *AB* until the solid is observed to melt at *B*. At *B* the vapor pressures of *both* the solid and liquid are identical. As the temperature continues to rise, the vapor pressure will increase along *BC* until the liquid is observed to boil at *C*. The description given is for the "normal" behavior expected for a solid substance. All three states (solid, liquid, and gas) are observed sequentially during the change in temperature.

In the second case (Fig. 17.1B), the substance develops enough vapor pressure to vaporize completely at a temperature below its melting point. The substance shows a solid-to-gas transition only. The dashed line is now located *below* the melting point *F* of this substance. Thus, the applied pressure (760 mm Hg) is *less* than the vapor pressure of the solid–liquid phase at the melting point. Starting at *D*, the vapor pressure of the solid rises as the temperature increases along line *DF*. However, the vapor pressure of the solid reaches atmospheric pressure (point *E*) *before* the melting point at *F* is attained. Therefore, sublimation occurs at *E*. No melting behavior will be observed at atmospheric pressure for this substance. For a melting point to be reached and the behavior along line *FG* to be observed, an applied pressure greater than the vapor pressure of the substance at point *F* would be required. This could be achieved by using a sealed pressure apparatus.

The sublimation behavior just described is relatively rare for substances at atmospheric pressure. Several compounds exhibiting this behavior—carbon dioxide, perfluorocyclohexane, and hexachloroethane—are listed in Table 17.1. Notice that these compounds have vapor pressures *above* 760 mm Hg at their melting points. In other words, their vapor pressures reach 760 mm Hg below their melting points, and they sublime rather than melt. Anyone trying to determine the melting point of hexachloroethane at atmospheric pressure will see vapor pouring from the end of the melting-point tube! With a sealed capillary tube, the melting point of 186°C is observed.

Table 17.1 *Vapor pressures of solids at their melting points*

Compound	Vapor Pressure of Solid at MP (mm Hg)	Melting Point (°C)
Carbon dioxide	3876 (5.1 atm)	−57
Perfluorocyclohexane	950	59
Hexachloroethane	780	186
Camphor	370	179
Iodine	90	114
Naphthalene	7	80
Benzoic acid	6	122
p-Nitrobenzaldehyde	0.009	106

17.2 Sublimation Behavior of Solids

Sublimation is usually a property of relatively nonpolar substances that also have highly symmetrical structures. Symmetrical compounds have relatively high melting points and high vapor pressures. The ease with which a substance can escape from the solid state is determined by the strength of intermolecular forces. Symmetrical molecular structures have a relatively uniform distribution of electron density and a small dipole moment. A smaller dipole moment means a higher vapor pressure because of lower electrostatic attractive forces in the crystal.

Solids sublime if their vapor pressures are greater than atmospheric pressure at their melting points. Some compounds with the vapor pressures at their melting points are listed in Table 17.1. The first three entries in the table were discussed in Section 17.1. At atmospheric pressure they would sublime rather than melt, as shown in Figure 17.1B.

The next four entries in Table 17.1 (camphor, iodine, naphthalene, and benzoic acid) exhibit typical change-of-state behavior (solid, liquid, and gas) at atmospheric pressure, as shown in Figure 17.1A. These compounds sublime readily under reduced pressure, however. Vacuum sublimation is discussed in Section 17.3.

Compared with many other organic compounds, camphor, iodine, and naphthalene have relatively high vapor pressures at relatively low temperatures. For example, they have a vapor pressure of 1 mm Hg at 42, 39, and 53°C, respectively. Although this vapor pressure does not seem very large, it is high enough to lead, after a time, to **evaporation** of the solid from an open container. Mothballs (naphthalene and 1,4-dichlorobenzene) show this behavior. When iodine stands in a closed container over a period of time, you can observe movement of crystals from one part of the container to another.

Although chemists often refer to any solid–vapor transition as sublimation, the process just described for camphor, iodine, and naphthalene is really an **evaporation** of a solid. Strictly speaking, a sublimation point is like a melting point or a boiling point. It is defined as the point at which the vapor pressure of the solid *equals* the applied pressure. Many liquids readily evaporate at temperatures far below their boiling points. It is, however, much less common for solids to evaporate. Solids that readily sublime (evaporate) must be stored in sealed containers. When the melting point of such a solid is being determined, some of the solid may sublime and collect toward the open end of the melting-point tube while the rest of the sample melts. To solve the sublimation problem, one seals the capillary tube or rapidly determines the melting point. It is possible to use the sublimation behavior to purify a substance. For example, at atmospheric pressure, camphor can be readily sublimed, just below its melting point at 175°C. At 175°C the vapor pressure of camphor is 320 mm Hg. The vapor solidifies on a cool surface.

17.3 Vacuum Sublimation

Many organic compounds sublime readily under reduced pressure. When the vapor pressure of the solid equals the applied pressure, sublimation occurs, and the behavior is identical to that shown in Figure 17.1B. The solid phase passes directly into the vapor phase. From the data given in Table 17.1, you should expect camphor, naphthalene, and benzoic acid to sublime at or below the respective applied pressures of 370, 7, and 6 mm Hg. In principle, you can sublime *p*-nitrobenzaldehyde (last entry in the table), but it would not be practical because of the low applied pressure required.

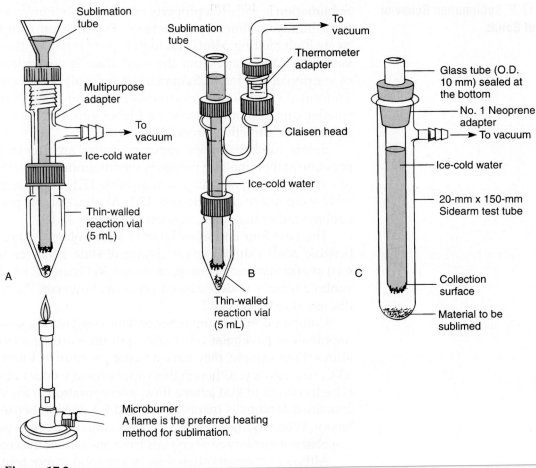

Figure 17.2
Sublimation apparatus.

17.4 Sublimation Methods

Sublimation can be used to purify solids. The solid is warmed until its vapor pressure becomes high enough for it to vaporize and condense as a solid on a cooled surface placed closely above. Several types of apparatus are illustrated in Figure 17.2. In each case, the cooled condensing surface is a tube filled with ice-cold water. The tube is filled from a beaker containing ice and water by using a Pasteur pipet. If the cooling water becomes warm before the sublimation is completed, the tube is emptied and refilled, once again by using a Pasteur pipet for these operations.

A flame is the preferred heating device for a sublimation. The burner can be held by its cool base (not the hot barrel!) and moved up and down the sides of the thin-walled outer vial or tube to "chase" any solid that has formed on the sides toward the cold tube in the center. With an aluminum block, a ring of solid often forms on the inside walls of the apparatus just where it leaves the heating block. If this happens, using the aluminum collars will improve the situation considerably. When using a conical vial, use a thin-walled conical vial instead of a regular conical vial, because the thicker glass can shatter when heated by a flame.

Many solids do not develop enough vapor pressure at atmospheric pressure (760 mm Hg) to be purified by sublimation, but they frequently can be sublimed at reduced pressure. Thus, most sublimation equipment has provision for connection to an aspirator or other vacuum source. Reduction of pressure also helps to prevent thermal decomposition of substances that would require high temperatures to sublime at ordinary pressures.

Remember that while performing a sublimation, it is important to keep the temperature below the melting point of the solid. After sublimation, the material that has collected on the cooled surface is recovered by removing the central tube (cold-finger) from the apparatus. Take care in removing this tube to avoid dislodging the crystals that have collected. The deposit of crystals is scraped from the inner tube with a spatula. If reduced pressure has been used, the pressure must be released carefully to keep a blast of air from dislodging the crystals.

17.5 Advantages of Sublimation

One advantage of sublimation is that no solvent is used, and therefore none needs to be removed later. Sublimation also removes occluded material, such as molecules of solvent, from the sublimed substance. For instance, caffeine (sublimes at 178°C, melts at 236°C) absorbs water gradually from the atmosphere to form a hydrate. During sublimation, this water is lost, and anhydrous caffeine is obtained. If too much solvent is present in a sample to be sublimed, however, instead of becoming lost, it condenses on the cooled surface and thus interferes with the sublimation.

Sublimation is a faster method of purification than crystallization but not as selective. Similar vapor pressures are often a factor in dealing with solids that sublime; consequently, little separation can be achieved. For this reason, solids are far more often purified by crystallization. Sublimation is most effective in removing a volatile substance from a nonvolatile compound, particularly a salt or other inorganic material. Sublimation is also effective in removing highly volatile bicyclic or other symmetrical molecules from less volatile reaction products. Examples of volatile bicyclic compounds are borneol, isoborneol, and camphor.

PROBLEMS

1. Why is solid carbon dioxide called dry ice? How does it differ from solid water in behavior?

2. Under what conditions can you have *liquid* carbon dioxide?

3. A solid substance has a vapor pressure of 800 mm Hg at its melting point (80°C). Describe how the solid behaves as the temperature is raised from room temperature to 80°C while as the atmospheric pressure is held constant at 760 mm Hg.

4. A solid substance has a vapor pressure of 100 mm Hg at the melting point (100°C). Assuming an atmospheric pressure of 760 mm Hg, describe the behavior of this solid as the temperature is raised from room temperature to its melting point.

5. A substance has a vapor pressure of 50 mm Hg at the melting point (100°C). Describe how you would experimentally sublime this substance.

18   TECHNIQUE 18

Steam Distillation

The simple, vacuum, and fractional distillations described in Techniques 14, 15, and 16 are applicable to completely soluble (miscible) mixtures only. When liquids are *not* mutually soluble (immiscible), they can also be distilled, but with a somewhat different result. A mixture of immiscible liquids will boil at a lower temperature than the boiling points of any of the separate components as pure compounds. When steam is used to provide one of the immiscible phases, the process is called **steam distillation.** The advantage of this technique is that the desired material distills at a temperature below 100°C. Thus, if unstable or very high-boiling substances are to be removed from a mixture, decomposition is avoided. Because all gases mix, the two substances can mix in the vapor and codistill. Once the distillate is cooled, the desired component, which is not miscible, separates from the water. Steam distillation is used widely in isolating liquids from natural sources. It is also used in removing a reaction product from a tarry reaction mixture.

18.1 Differences between Distillation of Miscible and Immiscible Mixtures

Two liquids, A and B, that are mutually soluble (miscible), and that do not interact, form an ideal solution and follow Raoult's Law, as shown in Equation 1.

$$\textbf{Miscible Liquids} \qquad P_{\text{total}} = P_A^{\circ} N_A + P_B^{\circ} N_B \qquad [1]$$

Note that the vapor pressures of pure liquids P_A° and P_B° are not added directly to give the total pressure P_{total} but are reduced by the respective mole fractions N_A and N_B. The total pressure above a miscible or homogeneous solution will depend on P_A° and P_B° and also N_A and N_B. Thus, the composition of the vapor will also depend on *both* the vapor pressures and the mole fractions of each component.

$$\textbf{Immiscible Liquids} \qquad P_{\text{total}} = P_A^{\circ} + P_B^{\circ} \qquad [2]$$

In contrast, when two mutually insoluble (immiscible) liquids are "mixed" to give a heterogeneous mixture, each exerts its own vapor pressure, independently of the other, as shown in Equation 2. The mole fraction term does not appear in this equation, because the compounds are not miscible. You simply add the vapor pressures of the pure liquids P_A° and P_B° at a given temperature to obtain the total pressure above the mixture. When the total pressure equals 760 mm Hg, the mixture boils. The composition of the vapor from an immiscible mixture, in contrast to the miscible mixture, is determined only by the vapor pressures of the two substances codistilling. Equation 3 defines the composition of the vapor from an immiscible mixture. Calculations involving this equation are given in Section 18.2.

$$\frac{\textbf{Moles A}}{\textbf{Moles B}} = \frac{P_A^{\circ}}{P_B^{\circ}} \qquad [3]$$

A mixture of two immiscible liquids boils at a lower temperature than the boiling points of either component. The explanation for this behavior is like that given for minimum-boiling-point azeotropes (Technique 15,

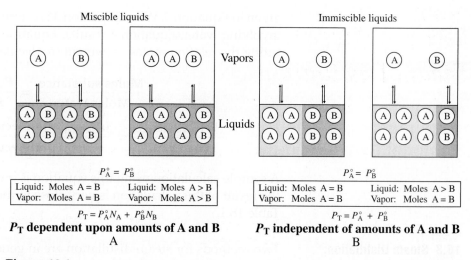

Figure 18.1

Total pressure behavior for miscible and immiscible liquids. (A) Ideal miscible liquids follow Raoult's Law: P_T depends on the mole fractions and vapor pressures of A and of B. (B) Immiscible liquids do not follow Raoult's Law: P_T depends only on the vapor pressures of A and B.

Section 15.7). Immiscible liquids behave as they do because an extreme incompatibility between the two liquids leads to higher combined vapor pressures than Raoult's Law would predict. The higher combined vapor pressures cause a lower boiling point for the mixture than for either single component. Thus, you may think of steam distillation as a special type of azeotropic distillation in which the substance is completely insoluble in water.

The differences in behavior of miscible and immiscible liquids, where it is assumed that P_A° equals P_B°, are shown in Figure 18.1. Note that with miscible liquids, the composition of the vapor depends on the relative amounts of A and B present (Fig. 18.1A). Thus, the composition of the vapor must change during a distillation. In contrast, the composition of the vapor with immiscible liquids is independent of the amounts of A and B present (Fig. 18.1B). Hence, the vapor composition must remain *constant* during the distillation of such liquids, as predicted by Equation 3. Immiscible liquids act as if they were being distilled simultaneously from separate compartments, as shown in Figure 18.1B, even though in practice they are "mixed" during a steam distillation. Because all gases mix, they do give rise to a homogeneous vapor and codistill.

18.2 Immiscible Mixtures: Calculations

The composition of the distillate is constant during a steam distillation, as is the boiling point of the mixture. The boiling points of steam-distilled mixtures will always be below the boiling point of water (bp 100°C), as well as the boiling point of any of the other substances distilled. Some representative boiling points and compositions of steam distillates are given in Table 18.1. Note that the higher the boiling point of a pure substance, the more closely the temperature of the steam distillate approaches, but does not exceed, 100°C. This is a reasonably low temperature, and it avoids the decomposition that might result at high temperatures with a simple distillation.

For immiscible liquids, the molar proportions of two components in a distillate equal the ratio of their vapor pressures in the boiling mixture, as given in Equation 3. When Equation 3 is rewritten for an immiscible mixture involving water, Equation 4 results. Equation 4 can be modified by substituting the relation moles = (weight/molecular weight) to give Equation 5.

$$\frac{\text{Moles substance}}{\text{Moles water}} = \frac{P^{\circ}_{\text{substance}}}{P^{\circ}_{\text{water}}} \qquad [4]$$

$$\frac{\text{Wt substance}}{\text{Wt water}} = \frac{(P^{\circ}_{\text{substance}})(\text{molecular weight}_{\text{substance}})}{(P^{\circ}_{\text{water}})(\text{molecular weight}_{\text{water}})} \qquad [5]$$

A sample calculation using this equation is given in Figure 18.2. Notice that the result of this calculation is very close to the experimental value given in Table 18.1.

18.3 Steam Distillation: Methods

Two methods for steam distillation are in general use in the laboratory: the **direct method** and the **live steam method**. In the first method, steam is generated *in situ* (in place) by heating a distillation flask containing the

---

Problem   How many grams of water must be distilled to steam distill 1.55 g of 1-octanol from an aqueous solution? What will be the composition (wt %) of the distillate? The mixture distills at 99.4°C.

Answer   The vapor pressure of water at 99.4°C must be obtained from the *CRC Handbook* (= 744 mm Hg).

(a) Obtain the partial pressure of 1-octanol.

$$P^{\circ}_{\text{1-octanol}} = P_{\text{total}} - P^{\circ}_{\text{water}}$$
$$P^{\circ}_{\text{1-octanol}} = (760 - 744) = 16 \text{ mm Hg}$$

(b) Obtain the composition of the distillate.

$$\frac{\text{wt 1-octanol}}{\text{wt water}} = \frac{(16)(130)}{(744)(18)} = 0.155 \text{ g/g-water}$$

(c) Clearly, 10 g of water must be distilled.

$$(0.155 \text{ g/g-water})(10 \text{ g-water}) = 1.55 \text{ g 1-octanol}$$

(d) Calculate the **weight** percentages.

$$\text{1-octanol} = 1.55 \text{ g}/(10\text{g} + 1.55 \text{ g}) = 13.4\%$$
$$\text{water} = 10 \text{ g}/(10 \text{ g} + 1.55 \text{ g}) = 86.6\%$$

Figure 18.2
Sample calculations for a steam distillation.

Table 18.1 *Boiling points and compositions of steam distillates*

Mixture	Boiling Point of Pure Substance (°C)	Boiling Point of Mixture (°C)	Composition (% Water)
Benzene–water	80.1	69.4	8.9%
Toluene–water	110.6	85.0	20.2%
Hexane–water	69.0	61.6	5.6%
Heptane–water	98.4	79.2	12.9%
Octane–water	125.7	89.6	25.5%
Nonane–water	150.8	95.0	39.8%
1-Octanol–water	195.0	99.4	90.0%

compound and water. In the second method, steam is generated outside and is passed into the distillation flask using an inlet tube.

A. Direct Method

Microscale. The direct method of steam distillation is the only one suitable for microscale reactions. Steam is produced in the conical vial or distillation flask (*in situ*) by heating water to its boiling point in the presence of the compound to be distilled. This method works well for small amounts of materials. A microscale steam distillation apparatus is shown in Figure 18.3. Water and the compound to be distilled are placed in the flask and heated. A stirring bar or a boiling stone should be used to prevent bumping. The vapors of the water and the desired compound codistill when they are heated. They are condensed and collect in the Hickman head. When the Hickman head fills, the distillate is removed with a Pasteur pipet and placed in another vial for storage. For the typical microscale experiment, it will be necessary to fill the well and remove the distillate three or four times. All these distillate fractions are placed in the same storage container. The efficiency in collecting the distillate can sometimes be improved if the inside walls of the Hickman head are rinsed several times into the well. A Pasteur pipet is used to perform the rinsing. Distillate is withdrawn from the well, and then it is used to wash the walls of the Hickman head all the way around the head. After the walls have been washed and when the well is full, the distillate can be withdrawn and transferred to the storage container. It may be necessary to add more water during the course of the distillation. More water is added (remove the condenser if used) through the center of the Hickman head by using a Pasteur pipet.

Semimicroscale. The apparatus shown in Figure 14.10, page 713, may also be used to perform a steam distillation at the microscale level or slightly above. This apparatus avoids the need to empty the collected distillate during the course of the distillation as is required when a Hickman head is used.

Macroscale. A larger-scale direct method steam distillation is illustrated in Figure 18.4. Although a heating mantle may be used, it is probably best to use a flame with this method because a large volume of water must be heated rapidly. A boiling stone must be used to prevent bumping. The separatory funnel allows more water to be added during the course of the distillation.

B. Live Steam Method

Macroscale. A large-scale steam distillation using the live steam method is shown in Figure 18.5. If steam lines are available in the laboratory, they may be attached directly to the steam trap (purge them first to drain water). If steam lines are not available, an external steam generator (see inset) must be prepared. The external generator usually will require a flame to produce steam at a rate fast enough for the distillation. When the distillation is first started, the clamp at the bottom of the steam trap is left open. The steam lines will have a large quantity of condensed water in them until they are well heated. When the lines become hot and condensation of steam ceases, the clamp may be closed. Occasionally, the clamp will have to be reopened to

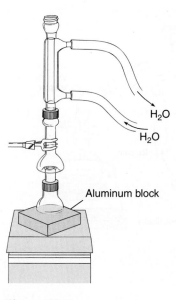

Figure 18.3
Microscale steam distillation.

H₂O
H₂O

Aluminum block

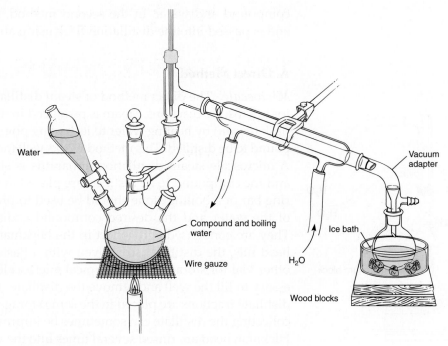

Figure 18.4
Macroscale direct steam distillation.

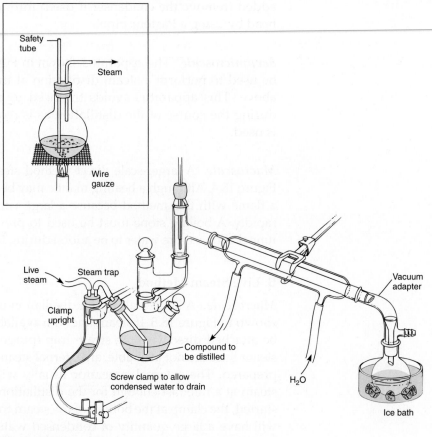

Figure 18.5
Macroscale steam distillation using live steam.

remove condensate. In this method, the steam agitates the mixture as it enters the bottom of the flask, and a stirrer or boiling stone is not required.

CAUTION

Hot steam can produce severe burns.

Sometimes it is helpful to heat the three-necked distilling flask with a heating mantle (or flame) to prevent excessive condensation at that point. Steam must be admitted fast enough for you to see the distillate condensing as a milky white fluid in the condenser. The vapors that codistill will separate on cooling to give this cloudiness. When the condensate becomes clear, the distillation is near the end. The flow of water through the condenser should be faster than in other types of distillation to help cool the vapors. Make sure the vacuum adapter remains cool to the touch. An ice bath may be used to cool the receiving flask if desired. When the distillation is to be stopped, the screw clamp on the steam trap should be opened, and the steam inlet tube must be removed from the three-necked flask. If this is not done, liquid will back up into the tube and steam trap.

PROBLEMS

1. Calculate the weight of benzene codistilled with each gram of water and the percentage composition of the vapor produced during a steam distillation. The boiling point of the mixture is 69.4°C. The vapor pressure of water at 69.4°C is 227.7 mm Hg. Compare the result with the data in Table 18.1.

2. Calculate the approximate boiling point of a mixture of bromobenzene and water at atmospheric pressure. A table of vapor pressures of water and bromobenzene at various temperatures is given.

| | Vapor Pressures (mm Hg) | |
Temperature (°C)	Water	Bromobenzene
93	588	110
94	611	114
95	634	118
96	657	122
97	682	127
98	707	131
99	733	136

3. Calculate the weight of nitrobenzene that codistills (bp 99°C) with each gram of water during a steam distillation. You may need the data given in Problem 2.

4. A mixture of *p*-nitrophenol and *o*-nitrophenol can be separated by steam distillation. The *o*-nitrophenol is steam volatile, and the *para* isomer is not volatile. Explain. Base your answer on the ability of the isomers to form hydrogen bonds internally.

19   TECHNIQUE 19

Column Chromatography

The most modern and sophisticated methods of separating mixtures that the organic chemist has available all involve **chromatography.** Chromatography is defined as the separation of a mixture of two or more compounds or ions by distribution between two phases, one of which is stationary and the other moving. Various types of chromatography are possible, depending on the nature of the two phases involved: **solid–liquid** (column, thin-layer, and paper), **liquid–liquid,** (high-performance liquid), and **gas–liquid** (vapor–phase) chromatographic methods are common.

All chromatography works on much the same principle as solvent extraction (Technique 12). Basically, the methods depend on the differential solubilities or adsorptivities of the substances to be separated relative to the two phases between which they are to be partitioned. In this chapter, column chromatography, a solid–liquid method, is considered. Thin-layer chromatography is examined in Technique 20; high-performance liquid chromatography is discussed in Technique 21; and gas chromatography, a gas–liquid method, is discussed in Technique 22.

19.1 Adsorbents

Column chromatography is a technique based on both adsorptivity and solubility. It is a solid–liquid phase-partitioning technique. The solid may be almost any material that does not dissolve in the associated liquid phase; the solids used most commonly are silica gel, $SiO_2 \cdot xH_2O$, also called silicic acid, and alumina, $Al_2O_3 \cdot xH_2O$. These compounds are used in their powdered or finely ground forms (usually 200 to 400 mesh).[1]

Most alumina used for chromatography is prepared from the impure ore bauxite $Al_2O_3 \cdot xH_2O + Fe_2O_3$. The bauxite is dissolved in hot sodium hydroxide and filtered to remove the insoluble iron oxides; the alumina in the ore forms the soluble amphoteric hydroxide $Al(OH)_4^-$. The hydroxide is precipitated by CO_2, which reduces the pH, as $Al(OH)_3$. When heated, the $Al(OH)_3$ loses water to form pure alumina Al_2O_3.

$$\text{Bauxite (crude)} \xrightarrow{\text{hot NaOH}} Al(OH)_4^-(aq) + Fe_2O_3 \text{ (insoluble)}$$

$$Al(OH)_4^-(aq) + CO_2 \longrightarrow Al(OH)_3 + HCO_3^-$$

$$2\ Al(OH)_3 \xrightarrow{\text{heat}} Al_2O_3(s) + 3\ H_2O$$

Alumina prepared in this way is called **basic alumina** because it still contains some hydroxides. Basic alumina cannot be used for chromatography of compounds that are base sensitive. Therefore, it is washed with acid to neutralize the base, giving **acid-washed alumina.** This material is unsatisfactory unless it has been washed with enough water to remove *all* the acid; on being so washed, it becomes the best chromatographic material, called

---

[1] The term *mesh* refers to the number of openings per linear inch found in a screen. A large number refers to a fine screen (finer wires more closely spaced). When particles are sieved through a series of these screens, they are classified by the smallest mesh screen that they will pass through. Mesh 5 would represent a coarse gravel, and mesh 800 would be a fine powder.

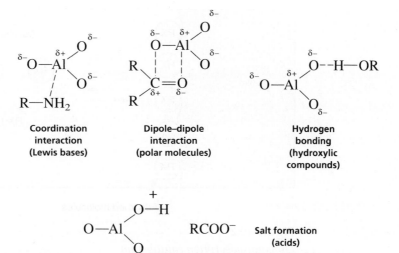

Figure 19.1
Possible interactions of organic compounds with alumina.

neutral alumina. If a compound is acid-sensitive, either basic or neutral alumina must be used. You should be careful to ascertain what type of alumina is being used for chromatography. Silica gel is not available in any form other than that suitable for chromatography.

19.2 Interactions

If powdered or finely ground alumina (or silica gel) is added to a solution containing an organic compound, some of the organic compound will **adsorb** onto or adhere to the fine particles of alumina. Many kinds of intermolecular forces cause organic molecules to bind to alumina. These forces vary in strength according to their type. Nonpolar compounds bind to the alumina using only van der Waals forces. These are weak forces, and nonpolar molecules do not bind strongly unless they have extremely high molecular weights. The most important interactions are those typical of polar organic compounds. Either these forces are of the dipole–dipole type or they involve some direct interaction (coordination, hydrogen bonding, or salt formation). These types of interactions are illustrated in Figure 19.1, which for convenience shows only a portion of the alumina structure. Similar interactions occur with silica gel. The strengths of such interactions vary in the approximate order:

Salt formation > coordination > hydrogen-bonding > dipole–dipole > van der Waals

Strength of interaction varies among compounds. For instance, a strongly basic amine would bind more strongly than a weakly basic one (by coordination). In fact, strong bases and strong acids often interact so strongly that they *dissolve* alumina to some extent. You can use the following rule of thumb:

NOTE: The more polar the functional group, the stronger the bond to alumina (or silica gel).

A similar rule holds for solubility. Polar solvents dissolve polar compounds more effectively than nonpolar solvents; nonpolar compounds are dissolved best by nonpolar solvents. Thus, the extent to which any given solvent can wash an adsorbed compound from alumina depends almost directly on the relative polarity of the solvent. For example, although a ketone adsorbed on alumina might not be removed by hexane, it might be removed

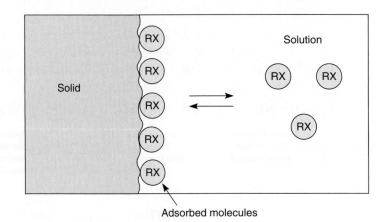

Figure 19.2
Dynamic adsorption equilibrium.

completely by chloroform. For any adsorbed material, a kind of **distribution** equilibrium can be envisioned between the adsorbent material and the solvent. This is illustrated in Figure 19.2.

The distribution equilibrium is *dynamic,* with molecules constantly *adsorbing* from the solution and *desorbing* into it. The average number of molecules remaining adsorbed on the solid particles at equilibrium depends both on the particular molecule (RX) involved and the dissolving power of the solvent with which the adsorbent must compete.

19.3 Principle of Column Chromatographic Separation

The dynamic equilibrium mentioned previously, and the variations in the extent to which different compounds adsorb on alumina or silica gel, underlie a versatile and ingenious method for *separating* mixtures of organic compounds. In this method, the mixture of compounds to be separated is introduced onto the top of a cylindrical glass column (Fig. 19.3) *packed,* or filled, with fine alumina particles (stationary solid phase). The adsorbent is continuously washed by a flow of solvent (moving phase) passing through the column.

Initially, the components of the mixture adsorb onto the alumina particles at the top of the column. The continuous flow of solvent through the column **elutes,** or washes, the solutes off the alumina and sweeps them down the column. The solutes (or materials to be separated) are called **eluates** or **elutants,** and the solvents are called **eluents.** As the solutes pass down the column to fresh alumina, new equilibria are established among the adsorbent, the solutes, and the solvent. The constant equilibration means that different compounds will move down the column at differing rates, depending on their relative affinity for the adsorbent on one hand and for the solvent on the other. Because the number of alumina particles is large, because they are closely packed, and because fresh solvent is being added continuously, the number of equilibrations between adsorbent and solvent that the solutes experience is enormous.

As the components of the mixture are separated, they begin to form moving bands (or zones), each band containing a single component. If the column is long enough and the other parameters (column diameter, adsorbent, solvent, and flow rate) are correctly chosen, the bands separate from one another, leaving gaps of pure solvent in between. As each band (solvent and solute) passes out the bottom of the column, it can be collected

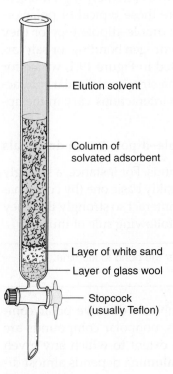

— Elution solvent

— Column of solvated adsorbent

— Layer of white sand
— Layer of glass wool

— Stopcock (usually Teflon)

Figure 19.3
Chromatographic column.

before the next band arrives. If the parameters mentioned are poorly chosen, the various bands either overlap or coincide, in which case either a poor separation or no separation at all is the result. A successful chromatographic separation is illustrated in Figure 19.4.

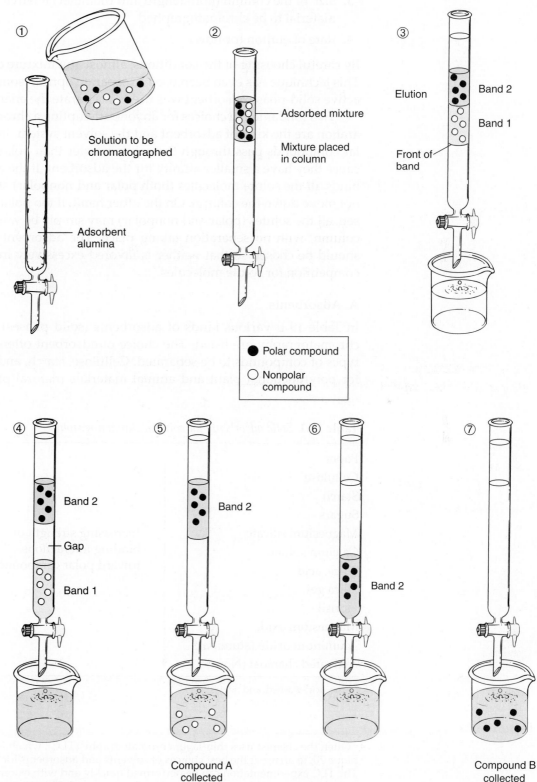

Figure 19.4

Sequence of steps in a chromatographic separation.

19.4 Parameters Affecting Separation

The versatility of column chromatography results from the many factors that can be adjusted. These include

1. Adsorbent chosen
2. Polarity of the solvents chosen
3. Size of the column (both length and diameter) relative to the amount of material to be chromatographed
4. Rate of elution (or flow)

By careful choosing of the conditions, almost any mixture can be separated. This technique has even been used to separate optical isomers. An optically active solid-phase adsorbent was used to separate the enantiomers.

Two fundamental choices for anyone attempting a chromatographic separation are the kind of adsorbent and the solvent system. In general, nonpolar compounds pass through the column faster than polar compounds because they have a smaller affinity for the adsorbent. If the adsorbent chosen binds all the solute molecules (both polar and nonpolar) strongly, they will not move down the column. On the other hand, if too polar a solvent is chosen, all the solutes (polar and nonpolar) may simply be washed through the column, with no separation taking place. The adsorbent and the solvent should be chosen so that neither is favored excessively in the equilibrium competition for solute molecules.[2]

A. Adsorbents

In Table 19.1, various kinds of adsorbents (solid phases) used in column chromatography are listed. The choice of adsorbent often depends on the types of compounds to be separated. Cellulose, starch, and sugars are used for polyfunctional plant and animal materials (natural products) that are

Table 19.1 *Solid adsorbents for column chromatography*

Paper	
Cellulose	
Starch	
Sugars	
Magnesium silicate	Increasing strength of
Calcium sulfate	binding interactions
Silicic acid	toward polar compounds
Silica gel	
Florisil	
Magnesium oxide	
Aluminum oxide (alumina)*	
Activated charcoal (Norit)	

*Basic, acid-washed, and neutral.

---

[2] Often the chemist uses thin-layer chromatography (TLC), which is described in Technique 20, to arrive at the best choices of solvents and adsorbents for the best separation. The TLC experimentation can be performed quickly and with extremely small amounts (microgram quantities) of the mixture to be separated. This saves significant time and materials. Technique 20 describes this use of TLC.

very sensitive to acid–base interactions. Magnesium silicate is often used for separating acetylated sugars, steroids, and essential oils. Silica gel and Florisil are relatively mild toward most compounds and are widely used for a variety of functional groups—hydrocarbons, alcohols, ketones, esters, acids, azo compounds, amines. Alumina is the most widely used adsorbent and is obtained in the three forms mentioned in Section 19.1: acidic, basic, and neutral. The pH of acidic or acid-washed alumina is approximately 4. This adsorbent is particularly useful for separating acidic materials such as carboxylic acids and amino acids. Basic alumina has a pH of 10 and is useful in separating amines. Neutral alumina can be used to separate a variety of nonacidic and nonbasic materials.

The approximate strength of the various adsorbents listed in Table 19.1 is also given. The order is only approximate, and therefore it may vary. For instance, the strength, or separating abilities, of alumina and silica gel largely depend on the amount of water present. Water binds tightly to either adsorbent, taking up sites on the particles that could otherwise be used for equilibration with solute molecules. If one adds water to the adsorbent, it is said to have been **deactivated.** Anhydrous alumina or silica gel is said to be highly **activated.** High activity is usually avoided with these adsorbents. Use of the highly active forms of either alumina or silica gel, or of the acidic or basic forms of alumina, can often lead to molecular rearrangement or decomposition in certain types of solute compounds.

The chemist can select the degree of activity that is appropriate to carry out a particular separation. To accomplish this, highly activated alumina is mixed thoroughly with a precisely measured quantity of water. The water partially hydrates the alumina and thus reduces its activity. By carefully determining the amount of water required, the chemist can have available an entire spectrum of possible activities.

B. Solvents

In Table 19.2, some common chromatographic solvents are listed, along with their relative ability to dissolve polar compounds. Sometimes a single

Table 19.2 *Solvents (eluents) for chromatography*

Petroleum ether	
Cyclohexane	
Carbon tetrachloride*	
Toluene	
Chloroform*	
Methylene chloride	Increasing polarity and
Diethyl ether	"solvent power" toward
Ethyl acetate	polar functional groups
Acetone	
Pyridine	
Ethanol	
Methanol	
Water	
Acetic acid	

*Suspected carcinogens.

Table 19.3 *Elution sequence for compounds*

Hydrocarbons	Fastest (will elute with nonpolar solvent)
Olefins	
Ethers	
Halocarbons	
Aromatics	
Ketones	Order of elution
Aldehydes	
Esters	
Alcohols	
Amines	
Acids, strong bases	Slowest (need a polar solvent)

solvent can be found that will separate all the components of a mixture. Sometimes a mixture of solvents can be found that will achieve separation. More often, you must start elution with a nonpolar solvent to remove relatively nonpolar compounds from the column and then gradually increase the solvent polarity to force compounds of greater polarity to come down the column, or elute. The approximate order in which various classes of compounds elute by this procedure is given in Table 19.3. In general, nonpolar compounds travel through the column faster (elute first), and polar compounds travel more slowly (elute last). However, molecular weight is also a factor in determining the order of elution. A nonpolar compound of high molecular weight travels more slowly than a nonpolar compound of low molecular weight, and it may even be passed by some polar compounds.

Solvent polarity functions in two ways in column chromatography. First, a polar solvent will better dissolve a polar compound and move it down the column faster. Therefore, as already mentioned, one usually increases the polarity of the solvent during column chromatography to wash down compounds of increasing polarity. Second, as the polarity of the solvent increases, the solvent itself will displace adsorbed molecules from the alumina or silica and take their place on the column. Because of this second effect, a polar solvent will move *all* types of compounds, both polar and nonpolar, down the column at a faster rate than a nonpolar solvent.

When the polarity of the solvent has to be changed during a chromatographic separation, some precautions must be taken. Rapid changes from one solvent to another are to be avoided (especially when silica gel or alumina is involved). Usually, small percentages of a new solvent are mixed slowly into the one in use until the percentage reaches the desired level. If this is not done, the column packing often "cracks" as a result of the heat liberated when alumina or silica gel is mixed with a solvent. The solvent solvates the adsorbent, and the formation of a weak bond generates heat.

$$\textbf{Solvent + alumina} \longrightarrow \textbf{(alumina} \cdot \textbf{solvent) + heat}$$

Often, enough heat is generated locally to evaporate the solvent. The formation of vapor creates bubbles, which forces a separation of the column packing; this is called **cracking.** A cracked column does not produce a good separation, because it has discontinuities in the packing. The way in which a column is packed or filled is also very important in preventing cracking.

Certain solvents should be avoided with alumina or silica gel, especially with the acidic, basic, and highly active forms. For instance, with any of these adsorbents, acetone dimerizes via an aldol condensation to give diacetone alcohol. Mixtures of esters *transesterify* (exchange their alcoholic portions) when ethyl acetate or an alcohol is the eluent. Finally, the most active solvents (pyridine, methanol, water, and acetic acid) dissolve and elute some of the adsorbent itself. Generally, try to avoid going to solvents more polar than diethyl ether or methylene chloride in the eluent series (Table 19.2).

C. Column Size and Adsorbent Quantity

The column size and the amount of adsorbent must also be selected correctly to separate a given amount of sample well. As a rule of thumb, the amount of adsorbent should be 25 to 30 times, by weight, the amount of material to be separated by chromatography. Furthermore, the column should have a height-to-diameter ratio of about 8:1. Some typical relations of this sort are given in Table 19.4.

Note, as a caution, that the difficulty of the separation is also a factor in determining the size and length of the column to be used and in the amount of adsorbent needed. Compounds that do not separate easily may require longer columns and more adsorbent than specified in Table 19.4. For easily separated compounds, a shorter column and less adsorbent may suffice.

D. Flow Rate

The rate at which solvent flows through the column is also significant in the effectiveness of a separation. In general, the time the mixture to be separated remains on the column is directly proportional to the extent of equilibration between stationary and moving phases. Thus, similar compounds eventually separate if they remain on the column long enough. The time a material remains on the column depends on the flow rate of the solvent. If the flow is too slow, however, the dissolved substances in the mixture may diffuse faster than the rate at which they move down the column. Then the bands grow wider and more diffuse, and the separation becomes poor.

Table 19.4 *Size of column and amount of adsorbent for typical sample sizes*

Amount of Sample (g)	Amount of Adsorbent (g)	Column Diameter (mm)	Column Height (mm)
0.01	0.3	3.5	30
0.10	3.0	7.5	60
1.00	30.0	16.0	130
10.00	300.0	35.0	280

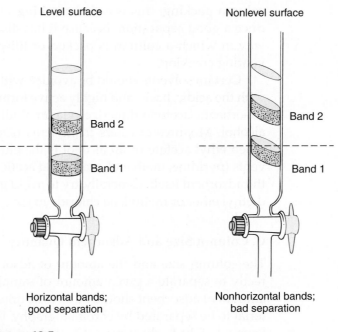

Figure 19.5
Comparison of horizontal and nonhorizontal band fronts.

19.5 Packing the Column: Typical Problems

The most critical operation in column chromatography is packing (filling) the column with adsorbent. The **column packing** must be evenly packed and free of irregularities, air bubbles, and gaps. As a compound travels down the column, it moves in an advancing zone, or **band.** It is important that the leading edge, or **front,** of this band be horizontal, or perpendicular to the long axis of the column. If two bands are close together and do not have horizontal band fronts, it is impossible to collect one band while completely excluding the other. The leading edge of the second band begins to elute before the first band has finished eluting. This condition can be seen in Figure 19.5. There are two main reasons for this problem. First, if the top surface edge of the adsorbent packing is not level, nonhorizontal bands result. Second, bands may be nonhorizontal if the column is not held in an exactly vertical position in both planes (front to back and side to side). When you are preparing a column, you must watch both these factors carefully.

Another phenomenon, called **streaming** or **channeling,** occurs when part of the band front advances ahead of the major part of the band. Channeling occurs if there are any cracks or irregularities in the adsorbent surface or any irregularities caused by air bubbles in the packing. A part of the advancing front moves ahead of the rest of the band by flowing through the channel. Two examples of channeling are shown in Figure 19.6.

19.6 Packing the Column: Microscale Methods

The following methods are used to avoid problems resulting from uneven packing and column irregularities. These procedures should be followed carefully in preparing a chromatography column. Failure to pay close attention to the preparation of the column may well affect the quality of the separation.

Preparation of a column involves two distinct stages. In the first stage, a support base on which the packing will rest is prepared. This must be done so that the packing, a finely divided material, does not wash out of

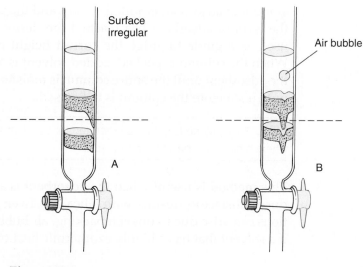

Figure 19.6
Channeling complications.

the bottom of the column. In the second stage, the column of adsorbent is deposited on top of the supporting base.

A. Preparing the Support Base

For microscale applications, select a Pasteur pipet (5¾-inch) and clamp it upright (vertically). To reduce the amount of solvent needed to fill the column, break off most of the tip of the pipet. Place a small ball of cotton in the pipet and tamp it into position using a glass rod or a piece of wire. Take care not to plug the column totally by tamping the cotton too hard. The correct position of the cotton is shown in Figure 19.7. A microscale chromatography column is packed by one of the dry pack methods described in Part B of this section.

B. Depositing the Adsorbent

Dry Pack Method 1. To fill a microscale column, fill the Pasteur pipet (with the cotton plug, prepared as described in Section A) about half full with solvent. Using a microspatula, add the solid adsorbent slowly to the solvent in the column. As you add the solid, tap the column *gently* with a pencil, a finger, or a glass rod. The tapping promotes even settling and mixing and gives an evenly packed column free of air bubbles. As the adsorbent is added, solvent flows out of the Pasteur pipet. Because the adsorbent must not be allowed to dry during the packing process, you must use a means of controlling the solvent flow. If a piece of small-diameter plastic tubing is available, it can be fitted over the narrow tip of the Pasteur pipet. The flow rate can then be controlled using a screw clamp. A simple approach to controlling the flow rate is to use a finger over the top of the Pasteur pipet, much as you control the flow of liquid in a volumetric pipet. Continue adding the adsorbent slowly, with constant tapping, until the level of the adsorbent has reached the desired level. As you pack the column, be careful not to let the column run dry. The final column should appear as shown in Figure 19.7.

Dry Pack Method 2. An alternative dry pack method for microscale columns is to fill the Pasteur pipet with *dry* adsorbent, without any solvent. Position a plug of cotton in the bottom of the Pasteur pipet. The desired

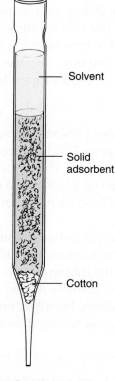

— Solvent

— Solid
adsorbent

— Cotton

Figure 19.7
Microscale chromatography column.

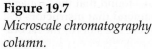

amount of adsorbent is added slowly, and the pipet tapped constantly, until the level of adsorbent has reached the desired height. Figure 19.7 can be used as a guide to judge the correct height of the column of adsorbent. When the column is packed, added solvent is allowed to percolate through the adsorbent until the entire column is moistened. The solvent is not added until just before the column is to be used.

NOTE: This method is not recommended for use with silica gel or for experiments where a careful separation is required.

This method is useful when the adsorbent is alumina, but it does not produce satisfactory results with silica gel. Even with alumina, poor separations can arise due to uneven packing, air bubbles, and cracking, especially if a solvent that has a highly exothermic heat of solvation is used.

19.7 Packing the Column: Semimicroscale and Macroscale Methods

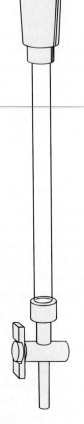

Figure 19.8
Commercial semimicroscale chromatography column. (The column is shown equipped with an optional solvent reservoir.)

As with microscale columns, the procedures described in this section should be followed carefully in preparing a semimicroscale or conventional-scale chromatography column. Failure to pay close attention to the details of these procedures may adversely affect the quality of the separation.

Again, preparation of a column involves two distinct stages: preparation of the support base and filling the column with adsorbent.

A. Preparing the Support Base

Semimicroscale Columns. An alternative apparatus for small-scale column chromatography is a commercial column, such as the one shown in Figure 19.8. This type of column is made of glass and has a solvent-resistant plastic stopcock at the bottom.[3] The stopcock assembly contains a filter disc to support the adsorbent column. An optional upper fitting, also made of solvent-resistant plastic, serves as a solvent reservoir. The column shown in Figure 19.8 is equipped with the solvent reservoir. This type of column is available in a variety of lengths, ranging from 100 to 300 mm. Because the column has a built-in filter disc, it is not necessary to prepare a support base before the adsorbent is added.

Macroscale Columns. For large-scale applications, clamp a chromatography column upright (vertically). The column (Figure 19.3, page 758) is a piece of cylindrical glass tubing with a stopcock attached at one end. The stopcock usually has a Teflon plug because stopcock grease (used on glass plugs) dissolves in many of the organic solvents used as eluents. Stopcock grease in the eluent will contaminate the eluates.

Instead of a stopcock, attach a piece of flexible tubing to the bottom of the column, with a screw clamp used to stop or regulate the flow (Fig. 19.9). When a screw clamp is used, care must be taken that the tubing used is not dissolved by the solvents that will pass through the column during the experiment. Rubber, for instance, dissolves in chloroform, benzene, methylene chloride, toluene, or tetrahydrofuran (THF). Tygon tubing dissolves

---

[3] *Note to the Instructor:* With certain organic solvents, we have found that the "solvent-resistant" plastic stopcock may tend to dissolve! We recommend that instructors test their equipment with the solvent that they intend to use before the start of the laboratory class.

Figure 19.9

Tubing with screw clamp to regulate solvent flow on a chromatography column.

(actually, the plasticizer is removed) in many solvents, including benzene, methylene chloride, chloroform, ether, ethyl acetate, toluene, and THF. Polyethylene tubing is the best choice for use at the end of a column because it is inert with most solvents.

Next, the column is partially filled with a quantity of solvent, usually a nonpolar solvent like hexane, and a support for the finely divided adsorbent is prepared in the following way. A loose plug of glass wool is tamped down into the bottom of the column with a long glass rod until all entrapped air is forced out as bubbles. Take care not to plug the column totally by tamping the glass wool too hard. A small layer of clean white sand is formed on top of the glass wool by pouring sand into the column. The column is tapped to level the surface of the sand. Any sand adhering to the side of the column is washed down with a small quantity of solvent. The sand forms a base that supports the column of adsorbent and prevents it from washing through the stopcock. The column is packed in one of two ways: by the slurry method or by the dry pack method.

B. Depositing the Adsorbent

Slurry Method. The slurry method is not recommended as a microscale method for use with Pasteur pipets. On a very small scale, it is too difficult to pack the column with the slurry without losing the solvent before the packing has been completed. Microscale columns should be packed by the dry pack method, as described in Section 19.6.

In the slurry method, the adsorbent is packed into the column as a mixture of a solvent and an undissolved solid. The slurry is prepared in a separate container (Erlenmeyer flask) by adding the solid adsorbent, a little at a time, to a quantity of the solvent. This order of addition (adsorbent added to solvent) should be followed strictly because the adsorbent solvates and liberates heat. If the solvent is added to the adsorbent, it may boil away almost as fast as it is added due to heat evolved. This will be especially true if ether or another low-boiling solvent is used. When this happens, the final mixture will be uneven and lumpy. Enough adsorbent is added to the solvent, and mixed by swirling the container, to form a thick but flowing slurry. The container should be swirled until the mixture is homogenous and relatively free of entrapped air bubbles.

For a standard-sized column, the procedure is as follows. When the slurry has been prepared, the column is filled about half full with solvent, and the stopcock is opened to allow solvent to drain slowly into a large beaker. The slurry is mixed by swirling and is then poured in portions into the top of the draining column (a wide-necked funnel may be useful here). Be sure to swirl the slurry thoroughly before each addition to the column. The column is tapped constantly and *gently* on the side during the pouring operation with the fingers or with a pencil fitted with a rubber stopper. A short piece of large-diameter pressure tubing may also be used for tapping. The tapping promotes even settling and mixing and gives an evenly packed column free of air bubbles. Tapping is continued until all the material has settled, showing a well-defined level at the top of the column. Solvent from the collecting beaker may be re-added to the slurry if it becomes too thick to be poured into the column at one time. In fact, the collected solvent should be cycled through the column several times to ensure that settling is complete and that the column is firmly packed. The downward flow of solvent tends to compact the adsorbent. You should take care never to let the column "run dry" during packing. There should always be solvent on top of the absorbent column.

Dry Pack Method 1. In the first of the dry pack methods introduced here, the column is filled with solvent and allowed to drain *slowly.* The dry adsorbent is added, a little at a time, while the column is tapped gently with a pencil, finger, or glass rod.

Semimicroscale Columns The procedure to fill a commercial semimicroscale column is essentially the same as that used to fill a Pasteur pipet (Section 19.6). The commercial column has the advantage that it is much easier to control the flow of solvent from the column during the filling process, because the stopcock can be adjusted appropriately. It is not necessary to use a cotton plug or to deposit a layer of sand before adding the adsorbent. The presence of the fritted disc at the base of the column acts to prevent adsorbent from escaping from the column.

Macroscale Columns A plug of cotton is placed at the base of the column, and an even layer of sand is formed on top (see p. 758). The column is filled about half full with solvent, and the solid adsorbent is added carefully from a beaker while the solvent is allowed to flow slowly from the column. As the solid is added, the column is tapped as described for the slurry method in order to ensure that the column is packed evenly. When the column has the desired length, no more adsorbent is added. This method also produces an evenly packed column. Solvent should be cycled through this column (for macroscale applications) several times before each use. The same portion of solvent that has drained from the column during the packing is used to cycle through the column.

Dry Pack Method 2. In this method, the column is filled with dry adsorbent without any solvent. When the desired amount of adsorbent has been added, solvent is allowed to percolate through the column.

Semimicroscale Columns The Dry Pack Method 2 is similar to that described for Pasteur pipets (Section 19.6), except that the plug of cotton is not required. The flow rate of solvent through the column can be controlled using the stopcock, which is part of the column assembly (see Fig. 19.8).

Macroscale Columns Macroscale columns can also be packed by a dry pack method that is similar to the microscale methods described in Section 19.6. The disadvantages described for the microscale method also apply to the macroscale method. This method is not recommended for use with silica gel or alumina, because the combination leads to uneven packing, air bubbles, and cracking, especially if a solvent that has a highly exothermic heat of solvation is used.

19.8 Applying the Sample to the Column

The solvent (or solvent mixture) used to pack the column is normally the least polar elution solvent that can be used during chromatography. The compounds to be chromatographed are not highly soluble in the solvent. If they were, they would probably have a greater affinity for the solvent than for the adsorbent and would pass right through the column without equilibrating with the stationary phase.

The first elution solvent, however, is generally not a good solvent to use in preparing the sample to be placed on the column. Because the compounds are not highly soluble in nonpolar solvents, it takes a large amount of the initial solvent to dissolve the compounds, and it is difficult to get the mixture to form a narrow band on top of the column. A narrow band is ideal for an optimum separation of components. For the best separation, therefore, the compound is applied to the top of the column undiluted if it is a liquid or in a *very small* amount of highly polar solvent if it is a solid. Water must not be used to dissolve the initial sample being chromatographed, because it reacts with the column packing.

In adding the sample to the column, use the following procedure. Lower the solvent level to the top of the adsorbent column by draining the solvent from the column. Add the sample (either a pure liquid or a solution) to form a small layer on top of the adsorbent. A Pasteur pipet is convenient for adding the sample to the column. Take care not to disturb the surface of the adsorbent. This is best accomplished by touching the pipet to the inside of the glass column and slowly draining it so as to allow the sample to spread into a thin film, which slowly descends to cover the entire adsorbent surface. Drain the pipet close to the surface of the adsorbent. When all the sample has been added, drain this small layer of liquid into the column until the top surface of the column *just begins* to dry. Then add a small layer of the chromatographic solvent carefully with a Pasteur pipet, again being careful not to disturb the surface. Drain this small layer of solvent into the column until the top surface of the column just dries. Add another small layer of fresh solvent, if necessary, and repeat the process until it is clear that the sample is strongly adsorbed on the top of the column. If the sample is colored and the fresh layer of solvent acquires some of this color, the sample has not been properly adsorbed. Once the sample has been properly applied, you can protect the level surface of the adsorbent by carefully filling the top of the column with solvent and sprinkling clean, white sand into the column so as to form a small protective layer on top of the adsorbent. For microscale applications, this layer of sand is not required.

Separations are often better if the sample is allowed to stand a short time on the column before elution. This allows a true equilibrium to be established. In columns that stand for too long, however, the adsorbent often compacts or even swells, and the flow can become annoyingly slow. Diffusion of the sample to widen the bands also becomes a problem if a column is allowed to stand over an extended period. For small-scale chromatography, using Pasteur pipets, there is no stopcock, and it is not possible to stop the flow. In this case, it is not considered necessary to allow the column to stand.

19.9 Elution Techniques

Solvents for analytical and preparative chromatography should be pure reagents. Commercial-grade solvents often contain small amounts of residue, which remains when the solvent is evaporated. For normal work, and for relatively easy separations that take only small amounts of solvent, the residue usually presents few problems. For large-scale work, commercial-grade solvents may have to be redistilled before use. This is especially true for hydrocarbon solvents, which tend to have more residue than other solvent types. Most of the experiments in this laboratory manual have been designed to avoid this particular problem.

One usually begins elution of the products with a nonpolar solvent, such as hexane or petroleum ether. The polarity of the elution solvent can be increased gradually by adding successively greater percentages of either ether or toluene (for instance, 1, 2, 5, 10, 15, 25, 50, 100%) or some other solvent of greater solvent power (polarity) than hexane. The transition from one solvent to another should not be too rapid in most solvent changes. If the two solvents to be changed differ greatly in their heats of solvation in binding to the adsorbent, enough heat can be generated to crack the column. Ether is especially troublesome in this respect because it has both a low boiling point and a relatively high heat of solvation. Most organic compounds can be separated on silica gel or alumina using hexane–ether or hexane–toluene combinations for elution and following these by pure methylene chloride. Solvents of greater polarity are usually avoided for the various reasons mentioned previously. In microscale work, the usual procedure is to use only one solvent for the chromatography.

The flow of solvent through the column should not be too rapid or the solutes will not have time to equilibrate with the adsorbent as they pass down the column. If the rate of flow is too low or stopped for a period, diffusion can become a problem—the solute band will diffuse, or spread out, in all directions. In either of these cases, separation will be poor. As a general rule (and only an approximate one), most macroscale columns are run with flow rates ranging from 5 to 50 drops of effluent per minute; a steady flow of solvent is usually avoided. Microscale columns made from Pasteur pipets do not have a means of controlling the solvent flow rate, but commercial microscale columns are equipped with stopcocks. The solvent flow rate in this type of column can be adjusted in a manner similar to that used with larger columns. To avoid diffusion of the bands, do not stop the column and do not set it aside overnight.

19.10 Reservoirs

For a microscale chromatography, the portion of the Pasteur pipet above the adsorbent is used as a reservoir of solvent. Fresh solvent, as needed, is added by means of another Pasteur pipet. When it is necessary to change solvent, the new solvent is also added in this manner. In some cases, the chromatography may proceed too slowly; the rate of solvent flow can be accelerated by attaching a rubber dropper bulb to the top of the Pasteur pipet column and squeezing *gently*. The additional air pressure forces the solvent through the column more rapidly. If this technique is used, however, care must be taken to remove the rubber bulb from the column before releasing it. Otherwise, air may be drawn up through the bottom of the column, destroying the column packing.

When large quantities of solvent are used in a chromatographic separation, it is often convenient to use a solvent reservoir to forestall having to add small portions of fresh solvent continually. The simplest type of reservoir, a feature of many columns, is created by fusing the top of the column to a round-bottom flask (Fig. 19.10A). If the column has a standard-taper joint at its top, a reservoir can be created by joining a standard-taper separatory funnel to the column (Fig. 19.10B). In this arrangement, the stopcock is left open, and no stopper is placed in the top of the separatory funnel. A third common arrangement is shown in Figure 19.10C. A separatory funnel is filled with solvent; its stopper is wetted with solvent and put *firmly* in place. The funnel is inserted into the empty filling space at the top of the chromatographic column, and the stopcock is opened. Solvent flows out of

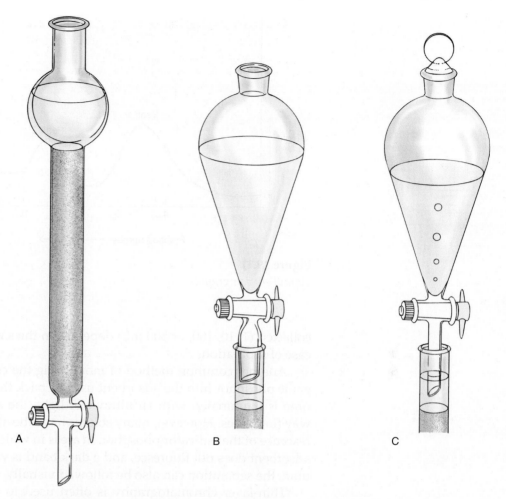

Figure 19.10
Various types of solvent-reservoir arrangements for chromatographic columns.

the funnel, filling the space at the top of the column until the solvent level is well above the outlet of the separatory funnel. As solvent drains from the column, this arrangement automatically refills the space at the top of the column by allowing air to enter through the stem of the separatory funnel. Some microscale columns, such as that shown in Figure 19.8, are equipped with a solvent reservoir that fits onto the top of the column. It functions just like the reservoirs described in this section.

19.11 Monitoring the Column

It is a happy instance when the compounds to be separated are colored. The separation can then be followed visually and the various bands collected separately as they elute from the column. For the majority of organic compounds, however, this lucky circumstance does not exist, and other methods must be used to determine the positions of the bands. The most common method of following a separation of colorless compounds is to collect **fractions** of constant volume in preweighed flasks, to evaporate the solvent from each fraction, and to reweigh the flask plus any residue. A plot of fraction number versus the weight of the residues after evaporation of solvent gives a plot like that in Figure 19.11. Clearly, fractions 2 through 7 (Peak 1) may be combined as a single compound, and so can fractions 8 through 11 (Peak 2) and 12 through 15 (Peak 3). The size of the fractions

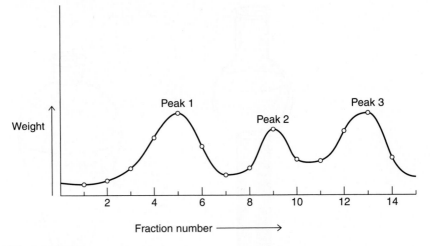

Figure 19.11
Typical elution graph.

collected (1, 10, 100, or 500 mL) depends on the size of the column and the ease of separation.

Another common method of monitoring the column is to mix an inorganic phosphor into the adsorbent used to pack the column. When the column is illuminated with an ultraviolet light, the adsorbent treated in this way fluoresces. However, many solutes have the ability to **quench** the fluorescence of the indicator phosphor. In areas in which solutes are present, the adsorbent does not fluoresce, and a dark band is visible. In this type of column, the separation can also be followed visually.

Thin-layer chromatography is often used to monitor a column. This method is described in Technique 20 (Section 20.10, p. 789). Several sophisticated instrumental and spectroscopic methods, which we shall not detail, can also monitor a chromatographic separation.

19.12 Tailing

When a single solvent is used for elution, an elution curve (weight versus fraction) like that shown as a solid line in Figure 19.12 is often observed. An ideal elution curve is shown by dashed lines. In the nonideal curve, the compound is said to be **tailing.** Tailing can interfere with the beginning of a curve or a peak of a second component and lead to a poor separation. One way to avoid this is to increase the polarity of the solvent constantly while eluting. In this way, at the tail of the peak, where the solvent polarity is increasing, the compound will move slightly faster than at the front and allow the tail to squeeze forward, forming a more nearly ideal band.

19.13 Recovering the Separated Compounds

In recovering each of the separated compounds of a chromatographic separation when they are solids, the various correct fractions are combined and evaporated. If the combined fractions contain sufficient material, they may be purified by recrystallization. If the compounds are liquids, the correct fractions are combined, and the solvent is evaporated. If sufficient material has been collected, liquid samples can be purified by distillation. The combination of chromatography–crystallization or chromatography–distillation usually yields very pure compounds. For microscale applications, the amount of sample collected is too small to allow a purification by crystallization or distillation. The samples that are obtained after the solvent has

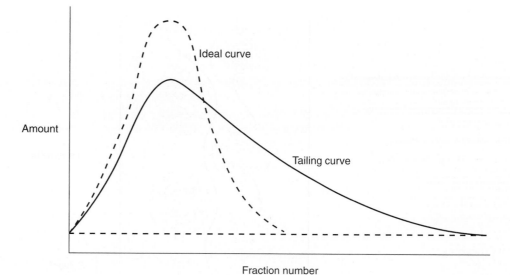

Figure 19.12
Elution curves: One ideal and one that "tails."

been evaporated are considered to be sufficiently pure, and no additional purification is attempted.

19.14 Decolorization by Column Chromatography

A common outcome of organic reactions is the formation of a product that is contaminated by highly colored impurities. Very often these impurities are highly polar, and they have a high molecular weight, as well as being colored. The purification of the desired product requires that these impurities be removed. Section 11.7 of Technique 11 (pp. 663–664) details methods of decolorizing an organic product. In most cases, these methods involve the use of a form of activated charcoal, or Norit.

An alternative, which is applied conveniently in microscale experiments, is to remove the colored impurity by column chromatography. Because of the polarity of the impurities, the colored components are strongly adsorbed on the stationary phase of the column, and the less polar desired product passes through the column and is collected.

Microscale decolorization of a solution on a chromatography column requires that a column be prepared in a Pasteur pipet, using either alumina or silica gel as the adsorbent (Section 19.6). The sample to be decolorized is diluted to the point where crystallization within the column will not take place, and it is then passed through the column in the usual manner. The desired compound is collected as it exits the column, and the excess solvent is removed by evaporation (Technique 7, Section 7.10, p. 611).

19.15 Gel Chromatography

The stationary phase in gel chromatography consists of a cross-linked polymeric material. Molecules are separated according to their *size* by their ability to penetrate a sieve-like structure. Molecules permeate the porous stationary phase as they move down the column. Small molecules penetrate the porous structure more easily than large ones. Thus, the large molecules move through the column faster than the smaller ones and elute first. The separation of molecules by gel chromatography is depicted in Figure 19.13. With adsorption chromatography using materials such as alumina or silica, the order is usually the reverse. Small molecules (of low molecular weight)

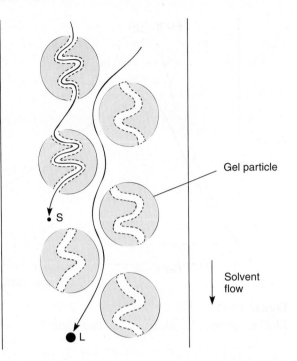

Figure 19.13
*Gel chromatography: Comparison of the paths
of large (L) and small (S) molecules through the
column during the same interval.*

pass through the column *faster* than large molecules (of high molecular weight) because large molecules are more strongly attracted to the polar stationary phase.

Equivalent terms used by chemists for the gel-chromatography technique are **gel filtration** (biochemistry term), **gel-permeation chromatography** (polymer chemistry term), and **molecular sieve chromatography. Size-exclusion chromatography** is a general term for the technique, and it is perhaps the most descriptive term for what occurs on a molecular level.

Sephadex is one of the most popular materials for gel chromatography. It is widely used by biochemists for separating proteins, nucleic acids, enzymes, and carbohydrates. Most often, water or aqueous solutions of buffers are used as the moving phase. Chemically, Sephadex is a polymeric carbohydrate that has been cross-linked. The degree of cross-linking determines the size of the "holes" in the polymer matrix. In addition, the hydroxyl groups on the polymer can adsorb water, which causes the material to swell. As it expands, "holes" are created in the matrix. Several different gels are available from manufacturers, each with its own set of characteristics. For example, a typical Sephadex gel, such as G-75, can separate molecules in the molecular weight (MW) range 3000 to 70,000. Assume for the moment that one has a four-component mixture containing compounds with molecular weights of 10,000, 20,000, 50,000, and 100,000. The 100,000-MW compound would pass through the column first because it cannot penetrate the polymer matrix. The 50,000-, 20,000-, and 10,000-MW compounds penetrate the matrix to various degrees and would be separated. The molecules would elute in the order given (decreasing order of molecular weights). The gel separates on the basis of molecular size and configuration, rather than molecular weight.

Sephadex LH-20 has been developed for nonaqueous solvents. Some of the hydroxyl groups have been alkylated, and thus the material can swell under both aqueous and nonaqueous conditions (it now has "organic" character). This material can be used with several organic solvents, such as alcohol, acetone, methylene chloride, and aromatic hydrocarbons.

Another type of gel is based on a polyacrylamide structure (Bio-Gel P and Poly-Sep AA). A portion of a polyacrylamide chain is shown here:

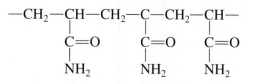

Gels of this type can also be used in water and some polar organic solvents. They tend to be more stable than Sephadex, especially under acidic conditions. Polyacrylamides can be used for many biochemical applications involving macromolecules. For separating synthetic polymers, cross-linked polystyrene beads (copolymer of styrene and divinylbenzene) find common application. Again, the beads are swollen before use. Common organic solvents can be used to elute the polymers. As with other gels, the higher-molecular-weight compounds elute before the lower-molecular-weight compounds.

19.16 Flash Chromatography

One of the drawbacks to column chromatography is that, for large-scale preparative separations, the time required to complete a separation may be very long. Furthermore, the resolution that is possible for a particular experiment tends to deteriorate as the time for the experiment grows longer. This latter effect arises because the bands of compounds that move very slowly through a column tend to "tail."

A technique that can be useful in overcoming these problems has been developed. This technique, called **flash chromatography,** is actually a simple modification of an ordinary column chromatography. In flash chromatography, the adsorbent is packed into a relatively short glass column, and air pressure is used to force the solvent through the adsorbent.

The apparatus used for flash chromatography is shown in Figure 19.14. The glass column is fitted with a Teflon stopcock at the bottom to control the flow rate of solvent. A plug of glass wool is placed in the bottom of the column to act as a support for the adsorbent. A layer of sand may also be added on top of the glass wool. The column is filled with adsorbent using the dry pack method. When the column has been filled, a fitting is attached to the top of the column, and the entire apparatus is connected to a source of high-pressure air or nitrogen. The fitting is designed so that the pressure applied to the top of the column can be adjusted precisely. The source of the high-pressure air is often a specially adapted air pump.

A typical column would use silica gel adsorbent (particle size = 40 to 63 μm) packed to a height of 5 inches in a glass column of 20-mm diameter. The pressure applied to the column would be adjusted to achieve a solvent flow rate such that the solvent level in the column would decrease by about 2 in./min. This system would be appropriate to separate the components of a 250-mg sample.

The high-pressure air forces the solvent through the column of adsorbent at a rate that is much greater than what would be achieved if the

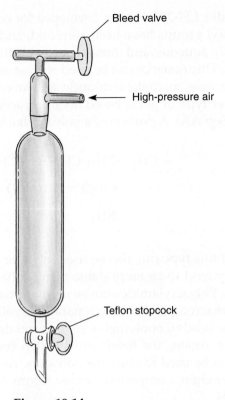

Figure 19.14
Apparatus for flash chromatography.

solvent flowed through the column under the force of gravity. Because the solvent is caused to flow faster, the time required for substances to pass through the column is reduced. By itself, simply applying air pressure to the column might reduce the clarity of the separation because the components of the mixture would not have time to establish themselves into distinctly separate bands. However, in flash chromatography, you can use a much finer adsorbent than would be used in ordinary chromatography. With a much smaller particle size for the adsorbent, the surface area is increased, and the resolution possible thereby improves.

A simple variation on this idea does not use air pressure. Instead, the lower end of the column is inserted into a stopper, which is fitted into the top of a suction flask. Vacuum is applied to the system, and the vacuum acts to draw the solvent through the adsorbent column. The overall effect of this variation is similar to that obtained when air pressure is applied to the top of the column.

REFERENCES

Deyl, Z., Macek, K., and Janák, J. *Liquid Column Chromatography.* Amsterdam: Elsevier, 1975.

Heftmann, E. *Chromatography,* 3rd ed. New York: Van Nostrand Reinhold, 1975.

Jacobson, B. M. "An Inexpensive Way to Do Flash Chromatography." *Journal of Chemical Education,* 65 (May 1988): 459.

Still, W. C., Kahn, M., and Mitra, A. "Rapid Chromatographic Technique for Preparative Separations with Moderate Resolution." *Journal of Organic Chemistry,* 43 (1978): 2923.

PROBLEMS

1. A sample was placed on a chromatography column. Methylene chloride was used as the eluting solvent. No separation of the components in the sample was observed. What must have been happening during this experiment? How would you change the experiment to overcome this problem?

2. You are about to purify an impure sample of naphthalene by column chromatography. What solvent should you use to elute the sample?

3. Consider a sample that is a mixture composed of biphenyl, benzoic acid, and benzyl alcohol. Predict the order of elution of the components in this mixture. Assume that the chromatography uses a silica column, and the solvent system is based on cyclohexane, with an increasing proportion of methylene chloride being added as a function of time.

4. An orange compound was added to the top of a chromatography column. Solvent was added immediately, with the result that the entire volume of solvent in the solvent reservoir turned orange. No separation could be obtained from the chromatography experiment. What went wrong?

5. A yellow compound, dissolved in methylene chloride, is added to a chromatography column. The elution is begun using petroleum ether as the solvent. After 6 L of solvent had passed through the column, the yellow band still had not traveled down the column appreciably. What should be done to make this experiment work better?

6. You have 0.50 g of a mixture that you wish to purify by column chromatography. How much adsorbent should you use to pack the column? Estimate the appropriate column diameter and height.

7. In a particular sample, you wish to collect the component with the *highest* molecular weight as the *first* fraction. What chromatographic technique should you use?

8. A colored band shows an excessive amount of tailing as it passes through the column. What can you do to rectify this problem?

9. How would you monitor the progress of a column chromatography when the sample is colorless? Describe at least two methods.

---

20 TECHNIQUE 20

Thin-Layer Chromatography

Thin-layer chromatography (TLC) is a very important technique for the rapid separation and qualitative analysis of small amounts of material. It is ideally suited for the analysis of mixtures and reaction products in microscale experiments. The technique is closely related to column chromatography. In fact, TLC can be considered simply column chromatography in reverse, with the solvent ascending the adsorbent, rather than descending. Because of this close relationship to column chromatography, and because the principles governing the two techniques are similar, Technique 19, on column chromatography, should be read first.

20.1 Principles of Thin-Layer Chromatography

Like column chromatography, TLC is a solid–liquid partitioning technique. However, the moving liquid phase is not allowed to percolate down the adsorbent; it is caused to ascend a thin layer of adsorbent coated onto a backing support. The most typical backing is a glass plate, but other materials are also used. A thin layer of the adsorbent is spread onto the plate and allowed to dry. A coated and dried plate of glass is called a **thin-layer plate** or a **thin-layer slide.** (The reference to *slide* comes about because microscope slides are often used to prepare small thin-layer plates.) When a thin-layer plate is placed upright in a vessel that contains a shallow layer of solvent, the solvent ascends the layer of adsorbent on the plate by capillary action.

In TLC, the sample is applied to the plate before the solvent is allowed to ascend the adsorbent layer. The sample is usually applied as a small spot near the base of the plate; this technique is often referred to as **spotting.** The plate is spotted by repeated applications of a sample solution from a small capillary pipet. When the filled pipet touches the plate, capillary action delivers its contents to the plate, and a small spot is formed.

As the solvent ascends the plate, the sample is partitioned between the moving liquid phase and the stationary solid phase. During this process, you are **developing,** or **running,** the thin-layer plate. In development, the various components in the applied mixture are separated. The separation is based on the many equilibrations the solutes experience between the moving and the stationary phases. (The nature of these equilibrations was thoroughly discussed in Technique 19, Sections 19.2 and 19.3, pp. 757–759.) As in column chromatography, the least polar substances advance faster than the most polar substances. A separation results from the differences in the rates at which the individual components of the mixture advance upward on the plate. When many substances are present in a mixture, each has its own characteristic solubility and adsorptivity properties, depending on the functional groups in its structure. In general, the stationary phase is strongly polar and strongly binds polar substances. The moving liquid phase is usually less polar than the adsorbent and most easily dissolves substances that are less polar or even nonpolar. Thus, substances that are the most polar travel slowly upward, or not at all, and nonpolar substances travel more rapidly if the solvent is sufficiently nonpolar.

When the thin-layer plate has been developed, it is removed from the developing tank and allowed to dry until it is free of solvent. If the mixture that was originally spotted on the plate was separated, there will be a vertical series of spots on the plate. Each spot corresponds to a separate component or compound from the original mixture. If the components of the mixture are colored substances, the various spots will be clearly visible after development. More often, however, the "spots" will not be visible because they correspond to colorless substances. If spots are not apparent, they can be made visible only if a **visualization method** is used. Often, spots can be seen when the thin-layer plate is held under ultraviolet light; the ultraviolet lamp is a common visualization method. Also common is the use of iodine vapor. The plates are placed in a chamber containing iodine crystals and left to stand for a short time. The iodine reacts with the various compounds adsorbed on the plate to give colored complexes that are clearly visible. Because iodine often changes the compounds by reaction, the components of the mixture cannot be recovered from the plate when the iodine method is used. (Other methods of visualization are discussed in Section 20.7.)

20.2 Commercially Prepared TLC Plates

The most convenient type of TLC plate is prepared commercially and sold in a ready-to-use form. Many manufacturers supply glass plates precoated with a durable layer of silica gel or alumina. More conveniently, plates are also available that have either a flexible plastic backing or an aluminum backing. The most common types of commercial TLC plates are composed of plastic sheets that are coated with silica gel and polyacrylic acid, which serves as a binder. A fluorescent indicator may be mixed with the silica gel. The indicator renders the spots due to the presence of compounds in the sample visible under ultraviolet light (see Section 20.7). Although these plates are relatively expensive when compared with plates prepared in the laboratory, they are far more convenient to use, and they provide more consistent results. The plates are manufactured quite uniformly. Because the plastic backing is flexible, they have the additional advantage that the coating does not flake off the plates easily. The plastic sheets (usually 8 in. by 8 in. square) can also be cut with a pair of scissors or paper cutter to whatever size may be required.

If the package of commercially prepared TLC plates has been opened previously, or if the plates have not been purchased recently, they should be dried before use. Dry the plates by placing them in an oven at 100°C for 30 minutes, and store them in a desiccator until they are to be used.

20.3 Preparation of Thin-Layer Slides and Plates

Commercially prepared plates (Section 20.2) are the most convenient to use, but if you must prepare your own slides or plates, this section gives the directions for preparing them. The two adsorbent materials used most often for TLC are alumina G (aluminum oxide) and silica gel G (silicic acid). The G designation stands for gypsum (calcium sulfate). Calcined gypsum, $CaSO_4 \cdot \frac{1}{2}H_2O$, is better known as plaster of Paris. When exposed to water or moisture, gypsum sets in a rigid mass, $CaSO_4 \cdot 2H_2O$, which binds the adsorbent together and to the glass plates used as a backing support. In the adsorbents used for TLC, about 10–13% by weight of gypsum is added as a binder. The adsorbent materials are otherwise like those used in column chromatography; the adsorbents used in column chromatography have a larger particle size, however. The material for thin-layer work is a fine powder. The small particle size, along with the added gypsum, makes it impossible to use silica gel G or alumina G for column work. In a column, these adsorbents generally set so rigidly that solvent virtually stops flowing through the column.

A. Microscope Slide TLC Plates

For qualitative work such as identifying the number of components in a mixture or trying to establish that two compounds are identical, small TLC plates made from microscope slides are especially convenient. Coated microscope slides are easily made by dipping the slides into a container holding a slurry of the adsorbent material. Although numerous solvents can be used to prepare a slurry, methylene chloride is probably the best choice. It has the two advantages of low boiling point (40°C) and inability to cause the adsorbent to set or form lumps. The low boiling point means that it is not necessary to dry the coated slides in an oven. Its inability to cause the gypsum binder to set means that slurries made with it are stable for several days. The layer of adsorbent formed is fragile, however, and must be treated carefully. For this reason, some people prefer to add a small amount of methanol to the methylene chloride to enable the gypsum to set more firmly. The methanol solvates the calcium sulfate much as water does. More durable plates can be made by dipping plates into a slurry prepared from water. These plates must be oven dried before use. Also, a slurry prepared from water must be used soon after its preparation. If it is not, it begins to set and form lumps. Thus, an aqueous slurry must be prepared immediately before use; it cannot be used after it has stood for any length of time. For microscope slides, a slurry of silica gel G in methylene chloride is not only convenient but also adequate for most purposes.

Preparing the Slurry. The slurry is most conveniently prepared in a 4-oz wide-mouthed screw-cap jar. About 3 mL of methylene chloride are required for each gram of silica gel G. For a smooth slurry without lumps, the silica gel should be added to the solvent while the mixture is being either stirred or swirled. Adding solvent to the adsorbent usually causes lumps to form in the mixture. When the addition is complete, the cap should be placed on the jar tightly and the jar shaken vigorously to ensure thorough mixing. The slurry may be stored in the tightly capped jar until it is to be used. More methylene chloride may have to be added to replace evaporation losses.

CAUTION ⚠

Avoid breathing silica dust or methylene chloride, prepare and use the slurry in a hood, and avoid getting methylene chloride or the slurry mixture on your skin.

Preparing the Slides. If new microscope slides are available, you can use them without any special treatment. However, it is more economical to reuse or recycle microscope slides. Wash the slides with soap and water, rinse with water, and then rinse with 50% aqueous methanol. Allow the plates to dry thoroughly on paper towels. They should be handled by the edges because fingerprints on the plate surface will make it difficult for the adsorbent to bind to the glass.

Coating the Slides. The slides are coated with adsorbent by dipping them into the container of slurry. You can coat two slides simultaneously by sandwiching them together before dipping them in the slurry.

NOTE: Perform the coating operation under a hood.

Shake the slurry vigorously just before dipping the slides. Because the slurry settles on standing, it should be mixed in this way before each set of slides is dipped. The depth of the slurry in the jar should be about 3 inches, and the plates should be dipped into the slurry until only about 0.25 inches at the top remains uncoated. The dipping operation should be performed smoothly. The plates may be held at the top (see Fig. 20.1), where they will not be coated. They are dipped into the slurry and withdrawn with a slow and steady motion. The dipping operation takes about 2 seconds. Some practice may be required to get the correct timing. After dipping, replace the cap on the jar and hold the plates for a minute until most of the solvent has evaporated. Separate the plates and place them on paper towels to complete the drying. The plates should have an even coating; there should be no streaks and no thin spots where glass shows through the adsorbent. The plates should not have a thick and lumpy coating.

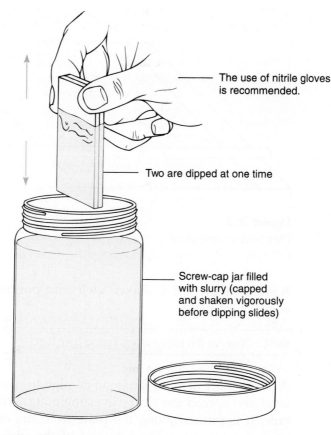

The use of nitrile gloves is recommended.

Two are dipped at one time

Screw-cap jar filled with slurry (capped and shaken vigorously before dipping slides)

Figure 20.1
Dipping slides to coat them.

Two conditions cause thin and streaked plates. First, the slurry may not have been mixed thoroughly before the dipping operation; the adsorbent might then have settled to the bottom of the jar, and the thin slurry at the top would not have coated the slides properly. Second, the slurry simply may not have been thick enough; more silica gel G must then be added to the slurry until the consistency is correct. If the slurry is too thick, the coating on the plates will be thick, uneven, and lumpy. To correct this, dilute the slurry with enough solvent to achieve the proper consistency.

Plates with an unsatisfactory coating may be wiped clean with a paper towel and redipped. Take care to handle the plates only from the top or by the sides to avoid getting fingerprints on the glass surface.

B. Larger Thin-Layer Plates

For separations involving large amounts of material, or for difficult separations, it may be necessary to use larger thin-layer plates. Plates with dimensions up to 20–25 cm² are common. With larger plates, it is desirable to have a more durable coating, and a water slurry of the adsorbent should be used to prepare them. If silica gel is used, the slurry should be prepared in the ratio of about 1 g silica gel G to each 2 mL of water. The glass plate used for the thin-layer plate should be washed, dried, and placed on a sheet of newspaper. Place two strips of masking tape along two edges of the plate. Use more than one layer of masking tape if a thicker coating is desired on the plate.

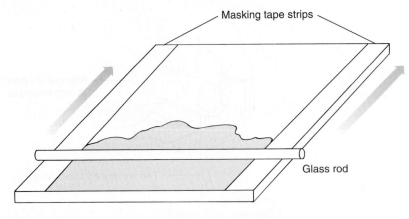

Masking tape strips

Glass rod

Figure 20.2
Preparing a large plate.

A slurry is prepared, shaken well, and poured along one of the untaped edges of the plate.

NOTE: Observe the precautions stated on p. 780.

A heavy piece of glass rod, long enough to span the taped edges, is used to level and spread the slurry over the plate. While the rod is resting on the tape, it is pushed along the plate from the end at which the slurry was poured toward the opposite end of the plate. This is illustrated in Figure 20.2. After the slurry is spread, the masking tape strips are removed, and the plates are dried in a 110°C oven for about 1 hour. Plates of 10–25 cm² are easily prepared by this method. Larger plates present more difficulties. Many laboratories have a commercially manufactured spreading machine that makes the entire operation simpler.

20.4 Sample Application: Spotting the Plates

Preparing a Micropipet

To apply the sample that is to be separated to the thin-layer plate, one uses a micropipet. A micropipet is easily made from a short length of thin-walled capillary tubing like that used for melting-point determinations, but open at both ends. The capillary tubing is heated at its midpoint with a microburner and rotated until it is soft. When the tubing is soft, the heated portion of the tubing is drawn out until a constricted portion of tubing 4–5 cm long is formed. After cooling, the constricted portion of tubing is scored at its center with a file or scorer and broken. The two halves yield two capillary micropipets. Figure 20.3 shows how to make such pipets.

Spotting the Plate

To apply a sample to the plate, begin by placing about 1 mg of a solid test substance, or one drop of a liquid test substance, in a small container such as a watch glass or a test tube. Dissolve the sample in a few drops of a volatile solvent. Acetone or methylene chloride is usually a suitable solvent. If a solution is to be tested, it can often be used directly. The small capillary pipet, prepared as described, is filled by dipping the pulled end into the solution to be examined. Capillary action fills the pipet. Empty the pipet by touching it *lightly* to the thin-layer plate at a point about 1 cm from the

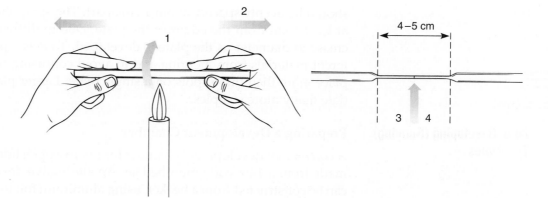

① Rotate in flame until soft.
② Remove from flame and pull.

③ Score lightly in center of pulled section.
④ Break in half to give two pipets.

Figure 20.3
Construction of two capillary micropipets.

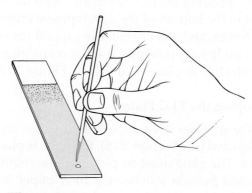

Figure 20.4
Spotting the plate with a drawn capillary pipet.

bottom (Fig. 20.4). The spot must be high enough that it does not dissolve in the developing solvent. It is important to touch the plate very lightly and not to gouge a hole in the adsorbent. When the pipet touches the plate, the solution is transferred to the plate as a small spot. The pipet should be touched to the plate *very briefly* and then removed. If the pipet is held to the plate, its entire contents will be delivered to the plate. Only a small amount of material is needed. It is often helpful to blow gently on the plate as the sample is applied. This helps to keep the spot small by evaporating the solvent before it can spread out on the plate. The smaller the spot formed, the better the separation obtainable. If needed, additional material can be applied to the plate by repeating the spotting procedure. You should repeat the procedure with several small amounts, rather than apply one large amount. The solvent should be allowed to evaporate between applications. If the spot is not small (about 2 mm in diameter), a new plate should be prepared. The capillary pipet may be used several times if it is rinsed between uses. It is repeatedly dipped into a small portion of solvent to rinse it and touched to a paper towel to empty it.

As many as three spots may be applied to a microscope-slide TLC plate. Each spot should be about 1 cm from the bottom of the plate, and all spots should be evenly spaced, about 1 cm apart. The spots should be positioned at least 1 cm from the edges of the plate. Due to diffusion, spots often increase in diameter as the plate is developed. To keep spots containing different materials from merging and to avoid confusing the samples, do not place more than three spots on a single plate. Larger plates can accommodate many more samples.

20.5 Developing (Running) TLC Plates

Preparing a Development Chamber

A convenient development chamber for microscope-slide TLC plates can be made from a 4-oz wide-mouthed jar. An alternative development chamber can be constructed from a beaker, using aluminum foil to cover the opening. The inside of the jar or beaker should be lined with a piece of filter paper, cut so that it does not quite extend around the inside of the jar. A small vertical opening (2–3 cm) should be left in the filter paper for observing the development. Before development, the filter paper inside the jar or beaker should be thoroughly moistened with the development solvent. The solvent-saturated liner helps to keep the chamber saturated with solvent vapors, thereby speeding the development. Once the liner is saturated, the level of solvent in the bottom of the development chamber is adjusted to a depth of about 5 mm, and the chamber is capped (or covered with aluminum foil) and set aside until it is to be used. A correctly prepared development chamber (with slide in place) is shown in Figure 20.5.

Developing the TLC Plate

Once the spot has been applied to the thin-layer plate and the solvent has been selected (see Section 20.5), the plate is placed in the chamber for development. The plate must be placed in the chamber carefully so that none of the coated portion touches the filter paper liner. In addition, the solvent level in the bottom of the chamber must not be above the spot that was applied to the plate, or the spotted material will dissolve in the pool of solvent instead of undergoing chromatography. Once the plate has been placed correctly, replace the cap on the developing chamber and wait for the solvent to

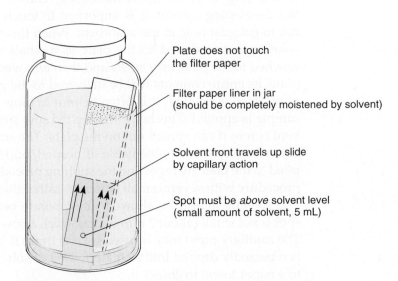

Figure 20.5
Development chamber with thin-layer plate undergoing development.

advance up the plate by capillary action. This generally occurs rapidly, and you should watch carefully. As the solvent rises, the plate becomes visibly moist. When the solvent has advanced to within 5 mm of the end of the coated surface, the plate should be removed, and the position of the solvent front should be marked *immediately* by scoring the plate along the solvent line with a *pencil.* The solvent front must not be allowed to travel beyond the end of the coated surface. The plate should be removed before this happens. The solvent will not actually advance beyond the end of the plate, but spots allowed to stand on a completely moistened plate on which the solvent is not in motion expand by diffusion. Once the plate has dried, any visible spots should be outlined on the plate with a pencil. If no spots are apparent, a visualization method (Section 20.7) may be needed.

20.6 Choosing a Solvent for Development

The development solvent used depends on the materials to be separated. You may have to try several solvents before a satisfactory separation is achieved. Because microscope slides can be prepared and developed rapidly, an empirical choice is usually not hard to make. A solvent that causes all the spotted material to move with the solvent front is too polar. One that does not cause any of the material in the spot to move is not polar enough. As a guide to the relative polarity of solvents, consult Table 19.2 in Technique 19 (p. 761). Figure 20.6 shows three TLC plates run with different solvents. As can be seen, if the solvent is too polar, the spots tend to run near the top of the plate, and the separation is poor. If the solvent is not sufficiently polar, the spots do not travel very far; again, the separation is poor. The ideal solvent choice allows the spots to travel well up the plate but allows for a clean separation.

Methylene chloride and toluene are solvents of intermediate polarity and good choices for a wide variety of functional groups to be separated. For hydrocarbon materials, good first choices are hexane, petroleum ether (ligroin), or toluene. Hexane or petroleum ether with varying proportions of toluene or ether gives solvent mixtures of moderate polarity that are useful

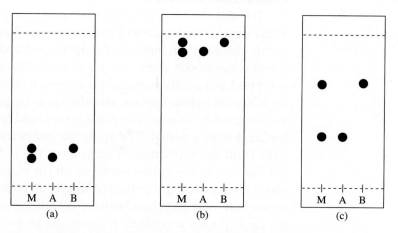

Figure 20.6
*TLC plates showing the effects of solvent polarity on a separation. The three spots on each plate are indicated by **M** for a two-component mixture, **A** for a standard sample of substance A, and **B** for a standard sample of substance B. **(a)** Solvent of low polarity: poor separation. **(b)** Solvent of high polarity: poor separation. **(c)** Solvent of intermediate polarity: clean separation.*

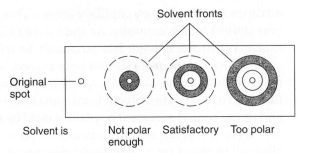

Figure 20.7
Concentric ring method of testing solvents.

for many common functional groups. Polar materials may require ethyl acetate, acetone, or methanol.

A rapid way to determine a good solvent is to apply several sample spots to a single plate. The spots should be placed a minimum of 1 cm apart. A capillary pipet is filled with a solvent and gently touched to one of the spots. The solvent expands outward in a circle. The solvent front should be marked with a pencil. A different solvent is applied to each spot. As the solvents expand outward, the spots expand as concentric rings. From the appearance of the rings, you can judge approximately the suitability of the solvent. Several types of behavior experienced with this method of testing are shown in Figure 20.7.

20.7 Visualization Methods

It is fortunate when the compounds separated by TLC are colored because the separation can be followed visually. More often than not, however, the compounds are colorless. In that case, the separated materials must be made visible by some reagent or some method that makes the separated compounds visible. Reagents that give rise to colored spots are called **visualization reagents.** Methods of viewing that make the spots apparent are **visualization methods.**

The visualization reagent used most often is iodine. Iodine reacts with many organic materials to form complexes that are either brown or yellow. In this visualization method, the developed and dried TLC plate is placed in a 4-oz wide-mouth screw-cap jar along with a few crystals of iodine. The jar is capped and gently warmed on a steam bath or a hot plate at low heat. The jar fills with iodine vapors, and the spots begin to appear. When the spots are sufficiently intense, the plate is removed from the jar and the spots are outlined with a pencil. The spots are not permanent. Their appearance results from the formation of complexes the iodine makes with the organic substances. As the iodine sublimes off the plate, the spots fade. Hence, they should be marked immediately. Nearly all compounds except saturated hydrocarbons and alkyl halides form complexes with iodine. The intensities of the spots do not accurately indicate the amount of material present, except in the crudest way.

The second most common method of visualization is by an ultraviolet (UV) lamp. Under UV light, compounds often look like bright spots on the plate. This often suggests the structure of the compound: Certain types of compounds shine very brightly under UV light because they fluoresce.

Another method that provides good results involves adding a fluorescent indicator to the adsorbent used to coat the plates. A mixture of zinc and cadmium sulfides is often used. When treated in this way and held under UV light, the entire plate fluoresces. However, dark spots appear on the plate where the separated compounds are seen to quench this fluorescence.

In addition to the preceding methods, several chemical methods are available that either destroy or permanently alter the separated compounds through reaction. Many of these methods are specific for particular functional groups.

Alkyl halides can be visualized if a dilute solution of silver nitrate is sprayed on the plates. Silver halides are formed. These halides decompose if exposed to light, giving rise to dark spots (free silver) on the TLC plate.

Most organic functional groups can be made visible if they are charred with sulfuric acid. Concentrated sulfuric acid is sprayed on the plate, which is then heated in an oven at 110°C to complete the charring. Permanent spots are thus created.

Colored compounds can be prepared from colorless compounds by making derivatives before spotting them on the plate. An example of this is the preparation of 2,4-dinitrophenylhydrazones from aldehydes and ketones to produce yellow and orange compounds. You may also spray the 2,4-dinitrophenylhydrazine reagent on the plate after the ketones or aldehydes have separated. Red and yellow spots form where the compounds are located. Other examples of this method are using ferric chloride for visualizing phenols and using bromocresol green for detecting carboxylic acids. Chromium trioxide, potassium dichromate, and potassium permanganate can be used for visualizing compounds that are easily oxidized. *p*-Dimethylaminobenzaldehyde easily detects amines. Ninhydrin reacts with amino acids to make them visible. Numerous other methods and reagents available from various supply outlets are specific for certain types of functional groups. These visualize only the class of compounds of interest.

20.8 Preparative Plates

If you use large plates (Section 20.3B), materials can be separated, and the separated components can be recovered individually from the plates. Plates used in this way are called **preparative plates.** For preparative plates, a thick layer of adsorbent is generally used. Instead of being applied as a spot or a series of spots, the mixture to be separated is applied as a line of material about 1 cm from the bottom of the plate. As the plate is developed, the separated materials form bands. After development, you can observe the separated bands, usually by UV light, and outline the zones in pencil. If the method of visualization is destructive, most of the plate is covered with paper to protect it, and the reagent is applied only at the extreme edge of the plate.

Once the zones have been identified, the adsorbent in those bands is scraped from the plate and extracted with solvent to remove the adsorbed material. Filtration removes the adsorbent, and evaporation of the solvent gives the recovered component from the mixture.

20.9 The R_f Value

Thin-layer chromatography conditions include

1. Solvent system
2. Adsorbent
3. Thickness of the adsorbent layer
4. Relative amount of material spotted

Under an established set of such conditions, a given compound always travels a fixed distance relative to the distance the solvent front travels. This ratio of the distance the compound travels to the distance the solvent travels is called the R_f **value.** The symbol R_f stands for "retardation factor," or "ratio-to-front," and it is expressed as a decimal fraction:

$$R_f = \frac{\text{distance traveled by substance}}{\text{distance traveled by solvent front}}$$

When the conditions of measurement are completely specified, the R_f value is constant for any given compound, and it corresponds to a physical property of that compound.

The R_f value can be used to identify an unknown compound, but like any other identification based on a single piece of data, the R_f value is best confirmed with some additional data. Many compounds can have the same R_f value, just as many compounds have the same melting point.

It is not always possible, in measuring an R_f value, to duplicate exactly the conditions of measurement another researcher has used. Therefore, R_f values tend to be of more use to a single researcher in one laboratory than they are to researchers in different laboratories. The only exception to this is when two researchers use TLC plates from the same source, as in commercial plates, or know the *exact* details of how the plates were prepared. Nevertheless, the R_f value can be a useful guide. If exact values cannot be relied on, the relative values can provide another researcher with useful information about what to expect. Anyone using published R_f values will find it a good idea to check them by comparing them with standard substances whose identity and R_f values are known.

To calculate the R_f value for a given compound, measure the distance that the compound has traveled from the point at which it was originally spotted. For spots that are not too large, measure to the center of the migrated spot. For large spots, the measurement should be repeated on a new plate, using less material. For spots that show tailing, the measurement is made to the "center of gravity" of the spot. This first distance measurement is then divided by the distance the solvent front has traveled from the same original spot. A sample calculation of the R_f values of two compounds is illustrated in Figure 20.8.

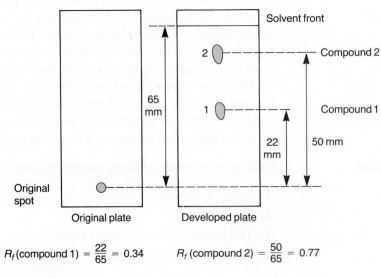

$$R_f(\text{compound 1}) = \frac{22}{65} = 0.34 \qquad R_f(\text{compound 2}) = \frac{50}{65} = 0.77$$

Figure 20.8
Sample calculation of R_f values.

20.10 Thin-Layer Chromatography Applied in Organic Chemistry

Thin-layer chromatography has several important uses in organic chemistry. It can be used in the following applications:

1. To establish that two compounds are identical
2. To determine the number of components in a mixture
3. To determine the appropriate solvent for a column-chromatographic separation
4. To monitor a column-chromatographic separation
5. To check the effectiveness of a separation achieved on a column, by crystallization or by extraction
6. To monitor the progress of a reaction

In all these applications, TLC has the advantage that only small amounts of material are necessary. Material is not wasted. With many of the visualization methods, less than a tenth of a microgram (10^{-7} g) of material can be detected. On the other hand, samples as large as a milligram may be used. With preparative plates that are large (about 9 in. on a side) and have a relatively thick coating of adsorbent (>500 μm), it is often possible to separate from 0.2 to 0.5 g of material at one time. The main disadvantage of TLC is that volatile materials cannot be used, because they evaporate from the plates.

Thin-layer chromatography can establish that two compounds suspected to be identical are in fact identical. Simply spot both compounds side by side on a single plate and develop the plate. If both compounds travel the same distance on the plate (have the same R_f value), they are probably identical. If the spot positions are not the same, the compounds are definitely not identical. It is important to spot compounds *on the same plate.* This is especially important with hand-dipped microscope slides. Because they vary widely from plate to plate, no two plates have exactly the same thickness of adsorbent. If you use commercial plates, this precaution is not necessary, although it is nevertheless a good idea.

Thin-layer chromatography can establish whether a sample is a single substance or a mixture. A single substance gives a single spot no matter what solvent is used to develop the plate. On the other hand, the number of components in a mixture can be established by trying various solvents on a mixture. A word of caution should be given. It may be difficult, in dealing with compounds of very similar properties, isomers for example, to find a solvent that will separate the mixture. Inability to achieve a separation is not absolute proof that a sample is a single pure substance. Many compounds can be separated only by *multiple developments* of the TLC slide with a fairly nonpolar solvent. In this method, you remove the plate after the first development and allow it to dry. After being dried, it is placed in the chamber again and developed once more. This effectively doubles the length of the slide. At times, several developments may be necessary.

When a mixture is to be separated, you can use TLC to choose the best solvent to separate it if column chromatography is contemplated. You can try various solvents on a plate coated with the same adsorbent as will be used in the column. The solvent that resolves the components best will probably work well on the column. These small-scale experiments are quick, use very little material, and save time that would be wasted by attempting to separate the entire mixture on the column. Similarly, TLC plates can **monitor** a column. A hypothetical situation is shown in Figure 20.9.

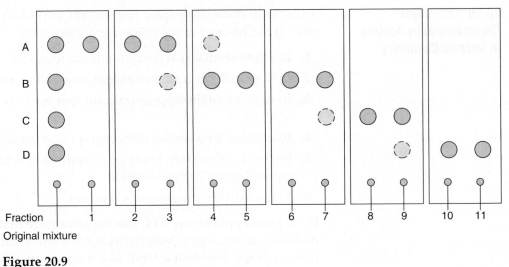

Figure 20.9
Monitoring a column.

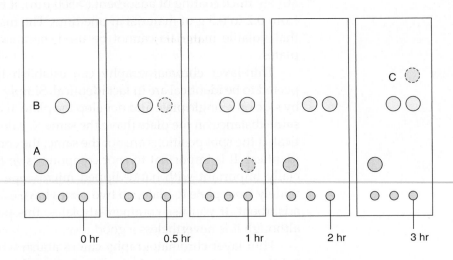

Figure 20.10
Monitoring a reaction.

A solvent was found that would separate the mixture into four components (A–D). A column was run using this solvent, and 11 fractions of 15 mL each were collected. Thin-layer analysis of the various fractions showed that Fractions 1–3 contained Component A; Fractions 4–7, Component B; Fractions 8–9, Component C; and Fractions 10–11, Component D. A small amount of cross-contamination was observed in Fractions 3, 4, 7, and 9.

In another TLC example, a researcher found a product from a reaction to be a mixture. It gave two spots, A and B, on a TLC slide. After the product was crystallized, the crystals were found by TLC to be pure A, whereas the mother liquor was found to have a mixture of A and B. The crystallization was judged to have purified A satisfactorily.

Finally, it is often possible to monitor the progress of a reaction by TLC. At various points during a reaction, samples of the reaction mixture are taken and subjected to TLC analysis. An example is given in Figure 20.10. In this case, the desired reaction was the conversion of A to B. At the beginning of the reaction (0 hr), a TLC slide was prepared that was spotted with pure A, pure B, and the reaction mixture. Similar slides were prepared at 0.5, 1, 2, and 3 hours after the start of the reaction. The slides showed that the reaction was complete in 2 hours. When the reaction was run longer than 2 hours, a new compound, side product C, began to appear. Thus, the optimum reaction time was judged to be 2 hours.

20.11 Paper Chromatography

Paper chromatography is often considered to be related to thin-layer chromatography. The experimental techniques are somewhat like those of TLC, but the principles are more closely related to those of extraction. Paper chromatography is actually a liquid–liquid partitioning technique, rather than a solid–liquid technique. For paper chromatography, a spot is placed near the bottom of a piece of high-grade filter paper (Whatman No. 1 is often used). Then the paper is placed in a developing chamber. The development solvent ascends the paper by capillary action and moves the components of the spotted mixture upward at differing rates. Although paper consists mainly of pure cellulose, the cellulose itself does not function as the stationary phase. Rather, the cellulose absorbs water from the atmosphere, especially from an atmosphere saturated with water vapor. Cellulose can absorb up to about 22% of water. It is this water adsorbed on the cellulose that functions as the stationary phase. To ensure that the cellulose is kept saturated with water, many development solvents used in paper chromatography contain water as a component. As the solvent ascends the paper, the compounds are partitioned between the stationary water phase and the moving solvent. Because the water phase is stationary, the components in a mixture that are most highly water-soluble, or those that have the greatest hydrogen-bonding capacity, are the ones that are held back and move most slowly. Paper chromatography applies mostly to highly polar compounds or to those that are polyfunctional. The most common use of paper chromatography is for sugars, amino acids, and natural pigments. Because filter paper is manufactured consistently, R_f values can often be relied on in paper chromatographic work. However, R_f values are customarily measured from the leading edge (top) of the spot—not from its center, as is customary in TLC.

PROBLEMS

1. A student spots an unknown sample on a TLC plate and develops it in dichloromethane solvent. Only one spot, for which the R_f value is 0.95, is observed. Does this indicate that the unknown material is a pure compound? What can be done to verify the purity of the sample?

2. You and another student were each given an unknown compound. Both samples contained colorless material. You each used the same brand of commercially prepared TLC plate and developed the plates using the same solvent. Each of you obtained a single spot of $R_f = 0.75$. Were the two samples necessarily the same substance? How could you prove unambiguously that they were identical using TLC?

3. Consider a sample that is a mixture composed of biphenyl, benzoic acid, and benzyl alcohol. The sample is spotted on a TLC plate and developed in a dichloromethane–cyclohexane solvent mixture. Predict the *relative* R_f values for the three components in the sample. *Hint:* See Table 19.3.

4. Calculate the R_f value of a spot that travels 5.7 cm, with a solvent front that travels 13 cm.

5. A student spots an unknown sample on a TLC plate and develops it in pentane solvent. Only one spot, for which the R_f value is 0.05, is observed. Is the unknown material a pure compound? What can be done to verify the purity of the sample?

6. A *colorless* unknown substance is spotted on a TLC plate and developed in the correct solvent. The spots do not appear when visualization with a UV lamp or iodine vapors is attempted. What could you do in order to visualize the spots if the compound is

   a. An alkyl halide

   b. A ketone

   c. An amino acid

   d. A sugar

---

21   TECHNIQUE 21

High-Performance Liquid Chromatography (HPLC)

The separation that can be achieved is greater if the column packing used in column chromatography is made more dense by using an adsorbent that has a smaller particle size. The solute molecules encounter a much larger surface area on which they can be adsorbed as they pass through the column packing. At the same time, the solvent spaces between the particles are reduced in size. As a result of this tight packing, equilibrium between the liquid and solid phases can be established rapidly with a fairly short column, and the degree of separation is markedly improved. The disadvantage of making the column packing more dense is that the solvent flow rate becomes very slow or even stops. Gravity is not strong enough to pull the solvent through a tightly packed column.

A recently developed technique can be applied to obtain much better separations with tightly packed columns. A pump forces the solvent through the column packing. As a result, solvent flow rate is increased and the advantage of better separation is retained. This technique, called **high-performance liquid chromatography (HPLC),** is becoming widely applied to problems where separations by ordinary column chromatography are unsatisfactory. Because the pump often provides pressures in excess of 1000 pounds per square inch (psi), this method is also known as **high-pressure liquid chromatography.** High pressures are not required, however, and satisfactory separations can be achieved with pressures as low as 100 psi.

The basic design of an HPLC instrument is shown in Figure 21.1. The instrument contains the following essential components:

1. Solvent reservoir

2. Solvent filter and degasser

3. Pump

4. Pressure gauge

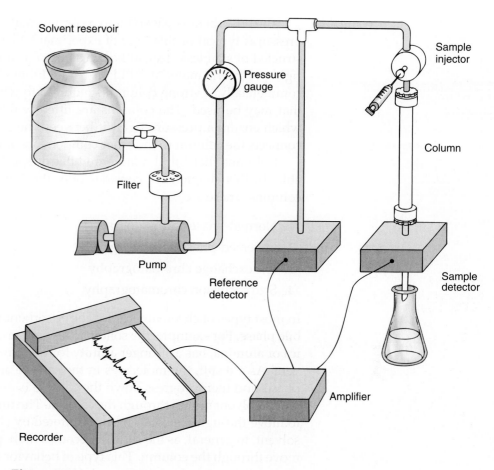

Figure 21.1
Schematic diagram of a high-performance liquid chromatograph.

5. Sample injection system
6. Column
7. Detector
8. Amplifier and electronic controls
9. Chart recorder

There may be other variations on this simple design. Some instruments have heated ovens in order to maintain the column at a specified temperature, fraction collectors, and microprocessor-controlled data-handling systems. Additional filters for the solvent and sample may also be included. You may find it interesting to compare this schematic diagram with that shown in Technique 22 (Fig. 22.2, p. 799) for a gas-chromatography instrument. Many of the essential components are common to both types of instruments.

21.1 Adsorbents and Columns

The most important factor to consider when choosing a set of experimental conditions is the nature of the material packed into the column. You must also consider the size of the column that will be selected. The chromatography column is generally packed with silica or alumina adsorbents. Unlike column chromatography, however, the adsorbents used for HPLC have a much smaller particle size. Typically, particle size ranges from 5 to 20 μm in diameter for HPLC; it is on the order of 100 μm for column chromatography.

The adsorbent is packed into a column that can withstand the elevated pressures typical of this type of experiment. Generally, the column is constructed of stainless steel, although some columns that are constructed of a rigid polymeric material ("PEEK"—Poly Ether Ether Ketone) are available commercially. A strong column is required to withstand the high pressures that may be used. The columns are fitted with stainless steel connectors, which ensure a pressure-tight fit between the column and the tubing that connects the column to the other components of the instrument.

Columns that fulfill a large number of specialized purposes are available. In this chapter, we consider only the four most important types of columns. These are

1. Normal-phase chromatography
2. Reversed-phase chromatography
3. Ion-exchange chromatography
4. Size exclusion chromatography

In most types of chromatography, the adsorbent is more polar than the mobile phase. For example, the solid packing material, which may be either silica or alumina, has a stronger affinity for polar molecules than does the solvent. As a result, the molecules in the sample adhere strongly to the solid phase, and their progress down the column is much slower than the rate at which solvent moves through the column. The time required for a substance to move through the column can be altered by changing the polarity of the solvent. In general, as the solvent becomes more polar, the faster substances move through the column. This type of behavior is known as **normal-phase chromatography.** In HPLC, you inject a sample onto a normal-phase column and elute it by varying the polarity of the solvent, much as you do with ordinary column chromatography. Disadvantages of normal phase chromatography are that retention times tend to be long, and bands have a tendency to "tail."

These disadvantages can be ameliorated by selecting a column in which the solid support is *less polar* than the moving solvent phase. This type of chromatography is known as **reversed-phase chromatography.** In this type of chromatography, the silica column packing is treated with alkylating agents. As a result, nonpolar alkyl groups are bonded to the silica surface, making the adsorbent nonpolar. The alkylating agents that are used most commonly can attach methyl ($—CH_3$), octyl ($—C_8H_{17}$), or octadecyl ($—C_{18}H_{37}$) groups to the silica surface. The latter variation, where an 18-carbon chain is attached to the silica, is the most popular. This type of column is known as a **C-18 column.** The bonded alkyl groups have an effect similar to that which would be produced by an extremely thin organic solvent layer coating the surface of the silica particles. The interactions that take place between the substances dissolved in the solvent and the stationary phase thus become more like those observed in a liquid–liquid extraction. The solute particles distribute themselves between the two "solvents"—that is, between the moving solvent and the organic coating on the silica. The longer the chains of the alkyl groups that are bonded to the silica, the more effective the alkyl groups are as they interact with solute molecules.

Reversed-phase chromatography is widely used because the rate at which solute molecules exchange between moving phase and stationary phase is rapid, which means that substances pass through the column relatively quickly. Furthermore, problems arising from the "tailing" of peaks are reduced. A disadvantage of this type of column, however, is that the chemically bonded solid phases tend to decompose. The organic groups are slowly hydrolyzed from the surface of the silica, which leaves a normal silica surface exposed. Thus, the chromatographic process that takes place on the column slowly shifts from a reversed-phase to a normal-phase separation mechanism.

Another type of solid support that is sometimes used in reversed-phase chromatography is organic polymer beads. These beads present a surface to the moving phase which is largely organic in nature.

For solutions of ions, select a column that is packed with an ion-exchange resin. This type of chromatography is known as **ion-exchange chromatography.** The ion-exchange resin that is chosen can be either an anion-exchange resin or a cation-exchange resin, depending upon the nature of the sample being examined.

A fourth type of column is known as a **size-exclusion column** or a **gel-filtration column.** The interaction that takes place on this type of column is similar to that described in Technique 19, Section 19.15, page 773.

21.2 Column Dimensions

The dimensions of the column that you use depend upon the application. For analytical applications, a typical column is constructed of tubing that has an inside diameter of between 4 and 5 mm, although analytical columns with inside diameters of 1 or 2 mm are also available. A typical analytical column has a length of about 7.5 to 30 cm. This type of column is suitable for the separation of a 0.1- to 5-mg sample. With columns of smaller diameter, it is possible to perform an analysis with samples smaller than 1 *micro*gram.

High-performance liquid chromatography is an excellent analytical technique, but the separated compounds may also be isolated. The technique can be used for preparative experiments. Just as in column chromatography, the fractions can be collected into individual receiving containers as they pass through the column. The solvents can be evaporated from these fractions, allowing you to isolate separated components of the original mixture. Samples that range in size from 5 to 100 mg can be separated on a semi-preparative, or **semiprep,** column. The dimensions of a semiprep column are typically 8 mm inside diameter and 10 cm in length. A semiprep column is a practical choice when you wish to use the same column for both analytical and preparative separations. A semiprep column is small enough to provide reasonable sensitivity in analyses, but it is also capable of handling moderate-size samples when you need to isolate the components of a mixture. Even larger samples can be separated using a **preparative column.** This type of column is useful when you wish to collect the components of a mixture and then use the pure samples for additional study (e.g., for a subsequent chemical reaction or for spectroscopic analysis). A preparative column may be as large as 20 mm in inside diameter and 30 cm in length. A preparative column can handle samples as large as 1 g per injection.

21.3 Solvents

The choice of solvent used for an HPLC separation depends on the type of chromatographic process selected. For a normal-phase separation, the solvent is selected based on its polarity. The criteria described in Technique 19, Section 19.4B, page 761, are used. A solvent of very low polarity might be pentane, petroleum ether, hexane, or carbon tetrachloride; a solvent of very high polarity might be water, acetic acid, methanol, or 1-propanol. For a reversed-phase experiment, a less polar solvent causes solutes to migrate *faster*. For example, for a mixed methanol–water solvent, as the percentage of methanol in the solvent increases (solvent becomes less polar), the time required to elute the components of a mixture from a column decreases. The behavior of solvents as eluents in a reversed-phase chromatography would be the reverse of the order shown in Table 19.2 on page 761.

If a single solvent (or solvent mixture) is used for the entire separation, the chromatogram is said to be **isochratic.** Special electronic devices are available with HPLC instruments that allow you to program changes in the solvent composition from the beginning to the end of the chromatography. These are called **gradient elution systems.** With gradient elution, the time required for a separation may be shortened considerably.

The need for pure solvents is especially acute with HPLC. The narrow bore of the column and the very small particle size of the column packing require that solvents be particularly pure and free of insoluble residue. In most cases, the solvents must be filtered through ultrafine filters and **degassed** (have dissolved gases removed) before they can be used.

The solvent gradient is chosen so that the eluting power of the solvent increases over the duration of the experiment. The result is that components of the mixture that tend to move very slowly through the column are caused to move faster as the eluting power of the solvent gradually increases. The instrument can be programmed to change the composition of the solvent following a linear gradient or a nonlinear gradient, depending on the specific requirements of the separation.

21.4 Detectors

A flow-through **detector** must be provided to determine when a substance has passed through the column. In most applications, the detector detects either the change in index of refraction of the liquid as its composition changes or the presence of solute by its absorption of ultraviolet or visible light. The signal generated by the detector is amplified and treated electronically in a manner similar to that found in gas chromatography (Technique 22, Section 22.6, p. 804).

A detector that responds to changes in the index of refraction of the solution may be considered the most universal of the HPLC detectors. The refractive index of the liquid passing through the detector changes slightly, but significantly, as the liquid changes from pure solvent to a liquid where the solvent contains some type of organic solute. This change in refractive index can be detected and compared to the refractive index of pure solvent. The difference in index values is then recorded as a peak on a chart. A disadvantage of this type of detector is that it must respond to very small changes in refractive index. As a result, the detector tends to be unstable and difficult to balance.

When the components of the mixture have some type of absorption in the ultraviolet or visible regions of the spectrum, a detector that is adjusted to detect absorption at a particular wavelength of light can be used. This type of detector is much more stable, and the readings tend to be more reliable. Unfortunately, many organic compounds do not absorb ultraviolet light, and this type of detector cannot be used.

21.5 Presentation of Data

The data produced by an HPLC instrument appear in the form of a chart, where detector response is the vertical axis and time is represented on the horizontal axis. These are recorded on a continuously moving strip of chart paper, although they may also be observed in graphic form on a computer display. In virtually all respects, the form of the data is identical to that produced by a gas chromatograph; in fact, in many cases, the data-handling system for the two types of instruments is essentially identical. To understand how to analyze the data from an HPLC instrument, read Sections 22.12 and 22.13 in Technique 22.

REFERENCE

Rubinson, K. A. *Chemical Analysis.* Boston: Little, Brown and Co., 1987. Chapter 14, "Liquid Chromatography."

PROBLEMS

1. For a mixture of biphenyl, benzoic acid, and benzyl alcohol, predict the order of elution and describe any differences that you would expect for a normal-phase HPLC experiment (in hexane solvent) compared with a reversed-phase experiment (in tetrahydrofuran–water solvent).

2. How would the *gradient elution program* differ between normal-phase and reversed-phase chromatography?

22   TECHNIQUE 22

Gas Chromatography

Gas chromatography is one of the most useful instrumental tools for separating and analyzing organic compounds that can be vaporized without decomposition. Common uses include testing the purity of a substance and separating the components of a mixture. The relative amounts of the components in a mixture may also be determined. In some cases, gas chromatography can be used to identify a compound. In microscale work, it can also be used as a preparative method to isolate pure compounds from a small amount of a mixture.

Gas chromatography resembles column chromatography in principle, but it differs in three respects. First, the partitioning processes for the compounds to be separated are carried out between a **moving gas phase** and a **stationary liquid phase.** (Recall that in column chromatography the moving phase is a liquid and the stationary phase is a solid adsorbent.) A second difference is that the temperature of the gas system can be controlled because the column is contained in an insulated oven. And third, the concentration of any given compound in the gas phase is a function of its vapor pressure only. Because gas chromatography separates the components of a mixture primarily on the basis of their vapor pressures (or boiling points), this technique is also similar in principle to fractional distillation. In microscale work, it is sometimes used to separate and isolate compounds from a mixture; fractional distillation would normally be used with larger amounts of material.

Gas chromatography (GC) is also known as vapor-phase chromatography (VPC) and as gas–liquid partition chromatography (GLPC). All three

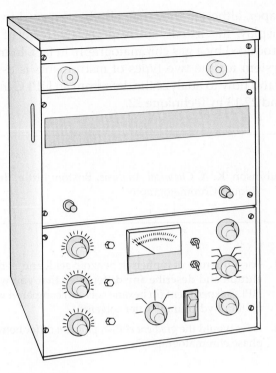

Figure 22.1
Gas chromatograph.

names, as well as their indicated abbreviations, are often found in the literature of organic chemistry. In reference to the technique, the last term, GLPC, is the most strictly correct and is preferred by most authors.

22.1 The Gas Chromatograph

The apparatus used to carry out a gas–liquid chromatographic separation is generally called a **gas chromatograph.** A typical student-model gas chromatograph, the GOW-MAC model 69-350, is illustrated in Figure 22.1. A schematic block diagram of a basic gas chromatograph is shown in Figure 22.2. The basic elements of the apparatus are apparent. In short, the sample is injected into the chromatograph, and it is immediately vaporized in a heated injection chamber and introduced into a moving stream of gas, called the **carrier gas.** The vaporized sample is then swept into a column filled with particles coated with a liquid adsorbent. The column is contained in a temperature-controlled oven. As the sample passes through the column, it is subjected to many gas–liquid partitioning processes, and the components are separated. As each component leaves the column, its presence is detected by an electrical detector that generates a signal that is recorded on a strip chart recorder.

Many modern instruments are also equipped with a microprocessor, which can be programmed to change parameters, such as the temperature of the oven, while a mixture is being separated on a column. With this capability, it is possible to optimize the separation of components and to complete a run in a relatively short time.

22.2 The Column

The heart of the gas chromatograph is the column. There are two types in common use: packed and capillary columns.

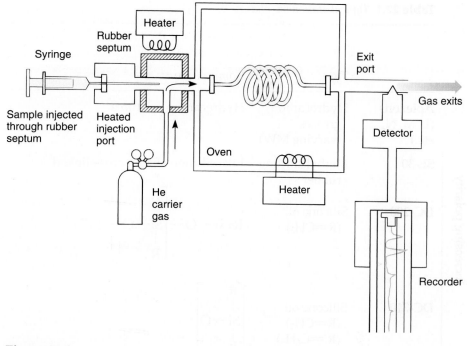

Figure 22.2
Schematic diagram of a gas chromatograph.

Packed Columns. These columns are usually constructed of stainless steel tubing with diameters of ⅛ in. (3 mm) or ¼ in. (6 mm) and lengths of from 4 to 12 feet. The column is packed with a liquid or low melting solid as the **stationary phase** distributed on a solid **support material.** The stationary phase must be relatively nonvolatile, that is, it should have a low vapor pressure and a high boiling point. Some typical stationary phases used with packed columns are listed in Table 22.1. Typical support materials are shown in Table 22.2. The most common support material consists of diatomaceous earth (Chromosorb).

Packed columns are bought from commercial sources or may sometimes be made in the laboratory by researchers. Basically, you dissolve one of the stationary phases listed in Table 22.1 in methylene chloride. Then you add the support material to the solution followed by removal of the solvent on a rotary evaporator (see Technique 7, Section 7.11, page 615 and Figure 7.19, p. 614). The evaporation process evenly distributes the stationary phase onto the support material and yields a dry solid. In the final step, the solid consisting of the stationary phase coated on the support material, is packed into the stainless steel tubing as evenly as possible. After plugging the ends of the tubing with glass wool to prevent the solid from coming out, the tubing is rolled into a coil that will fit into the oven of the gas chromatograph with the two ends connected to the gas entrance and exit ports (see Figure 22.2).

The selection of the packed column depends on the application. If one wants to separate nonpolar compounds that vary only by boiling point, one often uses one of the polydimethyl siloxanes (methyl silicones) columns such as SE-30 or DC-200. For more polar compounds, chemists will select a silicone column that has attached methyl and phenyl groups on the silicone polymer (DC-710). Separating even more polar compounds calls for a polyethylene glycol (carbowax) column or a column packed with diethylene glycol

Table 22.1 *Typical stationary phases*

Increasing polarity (↓)

	Type	Composition	Maximum Temperature (°C)	Typical Use
Apiezons (L, M, N, etc.)	Hydrocarbon greases (varying MW)	Hydrocarbon mixtures	250–300	Hydrocarbons
SE-30	Methyl silicone rubber	Like silicone oil, but cross-linked	350	General applications
DC-200	Silicone oil ($R=CH_3$)	$R_3Si-O-\left[\begin{array}{c} R \\ \| \\ Si-O \\ \| \\ R \end{array}\right]_n-SiR_3$	225	Aldehydes, ketones, halocarbons
DC-710	Silicone oil ($R=CH_3$) ($R'=C_6H_5$)	$\left[\begin{array}{c} R' \\ \| \\ Si-O \\ \| \\ R \end{array}\right]_n$	300	General applications
Carbowaxes (400–20M)	Polyethylene glycols (varying chain lengths)	Polyether $HO-(CH_2CH_2-O)_n-CH_2CH_2OH$	Up to 250	Alcohols, ethers, halocarbons
DEGS	Diethylene glycol succinate	Polyester $\left[CH_2CH_2-O-\underset{\underset{O}{\|}}{C}-(CH_2)_2-\underset{\underset{O}{\|}}{C}-O\right]_n$	200	General applications

Table 22.2 *Typical solid supports*

Crushed firebrick	Chromosorb T (Teflon beads)
Nylon beads	
Glass beads	Chromosorb P (pink diatomaceous earth, highly absorptive, pH 6–7)
Silica	
Alumina	
Charcoal	Chromosorb W (White diatomaceous earth, medium absorptivity, pH 8–10)
Molecular sieves	
	Chromosorb G (Like the above, low absorptivity, pH 8.5)

succinate (DEGS). Chemists will need to carefully note the maximum temperature that can be employed with the columns (Table 22.1). Above the specified temperature, the liquid phase itself will begin to "bleed" off the column.

Packed columns are cheaper to buy and can separate larger quantities of material than capillary columns. Packed columns, however, are not as efficient in separating materials, especially with compounds with very similar polarities or boiling points.

Capillary Columns. Many gas chromatographs sold today use capillary columns rather than packed columns. Capillary columns are made of very thin fused-silica with an inner diameter of about 0.25 mm. Typically the columns are very long, often 25 meters up to 100 meters in length. The stationary phase is coated as a film on the inner surface with a thickness of about 0.25 μm. The film is bonded to the silica and cross-linked to improve thermal stability and to help prevent bleeding of the stationary phase from the column. Most of the capillary columns do not have any support material in them.

The liquid stationary phases used with capillary columns are similar to those used in packed columns. The most common stationary phases are polysiloxanes (silicones) that contain various substituents that modify the polarity of the phase. Polydimethyl siloxane (methyl silicone) is nonpolar. Replacing methyl groups with increasing numbers of phenyl substituents increases the polarity of the silicone. For example, J&W DB-l, or a similar stationary phase sold by other companies,[1] has the same properties as methyl silicone (SE-30 or DC-200) and is used with nonpolar compounds that separate by boiling point differences. For other applications, a J&W DB-5 may be used. This silicone consists of a material with 95% methyl and 5% phenyl substituents on the silicone. A DB-17 column has 50% phenyl groups replacing methyl groups, and would have a medium polarity (similar to DC-710). J&W DB-wax is used to separate much more polar compounds and is similar to diethylene glycols (carbowax). Chiral capillary columns are also available that can separate enantiomers (Section 22.8).

Because of the length and small diameter of capillary columns, there is increased interaction between the compounds in the mixture and the stationary phase. Capillary columns are, therefore, much more efficient in separating compounds with similar properties than with packed columns, but only if a dilute solution of a mixture of compounds is injected into the column. In order to obtain a satisfactory separation, you must dissolve about 1 drop of a mixture in about 2 mL of a solvent such as methylene chloride or pentane. About 1 μL of this dilute sample is injected onto the column. In contrast, 1 to 10 μL of an undiluted sample can often be analyzed on a packed column without overloading the column. These expensive capillary columns must be purchased from commercial companies specializing in their manufacture. Sensible researchers would never try to make their own capillary columns!

In the final step, the liquid-phase-coated support material is packed into the tubing as evenly as possible. The tubing is bent or coiled so that it fits into the oven of the gas chromatograph with its two ends connected to the gas entrance and exit ports.

Selection of a liquid phase usually revolves about two factors. First, most of them have an upper temperature limit above which they cannot be used. Above the specified limit of temperature, the liquid phase itself will

---

[1] Many companies supply capillary columns: J&W (Alltech), Supelco, HP, Chrompack, and Quadrex are some of the suppliers. Comparison of chemical compositions, polarities and applications may be made by consulting: http://www.quadrexcorp.com/new/columns.htm

begin to "bleed" off the column. Second, the materials to be separated must be considered. For polar samples, it is usually best to use a polar liquid phase; for nonpolar samples, a nonpolar liquid phase is indicated. The liquid phase performs best when the substances to be separated *dissolve* in it.

Most researchers today buy packed columns from commercial sources, rather than pack their own. A wide variety of types and lengths is available. Alternatives to packed columns are Golay or glass capillary columns of diameters 0.1–0.2 mm. With these columns, no solid support is required, and the liquid is coated directly on the inner walls of the tubing. Liquid phases commonly used in glass capillary columns are similar in composition to those used in packed columns. They include DB-1 (similar to SE-30), DB-17 (similar to DC-710), and DB-WAX (similar to Carbowax 20M). The length of a capillary column is usually very long, typically 50–100 ft. Because of the length and small diameter, there is increased interaction between the sample and the stationary phase. Gas chromatographs equipped with these small-diameter columns are able to separate components more effectively than instruments using larger packed columns.

22.3 Principles of Separation

After a column is selected and installed, the **carrier gas** (usually helium, argon, or nitrogen) is allowed to flow through the column supporting the liquid phase. The mixture of compounds to be separated is introduced into the carrier gas stream, where its components are equilibrated (or partitioned) between the moving gas phase and the stationary liquid phase (Figure 22.3). The latter is held stationary because it is adsorbed onto the surfaces of the support material.

The sample is introduced into the gas chromatograph by a microliter syringe. It is injected as a liquid or as a solution through a rubber septum into a heated chamber, called the **injection port,** where it is vaporized and mixed with the carrier gas. As this mixture reaches the column, which is heated in a controlled oven, it begins to equilibrate between the liquid and

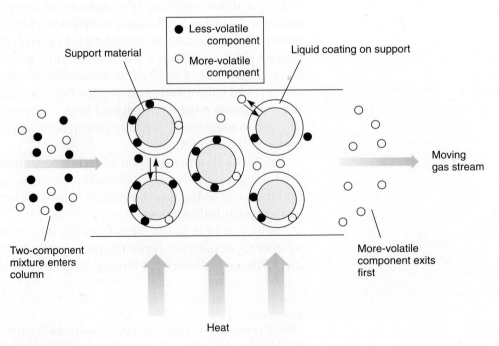

Figure 22.3
The separation process.

gas phases. The length of time required for a sample to move through the column is a function of how much time it spends in the vapor phase and how much time it spends in the liquid phase. The more time it spends in the vapor phase, the faster it gets to the end of the column. In most separations, the components of a sample have similar solubilities in the liquid phase. Therefore, the time the different compounds spend in the vapor phase is primarily a function of the vapor pressure, and the more volatile component arrives at the end of the column first, as illustrated in Figure 22.3. By selecting the correct temperature of the oven and the correct liquid phase, the compounds in the injected mixture travel through the column at different rates and are separated.

22.4 Factors Affecting Separation

Several factors determine the rate at which a given compound travels through a gas chromatograph. First of all, compounds with low boiling points will generally travel through the gas chromatograph faster than compounds of higher boiling points. This because the column is heated, and low-boiling compounds always have higher vapor pressures than compounds of higher boiling point. In general, therefore, for compounds with the same functional group, the higher the molecular weight, the longer the retention time. For most molecules, the boiling point increases as the molecular weight increases. If the column is heated to a temperature that is too high, however, the entire mixture to be separated is flushed through the column at the same rate as the carrier gas, and no equilibration takes place with the liquid phase. On the other hand, at too low a temperature, the mixture dissolves in the liquid phase and never revaporizes. Thus, it is retained on the column.

The second factor, the rate of flow of the carrier gas, is important. The carrier gas must not move so rapidly that molecules of the sample in the vapor phase cannot equilibrate with those dissolved in the liquid phase. This may result in poor separation between components in the injected mixture. If the rate of flow is too slow, however, the bands broaden significantly, leading to poor resolution (see Section 22.8).

The third factor is the choice of liquid phase used in the column. The molecular weights, functional groups, and polarities of the component molecules in the mixture to be separated must be considered when a liquid phase is being chosen. One generally uses a different type of material for hydrocarbons, for instance, than for esters. The materials to be separated should *dissolve* in the liquid. The useful temperature limit of the liquid phase selected must also be considered.

The fourth factor, the length of the column, is also important. Compounds that resemble one another closely, in general, require longer columns than dissimilar compounds. Many kinds of isomeric mixtures fit into the "difficult" category. The components of isomeric mixtures are so much alike that they travel through the column at very similar rates. You need a longer column, therefore, to take advantage of any differences that may exist.

22.5 Advantages of Gas Chromatography

All factors that have been mentioned must be adjusted by the chemist for any mixture to be separated. Considerable preliminary investigation is often required before a mixture can be separated successfully into its components by gas chromatography. Nevertheless, the advantages of the technique are many.

First, many mixtures can be separated by this technique when no other method is adequate. Second, as little as 1–10 μL (1 μL = 10^{-6} L) of a mixture can be separated by this technique. This advantage is particularly important when working at the microscale level. Third, when gas chromatography is coupled with an electronic recording device (see following discussion), the amount of each component present in the separated mixture can be estimated quantitatively.

The range of compounds that can be separated by gas chromatography extends from gases, such as oxygen (bp −183°C) and nitrogen (bp −196°C), to organic compounds with boiling points over 400°C. The only requirement for the compounds to be separated is that they have an appreciable vapor pressure at a temperature at which they can be separated and that they be thermally stable at this temperature.

22.6 Monitoring the Column (The Detector)

To follow the separation of the mixture injected into the gas chromatograph, it is necessary to use an electrical device called a **detector.** Two types of detectors in common use are the **thermal conductivity detector (TCD)** and the **flame-ionization detector (FID).**

The thermal conductivity detector is simply a hot wire placed in the gas stream at the column exit. The wire is heated by constant electrical voltage. When a steady stream of carrier gas passes over this wire, the rate at which it loses heat and its electrical conductance have constant values. When the composition of the vapor stream changes, the rate of heat flow from the wire, and hence its resistance, changes. Helium, which has a higher thermal conductivity than most organic substances, is a common carrier gas. Thus, when a substance elutes in the vapor stream, the thermal conductivity of the moving gases will be lower than with helium alone. The wire then heats up, and its resistance decreases.

A typical TCD operates by difference. Two detectors are used: one exposed to the actual effluent gas and the other exposed to a reference flow of carrier gas only. To achieve this situation, a portion of the carrier gas stream is diverted *before* it enters the injection port. The diverted gas is routed through a reference column into which no sample has been admitted. The detectors mounted in the sample and reference columns are arranged so as to the form the arms of a Wheatstone bridge circuit, as shown in Figure 22.4.

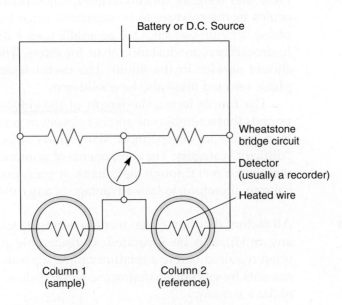

Figure 22.4
Typical thermal conductivity detector.

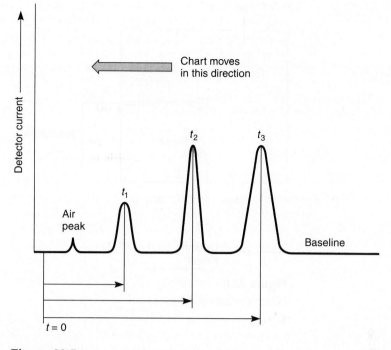

Figure 22.5
Typical chromatogram.

As long as the carrier gas alone flows over both detectors, the circuit is in balance. However, when a sample elutes from the sample column, the bridge circuit becomes unbalanced, creating an electrical signal. This signal can be amplified and used to activate a strip chart recorder. The recorder is an instrument that plots, by means of a moving pen, the unbalanced bridge current versus time on a continuously moving roll of chart paper. This record of detector response (current) versus time is called a **chromatogram.** A typical gas chromatogram is illustrated in Figure 22.5. Deflections of the pen are called **peaks.**

When a sample is injected, some air (CO_2, H_2O, N_2, and O_2) is introduced along with the sample. The air travels through the column almost as rapidly as the carrier gas; as it passes the detector, it causes a small pen response, thereby giving a peak, called the **air peak.** At later times (t_1, t_2, t_3), the components also give rise to peaks on the chromatogram as they pass out of the column and past the detector.

In a flame-ionization detector, the effluent from the column is directed into a flame produced by the combustion of hydrogen, as illustrated in Figure 22.6. As organic compounds burn in the flame, ion fragments are produced that collect on the ring above the flame. The resulting electrical signal is amplified and sent to a recorder in a similar manner to a TCD, except that a FID does not produce an air peak. The main advantage of the FID is that it is more sensitive and can be used to analyze smaller quantities of sample. Also, because a FID does not respond to water, a gas chromatograph with this detector can be used to analyze aqueous solutions. Two disadvantages are that it is more difficult to operate and the detection process destroys the sample. Therefore, an FID gas chromatograph cannot be used to do preparative work, which is often desired in the microscale laboratory.

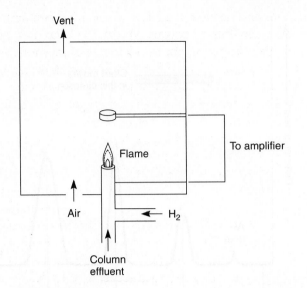

Figure 22.6
Flame-ionization detector.

22.7 Retention Time

The period following injection that is required for a compound to pass through the column is called the **retention time** of that compound. For a given set of constant conditions (flow rate of carrier gas, column temperature, column length, liquid phase, injection port temperature, carrier), the retention time of any compound is always constant (much like the R_f value in thin-layer chromatography, as described in Technique 20, Section 20.9, p. 787). The retention time is measured from the time of injection to the time of maximum pen deflection (detector current) for the component being observed. This value, when obtained under controlled conditions, can identify a compound by a direct comparison of it with values for known compounds determined under the same conditions. For easier measurement of retention times, most strip chart recorders are adjusted to move the paper at a rate that corresponds to time divisions calibrated on the chart paper. The retention times (t_1, t_2, t_3) are indicated in Figure 22.5 for the three peaks illustrated.

Most modern gas chromatographs are attached to a "data station," which uses a computer or a microprocessor to process the data. With these instruments, the chart often does not have divisions. Instead, the computer prints the retention time, usually to the nearest 0.01 minute, above each peak.

22.8 Chiral Stationary Phases

A recent innovation in gas chromatography is to use chiral adsorbent materials to achieve separations of stereoisomers. The interaction between a particular stereoisomer and the chiral adsorbent may be different from the interaction between the opposite stereoisomer and the same chiral adsorbent. As a result, retention times for the two stereoisomers are likely to be sufficiently different to allow for a clean separation. The interactions between a chiral substance and the chiral adsorbent will include hydrogen-bonding and dipole–dipole attraction forces, although other properties may also be involved. One enantiomer should interact more strongly with the adsorbent than its opposite form. Thus, one enantiomer should pass through the gas chromatography column more slowly than its opposite form.

The ability of chiral adsorbents to separate stereoisomers is rapidly finding many useful applications, particularly in the synthesis of pharmaceutical agents. The biological activity of chiral substances often depends upon their stereochemistry because the living body is a highly chiral environment. A large number of pharmaceutical compounds have two enantiomeric forms that in many cases show significant differences in their behavior and activity. The ability to prepare enantiomerically pure drugs is very important because these pure substances are much more potent (and often have fewer side effects) than their racemic analogues.

Another type of stationary phase in gas chromatography is based on molecules such as the **cyclodextrins.** With these materials, the discrimination between enantiomers depends on the interactions between the stereoisomers and the chiral cavity that is formed within these materials. Because enantiomers differ in shape, they will fit differently within the chiral cavity. The result will be that the enantiomers will pass through the cyclodextrin stationary phase at different rates, thus leading to a separation.

The cyclodextrins owe their specificity to their structure, which is based on polymers of D-(+)-glucose. The hydroxyl groups of the glucose have been alkylated, so that the cavity is relatively nonpolar. The exterior hydroxyl groups of the cyclodextrins have also been substituted with *tert*-butyldimethylsilyl groups. The result is a material that can also utilize differences in hydrogen-bonding and dipole–dipole interactions to separate stereoisomers.

The structure of one important cyclodextrin-based chiral adsorbent is shown in Figure 22.7. Gas chromatography using this chiral adsorbent as a stationary phase has been used to separate a wide variety of stereoisomers. In one recent publication, this method was used to isolate a pure sample of (S)-(+)-2-methyl-4-octanol, a male-specific compound released by the sugarcane weevil, *Sphenophorus levis.*[1]

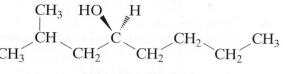

(S)-(+)-2-Methyl-4-octanol

22.9 Poor Resolution and Tailing

The peaks in Figure 22.5 are well **resolved.** That is, the peaks are separated from one another, and between each pair of adjacent peaks the tracing returns to the **baseline.** In Figure 22.8, the peaks overlap, and the resolution is not good. Poor resolution is often caused by using too much sample, too high a column temperature, too short a column, a liquid phase that does not discriminate well between the two components, a column with too large a diameter, or, in short, almost any wrongly adjusted parameter. When peaks are poorly resolved, it is more difficult to determine the relative amount of each component. Methods for determining the relative percentages of each component are given in Section 22.12.

Another desirable feature illustrated by the chromatogram in Figure 22.5 is that each peak is symmetrical. A common example of an unsymmetrical peak is one in which **tailing** has occurred, as shown in Figure 22.9. Tailing

[1] Zarbin, P. H. G., Princival, J. L., dos Santos, A. A., and de Oliveira, A. R. M. "Synthesis of (S)-(+)-2-Methyl-4-octanol: Male-Specific Compound Released by Sugarcane Weevil *Sphenophorus levis." Journal of the Brazilian Chemical Society,* 15 (2004): 331–334.

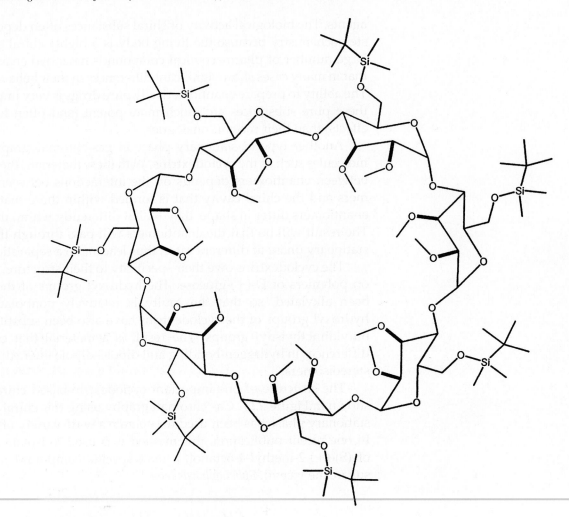

Figure 22.7
Cyclodextrin derivative used as a chiral adsorbent in gas chromatography.

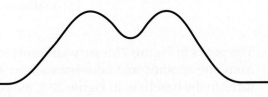

Figure 22.8
Poor resolution or peaks overlap.

usually results from injecting too much sample into the gas chromatograph. Another cause of tailing occurs with polar compounds, such as alcohols and aldehydes. These compounds may be temporarily adsorbed on column walls or areas of the support material that are not adequately coated by the liquid phase. Therefore, they do not leave in a band, and tailing results.

22.10 Qualitative Analysis   A disadvantage of the gas chromatograph is that it gives no information whatever about the identities of the substances it has separated. The little information it does provide is given by the retention time. It is hard to reproduce this quantity from day to day, however, and exact duplications of

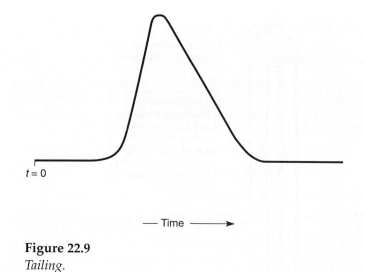

$t = 0$

— Time —→

Figure 22.9
Tailing.

separations performed last month may be difficult to make this month. It is usually necessary to **calibrate** the column each time it is used. That is, you must run pure samples of all known and suspected components of a mixture individually, just before chromatographing the mixture, to obtain the retention time of each known compound. As an alternative, each suspected component can be added, one by one, to the unknown mixture while the operator looks to see which peak has its intensity increased relative to the unmodified mixture. Another solution is to *collect* the components individually as they emerge from the gas chromatograph. Each component can then be identified by other means, such as by infrared or nuclear magnetic resonance spectroscopy or by mass spectrometry.

22.11 Collecting the Sample

For gas chromatographs with a thermal conductivity detector, it is possible to collect samples that have passed through the column. One method uses a gas-collection tube (see Figure 22.10), which is included in most microscale glassware kits. A collection tube is joined to the exit port of the column by inserting the ⊤5/5 inner joint into a metal adapter, which is connected to the exit port. When a sample is eluted from the column in the vapor state, it is cooled by the connecting adapter and the gas-collection tube and condenses in the collection tube. The gas-collection tube is removed from the adapter when the recorder indicates that the desired sample has completely passed through the column. After the first sample has been collected, the process can be repeated with another gas-collection tube.

To isolate the liquid, the tapered joint of the collection tube is inserted into a 0.1-mL conical vial, which has a ⊤5/5 outer joint. The assembly is placed into a test tube, as illustrated in Figure 22.11. During centrifugation, the sample is forced into the bottom of the conical vial. After the apparatus is disassembled, the liquid can be removed from the vial with a syringe for a boiling-point determination or analysis by infrared spectroscopy. If a determination of the sample weight is desired, the empty conical vial and cap should be tared and reweighed after the liquid has been collected. It is advisable to dry the gas collection tube and the conical vial in an oven before use to prevent contamination by water or other solvents used in cleaning this glassware.

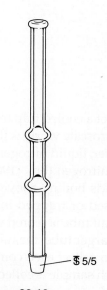

⊤ 5/5

Figure 22.10
Gas-chromatography collection tube.

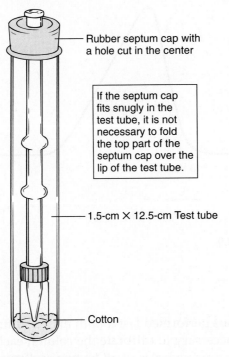

Rubber septum cap with a hole cut in the center

If the septum cap fits snugly in the test tube, it is not necessary to fold the top part of the septum cap over the lip of the test tube.

1.5-cm × 12.5-cm Test tube

Cotton

Figure 22.11
Gas-chromatography collection tube and 0.1-mL conical vial.

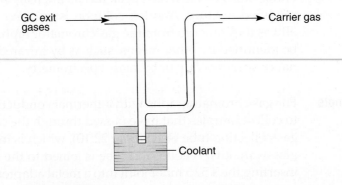

GC exit gases

Carrier gas

Coolant

Figure 22.12
Collection trap.

Another method for collecting samples is to connect a cooled trap to the exit port of the column. A simple trap, suitable for microscale work, is illustrated in Figure 22.12. Suitable coolants include ice water, liquid nitrogen, or dry ice–acetone. For instance, if the coolant is liquid nitrogen (bp −196°C) and the carrier gas is helium (bp −269°C), compounds boiling above the temperature of liquid nitrogen generally are condensed or trapped in the small tube at the bottom of the U-shaped tube. The small tube is scored with a file just below the point where it is connected to the larger tube, the tube is broken off, and the sample is removed for analysis. To collect each component of the mixture, you must change the trap after each sample is collected.

22.12 Quantitative Analysis   The area under a gas-chromatograph peak is proportional to the amount (moles) of compound eluted. Hence, the molar percentage composition of a mixture can be approximated by comparing relative peak areas. This method

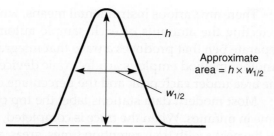

Figure 22.13
Triangulation of a peak.

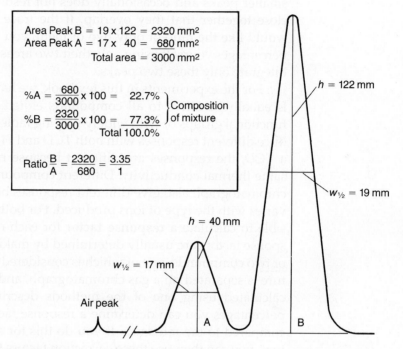

Area Peak B = 19 x 122 = 2320 mm²
Area Peak A = 17 x 40 = 680 mm²
                    Total area = 3000 mm²

$\%A = \dfrac{680}{3000} \times 100 = 22.7\%$ ⎫ Composition
$\%B = \dfrac{2320}{3000} \times 100 = 77.3\%$ ⎬ of mixture
                        Total 100.0% ⎭

$\text{Ratio}\ \dfrac{B}{A} = \dfrac{2320}{680} = \dfrac{3.35}{1}$

Figure 22.14
Sample percentage composition calculation.

of analysis assumes that the detector is equally sensitive to all compounds eluted and that it gives a linear response with respect to amount. Nevertheless, it gives reasonably accurate results.

The simplest method of measuring the area of a peak is by geometrical approximation, or triangulation. In this method, you multiply the height h of the peak above the baseline of the chromatogram by the width of the peak at half of its height $w_{1/2}$. This is illustrated in Figure 22.13. The baseline is approximated by drawing a line between the two side arms of the peak. This method works well only if the peak is symmetrical. If the peak has tailed or is unsymmetrical, it is best to cut out the peaks with scissors and weigh the pieces of paper on an **analytical balance.** Because the weight per area of a piece of good chart paper is reasonably constant from place to place, the ratio of the areas is the same as the ratio of the weights. To obtain a percentage composition for the mixture, first add all the peak areas (weights). Then, to calculate the percentage of any component in the mixture, divide its individual area by the total area and multiply the result by 100. A sample calculation is illustrated in Figure 22.14. If peaks overlap (see Fig. 22.8), either the gas-chromatographic conditions must be readjusted to achieve better resolution of the peaks or the peak shape must be estimated.

There are various instrumental means, which are built into recorders, of detecting the amounts of each sample automatically. One method uses a separate pen that produces a trace that integrates the area under each peak. Another method employs an electronic device that automatically prints out the area under each peak and the percentage composition of the sample.

Most modern data stations label the top of each peak with its retention time in minutes. When the trace is completed, the computer prints a table of all the peaks with their retention times, areas, and the percentage of the total area (sum of all the peaks) that each peak represents. Some caution should be used with these results because the computer often does not include smaller peaks and occasionally does not resolve narrow peaks that are so close together that they overlap. If the trace has several peaks and you would like the ratio of only two of them, you will have to determine their percentages yourself using only their two areas or instruct the instrument to integrate only these two peaks.

For the experiments in this textbook, we have assumed that the detector is equally sensitive to all compounds eluted. Compounds with different functional groups or with widely varying molecular weights, however, produce different responses with both TCD and FID gas chromatographs. With a TCD, the responses are different because not all compounds have the same thermal conductivity. Different compounds analyzed with a FID gas chromatograph also give different responses because the detector response varies with the type of ions produced. For both types of detectors, it is possible to calculate a **response factor** for each compound in a mixture. Response factors are usually determined by making up an equimolar mixture of two compounds, one of which is considered to be the reference. The mixture is separated on a gas chromatograph, and the relative percentages are calculated using one of the methods described previously. From these percentages you can determine a response factor for the compound being compared to the reference. If you do this for all the components in a mixture, you can then use these correction factors to make more accurate calculations of the relative percentages for the compounds in the mixture.

To illustrate how response factors are determined, consider the following example. An equimolar mixture of benzene, hexane, and ethyl acetate is prepared and analyzed using a flame-ionization gas chromatograph. The peak areas obtained are

Hexane 831158

Ethyl acetate 1449695

Benzene 966463

In most cases, benzene is taken as the standard, and its response factor is defined to be equal to 1.00. Calculation of the response factors for the other components of the test mixture proceeds as follows:

Hexane	$831158/966463 = 0.86$
Ethyl acetate	$1449695/966463 = 1.50$
Benzene	$966463/966463 = 1.00$ (by definition)

Notice that the response factors calculated in this example are molar response factors. It is necessary to correct these values by the relative molecular weights of each substance to obtain weight response factors.

When you use a flame-ionization gas chromatograph for quantitative analysis, it is first necessary to determine the response factors for each component of the mixture being analyzed, as just shown. For a quantitative analysis, it is likely that you will have to convert molar response factors into weight response factors. Next, the chromatography experiment using the unknown samples is performed. The observed peak areas for each component are corrected using the response factors in order to arrive at the correct weight percentage of each component in the sample. The application of response factors to correct the original results of a quantitative analysis will be illustrated in the following section.

22.13 Treatment of Data: Chromatograms Produced by Modern Data Stations

A. Gas Chromatograms and Data Tables

Most modern gas chromatography instruments are equipped with computer-based data stations. Interfacing the instrument with a computer allows the operator to display and manipulate the results in whatever manner might be desired. The operator thus can view the output in a convenient form. The computer can display the actual gas chromatogram and display the integration results. It can even display the result of two experiments simultaneously, making a comparison of parallel experiments convenient.

Figure 22.15 shows a gas chromatogram of a mixture of hexane, ethyl acetate, and benzene. The peaks corresponding to each substance can be seen; the peaks are labeled with their respective retention times.

	Retention Time (minutes)
Hexane	2.959
Ethyl acetate	3.160
Benzene	3.960

We can also see that there is a very small amount of an unspecified impurity, with a retention time of about 3.4 minutes.

Figure 22.16 shows part of the printed output that accompanies the gas chromatogram. It is this information that is used in the quantitative analysis of the mixture. According to the printout, the first peak has a retention time of 2.954 minutes (the difference between the retention times that appear as labels on the graph and those that appear in the data table are not significant). The computer has also determined the area under this peak (422,373 counts). Finally, the computer has calculated the percentage of the first substance (hexane) by determining the total area of all the peaks in the chromatogram (1,227,054 counts) and dividing that into the area for the hexane peak. The result is displayed as 34.4217%. In a similar manner, the data table shows the retention times and peak areas for the other two peaks in the sample, along with a determination of the percentage of each substance in the mixture.

B. Application of Response Factors

If the detector responded with equal sensitivity to each of the components of the mixture, the data table shown in Figure 22.16 would contain the complete quantitative analysis of the sample. Unfortunately, as we have seen (Section 22.12), gas chromatography detectors respond more sensitively to some substances than they do to others. To correct for this discrepancy, it is necessary to apply corrections that are based on the **response factors** for each component of the mixture.

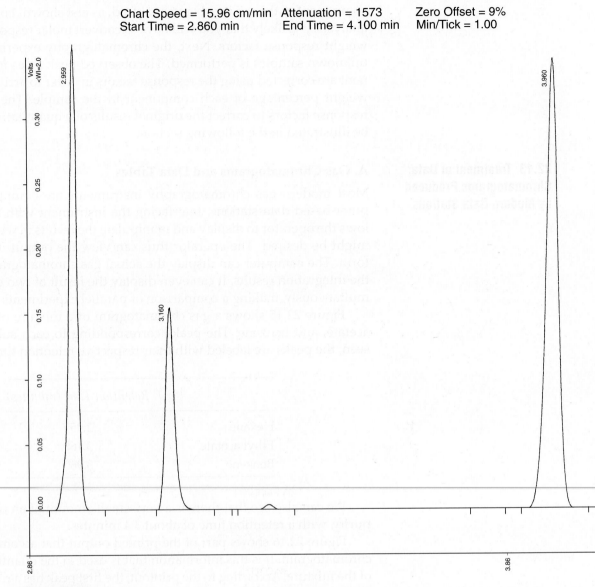

Chart Speed = 15.96 cm/min   Attenuation = 1573   Zero Offset = 9%
Start Time = 2.860 min   End Time = 4.100 min   Min/Tick = 1.00

Figure 22.15
A sample gas chromatogram obtained from a data station.

The method for determining the response factors was introduced in Section 22.12. In this section, we will see how this information is applied in order to obtain a correct analysis. This example should serve to demonstrate the procedure for correcting raw gas chromatography results when response factors are known. According to the data table, the reported peak area for the first (hexane) peak is 422,373 counts. The response factor for hexane was previously determined to be 0.86. The area of the hexane peak is thus corrected as follows:

$$422{,}373/0.86 = 491{,}000$$

Notice that the calculated result has been adjusted to reflect a reasonable number of significant figures.

```
Run Mode        : Analysis
Peak Measurement: Peak Area
Calculation Type: Percent
```

Peak No.	Peak Name	Result ()	Ret. Time (min)	Time Offset (min)	Area (counts)	Sep. Code	Width 1/2 (sec)	Status Codes
1		34.4217	2.954	0.000	422373	BB	1.0	
2		16.6599	3.155	0.000	204426	BB	1.2	
3		48.9184	3.954	0.000	600255	BB	1.6	
	Totals:	100.0000		0.000	1227054			

Total Unidentified Counts :   1227054 counts

Detected Peaks: 8   Rejected Peaks: 5   Identified Peaks: 0

Multiplier: 1   Divisor: 1   Unidentified Peak Factor: 0

Baseline Offset: 1 microVolts

Noise (used): 28 microVolts — monitored before this run

Manual injection

* *

Figure 22.16
A data table to accompany the gas chromatogram shown in Figure 22.15.

The areas for the other peaks in the gas chromatogram are corrected in a similar manner:

Hexane	422,373/0.86 =	491,000
Ethyl acetate	204,426/1.50 =	136,000
Benzene	600,255/1.00 =	600,000
Total peak area		1,227,000

Using these corrected areas, the true percentages of each component can be easily determined:

		Composition
Hexane	491,000/1,227,000	40.0%
Ethyl acetate	136,000/1,227,000	11.1%
Benzene	600,000/1,227,000	48.9%
Total		100.0%

C. Determination of Relative Percentages of Components in a Complex Mixture

In some circumstances, one may wish to determine the relative percentages of two components when the mixture being analyzed may be more complex and may contain more than two components. Examples of this situation might include the analysis of a reaction product where the laboratory worker might be interested in the relative percentages of two isomeric products when the sample might also contain peaks arising from the solvent, unreacted starting material, or some other product or impurity.

The example provided in Figures 22.15 and 22.16 can be used to illustrate the method of determining the relative percentages of some, but not all, of the components in the sample. Assume we are interested in the relative percentages of hexane and ethyl acetate in the sample but not in the percentage of benzene, which may be a solvent or an impurity. We know from the previous discussion that the *corrected* relative areas of the two peaks of interest are as follows:

	Relative Area
Hexane	491,000
Ethyl acetate	136,000
Total	627,000

We can determine the relative percentages of the two components simply by dividing the area of each peak by the total area of the two peaks:

		Percentage
Hexane	491,000/627,000	78.3%
Ethyl acetate	136,000/627,000	21.7%
Total		100.0%

22.14 Gas Chromatography–Mass Spectrometry (GC–MS)

A recently developed variation on gas chromatography is **gas chromatography–mass spectrometry,** also known as **GC–MS.** In this technique, a gas chromatograph is coupled to a mass spectrometer (see Technique 28). In effect, the mass spectrometer acts as a detector. The gas stream emerging from the gas chromatograph is admitted through a valve into a tube, where it passes over the sample inlet system of the mass spectrometer. Some of the gas stream is thus admitted into the ionization chamber of the mass spectrometer.

The molecules in the gas stream are converted into ions in the ionization chamber, and thus the gas chromatogram is actually a plot of time versus **ion current,** a measure of the number of ions produced. At the same time that the molecules are converted into ions, they are also accelerated and passed through the **mass analyzer** of the instrument. The instrument, therefore, determines the mass spectrum of each fraction eluting from the gas chromatography column.

A drawback of this method involves the need for rapid scanning by the mass spectrometer. The instrument must determine the mass spectrum of each component in the mixture before the next component exits from the column so that the spectrum of one substance is not contaminated by the spectrum of the next fraction.

Because high-efficiency capillary columns are used in the gas chromatograph, in most cases compounds are completely separated before the gas stream is analyzed. The typical GC–MS instrument has the capability of obtaining at least one scan per second in the range of 10–300 amu. Even more scans are possible if a narrow range of masses is analyzed. Using capillary columns, however, requires the user to take particular care to ensure that the sample does not contain any particles that might obstruct the flow of gases through the column. For this reason, the sample is carefully filtered through a very fine filter before the sample is injected into the chromatograph.

With a GC–MS system, a mixture can be analyzed and results obtained that resemble very closely those shown in Figures 22.15 and 22.16. A library search on each component of the mixture can also be conducted. The data stations of most instruments contain a library of standard mass spectra in their computer memory. If the components are known compounds, they can be identified tentatively by a comparison of their mass spectrum with the spectra of compounds found in the computer library. In this way, a "hit list" can be generated that reports on the probability that the compound in the library matches the known substance. A typical printout from a GC–MS instrument will list probable compounds that fit the mass spectrum of the component, the names of the compounds, their CAS Nos. (see Technique 29, Section 29.11, p. 952), and a "quality" or "confidence" number. This last number provides an estimate of how closely the mass spectrum of the component matches the mass spectrum of the substance in the computer library.

A variation on the GC–MS technique includes coupling a Fourier-transform infrared spectrometer (FT–IR) to a gas chromatograph. The substances that elute from the gas chromatograph are detected by determining their infrared spectra rather than their mass spectra. A new technique that also resembles GC–MS is **high-performance liquid chromatography–mass spectrometry (HPLC–MS).** An HPLC instrument is coupled through a special interface to a mass spectrometer. The substances that elute from the HPLC column are detected by the mass spectrometer, and their mass spectra can be displayed, analyzed, and compared with standard spectra found in the computer library built into the instrument.

PROBLEMS

1. **a.** A sample consisting of 1-bromopropane and 1-chloropropane is injected into a gas chromatograph equipped with a nonpolar column. Which compound has the shorter retention time? Explain your answer.

    b. If the same sample were run several days later with the conditions as nearly the same as possible, would you expect the retention times to be identical to those obtained the first time? Explain.

2. Using triangulation, calculate the percentage of each component in a mixture composed of two substances, A and B. The chromatogram is shown in Figure 22.17.

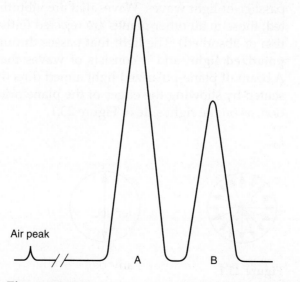

Figure 22.17
A chromatogram for problem 2.

3. Make a photocopy of the chromatogram in Figure 22.17. Cut out the peaks and weigh them on an analytical balance. Use the weights to calculate the percentage of each component in the mixture. Compare your answer to what you calculated in problem 2.

4. What would happen to the retention time of a compound if the following changes were made?

   a. Decrease the flow rate of the carrier gas

   b. Increase the temperature of the column

   c. Increase the length of the column

---

23   TECHNIQUE 23

Polarimetry

23.1 Nature of Polarized Light

Light has a dual nature because it shows properties of both waves and particles. The wave nature of light can be demonstrated by two experiments: polarization and interference. Of the two, polarization is the more interesting to organic chemists because they can take advantage of polarization experiments to learn something about the structure of an unknown molecule.

Ordinary white light consists of wave motion in which the waves have a variety of wavelengths and vibrate in all possible planes perpendicular to the direction of propagation. Light can be made to be **monochromatic** (of one wavelength or color) by using filters or special light sources. Frequently, a sodium lamp (sodium D line = 5893 Å) is used. Although the light from this lamp consists of waves of only one wavelength, the individual light waves still vibrate in all possible planes perpendicular to the beam. If we imagine that the beam of light is aimed directly at the viewer, ordinary light can be represented by showing the edges of the planes oriented randomly around the path of the beam, as on the left side of Figure 23.1.

A Nicol prism, which consists of a specially prepared crystal of Iceland spar (or calcite), has the property of serving as a screen that can restrict the passage of light waves. Waves that are vibrating in one plane are transmitted; those in all other planes are rejected (either refracted in another direction or absorbed). The light that passes through the prism is called **plane-polarized light,** and it consists of waves that vibrate in only one plane. A beam of plane-polarized light aimed directly at the viewer can be represented by showing the edges of the plane oriented in one particular direction, as on the right side of Figure 23.1.

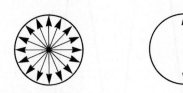

Figure 23.1
Ordinary versus plane-polarized light.

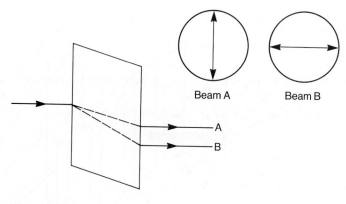

Figure 23.2
Double refraction.

Iceland spar has the property of **double refraction;** that is, it can split, or doubly refract, an entering beam of ordinary light into two separate emerging beams of light. Each of the two emerging beams (labeled A and B in Figure 23.2) has only a single plane of vibration, and the plane of vibration in beam A is perpendicular to the plane of beam B. In other words, the crystal has separated the incident beam of ordinary light into two beams of plane-polarized light, with the plane of polarization of beam A perpendicular to the plane of beam B.

To generate a single beam of plane-polarized light, one can take advantage of the double-refracting property of Iceland spar. A Nicol prism, invented by the Scottish physicist William Nicol, consists of two crystals of Iceland spar cut to specified angles and cemented by Canada balsam. This prism transmits one of the two beams of plane-polarized light while reflecting the other at a sharp angle so that it does not interfere with the transmitted beam. Plane-polarized light can also be generated by a Polaroid filter, a device invented by E. H. Land, an American physicist. Polaroid filters consist of certain types of crystals embedded in transparent plastic and capable of producing plane-polarized light.

After passing through a first Nicol prism, plane-polarized light can pass through a second Nicol prism, but only if the second prism has its axis oriented so that it is *parallel* to the incident light's plane of polarization. Plane-polarized light is *absorbed* by a Nicol prism that is oriented so that its axis is *perpendicular* to the incident light's plane of polarization. These situations can be illustrated by the picket-fence analogy, as shown in Figure 23.3. Plane-polarized light can pass through a fence whose slats are oriented in the proper direction but is blocked out by a fence whose slats are oriented perpendicularly.

An **optically active substance** is one that interacts with polarized light to rotate the plane of polarization through some angle α. Figure 23.4 illustrates this phenomenon.

23.2 The Polarimeter

An instrument called a **polarimeter** is used to measure the extent to which a substance interacts with polarized light. A schematic diagram of a polarimeter is shown in Figure 23.5. The light from the source lamp is polarized by being passed through a fixed Nicol prism, called a **polarizer.** This light passes through the sample, with which it may or may not interact to have its plane of polarization rotated in one direction or the other. A second,

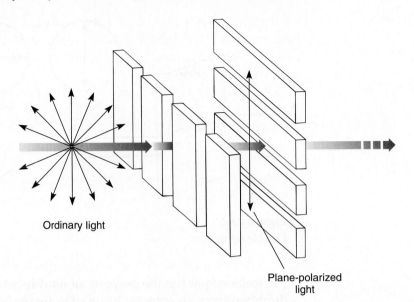

Figure 23.3
The picket-fence analogy.

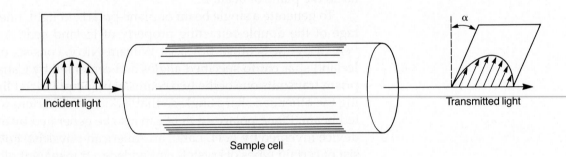

Figure 23.4
Optical activity.

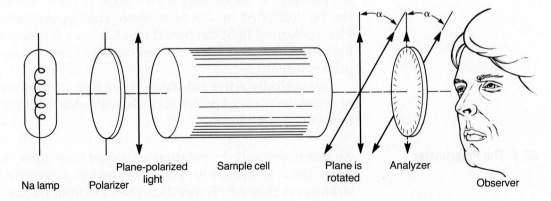

Figure 23.5
Schematic diagram of a polarimeter.

rotatable Nicol prism, called the **analyzer,** is adjusted to allow the maximum amount of light to pass through. The number of degrees and the direction of rotation required for this adjustment are measured to give the **observed rotation** α.

So that data determined by several persons under different conditions can be compared, a standardized means of presenting optical rotation data is necessary. The most common way of presenting such data is by recording the **specific rotation** $[\alpha]_\lambda^t$, which has been corrected for differences in concentration, cell path length, temperature, solvent, and wavelength of the light source. The equation defining the specific rotation of a compound in solution is

$$[\alpha]_\lambda^t = \frac{\alpha}{cl}$$

where α = observed rotation in degrees, c = concentration in grams per milliliter of solution, l = length of sample tube in decimeters, λ = wavelength of light (usually indicated as "D," for the sodium D line), and t = temperature in degrees Celsius. For pure liquids, the density d of the liquid in grams per milliliter replaces c in the preceding formula. You may occasionally want to compare compounds of different molecular weights, so a **molecular rotation,** based on moles instead of grams, is more convenient than a specific rotation. The molecular rotation M_λ^t is derived from the specific rotation $[\alpha]_\lambda^t$ by

$$M_\lambda^t = \frac{[\alpha]_\lambda^t \times \text{Molecular weight}}{100}$$

Usually, measurements are made at 25°C with the sodium D line as a light source; consequently, specific rotations are reported as $[\alpha]_D^{25}$.

Polarimeters that are now available incorporate electronics to determine the angle of rotation of chiral molecules. These instruments are essentially automatic. The only real difference between an automatic polarimeter and a manual one is that a light detector replaces the eye. No visual observation of any kind is made with an automatic instrument. A microprocessor adjusts the analyzer until the light reaching the detector is at a minimum. The angle of rotation is displayed digitally in an LCD window, including the sign of rotation. The simplest instrument is equipped with a sodium lamp that gives rotations based on the sodium D line (589 nm). More expensive instruments use a tungsten lamp and filters so that wavelengths can be varied over a range of values. Using the latter instrument, a chemist can observe rotations at different wavelengths.

23.3 Sample Preparation: The Sample Cell

It is important for the solution whose optical rotation is to be determined to contain no suspended particles of dust, dirt, or undissolved material that might disperse the incident polarized light. Therefore, you must clean the sample cell carefully, and your sample must be free of suspended particles. You must also prevent the presence of any air bubbles in the bore when you fill the cell. Most cells have a stem in the center or an area at one end of the cell where the diameter of the tube is increased. These features are designed to help you catch any bubbles in an area that is above the path that the light takes through the main bore.

Two modern **polarimetry cells** are shown in Figure 23.6. In the first case, the cell is filled until the liquid completely fills the bore and a small

Figure 23.6
Two modern polarimetry cells (Rudolph Research).

portion of the center stem. Then, if one gently rocks the cell back and forth along its axis, bubbles will rise and collect in the stem where they are above the light path. A stopper is placed in the stem when you are finished. In the second case, the cell is filled vertically, and the end is screwed on. Bubbles are trapped at the raised end when the cell is turned horizontally.

Sample cells are available in various lengths, with 0.5 dm and 1.0 dm being the most common. A typical 0.5-dm cell holds about 3–5 mL of solution, but many companies sell **microcells** that have a very narrow-diameter bore and require much less solution. Polarimeter cells are quite expensive because the windows must be made out of quartz rather than ordinary glass. Be sure to handle them carefully and to avoid getting fingerprints on the end windows because this will also disperse the polarized light.

With liquid samples, it is often possible to use the **neat** (undiluted) liquid as your sample. In this case, the concentration of the sample is just the density of the liquid (g/mL). If you have a solid sample or if you have too little of a liquid to fill the cell, you will have to either dissolve or dilute the sample with a solvent. In this case, you must weigh (grams) the amount of material you use and divide by the total volume (mL) to obtain the concentration in g/mL. Water, methanol, and ethanol are the best solvents to use because they are unlikely to attack the cell you are using. Many cells have rubber parts or use a cement to attach the windows to the ends of the bore. Rubber and cements will often dissolve in stronger solvents such as acetone or methylene chloride, thereby damaging the cell. Check with your instructor before using any solvent stronger than water, methanol, or ethanol. These are also the preferred solvents to use for cleaning the cells.

23.4 Operation of the Polarimeter

A. The Zeiss Polarimeter, a Classic Instrument

The procedures given here are for the operation of the Zeiss polarimeter (Figure 23.7), a classic analog instrument with a circular scale and a sodium lamp. Many other older models of polarimeter are operated in a similar fashion.

Before taking any measurements, turn on the sodium lamp and wait 5–10 minutes for the lamp to warm up and stabilize. After the warm-up period is complete, you should make an initial check of the instrument by taking a zero reading with a sample cell filled only with solvent. If the zero reading does not correspond with the zero-degree (0°) calibration mark, then the difference in readings must be used to correct all subsequent readings.

To take the zero measurement, place the polarimeter cell with the sample in the sloped cradle or rack inside the instrument. If you are using a cell

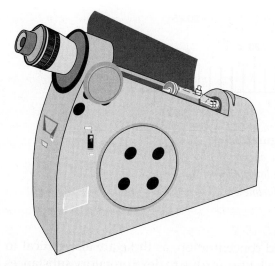

Figure 23.7
The Zeiss polarimeter.

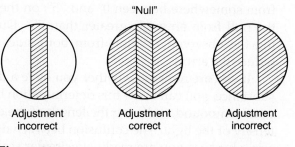

Figure 23.8
Image field sectors in the polarimeter.

with an enlarged end, that end must be placed at the high end of the cradle, making sure that no bubbles are in the bore of the cell. After closing the cover and while watching through the eyepiece, turn the analyzer knob or ring until the proper angle of the analyzer is reached (the angle that allows no light to pass through the instrument). Most analog instruments, including the Zeiss polarimeter, are of the split-field type. When you look upward through the eyepiece, you see a circle split into three sectors (Figure 23.8), with the center sector either lighter or darker than those on either side. The analyzer prism is rotated until all of the sectors are matched in intensity, usually the darker color (see Figure 23.8). This is called the **null** reading.

When you look downward in the eyepiece, you see the value of the angle through which the plane of the polarized light has been rotated (if any) indicated on a vernier degree scale (Figure 23.9). Some polarimeters, such as the original Rudolph polarimeter, have instead a large circular scale, like a halo, attached directly to the knob you turn.

After determining the zero setting on the blank solution, place the polarimeter cell containing your sample into the polarimeter and measure the observed angle of rotation in the same way as described for the zero measurement. Be sure to record not only the numerical value of the reading but also the direction of rotation. Also record the solvent, temperature,

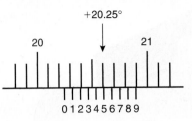

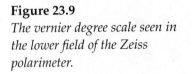

Figure 23.9
*The vernier degree scale seen in
the lower field of the Zeiss
polarimeter.*

and concentration, as these are also critical to the measurement. Rotations clockwise are due to dextrorotatory substances and are indicated by the "+" sign. Rotations counterclockwise are due to levorotatory substances and are indicated by the "−" sign. You should take several readings, including readings for which the value was approached from both sides. In other words, where the actual reading might be +75°, first approach this reading upward from somewhere between 0° and 75°; on the next measurement, approach the null from an angle greater than 75°. Duplicating readings, approaching the observed rotation from both sides, and averaging the readings reduce the error.

If you are not sure whether you have a dextrorotatory or a levorotaory substance, you can make this determination by halving the concentration of your compound, reducing the length of the cell by half, or reducing the intensity of the light. The confusion between dextrorotatory and levorotatory arises because you are reading a circular scale. The null reading can be approached from either direction (clockwise or counterclockwise), starting from zero (see Figure 23.10). For instance, is your null at +120°, or is it at −240°? Both readings are at the same point on the scale. Figure 23.10 shows that by reducing the concentration, the cell length, or the light intensity in half (any one of these), the reading will change, and it will move in a different direction for levorotatory substances than for dextrorotatory substances. The direction of rotation is most often determined by making measurements at different dilutions.

Once you have determined the value and direction of the observed rotation α, you must correct it by the zero value and then use the formulas in Section 23.2 to convert it to the specific rotation $[\alpha]_D$. The specific rotation is always reported as a function of temperature, indicating the wavelength by "D" if a sodium lamp was used, and the solvent and concentration used are reported. For example:

$$[\alpha]_D = +43.8 \ (c = 7.5 \text{ g}/100 \text{ mL, in absolute ethanol})$$

B. The Modern Digital Polarimeter

A modern digital polarimeter, such as the one shown in Figure 23.11, is much easier to operate than the older analog instruments. The modern instrument will store the zero reading for you, subtract it from every subsequent reading automatically, determine the direction of rotation, and calculate the specific rotation from the reading obtained on your sample.

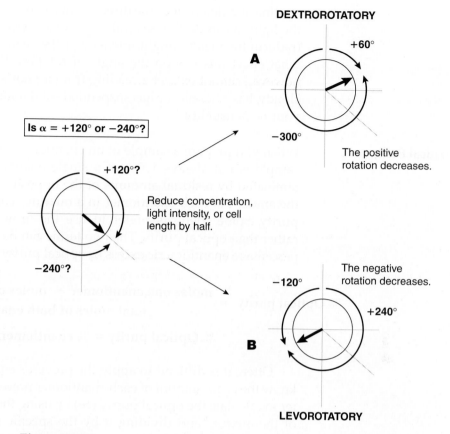

DEXTROROTATORY

A
+60°
−300°
The positive rotation decreases.

+120°?
−240°?
Reduce concentration, light intensity, or cell length by half.

B
−120°
+240°
The negative rotation decreases.

LEVOROTATORY

Figure 23.10

How to determine the direction of rotation. This diagram shows the effect on observed rotation if you reduce by half the concentration of the compound, the light intensity, or the length of the cell. By this method, it is easy to determine if the compound is dextrorotatory (A) or levorotatory (B).

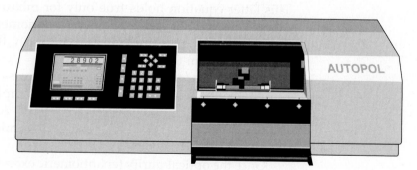

Figure 23.11

The Autopol IV (Rudolph Research), a modern digital polarimeter.

When finished, it can print everything on a sheet of paper for you to take with you. In a typical instrument, you first determine the zero reading and then store it in electronic memory. Once the zero reading is determined, you place your sample in the instrument. The instrument automatically finds the null angle and the direction of rotation and displays it on an LED readout. The instrument approaches the null several times to be sure of its reading and determines the direction of rotation by reducing the intensity of the light. It can do this several ways. One common method is to attenuate (reduce) the incident light intensity of the beam of polarized light and see what effect this has on the angle of rotation. Even a digital polarimeter, however, cannot extract a reading from a poor sample, such as one that is cloudy, has a bubble, or has suspended solid material. A good sample is still your responsibility.

23.5 Optical Purity

When you prepare a sample of an enantiomer by a resolution method, the sample is not always 100% of a single enantiomer. It frequently is contaminated by residual amounts of the opposite stereoisomer. If you know the amount of each enantiomer in a mixture, you can calculate the **optical purity**. Some chemists prefer to use the term **enantiomeric excess (ee)** rather than optical purity. The two terms can be used interchangeably. The percentage enantiomeric excess or optical purity is calculated as follows:

$$\% \text{ Optical purity} = \frac{\text{moles one enantiomer } - \text{ moles of other enantiomer}}{\text{total moles of both enantiomers}} \times 100$$

$$\% \text{ Optical purity} = \% \text{ enantiomeric excess (ee)}$$

Often, it is difficult to apply the previous equation because you do not know the exact amount of each enantiomer present in a mixture. It is far easier to calculate the optical purity (ee) by using the observed specific rotation of the mixture and dividing it by the specific rotation of the pure enantiomer. Values for the pure enantiomers can sometimes be found in literature sources.

$$\% \text{ Optical purity} = \% \text{ enantiomeric excess} = \frac{\text{observed specific rotation}}{\text{specific rotation of pure enantiomer}} \times 100$$

This latter equation holds true only for mixtures of two chiral molecules that are mirror images of each other (enantiomers). If some other chiral substance is present in the mixture as an impurity, then the actual optical purity will deviate from the value calculated.

In a racemic ($\pm$) mixture, there is no excess enantiomer, and the optical purity (enantiomeric excess) is zero; in a completely resolved material, the optical purity (enantiomeric excess) is 100%. A compound that is $x\%$ optically pure contains $x\%$ of one enantiomer and $(100 - x)\%$ of a racemic mixture.

Once the optical purity (enantiomeric excess) is known, the relative percentages of each of the enantiomers can be calculated easily. If the predominant form in the impure, optically active mixture is assumed to be the (+) enantiomer, the percentage of the (+) enantiomer is

$$\left[x + \left(\frac{100 - x}{2}\right)\right]\%$$

and the percentage of the (−) enantiomer is $[(100 - x)/2]\%$. The relative percentages of (+) and (−) forms in a partially resolved mixture of enantiomers can be calculated as shown next. Consider a partially resolved mixture of camphor enantiomers. The specific rotation for pure (+)-camphor is +43.8° in absolute ethanol, but the mixture shows a specific rotation of +26.3°.

$$\text{Optical purity} = \frac{+26.3°}{+43.8°} \times 100 = 60\% \text{ optically pure}$$

$$\% \, (+) \text{ enantiomer} = 60 + \left(\frac{100 - 60}{2}\right) = 80\%$$

$$\% \, (-) \text{ enantiomer} = \left(\frac{100 - 60}{2}\right) = 20\%$$

Notice that the difference between these two calculated values equals the optical purity or enantiomeric excess.

PROBLEMS

1. Calculate the specific rotation of a substance that is dissolved in a solvent (0.4 g/mL) and that has an observed rotation of −10° as determined with a 0.5-dm cell.

2. Calculate the observed rotation for a solution of a substance (2.0 g/mL) that is 80% optically pure. A 2-dm cell is used. The specific rotation for the optically pure substance is +20°.

3. What is the optical purity of a partially racemized product if the calculated specific rotation is −8° and the pure enantiomer has a specific rotation of −10°? Calculate the percentage of each of the enantiomers in the partially racemized product.

---

24   **TECHNIQUE 24**

Refractometry

The **refractive index** is a useful physical property of liquids. Often, a liquid can be identified from a measurement of its refractive index. The refractive index can also provide a measure of the purity of the sample being examined. This is accomplished by comparing the experimentally measured refractive index with the value reported in the literature for an ultrapure sample of the compound. The closer the measured sample's value to the literature value, the purer the sample.

24.1 The Refractive Index

The refractive index has as its basis the fact that light travels at a different velocity in condensed phases (liquids, solids) than in air. The refractive index n is defined as the ratio of the velocity of light in air to the velocity of light in the medium being measured:

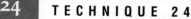

$$n = \frac{V_{\text{air}}}{V_{\text{liquid}}} = \frac{\sin \theta}{\sin \phi}$$

It is not difficult to measure the ratio of the velocities experimentally. It corresponds to $(\sin \theta / \sin \phi)$, where θ is the angle of incidence for a beam of light striking the surface of the medium and ϕ is the angle of refraction of the beam of light *within* the medium. This is illustrated in Figure 24.1.

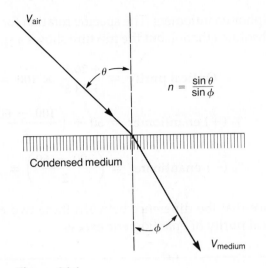

Figure 24.1
The refractive index.

The refractive index for a given medium depends on two variable factors. First, it is *temperature* dependent. The density of the medium changes with temperature; hence, the speed of light in the medium also changes. Second, the refractive index is *wavelength* dependent. Beams of light with different wavelengths are refracted to different extents in the same medium and give different refractive indices for that medium. It is usual to report refractive indices measured at 20°C, with a sodium discharge lamp as the source of illumination. The sodium lamp gives off yellow light of 589-nm wavelength, the so-called sodium D line. Under these conditions, the refractive index is reported in the following form:

$$n_D^{20} = 1.4892$$

The superscript indicates the temperature, and the subscript indicates that the sodium D line was used for the measurement. If another wavelength is used for the determination, the D is replaced by the appropriate value, usually in nanometers (1 nm = 10^{-9} m).

Notice that the hypothetical value reported has four decimal places. It is easy to determine the refractive index to within several parts in 10,000. Therefore, n_D is a very accurate physical constant for a given substance and can be used for identification. However, it is sensitive to even small amounts of impurity in the substance measured. Unless the substance is purified *extensively,* you will not usually be able to reproduce the last two decimal places given in a handbook or other literature source. Typical organic liquids have refractive index values between 1.3400 and 1.5600.

24.2 The Abbé Refractometer   The instrument used to measure the refractive index is called a **refractometer.** Although many styles of refractometer are available, by far the most common instrument is the Abbé refractometer. This style of refractometer has the following advantages:

1. White light may be used for illumination; the instrument is compensated, however, so that the index of refraction obtained is actually that for the sodium D line.

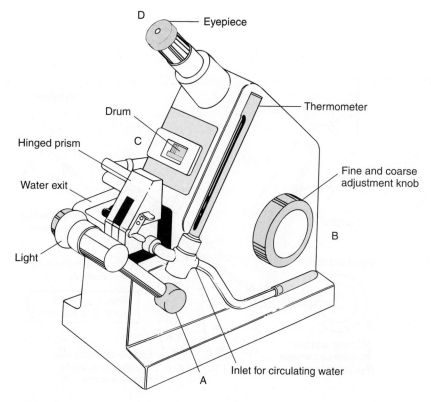

Figure 24.2
Abbé refractometer (Bausch and Lomb Abbé 3L).

2. The prisms can be temperature controlled.

3. Only a small sample is required (a few drops of liquid using the standard method or about 5 μL using a modified technique).

A common type of Abbé refractometer is shown in Figure 24.2.

The optical arrangement of the refractometer is complex; a simplified diagram of the internal workings is given in Figure 24.3. The letters *A*, *B*, *C*, and *D* label corresponding parts in both Figures 24.2 and 24.3. A complete description of refractometer optics is too difficult to attempt here, but Figure 24.3 gives a simplified diagram of the essential operating principles.

Using the standard method, introduce the sample to be measured between the two prisms. If it is a free-flowing liquid, it may be introduced into a channel along the side of the prisms, injected from a Pasteur pipet. If it is a viscous sample, the prisms must be opened (they are hinged) by lifting the upper one; a few drops of liquid are applied to the lower prism with a Pasteur pipet or a wooden applicator. If a Pasteur pipet is used, take care not to touch the prisms because they become scratched easily. When the prisms are closed, the liquid should spread evenly to make a thin film. With highly volatile samples, the remaining operations must be performed rapidly. Even when the prisms are closed, evaporation of volatile liquids can readily occur.

Next, turn on the light and look into the eyepiece *D*. The hinged lamp is adjusted to give the maximum illumination to the visible field in the eyepiece. The light rotates at pivot *A*.

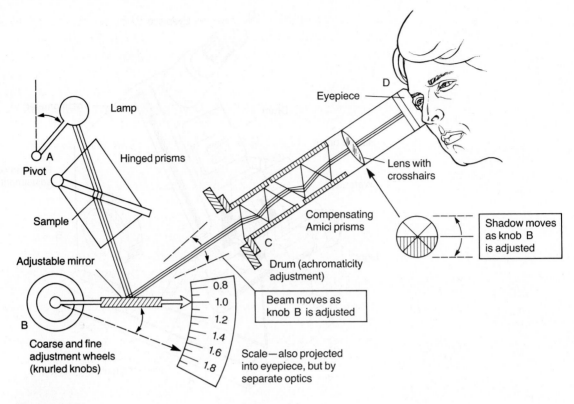

Figure 24.3
Simplified diagram of a refractometer.

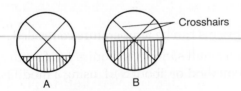

Figure 24.4
(A) Refractometer incorrectly adjusted.
(B) Correct adjustment.

Figure 24.5
Refractometer showing chromatic aberration (color dispersion). The dispersion is incorrectly adjusted.

Rotate the coarse and fine adjustment knobs at *B* until the dividing line between the light and dark halves of the visual field coincides with the center of the crosshairs (Figure 24.4). If the crosshairs are not in sharp focus, adjust the eyepiece to focus them. If the horizontal line dividing the light and dark areas appears as a colored band, as in Figure 24.5, the refractometer shows **chromatic aberration** (color dispersion). This can be adjusted with knob *C* drum (Figure 24.3). This knurled knob rotates a series of prisms, called Amici prisms, that color-compensate the refractometer and cancel out dispersion. Adjust the knob to give a sharp, uncolored division between the light and dark segments. When you have adjusted everything correctly (as in Figure 24.4B), read the refractive index. In the instrument described here, press a small button on the left side of the housing to make the scale visible in the eyepiece. In other refractometers, the scale is visible at all times, frequently through a separate eyepiece.

Occasionally, the refractometer will be so far out of adjustment that it may be difficult to measure the refractive index of an unknown. When this happens, it is wise to place a pure sample of known refractive index in the instrument, set the scale to the correct value of refractive index, and adjust the controls for the sharpest line possible. Once this is done, it is easier to measure an unknown sample. It is especially helpful to perform this procedure prior to measuring the refractive index of a highly volatile sample.

NOTE: There are many styles of refractometer, but most have adjustments similar to those described here.

In the procedure just described, several drops of liquid are required to obtain the refractive index. In some experiments, you may not have enough sample to use this standard method. It is possible to modify the procedure so that a reasonably accurate refractive index can be obtained on about 5 μL of liquid. Instead of placing the sample directly onto the prism, you apply the sample to a small piece of lens paper. The lens paper can be conveniently cut with a handheld paper punch,[1] and the paper disc (0.6-cm diameter) is placed in the center of the bottom prism of the refractometer. To avoid scratching the prism, use forceps or tweezers with plastic tips to handle the disc. About 5 μL of liquid is carefully placed on the lens paper using a microliter syringe. After closing the prisms, adjust the refractometer as described previously and read the refractive index. With this method, the horizontal line dividing the light and dark areas may not be as sharp as it is in the absence of the lens paper. It may also be impossible to eliminate color dispersion completely. Nonetheless, the refractive index values determined by this method are usually within 10 parts in 10,000 of the values determined by the standard procedure.

24.3 Cleaning the Refractometer

In using the refractometer, you should always remember that if the prisms are scratched, the instrument will be ruined.

NOTE: Do not touch the prisms with any hard object.

This admonition includes Pasteur pipets and glass rods.

When measurements are completed, the prisms should be cleaned with ethanol or petroleum ether. Moisten *soft* tissues with the solvent and wipe the prisms *gently.* When the solvent has evaporated from the prism surfaces, the prisms should be locked together. The refractometer should be left with the prisms closed to avoid collection of dust in the space between them. The instrument should also be turned off when it is no longer in use.

24.4 The Digital Refractometer

Today, there are modern digital refractometers available that determine the refractive index of a liquid electronically (Figure 24.6). Once the instrument has been calibrated, it is only necessary to place a drop of your liquid between the prisms (see the inset in Figure 24.6), close the lid, and read the

---

[1] In order to cut the lens paper more easily, place several sheets between two pieces of heavier paper, such as that used for file folders.

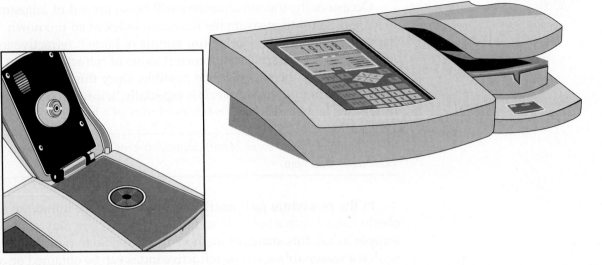

Figure 24.6
The Rudolph J-series, a modern digital refractometer. To make a measurement, place the sample on the lower prism (see the inset) and close the lid.

display. The instrument can make temperature corrections and store the values of your readings in its microprocessor memory. Once again, these instruments must be treated with respect, taking care not to scratch the prisms and to clean them after use.

24.5 Temperature Corrections

Most refractometers are designed so that circulating water at a constant temperature can maintain the prisms at 20°C. If this temperature-control system is not used or if the water is not at 20°C, a temperature correction must be made. Although the magnitude of the temperature correction may vary from one class of compound to another, a value of 0.00045 per degree Celsius is a useful approximation for most substances. The index of refraction of a substance *decreases* with *increasing* temperature. Therefore, add the correction to the observed n_D value for temperatures higher than 20°C and subtract it for temperatures lower than 20°C. For example, the reported n_D value for nitrobenzene is 1.5529. One would observe a value at 25°C of 1.5506. The temperature correction would be made as follows:

$$n_D^{20} = 1.5506 + 5(0.00045) = 1.5529$$

PROBLEMS

1. A solution consisting of isobutyl bromide and isobutyl chloride is found to have a refractive index of 1.3931 at 20°C. The refractive indices at 20°C of isobutyl bromide and isobutyl chloride are 1.4368 and 1.3785, respectively. Determine the molar composition (in percent) of the mixture by assuming a linear relation between the refractive index and the molar composition of the mixture.

2. The refractive index of a compound at 16°C is found to be 1.3982. Correct this refractive index to 20°C.

25 **TECHNIQUE 25**

Infrared Spectroscopy

Almost any compound having covalent bonds, whether organic or inorganic, will be found to absorb frequencies of electromagnetic radiation in the infrared region of the spectrum. The infrared region of the electromagnetic spectrum lies at wavelengths longer than those associated with visible light, which includes wavelengths from approximately 400 nm to 800 nm (1 nm = 10^{-9} m) but at wavelengths shorter than those associated with radio waves, which have wavelengths longer than 1 cm. For chemical purposes, we are interested in the *vibrational* portion of the infrared region. This portion includes radiations with wavelengths (λ) between 2.5 μm and 15 μm (1 μm = 10^{-6} m). The relation of the infrared region to other regions included in the electromagnetic spectrum is illustrated in Figure 25.1.

As with other types of energy absorption, molecules are excited to a higher energy state when they absorb infrared radiation. The absorption of the infrared radiation is, like other absorption processes, a quantized process. Only selected frequencies (energies) of infrared radiation are absorbed by a molecule. The absorption of infrared radiation corresponds to energy changes on the order of 8–40 kJ/mole (2–10 kcal/mole). Radiation in this energy range corresponds to the range encompassing the stretching and bending vibrational frequencies of the bonds in most covalent molecules. In the absorption process, those frequencies of infrared radiation that match the natural vibrational frequencies of the molecule in question are absorbed, and the energy absorbed increases the *amplitude* of the vibrational motions of the bonds in the molecule.

Most chemists refer to the radiation in the vibrational infrared region of the electromagnetic spectrum by units called **wavenumbers** ($\bar{\nu}$). Wavenumbers are expressed in reciprocal centimeters (cm^{-1}) and are easily computed by taking the reciprocal of the wavelength (λ) expressed in centimeters. This unit has the advantage, for those performing calculations, of being directly proportional to energy. Thus, the vibrational infrared region of the spectrum extends from about 4000 cm^{-1} to 650 cm^{-1} (or wavenumbers).

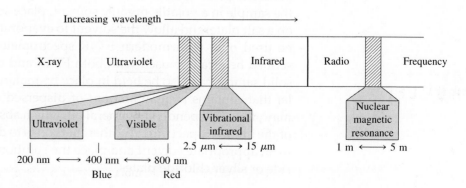

Figure 25.1
A portion of the electromagnetic spectrum showing the relation of vibrational infrared radiation to other types of radiation.

Wavelengths (μm) and wavenumbers (cm^{-1}) can be interconverted by the following relationships:

$$cm^{-1} = \frac{1}{(\mu m)} \times 10,000$$

$$\mu m = \frac{1}{(cm)^{-1}} \times 10,000$$

PART A. SAMPLE PREPARATION AND RECORDING THE SPECTRUM

25.1 Introduction

To determine the infrared spectrum of the compound, one must place the compound in a sample holder or cell. In infrared spectroscopy, this immediately poses a problem. Glass, quartz, and plastics absorb strongly throughout the infrared region of the spectrum (any compound with covalent bonds usually absorbs) and cannot be used to construct sample cells. Ionic substances must be used in cell construction. Metal halides (sodium chloride, potassium bromide, silver chloride) are commonly used for this purpose.

Sodium Chloride Cells. Single crystals of sodium chloride are cut and polished to give plates that are transparent throughout the infrared region. These plates are then used to fabricate cells that can be used to hold *liquid* samples. Because sodium chloride is water soluble, samples must be *dry* before a spectrum can be obtained. In general, sodium chloride plates are preferred for most applications involving liquid samples. Potassium bromide plates may also be used in place of sodium chloride.

Silver Chloride Cells. Cells may be constructed of silver chloride. These plates may be used for *liquid* samples that contain small amounts of water, because silver chloride is water insoluble. However, because water absorbs in the infrared region, as much water as possible should be removed, even when silver chloride is used. Silver chloride plates must be stored in the dark. They darken when exposed to light, and they cannot be used with compounds that have an amino functional group. Amines react with silver chloride.

Solid Samples. The easiest way to hold a *solid* sample in place is to dissolve the sample in a volatile organic solvent, place several drops of this solution on a salt plate, and allow the solvent to evaporate. This dry film method can be used only with modern FT-IR spectrometers. The other methods described here can be used with both FT-IR and dispersion spectrometers. A solid sample can also be held in place by making a potassium bromide pellet that contains a small amount of dispersed compound. A solid sample may also be suspended in mineral oil, which absorbs only in specific regions of the infrared spectrum. Another method is to dissolve the solid compound in an appropriate solvent and place the solution between two sodium chloride or silver chloride plates.

25.2 Liquid Samples— NaCl Plates

The simplest method of preparing the sample, if it is a liquid, is to place a thin layer of the liquid between two sodium chloride plates that have been ground flat and polished. This is the method of choice when you need to

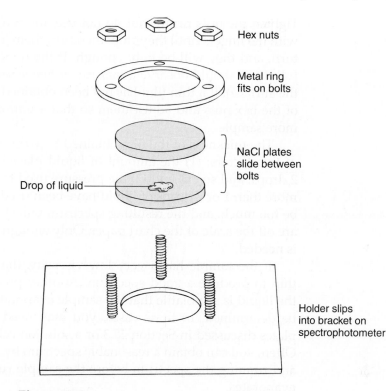

Hex nuts

Metal ring
fits on bolts

NaCl plates
slide between
bolts

Drop of liquid

Holder slips
into bracket on
spectrophotometer

Figure 25.2
Salt plates and holder.

determine the infrared spectrum of a pure liquid. A spectrum determined by this method is referred to as a **neat** spectrum. No solvent is used. The polished plates are expensive because they are cut from a large, single crystal of sodium chloride. Salt plates break easily, and they are water soluble.

Preparing the Sample. Obtain two sodium chloride plates and a holder from the desiccator where they are stored. Moisture from fingers will mar and occlude the polished surfaces. Samples that contain water will destroy the plates.

NOTE: The plates should be touched only on their edges. Be certain to use a sample that is dry or free from water.

Add 1 or 2 drops of the liquid to the surface of one plate and then place the second plate on top.[1] The pressure of this second plate causes the liquid to spread out and form a thin capillary film between the two plates. As shown in Figure 25.2, set the plates between the bolts in a holder and place the metal ring carefully on the salt plates. Use the hex nuts to hold the salt plates in place.

NOTE: Do not overtighten the nuts or the salt plates will cleave or split.

---

[1] Use a Pasteur pipet or a short length of microcapillary tubing. If you use the microcapillary tubing, it can be filled by touching it into the liquid sample. When you touch it (lightly) to the salt plate, it will empty. Be careful not to scratch the plate.

Tighten the nuts firmly, but do not use any force to turn them. Spin them with the fingers until they stop; then turn them just another fraction of a full turn, and they will be tight enough. If the nuts have been tightened carefully, you should observe a *transparent film of sample* (a uniform wetting of the surface). If a thin film has not been obtained, either loosen one or more of the hex nuts and adjust them so that a uniform film is obtained or add more sample.

The thickness of the film obtained between the two plates is a function of two factors: (1) the amount of liquid placed on the first plate (1 drop, 2 drops, and so on) and (2) the pressure used to hold the plates together. If more than 1 or 2 drops of liquid have been used, the amount will probably be too much, and the resulting spectrum will show strong absorptions that are off the scale of the chart paper. Only enough liquid to wet both surfaces is needed.

If the sample has a very low viscosity, the capillary film may be too thin to produce a good spectrum. Another problem you may find is that the liquid is so volatile that the sample evaporates before the spectrum can be determined. In these cases, you may need to use the silver chloride plates discussed in Section 25.3 or a solution cell described in Section 25.6. Often, you can obtain a reasonable spectrum by assembling the cell quickly and running the spectrum before the sample runs out of the salt plates or evaporates.

Determining the Infrared Spectrum. Slide the holder into the slot in the sample beam of the spectrophotometer. Determine the spectrum according to the instructions provided by your instructor. In some cases, your instructor may ask you to calibrate your spectrum. If this is the case, refer to Section 25.8.

Cleaning and Storing the Salt Plates. Once the spectrum has been determined, demount the holder and rinse the salt plates with methylene chloride (or *dry* acetone). (Keep the plates away from water!) Use a soft tissue, moistened with the solvent, to wipe the plates. If some of your compound remains on the plates, you may observe a shiny surface. Continue to clean the plates with solvent until no more compound remains on the surfaces of the plates.

CAUTION	

Avoid direct contact with methylene chloride. Return the salt plates and holder to the desiccator for storage.

25.3 Liquid Samples— AgCl Plates

The minicell shown in Figure 25.3 may also be used with liquids.[2] The cell assembly consists of a two-piece threaded body, an O-ring, and two silver chloride plates. The plates are flat on one side, and there is a circular depression (0.025 mm or 0.10 mm deep) on the other side of the plate. An advantage of using silver chloride plates is that they may be used with wet samples or

---

[2] The Wilks Mini-Cell liquid sample holder is available from the Foxboro Company, 151 Woodward Avenue, South Norwalk, CT 06856. We recommend the AgCl cell windows with 0.10-mm depression rather than the 0.025-mm depression.

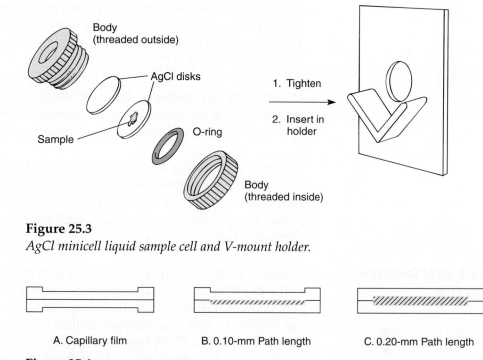

Figure 25.3
AgCl minicell liquid sample cell and V-mount holder.

A. Capillary film   B. 0.10-mm Path length   C. 0.20-mm Path length

Figure 25.4
Path-length variations for AgCl plates.

solutions. A disadvantage is that silver chloride darkens when exposed to light for extended periods. Silver chloride plates also scratch more easily than salt plates and react with amines.

Preparing the Sample. Silver chloride plates should be handled in the same way as salt plates. Unfortunately, they are smaller and thinner (about like a contact lens) than salt plates, and care must be taken not to lose them! Remove them from the light-tight container with care. It is difficult to tell which side of the plate has the slight circular depression. Your instructor may have etched a letter on each plate to indicate which side is the flat one. To determine the infrared spectrum of a pure liquid (neat spectrum), select the flat side of each silver chloride plate. Insert the O-ring into the cell body as shown in Figure 25.3, place the plate into the cell body with the flat surface up, and add 1 drop or less of liquid to the plate.

NOTE: Do not use amines with AgCl plates.

Place the second plate on top of the first with the flat side down. The orientation of the silver chloride plates is shown in Figure 25.4A. This arrangement is used to obtain a capillary film of your sample. Screw the top of the minicell into the body of the cell so that the silver chloride plates are held firmly together. A tight seal forms because AgCl deforms under pressure.

Other combinations may be used with these plates. For example, you may vary the sample path length by using the orientations shown in Figures 25.4B and C. If you add your sample and the 0.10-mm depression of one plate and cover it with the flat side of the other one, you obtain a path length of 0.10 mm (Figure 25.4B). This arrangement is useful for analyzing volatile or low-viscosity liquids. Placement of the two plates with their depressions toward each other gives a path length of 0.20 mm (Figure 25.4C). This orientation may be used for a solution of a solid (or liquid) in carbon tetrachloride (Section 25.6B).

Determining the Spectrum. Slide the V-mount holder shown in Figure 25.3 into the slot on the infrared spectrophotometer. Set the cell assembly in the V-mount holder and determine the infrared spectrum of the liquid.

Cleaning and Storing the AgCl Plates. Once the spectrum has been determined, the cell assembly holder should be demounted and the AgCl plates rinsed with methylene chloride or acetone. Do not use tissue to wipe the plates, because they scratch easily. AgCl plates are light sensitive. Store the plates in a light-tight container.

25.4 Solid Samples— Dry Film

A simple method for determining the infrared spectrum of a solid sample is the **dry film** method. This method is easier than the other methods described here, it does not require any specialized equipment, and the spectra are excellent.[3] The disadvantage is that the dry film method can be used only with modern FT-IR spectrometers.

To use this method, place about 5 mg of your solid sample in a small, clean test tube. Add about 5 drops of methylene chloride (or diethyl ether, pentane, or dry acetone), and stir the mixture to dissolve the solid. Using a Pasteur pipet (not a capillary tube), place several drops of the solution on the face of a salt plate. Allow the solvent to evaporate; a uniform deposit of your product will remain as a dry film coating the salt plate. Mount the salt plate on a V-shaped holder in the infrared beam. Note that only one salt plate is used; the second salt plate is not used to cover the first. Once the salt plate is positioned properly, you may determine the spectrum in the normal manner. With this method, it is *very important* that you clean your material off the salt plate. When you are finished, use methylene chloride or dry acetone to clean the salt plate.

25.5 Solid Samples—KBr Pellets and Nujol Mulls

The methods described in this section can be used with both FT-IR and dispersion spectrometers.

A. KBr Pellets

One method of preparing a solid sample is to make a **potassium bromide (KBr) pellet.** When KBr is placed under pressure, it melts, flows, and seals the sample into a solid solution, or matrix. Because potassium bromide does not absorb in the infrared spectrum, a spectrum can be obtained on a sample without interference.

Preparing the Sample. Remove the agate mortar and pestle from the desiccator for use in preparing the sample. (Take care of them; they are expensive.) Grind 1 mg (0.001 g) of the solid sample for 1 minute in the agate mortar. At this point, the particle size will become so small that the surface of the

---

[3] Feist, P. L. "Sampling Techniques for Organic Solids in IR Spectroscopy: Thin Solid Films as the Method of Choice in Teaching Laboratories." *Journal of Chemical Education,* 78 (2001): 351.

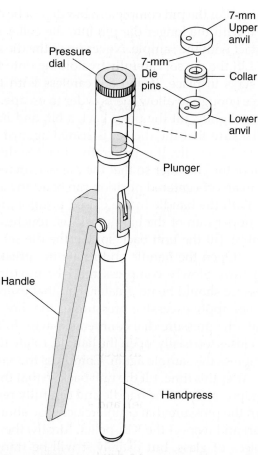

Figure 25.5
Making a KBr pellet with a handpress.

solid appears shiny. Add 80 mg (0.080 g) of *powdered* potassium bromide and grind the mixture for about 30 seconds with the pestle. Scrape the mixture into the middle with a spatula and grind the mixture again for about 15 seconds. This grinding operation helps to mix the sample thoroughly with the KBr. You should work as rapidly as possible because KBr absorbs water. The sample and KBr must be finely ground, or the mixture will scatter the infrared radiation excessively. Using your spatula, heap the mixture in the center of the mortar. Return the bottle of potassium bromide to the desiccator where it is stored when it is not in use.

The sample and potassium bromide should be weighed on an analytical balance the first few times that a pellet is prepared. After some experience, you can estimate these quantities quite accurately by eye.

Making a Pellet Using a KBr Handpress. Two methods are commonly used to prepare KBr pellets. The first method uses the handpress apparatus shown in Figure 25.5.[4] Remove the die set from the storage container. Take extreme care to avoid scratching the polished surfaces of the die set. Place the anvil with the shorter die pin (lower anvil in Figure 25.5) on a bench. Slip the collar over the pin. Remove about one fourth of your KBr mixture with a spatula and transfer it into the collar. The powder may not cover

---

[4] KBr Quick Press unit is available from Wilmad Glass Company, Inc., Route 40 and Oak Road, Buena, NJ 08310.

the head of the pin completely, but do not be concerned about this. Place the anvil with the longer die pin into the collar so that the die pin comes into contact with the sample. Never press the die set unless it contains a sample.

Lift the die set carefully by holding onto the lower anvil so that the collar stays in place. If you are careless with this operation, the collar may move enough to allow the powder to escape. Open the handle of the hand-press slightly, tilt the press back a bit, and insert the die set into the press. Make sure that the die set is seated against the side wall of the chamber. Close the handle. It is imperative that the die set be seated against the side wall of the chamber so that the die is centered in the chamber. Pressing the die in an off-centered position can bend the anvil pins.

With the handle in the closed position, rotate the pressure dial so that the upper ram of the handpress just touches the upper anvil of the die assembly. Tilt the unit back so that the die set does not fall out of the hand-press. Open the handle and rotate the pressure dial clockwise about one-half turn. Slowly compress the KBr mixture by closing the handle. The pressure should be no greater than that exerted by a very firm handshake. Do not apply excessive pressure or the dies may be damaged. If in doubt, rotate the pressure dial counterclockwise to lower the pressure. If the handle closes too easily, open the handle, rotate the pressure dial clockwise, and compress the sample again. Compress the sample for about 60 seconds.

After this time, tilt the unit back so that the die set does not fall out of the handpress. Open the handle and carefully remove the die set from the unit. Turn the pressure dial counterclockwise about one full turn. Pull the die set apart and inspect the KBr pellet. Ideally, the pellet should appear clear like a piece of glass, but usually it will be translucent or somewhat opaque. There may be some cracks or holes in the pellet. The pellet will produce a good spectrum, even with imperfections, as long as light can travel through the pellet. Clean the dies using the procedure outlined on page 841.

Making a Pellet with a KBr Minipress. The second method of preparing a pellet uses the minipress apparatus shown in Figure 25.6. Obtain a ground KBr mixture as described in "Preparing the Sample" and transfer a portion of the finely ground powder (usually not more than half) into a die that compresses it into a translucent pellet. As shown in Figure 25.6, the die consists of two stainless steel bolts and a threaded barrel. The bolts have their ends ground flat. To use this die, screw one of the bolts into the barrel, but not all the way; leave one or two turns. Carefully add the powder with a

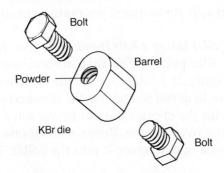

Figure 25.6
Making a KBr pellet with a minipress.

spatula into the open end of the partly assembled die and tap it lightly on the benchtop to give an even layer on the face of the bolt. While keeping the barrel upright, carefully screw the second bolt into the barrel until it is finger tight. Insert the head of the bottom bolt into the hexagonal hole in a plate bolted to the benchtop. This plate keeps the head of one bolt from turning. The top bolt is tightened with a torque wrench to compress the KBr mixture. Continue to turn the torque wrench until you hear a loud click (the ratchet mechanism makes softer clicks) or until you reach the appropriate torque value (20 ft-lb). If you tighten the bolt beyond this point, you may twist the head off one of the bolts. Leave the die under pressure for about 60 seconds; then reverse the ratchet on the torque wrench or pull the torque wrench in the opposite direction to open the assembly. When the two bolts are loose, hold the barrel horizontally and carefully remove the two bolts. You should observe a clear or translucent KBr pellet in the center of the barrel. Even if the pellet is not totally transparent, you should be able to obtain a satisfactory spectrum as long as light passes through the pellet.

Determining the Infrared Spectrum. To obtain the spectrum, slide the holder appropriate for the type of die that you are using into the slot on the infrared spectrophotometer. Set the die containing the pellet in the holder so that the sample is centered in the optical path. Obtain the infrared spectrum. If you are using a double-beam instrument, you may be able to compensate (at least partially) for a marginal pellet by placing a wire screen or attenuator in the reference beam, thereby balancing the lowered transmittance of the pellet. An FT-IR instrument will automatically deal with the low intensity if you select the "autoscale" option.

Problems with an Unsatisfactory Pellet. If the pellet is unsatisfactory (too cloudy to pass light), one of several things may have been wrong:

1. The KBr mixture may not have been ground finely enough, and the particle size may be too big. The large particle size creates too much light scattering.
2. The sample may not be dry.
3. Too much sample may have been used for the amount of KBr taken.
4. The pellet may be too thick; that is, too much of the powdered mixture was put into the die.
5. The KBr may have been "wet" or have acquired moisture from the air while the mixture was being ground in the mortar.
6. The sample may have a low melting point. Low-melting solids not only are difficult to dry but also melt under pressure. You may need to dissolve the compound in a solvent and run the spectrum in solution (Section 25.6).

Cleaning and Storing the Equipment. After you have determined the spectrum, punch the pellet out of the die with a wooden applicator stick (a spatula should not be used as it may scratch the dies). Remember that the polished faces of the die set must not be scratched or they become useless. After the pellet has been punched out, wash all parts of the die set or minipress with warm water. Then rinse the parts with acetone and dry them using a Kimwipe. Check with your instructor to see if there are additional instructions for cleaning the die set. Return the dies to the storage container. Wash the mortar and pestle with water, dry them carefully with paper towels, and return them to the desiccator. Return the KBr powder to its desiccator.

B. Nujol Mulls

If an adequate KBr pellet cannot be obtained or if the solid is insoluble in a suitable solvent, the spectrum of a solid may be determined as a **Nujol mull.** In this method, finely grind about 5 mg of the solid sample in an agate mortar with a pestle. Then add 1 or 2 drops of Nujol mineral oil (white) and grind the mixture to a very fine dispersion. The solid is not dissolved in the Nujol; it is actually a suspension. This mull is then placed between two salt plates using a rubber policeman. Mount the salt plates in the holder in the same way as for liquid samples (Section 25.2).

Nujol is a mixture of high-molecular-weight hydrocarbons. Hence, it has absorptions in the C—H stretch and CH_2 and CH_3 bending regions of the spectrum (Figure 25.7). Clearly, if Nujol is used, no information can be obtained in these portions of the spectrum. In interpreting the spectrum, you must ignore these Nujol peaks. It is important to label the spectrum immediately after it was determined, noting that it was determined as a Nujol mull. Otherwise, you might forget that the C—H peaks belong to Nujol and not to the dispersed solid.

25.6 Solid Samples— Solution Spectra

A. Method A—Solution between Salt (NaCl) Plates

For substances that are soluble in carbon tetrachloride, a quick and easy method for determining the spectra of solids is available. Dissolve as much solid as possible in 0.1 mL of carbon tetrachloride. Place 1 or 2 drops of the solution between sodium chloride plates in precisely the same manner as used for pure liquids (Section 25.2). The spectrum is determined as described for pure liquids using salt plates (Section 25.2). You should work as

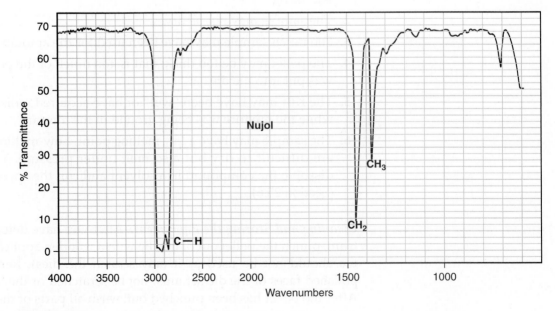

Figure 25.7
Infrared spectrum of Nujol (mineral oil).

quickly as possible. If there is a delay, the solvent will evaporate from between the plates before the spectrum is recorded. Because the spectrum contains the absorptions of the solute superimposed on the absorptions of carbon tetrachloride, it is important to remember that any absorption that appears near 800 cm^{-1} may be due to the stretching of the C—Cl bond of the solvent. Information contained to the right of about 900 cm^{-1} is not usable in this method. There are no other interfering bands for this solvent (see Figure 25.8), and any other absorptions can be attributed to your sample. Chloroform solutions should not be studied by this method because the solvent has too many interfering absorptions (see Figure 25.9).

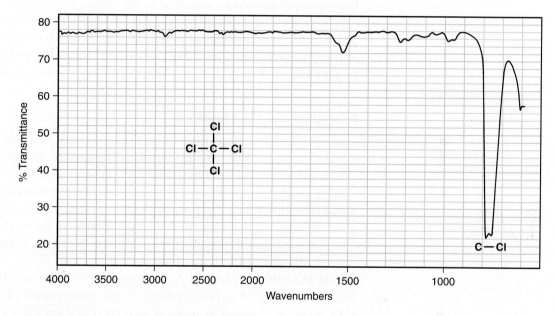

Figure 25.8
Infrared spectrum of carbon tetrachloride.

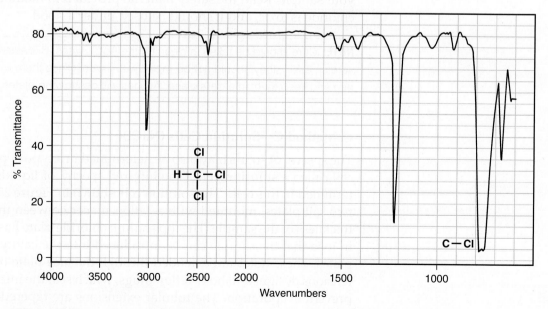

Figure 25.9
Infrared spectrum of chloroform.

> **CAUTION**   ⚠
>
> **Carbon tetrachloride is a hazardous solvent. Work under the hood!**

Carbon tetrachloride, besides being toxic, is suspected of being a carcinogen. In spite of the health problems associated with its use, there is no suitable alternative solvent for infrared spectroscopy. Other solvents have too many interfering infrared absorption bands. Handle carbon tetrachloride carefully to minimize the adverse health effects. The spectroscopic-grade carbon tetrachloride should be stored in a glass-stoppered bottle in a hood. A Pasteur pipet should be attached to the bottle, possibly by storing it in a test tube taped to the side of the bottle. All sample preparation should be conducted in a hood. Rubber or plastic gloves should be worn. The cells should also be cleaned in the hood. All carbon tetrachloride used in preparing samples should be disposed of in an appropriately marked waste container.

B. Method B—AgCl Minicell

The AgCl minicell described in Section 25.3 may be used to determine the infrared spectrum of a solid dissolved in carbon tetrachloride. Prepare a 5–10% solution (5–10 mg in 0.1 mL) in carbon tetrachloride. If it is not possible to prepare a solution of this concentration because of low solubility, dissolve as much solid as possible in the solvent. Following the instructions given in Section 25.3, position the AgCl plates as shown in Figure 25.4C to obtain the maximum possible path length of 0.20 mm. When the cell is tightened firmly, the cell will not leak.

As indicated in method A, the spectrum will contain the absorptions of the dissolved solid superimposed on the absorptions of carbon tetrachloride. A strong absorption appears near 800 cm^{-1} for C—Cl stretch in the solvent. No useful information may be obtained for the sample to the right of about 900 cm^{-1}, but other bands that appear in the spectrum will belong to your sample. Read the safety material provided in method A. Carbon tetrachloride is toxic, and it should be used under a hood.

NOTE: Care should be taken in cleaning the AgCl plates. Because AgCl plates scratch easily, they should not be wiped with tissue. Rinse them with methylene chloride and keep them in a dark place. Amines will destroy the plates.

C. Method C—Solution Cells (NaCl)

The spectra of solids may also be determined in a type of permanent sample cell called a **solution cell.** (The infrared spectra of liquids may also be determined in this cell.) The solution cell, shown in Figure 25.10, is made from two salt plates, mounted with a Teflon spacer between them to control the thickness of the sample. The top sodium chloride plate has two holes drilled in it so that the sample can be introduced into the cavity between the two plates. These holes are extended through the face plate by two tubular extensions designed to hold Teflon plugs, which seal the internal chamber and prevent evaporation. The tubular extensions are tapered so that a syringe body (Luer lock without a needle) will fit snugly into them from the outside. The cells are thus filled from a syringe; usually, they are held upright and filled from the bottom entrance port.

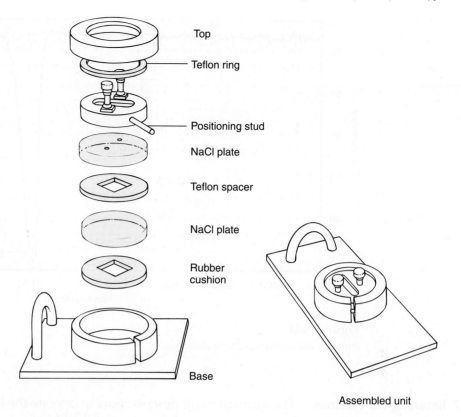

Top
Teflon ring
Positioning stud
NaCl plate
Teflon spacer
NaCl plate
Rubber cushion
Base

Assembled unit

Figure 25.10
A solution cell.

These cells are expensive, and you should try either method A or B before using solution cells. If you do need them, obtain your instructor's permission and receive instruction before using the cells. The cells are purchased in matched pairs, with identical path lengths. Dissolve a solid in a suitable solvent, usually carbon tetrachloride, and add the solution to one of the cells (**sample cell**) as described in the previous paragraph. The pure solvent, identical to that used to dissolve the solid, is placed in the other cell (**reference cell**). The spectrum of the solvent is subtracted from the spectrum of the solution (not always completely), and a spectrum of the solute is thus provided. For the solvent compensation to be as exact as possible and to avoid contamination of the reference cell, it is essential that one cell be used as a reference and that the other cell be used as a sample cell without ever being interchanged. After the spectrum is determined, it is important to clean the cells by flushing them with clean solvent. They should be dried by passing dry air through the cell.

Solvents most often used in determining infrared spectra are carbon tetrachloride (Figure 25.8), chloroform (Figure 25.9), and carbon disulfide (Figure 25.11). A 5–10% solution of solid in one of these solvents usually gives a good spectrum. Carbon tetrachloride and chloroform are suspected carcinogens; however, because there are no suitable alternative solvents, these compounds must be used in infrared spectroscopy. The procedure outlined on page 844 for carbon tetrachloride should be followed. This procedure serves equally well for chloroform.

NOTE: Before you use the solution cells, you must obtain the instructor's permission and instruction on how to fill and clean the cells.

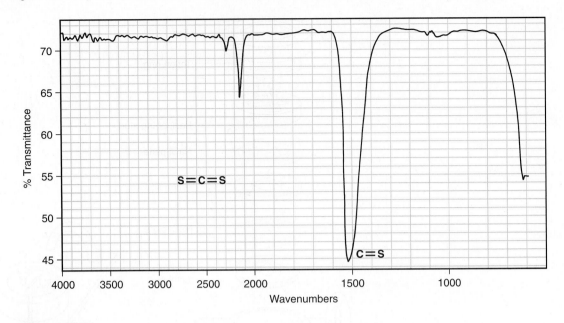

Figure 25.11
Infrared spectrum of carbon disulfide.

25.7 Recording the Spectrum

The instructor will describe how to operate the infrared spectrophotometer because the controls vary considerably, depending on the manufacturer, model of the instrument, and type. For example, some instruments involve pushing only a few buttons, whereas others use a more complicated computer interface system.

In all cases, it is important that the sample, the solvent, the type of cell or method used, and any other pertinent information be written on the spectrum immediately after the determination. This information may be important, and it is easily forgotten if not recorded. You may also need to calibrate the instrument (Section 25.8).

25.8 Calibration

For some instruments, the frequency scale of the spectrum must be calibrated so that you know the position of each absorption peak precisely. You can recalibrate by recording a very small portion of the spectrum of polystyrene over the spectrum of your sample. The complete spectrum of polystyrene is shown in Figure 25.12. The most important of these peaks is at 1603 cm^{-1}; other useful peaks are at 2850 cm^{-1} and 906 cm^{-1}. After you record the spectrum of your sample, substitute a thin film of polystyrene for the sample cell and record the **tips** (not the entire spectrum) of the most important peaks over the sample spectrum.

It is always a good idea to calibrate a spectrum when the instrument uses chart paper with a preprinted scale. It is difficult to align the paper properly so that the scale matches the absorption lines precisely. You often need to know the precise values for certain functional groups (for example, the carbonyl group). Calibration is essential in these cases.

With computer-interfaced instruments, the instrument does not need to be calibrated. With this type of instrument, the spectrum and scale are printed on blank paper at the same time. The instrument has an internal calibration that ensures that the positions of the absorptions are known precisely and that they are placed at the proper positions on the scale. With this

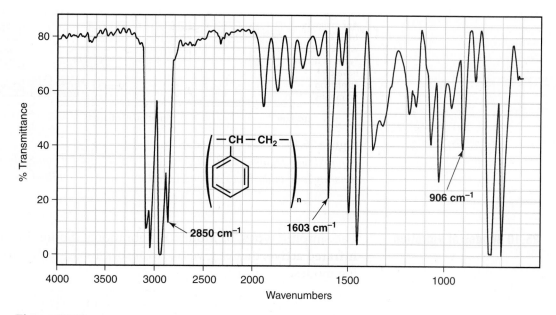

Figure 25.12
Infrared spectrum of polystyrene (thin film).

type of instrument, it is often possible to print a list of the locations of the major peaks, as well as to obtain the complete spectrum of your compound.

PART B. INFRARED SPECTROSCOPY

25.9 Uses of the Infrared Spectrum

Because every type of bond has a different natural frequency of vibration and because the same type of bond in two different compounds is in a slightly different environment, no two molecules of different structure have exactly the same infrared absorption pattern, or **infrared spectrum.** Although some of the frequencies absorbed in the two cases might be the same, in no case of two different molecules will their infrared spectra (the patterns of absorption) be identical. Thus, the infrared spectrum can be used to identify molecules much as a fingerprint can be used to identify people. Comparing the infrared spectra of two substances thought to be identical will establish whether or not they are in fact identical. If the infrared spectra of two substances coincide peak for peak (absorption for absorption), in most cases the substances are identical.

A second and more important use of the infrared spectrum is that it gives structural information about a molecule. The absorptions of each type of bond (N—H, C—H, O—H, C—X, C=O, C—O, C—C, C=C, C≡C, C≡N, and so on) are regularly found only in certain small portions of the vibrational infrared region. A small range of absorption can be defined for each type of bond. Outside this range, absorptions will normally be due to some other type of bond. Thus, for instance, any absorption in the range 3000 ± 150 cm^{-1} will almost always be due to the presence of a CH bond in the molecule; an absorption in the range 1700 ± 100 cm^{-1} will normally be due to the presence of a C=O bond (carbonyl group) in the molecule. The same type of range applies to each type of bond. The way these are spread out over the vibrational infrared is illustrated schematically in Figure 25.13. It is a good idea to remember this general scheme for future convenience.

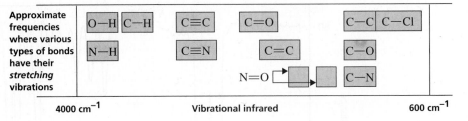

Figure 25.13
Approximate regions in which various common types of bonds absorb. (Bending, twisting, and other types of bond vibration have been omitted for clarity.)

25.10 Modes of Vibration

The simplest types, or **modes,** of vibrational motion in a molecule that are **infrared active,** that is, give rise to absorptions, are the stretching and bending modes.

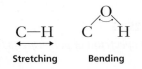

Other, more complex types of stretching and bending are also active, however. To introduce several words of terminology, the normal modes of vibration for a methylene group are shown below.

In any group of three or more atoms—at least two of which are identical—there are *two* modes of stretching or bending: the symmetric mode and asymmetric mode. Examples of such groupings are —CH$_3$, —CH$_2$—, —NO$_2$, —NH$_2$, and anhydrides (CO)$_2$O. For the anhydride, owing to asymmetric and symmetric modes of stretch, this functional group gives *two* absorptions in the C=O region. A similar phenomenon is seen for amino groups, where primary amines usually have *two* absorptions in the NH stretch region, whereas secondary amines R$_2$NH have only one absorption peak. Amides show similar bands. There are two strong N=O stretch peaks for a nitro group, which are caused by asymmetric and symmetric stretching modes.

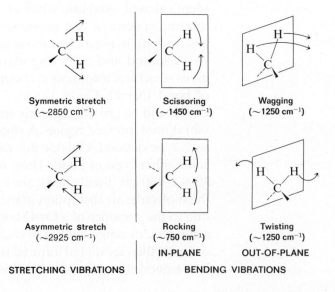

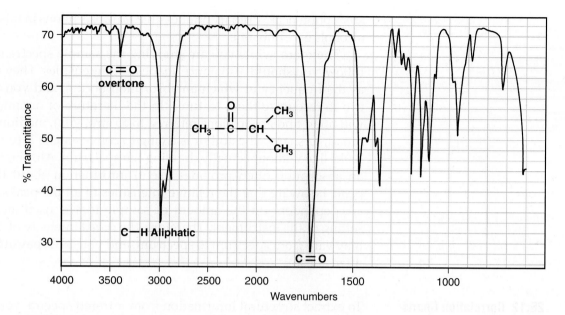

Figure 25.14
Infrared spectrum of methyl isopropyl ketone (neat liquid, salt plates).

25.11 What to Look for in Examining Infrared Spectra

The instrument that determines the absorption spectrum for a compound is called an **infrared spectrophotometer.** The spectrophotometer determines the relative strengths and positions of all the absorptions in the infrared region and plots this information on a piece of paper. This plot of absorption intensity versus wavenumber or wavelength is referred to as the **infrared spectrum** of the compound. A typical infrared spectrum, that of methyl isopropyl ketone, is shown in Figure 25.14.

The strong absorption in the middle of the spectrum corresponds to C=O, the carbonyl group. Note that the C=O peak is intense. In addition to the characteristic position of absorption, the **shape** and **intensity** of this peak are also unique to the C=O bond. This is true for almost every type of absorption peak; both shape and intensity characteristics can be described, and these characteristics often make it possible to distinguish the peak in a confusing situation. For instance, to some extent both C=O and C=C bonds absorb in the same region of the infrared spectrum:

$$\text{C=O} \quad 1850–1630 \text{ cm}^{-1}$$

$$\text{C=C} \quad 1680–1620 \text{ cm}^{-1}$$

However, the C=O bond is a strong absorber, whereas the C=C bond generally absorbs only weakly. Hence, a trained observer would not normally interpret a strong peak at 1670 cm^{-1} to be a carbon–carbon double bond or a weak absorption at this frequency to be due to a carbonyl group.

The shape of a peak often gives a clue to its identity as well. Thus, although the NH and OH regions of the infrared overlap,

$$\text{OH} \quad 3650–3200 \text{ cm}^{-1}$$

$$\text{NH} \quad 3500–3300 \text{ cm}^{-1}$$

NH usually gives a **sharp** absorption peak (absorbs a very narrow range of frequencies), and OH, when it is in the NH region, usually gives a **broad** absorption peak. Primary amines give *two* absorptions in this region, whereas alcohols give only one.

Therefore, while you are studying the sample spectra in the pages that follow, you should also notice shapes and intensities. They are as important as the frequency at which an absorption occurs, and you must train your eye to recognize these features. In the literature of organic chemistry, you will often find absorptions referred to as strong (s), medium (m), weak (w), broad, or sharp. The author is trying to convey some idea of what the peak looks like without actually drawing the spectrum. Although the intensity of an absorption often provides useful information about the identity of a peak, be aware that the relative intensities of all the peaks in the spectrum are dependent on the amount of sample that is used and the sensitivity setting of the instrument. Therefore, the *actual* intensity of a particular peak may vary from spectrum to spectrum, and you must pay attention to *relative* intensities.

25.12 Correlation Charts and Tables

To extract structural information from infrared spectra, you must know the frequencies or wavelengths at which various functional groups absorb. Infrared **correlation tables** present as much information as is known about where the various functional groups absorb. The books listed at the end of this chapter present extensive lists of correlation tables. Sometimes, the absorption information is given in a chart, called a **correlation chart.** A simplified correlation table is given in Table 25.1.

Although you may think assimilating the mass of data in Table 25.1 will be difficult, it is not if you make a modest start and then gradually increase your familiarity with the data. An ability to interpret the fine details of an infrared spectrum will follow. This is most easily accomplished by first establishing the broad visual patterns of Figure 25.13 firmly in mind. Then, as a second step, a "typical absorption value" can be memorized for each of the functional groups in this pattern. This value will be a single number that can be used as a pivot value for the memory. For instance, start with a simple aliphatic ketone as a model for all typical carbonyl compounds. The typical aliphatic ketone has carbonyl absorption of 1715 ± 10 cm^{-1}. Without worrying about the variation, memorize 1715 cm^{-1} as the base value for carbonyl absorption. Then learn the extent of the carbonyl range and the visual pattern of how the different kinds of carbonyl groups are arranged throughout this region. See, for instance, Figure 25.27 (p. 860), which gives typical values for carbonyl compounds. Also learn how factors such as ring size (when the functional group is contained in a ring) and conjugation affect the base values (that is, in which direction the values are shifted). Learn the trends—always remembering the base value (1715 cm^{-1}). It might prove useful as a beginning to memorize the base values in Table 25.2 for this approach. Notice that there are only eight values.

25.13 Analyzing a Spectrum (or What You Can Tell at a Glance)

In analyzing the spectrum of an unknown, concentrate first on establishing the presence (or absence) of a few major functional groups. The most conspicuous peaks are C=O, O—H, N—H, C—O, C=C, C≡C, C≡N, and NO$_2$. If they are present, they give immediate structural information. Do not try to analyze in detail the CH absorptions near 3000 cm^{-1}; almost all compounds have these absorptions. Do not worry about subtleties of the exact

Table 25.1 *A simplified correlation table*

	Type of Vibration		Frequency (cm^{-1})	Intensity[a]
C—H	Alkanes	(stretch)	3000–2850	s
	—CH$_3$	(bend)	1450 and 1375	m
	—CH$_2$—	(bend)	1465	m
	Alkenes	(stretch)	3100–3000	m
		(bend)	1700–1000	s
	Aromatics	(stretch)	3150–3050	s
		(out-of-plane bend)	1000–700	s
	Alkyne	(stretch)	ca. 3300	s
	Aldehyde		2900–2800	w
			2800–2700	w
C—C	Alkane	Not interpretatively useful		
C=C	Alkene		1680–1600	m–w
	Aromatic		1600–1400	m–w
C≡C	Alkyne		2250–2100	m–w
C=O	Aldehyde		1740–1720	s
	Ketone (acyclic)		1725–1705	s
	Carboxylic acid		1725–1700	s
	Ester		1750–1730	s
	Amide		1700–1640	s
	Anhydride		ca. 1810	s
			ca. 1760	s
C—O	Alcohols, ethers, esters, carboxylic acids		1300–1000	s
O—H	Alcohol, phenols			
	Free		3650–3600	m
	H-Bonded		3400–3200	m
	Carboxylic acids		3300–2500	m
N—H	Primary and secondary amines		ca. 3500	m
C≡N	Nitriles		2260–2240	m
N=O	Nitro (R—NO$_2$)		1600–1500	s
			1400–1300	s
C—X	Fluoride		1400–1000	s
	Chloride		800–600	s
	Bromide, iodide		<600	s

[a]s, strong; m, medium; w, weak.

Table 25.2 *Base values for absorptions of bonds*

O—H	3400 cm^{-1}	C≡C	2150 cm^{-1}
N—H	3500 cm^{-1}	C=O	1715 cm^{-1}
C—H	3000 cm^{-1}	C=C	1650 cm^{-1}
C≡N	2250 cm^{-1}	C—O	1100 cm^{-1}

type of environment in which the functional group is found. A checklist of the important gross features follows:

1. Is a carbonyl group present?

   The C=O group gives rise to a strong absorption in the region 1820–1600 cm^{-1}. The peak is often the strongest in the spectrum and of medium width. You can't miss it.

2. If C=O is present, check the following types. (If it is absent, go to item 3.)

Acids	Is O—H also present?
	Broad absorption near 3300–2500 cm^{-1} (usually overlaps C—H).
Amides	Is N—H also present?
	Medium absorption near 3500 cm^{-1}, sometimes a double peak, equivalent halves.
Esters	Is C—O also present?
	Medium intensity absorptions near 1300–1000 cm^{-1}.
Anhydrides	Have *two* C=O absorptions near 1810 and 1760 cm^{-1}.
Aldehydes	Is aldehyde C—H present?
	Two weak absorptions near 2850 cm^{-1} and 2750 cm^{-1} on the right side of C—H absorptions.
Ketones	The preceding five choices have been eliminated.

3. If C=O is absent

Alcohols	Check for O—H.
or Phenols	**Broad** absorption near 3600–3300 cm^{-1}.
	Confirm this by finding C—O near 1300–1000 cm^{-1}.
Amines	Check for N—H.
	Medium absorption(s) near 3500 cm^{-1}.
Ethers	Check for C—O (and absence of O—H) near 1300–1000 cm^{-1}.

4. Double bonds or aromatic rings or both

   C=C is a **weak** absorption near 1650 cm^{-1}.

   Medium to strong absorptions in the region 1650–1450 cm^{-1} often imply an aromatic ring.

   Confirm the above by consulting the C—H region. Aromatic and vinyl C—H occur to the left of 3000 cm^{-1} (aliphatic C—H occurs to the right of this value).

5. Triple bonds   C≡N is a medium, sharp absorption near 2250 cm^{-1}.

   C≡C is a weak but sharp absorption near 2150 cm^{-1}.

   Check also for acetylenic C—H near 3300 cm^{-1}.

6. Nitro groups   *Two* strong absorptions near 1600–1500 cm^{-1} and 1390–1300 cm^{-1}.

7. Hydrocarbons   None of the above is found.

   Main absorptions are in the C—H region near 3000 cm^{-1}.

   Very simple spectrum, only other absorptions are near 1450 cm^{-1} and 1375 cm^{-1}.

   The beginning student should resist the idea of trying to assign or interpret *every* peak in the spectrum. You simply will not be able to do this. Concentrate first on learning the principal peaks and recognizing their presence or absence. This is best done by carefully studying the illustrative spectra in the section that follows.

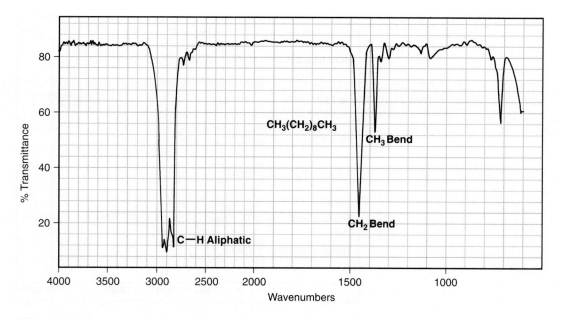

Figure 25.15
Infrared spectrum of decane (neat liquid, salt plates).

NOTE: In describing the shifts of absorption peaks or their relative positions, we have used the phrases "to the left" and "to the right." This was done to simplify descriptions of peak positions. The meaning is clear because all spectra are conventionally presented left to right from 4000 to 600 cm^{-1}.

25.14 Survey of the Important Functional Groups

A. Alkanes

The spectrum is usually simple, with a few peaks.

C—H   Stretch occurs around 3000 cm^{-1}.
   1. In alkanes (except strained ring compounds), absorption always occurs to the right of 3000 cm^{-1}.
   2. If a compound has vinylic, aromatic, acetylenic, or cyclopropyl hydrogens, the CH absorption is to the left of 3000 cm^{-1}.

CH$_2$ Methylene groups have a characteristic absorption at approximately 1450 cm^{-1}.

CH$_3$ Methyl groups have a characteristic absorption at approximately 1375 cm^{-1}.

C—C   Stretch—not interpretatively useful—has many peaks.

The spectrum of decane is shown in Figure 25.15.

B. Alkenes

=C—H   Stretch occurs to the left of 3000 cm^{-1}.
=C—H   Out-of-plane (oop) bending occurs at 1000–650 cm^{-1}.
   The C—H out-of-plane absorptions often allow you to determine the type of substitution pattern on the double bond, according to the number of absorptions and their positions. The correlation chart in Figure 25.16 shows the positions of these bands.

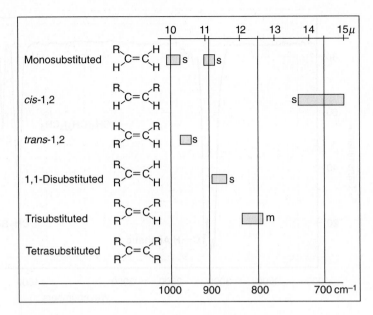

Figure 25.16
The C—H out-of-plane bending vibrations for substituted alkenes.

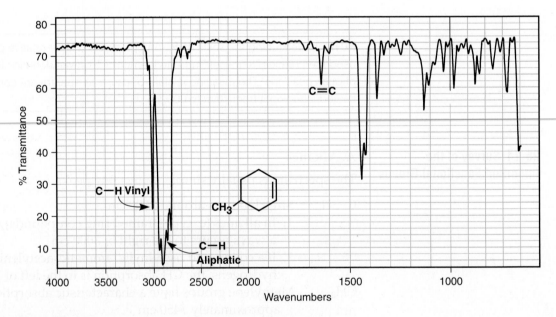

Figure 25.17
Infrared spectrum of 4-methylcyclohexene (neat liquid, salt plates).

C=C   Stretch 1675–1600cm^{-1}, often weak.
       Conjugation moves C=C stretch to the right.
       Symmetrically substituted bonds, as in 2,3-dimethyl-2-butene, do
          not absorb in the infrared region (no dipole change). Highly
          substituted double bonds are often vanishingly weak in
          absorption.

The spectra of 4-methylcyclohexene and styrene are shown in Figures 25.17
and 25.18.

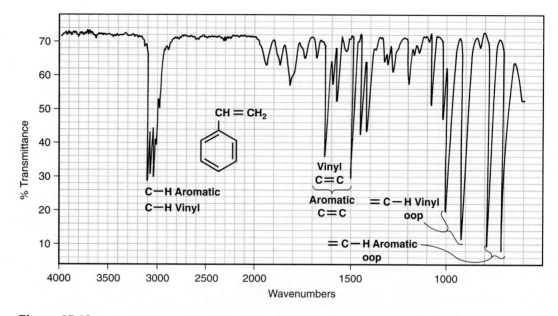

Figure 25.18
Infrared spectrum of styrene (neat liquid, salt plates).

C. Aromatic Rings

=C—H Stretch is always to the left of 3000 cm^{-1}.

=C—H Out-of-plane (oop) bending occurs at 900 to 690 cm^{-1}.

         The C—H out-of-plane absorptions often allow you to determine the type of ring substitution by their numbers, intensities, and positions. The correlation chart in Figure 25.19A indicates the positions of these bands.

         The patterns are generally reliable—they are most reliable for rings with alkyl substituents and least reliable for polar substituents.

Ring Absorptions (C=C). There are often four sharp absorptions that occur in pairs at 1600 cm^{-1} and 1450 cm^{-1} and are characteristic of an aromatic ring. See, for example, the spectra of anisole (Figure 25.23), benzonitrile (Figure 25.26), and methyl benzoate (Figure 25.35).

    There are many weak combination and overtone absorptions that appear between 2000 cm^{-1} and 1667 cm^{-1}. The relative shapes and numbers of these peaks can be used to determine whether an aromatic ring is monosubstituted or di-, tri-, tetra-, penta-, or hexa-substituted. Positional isomers can also be distinguished. Because the absorptions are weak, these bands are best observed by using neat liquids or concentrated solutions. If the compound has a high-frequency carbonyl group, this absorption overlaps the weak overtone bands, so no useful information can be obtained from analyzing this region. The various patterns that are obtained in this region are shown in Figure 25.19B.

    The spectra of styrene and *o*-dichlorobenzene are shown in Figures 25.18 and 25.20.

D. Alkynes

≡C—H Stretch is usually near 3300 cm^{-1}, sharp peak.

C≡C Stretch is near 2150 cm^{-1}, sharp peak.

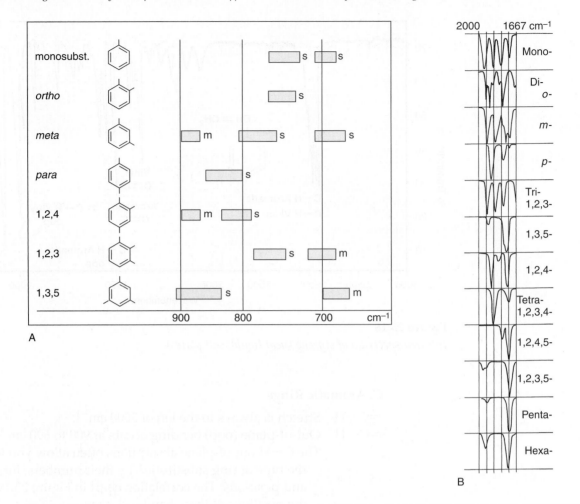

Figure 25.19

(A) The C—H out-of-plane bending vibrations for substituted benzenoid compounds. (B) The 2000–1667 cm⁻¹ region for substituted benzenoid compounds. (From Dyer, J. R., Applications of Absorption Spectroscopy of Organic Compounds, *Englewood Cliffs, NJ: Prentice Hall, 1965.)*

Conjugation moves C≡C stretch to the right.
Disubstituted or symmetrically substituted triple bonds give either no absorption or weak absorption.

E. Alcohols and Phenols

O—H   Stretch is a sharp peak at 3650–3600 cm⁻¹ if no hydrogen bonding takes place. (This is usually observed only in dilute solutions.)
If there is hydrogen bonding (usual in neat or concentrated solutions), the absorption is *broad* and occurs more to the right at 3500–3200 cm⁻¹, sometimes overlapping C—H stretch absorptions.

C—O   Stretch is usually in the range of 1300–1000 cm⁻¹.
Phenols are like alcohols. The 2-naphthol shown in Figure 25.21 has some molecules hydrogen bonded and some free. The spectrum of 4-methylcyclohexanol is shown in Figure 25.22. This alcohol, which was determined neat, would also have had a free OH spike to the left of this hydrogen-bonded band if it had been determined in dilute solution.

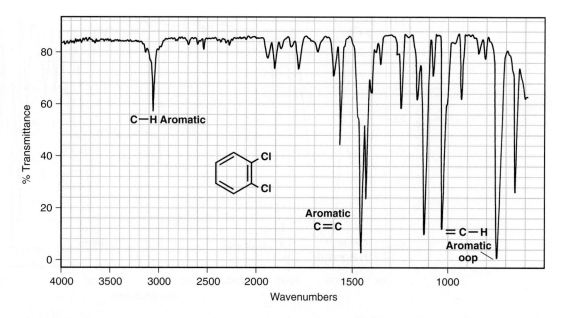

Figure 25.20
Infrared spectrum of o-dichlorobenzene (neat liquid, salt plates).

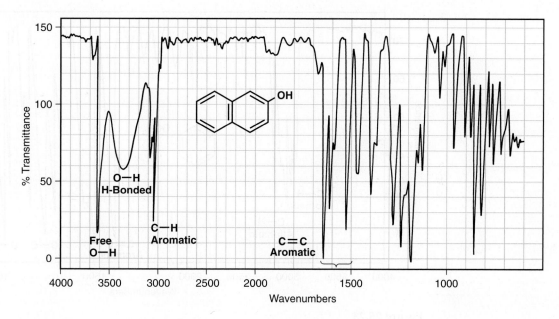

Figure 25.21
Infrared spectrum of 2-naphthol showing both free and hydrogen-bonded OH (CHCl₃ solution).

F. Ethers

C—O   The most prominent band is due to C—O stretch at 1300–1000 cm^{-1}. Absence of C=O and O—H bands is required to be sure C—O stretch is not due to an alcohol or ester. Phenyl and vinyl ethers are found in the left portion of the range, aliphatic ethers in the right. (Conjugation with the oxygen moves the absorption to the left.)

The spectrum of anisole is shown in Figure 25.23.

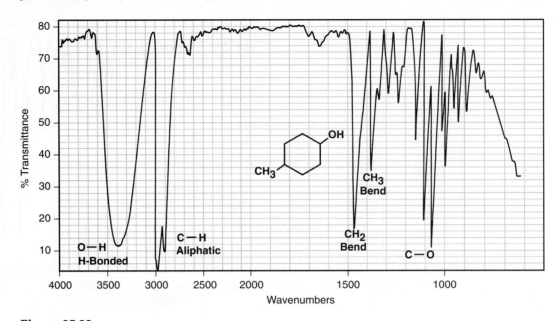

Figure 25.22
Infrared spectrum of 4-methylcyclohexanol (neat liquid, salt plates).

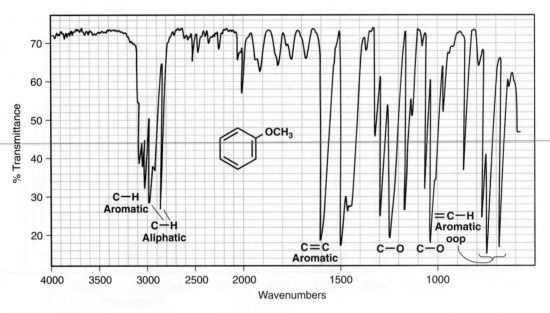

Figure 25.23
Infrared spectrum of anisole (neat liquid, salt plates).

G. Amines

N—H   Stretch occurs in the range of 3500–3300 cm^{-1}.
         Primary amines have *two* bands typically 30 cm^{-1} apart.
         Secondary amines have one band, often vanishingly weak.
         Tertiary amines have no NH stretch.

C—N   Stretch is weak and occurs in the range of 1350–1000 cm^{-1}.

N—H   Scissoring bending mode occurs in the range of 1640–1560 cm^{-1} (broad).
         An out-of-plane bending absorption can sometimes be observed at about 800 cm^{-1}.

The spectrum of *n*-butylamine is shown in Figure 25.24.

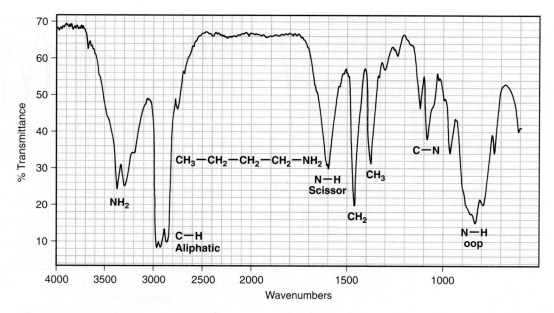

Figure 25.24
Infrared spectrum of n-*butylamine (neat liquid, salt plates).*

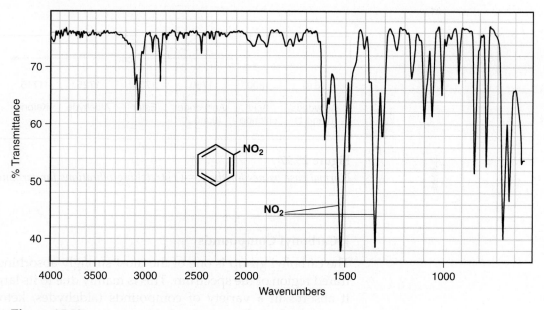

Figure 25.25
Infrared spectrum of nitrobenzene (neat liquid, salt plates).

H. Nitro Compounds

N=O Stretch is usually two strong bands at 1600–1500 cm^{-1} and 1390–1300 cm^{-1}.

The spectrum of nitrobenzene is shown in Figure 25.25.

I. Nitriles

C≡N Stretch is a sharp absorption near 2250 cm^{-1}.
Conjugation with double bonds or aromatic rings moves the absorption to the right.

The spectrum of benzonitrile is shown in Figure 25.26.

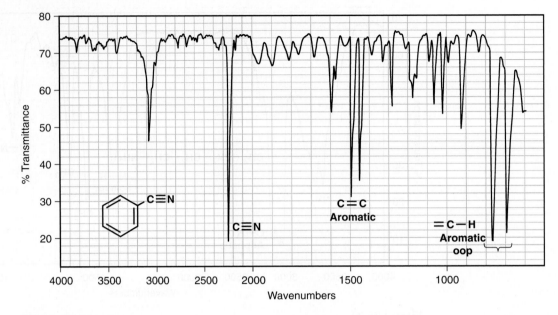

Figure 25.26
Infrared spectrum of benzonitrile (neat liquid, salt plates).

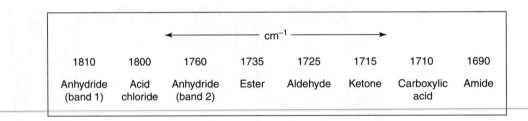

Figure 25.27
Normal base values for the C=O stretching vibrations for carbonyl groups.

J. Carbonyl Compounds

The carbonyl group is one of the most strongly absorbing groups in the infrared region of the spectrum. This is mainly due to its large dipole moment. It absorbs in a variety of compounds (aldehydes, ketones, acids, esters, amides, anhydrides, and acid chlorides) in the range of 1850–1650 cm^{-1}. In Figure 25.27, the normal values for the various types of carbonyl groups are compared. In the sections that follow, each type is examined separately.

K. Aldehydes

C=O   Stretch at approximately 1725 cm^{-1} is normal.
Aldehydes *seldom* absorb to the left of this value.
Conjugation moves the absorption to the right.

C—H   Stretch, aldehyde hydrogen (—CHO), consists of *weak* bands at about 2750 cm^{-1} and 2850 cm^{-1}. Note that the CH stretch in alkyl chains does not usually extend this far to the right.

The spectrum of an unconjugated aldehyde, nonanal, is shown in Figure 25.28, and the conjugated aldehyde, benzaldehyde, is shown in Figure 25.29.

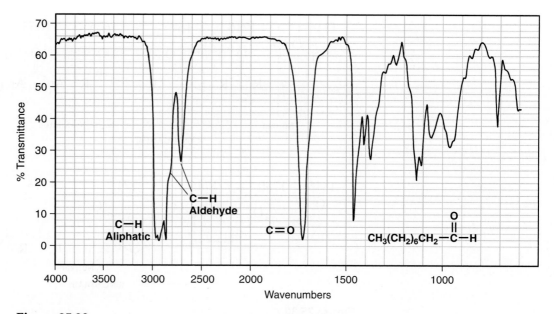

Figure 25.28
Infrared spectrum of nonanal (neat liquid, salt plates).

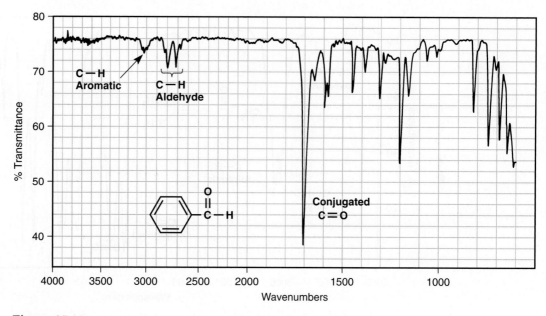

Figure 25.29
Infrared spectrum of benzaldehyde (neat liquid, salt plates).

L. Ketones

C=O   Stretch at approximately at 1715 cm^{-1} is normal.
      Conjugation moves the absorption to the right.
      Ring strain moves the absorption to the left in cyclic ketones
         (see Figure 25.30).

The spectra of methyl isopropyl ketone and mesityl oxide are shown in Figures 25.14 and 25.31. The spectrum of camphor, shown in Figure 25.32, has a carbonyl group that has been shifted to a higher frequency because of ring strain (1745 cm^{-1}).

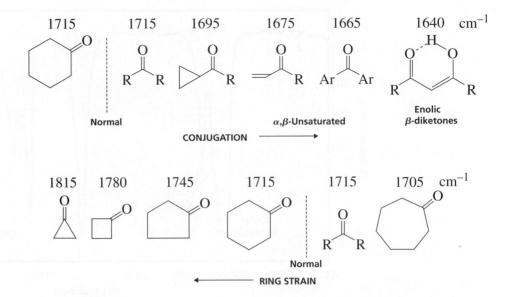

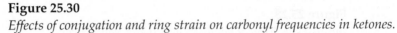

Figure 25.30
Effects of conjugation and ring strain on carbonyl frequencies in ketones.

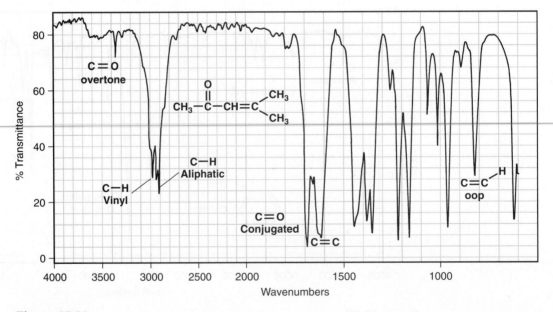

Figure 25.31
Infrared spectrum of mesityl oxide (neat liquid, salt plates).

M. Acids

O—H   Stretch, usually *very broad* (strongly hydrogen bonded) at
          $3300–2500$ cm^{-1}, often interferes with C—H absorptions.

C=O   Stretch, broad, $1730–1700$ cm^{-1}.
          Conjugation moves the absorption to the right.

C—O   Stretch, in the range of $1320–1210$ cm^{-1}, is strong.

The spectrum of benzoic acid is shown in Figure 25.33.

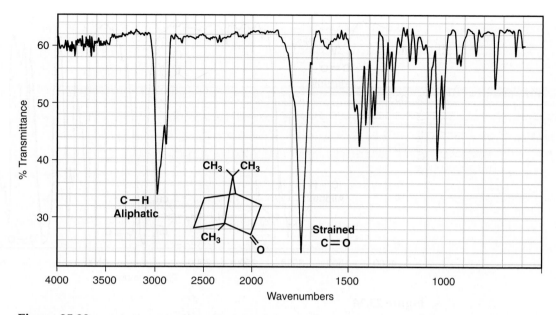

Figure 25.32
Infrared spectrum of camphor (KBr pellet).

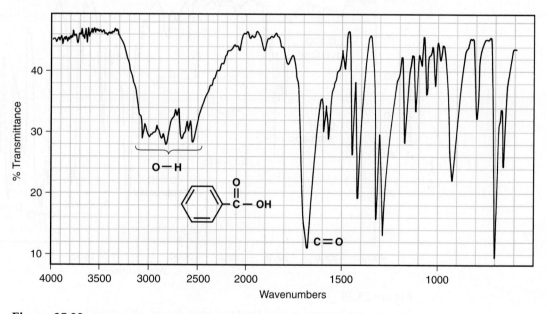

Figure 25.33
Infrared spectrum of benzoic acid (KBr pellet).

N. Esters (R—C—OR′)

C=O Stretch occurs at about 1735 cm^{-1} in normal esters.
1. Conjugation in the R part moves the absorption to the right.
2. Conjugation with the O in the R′ part moves the absorption to the left.
3. Ring strain (lactones) moves the absorption to the left.

C—O Stretch, two bands or more, one stronger than the others, is in the range of 1300–1000 cm^{-1}.

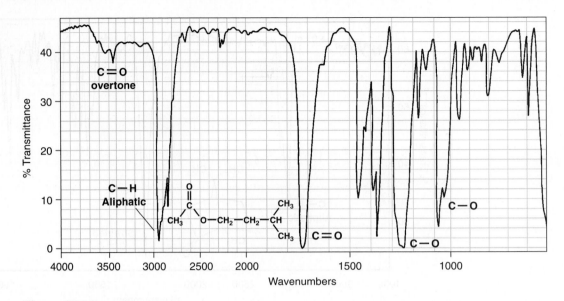

Figure 25.34
Infrared spectrum of isopentyl acetate (neat liquid, salt plates).

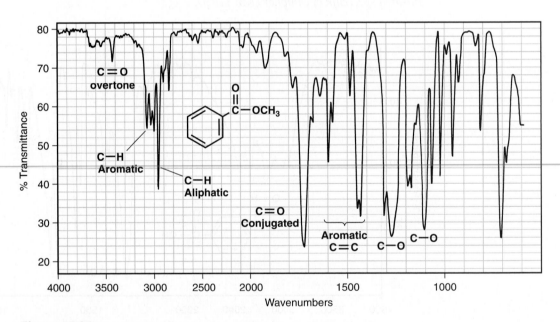

Figure 25.35
Infrared spectrum of methyl benzoate (neat liquid, salt plates).

The spectrum of an unconjugated ester, isopentyl acetate, is shown in Figure 25.34 (C=O appears at 1740 cm^{-1}). A conjugated ester, methyl benzoate, is shown in Figure 25.35 (C=O appears at 1720 cm^{-1}).

O. Amides

C=O   Stretch is at approximately 1670–1640 cm^{-1}.
         Conjugation and ring size (lactams) have the usual effects.

N—H   Stretch (if monosubstituted or unsubstituted) is at 3500–3100 cm^{-1}.
         Unsubstituted amides have two bands (—NH$_2$) in this region.

N—H   Bending around 1640–1550 cm^{-1}.

The spectrum of benzamide is shown in Figure 25.36.

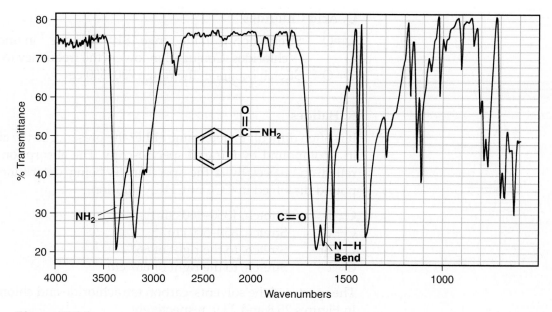

Figure 25.36
Infrared spectrum of benzamide (solid phase, KBr).

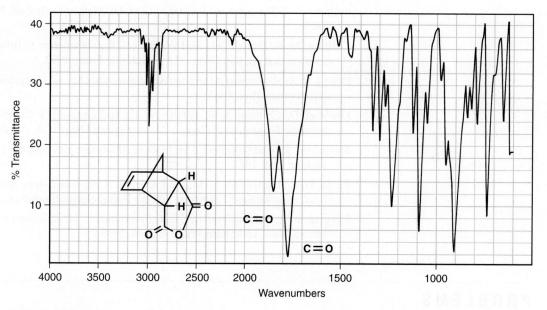

Figure 25.37
Infrared spectrum of cis-*norbornene-5,6-endo-dicarboxylic anhydride (KBr pellet).*

P. Anhydrides

C=O   Stretch always has *two* bands: 1830–1800 cm^{-1} and 1775–1740 cm^{-1}.
Unsaturation moves the absorptions to the right.
Ring strain (cyclic anhydrides) moves the absorptions to the left.

C—O   Stretch is at 1300–900 cm^{-1}. The spectrum of *cis*-norbornene-5,6-*endo*-dicarboxylic anhydride is shown in Figure 25.37.

Q. Acid Chlorides

C=O   Stretch occurs in the range 1810–1775 cm^{-1} in unconjugated chlorides. Conjugation lowers the frequency to 1780–1760 cm^{-1}.

C—O   Stretch occurs in the range 730–550 cm^{-1}.

R. Halides

It is often difficult to determine either the presence or the absence of a halide in a compound by infrared spectroscopy. The absorption bands cannot be relied on, especially if the spectrum is being determined with the compound dissolved in CCl_4 or $CHCl_3$ solution.

C—F   Stretch, 1350–960 cm^{-1}.

C—Cl   Stretch, 850–500 cm^{-1}.

C—Br   Stretch, to the right of 667 cm^{-1}.

C—I   Stretch, to the right of 667 cm^{-1}.

The spectra of the solvents, carbon tetrachloride and chloroform, are shown in Figures 25.8 and 25.9, respectively.

REFERENCES

Bellamy, L. J. *The Infra-red Spectra of Complex Molecules,* 3rd ed. New York: Methuen, 1975.

Colthup, N. B., Daly, L. H., and Wiberly, S. E. *Introduction to Infrared and Raman Spectroscopy,* 3rd ed. San Diego, CA: Academic Press, 1990.

Dyer, J. R. *Applications of Absorption Spectroscopy of Organic Compounds.* Englewood Cliffs, NJ: Prentice-Hall, 1965.

Lin-Vien, D., Colthup, N. B., Fateley, W. G., and Grasselli, J. G. *Infrared and Raman Characteristic Frequencies of Organic Molecules.* San Diego, CA: Academic Press, 1991.

Nakanishi, K., and Soloman, P. H. *Infrared Absorption Spectroscopy,* 2nd ed. San Francisco: Holden-Day, 1977.

Pavia, D. L., Lampman, G. M., and Kriz, G. S. *Introduction to Spectroscopy: A Guide for Students of Organic Chemistry,* 3rd ed. Philadelphia: Saunders, 2001.

Silverstein, R. M., and Webster, F. X. *Spectrometric Identification of Organic Compounds,* 6th ed. New York: John Wiley & Sons, 1998.

PROBLEMS

1. Comment on the suitability of running the infrared spectrum under each of the following conditions. If there is a problem with the conditions given, provide a suitable alternative method.

   a. A neat spectrum of liquid with a boiling point of 150°C is determined using salt plates.

   b. A neat spectrum of a liquid with a boiling point of 35°C is determined using salt plates.

   c. A KBr pellet is prepared with a compound that melts at 200°C.

   d. A KBr pellet is prepared with a compound that melts at 30°C.

   e. A solid aliphatic hydrocarbon compound is determined as a Nujol mull.

f. Silver chloride plates are used to determine the spectrum of aniline.

g. Sodium chloride plates are selected to run the spectrum of a compound that contains some water.

2. Indicate how you could distinguish between the following pairs of compounds by using infrared spectroscopy.

a. $CH_3CH_2CH_2\overset{\overset{O}{\|}}{C}-H$ $\qquad\qquad$ $CH_3CH_2\overset{\overset{O}{\|}}{C}CH_3$

b. 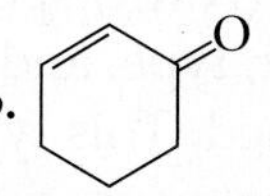 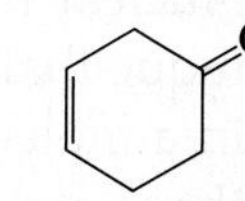 $\qquad\qquad$

c. $CH_3CH_2\overset{\overset{H}{|}}{N}CH_2CH_3$ $\qquad\qquad$ $CH_3CH_2CH_2CH_2NH_2$

d. $CH_3CH_2\overset{\overset{O}{\|}}{C}OCH_2CH_3$ $\qquad\qquad$ $CH_3CH_2\overset{\overset{O}{\|}}{C}CH_2OCH_3$

e. $CH_3CH_2\overset{\overset{O}{\|}}{C}OH$ $\qquad\qquad$ $CH_3CH_2CH_2OH$

f. 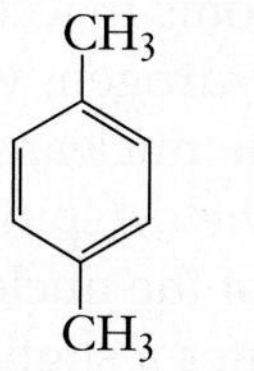 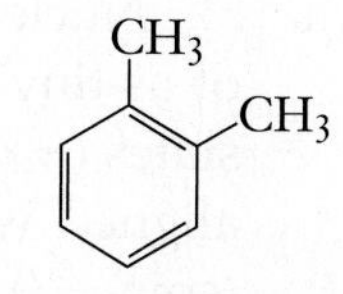 $\qquad\qquad$

g. $CH_3CH_2CH=CH_2$ $\qquad\qquad$ $CH_3CH=CHCH_3$ (trans)

h. $CH_3CH_2CH_2C\equiv CH$ $\qquad\qquad$ $CH_3CH_2CH_2CH=CH_2$

i. 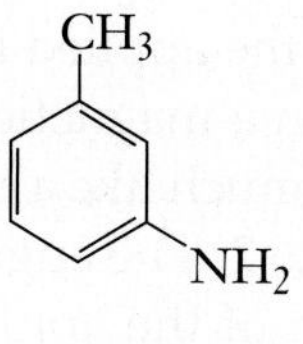 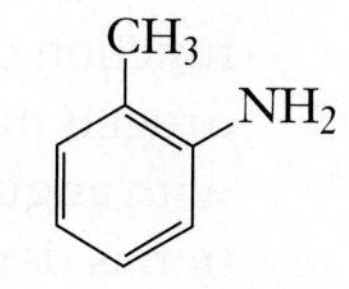 $\qquad\qquad$

j. $CH_3CH_2CH_2CH_2\overset{\overset{O}{\|}}{C}-OH$ $\qquad\qquad$ $CH_3CH_2CH_2\overset{\overset{O}{\|}}{C}OCH_3$

k. $CH_3CH_2CH_2CH_2CH_3$ $\qquad\qquad$ $CH_2=CHCH_2CH_2CH_2CH_3$

l. $CH_3CH_2CH_2CH_2C\equiv CH$ $\qquad\qquad$ $CH_3CH_2CH_2C\equiv CCH_3$

26   TECHNIQUE 26

Nuclear Magnetic Resonance Spectroscopy (Proton NMR)

Nuclear magnetic resonance (NMR) spectroscopy is an instrumental technique that allows the number, type, and relative positions of certain atoms in a molecule to be determined. This type of spectroscopy applies only to those atoms that have nuclear magnetic moments because of their nuclear spin properties. Although many atoms meet this requirement, hydrogen atoms (^1_1H) are of the greatest interest to the organic chemist. Atoms of the ordinary isotopes of carbon ($^{12}_6C$) and oxygen ($^{16}_8O$) do not have nuclear magnetic moments, and ordinary nitrogen atoms ($^{14}_7N$), although they do have magnetic moments, generally fail to show typical NMR behavior for other reasons. The same is true of the halogen atoms, except for fluorine ($^{19}_9F$), which does show active NMR behavior. Of the atoms mentioned here, the hydrogen nucleus (^1_1H) and carbon-13 nucleus ($^{13}_6C$) are the most important to organic chemists. Proton (^1H) NMR is discussed here and carbon (^{13}C) NMR is described in Technique 27.

Nuclei of NMR-active atoms placed in a magnetic field can be thought of as tiny bar magnets. In hydrogen, which has two allowed nuclear spin states ($+\frac{1}{2}$ and $-\frac{1}{2}$), either the nuclear magnets of individual atoms can be aligned with the magnetic field (spin $+\frac{1}{2}$) or they can be opposed to it (spin $-\frac{1}{2}$). A slight majority of the nuclei are aligned with the field, because this spin orientation constitutes a slightly lower-energy spin state. If radio-frequency waves of the appropriate energy are supplied, nuclei aligned with the field can absorb this radiation and reverse their direction of spin or become reoriented so that the nuclear magnet opposes the applied magnetic field (Figure 26.1).

The frequency of radiation required to induce spin conversion is a direct function of the strength of the applied magnetic field. When a spinning hydrogen nucleus is placed in a magnetic field, the nucleus begins to precess with angular frequency ω, much like a child's toy top. This precessional motion is depicted in Figure 26.2. The angular frequency of nuclear precession ω increases as the strength of the applied magnetic field is increased. The radiation that must be supplied to induce spin conversion in a hydrogen nucleus of spin $+\frac{1}{2}$ must have a frequency that just matches the angular precessional frequency ω. This is called the resonance condition, and spin conversion is said to be a resonance process.

For the average proton (hydrogen atom), if a magnetic field of approximately 1.4 tesla is applied, radio-frequency radiation of 60 MHz is required

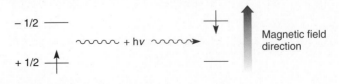

Figure 26.1
The NMR absorption process.

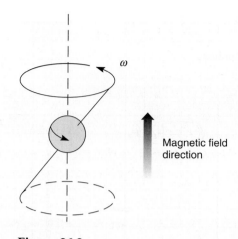

Figure 26.2
*Precessional motion of a spinning
nucleus in an applied magnetic field.*

to induce a spin transition.[1] Fortunately, the magnetic field strength required to induce the various protons in a molecule to absorb 60-MHz radiation varies from proton to proton within the molecule and is a sensitive function of the immediate *electronic* environment of each proton. The proton nuclear magnetic resonance spectrometer supplies a basic radio-frequency radiation of 60 MHz to the sample being measured and *increases* the strength of the applied magnetic field over a range of several parts per million from the basic field strength. As the field increases, various protons come into resonance (absorb 60-MHz energy), and a resonance signal is generated for each proton. An NMR spectrum is a plot of the strength of the magnetic field versus the intensity of the absorptions. A typical 60-MHz NMR spectrum is shown in Figure 26.3.

Modern FT–NMR instruments produce the same type of NMR spectrum just described, even though they do it by a different method. See your lecture textbook for a discussion of the differences between classic CW instruments and modern FT–NMR instruments. Fourier transform spectrometers operating at magnetic field strengths of at least 7.1 tesla and at spectrometer frequencies of 300 MHz and above allow chemists to obtain both the proton and carbon NMR spectra on the same sample.

PART A. PREPARING A SAMPLE FOR NMR SPECTROSCOPY

The NMR sample tubes used in most instruments are approximately 0.5 cm × 18 cm in overall dimension and are fabricated of uniformly thin glass tubing. These tubes are fragile and expensive, so care must be taken to avoid breaking the tubes.

CAUTION

NMR tubes are made out of very thin glass and break easily. Never place the cap on tightly, and take special care when removing it.

---

[1] Most modern instruments (FT–NMR instruments) use higher fields than described here and operate differently. The classical 60-MHz continuous wave (CW) instrument is used here as a simple example.

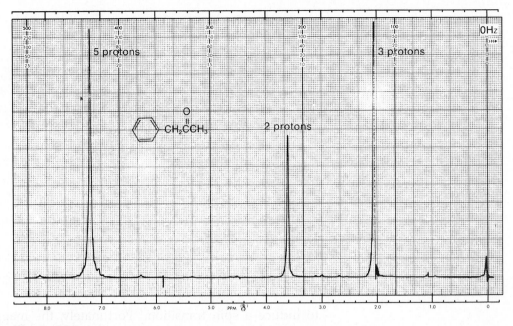

Figure 26.3
Nuclear magnetic resonance spectrum of phenylacetone (the absorption peak at the far right is caused by the added reference substance tetramethylsilane).

To prepare the solution, you must first choose the appropriate solvent. The solvent should not have NMR absorption peaks of its own, that is, it should contain no protons. Carbon tetrachloride (CCl_4) fits this requirement and can be used in some instruments. However, because FT–NMR spectrometers require deuterium to stabilize (lock) the field, organic chemists usually use deuterated chloroform ($CDCl_3$) as a solvent. This solvent dissolves most organic compounds and is relatively inexpensive. You can use this solvent with any NMR instrument. You should not use normal chloroform $CHCl_3$, because the solvent contains a proton. Deuterium ^2H does not absorb in the proton region and is thus "invisible," or not seen, in the proton NMR spectrum. Use deuterated chloroform to dissolve your sample, unless you are instructed to use another solvent, such as deuterated derivatives of water, acetone, or dimethylsulfoxide.

26.1 Routine Sample Preparation Using Deuterated Chloroform

1. Most organic liquids and low-melting solids will dissolve in deuterated chloroform. However, you should first determine whether your sample will dissolve in ordinary $CHCl_3$ before using the deuterated solvent. If your sample does not dissolve in chloroform, consult your instructor about a possible alternative solvent, or consult Section 26.2.

CAUTION   ⚠️

Chloroform, deuterated chloroform, and carbon tetrachloride are all toxic solvents. In addition, they may be carcinogenic substances.

---

2. If you are using an FT–NMR spectrometer, add 30 mg (0.030 g) of your liquid or solid sample to a tared conical vial or test tube. Use a Pasteur pipet to transfer a liquid or a spatula to transfer a solid. Non-FT instruments usually require a more concentrated solution in order to

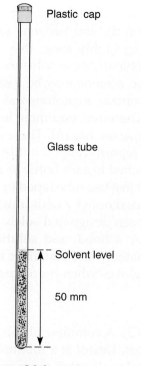

Figure 26.4
An NMR sample tube.

obtain an adequate spectrum. Typically, a 10–30% sample concentration (weight/weight) is used.

3. Transfer about 0.5 mL of the deuterated chloroform with a clean, dry Pasteur pipet to your sample. Swirl the test tube or conical vial to help dissolve the sample. At this point, the sample should have completely dissolved. Add a little more solvent, if necessary, to dissolve the sample fully.

4. Transfer the solution to the NMR tube using a clean, dry Pasteur pipet. Be careful when transferring the solution to avoid breaking the edge of the fragile NMR tube. It is best to hold the NMR tube and the container with the solution in the same hand when making the transfer.

5. Once the solution has been transferred to the NMR tube, use a clean pipet to add enough deuterated chloroform to bring the total solution height to about 50 mm (Figure 26.4). In some cases, you will need to add a small amount of tetramethylsilane (TMS) as a **reference substance** (Section 26.3). Check with your instructor to see if you need to add TMS to your sample. Deuterated chloroform has a small amount of $CHCl_3$ impurity, which gives rise to a low-intensity peak in the NMR spectrum at 7.27 parts per million (ppm). This impurity may also help you to "reference" your spectrum.

6. Cap the NMR tube. Do this firmly but not too tightly. If you jam the cap on, you may have trouble removing it later without breaking the end off of the very thin glass tube. Make sure that the cap is on straight. Invert the NMR tube several times to mix the contents.

7. You are now ready to record the NMR spectrum of your sample. Insert the NMR tube into its holder and adjust its depth by using the gauge provided to you.

Cleaning the NMR Tube

1. Carefully uncap the tube so that you do not break it. Turn the tube upside down and hold it vertically over a beaker. Shake the tube up and down gently so that the contents of the tube empty into the beaker.

2. Partially refill the NMR tube with acetone using a Pasteur pipet. Carefully replace the cap and invert the tube several times to rinse it.

3. Remove the cap and drain the tube as before. Place the open tube upside down in a beaker with a Kimwipe or paper towel placed in the bottom of the beaker. Leave the tube standing in this position for at least one laboratory period so that the acetone completely evaporates. Alternatively, you may place the beaker and NMR tube in an oven for at least 2 hours. If you need to use the NMR tube before the acetone has fully evaporated, attach a piece of pressure tubing to the tube and pull a vacuum with an aspirator. After several minutes, the acetone should have fully evaporated. Because acetone contains protons, you must not use the NMR tube until the acetone has evaporated completely.[2]

4. Once the acetone is evaporated, place the clean tube and its cap (do not cap the tube) in its storage container and place it in your desk. The storage container will prevent the tube from being crushed.

---

[2] If you can't wait to be sure all of the acetone has evaporated, you may rinse the tube once or twice with a *very small* amount of $CDCl_3$ before using it.

Health Hazards Associated with NMR Solvents

Carbon tetrachloride, chloroform (and chloroform-d), and benzene (and benzene-d_6) are hazardous solvents. Besides being highly toxic, they are suspected carcinogens. In spite of these health problems, these solvents are commonly used in NMR spectroscopy. Deuterated acetone may be a safer alternative. These solvents are used because they contain no protons and are excellent solvents for most organic compounds. Therefore, you must learn to handle these solvents with great care to minimize the hazard. These solvents should be stored either under a hood or in septum-capped bottles. If the bottles have screw caps, a pipet should be attached to each bottle. A recommended way of attaching the pipet is to store it in a test tube taped to the side of the bottle. Septum-capped bottles can be used only by withdrawing the solvent with a hypodermic syringe that has been designated solely for this use. All samples should be prepared under a hood, and solutions should be disposed of in an appropriately designated waste container that is stored under the hood. Wear rubber or plastic gloves when preparing or discarding samples.

26.2 Nonroutine Sample Preparation

Some compounds do not dissolve readily in $CDCl_3$. A commercial solvent called **Unisol** will often dissolve the difficult cases. Unisol is a mixture of $CDCl_3$ and DMSO-d_6. Deuterated acetone may also dissolve more polar substances.

With highly polar substances, you may find that your sample will not dissolve in deuterated chloroform or Unisol. If this is the case, you may be able to dissolve the sample in deuterium oxide D_2O. Spectra determined in D_2O often show a small peak at about 5 ppm because of OH impurity. If the sample compound has acidic hydrogens, they may *exchange* with D_2O, leading to the appearance of an OH peak in the spectrum and the *loss* of the original absorption from the acidic proton, owing to the exchanged hydrogen. In many cases, this will also alter the splitting patterns of a compound.

Many solid carboxylic acids do not dissolve in $CDCl_3$ or even D_2O. In such cases, add a small piece of sodium metal to about 1 mL of D_2O. The acid is then dissolved in this solution. The resulting basic solution enhances the solubility of the carboxylic acid. In such a case, the hydroxyl proton of the carboxylic acid cannot be observed in the NMR spectrum because it exchanges with the solvent. A large DOH peak is observed, however, due to the exchange and the H_2O impurity in the D_2O solvent.

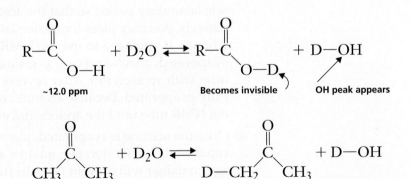

When the above solvents fail, other special solvents can be used. Acetone, acetonitrile, dimethylsulfoxide, pyridine, benzene, and dimethylformamide can be used if you are not interested in the region or regions of the NMR spectrum in which they give rise to absorption. The deuterated (but expensive) analogs of these compounds are also used in special instances (for example, acetone-d_6, dimethylsulfoxide-d_6, dimethylformamide-d_7, and benzene-d_6). If the sample is not sensitive to acid, trifluoroacetic acid (which has no protons with $\delta < 12$) can be used. You must be aware that these solvents often lead to chemical shift values different from those determined in CCl_4 or $CDCl_3$. Variations of as much as 0.5–1.0 ppm have been observed. In fact, it is sometimes possible, by switching to pyridine, benzene, acetone, or dimethylsulfoxide as solvents, to separate peaks that overlap when CCl_4 or $CDCl_3$ solutions are used.

26.3 Reference Substances

To provide the internal reference standard, tetramethylsilane (TMS) must be added to the sample solution. This substance has the formula $(CH_3)_4Si$. By universal convention, the chemical shifts of the protons in this substance are defined as 0.00 ppm. The spectrum should be shifted so that the TMS signal appears at this position on precalibrated paper.

The concentration of TMS in the sample should range from 1% to 3%. Some people prefer to add 1 to 2 drops of TMS to the sample just before determining the spectrum. Because TMS has 12 equivalent protons, not much of it needs to be added. A Pasteur pipet or a syringe may be used for the addition. It is far easier to have available in the laboratory a prepared solvent that already contains TMS. Deuterated chloroform and carbon tetrachloride often have TMS added to them. Because TMS is highly volatile (bp 26.5°C), such solutions should be stored, tightly stoppered, in a refrigerator. Tetramethylsilane itself is best stored in a refrigerator as well.

Tetramethylsilane does not dissolve in D_2O. For spectra determined in D_2O, a different internal standard, sodium 2,2-dimethyl-2-silapentane-5-sulfonate, must be used. This standard is water soluble and gives a resonance peak at 0.00 ppm.

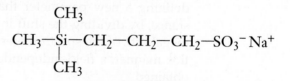

Sodium 2,2-dimethyl-2-silapentane-5-sulfonate (DSS)

PART B. NUCLEAR MAGNETIC RESONANCE (^{1}H NMR)

26.4 The Chemical Shift

The differences in the applied field strengths at which the various protons in a molecule absorb 60-MHz radiation are extremely small. The different absorption positions amount to a difference of only a few parts per million (ppm) in the magnetic field strength. Because it is experimentally difficult to measure the precise field strength at which each proton absorbs to less than one part in a million, a technique has been developed whereby the *difference* between two absorption positions is measured directly. A standard reference substance is used to achieve this measurement, and the positions of the absorptions of all other protons are measured relative to the values for the reference substance. The reference substance that has been universally accepted is **tetramethylsilane** $(CH_3)_4Si$, which is also called **TMS**. The proton resonances in this molecule appear at a higher field strength than the proton resonances in most other molecules, and all the protons of TMS have resonance at the same field strength.

To give the position of absorption of a proton, a quantitative measurement, a parameter called the **chemical shift** (δ), has been defined. One δ unit corresponds to a one-ppm change in the magnetic field strength. To determine the chemical shift value for the various protons in a molecule, the operator determines an NMR spectrum of the molecule with a small quantity of TMS added directly to the sample. That is, both spectra are determined *simultaneously*. The TMS absorption is adjusted to correspond to the $\delta = 0$ ppm position on the recording chart, which is calibrated in δ units, and the δ values of the absorption peaks for all other protons can be read directly from the chart.

Because the NMR spectrometer increases the magnetic field as the pen moves from left to right on the chart, the TMS absorption appears at the extreme right edge of the spectrum ($\delta = 0$ ppm) or at the *upfield* end of the spectrum. The chart is calibrated in δ units (or ppm), and most other protons absorb at a lower field strength (or *downfield*) from TMS.

The shift from TMS for a given proton depends on the strength of the applied magnetic field. In an applied field of 1.41 tesla, the resonance of a proton is approximately 60 MHz, whereas in an applied field of 2.35 tesla (23,500 gauss), the resonance appears at approximately 100 MHz. The ratio of the resonance frequencies is the same as the ratio of the two field strengths:

$$\frac{100 \text{ MHz}}{60 \text{ MHz}} = \frac{2.35 \text{ Tesla}}{1.41 \text{ Tesla}} = \frac{23,500 \text{ Gauss}}{14,100 \text{ Gauss}} = \frac{5}{3}$$

Hence, for a given proton, the shift (in hertz) from TMS is five-thirds larger in the 100-MHz range than in the 60-MHz range. This can be confusing for workers trying to compare data if they have spectrometers that differ in the strength of the applied magnetic field. The confusion is easily overcome by defining a new parameter that is independent of field strength—for instance, by dividing the shift in hertz of a given proton by the frequency in megahertz of the spectrometer with which the shift value was obtained. In this manner, a field-independent measure called the **chemical shift** (δ) is obtained

$$\delta = \frac{\text{(shift in Hz)}}{\text{(spectrometer frequency in MHz)}} \qquad [1]$$

The chemical shift in δ units expresses the amount by which a proton resonance is shifted from TMS, in parts per million (ppm), of the spectrometer's basic operating frequency. Values of δ for a given proton are always the same irrespective, of whether the measurement was made at 60 MHz, 100 MHz, or 300 MHz. For instance, at 60 MHz, the shift of the protons in CH_3Br is 162 Hz from TMS; at 100 MHz, the shift is 270 Hz; and at 300 MHz, the shift is 810 Hz. However, all three correspond to the same value of $\delta = 2.70$ ppm:

$$\delta = \frac{162 \text{ Hz}}{60 \text{ MHz}} = \frac{270 \text{ Hz}}{100 \text{ MHz}} = \frac{810 \text{ Hz}}{300 \text{ MHz}} = 2.70 \text{ ppm}$$

26.5 Chemical Equivalence—Integrals

All the protons in a molecule that are in chemically identical environments often exhibit the same chemical shift. Thus, all the protons in TMS or all the protons in benzene, cyclopentane, or acetone have their own respective resonance values all at the same δ value. Each compound gives rise to a single absorption peak in its NMR spectrum. The protons are said to be **chemically equivalent**. On the other hand, molecules that have sets of protons that are chemically distinct from one another may give rise to an absorption peak from each set.

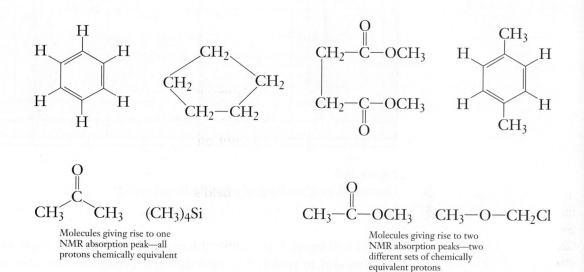

Molecules giving rise to one NMR absorption peak—all protons chemically equivalent

Molecules giving rise to two NMR absorption peaks—two different sets of chemically equivalent protons

The NMR spectrum given in Figure 26.3 is that of phenylacetone, a compound having *three* chemically distinct types of protons:

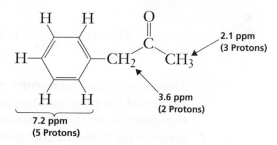

2.1 ppm (3 Protons)

3.6 ppm (2 Protons)

7.2 ppm (5 Protons)

You can immediately see that the NMR spectrum furnishes valuable information on this basis alone. In fact, the NMR spectrum not only can distinguish how many types of protons a molecule has but also can reveal *how many* of each type are contained within the molecule.

In the NMR spectrum, the area under each peak is proportional to the number of hydrogens generating that peak. Hence, in the case of phenylacetone, the area ratio of the three peaks is 5:2:3, the same as the ratio of the numbers of each type of hydrogen. The NMR spectrometer can electronically "integrate" the area under each peak. It does this by tracing over each peak a vertically rising line, which rises in height by an amount proportional to the area under the peak. Shown in Figure 26.5 is an NMR spectrum of benzyl acetate, with each of the peaks integrated in this way.

It is important to note that the height of the integral line does not give the absolute number of hydrogens; it gives the *relative* numbers of each type

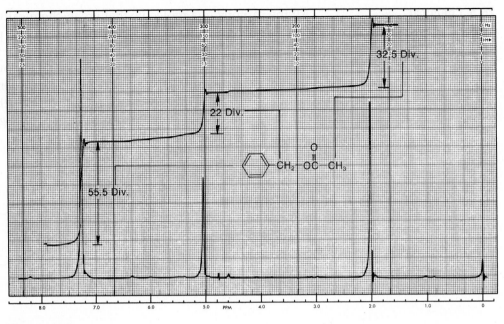

Figure 26.5
Determination of the integral ratios for benzyl acetate.

of hydrogen. For a given integral to be of any use, there must be a second integral to which it is referred. The benzyl acetate case provides a good example of this. The first integral rises for 55.5 divisions on the chart paper, the second for 22.0 divisions, and the third for 32.5 divisions. These numbers are relative and give the *ratios* of the various types of protons. You can find these ratios by dividing each of the larger numbers by the smallest number:

$$\frac{55.5 \text{ div}}{22.0 \text{ div}} = 2.52 \qquad \frac{22.0 \text{ div}}{22.0 \text{ div}} = 1.00 \qquad \frac{32.5 \text{ div}}{22.0 \text{ div}} = 1.48$$

Thus, the number ratio of the protons of each type is 2.52:1.00:1.48. If you assume that the peak at 5.1 ppm is really caused by two hydrogens and that the integrals are slightly in error (this can be as much as 10%), then you can arrive at the true ratios by multiplying each figure by 2 and rounding off; we then get 5:2:3. Clearly, the peak at 7.3 ppm, which integrates for 5, arises from the resonance of the aromatic ring protons, and the peak at 2.0 ppm, which integrates for 3, is caused by the methyl protons. The two-proton resonance at 5.1 ppm arises from the benzyl protons. Notice then that the integrals give the simplest ratios, but not necessarily the true ratios, of the number of protons in each type.

In addition to the rising integral line, modern instruments usually give digitized numerical values for the integrals. Like the heights of the integral lines, these digitized integral values are not absolute but relative, and they should be treated as explained in the preceding paragraph. These digital values are also not exact; like the integral lines, they have the potential for a small degree of error (up to 10%). Figure 26.6 is an example of an integrated spectrum of benzyl acetate determined on a 300-MHz pulsed FT–NMR instrument. The digitized values of the integrals appear under the peaks.

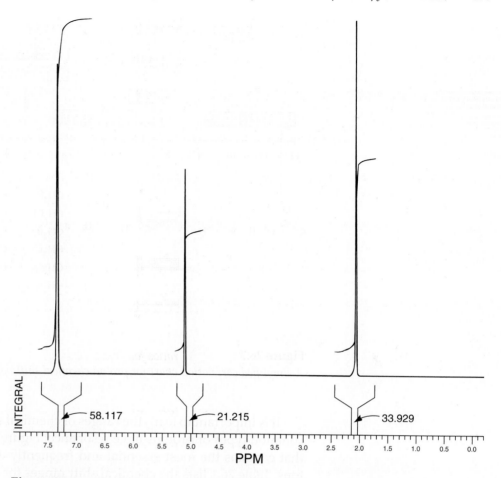

Figure 26.6

An integrated spectrum of benzyl acetate determined on a 300-MHz FT-NMR.

26.6 Chemical Environment and Chemical Shift

If the resonance frequencies of all protons in a molecule were the same, NMR would be of little use to the organic chemist. However, not only do different types of protons have different chemical shifts but they also have a value of chemical shift that characterizes the type of proton they represent. Every type of proton has only a limited range of δ values over which it gives resonance. Hence, the numerical value of the chemical shift for a proton indicates the *type of proton* originating the signal, just as the infrared frequency suggests the type of bond or functional group. Notice, for instance, that the aromatic protons of both phenylacetone (Figure 26.3) and benzyl acetate (Figure 26.5) have resonance near 7.3 ppm and that both methyl groups attached directly to a carbonyl group have a resonance of approximately 2.1 ppm. Aromatic protons characteristically have resonance near 7–8 ppm, and acetyl groups (the methyl protons) have their resonance near 2 ppm. These values of chemical shift are diagnostic. Notice also how the resonance of the benzyl (—CH₂—) protons comes at a higher value of chemical shift (5.1 ppm) in benzyl acetate than in phenylacetone (3.6 ppm). Being attached to the electronegative element, oxygen, these protons are more deshielded (see Section 26.7) than the protons in phenylacetone. A trained chemist would have readily recognized the probable presence of the oxygen by the chemical shift shown by these protons.

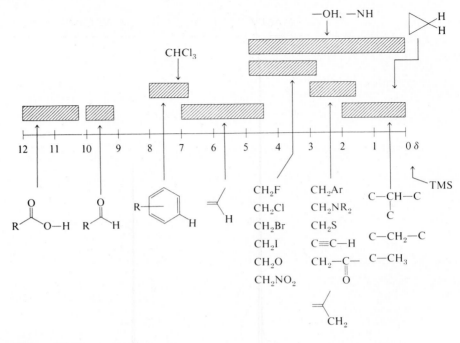

Figure 26.7
A simplified correlation chart for proton chemical shift values.

It is important to learn the ranges of chemical shifts over which the most common types of protons have resonance. Figure 26.7 is a correlation chart that contains the most essential and frequently encountered types of protons. Table 26.1 lists the chemical shift ranges for selected types of protons. For the beginner, it is often difficult to memorize a large body of numbers relating to chemical shifts and proton types. However, this needs to be done only crudely. It is more important to "get a feel" for the regions and the types of protons than to know a string of actual numbers. To do this, study Figure 26.7 carefully.

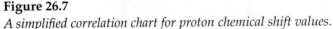

The values of chemical shift given in Figure 26.7 and in Table 26.1 can be easily understood in terms of two factors: local diamagnetic shielding and anisotropy. These two factors are discussed in Sections 26.7 and 26.8.

26.7 Local Diamagnetic Shielding

The trend of chemical shifts that is easiest to explain is that involving electronegative elements substituted on the same carbon to which the protons of interest are attached. The chemical shift simply increases as the electronegativity of the attached element increases. This is illustrated in Table 26.2 for several compounds of the type CH_3X.

Multiple substituents have a stronger effect than a single substituent. The influence of the substituent drops off rapidly with distance. An electronegative element has little effect on protons that are more than three carbons away from it. These effects are illustrated in Table 26.3.

Electronegative substituents attached to a carbon atom, because of their electron-withdrawing effects, reduce the valence electron density around the protons attached to that carbon. These electrons *shield* the proton from the applied magnetic field. This effect, called **local diamagnetic shielding,** occurs because the applied magnetic field induces the valence electrons to circulate. This circulation generates an induced magnetic field, which *opposes* the applied field. This is illustrated in Figure 26.8. Electronegative substituents on

Table 26.1 *Approximate chemical shift ranges (ppm) for selected types of protons*

$R-CH_3$		0.7–1.3	$R-\overset{\mid}{\underset{\mid}{N}}-\overset{\mid}{\underset{\mid}{C}}-H$	2.2–2.9
$R-CH_2-R$		1.2–1.4		
R_3CH		1.4–1.7	$R-S-\overset{\mid}{\underset{\mid}{C}}-H$	2.0–3.0
$R-\overset{\mid}{C}=\overset{\mid}{C}-\overset{\mid}{\underset{\mid}{C}}-H$		1.6 – 2.6	$I-\overset{\mid}{\underset{\mid}{C}}-H$	2.0–4.0
$R-\overset{O}{\overset{\|}{C}}-\overset{\mid}{\underset{\mid}{C}}-H,\ H-\overset{O}{\overset{\|}{C}}-\overset{\mid}{\underset{\mid}{C}}-H$		2.1 – 2.4	$Br-\overset{\mid}{\underset{\mid}{C}}-H$	2.7–4.1
			$Cl-\overset{\mid}{\underset{\mid}{C}}-H$	3.1–4.1
$RO-\overset{O}{\overset{\|}{C}}-\overset{\mid}{\underset{\mid}{C}}-H,\ HO-\overset{O}{\overset{\|}{C}}-\overset{\mid}{\underset{\mid}{C}}-H$		2.1 – 2.5	$R-\overset{O}{\underset{O}{\overset{\|}{\underset{\|}{S}}}}-O-\overset{\mid}{\underset{\mid}{C}}-H$	ca. 3.0
$N\equiv C-\overset{\mid}{\underset{\mid}{C}}-H$		2.1 – 3.0	$RO-\overset{\mid}{\underset{\mid}{C}}-H,\ HO-\overset{\mid}{\underset{\mid}{C}}-H$	3.2–3.8
⬡$-\overset{\mid}{\underset{\mid}{C}}-H$		2.3 – 2.7	$R-\overset{O}{\overset{\|}{C}}-O-\overset{\mid}{\underset{\mid}{C}}-H$	3.5–4.8
$R-C\equiv C-H$		1.7 – 2.7	$O_2N-\overset{\mid}{\underset{\mid}{C}}-H$	4.1–4.3
$R-S-H$	var	1.0 – 4.0[a]	$F-\overset{\mid}{\underset{\mid}{C}}-H$	4.2–4.8
$R-\underset{\mid}{N}-H$	var	0.5 – 4.0[a]		
$R-O-H$	var	0.5 – 5.0[a]		
⬡$-O-H$	var	4.0 – 7.0[a]	$R-\overset{\mid}{C}=\overset{\mid}{C}-H$	4.5–6.5
			⬡$-H$	6.5–8.0
⬡$-\underset{\mid}{N}-H$	var	3.0 – 5.0[a]	$R-\overset{O}{\overset{\|}{C}}-H$	9.0–10.0
$R-\overset{O}{\overset{\|}{C}}-\underset{\mid}{N}-H$	var	5.0 – 9.0[a]	$R-\overset{O}{\overset{\|}{C}}-OH$	11.0–12.0

Note: For those hydrogens shown as $-\overset{\mid}{\underset{\mid}{C}}-H$, if that hydrogen is part of a methyl group (CH_3), the shift is generally at the low end of the range given; if the hydrogen is in a methylene group ($-CH_2-$), the shift is intermediate; and if the hydrogen is in a methine group ($-\underset{\mid}{\overset{\mid}{C}}H-$), the shift is typically at the high end of the range given.

[a]The chemical shift of these groups is variable, depending on the chemical environment in the molecule and on concentration, temperature, and solvent.

Table 26.2 *Dependence of chemical shift of CH_3X on the element X*

Compound CH_3X	CH_3F	CH_3OH	CH_3Cl	CH_3Br	CH_3I	CH_4	$(CH_3)_4Si$
Element X	F	O	Cl	Br	I	H	Si
Electronegativity of X	4.0	3.5	3.1	2.8	2.5	2.1	1.8
Chemical shift (ppm)	4.26	3.40	3.05	2.68	2.16	0.23	0

Table 26.3 *Substitution effects*

	$CHCl_3$	CH_2Cl_2	CH_3Cl	$—CH_2Br$	$—CH_2—CH_2Br$	$—CH_2—CH_2CH_2Br$
δ (ppm)	7.27	5.30	3.05	3.3	1.69	1.25

Note: Values apply to underlined hydrogens.

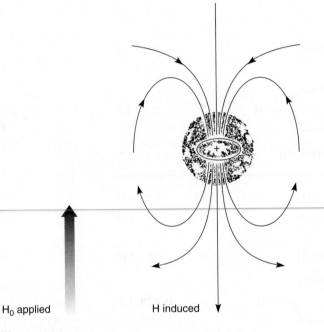

Figure 26.8
Local diamagnetic shielding of a proton due to its valence electrons.

carbon reduce the local diamagnetic shielding in the vicinity of the attached protons because they reduce the electron density around those protons. Substituents that produce this effect are said to *deshield* the proton. The greater the electronegativity of the substituent, the more the deshielding of the protons and, hence, the greater the chemical shift of those protons.

26.8 Anisotropy

Figure 26.7 clearly shows that several types of protons have chemical shifts not easily explained by a simple consideration of the electronegativity of the attached groups. Consider, for instance, the protons of benzene or other aromatic systems. Aryl protons generally have a chemical shift that is as large as that for the proton of chloroform. Alkenes, alkynes, and aldehydes also have protons whose resonance values are not in line with the expected magnitude

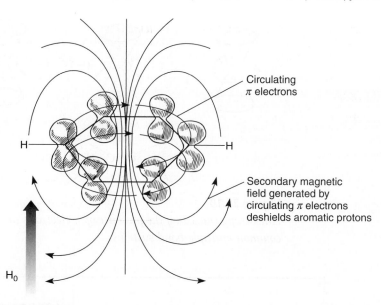

Figure 26.9
Diamagnetic anisotropy in benzene.

of any electron-withdrawing effects. In each of these cases, the effect is due to the presence of an unsaturated system (π electrons) in the vicinity of the proton in question. In benzene, for example, when the π electrons in the aromatic ring system are placed in a magnetic field, they are induced to circulate around the ring. This circulation is called a **ring current.** Moving electrons (the ring current) generate a magnetic field much like that generated in a loop of wire through which a current is induced to flow. The magnetic field covers a spatial volume large enough to influence the shielding of the benzene hydrogens. This is illustrated in Figure 26.9. The benzene hydrogens are deshielded by the **diamagnetic anisotropy** of the ring. An applied magnetic field is nonuniform (anisotropic) in the vicinity of a benzene molecule because of the labile electrons in the ring that interact with the applied field. Thus, a proton attached to a benzene ring is influenced by *three* magnetic fields: the strong magnetic field applied by the magnets of the NMR spectrometer and two weaker fields, one due to the usual shielding by the valence electrons around the proton and the other due to the anisotropy generated by the ring system electrons. It is this anisotropic effect that gives the benzene protons a greater chemical shift than is expected. These protons just happen to lie in a **deshielding** region of this anisotropic field. If a proton were placed in the center of the ring rather than on its periphery, the proton would be shielded because the field lines would have the opposite direction.

All groups in a molecule that have π electrons generate secondary anisotropic fields. In acetylene, the magnetic field generated by induced circulation of π electrons has a geometry such that the acetylene hydrogens are **shielded.** Hence, acetylenic hydrogens come at a higher field than expected. The shielding and deshielding regions due to the various π electron functional groups have characteristic shapes and directions; they are illustrated in Figure 26.10. Protons falling within the cones are shielded, and those falling outside the conical areas are deshielded. Because the magnitude of the anisotropic field diminishes with distance, beyond a certain distance anisotropy has essentially no effect.

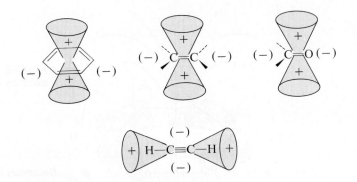

Figure 26.10
Anisotropy caused by the presence of π electrons in some common multiple-bond systems.

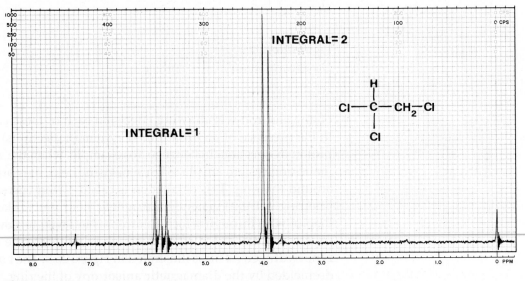

Figure 26.11
NMR spectrum of 1,1,2-trichloroethane. (Courtesy of Varian Associates.)

26.9 Spin–Spin Splitting (*n* + 1 Rule)

We have already considered how the chemical shift and the integral (peak area) can give information about the numbers and types of hydrogens contained in a molecule. A third type of information available from the NMR spectrum is derived from spin–spin splitting. Even in simple molecules, each type of proton rarely gives a single resonance peak. For instance, in 1,1,2-trichloroethane there are two chemically distinct types of hydrogen:

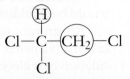

From information given thus far, you would predict *two* resonance peaks in the NMR spectrum of 1,1,2-trichloroethane with an area ratio (integral ratio) of 2:1. In fact, the NMR spectrum of this compound has *five* peaks. A group of three peaks (called a **triplet**) exists at 5.77 ppm, and a group of two peaks (called a **doublet**) is found at 3.95 ppm. The spectrum is shown in Figure 26.11. The methine (CH) resonance (5.77 ppm) is split into a triplet, and the methylene resonance (3.95 ppm) is split into a doublet. The area under the three triplet peaks is *one*, relative to an area of *two* under the two doublet peaks.

This phenomenon is called **spin–spin splitting.** Empirically, spin–spin splitting can be explained by the "$n + 1$ rule." Each type of proton "senses" the number of equivalent protons (n) on the carbon atom or atoms next to the one to which it is bonded, and its resonance peak is split into $n + 1$ components.

Let's examine the case at hand, 1,1,2-trichloroethane, using the $n + 1$ rule. First, the lone methine hydrogen is situated next to a carbon bearing two methylene protons. According to the rule, it has two equivalent neighbors ($n = 2$) and is split into $n + 1 = 3$ peaks (a triplet). The methylene protons are situated next to a carbon bearing only one methine hydrogen. According to the rule, they have one neighbor ($n = 1$) and are split into $n + 1 = 2$ peaks (a doublet).

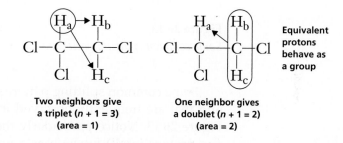

The spectrum of 1,1,2-trichloroethane can be explained easily by the interaction, or coupling, of the spins of protons on adjacent carbon atoms. The position of absorption of proton H_a is affected by the spins of protons H_b and H_c attached to the neighboring (adjacent) carbon atom. If the spins of these protons are aligned with the applied magnetic field, the small magnetic field generated by their nuclear spin properties will augment the strength of the field experienced by the first-mentioned proton H_a. The proton H_a will thus be *deshielded*. If the spins of H_b and H_c are opposed to the applied field, they will decrease the field experienced by proton H_a. It will then be *shielded*. In each of these situations, the absorption position of H_a will be altered. Among the many molecules in the solution, you will find all the various possible spin combinations for H_b and H_c; hence, the NMR spectrum of the molecular solution will give *three* absorption peaks (a triplet) for H_a because H_b and H_c have three different possible spin combinations (Figure 26.12). By a similar analysis, it can be seen that protons H_b and H_c should appear as a doublet.

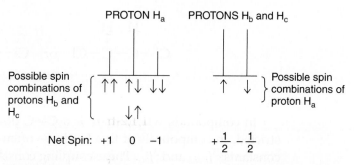

Figure 26.12

Analysis of spin–spin splitting pattern for 1,1,2-trichloroethane.

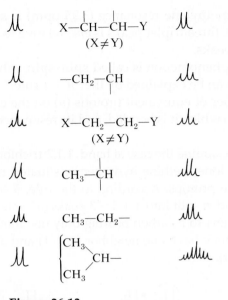

Figure 26.13
Some common splitting patterns.

Some common splitting patterns that can be predicted by the $n + 1$ rule and that are frequently observed in a number of molecules are shown in Figure 26.13. Notice particularly the last entry, where *both* methyl groups (six protons in all) function as a unit and split the methine proton into a septet ($6 + 1 = 7$).

26.10 The Coupling Constant

The quantitative amount of spin–spin interaction between two protons can be defined by the **coupling constant.** The spacing between the component peaks in a single multiplet is called the coupling constant J. This distance is measured on the same scale as the chemical shift and is expressed in hertz (Hz).

Coupling constants for protons on adjacent carbon atoms have magnitudes of from about 6 Hz to 8 Hz (see Table 26.4). You should expect to see a coupling constant in this range for compounds where there is free rotation about a single bond. Because three bonds separate protons from each other on adjacent carbon atoms, we label these coupling constants as ^3J. For example, the coupling constant for the compound shown in Figure 26.11 would be written as $^3J = 6$ Hz. The boldfaced lines in the following diagram show how the protons on adjacent carbon atoms are three bonds away from each other.

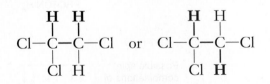

In compounds where there is a C=C double bond, free rotation is restricted. In compounds of this kind, we often find two types of ^3J coupling constants: $^3J_{trans}$ and $^3J_{cis}$. These coupling constants vary in value as shown in Table 26.4, but $^3J_{trans}$ is almost always larger than $^3J_{cis}$. The magnitudes of these ^3Js often provide important structural clues. You can distinguish, for

Table 26.4 *Representative coupling constants and approximate values (Hz)*

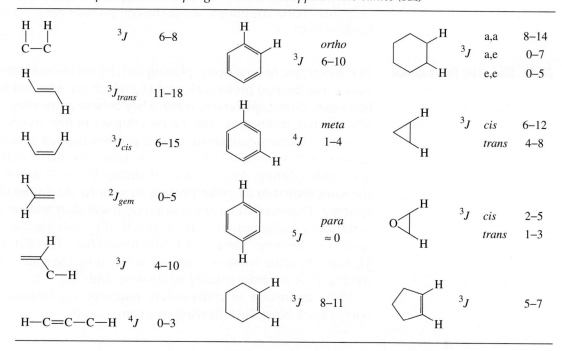

example, between a *cis* alkene and a *trans* alkene on the basis of the observed coupling constants for the two vinyl protons on disubstituted alkenes. Most of the coupling constants shown in the first column of Table 26.4 are three bond couplings, but you will notice that there is a two-bond (^2J) coupling constant listed. These protons that are bonded to a common carbon atom are often referred to as *geminal* protons and can be labeled as $^2J_{gem}$. Notice that the coupling constants for *geminal* protons are quite small for alkenes. The ^2J couplings are observed only when the protons on a methylene group are in a different environment (see Section 26.11). The following structure shows the various types of couplings that you observe for protons on a C=C double bond in a typical alkene, vinyl acetate. The spectrum for this compound is described in detail in Section 26.11.

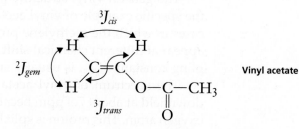

Vinyl acetate

Longer-range couplings that occur over four or more bonds are observed in some alkenes and also in aromatic compounds. Thus, in Table 26.4, we see that it is possible to observe a small H—H coupling (^4J = 0–3 Hz) occurring over four bonds in an alkene. In an aromatic compound, you often observe a small but measurable coupling between *meta* protons that are four bonds away from each other (^4J = 1–4 Hz). Couplings over five bonds are usually quite small, with values close to 0 Hz. The long-range couplings are usually observed only in *unsaturated* compounds. The spectra of saturated compounds are often more easily interpreted because they usually have only three bond couplings. Aromatic compounds are discussed in detail in Section 26.13.

26.11 Magnetic Equivalence

In the example of spin–spin splitting in 1,1,2-trichloroethane (Figure 26.11), notice that the two protons H_b and H_c, which are attached to the same carbon atom, do not split one another. They behave as an integral group. Actually, the two protons H_b and H_c *are* coupled to one another; however, for reasons we cannot explain fully here, protons that are attached to the same carbon and both of which have the same chemical shift do not show spin–spin splitting. Another way of stating this is that protons coupled to the same extent to *all* other protons in a molecule do not show spin–spin splitting. Protons that have the same chemical shift and are coupled equivalently to all other protons are magnetically equivalent and do not show spin–spin splitting. Thus, in 1,1,2-trichloroethane (p. 882), protons H_b and H_c have the same value of δ and are coupled by the same value of J to proton H_a. They are magnetically equivalent, and $^2J_{gem} = 0$.

It is important to differentiate magnetic equivalence and chemical equivalence. Note the following two compounds.

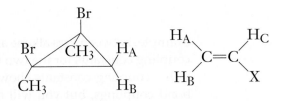

In the cyclopropane compound, the two geminal hydrogens H_A and H_B are chemically equivalent; however, they are not magnetically equivalent. Proton H_A is on the same side of the ring as the two halogens. Proton H_B is on the same side of the ring as the two methyl groups. Protons H_A and H_B will have different chemical shifts, will couple to one another, and will show spin–spin splitting. Two doublets will be seen for H_A and H_B. For cyclopropane rings, $^2J_{gem}$ is usually around 5 Hz.

The general vinyl structure (alkene) shown in the previous figure and the specific example of vinyl acetate shown in Figure 26.14 are examples of cases in which the methylene protons H_A and H_B are nonequivalent. They appear at different chemical shift values and will split each other. This coupling constant, $^2J_{gem}$, is usually small with vinyl compounds (about 2 Hz).

The spectrum of vinyl acetate is shown in Figure 26.14. H_C appears downfield at about 7.3 ppm because of the electronegativity of the attached oxygen atom. This proton is split by H_B into a doublet ($^3J_{trans} = {}^3J_{BC} = 15$ Hz), and then each leg of the doublet is split by H_A into a doublet ($^3J_{cis} = {}^3J_{AC} = 7$ Hz). Notice that the $n + 1$ rule is applied individually to each adjacent proton. The pattern that results is usually referred to as a doublet of doublets (dd). The graphic analysis shown in Figure 26.15 should help you understand the pattern obtained for proton H_C.

Now look at the pattern shown in Figure 26.14 for proton H_B at 4.85 ppm. It is also a doublet of doublets. Proton H_B is split by proton H_C into a doublet ($^3J_{trans} = {}^3J_{BC} = 15$ Hz), and then each leg of the doublet is split by the geminal proton H_A into doublets ($^2J_{gem} = {}^2J_{AB} = 2$ Hz).

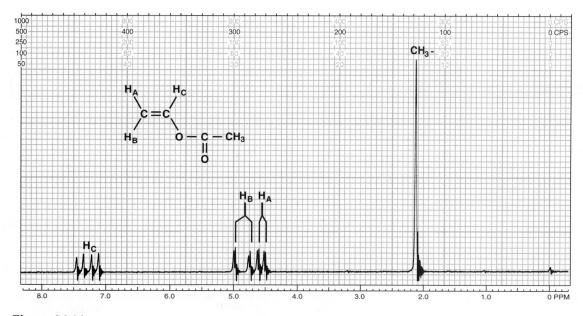

Figure 26.14

NMR spectrum of vinyl acetate. (Courtesy of Varian Associates.)

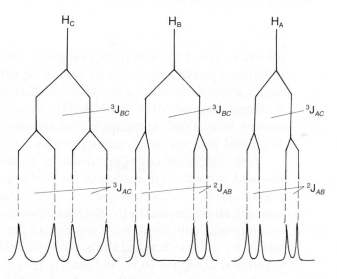

Figure 26.15

Analysis of the splittings in vinyl acetate.

Proton H_A, shown in Figure 26.14, appears at 4.55 ppm. This pattern is also a doublet of doublets. Proton H_A is split by proton H_C into a doublet ($^3J_{cis} = {}^3J_{AC} = 7$ Hz), and then each leg of the doublet is split by the geminal proton H_B into doublets ($^2J_{gem} = {}^2J_{AB} = 2$ Hz). For each proton shown in Figure 26.14, the NMR spectrum must be analyzed graphically, splitting by splitting. This complete graphic analysis is shown in Figure 26.15.

26.12 Spectra at Higher Field Strength

Occasionally, the 60-MHz spectrum of an organic compound, or a portion of it, is almost undecipherable because the chemical shifts of several groups of protons are all very similar. In these cases, all the proton resonances occur in

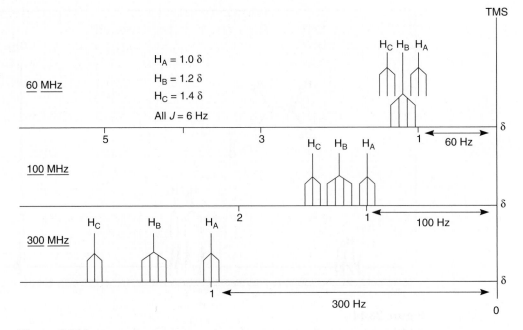

Figure 26.16
A comparison of the spectrum of a compound with overlapping multiplets at 60 MHz, with spectra of the same compound also determined at 100 MHz and 300 MHz.

the same area of the spectrum, and peaks often overlap so extensively that individual peaks and splittings cannot be extracted. One way to simplify such a situation is to use a spectrometer that operates at a higher frequency. Although both 60-MHz and 100-MHz instruments are still in use, it is becoming increasingly common to find instruments operating at much higher fields and with spectrometer frequencies 300, 400, or 500 MHz.

Although NMR coupling constants do not depend on the frequency or the field strength of operation of the NMR spectrometer, chemical shifts in *hertz* depend on these parameters. This circumstance can often be used to simplify an otherwise undecipherable spectrum. Suppose, for instance, that a compound contained three multiplets derived from groups of protons with very similar chemical shifts. At 60 MHz, these peaks might overlap, as illustrated in Figure 26.16, and simply give an unresolved envelope of absorption. It turns out that the $n + 1$ rule fails to make the proper predictions when chemical shifts are similar for the protons in a molecule. The spectral patterns that result are said to be **second order,** and what you end up seeing is an amorphous blob of unrecognizable patterns!

Figure 26.16 also shows the spectrum of the same compound at two higher frequencies (100 MHz and 300 MHz). When the spectrum is redetermined at a higher frequency, the coupling constants (J) do not change, but the chemical shifts in *hertz* (not ppm) of the proton groups (H_A, H_B, H_C) responsible for the multiplets do increase. It is important to realize, however, that the chemical shift in *ppm* is a constant, and it will not change when the frequency of the spectrometer is increased (see equation 1 on p. 874).

Notice that at 300 MHz, the individual multiplets are cleanly separated and resolved. At high frequency, the chemical shift differences of each proton increase, resulting in more clearly recognizable patterns (that is, triplets, quartets, and so on) and less overlap of proton patterns in the spectrum. At high frequency, the chemical shift differences are large, and the $n + 1$ rule will more likely correctly predict the patterns. Thus, it is a clear advantage to use NMR spectrometers operating at high frequency (300 MHz or above) because the resulting spectra are more likely to provide nonoverlapped and well-resolved peaks. When the protons in a spectrum follow the $n + 1$ rule, the spectrum is said to be **first order.** The result is that you will obtain a spectrum with much more recognizable patterns, as shown in Figure 26.16.

26.13 Aromatic Compounds—Substituted Benzene Rings

Phenyl rings are so common in organic compounds that it is important to know a few facts about NMR absorptions in compounds that contain them. In general, the ring protons of a benzenoid system have resonance near 7.3 ppm; however, electron-withdrawing ring substituents (for example, nitro, cyano, carboxyl, or carbonyl) move the resonance of these protons downfield (larger ppm values), and electron-donating ring substituents (for example, methoxy or amino) move the resonance of these protons upfield (smaller ppm values). Table 26.5 shows these trends for a series of symmetrically *p*-disubstituted benzene compounds. The *p*-disubstituted compounds were chosen because their two planes of symmetry render all of the hydrogens equivalent. Each compound gives only one aromatic peak (a singlet) in the proton NMR spectrum. Later, you will see that some positions are affected more strongly than others in systems with substitution patterns different from this one.

In the sections that follow, we will attempt to cover some of the most important types of benzene ring substitution. In some cases, it will be necessary to examine sample spectra taken at both 60 MHz and 300 MHz. Many benzenoid rings show second-order splittings at 60 MHz but are essentially first order at 300 MHz.

A. Monosubstituted Rings

Alkylbenzenes. In monosubstituted benzenes in which the substituent is neither a strongly electron-withdrawing nor a strongly electron-donating group, all the ring protons give rise to what appears to be a *single resonance* when the spectrum is determined at 60 MHz. This is a particularly common occurrence in alkyl-substituted benzenes. Although the protons *ortho*, *meta*,

Table 26.5 *Proton chemical shifts in p-disubstituted benzene compounds*

Substituent X		δ (ppm)	
—OCH$_3$		6.80	Electron donating (shielding)
—OH		6.60	
—NH$_2$		6.36	
—CH$_3$		7.05	
—H		7.32	
—COOH		8.20	Electron withdrawing (deshielding)
—NO$_2$		8.48	

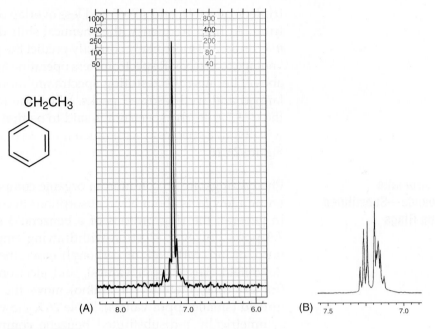

Figure 26.17
The aromatic ring portions of the ^{1}H NMR spectra of ethylbenzene at (A) 60 MHz and (B) 300 MHz.

and *para* to the substituent are not chemically equivalent, they generally give rise to a single unresolved absorption peak. A possible explanation is that the chemical shift differences, which should be small in any event, are somehow eliminated by the presence of the ring current, which tends to equalize them. All of the protons are nearly equivalent under these conditions. The NMR spectra of the aromatic portions of alkylbenzene compounds are good examples of this type of circumstance. Figure 26.17A is the 60-MHz ^{1}H spectrum of ethylbenzene.

The 300-MHz spectrum of ethylbenzene, shown in Figure 26.17B, presents quite a different picture. With the increased frequency shifts at 300 MHz, the nearly equivalent (at 60 MHz) protons are neatly separated into two groups. The *ortho* and *para* protons appear upfield from the *meta* protons. The splitting pattern is clearly second order.

Electron-Donating Groups. When electron-donating groups are attached to the ring, the ring protons are not equivalent, even at 60 MHz. A highly activating substituent such as methoxy clearly increases the electron density at the *ortho* and *para* positions of the ring (by resonance) and helps to give these protons greater shielding than those in the *meta* positions and, thus, a substantially different chemical shift.

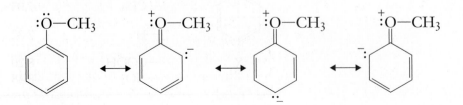

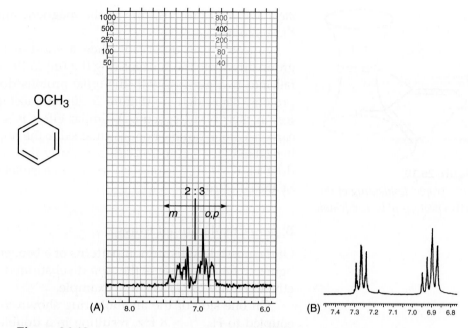

Figure 26.18
The aromatic ring portions of the ¹H NMR spectra of anisole at (A) 60 MHz and (B) 300 MHz.

At 60 MHz, this chemical shift difference results in a complicated second-order splitting pattern for anisole (methoxybenzene), but the protons do fall clearly into two groups, the *ortho/para* protons and the *meta* protons. The 60-MHz NMR spectrum of the aromatic portion of anisole (Figure 26.18A) has a complex multiplet for the *o,p* protons (integrating for three protons) that is upfield from the *meta* protons (integrating for two protons), with a clear distinction (gap) between the two types. Aniline (aminobenzene) provides a similar spectrum, also with a 3:2 split, owing to the electron-releasing effect of the amino group.

The 300-MHz spectrum of anisole (Figure 26.18B) shows the same separation between the *ortho/para* hydrogens (upfield) and the *meta* hydrogens (downfield). However, because the actual shift in hertz between the two types of hydrogens is greater, there is less second-order interaction, and the lines in the pattern are sharper at 300 MHz. In fact, it might be tempting to try to interpret the observed pattern as if it were first order, a triplet at 7.25 ppm (*meta*, 2 H) and an overlapping triplet (*para*, 1 H) with a doublet (*ortho*, 2 H) at about 6.9 ppm.

Anisotropy—Electron-Withdrawing Groups. A carbonyl or a nitro group would be expected to show (aside from anisotropy effects) a reverse effect, because these groups are electron withdrawing. It would be expected that the group would act to decrease the electron density around the *ortho* and *para* positions, thus deshielding the *ortho* and *para* hydrogens and providing a pattern exactly the reverse of the one shown for anisole (3:2 ratio, down-field:upfield). Convince yourself of this by drawing resonance structures. Nevertheless, the actual NMR spectra of nitrobenzene and benzaldehyde do not have the appearances that would be predicted on the basis of resonance structures. Instead, the *ortho* protons are much more deshielded than the

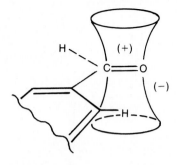

Figure 26.19
Anisotropic deshielding of the ortho *protons of benzaldehyde.*

meta and *para* protons, due to the magnetic anisotropy of the π bonds in these groups.

Anisotropy is observed when a substituent group bonds a carbonyl group directly to the benzene ring (Figure 26.19). Once again, the ring protons fall into two groups, with the *ortho* protons downfield from the *meta/para* protons. Benzaldehyde (Figure 26.20) and acetophenone both show this effect in their NMR spectra. A similar effect is sometimes observed when a carbon–carbon double bond is attached to the ring. The 300-MHz spectrum of benzaldehyde (Figure 26.20B) is a nearly first-order spectrum and shows a doublet (H_C, 2 H), a triplet (H_B, 1 H), and a triplet (H_A, 2 H). It can be analyzed by the $n + 1$ rule.

B. *para*-Disubstituted Rings

Of the possible substitution patterns of a benzene ring, some are easily recognized. One of these is the *para*-disubstituted benzene ring. Examine anethole (Figure 26.21) as a first example.

On one side of the anethole ring shown in Figure 26.21, proton H_a is coupled to H_b, $^3J = 8$ Hz, resulting in a doublet at about 6.80 ppm in the spectrum. Proton H_a appears upfield (smaller ppm value) relative to H_b because of shielding by the electron-releasing effect of the methoxy group (see p. 890). Likewise, H_b is coupled to H_a, $^3J = 8$ Hz, producing another doublet at 7.25 ppm for this proton. Because of the plane of symmetry, both halves of the ring are equivalent. Thus, H_a and H_b on the other side of the ring also appear at 6.80 ppm and 7.25 ppm, respectively. Each doublet, therefore, integrates for two protons each. A *para*-disubstituted ring, with two different substituents attached, is easily recognized by the appearance of two doublets, each integrating for two protons each.

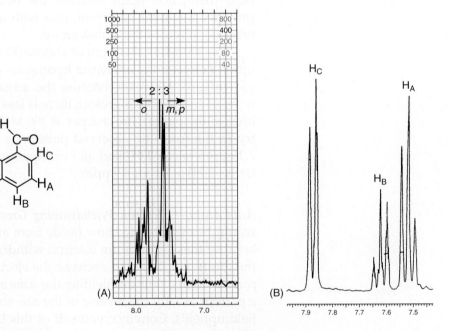

Figure 26.20
The aromatic ring portions of the ^1H *NMR spectra of benzaldehyde at (A) 60 MHz and (B) 300 MHz.*

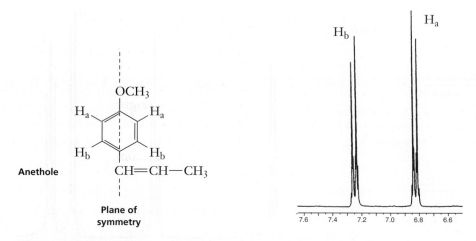

Figure 26.21

The aromatic ring protons of the 300-MHz ^{1}H NMR spectrum of anethole showing a para-*disubstituted pattern.*

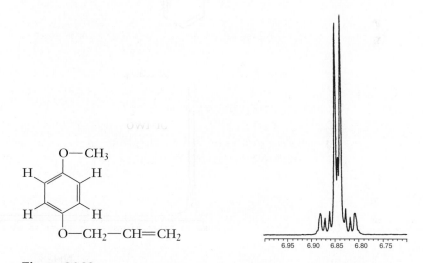

Figure 26.22

The aromatic ring protons of the 300-MHz ^{1}H NMR spectrum of 4-allyloxyanisole.

As the chemical shifts of H_a and H_b approach each other in value, the *para*-disubstituted pattern becomes similar to that of 4-allyloxyanisole (Figure 26.22). The inner peaks move closer together, and the outer ones become smaller or even disappear. Ultimately, when H_a and H_b approach each other closely enough in chemical shift, the outer peaks disappear, and the two inner peaks merge into a *singlet*; 1,4-dimethylbenzene (*para*-xylene), for instance, gives a singlet at 7.05 ppm. Hence, a single aromatic resonance integrating for four protons could easily represent a *para*-disubstituted ring, but the substituents would obviously be either identical or very similar.

C. Other Substitution

Figure 26.23 shows the 300-MHz ^{1}H spectra of the aromatic ring portions of 2-, 3-, and 4-nitroaniline (the *ortho, meta,* and *para* isomers). The characteristic

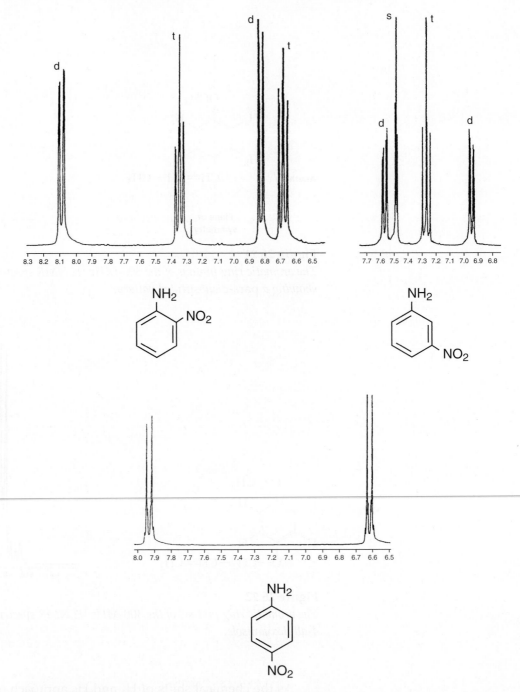

Figure 26.23
The 300-MHz ^{1}H NMR spectra of the aromatic ring portions of 2-, 3-, and 4-nitroaniline (s, singlet; d, doublet; t, triplet). The NH$_2$ group is not shown.

pattern of a *para*-disubstituted ring, with its pair of doublets, makes it easy to recognize 4-nitroaniline. The splitting patterns for 2- and 3-nitroaniline are first order, and they can be analyzed by the $n + 1$ rule. As an exercise, see if you can analyze these patterns, assigning the multiplets to specific protons on the ring. Use the indicated multiplicities (s, d, t) and expected chemical shifts to help your assignments. Remember that the amino group releases

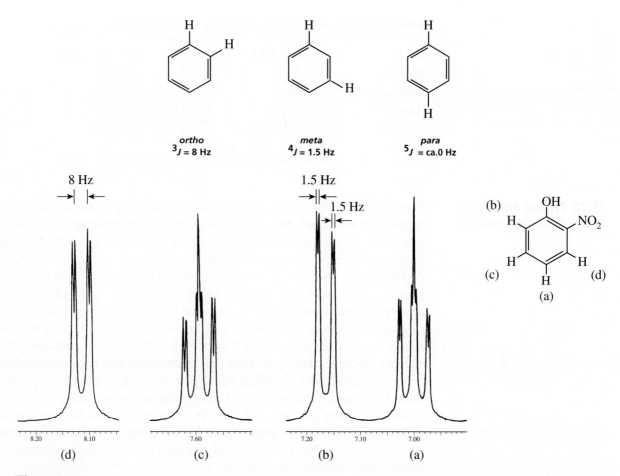

Figure 26.24

Expansions of the aromatic ring proton multiplets from the 300-MHz ^{1}H spectrum of 2-nitrophenol. The accompanying hydroxyl absorption (OH) is not shown. Coupling constants are indicated on some of the peaks of the spectrum to give an idea of scale.

electrons by resonance, and the nitro group shows a significant anisotropy toward *ortho* protons. You may ignore any *meta* and *para* couplings, remembering that these long-range couplings will be too small in magnitude to be observed on the scale on which these figures are presented. If the spectra were expanded, you would be able to observe ^4J couplings.

The spectrum shown in Figure 26.24 is of 2-nitrophenol. It is helpful to look also at the coupling constants for the benzene ring found in Table 26.4. Because the spectrum is expanded, it is now possible to see ^3J couplings (about 8 Hz), as well as ^4J couplings (about 1.5 Hz). ^5J couplings are not observed ($^5J \approx 0$). Each of the protons on this compound is assigned on the spectrum. Proton H_d appears downfield at 8.11 ppm as a doublet of doublets ($^3J_{ad}$ = 8 Hz and $^4J_{cd}$ = 1.5 Hz); H_c appears at 7.6 ppm as a triplet of doublets ($^3J_{ac}$ = $^3J_{bc}$ = 8 Hz and $^4J_{cd}$ = 1.5 Hz); H_b appears at 7.17 ppm as a doublet of doublets ($^3J_{bc}$ = 8 Hz and $^4J_{ab}$ = 1.5 Hz); and H_a appears at 7.0 ppm as a triplet of doublets ($^3J_{ac}$ = $^3J_{ad}$ = 8 Hz and $^4J_{ab}$ = 1.5 Hz). H_d appears the furthest downfield because of the anisotropy of the nitro group. H_a and H_b are relatively shielded because of the resonance-releasing effect of the hydroxyl group, which shields these two protons. H_c is assigned by a process of elimination in the absence of these two effects.

Table 26.6 *Typical ranges for groups with variable chemical shift*

Acids	RCOOH	10.5–12.0 ppm
Phenols	ArOH	4.0–7.0
Alcohols	ROH	0.5–5.0
Amines	RNH_2	0.5–5.0
Amides	$RCONH_2$	5.0–8.0
Enols	$CH=CH-OH$	≥ 15

26.14 Protons Attached to Atoms Other Than Carbon

Protons attached to atoms other than carbon often have a widely variable range of absorptions. Several of these groups are tabulated in Table 26.6. In addition, under the usual conditions of determining an NMR spectrum, protons on heteroelements normally do not couple with protons on adjacent carbon atoms to give spin–spin splitting. The primary reason is that such protons often exchange rapidly with those of the solvent medium. The absorption position is variable because these groups also undergo various degrees of hydrogen bonding in solutions of different concentrations. The amount of hydrogen bonding that occurs with a proton radically affects the valence electron density around that proton and produces correspondingly large changes in the chemical shift. The absorption peaks for protons that have hydrogen bonding or are undergoing exchange are frequently broad relative to other singlets and can often be recognized on that basis. For a different reason, called **quadrupole broadening,** protons attached to nitrogen atoms often show an extremely broad resonance peak, often almost indistinguishable from the baseline.

26.15 Chemical Shift Reagents

Researchers have known for some time that interactions between molecules and solvents, such as those due to hydrogen bonding, can cause large changes in the resonance positions of certain types of protons (for example, hydroxyl and amino). They have also known that the resonance positions of some groups of protons can be greatly affected by changing from the usual NMR solvents such as CCl_4 and $CDCl_3$ to solvents such as benzene, which impose local anisotropic effects on surrounding molecules. In many cases, it is possible to resolve partially overlapping multiplets by such a solvent change. The use of **chemical shift reagents** for this purpose dates from about 1969. Most of these chemical shift reagents are organic complexes of paramagnetic rare earth metals from the lanthanide series of elements. When these metal complexes are added to the compound whose spectrum is being determined, profound shifts in the resonance positions of the various groups of protons are observed. The direction of the shift (upfield or downfield) depends primarily on which metal is being used. Complexes of europium, erbium, thulium, and ytterbium shift resonances to lower field; complexes of cerium, praseodymium, neodymium, samarium, terbium, and holmium generally shift resonances to higher field. The advantage of using such reagents is that shifts similar to those observed at higher field can be induced without the purchase of an expensive higher-field instrument.

Of the lanthanides, europium is probably the most commonly used metal. Two of its widely used complexes are *tris*-(dipivalomethanato)europium and *tris*-(6,6,7,7,8,8,8-heptafluoro-2,2-dimethyl-3,5-octanedionato)europium. These are frequently abbreviated Eu(dpm)$_3$ and Eu(fod)$_3$, respectively.

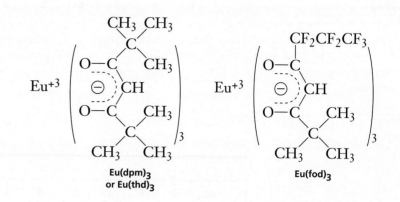

Eu(dpm)$_3$
or Eu(thd)$_3$

Eu(fod)$_3$

These lanthanide complexes produce spectral simplifications in the NMR spectrum of any compound that has a relatively basic pair of electrons (unshared pair) that can coordinate with Eu^{3+}. Typically, aldehydes, ketones, alcohols, thiols, ethers, and amines will all interact:

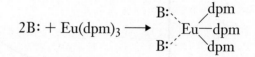

The amount of shift that a given group of protons will experience depends (1) on the distance separating the metal (Eu^{3+}) and that group of protons, and (2) on the concentration of the shift reagent in the solution. Because of the latter dependence, it is necessary when reporting a lanthanide-shifted spectrum to report the number of mole equivalents of shift reagent used or its molar concentration.

The distance factor is illustrated in the spectra of hexanol, which are given in Figures 26.25 and 26.26. In the absence of shift reagent, the normal spectrum is obtained (Figure 26.25). Only the triplet of the terminal methyl group and the triplet of the methylene group next to the hydroxyl are resolved in the spectrum. The other protons (aside from OH) are found together in a broad unresolved group. With shift reagent added (Figure 26.26), each of the methylene groups is clearly separated and resolved into the proper multiplet structure. The spectrum is first order and simplified; all the splittings are explained by the $n + 1$ rule.

One final consequence of using a shift reagent should be noted. Notice in Figure 26.26 that the multiplets are not as nicely resolved into sharp peaks as you might expect. This is due to the fact that shift reagents cause a small amount of peak broadening. At high-shift reagent concentrations, this problem becomes serious, but at most useful concentrations the amount of broadening experienced is tolerable.

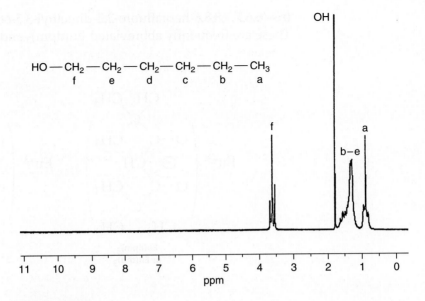

Figure 26.25
90-MHz ^{1}H NMR spectrum of hexanol determined without Eu(dpm)$_3$
© *National Institute of Advanced Industrial Science and Technology.*

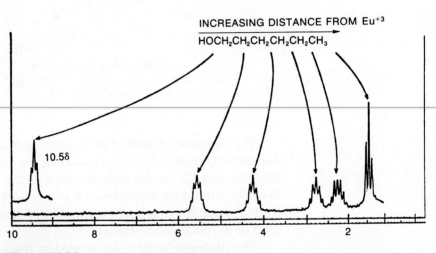

Figure 26.26
The 100-MHz ^{1}H NMR spectrum of hexanol with 0.29 mole equivalents of Eu(dpm)$_3$ added. From Sanders, J. K. M., and Williams, D. H. Chemical Communications, (1970): 422. Reproduced by permission of The Royal Society of Chemistry.

REFERENCES

Textbooks   Friebolin, H. *Basic One- and Two-Dimensional NMR Spectroscopy,* 3rd ed. New York: VCH Publishers, 1998.

Gunther, H. *NMR Spectroscopy,* 2nd ed. New York: John Wiley & Sons, 1995.

Jackman, L. M., and Sternhell, S. *Nuclear Magnetic Resonance Spectroscopy in Organic Chemistry,* 2nd ed. New York: Pergamon Press, 1969.

Macomber, R. S. *A Complete Introduction to Modern NMR Spectroscopy.* New York: John Wiley & Sons, 1997.

Macomber, R. S. *NMR Spectroscopy: Essential Theory and Practice.* New York: College Outline Series, Harcourt Brace Jovanovich, 1988.

Pavia, D. L., Lampman, G. M., and Kriz, G. S. *Introduction to Spectroscopy,* 3rd ed. Philadelphia: Harcourt College Publishers, 2001.

Sanders, J. K. M., and Hunter, B. K. *Modern NMR Spectroscopy—A Guide for Chemists,* 2nd ed. Oxford: Oxford University Press, 1993.

Silverstein, R. M., Webster, F. X. and Kiemle, D. *Spectrometric Identification of Organic Compounds,* 7th ed. New York: John Wiley & Sons, 2004.

Compilations of Spectra

Pouchert, C. J. *The Aldrich Library of NMR Spectra, 60 MHz,* 2nd ed. Milwaukee, WI: Aldrich Chemical Company, 1983.

Pouchert, C. J., and Behnke, J. *The Aldrich Library of ^{13}C and ^{1}H FT–NMR Spectra, 300 MHz.* Milwaukee, WI: Aldrich Chemical Company, 1993.

Pretsch, E., Clerc, T., Seibl, J., and Simon, W. *Tables of Spectral Data for Structure Determination of Organic Compounds,* 2nd ed. Berlin and New York: Springer-Verlag, 1989. Translated from the German by K. Biemann.

Web Sites

http://www.aist.go.jp/RIODB/SDBS/menu-e.html
Integrated Spectral DataBase System for Organic Compounds, National Institute of Materials and Chemical Research, Tsukuba, Ibaraki 305-8565, Japan. This database includes infrared, mass spectra, and NMR data (proton and carbon-13) for a large number of compounds.

http://www.chem.ucla.edu/~webspectra/
UCLA Department of Chemistry and Biochemistry in connection with Cambridge University Isotope Laboratories maintains a Web site, WebSpectra, that provides NMR and IR spectroscopy problems for students to interpret. They provide links to other sites with problems for students to solve.

PROBLEMS

1. Describe the method that you should use to determine the proton NMR spectrum of a carboxylic acid, which is insoluble in *all* the common organic solvents that your instructor is likely to make available.

2. To save money, a student uses chloroform instead of deuterated chloroform to run a proton NMR spectrum. Is this a good idea?

3. Look up the solubilities for the following compounds and decide whether you would select deuterated chloroform or deuterated water to dissolve the substances for NMR spectroscopy.

   a. Glycerol (1,2,3-propanetriol)

   b. 1,4-Diethoxybenzene

   c. Propyl pentanoate (propyl ester of pentanoic acid)

4. Assign each of the proton patterns in the spectra of 2-, 3-, and 4-nitroaniline as shown in Figure 26.23.

5. The following two compounds are isomeric esters derived from acetic acid, each with formula $C_5H_{10}O_2$. These expanded spectra clearly show the splitting patterns: singlet, doublet, triplet, quartet, etc. Integral curves are drawn on the spectra, along with relative integration values provided just above the scale and under each set of peaks. These numbers indicate the number of protons assigned to each pattern. Remember that these integral values are approximate. You will need to round the values off to the nearest whole number. Draw the structure of each compound.

a.

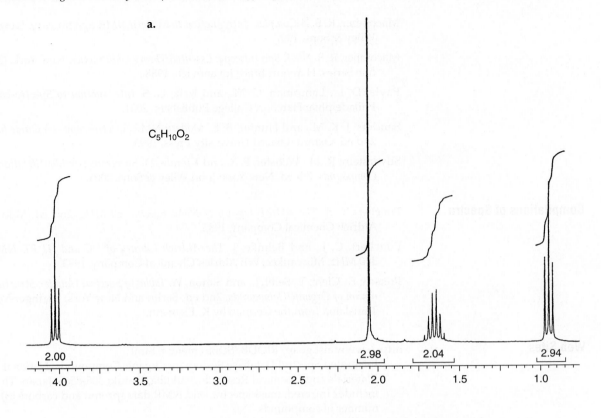

$C_5H_{10}O_2$

b. The set of peaks centering on 5 ppm is expanded in both the *x* and *y* directions in order to show the pattern more clearly. This expanded pattern is shown as an inset on the full spectrum.

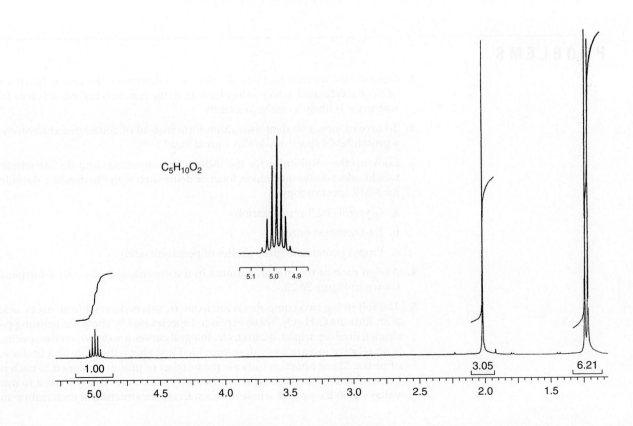

$C_5H_{10}O_2$

6. The compound that gives the following NMR spectrum has the formula $C_3H_6Br_2$. Draw the structure.

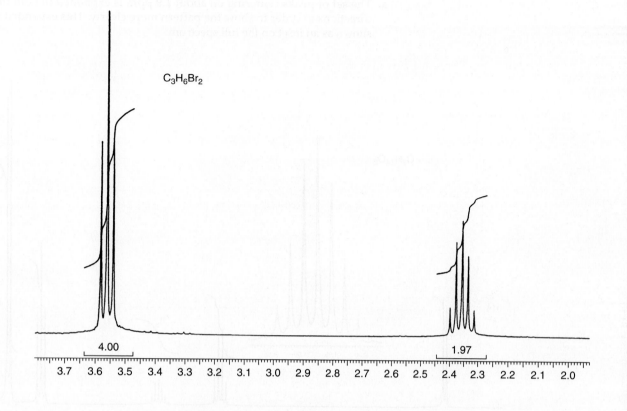

7. Draw the structure of an ether with formula $C_5H_{12}O_2$ that fits the following NMR spectrum.

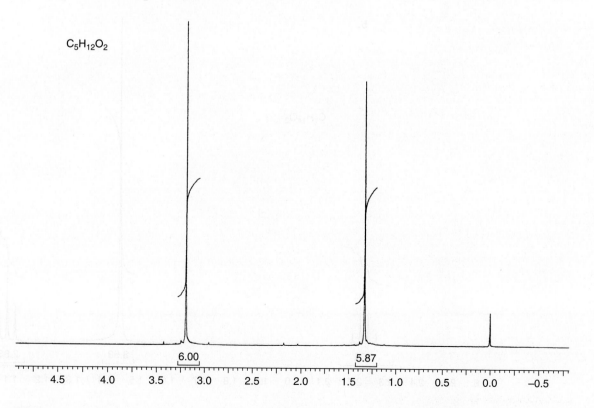

8. Following are the NMR spectra of three isomeric esters with the formula $C_7H_{14}O_2$, all derived from propanoic acid. Provide a structure for each.

a. The set of peaks centering on about 1.9 ppm is expanded in both the x and y directions in order to show the pattern more clearly. This expanded pattern is shown as an inset on the full spectrum.

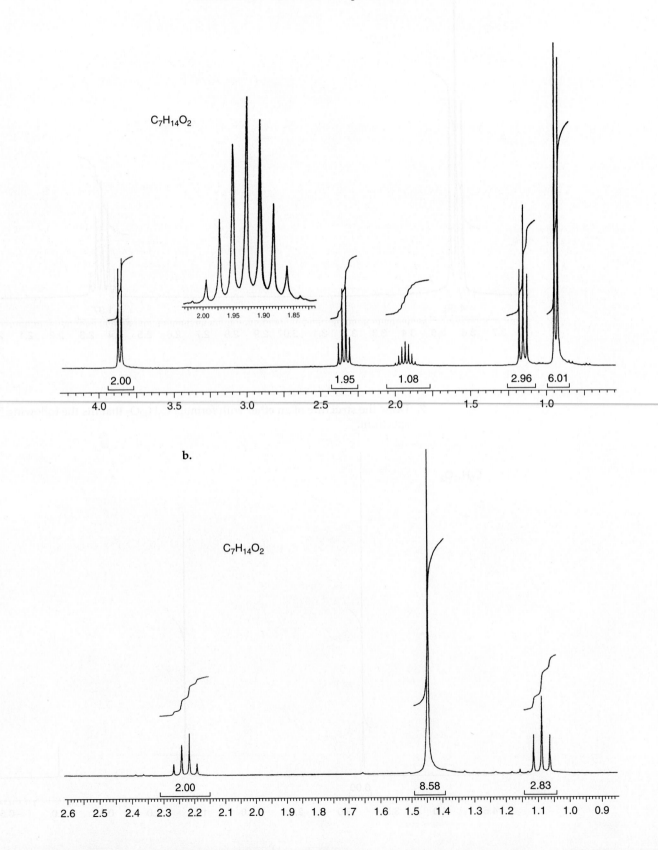

c.

$C_7H_{14}O_2$

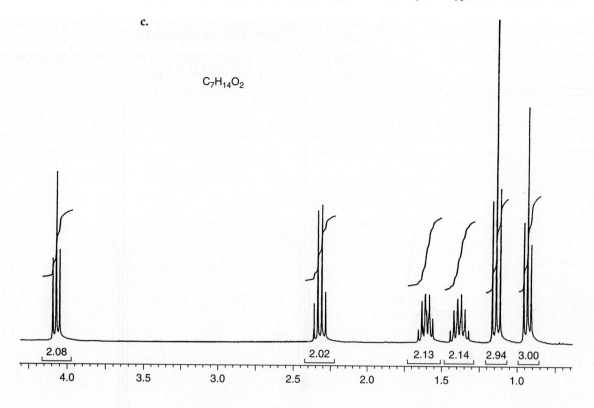

2.08   2.02   2.13   2.14   2.94   3.00

4.0   3.5   3.0   2.5   2.0   1.5   1.0

9. The two isomeric carboxylic acids that give the following NMR spectra both have the formula $C_3H_5ClO_2$. Draw their structures.

   a. The broad singlet integrating for one proton that is shown as an inset on the spectrum appears downfield at 11.5 ppm.

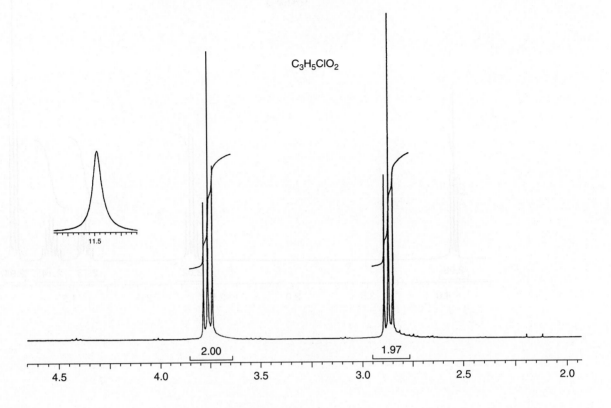

   b. The singlet integrating for one proton that is shown as an inset on the spectrum appears downfield at 12.0 ppm.

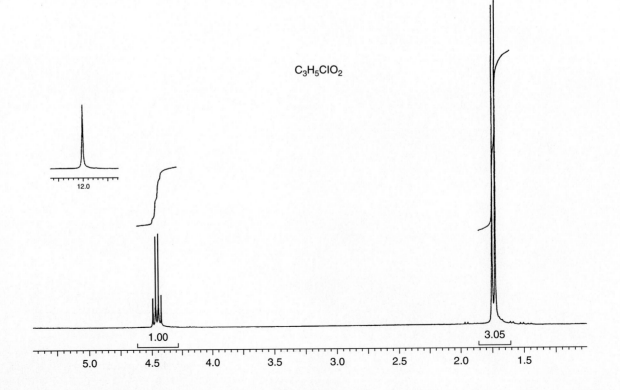

10. The following compounds are isomers with formula $C_{10}H_{12}O$. Their infrared spectra show strong bands near 1715 cm^{-1} and in the range from 1600 cm^{-1} to 1450 cm^{-1}. Draw their structures.

a.

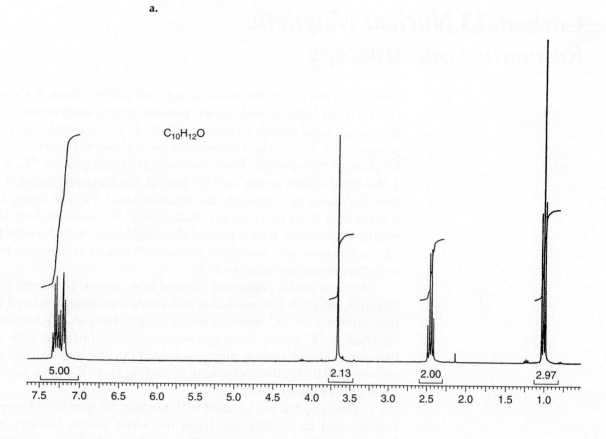

b.

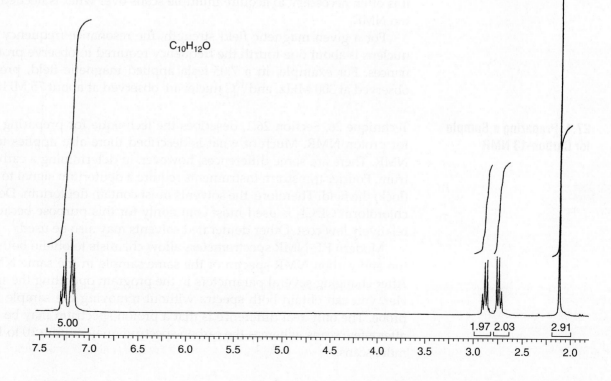

TECHNIQUE 27

Carbon-13 Nuclear Magnetic Resonance Spectroscopy

Carbon-12, the most abundant isotope of carbon, does not possess spin ($I = 0$); it has both an even atomic number and an even atomic weight. The second principal isotope of carbon, ^{13}C, however, does have the nuclear spin property ($I = \frac{1}{2}$). ^{13}C atom resonances are not easy to observe, due to a combination of two factors. First, the natural abundance of ^{13}C is low; only 1.08% of all carbon atoms are ^{13}C. Second, the magnetic moment μ of ^{13}C is low. For these two reasons, the resonances of ^{13}C are about 6000 times weaker than those of hydrogen. With special Fourier transform (FT) instrumental techniques, which are not discussed here, it is possible to observe ^{13}C nuclear magnetic resonance (carbon-13) spectra on samples that contain only the natural abundance of ^{13}C.

The most useful parameter derived from carbon-13 spectra is the chemical shift. Integrals are unreliable and are not necessarily related to the relative numbers of ^{13}C atoms present in the sample. Hydrogens that are attached to ^{13}C atoms cause spin–spin splitting, but spin–spin interaction between adjacent carbon atoms is rare. With the low natural abundance carbon-13 (0.0108), the probability of finding two ^{13}C atoms adjacent to one another is extremely low.

Carbon spectra can be used to determine the number of nonequivalent carbons and to identify the types of carbon atoms (methyl, methylene, aromatic, carbonyl, and so on) that may be present in a compound. Thus, carbon NMR provides direct information about the carbon skeleton of a molecule. Because of the low natural abundance of carbon-13 in a sample, it is often necessary to acquire multiple scans over what is needed for proton NMR.

For a given magnetic field strength, the resonance frequency of a ^{13}C nucleus is about one fourth the frequency required to observe proton resonances. For example, in a 7.05-tesla applied magnetic field, protons are observed at 300 MHz, and ^{13}C nuclei are observed at about 75 MHz.

27.1 Preparing a Sample for Carbon-13 NMR

Technique 26, Section 26.1, describes the technique for preparing samples for proton NMR. Much of what is described there also applies to carbon NMR. There are some differences, however, in determining a carbon spectrum. Fourier transform instruments require a deuterium signal to stabilize (lock) the field. Therefore, the solvents must contain deuterium. Deuterated chloroform $CDCl_3$ is used most commonly for this purpose because of its relatively low cost. Other deuterated solvents may also be used.

Modern FT–NMR spectrometers allow chemists to obtain both the proton and carbon NMR spectra of the same sample in the same NMR tube. After changing several parameters in the program operating the spectrometer, you can obtain both spectra without removing the sample from the probe. The only real difference is that a proton spectrum may be obtained after a few scans, whereas the carbon spectrum may require 10 to 100 times more scans.

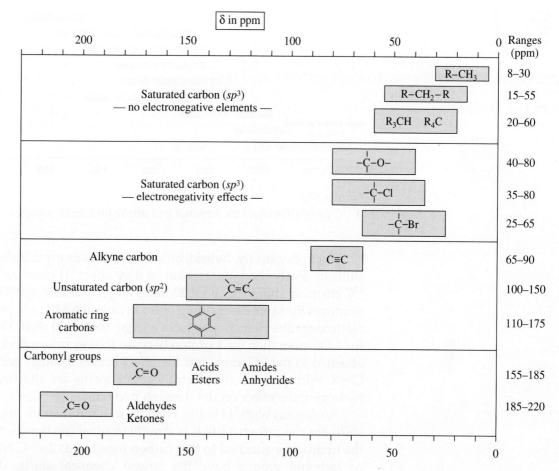

Figure 27.1

A correlation chart for ^{13}C *chemical shifts (chemical shifts are listed in parts per million from tetramethylsilane).*

Tetramethylsilane (TMS) may be added as an internal reference standard, where the chemical shift of the methyl carbon is defined as 0.00 ppm. Alternatively, you may use the center peak of the $CDCl_3$ pattern, which is found at 77.0 ppm. This pattern can be observed as a small "triplet" near 77.0 ppm in a number of the spectra given in this chapter.

27.2 Carbon-13 Chemical Shifts

An important parameter derived from carbon-13 spectra is the chemical shift. The correlation chart in Figure 27.1 shows typical ^{13}C chemical shifts, listed in parts per million (ppm) from TMS, where the carbons of the methyl groups of TMS (not the hydrogens) are used for reference. Notice that the chemical shifts appear over a range (0–220 ppm) much larger than that observed for protons (0–12 ppm). Because of the very large range of values, nearly every nonequivalent carbon atom in an organic molecule gives rise to a peak with a different chemical shift. Peaks rarely overlap as they often do in proton NMR.

The correlation chart is divided into four sections. Saturated carbon atoms appear at the highest field, nearest to TMS (8–60 ppm). The next section of the chart demonstrates the effect of electronegative atoms (40–80 ppm). The third section includes alkene and aromatic-ring carbon atoms (100–175 ppm). Finally, the fourth section contains carbonyl carbons, which appear at the lowest field values (155–220 ppm).

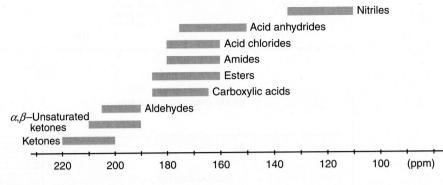

Figure 27.2
A ^{13}C correlation chart for carbonyl and nitrile functional groups.

Electronegativity, hybridization, and anisotropy all affect ^{13}C chemical shifts in nearly the same fashion as they affect ^1H chemical shifts; however ^{13}C chemical shifts are about 20 times larger. Electronegativity (Section 26.7) produces the same deshielding effect in carbon NMR as in proton NMR—the electronegative element produces a large downfield shift. The shift is greater for a ^{13}C atom than for a proton because the electronegative atom is directly attached to the ^{13}C atom and the effect occurs through only a single bond, C—X. With protons, the electronegative atoms are attached to carbon, not hydrogen; the effect occurs through two bonds, H—C—X, rather than one.

Analogous with ^1H shifts, changes in hybridization also produce larger shifts for the carbon-13 that is *directly involved* (no bonds) than they do for the hydrogens attached to that carbon (one bond). In ^{13}C NMR, the carbons of carbonyl groups have the largest chemical shifts, due both to sp^2 hybridization and to the fact that an electronegative oxygen is directly attached to the carbonyl carbon, deshielding it even further. Anisotropy (Section 26.8) is responsible for the large chemical shifts of the carbons in aromatic rings and alkenes.

Notice that the range of chemical shifts is larger for carbon atoms than for hydrogen atoms. Because the factors affecting carbon shifts operate either through one bond or directly on carbon, they are greater than those for hydrogen, which operate through more bonds. As a result, the entire range of chemical shifts becomes larger for ^{13}C (0 to 220 ppm) than for ^1H (0–12 ppm).

Many of the important functional groups of organic chemistry contain a carbonyl group. In determining the structure of a compound containing a carbonyl group, it is frequently helpful to have some idea of the type of carbonyl group in the unknown. Figure 27.2 illustrates the typical ranges of ^{13}C chemical shifts for some carbonyl-containing functional groups. Although there is some overlap in the ranges, ketones and aldehydes are easy to distinguish from the other types. Chemical shift data for carbonyl carbons are particularly powerful when combined with data from an infrared spectrum.

27.3 Proton-Coupled ^{13}C Spectra—Spin–Spin Splitting of Carbon-13 Signals

Unless a molecule is artificially enriched by synthesis, the probability of finding two ^{13}C atoms in the same molecule is low. The probability of finding two ^{13}C atoms adjacent to each other in the same molecule is even lower. Therefore, we rarely observe **homonuclear** (carbon–carbon) spin–spin splitting patterns where the interaction occurs between two ^{13}C atoms. However, the spins of protons attached directly to ^{13}C atoms do interact with the spin of carbon and cause the carbon signal to be split according to the $n + 1$ rule. This is **heteronuclear** (carbon–hydrogen) coupling involving two

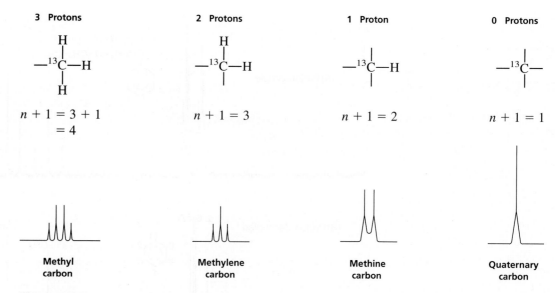

Figure 27.3

The effect of attached protons on ^{13}C resonances.

different types of atoms. With ^{13}C NMR, we generally examine splitting that arises from the protons *directly attached* to the carbon atom being studied. This is a one-bond coupling. In proton NMR, the most common splittings are *homonuclear* (hydrogen–hydrogen), which occur between protons attached to *adjacent* carbon atoms. In these cases, the interaction is a three-bond coupling, H—C—C—H.

Figure 27.3 illustrates the effect of protons directly attached to a ^{13}C atom. The $n + 1$ rule predicts the degree of splitting in each case. The resonance of a ^{13}C atom with three attached protons, for instance, is split into a quartet ($n + 1 = 3 + 1 = 4$). Because the hydrogens are directly attached to the carbon-13 (one-bond couplings), the coupling constants for this interaction are quite large, with J values of about 100 Hz to 250 Hz. Compare the typical three-bond H—C—C—H couplings that are common in NMR spectra, which have J values of about 4 Hz to 18 Hz.

It is important to note while examining Figure 27.3 that you are not "seeing" protons directly when looking at a ^{13}C spectrum (proton resonances occur at frequencies outside the range used to obtain ^{13}C spectra); you are observing only the effect of the protons on ^{13}C atoms. Also remember that we cannot observe ^{12}C, because it is NMR inactive.

Spectra that show the spin–spin splitting, or coupling, between carbon-13 and the protons directly attached to it are called **proton-coupled spectra.** Figure 27.4A is the proton-coupled ^{13}C NMR spectrum of ethyl phenylacetate. In this spectrum, the first quartet downfield from TMS (14.2 ppm) corresponds to the carbon of the methyl group. It is split into a quartet ($J = 127$ Hz) by the three attached hydrogen atoms (^{13}C—H, one-bond couplings). In addition, although it cannot be seen on the scale of this spectrum (an expansion must be used), each of the quartet lines is split into a closely spaced triplet ($J = $ ca. 1 Hz). This additional fine splitting is caused by the two protons on the adjacent —CH$_2$— group. These are two-bond couplings (H—C—^{13}C) of a type that occurs commonly in ^{13}C spectra, with coupling constants that are generally quite small ($J = 0$–2 Hz) for systems with carbon atoms in an aliphatic chain. Because of their small size, these couplings are frequently ignored in the routine analysis of spectra, with

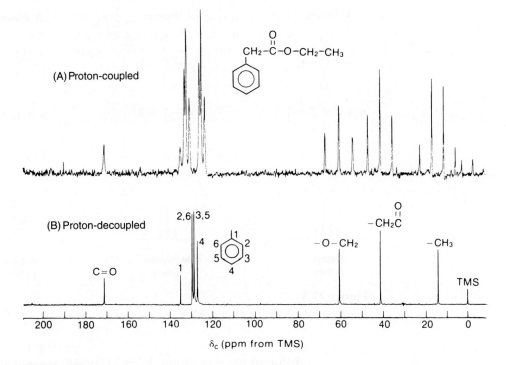

Figure 27.4

Ethyl phenylacetate. (A) The proton-coupled ^{13}C *NMR spectrum (20 MHz). (B) The proton-decoupled* ^{13}C *spectrum (20 MHz). (From Moore, J. A., Dalrymple, D. L., and Rodig, O. R.* Experimental Methods in Organic Chemistry, *3rd ed. Philadelphia: W. B. Saunders, 1982.)*

greater attention being given to the larger one-bond splittings seen in the quartet itself.

There are two —CH$_2$— groups in ethyl phenylacetate. The one corresponding to the ethyl —CH$_2$— group is found farther downfield (60.6 ppm), as this carbon is deshielded by the attached oxygen. It is a triplet because of the two attached hydrogens (one-bond couplings). Again, although it is not seen in this unexpanded spectrum, the three hydrogens on the adjacent methyl group finely split each of the triplet peaks into a quartet. The benzyl —CH$_2$— carbon is the intermediate triplet (41.4 ppm). Farthest downfield is the carbonyl-group carbon (171.1 ppm). On the scale of this presentation, it is a singlet (no directly attached hydrogens), but because of the adjacent benzyl —CH$_2$— group, it is actually split finely into a triplet. The aromatic ring carbons also appear in the spectrum, and they have resonances in the range from 127 ppm to 136 ppm. Section 27.7 will discuss aromatic ring ^{13}C resonances.

Proton-coupled spectra for large molecules are often difficult to interpret. The multiplets from different carbons commonly overlap because the ^{13}C—H coupling constants are frequently larger than the chemical shift differences of the carbons in the spectrum. Sometimes, even simple molecules such as ethyl phenylacetate (Figure 27.4A) are difficult to interpret. Proton decoupling, which is discussed in the next section, avoids this problem.

27.4 Proton-Decoupled ^{13}C Spectra

By far, the great majority of ^{13}C NMR spectra are obtained as **proton-decoupled spectra.** The decoupling technique obliterates all interactions between protons and ^{13}C nuclei; therefore, only **singlets** are observed in a decoupled ^{13}C NMR spectrum. Although this technique simplifies the spectrum and avoids overlapping multiplets, it has the disadvantage that the information on attached hydrogens is lost.

Proton **decoupling** is accomplished in the process of determining a ^{13}C NMR spectrum by simultaneously irradiating all of the protons in the molecule with a broad spectrum of frequencies in the proper range for protons. Modern NMR spectrometers provide a second, tunable radio-frequency generator, the **decoupler,** for this purpose. Irradiation causes the protons to become saturated, and they undergo rapid upward and downward transitions, among all their possible spin states. These rapid transitions decouple any spin–spin interactions between the hydrogens and the ^{13}C nuclei being observed. In effect, all spin interactions are averaged to zero by the rapid changes. The carbon nucleus "senses" only one average spin state for the attached hydrogens rather than two or more distinct spin states.

Figure 27.4B is a proton-decoupled spectrum of ethyl phenylacetate. The proton-coupled spectrum (Figure 27.4A) was discussed in Section 27.3. It is interesting to compare the two spectra to see how the proton-decoupling technique simplifies the spectrum. Every chemically and magnetically distinct carbon gives only a single peak. Notice, however, that the two *ortho* ring carbons (carbons 2 and 6) and the two *meta* ring carbons (carbons 3 and 5) are equivalent by symmetry and that each pair gives only a single peak.

Figure 27.5 is a second example of a proton-decoupled spectrum. Notice that the spectrum shows three peaks corresponding to the exact number of

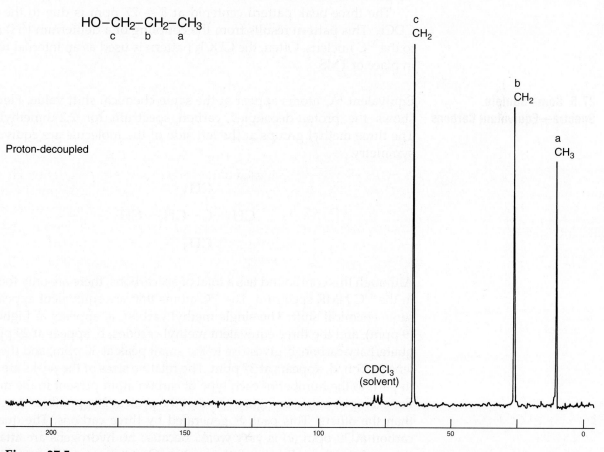

Figure 27.5
The proton-decoupled ^{13}C NMR spectrum of 1-propanol (22.5 MHz).

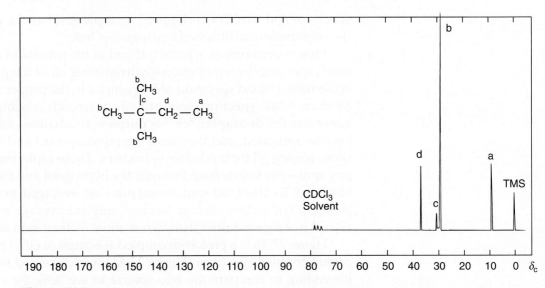

Figure 27.6
The proton-decoupled ^{13}C NMR spectrum of 2,2-dimethylbutane.

carbon atoms in 1-propanol. If there are no equivalent carbon atoms in a molecule, a ^{13}C peak will be observed for *each* carbon. Notice also that the assignments given in Figure 27.5 are consistent with the values in the chemical shift chart (Figure 27.1). The carbon atom closest to the electronegative oxygen is farthest downfield, and the methyl carbon is at highest field.

The three-peak pattern centered at δ = 77 ppm is due to the solvent $CDCl_3$. This pattern results from the coupling of a deuterium (^2H) nucleus to the ^{13}C nucleus. Often, the $CDCl_3$ pattern is used as an internal reference in place of TMS.

27.5 Some Sample Spectra—Equivalent Carbons

Equivalent ^{13}C atoms appear at the same chemical shift value. Figure 27.6 shows the proton-decoupled carbon spectrum for 2,2-dimethylbutane. The three methyl groups at the left side of the molecule are equivalent by symmetry.

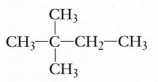

Although this compound has a total of six carbons, there are only four peaks in the ^{13}C NMR spectrum. The ^{13}C atoms that are equivalent appear at the same chemical shift. The single methyl carbon, a, appears at highest field (9 ppm), and the three equivalent methyl carbons, b, appear at 29 ppm. The quaternary carbon, c, gives rise to the small peak at 30 ppm, and the methylene carbon, d, appears at 37 ppm. The relative sizes of the peaks are related, in part, to the number of each type of carbon atom present in the molecule. For example, notice in Figure 27.6 that the peak at 29 ppm (b) is much larger than the others. This peak is generated by three carbons. The quaternary carbon at 30 ppm (c) is very weak. Because no hydrogens are attached to this carbon, there is very little nuclear Overhauser enhancement (NOE) (Section 27.6). Without attached hydrogen atoms, relaxation times are also

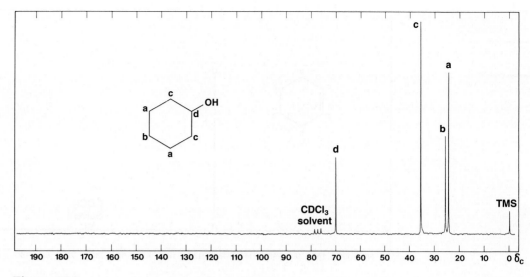

Figure 27.7

The proton-decoupled ^{13}C NMR spectrum of cyclohexanol.

longer than for other carbon atoms. Quaternary carbons, those with no hydrogens attached, frequently appear as weak peaks in proton-decoupled ^{13}C NMR spectra (see Section 27.6).

Figure 27.7 is a proton-decoupled ^{13}C spectrum of cyclohexanol. This compound has a plane of symmetry passing through its hydroxyl group, and it shows only four carbon resonances. Carbons a and c are doubled due to symmetry and give rise to larger peaks than carbons b and d. Carbon d, bearing the hydroxyl group, is deshielded by oxygen and has its peak at 70.0 ppm. Notice that this peak has the lowest intensity of all of the peaks. Its intensity is lower than that of carbon b in part because the carbon d peak receives the least amount of NOE; there is only one hydrogen attached to the hydroxyl carbon, whereas each of the other carbons has two hydrogens.

A carbon attached to a double bond is deshielded due to its sp^2 hybridization and some diamagnetic anisotropy. This effect can be seen in the ^{13}C NMR spectrum of cyclohexene (Figure 27.8). Cyclohexene has a plane of symmetry that runs perpendicular to the double bond. As a result, we observe only three absorption peaks. There are two of each type of sp^3 carbon. Each of the double-bond carbons c has only one hydrogen, whereas each of the remaining carbons has two. As a result of a reduced NOE, the double-bond carbons (127 ppm) have a lower-intensity peak in the spectrum.

In Figure 27.9, the spectrum of cyclohexanone, the carbonyl carbon has the lowest intensity. This is due not only to reduced NOE (no hydrogen attached) but also to the long relaxation time of the carbonyl carbon (Section 27.6). Notice also that Figure 27.2 predicts the large chemical shift for this carbonyl carbon (211 ppm).

27.6 Nuclear Overhauser Enhancement (NOE)

When we obtain a proton-decoupled ^{13}C spectrum, the intensities of many of the carbon resonances increase significantly above those observed in a proton-coupled experiment. Carbon atoms with hydrogen atoms directly attached are enhanced the most, and the enhancement increases (but not always linearly) as more hydrogens are attached. This effect is known as the nuclear Overhauser effect, and the degree of increase in the signal is called

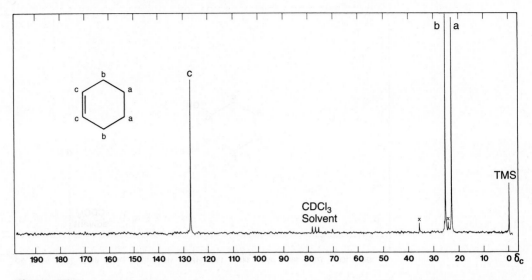

Figure 27.8
The proton-decoupled ^{13}C NMR spectrum of cyclohexene. (The peaks marked with an x are impurities.)

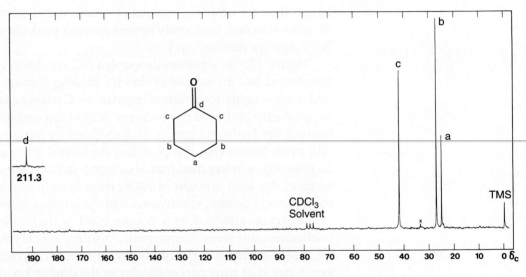

Figure 27.9
The proton-decoupled ^{13}C NMR spectrum of cyclohexanone. (The peak marked with an x is an impurity.)

the **nuclear Overhauser enhancement (NOE).** Thus, we expect that the intensity of the carbon peaks should increase in the following order in a typical carbon-13 NMR spectrum:

$$CH_3 > CH_2 > CH > C$$

Carbon atom relaxation times influence the intensity of peaks in a spectrum. When more protons are attached to a carbon atom, relaxation times become shorter, resulting in more intense peaks. Thus, we expect methyl and methylene groups to be relatively more intense than the intensity observed for quaternary carbon atoms where there are no attached protons. Thus, a weak-intensity peak is observed for the quaternary carbon atom at 30 ppm in 2,2-dimethylbutane (Figure 27.6). In addition, weak carbonyl

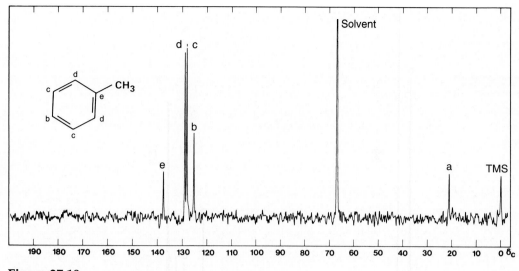

Figure 27.10
The proton-decoupled ^{13}C NMR spectrum of toluene.

carbon peaks are observed at 171 ppm in ethyl phenylacetate (Figure 27.4) and at 211 ppm in cyclohexanone (Figure 27.9).

27.7 Compounds with Aromatic Rings

Compounds with carbon–carbon double bonds or aromatic rings give rise to chemical shifts from 100 ppm to 175 ppm. Because relatively few other peaks appear in this range, a great deal of useful information is available when peaks appear here.

A **monosubstituted** benzene ring shows *four* peaks in the aromatic carbon area of a proton-decoupled ^{13}C spectrum, because the *ortho* and *meta* carbons are doubled by symmetry. Often the carbon with no protons attached, the *ipso* carbon, has a very weak peak due to a long relaxation time and a weak NOE. In addition, there are two larger peaks for the doubled *ortho* and *meta* carbons and a medium-sized peak for the *para* carbon. In many cases, it is not important to be able to assign all of the peaks precisely. In the example of toluene, shown in Figure 27.10, notice that carbons c and d are not easy to assign by inspection of the spectrum.

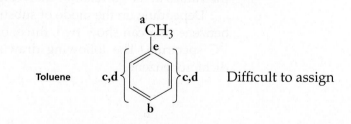

In a proton-coupled ^{13}C spectrum, a monosubstituted benzene ring shows three doublets and one singlet. The singlet arises from the *ipso* carbon, which has no attached hydrogen. Each of the other carbons in the ring (*ortho*, *meta*, and *para*) has one attached hydrogen and yields a doublet.

Figure 27.4B is the proton-decoupled spectrum of ethyl phenylacetate, with the assignments noted next to the peaks. Notice that the aromatic ring region shows four peaks between 125 ppm and 135 ppm, consistent with a monosubstituted ring. There is one peak for the methyl carbon (13 ppm)

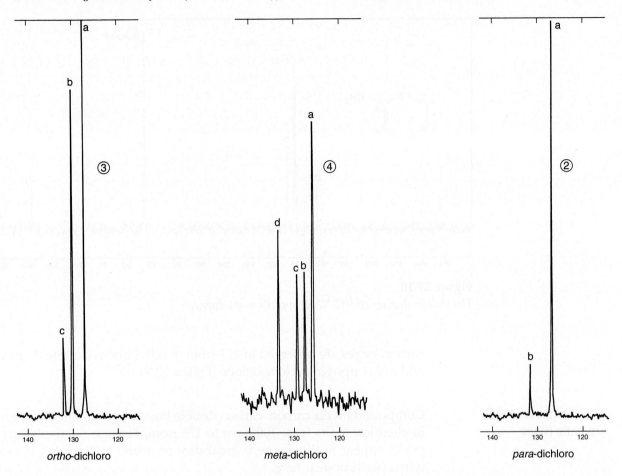

Figure 27.11
The proton-decoupled ^{13}C NMR spectra of the three isomers of dichlorobenzene (25 MHz).

and there are two peaks for the methylene carbons. One of the methylene carbons is directly attached to an electronegative oxygen atom and appears at 61 ppm, and the other is more shielded (41 ppm). The carbonyl carbon (an ester) has resonance at 171 ppm. All of the carbon chemical shifts agree with the values in the correlation chart (Figure 27.1).

Depending on the mode of substitution, a symmetrically **disubstituted** benzene ring can show two, three, or four peaks in the proton-decoupled ^{13}C spectrum. The following drawings illustrate this for the isomers of dichlorobenzene.

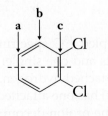

Three unique carbon atoms

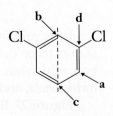

Four unique carbon atoms

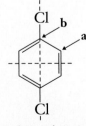

Two unique carbon atoms

Figure 27.11 shows the spectra of all three dichlorobenzenes, each of which has the number of peaks consistent with the analysis just given. You can see that ^{13}C NMR spectroscopy is very useful in the identification of isomers.

Most other polysubstitution patterns on a benzene ring yield six peaks in the proton-decoupled ^{13}C NMR spectrum, one for each carbon. However, when identical substituents are present, watch carefully for planes of symmetry that may reduce the number of peaks.

REFERENCES

Textbooks

Friebolin, H. *Basic One- and Two-Dimensional NMR Spectroscopy,* 3rd ed. New York: VCH Publishers, 1998.

Gunther, H. *NMR Spectroscopy,* 2nd ed. New York: John Wiley & Sons, 1995.

Levy, G. C. *Topics in Carbon-13 Spectroscopy.* New York: John Wiley & Sons, 1984.

Levy, G. C., Lichter, R. L., and Nelson, G. L. *Carbon-13 Nuclear Magnetic Resonance Spectroscopy,* 2nd ed. New York: John Wiley & Sons, 1980.

Macomber, R. S. *A Complete Introduction to Modern NMR Spectroscopy.* New York: John Wiley & Sons, 1997.

Macomber, R. S. *NMR Spectroscopy—Essential Theory and Practice.* New York: College Outline Series, Harcourt Brace Jovanovich, 1988.

Pavia, D. L., Lampman, G. M., and Kriz, G. S. *Introduction to Spectroscopy,* 3rd ed. Philadelphia: Harcourt College Publishers, 2001.

Sanders, J. K. M., and Hunter, B. K. *Modern NMR Spectroscopy—A Guide for Chemists,* 2d ed. Oxford, England: Oxford University Press, 1993.

Silverstein, R. M., Webster, F. X., and Kiemle, D. *Spectrometric Identification of Organic Compounds,* 7th ed. New York: John Wiley & Sons, 2004.

Compilations of Spectra

Johnson, L. F., and Jankowski, W. C. *Carbon-13 NMR Spectra: A Collection of Assigned, Coded, and Indexed Spectra, 25 MHz.* New York: Wiley-Interscience, 1972.

Pouchert, C. J., and Behnke, J. *The Aldrich Library of ^{13}C and ^{1}H FT–NMR Spectra, 75 and 300 MHz.* Milwaukee, WI: Aldrich Chemical Company, 1993.

Pretsch, E., Clerc, T., Seibl, J., and Simon, W. *Tables of Spectral Data for Structure Determination of Organic Compounds,* 2nd ed. Berlin and New York: Springer-Verlag, 1989. Translated from the German by K. Biemann.

Web Sites

http://www.aist.go.jp/RIODB/SDBS/menu-e.html
Integrated Spectral DataBase System for Organic Compounds, National Institute of Materials and Chemical Research, Tsukuba, Ibaraki 305-8565, Japan. This database includes infrared, mass spectra, and NMR data (proton and carbon-13) for a number of compounds.

http://www.chem.ucla.edu/~webspectra
UCLA Department of Chemistry and Biochemistry in connection with Cambridge University Isotope Laboratories maintains a Web site, WebSpectra, that provides NMR and IR spectroscopy problems for students to interpret. They provide links to other sites with problems for students to solve.

PROBLEMS

1. Predict the number of peaks that you would expect in the proton-decoupled ^{13}C spectrum of each of the following compounds. Problems 1a and 1b are provided as examples. Dots are used to show the nonequivalent carbon atoms in these two examples.

a.

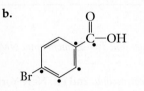   **Four peaks**

b.

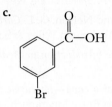   **Five peaks**

c.

d.

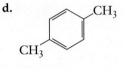

e.

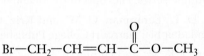

f.

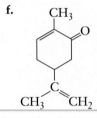

g.

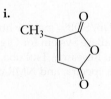

h.

i.

CH$_3$

j.

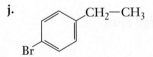

k.

CH$_2$—CH$_3$

Br

2. Following are the ^{1}H and ^{13}C spectra for two isomeric bromoalkanes (**A** and **B**) with formula C_4H_9Br. Integral curves are drawn on the spectra, along with relative integral values provided just above the scale and under each set of peaks. These numbers indicate the relative number of protons assigned to each pattern. Remember that these integral values are approximate. You will need to round the values off to the nearest whole number. Also, in some cases, the lowest whole-number ratios are given. In that case, the values provided may need to be multiplied by two or three in order to obtain the actual number of protons in each pattern.

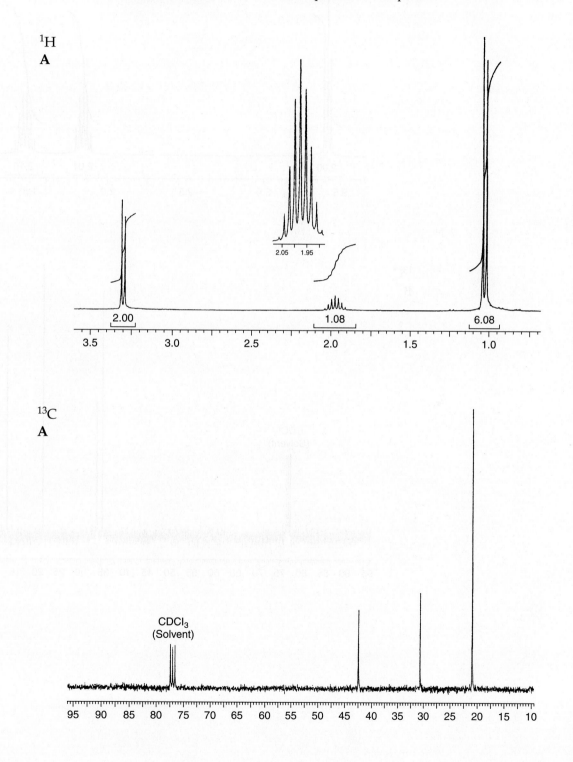

^{1}H
A

^{13}C
A

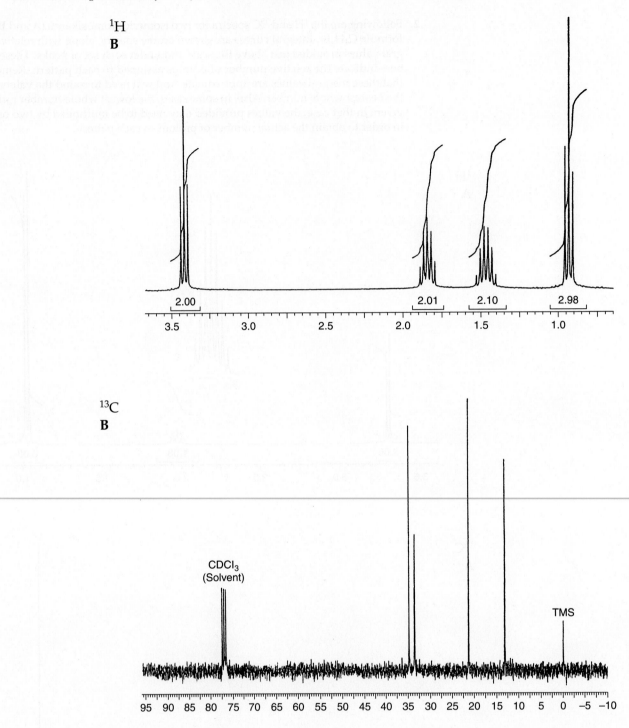

3. Following are the ^1H and ^{13}C spectra for each of three isomeric ketones (**A**, **B**, and **C**) with formula $C_7H_{14}O$. Integral curves are drawn on the spectra, along with relative integral values provided just above the scale and under each set of peaks. These numbers indicate the relative number of protons assigned to each pattern. Remember that these integral values are approximate. You will need to round the values off to the nearest whole number. Also, in some cases, the lowest whole-number ratios are given. In that case, the values provided may need to be multiplied by two or three in order to obtain the actual number of protons in each pattern.

^1H
A

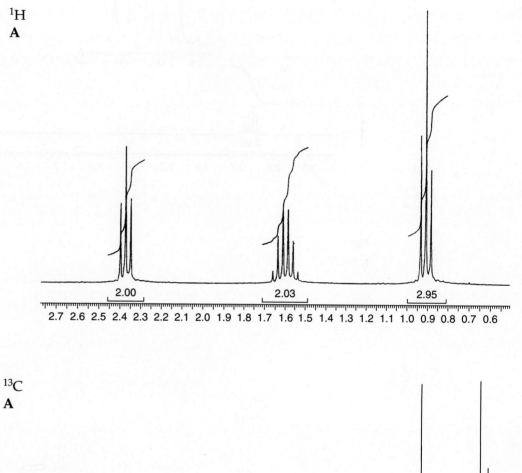

^{13}C
A

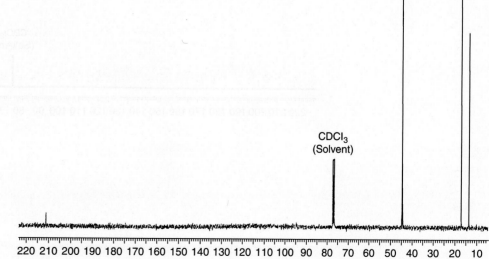

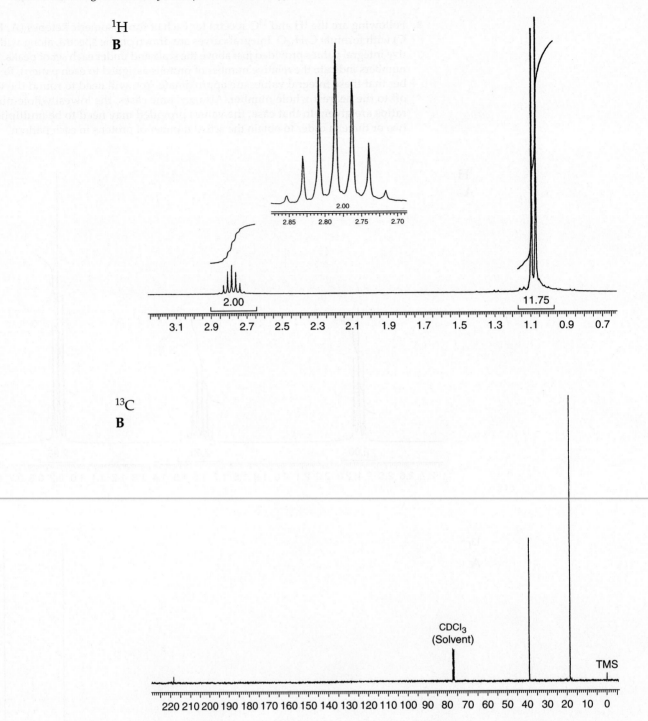

¹H
B

2.00

2.85   2.80   2.75   2.70

2.00

11.75

3.1   2.9   2.7   2.5   2.3   2.1   1.9   1.7   1.5   1.3   1.1   0.9   0.7

¹³C
B

CDCl₃
(Solvent)

TMS

220 210 200 190 180 170 160 150 140 130 120 110 100 90 80 70 60 50 40 30 20 10 0

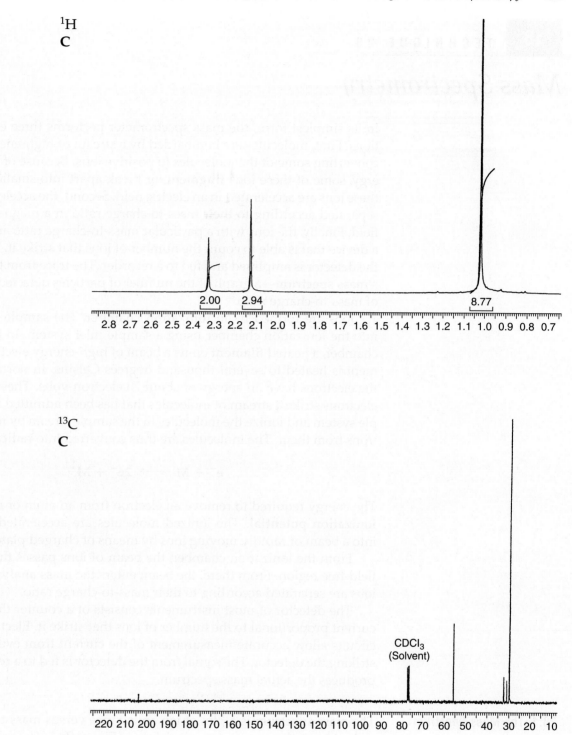

Mass Spectrometry

In its simplest form, the mass spectrometer performs three essential functions. First, molecules are bombarded by a stream of high-energy electrons, converting some of the molecules to positive ions. Because of their high energy, some of these ions **fragment,** or break apart into smaller ions. All of these ions are accelerated in an electric field. Second, the accelerated ions are separated according to their mass-to-charge ratio in a magnetic or electric field. Finally, the ions with a particular mass-to-charge ratio are detected by a device that is able to count the number of ions that strike it. The output of the detector is amplified and fed to a recorder. The trace from the recorder is a **mass spectrum**—a graph of the number of particles detected as a function of mass-to-charge ratio.

Ions are formed in an **ionization chamber.** The sample is introduced into the ionization chamber using a sample inlet system. In the ionization chamber, a heated **filament** emits a beam of high-energy electrons. The filament is heated to several thousand degrees Celsius. In normal operation, the electrons have an energy of about 70 electron-volts. These high-energy electrons strike a stream of molecules that has been admitted from the sample system and ionize the molecules in the sample stream by removing electrons from them. The molecules are thus converted into **radical-cations.**

$$e^- + M \longrightarrow 2e^- + M^{\overset{+}{\bullet}}$$

The energy required to remove an electron from an atom or molecule is its **ionization potential.** The ionized molecules are accelerated and focused into a beam of rapidly moving ions by means of charged plates.

From the ionization chamber, the beam of ions passes through a short field-free region. From there, the beam enters the **mass analyzer,** where the ions are separated according to their mass-to-charge ratio.

The detector of most instruments consists of a counter that produces a current proportional to the number of ions that strike it. Electron multiplier circuits allow accurate measurement of the current from even a single ion striking the detector. The signal from the detector is fed to a **recorder,** which produces the actual mass spectrum.

28.1 The Mass Spectrum

The **mass spectrum** is a plot of ion abundance versus mass-to-charge (m/e) ratio. A typical mass spectrum is shown in Figure 28.1. The spectrum shown is that of dopamine, a substance that acts as a neurotransmitter in the central nervous system. The spectrum is displayed as a bar graph of percentage ion abundance (relative abundance) plotted against m/e.

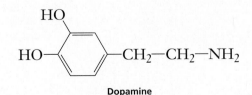

Dopamine

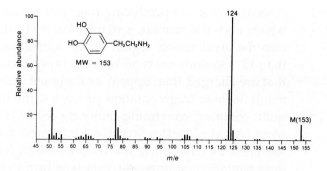

Figure 28.1
The mass spectrum of dopamine.

The most abundant ion formed in the ionization chamber gives rise to the tallest peak in the mass spectrum, called the **base peak.** For dopamine, the base peak appears at $m/e = 124$. The relative abundances of all the other peaks in the spectrum are reported as percentages of the abundance of the base peak.

The beam of electrons in the ionization chamber converts some of the sample molecules into positive ions. Removal of a single electron from a molecule yields an ion whose weight is the actual molecular weight of the original molecule. This ion is the **molecular ion,** frequently symbolized as M^+. The value of m/e at which the molecular ion appears on the mass spectrum, assuming that the ion has only one electron removed, gives the molecular weight of the original molecule. In the mass spectrum of dopamine, the molecular ion appears at $m/e = 153$, the molecular weight of dopamine. If you can identify the molecular ion peak in the mass spectrum, you can use the spectrum to determine the molecular weight of an unknown substance. If the presence of heavy isotopes is ignored for the moment, the molecular ion peak corresponds to the heaviest particle observed in the mass spectrum.

Molecules do not occur in nature as isotopically pure species. Virtually all atoms have heavier isotopes that occur in varying natural abundances. Hydrogen occurs largely as ^1H, but a few percent of hydrogen atoms occur as the isotope ^2H. Further, carbon normally occurs as ^{12}C, but a few percent of carbon atoms are the heavier isotope, ^{13}C. With the exception of fluorine, most other elements have a certain percentage of heavier isotopes that occur naturally. Peaks caused by ions bearing these heavier isotopes are also found in the mass spectrum. The relative abundances of these isotopic peaks are proportional to the abundances of the isotopes in nature. Most often, the isotopes occur at one or two mass units above the mass of the "normal" atom. Therefore, besides looking for the molecular ion (M^+) peak, you should also attempt to locate the M+1 and M+2 peaks. As will be demonstrated later, you can use the relative abundances of these M+1 and M+2 peaks to determine the molecular formula of the substance being studied.

The beam of electrons in the ionization chamber can produce the molecular ion. This beam also has sufficient energy to break some of the bonds in the molecule, producing a series of molecular fragments. Fragments that are positively charged are also accelerated in the ionization chamber, sent through the analyzer, detected, and recorded on the mass spectrum. These **fragment ion peaks** appear at m/e values corresponding to their individual masses. Very often, a fragment ion rather than the molecular ion will be the most abundant ion produced in the mass spectrum (the base peak).

A second means of producing fragment ions occurs with the molecular ion, which, once it is formed, is so unstable that it disintegrates before it can pass into the accelerating region of the ionization chamber. Lifetimes shorter than 10^{-5} seconds are typical in this type of fragmentation. Those fragments that are charged then appear as fragment ions in the mass spectrum. As a result of these fragmentation processes, the typical mass spectrum can be quite complex, containing many more peaks than the molecular ion and M+1 and M+2 peaks. Structural information about a substance can be determined by examining the fragmentation pattern in the mass spectrum. Fragmentation patterns are discussed further in Section 28.3.

28.2 Molecular Formula Determination

Mass spectrometry can be used to determine the molecular formulas of molecules that provide reasonably abundant molecular ions. Although there are at least two principal techniques for determining a molecular formula, only one will be described here.

The molecular formula of a substance can be determined through the use of **precise atomic masses.** High-resolution mass spectrometers are required for this method. Atoms are normally thought of as having integral atomic masses; for example, H = 1, C = 12, and O = 16. If you can determine atomic masses with sufficient precision, however, you find that the masses do not have values that are exactly integral. The mass of each atom actually differs from a whole mass number by a small fraction of a mass unit. The actual masses of some atoms are given in Table 28.1.

Table 28.1 *Precise masses of some common elements*

Element	Atomic Weight	Nuclide	Precise Mass
Hydrogen	1.00797	^1H	1.00783
		^2H	2.01410
Carbon	12.01115	^{12}C	12.0000
		^{13}C	13.00336
Nitrogen	14.0067	^{14}N	14.0031
		^{15}N	15.0001
Oxygen	15.9994	^{16}O	15.9949
		^{17}O	16.9991
		^{18}O	17.9992
Fluorine	18.9984	^{19}F	18.9984
Silicon	28.086	^{28}Si	27.9769
		^{29}Si	28.9765
		^{30}Si	29.9738
Phosphorus	30.974	^{31}P	30.9738
Sulfur	32.064	^{32}S	31.9721
		^{33}S	32.9715
		^{34}S	33.9679
Chlorine	35.453	^{35}Cl	34.9689
		^{37}Cl	36.9659
Bromine	79.909	^{79}Br	78.9183
		^{81}Br	80.9163
Iodine	126.904	^{127}I	126.9045

Depending on the atoms that are contained within a molecule, it is possible for particles of the same nominal mass to have slightly different measured masses when precise mass determinations are possible. To illustrate, a molecule whose molecular weight is 60 could be C_3H_8O, $C_2H_8N_2$, $C_2H_4O_2$, or CH_4N_2O. The species have the following precise masses:

C_3H_8O	60.05754
$C_2H_8N_2$	60.06884
$C_2H_4O_2$	60.02112
CH_4N_2O	60.03242

Observing a molecular ion with a mass of 60.058 would establish that the unknown molecule was C_3H_8O. Distinguishing among these possibilities is well within the capability of a modern high-resolution instrument.

In another method, these four compounds may also be distinguished by differences in the relative intensities of their M, M+1, and M+2 peaks. The predicted intensities are either calculated by formula or looked up in tables. Details of this method may be found in the References (p. 942).

28.3 Detecting Halogens

When chlorine or bromine is present in a molecule, the isotope peak that is two mass units heavier than the molecular ion (the M+2 peak) becomes very significant. The heavy isotope of each of these elements is two mass units heavier than the lighter isotope. The natural abundance of ^{37}Cl is 32.5% that of ^{35}Cl; the natural abundance of ^{81}Br is 98.0% that of ^{79}Br. When these elements are present, the M+2 peak becomes intense, and the pattern is characteristic of the particular halogen present. If a compound contains two chlorine or bromine atoms, a distinct M+4 peak should be observed, as well as an intense M+2 peak. In these cases, you should exercise caution in identifying the molecular ion peak in a mass spectrum, but the pattern of peaks is characteristic of the nature of the halogen substitution in the molecule. Table 28.2 gives the relative intensities of isotope peaks for various combinations of bromine and chlorine atoms. The patterns of molecular ion and isotopic peaks observed with halogen substitution are shown in Figure 28.2. Examples of these patterns can be seen in the mass spectra of chloroethane (Figure 28.3) and bromoethane (Figure 28.4).

Table 28.2 *Relative intensities of isotope peaks for various combinations of bromine and chlorine*

Halogen	M	M+2	M+4	M+6
Br	100	97.7	—	—
Br_2	100	195.0	95.4	—
Br_3	100	293.0	286.0	93.4
Cl	100	32.6	—	—
Cl_2	100	65.3	10.6	—
Cl_3	100	97.8	31.9	3.47
BrCl	100	130.0	31.9	—
Br_2Cl	100	228.0	159.0	31.2
$BrCl_2$	100	163.0	74.4	10.4

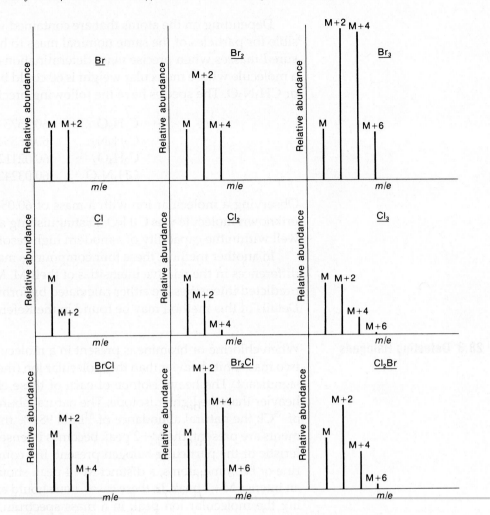

Figure 28.2
Mass spectra expected for various combinations of bromine and chlorine.

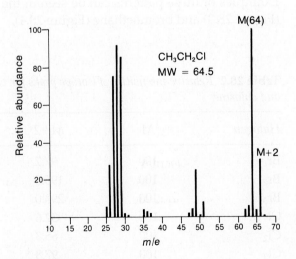

Figure 28.3
The mass spectrum of chloroethane.

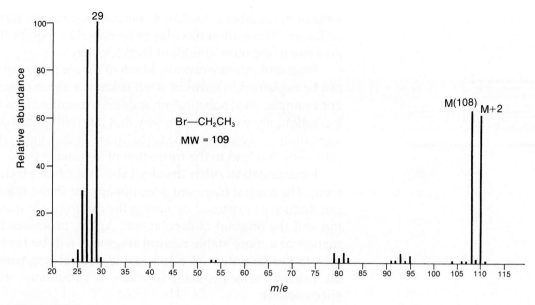

Figure 28.4
The mass spectrum of bromoethane.

28.4 Fragmentation Patterns

When the molecule has been bombarded by high-energy electrons in the ionization chamber of a mass spectrometer, besides losing one electron to form an ion, the molecule also absorbs some of the energy transferred in the collision between the molecule and the incident electrons. This extra energy puts the molecular ion in an excited vibrational state. The vibrationally excited molecular ion is often unstable and may lose some of this extra energy by breaking apart into fragments. If the lifetime of an individual molecular ion is longer than 10^{-5} seconds, a peak corresponding to the molecular ion will be observed in the mass spectrum. Those molecular ions with lifetimes shorter than 10^{-5} seconds will break apart into fragments before they are accelerated within the ionization chamber. In such cases, peaks corresponding to the mass-to-charge ratios for these fragments will also appear in the mass spectrum. For a given compound, not all the molecular ions formed by ionization have precisely the same lifetime. The ions have a range of lifetimes; some individual ions may have shorter lifetimes than others. As a result, peaks are usually observed arising from both the molecular ion and the fragment ions in a typical mass spectrum.

For most classes of compounds, the mode of fragmentation is somewhat characteristic. In many cases, it is possible to predict how a molecule will fragment. Remember that the ionization of the sample molecule forms a molecular ion that not only carries a positive charge but also has an unpaired electron. The molecular ion, then, is actually a **radical–cation,** and it contains an odd number of electrons. In the structural formulas that follow, the radical–cation is indicated by enclosing the structure in square brackets. The positive charge and the unshared electron are shown as superscripts.

$$[\mathbf{R{-}CH_3}]^{\overset{+}{\cdot}}$$

When fragment ions form in the mass spectrometer, they almost always form by means of unimolecular processes. The pressure of the sample in the ionization chamber is too low to permit a significant number of bimolecular collisions. Those unimolecular processes that require the least energy will give rise to the most abundant fragment ions.

Fragment ions are cations. Much of the chemistry of these fragment ions can be explained in terms of what is known about carbocations in solution. For example, alkyl substitution stabilizes fragment ions (and promotes their formation) in much the same way that it stabilizes carbocations. Those fragmentation processes that lead to more stable ions will be favored over processes that lead to the formation of less stable ions.

Fragmentation often involves the loss of an electrically neutral fragment. The neutral fragment does not appear in the mass spectrum, but you can deduce its existence by noting the difference in masses of the fragment ion and the original molecular ion. Again, processes that lead to the formation of a more stable neutral fragment will be favored over those that lead to the formation of a less stable neutral fragment. The loss of a stable neutral molecule, such as water, is commonly observed in the mass spectrometer.

A. Cleavage of One Bond

The most common mode of fragmentation involves the cleavage of one bond. In this process, the odd-electron molecular ion yields an odd-electron neutral fragment and an even-electron fragment ion. The neutral fragment that is lost is a **free radical,** whereas the ionic fragment is of the carbocation type. Cleavages that lead to the formation of more stable carbocations will be favored. Thus, the ease of fragmentation to form ions increases in the following order:

$$CH_3^+ < RCH_2^+ < R_2CH^+ < R_3C^+ < CH_2\!=\!CH\!-\!CH_2^+ < C_6H_5\!-\!CH_2^+$$

Increasing ease of formation →

The following reactions show examples of fragmentation that take place with the cleavage of one bond:

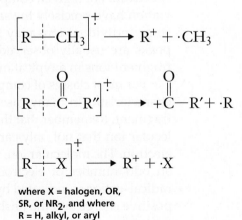

where X = halogen, OR,
SR, or NR$_2$, and where
R = H, alkyl, or aryl

B. Cleavage of Two Bonds

The next most important type of fragmentation involves the cleavage of two bonds. In this type of process, the odd-electron molecular ion yields an odd-electron fragment ion and an even-electron neutral fragment, usually a small, stable molecule. Examples of this type of cleavage are shown next:

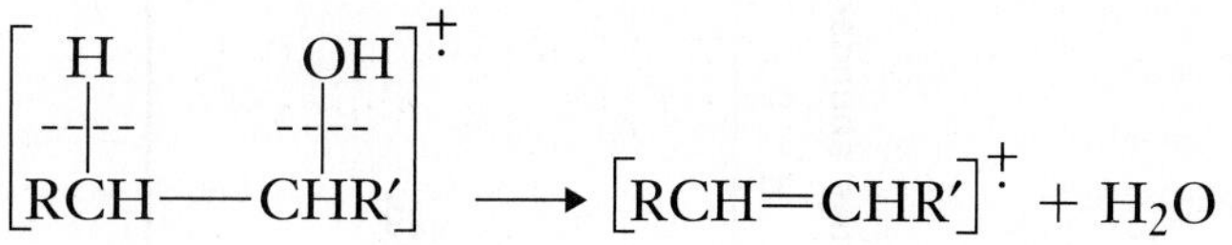

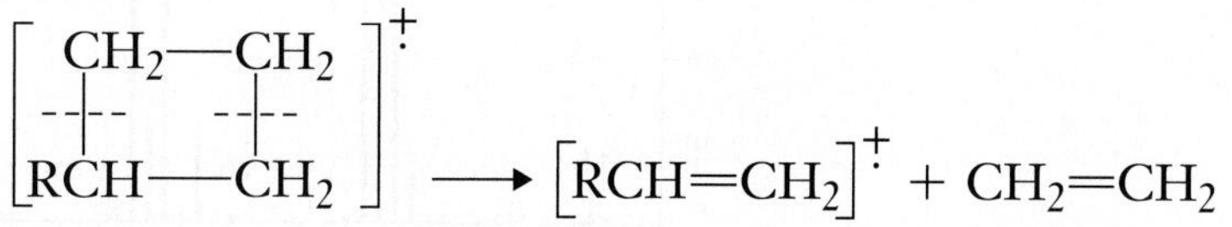

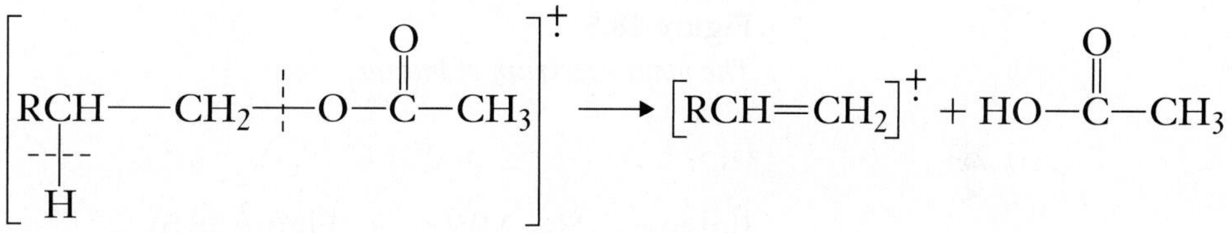

$$\left[\begin{array}{cc} H & OH \\ \hline RCH & CHR' \end{array}\right]^{+\cdot} \longrightarrow \left[RCH{=}CHR'\right]^{+\cdot} + H_2O$$

$$\left[\begin{array}{cc} CH_2 & CH_2 \\ \hline RCH & CH_2 \end{array}\right]^{+\cdot} \longrightarrow \left[RCH{=}CH_2\right]^{+\cdot} + CH_2{=}CH_2$$

$$\left[\begin{array}{c} RCH{-}CH_2 \!\mid\! O{-}\overset{\displaystyle O}{\overset{\|}{C}}{-}CH_3 \\ H \end{array}\right]^{+\cdot} \longrightarrow \left[RCH{=}CH_2\right]^{+\cdot} + HO{-}\overset{\displaystyle O}{\overset{\|}{C}}{-}CH_3$$

C. Other Cleavage Processes

In addition to the processes just mentioned, fragmentation reactions involving rearrangements, migrations of groups, and secondary fragmentations of fragment ions are also possible. These processes occur less often than the types of processes just described. Nevertheless, the pattern of molecular ion and fragment ion peaks observed in the typical mass spectrum is quite complex and unique for each particular molecule. As a result, the mass spectral pattern observed for a given substance can be compared with the mass spectra of known compounds as a means of identification. The mass spectrum is like a fingerprint. For a treatment of the specific modes of fragmentation characteristic of particular classes of compounds, refer to more advanced textbooks (see p. 942). The unique appearance of the mass spectrum for a given compound is the basis for identifying the components of a mixture in the **gas chromatography–mass spectrometry (GC–MS)** technique (see Technique 22, Section 22.14, p. 816). The mass spectrum of every component in a mixture is compared with standard spectra stored in the computer memory of the instrument. The printed output produced by a GC–MS instrument includes an identification based on the results of the computer matching of mass spectra.

28.5 Interpreted Mass Spectra

In this section, the mass spectra of some representative organic compounds are presented. The important fragment ion peaks in each mass spectrum are identified. In some of the examples, identification of the fragments is presented without explanation, although some interpretation is provided where an unusual or interesting process takes place. In the first example, that of butane, a more complete explanation of the symbolism used is offered.

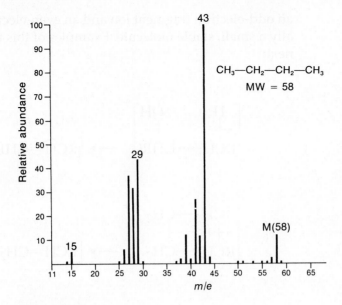

Figure 28.5
The mass spectrum of butane.

Butane; C_4H_{10}, MW = 58 (Figure 28.5)

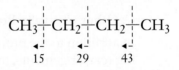

In the structural formula of butane, the dashed lines represent the location of bond-breaking processes that occur during fragmentation. In each case, the fragmentation process involves the breaking of one bond to yield a neutral radical and a cation. The arrows point toward the fragment that bears the positive charge. This positive fragment is the ion that appears in the mass spectrum. The mass of the fragment ion is indicated beneath the arrow.

The mass spectrum shows the molecular ion at $m/e = 58$. Breaking of the C1—C2 bond yields a three-carbon fragment with a mass of 43.

$$CH_3—CH_2—CH_2{\mid}CH_3 \longrightarrow CH_3—CH_2—CH_2^+ + \cdot CH_3$$
$$m/e = 43$$

Cleavage of the central bond yields an ethyl cation, with a mass of 29.

$$CH_3—CH_2{\mid}CH_2—CH_3 \longrightarrow CH_3—CH_2^+ + \cdot CH_2—CH_3$$
$$m/e = 29$$

The terminal bond can also break to yield a methyl cation, which has a mass of 15.

$$CH_3{\mid}CH_2—CH_2—CH_3 \longrightarrow CH_3^+ + \cdot CH_2—CH_2—CH_3$$
$$m/e = 15$$

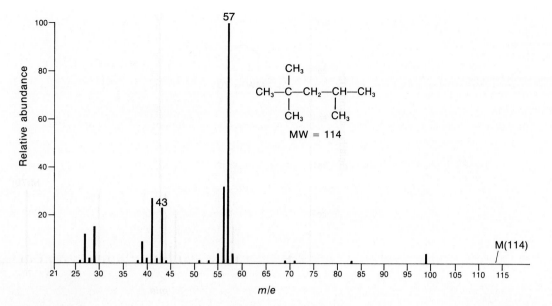

Figure 28.6
The mass spectrum of 2,2,4-trimethylpentane ("isooctane").

Each of these fragments appears in the mass spectrum of butane and has been identified.

2,2,4-Trimethylpentane; C_8H_{18}, MW = 114 (Figure 28.6)

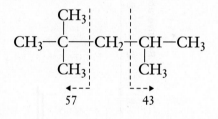

Notice that in the case of 2,2,4-trimethylpentane, by far the most abundant fragment is the *tert*-butyl cation (m/e = 57). This result is not surprising when one considers that the *tert*-butyl cation is a particularly stable carbocation.

Cyclopentane; C_5H_{10}, MW = 70 (Figure 28.7)

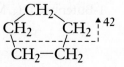

In the case of cyclopentane, the most abundant fragment results from the simultaneous cleavage of two bonds. This mode of fragmentation eliminates a neutral molecule of ethene (MW = 28) and results in the formation of a cation at m/e = 42.

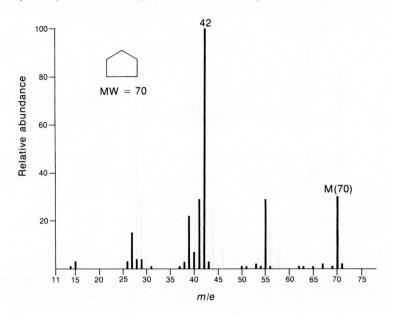

Figure 28.7
The mass spectrum of cyclopentane.

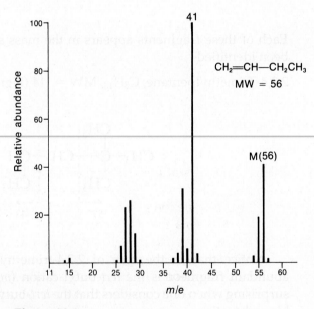

Figure 28.8
The mass spectrum of 1-butene.

1-Butene; C_4H_8, MW = 56 (Figure 28.8)

$$CH_2{=}CH{-}CH_2{\dashv}CH_3$$

41

An important fragment in the mass spectra of alkenes is the allyl cation ($m/e = 41$). This cation is particularly stable due to resonance.

$$[{}^+CH_2{-}CH{=}CH_2 \longleftrightarrow CH_2{=}CH{-}CH_2{}^+]$$

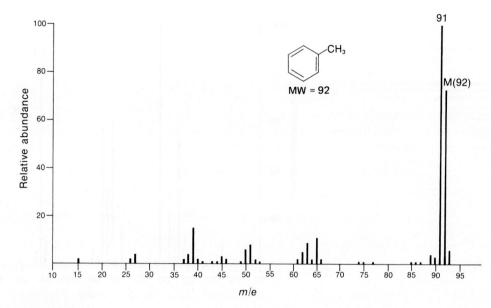

Figure 28.9
The mass spectrum of toluene.

Toluene; C_7H_8, MW = 92 (Figure 28.9)

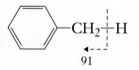

When an alkyl group is attached to a benzene ring, preferential fragmentation occurs at a benzylic position to form a fragment ion of the formula $C_7H_7^+$ (m/e = 91). In the mass spectrum of toluene, loss of hydrogen from the molecule ion gives a strong peak at m/e = 91. Although it may be expected that this fragment ion peak is due to the benzyl carbocation, evidence suggests the benzyl carbocation actually rearranges to form the **tropylium ion.** Isotope-labeling experiments tend to confirm the formation of the tropylium ion. The tropylium ion is a seven-carbon ring system that contains six electrons in π-molecular orbitals and hence is resonance stabilized in a manner similar to that observed in benzene.

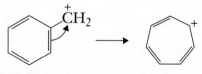

1-Butanol; $C_4H_{10}O$, MW = 74 (Figure 28.10)

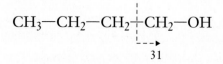

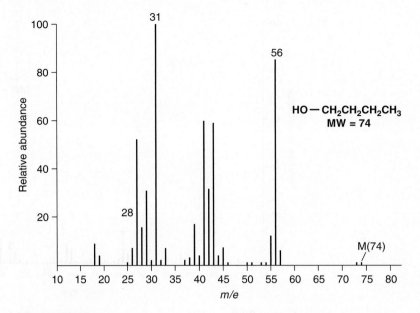

Figure 28.10
The mass spectrum of 1-butanol.

The most important fragmentation reaction for alcohols is loss of an alkyl group:

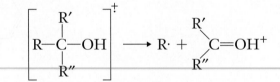

The largest alkyl group is the one that is lost most readily. In the spectrum of 1-butanol, the intense peak at $m/e = 31$ is due to the loss of a propyl group to form

A second common mode of fragmentation involves dehydration. Loss of a molecule of water from 1-butanol leaves a cation of mass 56.

$$CH_3-CH_2-CH-CH_2^{\uparrow 56}$$
$$\underset{H \quad\quad OH}{|\;\;\;\;\;\;\;|}$$

Benzaldehyde; C_7H_6O, MW = 106 (Figure 28.11)

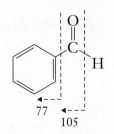

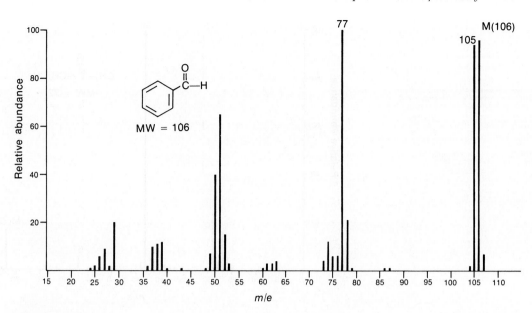

Figure 28.11
The mass spectrum of benzaldehyde.

The loss of a hydrogen atom from an aldehyde is a favorable process. The resulting fragment ion is a benzoyl cation, a particularly stable type of carbocation.

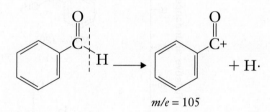

Loss of the entire aldehyde functional group leaves a phenyl cation. This ion can be seen in the spectrum of an *m/e* value of 77.

2-Butanone; C_4H_8O, MW = 72 (Figure 28.12)

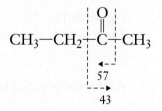

If the methyl group is lost as a neutral fragment, the resulting cation, an **acylium ion,** has an *m/e* value of 57. If the ethyl group is lost, the resulting acylium ion appears at an *m/e* value of 43.

$$CH_3{-}CH_2{-}\overset{\overset{\displaystyle O}{\|}}{C}{-}CH_3 \longrightarrow CH_3{-}CH_2{-}\overset{\overset{\displaystyle O}{\|}}{C^+} + \cdot CH_3$$
$$m/e = 57$$

$$CH_3{-}CH_2{-}\overset{\overset{\displaystyle O}{\|}}{C}{-}CH_3 \longrightarrow CH_3{-}\overset{\overset{\displaystyle O}{\|}}{C^+} + \cdot CH_2CH_3$$
$$m/e = 43$$

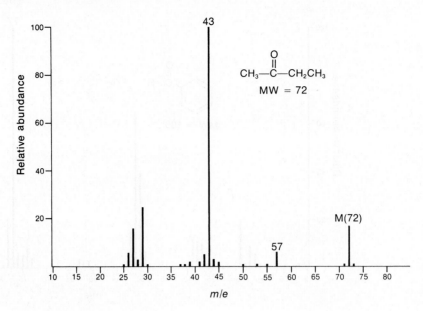

Figure 28.12
The mass spectrum of 2-butanone.

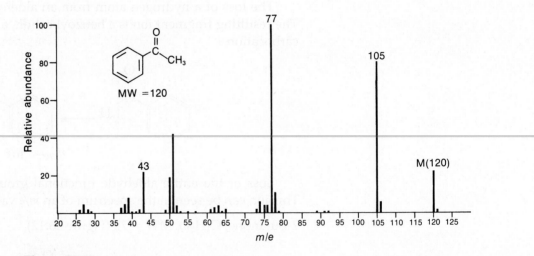

Figure 28.13
The mass spectrum of acetophenone.

Acetophenone; C_8H_8O, MW = 120 (Figure 28.13)

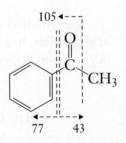

Aromatic ketones undergo α-cleavage to lose the alkyl group and form the benzoyl cation (m/e = 105). This ion subsequently loses carbon monoxide to form the phenyl cation (m/e = 77). Aromatic ketones also undergo

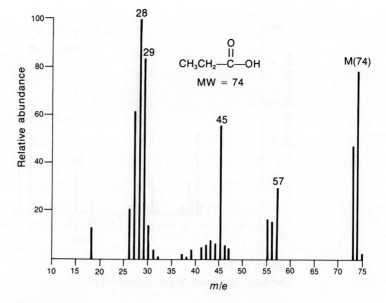

Figure 28.14
The mass spectrum of propanoic acid.

α-cleavage on the other side of the carbonyl group, forming an alkyl acylium ion. In the case of acetophenone, this ion appears at an *m/e* value of 43.

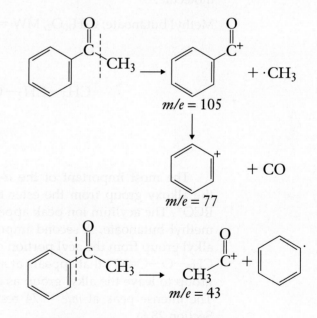

Propanoic acid; $C_3H_6O_2$, MW = 74 (Figure 28.14)

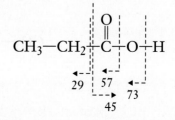

With short-chain carboxylic acids, the loss of OH and COOH through α-cleavage on either side of the C=O group may be observed. In the mass

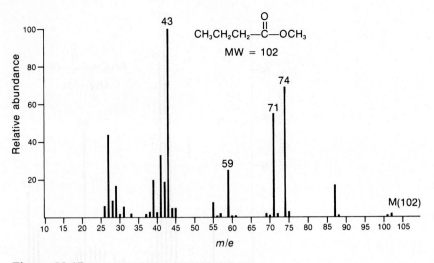

Figure 28.15
The mass spectrum of methyl butanoate.

spectrum of propanoic acid, loss of OH gives rise to a peak at $m/e = 57$. Loss of COOH gives rise to a peak at $m/e = 29$. Loss of the alkyl group as a free radical, leaving the COOH$^+$ ion ($m/e = 45$), also occurs. The intense peak at $m/e = 28$ is due to additional fragmentation of the ethyl portion of the acid molecule.

Methyl butanoate; $C_5H_{10}O_2$, MW = 102 (Figure 28.15)

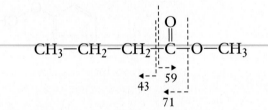

The most important of the α-cleavage reactions involves the loss of the alkoxy group from the ester to form the corresponding acylium ion, RCO$^+$. The acylium ion peak appears at $m/e = 71$ in the mass spectrum of methyl butanoate. A second important peak results from the loss of the alkyl group from the acyl portion of the ester molecule, leaving a fragment CH$_3$—O—C=O$^+$ that appears at $m/e = 59$. Loss of the carboxylate function group to leave the alkyl group as a cation gives rise to a peak at $m/e = 43$. The intense peak at $m/e = 74$ results from a rearrangement process (see Section 28.6).

1-Bromohexane; $C_6H_{13}Br$, MW = 165 (Figure 28.16)

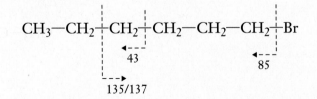

The most interesting characteristic of the mass spectrum of 1-bromo-hexane is the presence of the doublet in the molecular ion. These two peaks, of equal height, separated by two mass units, are strong evidence that

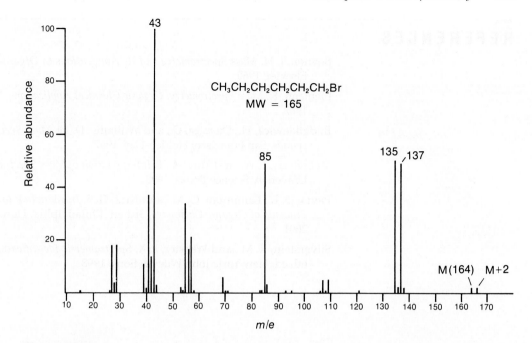

Figure 28.16
The mass spectrum of 1-bromohexane.

bromine is present in the substance. Notice also that loss of the terminal ethyl group yields a fragment ion that still contains bromine (m/e = 135 and 137). The presence of the doublet demonstrates that this fragment contains bromine.

28.6 Rearrangement Reactions

Because the fragment ions that are detected in a mass spectrum are cations, we can expect that these ions will exhibit behavior we are accustomed to associate with carbocations. It is well known that carbocations are prone to rearrangement reactions, converting a less stable carbocation into a more stable one. These types of rearrangements are also observed in the mass spectrum. If the abundance of a cation is especially high, it is assumed that a rearrangement to yield a longer-lived cation must have occurred.

Other types of rearrangements are also known. An example of a rearrangement that is not normally observed in solution chemistry is the rearrangement of a benzyl cation to a tropylium ion. This rearrangement is seen in the mass spectrum of toluene (Figure 28.9).

A particular type of rearrangement process that is unique to mass spectrometry is the **McLafferty rearrangement.** This type of rearrangement occurs when an alkyl chain of at least three carbons in length is attached to an energy-absorbing structure such as a phenyl or carbonyl group that can accept the transfer of a hydrogen ion. The mass spectrum of methyl butanoate (Figure 28.15) contains a prominent peak at m/e = 74. This peak arises from a McLafferty rearrangement of the molecular ion.

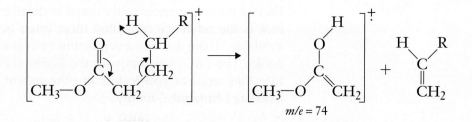

REFERENCES

Beynon, J. H. *Mass Spectrometry and Its Applications to Organic Chemistry.* Amsterdam: Elsevier, 1960.

Biemann, K. *Mass Spectrometry: Organic Chemical Applications.* New York: McGraw-Hill, 1962.

Budzikiewicz, H., Djerassi, C., and Williams, D. H. *Mass Spectrometry of Organic Compounds.* San Francisco: Holden-Day, 1967.

McLafferty, F. W., and Tureček, F. *Interpretation of Mass Spectra,* 4th ed. Mill Valley, CA: University Science Books, 1993.

Pavia, D. L., Lampman, G. M., and Kriz, G. S. *Introduction to Spectroscopy: A Guide for Students of Organic Chemistry,* 3rd ed. Philadelphia: Harcourt College Publishers, 2001.

Silverstein, R. M., and Webster, F. X. *Spectrometric Identification of Organic Compounds,* 6th ed. New York: John Wiley & Sons, 1998.

29   TECHNIQUE 29

Guide to the Chemical Literature

Often, you may need to go beyond the information contained in the typical organic chemistry textbook and use reference material in the library. At first glance, using library materials may seem formidable because of the numerous sources the library contains. If, however, you adopt a systematic approach, the task can prove rather useful. This description of various popular sources and an outline of logical steps to follow in the typical literature search should be helpful.

29.1 Locating Physical Constants: Handbooks

To find information on routine physical constants, such as melting points, boiling points, indices of refraction, and densities, you should first consider a handbook. Examples of suitable handbooks are

Aldrich Handbook of Fine Chemicals. Milwaukee, WI: Sigma-Aldrich, 2005–2006.

Budavari, S., ed. *The Merck Index,* 13th ed. Whitehouse Station, NJ: Merck, 2002.

Dean, J. A., ed. *Lange's Handbook of Chemistry,* 14th ed. New York: McGraw-Hill, 1999.

Lide, D. R., ed. *CRC Handbook of Chemistry and Physics,* 85th ed. Boca Raton, FL: CRC Press, 2004.

Each of these references is discussed in detail in Technique 4. The *CRC Handbook* is the reference consulted most often because the book is so widely available. There are, however, distinct advantages to using the other handbooks. The *CRC Handbook* uses the *Chemical Abstracts* system of nomenclature that requires you to identify the parent name; 3-methyl-1-butanol is listed as 1-butanol, 3-methyl.

The Merck Index has fewer compounds listed, but for the ones listed, there is far more information provided. If the compound is a medicinal or natural product, this is the reference of choice. This handbook contains literature references on the isolation and synthesis of a compound, along with certain properties of medicinal interest, such as toxicity. *Lange's Handbook* and the *Aldrich Handbook* list compounds in alphabetical order; 3-methyl-1-butanol is listed as 3-methyl-1-butanol.

A more complete handbook that is usually housed in the library is

Buckingham, J., ed. *Dictionary of Organic Compounds.* New York: Chapman & Hall/Methuen, 1982–1992.

This is a revised version of an earlier four-volume handbook edited by I. M. Heilbron and H. M. Bunbury. In its present form, it consists of seven volumes with 10 supplements.

29.2 General Synthetic Methods

Many standard introductory textbooks in organic chemistry provide tables that summarize most of the common reactions, including side reactions, for a given class of compounds. These books also describe alternative methods of preparing compounds.

Brown, W. H., and Foote, C. *Organic Chemistry,* 3rd ed. Pacific Grove, CA: Brooks/Cole, 2002.

Carey, F. A. *Organic Chemistry,* 6th ed. New York: McGraw-Hill, 2006.

Ege, S. *Organic Chemistry,* 5th ed. Boston: Houghton-Mifflin, 2004.

Fessenden, R. J., and Fessenden, J. S. *Organic Chemistry,* 6th ed. Pacific Grove, CA: Brooks/Cole, 1998.

Fox, M. A., and Whitesell, J. K. *Organic Chemistry,* 3rd ed. Boston: Jones & Bartlett, 2004.

Hornback, Joe, *Organic Chemistry.* Pacific Grove, CA: Brooks/Cole, 1998.

Jones, M., Jr. *Organic Chemistry,* 3rd ed. New York: W. W. Norton, 2003.

Loudon, G. M. *Organic Chemistry,* 4th ed. Menlo Park, CA: Benjamin/Cummings, 2004.

McMurry, J. *Organic Chemistry,* 6th ed. Pacific Grove, CA: Brooks/Cole, 2004.

Morrison, R. T., and Boyd, R. N. *Organic Chemistry,* 7th ed. Englewood Cliffs, NJ: Prentice-Hall, 1999.

Smith, M. B., and March, J. *Advanced Organic Chemistry,* 5th ed. New York: John Wiley & Sons, 2000.

Solomons, T. W. G., and Fryhle, C. *Organic Chemistry,* 8th ed. New York: John Wiley & Sons, 2003.

Streitwieser, A., Heathcock, C. H., and Kosower, E. M. *Introduction to Organic Chemistry,* 4th ed. New York: Prentice-Hall, 1992.

Vollhardt, K. P. C., and Schore, N. E. *Organic Chemistry,* 4th ed. New York: W. H. Freeman, 2003.

Wade, L. G., Jr. *Organic Chemistry,* 5th ed. Englewood Cliffs, NJ: Prentice-Hall, 2003.

29.3 Searching the Chemical Literature

If the information you are seeking is not available in any of the handbooks mentioned in Section 29.1 or if you are searching for more detailed information than they can provide, then a proper literature search is in order. Although an examination of standard textbooks can provide some help, you often must use all the resources of the library, including journals, reference collections, and abstracts. The following sections outline how the various types of sources should be used and what sort of information can be obtained from them.

The methods discussed for searching the literature use mainly printed materials. Modern search methods also make use of computerized databases and are discussed in Section 29.11. These are vast collections of data and bibliographic materials that can be scanned rapidly from remote computer terminals. Although computerized searching is widely available, it may not be readily accessible to undergraduate students. The following references provide excellent introductions to the literature of organic chemistry:

Carr, C. "Teaching and Using Chemical Information." *Journal of Chemical Education, 70* (September 1993): 719.

Maizell, R. E. *How to Find Chemical Information*, 3rd ed. New York: John Wiley & Sons, 1998.

Smith, M. B., and March, J. *Advanced Organic Chemistry*, 5th ed. New York: John Wiley & Sons, 2001.

Somerville, A. N. "Information Sources for Organic Chemistry, 1: Searching by Name Reaction and Reaction Type." *Journal of Chemical Education, 68* (July 1991): 553.

Somerville, A. N. "Information Sources for Organic Chemistry, 2: Searching by Functional Group." *Journal of Chemical Education, 68* (October 1991): 842.

Somerville, A. N. "Information Sources for Organic Chemistry, 3: Searching by Reagent." *Journal of Chemical Education, 69* (May 1992): 379.

Wiggins, G. *Chemical Information Sources*. New York: McGraw-Hill, 1991. Integrates printed materials and computer sources of information.

29.4 Collections of Spectra

Collections of infrared, nuclear magnetic resonance, and mass spectra can be found in the following catalogues of spectra:

Cornu, A., and Massot, R. *Compilation of Mass Spectral Data*, 2nd ed. London: Heyden and Sons, 1975.

High-Resolution NMR Spectra Catalog. Palo Alto, CA: Varian Associates. Vol. 1, 1962; Vol. 2, 1963.

Johnson, L. F., and Jankowski, W. C. *Carbon-13 NMR Spectra*. New York: John Wiley & Sons, 1972.

Pouchert, C. J. *Aldrich Library of Infrared Spectra*, 3rd ed. Milwaukee: Aldrich Chemical Co., 1981.

Pouchert, C. J. *Aldrich Library of FT-IR Spectra*, 2nd ed. Milwaukee: Aldrich Chemical Co., 1997.

Pouchert, C. J. *Aldrich Library of NMR Spectra*, 2nd ed. Milwaukee: Aldrich Chemical Co., 1983.

Pouchert, C. J., and Behnke, J. *Aldrich Library of ^{13}C and ^1H FT NMR Spectra*. Milwaukee: Aldrich Chemical Co., 1993.

Sadtler Standard Spectra. Philadelphia: Sadtler Research Laboratories. Continuing collection.

Stenhagen, E., Abrahamsson, S., and McLafferty, F. W. *Registry of Mass Spectral Data*, 4 vols. New York: Wiley-Interscience, 1974.

The American Petroleum Institute has also published collections of infrared, nuclear magnetic resonance, and mass spectra.

29.5 Advanced Textbooks

Much information about synthetic methods, reaction mechanisms, and reactions of organic compounds is available in any of the many current advanced textbooks in organic chemistry. Examples of such books are

Carey, F. A., and Sundberg, R. J. *Advanced Organic Chemistry. Part A. Structure and Mechanisms; Part B. Reactions and Synthesis,* 4th ed. New York: Kluwer Academic, 2000.

Carruthers, W. *Some Modern Methods of Organic Synthesis,* 4th ed. Cambridge, UK: Cambridge University Press, 2004.

Corey, E. J., and Cheng, X.-M. *The Logic of Chemical Synthesis.* New York: John Wiley & Sons, 1995.

Fieser, L. F., and Fieser, M. *Advanced Organic Chemistry.* New York: Reinhold, 1961.

Finar, I. L. *Organic Chemistry,* 6th ed. London: Longman Group, 1986.

House, H. O. *Modern Synthetic Reactions,* 2nd ed. Menlo Park, CA: W. H. Benjamin, 1972.

Noller, C. R. *Chemistry of Organic Compounds,* 3rd ed. Philadelphia: W. B. Saunders, 1965.

Smith, M. B. *Organic Synthesis,* 2nd ed. New York: McGraw-Hill, 2002.

Smith, M. B., and March, J. *Advanced Organic Chemistry,* 5th ed. New York: John Wiley & Sons, 2000.

Stowell, J. C. *Intermediate Organic Chemistry,* 2nd ed. New York: John Wiley & Sons, 1993.

Warren, S. *Organic Synthesis: The Disconnection Approach.* New York: John Wiley & Sons, 1982.

These books often contain references to original papers in the literature for students wanting to follow the subject further. Consequently, you obtain not only a review of the subject from such a textbook but also a key reference that is helpful toward a more extensive literature search. The textbook by Smith and March is particularly useful for this purpose.

29.6 Specific Synthetic Methods

Anyone interested in locating information about a particular method of synthesizing a compound should first consult one of the many general textbooks on the subject. Useful ones are

Anand, N., Bindra, J. S., and Ranganathan, S. *Art in Organic Synthesis,* 2nd ed. New York: John Wiley & Sons, 1988.

Barton, D., and Ollis, W. D., eds. *Comprehensive Organic Chemistry,* 6 vols. Oxford: Pergamon Press, 1979.

Buehler, C. A., and Pearson, D. E. *Survey of Organic Synthesis.* New York: Wiley-Interscience, 1970, 2 vols., 1977.

Carey, F. A., and Sundberg, R. J. *Advanced Organic Chemistry. Part B. Reactions and Synthesis,* 4th ed. New York: Kluwer, 2000.

Compendium of Organic Synthetic Methods. New York: Wiley-Interscience, 1971–2002. This is a continuing series, now in 10 volumes.

Fieser, L. F., and Fieser, M. *Reagents for Organic Synthesis.* New York: Wiley-Interscience, 1967–1999. This is a continuing series, now in 21 volumes.

Greene, T. W., and Wuts, P. G. M. *Protective Groups in Organic Synthesis,* 3rd ed. New York: John Wiley & Sons, 1999.

House, H. O. *Modern Synthetic Reactions,* 2nd ed. Menlo Park, CA: W. H. Benjamin, 1972.

Larock, R. C. *Comprehensive Organic Transformations,* 2nd ed. New York: Wiley-VCH, 1999.

Mundy, B. P., and Ellerd, M. G. *Name Reactions and Reagents in Organic Synthesis.* New York: John Wiley & Sons, 1988.

Patai, S., ed. *The Chemistry of the Functional Groups.* London: Interscience, 1964–present. This series consists of many volumes, each one specializing in a particular functional group.

Smith, M. B., and March, J. *Advanced Organic Chemistry,* 5th ed. New York: John Wiley & Sons, 2000.

Trost, B. M., and Fleming, I. *Comprehensive Organic Synthesis.* Amsterdam: Pergamon/Elsevier Science, 1992. This series consist of 9 volumes plus supplements.

Vogel, A. I. *Vogel's Textbook of Practical Organic Chemistry, Including Qualitative Organic Analysis,* 5th ed. London: Longman Group, 1989. Revised by members of the School of Chemistry, Thames Polytechnic.

Wagner, R. B., and Zook, H. D. *Synthetic Organic Chemistry.* New York: John Wiley & Sons, 1956.

More specific information, including actual reaction conditions, exists in collections specializing in organic synthetic methods. The most important of these are

Organic Syntheses. New York: John Wiley & Sons, 1921–present. Published annually.

Organic Syntheses, Collective Volumes. New York: John Wiley & Sons, 1941–2004.

Vol. 1, 1941, Annual Volumes 1–9
Vol. 2, 1943, Annual Volumes 10–19
Vol. 3, 1955, Annual Volumes 20–29
Vol. 4, 1963, Annual Volumes 30–39
Vol. 5, 1973, Annual Volumes 40–49
Vol. 6, 1988, Annual Volumes 50–59
Vol. 7, 1990, Annual Volumes 60–64
Vol. 8, 1993, Annual Volumes 65–69
Vol. 9, 1998, Annual Volumes 70–74
Vol. 10, 2004, Annual Volumes 75–79

It is much more convenient to use the collective volumes where the earlier annual volumes of *Organic Syntheses* are combined in groups of nine or 10 in the first six collective volumes (Volumes 1–6), and then in groups of five for the next three volumes (Volumes 7, 8, 9, and 10). Useful indices are included at the end of each of the collective volumes that classify methods according to the type of reaction, type of compound prepared, formula of compound prepared, preparation or purification of solvents and reagents, and use of various types of specialized apparatus.

The main advantage of using one of the *Organic Syntheses* procedures is that they have been tested to make sure that they work as written. Often, an organic chemist will adapt one of these tested procedures to the preparation of another compound. One of the features of the advanced organic textbook by Smith and March is that it includes references to specific preparative methods contained in *Organic Syntheses.*

More advanced material on organic chemical reactions and synthetic methods may be found in any one of a number of annual publications that review the original literature and summarize it. Examples include

Advances in Organic Chemistry: Methods and Results. New York: John Wiley & Sons, 1960–present.

Annual Reports in Organic Synthesis. Orlando, FL: Academic Press, 1985–1995.

Annual Reports of the Chemical Society, Section B. London: Chemical Society, 1905–present. Specifically, the section "Synthetic Methods."

Organic Reactions. New York: John Wiley & Sons, 1942–present.

Progress in Organic Chemistry. New York: John Wiley & Sons, 1952–1973.

Each of these publications contains a great many citations to the appropriate articles in the original literature.

29.7 Advanced Laboratory Techniques

The student who is interested in reading about techniques more advanced than those described in this textbook, or in more complete descriptions of techniques, should consult one of the advanced textbooks specializing in organic laboratory techniques. Besides focusing on apparatus construction and the performance of complex reactions, these books provide advice on purifying reagents and solvents. Useful sources of information on organic laboratory techniques include

Bates, R. B., and Schaefer, J. P. *Research Techniques in Organic Chemistry.* Englewood Cliffs, NJ: Prentice-Hall, 1971.

Krubsack, A. J. *Experimental Organic Chemistry.* Boston: Allyn & Bacon, 1973.

Leonard, J., Lygo, B., and Procter, G. *Advanced Practical Organic Chemistry,* 2nd ed. London: Chapman & Hall, 1995.

Monson, R. S. *Advanced Organic Synthesis: Methods and Techniques.* New York: Academic Press, 1971.

Techniques of Chemistry. New York: John Wiley & Sons, 1970–present. Currently 23 volumes. The successor to *Technique of Organic Chemistry,* this series covers experimental methods of chemistry, such as purification of solvents, spectral methods, and kinetic methods.

Weissberger, A., et al., eds. *Technique of Organic Chemistry,* 3rd ed., 14 vols. New York: Wiley-Interscience, 1959–1969.

Wiberg, K. B. *Laboratory Technique in Organic Chemistry.* New York: McGraw-Hill, 1960.

Numerous works and some general textbooks specialize in particular techniques. The preceding list is representative only of the most common books in this category. The following books deal specifically with microscale and semimicroscale techniques.

Cheronis, N. D. "Micro and Semimicro Methods." In A. Weissberger, ed., *Technique of Organic Chemistry,* Vol. 6. New York: Wiley-Interscience, 1954.

Cheronis, N. D., and Ma, T. S. *Organic Functional Group Analysis by Micro and Semimicro Methods.* New York: Wiley-Interscience, 1964.

Ma, T. S., and Horak, V. *Microscale Manipulations in Chemistry.* New York: Wiley-Interscience, 1976.

29.8 Reaction Mechanisms

As with the case of locating information on synthetic methods, you can obtain a great deal of information about reaction mechanisms by consulting one of the common textbooks on physical organic chemistry. The textbooks listed here provide a general description of mechanisms, but they do not contain specific literature citations. Very general textbooks include

Bruckner, R. *Advanced Organic Chemistry: Reaction Mechanisms.* New York: Academic Press, 2001.

Miller, A., and Solomon, P. *Writing Reaction Mechanisms in Organic Chemistry,* 2nd ed. San Diego, CA: Academic Press, 1999.

Sykes, P. *A Primer to Mechanisms in Organic Chemistry.* Menlo Park, CA: Benjamin/Cummings, 1995.

More advanced textbooks include

Carey, F. A., and Sundberg, R. J. *Advanced Organic Chemistry. Part A. Structure and Mechanisms,* 4th ed. New York: Kluwer, 2000.

Hammett, L. P. *Physical Organic Chemistry: Reaction Rates, Equilibria, and Mechanisms,* 2nd ed. New York: McGraw-Hill, 1970.

Hine, J. *Physical Organic Chemistry,* 2nd ed. New York: McGraw-Hill, 1962.

Ingold, C. K. *Structure and Mechanism in Organic Chemistry,* 2nd ed. Ithaca, NY: Cornell University Press, 1969.

Isaacs, N. S. *Physical Organic Chemistry,* 2nd ed. New York: John Wiley & Sons, 1995.

Jones, R. A. Y. *Physical and Mechanistic Organic Chemistry,* 2nd ed. Cambridge: Cambridge University Press, 1984.

Lowry, T. H., and Richardson, K. S. *Mechanism and Theory in Organic Chemistry,* 3rd ed. New York: Harper & Row, 1987.

Moore, J. W., and Pearson, R. G. *Kinetics and Mechanism,* 3rd ed. New York: John Wiley & Sons, 1981.

Smith, M. B., and March, J. *Advanced Organic Chemistry,* 5th ed. New York: John Wiley & Sons, 2001.

These books include extensive bibliographies that permit the reader to delve more deeply into the subject.

Most libraries also subscribe to annual series of publications that specialize in articles dealing with reaction mechanisms. Among these are

Advances in Physical Organic Chemistry. London: Academic Press, 1963–present.

Annual Reports of the Chemical Society. Section B. London: Chemical Society, 1905–present. Specifically, the section "Reaction Mechanisms."

Organic Reaction Mechanisms. Chichester: John Wiley & Sons, 1965–present.

Progress in Physical Organic Chemistry. New York: Interscience, 1963–present.

These publications provide the reader with citations from the original literature that can be very useful in an extensive literature search.

29.9 Organic Qualitative Analysis

Many laboratory manuals provide basic procedures for identifying organic compounds through a series of chemical tests and reactions. Occasionally, you might require a more complete description of analytical methods or a more complete set of tables of derivatives. Textbooks specializing in organic qualitative analysis should fill this need. Examples of sources for such information include

Cheronis, N. D., and Entriken, J. B. *Identification of Organic Compounds: A Student's Text Using Semimicro Techniques.* New York: Interscience, 1963.

Pasto, D. J., and Johnson, C. R. *Laboratory Text for Organic Chemistry: A Source Book of Chemical and Physical Techniques.* Englewood Cliffs, NJ: Prentice-Hall, 1979.

Rappoport, Z. ed. *Handbook of Tables for Organic Compound Identification,* 3rd ed. Boca Raton, FL: CRC Press, 1967.

Shriner, R. L., Hermann, C. K. F., Merrill, T. C., Curtin, D. Y., and Fuson, R. C. *The Systematic Identification of Organic Compounds,* 7th ed. New York: John Wiley & Sons, 1998.

Vogel, A. I. *Elementary Practical Organic Chemistry. Part 2. Qualitative Organic Analysis,* 2nd ed. New York: John Wiley & Sons, 1966.

Vogel, A. I. *Vogel's Textbook of Practical Organic Chemistry, Including Qualitative Organic Analysis,* 5th ed. London: Longman Group, 1989. Revised by members of the School of Chemistry, Thames Polytechnic.

29.10 *Beilstein* and *Chemical Abstracts*

One of the most useful sources of information about the physical properties, synthesis, and reactions of organic compounds is *Beilsteins Handbuch der Organischen Chemie.* This is a monumental work, initially edited by Friedrich Konrad Beilstein and updated through several revisions by the Beilstein Institute in Frankfurt am Main, Germany. The original edition (the *Hauptwerk,* abbreviated H) was published in 1918 and covers completely the literature to 1909. Five supplementary series (*Ergänzungswerken*) have been published since that time. The first supplement (*Erstes Ergänzungswerk,* abbreviated E I) covers the literature from 1910 to 1919; the second supplement (*Zweites Ergänzungswerk,* E II) covers 1920–1929; the third supplement (*Drittes Ergänzungswerk,* E III) covers 1930–1949; the fourth supplement (*Viertes Ergänzungswerk,* E IV) covers 1950–1959; and the fifth supplement (in English) covers 1960–1979. Volumes 17–27 of supplementary series III and IV, covering heterocyclic compounds, are combined in a joint issue, E III/IV. Supplementary series III, IV, and V are not complete, so the coverage of *Handbuch der Organischen Chemie* can be considered complete to 1929, with partial coverage to 1979.

Beilsteins Handbuch der Organischen Chemie, usually referred to simply as *Beilstein,* also contains two types of cumulative indices. The first of these is a name index (*Sachregister*), and the second is a formula index (*Formelregister*). These indices are particularly useful for a person wishing to locate a compound in *Beilstein.*

The principal difficulty in using *Beilstein* is that it is written in German through the fourth supplement. The fifth supplement is in English. Although some reading knowledge of German is useful, you can obtain information from the work by learning a few key phrases. For example, *Bildung* is "formation" or "structure." *Darst* or *Darstellung* is "preparation," K_P or *Siedepunkt* is "boiling point," and *F* or *Schmelzpunkt* is "melting point." Furthermore, the names of some compounds in German are not cognates of the English names. Some examples are *Apfelsäure* for "malic acid" (*säure* means "acid"), *Harnstoff* for "urea," *Jod* for "iodine," and *Zimtsäure* for "cinnamic acid." If you have access to a German–English dictionary for chemists, many of these difficulties can be overcome. The best such dictionary is

Patterson, A. M. *German–English Dictionary for Chemists,* 4th ed. New York: John Wiley & Sons, 1991.

Beilstein is organized according to a very sophisticated and complicated system. However, most students do not wish to become experts on

Beilstein to this extent. A simpler, though slightly less reliable, method is to look for the compound in the formula index that accompanies the second supplement. By looking under the molecular formula, you will find the names of compounds that have that formula. After that name will be a series of numbers that indicate the pages and volume in which that compound is listed. Suppose, as an example, that you are searching for information on *p*-nitroaniline. This compound has the molecular formula $C_6H_6N_2O_2$. Searching for this formula in the formula index to the second supplement, you find

4-Nitro-anilin **12** 711, **I** 349, **II** 383

This information tells you that *p*-nitroaniline is listed in the main edition, *Hauptwerk,* in Volume 12, page 711. Locate this particular volume, which is devoted to isocyclic monoamines and turn to page 711 to find the beginning of the section on *p*-nitroaniline. At the left side of the top of this page is "Syst. No. 1671." This is the system number given to compounds in this part of Volume 12. The system number is useful, as it can help you find entries for this compound in subsequent supplements. The organization of *Beilstein* is such that all entries on *p*-nitroaniline in each of the supplements will be found in Volume 12. The entry in the formula index also indicates that material on this compound may be found in the first supplement on page 349 and in the second supplement on page 383. On page 349 of Volume 12 of the first supplement, there is a heading, "XII, 710–712," and on the left is "Syst. No. 1671." Material on *p*-nitroaniline is found in each supplement on a page that is headed with the volume and page of the *Hauptwerk* in which the same compound is found. On page 383 of Volume 12 of the second supplement, the heading in the center of the top of the page is "H12, 710–712." On the left, you find "Syst. No. 1671." Again, because *p*-nitroaniline appeared in Volume 12, page 711, of the main edition, you can locate it by searching through Volume 12 of any supplement until you find a page with the heading corresponding to Volume 12, page 711.

Because the third and fourth supplements are not complete, there is no comprehensive formula index for these supplements. However, you can still find material on *p*-nitroaniline by using the system number and the volume and page in the main work. In the third supplement, because the amount of information available has grown so much since the early days of Beilstein's work, Volume 12 has now expanded so that it is found in several bound parts. However, you select the part that includes system number 1671. In this part of Volume 12, you look through the pages until you find a page headed "Syst. No. 1671/H711." The information on *p*-nitroaniline is found on this page (page 1580). If Volume 12 of the fourth supplement were available, you would go on in the same way to locate more recent data on *p*-nitroaniline. This example is meant to illustrate how you can locate information on particular compounds without having to learn the *Beilstein* system of classification. You might do well to test your ability at finding compounds in *Beilstein* as we have described here.

Guidebooks to using *Beilstein,* which include a description of the *Beilstein* system, are recommended for anyone who wants to work extensively with *Beilstein.* Among such sources are

Heller, S. R. *The Beilstein System: Strategies for Effective Searching.* New York: Oxford University Press, 1997.

How to Use Beilstein. Beilstein Institute, Frankfurt am Main. Berlin: Springer-Verlag.

Huntress, E. H. *A Brief Introduction to the Use of* Beilsteins Handbuch der Organischen Chemie, 2nd ed. New York: John Wiley & Sons, 1938.

Weissbach, O. *The Beilstein Guide: A Manual for the Use of* Beilsteins Handbuch der Organischen Chemie. New York: Springer-Verlag, 1976.

Beilstein reference numbers are listed in such handbooks as *CRC Handbook of Chemistry and Physics* and *Lange's Handbook of Chemistry.* Additionally, *Beilstein* numbers are included in the *Aldrich Handbook of Fine Chemicals,* issued by the Aldrich Chemical Company. If the compound you are seeking is listed in one of these handbooks, you will find that using *Beilstein* is simplified.

Another very useful publication for finding references for research on a particular topic is *Chemical Abstracts,* published by the Chemical Abstracts Service of the American Chemical Society. *Chemical Abstracts* contains abstracts of articles appearing in more than 10,000 journals from virtually every country conducting scientific research. These abstracts list the authors, the journal in which the article appeared, the title of the paper, and a short summary of the contents of the article. Abstracts of articles that appeared originally in a foreign language are provided in English, with a notation indicating the original language.

To use *Chemical Abstracts,* you must know how to use the various indices that accompany it. At the end of each volume, there appears a set of indices, including a formula index, a general subject index, a chemical substances index, an author index, and a patent index. The listings in each index refer the reader to the appropriate abstract according to the number assigned to it. There are also collective indices that combine all the indexed material appearing in a 5-year period (10-year period before 1956). In the collective indices, the listings include the volume number as well as the abstract number.

For material after 1929, *Chemical Abstracts* provides the most complete coverage of the literature. For material before 1929, use *Beilstein* before consulting *Chemical Abstracts. Chemical Abstracts* has the advantage that it is written entirely in English. Nevertheless, most students perform a literature search to find a relatively simple compound. Finding the desired entry for a simple compound is much easier in *Beilstein* than in *Chemical Abstracts.* For simple compounds, the indices in *Chemical Abstracts* are likely to contain very many entries. To locate the desired information, you must comb through this multitude of listings—potentially a very time-consuming task.

The opening pages of each index in *Chemical Abstracts* contain a brief set of instructions on using that index. If you want a more complete guide to *Chemical Abstracts,* consult a textbook designed to familiarize you with these abstracts and indices. Two such books are

CAS Printed Access Tools: A Workbook. Washington, DC: Chemical Abstracts Service, American Chemical Society, 1977.

How to Search Printed CA. Washington, DC: Chemical Abstracts Service, American Chemical Society, 1989.

Chemical Abstracts Service maintains a computerized database that permits users to search through *Chemical Abstracts* rapidly and thoroughly. This service, which is called *CA Online,* is described in Section 29.11. *Beilstein* is also available for online searching by computer.

29.11 Computer Online Searching

You can search a number of chemistry databases online by using a computer and modem or a direct Internet connection. Many academic and industrial libraries can access these databases through their computers. One organization that maintains a large number of databases is the Scientific and Technical Information Network (STN International). The fee charged to the library for this service depends on the total time used in making the search, the type of information being asked for, the time of day when the search is being conducted, and the type of database being searched.

The Chemical Abstracts Service database (*CA Online*) is one of many databases available on STN. It is particularly useful to chemists. Unfortunately, this database extends back only to about 1967, although some earlier references are available. Searches for references earlier than 1967 must be made with printed abstracts (Section 29.10). Searching online is much faster than searching in the printed abstracts. In addition, you can tailor the search in a number of ways by using keywords and the Chemical Abstracts Service Registry Number (CAS Number) as part of the search routine. The CAS Number is a specific number assigned to every compound listed in the *Chemical Abstracts* database. The CAS Number is used as a key in an online search to locate information about the compound. For the more common organic compounds, you can easily obtain CAS Numbers from the catalogs of most of the companies that supply chemicals. Another advantage of performing an online search is that the *Chemical Abstracts* files are updated much more quickly than the printed versions of abstracts. This means that your search is more likely to reveal the most current information available.

Other useful databases available from STN include *Beilstein* and *CASREACTS*. As described in Section 29.10, *Beilstein* is very useful to organic chemists. Currently, there are over 3.5 million compounds listed in the database. You can use the CAS Numbers to help in a search that has the potential of going back to 1830. *CASREACTS* is a chemical reactions database derived from over 100 journals covered by *Chemical Abstracts,* starting in 1985. With this database, you can specify a starting material and a product using the CAS Numbers. Further information on *CA Online, Beilstein, CASREACTS,* and other databases can be obtained from the following references:

Heller, S. R., ed. *The Beilstein Online Database: Implementation, Content and Retrieval.* Washington, DC: American Chemical Society, 1990.

Smith, M. B., and March, J. *Advanced Organic Chemistry,* 5th ed. New York: John Wiley & Sons, 2001.

Somerville, A. N. "Information Sources for Organic Chemistry, 2: Searching by Functional Group." *Journal of Chemical Education, 68* (October 1991): 842.

Somerville, A. N. "Subject Searching of Chemical Abstracts Online." *Journal of Chemical Education, 70* (March 1993): 200.

Wiggins, G. *Chemical Information Sources.* New York: McGraw-Hill, 1990. Integrates printed materials and computer sources of information.

SciFinder and SciFinder Scholar

The newest tools for online searching are SciFinder and SciFinder Scholar, the latter being the academic version of the software. This online service requires a yearly subscription and is available for use at many colleges and universities. SciFinder allows you to search several multidisciplinary CAS databases that contain information from as far back as 1907 to the present. The databases may be searched in a variety of ways: by name, chemical substance, reaction, research topic, CAS number, or author. The program has drawing tools similar to ChemDraw, and what makes the program extremely useful is the ability to draw a structure on the screen and search for it. This avoids the need to name the structure first. In addition, substructure searching is allowed, which means that you may enter a partial structure and the program will find all references having compounds with the features you have indicated. Once you retrieve literature references, hyperlinks allow you view abstracts of the papers or retrieve physical property information. SciFinder is easy to use and requires minimal training. A recent book explains the program thoroughly:

Ridley, D. D. *Information Retrieval: SciFinder and SciFinder Scholar*. New York: Wiley, 2002.

For those who are at a university that subscribes to the service, there is an online tutorial at www.cas.org/SCIFINDER/SCHOLAR.

29.12 Scientific Journals

Ultimately, someone wanting information about a particular area of research will be required to read articles from the scientific journals. These journals are of two basic types: review journals and primary scientific journals. Journals that specialize in review articles summarize all the work that bears on the particular topic. These articles may focus on the contributions of one particular researcher but often consider the contributions of many researchers to the subject. These articles also contain extensive bibliographies, which refer you to the original research articles. Among the important journals devoted, at least partly, to review articles are

Accounts of Chemical Research
Angewandte Chemie (International Edition, in English)
Chemical Reviews
Chemical Society Reviews (formerly known as *Quarterly Reviews*)
Nature
Science

The details of the research of interest appear in the primary scientific journals. Although there are thousands of journals published in the world, a few important journals specializing in articles dealing with organic chemistry include

Canadian Journal of Chemistry
European Journal of Organic Chemistry (formerly known as *Chemische Berichte*)
Journal of Organic Chemistry
Journal of the American Chemical Society
Journal of the Chemical Society, Chemical Communications
Journal of the Chemical Society, Perkin Transactions (Parts I and II)
Journal of Organometallic Chemistry
Organic Letters
Organometallics
Synlett
Synthesis
Tetrahedron
Tetrahedron Letters

29.13 Topics of Current Interest

The following journals and magazines are good sources for topics of educational and current interest. They specialize in news articles and focus on current events in chemistry or in science in general. Articles in these journals (magazines) can be useful in keeping you abreast of developments in science that are not part of your normal specialized scientific reading.

American Scientist
Chemical and Engineering News
Chemistry and Industry
Chemistry in Britain
Chemtech
Discover
Journal of Chemical Education
Nature
Omni
Science
Scientific American

Other sources for topics of current interest include the following:

Encyclopedia of Chemical Technology, 4th ed., 25 vols. plus index and supplements, 1992. Also called *Kirk-Othmer Encyclopedia of Chemical Technology*.
McGraw-Hill Encyclopedia of Science and Technology, 20 volumes and supplements, 1997.

29.14 How to Conduct a Literature Search

The easiest method to follow in searching the literature is to begin with secondary sources and then go to the primary sources. In other words, you would try to locate material in a textbook, *Beilstein*, or *Chemical Abstracts*. From the results of that search, you would then consult one of the primary scientific journals.

A literature search that ultimately requires you to read one or more papers in the scientific journals is best conducted if you can identify a particular paper central to the study. Often, you can obtain this reference from a textbook or a review article on the subject. If this is not available, a search through *Beilstein* is required. A search through one of the handbooks that provides *Beilstein* reference numbers (see Section 29.10) may be helpful. Searching through *Chemical Abstracts* would be considered the next logical step. From these sources, you should be able to identify citations from the original literature on the subject.

Additional citations may be found in the references cited in the journal article. In this way, the background leading to the research can be examined. It is also possible to conduct a search forward in time from the date of the journal article through the *Science Citation Index*. This publication provides the service of listing articles and the papers in which these articles were cited. Although the *Science Citation Index* consists of several types of indices, the *Citation Index* is most useful for the purposes described here. A person who knows of a particular key reference on a subject can examine the *Science Citation Index* to obtain a list of papers that have used that seminal reference in support of the work described. The *Citation Index* lists papers by their senior author, journal, volume, page, and date, followed by citations of papers that have referred to that article, author, journal, volume, page, and date of each. The *Citation Index* is published in annual volumes, with quarterly supplements issued during the current year. Each volume contains a complete list of the citations of the key articles made during that year. A disadvantage is that *Science Citation Index* has been available only since 1961. An additional disadvantage is that you may miss journal articles on the subject of interest if *Citation Index* failed to cite that particular key reference in its bibliographies—a reasonably likely possibility.

You can, of course, conduct a literature search by a "brute force" method, by beginning the search with *Beilstein* or even with the indices in *Chemical Abstracts*. However, the task can be made much easier by performing a computer search (Section 29.11) or by starting with a book or an article of general and broad coverage, which can provide a few citations for starting points in the search.

The following guides to using the chemical literature are provided for the reader who is interested in going farther into this subject.

Bottle, R. T., and Rowland, J. F. B., eds. *Information Sources in Chemistry*, 4th ed. New York: Bowker-Saur, 1992.

Maizell, R. E. *How to Find Chemical Information: A Guide for Practicing Chemists, Educators, and Students*, 3rd ed. New York: John Wiley & Sons, 1998.

Mellon, M. G. *Chemical Publications*, 5th ed. New York: McGraw-Hill, 1982.

Wiggins, G. *Chemical Information Sources*. New York: McGraw-Hill, 1991. Integrates printed materials and computer sources of information.

PROBLEMS

1. Find the following compounds in the formula index for the *Second Supplement of Beilstein* (Section 29.10). (1) List the page numbers from the main work and the supplements (first and second). (2) Using these page numbers, look up the system number (Syst. No.) and the main work number (*Hauptwerk* number, H) for each compound in the main work and the first and second supplements. In some cases, a compound may not be found in all three places. (3) Now use the system number and main work number to find each of these compounds in the third and fourth supplements. List the page numbers where these compounds are found.

   a. 2,5-hexanedione (acetonylacetone)

   b. 3-nitroacetophenone

   c. 4-*tert*-butylcyclohexanone

   d. 4-phenylbutanoic acid (4-phenylbutyric acid, γ-phenylbuttersäure)

2. Using the *Science Citation Index* (Section 29.14), list five research papers by complete title and journal citation for each of the following chemists who have been awarded the Nobel Prize. Use the *Five-Year Cumulative Source Index* for the years 1980–1984 as your source.

   a. H. C. Brown

   b. R. B. Woodward

   c. D. J. Cram

   d. G. Olah

3. The reference book by Smith and March is listed in Section 29.2. Using Appendix 2 in this book, give two methods for preparing the following functional groups. You will need to provide equations.

   a. carboxylic acids

   b. aldehydes

   c. esters (carboxylic esters)

4. *Organic Syntheses* is described in Section 29.6. There are currently nine collective volumes in the series, each with its own index. Find the compounds listed below and provide the equations for preparing each compound.

   a. 2-methylcyclopentane-1,3-dione

   b. *cis*-Δ⁴-tetrahydrophthalic anhydride (listed as tetrahydrophthalic anhydride)

5. Provide four ways that may be used to oxidize an alcohol to an aldehyde. Give complete literature references for each method, as well as equations. Use the *Compendium of Organic Synthetic Methods* or *Survey of Organic Syntheses* by Buehler and Pearson (Section 29.6).

Appendices

1   **APPENDIX 1**

Tables of Unknowns and Derivatives

More extensive tables of unknowns may be found in Z. Rappoport, ed. *Handbook of Tables for Organic Compound Identification*, 3rd ed. Boca Raton, FL: CRC Press, 1967.

ALDEHYDES

Compound	BP	MP	Semi-carbazone*	2,4-Dinitro-phenyl-hydrazone*
Ethanal (acetaldehyde)	21	—	162	168
Propanal (propionaldehyde)	48	—	89	148
Propenal (acrolein)	52	—	171	165
2-Methylpropanal (isobutyraldehyde)	64	—	125	187
Butanal (butyraldehyde)	75	—	95	123
3-Methylbutanal (isovaleraldehyde)	92	—	107	123
Pentanal (valeraldehyde)	102	—	—	106
2-Butenal (crotonaldehyde)	104	—	199	190
2-Ethylbutanal (diethylacetaldehyde)	117	—	99	95
Hexanal (caproaldehyde)	130	—	106	104
Heptanal (heptaldehyde)	153	—	109	108
2-Furaldehyde (furfural)	162	—	202	212
2-Ethylhexanal	163	—	254	114
Octanal (caprylaldehyde)	171	—	101	106
Benzaldehyde	179	—	222	237
Nonanal (nonyl aldehyde)	185	—	100	100
Phenylethanal (phenylacetaldehyde)	195	33	153	121
2-Hydroxybenzaldehyde (salicylaldehyde)	197	—	231	248
4-Methylbenzaldehyde (*p*-tolualdehyde)	204	—	234	234
3,7-Dimethyl-6-octenal (citronellal)	207	—	82	77
Decanal (decyl aldehyde)	207	—	102	104
2-Chlorobenzaldehyde	213	11	229	213
3-Chlorobenzaldehyde	214	18	228	248
3-Methoxybenzaldehyde (*m*-anisaldehyde)	230	—	233 d.	—
3-Bromobenzaldehyde	235	—	205	—
4-Methoxybenzaldehyde (*p*-anisaldehyde)	248	2.5	210	253
trans-Cinnamaldehyde	250 d.	—	215	255
3,4-Methylenedioxybenzaldehyde (piperonal)	263	37	230	266 d.
2-Methoxybenzaldehyde (*o*-anisaldehyde)	245	38	215 d.	254
3,4-Dimethoxybenzaldehyde	—	44	177	261
2-Nitrobenzaldehyde	—	44	256	265

ALDEHYDES *(Cont.)*

Compound	BP	MP	Semi-carbazone*	2,4-Dinitro-phenyl-hydrazone*
4-Chlorobenzaldehyde	—	48	230	254
4-Bromobenzaldehyde	—	57	228	257
3-Nitrobenzaldehyde	—	58	246	293
2,4-Dimethoxybenzaldehyde	—	71	—	—
2,4-Dichlorobenzaldehyde	—	72	—	—
4-Dimethylaminobenzaldehyde	—	74	222	325
4-Hydroxy-3-methoxybenzaldehyde (vanillin)	—	82	230	271
3-Hydroxybenzaldehyde	—	104	198	259
5-Bromo-2-hydroxybenzaldehyde (5-bromosalicylaldehyde)	—	106	297 d.	—
4-Nitrobenzaldehyde	—	106	221	320 d.
4-Hydroxybenzaldehyde	—	116	224	280 d.
(±)-Glyceraldehyde	—	142	160 d.	167

Note: "d" indicates "decomposition."
*See Appendix 2, "Procedures for Preparing Derivatives."

KETONES

Compound	BP	MP	Semi-carbazone*	2,4-Dinitro-phenyl-hydrazone*
2-Propanone (acetone)	56	—	187	126
2-Butanone (methyl ethyl ketone)	80	—	146	117
3-Buten-2-one (methyl vinyl ketone)	81	—	140	—
3-Methyl-2-butanone (isopropyl methyl ketone)	94	—	112	120
2-Pentanone (methyl propyl ketone)	102	—	112	143
3-Pentanone (diethyl ketone)	102	—	138	156
3,3-Dimethyl-2-butanone (pinacolone)	106	—	157	125
4-Methyl-2-pentanone (isobutyl methyl ketone)	117	—	132	95
2,4-Dimethyl-3-pentanone (diisopropyl ketone)	124	—	160	86
3-Hexanone	125	—	113	130
2-Hexanone (methyl butyl ketone)	128	—	121	106
4-Methyl-3-penten-2-one (mesityl oxide)	130	—	164	200
Cyclopentanone	131	—	210	146
5-Hexen-2-one	131	—	102	108
2,3-Pentanedione	134	—	122 (mono) 209 (di)	209
5-Methyl-3-hexanone	136	—	—	—
2,4-Pentanedione (acetylacetone)	139	—	122 (mono) 209 (di)	209
4-Heptanone (dipropyl ketone)	144	—	132	75

KETONES *(Cont.)*

Compound	BP	MP	Semi-carbazone*	2,4-Dinitro-phenyl-hydrazone*
5-Methyl-2-hexanone	145	—	—	—
1-Hydroxy-2-propanone (hydroxyacetone, acetol)	146	—	196	129
3-Heptanone	148	—	101	—
2-Heptanone (methyl amyl ketone)	151	—	123	89
Cyclohexanone	156	—	166	162
2-Methylcyclohexanone	165	—	191	136
3-Octanone	167	—	—	—
2,6-Dimethyl-4-heptanone (diisobutyl ketone)	168	—	122	66
2-Octanone	173	—	122	92
Cycloheptanone	181	—	163	148
Ethyl acetoacetate	181	—	129 d.	93
5-Nonanone	186	—	90	—
3-Nonanone	187	—	112	—
2,5-Hexanedione (acetonylacetone)	191	−9	185 (mono) 224 (di)	257 (di)
2-Nonanone	195	−8	118	—
Acetophenone (methyl phenyl ketone)	202	20	198	238
2-Hydroxyacetophenone	215	28	210	212
1-Phenyl-2-propanone (phenylacetone)	216	27	198	156
Propiophenone (1-phenyl-1-propanone)	218	21	173	191
Isobutyrophenone (2-methyl-1-phenyl-1-propanone)	222	—	181	163
1-Phenyl-2-butanone	226	—	135	—
4-Methylacetophenone	226	28	205	258
3-Chloroacetophenone	228	—	232	—
2-Chloroacetophenone	229	—	160	—
Butyrophenone (1-phenyl-1-butanone)	230	12	187	190
2-Undecanone	231	12	122	63
4-Chloroacetophenone	232	12	204	231
4-Phenyl-2-butanone (benzylacetone)	235	—	142	127
2-Methoxyacetophenone	239	—	183	—
3-Methoxyacetophenone	240	—	196	—
Valerophenone (1-phenyl-1-pentanone)	248	—	160	166
4-Chloropropiophenone	—	36	176	—
4-Phenyl-3-buten-2-one (benzalacetone)	—	37	187	227
4-Methoxyacetophenone	—	38	198	220
3-Bromopropiophenone	—	40	183	—
1-Indanone	—	41	233	258
Benzophenone	—	48	164	238
4-Bromoacetophenone	—	51	208	230
3,4-Dimethoxyacetophenone	—	51	218	207
2-Acetonaphthone (methyl 2-naphthyl ketone)	—	53	234	262 d.
Desoxybenzoin (benzyl phenyl ketone)	—	60	148	204
1,1-Diphenylacetone	—	61	170	—

KETONES *(Cont.)*

Compound	BP	MP	Semi-carbazone*	2,4-Dinitro-phenyl-hydrazone*
4-Chlorobenzophenone	—	76	—	185
3-Nitroacetophenone	—	80	257	228
4-Nitroacetophenone	—	80	—	—
4-Bromobenzophenone	—	82	350	230
Fluorenone	—	83	—	283
4-Hydroxyacetophenone	—	109	199	210
Benzoin	—	136	206	245
4-Hydroxypropiophenone	—	148	—	229
(±)-Camphor	—	179	237	164

Note: "d" indicates "decomposition."

*See Appendix 2, "Procedures for Preparing Derivatives."

CARBOXYLIC ACIDS

Compound	BP	MP	p-Toluidide*	Anilide*	Amide*
Methanoic acid (formic acid)	101	8	53	47	43
Ethanoic acid (acetic acid)	118	17	148	114	82
Propenoic acid (acrylic acid)	139	13	141	104	85
Propanoic acid (propionic acid)	141	—	124	103	81
2-Methylpropanoic acid (isobutyric acid)	154	—	104	105	128
Butanoic acid (butyric acid)	162	—	72	95	115
3-Butenoic acid (vinylacetic acid)	163	—	—	58	73
2-Methylpropenoic acid (methacrylic acid)	163	16	—	87	102
Pyruvic acid	165 d.	14	109	104	124
3-Methylbutanoic acid (isovaleric acid)	176	—	106	109	135
3,3-Dimethylbutanoic acid	185	—	134	132	132
Pentanoic acid (valeric acid)	186	—	74	63	106
2-Chloropropanoic acid	186	—	124	92	80
Dichloroacetic acid	194	6	153	118	98
2-Methylpentanoic acid	195	—	80	95	79
Hexanoic acid (caproic acid)	205	—	75	95	101
2-Bromopropanoic acid	205 d.	24	125	99	123
Heptanoic acid	223	—	81	70	96
2-Ethylhexanoic acid	228	—	—	—	102
Cyclohexanecarboxylic acid	233	31	—	146	186
Octanoic acid (caprylic acid)	237	16	70	57	107
Nonanoic acid	254	12	84	57	99
Decanoic acid (capric acid)	—	32	78	70	108
4-Oxopentanoic acid (levulinic acid)	—	33	108	102	108 d.

CARBOXYLIC ACIDS *(Cont.)*

Compound	BP	MP	p-Toluidide*	Anilide*	Amide*
Trimethylacetic acid (pivalic acid)	—	35	120	130	155
3-Chloropropanoic acid	—	40	—	—	101
Dodecanoic acid (lauric acid)	—	43	87	78	100
3-Phenylpropanoic acid (hydrocinnamic acid)	—	48	135	98	105
Bromoacetic acid	—	50	—	131	91
4-Phenylbutanoic acid	—	52	—	—	84
Tetradecanoic acid (myristic acid)	—	54	93	84	103
Trichloroacetic acid	—	57	113	97	141
3-Bromopropanoic acid	—	61	—	—	111
Hexadecanoic acid (palmitic acid)	—	62	98	90	106
Chloroacetic acid	—	63	162	137	121
Cyanoacetic acid	—	66	—	198	120
Octadecanoic acid (stearic acid)	—	69	102	95	109
trans-2-Butenoic acid (crotonic acid)	—	72	132	118	158
Phenylacetic acid	—	77	136	118	156
α-Methyl-*trans*-cinnamic acid	—	81	—	—	128
4-Methoxyphenylacetic acid	—	87	—	—	189
3,4-Dimethoxyphenyl acetic acid	—	97	—	—	147
Pentanedioic acid (glutaric acid)	—	98	218 (di)	224 (di)	176 (di)
Phenoxyacetic acid	—	99	—	99	102
2-Methoxybenzoic acid (*o*-anisic acid)	—	100	—	131	129
2-Methylbenzoic acid (*o*-toluic acid)	—	104	144	125	142
Nonanedioic acid (azelaic acid)	—	106	201 (di)	107 (mono) 186 (di)	93 (mono) 175 (di)
3-Methoxybenzoic acid (*m*-anisic acid)	—	107	—	—	136
3-Methylbenzoic acid (*m*-toluic acid)	—	111	118	126	94
4-Bromophenylacetic acid	—	117	—	—	194
(±)-Phenylhydroxyacetic acid (mandelic acid)	—	118	172	151	133
Benzoic acid	—	122	158	163	130
2,4-Dimethylbenzoic acid	—	126	—	141	180
2-Benzoylbenzoic acid	—	127	—	195	165
Maleic acid	—	130	142 (di)	198 (mono) 187 (di)	172 (mono) 260 (di)
Decanedioic acid (sebacic acid)	—	133	201 (di)	122 (mono) 200 (di)	170 (mono) 210 (di)
3-Chlorocinnamic acid	—	133	142	135	76
2-Furoic acid	—	133	170	124	143
trans-Cinnamic acid	—	133	168	153	147
2-Acetylsalicylic acid (aspirin)	—	138	—	136	138
5-Chloro-2-nitrobenzoic acid	—	139	—	164	154
2-Chlorobenzoic acid	—	140	131	118	139
3-Nitrobenzoic acid	—	140	162	155	143
4-Chloro-2-nitrobenzoic acid	—	142	—	—	172
2-Nitrobenzoic acid	—	146	—	155	176

CARBOXYLIC ACIDS *(Cont.)*

Compound	BP	MP	p-*Toluidide**	*Anilide**	*Amide**
2-Aminobenzoic acid (anthranilic acid)	—	146	151	131	109
Diphenylacetic acid	—	148	172	180	167
2-Bromobenzoic acid	—	150	—	141	155
Benzilic acid	—	150	190	175	154
Hexanedioic acid (adipic acid)	—	152	239	151 (mono) 241 (di)	125 (mono) 220 (di)
Citric acid	—	153	189 (tri)	198 (tri)	210 (tri)
4-Nitrophenylacetic acid	—	153	—	198	198
2,5-Dichlorobenzoic acid	—	153	—	—	155
3-Chlorobenzoic acid	—	156	—	123	134
2,4-Dichlorobenzoic acid	—	158	—	—	194
4-Chlorophenoxyacetic acid	—	158	—	125	133
2-Hydroxybenzoic acid (salicylic acid)	—	158	156	136	142
5-Bromo-2-hydroxybenzoic acid (5-bromosalicylic acid)	—	165	—	222	232
3,4-Dimethylbenzoic acid	—	165	—	104	130
2-Chloro-5-nitrobenzoic acid	—	166	—	—	178
Methylenesuccinic acid (itaconic acid)	—	166 d.	—	152 (mono)	191 (di)
(+)-Tartaric acid	—	169	—	180 (mono) 264 (di)	171 (mono) 196 (di)
5-Chlorosalicylic acid	—	172	—	—	227
4-Methylbenzoic acid (*p*-toluic acid)	—	180	160	145	160
4-Chloro-3-nitrobenzoic acid	—	182	—	131	156
4-Methoxybenzoic acid (*p*-anisic acid)	—	184	186	169	167
Butanedioic acid (succinic acid)	—	188	180 (mono) 255 (di)	143 (mono) 230 (di)	157 (mono) 260 (di)
4-Ethoxybenzoic acid	—	198	—	170	202
Fumaric acid	—	200 s.	—	233 (mono) 314 (di)	270 (mono) 266 (di)
3-Hydroxybenzoic acid	—	201 s.	163	157	170
3,5-Dinitrobenzoic acid	—	202	—	234	183
3,4-Dichlorobenzoic acid	—	209	—	—	133
Phthalic acid	—	210 d.	150 (mono) 201 (di)	169 (mono) 253 (di)	144 (mono) 220 (di)
4-Hydroxybenzoic acid	—	214	204	197	162
3-Nitrophthalic acid	—	215	226 (di)	234 (di)	201 (di)
Pyridine-3-carboxylic acid (nicotinic acid)	—	236	150	132	128
4-Nitrobenzoic acid	—	240	204	211	201
4-Chlorobenzoic acid	—	242	—	194	179
4-Bromobenzoic acid	—	251	—	197	190

Note: "d" indicates "decomposition"; "s" indicates "sublimation."

*See Appendix 2, "Procedures for Preparing Derivatives."

PHENOLS†

Compound	BP	MP	α-Naphthyl-urethane*	Bromo Derivative* Mono	Di	Tri	Tetra
2-Chlorophenol	176	7	120	48	76	—	—
3-Methylphenol (*m*-cresol)	203	12	128	—	—	84	—
2-Ethylphenol	207	—	—	—	—	—	—
2,4-Dimethylphenol	212	23	135	—	—	—	—
2-Methylphenol (*o*-cresol)	191	32	142	—	56	—	—
2-Methoxyphenol (guaiacol)	204	32	118	—	—	116	—
4-Methylphenol (*p*-cresol)	202	35	146	—	49	—	198
3-Chlorophenol	214	35	158	—	—	—	—
4-Methyl-2-nitrophenol	—	35	—	—	—	—	—
2,4-Dibromophenol	238	40	—	95	—	—	—
Phenol	181	42	133	—	—	95	—
4-Chlorophenol	217	43	166	33	90	—	—
4-Ethylphenol	219	45	128	—	—	—	—
2-Nitrophenol	216	45	113	—	117	—	—
2-Isopropyl-5-methylphenol (thymol)	234	51	160	55	—	—	—
4-Methoxyphenol	243	56	—	—	—	—	—
3,4-Dimethylphenol	225	64	141	—	—	171	—
4-Bromophenol	238	64	169	—	—	—	—
4-Chloro-3-methylphenol	235	66	153	—	—	—	—
3,5-Dimethylphenol	220	68	—	—	—	166	—
2,6-Di-*tert*-butyl-4-methylphenol	—	70	—	—	—	—	—
2,4,6-Trimethylphenol	232	72	—	—	—	—	—
2,5-Dimethylphenol	212	75	173	—	—	178	—
1-Naphthol (α-naphthol)	278	94	152	—	105	—	—
2-Methyl-4-nitrophenol	186	96	—	—	—	—	—
2-Hydroxyphenol (catechol)	245	104	175	—	—	—	192
2-Chloro-4-nitrophenol	—	106	—	—	—	—	—
3-Hydroxyphenol (resorcinol)	—	109	—	—	—	112	—
4-Nitrophenol	—	112	150	—	142	—	—
2-Naphthol (β-naphthol)	—	123	157	84	—	—	—
3-Methyl-4-nitrophenol	—	129	—	—	—	—	—
1,2,3-Trihydroxybenzene (pyrogallol)	—	133	—	—	158	—	—
4-Phenylphenol	—	164	—	—	—	—	—

*See Appendix 2, "Procedures for Preparing Derivatives."

†Also check:

   Salicylic acid (2-hydroxybenzoic acid)

   Esters of salicylic acid (salicylates)

   Salicylaldehyde (2-hydroxybenzaldehyde)

   4-Hydroxybenzaldehyde

   4-Hydroxypropiophenone

   3-Hydroxybenzoic acid

   4-Hydroxybenzoic acid

   4-Hydroxybenzophenone

PRIMARY AMINES†

Compound	BP	MP	Benzamide*	Picrate*	Acetamide*
t-Butylamine	46	—	134	198	101
Propylamine	48	—	84	135	—
Allylamine	56	—	—	140	—
sec-Butylamine	63	—	76	139	—
Isobutylamine	69	—	57	150	—
Butylamine	78	—	42	151	—
Isopentylamine (ioamylamine)	96	—	—	138	—
Pentylamine (amylamine)	104	—	—	139	—
Ethylenediamine	118	—	244 (di)	233 (di)	172 (di)
Hexylamine	132	—	40	126	—
Cyclohexylamine	135	—	149	—	101
1,3-Diaminopropane	140	—	148 (di)	250	126 (di)
Furfurylamine	145	—	—	150	—
Heptylamine	156	—	—	121	—
Octylamine	180	—	—	112	—
Benzylamine	184	—	105	194	65
Aniline	184	—	163	180	114
2-Methylaniline (*o*-toluidine)	200	—	144	213	110
3-Methylaniline (*m*-toluidine)	203	—	125	200	65
2-Chloroaniline	208	—	99	134	87
2,6-Dimethylaniline	216	11	168	180	177
2,5-Dimethylaniline	216	14	140	171	139
3,5-Dimethylaniline	220	—	144	225	—
4-Isopropylaniline	225	—	162	—	102
2-Methoxyaniline (*o*-anisidine)	225	6	60	200	85
3-Chloroaniline	230	—	120	177	74
2-Ethoxyaniline (*o*-phenetidine)	231	—	104	—	79
4-Chloro-2-methylaniline	241	29	142	—	140
4-Ethoxyaniline (*p*-phenetidine)	250	2	173	69	137
3-Bromoaniline	251	18	120	180	87
2-Bromoaniline	250	31	116	129	99
2,6-Dichloroaniline	—	39	—	—	—
4-Methylaniline (*p*-toluidine)	200	43	158	182	147
2-Ethylaniline	210	47	147	194	111
2,5-Dichloroaniline	251	50	120	86	132
4-Methoxyaniline (*p*-anisidine)	—	58	154	170	130
2,4-Dichloroaniline	245	62	117	106	145
4-Bromoaniline	245	64	204	180	168
4-Chloroaniline	—	72	192	178	179
2-Nitroaniline	—	72	110	73	92
2,4,6-Trichloroaniline	262	75	174	83	204
Ethyl *p*-aminobenzoate	—	89	148	—	110
o-Phenylenediamine	258	102	301 (di)	208	185 (di)
2-Methyl-5-nitroaniline	—	106	186	—	151
4-Aminoacetophenone	—	106	205	—	167

PRIMARY AMINES† *(Cont.)*

Compound	BP	MP	Benzamide*	Picrate*	Acetamide*
2-Chloro-4-nitroaniline	—	108	161	—	139
3-Nitroaniline	—	114	157	143	155
4-Methyl-2-nitroaniline	—	116	148	—	99
4-Chloro-2-nitroaniline	—	118	133	—	104
2,4,6-Tribromoaniline	—	120	200	—	232
2-Methyl-4-nitroaniline	—	130	—	—	202
2-Methoxy-4-nitroaniline	—	138	149	—	153
p-Phenylenediamine	—	140	128 (mono)	—	162 (mono)
			300 (di)		304 (di)
4-Nitroaniline	—	148	199	100	215
4-Aminoacetanilide	—	162	—	—	304
2,4-Dinitroaniline	—	180	202	—	120

*See Appendix 2, "Procedures for Preparing Derivatives."
†Also check 4-aminobenzoic acid and its esters.

SECONDARY AMINES

Compound	BP	MP	Benzamide*	Picrate*	Acetamide*
Diethylamine	56	—	42	155	—
Diisopropylamine	84	—	—	140	—
Pyrrolidine	88	—	Oil	112	—
Piperidine	106	—	48	152	—
Dipropylamine	110	—	—	75	—
Morpholine	129	—	75	146	—
Diisobutylamine	139	—	—	121	86
N-Methylcyclohexylamine	148	—	85	170	—
Dibutylamine	159	—	—	59	—
Benzylmethylamine	184	—	—	117	—
N-Methylaniline	196	—	63	145	102
N-Ethylaniline	205	—	60	132	54
N-Ethyl-*m*-toluidine	221	—	72	—	—
Dicyclohexylamine	256	—	153	173	103
N-Benzylaniline	298	37	107	48	58
Indole	254	52	68	—	157
Diphenylamine	302	52	180	182	101
N-Phenyl-1-naphthylamine	335	62	152	—	115

*See Appendix 2, "Procedures for Preparing Derivatives."

TERTIARY AMINES[†]

Compound	BP	MP	Picrate*	Methiodide*
Triethylamine	89	—	173	280
Pyridine	115	—	167	117
2-Methylpyridine (α-picoline)	129	—	169	230
2,6-Dimethylpyridine (2,6-lutidine)	143	—	168	233
4-Methylpyridine (4-picoline)	143	—	167	—
3-Methylpyridine (β-picoline)	144	—	150	92
Tripropylamine	157	—	116	207
N,N-Dimethylbenzylamine	183	—	93	179
N,N-Dimethylaniline	193	—	163	228 d.
Tributylamine	216	—	105	186
N,N-Diethylaniline	217	—	142	102
Quinoline	237	—	203	72/133

Note: "d" indicates "decomposition."

*See Appendix 2, "Procedures for Preparing Derivatives."

[†]Also check nicotinic acid and its esters.

ALCOHOLS

Compound	BP	MP	3,5-Di-nitrobenzoate*	Phenyl-urethane*
Methanol	65	—	108	47
Ethanol	78	—	93	52
2-Propanol (isopropyl alcohol)	82	—	123	88
2-Methyl-2-propanol (t-butyl alcohol)	83	26	142	136
3-Buten-2-ol	96	—	54	—
2-Propen-1-ol (allyl alcohol)	97	—	49	70
1-Propanol	97	—	74	57
2-Butanol (sec-butyl alcohol)	99	—	76	65
2-Methyl-2-butanol (t-pentyl alcohol)	102	−8.5	116	42
2-Methyl-3-butyn-2-ol	104	—	112	—
2-Methyl-1-propanol (isobutyl alcohol)	108	—	87	86
3-Buten-1-ol	113	—	59	25
3-Methyl-2-butanol	114	—	76	68
2-Propyn-1-ol (propargyl alcohol)	114	—	—	—
3-Pentanol	115	—	101	48
1-Butanol	118	—	64	61
2-Pentanol	119	—	62	—
3,3-Dimethyl-2-butanol	120	—	107	77
2,3-Dimethyl-2-butanol	121	—	111	65
2-Methyl-2-pentanol	123	—	72	—
3-Methyl-3-pentanol	123	—	96	43
2-Methoxyethanol	124	—	—	(113)[†]
2-Methyl-3-pentanol	128	—	85	50
2-Chloroethanol	129	—	95	51

ALCOHOLS *(Cont.)*

Compound	BP	MP	3,5-Di-nitrobenzoate*	Phenyl-urethane*
3-Methyl-1-butanol (isoamyl alcohol)	132	—	61	56
4-Methyl-2-pentanol	132	—	65	143
2-Ethoxyethanol	135	—	75	(67)[†]
3-Hexanol	136	—	97	—
1-Pentanol	138	—	46	46
2-Hexanol	139	—	39	(61)[†]
2,4-Dimethyl-3-pentanol	140	—	—	95
Cyclopentanol	140	—	115	132
2-Ethyl-1-butanol	146	—	51	—
2,2,2-Trichloroethanol	151	—	142	87
1-Hexanol	157	—	58	42
2-Heptanol	159	—	49	(54)[†]
Cyclohexanol	160	—	113	82
3-Chloro-1-propanol	161	—	77	38
(2-Furyl)-methanol (furfuryl alcohol)	170	—	80	45
1-Heptanol	176	—	47	60
2-Octanol	179	—	32	114
2-Ethyl-1-hexanol	185	—	—	(61)[†]
1-Octanol	195	—	61	74
3,7-Dimethyl-1,6-octadien-3-ol (linalool)	196	—	—	66
2-Nonanol	198	—	43	(56)[†]
Benzyl alcohol	204	—	113	77
1-Phenylethanol	204	20	92	95
1-Nonanol	214	—	52	62
1,3-Propanediol	215	—	178 (di)	137 (di)
2-Phenylethanol	219	—	108	78
1-Decanol	231	7	57	59
3-Phenylpropanol	236	—	45	92
1-Dodecanol (lauryl alcohol)	—	24	60	74
3-Phenyl-2-propen-1-ol (cinnamyl alcohol)	250	34	121	90
α-Terpineol	221	36	78	112
1-Tetradecanol (myristyl alcohol)	—	39	67	74
(−)-Menthol	212	41	158	111
1-Hexadecanol (cetyl alcohol)	—	49	66	73
2,2-Dimethyl-1-propanol (neopentyl alcohol)	113	56	—	144
4-Methylbenzyl alcohol	217	59	117	79
1-Octadecanol (stearyl alcohol)	—	59	77	79
Diphenylmethanol (benzhydrol)	—	68	141	139
4-Nitrobenzyl alcohol	—	93	157	—
Benzoin	—	136	—	165
Cholesterol	—	147	—	168
Triphenylmethanol	—	161	—	—
(+)-Borneol	—	208	154	138

*See Appendix 2, "Procedures for Preparing Derivatives."

[†]α-Naphthylurethane.

ESTERS

Compound	BP	MP	Compound	BP	MP
Methyl formate	32	—	Isopentyl acetate	142	—
Ethyl formate	54	—	(3-methylbutyl acetate)		
Methyl acetate	57	—	2-Methoxyethyl acetate	145	—
Isopropyl formate	71	—	Ethyl chloroacetate	145	—
Vinyl acetate	72	—	Ethyl valerate (ethyl pentanoate)	146	—
Ethyl acetate	77	—	Ethyl α-chloropropanoate	146	—
Methyl propionate (methyl propanoate)	80	—	Pentyl acetate	147	—
Methyl acrylate	80	—	Methyl hexanoate	151	—
Propyl formate	81	—	Ethyl lactate	154	—
Isopropyl acetate	89	—	Butyl butyrate	167	—
Ethyl chloroformate	93	—	Ethyl hexanoate	168	—
Methyl isobutyrate	93	—	Hexyl acetate	169	—
(methyl 2-methylpropanoate)			Methyl acetoacetate	170	—
2-Propenyl acetate (isopropenyl acetate)	94	—	Methyl heptanoate (methyl enanthlate)	172	—
tert-Butyl acetate	98	—	Furfuryl acetate	176	—
(1,1-dimethylethyl acetate)			Methyl 2-furoate	181	—
Ethyl propionate (ethyl propanoate)	99	—	Dimethyl malonate	181	—
Methyl methacrylate	100	—	Ethyl acetoacetate	181	—
(methyl 2-methylpropenoate)			Diethyl oxalate	185	—
Methyl pivalate	101	—	Ethyl heptanoate	187	—
(methyl trimethyl acetate)			Heptyl acetate	192	—
Ethyl acrylate (ethyl propenoate)	101	—	Dimethyl succinate	196	—
Propyl acetate	102	—	Phenyl acetate	197	—
Methyl butyrate (methyl butanoate)	102	—	Diethyl malonate	199	—
Ethyl isobutyrate	110	—	Methyl benzoate	199	—
(ethyl 2-methylpropanoate)			Dimethyl maleate	204	—
Isopropyl propionate	110	—	Ethyl levulinate	206	—
(isopropyl propanoate)			Ethyl octanoate	208	—
2-Butyl acetate (*sec*-butyl acetate)	111	—	Ethyl cyanoacetate	208	—
Methyl isovalerate	117	—	Ethyl benzoate	212	—
(methyl 3-methylbutanoate)			Benzyl acetate	217	—
Isobutyl acetate	117	—	Diethyl succinate	217	—
(2-methylpropyl acetate)			Diethyl fumarate	219	—
Ethyl pivalate	118	—	Methyl phenylacetate	220	—
(ethyl 2,2-dimethylpropanoate)			Methyl salicylate	224	—
Methyl crotonate (methyl 2-butenoate)	119	—	Diethyl maleate	224	—
Ethyl butyrate (ethyl butanoate)	121	—	Ethyl phenylacetate	228	—
Propyl propionate (propyl propanoate)	123	—	Propyl benzoate	231	—
Butyl acetate	126	—	Ethyl salicylate	234	—
Methyl valerate (methyl pentanoate)	128	—	Dimethyl suberate	268	—
Methyl methoxyacetate	130	—	Ethyl cinnamate	271	—
Methyl chloroacetate	130	—	Dimethyl phthalate	284	—
Ethyl isovalerate	134	—	Diethyl phthalate	298	—
(ethyl 3-methylbutanoate)			Methyl cinnamate	—	36
Ethyl crotonate (ethyl 2-butenoate)	138	—			

ESTERS *(Cont.)*

Compound	BP	MP	Compound	BP	MP
Ethyl 2-furoate	—	36	Dimethyl isophthalate	—	68
Methyl stearate	—	39	Phenyl benzoate	—	69
Dimethyl itaconate	—	39	Methyl 3-nitrobenzoate	—	78
Phenyl salicylate	—	42	Methyl 4-bromobenzoate	—	81
Diethyl terephthalate	—	44	Ethyl 4-aminobenzoate	—	89
Methyl 4-chlorobenzoate	—	44	Methyl 4-nitrobenzoate	—	96
Ethyl 3-nitrobenzoate	—	47	Dimethyl fumarate	—	102
Methyl mandelate	—	53	Cholesterol acetate	—	114
Ethyl 4-nitrobenzoate	—	56	Ethyl 4-hydroxybenzoate	—	116

2 | APPENDIX 2

Procedures for Preparing Derivatives

> ### CAUTION
>
> Some of the chemicals used in preparing derivatives are suspected carcinogens. Before beginning any of these procedures, consult the list of suspected carcinogens on page 561. Exercise care in handling these substances.

ALDEHYDES AND KETONES

Semicarbazones

Place 0.5 mL of a 2 M stock solution of semicarbazide hydrochloride (or 0.5 mL of a solution prepared by dissolving 1.11 g of semicarbazide hydrochloride [$MW = 111.5$] in 5 mL of water) in a small test tube. Add 0.15 g of the unknown compound to the test tube. If the unknown does not dissolve in the solution or if the solution becomes cloudy, add enough methanol (maximum of 2 mL) to dissolve the solid and produce a clear solution. If a solid or cloudiness remains after adding 2 mL of methanol, do not add any more methanol and continue this procedure with the solid present. Using a Pasteur pipet, add 10 drops of pyridine and heat the mixture in a hot waterbath (about 60°C) for about 10–15 minutes. By that time, the product should have begun to crystallize. Collect the product by vacuum filtration. The product can be recrystallized from ethanol if necessary.

Semicarbazones (Alternative Method)

Dissolve 0.25 g of semicarbazide hydrochloride and 0.38 g of sodium acetate in 1.3 mL of water. Then dissolve 0.25 g of the unknown in 2.5 mL of ethanol. Mix the two solutions together in a 25-mL Erlenmeyer flask and heat the mixture to boiling for about 5 minutes. After heating the mixture, place the reaction flask in a beaker of ice and scratch the sides of the flask with a glass rod to induce crystallization of the derivative. Collect the derivative by vacuum filtration and recrystallize it from ethanol.

2,4-Dinitrophenylhydrazones

Place 10 mL of a solution of 2,4-dinitrophenylhydrazine (prepared as described for the classification test in Experiment 54D) in a test tube and add 0.15 g of the unknown compound. If the unknown is a solid, it should be dissolved in the minimum amount of 95% ethanol or 1,2-dimethoxyethane before it is added. If crystallization is not immediate, gently warm the solution for a minute in a hot waterbath (90°C) and then set it aside to crystallize. Collect the product by vacuum filtration.

CARBOXYLIC ACIDS

Working in a hood, place 0.50 g of the acid and 2 mL of thionyl chloride into a small round-bottom flask. Add a magnetic stir bar and attach a water-jacketed condenser and a drying tube packed with calcium chloride to the flask. While stirring, heat the reaction mixture to boiling for 30 minutes on a hot plate. Allow the mixture to cool to room temperature. Use this mixture to prepare the amide, anilide, or *p*-toluidide derivatives by one of the following three procedures.

Amides

Working in a hood, add the thionyl chloride/carboxylic acid mixture dropwise from a Pasteur pipet into a beaker containing 5 mL of ice-cold concentrated ammonium hydroxide. The reaction is very exothermic. Stir the mixture vigorously after the addition for about 5 minutes. When the reaction is complete, collect the product by vacuum filtration and recrystallize it from water or from water–ethanol, using the mixed-solvents method (Technique 11, Section 11.10).

Anilides

Dissolve 0.5 g of aniline in 13 mL of methylene chloride in a 50-mL Erlenmeyer flask. Using a Pasteur pipet, carefully add the mixture of thionyl chloride/carboxylic acid to this solution. Warm the mixture for an additional 5 minutes on a hot plate, unless a significant color change occurs. *If a color change occurs*, discontinue heating, add a magnetic stir bar, and stir the mixture for 20 minutes at room temperature. Then transfer the methylene chloride solution to a small separatory funnel and wash it sequentially with 2.5 mL of water, 2.5 mL of 5% hydrochloric acid, 2.5 mL of 5% sodium hydroxide, and a second 2.5-mL portion of water (the methylene chloride solution should be the bottom layer). Dry the methylene chloride layer over a small amount of anhydrous sodium sulfate. Decant the methylene chloride layer away from the drying agent into a small flask and evaporate the methylene chloride on a warm hot plate in the hood. Use a stream of air or nitrogen to speed up the evaporation. Recrystallize the product from water or from ethanol–water, using the mixed-solvents method (Technique 11, Section 11.10).

p-Toluidides

Use the same procedure as that described in preparing anilides but substitute *p*-toluidine for aniline.

PHENOLS

α-Naphthylurethanes

Follow the procedure given later for preparing phenylurethanes from alcohols but substitute α-naphthylisocyanate for phenylisocyanate.

Bromo Derivatives

First, if a stock brominating solution is not available, prepare one by dissolving 0.75 g of potassium bromide in 5 mL of water and adding 0.5 g of bromine. Dissolve 0.1 g of the phenol in 1 mL of methanol or 1,2-dimethoxyethane; then add 1 mL of water. Add 1 mL of the brominating mixture to the phenol solution and swirl the mixture vigorously. Then continue adding the brominating solution dropwise while swirling, until the color of the bromine reagent persists. Finally, add 3–5 mL of water and shake the mixture vigorously. Collect the precipitated product by vacuum filtration and wash it well with water. Recrystallize the derivative from methanol–water, using the mixed-solvents method (Technique 11, Section 11.10).

AMINES

Acetamides

Place 0.15 g of the amine and 0.5 mL of acetic anhydride in a small Erlenmeyer flask. Heat the mixture for about 5 minutes; then add 5 mL of water and stir the solution vigorously to precipitate the product and hydrolyze the excess acetic anhydride. If the product does not crystallize, it may be necessary to scratch the walls of the flask with a glass rod. Collect the crystals by vacuum filtration and wash them with several portions of cold 5% hydrochloric acid. Recrystallize the derivative from methanol–water, using the mixed-solvents method (Technique 11, Section 11.10).

Aromatic amines, or those amines that are not very basic, may require pyridine (2 mL) as a solvent and a catalyst for the reaction. If pyridine is used, a longer period of heating is required (up to 1 hour), and the reaction should be carried out in an apparatus equipped with a reflux condenser. After reflux, the reaction mixture must be extracted with 5–10 mL of 5% sulfuric acid to remove the pyridine.

Benzamides

Using a centrifuge tube, suspend 0.15 g of the amine in 1 mL of 10% sodium hydroxide solution and add 0.5 g of benzoyl chloride. Cap the tube and shake the mixture vigorously for about 10 minutes. After shaking the mixture, add enough dilute hydrochloric acid to bring the pH of the solution to pH 7 or 8. Collect the precipitate by vacuum filtration, wash it thoroughly with cold water, and recrystallize it from ethanol–water, using the mixed-solvents method (Technique 11, Section 11.10).

Benzamides (Alternative Method)

In a small round-bottom flask, dissolve 0.25 g of the amine in a solution of 1.2 mL of pyridine and 2.5 mL of toluene. Add 0.25 mL of benzoyl chloride to the solution and heat the mixture under reflux for about 30 minutes. Pour the cooled reaction mixture into 25 mL of water and stir the mixture vigorously to hydrolyze the excess benzoyl chloride. Separate the toluene layer and wash it, first with 1.5 mL of water and then with 1.5 mL of 5% sodium carbonate. Dry the toluene over granular anhydrous sodium sulfate, decant the toluene into a small Erlenmeyer flask, and remove the toluene by evaporation on a hot plate in the hood. Use a stream of air or nitrogen to speed up the evaporation. Recrystallize the benzamide from ethanol or ethanol–water, using the mixed-solvents method (Technique 11, Section 11.10).

Picrates

In an Erlenmeyer flask, dissolve 0.2 g of the unknown in about 5 mL of ethanol and add 5 mL of a saturated solution of picric acid in ethanol. Heat the solution to boiling and then allow it to cool slowly. Collect the product by vacuum filtration and rinse it with a small amount of cold ethanol.

CAUTION ⚠️

Great care must be taken when working with saturated solutions of picric acid. Picric acid may detonate when heated above 300°C! It is also known to explode when heated rapidly. For this reason, it is strongly recommended that you check with your instructor before preparing this derivative.

Methiodides

Mix equal-volume quantities of the amine and methyl iodide in a small round-bottom flask (about 0.25 mL is sufficient) and allow the mixture to stand for several minutes. Then heat the mixture gently under reflux for about 5 minutes. The methiodide should crystallize on cooling. If it does not, you can induce crystallization by scratching the walls of the flask with a glass rod. Collect the product by vacuum filtration and recrystallize it from ethanol or ethyl acetate.

ALCOHOLS

3,5-Dinitrobenzoates

Liquid Alcohols

Dissolve 0.25 g of 3,5-dinitrobenzoyl chloride in 0.25 mL of the alcohol and heat the mixture for about 5 minutes.[1] Allow the mixture to cool and add 1.5 mL of a 5% sodium carbonate solution and 1 mL of water. Stir the mixture vigorously and crush any solid that forms. Collect the product by vacuum filtration and wash it with cold water. Recrystallize the derivative from ethanol–water, using the mixed-solvents method (Technique 11, Section 11.10).

Solid Alcohols

Dissolve 0.25 g of the alcohol in 1.5 mL of dry pyridine and add 0.25 g of 3,5-dinitrobenzoyl chloride.[2] Heat the mixture under reflux for 15 minutes. Pour the cooled reaction mixture into a cold mixture of 2.5 mL of 5% sodium carbonate and 2.5 mL of water. Keep the solution cooled in an ice bath until the product crystallizes and stir it vigorously during the entire period. Collect the product by vacuum filtration, wash it with cold water, and recrystallize it from ethanol–water, using the mixed-solvents method (Technique 11, Section 11.10).

Phenylurethanes

Place 0.25 g of the *anhydrous* alcohol in a dry test tube and add 0.25 mL of phenylisocyanate (α-naphthylisocyanate for a phenol). If the compound is a phenol, add 1 drop of pyridine to catalyze the reaction. If the reaction is not spontaneous, heat the mixture in a hot waterbath (90°C) for 5–10 minutes. Cool the test tube in a beaker of ice and scratch the tube with a glass rod to induce crystallization. Decant the liquid from the solid product or, if necessary, collect the product by vacuum filtration. Dissolve the product in 2.5–3 mL of hot ligroin or hexane and filter the mixture by gravity (preheat funnel) to remove any unwanted and insoluble diphenylurea present. Cool the filtrate to induce crystallization of the urethane. Collect the product by vacuum filtration.

ESTERS

We recommend that esters be characterized by spectroscopic methods whenever possible. A derivative of the alcohol part of an ester can be prepared with the following procedure. For other derivatives, consult a comprehensive textbook. Several are listed in Experiment 54 (p. 491).

3,5-Dinitrobenzoates

Place 1.0 mL of the ester and 0.75 g of 3,5-dinitrobenzoic acid in a small round-bottom flask. Add 2 drops of concentrated sulfuric acid and a magnetic stir bar to the flask and attach a condenser. If the boiling point of the ester is below 150°C, heat at reflux while stirring for 30–45 minutes. If the boiling point of the ester is above 150°C, heat the mixture at about 150°C for

---

[1] 3,5-Dinitrobenzoyl chloride is an acid chloride and undergoes hydrolysis readily. The purity of this reagent should be checked before its use by determining its melting point (mp 69–71°C). When the carboxylic acid is present, the melting point will be high.

[2] See footnote 1.

30–45 minutes. Cool the mixture and transfer it to a small separatory funnel. Add 10 mL of ether. Extract the ether layer two times with 5 mL of 5% aqueous sodium carbonate (save the ether layer). Wash the organic layer with 5 mL of water and dry the ether solution over magnesium sulfate. Evaporate the ether in a hot waterbath in the hood. Use a stream of air or nitrogen to speed the evaporation. Dissolve the residue, usually an oil, in 2 mL of boiling ethanol and add water dropwise until the mixture becomes cloudy. Cool the solution to induce crystallization of the derivative.

Preparation of a Solid Carboxylic Acid from an Ester

An excellent derivative of an ester can be prepared by a basic hydrolysis of an ester when it yields a solid carboxylic acid. A procedure is provided in Experiment 54I, p. 491. Melting points for solid carboxylic acids are included in the Carboxylic Acids Table in Appendix 1 (pp. 961–963).

Index of Spectra

Infrared Spectra

2-Acetylcyclohexanone   358
Anisole   858
Benzaldehyde   308, 861
Benzamide   865
Benzil   311
Benzilic acid   313
Benzocaine   372
Benzoic acid   327, 863
Benzoin   308
Benzonitrile   860
Borneol   296
n-Butyl bromide   199
n-Butylamine   859
Caffeine   99
Camphor   296, 863
Carbon disulfide   846
Carbon tetrachloride   843
Carvone   128
Chloroform   843
Decane   853
1,2-Dichlorobenzene   857
7,7-Dichloronorcarane   222
N,N-Diethyl-*m*-toluamide   388
6-Ethoxycarbonyl-3,5-diphenyl-2-cyclohexenone   347
Isoborneol   297
Isopentyl acetate   107, 864
Limonene   128
Mesityl oxide   862
Methyl benzoate   864
Methyl isopropyl ketone   849
Methyl *m*-nitrobenzoate   233
Methyl salicylate   376
4-Methylcyclohexanol   216, 858
4-Methylcyclohexene   215, 854
Mineral oil   842
2-Naphthol   857
Nitrobenzene   859
Nonanal   861
cis-Norbornene-5,6-*endo*-dicarboxylic anhydride   422, 865
Nujol   842
Paraffin oil   842
t-Pentyl chloride   201
Polystyrene   847
Styrene   855

Sulfanilamide   396
Triphenylmethanol   324

¹H NMR Spectra

2-Acetylcyclohexanone   358
Benzocaine   372
Benzyl acetate (60-MHz NMR spectrum)   876
Benzyl acetate (300-MHz NMR spectrum)   877
Borneol   298
Camphor   297
Carvone   129
Ethyl 3-hydroxybutanoate   280, 282, 283
Eugenol   115
1-Hexanol   898
Isoborneol   298
Limonene   129
Methyl Salicylate   376
Phenylacetone   870
1,1,2-Trichloroethane   882
Vinyl acetate   887

¹³C NMR Spectra

Borneol   299
Camphor   299
Carvone   130
Cyclohexanol   913
Cyclohexanone   914
Cyclohexene   914
7,7-Dichloronorcarane   223
2,2-Dimethylbutane   912
Ethyl phenylacetate   910
Isoborneol   300
1-Propanol   911
Toluene   915

Mass Spectra

Acetophenone   938
Benzaldehyde   937
Bromoethane   929
1-Bromohexane   941
Butane   932
1-Butanol   936
2-Butanone   938
1-Butene   934
Chloroethane   928
Cyclopentane   934
Dopamine   925
Methyl butanoate   940
Propanoic acid   939
Toluene   935
2,2,4-Trimethylpentane   933

Ultraviolet-Visible Spectra

Benzophenone   429
Naphthalene   429

Mixtures

Borneol and isoborneol   300
t-Butyl chloride and *t*-butyl bromide   188
1-Chlorobutane and 1-bromobutane   188

Index

A

Ab initio calculations, 161–165

Abbé refractometer, 828–831

Acetamides, 972–973

p-Acetamidobenzenesulfonamide, 395–396

p-Acetamidobenzenesulfonyl chloride, 395

Acetaminophen, 68, 71, 75–80
  decolorization, 77
  microscale procedure, 76–78
  semimicroscale procedure, 78–80

Acetanilide, 67–68, 228, 394

Acetate esters, 498–501

Acetic acid, safety information, 554

Acetone
  alkyl halides reactivities, 176–178
  enolate ion, 342
  safety information, 554
  wash acetone, 566

Acetophenone, 938

Acetyl chloride test, 484, 487–488

2-Acetylcyclohexanone, 352, 354–358
  infrared spectrum, 358
  NMR spectrum, 358

Acetylsalicylic acid, 63–67, 71
  crystallization, 64–65
  ferric chloride test, 66
  melting point, 66
  vacuum filtration, 65–66

Acid chlorides, 865

Acids, infrared spectroscopy, 862

Acid-washed alumina, 756

Activators, 224

Actual yield, 562

Acylglycerols, 233

Acylium ion, 937

Addition funnel, 604

Addition polymers, 397–398, 399–401

Addition reaction, 467

Adjustable metal clamps, 598–600

Adsorbents
  chromatography, 756–757
  column chromatography, 760–761

Advance preparation, 559–561

Air condenser, 600

Air peak, 805

Alcohols
  derivatives, 490–491, 967–968, 974
  identification tests, 486–490

infrared spectroscopy, 856
  as solvent, 30
  spectroscopy, 490
  unknowns, 967–968

Aldehydes
  derivatives, 476, 958–959, 971
  enigma of, 527–529
  identification tests, 470–475
  infrared spectroscopy, 860
  spectroscopy, 475–476
  unknowns, 958–959

Alder, Kurt, 414

Aldol condensation reaction, 315, 339–343, 344, 533–536

Aldrich Handbook of Fine Chemicals, 560, 579–580, 628, 942–943

Alembic, 703

Alkaloid, 87

Alkanes, 853

Alkenes
  elimination reactions, 204–210
  infrared spectroscopy, 853–854

Alkyl chlorides
  hydrolysis, 190–195
  kinetic study, 193–194
  preparation, 193

Alkyl halides reactivities, 176–180

Alkylates, 258

Alkylation, 258

Alkylbenzenes, 889–890

Alkynes, 855–856

Allergens, 522

Allinger, Norman, 155–156

4-Allyloxyanisole, 893

Alumina, 664, 756–757

Aluminum block, 2–4, 590–593

Ambident nucleophile, 342

American Conference of Governmental Industrial Hygienists (ACGIH), 548

Amides, 383, 385–386
  infrared spectroscopy, 864
  preparation, 972

Amines
  derivatives, 485–486, 965–967, 972–973
  identification tests, 482–485
  infrared spectroscopy, 858
  spectroscopy, 485
  unknowns, 965–967

p-Aminophenol, 68

Amoore, J. E., 120

Analgesic, 61

Analgesics
  in common preparations, 69
  essay, 67–70
  isolating active ingredient, 71–74
  thin-layer chromatography, 82–87

Analytical balance, 6, 7, 588, 811

Analyzer, 821

Anesthetics, 367–370
  benzocaine, 370–373
  cocaine, 367–368

Anethole, 893

Angina pectoris, 88

Anhydrides, 865

Anilides, 972

Aniline, 228

Anisole
  aromatic ring portions, 891
  crystallization, 228
  infrared spectrum, 858
  nitration, 232

Anisotropy, 880–882, 891–892

Antihistamine drugs analysis, 522–523

Antihistamines, 522

Anti-inflammatory, 61

Antipyretic, 61

APC tablet, 68

Apparatus assembly, 569, 598–600

Aqueous solution, pH of, 477–478, 484

Aqueous-based organozinc reactions, 328–331, 524–527

Arachidonic acid, 61

Aromatic compounds
  Friedel–Crafts acylation, 515–521
  infrared spectroscopy, 855
  nitration, 284–288
  nuclear magnetic resonance (NMR) spectroscopy, 889–895
  relative reactivities, 224–228
  tests for, 469–470

Aspartame, 443

Aspirator, 610, 624–625

Aspirin, 61–67
  essay, 61–62
  preparation, 65–66

Atom economy, 270–271

Automatic pipets, 6–8, 10, 581

Axel, Richard, 121

Azeotropes, 724–732
  applications, 728–730
  maximum-boiling-point diagrams, 726–727
  minimum-boiling-point diagrams, 724–726

B

Back extraction, 688
Backwashing, 688
Bactericidal drugs, 390
Bacteriostatic drugs, 390
Baeyer test, 468
Baker SAF-T-DATA System, 548
Balances, 588–589
  analytical, 6, 7, 588, 811
  top-loading, 6, 7, 588
Banana oil. *See* Isopentyl acetate
Base peak, 925
Basic alumina, 756
Basis set orbitals, 162–164
Bayer, 61, 68
Beakers, 588
Beilstein, 949–951, 952
Beilstein test, 461
Bent vacuum adapter, 713
Benzalacetophenones. *See* Chalcones
Benzaldehyde
  aromatic ring portions, 892
  infrared spectrum, 308, 861
  mass spectrum, 937
  multiple reaction sequences, 302–314
  thiamine catalysis, 303–305, 307–308
Benzamides, 865, 973
Benzene
  safety information, 554
  as solvent, 30
Benzene ring substitution
  carbon-13 spectroscopy, 915–917
  NMR spectroscopy, 889–895
Benzenoid compounds, 856
Benzil
  infrared spectrum, 311
  preparation, 309
Benzilic acid
  infrared spectrum, 313
  multiple reaction sequences, 302–314
  preparation, 311–314
Benzocaine, 370–373
  infrared spectrum, 372
  NMR spectrum, 372
Benzoic acid, 318–319, 325
  infrared spectrum, 327, 863
Benzoin
  infrared spectrum, 308
  preparation, 303–305, 307–308
Benzonitrile, 860
Benzophenone, 424–431
Benzopinacolone, 424, 431–433

Benzpinacol, 424, 431–433
Benzyl acetate, 876, 877
Benzyltriphenylphosphonium chloride, 362
Beriberi, 304
Bernoulli effect, 624
Binding energy, 165
Biodegradability, 246
Bioluminescence, 433
Biomass, 270
Biphenyl, 319
Bleed valve, 743
Boiling point, 694–702
  barometric pressure, 694–695
  carvones, 130
  determination, 57–60
  distillation method, 699
  list of compounds, 60
  4-methylcyclohexene, 214, 215
  microscale inverted capillary method, 697–699
  semimicroscale direct method, 695–697
  semimicroscale inverted capillary method, 697
  thermometers and stem corrections, 700–702
  unknowns determined from, 450
Boiling stones, 603–604, 712
Boll, Franz, 132
Bond-density surface, 167
Borneol
  carbon-13 spectrum, 299
  infrared spectrum, 296
  NMR spectrum, 298
  oxidation, 289, 293–295
Born-Oppenheimer approximation, 160
Bottle labels, 550–554
Bromide, 465
Bromine
  isotope peaks, 927
  mass spectra, 928
  test, 466–467, 481
Bromo derivatives, 972
1-Bromobutane, 208–209
2-Bromobutane, 208–209
Bromoethane, 929
1-Bromohexane, 940–941
Bronchodilator, 88
Büchner funnel, 623
Buck, Linda, 121
Builder, 248
Bumping, 602, 735
Bunsen burner, 589, 595
Butane
  conformations, 157–158
  mass spectrum, 932
1-Butanol
  dehydration, 207–208

mass spectrum, 935–936
  nucleophilic substitution, 183–185
2-Butanol
  dehydration, 207–208
  nucleophilic substitution, 183–185
2-Butanone, 937–938
1-Butene, 934–935
Butenes, 159–160
n-Butyl bromide
  distillation, 199
  infrared spectrum, 199
  preparation, 195–196, 197–198
  semimicroscale procedure, 199
n-Butylamine, 859

C

Caffeine
  in common preparations, 69
  essay, 87–90
  extraction, 33–35
  extraction with methylene chloride, 93–95
  isolation, 90–99
  solid phase extraction, 95–99, 690–692
Calculations, 562–565
Calibration of pipets, 11
Camphor
  carbon-13 spectrum, 299
  extraction, 294
  gas chromatography, 296
  infrared spectrum, 296, 863
  molecular modeling, 301
  NMR spectrum, 297
  oxidation from borneol, 293–295
  reduction, 289–291, 295
Capacity, of drying agent, 680
Caps, for glassware, 569–570
Caraway oil, 124–131
Carbocations, 173
Carbon dioxide
  decaffeination process, 89–90
Carbon disulfide, 846
Carbon tetrachloride
  infrared spectroscopy, 843, 844
  safety information, 554–555
Carbon-13 nuclear magnetic resonance spectroscopy, 906–923
  aromatic compounds, 516, 915–917
  borneol, 299
  camphor, 299
  chemical shifts, 907–908
  dichloronorcarane, 223
  equivalent carbons, 912–913
  isoborneol, 300
  nuclear Overhauser effect, 913–915
  proton-coupled spectra and spin–spin splitting, 908–910
  proton-decoupled spectra, 910–912
  sample preparation, 906–907
  uses, 906

Carbonyl
  conjugation and ring strain effects, 862
  infrared spectroscopy, 860
  reactivities, 173–174
Carboxylic acids
  acidities, 172
  derivatives, 478–479, 961–963, 971–972, 975
  identification tests, 477–478
  spectroscopy, 478
  unknowns, 961–963
Carcinogens, 557
Carotenoids, 134
  column chromatography, 139–141
  definition, 137
  extraction, 138–139
  isolation, 136–142
  thin-layer chromatography, 141–142
Carrier gas, 798, 802
Cartesian coordinates, 161
Carvones, 124–131
  boiling point, 130
  gas chromatography, 126–127, 130–131
  infrared spectroscopy, 127
  polarimetry, 127
  refractive index, 130
CASREACTS, 952
Catalytic hydrogenation, 239–243
C-18 column, 794
Celite, 623
Centrifugation, 627
Centrifuge tube, 677
Cerium (IV) test, 480, 487
Chalcones
  aldol condensation reaction, 339–343
  synthesis of substituted, 529–532
Channeling, 764
Charcoal, 663–664
Chemical Abstracts, 951
Chemical Abstracts Service, 952
Chemical Abstracts Service Number (CAS No.), 548
Chemical Hygiene Plan, 547
Chemical literature, 574–580, 942–956
  advanced laboratory techniques, 947
  advanced textbooks, 945
  Aldrich Handbook of Fine Chemicals, 579–580
  Beilstein , 949–951
  Chemical Abstracts, 951
  CRC Handbook of Chemistry and Physics, 574–576
  current topics, 954
  general synthetic methods, 943
  handbooks, 942–943
  introductory textbooks, 943
  journals, 953–954

Lange's Handbook of Chemistry, 576–578
  The Merck Index, 578–579
  online searching, 952–953
  organic qualitative analysis, 948–949
  reaction mechanisms, 947–948
  searching, 943–944, 954–955
  specific synthetic methods, 945–946
  spectra catalogues, 944
  strategy for using, 580
Chemical shift
  carbon-13 spectroscopy, 907–908
  definition, 874
  NMR spectroscopy, 877–880
  reagents, 896–897
Chemiluminescence, 433, 440–441
Chemotherapy, 388
Chiral gas chromatography, 806–807
  ethyl acetoacetate, 278
  α-phenylethylamine, 336–337
Chiral recognition, 124
Chiral reduction, 274–284
Chloride, 465
Chlorine
  isotope peaks, 927
  mass spectra, 928
Chloroethane, 928
Chloroform
  infrared spectroscopy, 843
  infrared spectrum, 843
  NMR spectroscopy, 870–872
  safety information, 555
Chlorophyll
  column chromatography, 139–141
  definition, 137
  extraction, 138–139
  isolation, 136–142
  thin-layer chromatography, 141–142
Chlorosulfonic acid, 394
Chromatic aberration, 830
Chromatogram, 805, 813
Chromatography, 42–50. *See also* Chiral gas chromatography; Column chromatography; Gas chromatography; High-performance liquid chromatography; Thin-layer chromatography
  adsorbents, 756–757
  definition, 756
  high-performance liquid, 444–446
  ion-exchange, 795
  normal-phase, 96, 794
  reversed-phase, 96, 794
Chromic acid test, 473–474, 489
Clamps, 598–600
Clean Air Act, 259, 260
Clothing, and laboratory safety, 546–547

Cloves, 112–119
  liquid carbon dioxide, 116–119
  microscale procedure, 113–114
  semimicroscale procedure, 115–116
Cocaine, 367–368
Coffee. *See* Caffeine
Cold bath, 595
Cold pressing, 234
Column chromatography, 48–50, 756–777
  2-acetylcyclohexanone, 355
  adsorbents, 760–761
  analgesics, 73–74
  applying sample to column, 768–769
  chlorophyll and carotenoids, 139–141
  column size and adsorbent quantity, 763
  decolorization, 773
  N,N-diethyl-*m*-toluamide, 387
  elution techniques, 769–770
  flash chromatography, 775–776
  flow rate, 763
  gas versus, 797
  gel chromatography, 773–775
  interactions, 757–758
  monitoring, 771–772
  packing the column, 764–768
  recovering separated compounds, 772–773
  reservoirs, 770–771
  separation, 758–763
  solvents, 761–763
  tailing, 772, 773
  thin-layer chromatography and, 777
Completeness, of drying agent, 680
Compounds, polarity of, 639
Computational chemistry. *See also* Molecular modeling
  density-electrostatic potential maps, 172–173
  density-LUMO maps, 173–174
  essay, 160–168
  experiments, 168–174
  heats of formation, 169–171
  heats of reaction, 171–172
Condensation polymers, 398, 401–403
Condensers, 600–601, 708
Cones, 132
Conformation
  n-Butane, 157–158
  molecular mechanics, 154–155
Conical vials, 4–6, 588, 674–677
Connectivity matrix, 161
Continuous liquid–liquid extraction, 688–689
Continuous solid–liquid extraction, 688–689
Cooling methods, 595
Copolymers, 397

Coupling constant, 884–886
Cracking, 256, 763
Craig tube, 25–27, 614, 625–627, 656
CRC Handbook of Chemistry and Physics, 560, 574–576, 628, 942
Crown ether catalysis, 219–220
Crystallization, 21–32, 647–668
  acetaminophen, 77–78, 79
  acetylsalicylic acid, 64–65
  benzalacetophenones, 341
  benzil, 310
  benzilic acid, 313
  benzoic acid, 326
  benzoin, 307–308
  decolorization, 663–664
  definition, 647
  1,4-diphenyl-1,3-butadiene, 365
  drying crystals, 665–667
  6-ethoxycarbonyl-3,5-diphenyl-2-cyclohexenone, 346
  inducing, 664–665
  microscale, 25–27, 647, 656–659
  mixed solvents, 667
  mixture melting points, 28–29
  α-phenylethylamine, 335
  potassium benzilate, 312–313
  recrystallization, 650
  semimicroscale, 22–25, 647, 650–656
  solubility, 647–648
  solvent selection, 21, 27–28, 660–662
  solvent testing, 662–663
  sulfanilamide, 396
  theory, 649–650
  triphenylmethanol, 324
Curds, 245–246
Cyano group, 463–464
Cyclodextrins, 807
Cyclohexane, 158–159
  phase-transfer catalysis, 217–224
Cyclohexanol, 913
Cyclohexanone, 914
Cyclohexene, 914
Cyclo-oxygenase, 61
Cyclopentadiene, 419–423
Cyclopentane, 933–934

D

Data sources. *See* Chemical literature
Data tables, 813
DDT, 415–416
Dean–Stark water separator, 728
Decaffeination, 89–90
Decane, 853
Decantation, 620–621
Decolorization
  acetaminophen, 77, 79
  column chromatography, 773
  crystallization, 663–664
Decomposition, 634–635

Dehydration
  1-butanol and 2-butanol, 207–208
  4-methylcyclohexene, 213, 214–215
Dehydrobromination, 208–209
Density, 702
Density-electrostatic potential, 167, 172–173
Density-elpot, 167
Density-LUMO maps, 173–174
Derivatives
  alcohols, 490–491
  aldehydes and ketones, 476
  amines, 485–486
  carboxylic acids, 478–479
  essay, 494
  phenols, 481–482
  preparation, 971–975
  tables of unknowns and, 958–970
  unknowns identification, 453
Design for the Environment Program (EPA), 269
Detent, 7
Detergents. *See* Soaps and detergents
Detonation, 257
Developing the thin-layer plate, 778, 784–785
Dewar flask, 739–740
Diamagnetic anisotropy, 881
Diatomaceous earth, 623
o-Dichlorobenzene, 857
Dichlorobenzene, 916
Dichlorocarbene, 217–224
Dictionary of Organic Compounds, 943
Diels, Otto, 414
Diels–Alder reaction
  insecticides, 414–419
  cis-norbornene-5,6-*endo*-dicarboxylic anhydride formation, 419–423
Diesel fuel, 262
Diet soft drink analysis, 444–446
N,N-Diethyl-*m*-toluamide, 383–388
  infrared spectrum, 388
Dilation, 88
1,2-Dimethoxyethane, safety information, 555
3,5-Dinitrobenzoates, 974, 974–975
2,4-Dinitrophenylhydrazine, 471–472
2,4-Dinitrophenylhydrazones, 971
Dioxane, safety information, 555
1,4-Diphenyl-1,3-butadiene, 359–366, 364
  NMR spectrum, 366
  preparation, 362–365
  thin-layer chromatography, 363–364, 365–366
Dipole–dipole interaction, 638
Direct contact decaffeination, 89
Discoloration, 634
Disparlure, 379
Dispensing pumps, 8–9, 10, 582–583
Dispersion forces, 639

Disposable (Pasteur) pipets, 11
Distillate, 51, 707
Distillation, 51–57. *See also* Fractional distillation; Simple distillation; Steam distillation; Vacuum distillation
  acetate esters, 500
  n-butyl bromide, 199
  equipment evolution, 703–705
  isopentyl acetate, 105, 107
  methyl salicylate, 375
  oil of cloves, 113–114
  t-pentyl chloride, 201, 203
Distillation head, 713
Distribution coefficient, 670–671
Diuretic, 88
Domagk, Gerhard, 389
Dopamine, 925
Double refraction, 819
Drug identification, 80–81
Dry film method, 838
Drying agents, 680–684
Drying tubes, 605–606, 607–608
Dubois, Raphael, 433
Duisberg, Carl, 68

E

Ebulliator tube, 735
Eckert, Charles, 271
Ehrlich, Paul, 388
Electron-density surface, 167–168
Electrophilic addition, 229
Electrophilic aromatic substitution reactions, 230
Elemental analysis, 453, 460–465
Elimination reactions, 204–210
  apparatus, 206–207
  aromaticity, 230
  dehydration of 1-butanol and 2-butanol, 207–208
  dehydrobromination of 1-bromobutane and 2-bromobutane, 208–209
Eluates/elutants, 758
Eluents, 758
Emulsions, 684–685
Enamine reactions, 348–351, 353–354
Enantiomeric excess, 826
Enantioselective, 275
Enantiospecific, 274
Endo approach, 290
Energy transfer, 428–430
Environmental Protection Agency, 259, 269, 416
Equipment. *See also* Glassware; *specific pieces*
  common, 572–573
  description, 571
  microscale kit, 570, 704–705
  microscale optional, 571

Erlenmeyer flask, 22–25, 588, 614
Essential oils, 108, 112–119, 502–509
    identification by gas
        chromatography–mass
        spectrometry, 507–508
    investigation, 508–509
    isolation by steam distillation,
        504–507
Esterification, 63
    vanillin, 536–538
Esters
    derivatives, 494, 969–970, 974–975
    essay, 99–102
    fats and oils, 233
    identification tests, 491–493
    infrared spectroscopy, 863
    spectroscopy, 493
    unknowns, 969–970
Ethanol
    alkyl halides reactivities, 176
    essay, 142–145
    safety information, 555
    sucrose, 145–149
Ethers
    infrared spectroscopy, 857
    safety information, 555
6-Ethoxycarbonyl-3,5-diphenyl-
        2-cyclohexenone, 344–347
    infrared spectrum, 347
Ethyl acetoacetate, chiral reduction,
        274–284
Ethyl alcohol, 30
Ethyl phenylacetate, 910
Ethyl (S)-3-hydroxybutanoate,
        279–284
Ethylbenzene, 890
Eugenol, 112
Eutectic, 629, 630
Eutrophication, 248
Evaporation
    analgesics, 73–74
    caffeine, 94
    oil of cloves, 114
    sublimation versus, 747
Exo approach, 290
Extraction, 11–12, 28–29, 32–42,
        669–693
    acetate esters, 500
    2-acetylcyclohexanone, 354
    analgesics, 72–73, 72–73
    apparatus, 673
    caffeine, 33–35, 94
    camphor, 294
    centrifuge tube, 677
    conical vial, 674–677
    continuous liquid–liquid, 688–689
    continuous solid–liquid, 688–689
    distribution coefficient, 670–671
    distribution of solute, 36–37
    drying agents, 680–684
    emulsions, 684–685

essential oils, 502–509
    isopentyl acetate, 106
    method selection, 671–673
    neutral compound isolation, 38–40
    oil of cloves, 114, 118–119
    organic layer determination, 37,
        679–680
    t-pentyl chloride, 200, 202–203
    purification, 685–688
    separation, 685–688
    separatory funnel, 677–679
    solid phase, 95–99, 689–692
    solvent selection, 671–673
Eye. See Vision, chemistry of
Eye safety, 542

F

Fat replacers, 237–238
Fats and oils, 233–238. See also
        Essential oils
    catalytic hydrogenation, 239–243
    common fatty acids, 234
    diet and nutrition, 236–238
    fatty acid composition, 235
    hydrolysis, 249–251
    recovery, 234
    soaps, 244–245, 249–251
Feedstock, 270
Fermentation
    essay, 142–145
    sucrose, 147–148
Ferric chloride test, 66, 475, 480
Ferric hydroxamate test, 492
Ferrous hydroxide test, 463
Filter Aid, 623
Filter cones, 616–618
Filter flask, 621
Filter paper, 621
Filtering media, 623
Filtering pipets, 618–620
Filter-tip pipet, 625
Filtration, 616–627
    aspirator, 624–625
    centrifugation, 627
    Craig tubes, 625–627
    filter paper, 621
    filtering media, 623
    filter-tip pipet, 625
    gravity filtration, 616–621
    overview of methods, 617
    vacuum filtration, 621–623
Fire safety, 542, 549
Fireflies, 433–436
First aid, 547
First order patterns, 889
Flame front, 257
Flame-ionization detector, 805–806
Flames, use in laboratory, 546, 595
Flash chromatography, 775–776
Flavors, 99–102
Fluorescence, 426

Fluted filters, 618–619
Folic acid, 390
Food and Drug Administration, 444
Food, and laboratory safety, 546
Force constant, 153
Force field, 152, 153
Forerun, 707
Fossil fuels. See Petroleum and
        fossil fuels
Fractional distillation, 715–732
    column efficiency, 721
    fractionating columns and packings,
        721–723
    macroscale apparatus, 724
    methods and practice, 723–724
    microscale apparatus, 717, 723–724
    microscale procedure, 56–57
    need for, 706
    Raoult's Law, 719–720
    semimicroscale procedure, 53–56
    simple versus, 715–716
    sucrose, 148–149
    vapor–liquid composition dia-
        grams, 716–719
Fractionating column, 716, 721–723
Fractions, 707
Fragment ion peaks, 925
Fragrances, 99–102. See also Odor
Free radical, 930
Freud, Sigmund, 367
Friedel–Crafts acylation, 515–521
Frontier orbitals, 166
Frozen core approximation, 164

G

Gas chromatography, 797–818. See also
        Chiral gas chromatography
    advantages, 803–804
    antihistamine drugs analysis,
        522–523
    apparatus, 798, 799
    camphor reduction, 296
    carvones, 126–127, 130–131
    chiral stationary phases, 806–807
    chromatogram, 805, 813
    column chromatography versus, 797
    column used in, 798–802
    cyclodextrins, 807
    data treatment, 813–816
    detectors, 804–806
    distillation, 56
    essential oils, 507–508
    gasolines, 264–268
    mass spectrometry and, 816–817
    nucleophilic substitution, 187
    qualitative analysis, 808–809
    quantitative analysis, 810–813
    resolution, 807–808
    response factors, 812–814
    retention time, 806
    sample collection, 809–810

Gas chromatography (*continued*)
   separation, 802–803
   solid supports, 800
   stationary phases, 800
   tailing, 807–809
   uses, 797
Gas collection, 610–611
Gas–liquid partition chromatography.
   See Gas chromatography
Gasoline, 255–262
   gas chromatography, 264–268
   major components, 267
Gaussian-type orbitals, 163
Gel chromatography, 773–775
Gel-filtration column, 795
Geminal methyl groups, 290
Geometry optimization, 164
Glassware, 566–574
   apparatus assembly, 569
   capping vials or openings, 569–570
   cleaning, 566–567
   drying, 567
   etching, 569
   joints, 568–569
   microscale organic kit, 570, 704–705
   optional microscale pieces, 571
   tubing attachment, 571
Global minimum, 154
Glycerides, 233
Gossyplure, 379
Gradient elution systems, 796
Graduated cylinders, 587
Graduated pipets, 6, 9–10, 583–585
Grain alcohol, 143
Gravity filtration, 616–621
   decantation, 620–621
   filter cones, 616–618
   filtering pipets, 618–620
   fluted filters, 618–619
Green chemistry, 268–273
   applications, 269–273
   definition, 268
   principles, 269
Grignard reagent
   carbon-carbon bond, 328
   characteristics, 318–319
   preparation, 320–322
Ground-state singlet state, 426
Gyplure, 379

H
Half-life, 192
Halides
   alkyl halides reactivities, 176–180
   identification tests, 461–463, 465
   infrared spectroscopy, 865
Halogens, mass spectrometry
   detection of, 927–928
Hamiltonian operator, 161
Hammond Postulate, 171
Hartree, 165

Hazard Communication Act, 547
Heating mantles, 713
Heating methods
   aluminum block, 2–4, 590–593
   flames, 595
   microwave ovens, 596–597
   under reflux, 600–602
   sand bath, 3–4, 593–594
   steam bath, 595–596
   water bath, 4, 594
Heats of formation, 165, 169–171
Heats of reaction, 171–172
HETP (column efficiency), 722
Hexane, safety information, 555
Hexanol, 898
Hickman head, 56, 707
High-performance liquid chromatog-
      raphy (HPLC), 444–446, 792–797
   adsorbents and columns, 793–795
   column dimensions, 795
   data presentation, 796–797
   detectors, 796
   mass spectrometry and, 817
   solvent selection, 795–796
   spice analysis, 507
Hinsberg test, 484
Hirsch funnel, 22–25, 622–623
Hofmann, Felix, 61
Holdup, 722
HOMO (highest occupied molecular
      orbital), 166
Honey, 442
Hot plates, 590–591
Hot pressing, 234
Hubbert, Marion King, 261
Hydrochloric acid, 457
Hydrocracking, 258
Hydrolysis, 64, 91, 492–493
Hydroxyl group, 63
Hygroscopic substance, 666

I
I. G. Farbenindustrie, 389
Ibuprofen, 69–70, 71, 271
Ice bath, 595
Iceland spar, 818–819
Ideal solution, 716
Ignition test, 469–470
Immiscibility, 32, 638
   steam distillation, 750–752
In vitro, 389
In vivo, 389
Inert atmosphere, reactions in, 606–607
Inert compounds, 459
Infrared spectroscopy, 57–60, 833–867
   2-acetylcyclohexanone, 358
   acid chlorides, 865
   alcohols, 490, 856
   aldehydes and ketones, 475–476,
      860–861
   alkanes, 853

alkenes, 853–854
alkynes, 855–856
amides, 864
amines, 485, 858
anhydrides, 865
aromatic compounds, 515–516, 855
aromaticity, 470
benzaldehyde, 308
benzil, 311
benzilic acid, 313
benzocaine, 372
benzoic acid, 327
benzoin, 308
borneol, 296
n-butyl bromide, 199
calibration, 846–847
camphor, 296
carbonyl compounds, 860
carboxylic acids, 478
carvones, 127–128
correlation tables and charts, 850, 851
cyano group detection, 463
dichloronorcarane, 222
N,*N*-diethyl-*m*-toluamide, 388
esters, 493, 863
ethers, 857
6-ethoxycarbonyl-3,5-diphenyl-2-
      cyclohexenone, 347
ethyl acetoacetate, 278
halides, 865
isoborneol, 297
liquid samples, 834–838
methyl benzoate, 233
methyl salicylate, 376
4-methylcyclohexene, 214, 215–216
nitriles, 859
nitro compounds, 859
cis-norbornene-5,6-*endo*-dicarboxylic
      anhydride, 422
oxidation puzzle, 538–540
t-pentyl chloride, 201
phenols, 481, 856
polymers, 412–413
recording spectrum, 846
sample preparation, 834–846
simple multiple bonds test, 469
solid samples, 838–846
spectrophotometer, 849
spectrum analysis, 850–853
sulfanilamide, 396
triphenylmethanol, 324
uses, 847
vibration, 848
what to look for, 849–850
Injection port, 802
Insect repellents, 381–382
   N,*N*-diethyl-*m*-toluamide, 383–388
Insecticides, 414–419
   alternatives, 417–418
   hard, 415–416
   soft, 417

Internal conversion, 428
International Union of Pure and
    Applied Chemistry (IUPAC), 548
Intersystem crossing, 426
Invert sugar, 442
Invertase, 145, 442
Iodide, 465
Iodine, 786
Iodoform test, 474–475, 490
Ion current, 816
Ion-exchange chromatography, 795
Ionic liquids, 272
Ionization potential, 924
Isoborneol
    camphor reduction, 291
    carbon-13 spectrum, 300
    infrared spectrum, 297
    NMR spectrum, 298
Isochratic, 796
Isolation experiments, 560–561
Isomerism, 169–170
Isopentyl acetate (banana oil), 98,
    103–108
    infrared spectrum, 864
    microscale procedure, 104–105
    semimicroscale procedure, 106–107
Isoprene rule, 109
Isoprene units, 109

J

Joint clips, 598–600
*Journal of the American Chemical
    Society*, 565
Journals, 953–954
Juvenile hormone, 418

K

Ketones
    conjugation and ring strain
        effects, 862
    derivatives, 476, 959–961, 971
    identification tests, 470–475
    infrared spectroscopy, 861
    spectroscopy, 475–476
    unknowns, 959–961
    α,β-unsaturated, 344–347
Ketoprofen, 70
Knocking, 257
Koller, Karl, 368

L

Laboratory information management,
    558–565
    notebook, 559–565
    reports, 565
    sample submission, 565
Laboratory notebook, 559–565
    advance preparation, 559–561
    basics, 559
    calculations, 562–565
    records, 561–562

Laboratory records, 561–562
Laboratory reports, 565
Laboratory safety, 542–557
    carcinogenic substances, 557
    clothing, 546–547
    common solvents, 554–556
    eye safety, 542
    fires, 542–543
    first aid, 547
    flames, 546
    food, 546
    guidelines, 542–547
    inadvertently mixed chemicals, 546
    organic solvent hazards, 543–544
    right-to-know laws, 547–554
    unauthorized experiments, 546
    waste disposal, 544–545
Land, E. H., 819
Lange's Handbook of Chemistry, 560,
    576–578, 942–943
LeChâtelier's Principle, 176
Lethal Dose, 50% Mortality
    (LD 50), 549
Light, 818–819
Ligroin, 643–644. *See also* Hexane
Limiting reagent, 562, 580
Linear combination of atomic
    orbitals, 162
Liotta, Charles, 271
Liquid reagents, 604–605
Liquids, measurement of, 6–13
Literature. *See* Chemical literature
Local anesthetics, 367–370
Local diamagnetic shielding,
    878–880
Local minimum, 155
London forces, 639
Lucas test, 488
Luciferase, 434
Luciferin, 434
Lucretius, 120
Luminol, 437–441
LUMO (lowest unoccupied molecular
    orbital), 166, 173–174

M

Macroscale experiments, 2
Magnetic spin vane, 603
Magnetic stirrers, 602–603
Maleic anhydride, 419–423
Malt, 143
Maltase, 145
Maltose, 143
Manometers, 741–744
Mash, 143
Mass analyzer, 924
Mass spectrometry, 924–941
    acetophenone, 938
    antihistamine drugs analysis,
        522–523
    benzaldehyde, 937

1-bromohexane, 940
butane, 932–933
1-butanol, 935–936
2-butanone, 937–938
1-butene, 934–935
cyclopentane, 933–934
essential oils, 507–508
fragmentation patterns, 929–931
gas chromatography and,
    816–817
halogen detection, 927–928
high-performance liquid
    chromatography and, 817
mass spectrum, 924–926
methyl butanoate, 940
molecular formula determination,
    926–927
operation, 924
propanoic acid, 939
rearrangement reactions, 941
2,2,4-trimethylpentane, 933
uses, 924
Material Safety Data Sheets (MSDSs),
    547–549
Maximum-boiling-point diagrams,
    726–727
McElroy, W. D., 433
McLafferty rearrangement, 941
Measurement, 580–589
    automatic pipets, 581
    balances, 588–589
    dispensing pumps, 582–583
    graduated cylinders, 587
    graduated pipets, 583–585
    liquids, 6–13, 580
    Pasteur pipets, 586–587
    solids, 6, 580
    syringes, 587
    vials, beakers, and flasks, 588
Melting point, 627–637
    acetaminophen, 78, 79–80
    behaviors prior to, 634–635
    1,4-diphenyl-1,3-butadiene, 364
    electrical instruments, 632–634
    mixture, 28–29, 630–631
    packing the tube, 631
    range, 628–629
    standards, 636
    theory, 629–630
    thermometer calibration, 635–636
    Thiele tube, 631–632
    unknowns determined from, 28–29,
        450, 627
The Merck Index, 560, 578–579, 628,
    942–943
Mesityl oxide, 862
Methanol, 262
    safety information, 556
Methemoglobinemia, 68
Methiodides, 973
Methods. *See* Reaction methods

Methyl benzoate
  infrared spectrum, 864
  nitration, 228–233
Methyl butanoate, 940
Methyl isopropyl ketone, 849
Methyl oleate, 239–243
Methyl salicylate, 373–377
  infrared spectrum, 376
  NMR spectrum, 376
Methyl stearate, 239–243
2-Methyl-2-propanol, 186–187
4-Methylcyclohexanol, 858
4-Methylcyclohexene, 211–217
  infrared spectrum, 854
  microscale procedure, 213–214
  semimicroscale procedure, 214–215
  unsaturation tests, 216
Methylene chloride
  caffeine extraction, 93–95
  safety information, 556
Mevalonic acid, 109
Micelle, 246–247
Michael addition reaction, 344,
    533–536
Micropipets, 782
Microscale experiments, 2
Microspatulas, 13
Microwave ovens, 596–597
Minimization, in molecular mechan-
    ics, 154–155
Minimum-boiling-point diagrams,
    724–726
Miscibility, 638
  steam distillation, 750–751
Mixture melting point, 28–29, 630–631
Mole fraction, 717
Mole percentage, 717
Molecular ion, 925
Molecular mechanics, 152, 160
  conformation, 154–155
  current implementations, 155
  essay, 152–156
  limitations, 155
  minimization, 154–155
Molecular modeling, 156–160. *See also*
    Computational chemistry
  alkyl halides, 179–180
  butane conformations, 157–158
  butenes, 159–160
  camphor reduction, 301
  cyclohexane chair and boat confor-
    mations, 158–159
  enolate ion of acetone, 342
  molecular mechanics, 152–156
  nitration of anisole, 232
  nitration of methyl benzoate,
    231–232
  cis-norbornene-5,6-*endo*-dicarboxylic
    anhydride, 422–423
  quantum mechanics, 152
  substituted cyclohexane rings, 159

Molecular sieve chromatography, 774
Moncrieff, R. W., 120
Monomers, 397
Monoterpenes, 109
Mother liquor, 21, 626, 649
Mydriasis, 368
Myocardial, 88

N

Naphthalene, 429–431
Naphthenes, 255
2-Naphthol, 857
α-Naphthylurethanes, 972
Naproxen, 70
National Fire Protection
    Association, 549
Neutral alumina, 757
Neutraliztion equivalent, 478
Nicol prism, 818–819
Nicol, William, 819
Nitration
  anisole, 232
  aromatic compounds, 284–288
  methyl benzoate, 228–233
Nitriles, 859
Nitro compounds, infrared spec-
    troscopy, 859
Nitro groups, identification
    tests, 463
2-, 3-, and 4-Nitroaniline, 894
Nitrobenzene, 859
Nitrogen, 465
2-Nitrophenol, 895
Nitrous acid test, 483–484
NMR spectroscopy. *See* Nuclear
    magnetic resonance (NMR)
    spectroscopy
Nonanal, 861
Norbornanone, 301
cis-norbornene-5,6-*endo*-dicarboxylic
    anhydride, 419–423
  infrared spectrum, 422, 865
Norit, 663–664
Normal-phase chromatography,
    96, 794
Notebook. *See* Laboratory notebook
Noxious gases, capturing, 607–610
Nuclear magnetic resonance (NMR)
    spectroscopy, 128–129, 868–905.
    See also Carbon-13 nuclear mag-
    netic resonance spectroscopy
  2-acetylcyclohexanone, 358
  alcohols, 490
  aldehydes and ketones, 476
  amines, 485
  anisotropy, 880–882
  aromatic compounds, 516, 889–895
  aromaticity, 470
  benzene ring substitution, 889–895
  benzocaine, 372
  borneol, 298

borohydride reduction product, 300
camphor, 295, 297
carboxylic acids, 478
chemical environment and chemical
    shift, 877–880
chemical equivalence, 875
chemical shift, 874
chemical shift reagents, 896–897
coupling constant, 884–886
1,4-diphenyl-1,3-butadiene, 366
esterification of vanillin, 536–538
esters, 493
ethyl (S)-3-hydroxybutanoate,
    279–284
isoborneol, 298
local diamagnetic shielding,
    878–880
magnetic equivalence, 886–887
methyl salicylate, 376
nucleophilic substitution, 189
operation, 868–869
phenols, 481
α-phenylethylamine, 332, 338–339
protons attached to atoms other
    than carbon, 896
reference substances, 873–874
sample preparation, 869–873
simple multiple bonds test, 469
spectra at higher field strength,
    887–889
spin–spin splitting, 882–884
Nuclear Overhauser enhancement
    (NOE), 913–915
Nucleophilic substitution reactions,
    180–190, 510–514
  competitive nucleophiles with
    1-butanol or 2-butanol, 183–185
  competitive nucleophiles with
    2-methyl-2-propanol, 186–187
  gas chromatography, 187
  NMR spectroscopy, 189
  refractive index, 187–188
Nujol mull, 842
Nylon. *See* Polyamide

O

Occupational Safety and Health
    Administration (OSHA), 547
Odor, 119–123. *See also* Fragrances;
    Pheromones
OFF, 383–388
Oiling out, 661
Oils. *See* Essential oils; Fats and oils
Olestra, 237–238
Online searching, 952
Opsin, 132
Optical purity, 826–827
Optically active substance, 819, 820
Organic solvent hazards, 543–544
Organic Syntheses, 946
Organoleptic qualities, 100

Organozinc reagents, 328–331, 524–527
ortho, para directors, 224
Oxidation
  borneol, 289, 293–295
  puzzle, 538–540
Oxides of nitrogen, 261
Oxygenates, 259
Ozone, 261

P

Paper chromatography, 791
Paper factor, 418
Paraffins, 255
Parameterization, 154
Particulate matter, 261
Pasteur, Louis, 142–145
Pasteur pipets, 11, 586–587, 619–620
Peaks, 805, 925
Penicillin, 390
Pentane. *See* Hexane
t-Pentyl chloride
  distillation, 201, 203
  extraction, 200, 202–203
  infrared spectrum, 201
  macroscale procedure, 202
  microscale procedure, 200
  preparation, 196, 200, 202
  semimicroscale procedure, 201–202
Percentage yield, 562
Permissible Exposure Limit (PEL), 548
Peroxyacetyl nitrate, 261
Pesticides. *See* Insecticides
Petroleum and fossil fuels, 255–263
  alternate fuels and electricity, 262–263
  combustion, 256–257, 260
  fractions from crude oil distillation, 255
  health and environment, 258–262
  octane ratings, 257–259
  refining, 255–256
  supply, 261–262
Petroleum ether, 643–644.
  See also Hexane
Phases, 669
Phase-transfer catalysis, 217–224
  crown ether catalysis, 219–220
  quaternary ammonium salt catalysis, 218–219
Phenacetin, 68
Phenols
  derivatives, 481–482, 964, 972
  identification tests, 479–481
  infrared spectroscopy, 856
  salicylic acid, 64
  spectroscopy, 481
  unknowns, 964

Phenylacetone, 870
α-Phenylethylamine, 331–339
  NMR spectrum, 332
  resolution, 334–337
  resolution of enantiomers, 332
Phenylpropanoids, 110–111
Phenylurethanes, 974
Pheophytins, 137
Pheromones, 102, 377–382
  alarm, 379, 381
  insect repellents, 381–382
  other, 379–381
  recognition, 381
  recruiting, 381
  releaser versus primer, 377
  sex attractants, 377–379
Phosphorescence, 426
Photochemistry, 433–436
Photoreduction, 424–431
Picrates, 973
Pinacol rearrangement, 424, 431–433
Pipets
  automatic, 6–8, 10, 581
  calibration, 11
  disposable Pasteur, 11
  filtering, 618–620
  filter-tip, 625
  graduated, 6, 9–10, 583–585
  micropipets, 782
  Pasteur, 11, 586–587, 619–620
Plane-polarized light, 818
Plastic joint clips, 598–600
Plasticizers, 398
Plastics. *See* Polymers and plastics
Pliny the Elder, 244
Polarimeters, 819–826
  modern digital, 824–826
  null reading, 823
  operation, 822–824
  Zeiss, 822–824
Polarimetry, 818–827
  carvones, 127
  cells, 821–822
  ethyl acetoacetate, 278–279
  optical purity, 826–827
  α-phenylethylamine, 336
  polarimeter, 819–821
  polarimeter operation, 822–824
  polarized light, 818–819
  sample preparation, 821–822
Polarity, 639–642
Polarization basis sets, 163–164
Polarizer, 819
Polaroid filters, 819
Pollution Prevention Act, 269
Polyamide, 409–410
Polyesters, 237
  preparation, 407–409
Polymerization, 256
Polymers and plastics, 397–405
  addition, 397–398, 399–401

chemical structures, 397
condensation, 398, 401–403
cross-linked, 398
disposability problems, 403–405
infrared spectra, 412–413
polyamide, 409–410
polyesters, 407–409
polystyrene, 410–412
preparation, 406–414
thermal classification, 398
Polystyrene, 410–412
  infrared spectrum, 847
Potassium bromide, infrared spectroscopy and, 838–842
Potassium iodide, 176
Potassium permanganate, 468
Powerfully electrophilic reagents, 229
Precipitation, 647
Precise atomic masses, 926–927
Preparative column, 795
Preparative experiments, 560
Preparative plates, 787
Pressure tubing, 733
Pressure-equalizing addition funnel, 604
Product development control, 291
Propanoic acid, 939
Propylure, 379
Prostaglandins, 61
Proton NMR. *See* Nuclear magnetic resonance (NMR) spectroscopy
Purification
  extraction, 685–688
  separation and purification scheme, 496–498
  unknowns identification, 450–451
Purity, 628. *See also* Optical purity
Pyridine, safety information, 556

Q

Qualitative analysis, literature on, 948–949
Quantum mechanics, 152, 160–168
  ab initio calculations, 161–165
  basis set orbitals, 162–164
  graphic models and visualization, 165–166
  heats of formation, 165
  mapping properties onto density surface, 167–168
  Schrödinger equations, 161–162
  semiempirical calculations, 161, 164
  surfaces, 166–168
  terms and methods, 160–161
  wavefunction, 162
Quaternary ammonium salt catalysis, 218–219
Queen substance, 379
Quencher, 429

R

Radiationless transition, 426
Radical-cations, 924, 929
Raoult's Law, 719–720, 751
Rate of takeoff, 723
Reaction mechanisms, literature on, 947–948
Reaction methods, 598–616
    addition of liquid reagents, 604–605
    apparatus assembly, 598–600
    boiling stones, 603–604
    drying tubes, 605–606
    gas collection, 610–611
    heating under reflux, 600–602
    inert atmosphere, 606–607
    noxious gas capture, 607–610
    rotary evaporator, 614–615
    solvent evaporation, 611–614
    stirring, 602–603
Records, 561–562
Recrystallization, 650
Reduction, camphor, 289–291, 295
Reflux apparatus, 600–602
Reflux ratio, 723
Reflux ring, 602
Reformates, 258
Reforming (gasoline), 258
Reformulated gasoline, 260
Refractive index, 827–828
    carvones, 130
    nucleophilic substitution, 187–188
Refractometry, 827–832
    Abbé refractometer, 828–831
    cleaning refractometer, 831
    digital refractometers, 831–832
    refractive index, 827–828
    temperature corrections, 832
Regioselectivity, 170
Rendering, 234
Reports, 565
Resolution, 807–808
Response factor, 812–814
Retardation factor, 788
Retention time, 56, 806
Retinene, 132
Retort, 703
Reversed-phase chromatography, 96, 794
Rhodopsin, 132
Right-to-Know Laws, 547–554
Ring current, 881
Robinson annelation (ring-formation) reactions, 351–352
Robiquet, Pierre Jean, 87
Rods, 132
Rotary evaporator, 614–615
Rotary fraction collectors, 739, 740
Rubber tubing, 571
Rudolph polarimeter, 823
Running the thin-layer plate, 778, 784–785

S

S_N1 reaction rates, 171–172, 179
S_N2 reaction rates, 179–180
Saccharin, 443
Safety. *See* Laboratory safety
Salicylamide, 68
Salicylic acid, 61
Salting out, 528
Sample submission, 565
Sand bath, 3–4, 593–594
Saponification, 244
Saponins, 243–244
Schrödinger equations, 161–162
Science Citation Index, 954
Scientific and Technical Information Network (STN International), 952
SciFinder, 952–953
SciFinder Scholar, 952
Second order patterns, 888
Secondary containment, 544
Seeding, 665
Seliger, H. H., 433
Semicarbazones, 971
Semiempirical calculations, 161, 164
Semiprep column, 795
Separation
    column chromatographic, 758–763
    extraction, 685–688
    gas chromatography, 802–803
Separation scheme, 496–498, 560–561
Separatory funnel, 677–679
Sephadex, 774–775
Sex attractants, 377–379
Shikimic acid pathway, 111
Shrinkage, 635
Side products, 559
Side reactions, 559
Silica gel, 664
Silver chloride, infrared spectroscopy and, 834, 836–838, 844
Silver nitrate
    alkyl halides reactivities, 176, 178
    test, 461–462, 477
Simple distillation, 703–715
    equipment evolution, 703–705
    fractional versus, 715–716
    microscale equipment, 707–713
    microscale procedure, 56–57
    semimicroscale and macroscale equipment, 713–714
    semimicroscale procedure, 53–56
    temperature monitoring, 709–712
    theory, 705–707
Simple multiple bonds test, 466–469
Single-point calculations, 164
Single-point energy, 158
Singlet energy transfer, 429
Singlet state, 426
Size-exclusion chromatography, 774

Size-exclusion column, 795
Slater-type orbitals (STOs), 162–163
Slurry, 623
Smog, 261
Soaps and detergents, 243–248
    cleaning action, 246
    preparation, 244–246, 249–255
    problems, 248
Sodium bicarbonate, 458, 477
Sodium borohydride, 289–291, 295, 300
Sodium chloride, infrared spectroscopy and, 834–836, 844–845
Sodium cyclamate, 443
Sodium fusion tests, 464
Sodium hydroxide test, 458, 479
Sodium hypochlorite, 289, 293
Sodium iodide, 176–178, 462
Sodlium lauryl sulfate, 251–255
Softening, 635
Solid phase extraction (SPE), 95–99, 689–692
Solids
    measurement, 6
    sublimation behavior, 747
    vapor-pressure behavior, 745–746
Solubility, 13–20, 637–646
    chart of functional groups, 455
    crystallization, 647–648
    definition, 637–638
    in hydrochloric acid, 457
    inert compounds, 459
    organic solvents, 643–644
    predicting, 638–643
    in sodium bicarbonate, 458
    in sodium hydroxide, 458
    in sulfuric acid, 459
    tests for unknowns identification, 451, 454–459
    in water, 456–457
Solute, 637
Solution cell, 844–845
Solvent extraction, 234
Solvents
    column chromatography, 761–763
    crystallization, 660–663, 667
    definition, 637
    evaporation, 611–614
    for extraction process, 671–673
    green chemistry, 271
    high-performance liquid chromatography, 795–796
    organic, 643–644
    polarity, 642
    safety information, 554–556
    thin-layer chromatography, 785–786
Sorbent, 689
Sorbitol, 442
Soxhlet extractor, 689
Spearmint oil, 124–131
Spectra catalogues, 944

Spectrophotometer, 849
Spectroscopy, unknowns
    identification, 452–453. *See also*
    Infrared spectroscopy
Spinach, 136–142
Spinning-band column, 722–723
Spin–spin splitting, 882–884,
    908–910
Split-valence basis sets, 163
Spotting, 778, 782–784
Steam bath, 595–596
Steam distillation, 750–755
    boiling points and
        compositions, 753
    direct method, 752–754
    immiscible mixture calculations,
        751–752
    isolation of essential oils, 504–507
    live steam method, 753–755
    miscible versus immiscible
        mixtures, 750–751
Stem correction, 634, 700
Steric approach control, 291
Steric energy, 152
Stirring methods, 602–603, 712
STO-3G basis set, 163
Stock solution, 464
Stone, Edward, 61
Straight-run gasoline, 256
Strain energy, 152
Streaming, 764
Strokes, 256
Styrene, 855
Sublimation, 635, 745–749
    advantages, 749
    apparatus, 748
    caffeine, 95, 98
    methods, 748–749
    solids, 747
    vacuum sublimation, 747
    vapor-pressure behavior of solids
        and liquids, 745–746
Substitution reaction, 467
Sucrose, 145–149, 237, 442
    fermentation, 147–148
    fractional distillation, 148–149
Sudser, 248
Sulfa drugs
    essay, 388–391
    preparation, 392–396
    uses, 21
Sulfamates, 443
Sulfanilamide, 21, 389, 395–396
Sulfur, 465
Sulfuric acid, 459
Supercritical carbon dioxide, 271
Supercritical state, 90
Supernatant liquid, 73
Surfaces, 166–167
Sweeteners, 441–444
Syndets, 246

Synthetic methods, 943, 945–946
Syringes, 587

T

Tailing, 772, 773, 807–809
Tannin, 91
Tare, 6, 588
Taste, 101
Tautomerism, 170
Tea. *See* Caffeine
Terpenes, 108–109
Terpenoids, 108
Tetracycline, 390
Tetraethyllead, 258
Tetrahydrofuran, 556
Tetramethylsilane (TMS), 873–874
Tetraphenylcyclopentadienone,
    314–317
Theoretical plates, 721
Theoretical yield, 562
Thermal conductivity detector
    (TCD), 804
Thermometers
    aluminum block, 2–3, 590–592
    boiling point determination,
        700–702
    bulb immersion, 700
    calibration, 635–636
    mercury versus nonmercury, 700
    partial immersion, 700
    stem correction, 634, 700
    total immersion, 700
Thermoplastics, 398–399, 403–404
Thermoset plastics, 399, 403
Thiamine catalysis, 303–305,
    307–308
Thiele tube, 631–632
Thin-layer chromatography (TLC),
    43–45, 777–792
    analgesic drug analysis, 82–87
    applications, 789–791
    borneol oxidation, 293–294
    chlorophyll and carotenoids,
        141–142
    column chromatography and, 777
    commercially prepared plates,
        778–779
    developing plates, 784–785
    1,4-diphenyl-1,3-butadiene, 363,
        365–366
    drug identification, 81
    monitoring reaction, 46–48
    paper chromatography and, 791
    plate preparation, 44, 779–782
    preparative plates, 787
    principles, 777–778
    R_f value, 787–788
    solvent selection, 45–46, 785–786
    spotting plates, 782–784
    visualization methods, 786–787
Thin-layer plates, 778

commercially prepared, 778–779
    larger, 781–782
    microscopic slide, 779–781
    preparation of plates, 779–782
Threshold Limit Value (TLV), 548
Throughput, 723
Tollens test, 472–473
Toluene
    mass spectrum, 935
    NMR spectrum, 915
    as solvent, 30, 556
p-Toluidides, 972
Top-loading balance, 6, 7, 588
Total energy, 165
trans-fatty acids, 236
Transportation of chemicals, 544
Tréfouël, J., 389
Triacylglycerols, 244
Triangulation, 811
1,1,2-Trichloroethane, 882–883
Triglycerides, 244
2,2,4-Trimethylpentane, 933
Triphenylmethanol, 317, 322
    infrared spectrum, 324
Triplet energy transfer, 429
Triplet state, 426
Tropylium ion, 935
Tubing
    pressure, 733
    rubber, 571

U

Ultraviolet lamp, 786
Unisol, 872
Unknowns, identification of,
    448–453
    alcohols, 486–491
    aldehydes, 470–476
    amines, 482–486
    boiling/melting points, 60, 450
    carboxylic acids, 477–479
    chemical classification tests, 451–452
    confirmation of identity, 453
    derivatives, 453
    drug identification, 80–81
    elemental analysis, 453, 460–465
    esters, 491–494
    general unknowns, 448
    ketones, 470–476
    phenols, 479–482
    preliminary classification, 449
    preliminary tests, 451
    purification, 450–451
    solubility tests, 451, 454–459
    specific unknowns, 448
    spectroscopy, 452–453
    structure determination using
        tables, 452
    tables, 958–970
    unsaturation tests, 466–470
Unsaturation tests, 466

V

Vacuum distillation, 732–744
  apparatus, 735–736
  aromatic compounds, 519–520
  bulb-to-bulb, 739–740
  manometers, 741–744
  mechanical pump, 740–741
  microscale methods, 732–735
  rotary fraction collectors, 739, 740
  semimicroscale and macroscale
    equipment, 736–737
  simplified microscale apparatus,
    735–736
  step-by-step instructions,
    737–739
  uses, 732
Vacuum filtration, 621–623
  acetylsalicylic acid, 65–66
  analgesics, 73–74
van der Waals forces, 639
Vanillin, 536–538

Vapor–liquid composition diagrams,
    716–719
Vapor-phase chromatography.
    See Gas chromatography
Variation Principle, 161
Vasodilator, 88
Vermifuge, 67
Vinyl acetate, 887
Vision, chemistry of, 132–135
Visualization methods, 778,
    786–787
Volatility, 662
Vomeronasal organ, 378

W

Wald, George, 132, 134
Wash acetone, 566
Waste disposal, 544–545
Water bath, 3–4, 594
Water process decaffeination, 89
Water trap, 733, 735

Water-jacketed condenser, 600
Wavefunction, 162
Wavefunction Ψ, 160–161
Wavenumbers, 833
Wintergreen. *See* Methyl salicylate
Wittig reaction, 359, 364
Wittig salt, 359, 362
Wort, 143

X

Xanthines, 88
Xanthophylls, 136–142

Y

Yield, 562
Ylide, 304, 359–360, 363

Z

Zeiss polarimeter, 822–824
Zinc, 328–331
Zymase, 145

Infrared Absorption Bands

	Type of Vibration		Frequency (cm^{-1})	Intensity
C—H	Alkanes	(stretch)	3000–2850	s
	—CH$_3$	(bend)	1450 and 1375	m
	—CH$_2$—	(bend)	1465	m
	Alkenes	(stretch)	3100–3000	m
		(out-of-plane bend)	1000–650	s
	Aromatics	(stretch)	3150–3050	s
		(out-of-plane bend)	900–690	s
	Alkyne	(stretch)	ca. 3300	s
	Aldehyde		2900–2800	w
			2800–2700	w
O—H	Alcohol, phenols			
	Free		3650–3600	m
	H-bonded		3400–3200	m
	Carboxylic acids		3400–2400	m
N—H	Primary and secondary amines and amides			
	(stretch)		3500–3100	m
	(bend)		1640–1550	m–s
C≡C	Alkyne		2250–2100	m–w
C≡N	Nitriles		2260–2240	m
C=C	Alkene		1680–1600	m–w
	Aromatic		1600 and 1475	m–w
N=O	Nitro (R—NO$_2$)		1550 and 1350	s
C=O	Aldehyde		1740–1720	s
	Ketone		1725–1705	s
	Carboxylic acid		1725–1700	s
	Ester		1750–1730	s
	Amide		1680–1630	s
	Anhydride		1810 and 1760	s
	Acid chloride		1800	s
C—O	Alcohols, ethers, esters, carboxylic acids, anhydrides		1300–1000	s
C—N	Amines		1350–1000	m–s
C—X	Fluoride		1400–1000	s
	Chloride		785–540	s
	Bromide, iodide		< 667	s